ENCYCLOPEDIA OF
Physical Science
AND Technology

THIRD EDITION

Cr-Ela

Volume 4

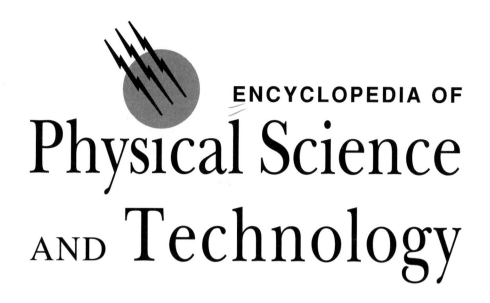

ENCYCLOPEDIA OF
Physical Science
AND Technology

THIRD EDITION

Cr-Ela

Volume 4

Editor-in-Chief

Robert A. Meyers, Ramtech, Inc.

ACADEMIC PRESS

A Harcourt Science and Technology Company

San Diego   San Francisco   Boston   New York   London   Sydney   Tokyo

This book is printed on acid-free paper.

Academic Press
A Harcourt Science and Technology Company
525 B Street, Suite 1900, San Diego, California 92101-4495, USA
http://www.academicpress.com

Academic Press
Harcourt Place, 32 Jamestown Road, London NW1 7BY, UK
http://www.academicpress.com

Library of Congress Catalog Card Number: 2001090661

International Standard Book Number:

0-12-227410-5 (Set)	0-12-227420-2 (Volume 10)
0-12-227411-3 (Volume 1)	0-12-227421-0 (Volume 11)
0-12-227412-1 (Volume 2)	0-12-227422-9 (Volume 12)
0-12-227413-X (Volume 3)	0-12-227423-7 (Volume 13)
0-12-227414-8 (Volume 4)	0-12-227424-5 (Volume 14)
0-12-227415-6 (Volume 5)	0-12-227425-3 (Volume 15)
0-12-227416-4 (Volume 6)	0-12-227426-1 (Volume 16)
0-12-227417-2 (Volume 7)	0-12-227427-X (Volume 17)
0-12-227418-0 (Volume 8)	0-12-227429-6 (Index)
0-12-227419-9 (Volume 9)	

PRINTED IN THE UNITED STATES OF AMERICA
01 02 03 04 05 06 MM 9 8 7 6 5 4 3 2 1

SUBJECT AREA EDITORS

Ruzena Bajcsy
University of Pennsylvania
Robotics

Allen J. Bard
University of Texas
Analytical Chemistry

Wai-Kai Chen
University of Illinois
Circuit Theory and Circuits

John T. Christian
Consultant, Waban, Massachusetts
Mining Engineering
Snow, Ice, and Permafrost
Soil Mechanics

Robert Coleman
U.S. Geological Survey
Earth Science
Geochemistry
Geomorphology
Geology and Mineralogy

Elias J. Corey
Harvard University
Organic Chemistry

Paul J. Crutzen
Max Planck Institute for Chemistry
Atmospheric Science

Gordon Day
National Institute of Standards and Technology
Lasers and Masers
Optical Materials

Jack Dongarra
University of Tennessee
Computer Hardware

Gerard M. Faeth
University of Michigan
Aeronautics

William H. Glaze
University of North Carolina
Environmental Chemistry, Physics, and Technology

Tomifumi Godai
National Space Development Agency of Japan
Space Technology

Barrie A. Gregory
University of Brighton
Instrumentation and Special Applications
for Electronics and Electrical Engineering

Gordon Hammes
Duke University
Biochemistry

Frederick Hawthorne
University of California, Los Angeles
Inorganic Chemistry

Leonard V. Interrante
Rensselaer Polytechnic Institute
Materials Science and Engineering

Bruce H. Krogh
Carnegie Mellon University
Control Technology

Joseph P. S. Kung
University of North Texas
Mathematics

William H. K. Lee
U.S. Geological Survey
Geodesy
Solid Earth Geophysics
Terrestrial Magnetism

Yuan T. Lee
Academica Sinica
Physical Chemistry
Radio, Radiation, and Photochemistry

Robert W. Lenz
University of Massachusetts
Polymer Chemistry

Irvin E. Liener
University of Minnesota
Agricultural and Food Chemistry

Larry B. Milstein
University of California, San Diego
Communications

Donald R. Paul
University of Texas
Applied Chemistry and Chemical
Engineering

S. S. Penner
University of California, San Diego
Energy Resources
Power Systems

George S. Philander
Princeton University
Dynamic Oceanography
Physical and Chemical Oceanography

Peter H. Rose
Orion Equipment, Inc
Materials Science for Electrical and
Electronic Engineering

Jerome S. Schultz
University of Pittsburgh
Biotechnology

Melvin Schwartz
Columbia University
Physics

Phillip A. Sharp
Massachusetts Institute of Technology
Molecular Biology

Martin Shepperd
Bournemouth University
Computer Software

Steven N. Shore
Indiana University
Astronomy

Leslie Smith
University of British Columbia
Hydrology and Limnology

Stein Sture
University of Colorado
Mechanical, Industrial, and Civil
Engineering

Robert Trappl
University of Vienna
Systems Theory and Cybernetics

Nicholas Tsoulfanidis
University of Missouri
Nuclear Technology

M. Van Rossum
Catholic University of Leuven
Components, Electronic Devices, and
Materials

R. H. Woodward Waesche
Science Applications International Corporation
Propulsion and Propellants

Contents

Foreword

The editors of the *Encyclopedia of Physical Science and Technology* had a daunting task: to make an accurate statement of the status of knowledge across the entire field of physical science and related technologies.

No such effort can do more than describe a rapidly changing subject at a particular moment in time, but that does not make the effort any less worthwhile. Change is inherent in science; science, in fact, seeks change. Because of its association with change, science is overwhelmingly the driving force behind the development of the modern world.

The common point of view is that the findings of basic science move in a linear way through applied research and technology development to production. In this model, all the movement is from science to product. Technology depends on science and not the other way around. Science itself is autonomous, undisturbed by technology or any other social forces, and only through technology does science affect society.

This superficial view is seriously in error. A more accurate view is that many complex connections exist among science, engineering, technology, economics, the form of our government and the nature of our politics, and literature, and ethics.

Although advances in science clearly make possible advances in technology, very often the movement is in the other direction: Advances in technology make possible advances in science. The dependence of radio astronomy and high-energy physics on progress in detector technology is a good example. More subtly, technology may stimulate science by posing new questions and problems for study.

The influence of the steam engine on the development of thermodynamics is the classic example. A more recent one would be the stimulus that the problem of noise in communications channels gave to the study of information theory.

As technology has developed, it has increasingly become the object of study itself, so that now much of science is focused on what we have made ourselves, rather than only on the natural world. Thus, the very existence of the computer and computer programming made possible the development of computer science and artificial intelligence as scientific disciplines.

The whole process of innovation involves science, technology, invention, economics, and social structures in complex ways. It is not simply a matter of moving ideas out of basic research laboratories, through development, and onto factory floors. Innovation not only requires a large amount of technical invention, provided by scientists and engineers, but also a range of nontechnical or "social" invention provided by, among others, economists, psychologists, marketing people, and financial experts. Each adds value to the process, and each depends on the others for ideas.

Beyond the processes of innovation and economic growth, science has a range of direct effects on our society.

Science affects government and politics. The U.S. Constitution was a product of eighteenth century rationalism and owes much to concepts that derived from the science of that time. To a remarkable extent, the Founding Fathers were familiar with science: Franklin, Jefferson, Madison, and Adams understood science, believed passionately in empirical inquiry as the source of truth, and felt that government should draw on scientific concepts for inspiration. The concept of "checks and balances" was borrowed from Newtonian physics, and it was widely believed that, like the orderly physical universe that science was discovering, social relations were subject to a series of natural laws as well.

Science also pervades modern government and politics. A large part of the Federal government is concerned either with stimulating research or development, as are the National Aeronautics and Space Administration (NASA) and the National Science Foundation, or with seeking to regulate technology in some way. The reason that science and technology have spawned so much government activity is that they create new problems as they solve old ones. This is true in the simple sense of the "side effects" of new technologies that must be managed and thus give rise to such agencies as the Environmental Protection Agency (EPA). More importantly, however, the availability of new technologies makes possible choices that did not exist before, and many of these choices can only be made through the political system.

Biotechnology is a good example. The Federal government has supported the basic science underlying biotechnology for many years. That science is now making possible choices that were once unimagined, and in the process a large number of brand new political problems are being created. For example, what safeguards are necessary before genetically engineered organisms are tested in the field? Should the Food and Drug Administration restrict the development of a hormone that will stimulate cows to produce more milk if the effect will be to put a large number of dairy farmers out of business? How much risk should be taken to develop medicines that may cure diseases that are now untreatable?

These questions all have major technical content, but at bottom they involve values that can only be resolved through the political process.

Science affects ideas. Science is an important source of our most basic ideas about reality, about the way the world is put together and our place in it. Such "world views" are critically important, for we structure all our institutions to conform with them.

In the medieval world view, the heavens were unchanging, existing forever as they were on the day of creation. Then Tycho Brahe observed the "new star"—the nova of 1572—and the inescapable fact of its existence forced a reconstruction of reality. Kepler, Galileo, and Newton followed and destroyed the earth- and human-centered universe of medieval Christianity.

Darwin established the continuity of human and animal, thus undermining both our view of our innate superiority and a good bit of religious authority. The germ theory of disease-made possible by the technology of the microscope destroyed the notion that disease was sent by God as a just retribution for unrepentant sinners.

Science affects ethics. Because science has a large part in creating our reality, it also has a significant effect on ethics. Once the germ theory of disease was accepted, it could no longer be ethical, because it no longer made sense, to berate the sick for their sins. Gulliver's voyage to the land of the Houynhyms made the point:

By inventing a society in which individuals' illnesses were acts of free will while their crimes were a result of outside forces, he made it ethical to punish the sick but not the criminal.

Knowledge—most of it created by science—creates obligations to act that did not exist before. An engineer, for instance, who designs a piece of equipment in a way that is dangerous, when knowledge to safely design it exists, has violated both an ethical and a legal precept. It is no defense that the engineer did not personally possess the knowledge; the simple existence of the knowledge creates the ethical requirement.

In another sense, science has a positive effect on ethics by setting an example that may be followed outside science. Science must set truth as the cardinal value, for otherwise it cannot progress. Thus, while individual scientists may lapse, science as an institution must continually reaffirm the value of truth. To that extent science serves as a moral example for other areas of human endeavor.

Science affects art and literature. Art, poetry, literature, and religion stand on one side and science on the other side of C. P. Snow's famous gulf between the "two cultures." The gulf is largely artificial, however; the two sides have more in common than we often realize. Both science and the humanities depend on imagination and the use of metaphor. Despite widespread belief to the contrary, science does not proceed by a rational process of building theories from undisputed facts. Scientific and technological advances depend on imagination, on some intuitive, creative vision of how reality might be constructed. As Peter Medawar puts it: [Medawar, P. (1969). Encounter 32(l), 15-23]: All advance of scientific understanding, at every level, begins with a speculative adventure, an imaginative preconception of what might be true—a preconception that always, and necessarily, goes a little way (sometimes a long way) beyond anything that we have logical or factual authority to believe in.

The difference between literature and science is that in science imagination is controlled, restricted, and tested by reason. Within the strictures of the discipline, the artist or poet may give free rein to imagination. Although we may critically compare a novel to life, in general, literature or art may be judged without reference to empirical truth. Scientists, however, must subject their imaginative construction to empirical test. It is not established as truth until they have persuaded their peers that this testing process has been adequate, and the truth they create is always tentative and subject to renewed challenge.

The genius of science is that it takes imagination and reason, which the Romantics and the modern counterculture both hold to be antithetical, and combines them in a synergistic way. Both science and art, the opposing sides of the "two cultures," depend fundamentally on the creative use of imagination. Thus, it is not surprising that many mathematicians and physicists are also accomplished musicians, or that music majors have often been creative computer programmers.

Science, technology, and culture in the future. We can only speculate about how science and technology will affect society in the future. The technologies made possible by an understanding of mechanics, thermodynamics, energy, and electricity have given us the transportation revolutions of this century and made large amounts of energy available for accomplishing almost any sort of physical labor. These technologies are now mature and will continue to evolve only slowly. In their place, however, we have the information revolution and soon will have the biotechnology revolution. It is beyond us to say where these may lead, but the implications will probably be as dramatic as the changes of the past century.

Computers affected first the things we already do, by making easy what was once difficult. In science and engineering, computers are now well beyond that. We can now solve problems that were only recently impossible. Modeling, simulation, and computation are rapidly be-

coming a way to create knowledge that is as revolutionary as experimentation was 300 years ago.

Artificial intelligence is only just beginning; its goal is to duplicate the process of thinking well enough so that the distinction between humans and machines is diminished. If this can be accomplished, the consequences may be as profound as those of Darwin's theory of evolution, and the working out of the social implications could be as difficult.

With astonishing speed, modern biology is giving us the ability to genuinely understand, and then to change, biological organisms. The implications for medicine and agriculture will be great and the results should be overwhelmingly beneficial.

The implications for our view of ourselves will also be great, but no one can foresee them. Knowledge of how to change existing forms of life almost at will confers a fundamentally new power on human beings. We will have to stretch our wisdom to be able to deal intelligently with that power.

One thing we can say with confidence: Alone among all sectors of society and culture, science and technology progress in a systematic way. Other sectors change, but only science and technology progress in such a way that today's science and technology can be said to be unambiguously superior to that of an earlier age. Because science progresses in such a dramatic and clear way, it is the dominant force in modern society.

Erich Bloch
National Science Foundation
Washington, D.C.

Preface

We are most gratified to find that the first and second editions of the *Encyclopedia of Physical Science and Technology* (1987 and 1992) are now being used in some 3,000 libraries located in centers of learning and research and development organizations world-wide. These include universities, institutes, technology based industries, public libraries, and government agencies. Thus, we feel that our original goal of providing in-depth university and professional level coverage of every facet of physical sciences and technology was, indeed, worthwhile.

The editor-in-chief (EiC) and the Executive Board determined in 1998 that there was now a need for a Third Edition. It was apparent that there had been a blossoming of scientific and engineering progress in almost every field and although the World Wide Web is a mighty river of information and data, there was still a great need for our articles, which comprehensively explain, integrate, and provide scientific and mathematical background and perspective. It was also determined that it would be desirable to add a level of perspective to our Encyclopedia team, by bringing in a group of eminent Section Editors to evaluate the existing articles and select new ones reflecting fields that have recently come into prominence.

The Third Edition Executive Board members, Stephen Hawking (astronomy, astrophysics, and mathematics), Daniel Goldin (space sciences), Elias Corey (chemistry), Paul Crutzen (atmospheric science), Yuan Lee (chemistry), George Olah (chemistry), Melvin Schwartz (physics), Edward Teller (nuclear technology), Frederick Seitz (environment), Benoit Mandelbrot (mathematics), Allen Bard (chemistry) and Klaus von Klitzing (physics)

concurred with the idea of expanding our coverage into molecular biology, biochemistry, and biotechnology in recognition of the fact that these fields are based on physical sciences. Military technology such as weapons and defense systems was eliminated in concert with present trends moving toward emphasis on peaceful uses of science and technology. Aaron Klug (molecular biology and biotechnology) and Phillip Sharp (molecular and cell biology) then joined the board to oversee their fields as well as the overall Encyclopedia. The Advisory Board was completed with the addition of John Bollinger (engineering), Michael Buckland (library sciences), Jean Carpentier (aerospace sciences), Ludwig Faddeev (physics), Herbert Friedman (space sciences), R. A. Mashelkar (chemical engineering), Karl Pister (engineering) and Gordon Slemon (engineering).

A 40 page topical outline of physical sciences and technology was prepared by the EiC and then reviewed by the board and modified according to their comments. This formed the basis for assuring complete coverage of the physical sciences and for dividing the science and engineering disciplines into 50 sections for selection of section editors. Six of the advisory board members decided to serve also as section editors (Allen Bard for analytical chemistry, Elias Corey for organic chemistry, Paul Crutzen for atmospheric sciences, Yuan Lee for physical chemistry, Phillip Sharp for molecular biology, and Melvin Schwartz for physics). Thirty-two additional section editors were then nominated by the EiC and the board for the remaining sections. A listing of the section editors together with their section descriptions is presented on p. v.

The section editors then provided lists of nominated articles and authors, as well as peer reviewers, to the EiC based on the section scopes given in the topical outline. These lists were edited to eliminate overlap. The Board was asked to help adjudicate the lists as necessary. Then, a complete listing of topics and nominated authors was assembled. This effort resulted in the deletion of about 200 of the Second Edition articles, the addition of nearly 300 completely new articles, and updating or rewrite of approximately 480 retained article topics, for a total of over 780 articles, which comprise the Third Edition. Examples of the new articles, which cover science or technology areas arising to prominence after the second edition, are: molecular electronics; nanostructured materials; image-guided surgery; fiber–optic chemical sensors; metabolic engineering; self-organizing systems; tissue engineering; humanoid robots; gravitational wave physics; pharmacokinetics; thermoeconomics, and superstring theory.

Over 1000 authors prepared the manuscripts at an average length of 17-18 pages. The manuscripts were peer reviewed, indexed, and published. The result is the eighteen volume work, of over 14,000 pages, comprising the Third Edition.

The subject distribution is: 17% chemistry; 5% molecular biology and biotechnology; 11% physics; 10% earth sciences; 3% environment and atmospheric sciences; 12% computers and telecommunications; 8% electronics, optics, and lasers; 7% mathematics; 8% astronomy, astrophysics, and space technology; 6% energy and power; 6% materials; 7% engineering, aerospace, and transportation. The relative distribution between basic and applied subjects is: 60% basic sciences, 7% mathematics, and 33% engineering and technology. It should be pointed out that a subject such as energy and power with just a 5% share of the topic distribution is about 850 pages in total, which corresponds to a book-length treatment.

We are saddened by the passing of six of the Board members who participated in previous editions of this Encyclopedia. This edition is therefore dedicated to the memory of S. Chandrasekhar, Linus Pauling, Vladimir Prelog, Abdus Salam, Glenn Seaborg, and Gian-Carlo Rota with gratitude for their contributions to the scientific community and to this endeavor.

Finally, I wish to thank the following Academic Press personnel for their outstanding support of this project: Robert Matsumura, managing editor, Carolan Gladden and Amy Covington, author relations; Frank Cynar, sponsoring editor; Nick Panissidi, manuscript processing; Paul Gottehrer and Michael Early, production; and Chris Morris, Major Reference Works director.

Robert A. Meyers, Editor-in-Chief
Ramtech, Inc.
Tarzana, California, USA

FROM THE PREFACE TO THE FIRST EDITION

In the summer of 1983, a group of world-renowned scientists were queried regarding the need for an encyclopedia of the physical sciences, engineering, and mathematics written for use by the scientific and engineering community. The projected readership would be endowed with a basic scientific education but would require access to authoritative information not in the reader's specific discipline. The initial advisory group, consisting of Subrahmanyan Chandrasekhar, Linus Pauling, Vladimir Prelog, Abdus Salam, Glenn Seaborg, Kai Siegbahn, and Edward Teller, encouraged this notion and offered to serve as our senior executive advisory board.

A survey of the available literature showed that there were general encyclopedias, which covered either all facets of knowledge or all of science including the biological sciences, but there were no encyclopedias specifically in the physical sciences, written to the level of the scientific community and thus able to provide the detailed information and mathematical treatment needed by the intended readership. Existing compendia generally limited their mathematical treatment to algebraic relationships rather than the in-depth treatment that can often be provided only by calculus. In addition, they tended either to fragment a given scientific discipline into narrow specifics or to present such broadly drawn articles as to be of little use to practicing scientists.

In consultation with the senior executive advisory board, Academic Press decided to publish an encyclopedia that contained articles of sufficient length to adequately cover a scientific or engineering discipline and that provided accuracy and a special degree of accessibility for its intended audience.

This audience consists of undergraduates, graduate students, research personnel, and academic staff in colleges and universities, practicing scientists and engineers in industry and research institutes, and media, legal, and management personnel concerned with science and engineering employed by government and private institutions. Certain advanced high school students with at least a year of chemistry or physics and calculus may also benefit from the encyclopedia.

Robert A. Meyers
TRW, Inc.

Guide to the Encyclopedia

eaders of the *Encyclopedia of Physical Science and Technology (EPST)* will find within these pages a comprehensive study of the physical sciences, presented as a single unified work. The encyclopedia consists of eighteen volumes, including a separate Index volume, and includes 790 separate full-length articles by leading international authors. This is the third edition of the encyclopedia published over a span of 14 years, all under the editorship of Robert Meyers.

Each article in the encyclopedia provides a comprehensive overview of the selected topic to inform a broad spectrum of readers, from research professionals to students to the interested general public. In order that you, the reader, will derive the greatest possible benefit from the *EPST*, we have provided this Guide. It explains how the encyclopedia was developed, how it is organized, and how the information within it can be located.

LOCATING A TOPIC

The *Encyclopedia of Physical Science and Technology* is organized in a single alphabetical sequence by title. Articles whose titles begin with the letter A are in Volume 1, articles with titles from B through Ci are in Volume 2, and so on through the end of the alphabet in Volume 17.

A reader seeking information from the encyclopedia has three possible methods of locating a topic. For each of these, the proper point of entry to the encyclopedia is the Index volume. The first method is to consult the alphabetical Table of Contents to locate the topic as an article title; the Index volume has a complete A-Z listing of all article titles with the appropriate volume and page number.

Article titles generally begin with the key term describing the topic, and have inverted word order if necessary to begin the title with this term. For example, "Earth Sciences, History of" is the article title rather than "History of Earth Sciences." This is done so that the reader can more easily locate a desired topic by its key term, and also so that related articles can be grouped together. For example, 12 different articles dealing with lasers appear together in the La- section of the encyclopedia.

The second method of locating a topic is to consult the Contents by Subject Area section, which follows the Table of Contents. This list also presents all the articles in the encyclopedia, in this case according to subject area rather than A-Z by title. A reader seeking information on nuclear technology, for example, will find here a list of more than 20 articles in this subject area.

The third method is to consult the detailed Subject Index that is the essence of the Index volume. This is the best starting point for a reader who wishes to refer to a relatively specific topic, as opposed to a more general topic that will be the focus of an entire article. For example, the Subject Index indicates that the topic of "biogas" is discussed in the article Biomass Utilization.

CONSULTING AN ARTICLE

The First Edition of the *Encyclopedia of Physical Science and Technology* broke new ground in scholarly reference publishing through its use of a special format for articles.

The purpose of this innovative format was to make each article useful to various readers with different levels of knowledge about the subject. This approach has been widely accepted by readers, reviewers, and librarians, so much so that it has not only been retained for subsequent editions of *EPST* but has also been adopted in many other Academic Press encyclopedias, such as the *Encyclopedia of Human Biology*. This format is as follows:

• Title and Author
• Outline
• Glossary
• Defining Statement
• Main Body of the Article
• Cross References
• Bibliography

Although it is certainly possible for a reader to refer only to the main body of the article for information, each of the other specialized sections provides useful material, especially for a reader who is not entirely familiar with the topic at hand.

USING THE OUTLINE

Entries in the encyclopedia begin with a topical outline that indicates the general content of the article. This outline serves two functions. First, it provides a preview of the article, so that the reader can get a sense of what is contained there without having to leaf through all the pages. Second, it serves to highlight important subtopics that are discussed within the article. For example, the article "Asteroid Impacts and Extinctions" includes subtopics such as "Cratering," "Environmental Catastophes," and "Extinctions and Speciation."

The outline is intended as an overview and thus it lists only the major headings of the article. In addition, extensive second-level and third-level headings will be found within the article.

USING THE GLOSSARY

The Glossary section contains terms that are important to an understanding of the article and that may be unfamiliar to the reader. Each term is defined in the context of the article in which it is used. The encyclopedia includes approximately 5,000 glossary entries. For example, the article "Image-Guided Surgery" has the following glossary entry:

Focused ultrasound surgery (FUS) Surgery that involves the use of extremely high frequency sound targeted to highly specific sites of a few millimeters or less.

USING THE DEFINING STATEMENT

The text of most articles in the encyclopedia begins with a single introductory paragraph that defines the topic under discussion and summarizes the content of the article. For example, the article "Evaporites" begins with the following statement:

EVAPORITES are rocks composed of chemically precipitated minerals derived from naturally occurring brines concentrated to saturation either by evaporation or by freeze-drying. They form in areas where evaporation exceeds precipitation, especially in a semiarid subtropical belt and in a subpolar belt. Evaporite minerals can form crusts in soils and occur as bedded deposits in lakes or in marine embayments with restricted water circulation. Each of these environments contains a specific suite of minerals.

USING THE CROSS REFERENCES

Though each article in the *Encyclopedia of Physical Science and Technology* is complete and self-contained, the topic list has been constructed so that each entry is supported by one or more other entries that provide additional information. These related entries are identified by cross references appearing at the conclusion of the article text. They indicate articles that can be consulted for further information on the same issue, or for pertinent information on a related issue. The encyclopedia includes a total of about 4,500 cross references to other articles. For example, the article "Aircraft Aerodynamic Boundary Layers" contains the following list of references:

Aircraft Performance and Design • Aircraft Speed and Altitude • Airplanes, Light • Computational Aerodynamics • Flight (Aerodynamics) • Flow Visualization • Fluid Dynamics

USING THE BIBLIOGRAPHY

The Bibliography section appears as the last element in an article. Entries in this section include not only relevant print sources but also Websites as well.

The bibliography entries in this encyclopedia are for the benefit of the reader and are not intended to represent a complete list of all the materials consulted by the author in preparing the article. Rather, the sources listed are the author's recommendations of the most appropriate materials for further research on the given topic. For example, the article "Chaos" lists as references (among others) the works *Chaos in Atomic Physics*, *Chaos in Dynamical Systems*, and *Universality in Chaos*.

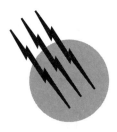

Critical Data in Physics and Chemistry

David R. Lide, Jr.
National Institute of Standards and Technology (Retired)

Bettijoyce B. Lide
National Institute of Standards and Technology

GLOSSARY

Data Factual information, usually expressed in numerical form, that is derived from an experiment, observation, or calculation.

Database An organized collection of data; the term generally implies that the data are expressed in a computer-readable form.

Evaluated data Data whose accuracy has been assessed through an independent review.

THE TERM CRITICAL DATA refers to measured properties of well-defined substances or materials that have been carefully evaluated and organized for convenient use by scientists and engineers. Such collections of data have traditionally been published as handbooks or tables, which have served as basic reference sources for the technical community. Modern computer technology makes it possible to express these collections as databases, which can be stored, retrieved, and accessed in a variety of ways.

I. HISTORY OF CRITICAL DATA PROGRAMS

As physics and chemistry developed into active scientific disciplines in the eighteenth and nineteenth centuries, it was recognized that the numerical results of experiments and observations were valuable to other researchers, often many years after the data were initially obtained. The archival research literature began to serve the function of a storage medium for these data. By the end of the nineteenth century, this literature had grown to the point that locating previously published data was time consuming and difficult. This led to the practice of compiling data from the primary literature and publishing this information in handbook format. An early example was the Landolt-Börnstein tables, *Numerical Data and Functional Relationships in Science and Technology*, which first appeared in 1883. Scientists came to depend on such tables and handbooks for quick access to data on physical and chemical properties.

The process of compiling data from the literature often revealed inconsistencies and discrepancies, which indicated errors in the original research. Thus, it became evident that some form of critical selection or evaluation of the published data was highly desirable. The first broad-coverage handbook to attempt this was the *International Critical Tables*, a seven-volume set of data books published in the 1920s. Experts from many nations evaluated the available data in their specialty areas and selected recommended values for the final publication. Further efforts of this nature were started in the 1930s and 1940s in such important areas of physical science as thermodynamics and atomic physics. In the 1950s, programs for the collection and evaluation of data in nuclear physics were established at Brookhaven and Oak Ridge National Laboratories. As scientific research has expanded and the technological applications of research findings have increased, it has become more and more evident that a critically evaluated base of physical and chemical data is essential for the orderly progress of science and technology.

II. ROLE OF THE NATIONAL INSTITUTE OF STANDARDS AND TECHNOLOGY

Scientists from the U.S. National Bureau of Standards (NBS), whose name was changed to the National Institute of Standards and Technology (NIST) in 1988, played a prominent part in these early critical data projects. In the mid-1960s, NBS was designated as the national focal point for such activities in the United States. It undertook the coordination of a set of activities, known as the National Standard Reference Data System (NSRDS), conducted at universities, industrial laboratories, and NIST itself. Some of these projects were long-term efforts, referred to as data centers, in which relevant data were systematically compiled from the scientific literature, evaluated, and organized into databases. Examples of such data centers include the Atomic Energy Levels Data Center and the Crystal Data Center at NIST and the Radiation Chemistry Data Center at Notre Dame University.

Other organizations have also been active in data compilation and evaluation. Such federal agencies as the De-partment of Energy, Department of Defense, and National Aeronautics and Space Agency have supported selected data projects relevant to their missions. Certain industrial trade associations (e.g., Gas Producers Association and International Copper Research Association) have sponsored data compilation projects of interest to the industry in question. Many professional societies take an active role in sponsoring or coordinating data projects. Examples include the American Institute of Chemical Engineers (Design Institute for Physical Property Data), ASM International (Alloy Phase Diagram Program), and American Society of Mechanical Engineers (steam properties and other data). The National Academy of Sciences–National Research Council has helped to assess needs for data and has examined several national issues associated with access by scientists to data needed in their research.

III. INTERNATIONAL ACTIVITIES

Like many other aspects of science, data compilation efforts can be carried out more efficiently if there is cooperation at an international level. This is particularly important when physical and chemical data affect technological issues, such as performance specifications for articles in international trade or rules for custody transfer of commodities. An early example of the need for international agreement on physical data is provided by the International Association for the Properties of Steam (IAPS). This group was established more than 60 years ago with the aim of reaching international agreement on the thermophysical properties of water and steam, which are crucial in specifying the performance of turbines, boilers, and pumps. Its international steam tables have been adopted as standards for trade and commerce, as well as for scientific applications.

Several international scientific unions have played a role in data compilation. In particular, the International Union of Pure and Applied Chemistry sponsors projects that deal with various types of chemical data. Unions in the geosciences are concerned with data such as terrestrial magnetism, where correlations with geographic location are important. There are also intergovernmental organizations such as the International Atomic Energy Agency (IAEA), which has evaluated data from nuclear and atomic physics that are important in energy research and development.

In 1966, the International Council of Scientific Unions established a standing Committee on Data for Science and Technology (CODATA), with a mandate to improve the quality, reliability, processing, management, and accessibility of data of importance to science and technology. CODATA has representation from the major countries and scientific unions and approaches data issues on both an

TABLE I   CODATA 1998 Recommended Values of the Fundamental Physical Constants

Quantity	Symbol	Value	Unit	Relative std. uncert. u_r
Speed of light in vacuum	c, c_0	299 792 458	$m\,s^{-1}$	(exact)
Magnetic constant	μ_0	$4\pi \times 10^{-7}$ $= 12.566\,370\,614\ldots \times 10^{-7}$	$N\,A^{-2}$ $N\,A^{-2}$	(exact)
Electric constant $1/\mu_0 c^2$	ε_0	$8.854\,187\,817\ldots \times 10^{-12}$	$F\,m^{-1}$	(exact)
Newtonian constant of gravitation	G	$6.673(10) \times 10^{-11}$	$m^3\,kg^{-1}\,s^{-2}$	1.5×10^{-3}
Planck constant $h/2\pi$	h	$6.626\,068\,76(52) \times 10^{-34}$	$J\,s$	7.8×10^{-8}
	$\hbar$	$1.054\,571\,596(82) \times 10^{-34}$	$J\,s$	7.8×10^{-8}
Elementary charge	e	$1.602\,176\,462(63) \times 10^{-19}$	C	3.9×10^{-8}
Magnetic flux quantum $h/2e$	Φ_0	$2.067\,833\,636(81) \times 10^{-15}$	Wb	3.9×10^{-8}
Conductance quantum $2e^2/h$	G_0	$7.748\,091\,696(28) \times 10^{-5}$	S	3.7×10^{-9}
Electron mass	m_e	$9.109\,381\,88(72) \times 10^{-31}$	kg	7.9×10^{-8}
Proton mass	m_p	$1.672\,621\,58(13) \times 10^{-27}$	kg	7.9×10^{-8}
Proton-electron mass ratio	m_p/m_e	$1\,836.152\,6675(39)$		2.1×10^{-9}
Fine-structure constant $e^2/4\pi\epsilon_0\hbar c$	α	$7.297\,352\,533(27) \times 10^{-3}$		3.7×10^{-9}
Inverse fine-structure constant	α^{-1}	$137.035\,999\,76(50)$		3.7×10^{-9}
Rydberg constant $\alpha^2 m_e c/2h$	R_∞	$10\,973\,731.568\,548(83)$	m^{-1}	7.6×10^{-12}
Avogadro constant	N_A, L	$6.022\,141\,99(47) \times 10^{23}$	mol^{-1}	7.9×10^{-8}
Faraday constant $N_A e$	F	$96\,485.3415(39)$	$C\,mol^{-1}$	4.0×10^{-8}
Molar gas constant	R	$8.314\,472(15)$	$J\,mol^{-1}\,K^{-1}$	1.7×10^{-6}
Boltzmann constant R/N_A	k	$1.380\,650\,3(24) \times 10^{-23}$	$J\,K^{-1}$	1.7×10^{-6}
Stefan-Boltzmann constant $(\pi^2/60)k^4/\hbar^3 c^2$	σ	$5.670\,400(40) \times 10^{-8}$	$W\,m^{-2}\,K^{-4}$	7.0×10^{-6}
Non-SI units accepted for use with the SI				
Electron volt: (e/C) J	eV	$1.602\,176\,462(63) \times 10^{-19}$	J	3.9×10^{-8}
(Unified) atomic mass unit $1u = m_u = \frac{1}{12}m(^{12}C)$ $= 10^{-3}\,kg\,mol^{-1}/N_A$	u	$1.660\,538\,73(13) \times 10^{-27}$	kg	7.9×10^{-8}

Source: Mohr, P. J., and Taylor, B. N., *J. Phys. Chem. Ref. Data*, in press.

international and an interdisciplinary basis. It has provided recommended sets of certain key values, such as fundamental physical constants and important thermodynamic properties, which have been generally accepted for international use. CODATA also serves as a forum for reaching consensus on standards and formats for presenting data, and it carries out several educational activities such as conferences, training courses, and preparation of tutorial publications.

The best current values of some frequently used physical and chemical data published by various data evaluation groups are presented in Tables I to VIII.

IV. METHODS OF EVALUATING DATA

The question of how to evaluate published data is not easy to answer in a general way. Specific methodologies have been developed in some fields, and these have certain ele-

ments in common. However, a technique effective for one physical property may be entirely unsuitable for another.

A common feature of most evaluation efforts is the reduction of all published data to the same basis. Corrections for changes in temperature scale, atomic weights, fundamental constants, conversion relations, and other factors must be made before a true evaluation can be started. This often requires considerable effort to deduce the subsidiary data used by the original authors.

Critical evaluation implies a process of independent assessment of the reliability of data appearing in the literature. This process should be conducted by scientists who are familiar with the type of data in question and who have had experience in the measurement techniques that produced the data. There are usually some subjective elements of the evaluation process. For example, the evaluator will generally have a feeling for the accuracy of each measurement technique and for the pitfalls that can lead to unsuspected errors. The reputation of the researcher or

TABLE II   IUPAC Atomic Weights (1995)

Atomic number	Name	Symbol	Atomic weight	Atomic number	Name	Symbol	Atomic weight
1	Hydrogen	H	1.00794(7)	56	Barium	Ba	137.327(7)
2	Helium	He	4.002602(2)	57	Lanthanum	La	138.9055(2)
3	Lithium	Li	6.941(2)	58	Cerium	Ce	140.116(1)
4	Beryllium	Be	9.012182(3)	59	Praseodymium	Pr	140.90765(2)
5	Boron	B	10.811(7)	60	Neodymium	Nd	144.24(3)
6	Carbon	C	12.0107(8)	61	Promethium	Pm	[145]
7	Nitrogen	N	14.00674(7)	62	Samarium	Sm	150.36(3)
8	Oxygen	O	15.9994(3)	63	Europium	Eu	151.964(1)
9	Fluorine	F	18.9984032(5)	64	Gadolinium	Gd	157.25(3)
10	Neon	Ne	20.1797(6)	65	Terbium	Tb	158.92534(2)
11	Sodium	Na	22.989770(2)	66	Dysprosium	Dy	162.50(3)
12	Magnesium	Mg	24.3050(6)	67	Holmium	Ho	164.93032(2)
13	Aluminum	Al	26.981538(2)	68	Erbium	Er	167.26(3)
14	Silicon	Si	28.0855(3)	69	Thulium	Tm	168.93421(2)
15	Phosphorus	P	30.973761(2)	70	Ytterbium	Yb	173.04(3)
16	Sulfur	S	32.066(6)	71	Lutetium	Lu	174.967(1)
17	Chlorine	Cl	35.4527(9)	72	Hafnium	Hf	178.49(2)
18	Argon	Ar	39.948(1)	73	Tantalum	Ta	180.9479(1)
19	Potassium	K	39.0983(1)	74	Tungsten	W	183.84(1)
20	Calcium	Ca	40.078(4)	75	Rhenium	Re	186.207(1)
21	Scandium	Sc	44.955910(8)	76	Osmium	Os	190.23(3)
22	Titanium	Ti	47.867(1)	77	Iridium	Ir	192.217(3)
23	Vanadium	V	50.9415(1)	78	Platinum	Pt	195.078(2)
24	Chromium	Cr	51.9961(6)	79	Gold	Au	196.96655(2)
25	Manganese	Mn	54.938049(9)	80	Mercury	Hg	200.59(2)
26	Iron	Fe	55.845(2)	81	Thallium	Tl	204.3833(2)
27	Cobalt	Co	58.933200(9)	82	Lead	Pb	207.2(1)
28	Nickel	Ni	58.6934(2)	83	Bismuth	Bi	208.98038(2)
29	Copper	Cu	63.546(3)	84	Polonium	Po	[209]
30	Zinc	Zn	65.39(2)	85	Astatine	At	[210]
31	Gallium	Ga	69.723(1)	86	Radon	Rn	[222]
32	Germanium	Ge	72.61(2)	87	Francium	Fr	[223]
33	Arsenic	As	74.92160(2)	88	Radium	Ra	[226]
34	Selenium	Se	78.96(3)	89	Actinium	Ac	[227]
35	Bromine	Br	79.904(1)	90	Thorium	Th	232.0381(1)
36	Krypton	Kr	83.80(1)	91	Protactinium	Pa	231.03588(2)
37	Rubidium	Rb	85.4678(3)	92	Uranium	U	238.0289(1)
38	Strontium	Sr	87.62(1)	93	Neptunium	Np	[237]
39	Yttrium	Y	88.90585(2)	94	Plutonium	Pu	[244]
40	Zirconium	Zr	91.224(2)	95	Americium	Am	[243]
41	Niobium	Nb	92.90638(2)	96	Curium	Cm	[247]
42	Molybdenum	Mo	95.94(1)	97	Berkelium	Bk	[247]
43	Technetium	Tc	[98]	98	Californium	Cf	[251]
44	Ruthenium	Ru	101.07(2)	99	Einsteinium	Es	[252]
45	Rhodium	Rh	102.90550(2)	100	Fermium	Fm	[257]
46	Palladium	Pd	106.42(1)	101	Mendelevium	Md	[258]
47	Silver	Ag	107.8682(2)	102	Nobelium	No	[259]
48	Cadmium	Cd	112.411(8)	103	Lawrencium	Lr	[262]
49	Indium	In	114.818(3)	104	Rutherfordium	Rf	[261]
50	Tin	Sn	118.710(7)	105	Dubnium	Db	[262]
51	Antimony	Sb	121.760(1)	106	Seaborgium	Sg	[266]
52	Tellurium	Te	127.60(3)	107	Bohrium	Bh	[264]
53	Iodine	I	126.90447(3)	108	Hassium	Hs	[269]
54	Xenon	Xe	131.29(2)	109	Meitnerium	Mt	[268]
55	Cesium	Cs	132.90545(2)				

Note: Numbers in parentheses represent the uncertainty in the last digit. Values in brackets are the mass numbers of the longest-lived isotope of elements for which a standard atomic weight cannot be defined.

Source: Pure Appl. Chem. **68**, 2339 (1996).

TABLE III Ground Levels and Ionization Energies for the Neutral Atoms

Z		Element	Ground-state configuration	Ground level	Ionization energy (eV)
1	H	Hydrogen	$1s$	$^2S_{1/2}$	13.5984
2	He	Helium	$1s^2$	$^1S_0$	24.5874
3	Li	Lithium	$1s^2\,2s$	$^2S_{1/2}$	5.3917
4	Be	Beryllium	$1s^2\,2s^2$	$^1S_0$	9.3227
5	B	Boron	$1s^2\,2s^2\,2p$	$^2P^o_{1/2}$	8.2980
6	C	Carbon	$1s^2\,2s^2\,2p^2$	$^3P_0$	11.2603
7	N	Nitrogen	$1s^2\,2s^2\,2p^3$	$^4S^o_{3/2}$	14.5341
8	O	Oxygen	$1s^2\,2s^2\,2p^4$	$^3P_2$	13.6181
9	F	Fluorine	$1s^2\,2s^2\,2p^5$	$^2P^o_{3/2}$	17.4228
10	Ne	Neon	$1s^2\,2s^2\,2p^6$	$^1S_0$	21.5646
11	Na	Sodium	[Ne] $3s$	$^2S_{1/2}$	5.1391
12	Mg	Magnesium	[Ne] $3s^2$	$^1S_0$	7.6462
13	Al	Aluminum	[Ne] $3s^2\,3p$	$^2P^o_{1/2}$	5.9858
14	Si	Silicon	[Ne] $3s^2\,3p^2$	$^3P_0$	8.1517
15	P	Phosphorus	[Ne] $3s^2\,3p^3$	$^4S^o_{3/2}$	10.4867
16	S	Sulfur	[Ne] $3s^2\,3p^4$	$^3P_2$	10.3600
17	Cl	Chlorine	[Ne] $3s^2\,3p^5$	$^2P^o_{3/2}$	12.9676
18	Ar	Argon	[Ne] $3s^2\,3p^6$	$^1S_0$	15.7596
19	K	Potassium	[Ar] $4s$	$^2S_{1/2}$	4.3407
20	Ca	Calcium	[Ar] $4s^2$	$^1S_0$	6.1132
21	Sc	Scandium	[Ar] $3d\,4s^2$	$^2D_{3/2}$	6.5615
22	Ti	Titanium	[Ar] $3d^2\,4s^2$	$^3F_2$	6.8281
23	V	Vanadium	[Ar] $3d^3\,4s^2$	$^4F_{3/2}$	6.7462
24	Cr	Chromium	[Ar] $3d^5\,4s$	$^7S_3$	6.7665
25	Mn	Manganese	[Ar] $3d^5\,4s^2$	$^6S_{5/2}$	7.4340
26	Fe	Iron	[Ar] $3d^6\,4s^2$	$^5D_4$	7.9024
27	Co	Cobalt	[Ar] $3d^7\,4s^2$	$^4F_{9/2}$	7.8810
28	Ni	Nickel	[Ar] $3d^8\,4s^2$	$^3F_4$	7.6398
29	Cu	Copper	[Ar] $3d^{10}\,4s$	$^2S_{1/2}$	7.7264
30	Zn	Zinc	[Ar] $3d^{10}\,4s^2$	$^1S_0$	9.3942
31	Ga	Gallium	[Ar] $3d^{10}\,4s^2\,4p$	$^2P^o_{1/2}$	5.9993
32	Ge	Germanium	[Ar] $3d^{10}\,4s^2\,4p^2$	$^3P_0$	7.8994
33	As	Arsenic	[Ar] $3d^{10}\,4s^2\,4p^3$	$^4S^o_{3/2}$	9.7886
34	Se	Selenium	[Ar] $3d^{10}\,4s^2\,4p^4$	$^3P_2$	9.7524
35	Br	Bromine	[Ar] $3d^{10}\,4s^2\,4p^5$	$^2P^o_{3/2}$	11.8138
36	Kr	Krypton	[Ar] $3d^{10}\,4s^2\,4p^6$	$^1S_0$	13.9996
37	Rb	Rubidium	[Kr] $5s$	$^2S_{1/2}$	4.1771
38	Sr	Strontium	[Kr] $5s^2$	$^1S_0$	5.6949
39	Y	Yttrium	[Kr] $4d\,5s^2$	$^2D_{3/2}$	6.2171
40	Zr	Zirconium	[Kr] $4d^2\,5s^2$	$^3F_2$	6.6339
41	Nb	Niobium	[Kr] $4d^4\,5s$	$^6D_{1/2}$	6.7589
42	Mo	Molybdenum	[Kr] $4d^5\,5s$	$^7S_3$	7.0924
43	Tc	Technetium	[Kr] $4d^5\,5s^2$	$^6S_{5/2}$	7.28
44	Ru	Ruthenium	[Kr] $4d^7\,5s$	$^5F_5$	7.3605
45	Rh	Rhodium	[Kr] $4d^8\,5s$	$^4F_{9/2}$	7.4589
46	Pd	Palladium	[Kr] $4d^{10}$	$^1S_0$	8.3369
47	Ag	Silver	[Kr] $4d^{10}\,5s$	$^2S_{1/2}$	7.5762
48	Cd	Cadmium	[Kr] $4d^{10}\,5s^2$	$^1S_0$	8.9938
49	In	Indium	[Kr] $4d^{10}\,5s^2\,5p$	$^2P^o_{1/2}$	5.7864
50	Sn	Tin	[Kr] $4d^{10}\,5s^2\,5p^2$	$^3P_0$	7.3439
51	Sb	Antimony	[Kr] $4d^{10}\,5s^2\,5p^3$	$^4S^o_{3/2}$	8.6084
52	Te	Tellurium	[Kr] $4d^{10}\,5s^2\,5p^4$	$^3P_2$	9.0096
53	I	Iodine	[Kr] $4d^{10}\,5s^2\,5p^5$	$^2P^o_{3/2}$	10.4513
54	Xe	Xenon	[Kr] $4d^{10}\,5s^2\,5p^6$	$^1S_0$	12.1298
55	Cs	Cesium	[Xe] $6s$	$^2S_{1/2}$	3.8939
56	Ba	Barium	[Xe] $6s^2$	$^1S_0$	5.2117
57	La	Lanthanum	[Xe] $5d\,6s^2$	$^2D_{3/2}$	5.5769
58	Ce	Cerium	[Xe] $4f\,5d\,6s^2$	$^1G^o_4$	5.5387

Continues

TABLE III (*Continued*)

Z		Element	Ground-state configuration	Ground level	Ionization energy (eV)
59	Pr	Praseodymium	[Xe] $4f^3 6s^2$	$^4I^o_{9/2}$	5.473
60	Nd	Neodymium	[Xe] $4f^4 6s^2$	$^5I_4$	5.5250
61	Pm	Promethium	[Xe] $4f^5 6s^2$	$^6H^o_{5/2}$	5.582
62	Sm	Samarium	[Xe] $4f^6 6s^2$	$^7F_0$	5.6436
63	Eu	Europium	[Xe] $4f^7 6s^2$	$^8S^o_{7/2}$	5.6704
64	Gd	Gadolinium	[Xe] $4f^7 5d 6s^2$	$^9D^o_2$	6.1501
65	Tb	Terbium	[Xe] $4f^9 6s^2$	$^6H^o_{15/2}$	5.8638
66	Dy	Dysprosium	[Xe] $4f^{10} 6s^2$	$^5I_8$	5.9389
67	Ho	Holmium	[Xe] $4f^{11} 6s^2$	$^4I^o_{15/2}$	6.0215
68	Er	Erbium	[Xe] $4f^{12} 6s^2$	$^3H_6$	6.1077
69	Tm	Thulium	[Xe] $4f^{13} 6s^2$	$^2F^o_{7/2}$	6.1843
70	Yb	Ytterbium	[Xe] $4f^{14} 6s^2$	$^1S_0$	6.2542
71	Lu	Lutetium	[Xe] $4f^{14} 5d 6s^2$	$^2D_{3/2}$	5.4259
72	Hf	Hafnium	[Xe] $4f^{14} 5d^2 6s^2$	$^3F_2$	6.8251
73	Ta	Tantalum	[Xe] $4f^{14} 5d^3 6s^2$	$^4F_{3/2}$	7.5496
74	W	Tungsten	[Xe] $4f^{14} 5d^4 6s^2$	$^5D_0$	7.8640
75	Re	Rhenium	[Xe] $4f^{14} 5d^5 6s^2$	$^6S_{5/2}$	7.8335
76	Os	Osmium	[Xe] $4f^{14} 5d^6 6s^2$	$^5D_4$	8.4382
77	Ir	Iridium	[Xe] $4f^{14} 5d^7 6s^2$	$^4F_{9/2}$	8.9670
78	Pt	Platinum	[Xe] $4f^{14} 5d^9 6s$	$^3D_3$	8.9587
79	Au	Gold	[Xe] $4f^{14} 5d^{10} 6s$	$^2S_{1/2}$	9.2255
80	Hg	Mercury	[Xe] $4f^{14} 5d^{10} 6s^2$	$^1S_0$	10.4375
81	Tl	Thallium	[Xe] $4f^{14} 5d^{10} 6s^2 6p$	$^2P^o_{1/2}$	6.1082
82	Pb	Lead	[Xe] $4f^{14} 5d^{10} 6s^2 6p^2$	$^3P_0$	7.4167
83	Bi	Bismuth	[Xe] $4f^{14} 5d^{10} 6s^2 6p^3$	$^4S^o_{3/2}$	7.2856
84	Po	Polonium	[Xe] $4f^{14} 5d^{10} 6s^2 6p^4$	$^3P_2$	8.417?
85	At	Astatine	[Xe] $4f^{14} 5d^{10} 6s^2 6p^5$	$^2P^o_{3/2}$	
86	Rn	Radon	[Xe] $4f^{14} 5d^{10} 6s^2 6p^6$	$^1S_0$	10.7485
87	Fr	Francium	[Rn] $7s$	$^2S_{1/2}$	4.0727
88	Ra	Radium	[Rn] $7s^2$	$^1S_0$	5.2784
89	Ac	Actinium	[Rn] $6d 7s^2$	$^2D_{3/2}$	5.17
90	Th	Thorium	[Rn] $6d^2 7s^2$	$3F_2$	6.3067
91	Pa	Protactinium	[Rn] $5f^2(^3H_4) 6d 7s^2$	$(4,\frac{3}{2})_{11/2}$	5.89
92	U	Uranium	[Rn] $5f^3(^4I^o_{9/2}) 6d 7s^2$	$(\frac{9}{2},\frac{3}{2})^o_6$	6.1941
93	Np	Neptunium	[Rn] $5f^4(^5I_4) 6d 7s^2$	$(4,\frac{3}{2})_{11/2}$	6.2657
94	Pu	Plutonium	[Rn] $5f^6 7s^2$	$^7F_0$	6.0262
95	Am	Americium	[Rn] $5f^7 7s^2$	$^8S^o_{7/2}$	5.9738
96	Cm	Curium	[Rn] $5f^7 6d 7s^2$	$^9D^o_2$	5.9915
97	Bk	Berkelium	[Rn] $5f^9 7s^2$	$^6H^o_{15/2}$	6.1979
98	Cf	Californium	[Rn] $5f^{10} 7s^2$	$^5I_8$	6.2817
99	Es	Einsteinium	[Rn] $5f^{11} 7s^2$	$^4I^o_{15/2}$	6.42
100	Fm	Fermium	[Rn] $5f^{12} 7s^2$	$^3H_6$	6.50
101	Md	Mendelevium	[Rn] $5f^{13} 7s^2$	$^2F^o_{7/2}$	6.58
102	No	Nobelium	[Rn] $5f^{14} 7s^2$	$^1S_0$	6.65
103	Lr	Lawrencium	[Rn] $5f^{14} 7s^2 7p$?	$^2P^o_{1/2}$?	4.9?
104	Rf	Rutherfordium	[Rn] $5f^{14} 6d^2 7s^2$?	$^3F_2$?	6.0?

Source: Martin, W. C., and Musgrove, A. (2001). NIST Physics Reference Data Web Site, www. physics.nist.gov/PhysRefData/.

laboratory from which the data came is also a factor, since some research groups are known to take greater care in their work than others.

When there is a high degree of interrelation among a set of independently measured quantities, a systematic correlation scheme can be devised. Thermodynamics provides the prime example. Here one may have available calorimetric measurements of enthalpy changes in chemical reactions, heat capacity measurements, equilibrium constants as a function of temperature, entropy calculated from molecular constants, and perhaps other pertinent experimental measurements. When reduced to standard temperature and pressure, all the data relevant to a given reaction must satisfy well-established thermodynamic

TABLE IV   Properties of Selected Nuclides

	Abundance or half-life	Atomic mass (u)	Mass excess (keV)	Spin	Magnetic moment (μ_N)	Quadrupole moment (fm^2)
^{1}n	10.3 m	1.008 664 916	8071.317	1/2	−1.91304272	
^{1}H	99.985%	1.007 825 032	7288.969	1/2	+2.7928473	
^{2}H	0.015%	2.014 101 778	13135.720	1	+0.8574382	+0.286
^{3}H	12.32 y	3.016 049 268	14949.794	1/2	+2.9789625	
^{3}He	0.000137%	3.016 029 310	14931.204	1/2	−2.1276248	
^{4}He	99.999863%	4.002 603 250	2424.911	0	0	
^{6}Li	7.5%	6.015 122 3	14086.312	1	+0.8220467	−0.082
^{7}Li	92.5%	7.016 004 0	14907.673	3/2	+3.256427	−4.01
^{9}Be	100%	9.012 182 1	11347.584	3/2	−1.1779	+5.288
^{10}B	19.9%	10.012 937 0	12050.761	3	+1.800645	+8.459
^{11}B	80.1%	11.009 305 5	8667.984	3/2	+2.688649	+4.059
^{12}C	98.90%	12	0	0	0	
^{13}C	1.10%	13.003 354 838	3125.011	1/2	+0.7024118	
^{14}C	5715 y	14.003 241 988	3019.892	0	0	
^{14}N	99.634%	14.003 074 005	2863.417	1	+0.4037610	+2.02
^{15}N	0.366%	15.000 108 898	101.438	1/2	−0.2831888	
^{16}O	99.762%	15.994 914 622	−4736.998	0	0	
^{19}F	100%	18.998 403 21	−1487.405	1/2	+2.628868	
^{23}Na	100%	22.989 769 7	−9529.485	3/2	+2.217522	+10.89
^{31}P	100%	30.973 761 5	−24440.991	1/2	+1.13160	
^{32}S	95.02%	31.972 070 7	−26015.981	0	0	
^{34}S	4.21%	33.967 866 8	−29931.850	0	0	
^{55}Fe	2.73 y	54.938 298 029	−57475.007	3/2		
^{60}Co	5.271 y	59.933 822 196	−61644.218	5	+3.799	+44
^{90}Sr	29.1 y	89.907 737 596	−85941.863	0		
^{131}I	8.040 d	130.906 124 168	−87444.761	7/2	+2.742	−40
^{222}Rn	3.8235 d	222.017 570	16366.787	0	0	
^{226}Ra	1599 y	226.025 403	23662.324	0	0	
^{235}U	0.7200%	235.043 923	40914.062	7/2	−0.38	+493.6
^{238}U	99.2745%	238.050 783	47303.664	0	0	
^{239}Pu	24110 y	239.052 157	48583.478	1/2	+0.203	

Source: Lide, D. R., ed. (1999). "CRC Handbook of Chemistry and Physics," CRC Press, Boca Raton, FL.

relations. Furthermore, the energy and entropy changes for a process must be independent of the path followed. These constraints enable one to check the internal consistency of large data sets whose individual values come from a variety of sources. In this way, faulty measurements are frequently recognized that would not be suspected if examined in isolation.

Chemical thermodynamic data and thermophysical properties of fluids are routinely evaluated in this manner. Computer programs have been developed to assess large data sets and select recommended values through a least-squares or similar fitting procedure. Other fields amenable to this approach are atomic and molecular spectroscopy, nuclear physics, and crystallography. In still other cases, such as chemical kinetics and atomic collision cross sections, theory can be used to place limits on data values.

Ideally, the aim of every evaluation effort is to present a "best" or "recommended" value plus a quantitative statement of its uncertainty. If the dominant errors are truly random, a standard deviation or 95% confidence interval can be quoted, which gives the user a sound basis for deciding the implication of this uncertainty for a given problem. However, this situation almost never applies; instead, the most significant errors are usually systematic in nature, deriving from either the initial measurement process or the model used in analyzing the data. The correlations of large data sets described above are very helpful in uncovering such systematic errors, but the judgment of an experienced researcher is also extremely important.

TABLE V Specific Heat, Thermal Conductivity, and Coefficient of Thermal Expansion of the Solid Elements at 25°C

Element	c_p (J g^{-1} K^{-1})	λ (W cm^{-1} K^{-1})	α (10^{-6} K^{-1})
Aluminum	0.897	2.37	23.1
Antimony	0.207	0.24	11.0
Arsenic	0.329	0.50	15.5
Barium	0.204	0.18	20.6
Beryllium	1.825	2.00	11.3
Bismuth	0.122	0.08	13.4
Boron	1.026	0.27	4.7
Cadmium	0.232	0.97	30.8
Calcium	0.647	2.00	22.3
Carbon (diamond)	0.509	9.00	1.1
Cerium	0.192	0.11	5.2
Cesium	0.242	0.36	—
Chromium	0.449	0.94	4.9
Cobalt	0.421	1.00	13.0
Copper	0.385	4.01	16.5
Dysprosium	0.173	0.11	9.9
Erbium	0.168	0.15	12.2
Europium	0.182	0.14	35.0
Gadolinium	0.236	0.11	9.4
Gallium	0.371	0.41	—
Germanium	0.320	0.60	5.8
Gold	0.129	3.17	14.2
Hafnium	0.144	0.23	5.9
Holmium	0.165	0.16	11.2
Indium	0.233	0.82	32.1
Iridium	0.131	1.47	6.4
Iron	0.449	0.8	11.8
Lanthanum	0.195	0.13	12.1
Lead	0.129	0.35	28.9
Lithium	3.582	0.85	46
Lutetium	0.154	0.16	9.9
Magnesium	1.023	1.56	24.8
Manganese	0.479	0.08	21.7
Mercury	0.140	0.08	—
Molybdenum	0.251	1.38	4.8
Neodymium	0.190	0.17	9.6
Nickel	0.444	0.91	13.4
Niobium	0.265	0.54	7.3
Osmium	0.130	0.88	5.1
Palladium	0.246	0.72	11.8
Phosphorus (white)	0.769	0.24	—
Platinum	0.133	0.72	8.8
Plutonium	—	0.07	46.7
Potassium	0.757	1.02	—
Praseodymium	0.193	0.13	6.7
Promethium	—	0.15	11
Rhenium	0.137	0.48	6.2

Continues

TABLE V *(Continued)*

Element	c_p (J g^{-1} K^{-1})	λ (W cm^{-1} K^{-1})	α (10^{-6} K^{-1})
Rhodium	0.243	1.50	8.2
Rubidium	0.363	0.58	—
Ruthenium	0.238	1.17	6.4
Samarium	0.197	0.13	12.7
Scandium	0.568	0.16	10.2
Silicon	0.705	1.48	2.6
Silver	0.235	4.29	18.9
Sodium	1.228	1.41	71
Strontium	0.301	0.35	22.5
Sulfur (rhombic)	0.710	0.27	—
Tantalum	0.140	0.58	6.3
Technetium	—	0.51	—
Terbium	0.182	0.11	10.3
Thallium	0.129	0.46	29.9
Thorium	0.113	0.540	11.0
Thulium	0.160	0.17	13.3
Tin	0.228	0.67	22.0
Titanium	0.523	0.22	8.6
Tungsten	0.132	1.74	4.5
Uranium	0.116	0.28	13.9
Vanadium	0.489	0.31	8.4
Ytterbium	0.155	0.39	26.3
Yttrium	0.298	0.17	10.6
Zinc	0.388	1.16	30.2
Zirconium	0.278	0.227	5.7

Source: Adapted from Anderson, H. L., ed. (1989). "A Physicist's Desk Reference," Springer-Verlag, New York; with updates from Lide, D. R., ed. (1999). "CRC Handbook of Chemistry and Physics," CRC Press, Boca Raton, FL.

V. DISSEMINATION OF CRITICAL DATA

Traditionally, books and journals have served as the major vehicles for disseminating critically evaluated data to scientists and engineers. Several widely used series of tables have already been mentioned. The *Journal of Physical and Chemical Reference Data*, published jointly by the American Institute of Physics and the National Institute of Standards and Technology, is one of the major vehicles for disseminating tables of recommended data and documenting the methodology used for their evaluation. This journal is published bimonthly, with supplements appearing on an irregular basis. More specialized data journals also exist—for example, *Atomic Data and Nuclear Data Tables* (Academic Press) and *Journal of Phase Equilibria* (ASM International). Finally, many technical publishers offer monographs and handbooks containing evaluated data on specialized subjects.

TABLE VI Vapor Pressure of the Elements

	Element	1 Pa	10 Pa	100 Pa	1 kPa	10 kPa	100 kPa
		\multicolumn{6}{c}{Temperature (°C) for the indicated pressure[a]}					

	Element	1 Pa	10 Pa	100 Pa	1 kPa	10 kPa	100 kPa
Ag	Silver	1010	1140	1302	1509	1782	2160
Al	Aluminum	1209	1359	1544	1781	2091	2517
Ar	Argon	—	−226.4 s	−220.3 s	−212.4 s	−201.7 s	−186.0
As	Arsenic	280 s	323 s	373 s	433 s	508 s	601 s
At	Astatine	88 s	119 s	156 s	202 s	258 s	334
Au	Gold	1373	1541	1748	2008	2347	2805
B	Boron	2075	2289	2549	2868	3272	3799
Ba	Barium	638 s	765	912	1115	1413	1897
Be	Beryllium	1189 s	1335	1518	1750	2054	2469
Bi	Bismuth	668	768	892	1052	1265	1562
Br_2	Bromine	−87.7 s	−71.8 s	−52.7 s	−29.3 s	2.5	58.4
C	Carbon (graphite)	—	2566 s	2775 s	3016 s	3299 s	3635 s
Ca	Calcium	591 s	683 s	798 s	954	1170	1482
Cd	Cadmium	257 s	310 s	381	472	594	767
Ce	Cerium	1719	1921	2169	2481	2886	3432
Cl_2	Chlorine	−145 s	−133.7 s	−120.2 s	−103.6 s	−76.1	−34.2
Co	Cobalt	1517	1687	1892	2150	2482	2925
Cr	Chromium	1383 s	1534 s	1718 s	1950	2257	2669
Cs	Cesium	144.5	195.6	260.9	350.0	477.1	667.0
Cu	Copper	1236	1388	1577	1816	2131	2563
Dy	Dysprosium	1105 s	1250 s	1431	1681	2031	2558
Er	Erbium	1231 s	1390 s	1612	1890	2279	2859
Eu	Europium	590 s	684 s	799 s	961	1179	1523
F_2	Fluorine	−235 s	−229.5 s	−222.9 s	−214.8	−204.3	−188.3
Fe	Iron	1455 s	1617	1818	2073	2406	2859
Fr	Francium	131	181	246	335	465	673
Ga	Gallium	1037	1175	1347	1565	1852	2245
Gd	Gadolinium	1563	1755	1994	2300	2703	3262
Ge	Germanium	1371	1541	1750	2014	2360	2831
H_2	Hydrogen	—	—	—	—	−258.6	−252.8
He	Helium	—	—	—	—	−270.6	−268.9
Hf	Hafnium	2416	2681	3004	3406	3921	4603
Hg	Mercury	42.0	76.6	120.0	175.6	250.3	355.9
Ho	Holmium	1159 s	1311 s	1502	1767	2137	2691
I_2	Iodine	−12.8 s	9.3 s	35.9 s	68.7 s	108 s	184.0
In	Indium	923	1052	1212	1417	1689	2067
Ir	Iridium	2440 s	2684	2979	3341	3796	4386
K	Potassium	200.2	256.5	328	424	559	756.2
Kr	Krypton	−214.0 s	−208.0 s	−199.4 s	−188.9 s	−174.6 s	−153.6
La	Lanthanum	1732	1935	2185	2499	2905	3453
Li	Lithium	524.3	612.3	722.1	871.2	1064.3	1337.1
Lu	Lutetium	1633 s	1829.8	2072.8	2380	2799	3390
Mg	Magnesium	428 s	500 s	588 s	698	859	1088
Mn	Manganese	955 s	1074 s	1220 s	1418	1682	2060
Mo	Molybdenum	2469 s	2721	3039	3434	3939	4606
N_2	Nitrogen	−236 s	−232 s	−226.8 s	−220.2 s	−211.1 s	−195.9
Na	Sodium	280.6	344.2	424.3	529	673	880.2

Continues

TABLE VI (*Continued*)

	Element	1 Pa	10 Pa	100 Pa	1 kPa	10 kPa	100 kPa
						Temperature (°C) for the indicated pressure[a]		
Nb	Niobium	2669	2934	3251	3637	4120	4740
Nd	Neodymium	1322.3	1501.2	1725.3	2023	2442	3063
Ne	Neon	−261 s	−260 s	−258 s	−255 s	−252 s	−246.1
Ni	Nickel	1510	1677	1881	2137	2468	2911
O₂	Oxygen	—	—	—	−211.9	−200.5	−183.1
Os	Osmium	2887 s	3150	3478	3875	4365	4983
P	Phosphorus (white)	6 s	34 s	69	115	180	276
P	Phosphorus (red)	182 s	216 s	256 s	303 s	362 s	431 s
Pb	Lead	705	815	956	1139	1387	1754
Pd	Palladium	1448 s	1624	1844	2122	2480	2961
Po	Polonium	—	—	—	573	730.2	963.3
Pr	Praseodymium	1497.7	1699.4	1954	2298	2781	3506
Pt	Platinum	2057	2277	2542	2870	3283	3821
Pu	Plutonium	1483	1680	1925	2238	2653	3226
Ra	Radium	546 s	633 s	764	936	1173	1526
Rb	Rubidium	160.4	212.5	278.9	368	496.1	685.3
Re	Rhenium	3030 s	3341	3736	4227	4854	5681
Rh	Rhodium	2015	2223	2476	2790	3132	3724
Rn	Radon	−163 s	−152 s	−139 s	−121.4 s	−97.6 s	−62.3
Ru	Ruthenium	2315 s	2538	2814	3151	3572	4115
S	Sulfur	102 s	135	176	235	318	444
Sb	Antimony	534 s	603 s	738	946	1218	1585
Sc	Scandium	1372 s	1531 s	1733	1993	2340	2828
Se	Selenium	227	279	344	431	540	685
Si	Silicon	1635	1829	2066	2363	2748	3264
Sm	Samarium	728 s	833 s	967 s	1148	1402	1788
Sn	Tin	1224	1384	1582	1834	2165	2620
Sr	Strontium	523 s	609 s	717 s	866	1072	1373
Ta	Tantalum	3024	3324	3684	4122	4666	5361
Tb	Terbium	1516.1	1706.1	1928	2232	2640	3218
Tc	Technetium	2454	2725	3051	3453	3961	4621
Te	Tellurium	—	—	502	615	768.8	992.4
Th	Thorium	2360	2634	2975	3410	3986	4782
Ti	Titanium	1709	1898	2130	2419	2791	3285
Tl	Thallium	609	704	824	979	1188	1485
Tm	Thulium	844 s	962 s	1108 s	1297 s	1548	1944
U	Uranium	2052	2291	2586	2961	3454	4129
V	Vanadium	1828 s	2016	2250	2541	2914	3406
W	Tungsten	3204 s	3500	3864	4306	4854	5550
Xe	Xenon	−190 s	−181 s	−170 s	−155.8 s	−136.6 s	−108.4
Y	Yttrium	1610.1	1802.3	2047	2354	2763	3334
Yb	Ytterbium	463 s	540 s	637 s	774 s	993	1192
Zn	Zinc	337 s	397 s	477	579	717	912
Zr	Zirconium	2366	2618	2924	3302	3780	4405

[a] An "s" following an entry indicates the substance is solid at that temperature.

Source: Lide, D. R., ed. (1999). "CRC Handbook of Chemistry and Physics," CRC Press, Boca Raton, FL.

TABLE VII   Properties of Some Common Fluids

Formula	Fluid	Normal melting point (t_m/°C)	Normal boiling point (t_b/°C)	Critical constants (t_c/°C)	(p_c/°C)
He	Helium	—	−268.93	−267.96	0.23
Ar	Argon	−189.36[a]	−185.85	−122.28	4.9
H_2	Hydrogen	−259.34	−252.87	−240.18	1.29
O_2	Oxygen	−218.79	−182.95	−118.56	5.04
N_2	Nitrogen	−210	−195.79	−146.94	3.39
CO	Carbon monoxide	−205.02	−191.5	−140.24	3.5
CO_2	Carbon dioxide	−56.56[a]	−78.4[b]	30.98	7.38
H_2O	Water	0.00	100.0	373.99	22.06
NH_3	Ammonia	−77.73	−33.33	132.4	11.35
N_2O	Nitrous oxide	−90.8	−88.48	36.42	7.26
CH_4	Methane	−182.47	−161.4	−82.59	4.6
C_2H_6	Ethane	−182.79	−88.6	32.17	4.87
C_3H_8	Propane	−187.63	−42.1	96.68	4.25
C_4H_{10}	Butane	−138.3	−0.5	151.97	3.8
C_2H_4	Ethylene	−169.15	−103.7	9.19	5.04
C_6H_6	Benzene	5.49	80.09	288.9	4.9
CH_4O	Methanol	−97.53	64.6	239.4	8.08
C_2H_6O	Ethanol	−114.14	78.29	240.9	6.14
C_3H_6O	Acetone	−94.7	56.05	235.0	4.700

[a] Solid–liquid–gas triple point.
[b] Sublimation point, where vapor pressure of solid reaches 1 atm.
Source: Lide, D. R., ed. (1999). "CRC Handbook of Chemistry and Physics," CRC Press, Boca Raton, FL.

TABLE VIII   CODATA Key Values for Thermodynamics

Substance	State	Relative molecular mass	$\Delta_r H°$(298.15 K) (kJ mol^{-1})	$S°$(298.15 K) (J K^{-1} mol^{-1})	$H°$(298.15 K) − $H°$(0) (kJ mol^{-1})
O	Gas	15.9994	249.18 ± 0.10	160.950 ± 0.003	6.725 ± 0.001
O_2	Gas	31.9988	0	205.043 ± 0.005	8.680 ± 0.002
H	Gas	1.00794	217.998 ± 0.006	114.608 ± 0.002	6.197 ± 0.001
H^+	Aqueous	1.0074	0	0	—
H_2	Gas	2.0159	0	130.571 ± 0.005	8.468 ± 0.001
OH^-	Aqueous	17.0079	−230.015 ± 0.040	−10.90 ± 0.20	—
H_2O	Liquid	18.0153	−285.830 ± 0.040	69.95 ± 0.03	13.273 ± 0.020
H_2O	Gas	18.0153	−241.826 ± 0.040	188.726 ± 0.010	9.905 ± 0.005
He	Gas	4.00260	0	126.044 ± 0.002	6.197 ± 0.001
Ne	Gas	20.179	0	146.219 ± 0.003	6.197 ± 0.001
Ar	Gas	39.948	0	154.737 ± 0.003	6.197 ± 0.001
Kr	Gas	83.80	0	163.976 ± 0.003	6.197 ± 0.001
Xe	Gas	131.29	0	169.576 ± 0.003	6.197 ± 0.001
F	Gas	18.99840	79.38 ± 0.30	158.642 ± 0.004	6.518 ± 0.001
F^-	Aqueous	18.9989	−335.35 ± 0.65	−13.8 ± 0.8	—
F_2	Gas	37.9968	0	202.682 ± 0.005	8.825 ± 0.001
HF	Gas	20.0063	−273.30 ± 0.70	173.670 ± 0.003	8.599 ± 0.001
Cl	Gas	35.453	121.301 ± 0.008	165.081 ± 0.004	6.272 ± 0.001
Cl^-	Aqueous	35.4535	−167.080 ± 0.10	56.60 ± 0.20	—

Source: Excerpted from Cox, J. D., Wagman, D. D., and Medvedev, V. A. (1989). "CODATA Key Values for Thermodynamics," Hemisphere Publishing, New York.

There are also many handbooks with a broad coverage of physical and chemical data; among the most familiar of these are the *CRC Handbook of Chemistry and Physics, The Merck Index,* and the *American Institute of Physics Handbook.* Such handbooks are very convenient data sources. While they cannot provide the backup documentation found in the data journals and monographs discussed above, the better ones carry references to more detailed publications.

The decade beginning in 1990 saw a major change in the manner of disseminating all types of information, and scientific data were no exception. There are many advantages associated with computerized data dissemination. One consideration is economic. While the costs incurred with composition and printing have continued to increase, computer costs for data storage and network communications have decreased sharply, thus making the electronic dissemination of critical data more attractive. Often the sheer volume of data makes a machine-readable format the only practical way of storage and distribution. Electronic databases lend themselves to easy updating, thus promoting the currency of the data, and search and retrieval are far more powerful. Having the data in electronic form also makes it easier for the user to carry out calculations and look for trends that may lead to new scientific insights.

Although these advantages were recognized much earlier, the transition to electronic dissemination of scientific data did not begin to accelerate until the mid-1990s. Two factors have contributed: the expanding availability of personal computers with CD ROM drives and the explosive growth of the Internet. The CD ROM has proved to be an efficient means for distributing physical and chemical databases and the accompanying software to individuals for use on their personal computers. The Internet provides an inexpensive way for users to access large databases maintained on institutional computers. The graphical capabilities of the World Wide Web have also contributed by making it easy to display special characters, chemical structures, and other non-text information. Finally, the growing use of computers for data analysis, process simulation, engineering design, and similar applications has created a demand for data in digital, as opposed to paper, format.

The ease with which information can be posted on the Internet has had one unfortunate consequence. There are a great many sites that purport to provide physical and chemical data, but the quality is highly variable. Data quality is a consideration even when dealing with printed compilations, but on the Internet the traditional filter of the publication process can no longer be relied upon. A search for a specific property on a standard Internet search engine is likely to turn up hundreds of sites, most of which have no documentation and provide no basis for confidence in the correctness of the data presented. It is important for all users of data taken from the Internet to evaluate the reliability of the source and assure that it is truly critical data.

Some of the important World Wide Web sites for evaluated physical and chemical data are listed below:

- NIST Physics Data, covering fundamental constants, atomic spectra, and X-ray data; <physics.nist.gov>
- Fundamental Particle Properties, prepared by the Particle Data Group at Lawrence Berkeley Laboratories; <pdg.lbl.gov>
- NIST Chemistry Webbook, whose topics include thermodynamics, ion energetics, infrared and mass spectra, fluid properties, etc; <webbook.nist.gov>
- Beilstein and Gmelin Databases, covering chemical properties of organic and inorganic compounds; <www.beilstein.com>
- Hazardous Substances Data Bank, maintained by the National Library of Medicine and containing physical property data as well as toxicity and safety data; <chem.sis.nlm.nih.gov/hsdb/>
- CRCnetBase, including the Web version of the CRC Handbook of Chemistry and Physics, The Merck Index, and other databases; <www.crcpress.com>

Crystallographic databases are maintained for different classes of materials:

- Organic compounds: <www.ccdc.cam.ac.uk>
- Inorganic compounds: <www.nist.gov/srd/> and <www.fiz-karlsruhe.de>
- Metals: <www.tothcanada.com>
- Proteins: <www.rcsb.org>
- Nucleic acids: <www.ndbserver.rutgers.edu>

SEE ALSO THE FOLLOWING ARTICLES

CHEMICAL THERMODYNAMICS • DATABASES • MECHANICS, CLASSICAL • PERIODIC TABLE (CHEMISTRY) • THERMAL ANALYSIS • THERMOMETRY

BIBLIOGRAPHY

Anderson, H. L., ed. (1989). "A Physicist's Desk Reference," 2nd ed., Amer. Inst. of Phys., New York.

Dubois, J.-E., and Gershon, N., eds. (1996). "The Information Revolution: Impact on Science and Technology," Springer-Verlag, Berlin.

Glaeser, P. S., ed. (1992). "Data for Discovery," Begell House, New York.

Lide, D. R. (1973). "Status report on critical compilation of physical chemical data." *Ann. Rev. Phys. Chem.* **24,** 135–158.

Lide, D. R. (1981). "Critical data for critical needs." *Science* **212,** 135–158.

Maizell, R. E. (1998). "How To Find Chemical Information," Wiley-Interscience, New York.

Molino, B. B. (1985). *In* "The Role of Data in Scientific Progress" (P. S. Glaser, ed.), North-Holland, Amsterdam.

Rumble, J. R., and Hampel, V. E. (1984). "Database Management in Science and Technology," North-Holland, Amsterdam.

Cryogenic Process Engineering

Klaus D. Timmerhaus
University of Colorado

GLOSSARY

Coefficient of performance Criterion used to compare refrigerator performance.

Cryobiology The use of cryogens in applications of biology.

Cryocooler Small cryogenic refrigerators and liquefiers.

Cryogen Any fluid that becomes a liquid below 125 K.

Cryogenics Term commonly associated with activities performed below 125 K.

Cryomedicine The use of cryogens in various medical and surgical applications.

Cryopumping Procedure for attaining ultrahigh vacuum by the freezing of residual gases in a chamber.

Dewar Any cryogenic container shielded by a vacuum-insulated space.

Figure of merit Criterion used to compare liquefier performance.

Helium I Normal form of helium-4 liquid existing above 2.17 K.

Helium II Superfluid form of helium-4 liquid existing below 2.17 K.

Joule–Thomson expansion Expression representing an isenthalpic throttling process.

n-Hydrogen Equilibrium hydrogen existing at ambient temperatures consisting of 75% orthohydrogen and 25% parahydrogen.

MLI A high-vacuum, multilayered insulation used in cryogen storage and transfer.

Orthohydrogen Higher energy form of hydrogen resulting from the nuclear spin of the two protons in the same direction.

Parahydrogen Lower energy form of hydrogen resulting from the nuclear spin of the two protons in the opposite direction.

Regenerator Heat exchanger that periodically stores and releases heat from a matrix of high heat capacity.

Superconductivity Simultaneous disappearance of electrical resistivity and the appearance of perfect diamagnetism in a material.

Encyclopedia of Physical Science and Technology, Third Edition, Volume 4

Superfluid Designation for helium II existing below 2.17 K and exhibiting negligible viscosity.

CRYOGENICS is a term commonly associated with low temperatures. However, the point on the temperature scale at which refrigeration in the ordinary sense of the term ends and cryogenic engineering begins is not well defined. Most scientists and engineers working in cryogenic engineering restrict this term to a temperature range below 125 K. This is a reasonable dividing line since the normal boiling points of the more permanent gases, such as helium, hydrogen, neon, nitrogen, oxygen, and air, lie below 125 K, while the more common refrigerants have boiling points above this temperature. Thus, cryogenic process engineering is concerned with the industrial development, utilization, and improvement of low-temperature techniques, processes, and equipment.

I. GENERAL APPLICATIONS

The industrial production and utilization of temperatures below 125 K are commonly referred to as *cryogenic engineering* or *cryogenic process engineering*. This field of endeavor has grown significantly since World War II. It is now a major business in the United States with a national value in excess of $2.5 billion annually, based on the previously defined temperature range. If the definition is broadened slightly to include the production of some petrochemicals that utilize low-temperature processing in their manufacture, such as ethylene, the annual value rapidly escalates to over $12 billion.

An examination of cryogenic engineering shows it to be a very diverse supporting technology, a means to an end and not an end in itself. For example, oxygen, one of the most important industrial gases, is obtained by the low-temperature separation of air. Fifty percent of the oxygen produced in this manner is used by the steel industry to reduce the cost of high-grade steel, while another 20% is used in the chemical process industry to produce a variety of oxygenated compounds. Liquid hydrogen production since the mid-1950s has risen from laboratory quantities to a level of more than 250 tons/day (227,000 kg/day). Similarly, the need for liquid helium has increased by more than a factor of 15, requiring the construction of large plants to separate helium from natural gas by cryogenic means. Demands for energy have likewise accelerated the construction of tonnage base load liquefied natural gas (LNG) plants around the world and have been responsible for the associated domestic LNG industry of today with its use of peak-shaving plants.

An introduction of cryogenics would be incomplete without brief mention of some of the many current applications. For example, the phenomenon of superconductivity occurring at low temperatures has been successfully exploited in the development of high-field magnets for various uses. Space simulation is another application using a low-temperature concept. In this case, cryopumping, or the freezing of residual gases in a chamber on a cold surface, is used to provide the ultrahigh vacuum representative of outer space. This concept has been encompassed in several commercial vacuum pumps.

Freezing as a means of preserving food dates back to 1840. However, today the food industry uses large amounts of liquid nitrogen for this purpose and as a refrigerant in frozen-food transport systems. The use of cryogenics in biology and medicine has generated such interest that work in these low-temperature areas is now identified as cryobiology and cryomedicine, respectively. For example, liquid nitrogen-cooled containers are routinely used to preserve whole blood, bone marrow, and animal semen for extended periods of time. Liquid helium is used to cool the magnets in the MRI units employed by most modern hospitals. Cryogenic surgery is an acceptable procedure for curing such involuntary disorders as Parkinson's disease. Finally, one must recognize the role of cryogenics in the chemical processing industry with the treatment of natural gas streams to recover valuable heavy components or upgrade the heat content of fuel gas, the recovery of useful components from air, and the purification of various process streams.

II. LOW-TEMPERATURE PROPERTIES

Familiarity with the properties and behavior of materials used in any system operating at low temperatures is essential for proper design considerations. Since there are several significant effects among materials that become evident only at low temperatures, it is risky to obtain needed properties by an extrapolation of the variation in properties observed at ambient conditions. For example, the vanishing of specific heats, the phenomenon of superconductivity, and the onset of ductile–brittle transitions in carbon steel cannot be inferred from property measurements obtained at ambient temperatures. Accordingly, there is no substitute for test data on a truly representative sample specimen when designing for the limit of effectiveness of a cryogenic material or structure.

A. Fluid Properties

Numerous tabulations of thermodynamic property data are available in the literature. For example, a very recent tabulation of thermodynamic data by Jacobsen, et al (1997) covers all of the cryogenic fluids of interest. Sufficient detail on the models used for each fluid is available so

TABLE I Selected Properties of Cryogenic Liquids at Normal Boiling Point

Saturated liquid property at 0.1 MPa	Helium-4	Hydrogen[a]	Neon	Nitrogen	Air	Fluorine	Argon	Oxygen	Methane
Normal boiling point, K	4.2	20.4	27.1	77.3	78.9	85.3	87.3	90.2	111.7
Critical temperature, K	5.2	33.2	44.4	126.1	133.3	118.2	150.7	154.6	190.7
Critical pressure, MPa	0.23	1.31	2.71	3.38	3.90	5.55	4.87	5.06	4.63
Temperature at triple point, K	b	13.9	19.0	63.2	—	53.5	83.8	54.4	88.7
Pressure at triple point, MPa × 10³	c	7.2	43.0	12.8	—	0.22	68.6	0.15	10.1
Density, kg/m³	124.9	70.9	1204	810.8	874.0	1505	1403	1134	425.0
Heat of vaporization, kJ/kg	20.7	446.3	86.6	198.4	205.1	166.5	161.6	213.1	509.7
Specific heat, kJ/kg · K	4.56	9.78	1.84	2.04	1.97	1.55	1.14	1.70	3.45
Viscosity, (kg/m · sec) × 10⁶	3.57	13.06	124.0	157.9	80.6	244.7	252.1	188.0	118.6
Thermal conductivity, (kJ/m · sec · K) × 10³	0.027	0.118	0.130	0.139	—	0.135	0.123	0.148	0.111
Dielectric constant	1.0492	1.226	—	1.434	—	1.43	1.52	1.4837	1.6758

[a] Equilibrium hydrogen.

[b] λ-Point temperature, 2.17 K.

[c] λ-Point pressure, 5.02×10^{-3} MPa.

that the user may program the formulations in any appropriate computer language or format consistent with a particular application. Selected property data for some common cryogens are presented in Table I. Unique properties of several of these cryogens are noted below.

Liquid helium-4 has some very unusual properties since it can exist in two different liquid phases, namely, liquid helium I and liquid helium II (Fig. 1). The former is labeled the normal fluid, while the latter has been designated the superfluid since under certain conditions the fluid acts as if it had no viscosity. The phase transition between the two liquid phases is identified as the λ line. Intersection of the latter with the vapor-pressure curve is known as the λ point. Helium-4 has no triple point and requires a

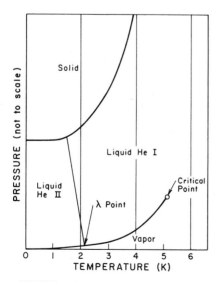

FIGURE 1 Phase diagram for helium-4.

pressure of 2.5 MPa or more even to exist as a solid below a temperature of 3 K.

Other properties of helium-4 show similar surprises. At the λ point, the specific heat of the liquid increases to a large value as the temperature is decreased through this point. Once below the λ point, the specific heat of helium II rapidly decreases to zero. The thermal conductivity of helium I, on the other hand, decreases with decreasing temperature. However, once the transition to helium II has been made, the thermal conductivity of the liquid can increase in value by as much as 10^6 that of helium I.

A unique property of hydrogen is that it can exist in two different molecular forms, namely, orthohydrogen and parahydrogen. The ortho and para forms differ in the relative orientation of the nuclear spins of the two atoms associated with the diatomic molecule. The thermodynamic equilibrium composition of the ortho and para varieties is temperature dependent. At ambient temperatures, the equilibrium mixture is 75% orthohydrogen and 25% parahydrogen and is designated as normal hydrogen. With decreasing temperatures, the thermodynamic equilibrium shifts to essentially 100% parahydrogen at 20.4 K. the normal boiling point of hydrogen. The conversion from normal hydrogen to parahydrogen is exothermic and evolves sufficient energy to vaporize ~1% of the stored liquid per hour. To minimize such losses in the commercial production of liquid hydrogen, a catalyst is used to effect the conversion from normal hydrogen to the thermodynamic equilibrium concentration during the liquefaction process.

The two forms of hydrogen have different specific heats. This difference, in turn, affects other thermal and transport properties of hydrogen. For example, parahydrogen gas has a higher thermal conductivity than orthohydrogen gas because of the higher specific heat of the parahydrogen gas.

In contrast to other cryogens, liquid oxygen is slightly magnetic. It is also chemically very reactive with hydrocarbon materials. It thus presents a safety problem and requires extra precautions in handling.

Fluorine is characterized by its high toxicity and extreme reactivity. The fatal concentration range for animals is 200 ppm × hr, while the maximum allowable dosage for humans is usually considered to be 1 ppm × hr.

B. Thermal Properties

The thermal properties of most interest at low temperatures are specific heat, thermal conductivity, and thermal expansivity.

1. Specific Heat

Specific heat can be predicted fairly accurately by mathematical models through statistical mechanics and quantum theory. For solids, the Debye model gives a satisfactory representation of the specific heat with temperature. Difficulties, however, are encountered when the Debye theory is applied to alloys and compounds. Plastics and glasses are other classes of solids that fail to follow this theory. In such cases, only experimental test data will provide sufficiently reliable specific heat values.

In general, the specific heat of cryogenic liquids decreases in a manner similar to that noted for crystalline solids as the temperature is lowered. At low pressures, the specific heat decreases with a decrease in temperature. However, at high pressures in the neighborhood of the critical, humps in the specific-heat curve are also observed for all normal cryogens.

2. Thermal Conductivity

Adequate predictions of thermal conductivity for pure metals can be made by means of the Wiedemann–Franz law, which states that the ratio of the thermal conductivity to the product of the electrical conductivity and the absolute temperature is a constant. High-purity aluminum and copper exhibit peaks in thermal conductivity between 20 and 50 K, but these peaks are rapidly suppressed with increased impurity levels and cold work of the metal. The aluminum alloys Inconel, Monel, and stainless steel show a steady decrease in thermal conductivity with a decrease in temperature. This behavior makes these structural materials useful in any cryogenic service that requires low thermal conductivity over an extended temperature range.

All cryogenic liquids except hydrogen and helium have thermal conductivities that increase as the temperature is decreased. For these two exceptions, the thermal conductivity decreases with a decrease in temperature. The kinetic theory of gases correctly predicts the decrease in thermal conductivity of all gases as the temperature is lowered.

3. Thermal Expansivity

The expansion coefficient of a solid can be estimated with the aid of an approximate thermodynamic equation of state for solids that equates the thermal expansion coefficient β with the quantity $\gamma C_v \rho / B$, where γ is the Grüneisen dimensionless ratio. C_v the specific heat of the solid, ρ the density of the material, and B the bulk modulus. For face-centered cubic (fcc) metals, the average value of the Grüneisen constant is $\sim$2.3. However, there is a tendency for this constant to increase with atomic number.

C. Electrical and Magnetic Properties

1. Electrical Resistivity

The electrical resistivity of most pure metallic elements at ambient and moderately low temperatures is approximately proportional to the absolute temperature. At very low temperatures, however, the resistivity (except that of superconductors) approaches a residual value almost independent of temperature. Alloys, on the other hand, have resistivities much higher than those of their constituent elements and resistance–temperature coefficients that are quite low. The electrical resistivity as a consequence is largely independent of temperature and may often be of the same magnitude as the room-temperature value.

The insulating quality of solid electrical conductors usually improves as the temperature is lowered. In fact, all the common cryogenic fluids are good electrical insulators.

2. Superconductivity

The phenomenon of superconductivity involving the simultaneous disappearance of all electrical resistance and the appearance of diamagnetism is undoubtedly the most distinguishing characteristic of cryogenics. The Bardeen–Cooper–Schriefer (BCS) theory has been successful in accounting for most of the basic features observed of the superconducting state for low-temperature superconductors (LTS) operating below 23 K. The advent of the ceramic high-temperature superconductors (HTS), operating between 77 and 125 K, has called for modifications to existing theories that still have not been finalized. The list of materials whose superconducting properties have been measured extends into the thousands.

Three important characteristics of the superconducting state are the critical temperature, the critical magnetic field, and the critical current. These parameters can be varied by using different materials or giving them special metallurgical treatments. For pure, unstrained metals, the normal (atmospheric) transition temperature from the superconducting state to the normal state is very sharp. For alloys, intermetallic compounds, and ceramics, the transition temperature can be quite large. Superconductivity in any of these materials, however, can be destroyed by subjecting the material either to an eternal or a self-induced magnetic field that exceeds a predetermined threshold field.

The alloy niobium–titanium (NbTi) and the intermetallic compound of niobium and tin (Nb_3Sn) are the most technologically advanced LTS materials presently available. Even though NbTi has a lower critical field and critical current density, it is often selected because its metallurgical properties favor convenient wire fabrication.

There are several families of high-temperature superconductors under investigation for practical magnet applications. Most of these HTS materials are copper oxide ceramics with varying oxygen contents. Becuase of their ceramic nature, HTS materials are quite brittle. This has introduced problems with some rather unique solutions relative to the fabrication of flexible wires that can be used in the windings of superconducting magnets.

Several of the low-temperature superconducting metals, such as lead, brass, and some solders (particularly lead–tin alloys), experience property changes when they become superconducting. Such changes can include specific heat, thermal conductivity, electrical resistance, magnetic permeability, and thermoelectric resistance. Consequently, the use of these superconducting metals in the construction of equipment for low-temperature operation must be evaluated carefully.

D. Mechanical Properties

A number of mechanical properties are of interest to the cryogenic engineer contemplating the design of a low-temperature facility. These properties include ultimate and yield strength, fatigue strength, impact strength, hardness, ductility, and elastic moduli.

1. Strength, Ductility, and Elastic Modulus

It is most convenient to classify metals by their lattice symmetry for low-temperature mechanical properties considerations. The fcc metals and their alloys are most often used in the construction of cryogenic equipment. Aluminum, copper, nickel, their alloys, and the austenitic stainless steels of the 18–8 type are fcc and do not exhibit an impact ductile-to-brittle transition at low temperatures. Generally, the mechanical properties of these metals im-

prove as the temperature is reduced. The yield strength at 20 K is considerably larger than at ambient temperature; Young's modulus is 5 to 20% larger at the lower temperatures, and fatigue properties, with the exception of 2024-T4 aluminum, are also improved at the lower temperatures. Since annealing of these metals and alloys can affect both the ultimate and yield strengths, care must be exercised under these conditions.

The body-centered cubic (bcc) metals and alloys are normally classified as undesirable for low-temperature construction. This class includes iron, the martensitic steels (low carbon and the 400 series of stainless steels), molybdenum, and niobium. If not brittle at room temperature, these materials exhibit a ductile-to-brittle transition at low temperatures. Cold working of some steels, in particular, can induce the austenite-to-martensite transition.

The hexagonal close-packed (hcp) metals exhibit mechanical properties intermediate between those of the fcc and bcc metals. For example, zinc suffers a ductile-to-brittle transition, whereas zirconium and pure titanium do not. The latter and its alloys have an hcp structure, remain reasonably ductile at low temperatures, and have been used for many applications where weight reduction and reduced heat leakage through the material have been important. However, small impurities of oxygen, nitrogen, hydrogen, and carbon can have a detrimental effect on the low-temperature ductility properties of titanium and its alloys.

Plastics increase in strength as the temperature is decreased, but this is also accompanied by a rapid decrease in elongation in a tensile test and a decrease in impact resistance. Teflon and glass-reinforced plastics retain appreciable impact resistance as the temperature is lowered. The glass-reinforced plastics also have high strength-to-weight and strength-to-thermal conductivity ratios. All elastomers, on the other hand, become brittle at low temperatures. Nevertheless, many of these materials, including rubber. Mylar, and nylon, can be used for static seal gaskets provided that they are highly compressed at room temperature before cooling.

The strength of glass under constant loading also increases with a decrease in temperature. Since failure occurs at a lower stress when the glass surface contains surface defects, the strength can be improved by tempering the surface.

III. REFRIGERATION AND LIQUEFACTION

Refrigeration in a thermodynamic process is accomplished when the process fluid absorbs heat at temperatures below that of the environment. Heat absorption at refrigerating temperatures can take place in totally different ways. If a low-temperature liquid is formed in the process, the heat that is absorbed evaporates the liquid, and refrigeration is accomplished at constant temperature. If the refrigerator is designed to reduce the process fluid to a cold gaseous state, the heat absorbed changes the sensible heat and consequently the temperature of the fluid.

In a continuous refrigeration process, there is no accumulation of refrigerant in any part of the system. This contrasts with a gas-liquefying system, where liquid accumulates and is withdrawn. Thus, in a liquefying system, the total mass of gas that is warmed and returned to the low-pressure side of the compressor is less than the gas to be cooled by the amount liquefied, creating an unbalanced flow in the heat exchangers. In a refrigerator, the warm and cool gas flows are usually equal in the heat exchangers, except where a portion of the flow is diverted through a work-producing expander. This results in what is usually referred to as "balanced flow condition" in a heat exchanger.

A process for producing refrigeration at cryogenic temperatures usually involves equipment at ambient temperature in which the process fluid is compressed and heat is rejected to a coolant. During the ambient temperature compression process, the enthalpy and entropy are decreased. At the cryogenic temperature where heat is absorbed, the enthalpy and entropy are increased. The reduction in temperature of the process fluid is usually accomplished by heat exchange between the cooling and warming fluid followed by an expansion. This expansion may take place either through a throttling device (isenthalpic expansion), where there is only a reduction in temperature, or in a work-producing device (isentropic expansion), where both temperature and enthalpy are decreased.

A. Isenthalpic Expansion

The simple Linde cycle shown in Fig. 2a provides a good example of an isenthalpic expansion process. In this process the gaseous refrigerant is initially compressed to approximate isothermal conditions by rejecting heat to a coolant. The compressed refrigerant is cooled in a heat exchanger by the stream returning to the compressor intake until it reaches the throttling valve. Joule–Thomson cooling on expansion further reduces the temperature until a portion of the refrigerant is liquefied. For a refrigerator, the unliquefied fraction and the vapor formed by liquid evaporation from the absorbed heat Q are warmed in the heat exchanger as they are returned to the compressor intake. Figure 2b shows this process on a temperature–entropy diagram. If one applies the first law to this refrigeration cycle and assumes no heat inleaks as well as negligible kinetic and potential energy changes in the re-

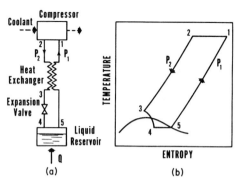

FIGURE 2 (a) Schematic for simple Linde-cycle refrigerator; (b) temperature–entropy diagram for cycle.

frigerant fluid, the refrigeration effect per unit mass of refrigerant compressed will simply be the difference in enthalpies of streams 1 and 2 of Fig. 2a. Thus, the coefficient of performance (COP) of the ideal simple Linde cycle is given by:

$$COP = \frac{Q_{ref}}{W} = \frac{h_1 - h_2}{T_1(s_1 - s_2) - (h_1 - h_2)} \quad (1)$$

where Q_{ref} is the refrigeration effect; W the work of compression; h_1 and h_2 the enthalpies at points 1 and 2, respectively; and s_1 and s_2 the entropies at points 1 and 2, respectively, of Fig. 2a.

For a simple Linde liquefier, the liquefied portion is continuously withdrawn from the reservoir, and only the unliquefied portion of the fluid is warmed in the countercurrent heat exchanger and returned to the compressor. The fraction y that is liquefied is obtained by applying the first law to the heat exchanger, throttling valve, and liquid reservoir. This results in:

$$y = (h_1 - h_2)/(h_1 - h_f) \quad (2)$$

where h_f is the specific enthalpy of the liquid being withdrawn. Note maximum liquefaction occurs when the difference between h_1 and h_2 is maximized. To account for heat inleak q_L, the relation is modified to:

$$y = (h_1 - h_2 - q_L)/(h_1 - h_f) \quad (3)$$

with a resultant decrease in the fraction liquefied. The work of compression is identical to that for the simple Linde refrigerator. The figure of merit (FOM), defined as $(W/m_f)_i/(W/m_f)$, where $(W/m_f)_i$ is the work of compression per unit mass liquefied for the ideal liquefier and (W/m_f) the work of compression per unit mass liquefied for the simple Linde cycle, reduces to the expression:

$$FOM = \left[\frac{T_1(s_1 - s_f) - (h_1 - h_f)}{T_1(s_1 - s_2) - (h_1 - h_2)} \right] \left(\frac{h_1 - h_2}{h_1 - h_f} \right) \quad (4)$$

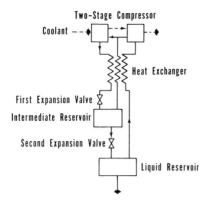

FIGURE 3 Liquefier using dual-pressure process.

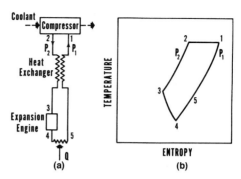

FIGURE 4 (a) Schematic for isentropic gas expansion refrigerator; (b) temperature–entropy diagram for cycle.

Liquefaction by this cycle requires that the inversion temperature of the refrigerant be above the ambient temperature to provide cooling as the process is started. Auxiliary refrigeration is required if the simple Linde cycle is to be used to liquefy fluids whose inversion temperature is below ambient. Liquid nitrogen is the optimum auxiliary refrigerant for hydrogen and neon liquefaction systems, while liquid hydrogen is the normal auxiliary refrigerant for helium liquefaction systems.

To reduce the work of compression, a two-stage, or dual-pressure, process can be used whereby the pressure is reduced by two successive isenthalpic expansions (Fig. 3). Since the work of compression is approximately proportional to the logarithm of the pressure ratio and the Joule–Thomson cooling is roughly proportional to the pressure difference, there is a much greater reduction in compressor work than in refrigerating performance. Hence, the dual-pressure process produces a given amount of refrigeration with less energy input than the simple Linde cycle refrigerator in Fig. 2.

B. Isentropic Expansion

Refrigeration can always be produced by expanding the process fluid in an engine and causing it to do work. A schematic of a simple gas refrigerator using this principle and the corresponding temperature–entropy diagram are shown in Fig. 4. Gas compressed isothermally at ambient temperature is cooled countercurrently in a heat exchanger by the low-pressure gas being returned to the compressor intake. Further cooling takes place during the work-producing expansion. In practice, this expansion is never truly isentropic, and this is reflected by path 3–4 on the temperature–entropy diagram (Fig. 4b).

Since the temperature in a work-producing expansion is always reduced, cooling does not depend on being below the inversion temperature before expansion. In large machines, the work produced during expansion is conserved.

In small refrigerators, the energy from the expansion is usually expended in a gas or hydraulic pump or other suitable work-absorbing device.

The refrigerator in Fig. 4a produces a cold gas, which absorbs heat from 4–5 and provides a method of refrigeration for obtaining temperatures other than those at the boiling points of cryogenic fluids.

C. Combined Isenthalpic and Isentropic Expansion

It is not uncommon to combine the isentropic and isenthalpic expansions to allow the formation of liquid in the refrigerator. This is done because of the technical difficulties associated with forming liquid in the engine. The Claude cycle is an example of a combination of these methods and is shown in Fig. 5a along with the corresponding temperature–entropy diagram (Fig. 5b).

One modification of the Claude cycle that has been used extensively in high-pressure liquefaction plants for air is the Heylandt cycle. In this cycle, the first warm heat exchanger in Fig. 5a has been eliminated, permitting the inlet of the expander to operate with ambient temperature

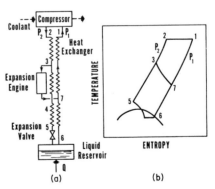

FIGURE 5 (a) Schematic for combined isenthalpic and isentropic expansion refrigerator; (b) temperature–entropy diagram for cycle.

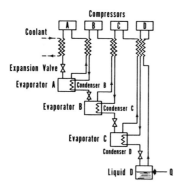

FIGURE 6 Cascade compressed vapor refrigerator.

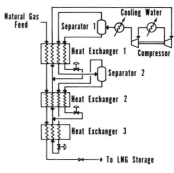

FIGURE 8 Mixed refrigerant cycle for liquefying natural gas.

seals, thereby minimizing lubrication problems. Another modification of the basic Claude cycle is the dual-pressure cycle utilizing the same principle as shown for the simple Linde cycle in Fig. 3. Still another extension of the Claude cycle is the Collins helium liquefier. Depending on the helium inlet pressure, from two to five expansion engines are used to provide the cooling needed in the system.

D. Mixed Refrigerant Cycle

Another cycle that has been used exclusively for large natural gas liquefaction plants is the mixed refrigerant cycle. Since this cycle resembles the classic cascade cycle in principle, it can best be understood by reference to a simplified flow sheet of that cycle presented in Fig. 6.

After purification, the natural gas stream is cooled successively by vaporization of propane, ethylene, and methane. Each of these gases, in turn, has been liquefied in a conventional refrigeration loop. Each refrigerant may be vaporized at two or three pressure levels to increase the natural gas cooling efficiency, but at a cost of considerably increased process complexity.

Cooling curves for natural gas liquefaction by the cascade process are shown in Fig. 7a,b. It is evident that the cascade cycle efficiency can be considerably improved by increasing the number of refrigerants or the number

of pressure levels employed. The actual work required for the nine-level cascade cycle depicted in Fig. 7b is ~80% of that required by the three-level cascade cycle depicted in Fig. 7a for the same throughput. The cascade system can be adapted to any cooling curve; that is, the quantity of refrigeration supplied at the various temperature levels can be chosen so that the temperature differences in the evaporators and heat exchangers approach a practical minimum (smaller temperature differences result in lower irreversibility and therefore lower power consumption).

The mixed refrigerant cycle (Fig. 8) is a variation of the cascade cycle just described and involves the circulation of a single mixed refrigerant stream, which is repeatedly condensed, vaporized, separated, and expanded. As a result, such processes require more sophisticated design methods and more complete knowledge of the thermodynamic properties of gaseous mixtures than expander or cascade cycles. Also, such processes must handle two-phase mixtures in heat exchangers. Nevertheless, simplification of the compression and heat exchange services in such cycles generally offers potential for reduced capital expenditure over conventional cascade cycles.

E. Cryocoolers

Mechanical coolers are generally classified as regenerative or recuperative. Regenerative coolers use reciprocating components that periodically move the working fluid back and forth in a regenerator. The recuperative coolers, on the other hand, use countercurrent heat exchangers to perform the heat-transfer operation. The Stirling and Gifford–McMahon cycles are typically regenerative coolers, while the Joule–Thomson and Brayton cycles are associated with recuperative coolers.

The past few years have witnessed an enhanced interest in pulse tube cryocoolers following the achievement by TRW of high-efficiency, long-life pulse tube cryocoolers based on the flexure-bearing, Stirling-cooler compressors developed at Oxford University. This interest has initiated

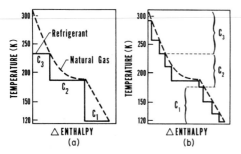

FIGURE 7 Three-level (a) and nine-level (b) cascade cycle cooling curves for natural gas.

the development of long-life, low-cost cryocoolers for the emerging high-temperature superconductor electronic market.

During this same time period, hydrogen sorption cryocoolers have achieved their first successful operation in space, and closed-cycle, helium, Joule–Thomson cryocoolers have continued to make progress in promising long-life space applications in the 4 K temperature range. In the commercial area, Gifford–McMahon cryocoolers with rare earth regenerators have made significant progress in opening up the 4 K market.

Mixtures of highly polar gases are receiving considerable attention as refrigerants for Joule–Thomson (J–T) cycles since the magnitude of the J–T coefficient increases with nonideality of the gas. New closed-cycle J–T or throttle-cycle refrigerators have taken advantage of these mixed refrigerants to achieve low-cost cryocooler systems in the 65 to 80 K temperature range. Microminiature J–T cryocoolers have also been developed over the past decade using these mixed refrigerants. Fabrication of these cryocoolers uses a photolithography process in which gas channels for the heat exchangers, expansion capillary, and liquid reservoir are etched on planar glass substrates that are fused together to form the sealed refrigerator. These microminiature refrigerators have been fabricated in a wide range of sizes and capacities.

Because of the rapidly increasing availability of cryocoolers, numerous new applications have become possible; many of these involve infrared imaging systems, spectroscopy, and high-temperature superconductors in the medical and communication fields. Many of these applications have required additional control of cryocooler-generated vibration and EMI susceptibility.

F. Comparison of Refrigeration and Liquefaction Systems

A thermodynamic measure of the quality of a low-temperature refrigeration and liquefaction system is its reversibility. The second law, or more precisely the entropy increase, is an effective guide to the degree of irreversibility associated with such a system. However, to obtain a clearer picture of what these entropy increases mean, it has become convenient to relate such an analysis to the additional work required to overcome these irreversibilities. The fundamental equation for such an analysis is

$$W = W_{rev} + T_0 \sum \dot{m} \Delta s \qquad (5)$$

where the total work W is the sum of the reversible work W_{rev} plus a summation of the losses in availability for vari-

ous steps in the analysis. Here, T_0 is the reference temperature (normally ambient), $\dot{m}$ the mass flow rate through the system, and Δs the change in entropy through the system.

Numerous analyses and comparisons of refrigeration and liquefaction cycles are presented in the literature. Great care must be exercised in accepting these comparisons since it is quite difficult to place all processes on a strictly comparable basis. Many assumptions are generally made in the course of these calculations, and these can have considerable effect on the conclusions. Assumptions that generally have to be made include heat leak, temperature differences in the exchangers, efficiencies of compressors and expanders, number of stages of compression, fraction of expander work recovered, state of expander exhaust, purity and condition of inlet gases, and pressure drop in the various streams. In view of this fact, differences in power requirements of 10 to 20% can be due to differences in assumed variables and can negate the advantage of one cycle over another. A comparison that demonstrates this point rather well is shown in Table II, which lists some common liquefaction systems described earlier using air as the working fluid and based on an inlet gas temperature and pressure of 294.4 K and 0.1 MPa, respectively.

IV. SEPARATION AND PURIFICATION OF GASES

The major industrial application of low-temperature processes involves the separation and purification of gases. Much of the commercial oxygen and nitrogen, and all the neon, argon, krypton, and xenon, are obtained by the distillation of liquid air. Commercial helium is separated from helium-bearing natural gas by a well-established low-temperature process. Cryogenics has also been used commercially to separate hydrogen from various sources of impure hydrogen. The low-boiling, valuable components of natural gas—namely, ethane, ethylene, propane, propylene, and others—are recovered and purified by various low-temperature schemes.

The separation of these gases is dictated by the Gibbs phase rule. The degree to which they separate is based on the physical behavior of the liquid and vapor phases. This behavior is governed, as at ambient temperatures, by Raoult's and Dalton's laws.

A. Air Separation

The simplest air separation device is the Linde single-column system, which utilizes the simple Linde

TABLE II Comparison of Several Liquefaction Systems Using Air as the Working Fluid[a]

Air liquefaction system[b]	Liquid yield ($y = \dot{m}_f/\dot{m}$)	Work per unit mass liquefied, (kJ/kg)	Figure of merit
Ideal reversible system	1.000	715	1.000
Simple Linde system, $p_2 = 20$ MPa, $\eta_c = 100\%$, $\varepsilon = 1.0$	0.086	5240	0.137
Simple Linde system, $p_2 = 20$ MPa, $\eta_c = 70\%$, $\varepsilon = 0.95$	0.061	10620	0.068
Simple Linde system observed	—	10320	0.070
Precooled simple Linde system, $p_2 = 20$ MPa, $T_3 = 228$ K, $\eta_c = 100\%$, $\varepsilon = 1.00$	0.179	2240	0.320
Precooled simple Linde system, $p_2 = 20$ MPa, $T_3 = 228$ K, $\eta_c = 70\%$, $\varepsilon = 0.95$	0.158	3700	0.194
Precooled simple Linde system, observed	—	5580	0.129
Linde dual-pressure system, $p_3 = 20$ MPa, $p_2 = 6$ MPa, $i = 0.8$, $\eta_c = 100\%$, $\varepsilon = 1.00$	0.060	2745	0.261
Linde dual-pressure system, $p_3 = 20$ MPa, $P_2 = 6$ MPa, $i = 0.8$, $\eta_c = 70\%$, $\varepsilon = 0.95$	0.032	8000	0.090
Linde dual-pressure system, observed	—	6340	0.113
Linde dual-pressure system, precooled to 228 K, observed	—	3580	0.201
Claude system, $p_2 = 4$ MPa, $x = \dot{m}_e/\dot{m} = 0.7$ $\eta_c = \eta_e = 100\%$, $\varepsilon = 1.00$	0.260	890	0.808
Claude system, $p_2 = 4$ MPa, $x = \dot{m}_e/\dot{m} = 0.7$, $\eta_c = 70\%$, $\eta_{e,ad} = 80\%$, $\eta_{e,m} = 90\%$, $\varepsilon = 0.95$	0.189	2020	0.356
Claude system, observed	—	3580	0.201
Cascade system, observed	—	3255	0.221

[a] Inlet conditions of 294.4 K and 0.1 MPa.

[b] η_c denotes compressor overall efficiency; η_c expander overall efficiency; $\eta_{e,ad}$ expander adiabatic efficiency; $\eta_{e,m}$ expander mechanical efficiency; and ε heat exchanger effectiveness. $i = m_1/m$ is mass in intermediate stream divided by mass through compressor, and $x = m_e/m$ is mass through expander divided by mass through compressor.

liquefaction cycle considered earlier but with a rectification column substituted for the liquid reservoir. (Since it is immaterial how the liquid is to be furnished to the column, any of the other liquefaction cycles could have been used in place of the simple Linde cycle.)

Although the oxygen product purity is high from a simple single-column separation scheme, the nitrogen effluent stream always contains about 6 to 7% oxygen. In other words, approximately one-third of the oxygen liquefied as feed to the column is lost in the nitrogen stream. This inherent loss of a valuable product in the single-column operation is not only undesirable but highly wasteful in terms of compression requirements.

This problem was solved by the introduction of the Linde double-column system. Two rectification columns are placed one on top of the other (hence the name double-column system). In this system, liquid air is introduced at an intermediate point in the lower column. A condenser–evaporator at the top of the lower column provides the reflux needed for the rectification process to obtain essentially pure nitrogen at this point. In order for the column to also deliver pure oxygen, the oxygen-rich liquid (~45% oxygen), from the boiler in the lower column is introduced at an intermediate level in the upper column. The reflux and the rectification process in the upper column produce pure oxygen at the bottom and pure nitrogen at the top (provided that argon and the rare gases have been previously removed). More than enough liquid nitrogen is produced in the lower column, so that some may be withdrawn and introduced in the upper column as needed reflux. Since the condenser must condense nitrogen vapor in the lower column by evaporating liquid oxygen in the upper column, it is necessary to operate the lower column at a higher pressure, ~0.5 MPa, while the upper column is operated at ~0.1 MPa. This requires throttling to reduce the pressure of the fluids from the lower column as they are transferred to the upper column.

The processes used in industrial air-separation plants have changed very little in basic principle during the past 25 years. After cooling the compressed air to its dew point in a main heat exchanger by flowing counter current to the products of separation, the air feed, at an absolute pressure of about 6 MPa, is separated in a double distillation column. This unit is kept cold by refrigeration developed in a turbine, which expands a flow equivalent to between 8 and 15% of the air-feed stream down to approximately atmospheric pressure.

Figure 9 shows a modern air-separation plant with front-end cleanup and product liquefaction. Production of such plants can exceed 2800 tons per day of liquid oxygen with an overall efficiency of about 15 to 20% of the theoretical optimum. The recent introduction of molecular sieve technology has provided an arrangement that increases the product to about 85% of the air input to the compressor. Thus, there has been a strong tendency over the past decade to retrofit older air-separation plants with this new arrangement to improve the process.

B. Rare Gas Recovery

Argon, neon, krypton, and xenon are recovered as products in commercial air separation plants. Since atmospheric air contains 0.93% argon with a boiling point intermediate between those of nitrogen and oxygen, the argon will appear as an impurity in either or both the nitrogen and the oxygen product of an air separation plant. Thus, removal of the argon is necessary if pure oxygen and nitrogen are desired from the air separation.

Figure 10 illustrates the scheme for removing and concentrating the argon. The upper column is tapped at the level where the argon concentration is highest in the column. Gas rich in argon is fed to an auxiliary column, where the argon is separated, and the remaining oxygen and nitrogen mixture is returned to the appropriate level in the primary column. The yield for this type of plant is about 50% of the atmospheric argon. The crude argon product generally contains 45% argon, 50% oxygen, and 5% nitrogen. The oxygen is readily removed by chemical reduction or adsorption. The remaining nitrogen impurity is of no consequence if the argon is to be used for filling incandescent lamps. However, for shielded-arc welding, the nitrogen must be removed by another rectification column.

Since helium and neon have boiling points considerably below that of nitrogen, these gases will collect on the nitrogen side of the condenser–reboiler associated with the double-column air separation system. Recovery of these gases is accomplished by periodic venting of a small portion of the gas from the dome of the condenser and transfer to a small condenser–rectifier refrigerated with

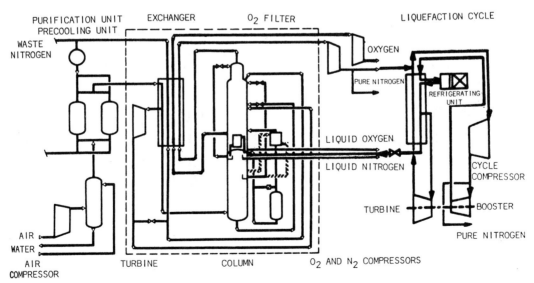

FIGURE 9 Schematic of a modern air-separation plant.

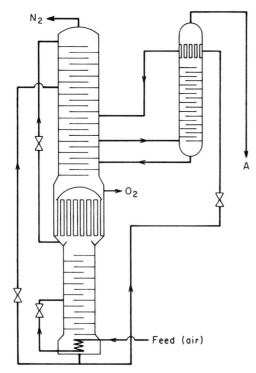

FIGURE 10 Air-separation plant with auxiliary argon separation column.

liquid nitrogen from the condenser of the double column. The resulting crude helium and neon are further purified by a series of charcoal adsorption units to provide high-purity neon.

The concentrations of krypton and xenon in atmospheric air are quite small. Thus, a very large amount of air has to be processed to produce an appreciable amount of these rare gases. Since the boiling points of krypton and xenon are higher than those of oxygen, these two components in atmospheric air tend to collect in the oxygen product of the double column. To recover these rare gases, liquid oxygen from the reboiler of the upper column is first sent to an auxiliary condenser–boiler to increase the concentration of the krypton and xenon. The product is further concentrated in another auxiliary rectification column before being vaporized and passed through a catalytic furnace to combine any remaining hydrocarbons with oxygen. The resulting water vapor and carbon dioxide are removed by a caustic trap, and the krypton and xenon are adsorbed in a silica gel trap. The krypton and xenon are then separated either in a small rectification column or by a series of adsorptions and desorptions on activated charcoal.

C. Helium Recovery

Most of the helium produced in the United States is obtained by recovering this component from helium-rich natural gas. Fortunately, since the major constituents of natural gas have boiling points considerably higher than that of helium, the separation can be accomplished with condenser–evaporators rather than with the more expensive rectification columns.

A typical scheme for separating helium from natural gas was pioneered by the U.S. Bureau of Mines. In this scheme, the natural gas is compressed to 4.13 MPa and treated to remove water vapor, carbon dioxide, and hydrogen sulfide. The purified stream is then partially condensed by the returning low-pressure, cold natural gas stream, throttled to a pressure of 1.72 MPa, and further cooled with cold nitrogen vapor in a heat exchanger–separator. where 98% of the gas is liquefied. The cold nitrogen vapor, supplied by an auxiliary refrigeration system, not only provides the necessary cooling but also causes some rectification of the gas phase in the heat exchanger, thereby increasing the helium content. The remaining vapor phase, consisting of about 60% helium and 40% nitrogen with a very small amount of methane, is warmed to ambient temperatures and sent to temporary storage pending further purification. The liquid phase, having been depleted of helium, is used to furnish the refrigeration required to cool and condense the incoming high-pressure gas. The process is completed by recompressing the stripped natural gas and returning it to the natural gas pipeline.

Purification of the crude helium is accomplished by compressing the gas to 18.6 MPa and cooling it first in a heat exchanger and then in a separator that is immersed in a bath of liquid nitrogen. In the separator nearly all of the nitrogen in the crude helium is condensed and removed as liquid. This liquid contains some dissolved helium, which is largely removed and returned to the process gas by reducing the pressure to 1.7 MPa and separating the resultant liquid and vapor phases in a nitrogen maker. Helium from the separator has a purity of ~98.5%. The final purification is accomplished by passing the cold helium through charcoal adsorption purifiers to remove the remaining nitrogen.

D. Natural Gas Processing

The need to recover increasing amounts of valuable hydrocarbon feedstocks from natural gas streams has resulted in expanded use of low-temperature processing of these streams. The majority of such natural gas processing is now accomplished using a turboexpander in a modified isentropic expansion cycle with feed gas normally available from 1 to 10 MPa. The first step is to dry the gas to dew points of 200 K and lower. After drying, the feed gas is cooled with cold residue gas. Liquid produced at this point is separated before entering the expander and sent to the condensate stabilizer. The gas from the separator flows to the expander. The expander exhaust stream can contain as much as 20 wt% liquid. This two-phase mixture is sent to the top section of the stabilizer, which separates the two phases. The liquid is used as reflux in this unit, while the cold gas exchanges heat with fresh feed and is recompressed by the expander-driven compressor. Many variations of this cycle are possible and have been used in actual plants.

E. Purification Schemes

The type and amount of impurities to be removed from a gas stream depend entirely on the type of process involved. For example, in the production of tonnage oxygen, various impurities must be removed to avoid plugging of the cold-process lines or to avoid buildup of hazardous contaminants. The impurities in air that contribute most to plugging are water and carbon dioxide. Helium, hydrogen, and neon, on the other hand, accumulate on the condensing side of the oxygen reboiler and reduce the rate of heat transfer unless removed by intermittent purging. The buildup of acetylene, on the other hand, can prove to be dangerous even if its feed concentration in the air does not exceed 0.04 ppm.

Refrigeration purification is a relatively simple method for removing water, carbon dioxide, and certain other contaminants from a process stream by condensation or freezing. (Either regenerators or reversing heat exchangers can be used for this purpose since a flow reversal is periodically necessary to reevaporate and remove the solid deposits.) The effectiveness of this method depends on the vapor pressure of the impurities relative to that of the major components of the process stream at the refrigeration temperature. Thus, assuming ideal gas behavior, the maximum impurity content in a gas stream after refrigeration would be inversely proportional to its vapor pressure. However, due to the departure from ideality at higher pressures, the impurity level generally will be considerably higher than that predicted for the ideal situation. Familiarity with these deviations is necessary if problems are to be avoided with this purification method.

One of the most common low-temperature methods for removing impurities involves the use of selective solid adsorbents. Such materials as silica gel, carbon, and synthetic zeolites (molecular sieves) are widely used as adsorbents because of their extremely large effective surface areas. Most of the gels and carbons have pores of various sizes in a given sample, but the synthetic zeolites can be manufactured with closely controlled pore size openings ranging from 0.4 to 1.3 nm. This additional selectivity is useful because it permits separation of gases on the basis of molecular size.

The design of low-temperature adsorbers requires knowledge of the rate of adsorption and the equilibrium conditions that exist between the solid and the gas as a function of temperature. The data for the latter are generally available from the suppliers of such adsorbents. The rate of adsorption is generally very rapid, and the adsorption is essentially complete in a relatively narrow zone of the adsorber. In usual plant operation at least two adsorption purifiers are employed—one in service while the other is being desorbed of impurities. In some cases there is an advantage in using three purifiers—one adsorbing, one desorbing, and one being cooled, with the latter two units being in series. The cooling of the purifier is generally performed using some of the purified gas to avoid adsorption during this period.

Low-temperature adsorption systems continue to find an increasing number of applications. For example, systems are used to remove the last traces of carbon dioxide and hydrocarbons in many air-separation plants. Adsorbents are also used in hydrogen liquefaction to remove oxygen, nitrogen, methane, and other trace impurities. They are also used in the purification of helium suitable for liquefaction (grade A) and for ultrapure helium (grade AAA, 99.999% purity). Adsorption at 35 K will, in fact, yield a helium with less than 2 ppb of neon, which is the only detectible impurity in helium after this treatment.

Even though most chemical purification methods are not carried out at low temperatures, they are useful in several cryogenic gas separation systems. Ordinarily water vapor is removed by refrigeration and adsorption methods. However, for small-scale purification, the gas can be passed over a desiccant, which removes the water vapor as water of crystallization. In the krypton–xenon purification system, carbon dioxide is removed by passage of the gas through a caustic, such as sodium hydroxide, to form sodium carbonate.

When oxygen is an impurity, it can be removed by reaction of the oxygen in the presence of a catalyst with hydrogen to form water. The latter then is removed by refrigeration or adsorption. Palladium and metallic nickel have proved to be effective catalysts for the hydrogen–oxygen reaction.

V. EQUIPMENT FOR CRYOGENIC PROCESSING

The achievement and utilization of low temperatures require the use of various specialized pieces of equipment including compressors, expanders, heat exchangers, pumps, transfer lines, and storage tanks. As a general rule, design principles applicable at ambient temperatures are also valid for low-temperature design. However, underlying each aspect of such a design must be a thorough understanding of the effect of temperature on the properties of the fluids being handled and the materials of construction being selected.

A. Compression Systems

Compression power accounts for more than 80% of the total energy required in the production of industrial gases and the liquefaction of natural gas. In order to minimize the cost and maintenance of cryogenic facilities, special care must be exercised to select the appropriate compression system. The three major types of compressors widely used today are reciprocating, centrifugal, and screw. Currently, there is no particular type of compressor that is generally preferred for all applications. The final selection will ultimately depend on the specific application as well as the effect of plant site and existing facilities.

1. Reciprocating Compressors

The key feature of reciprocating compressors is their adaptability to a wide range of volumes and pressures with high efficency. Some of the largest units for cryogenic gas production range up to 15,000 bhp (1.12×10^4 kW) and use the balanced–opposed machine concept in multistage designs with synchronous motor drive. When designed for multistage, multiservice operation, these units incorporate manual or automatic, fixed or variable, volume clearance pockets and externally actuated unloading devices where required. Balanced–opposed units not only minimize vibrations, resulting in smaller foundations, but also allow compact installation of coolers and piping, further increasing the savings.

Operating speeds of larger units are as high as 277 rpm with piston speeds for air service up to 4.3 m/sec. The larger compressors with provision for multiple services reduce the number of motors or drivers and minimize the accessory equipment, resulting in lower maintenance cost.

Compressors for oxygen service are characteristically operated at lower piston speeds, of the order of 3.3 m/sec. Maintenance of these machines requires rigid control of cleaning procedures and inspection of parts to ensure the absence of oil in the working cylinder and valve assemblies.

Engine drivers of the variable-speed type can generally operate over a 100 to 50% variation in the design speed with little loss in operating efficiency since compressor fluid friction losses decrease at the lower revolutions per minute.

2. Centrifugal Compressors

Technological advances achieved in centrifugal compressor design have resulted in improved high-speed compression equipment with capacities exceeding 280 m^3/sec in a single unit. As a consequence of their high efficiency, better reliability, and design upgrading, centrifugal compressors have become accepted for low-pressure cryogenic processes such as air-separation and base-load LNG plants.

Separately driven centrifugal compressors are adaptable to low-pressure cryogenic systems because they can be coupled directly to steam turbine drives, are less critical from the standpoint of foundation design criteria, and lend themselves to gas turbine or combined cycle applications. Isentropic efficiencies of 80 to 85% are usually obtained.

3. Screw Compressors

Most screw compressors are of the oil-lubricated type. There are two types—the semihermetic and the open-drive type. In the former, the motor is located in the same housing as the compressor, while in the latter the motor is located outside of the compressor housing and thus requires a shaft seal. The only moving parts in screw compressors are two intermeshing helical rotors. The rotors consist of one male lobe, which functions as a rolling piston, and a female flute, which acts as a cylinder. Since rotary screw compression is a continuous positive-displacement process, no surges are created in the system.

Screw compressors require very little maintenance because the rotors turn at conservative speeds and are well lubricated with coolant oil. Fortunately, most of the oil can easily be separated from the gas in screw compressors. Typically, only small levels of impurities of between 1.0 and 2.0 ppm by weight remain in the gas after compression. Charcoal filters can be used to reduce the impurities below this level.

A major advantage of screw compressors is that they permit the attainment of high-pressure ratios in a single mode. To handle these same large volumes with a reciprocating compressor would require a double-stage unit. Because of this and other advantages, screw compressors are now preferred over reciprocating compressors for helium refrigeration and liquefaction applications. They are competitive with centrifugal compressors in other applications as well.

B. Expansion Devices

The primary function of a cryogenic expansion device is to reduce the temperature of the gas to provide useful refrigeration for the process. In expansion engines, this is accomplished by converting part of the energy of the high-pressure gas stream into mechanical work. This work in large cryogenic facilities is recovered and utilized to reduce the overall compression requirements of the process. On the other hand, cooling of a gas can also be achieved by expanding the gas through an expansion valve (provided that its initial temperature is below the inversion temperature of the gas). The cooling here is accomplished by converting part of the energy of the high-pressure gas stream into kinetic energy. No mechanical work is obtained from such an expansion.

Expanders are of either the reciprocating or the centrifugal type. With the rapid growth of tonnage in cryogenic processes, centrifugal expanders have gradually displaced the reciprocating type in large plants. However, the reciprocating expander is still popular for those processes where the inlet temperature is very low, such as for hydrogen or helium gas. Units up to 3600 hp (2685 kW) have been put in service for nitrogen expansion in liquid hydrogen plants, while nonlubricated expanders with exhausts well below 33.3 K are being used in liquid hydrogen plants developed for the space program.

1. Reciprocating Expanders

Generally, reciprocating expanders are selected when the inlet pressure and pressure ratio are high and when the volume of gas handled is low. The inlet pressure to expansion engines used in air-separation plants varies between 4 and 20 MPa, while capacities range from 0.1 to 3 m^3/sec. Isentropic efficiencies achieved are from 70 to 80%.

The design features of reciprocating expanders employed in low-temperature processes include rigid, guided cam-actuated valve gears, renewable hardened valve seats, helical steel or air-springs, and special valve packings that eliminate leakage. Cylinders are normally steel forgings effectively insulated from the rest of the structure. Removable cylinder liners of Micarta or similar nonmetallic material and floating piston design offer wear resistance and good alignment in operation. Piston rider rings serve as guides for the piston. Nonmetallic rings are used for nonlubricated service. Both horizontal and vertical design, and one and two cylinder versions, have been used successfully.

Reciprocating expanders, in normal operation, should not accept liquid in any form during the expansion cycle. However, the reciprocating device can tolerate some liquid for short periods of time provided that none of the constituents freeze out in the expander cylinder and cause serious mechanical problems. If selected design conditions indicate possibilities of entering the liquid and especially the triple point range on expansion during normal operation, then inlet pressure and temper-

ature must be revised or thermal efficiency modified accordingly.

C. Centrifugal Expanders

Turboexpanders are classified as either axial or radial. Most turboexpanders built today are of the radial type because of their generally lower cost and reduced stresses for a given tip speed. This permits them to run at higher speeds with higher efficiencies and lower operating costs. On the other hand, axial flow expanders are more suitable for multistage expanders since they permit a much easier flow path from one stage to the next. Where low flow rates and high enthalpy reductions are required, an axial flow two-stage expander is generally used with nozzle valves controlling the flow. For example, in the processing of ethylene, gas leaving the demethanizer is normally saturated, and expansion conditions result in a liquid product coming out of the expander. Since up to 15 to 20% liquid at the isentropic end point can be handled in actual flow impulse turbine expanders, recovery of ethylene is feasible by the procedure. Depending on the initial temperature and pressure into the expander and the final exit pressure, good flow expanders are capable of reducing the enthalpy of an expanded fluid by between 175 and 350 kJ/kg. and this may be multistaged. The change in enthalpy drop can be automatically regulated by turbine speed. The development of highly reliable and efficient turboexpanders has made today's large-tonnage air-separation plants and baseload LNG facilities a reality. Notable advances in turboexpander design for these applications center on improved bearings, lubrication, and wheel and rotor design to permit nearly ideal rotor assembly speeds with good reliability. Pressurized labyrinth sealing systems use dry seal gas under pressure mixed with cold gas from the process to provide seal output temperatures above the frost point. Seal systems for oxygen compressors are more complex than for air or nitrogen and must prevent oil carryover to the processed gas. By the combination of variable-area nozzle grouping or partial admission of multiple nozzle grouping, efficiencies up to 85% have been obtained with radial turboexpanders.

Turboalternators were developed in the 1960s to improve the efficiency of small cryogenic refrigeration systems. This is accomplished by converting the kinetic energy in the expanding fluid to electrical energy, which in turn is transferred outside of the system, where it can be converted to heat and dissipated to an ambient heat sink.

D. Expansion Valve

The expansion valve or Joule–Thomson valve, as it is often called, is an important component in any liquefaction system, although not as critical a component as the others mentioned in this section. In simplest terms the expansion valve resembles a normal valve that has been modified to handle the flow of cryogenic fluids. These modifications include exposing the high-pressure stream to the lower part of the valve seat to reduce sealing problems, a valve stem that has been lengthened and constructed of a thin-walled tube to reduce heat transfer, and a stem seal that is accomplished at ambient temperatures.

E. Heat Exchangers and Regenerators

One of the more critical components of any low-temperature liquefaction and refrigeration system is the heat exchanger. This point is readily demonstrated by considering the effect of the heat exchanger effectiveness on the liquid yield of nitrogen in a simple Linde-cycle liquefaction process operating between a lower and upper pressure of 0.1 and 10 MPa, respectively. The liquid yield under these conditions will be zero whenever the effectiveness of the heat exchanger falls below 90%. (Heat exchanger effectiveness is defined as the ratio of the actual heat transfer to the maximum possible heat transfer.)

Fortunately, most cryogens, with the exception of helium II, behave as "classical" fluids. As a result, it has been possible to predict their behavior by using well-established principles of mechanics and thermodynamics applicable to many room-temperature fluids. In addition, this has permitted the formulation of convective heat transfer correlations for low-temperature designs of simple heat exchangers that are similar to those used at ambient conditions and utilize such well-known dimensionless quantities as the Nusselt, Reynolds, Prandtl, and Grashof numbers.

However, the requirements imposed by the need to operate more efficiently at low temperatures has made the use of simple exchangers impractical in many cryogenic applications. In fact, some of the important advances in cryogenic technology are directly related to the development of rather complex but very efficient types of heat exchangers. Some of the criteria that have guided the development of these units for low-temperature service are (1) a small temperature difference at the cold end of the exchanger to enhance efficiency, (2) a large ratio of heat-exchange surface area to heat-exchanger volume to minimize heat leak, (3) a high heat-transfer rate to reduce surface area, (4) a low mass to minimize start-up time, (5) multichannel capability to minimize the number of exchangers, (6) high-pressure capability to provide design flexibility, (7) a low or reasonable pressure drop to minimize compression requirements, and (8) minimum maintenance to minimize shutdowns.

The selection of an exchanger for low-temperature operation is normally determined by process design requirements, mechanical design limitations, and economic considerations. The principal industrial exchangers finding use in cryogenic applications are coiled-tube, plate–fin, reversing, and regenerator types.

1. Coiled-Tube Exchangers

Construction of these widely used heat exchangers involves winding a large number of tubes in helix fashion around a central core mandrel with each exchanger containing many layers of tubes, both along the principal and radial axes. Pressure drops in the coiled tubes are equalized for each specific stream by using tubes of equal length and carefully varying the spacing of these tubes in the different layers. A shell is fitted over the outermost tube layer, and this shell together with the outside surface of the core mandrel form the annular space in which the tubes are nested. Coiled-tube heat exchangers offer unique advantages, especially for those low-temperature design conditions where simultaneous heat transfer between more than two streams is desired, a large number of heat transfer units is required, and high operating pressures in various streams are encountered. The geometry of these exchangers can be varied widely to obtain optimum flow conditions for all streams and still meet heat transfer and pressure drop requirements.

Optimization of the coiled-tube heat exchanger is quite complex. There are numerous variables, such as tube and shell flow velocities, tube diameter, tube pitch, and layer spacing. Other considerations include single-phase and two-phase flow, condensation on either the tube or shell side, and boiling or evaporation on either the tube or shell side. Additional complications come into play when multicomponent streams are present, as in natural gas liquefaction, since mass transfer accompanies the heat transfer in the two-phase region.

Many empirical relationships have been developed to aid the optimization of coiled-tube exchangers under ambient conditions. Many of the same relationships are currently being used in low-temperature applications as well. A number of these relationships are tabulated in readily available cryogenic texts. However, no claim is made that these relationships will be more suitable than others for a specific design. This can be verified only by experimental measurements on the heat exchanger.

2. Plate–Fin Exchangers

These types of heat exchangers normally consist of heat-exchange surfaces obtained by stacking alternate layers of corrugated, high-uniformity, die-formed aluminum sheets (fins) between flat aluminum separator plates to form individual flow passages. Each layer is closed at the edge with solid aluminum bars of appropriate shape and size.

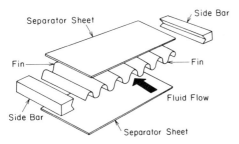

FIGURE 11 Exploded view of one layer of a plate–fin exchanger before brazing.

Figure 11 illustrates by exploded view the elements of one layer, in relative position, before being joined by a brazing operation to yield an integral rigid structure with a series of fluid flow passages. The latter normally have integral welded headers. Several sections can be connected to form one large exchanger. The main advantages of this type of exchanger are that it is compact (about nine times as much surface area per unit volume as conventional shell and tube exchangers), yet permits wide design flexibility, involves minimum weight, and allows design pressures to 6 MPa from 4.2 to 340 K.

The fins for these heat exchangers can be manufactured in a variety of configurations that can significantly alter the heat transfer and pressure drop characteristics of the exchanger. Various flow patterns can be developed to provide multipass or multistream arrangements by incorporating suitable internal seals, distributors, and external headers. The type of headers used depends on the operating pressures, the number of separate streams involved, and, in the case of a counterflow exchanger, whether reversing duty is required.

Design of the plate–fin exchanger involves selecting a geometry and surface arrangement that will give a product UA of the correct magnitude to satisfy the relation:

$$Q = UA\Delta T_e \qquad (6)$$

where Q is the heat transfer rate, U the overall heat-transfer coefficient, A the area of heat transfer, and ΔT_e the equivalent temperature difference between the two streams exchanging energy.

Plate–fin exchangers can be supplied as single units or as manifolded assemblies that consist of multiple units connected in parallel or in series. Sizes of single units are currently limited by manufacturing capabilities and assembly tolerances. Nevertheless, the compact design of brazed aluminum plate–fin exchangers makes it possible even now to furnish more than 35,000 m² of heat-transfer surface in one manifolded assembly. These exchangers are finding application throughout the world in such specific processes as helium liquefaction, helium extraction from natural gas, hydrogen purification and liquefaction,

air separation, and low-temperature hydrocarbon processing. Additional design details for plate–fin exchangers are available in most heat-exchanger texts.

3. Reversing Exchangers

Operation of low-temperature processes on a continuous basis necessitates the removal of all impurities that would solidify on cooling to very low temperatures. This cleanup is necessary because an accumulation of impurities in certain parts of the system can create operational difficulties or constitute potential hazards. Under certain conditions, the necessary purification steps can be carried out with the aid of reversing heat exchangers.

A typical arrangement of a reversing exchanger for an air-separation plant is shown in Fig. 12. Channels A and B constitute the two main reversing streams. Operation of such an exchanger is characterized by the cyclical changeover of one of these streams from one channel to the other. The reversal normally is accomplished by pneumatically operated valves on the warm end and by check valves on the cold end of the exchanger. Feed enters the warm end of the exchanger, and as it is progressively cooled, impurities are deposited on the cold surface of the exchanger. When the flows are reversed, the waste stream reevaporates the deposited impurities and removes them from the system. Pressure differences of the two streams, which in turn affect the saturation concentrations of impurities in those streams, permit impurities to be deposited during the warming period and reevaporated during the cooling period. Temperature differences, particularly at the cold end

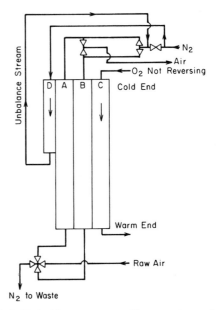

FIGURE 12 Typical flow arrangement for reversing exchanger in air separation plant.

of the reversing heat exchanger, are critical to the proper functioning of these types of exchangers.

4. Regenerators

Another method for the simultaneous cooling and purification of gases in low-temperature processes is based on the use of regenerators, first suggested by Fränkl in the 1920s. Whereas in the reversing exchanger the flows of the two fluids are continuous and countercurrent during any one period, the regenerator operates periodically, storing heat in a high heat-capacity packing in one-half of the cycle and then giving up the stored heat to the fluid in the other half of the cycle.

Such an exchanger normally consists of two identical columns packed with a material of high heat capacity and high heat-transfer area through which the gases that are to be cooled or warmed flow. Such regenerator materials and geometries generally fall into three groups, based on the temperature range over which they are to be used. The first group includes woven screen materials of stainless steel, bronze, or copper used over the temperature range from 30 to 300 K. In the range between 10 and 30 K, lead and antimony spheres are used becauase their heat capacity is higher than any of the screen materials. However, below 10 K, lead loses 89% of its room-temperature specific heat, and its volumetric heat capacity is less than that of helium at a pressure of 1 MPa. In the late 1980s, a third category of essentially heavy rere-earth intermetallic compounds was developed with the potential for enhancing the heat capacity at temperatures below 10 K. The increase in specific heat of two of these rare-earth compounds is shown in Fig. 13.

In the process of cooldown, the warm feed stream deposits impurities on the cold surface of the packing. When the streams are switched, the impurities are reevaporated in the cold stream while simultaneously cooling the packing. Thus, the purifying action of the regenerator is based on the same principles as for the reversing exchanger, and the same limiting critical temperature differences must be observed if complete reevaporation of the impurities is to take place.

Regenerators quite frequently are chosen for applications where the heat-transfer effectiveness, defined as Q_{actual}/Q_{ideal}, must approach values of 0.98 to 0.99. It is clear that a high regenerator effectiveness requires a high heat capacity per unit volume and a large surface area per unit volume.

The low cost of the heat-transfer surface along with the low pressure drop are the principal advantages of the regenerator. However, the intercontamination of fluid streams by mixing due to periodic flow reversals and the difficulty of regenerator design to handle three or more fluids have restricted its use and favored the adoption of brazed aluminum exchangers.

VI. STORAGE AND TRANSFER SYSTEMS

Once a cryogen has been produced, it must be stored, transferred, or transported to its end use. The effectiveness of the cryogenic storage transfer or transport system depends on how well it reduces the loss of the cryogen due to unavoidable heat leak into the system and how well it maintains the purity of the cryogen. Good design, with a knowledge of the heat-transfer mechanisms and the properties of available insulations, is essential in minimizing the boil-off losses due to heat leak. Proper operating procedures, on the other hand, are necessary if product purity is to be maintained.

A. Insulation Concepts

Since heat leak is a major concern in storage and transfer systems of cryogenic liquids, selecting the proper insulation to use in such systems is vitally important. The normal design strategy is to minimize radiative and convective heat transfer while introducing a minimum of solid conductance media. The choice of insulation, however, is generally governed by an attempt to balance the cost of installed insulation with the savings anticipated by lowered boil-off losses.

The various types of insulation used in the storage and transfer of cryogenic liquids can be divided into five categories: (1) vacuum, (2) multilayer, (3) powder and fibrous, (4) foam, and (5) special. The boundaries between these

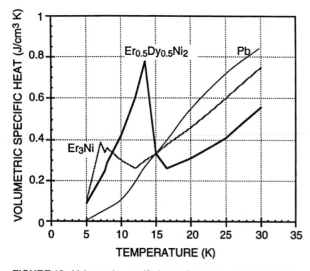

FIGURE 13 Volumetric specific heat of two rare-earth intermetallic compounds and lead.

TABLE III   Characteristics of Selected Insulation

Type of insulation	Apparent thermal conductivity (k_a, J/sec · m · K) (between 77 and 300 K)	Bulk density (k/gm³)
Pure gas at 0.1 MPa, 180 K		
H_2	34.07×10^{-2}	0.080
N_2	5.67×10^{-2}	1.21
Pure vacuum, 0.13 mPa or less	1.70×10^{-2}	Nil
Straight insulation		
Polystyrene foam	8.52×10^{-2}	32–48
Polyurethane foam	10.79×10^{-2}	80–128
Glass foam	11.36×10^{-2}	144
Evacuated powder		
Perlite (133 mPa)	0.34–0.68×10^{-2}	144–64
Silica (133 mPa)	0.57–0.68×10^{-2}	64–96
Combination insulation		
Aluminum foil and fiberglass		
(12–28 layers/cm, 1.33 mPa)	1.14–2.27×10^{-4}	64–112
(30–60 layers/cm, 1.33 mPa)	0.57×10^{-4}	120
Aluminum foil and nylon net		
(32 layers/cm, 1.33 mPa)	5.68×10^{-4}	89

general categories are by no means distinct. However, the classification scheme does offer a framework by which the widely varying types of insulation can be discussed.

Since heat transfer through these insulations can occur by several different mechanisms, the apparent thermal conductivity k_a of an insulation that incorporates all of these heat-transfer possibilities offers the best means of comparing these difference types. Table III provides a listing of some accepted k_a values for popular insulations used in cryogenic storage and transfer systems.

1. Vacuum Insulation

The mechanism of heat transfer prevailing across an evacuated space (0.13 mPa or less) is by radiation and conduction through the residual gas. Radiation is generally the more predominant mechanism and can be approximated by:

$$\frac{Q_r}{A_1} = \sigma \left(T_2^4 - T_1^4 \right) \left[\frac{1}{\varepsilon_1} + \frac{A_1}{A_2} \left(\frac{1}{\varepsilon_2} - 1 \right) \right]^{-1} \quad (7)$$

where Q_r/A_1 is the radiant heat flux, σ the Stefan–Boltzmann constant, and ε the emissivity of the surface. The subscripts 1 and 2 refer to the cold and warm surfaces, respectively. The bracketed term on the right is generally referred to as the emissivity factor.

When the mean free path of gas molecules becomes large relative to the distance between the walls of the

evacuated space as the pressure is reduced, free molecular conduction is encountered. The gaseous heat conduction under free molecular conditions for most cryogenic applications is given by:

$$\frac{Q_{gc}}{A_1} = \frac{\gamma + 1}{\gamma - 1} \left(\frac{R}{8\pi MT} \right)^{1/2} \alpha p (T_2 - T_1) \quad (8)$$

where α, the overall accommodation coefficient, is defined by:

$$\alpha = \frac{\alpha_1 \alpha_2}{\alpha_2 + \alpha_1 (1 - \alpha_2)(A_1/A_2)} \quad (9)$$

and γ is the ratio of the heat capacities, R the molar gas constant, M the molecular weight of the gas, and T the temperature of the gas at the point where the pressure p is measured. The subscripted A_1 and A_2, T_1 and T_2, and α_1 and α_2 are the areas, temperatures, and accommodation coefficients of the cold and warm surfaces, respectively. The accommodation coefficient depends on the specific gas–surface combination and the surface temperature.

Heat transport across an evacuated space by radiation can be reduced significantly by inserting one or more low-emissivity floating shields within the evacuated space. Such shields provide a reduction in the emissivity factor. The only limitation on the number of floating shields used is one of complexity and cost.

2. Multilayer Insulation

Multilayer insulation provides the most effective thermal protection available for cryogenic storage and transfer systems. It consists of alternating layers of highly reflecting material, such as aluminum foil or aluminized Mylar, and a low-conductivity spacer material or insulator, such as fiberglass mat or paper, glass fabric, or nylon net, all under high vacuum. When properly applied at the optimum density, this type of insulation can have an apparent thermal conductivity as low as 10 to 50 μW/m · K between 20 and 300 K. The very low thermal conductivity of multilayer insulations can be attributed to the fact that all modes of heat transfer are reduced to a bare minimum.

The apparent thermal conductivity of a highly evacuated (pressures on the order of 0.13 mPa or less) multilayer insulation can be determined from:

$$k_a = \frac{1}{N/\Delta x} \left[h_s + \frac{\sigma e T_2^3}{2 - e} \left(1 + \frac{T_1}{T_2} \right)^2 \left(1 + \frac{T_1}{T_2} \right) \right] \quad (10)$$

where $N/\Delta x$ is the number of complete layers (reflecting shield plus spacer) of insulation per unit thickness, h_s the solid conductance for the spacer material, σ the Stefan–Boltzmann constant. e the effective emissivity of the reflecting shield, and T_2 and T_1 the temperatures of the warm and cold sides of the insulaion, respectively. It is evident that the apparent thermal conductivity can be reduced by increasing the layer density up to a certain point.

Unfortunately, the effective thermal conductivity values generally obtained with actual cryogenic storage and transfer systems are at least a factor of 2 greater than the thermal conductivity values measured in the laboratory with carefully controlled techniques. This degradation in insulation thermal performance is caused by the combined presence of edge exposure to isothermal boundaries, gaps, joints, or penetrations in the insulation blanket required for structural supports, fill and vent lines, and the high lateral thermal conductivity of these insulation systems.

3. Powder Insulation

The difficulties encountered with the use of multilayer insulation for complex structural storage and transfer systems can be minimized by the use of evacuated powder insulation. This substitution in insulation materials, however, incurs a 10-fold decrease in overall thermal effectiveness of the insulation system. Nevertheless, in applications where this is not a serious factor and investment cost is a major factor, even unevacuated powder insulation with still a lower thermal effectiveness may be the proper choice of insulating material. Such is the case for large LNG storage facilities.

A powder insulation system consists of a finely divided particulate material such as perlite, expanded SiO_2, calcium silicate, diatomaceous earth, or carbon black packed between the surfaces to be insulaed. When used at 0.1 MPa gas pressure (generally with an inert substance), the powder reduces both convection and radiation and, if the particle size is sufficiently small, can also reduce the mean free path of the gas molecules.

The radiation contribution for highly evacuated powders near room temperature is larger than the solid-conduction contribution to the total heat transfer rate. On the other hand, the radiant contribution is smaller than the solid-conduction contribution for temperatures between 77 and 20 or 4 K. Thus, evacuated powders can be superior to vacuum alone (for insulation thicknesses greater than ~0.1 m) for heat transfer between ambient and liquid nitrogen temperatures. Conversely, since solid conduction becomes predominant at lower temperatures, it is usually more advantageous to use vacuum alone for reducing heat transfer between two cryogenic temperatures.

4. Foam Insulation

The apparent thermal conductivity of foams is dependent on the bulk density of the foamed material, the gas used as the foaming agent, and the temperature levels to which the insulation is exposed. Heat transport across a foam is determined by convection and radiation within the cells of the foam and by conduction in the solid structure. Evacuation of a foam is effective in reducing its thermal conductivity, indicating a partially open cellular structure, but the resulting values are still considerably higher than either multilayer or evacuated powder insulations. The opposite effect, diffusion of atmospheric gases into the cells, can cause an increase in the apparent thermal conductivity. This is particularly true with the diffusion of hydrogen and helium into the cells. Of all the foams, polyurethane and polystyrene have received the widest use at low temperatures.

The major disadvantage of foams is not their relatively high thermal conductivity compared with that of other insulations, but rather their poor thermal behavior. When applied to cryogenic systems, they tend to crack on repeated cycling and lose their insulation value.

5. Special Insulations

No single insulation has all the desirable thermal and strength characteristics required in many cryogenic applications. Consequently, numerous composite insulations have been developed. One such insulation consists of a polyurethane foam, reinforcement of the foam to provide adequate compressive strength, adhesives for sealing and securing the foam to the container, enclosures to prevent damage to the foam from external sources, and vapor barriers to maintain a separation between the foam and atmospheric gases. Another external insulation system for space applications uses honeycomb structures. Phenolic resin-reinforced fiberglass-cloth honeycomb is most commonly used. Filling the cells with a low-density polyurethane foam further improves the thermal effectiveness of the insulation.

B. Storage Systems

Storage vessels range from low-performance containers where the liquid in the container boils away in a few hours to high-performance containers and dewars where less than 0.1% of the fluid contents is evaporated per day. Since storage and transfer systems are important components of any cryogenic support facility, many examples of storage vessel design have appeared in the literature. The essential elements of a storage vessel consist of an inner vessel, which encloses the cryogenic fluid to be stored, and an outer vessel, which contains the appropriate insulation and serves as a vapor barrier to prevent water and other condensables from reaching the cold inner vessel. The value of the cryogenic liquid stored will dictate whether or not the insulation space is evacuated. In small laboratory dewars, the insulation is obtained by coating the two surfaces facing the insulation space with low-emissivity materials and then evacuating the space to a pressure of 0.13 mPa or lower. In larger vessels, insulations such as powders, fibrous materials, or multilayer insulations are used.

Several requirements must be met in the design of the inner vessel. The material of construction must be compatible with the stored cryogen. Nine percent nickel steels are acceptable for high-boiling cryogens ($T > 75$ K), while many aluminum alloys and austenitic steels are usually structurally acceptable throughout the entire temperature range. Economic and cooldown considerations dictate that the inner shell be as thin as possible. Accordingly, the inner container is designed to withstand only the internal pressure and bending forces, while stiffening rings are used to support the weight of the fluid. The minimum thickness of the inner shell for a cylindrical vessel under such a design arrangement is given in Section VIII of the American Society of Mechanical Engineers' (ASME) *Boiler and Pressure Vessel Code.*

The outer shell of the stroage vessel, on the other hand, is subjected to atmospheric pressure on the outside and evacuated conditions on the inside. Such a pressure difference requires an outer shell of sufficient material thickness with appropriately placed stiffening rings to withstand collapsing or buckling. Here again, specific design charts addressing this situation can be found in the ASME code.

Heat leak into a storage system for cryogens generally occurs by radiation and conduction through the insulation and conduction through the inner shell supports, piping, instrumentation leads, and access ports. Conduction losses are reduced by introducing long heat-leak paths, by making the cross-sections for heat flow small, and by using materials with low thermal conductivity. Radiation losses, a major factor in the heat leak through insulations, are reduced with the use of radiation shields, such as multilayer insulation, boil-off vapor-cooled shields, and opacifiers in powder insulation.

Most storage vessels for cryogens are designed for a 90% liquid volume and a 10% vapor or ullage volume. The latter permits reasonable vaporization of the liquid contents due to heat leak without incurring too rapid a buildup of pressure in the vessel. This, in turn, permits closure of the container for short periods either to avoid partial loss of the contents or to permit the safe transport of flammable or hazardous cryogens.

C. Transfer Systems

Three methods are commonly used to transfer a cryogen from the storage vessel. These are self-pressurization of the container, external gas pressurization, and mechanical pumping. Self-pressurization involves removing some of the fluid from the container, vaporizing the extracted fluid, and then reintroducing the vapor into the ullage space, thereby displacing the contents of the container. The external gas pressurization method utilizes an external gas to accomplish the desired displacement of the container contents. In the mechanical pumping method, the contents of the stroage vessel are removed by a cryogenic pump located in the liquid drain line.

Several different types of pumps have been used with cryogenic fluids. In general, the region of low flow rates at high pressures is best suited to positive displacement pumps, while the high-flow applications are generally best served by the use of centrifugal or axial flow pumps. The latter have been built and used for liquid hydrogen with flow rates of up to 3.8 m^3/sec and pressures of more than 6.9 MPa. For successful operation, cryogen subcooling, thermal contraction, lubrication, and compatibility of materials must be carefully considered.

Cryogenic fluid transfer lines are generally classified as one of three types: uninsulated, foam-insulated lines, and vacuum-insulated lines. The latter may entail vacuum insulation alone, evacuated powder insulation, or multilayer insulation. A vapor barrier must be applied to the outer surface of foam-insulated transfer lines to minimize the degradation of the insulation that occurs when water vapor and other condensables are permitted to diffuse through the insulation to the cold surface of the lines.

Two-phase flow is always involved in the cooldown of a transfer line. Since this process is a transient one, several different types of two-phase flow will exist simultaneously along the inlet of the transfer line. Severe pressure and flow oscillations occur as the cold liquid comes in contact with successive warm sections of the line. Such instability continues until the entire transfer line is cooled down and filled with liquid cryogen.

The transport of cryogens for more than a few hundred meters generally requires specially built transport systems for truck, railroad, or airline delivery. Volumes from 0.02 to more than 100 m^3 have been transported successfully by these carriers. The use of large barges and ships built specifically for cryogen shipment has increased the volume transported manyfold. This has been particularly true for the worldwide transport of LNG.

VII. INSTRUMENTATION

Once low temperatures have been attained and cryogens have been produced, property measurements must often be made at these temperatures. Such measurements as temperature and pressure are typically required for process optimization and control. In addition, as cryogenic fluids have acquired greater commercial importance, questions have arisen relative to the quantities of these fluids transferred or delivered. Accordingly, the instrumentation used must be able to indicate liquid level, density, and flow rate accurately.

A. Thermometry

Most low-temperature engineering temperature measurements are made with metallic resistance thermometers, nonmetallic resistance thermometers, or thermocouples. In the selection of a thermometer for a specific application one must consider such factors as absolute accuracy, reproducibility, sensitivity, heat capacity, self-heating, heat conduction, stability, simplicity and convenience of operation, ruggedness, and cost. Other characteristics may be of importance in certain applications.

B. Fluid Measurements

Liquid level is one of several measurements needed to establish the contents of a cryogenic container. Other measurements may include volume as a function of depth, density as a function of physical storage conditions, and sometimes discerning useful contents from total contents. Of these measurements, the liquid-level determination is presently the most advanced and can be made with an accuracy and precision comparable to that of thermometry and often with greater simplicity.

There are as many ways of classifying liquid-level sensors as there are developers who have described them. A convenient way to classify such devices is according to whether the output is discrete (point sensors) or continuous.

C. Density Measurements

Measurements of liquid density are closely related to quantity and liquid-level measurements since both are often required simultaneously to establish the mass contents of a tank, and the same physical principle may often be used for either measurement, since liquid-level detectors sense the steep density gradient at the liquid–vapor interface. Thus, the methods of density determination include the following techniques: direct weighing, differential pressure, capacitance, optical, acoustic, and nuclear radiation attenuation. In general, the various liquid level principles apply to density measurement techniques as well.

Two exceptions are noteworthy. In the case of homogeneous pure fluids, density can usually be determined more accurately by an indirect measurement, namely, the measurement of pressure and temperature which is then coupled with the analytical relationship between these intensive properties and density through accurate thermophysical properties data.

The case of nonhomogeneous fluids is quite different. LNG is often a mixture of five or more components whose composition and, hence, density vary with time and place. Accordingly, temperature and pressure measurements alone will not suffice. A dynamic, direct measurement is required, embodying one or more of the liquid-level principles used in liquid-level measurements.

D. Flow Measurements

Three basic types of flow meters are useful for liquid cryogens. These are the pressure drop or "head" type, the turbine type, and the momentum type.

VIII. SAFETY

No discussion of cryogenic systems would be complete without a review of some of the safety aspects associated with either laboratory or industrial use of cryogenic fluids. Ealier discussion of the properties of cryogenic fluids and the behavior of materials at low temperatures revealed that there are a number of unique hazards associated with cryogenic fluids. These hazards can best be classified as those associated with the response of the human body and the surroundings to cryogenic fluids and their vapors, and those associated with reactions between certain of the cryogenic fluids and their surroundings.

A. Human Hazards

It is well known that exposure of the human body to cryogenic fluids or to surfaces cooled by cryogenic fluids can result in severe "cold burns" since damage to the skin or tissue is similar to that caused by an ordinary burn. The severity of the burn depends on the contact area and the contact time; prolonged contact results in deeper burns. Severe burns are seldom sustained if rapid withdrawal is possible.

Protective clothing is mandatory to insulate the body from these low temperatures and prevent "frostbite." Safety goggles, gloves, and boots are imperative for personnel involved in the transfer of liquid cryogens. Such transfers, in the interest of good safety practices, should be attempted only when sufficient personnel are available to monitor the activity. Since nitrogen is a colorless, odorless, inert gas, personnel must be aware of the associated respiratory and asphyxiation hazards. Whenever the oxygen content of the atmosphere is diluted due to spillage or leakage of nitrogen, there is danger of nitrogen asphyxiation. In general, the oxygen content of air for breathing purposes should never be below 16%. Whenever proper air ventilation cannot be ensured, air-line respirators or a self-contained breathing apparatus should be used.

An oxygen-enriched atmosphere, on the other hand, produces exhilarating effects when breathed. However, lung damage can occur if the oxygen concentration in the air exceeds 60%, and prolonged exposure to an atmosphere of pure oxygen may initiate bronchitis, pneumonia, or lung collapse. An additional threat of oxygen-enriched air can come from the increased flammability and explosion hazards.

B. Materials Compatibility

Most failures of cryogenic systems can generally be traced to an improper selection of construction materials or a disregard for the change of some material property from ambient to low temperatures. For example, the ductility property of a material requires careful consideration since low temperatures have the effect of making some construction materials brittle or less ductile. This behavior is further complicated because some materials become brittle at low temperatures but still can absorb considerable impact, while others become brittle and lose their impact strength. Brittle fracture can occur very rapidly, resulting in almost instantaneous failure. Such failure can cause shrapnel damage if the system is under pressure, while release of a fluid such as oxygen can result in fire or explosions.

Low-temperature equipment can also fail because of thermal stresses caused by thermal contraction of the materials used. In solder joints, the solder must be able to withstand stresses caused by differential contraction where two dissimilar metals are joined. Contraction in long pipes is also a serious problem; a stainless-steel pipeline 30 m long will contract ~0.085 m when filled with liquid oxygen or nitrogen. Provisions must be made for this change in length during both cooling and warming of the pipeline by using bellows, expansion joints, or flexible hose. Pipe anchors, supports, and so on likewise must be carefully designed to permit contraction and expansion to take place. The primary hazard of failure due to thermal contraction is spillage of the cryogen and the possibility of fire or explosion.

All cryogenic systems should be protected against overpressure due to phase change from liquid to gas. Systems containing liquid cryogens can reach bursting pressures, if not relieved, simply by trapping the liquid in an enclosure. The rate of pressure rise depends on the rate of heat transfer into the liquid. In uninsulated systems, the liquid is vaporized rapidly and pressure in the closed system can rise very rapidly. The more liquid there is originally in the tank before it is sealed off, the greater will be the resulting final pressure. Relief valves and burst disks are normally used to relieve piping systems at a pressure near the design pressure of the equipment. Such relief should be provided between valves, on tanks, and at all points of possible (though perhaps unintentional) pressure rise in a piping system.

Overpressure in cryogenic systems can also occur in a more subtle way. Vent lines without appropriate rain traps can collect rainwater. Which when frozen can block the line. Exhaust tubes on relief valves and burst disks likewise can become inoperable. Small-necked, open-mouth dewars can collect moisture from the air and freeze closed. Entrapment of cold liquids or gases can occur by freezing water or other condensables in some portion of the cold system. If this occurs in an unanticipated location, the relief valve or burst disk may be isolated and afford no protection. Such a situation usually arises from improper operating procedures and emphasizes the importance of good operating practices.

Another source of system overpressure that is frequently overlooked results from cooldown surges. If a liquid cryogen is admitted to a warm line for the purpose of transfer of the liquid from one point to another, severe pressure surges will occur. These pressure surges can be up to 10 times the operating or transfer pressure and can even cause backflow into the storage container. Protection against such overpressure must be included in the overall design and operating procedures for the transfer system.

In making an accident or safety analysis, it is always wise to consider the possibility of encountering even more serious secondary effects from any cryogenic accident. For example, any one of the failures discussed previously (brittle fracture, contraction, overpressure, etc.) may release sizable quantities of cryogenic liquids, causing a severe fire or explosion hazard, asphyxiation possibilities, further brittle fracture problems, or sharpnel damage to other flammable or explosive materials. In this way the situation can rapidly and progressively become much more serious.

C. Flammability and Detonability

Almost any flammable mixture will, under favorable conditions of confinement, support an explosive flame propagation or even a detonation. When a fuel–oxidant mixture of a composition favorable for high-speed combustion is weakened by dilution with an oxidant, fuel, or an inert substance, it will first lose its capacity to detonate. Further dilution will then cause it to lose its capacity to burn explosively. Eventually, the lower or upper flammability limits will be reached and the mixture will not maintain its combustion temperature and will automatically extinguish itself. These principles apply to the combustible cryogens hydrogen and methane. The flammability and detonability

TABLE IV Flammability and Detonability Limits of Hydrogen and Methane Gas

Mixture	Flammability limits (mol%)	Detonability limits (mol%)
H_2–air	4–75	20–65
H_2–O_2	4–95	15–90
CH_4–air	5–15	6–14
CH_4–O_2	5–61	10–50

limits for these two cryogens with either air or oxygen are presented in Table IV. Since the flammability limits are rather broad, great care must be exercised to exclude oxygen from these cryogens. This is particularly true with hydrogen since even trace amounts of oxygen will condense, solidify, and build up with time in the bottom of the liquid hydrogen storage container and eventually attain the upper flammability limits. Then it is just a matter of time until some ignition source, such as a mechanical or electrostatic spark, accidentally initiates a fire or possibly an explosion.

Because of its chemical activity, oxygen also presents a safety problem in its use. Liquid oxgen is chemically reactive with hydrocarbon materials. Ordinary hydrocarbon lubricants are even dangerous to use in oxygen compressors and vacuum pumps exhausting gaseous oxygen. In fact, valves, fittings, and lines used with oil-pumped gases should never be used with oxygen. Serious explosions have resulted from the combination of oxygen and hydrocarbon lubricants.

To ensure against such unwanted chemical reactions, systems using liquid oxygen must be kept scrupulously clean of any foreign matter. The phrase "LOX clean" in the space industry has come to be associated with a set of elaborate cleaning and inspection specifications nearly representing the ultimate in large-scale equipment cleanliness.

Liquid oxygen equipment must also be constructed of materials incapable of initiating or sustaining a reaction. Only a few polymeric materials can be used in the design of such equipment since most will react violently with oxygen under mechanical impact. Also, reactive metals such as titanium and aluminum should be used cautiously, since they are potentially hazardous. Once the reaction is started, an aluminum pipe containing oxygen burns rapidly and intensely. With proper design and care, however, liquid oxygen systems can be operated safely.

Even though nitrogen is an inert gas and will not support combustion, there are some subtle means whereby a flammable or explosive hazard may develop. Cold traps or open-mouth dewars containing liquid nitrogen can con-

dense air and cause oxygen enrichment of the liquid nitrogen. The composition of air as it condenses into the liquid nitrogen container is about 50% oxygen and 50% nitrogen. As the liquid nitrogen evaporates, the liquid oxygen content steadily increases so that the last portion of liquid to evaporate will have a relatively high oxygen concentration. The nitrogen container must then be handled as if it contained liquid oxygen. Explosive hazards all apply to this oxygen-enriched liquid nitrogen.

Since air condenses at temperatures below ~82 K, uninsulated pipelines transferring liquid nitrogen will condense air. This oxygen-enriched condensate can drip on combustible materials, causing an extreme fire hazard or explosive situation. The oxygen-rich air condensate can saturate clothing, rags, wood, asphalt pavement, and so on and cause the same problems associated with the handling and spillage of liquid oxygen.

D. Summary

It is obvious that the best designed cryogenic facility is no better than the attention paid to every potential hazard. Unfortunately, the existence of such potential hazards cannot be considered once and then forgotten. Instead, there must be an ongoing safety awareness that focuses on every conceivable hazard that might be encountered. Assistance with identifying these safety hazards is adequately covered by Edeskuty and Stewart (1996).

SEE ALSO THE FOLLOWING ARTICLES

CHEMICAL ENGINEERING THERMODYNAMICS • CRYOGENICS • HEAT EXCHANGERS • METALLURGY, MECHANICAL • SUPERCONDUCTIVITY MECHANISMS • VACUUM TECHNOLOGY

BIBLIOGRAPHY

Barron R. F. (1986). "Cryogenic Systems," Oxford Univ. Press, London.

Edeskuty, F. J., and Stewart, W. F. (1996). "Safety in the Handling of Cryogenic Fluids," Plenum Press, New York.

Flynn, T. M. (1996). "Cryogenic Engineering," Dekker, New York.

Jacobsen, R. T., Penoncello, S. G., and Lemmon, E. W. (1997). "Thermodynamic Properties of Cryogenic Fluids," Plenum Press, New York.

Ross, R. G., Jr. (1999). "Cryocoolers 10," Kluwer Academic/Plenum Publishers, New York.

Timmerhaus, K. D., and Flynn, T. M. (1989). "Cryogenic Process Engineering," Plenum Press, New York.

Van Sciver, S. W. (1986). "Helium Cryogenics," Plenum Press, New York.

Weisend, J. G., II (1998). "Handbook of Cryogenic Engineering," Taylor & Francis, London.

Cryogenics

P. V. E. McClintock

University of Lancaster

GLOSSARY

Boson Entity whose intrinsic spin is an even multiple of $\hbar$, examples being photons, phonons, and atoms made up of an even number of fundamental particles.

Bose–Einstein statistics Form of quantum statistics that must be used to describe a gaseous assembly of noninteracting bosons at low temperatures.

Cryostat Apparatus used to achieve and maintain cryogenic temperatures.

Debye cut-off frequency Maximum possible frequency for a vibrational wave in a crystal: the Debye characteristic temperature is the Debye cut-off frequency multiplied by Planck's constant and divided by Boltzmann's constant.

Dewar Vacuum-insulated vessel, of a type commonly used for cryostats or as containers for liquefied gases.

Dispersion curve Frequencies (energies) of the permitted excitations of a system plotted as a function of their momenta (wave vectors).

Fermi–Dirac statistics Form of quantum statistics that must be used to describe a gaseous assembly of noninteracting fermions at low temperatures.

Fermi sphere Sphere of filled states in momentum space, characteristic of an ideal Fermi–Dirac gas at a temperature very much less than the Fermi temperature.

Fermi surface Surface of the Fermi sphere, which is a region of particular importance since it relates to the only particles that can readily undergo interactions; particles on the Fermi surface at very low temperatures have the Fermi momentum and Fermi energy; the Fermi temperature is the Fermi energy divided by Boltzmann's constant.

Fermion Entity whose intrinsic spin is a halfintegral multiple of $\hbar$, examples being electrons, protons, neutrons, and atoms made up of an odd number of fundamental particles.

He I Nonsuperfluid, higher temperature phase of liquid ^{4}He.

He II Superfluid, lower temperature phase of liquid ^{4}He.

^{3}He-A, ^{3}He-B Superfluid phases of liquid ^{3}He.

Phonon Quantum of vibrational energy in a crystal or in He II; phonons can be envisaged as quasi-particles traveling at the velocity of sound.

Roton Excitation, somehow analogous to, but different from, a phonon, near the minimum of the dispersion curve of He II.

Superconductivity State of zero electrical resistivity and perfect diamagnetism in a material.

Superfluidity State of a fluid in which it has zero viscosity, examples being He II, ^{3}He-A, ^{3}He-B, and the electron gas in a superconductor.

Superleak Material, usually a tightly packed powder or a medium with very fine pores, through which a superfluid can pass easily but which is impermeable to any normal (viscous) fluid.

CRYOGENICS is the science and technology of very low temperatures. Although there is no hard and fast definition of where ordinary refrigeration gives way to cryogenics, it is customary to consider the dividing line as being at ~100 K. Of particular scientific interest and importance are the cryogenic phenomena of superconductivity (below ~120 K) and superfluidity (below ~2 K), for which no known analogues exist in the everyday world at room temperature.

Cryogenic techniques are now being employed in applications as diverse as medicine, rocketry, archeology, and metrology.

I. NATURE AND SIGNIFICANCE OF LOW TEMPERATURES

A. Temperature

It is now possible to study the properties of matter experimentally over some 15 decades in temperature. That is, the highest temperature attainable is about 10^{15} times larger than the lowest one. As indicated in Fig. 1, it is feasible to reach temperatures as low as 10^{-7} K (for copper nuclei) with state-of-the-art cryogenic technology, or as high as 10^8 K in the case of hydrogen fusion experiments in a tokamak, starting in each case from the ambient temperature of 3×10^2 K. Important scientific insights into the fundamental nature of matter, with corresponding innovations and applications in technology, have arisen from continued progress at both ends of the temperature scale (and it is interesting to note that low-temperature physicists and engineers have already traveled further from ambient than their high-temperature colleagues by a factor of 10^4). In this article, we concentrate on what happens in the lower temperature two-thirds of Fig. 1.

The Kelvin temperature scale is absolute in the sense that it does not depend on any particular property of any particular material (the ratio of two temperatures being formally defined as the ratio of the heats accepted and rejected by an ideal engine operating between thermal reservoirs at those temperatures). The size of the degree Kelvin (K)

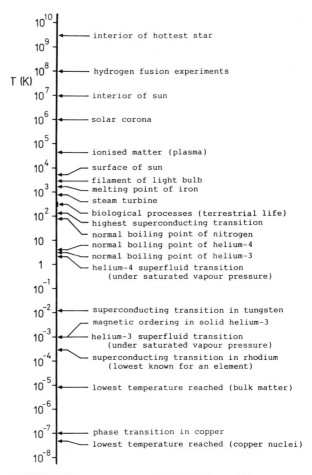

FIGURE 1 The temperatures T (in Kelvin) at which some selected phenomena occur. Note that the scale is logarithmic.

has been chosen, however, so as to be consistent with the degree Celsius of the earlier ideal gas temperature scale.

There is an absolute zero of temperature: 0 K on the Kelvin scale. In terms of classical concepts this would correspond to a complete cessation of all atomic motion. In reality, however, this is far from being the case. Real systems are governed by quantum mechanics for which the ground state does not have zero energy. The quantity that does approach zero as $T \rightarrow 0$ is the *entropy*, which gives a measure of the disorder of the system. The third law of thermodynamics, embodying this observation, is of particular importance for cryogenics. It can be stated in two forms. The first of these is

The entropy of all systems and of all states of a system is zero at absolute zero,

which, it should be noted, is subject to certain restrictions concerning the thermodynamic reversibility of connecting paths between the states of the system. The second form of the law is of more immediately obvious relevance:

It is impossible to reach the absolute zero of temperature by any finite number of processes.

In other words, the absolute zero is an unattainable goal to which the cryogenic engineer may aspire but which he or she can, by definition, never quite reach. The logarithmic temperature scale of Fig. 1 provides a convenient reminder of the restrictive nature of the third law (since 0 K would be situated an infinite distance downwards on such a plot). It is also appropriate, in the sense that the number of interesting changes in properties that occur when the temperature is altered often seems to depend more on the *factor* by which T changes than on the absolute magnitude of any change.

B. Matter at Low Temperatures

It is scarcely surprising that the properties of all materials undergo enormous changes as they are cooled through the huge range of low temperatures that is accessible. We discuss some particular cryogenic phenomena below; in this section we will try to provide a general overview of the sorts of changes in properties that occur in practice between, say, room temperature and 1 K.

The first and most striking difference between materials at room temperature and at 1 K is that (with the sole exception of helium) bulk matter at 1 K exists only in solid form. There are no liquids or gases.

The specific heats of most materials at 1 K are very small compared to their values at room temperature. This is, in fact, only to be expected because the specific heat at constant volume may be written in terms of the entropy S as:

$$C_v = T(\partial S / \partial T)_v$$

and the third law tells us that entropy changes all tend to zero as the temperature T tends to zero. Thus, at a sufficiently low temperature, all specific heats must become negligible.

The coefficient of thermal conductivity κ usually varies with T in quite a complicated way (see below) and at 1 K it may be larger or smaller than it is at 300 K, depending on the material. The coefficient of thermal diffusion, given by:

$$k_D = \kappa / C_v \rho$$

where ρ is the density, is usually much larger than at 300 K, however, which causes thermal equilibrium times at 1 K to be very short.

The electrical resistivity of pure metals usually falls by orders of magnitude on their being cooled from 300 K to 1 K. That of alloys, on the other hand, tends to be relatively temperature independent. In both cases, there is a possible exception. At a range of critical temperatures below about

20 K for metals and alloys, or below about 120 K for the high T_c oxide systems (see Fig. 1), the electrical resistivity of certain materials drops discontinuously to zero; they are then said to have become *superconductors*.

The coefficient of thermal expansion β decreases with T and, in the design of cryostats and other low-temperature apparatus, it can usually be ignored for temperatures below the normal boiling point of nitrogen (77 K). At 1 K, β is negligible. It can be shown that the third law requires that $\beta \to 0$ as $T \to 0$.

The behavior of helium at low temperatures is quite different from that of all other materials. When cooled under atmospheric pressure, helium liquefies at 4.2 K, but it never solidifies, no matter how cold it is made. This behavior is inexplicable in terms of classical physics where, at a sufficiently low temperature, even the very weak interatomic forces that exist between helium atoms should be sufficient to draw the material together into a crystalline solid. Furthermore, the liquid in question has many extraordinary properties. In particular, it undergoes a transition at ~2 K to a state of *superfluidity*, such that it has zero viscosity and can flow without dissipation of energy, even through channels of vanishingly small dimensions. This frictionless flow of the liquid is closely analogous to the frictionless flow of the electrons in a superconductor. On Earth, superfluidity and superconductivity are exclusively low-temperature phenomena, but it is inferred that they probably also arise in the proton and neutron fluids within neutron stars.

We have been referring, thus far, to the common isotope of helium, ^{4}He. The rare isotope. ^{3}He, behaves in an equally extraordinary manner at low temperatures but with some interesting and important differences to which we will return, below. When cooled under atmospheric pressure, ^{3}He liquefies at 3.2 K but it does not undergo a superfluid transition until the temperature has fallen to 1 mK.

Superfluidity and superconductivity represent particularly dramatic manifestations of the quantum theory on a grand scale and, in common also with virtually all other cryogenic phenomena, they can only be understood satisfactorily in terms of quantum statistical mechanics.

C. Quantum Statistics

The behavior of materials at cryogenic temperatures is dominated by quantum effects. It matters, to a very much greater extent than at room temperature, that the energy levels available to a system are in reality discrete and not continuous. The symmetry of the wave functions of the constituent particles of the system is also of crucial importance since it is this that determines whether or not the occupation of a given quantum state by more than one particle is possible.

Many cryogenic systems can be modeled successfully as gases, even including some that appear at first sight to be solids or liquids, as we shall see. In gases, the constituent particles are indistinguishable; in terms of quantum mechanics, each of them can be regarded as a wave that fills the entire container, with permitted energy states

$$E = \frac{1}{2m}\left(p_x^2 + p_y^2 + p_z^2\right)$$

where p_x, p_y, and p_z are the x, y, and z components of its momentum and L is the length of a side of the container (which is assumed to be a cube). The momenta take discrete values

$$p_x = hn_1/L, \qquad p_y = hn_2/L, \qquad p_z = hn_3/L$$

where h is Planck's constant and n_1, n_2, and n_3 can be any positive or negative integers, including zero. The magnitude of the momentum vector $|\mathbf{p}| = (p_x^2 + p_y^2 + p_z^2)^{1/2}$; $\mathbf{p} = \hbar\mathbf{k}$ where $\mathbf{k}$ is the wave vector.

Particles come in two kinds: (1) bosons, with symmetric wave functions, in which case any number of them can occupy the same quantum state (i.e., having the same values of n_1, n_2, and n_3); and (2) fermions, with antisymmetric wave functions such that there is a rigid restriction that (neglecting spin) precludes the occupation of a given state by more than one particle. Examples of bosons include photons, phonons, and entities, such as ^{4}He atoms, that are made up from an even number of fundamental particles. Examples of fermions include electrons, protons, neutrons, and entities, such as ^{3}He atoms, that are made up from an odd number of fundamental particles. Bosons are described by *Bose–Einstein statistics*; fermions by *Fermi–Dirac statistics*.

At extremely low temperatures, therefore, the particles of a boson assembly would almost all be in the same zero momentum state with $n_1 = n_2 = n_3 = 0$. A fermion assembly would also be in a state close to its lowest possible energy, but this would be very different from zero, since only one particle can occupy the zero momentum state and the others must therefore be in excited states of finite momentum and energy.

In practice, one deals with very large numbers of particles, the levels are very closely spaced, and it is reasonable to envisage a fermion system at very low temperatures as a sphere of filled states in momentum space as sketched in Fig. 2; that is, in a space whose axes represent the components of momenta in the x, y, and z directions. As the temperature is gradually raised, the surface of the sphere becomes "fuzzy" as a few particles just under the surface get promoted to unfilled states just above the surface. At higher temperatures the sphere progressively becomes less

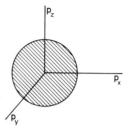

FIGURE 2 Sketch of the Fermi sphere of filled states in momentum space for an ideal gas of fermions at a temperature $T \ll T_F$. All states with momenta $p \le (p_x^2 + p_y^2 + p_z^2)^{1/2}$ are filled. Those for larger momenta are all empty.

well defined and finally "evaporates." For high temperatures, the average energy of the particles is large enough that the likelihood of two of them wanting to be in the same state is negligible, since they are then well spread out in momentum space; it has consequently become irrelevant whether they are bosons or fermions, and their behavior is classical. The sphere of filled states at low temperatures is known as the *Fermi sphere*, and the particles at its surface, the *Fermi surface*, possess the *Fermi energy* E_F and the *Fermi momentum* p_F. The criterion as to whether the temperature T is to be regarded as "high" or "low" is whether $T > T_F$ or $T < T_F$, where $E_F = k_B T_F$, k_B is Boltzmann's constant, and

$$T_F = (3\pi^2 N/V)^{2/3}(h^2/4\pi^2 m k_B)$$

where N/V is the number of particles of mass m per unit volume. T_F is known as the *Fermi temperature*.

An interesting phenomenon predicted for boson systems where the number of particles is fixed is that of *Bose–Einstein condensation*. This implies that, as the temperature is gradually reduced. there is a *sudden* occupation of the zero momentum state by a macroscopic number of particles. The number of particles per increment range of energy is sketched in Fig. 3 for temperatures above and just below the Bose–Einstein condensation temperature T_b. The distribution retains, at least approximately. the classical Maxwellian form until T_b is reached. Below T_b there are two classes of particles: those in the condensate (represented by the "spike" at $E = 0$) and those in excited states. The criterion for a high or low temperature in a boson system is simply that of whether $T > T_b$ or $T < T_b$, where T_b is given by:

$$T_b = \left(\frac{N/V}{2.612}\right)^{2/3} \frac{h^2}{2\pi m k_B}$$

In the case of particles that are localized, such as paramagnetic ions within a crystal lattice, it is largely irrelevant whether they are bosons or fermions. They are distinguishable by virtue of their positions in the lattice and, for the

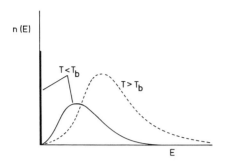

FIGURE 3 Sketch of the energy distribution function $n(E)$ of an ideal Bose–Einstein gas above, and very slightly below, the Bose–Einstein condensation temperature T_b.

same reason, there is no way in which two of them could ever be in the same state. Such assemblies are described by Boltzmann statistics. In a system containing N localized particles, the number n of them occupying a particular energy level E_i (each on its own separate site) is given by:

$$n = Ne^{-E_i/kT} \bigg/ \sum_j e^{-E_j/kT}$$

where the summation is taken over all the permitted energy levels.

II. SOLIDS AT LOW TEMPERATURES

A. Insulators

The thermal energy in an insulating crystal takes the form of vibrations of the atoms about their equilibrium positions. The atoms do not vibrate independently, however, since the movements of one atom directly modulate the potential wells occupied by its neighbors. The result is coupled modes of vibration of the crystal as a whole, known as *phonons*. The phonon dispersion curve for an idealized one-dimensional lattice is sketched in Fig. 4. It is periodic, and it may be seen that there is a definite maximum phonon frequency, known as the Debye cut-

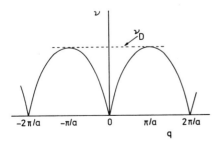

FIGURE 4 The phonon dispersion curve for an ideal one-dimensional lattice; the phonon frequency ν is plotted as a function of its wave vector **q**. The phonon energy $E = h\nu$.

off frequency ν_D. This corresponds physically to the fact that the shortest meaningful wavelength of a vibration in a discrete lattice is equal to twice the nearest neighbor separation; vibrations with apparently shorter wavelengths than this are indistinguishable from equivalent vibrations of longer wavelength. corresponding to the bending over of the dispersion curves at wave vectors of $\pm\pi/a$. It is also an expression of the fact there are only a limited number of normal modes of vibration of the crystal, equal to $3N$ where N is the total number of atoms in it. All the physical information about the dispersion curve is therefore contained in the range of wave vectors between $\pm\pi/a$, which is known as the *first Brillouin zone*. Dispersion curves for real three-dimensional solids are considerably more complicated than this, and there are longitudinal and transverse modes of vibration to be considered, but the essential physics is still as summarized in Fig. 4.

Phonons in a crystal are closely analogous to the photons of electromagnetic radiation (although the latter do not, of course, have any known upper limit on their frequency). Just as in the case of photons, it is possible to envisage phonons as traveling, particle-like packets of energy. They move through the crystal at the velocity of sound (except for very high-energy phonons whose group velocities, given by the gradient of the dispersion curve, fall towards zero on the Brillouin zone boundary).

An excellent way of deducing the specific heat of a crystal is to model it as an empty box containing a gas of phonons. The phonons are bosons so Bose–Einstein statistics must be used. It should be noted, however, that the total number of phonons is not conserved (so that a Bose–Einstein condensation is not to be expected), quite unlike a gas of, for example, helium atoms. This approach yields the famous Debye specific heat curve shown in Fig. 5. It is plotted, not as a function of T directly, but as a function of T/θ_D where $\theta_D = h\nu_D/k_B$ is the Debye characteristic temperature, thereby yielding a universal curve that turns out to be in excellent agreement with experiment for a very wide range of materials. Whether or not the temperature should be considered to be low in the context of any given material depends on whether or not $T \ll \theta_D$. Representative values of θ_D for a few selected materials are given in Table I.

It can be seen from Fig. 5 that the specific heat approaches the classical (Dulong and Petit) value of $3R$ when $T \gg \theta_D$, as predicted by the theorem of the Equipartition of Energy. At lower temperatures, however, the specific heat decreases rapidly towards zero in accordance with the third law. For $T \ll \theta_D$, it is found, both experimentally and theoretically, that $C_v \propto T^3$.

The magnitude of the thermal conductivity coefficient κ of an insulator is determined both by the number of phonons it contains and also by the ease with which they

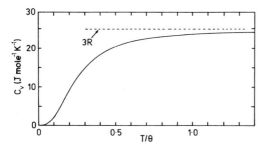

FIGURE 5 The molar specific heat C_v of an insulating solid, according to Debye theory, plotted as a function of T/θ_D where T is the absolute temperature and θ_D is the Debye characteristic temperature for any given material (see Table I).

can travel through the lattice. In fact, the process can accurately be modeled by analogy with heat conduction in a classical gas, for which

$$\kappa = C_V v \lambda / 3$$

where v is the velocity and λ the mean free path of the particles. In the present case, C_V is given by the Debye curve of Fig. 5, v is the (approximately constant) velocity of sound, and λ is determined by a variety of scattering processes. A typical $\kappa(T)$ curve for a crystal is shown by the upper curve of Fig. 6. At high temperatures, where C_V is constant, κ falls with increasing T because of *Umklapp processes*: a pair of phonons interacts to produce a

third phonon of such large momentum that it lies outside the first Brillouin zone or, equivalently, lies within the first Brillouin zone but with a momentum of opposite sign from those of the initial phonons. The result is that the group velocity and direction of energy flow are reversed. Umklapp processes are very effective in limiting the thermal conductivity at high temperatures. As the temperature is reduced, however, there are fewer and fewer phonons of sufficient energy for the process to occur, so λ increases rapidly and so does κ. Eventually, for a good-quality single crystal, λ becomes equal to the dimensions of the sample and (usually) can increase no further; the mean free path is then limited by *boundary scattering* in which the phonons are scattered by the walls and not to any important extent by each other. In this regime, λ is constant and approximately equal to the diameter of the crystal, v is constant, and C_V is decreasing with falling T since $T < \theta_D$ (see Fig. 5), so that κ passes through a maximum and also decreases, varying at T^3, like C_V, in the low temperature limit.

There are many other scattering processes that also influence κ. Phonons may, for example, be scattered by impurities, defects, and grain boundaries. Furthermore, boundary scattering at low temperatures can be either specular (equal angles of incidence and reflection) or, more commonly, diffuse (random angle of reflection). The former process does not limit λ, which can consequently become considerably larger than the diameter of the crystal; it occurs when the phonon wavelength is much longer than the scale of the surface roughness, for example in the

TABLE I Debye Characteristic Temperatures θ_D for Some Common Elements[a]

Element	θ_D(K)
Ag	225
Al	426
Au	164
C (diamond)	2065
Co	443
Cr	585
Fe	464
H (para)	116
He[b]	28–36
Hg	75
N	68
Ni	440
O	91
Pb	108
Si	636
Sn[c]	212
Zn	310

[a] From Rosenberg. H. M. (1963). "Low Temperature Solid State Physics," Clarendon Press, Oxford.

[b] Precise value dependent on pressure.

[c] Value given for grey tin.

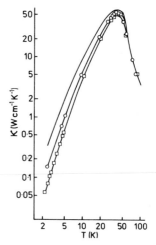

FIGURE 6 The thermal conductivity κ of a synthetic sapphire crystal of diameter 3 mm (upper curve), plotted as a function of temperature T. The circles and the associated curve represent measurements of κ for the same sample after it had been ground down to a diameter of 1.55 mm; and the squares, a diameter of 1.02 mm. [From Berman, R., Foster, E. L., and Ziman, J. M. (1955). *Proc. R. Soc. A* **231**, 130.]

case of a flame-polished sapphire. Thermal conduction in the boundary scattering regime is strange, at first sight, in that the magnitude of κ is geometry and size dependent, becoming smaller if the physical dimensions of the sample are reduced.

The addition of impurities to a crystal drastically reduces κ by providing an additional phonon scattering mechanism, thereby reducing λ. Similarly, κ for disordered or glassy insulators is usually very small because of the greatly enhanced phonon scattering.

B. Metals

In relation to its low temperature properties, a metal can be viewed as being like the insulating crystal discussed above but where, in addition, there is a gas of highly mobile quasi-free conduction electrons. The lattice has remarkably little effect on their motion, apart from opening up energy gaps, or forbidden energy ranges, at Brillouin zone boundaries. The electrons can be treated as an ideal Fermi gas.

Measurements of the specific heats of metals yield results almost indistinguishable from the Debye specific heat curve of Fig. 5, and it can be concluded, therefore, that the electrons make a negligible contribution to the specific heat. This is at first sight astonishing since, according to the classical Equipartition of Energy theorem, there should on average be a thermal energy of $\frac{3}{2}k_BT$ per electron ($\frac{1}{2}k_BT$ per degree of freedom, of which there are three for a gas), leading to an electronic molar specific heat contribution of $\frac{3}{2}R$. The result is easily explicable, however, in terms of Fermi–Dirac statistics. The Fermi temperature of the electron gas in a typical metal turns out to be $T_F \simeq 5 \times 10^4$ K. Thus, even at room temperature, $T \ll T_F$, and the electrons must be regarded as being effectively at a very low temperature. There is consequently a well-defined Fermi sphere in momentum space (see Fig. 2). Most of the electrons, being deep inside the sphere, cannot change state because there are no empty adjacent states into which to move and so are unresponsive to an alteration of temperature and do not contribute to the specific heat. The only electrons that contribute are those within an energy range of $\sim \pm k_BT$ of the Fermi surface, which can be shown to lead to a linear dependence of C_V on T. Such behavior is characteristic of highly degenerate ($T \ll T_F$) Fermi systems and is also seen in liquid ^{3}He and in liquid ^{3}He–^{4}He solutions (see below).

With a lattice (phonon) specific heat varying as T^3 at low temperatures, and an electron specific heat varying as T, it is evident that the latter contribution will eventually dominate if the temperature is reduced sufficiently. For a typical metal, it is necessary to go down at least to temperatures near 1 K in order to measure the electronic contribution to the specific heat.

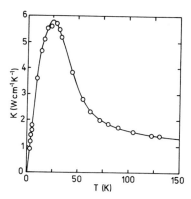

FIGURE 7 The thermal conductivity κ of chromium plotted as a function of temperature T. [From Harper. A. F. A., Kemp, W. R. G., Klements, P. G., Tainsh, R. J., and White, G. K. (1957). *Philos. Mag.* **2**, 577.]

Thermal conduction in metals can in principle occur through both the phonon and the electron gases. However, the phonons tend to scatter the electrons, and *vice versa*; in practice, it turns out that most of the conduction is due to the electrons. A typical $\kappa(T)$ characteristic for a pure metal is shown in Fig. 7. At high temperatures electron-phonon scattering limits the electronic mean free path, so κ rises with decreasing T because of the corresponding decrease in the number (and average energy) of phonons. Eventually, the phonons become so few in number that this scattering process becomes unimportant. The electronic mean free path then becomes temperature independent, being limited by defect scattering or, for a very pure and perfect crystal, by boundary scattering. The heat capacity falls linearly with decreasing T; the relevant velocity for the electrons is the Fermi velocity v_F, which is constant; the mean free path is constant; and κ, too, falls linearly with T at very low temperatures.

The electron gas can also, of course, support the passage of an electrical current. The electrical resistivity ρ is governed by electron–phonon scattering at high temperatures. It consequently falls as T is decreased; a typical result is as sketched in Fig. 8. At a sufficiently low temperature, electron–phonon scattering becomes negligible and the resistivity approaches a temperature-independent value, ρ_0, known as the *residual resistivity*. The magnitude of ρ_0 provides an excellent measure of the purity and crystalline quality of the sample and is usually quoted in terms of the *resistivity ratio*, the ratio by which the resistivity changes on cooling from room temperature. In extreme cases of purity and crystal perfection, resistivity ratios as large as 10^5 have been recorded.

Both the electrical and thermal conductivities of metals are greatly reduced by impurities or imperfections in the lattice. Alloys, for example, are an extreme case and usually have conductivities that are many orders of magnitude

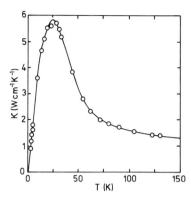

FIGURE 8 Sketch to indicate the variation of electrical resistivity ρ with temperature T in a metal. In the limit of very low temperatures, ρ usually approaches a constant value, ρ_0, known as the *residual resistivity*.

lower than those of pure metals. For this reason, they are especially valuable in the construction of cryostats since less heat will then flow from the hotter to the colder parts of the system. Stainless steel and, to a lesser extent, brass and German silver, are frequently used for such purposes.

C. Superconductivity

There are a large number of materials whose electrical resistance vanishes abruptly at a critical temperature T_c somewhere in the range below 120 K, corresponding to the onset of super conductivity. The phenomenon is a remarkable one, for which no known analog exists at room temperature. The transition itself is extremely sharp for a pure metal (with a width of 10^{-5} K in gallium, for example), and the electrical resistance at very low temperatures does actually appear to be zero. A circulating current in a closed loop of the superconductor, which is easily detectable by the external magnetic field that it creates, will flow without measurable diminution for as long as the experiment is continued.

As well as being perfect conductors, superconductors are also perfect diamagnets. An applied magnetic field is always excluded from the bulk of the material, regardless of whether the field was applied before or after cooling through the superconducting transition at T_c. This phenomenon, known as the *Meissner effect*, involves the appearance of circulating currents at the surface of the superconductor which create a magnetic field that exactly cancels the applied one, inside the material.

The basic mechanism of superconductivity was accounted for in considerable detail by the theory of Bardeen, Cooper, and Schrieffer (BCS theory), which discusses the phenomenon in terms of the formation of *Cooper pairs* of electrons through an attractive interaction mediated by the lattice. Unlike individual electrons,

the Cooper pairs are bosons and form a condensate analogous to those for liquid ^{4}He or liquid ^{3}He below their superfluid transition temperatures. The condensate possesses a macroscopic wave function that extends throughout the sample and implies the coherent collective motion of a very large number of particles. To change the state of any single particle would involve making simultaneous changes in the states of all the others, which is a highly improbable event and helps to explain the resistanceless flow of a current.

III. LIQUID HELIUM

A. Liquid ^{4}He and Liquid ^{3}He

The low-temperature phase diagrams of ^{4}He and ^{3}He are shown in Fig. 9(a) and (b), drawn on the same scale for convenient comparison. It can be seen immediately that, although the two phase diagrams clearly have much in common, there are also some interesting differences. Both isotopes remain liquid to the lowest attainable temperatures, but each can be solidified by the application of sufficient external pressure, greater pressure being needed in the case of ^{3}He (35 bar in the low-temperature limit) than in the case of ^{4}He (25 bar). A pronounced minimum is evident at $\sim$0.3 K in the melting curve of ^{3}He. Both liquids undergo superfluid transitions, although, as remarked above, that for ^{3}He takes place at a temperature (T_c) a thousand times lower than that (T_λ) for ^{4}He.

The non-solidification of the liquid heliums at very low temperatures is, in part, a consequence of the extreme weakness of their interatomic forces. The helium atom

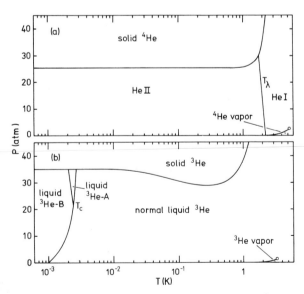

FIGURE 9 The low-temperature phase diagrams of (a) ^{4}He and (b) ^{3}He in zero magnetic field.

possesses an inert gas structure and is chemically inert. It can be envisaged as a small, rigid sphere, much like the hard sphere conventionally assumed by classical kinetic theory. The only attractive force that exists between helium atoms is that due to the Van der Waals interaction, which is exceptionally weak owing to the tightness with which the two electrons are bound to the nucleus (helium having the highest ionization potential of any atom). The weakness of the interatomic forces would not, in itself, prevent the solidification of helium under its saturated vapor pressure, though it would imply, of course, that a very low temperature would be needed for the process to occur.

The other vital consideration affecting the possibility of solidification and influencing also almost all of the other properties of liquid helium is its very high zero-point energy. In quantum mechanics, the lowest permitted kinetic energy for a particle is not necessarily zero. If a particle of mass m is confined to a volume V, then its minimum kinetic energy, or zero-point energy, is

$$E_z = h^2/8m(4\pi/3V)^{2/3}$$

where h is Planck's constant. The zero point energy of helium is particularly large because of its relatively small atomic mass.

In order to minimize its total energy, the liquid tends to expand, thereby increasing V and decreasing E_z. Because of the weakness of the interatomic forces, the effect of zero point energy on the density of the liquid is very large. As a result, liquid ^{4}He has a density only one-third of the value that would be obtained if the atoms were touching each other. Liquid ^{3}He is affected even more strongly on account of its smaller atomic mass, its density being reduced by a factor of about one-quarter. Both isotopes of helium can thus be envisaged as forming liquids in which the atoms are exceptionally widely spaced and therefore able to slip past each other particularly easily; liquid helium, therefore, has rather gas-like properties.

The key to an understanding of the liquid heliums lies, in fact, in the realization that they are often best considered as though they were gases. As a first approximation, therefore, they can be treated as ideal gases. Liquid ^{4}He, being composed of atoms that are bosons, should thus be subject to Bose–Einstein statistics; liquid ^{3}He, composed of fermions, should be subject to Fermi–Dirac statistics. Any deviations of the properties of the actual liquids from those predicted for ideal gases of the same density can then be attributed to the finite sizes of the atoms and to the attractive interactions between them.

At temperatures above 2.17 K, liquid ^{3}He and liquid ^{4}He are very similar to each other and both behave like rather non-ideal dense classical gases. At 2.17 K under its natural vapor pressure, however, ^{4}He undergoes the so-called

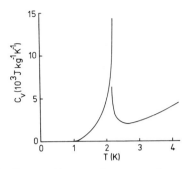

FIGURE 10 The specific heat C_V of liquid ^{4}He as a function of temperature T [From Atkins, K. R. (1959). "Liquid Helium." Cambridge Univ. Press, London.

lambda transition and acquires a whole new set of properties, many of which are quite unexpected in character.

B. Superfluid Properties of Liquid ^{4}He

The properties of liquid ^{4}He above and below the lambda transition temperature T_λ are so completely different that it is almost as though one were dealing with two separate liquids; to emphasize the distinction they are referred to as He I and He II respectively. The name of the transition derives from the characteristic shape of the logarithmic singularity that occurs in the specific heat (Fig. 10) at that temperature, which is strongly reminiscent of a Greek lambda. As can be seen from Fig. 9, T_λ is weakly pressure dependent, falling towards lower temperatures as the pressure is increased.

One of the most striking properties of He II is its ability to flow effortlessly through very narrow channels. An interesting demonstration can be seen in superfluid film flow, illustrated in Fig. 11, where a vessel containing He II tends to empty; and an empty one that is partially immersed in He II tends to fill up until the inside and outside levels

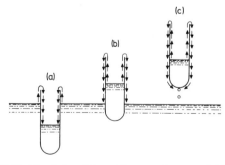

FIGURE 11 Gravitational flow of the creeping superfluid film. (a) and (b) In the case of a partially immersed vessel, film flow equalizes the inside and outside levels. (c) A suspended vessel of He II will eventually empty completely via the film. [From Daunt, J. G., and Mendelssohn, K. (1939). *Proc. R. Soc. (London)* **A170**, 423.]

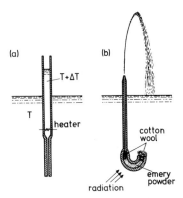

FIGURE 12 (a) When the temperature of the He II inside the vessel is raised slightly above that of the bath, liquid flows in through the capillary tube, which acts as a superleak. (b) The same phenomenon (using radiative heating in the case illustrated) can be used to create a dramatic fountain of liquid helium. [From Allen, J. F., and Jones, H. (1938). *Nature* **141**, 243.]

equalize. What happens is that the adsorbed film of helium on the walls of the vessel acts as a syphon through which the liquid can pass under the influence of the gravitational potential. The same phenomenon would doubtless take place for other liquids, too, such as water, were it not that their viscosities effectively immobilize their films on the walls, so that flow at an appreciable rate cannot take place. In the case of He II, rapid nonviscous flow occurs within the film which, despite its narrowness (∼10 nm), can still carry an appreciable volume flow rate of liquid.

It would be a gross oversimplification, however, to say that He II is inviscid. The situation is much more complicated than this, as can be seen from the fact there are often thermal effects associated with superflow. For example, when He II is allowed to drain partially from a vessel whose exit is blocked with a superleak of tightly packed power (impenetrable to conventional fluids) it is found that the remaining liquid is at a higher temperature than it was at the start. The inverse experiment can also be carried out. When an open-topped cylinder is connected to a reservoir of He II via a superleak (e.g., a very fine capillary tube) as shown in Fig. 12a, and the He II inside the cylinder is warmed to a temperature slightly above that of the reservoir, its level rises. If the top of the vessel is drawn out into a fine jet and the temperature of the He II inside is raised sufficiently, the liquid squirts to form a fountain as shown in Fig. 12(b), hence the common apellation *fountain effect* for this phenomenon.

Attempts to measure the viscosity of He II directly can lead to ambiguous results. The liquid can flow through a narrow tube or orifice, with zero pressure head needed to induce the flow, implying that the viscosity is itself zero. On the other hand, finite viscosity values can be measured

through investigations of the damping of oscillating disks or spheres immersed in the liquid.

Two seemingly quite different models have been developed to describe the strange modes of behavior of He II: the *two-fluid model* and the *excitation model*, each of which has its own range of utility and appliability. It can be demonstrated theoretically that the two pictures are formally equivalent, in that the latter model implies the former.

We discuss first the phenomenological two-fluid model, which is generally the more useful in describing the behavior of the liquid between 0.5 K and T_λ. The model postulates that He II consists of an intimate mixture of two separate but freely interpenetrating fluids, each of which fills the entire container. There is a superfluid component of density ρ_s that has zero viscosity and carries no entropy, and there is a normal fluid component of density ρ_n that has a finite viscosity, carries the whole entropy of the liquid, and behaves much like any ordinary fluid. The ratios ρ_s/ρ and ρ_n/ρ are not predicted by the model, but can readily be determined by experiment, when it is found that $\rho_s/\rho \to 1$ as $T \to 0$ and $\rho_s/\rho \to 0$ as $T \to T_\lambda$ (with opposite behavior occurring in ρ_n/ρ so as to keep the total density ρ equal to $\rho_n + \rho_s$), as shown in Fig. 13. Any transfer of heat to the liquid transforms some of the superfluid into normal fluid in such a way that the total density $(\rho_n + \rho_s)$ remains constant. A form of two-fluid hydrodynamics, taking explicit account of these thermal effects, has been developed and gives an excellent description of most of the properties of the liquid. It is found that, unlike ordinary fluids, the superfluid may be accelerated, not only by a gradient in pressure but also by a gradient in temperature.

The two-fluid model provides a conceptual framework within which most of the phenomena associated with He II can be understood. For example, a viscosity of zero is found for flow-through fine channels, because only the superfluid component then moves. When a sphere or disk oscillates in bulk He II, however, its motion is damped by the normal fluid component which surrounds it, and the

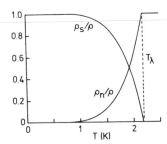

FIGURE 13 The normal and superfluid densities, ρ_n and ρ_s, of He II as functions of temperature T, divided in each case by the total density ρ of the liquid.

viscosity appears to be finite. The rise in the temperature of the residual liquid observed when He II is allowed to drain away through a superleak occurs because only the superfluid component leaves so the entropy of the liquid left behind must remain constant; its entropy density, and hence its temperature, therefore increases.

The liquid is effectively a superconductor of heat, so it is extremely difficult to create a temperature gradient. If a heater is switched on in the liquid, it converts superfluid component to normal fluid, which flows away while more superfluid moves through it towards the heater—an extremely efficient heat-removal process. This *superfluid heat flush* effect carries all suspended impurities away from the heater, because they are part of the normal fluid, and it can therefore be used as the basis of a technique for isotopic purification.

New phenomena have been predicted on the basis of the two-fluid equations of motion—for example, the existence of a novel wave mode known as *second sound* in which the normal and superfluid components oscillate in antiphase such that the density remains constant. The characteristic wave velocity (typically 20 m sec^{-1} but depending on temperature) is much slower than that of ordinary or first sound (typically 240 m sec^{-1}), a pressure-density wave in which the two components oscillate in phase. Second sound is a temperature-entropy wave at constant pressure and density; it may be created by means of an oscillating voltage applied to a heater and may be detected by means of a thermometer.

It is tempting to identify the superfluid and normal fluid components of He II with the atoms of the condensate and excited states, respectively, of an ideal Bose–Einstein gas, but this, too, turns out to be a gross oversimplification. Nonetheless, it is clear that Bose–Einstein statistics do play an essential role in determining the properties of the liquid. If one calculates the Bose–Einstein condensation temperature of an ideal gas with the same density and atomic mass as liquid ^{4}He, one finds $T_c \simeq 3.1$ K, and the difference between this value and T_λ may plausibly be ascribed to the influence of the interatomic forces. The presence of a condensate of ^{4}He atoms in the zero momentum state has recently been confirmed by inelastic neutron scattering measurements. As shown in Fig. 14, the condensate fraction rises from zero at T_λ towards an asymptotic limit of about 14% at low temperatures. This departure from the approximately 100% condensate characteristic of an ideal gas for $T \ll T_c$ is an expected consequence of the non-negligible interatomic forces. Of couse, as already noted in Fig. 13, the superfluid component does form 100% of the liquid at $T = 0$ and it cannot, therefore, be directly identified with the condensate fraction.

The excitation model of He II perceives the liquid as being somewhat akin to a crystal at low temperatures. That

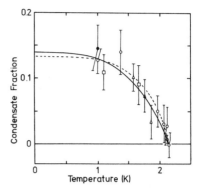

FIGURE 14 The Bose–Einstein condensate fraction of liquid ^{4}He as a function of temperature T. The points and the dashed curve represent experimental data from various sources. The full curve is a theoretical temperature dependence fitted to the data, enbling a $T = 0$ K value of $14 \pm 2\%$ to be deduced. [From Sears, V. F., Svensson, E. C., Martel, P., and Woods, A. D. B. (1982). *Phys. Rev. Lett.* **49**, 279; Campbell, L. J. (1983). *Phys. Rev. B* **27**, 1913.]

is, the liquid is envisaged as an inert background "either" in which there can move a gas of particle-like excitations which carry all the entropy of the system. The dispersion curve of the excitations as determined by inelastic neutron scattering is shown in Fig. 15. Usually, only those excitations within the thickened portions of the curve need be considered, corresponding to *phonons* at low momentum and to *rotons* at higher momentum. The phonons are quantized, longituadinal sound waves, much like those found in crystals; the physical nature of the rotons remains something of an enigma.

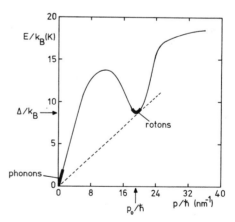

FIGURE 15 The dispersion curve for elementary excitations in He II under its saturated vapor pressure. The energy E of an excitation (divided by Boltzmann's constant to yield temperature units) is plotted against its momentum p (divided by h to yield the wave vector in units of reciprocal length). Only those states within the thickened portions of the curve, *phonons* and *rotons*, are populated to any significant extent in thermal equilibrium. The graident of the dashed line drawn from the origin to make a tangent near the roton minimum is equal to the Landau critical velocity v_L. [From Cowley, R. A., and Woods, A. D. B. (1971). *Can. J. Phys.* **49**, 177.]

The entropy and specific heat of the liquid are readily calculated by standard statistical mechanical methods in terms of the shape of the dispersion curve; that is, in terms of the velocity of sound c, which specifies the gradient dE/dP at small values of p, and of the roton parameters Δ, p_0, and μ, which specify, respectively, the energy, momentum, and effective mass of a roton at the minimum. Good agreement is obtained between calculated and measured values, provided that the temperature is not too high. Close to T_λ, this simple picture breaks down, for two reasons. First, the roton parameters themselves become temperature dependent, and, second, the short lifetimes of the excitations cause a broadening of their energies. Again, provided that the temperature is not too high, it is possible to calculate the momentum carried by the gas of excitations. This quantity may be identified with the momentum carried by the normal fluid component of the two-fluid model. Indeed, it emerges that, in a very real sense, the excitations *are* the normal fluid component. It is then straightforward to calculate ρ_s and ρ_n as functions of T, in terms of c, Δ, p_0, and μ.

The excitation model also provides a natural explanation for the superfluidity of He II. If it is assumed that a moving object in the liquid can dissipate energy only through the creation of excitations of energy E and momentum p and that energy and momentum must be locally conserved, then it can easily be demonstrated that the minimum velocity at which such processes can occur is given by E/p. For most liquids, where the thermal energy is associated with individual atoms or molecules, $E = p^2/2m$ and E/p can take any value down to zero. The same argument would apply to the Bose–condensate for an ideal gas. In the case of He II, however, E/p can never fall below the value given by the gradient of a tangent drawn from the origin to the roton minimum of the dispersion curve, as shown in Fig. 15. This is the minimum velocity at which dissipation can occur in the superfluid component and is known as the Landau critical velocity, v_L. Values of v_L have been determined from measurements of the drag on a small moving object (a negative ion) in He II as a function of its speed v. As shown in Fig. 16, the results show that there is no drag on the object until a speed of v_L has been attained, in excellent agreement with theoretical predictions based on the dispersion curve.

Fundamental (quantum mechanical, manybody) theories of He II usually seek to calculate the dispersion curve of Fig. 15 starting from a knowledge of the radii and massess of ^{4}He atoms and of the interactions between them. Once the dispersion curve has been correctly derived, superfluidity and the two-fluid model follow automatically.

In actual fact, critical velocities measured for He II, for example in flow experiments, are usually a great deal

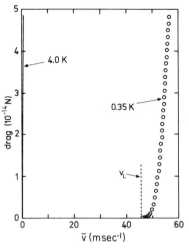

FIGURE 16 The drag force on a negative ion moving through He II at 0.35 K, as a function of its speed $\bar{v}$. The drag remains immeasurably small until $\bar{v}$ has reached the Landau critical velocity for roton creation, v_L, but then rises rapidly with further increase of $\bar{v}$. The behavior of a negative ion in He I at 4.0 K is also plotted, in order to emphasize the profound qualitative difference between the two cases. [From Allum, D. R., McClintock, P. V. E., Phillips, A., and Bowley, R. M. (1977). *Phil. Trans. R. Soc. (London)* **A284**, 179.]

smaller than v_L. This is not because there is anything wrong with the physical arguments leading to v_L; rather, it is an indication that other kinds of excitation can exist in the liquid, in addition to phonons, rotons, and related regions of the dispersion curve. In particular, *quantized vortices* can exist in the superfluid component. A quantized vortex in He II is a linear singularity, with a core region of atomic or subatomic dimensions, around which the superfluid flows tangentially with a velocity $\mathbf{v}_s$, that varies inversely with radial distance from the core, so that the circulation:

$$\kappa_c = \int \mathbf{v}_s \cdot \mathbf{dl} = nh/m_4$$

where d is a superfluid order parameter, l the intrinsic angular momentum, h Planck's constant, m_4 the mass of a ^{4}He atom, and n an integer. In practice, for free vortices n is always equal to unity. Samples of He II, regardless of their history, always seem to contain at least a few quantized vortices. When a characteristic critical flow velocity is exceeded (typically of a few mm sec^{-1} or cm sec^{-1}, depending on the channel geometry and dimensions), these grow into a dense disordered tangle, somewhat akin to a mass of spaghetti but with no free ends, thereby giving rise to a dissipation of the kineitc energy of the flowing liquid.

Quantized vortices also play a central role in the "rotation" of the He II within a rotating vessel. Strictly, the superfluid component of He II cannot rotate at all because of a requirement that:

$$\text{curl } \mathbf{v}_s = 0$$

which, by use of Stokes' theorem, quickly leads to the conclusion that the circulation $\kappa_c = 0$ in any simply connected geometry (i.e., where any closed loop can in principle be shrunk to a point without intercepting boundaries). This condition implies that the superfluid component of He II in a vessel on the earth's surface does not rotate with the vessel but remains at rest relative to the fixed stars. There is convincing experimental evidence that this is indeed the case for small enough angular velocites. For sufficiently large angular velocities of the container, though, both components of the He II *seem* to come into rotation with the vessel. What is actually happening, however, is that a number of quantized vortices are formed parallel to the axis of rotation of the vessel; these cause the superfluid component to simulate solid-body-like rotation, because each vortex core moves in the combined flow fields of all the other vortices. Once vortices have been formed, the geometry is no longer simply connected, and the form of the flow field of a vortex is such that, except in the vortex cores, the liquid still remains curl-free, as required.

The inability of the superfluid component to rotate in a conventional manner is one of the most striking demonstrations of a macroscopic quantum phenomenon. The quantization condition for the vortices (above) emerges naturally from the postualte that the superfluid must be described by a *macroscopic wave function*, exactly analogous to the microscopic wave functions used to describe electrons in atoms, but on the scale of the containing vessel. The existence of macroscopic quantized vortices may seem surprising, but there is overwhelming evidence for their reality, culminating in the development of an imaging system that achieved the vortex photographs shown in Fig. 17.

Although quantized vortices are central to an understanding of He II, and most of their properties are now reasonably well understood, the mechanism by which they are created in the first place remains something of a mystery. As mentioned above, all samples of He II invariably seem to contain a few vortices. and it is now believed that these "background vortices" are formed as the liquid is being cooled through the lambda transition. It is also clear that vortices can be created in the superfluid at lower temperatures (for example, by a moving object), provided that a characteristic critical velocity is exceeded, but the mechanism is difficult to study because of the interfering effect of the background vortices.

The only reliable experiments on the *creation* of quantized vortices in the cold superfluid are those based on the

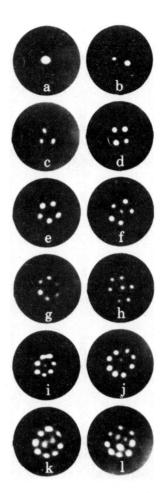

FIGURE 17 Quantized vortex lines in a rotating bucket of He II for various angular velocities, photographed from above. [From Yarmchuk, E. J., Gordon, M. J. V., and Packard. R. E. (1979). *Phys. Rev. Lett.* **43**, 214. Photograph by courtesy of R. E. Packard.]

use of negative ions. They have revealed that the mechanism involves a form of *macroscoopic quantum tunneling* (MQT); the whole system undergoes a discontinuous quantum transition, though an energy barrier, to a state with a vortex loop of finite length attached to the ion (there being no intermediate states for which momentum and energy can be conserved). Figure 18 shows measurements of the rate v at which negative ions nucleate vortices in isotopically pure He II for three electric fields, as a function of reciprocal temperature T^{-1}. The flat regions of the curves on the right-hand side of the figure correspond to MQT through the energy barrier; the rapidly rising curves at higher temperatures (left-hand side) correspond to thermally activated jumps over the top of the barrier attributable to phonons.

Except for low velocities (usually much smaller than v_L) and small heat currents, the flow and thermal properties of He II are dominated by quantized vortices. If

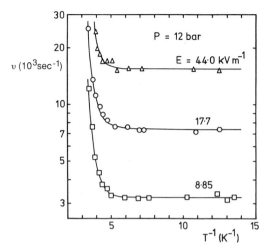

FIGURE 18 Measurements of the rate ν at which quantized vortices are created by negative ions moving through isotopically pure He II, plotted as a function of reciprocal temperature T^{-1} for three electic fields E. In stronger fields, the ions move faster and ν is correspondingly larger. The temperature-independent mechanism operative at low temperatures-independent mechanism operative at low temperatures (right-hand side of figure) is attributable to macroscopic quantum tunneling through an energy barrier; the rapid rise of ν with temperature at higher temperatures (left-hand side) is attributable to thermal activation over the barrier. [From Hendry, P. C., Lawson, N. S., McClintock, P. V. E., and Williams, C. D. H. (1988). *Phys. Rev. Lett.* **60,** 604.]

a critical velocity or heat current is exceeded, then the superfluidity breaks down and a self-sustained tangle of vortex lines builds up within the liquid. Being a metastable state, it continuously decays and can only be maintained by drawing energy from the flow field or heat current. Recent studies have shown that, despite its quantization, such turbulence can often behave in a surprisingly classical manner.

Quantized vortices in He II are the analog of the quantized flux lines found in superconductors, where magnetic field plays the same role as angular velocity in He II. From this viewpoint, He II can be regarded as analogous to an extreme Type II semiconductor. The very small core dimensions of the helium vortices implies, however, that the equivalent of the "second critical field" can never be attained in practice. Corresponding to that angular velocity of the system at which the vortex cores touched each other, it would require angular velocities far beyond the strengths of available container materials.

C. Normal Liquid ^{3}He

Liquid ^{3}He above 1 K is much like a dense classical gas and has properties very similar to those of He I. On further cooling, however, there is a *gradual* change in its behavior to a quite different set of properties, even for

temperatures well above that of the superfluid transition, $T_c \simeq 1$ mK. Above 1 K, for example, the viscosity of the liquid is almost temperature independent and its thermal conductivity κ decreases with decreasing temperature, but at 50 mK, on the other hand, η varies as T^{-2} and κ as T^{-1}.

Qualitatively, this is very much the kind of behavior to be expected of an ideal Fermi gas. The Fermi temperature calculated for an ideal gas of the same density and atomic mass as liquid ^{3}He is $T_F \simeq 5$ K, but one has to bear in mind that a consequence of the interactions between the atoms is to increase the effective mass of each atom, thereby reducing T_F. The gradual change in properties of the liquid ^{3}He on cooling can therefore be attributed to the change between classical and Fermi–Dirac behavior as the liquid enters the degenerate ($T \ll T_F$) regime where a well-defined Fermi-sphere exists in momentum space. The specific heat of liquid ^{3}He becomes linear in T at very low temperatures, just as expected for an ideal Fermi gas, and is thus closely analogous to the specific heat of the electron gas in a metal. The paramagnetic spin susceptibility of the liquid, which varies as T^{-1} at high temperatures, becomes temperature independent below ~ 100 mK. Again, this parallels the properties of the electron gas in a normal metal and is the behavior predicted theoretically for the ideal Fermi gas.

Although the properties of liquid ^{3}He are qualitatively almost identical with those of the ideal Fermi gas, it is not surprising that there are large quantitative differences; the system is, after all, a liquid and not a gas, so the interatomic forces cannot be left completely out of account. As already remarked, one effect of the interatomic forces is to raise the effective mass from m_3 for a bare ^{3}He atom to a larger value, m_3^*. It is not, however, possible on this basis alone to reconcile the departures from ideal behavior for all the properties of ^{3}He. That is to say, the values of m_3^* derived on this basis from measurements of different properties would themselves be widely different.

A detailed quantitative understanding of liquid ^{3}He at low temperatures requires the application of Landau's *Fermi liquid theory*, which takes explicit account of the interactions, and parameterises them in the form of a small number of dimensionless constants known as *Landau parameters*. For most purposes, only three of these parameters are needed (usually written as F_0, F_1 and G_1) and almost all of the properties of the interacting Fermi liquid can be calculated in terms of them. Numerical values of the Landau parameters are not predicted by the theory but are to be found by experiment. The crucial test of the theory—a requirement that consistent values of the Landau parameters should be obtained from widely differing kinds of experiment—is convincingly fulfilled.

An interesting situation arises in a Fermi system at very low temperatures when the average time τ between collisions becomes very long. For any given sound frequency ω, there will be a temperature at which ω and τ^{-1} become equal. At lower temperatures than this, where $\omega\tau \gg 1$, the propagation of ordinary sound in an ideal Fermi gas clearly becomes impossible because the collisions simply do not occur fast enough. For the interacting Fermi liquid, however, Landau predicted that a novel, "collisionless" form of sound propagation known as *zero sound* should still be possible corresponding to a characteristic oscillating distortion in the shape of the Fermi sphere; zero sound has indeed been observed in liquid ^{3}He.

The curious minimum in the melting curve of ^{3}He near 0.3 K (Fig. 9) may be understood in terms of relative entropies of the two phases and of the Clausius–Clapeyron equation:

$$(dp/dT)_m = \Delta S_m / \Delta V_m$$

where m the subscript refers to the melting curve and ΔS and ΔV are the changes in entropy and volume that occur on melting. The negative sign of $(dP/dT)_m$ below 0.3 K (Fig. 9) shows that the solid entropy must then be *larger* than that of the liquid. In fact, this is only to be expected. Atoms in solid are localized on lattice sites and therefore distinguishable, so they will be described by Boltzmann statistics. The entropy of the solid below 1 K arises almost entirely from the nuclear spins, which can exist in either of two permitted states, and will therefore take the (temperature-independent) value of $R \ln 2$ per mole in accordance with the Boltzmann–Planck equation. Given that the specific heat of the liquid, and therefore its entropy, varies approximately linearly with T, there is bound to be a characteristic temperature where the entropy of the liquid falls below that of the solid. Experimental values of the solid and liquid entropies are plotted in Fig. 19, where it can be seen that the two curves do indeed cross at exactly the temperature of the minimum in the melting curve. (The sudden descent toward zero of the solid entropy, in accordance with the third law, corresponds to an antiferromagnetic ordering transition at about 1 mK.)

The observation that the entropy of solid ^{3}He can sometimes be lower than that of the liquid appears at first sight to be quite astonishing considering that entropy gives a measure of the disorder in a system and that a liquid is usually considered to be in a highly disordered state compared to a crystalline solid. The resolution of this apparent paradox lies in the fact that, actually, liquid ^{3}He really is a highly ordered system but one in which the ordering takes place in momentum space, rather than in ordinary Cartesian space.

The existence of the minimum in the melting curve of ^{3}He is of considerable importance to cryogenic technology, in addition to its purely scientific interest, in that it provides the basis for the cooling method known

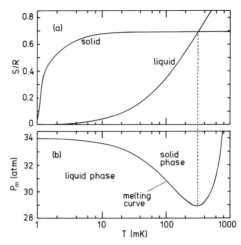

FIGURE 19 (a) The entropy S of liquid and solid ^{3}He (upper diagram) measured in units of the gas constant R, plotted as functions of temperature T and the melting pressure P_m of ^{3}He as a function of T. It may be noted that the minimum in the melting curve occurs at the temperature where the difference between the solid and liquid entropies changes sign. [From values tabulated by Betts, D. S. (1976). "Refrigeration and Thermometry Below One Kelvin," Sussex Univ. Press, London.]

as *Pomerancluk refrigeration*, a topic to which we will return.

D. Superfluid Properties of Liquid ^{3}He

Very soon after the publication of the BCS theory of superconductivity in 1957, it was realized that a similar phenomenon could perhaps occur in liquid ^{3}He. That is, the system might be able to reduce its energy if the ^{3}He atoms formed Cooper pairs. It was quickly appreciated, however, that the hard-core repulsion of the relatively large-diameter ^{3}He atoms would preclude the formation of pairs with zero relative orbital angular momentum ($L = 0$). Cooper paris with finite orbital angular momenta might, however, be energetically fovored but the temperature of the transition was difficult to predict with confidence. The phenomenon was eventually discovered at Cornell University in 1972 by Osheroff, who noted some unexpected anomalies in the otherwise smooth pressurization characteristics of a Pomeranchuk (compressional cooling) cryostat. Follwing that initial discovery, an immense amount of research has taken place and the superfluid phases of the liquid are now understood in remarkable detail. They share many features in common with He II but are vastly more complicted on account of their inherent anisotropy and their magnetic properties.

The low-temperature part of the ^{3}He phase diagram is shown in Fig. 20. It is very strongly influenced by the application of a magnetic field. In Fig. 20(a), with zero magnetic filed, there are two distinct superfluid phases

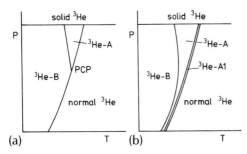

FIGURE 20 Schematic diagram illustrating the effect of a magnetic field on the low-temperature region of the ^{3}He phase diagram. In (a) there is no magnetic field and in (b) the magnetic field has been applied.

designated ^{3}He-A and ^{3}He-B, and there is a unique point, the polycritical point (PCP), where ^{3}He-A, ^{3}He-B and normal ^{3}He all co-exist in equilibrium. The application of a magnetic field causes this picture to change dramatically, as shown in Fig. 20(b). The PCP has then disappeared and, instead, a narrow slice of A-phase extends right down to meet the temperature axis at zero pressure. Simultaneously, a very narrow region of a third superfluid phase, the Al-phase, has appeared separating the A-phase from the normal Fermi liquid. As the magnetic field is further increased. the region of B-phase retreats towards lower temperatures until, at a field of $\sim$0.6T, it has disappeared entirely.

The specific heat of liquid ^{3}He undergoes a finite discontinuity at the superfluid transition. This behavior is utterly different from the lambda transition (a logarithmic singularity) observed for liquid ^{3}He, shown in Fig. 10. It is, however, precisely the behavior expected at the transition to a BCS-like state; it is a beautiful example of a second-order phase transformation and is closely similar to the specific heat anomaly observed at the normal/superconducting transition for a metal in a zero magnetic field. When the liquid is cooled in a finite magnetic field, a pair of second-order transitions is seen, one for normal/Al and the other for the Al/A transition. The A/B transition is completely different in character, however, being a first-order phase transition (like boiling or freezing) in which a significant degree of supercooling can occur as the temperature of the liquid is being reduced.

Numerous experiments, and particularly those based on nuclear magnetic resonance (NMR) have shown that the superfluid phases of ^{3}He involve Cooper pairs in which the nuclear spins are parallel (equal spin pairing, or ESP, states with net $\mathbf{S} = 1$) and for which the orbital angular momentum quantum number $L = 1$. The Cooper pairs are thus rather like diatonic molecules, with atoms orbiting each other in states of finite relative angular momentum.

For Cooper pairs with $\mathbf{S} = 1$, $L = 1$ there are a very large number of possible states for the liquid as a whole. It has turned out that of the three possible spin projection values $\mathbf{S}_z = 0$ or ± 1, on the quantization axis, only $\mathbf{S}_z = \pm 1$ pairs are present in the A-phase; all three types of pairs are present in the B-phase; and only $\mathbf{S}_z = \pm 1$ pairs are present in the Al-phase. In each case, the relative orientations of the spin and orbital angular momenta adjust themselves so as to minimize the free energy of the liquid as a whole, and they do so in quite different ways for the A- and B-phases.

In the A-phase (and the Al-phase), the orbital angular momentum vector for every pair is orientated in the same direction, so the liquid as a whole must carry a resultant intrinsic angular momentum $\mathbf{I}$. The A-phase is strongly anisotropic, with an energy gap $\Delta(T)$ that varies from zero in the direction of $\mathbf{I}$ to a maximum value in the direction of $\mathbf{S}$, perpendicular to $\mathbf{I}$. There is a superfluid order parameter $\mathbf{d}$, analogous to the (scalar) macroscopic wave function in He II, that points in the same direction as $\mathbf{I}$. The vector $\mathbf{I}$ intersects boundaries at right angles, but it also tends to orientate perpendicular to $\mathbf{S}$ and thus perpendicular to an external magnetic field. The liquid therefore acquires *textures* in much the same way as liquid crystals.

The B-phase has an isotropic energy gap, but the direction of $\mathbf{d}$ varies for different points on the Fermi surface. The liquid itself is not isotropic, however, and possesses a symmetry axis $\mathbf{N}$ that tends to align with external magnetic fields and is analogous to the directorix vector of a liquid crystal.

There is overwhelming evidence from numerous experiments for the superfluidity of both the A- and B-phases. One convincing demonstration is shown in Fig. 21, which plots the damping of a wire vibrating in the liquid as a function of temperature. Just above T_c, the viscosity of the liquid is comparable with that of light machine oil and rising so fast as T is reduced that the damping at T_c itself was too large to measure. Below T_c, however, the damping fell steadily with T to reach a value that was actually *smaller* than had been measured in a good vacuum at room temperature.

Most of the superfluid properties of He II (such as the creeping film, second-sound, and so on) have now also been demonstrated for superfluid ^{3}He. Like He II, the liquid can be described in terms of a two-fluid hydrodynamics. The ^{3}He superfluid component is associated with the condensate of Cooper pairs, and the normal fluid component with unpaired fermions in excited states. There is a Landau critical velocity representing the onset condition for pair-breaking, about 10^{-3} of the size of v_L for He II. The two-fluid properties of superfluid ^{3}He are greatly complicated by its anisotropy, particularly in the case of the

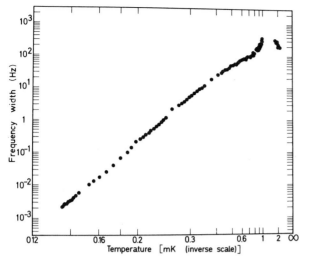

FIGURE 21 The damping experienced by a vibrating wire immersed in liquid ^{3}He under a pressure of 0.0 atm. The measured frequency width Δv of the resonance, which gives a measure of the damping, fell by more than five orders of magnitude as the sample was cooled from T_c (1.04 mK) to the minimum temperature reached (0.14 mK). [From Guénault, A. M., Keith, V., Kennedy, C. J., Miller, I. E., and Pickett, G. R. (1983). *Nature* **302**, 695.]

A-phase where quantities such as the superfluid density vary with direction.

Superfluid ^{3}He has a wide range of interesting and important magnetic properties (which were of particular importance in the original identification of the physical nature of the phases). For example, both the A and B phases possess characteristic frequencies corresponding to oscillations of the **d** vector about its equilibrium orientation. They may be measured either by NMR, when the A-phase exhibits an extraordinary longitudinal resonance (with parallel static and radiofrequency fields), or by measurement of the frequency of parallel ringing. The latter phenomenon, known as the *Leggett effect*, is a very weakly damped magnetic oscillation of the liquid that occurs when the external magnetic field is stepped up or down by a small increment. It can be understood in terms of a transfer, by tunneling, of Cooper pairs between what are effectively separate but weakly coupled superfluids composed respectively of $\mathbf{S}_z = +1$ or $\mathbf{S}_z = -1$ pairs; that is, a kind of internal Josephson effect within the liquid.

IV. ACHIEVEMENT AND MEASUREMENT OF LOW TEMPERATURES

A. Evaporation Cryostats

Cryogenic temperatures are most easily achieved and maintained with the aid of liquefied gases, or *cryogens*,

the most commonly used being nitrogen and helium. They may be liquefied on site or may be purchased as liquid. In either case, they are stored in highly insulated vessels prior to their use for cooling devices or samples of material in cryostats. For temperatures between the normal boiling points of N_2 and ^{4}He (77 K and 4.2 K, respectively), continuous-flow helium cryostats are often employed, with the temperature being adjusted by controlling the rate at which cold helium vapor removes the heat flowing in to the sample from room temperature. Table II summarizes the main cooling methods available in the temperature range below 4 K.

To reach temperatures below its normal boiling point, any given cryogen can be pumped so that it boils under reduced pressure. Nitrogen becomes solid at 63 K, however; its vapor pressure is so small that this temperature is the effective lower limit for a nitrogen cryostat. Liquid hydrogen solidifies at about 14 K, but the solid has a remarkably high vapor pressure and the temperature can be further reduced to about 10 K by sublimation under reduced pressure; on safety grounds, however, liquid hydrogen is seldom used in the laboratory. Liquid ^{4}He, as already mentioned, never solidifies under its own vapor pressure. The temperature reached is limited, instead, very largely by superfluid film flow; the creeping film tends to climb up the walls or up the pumping tube, evaporates in the warmer regions, and consequently saturates the vacuum pump. For this reason, the practical lower limit for a pumped ^{4}He cryostat is about 1 K, although with a diaphragm to reduce the perimeter over which the film flows and with the aid of very powerful pumps, lower temperatures may be achieved.

A typical liquid ^{4}He cryostat dewar is sketched in Fig. 22. Particular care is taken to minimize heat fluxes flowing into the liquid ^{4}He on account of its tiny latent heat; 1 watt is sufficient to evaporate 1 liter of liquid in approximately 1 hour. The inner vessel of liquid ^{4}He (and

TABLE II Some Cooling Techniques Available for Use Below 4 K

Technique	Temperature limits (K)	
	Upper	Lower
^{4}He evaporation[a]	4	1
^{3}He evaporation	1	3×10^{-1}
Dilution refrigeration	5×10^{-1}	3×10^{-3}
Pomeranchuk cooling	3×10^{-1}	10^{-3}
Adiabatic demagnetization[b] (electronic)	1	10^{-3}
Adiabatic demagnetization (nuclear)	10^{-2}	10^{-7}

[a] The lower limit can be pushed below 0.5 K by use of huge pumping speeds, but the rate at which heat can be extracted becomes too small to be useful.

[b] Limits shown are for commonly used salts.

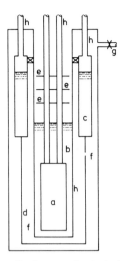

FIGURE 22 Schematic diagram of a typical metal helium cryostat dewar: a—insert with sample/device to be cooled; b—liquid ^{4}He at 4.2 K; c—liquid N$_2$ at 77 K in annular can; d—copper radiation shield at 77 K; e—gas-cooled copper radiation shields; f—vacuum; g—pump-out port for vacuum spaces; h—thin-walled stainless-steel tubes. The insert is usually sealed to the top of the dewar with a top plate and rubber O-ring (not illustrated).

most of the other vertical tubes, also) is made of thin-walled stainless steel in order to minimize the inflow of heat through conduction. For the same reason, it is surrounded by a vacuum space. There is then an annular liquid nitrogen vessel at 77 K, in order to intercept radiation from room temperature, and an outer vacuum space to protect the liquid nitrogen from thermal conduction from room temperature. Some radiation shields placed in the cold outgoing gas are used to prevent room-temperature radiation from shining directly on the cold space at the bottom. The whole helium bath may be pumped in order to adjust the temperature; more commonly, a dewar such as the one in Fig. 22 is used as the starting point for the other refrigeration methods described below. A ^{3}He evaporator or dilution refrigerator, for example, may be immersed in liquid helium in the dewar, boiling under atmospheric pressure at 4.2 K.

For temperatures between 0.3 K and 1.0 K it is convenient to use a ^{3}He evaporation cryostat. Liquid ^{3}He can be pumped to a lower temperature than ^{4}He, partly because of its larger vapor pressure at any given temperature, on account of its lower atomic mass and partly because of the absence of a creeping superfluid film. The material is expensive (at the time of writing, ~\$350 for 1 liter of gas at STP, which yields ~1 cm^3 of liquid), and so is usually used in rather small quantities. A typical "single-shot" ^{3}He cryostat is sketched in Fig. 23. The ^{3}He gas, in total usually 1–5 liters at STP, passes down the ^{3}He pumping tube. It cools as it goes through the region where the tube is im-

mersed in liquid helium at 4.2 K; it condenses in thermal contact with a pumped pot of liquid ^{4}He at ~1.1 K and then runs down into the ^{3}He pot. When condensation is complete, the ^{3}He pot is pumped. Its temperature can readily be reduced to ~0.3 K by use of an oil-diffusion pump. Depending on the influx of heat, the ^{3}He will typically last for several hours before recondensation is needed. Alternatively, the ^{3}He can be returned continuously to the pot from the outlet of the pump, via a capillary tube, with a restriction in the capillary to support the pressure difference of about 1 atmosphere between its top and bottom. In this mode, the cryostat has a higher base temperature which it can, however, maintain for an indefinite period.

B. Helium Dilution Refrigerators

Helium dilution refrigerators can provide continuous cooling within the remarkably large temperature range from about 0.5 K down to about 3 mK. The technique is based on the fact of there being a negative heat of solution when liquid ^{3}He dissolves in liquid ^{4}He. That the method can be carried to such extremely low temperatures is due to the fact that the limiting solubility of ^{3}He in ^{4}He, on the left-hand side of the phase separation diagram (Fig. 24), approaches 6% as $T \to 0$ rather than approaching zero (as happens for ^{4}He in ^{3}He on the right-hand side). There is no conflict with the third law, however, because the ^{3}He atoms have become highly ordered in momentum space, being a degenerate Fermi system with a well-defined Fermi sphere.

In fact, for concentrations of 6% and below, the properties of a ^{3}He–^{4}He solution at low temperatures are

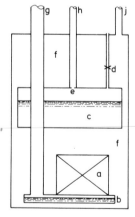

FIGURE 23 Schematic diagram of lower part of a typical single-shot ^{3}He cryostat: a—sample to be cooled; b—liquid ^{3}He at 0.3 K in copper pot; c—liquid ^{4}He at 1.1 K; d—needle-valve on stainless-steel filling tube for liquid ^{4}He; e—restriction to reduce superfluid film flow; f—vacuum; g, h, j—stainless-steel pumping tubes for ^{3}He, ^{4}He, and high vacuum, respectively.

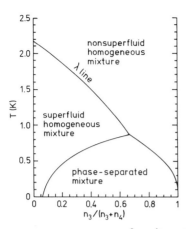

FIGURE 24 Phase-separation for liquid ^{3}He–^{4}He mixtures under their saturated vapor pressure, where n_3 and n_4 are the number densities of ^{3}He and ^{4}He, respectively, and T is the temperature. The finite solubility of ^{3}He in ^{4}He at $T = 0$ K should be particularly noted.

dominated by the ^{3}He. The liquid ^{4}He is almost 100% superfluid component through which the ^{3}He atoms can move freely, quite unhindered by any viscous effects. The ^{4}He can therefore be considered as a kind of background or ether whose only significant influences on the ^{3}He atoms are to increase their effective mass by a factor of about 2.5 and, of course, to prevent them from condensing. The properties of the ^{3}He are almost identical to those calculated for an ideal Fermi gas of the same density. When ^{3}He dissolves in pure liquid ^{4}He at a very low temperature it is as though it were "evaporating" in a "vacuum"; there is a corresponding latent heat, and it is this that is exploited in the dilution refrigerator.

Dilution refrigerators have been built successfully in a variety of different configurations, but the arrangement sketched in Fig. 25 is by far the most common one used in practice. In operation, the cooling effect occurs in the mixing chamber as ^{3}He atoms cross the interface between the concentrated ^{3}He-rich phase floating on top of the dilute ^{4}He-rich phase; the "evaporation" therefore takes place downwards. The dilute phase is connected to a still, which is usually held at about 0.7 K. When the still is pumped, the vapor that is removed is predominantly ^{3}He, because its vapor pressure is so much greater than that of ^{4}He. The depletion of ^{3}He within the still gives rise to an osmotic pressure, driving ^{3}He toward the still from the mixing chamber, where more ^{3}He can then dissolve across the phase boundary. The ^{3}He pumped off from the still is subsequently returned to the mixing chamber via a heat exchanger, so that it is cooled by the cold outgoing stream of ^{3}He on the dilute side. The art of designing efficient dilution refrigerators lies principally in the optimization of the heat exchanger.

The dilution refrigerator has become the work-horse of ultralow temperature physics. Its great virtues are, first, the huge temperature range that can be covered and, second, the fact that the cooling effect is continuous. Experiments can thus be kept at a temperature of a few mK for weeks or months at a time. In practice, the dilution refrigerator provides the usual starting point for additional "one shot" cooling by adiabatic nuclear demagnetization (see below), enabling experiments to be carried down to temperatures deep in the microkelvin range.

C. Pomeranchuk Refrigeration

Pomeranchuk cooling enables temperatures as low as 2 mK to be achieved and maintained, typically for a few hours. The technique exploits the negative latent heat of solidification of the ^{3}He that arises from the Clausius–Clapeyron equation and the fact that $(dP/dT)_m$ is negative for ^{3}He when $T < 0.3$ K (see above). If a volume of liquid ^{3}He is progressively compressed, therefore, it will cool as soon as solid starts to form, provided $T < 0.3$ K. When all of the liquid has been solidified, then the experiment is over and the cell must be allowed to warm

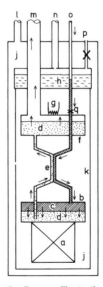

FIGURE 25 Schematic diagram illustrating the mode of operation of the standard dilution refrigerator. The concentrated (^{3}He-rich) phase is diagonally shaded and the dilute (^{4}He-rich) phase is dotted, the arrows indicating the direction of the ^{3}He circulation. a—sample to be cooled; b—mixing chamber between approximately 3 mK and 0.8 K; c—almost pure ^{3}He; d—dilute (approximately 6%) solution of ^{3}He in ^{4}He; e—heat exchanger; f—still at approximately 0.8 K; g—heater; h—pumped pot of liquid ^{4}He at approximately 1.2 K; j—vacuum; k—radiation shield at 1.2 K; l, m, n—stainless-steel pumping tubes for high vacuum, ^{3}He and ^{4}He, respectively; o—stainless-steel condensing-in tube for ^{3}He returned from pump; p—stainless-steel filling tube for ^{4}He pot, with needle-valve; q—restriction.

up again as the solid decompresses before a new cycle can start.

The main problem in practice is to compress the liquid/solid mixture smoothly, without jerking, which would introduce heat. This can be accomplished through the use of liquid ^{4}He as a hydraulic medium to effect compression of the ^{3}He on the other side of a bellows. Because the solidification pressure of ^{4}He is considerably lower than that of ^{3}He (see Fig. 9), it is actually necessary to use, for example, two sets of bellows to form a hydraulic amplifier.

The great merit of Pomeranchuk refrigeration is the relatively large amount of heat that can be removed, being about ten times greater than in dilution refrigeration for any given ^{3}He circulation or solidification rate. It is particularly suitable for studies of ^{3}He itself, given the difficulties of heat transfer at very low temperatures and the fact that the cooling effect occurs within the sample under investigation. As already mentioned above, the original discovery of the superfluid phases of liquid ^{3}He took place in a Pomeranchuk cell. An inherent limitation of such studies, however, is that they are restricted to the melting pressure.

D. Magnetic Cooling

The technique of adiabatic demagnetization has been used for temperatures from 1 K down to the lowest attainable. Following the successful development of dilution refrigerators capable of providing useful continuous cooling powers below 10 mK the method is now used principally to cover the temperature range below ∼5 mK. It is inherently a single-shot procedure. A variety of materials have been used as working media for demagnetization, but the basis of the technique is the same. The medium must be paramagnetic; that is, it must contain magnetic elements that can be orientated by an externally applied magnetic field thereby increasing the order and decreasing the entropy of the system.

The cooling procedure is sketched in Fig. 26, where it can be seen that, for any given temperature, the entropy is smaller when the magnetic field has been applied. The first stage is (ideally) an isothermal magnetization of the sample; the heat liberated as a result must be removed by, for example, the dilution refrigerator used for precooling the system. Once the magnetic field has fully been applied, the thermal link between it and the dilution refrigerator is broken, and the magnetic field is gradually reduced. Because the sample is now thermally isolated, this process is adiabatic and, hence, ideally, isentropic. The sample must therefore cool in order to get back to the B = 0 entropy curve again.

The temperature that can be reached must be limited ultimately by a ferromagnetic (or antiferromagnetic) or-

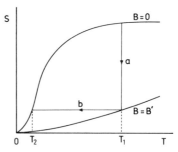

FIGURE 26 Schematic diagram illustrating the working principles of cooling by adiabatic demagnetization. The entropy S of the paramagnetic material is plotted against temperature T for zero applied magnetic field B, and for the large value $B = B'$. The system is precooled (e.g., by means of a dilution refrigerator) to $T = T_1$ and the external field is applied isothermally (a). The thermal link to the dilution refrigerator is then broken and the magnetic field is removed adiabatically, i.e., isentropically (b), when the system cools to the reduced final temperature T_2.

dering transition among the magnetic elements. It is not possible to cool the sample very far below the transition. Materials whose paramagnetism arises from electron spins, such as cerium magnesium nitrate (CMN), can be used to cool experiments to temperatures in the mK range, CMN itself being one of the best known and allowing temperatures near 1 mK to be attained. For lower temperatures than this, it is essential to choose a cryogenic medium whose magnetic elements interact less strongly with each other. The interactions between the magnetic C_e^{3+} ions in the CMN may, for example, be reduced by increasing their average separation. This can be accomplished by dilution of the CMN with the (nonmagnetic) lanthanum magnesium nitrate to produce a mixture known as CLMN. Although lower temperatures may thereby be attained, in principle, because of the resultant depression of the ordering temperature to well below 1 mK there is less available cooling power because of the reduced number of magnetic ions. A better approach is to use a nuclear paramagnetic system, such as the nuclei of copper atoms. The magnetic moments of the nuclei are so small that the interactions between them are negligible near 1 mK; the ordering transition takes place at ∼100 nK. On the other hand, precisely because of the small nuclear moment, it is relatively difficult to align the nuclei with an external magnetic field. Successful nuclear adiabatic demagnetization therefore requires the largest attainable fields (in practice ∼10 T) and the lowest possible starting temperatures (in practice ∼7 mK, if the precooling time is not to be too long).

The main experimental problem in the μK and low mK ranges lies not so much in providing sufficient cooling power, for which the cryogenic technology is now readily available, as in promoting heat flow between the system under investigation and the refrigerant. For example, the nuclear spin system of copper is relatively easily and quickly cooled to 10^{-6} K but it is very much more difficult to cool even the conduction electrons in the same piece of copper and, more importantly, via the electrons and the lattice, to cool the item under study. This is because there is a thermal boundary resistance between all materials which becomes enormous at ultralow temperatures and implies the existence of a temperature discontinuity across any boundary across which heat is flowing. Thus, it is often the heat leak into the sample in the nuclear demagnetization cryostar that determines the ultimate temperature to which it can be cooled. By very careful vibration isolation, the screening of external electromagnetic fields, and judicious design and choice of materials, it is possible to reduce this heat leak to well below 1 nW. Such precautions have enabled liquid ^{3}He, for example, to be studied down to a temperature of approximately 120 μK (see Fig. 21).

An essential feature of any adiabatic demagnetization cryostat is a breakable thermal link between the refrigerant and the mixing chamber of the dilution refrigerator (or other precooler). A superconducting link of, for example, tin has been found particularly convenient and effective. It has a high thermal conductivity when driven into the normal state by an applied magnetic field but an extremely low thermal conductivity when in the superconducting state. The refrigerant can therefore be connected thermally to the mixing chamber or thermally isolated, simply by switching the magnetic field at the tin link on or off.

The most common nuclear refrigerant is copper, usually powdered or in the form of a bundle of wires, in order to reduce eddy current heating during demagnetization. Other nuclear refrigerants have also been used including, particularly, enhanced hyperfine interaction systems such as $PrNi_5$, which provides a much larger cooling power but cannot reach as low a final temperature as copper.

E. Thermometry

A very wide range of thermometric devices has been developed for the measurement of cryogenic temperatures. Thermometry between 4 K and room temperature is dealt with under *Cryogenic Process Engineering*; in this article, we discuss thermometry in the temperature range below 4 K. The principal thermometers and their approximate ranges of useful application are shown in Table III.

Vapor-pressure thermometry can only be based on ^{3}He or ^{4}He, being the only non-solid materials below 4 K. The pressure may be measured relatively easily by means of a manometer or McLeod gauge and then compared with international vapor pressure tables in order to find the

TABLE III   Some Cryogenic Thermometers Suitable for Use Below 4 K

Thermometer	Temperature limits (K)	
	Upper	Lower
^{4}He vapor pressure	4	1
^{3}He vapor pressure	1.5	5×10^{-1}
Carbon resistance	>4	10^{-3}
Germanium resistance	>4	5×10^{-2}
Thermocouples	>4	5×10^{-1}
Paramagnetic susceptibility[a] (electronic)	2	10^{-3}
Paramagnetic susceptibility (nuclear)	10^{-2}	$<10^{-6}$
Nuclear orientation	10^{-1}	10^{-3}
Noise thermometry	>4	10^{-2}

[a] For materials used in practice.

temperature. The rapid variation of vapor pressure with temperature ensures that the latter can be determined to high precision from pressure measurements of modest accuracy.

Resistance thermometry below 4 K is usually based on commercial carbon (radio) resistors or on germanium resistors. The measuring power must be kept small in order to prevent internal self-heating and varies in practice from about 10^{-7} W at 4 K down to 10^{-15} W in the low mK range. Both forms of resistor provide for rapid, convenient, and sensitive measurements, but both must be individually calibrated prior to use. Carbon resistors are usually recalibrated after each thermal cycle to room temperature but have the merits of cheapness and ready availability. Germanium resistors are much more stable over long periods and many thermal cycles, but are relatively expensive. In each case, more than one device is needed to cover the temperature range indicated with adequate sensitivity but without the resistance becoming too large for reliable measurement.

Thermocouples are generally best for temperature measurements above 4 K, where they are more sensitive, but some recently developed rhodium/iron alloys now permit measurements of useful sensitivity down to about 0.5 K. Once a given batch of material has been characterized, no further calibration is needed.

Paramagnetic susceptibility thermometers using salts such as CMN permit accurate temperature measurements down to about 1 mK and have the advantage that the technique becomes more sensitive as the temperature decreases. To a good approximation, over much of the range, a CMN thermometer follows Curie's Law and its susceptibility is inversely proportional to the absolute temperature; a single calibration point—for example, against ^{3}He vapor pressure at 0.6 K—is therefore in principle sufficient to calibrate the device for measurements down to temperatures 50 times colder. At temperatures below 10 mK, however, due allowance must be made for the fact that Curie's Law is no longer accurately followed. The onset of the ordering transition provides a lower limit below which that particular susceptibility thermometer cannot be used. Thermometers based on the nuclear susceptibility of a metal can be employed to below 1 μK but require particularly sensitive measurement techniques because of the small magnitude of this quantity. Superconducting quantum interferometer devices (SQUIDs) can be employed to particular advantage in such measurements.

Nuclear orientation thermometry depends on the anisotropic emission of γ-rays from polarized radioactive nuclei. The extent of the anisotropy is detemined by the degree of polarization which, in turn, varies inversely with the absolute temperature. Since the nuclear hyperfine splittings are usually known to high precision, the thermometer is in principle absolute and does not require calibration. The useful temperature range is relatively narrow. however, being limited by radioactive self-heating at the lower end and insensitivity at the upper end.

Noise thermometry is another technique that, in principle, provides absolute measurements of temperature. The Johnson noise voltage V_N generated by the random (Brownian) motion of electrons in an element of resistance R is measured over a frequency bandwidth Δf. The mean-squared noise voltage $\langle V_n^2 \rangle = 4k_B R T \Delta f$, from which T may be deduced if ΔF can also be measured. Unfortunately, a lack of sensitivity prevents the extension of this type of measurement below 10 mK where it would be most useful.

V. CRYOGENIC APPLICATIONS

Cryogenics is finding useful applications over an extraordinarily diverse range of engineering and technology. One of the most important and most widely exploited of all low-temperature phenomena is that of superconductivity, which is being applied to the construction of powerful magnets used for particle accelerators, for power storage, in medicine, and in superconducting motors and generators. It also holds out the promise of loss-free power transmission along superconducting cables. Superconducting instrumentation based on SQUIDs enables extraordinarily sensitive measurements to be made of very weak magnetic fields and is being widely used in archaeology, geology, and medicine.

The reduction of thermal (Johnson) noise in electronic circuits by cooling to cryogenic temperatures can yield ex-

tremely valuable improvements in signal/noise, thus facilitating the detection of very weak signals. Cooled infrared detectors are regularly used in astronomy, particularly for space and upper-atmosphere experiments.

The liquid heliums are being used to model phenomena that occurred in the very early universe, 10^{-35} s after the "big bang." There are theoretical reasons to expect that topological defects in space–time (e.g., cosmic strings) were formed as the universe fell through the critical temperature of 10^{27} K at which the GUT (grand unified theory) phase transition occurred. Cosmic strings are of particular interest and importance because, in some cosmologies, they are believed to have provided the primordial density inhomogeneities on which the galaxies later condensed. Unfortunately, defect formation at the GUT transition cannot be studied experimentally on account of the enormous energies involved, many orders of magnitude larger than available in even the most powerful particle acclerators, but the physics of the process can be studied by exploiting the close mathematical analogy that exists between the GUT and superfluid transitions. In practice, liquid helium is taken through the transition quickly and evidence is sought for the production of quantized vortices, which are the superfluid analogues of cosmic strings. At the time of writing, this process seems to have been reliably confirmed at the superfluid transition in ^{3}He but, surprisingly, not yet at the lambda transition in ^{4}He.

Isotopically pure liquid ^{4}He is being used as a down-scattering medium for the production and storage of ultracold neutrons (UCN)—neutrons of such low energy (and thus long de Broglie wavelength) that they can be totally reflected from certain materials. UCN can be held inside a neutrons, bottle and a range of experiments conducted on them. Because ^{3}He is a strong absorber of neutrons, the required isotopic ratio is extremely demanding, being $<10^{-12}$ compared with the $\sim 10^{-7}$ ratio typically found in natural well helium. Fortunately, purities of this order and better are readily achieved through use of superfluid heat flush to sweep away the ^{3}He atoms, which move with the normal fluid component.

Liquefied gases (O_2, N_2, H_2, He) find a wide range of applications in the steel industry, biology and medicine, the space program, and the food industry.

An interesting phenomenon known as the *quantum Hall effect*, which occurs in the two-dimensional electron gas in, for example, MOSFETs (metal-oxide-semiconductor-field-effect transistors), is becoming extremely valuable in metrology. At very low temperatures and strong magnetic fields, it is possible to exploit the effect to make absolute measurements of e^2/h (in effect, the atomic fine-structure constant) to very high precision.

SEE ALSO THE FOLLOWING ARTICLES

CHEMICAL THERMODYNAMICS • CRYOGENIC PROCESS ENGINEERING • MAGNETIC MATERIALS • QUANTUM THEORY • STATISTICAL MECHANICS • SUPERCONDUC-TIVITY • THERMOMETRY

BIBLIOGRAPHY

Barone, A., and Paterno, G. (1982). "Physics and Applications of the Josephson Effect," Wiley, New York.

Barron, R. (1985). "Cryogenic Systems," 2nd ed. Oxford Univ. Press, London.

Betts, D. S. (1989). "An Introduction to Millikelvin Technology," Cambridge Univ. London.

Donnelly, R. J. (1990). "Quantized Vortices in He II," Cambridge University Press, London.

Hands, B. A., ed. (1986). "Cryogenic Engineering," Academic Press, London.

Lounasmaa, O. (1974). "Experimental Principles and Methods below 1 K," Academic Press, New York.

McClintock, P. V. E., Meredith, D. J., and Wigmore, J. K. (1984). "Matter at Low Temperatures," Blackie, Glasgow; Wiley-Interscience, New York.

Mendelssohn, K. (1977). "The Quest for Absolute Zero," 2nd ed. Taylor & Francis, London.

Shachtman, T. (1999). "Absolute Zero and the Conquest of Cold," Houghton Mifflin, Boston, MA.

Tilley, D. R., and Tilley, J. (1986). "Superfluidity and Superconductivity," 2nd ed. Hilger, London.

Vollhardt, D., and Wölfle, P. (1990). "The Superfluid Phases of Helium 3," Taylor & Francis, London.

Weisend, J. G. (1998). "Handbook of Cryogenic Engineering," Taylor & Francis, London.

White, G. K. (1979). "Experimental Techniques in Low Temparature Physics, " 3rd ed. Clarendon Press, Oxford.

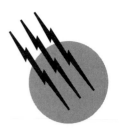

Cryptography

Rebecca N. Wright

AT&T Labs—Research

GLOSSARY

Ciphertext All or part of an encrypted message or file.

Computationally infeasible A computation is *computationally infeasible* if it is not practical to compute, for example if it would take millions of years for current computers.

Cryptosystem A *cryptosystem* or *cryptographic system* consists of an encryption algorithm and the corresponding decryption algorithm.

Digital signature Cryptographic means of authenticating the content and sender of a message, like a handwritten signature is to physical documents.

Encryption The process of transforming information to hide its meaning. An encryption algorithm is also called a *cipher* or *code*.

Decryption The process of recovering information that has been encrypted.

Key A parameter to encryption and decryption that controls how the information is transformed.

Plaintext The *plaintext* or *cleartext* is the data to be encrypted.

Public key cryptosystem A two-key cryptosystem in which the encryption key is made public, and the decryption key is kept secret.

Symmetric key cryptosystem A traditional single-key cryptosystem, in which the same key is used for encryption and decryption.

CRYPTOGRAPHY, from the Greek *krýpt-*, meaning hidden or secret, and *gráph-*, meaning to write, is the science of secret communication. Cryptography consists of encryption or enciphering, in which a *plaintext* message is transformed using an encryption key and an encryption algorithm into a *ciphertext* message, and decryption or deciphering, in which the ciphertext is transformed using the decryption key and the decryption algorithm back into the original plaintext. Cryptography protects the privacy and sometimes the authenticity of messages in a hostile environment. For an encryption algorithm to be considered secure, it should be difficult or impossible to determine the plaintext from the ciphertext without knowledge of the decryption key.

Historically, cryptography has been used to safeguard military and diplomatic communications and has therefore been of interest mainly to the government. Now, as the use of computers and computer networks grows, there is an increasing amount of information that is stored electronically, on computers that can be accessed from around the world via computer networks. As this happens, businesses and individuals are finding more of a need for protection of information that is proprietary, sensitive or expensive to obtain.

Traditionally, encryption was used for a *sender* to send a message to a *receiver* in such a way that others could not read or undetectably tamper with the message. Today, encryption protects the privacy and authenticity of data in transit and stored data, prevents unauthorized access to computer resources, and more. Cryptography is commonly used by almost everyone, often unknowingly, as it is increasingly embedded into ubiquitous systems, such as automated bank teller machines, cellular telephones, and World Wide Web browsers.

I. INTRODUCTION

Cryptography probably dates back close to the beginnings of writing. One of the earliest known examples is the Caesar cipher, named for its purported use by Julius Caesar in ancient Rome. The Caesar cipher, which can not be considered secure today, replaced each letter of the alphabet with the letter occurring three positions later or 23 positions earlier in the alphabet: A becomes D, B becomes E, X becomes A, and so forth. A generalized version of the Caesar cipher is an alphabetic substitution cipher. As an example, a simple *substitution cipher* might use the following secret key to provide a substitution between characters of the original message and characters of the encrypted message.

```
A B C D E F G H I J K L M N O P Q R S T U V W X Y Z
F R O A H I C W T Z X L U Y N K E B P M V G D S Q J
```

Using this key, a sample encrypted message is:

```
IEQ IBQDXMBQ FX RMBFQW MOWQB IEQ QLT IBQQ.
```

As is evident from even such a short example, this simple method does not disguise patterns in the text such as repeated letters and common combinations of letters. In fact, if the encrypted message is known to be English text, it is usually quite easy to determine the original message, even without the knowledge of the secret key, by using letter frequency analysis, guessing and checking, and maybe a little intuition and luck. Such substitution ciphers are commonly used today as puzzles in newspapers and puzzle books, but are not secure when used alone as cryptosystems. Polyalphabetic substitution ciphers, developed by Len Battista in 1568, improved on regular substitution ciphers by changing the substitution scheme partway through a message. Although substitution is not secure when used alone, it can be useful when used in conjunction with other techniques, and in fact, many cryptosystems used today benefit from substitution when it is carefully used as part of their encryption algorithms.

Another simple technique that is not secure alone, but can be secure when used as part of a cryptosystem, is *transposition* or *permutation*. A simple transposition cipher might rearrange the message

```
WE MEET AT DAWN IN THE MUSEUM,
```

by removing spaces, writing it in columns of letters whose length is determined by the key, and then reading across the columns

```
W E D I E E
E T A N M U
M A W T U M
E T N H S,
```

yielding the ciphertext

```
WEDIEE ETANMU MAWTUM ETNHS.
```

In this simple example, it is easy to see that an attacker can fairly easily determine the column length by seeing which pairs of letters are most likely to be adjacent in English words.

Used together, repeated combinations of substitution and transpositions can make the job of an attacker who is trying to break the system without knowing the key harder by more thoroughly obscuring the relationship between the plaintext and the ciphertext, requiring an attacker to explore many more possibilities. Many of the mechanical and electrical ciphers used in World Wars I and II, such as the Enigma rotor machine, relied on various combinations of substitutions and permutations.

Cryptanalysis is the process of trying to determine the plaintext of a ciphertext message without the decryption key, as well as possibly trying to determine the decryption key itself. Together, cryptography and cryptanalysis comprise the field of cryptology. The job of the cryptographer is to devise systems faster than the cryptanalysts can break them. Until early the 20th century, cryptanalysts generally had a clear upper hand over cryptographers, and most known ciphers or cryptosystems could easily be broken.

The transition to modern cryptography began with the invention of computers, and continued with the development of ciphers such as the Data Encryption Standard (DES) and the exciting discovery of public key cryptography. The use of computers means more possibilities for ciphers, as sophisticated and lengthy computations that would have error-prone if done by hand and expensive if done by mechanical devices have now become possible.

A. Cryptosystems: A Mathematical Definition

Mathematically, a cryptosystem is defined as three algorithms—a (randomized) key generation algorithm *keyGen*, a (possibly randomized) encryption algorithm *Enc*, and a (usually deterministic) decryption algorithm *Dec*. More specifically, we define the following sets.

$\mathbf{M}$ = set of plaintext messages

$\mathbf{C}$ = set of encrypted messages

$\mathbf{K_E}$ = set of encryption keys

$\mathbf{K_D}$ = set of decryption keys

$\mathbf{R}$ = set of random values for encryption

$\mathbf{R'}$ = set of random values for key generation.

M is called the *message space*, and $\mathbf{K_E} \cup \mathbf{K_D}$ is called the *key space*. In computerized algorithms, the key space and message space are typically sets of all bit strings of particular lengths. As a notational convention, we denote sets by boldface letters and elements of a set by the same letter in italic typeface. The functions *KeyGen*, *Enc*, and *Dec* are defined as follows.

$$KeyGen: \mathbf{R'} \to \mathbf{K_E} \times \mathbf{K_D}$$

$$Enc: \quad \mathbf{M} \times \mathbf{K_E} \times \mathbf{R} \to \mathbf{C}$$

$$Dec: \quad \mathbf{C} \times \mathbf{K_D} \to \mathbf{M},$$

such that for every $r \in \mathbf{R}$ and $r' \in \mathbf{R'}$,

$$Dec(Enc(M, K_E, r), K_D) = M,$$

where $KeyGen(r') = (K_E, K_D)$. We usually suppress explicit mention of the randomization used by *Enc*, and instead write $Enc(M, K_E)$ to donate $Enc(M, K_E, r)$, where r is chosen randomly as specified by an algorithmic description of *Enc*.

Often, $K_E = K_D$; that is, the encryption and decryption keys are identical, and in this case we refer to it simply as the key or the secret key. This is called *symmetric* or *secret key* cryptography. In contrast, in *asymmetric* or *public key* cryptography, the encryption keys and decryption keys are different from each other, and only the decryption key needs to be kept secret. In public key cryptography, the decryption key is also sometimes called the *secret key*.

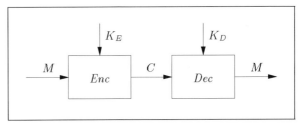

FIGURE 1 Diagram of a cryptosystem.

In order for a cryptosystem to be useful, all three functions *KeyGen*, *Enc*, and *Dec* must be efficiently computable. Since key generation is generally done only infrequently and can often be done in advance, it is acceptable for *KeyGen* to be somewhat less efficient, perhaps even taking many minutes to compute. In contrast, encryption and decryption are usually done more frequently and in real time, so *Enc* and *Dec* should be more efficient, measuring on the order of milliseconds or less.

We would also like additional requirements to capture the security of the cryptosystem—for example, that it is difficult to determine any information about K_D or M from $Enc(M, K_E,)$ alone. However, the specific meaning of this requirement depends the computational power available to an attacker, the abilities of the attacker to learn the encryptions of various messages, and other such factors, so there is not one single definition that can capture security in all settings. A rather strong, and desirable, notion of security is that of *semantic security*: from the ciphertext only, it should be computationally infeasible to learn *anything* about the plaintext except its length. The ciphertext should not reveal, for example, any of the bits of the ciphertext, nor should it suggest that some plaintext messages are more probably than others. Semantic security is much stronger than simply stating that an attacker does not learn the plaintext.

To see the importance of the key generation function in a cryptosystem, consider again the Caesar cipher presented previously. This can be thought of as encryption function that rotates characters in the alphabet according to the key, with a key generation function that always chooses the key 3. It is intuitively easy to see that an attacker who knows that the key is always 3, or infers it by seeing a number of plaintext/ciphertext pairs, clearly has an advantage over an attacker in a system where the key is chosen randomly from 1 to 26. In a system with a large key space, the key generation function can help to formally express the security of the cryptosystem by quantifying the *a priori* uncertainty the attacker has about the decryption key. In some implemented systems, key generation is left to the user, but this can be problematic because users are a bad source of randomness, and therefore this effectively reduces the size of the key space.

B. Goals of Cryptosystems: What Cryptography Can and Cannot Provide

Cryptography can be used to provide a variety of security-related properties. We will use the term "message" to refer either to a message or any other kind of data, either in transit or stored. Cryptography is often used to provide the following important properties.

Confidentiality: protects the contents of data from being read by unauthorized parties.

Authentication: allows the recipient of a message to positively determine the identity of the sender.

Integrity: ensures the recipient that a message has not been altered from its original contents.

Nonrepudiation: allows the recipient of a message to prove to a third party that the sender sent the message.

There are a number of additional security-related properties that cryptography does not directly provide, but for which cryptography can be part of a solution. These include anonymous communication, in which the receiver of a message is prevented from learning the identity of the sender; fair exchange, in which Alice should receive a valid message from Bob if and only if Bob receives a valid message from Alice; privacy from spam (unwanted bulk electronic mail); preventing the recipient of a message from further distributing the message; and protection against message traffic analysis.

Although cryptography is an important tool in securing computer systems, it alone is not sufficient. Even if a strong cryptosystem is used to authenticate users of a computer system before allowing them access, this authentication procedure can be easily subverted if there are other ways for attackers to access the system, whether through mistakenly installed, poorly configured, or just plain buggy software.

II. ATTACKS ON CRYPTOGRAPHIC SYSTEMS

It is generally assumed that an attacker on a cryptographic system knows everything about the specification of the system: that is, the key generation function, the encryption function, and the decryption function. Thus, the security of the system depends only on the fact that the attacker does not know the key, and on the degree to which the ciphertexts produced by the system hide the keys and plaintexts that were used. This is called Kerckhoff's principle, named for Auguste Kerckhoff who advocated it in a book he wrote in 1883. While the attacker may not in fact always know this much information, it is a conservative assumption and avoids "security by obscurity," or basing security on the assumption that the attacker does not know the encryption function. Captured equipment or documentation have frequently played a role in the breaking of military ciphers. In the industrial world, disgruntled employees can often be induced to reveal which ciphers their employers' systems use. If a cryptosystem is implemented in widely distributed software, the algorithm will almost certainly be discovered by reverse engineering, as happened with the RC4 cryptosystem, or leaked, as happened with the A5 cryptosystem. In practice, security by obscurity often turns out not to be security at all because cryptosystems that have not withstood public scrutiny are far more likely to have flaws in their design than those that were heavily scrutinized. Due to these factors, security by obscurity is generally frowned on by the scientific community.

In addition to knowing the specification of the system, an attacker may also know some additional information, such as what kinds of messages will be encrypted. The attacker may also have access to some pairs of plaintext and their corresponding ciphertexts, possibly chosen by the attacker.

A. Types of Attacks and Attackers

There are a number of types of attacks and attackers. One measurement of an attack is how much the attacker is able to learn, described here from weakest to strongest as categorized by Lars Knudsen.

Information deduction: The attacker learns partial information about the plaintext of an intercepted ciphertext or about the secret key.

Instance deduction: The attacker learns the plaintext of an intercepted ciphertext.

Global deduction: The attacker discovers an alternate algorithm for deciphering ciphertexts without the secret key.

Total break: The attacker learns the secret key.

In both global deductions and total breaks, the attacker can then decrypt any ciphertext.

The attacker may have different messages available to analyze in mounting an attack, again described from weakest to strongest.

Ciphertext-only attack: the attacker has access to a number of ciphertexts.

Known-plaintext attack: the attacker has access to a number of plaintext/ciphertext pairs.

Chosen-plaintext attack: the attacker can choose a number of plaintexts and learn their ciphertexts.

Adaptive chosen-plaintext attack: a chosen-plaintext attack in which the attacker can choose which plaintext message to see the ciphertext of next based on all the messages he has seen so far.

Chosen-ciphertext attack: the attacker can choose a number of ciphertexts and learn their plaintexts.

Adaptive chosen-ciphertext attack: a chosen-ciphertext attack in which the attacker can choose which ciphertext message to see the plaintext of next based on all the messages he or she has seen so far.

In the types of attacks just described, it is usually assumed that all the ciphertexts were generated with the same encryption key. In addition, some cryptosystems are susceptible to related-message and related-key attacks, in which the attacker has access to ciphertexts or plaintext/ciphertext pairs for keys or plaintext messages with certain known relationships.

One measure of the practicality of an attack is the number and type of ciphertexts or plaintext/ciphertext pairs it requires. Other measures include the computational complexity, also called the *work factor*, and the storage requirements of the attack.

In the case that encryption is being used to provide properties other than just secrecy, there are additional types of attacks. For example, if encryption is being used to provide authentication and integrity through digital signatures, an attacker may attempt to forge signatures. As above, successful attacks can range from an existential forgery, in which one signed message is forged, to a total break, and attacks can use any number and type of signed messages.

Cryptanalysis describes attacks that directly attack the cryptosystem itself. The main two classes of cryptanalytic attacks, described below, are brute force attacks and structural attacks. We also describe some non-cryptanalytic attacks.

B. Brute Force Attacks

Brute force attacks are ciphertext-only attacks or known-plaintext attacks in which the decryption algorithm is used as a "black box" to try decrypting a given ciphertext with all possible keys until, in the case of a ciphertext-only attack, a meaningful message is found (if here is a way to determine in the context under attack whether a message is "meaningful"), or in the case of known-plaintext attacks, until the ciphertext decrypts to the corresponding plaintext. On average, a brute force will have to check half the key space before finding the correct key. While such attacks are exponential in the length of the key, they can be successfully carried out if the key space is small enough.

C. Structural Attacks

Brute force attacks simply use then encryption algorithm as a black box. It is often possible to mount more effi-

cient attacks by exploiting the structure of the cipher. For example, in attacking the alphabetic substitution cipher, an efficient attacker makes use of the fact that different occurrences of the same letter is always substituted by the same substitute. More examples of structural attacks are given in the discussion of current cryptosystems.

D. Non-Cryptanalytic Attacks

When strong cryptographic systems with sufficiently long key lengths are used, brute force and other cryptanalytic attacks will not have a high probability of success. However, there a number of attacks that exploit the implementation and deployment of cryptosystems that can be tried instead and indeed are often successful in practice. These attacks usually have the goal of learning a user's secret key, though they also may be carried out only to learn a particular plaintext message. As in the case of cryptanalytic attacks, the attacker may or may not have access to plaintext/ciphertext pairs.

1. Social Attacks

Social attacks describe a broad range of attacks that use social factors to learn a user's secret key. These range from attempting to guess a secret key chosen by a user by using information known about the user to calling a user on the telephone pretending to be a system administrator and asking to be given information that will allow the attacker to gain access to the user's private keys. Alternately, the target of a social attack can be the contents of a particular sensitive message, again by fooling the sender or recipient into divulging information about it. Bribery and coercion are also considered social attacks. The best defense against social attacks is a combination of user education and legal remedies.

2. System Attacks

System attacks are attacks in which an attacker attempts to gain access to stored secret keys or stored unencrypted documents by attacking through non-cryptographic means the computer systems on which they are stored. Common ways that this is done are:

- Exploiting known, publicized holes in common programs such as *sendmail* or World Wide Web browsers.
- Computer viruses (usually distributed through e-mail).
- Trojan horse programs (usually downloaded from the Web by unsuspecting users).

In many cases, there are existing and easily available tools to carry out these types of attacks.

Edward Felten and others have described a number of strong attacks that are partially system attacks and partially social attacks, in which they take advantage of certain features in the way systems such as Web browsers are designed, combined with expected user behavior.

The best defenses against system attacks are prevention, detection, and punishment, achieved by a combination of good system administration, good firewalls, user education, and legal remedies.

3. Timing Attacks

Timing attacks were publicized by Paul Kocher in 1996. They attack the implementation of cryptosystems by measuring observable differences in the timing of the algorithm based on the particular value of the key. They then use statistical methods to determine the bits of key by observing many operations using the same key. Timing attacks typically require a significant number of chosen ciphertexts.

Related attacks can use any measure of differences in the performance of the encryption and decryption functions such as power consumption and heat dissipation.

Timing attacks and related attacks can be protected against to some degree by "blinding" the devices performing encryption and decryption computations so that all computations have the same performance, regardless of the particular key and message being used. However, this can have a substantial performance cost, as it requires all computations to have worst-case performance. Such attacks can also be protected against by designing systems so that they will not act as an "oracle" by decrypting and returning all and any messages that come their way, thereby preventing an attacker from obtaining the necessary data to carry out the attack. However, this is not always possible without interfering with the purpose of the system.

III. DESIGN AND USE OF CRYPTOGRAPHIC SYSTEMS

A good cryptosystem should satisfy several properties. It must be efficient to perform encryption and decryption, the encryption and decryption algorithms should be easy to implement, and the system should be resistant to attacks. Earlier, Kerckhoff's principle was noted, it says that the security of a cryptosystem should depend only on the secrecy of the secret key.

A. Provable Versus Heuristic Security

In some cases, it is possible to actually prove that a cryptosystem is secure, usually relative to certain *hardness* *assumptions* about the difficulty of breaking some of its components. In other cases, the security of a cryptosystem is only *heuristic*: the system appears to be secure based on resistance to known types of attacks and scrutiny by experts, but no proof of security is known. While provable security is certainly desirable, most of today's cryptosystems are not in fact provably secure.

B. Confusion and Diffusion

Two important properties that can be used to help in guiding the design of a secure cryptosystem, identified by Claude Shannon in 1949, are *confusion* and *diffusion*. Confusion measures the complexity of the relationship between the key and the ciphertext. Diffusion measures the degree to which small changes in the plaintext have large changes in the ciphertext.

For example, substitution creates confusion, while transpositions create diffusion. While confusion alone can be enough for a strong cipher, diffusion and confusion are most effective and efficient when used in conjunction.

C. Modes of Encryption

A *block cipher* encrypts messages of a small, fixed size, such as 128 bits. A *stream cipher* operates on a message stream, encrypting a small unit of data, say a bit or byte, at a time. While stream ciphers can be designed from scratch, it is also possible to use a block cipher as a stream cipher, as we will see below.

To encrypt a large message or data file using a block cipher, it must first be broken into blocks. A mode of encryption describes how the block cipher is applied to different blocks, usually by applying some sort of feedback so that, for example, repeated occurrences of the same block within a message do not encrypt to the same ciphertext. The main modes of block cipher encryption are *electronic codebook mode (ECB), cipher block chaining mode (CBC), cipher feedback mode (CFB), output feedback mode (OFB)*, and *counter mode (CTR)*.

1. Electronic Codebook Mode (ECB)

In electronic codebook mode, the block cipher is applied independently (with the same key) to each block of a message. While this is the easiest mode to implement, it does not obscure the existence of repeated blocks of plaintext. Furthermore, if an attacker can learn known plaintext/ciphertext pairs, these will allow him or her to decrypt any additional occurrences of the same block. Additionally, the attacker can replay selected blocks of previously sent ciphertexts, and an attacker may be able to use collected ciphertext blocks to replace blocks of a new ciphertext message and change its meaning. For example, a lucky attacker might be able to change an electronic funds transfer to a different amount or a different payee.

The other modes discussed as follows avoid this problem by incorporating some feedback that causes different occurrences of the same plaintext block to encrypt to different ciphertexts.

2. Cipher Block Chaining Mode (CBC)

In cipher block chaining mode, the plaintext of a block is combined with the ciphertext of the previous block via an exclusive or (xor) operation, and the result is encrypted. The result is the ciphertext of that block, and will also be used in the encryption of the following block. An initialization vector (IV) acts as the "previous ciphertext block" for the first plaintext block. The initialization vector can be made public (i.e., can be sent in the clear along with the ciphertext), but ideally should not be reused for encryption of different messages to avoid having the same ciphertext prefix for two messages with the same plaintext prefix.

Decryption reverses the process. The first block of ciphertext is decrypted and then xored with the initialization vector; the result is the first plaintext block. Subsequent ciphertext blocks are decrypted and then xored with the ciphertext of the previous block.

One concern in feedback modes is synchronization after transmission errors. Cipher block chaining is *self-synchronizing*: a transmission error in one block will result in an error in that block and the following block, but will not affect subsequent blocks.

Plaintext block chaining is also possible.

3. Cipher Feedback Mode (CFB)

Cipher feedback mode allows a block cipher with block size n bits to be used as a stream cipher with a data encryption unit of m bits, for any $m \leq n$.

In CFB mode, the block cipher operates on a register of n bits. The register is initially filled with an initialization vector. To encrypt m bits of data, the block cipher is used to encrypt the contents of the register, the leftmost m bits of the result are xored with the m bits of data, and the result is m bits of ciphertext. In addition, the register is shifted left by m bits, and those m ciphertext bits are inserted in the right-most m register bits to be used in processing the next m bits of plaintext.

Decryption reverses the process. The register initially contains the initialization vector. To decrypt m bits of ciphertext, the block cipher is used to encrypt the contents of the register, and the resulting leftmost m bits are xored with the m ciphertext bits to recover m plaintext bits. The m ciphertext bits are then shifted left into the register.

Note that the encryption function of the block cipher is used in encryption *and* decryption of CFB mode, and the decryption function of the block cipher is not used at all.

As in CBC mode, an initialization vector is needed to get things started, and can be made public. In CBC mode, however, the initialization vector *must* be unique for each message encrypted with the same key, or else an eavesdropper can recover the xor of the corresponding plaintexts.

A single transmission error in the ciphertext will cause an error in $n/m + 1$ blocks as the affected ciphertext block is shifted through the register, and then the system recovers.

4. Output Feedback Mode (OFB)

Output feedback mode is similar to CFB mode, except that instead of the leftmost m bits of the ciphertext being shifted left into the register, the leftmost m bits of the output of the block cipher are used. As in CBC mode, encryption proceeds by encrypting the contents of the register using the block cipher and xoring the leftmost m bits of the result with the current m plaintext bits. However, OFB mode introduces insecurity unless $m = n$. As with CFB mode, The initialization vector can be made public and must be unique for each message encrypted with the same key.

In OFB mode, the *key stream*—the sequences of m bits that will be xored with the plaintext (by the sender) or the ciphertext (by the receiver)—depend only on the initialization vector, not on the plaintext or the ciphertext. Hence OFB mode has the efficiency advantage that, provided the sender and receiver agree in advance about what initialization vector their next message will use, the key stream can be computed in advance, rather than having to be computed while a message is being encrypted or decrypted. Since xor is a much more efficient operation than most block ciphers, this can be a substantial gain in the time between the receipt of an encrypted message and its decryption.

In OFB mode, the key must be changed before the key stream repeats, on average after $2^m - 1$ bits are encrypted if $m = n$.

5. Counter Mode (CTR)

Counter mode is similar in structure to the feedback modes, CFB and OFB, except that the register contents are determined by a simple counter modulo 2^m, or some other method for generating unique registers for each application of the block cipher. As in OFB mode, the key stream can be computed in advance. Despite the apparent simplicity of CTR mode, it has been shown to be in some sense at least as secure as CBC mode.

IV. SYMMETRIC KEY CRYPTOGRAPHY

Traditionally, all cryptography was symmetric key cryptography. In a symmetric key cryptosystem, the encryption key K_E and the decryption key K_D are the same, denoted simply by K. The key K must be kept secret, and it is also important that an eavesdropper who sees repeated encryptions using the same key can not learn the key. The simple substitution cipher described earlier is an example of a symmetric key cryptosystem.

A. The One-Time Pad

Invented by Gilbert Vernam and Major Joseph Mauborgne in 1917, the one time pad is a provably secure cryptosystem. It is also *perfectly* secure, in the sense that the proof of security does not depend on *any* hardness assumptions. In the one-time pad, the message space M can be the set of all n-bit strings. The space of keys and ciphertexts are also the set of all n-bit strings. The key generation function chooses an n-bit key uniformly at random. Given a message M and a key K, the encryption $Enc(M, K) = M \oplus K$, the xor of M and K. The decryption of a ciphertext C is $Dec(C, K) = C \oplus K$. It follows that $Dec(Enc(M, K), K) = (M \oplus K) \oplus K = M \oplus (K \oplus K) = M$.

A given ciphertext C can correspond to *any* plaintext message M, specifically when the key is $K = M \oplus C$. Hence, since the K is random and is never reused, it is impossible to learn anything about M from C without the secret key. That is, the one-time pad is perfectly secure.

One-time pads are impractical in most settings because the secret key must be as long as the message that is to be sent and cannot be reused. If the key is reused for two different messages M and M', then the corresponding ciphertexts can be xored to learn $M \oplus M'$. If additional information is known about the plaintexts, such as they are English text, or that it is a bit string with a fixed header, this is usually sufficient to reveal M and M', which in turn also reveals the key K. Nonetheless, if it is possible to exchange a sufficiently long random key stream in advance, which can then be used to encrypt messages until it runs out, the one-time pad still can be useful.

B. Pseudorandom Number Generators

A pseudorandom number generator is a function that takes a short random *seed* and outputs a longer bit sequence that "appears random." To be cryptographically secure, the output of a pseudorandom number generator should be computationally indistinguishable from a random string. In particular, given a short prefix of the sequence, it should be computationally infeasible to predict the rest of the sequence without knowing the seed. Many so-called random number generators, such as those based on linear feedback shift registers (LFSR) or linear congruences, are not cryptographically secure, as it is possible to predict the sequence from a short prefix of the sequence. Despite the fact that LFSRs are not secure, a large number of stream ciphers have been developed using them. Most of these have themselves since been shown to be insecure.

The secure Blum–Blum–Shub generator, developed by Lenore Blum, Manuel Blum, and Michael Shub in 1986, is based on the believed computational difficulty of distinguishing quadratic residues modulo n from certain kinds of nonquadratic residues modulo n.

A cryptographically secure pseudorandom number generator can be used to make a stream cipher by using the seed as a key and treating the generator output as a long key for a pseudorandom one-time pad.

C. Data Encryption Standard (DES)

The Data Encryption Standard (DES) was issued by the United States National Bureau of Standards (NBS) in 1977 as a government standard to be used in the encryption of data. The DES algorithm is based on a cryptosystem called Lucifer that was proposed to the NBS by the IBM Corporation. The standard includes the data encryption algorithm itself, as well as instructions for its implementation in hardware and an order for U.S. federal government agencies to use it for protection of sensitive but non-classified information. Since its initial standardization, DES was heavily used for encryption both for government and non-government applications, both in the United States and elsewhere. Although its key length is short enough to now be susceptible to brute force attacks, DES lived a long lifetime and served its function well.

DES is a block cipher with a 56-bit key and 64-bit blocks. DES is an example of a *Feistel* cipher, so named for one of its designers, Horst Feistel. In the Feistel structure, encryption proceeds in rounds on an intermediate 64-bit result. Each round divides the intermediate result into a left half and a right half. The right half then undergoes a substitution followed by a permutation, chosen based on the key and the round. This output is xored with the left half to make the right half of the next intermediate result; the right half of the old intermediate result becomes the left half of the new intermediate result. In DES, there are 16 such rounds. An initial permutation of the plaintext is performed before the first round, and the output from the last round is transformed by the inverse permutation before the last round. This structure has the advantage that the encryption and decryption algorithms are almost identical, an important advantage for efficient hardware implementations.

As approved as a standard, DES can be used in any of ECB, CBC, CFB, and OFB modes of encryption.

1. Stronger Variants of DES

Although the key length of DES is now short enough to be susceptible to brute force attacks, there are variants of DES that effectively have a longer key length. It is important to note that even though these constructions are more resistant to brute force attacks, it is theoretically possible that they are more susceptible to structural attacks. However, no structural attacks are known on the two constructions presented here that are more efficient than structural attacks on DES itself.

a. Triple-DES.

Triple-DES uses two DES encryption keys, K_1 and K_2. The triple-DES encryption C of a plaintext block M is

$$C = Enc(Dec(Enc(M, K_1), K_2), K_1),$$

where Enc and Dec denote regular DES encryption and decryption. Decryption is

$$M = Dec(Enc(Dec(C, K_1), K_2), K_1).$$

The reason for the encrypt/decrypt/encrypt pattern is for compatibility with regular DES: triple-DES with $K_1 = K_2$ is identical to regular DES. With independently chosen keys, triple-DES has an effective key length of 128 bits.

b. DESX.

DESX, suggested by Ronald Rivest, has an effective key length of 184 bits, and is much more efficient that triple-DES because it only requires a single DES encryption. In DESX, K is a 56-bit key and K_1 and K_2 are 64-bit keys. The DESX encryption C of a plaintext block M is

$$C = K_2 \oplus Enc(K_1 \oplus M, K),$$

where Enc denotes regular DES encryption. Decryption is

$$M = K_1 \oplus Dec(K_2 \oplus C, K).$$

DES compatibility is obtained by taking $K_1 = K_2 = 0$. Joe Kilian and Phillip Rogaway proved that the DESX construction is sound, in that it is in fact more resistant to brute force attacks than DES.

2. Brute Force Attacks on DES

Even at the time of DES's standardization, there was some concern expressed about the relatively short key length, which was shortened from IBM's original proposal. Given a ciphertext/plaintext pair, or from several ciphertexts and a notion of a meaningful message, a brute force attack would on average need to try 2^{55} keys before finding the right key.

Since its introduction, a number of estimates have been given of the cost and speed of doing a brute force DES attack, and a handful of such attacks have actually been performed. Since computers tend to double in speed and halve in price every 18 months, both the cost and the time of these attacks has steadily declined. Furthermore, since DES key search is completely parallelizable, it is possible to find keys twice as fast by spending twice as much money.

When DES was standardized, Whitfield Diffie and Martin Hellman estimated that it would be possible to build a DES-cracking computer for $20 million that would crack a DES key in a day. In 1993, Michael Wiener designed on paper a special purpose brute force DES-cracking computer that he estimated could be built for $1 million and would crack an average DES key in about three and a half hours. In 1997, Wiener updated his analysis based on then-current computers, estimating that a $1 million machine would crack keys in 35 minutes.

In 1998, the Electronic Frontier Foundation (EFF) actually built a DES-cracking computer, at the cost of $200,000, consisting of an ordinary personal computer with a large array of custom-designed chips. It cracks a DES key in an average of four and a half days.

3. Differential and Linear Cryptanalysis of DES

a. Differential cryptanalysis.

In 1990, Eli Biham and Adi Shamir introduced *differential cryptanalysis*, a chosen-plaintext attack for cryptanalyzing ciphers based on substitutions and permutations. Applied to DES, the attack is more efficient than brute force, but it is a largely theoretical attack because of the large number of chosen plaintexts required. As compared to brute force, which requires a single known plaintext/ciphertext pair and takes time 2^{55}, differential cryptanalysis requires 2^{36} chosen plaintext/ciphertext pairs and takes time 2^{37}.

Differential cryptanalysis operates by taking many pairs of plaintexts with fixed xor difference, and looking at the differences in the resulting ciphertext pairs. Based on these differences, probabilities are assigned to possible keys. As more pairs are analyzed, the probability concentrates around a smaller number of keys. One can continue until the single correct key emerges as the most probable, or stop early and perform a reduced brute force search on the remaining keys.

Since the publication of differential cryptanalysis, Don Coppersmith, one of DES's designers at IBM, revealed that the DES design team in fact knew about differential cryptanalysis, but had to keep it secret for reasons of national security. He also revealed that they chose the specific substitution and permutation parameters of DES to provide as much resistance to differential cryptanalysis as possible.

b. Linear cryptanalysis. Linear cryptanalysis was invented by Mitsuru Matsui and Atsuhiro Yamagishi in 1992, and applied by Matsui to DES in 1993. Like differential cryptanalysis, linear cryptanalysis also requires a large number of plaintext/ciphertext pairs. Linear cryptanalysis uses plaintext/ciphertext pairs to generate a linear approximation to each round, that is, a function that approximates the key for each round as an xor of some of the rounds input bits and output bits. An approximation to DES can be obtained by combining the 16 1-round approximations. The more plaintext/ciphertext pairs that are used, the more accurate the approximation will be. With 2^{43} plaintext/ciphertext pairs, linear cryptanalysis requires time 2^{13} and has success probability .85 of recovering the key.

D. Advanced Encryption Standard

In 1997, the United States National Institute of Standards (NIST) began the process of finding a replacement for DES. The new advanced encryption standard (AES) would need to be an unpatented, publicly disclosed, symmetric key block cipher, operating on 128 bit blocks, and supporting key sizes of 128, 192, and 256 bits, large enough to resist brute force attacks well beyond the foreseeable future. Several candidates were submitted, and were considered for security, efficiency, and ease of implementation. Fifteen submitted algorithms from twelve countries were considered in the first round of the selection process, narrowed to five in the second round. On October 2, 2000, NIST announced that it had selected *Rijndael*, a block cipher developed by Belgian cryptographers Joan Daemen and Vincent Rijmen, as the proposed AES algorithm. Rijndael was chosen for its security, performance, efficiency, implementability, and flexibility. Before Rijndael can actually become the standard, it must first undergo a period of public review as a Draft Federal Information Processing Standard (FIPS) and then be officially approved by the United States Secretary of Commerce. This process is expected to be completed by the middle of 2001.

The Rijndael algorithm supports a variable key size and variable block size of 128, 192, or 256 bits, but the standard is expected to allow only block size 128, and key size 128, 192, or 256. Rijndael proceeds in rounds. For a 128-bit block, the total number of rounds performed is 10 if the key length is 128 bits, 12 if the key length is 192 bits, and 14 if the key length is 256 bits.

Unlike the Feistel structure of DES, Rijndael's rounds are divided into three "layers," in each of which each bit of an intermediate result is treated in the same way. (In contrast, the left half and the right half of intermediate results are treated differently in each round of a Fiestel cipher.) The layered structure is designed to resist linear and differential cryptanalysis, and consists of the following.

- Linear mixing layer: adds diffusion.
- non-Linear layer: adds confusion.
- Key addition layer: adds feedback between rounds by xoring a current round key with an intermediate encryption result.

The Rijndael algorithm considers bytes as elements of the finite field GF(2^8), represented as degree 8 polynomials. For example, the byte with decimal representation 105, or binary representation 01101001, is represented as

$$x^6 + x^5 + x^3 + 1.$$

The sum of two such elements is the polynomial obtained by summing the coefficients modulo 2. The product of two such elements is the multiplication of the polynomial modulo the irreducible polynomial

$$m(x) = x^8 + x^4 + x^3 + 1.$$

Let b be the block length in bits. Throughout the encryption process, a matrix of bytes containing containing a partial result is maintained. It is called the State, and has $b/32$ columns and four rows. The State initially consists of a block of plaintext, written "vertically" into the arrays, column by column. At the end, the ciphertext for the block will be taken by reading the State in the same order.

A diagram illustrating the structure of a round is given in Fig. 2. Each round (except the last) consists of four transformations on the State.

ByteSub: This is the non-linear layer. The ByteSub transformation consists of a non-linear byte substitution operating independently on each of the bytes of the State. The substitution is done using an "S-box" determined by taking for each element its inverse element in GF(2^8) followed by some additional algebraic operations. The resulting S-box satisfies several properties including invertibility, minimization of certain kinds of correlation to provide resistance to linear and differential cryptanalysis, and simplicity of description. In an implementation, the entire S-box can either be calculated once and stored as a table, or S-box transformations can be calculated as needed.

ShiftRow: Together, the ShiftRow transformation and the MixColumn operation are the linear mixing layer. The ShiftRow transformation cyclically shifts each rows of the State. For 128-bit blocks, the first row is not shifted, the second row is shifted left by 1 byte,

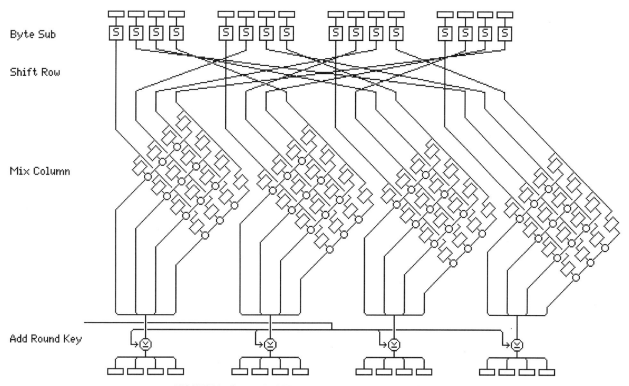

FIGURE 2 A round of Rijndael. (Illustration by John Savard.)

the third row is shifted by 2 bytes, and the last row shifted by 3 bytes.

MixColumn: In the MixColumn transformation, each column of the State is considered as the coefficients of a degree 4 polynomial over $GF(2^8)$, which is then multiplied by the fixed polynomial $c(x) = 3x^3 + x^2 + x + 2$ modulo $x^4 + 1$, where the sum and products of coefficients are as described above.

AddRoundKey: This is the key addition layer. In the AddRoundKey transformation, the State is transformed by being xored with the Round Key, which is obtained as follows. Initially, the encryption key is expanded into a longer key using a Key Expansion transformation. Each AddRoundKey uses the next b bits of the expanded key as its Round Key.

The final round eliminates the MixColumn transformation, consisting only of ByteSub, ShiftRow, and AddRoundKey. This is done to make decryption more structurally similar to encryption, which is desirable for ease of implementation. An additional AddRound-Key is performed before the first round.

Decryption in Rijndael consists of inverting each step. Due to the algebraic properties of Rijndael, the order of steps does not need to be changed for decryption. InvByteSub describes the application of the inverse S-box calculation. InvShiftRow shifts the rows right instead of

left, by the same offsets as ShiftRow. InvMixColumn replaces the polynomial by its inverse in $GF(2^8)$. Since xor is its own inverse, AddRoundKey remains the same, except that in decryption the Round keys must be used in the reverse order. Hence, Rijndael decryption starts with an AddRoundKey step, and then operates in rounds consisting of InvByteSub, InvShiftRow, InvMixColumn, and AddRoundKey, followed by a final round of InvByteSub, InvShiftRow, and InvMixColumn.

At the time of this writing, encryption modes for AES are being determined, and will probably consist of ECB, CBC, CFB, OFB, and counter modes, as well as possibly others.

V. PUBLIC KEY CRYPTOGRAPHY

The concept of public key cryptography was originally proposed in 1976 by Whitfield Diffie and Martin Hellman, and independently by Ralph Merkle. In 1997, Britain's Government Communications Headquarters (GCHQ) released previously classified documents revealing three British government employees, James Ellis, Clifford Cocks, and Malcolm Williamson, developed these same ideas several years earlier, but kept them secret for reasons of national security. There is some evidence that the United States' National Security Agency (NSA) also secretly developed similar ideas as early as the 1960's.

In public key systems, the encryption key K_E, also called the *public key*, and the decryption key K_D, also called the *secret key*, are different from each other. Furthermore, it is computationally infeasible to determine K_D from K_E. Therefore, the encryption key can be made public. This has two advantages for key management. First, instead of each pair of users requiring a different secret key, as in the case of symmetric key cryptography (for a total of $n^2 - n$ encryption keys for n users), each user can have a single encryption key that all the other users can use to send her encrypted messages (for a total of n encryption keys for n users). Second, keys no longer need to be exchanged privately before encrypted messages can be sent.

Although the public keys in a public key cryptosystem need not be communicated secretly, they still must be communicated in an authenticated manner. Otherwise, an attacker Marvin could try convince Bob into accepting Marvin's public key in place of Alice's public key. If Marvin succeeds, then encrypted messages from Bob to Alice will actually be readable by Marvin instead of by Alice. If Marvin has sufficient control over the communication network, he can even prevent detection by Alice by intercepting the messages from Bob and then reencrypting the messages with Alice's real public key and sending them to her. In order to avoid such "man-in-the-middle" attacks, public keys are usually *certified* by being digitally signed by other entities.

Assuming that Alice and Bob already know each other's public keys, they can communicate privately as follows. To send her message M_A to Bob, Alice encrypts it with Bob's public key and sends the resulting ciphertext to Bob. Bob uses his private key to decrypt the ciphertext and obtain M_A. To send his response M_B to Alice, Bob encrypts it with Alice's public key and sends the resulting ciphertext to Alice. Alice uses her private key to decrypt the ciphertext and obtain M_B. An eavesdropper who overhears the ciphertexts does not learn anything about M_A and M_B because she does not have the necessary decryption keys.

The fundamental mathematical idea behind public key cryptosystems are *trapdoor one-way functions*. A function is one-way if it is hard to invert it: that is, given a value y it is computationally infeasible to find x such that $f(x) = y$. A one-way function is said to have the *trapdoor* property if given the trapdoor information, it becomes easy to invert the function. To use a trapdoor one-way function as a public key cryptosystem, the one-way function is used as the encryption algorithm, parametrized by its public key. The trapdoor information is the secret key. Trapdoor one-way functions are conjectured, but not proven, to exist. As such, all known public key cryptosystems are in fact based on widely believed but unproven assumptions. The famous and widely believed conjecture in theoretical computer science, $P \neq NP$, is a necessary but not sufficient condition for most of these assumptions to be true. It is an active area of research in public key cryptography to determine minimal assumptions on which public key cryptosystems can be based.

The earliest proposed public key systems were based on NP-complete problems such as the knapsack problem, but these were quickly found to be insecure. Some variants are still considered secure, but are not efficient enough to be practical. The most widely used public key cryptosystems, the RSA and El Gamal systems, are based on number theoretic and algebraic properties. Some newer systems are based on elliptic curves and lattices. Recently, Ronald Cramer and Victor Shoup developed a public key cryptosystem that is both practical and provably secure against adaptive chosen ciphertext attacks, the strongest kind of attack. The RSA system is described in detail below.

A. Using Public Key Cryptosystems

1. Hybrid Systems

Since public key cryptosystems are considerable less efficient than comparably secure symmetric key systems, public key cryptosystems are almost always used in conjunction with symmetric key systems, called *hybrid systems*. In a hybrid system, when Alice and Bob are initiating a new communication session, they first use a public key cryptosystem to authenticate each other and privately exchange a new symmetric key, called a session key. For the remainder of the session, they use the symmetric session key to encrypt their messages. Hybrid systems enjoy the key management benefits of public key cryptography, most of the efficiency benefits of symmetric key cryptography, as well as gaining additional security from the use of a frequently changing session key.

2. Probabilistic Encryption

It is crucial to the security of a public key cryptosystem that messages are somehow randomized before encryption. To see this, suppose that encryption were deterministic and that an attacker knows that an encrypted message sent to Alice is either "Yes" or "No," but wants to learn which. Since Alice's public key K_E is public, the attacker can simply compute $Enc(\text{"Yes"}, K_E)$ and $Enc(\text{"No"}, K_E)$ to see which one actually corresponds to the ciphertext.

To avoid this problem, Shafi Goldwasser and Silvio Micali introduced *probabilistic encryption*. In probabilistic encryption, the encryption function is randomized rather than deterministic, and ciphertexts of one message should be computationally indistinguishable from ciphertexts of another message. The same plaintext will encrypt to different ciphertexts if different randomization is used, so an attacker who performs trial encryptions as above will only get the same result as the sender of a message if he also used the same randomness. Provably secure probabilistic encryption schemes have been developed based on the Blum–Blum–Shub pseudorandom number generator.

In practice, randomization is achieved by randomly padding the plaintext before encryption, which is then removed as part of decryption. However, this padding must be done carefully, as there have been attacks that successfully exploit padding.

3. Digital Signatures

Some public key cryptosystems can be used to digitally sign messages. Here, the private key is used to sign the message by applying the *Dec* function. The signature is appended to the plaintext message. Anyone can verify Alice's signature on a document by applying the *Enc* function with her public key to the signature part of the message and checking that the result matches the plaintext. If the public key system is secure, digital signatures provide the following.

- **Authenticity:** only Alice knows her private key and could create the signature.
- **Integrity:** if the plaintext is changed in transit, the signature will not match.
- **Non-repudiation:** a third party who knows Alice's public key can also verify that the signature is Alice's.

Unlike handwritten signatures, digital signatures are a function of the document they sign, so it is not possible to undetectably copy the signature from one document onto another.

If Alice and Bob wish to communicate using both encryption and digital signatures, Alice signs messages with her private signature key then encrypts with Bob's public encryption key. Bob decrypts with his private decryption key and then checks the digital signature using Alice's public signature verification key. If the encryption and decryption functions are commutative, as in the case with RSA, the same key pair could be used for both encryption and signatures. However, this is not recommended as it unnecessarily creates scenarios that provide chosen ciphertext and/or chosen plaintext to a potential attacker.

Since public key operations are expensive, often a hash function—a cryptographic compression function—is applied to a long message or file before it is signed. In this case, the signature consists of plaintext appended to the signature of the hash. The signature is verified by applying the hash to the plaintext and applying the decryption function to the signature, and checking whether these two results are identical.

Some public key systems, such as the Digital Signature Algorithm (DSA), can only be used for digital signatures, and not for encryption.

B. RSA

The first concrete public key system proposed with a security that has withstood the test of time was the RSA system, named for its inventors Ronald Rivest, Adi Shamir, and Leonard Adleman. RSA is based on the computational difficulty of factoring the product of large primes. The public key consists of an n-bit modulus N, which is the product of two primes p and q each of length $n/2$ bits, and an element e of the multiplicative group $\mathbb{Z}_n^*$. N is called the *RSA modulus* and e is called the encryption exponent. The decryption exponent is d such that $ed = 1 \mod \phi(n)$, where $\phi(N) = (p-1)(q-1)$ is the Euler totient function. The private key is the pair $\langle N, d \rangle$. Once the public and private keys have been generated, p and q are no longer needed. They can be discarded, but should not be disclosed.

Based on the current state of the art for factoring and other attacks on RSA, current security recommendations as of 2001 usually stipulate that n should be 1024 bits, or 309 decimal digits. Although any efficient algorithm for factoring translates into an efficient attack on RSA, the reverse is not known to be true. Indeed, Boneh and Venkatesan have given some evidence that factoring may be harder than breaking RSA.

Suppose we have a message M that we want to encrypt, and further suppose that it has already been padded with appropriate random padding (which is necessary for security reasons) and represented as an element $m \in \mathbb{Z}_n^*$. If M is too large to represent in $\mathbb{Z}_n^*$, it must first be broken into blocks, each of which will be encrypted as described here. To encrypt the resulting message $m \in \mathbb{Z}_n^*$, it is raised to the eth power modulo N, that is, $Enc(m, \langle N, e \rangle) = m^e \mod N$. Decryption is done by reversing the process: $Dec(c, \langle N, d \rangle) = c^d \mod N$. Therefore, $Dec(Enc(m, \langle N, e \rangle), \langle N, d \rangle) = m^{ed} \mod N = m$.

1. An RSA Example

We illustrate the use of RSA by an example. Let $p = 23$ and $q = 31$. While these values are much too small to produce a secure cryptosystem, they suffice to demonstrate the RSA algorithm. Then $n = pq = 713$ and

$$\phi(n) = \phi(713) = 22 \cdot 30 = 660.$$

The encryption exponent e must be chosen relatively prime to 660, say $e = 97$. The decryption exponent is $d = e^{-1}$ mod $\phi(n) = 313$. It can be found using the extended euclidean algorithm.

To encrypt the plaintext message $m = 542$, we have

$$c = m^e \bmod 437 = 302.$$

Note that if we attempt to compute m^e as an integer first, and then reduce modulo n, the intermediate result will be quite large, even for such small values of m and e. For this reason, it is important for RSA implementations to use modular exponentiation algorithms that reduce partial results as they go and to use more efficient techniques such as squaring and multiply rather than iterated multiplications. Even with these improvements, modular exponentiation is still somewhat inefficient, particularly for the large moduli that security demands. To speed up encryption, the encryption exponent is often chosen to be of the form $2^k + 1$, to allow for the most efficient use of repeated squarings. To speed up decryption, the Chinese Remainder Theorem can be used provided p and q are remembered as part of the private key.

2. Choosing the RSA Parameters and Attacks Against RSA

Despite a number of interesting attacks on it that have been discovered over the years, it is still generally believed secure today provided certain guidelines are followed in its implementation: the keys are large enough, certain kinds of keys are avoided, and messages are randomly padded prior to encryption in a "safe" way.

The key generation function for RSA specfies how to generate N, e, and d. The usual method is to first choose p and q, compute $N = pq$, choose e, and compute $d = e^{-1}$ mod N. There are several additional steps left to specify: how are p and q chosen, and how is e chosen. Both steps are influenced by the need to avoid certain attacks, which are described below. A careful choice of the RSA parameters proceeds as follows.

1. Choose the modulus size n large enough, and restrict p and q to be $n/2$ bits (to avoid factoring attacks).
2. Choose p and q to be "safe" primes of length $n/2$ (to avoid re-encryption attacks), and compute $N = pq$.
3. Choose e large enough (to avoid small public exponent attacks), either randomly or according to a specified calculation, and compute $d = e^{-1}$ mod N.
4. If the resulting d is too small, go back to the previous step and choose a new e and compute a new d (to avoid small private exponent attacks).

a. Breaking RSA by factoring N. If an attacker can factor N to obtain p and q, then he or she can compute $\phi(N)$ and use the extended euclidean algorithm to compute the private exponent $d = e^{-1} mod \phi(N)$. Since it is easy for a brute force search algorithm to find small factors of any integer by trial division, it is clear that p and q should be taken of roughly equal size.

When RSA was first introduced, the continued fraction factoring algorithm could factor numbers up to about 50 digits (around 200 bits). Since that time, spurred by the application of breaking RSA as well as by the inherent mathematical interest, factoring has been a much studied problem. By 1990, the quadratic sieve factoring algorithm could routinley factor numbers around 100 digits, the record being a 116-digit number (385 bits). In 1994, the quadratic sieve algorithm factored a 129-digit number (428 bits), and in 1996, the number field sieve algorithm factored a 130-digit number (431 bits)in less than a quarter of the time the quadratic sieve would have taken. The general number field sieve algorithm is currently the fastest factoring algorithm, with a running time less than $e^{3n^{1/3} \log^{2/3} n}$, where n is the length in bits.

At the time of this writing, security experts usually recommend taking $n = 1024$ (or 309 decimal digits) for general use of RSA. This recommendation usually increases every five to ten years as computing technology improves and factoring algorithm become more sophisticated.

b. Re-encryption attacks and safe primes. Gus Simmons and Robert Morris describe a general attack that can be applied to any deterministic public key cryptosystem, or as a chosen plaintext attack on any deterministic symmetric key cryptosystem. Given a ciphertext C, an attacker should re-encrypt C under the same key, re-encrypt that results, and so forth, until the result is the original ciphertext C. Then the previous result must be the original plaintext M. The success of the attack is determined by the length of such cycles. Although public key systems should not be, and are not, generally used without randomization, it is still desirable to avoid small cycles. Rivest recommends the following procedure for choosing *safe* primes.

1. Select a random $n/2$-bit number. Call it r.
2. Test $2r + 1$, $2r + 3$, ... for primality until a prime is found. Call it p''.
3. Test $2p'' + 1$, $4p'' + 1$, ... for primality until a prime is found. Call it p''.
4. Test $2p' + 1$, $4p' + 1$, ... for primality until a prime is found. This is p.
5. Repeat steps 1–4 to find q.

c. Small public exponent attacks.

In order to improve efficiency of encryption, it has been suggested to instead fix $e = 3$. However, certain attacks have been demonstrated when either e is too small. The most powerful of these attacks is based on lattice basis reduction and is due to Don Coppersmith. Coppersmith's attack is not a total break. However, if the public exponent is small enough and certain relationships between messages are known, it allows the attacker to succeed in learning the actual messages. If the encryption key is small enough and some bits of the decryption key are known, it allows the attacker to learn the complete decryption key. To avoid these attacks, it is important that the public exponent is chosen to be sufficiently large. It is still believed secure, and is desirable for efficiency reasons, to choose e to be of the form $2^k + 1$ for some $k \geq 16$.

d. Small private exponent attacks.

An attack of Michael Wiener shows that if $d < (1/3)N^{1/4}$, than attacker can efficiently recover the private exponent d from the public key $< N, e$). The attack is based on continued fraction approximations.

In addition to the attacks just described that relate to how the RSA parameters are chosen, there are also a number of attacks on RSA that relate to how RSA is used. As mentioned earlier, if the message space is small and no randomization is used, an attacker can learn the plaintext of a ciphertext C by encrypting each message in the message space and see which one gives the target ciphertext C. Some additional usage attacks on RSA are described below. RSA is also susceptible to timing attacks, described earlier.

e. Bleichenbacher's padding attack.

Daniel Bleichenbacher showed a adaptive chosen-ciphertext attack on RSA as implemented in the PKCS1 standard, which uses the approach of appending random bits to a short message M before encrypting it to make it n bits long. In PKCS1, a padded message looks like this:

| 02 | random pad | 00 | M |

which is then encrypted using RSA. The recipient of the message decrypts it, checks that the structure is correct, and strips of the random pad. However, some applications using PKCS1 then responded with an "invalid ciphertext" message if the initial "02" was not present. Given a target ciphertext C, the attacker sends related ciphertexts of unknown plaintexts to the recipient, and waits to see if the response indicates that the plaintexts start with "02" or not. Bleichenbacher showed how this information can be used to learn the target ciphertext C.

This attack demonstrates that the way randomization is added to a message before encryption is very important.

f. Multiplication attacks.

When used for signatures, the mathematical properties of exponentiation creates the possibilities for forgery. For example,

$$M_1^d \bmod N \cdot M_2^d \bmod N = (M_1 M_2)^d \bmod N,$$

so an attacker who sees the signature of M_1 and M_2 can compute the signature of $M_1 M_2$. Similarly, if the attacker wants to obtain Alice's signature on a message M that Alice is not willing to sign, he or she can try to "blind" it by producing a message that she would be willing to sign. To do this, the attacker chooses a random r and computes $M' = M \cdot r^e$. If Alice is willing to sign M', its signature is $M^d \cdot r \bmod N$, and the attacker divide by r to obtain the signature for M.

In practice, signatures are generated on hashes of messages, rather than the messages themselves, so this attack is not a problem. Furthermore, it is a useful property for allowing digital signatures where the signer does not learn the contents of a message, which can be useful in desiging systems that require both anonymity of participants and certification by a particular entity, such as anonymous digital cash systems and electronic voting systems.

g. Common modulus attacks.

In a system with many users, a system administrator might try to use the same modulus for all users, and give each user their own encryption and decryption exponents. However, Alice can use the Chinese Remainder theorem together with her private key d to factor the modulus. Once she has done that, she can invert other users public exponents to learn their decryption exponents.

VI. KEY DISTRIBUTION AND MANAGEMENT

In order for encryption to protect the privacy of a message, it is crucial that the secret keys remain secret. Similarly, in order for a digital signature to protect the authenticity and integrity of a message, it is important that the signing key remains secret and that the public key is properly identified as the public key of the reputed sender. Therefore is it of paramount importance that the distribution and management of public and private keys be done securely.

Historically, keys were hand delived, either directly or through a trusted courier. When keys had to be changed, replacement keys were also hand delivered. However, this is often difficult or dangerous, for the very same reasons that motivated the need for encryption in the first place. While the initial key may need to be communicated by hand, it is desirable to use encryption to communicate additional keys, rather than communicating them by hand. Method to do this are called key exchange protocols, and are described below.

With public key cryptography, some of the key management problems are solved. However, in order for Alice's public key to be useful, it is important that others know that it is her key, and not someone else masquerading as her for the purpose of receiving her secret messages. Hence, it is important that the binding between a public key and an identity is authenticated. Wide-scale methods for doing this are called public key infrastructures, and are also described below.

A. Key Exchange Protocols

The simplest key exchange protocol would be to use one secret key for a while, then use it to communicate a new secret key, and switch to that key. However, this is not a satisfactory solution because if one key is compromised (i.e., discovered by an attacker), then all future keys will be compromised as well. Instead, *session keys* are commonly used. A long-term key is exchanged securely (possibly by hand). A session key protocol is used to generate a short-term session key that will be used to encrypt messages for a period of time until the next time the session key protocol is run. Although exposure of the long-term key still results in compromise of all session keys, exposure of one session key does not reveal anything about past or future session keys. Long-term keys can be chosen to optimize security over efficiency, since they are only infrequently used, and long-term keys are less exposed because fewer messages are encrypted with them. Often the long-term key is the public key and private key of a public key cryptosystem, while the session keys are symmetric cryptosystem keys.

B. Diffie–Hellman Key Exchange

Diffie-Hellman key exchange is based on the assumed difficulty of the discrete logarithm problem modulo a prime number—that is, that it is difficult to compute z from $g^z \bmod p$. Diffie–Hellman allows to parties who have not previously exchanged any keys to agree on a secret key. Alice and Bob agree on a prime modulus p and a primitive element g. Alice picks a random number x and sends

$$a = g^x \bmod p$$

to Bob. Similarly, Bob picks a random number y and sends

$$b = g^y \bmod p$$

to Alice. Alice then computes $b^x \bmod p = g^{xy} \bmod p$ and Bob computes $a^y \bmod p = g^{xy} \bmod p$. The computed value $g^{xy} \bmod p$ is then used as a secret key.

Assuming that the discrete logarithm problem is computationally infeasible, an attacker overhearing the conversation between Alice and Bob can not learn $g^{xy} \bmod p$. However, it is subject to the kind of man-in-the-middle attack discussed earlier.

C. Key Distribution Centers

In a key distribution center (KDC) solution, a key distribution center shares a secret key with all participants and is trusted to communicate keys from one user to another. If Alice wants to exchange a key with Bob, she asks the KDC to choose a key for Alice and Bob to use and send it securely to each of them. While it may be possible to have such solutions within a particular business, they do not scale well to large systems or systems that cross administrative boundaries.

D. Public Key Infrastructures

In a public key infrastructure (PKI), any user Alice should be able to determine the public key of any other user Bob, and to be certain that it is really Bob's public key. This is done by having different entities digitally sign the pair: $\langle Bob, K_E \rangle$, consisting of Bob's identity and public key. In practice, a certificate will also contain other information, such as an expiration date, the algorithm used, and the identity of the signer. Now, Bob can present his certificate to Alice, and if she can verify the signature and trusts the signer to tell the truth, she knows K_E is Bob's public key. As with other key exchange solutions, this is simply moving the need for secrecy or authentication from one place to another, but can sometimes be useful.

The two main approaches to building a large-scale PKI are the hierarchical approach and the "web of trust" approach. In either model, a participant authenticates user/key bindings by determining one or more paths of certificates such that the user trusts the first entity in the path, certificates after the first are signed by the previous entity, and the final certificate contains the user/key binding in question. The difference between the two models is in the way trust is conveyed on the path.

In the hierarchical model, a certificate is signed by a certificate authority (CA). Besides a key binding, a CA certificate authorizes a role or privilege for the certified entity, by virtue of its status as an "authority" within its domain. For example, a company can certify its employees' keys because it hired those employees; a commercial certificate authority (CA) can certify its customer's keys because it generated them; a government or commercial CA can certify keys of hierarchically subordinate CAs by its powers of delegation; government agencies can certify keys of government agencies and licensed businesses, as empowered by law; and an international trade bureau can certify government keys by international agreement. An individual is assumed to know and trust the key of a CA within its domain. A CA is assumed to know and trust the key of the CA who certifies its own keys, and it has a responsibility for accuracy when signing certificates of a principal in its domain. In summary, the hierarchical model conveys trust transitively, but only within a prescribed domain of control and authority.

In the web of trust model, individuals act as *introducers*, by certifying the keys of other individuals whom they have personally authenticated. In order for Alice to determine whether a key K_E belongs to Bob, she considers the signature(s) certifying the binding of K_E to Bob, and must ask whether any of the users who signed Bob certificates are considered trusted to verify and sign someone else's certificate. In other words, trust is not conveyed along the path of certificates, but rather it is awarded by the user of the certificate. Belief in the final certificate is possible only if the user trusts all of the certifying users on a path.

VII. APPLICATIONS OF CRYPTOGRAPHY

Cryptography has found a wide range of applications. Many cryptographic tools use cryptography to create building blocks that provide privacy, authentication, anonymity, and other such properties. In turn, these tools can be used to create secure applications for users. One strong and very general tool, called *secure multiparty computation*, allows a group of parties each holding a private input to jointly compute an output dependent on their private inputs without revealing their private inputs to each other. Secure multiparty computation can be used to solve problems like electronic voting, electronic auctions, and many other such problems.

Cryptography and cryptographic tools are particularly important for providing security in communications networks and on computer systems. Link encryption, which encrypts along a single link of a communication network, and end-to-end encryption, which encrypts all the way from the start to the end of a path in a communication network, are both used to protect the privacy of messages in transit. In computer systems, cryptography can be used to provide access control and prevent unwanted intruders from reading files, changing files, or accessing other resources.

Cryptography can also be used to provide important security properties in electronic commerce. A now familiar example is the use of cryptography to authenticate a merchant and encrypt a credit card number when buying goods over the World Wide Web. Cryptography can also protect the provider of digital content such as music or video by ensuring that recipients cannot widely redistribute the content without being detected. In the future, more advanced applications of cryptography in electronic commerce may be seen, where credit cards are replaced by digital cash and automated agents securely participate in auctions on a user's behalf.

SEE ALSO THE FOLLOWING ARTICLES

COMPUTER ALGORITHMS • COMPUTER NETWORKS • COMPUTER VIRUSES • SOFTWARE RELIABILITY • WWW (WORLD-WIDE WEB)

BIBLIOGRAPHY

Boneh, D. (1999). "Twenty years of attacks on the RSA cryptosystem," *Notices Am. Math. Soc. (AMS)*, **46**(2), 203–213.

Daemen, J., and Rijmen V. (2000). The Block Cipher Rijndael. In "Proceedings of Smart Card Research and Applications" (CARDIS '98), Louvain-la-Neuve, Belgium, September 14–16, 1998, Lecture Notes in Computer Science, Vol. 1820, Springer, 277–284.

Diffie, W., and Hellman, M. E. (1976). "New directions in cryptography," *IEEE Trans. Inf. Theory*, **IT-22**(6), 644–654.

(1998). "electronic frontier foundation," (ed. John Gilmore) Cracking DES: Secrets of Encryption Research, Wiretap Politics and Chip Design, O'Reilly & Associates.

Goldreich, Oded (1999). "Modern Cryptography, Probabilistic Proofs and Pseudo-randomness," Springer-Verlag.

Kahn, David (1967). "The Codebreakers: The Story of Secret Writing," Macmillan Co., New York, New York.

Menenez, Alfred J., Van Oorschot, Paul C., and Vanstone, Scott A. (1996), "Handbook of Applied Cryptography," CRC Press Series on Discrete Mathematics and Its Applications, CRC Press.

Rivest, R. L., Shamir A., and Adleman, L. (1978). A method for obtaining digital signatures and public-key cryptosystems, *Comm. ACM*, February 1978, 120–126.

Schneier, Bruce (1996). "Applied Cryptography Second Edition: protocols, algorithms, and source code in C," John Wiley and Sons.

Shannon, C. E. (1949). "Communication theory of secrecy systems," *Bell Sys. Tech. J.* **1949,** 656–715.

Stinson, D. E. (1995). "Cryptography Theory and Practice," CRC Press, Boca Raton.

Crystal Growth

Lynn F. Schneemeyer

Bell Laboratories, Lucent Technologies

GLOSSARY

Convection Process or method of heat transport in a fluid described as heat flow patterns. In a heated melt, the warmer, less dense melt near the heater walls rises, cools at the surface and becomes more dense, then sinks near the cooler center.

Diffusion Movement of an elemental species (atom or ion) towards region where it has a lower concentration.

Dopant Impurity atom or ion added to a host crystal for the purpose of modifying selected properties.

Epitaxy Single-crystal growth of a material, usually as a thin layer on a seed (called the *substrate*) of the same material or an isomorphous material having nearly the same lattice constants.

Eutectic The invarient point in a simple binary phase diagram where two solids, A and B, are in equilibrium with a liquid. The eutectic composition is that combination of componenets that has the lowest melting point.

Existence region Phase field where a solid compound exists with variations in stoichiometry.

Flux A solvent, often a high-temperature melt, that dissolves a compound, producing a mixture with a lower liquid temperature than the melting point of the compound.

Isomorphous Term applied to isostructural compounds, compounds that share a common crystallographic structure.

Mass transport Movement of a dissolved species in a fluid medium.

Mixed crystal Crystal made up of two or more isomorphous components; also referred to as a solid solution.

Morphological stability Stability, or tendency to maintain its shape, of the fluid–solid interface during solidification.

Nucleation Initial deposition or precipitation of solid crystalline material which acts as a site for subsequent crystal growth.

Encyclopedia of Physical Science and Technology, Third Edition, Volume 4
Copyright © 2002 by Academic Press. All rights of reproduction in any form reserved.

Retrograde solubility Decreasing solubility of a species in a solid crystal with decreasing temperature; results in the formation of a second phase as precipitates within the crystal.

Seed crystal An isostructural crystal (sometimes the material that will be grown) introduced into the melt to act as a site for subsequent growth.

Stoichiometry Proportion of elemental constituents in a compound precisely as indicated in its chemical formula.

Sublimation Phase transition from the solid state directly to the vapor state.

CRYSTAL GROWTH is the controlled change of a substance to an ordered solid (condensed) state. This transformation may occur from the vapor, liquid, or even solid state. The objective of crystal growth is typically to obtain either bulk single crystals, which are macroscopic assemblies built from a repeated regular geometric network of atoms (the crystal lattice) or thin films grown on a suitable substrate. Crystals can be as small as fractions of a millimeter in physical dimensions to sizes measured in tens of centimeters to meters. In the case of a crystalline thin film, film thickness may be tens of angstroms to microns in thickness. Whatever its size, a single crystal is expected to contain no crystal grain boundaries.

I. INTRODUCTION

The growth of single crystals has been studied for hundreds of years in order to produce materials of esthetic value such as gemstones. Many also believed that gemstones had medical or mystical virtues, as well, which further added to their value. More recently, crystals of scientific or technological importance such as semiconductors or laser materials have received the greatest attention and have formed the basis of new industries. Today, single crystals and epitaxial layers are at the heart of computers, communication systems, medical devices, and many of the other miracles of the modern age.

Crystal growth is a multidisciplinary subject and draws from fields including chemistry, solid state physics, theoretical physics, crystallography, thermodynamics, kinetics, and fluid dynamics. The crystal grower uses information from various fields to develop techniques to produce the desired stoichiometry and structure, as well as crystalline form—bulk single crystal, epitaxial layer, or other—as nature permits.

This chapter focuses on practical aspects of crystal growth. Various crystal growth approaches will be presented together with the key experimental considerations. Despite important advances in crystal growth theory, its impact on experimental practice is still very limited. Thin-film growth approaches are of increasing importance and will be outlined here.

II. CRYSTAL GROWTH TECHNIQUES

A number of different crystal growth techniques have been developed for both bulk crystal growth and thin film growth. The different approaches are illustrated with actual crystal growth examples. The selection of a particular growth technique is discussed in terms of phase equilibria, with a brief discussion of the role of mass transport. The relevant chemical and thermodynamic factors are also mentioned as the choice of growth parameters is outlined. The standard techniques employed for the growth of bulk single crystals include melt growth, which is useful only for congruently melting materials; vapor phase transport, which requires volatile constituents and similar rates of transport for the various constituents; and flux growth, in which selection of the flux and growth condition must be tailored to the chemistry of a particular compound. A number of excellent books on crystal growth discuss in detail all of the techniques that have been developed to produce bulk crystals. This article will briefly outline only a few of the common techniques. Thin films can also be grown using a variety of techniques, the most common of which are molecular beam epitaxy (MBE), sputtering, and chemical vapor deposition (CVD). These will be outlined below.

There are a number of ways to classify crystal growth techniques. One is to separate them into techniques that involve only the chemical components of the desired crystal (the *direct* techniques) versus those that use additional foreign components or excesses of constituent components that act as mineralizers (the *indirect* techniques). Table I lists various crystal growth approaches according to this classification scheme.

Direct techniques, those approaches that involve only the chemical components of the desired crystal, include the bulk techniques Czochralski growth and Bridgeman–Stockbarger growth for the growth of congruently melting materials, as well as MBE for epitaxial thin films. The phase from which the crystal is growing, which may be gas, liquid, or solid, contains the same major atomic components as are present in the growing crystal (but excluding trace impurities, whether added to dope the crystal or present inadvertently). A simple example of direct crystal growth is the melt growth of single crystals of silicon from a pure silicon melt. Trace amounts of "dopant" atoms, species which, in very small amounts, modify the electrical and/or optical properties of the resulting crystals, may or may not be present in the melt.

Indirect techniques employ additional chemical components that act as fluxes, solvents, or carriers of the atoms

TABLE I Classification of Crystal Growth Techniques[a,b]

Phase	Direct		Indirect	
	Source	Growth technique	Source	Growth techniques
Liquid	Melt	Directional Solidification Bridgman Stockbarger Cooled seed Kyropoulos Pulling Czochralski Zoning techniques Float zone	Flux–solution (Off-stoichiometry; solid solution) Reaction	Slow cooling Temp differential Solvent evaporation Top seeded (isothermal) Hydrothermal Any solvent aqueous, molten, etc. Chemical, electrochemical
Vapor	Constituent gas (Epitaxy)	Sublimation condensation PVD, MBE, ALE	Compound gas	Reaction–condensation CVD, MOCVD
Solid	Solid	Recrystallization Strain Polycrystal	Solid solution	Exsolution Spinodal decomposition

[a] PVD, Physical vapor deposition; MBE, molecular beam epitaxy; ALE, atomic layer epitaxy; CVD, chemical vapor deposition; MOCVD, metal-organic chemical vapor deposition.

[b] Reprinted with permission from American Institute of Chemical Engineers, AIChE Symposium Series, *Tutorial Lect. Electrochem. Eng. Technol.*, *II 79*, 143, Copyright 1983.

forming the crystal. Such techniques can be used to grow crystals of materials that decompose on melting to form a liquid and a solid of another composition and are known as incongruently melting compounds. Most complex solids are incongruently melting. The term also applies to crystals grown from off-stoichiometric melts—that is, melts whose composition varies slightly from the precise stoichiometry of the crystal. Frequently, such compounds have a eutectic-melting composition as well as an "existence region" in which the solid compound exists with variations in stoichiometry. Indirect techniques employing high-temperature fluxes can also be used to grow crystals that can be grown directly. The use of a solvent may lower the growth temperature and avoid the formation of subsolidus (solid-to-solid) phase transitions as well as certain types of crystal defects.

Other classification schemes for crystal growth techniques are possible. One alternative crystal growth classification scheme divides techniques into those used in the growth of congruently melting materials, those used for incongruently melting materials, and those used for the growth of crystalline thin films. Congruently melting materials are those in which the composition of the melt exactly matches the crystal composition. These are typically grown using Czochralski growth, Bridgeman–Stockbarger growth, or float-zone growth. Incongruently melting materials, as discussed above, are typically grown using fluxes or other added components. In the case of thin films, incongruent materials are grown using chemical approaches such as CVD or sputtering.

III. BULK CRYSTAL GROWTH TECHNIQUES AND PARAMETERS

A. Techniques to Grow Congruently Melting Materials

Melt growth techniques are applied to congruently melting solids. The researcher must first verify that the material is truly congruently melting. This is best diagnosed in a crystal growth experiment in which a relatively large fraction, 50–70% or more, of the melt is crystallized. Then, by comparing the compositions of the top and bottom of the crystal, very small differences in the composition are magnified and the true congruent composition can be determined. Note that a congruent composition is not necessarily a stoichiometric composition. Typically, a bulk single crystal is grown from a melt by engineering the thermal gradient at the growth interface. Melt growth techniques are among the most important approaches for the growth of large, commercially important crystals, including silicon.

1. Directional Solidification: Bridgeman–Stockbarger Technique

Directional solidification is a melt growth technique that involves cooling a liquid melt contained in an ampoule

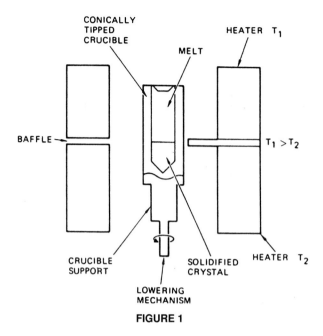

FIGURE 1

through its solidification temperature. The tip of the ampoule is often conically shaped to enhance nucleation of a single crystal. In favorable situations, only one or a few nuclei persist to the end of the narrow region of the ampoule so that a sample with large single-crystal domains is grown. In the Bridgeman technique, the furnace temperature is held constant and the ampoule lowered through a temperature gradient at a controlled rate. Because Stockbarger was interested in growing large alkali halide crystals, he created a steeper temperature gradient by adding baffles inside the furnace to assist in cooling a region of the furnace. Many other ways have been developed to produce a desired temperature gradient. Crystal growth involving a sample contained inside of an ampoule and cooled through a temperature gradient is generally referred to as the Bridgeman–Stockbarger technique (Fig. 1). The figure shows a vertical configuration. Horizontal directional solidification has also been extensively employed. Vertical growth systems can be designed with axial symmetry providing more symmetric heat flow and convective flow patterns and leading to a more homogeneous crystal.

The simplicity of the directional solidification approach and the ease of construction of suitable equipment prompts its wide application for the production of research samples of an enormous range of different materials. The details of the equipment setup are dictated by melting temperatures, chemical reactivity of the melt, vapor pressure of the material to be grown, etc. Growth can be seeded or unseeded. Often, initial experiments are unseeded. Preferred growth directions lead to large single-crystal regions in the sam-

ples that can be isolated and used as seeds in subsequent crystal growth experiments.

Another version of direction solidification is zone refining. In this approach, only a narrow zone of liquid is created in the sample. This zone is moved through the sample, and the width of the zone depends on the viscosity and surface tension of the melt. The technique, first developed by Pfann for purification, can also be used for single crystal growth. If, at the start of the zone melting, a single crystal is in contact with a polycrystalline ingot, then solidification can occur at the crystal surface so that a single crystal sample results after the molten zone has been moved through the ingot.

2. Czochralski Growth

Czochralski growth, or crystal pulling, is the dominant commercial process for the growth of single crystals of silicon, compound semiconductors, metals, oxides, and halides. The Czochralski technique essentially consists of a crucible containing the material to be crystallized surrounded by a heater to melt the charge. A pull rod, which is rotated and slowly raised, is mounted above the crucible and crystal growth takes place on the seed attached to that rod (or the rod itself if no seed is available), as shown in Fig. 2. Each Czochralski apparatus is tailored to the material it is used to grow.

In Czochralski growth, the seed acts as a heat leak that causes solidification of the growing crystal; heat transfer occurs through the growing crystal and pull rod and is enhanced by the steep thermal gradients at the interface

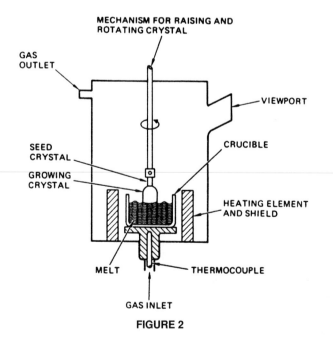

FIGURE 2

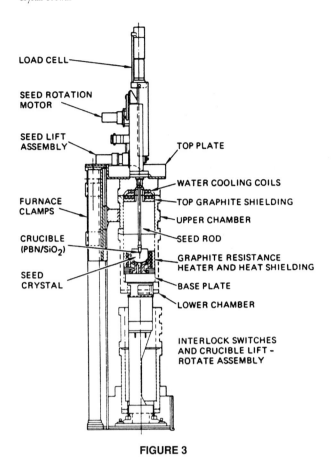

FIGURE 3

above the melt. Heating of the crucible can be accomplished by a resistance-heated furnace, radiofrequency induction heating using the crucible as a susceptor, or laser heating. Crystals are weighed during growth and the diameter controlled automatically through computer-controlled adjustments of the power to the furnace.

Numerous improvements to the Czochralski technique have been made, such as development of encapsulation techniques to control evaporation and techniques that allow control of the diameter of the growing crystal. Liquid-encapsulated Czochralski (LEC) is commonly used for the crystal growth of GaAs. Here, a layer of molten B_2O_3 lies above the GaAs melt and, together with an overpressure of inert gas, prevents the loss of the volatile arsenic during crystal growth (Fig. 3).

3. The Kyropoulos Technique

Related to the Czochralski technique is the Kyropoulos technique, which uses a cooled seed to initiate single crystal growth within the melt-containing crucible. Heat removal is continued by ramping down the furnace temperature, thus growing the crystal.

B. Techniques to Grow Incongruently Melting Materials

Many materials, particularly those with more complicated compositions or structures, do not melt congruently. Often, crystals of the phase of interest are precipitated from a high-temperature solution which can be formed below the decomposition temperature of the desired phase. Such approaches include flux growth, high-temperature solution growth; hydrothermal growth, solution growth at high pressures and temperatures; the traveling solvent method, a containerless process where the flux is moved along the length of the growing crystal; and electrochemical growth, where the insoluble species is produced as a result of an electrochemical oxidation or reduction. Variants of all of these involving the use of a seed to control nucleation are known.

1. Flux Growth

A very common way of producing crystals suitable for research studies is flux growth, generally high-temperature solution growth. The flux may have a different composition from the desired crystal (for example, the use of molten KF as a flux for the growth of $BaTiO^3$) or may contain one or more of the constituents of the desired phase (for example, the growth of $Ba_2YCu_3O_{7-x}$ from $BaCuO_2$–CuO mixtures). The latter is often referred to as a *eutectic* or *nonstoichiometric melt*. In flux growth, as in many solution growth approaches, crystals form as the solubility of component A, the desired phase, is lowered, typically by lowering the temperature of the melt, as indicated in the phase diagram of Fig. 4.

As a relatively low-temperature growth technique, flux growth can be used in the growth of materials containing volatile components and highly refractory materials in addition to the incongruently melting materials discussed

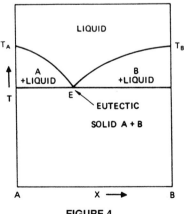

FIGURE 4

already. A wide variety of fluxes have been employed for the growth of crystals. Fluxes vary in characteristics, including acid–base properties and redox behavior. An empirical approach is usually used to identify suitable fluxes for the growth of a new compound. Beyond the initial selection of a flux material, appropriate choices for other growth parameters—including the melt composition, flux-to-compound ratio, and maximum heating temperature (T_{max}) being kept below the decomposition temperature—must be investigated in order to grow large, high-quality crystals of a desired phase free of defects such as flux inclusions.

2. Top-Seeded Solution Growth

A variation of flux growth involving the use of a seed to control the orientation of the product crystal is top-seeded solution growth (TSSG). A furnace is arranged to produce a well-defined temperature gradient with the temperature at the melt surface near that of the onset of solidification. In TSSG, as in Czochralski growth, the seed and its attached support rod act as a heat leak so that growth takes place on the seed in an experimental setup similar to that shown in Fig. 2. The furnace temperature is slowly lowered and the seed rotated and slowly raised, allowing additional growth to take place. Finally, the sample is raised out of the melt, usually leaving it largely free of the flux. Large crystals of the ferroelectric material, $BaTiO_3$, and other oxides have been grown by top-seeding.

3. Hydrothermal Growth

Hydrothermal growth uses supercritical water ($T > 373°C$) as the solvent at high pressures. A wide variety of complex oxides as well as chalcogenides and other materials have been grown hydrothermally. The approach depends on the greatly improved solvent properties and mobilities of supercritical water. Solvent other than water have been used as well. The mineralogy community has paid particular attention to this crystal growth approach which is reminiscent of common geological crystal growth processes. Also, electronic-grade quartz is grown by a hydrothermal technique.

4. Electrochemical Crystallization

Electrochemical crystallization involves the use of an electrochemical oxidation or reduction to convert a soluble solution species into an insoluble species at one of the electrodes. This approach is useful only for the growth of conducting crystals. If product material does not conduct, then the electrode at which precipitation is occurring becomes covered with an insulating coat and current flow

ceases, a situation called *passivation*. Metals such as copper and silver and conducting oxides such as $LiTi_2O_4$ and various tungsten bronze phases have been grown electrochemically.

5. Floating Zone Technique

A melt growth technique that does not require the use of a crucible, thus eliminating possible contamination by the crucible, is the floating zone technique. As indicated in Fig. 5, a vertical sample is heated with an rf or infared light source or other heat source to produce a molten zone of liquid that is held in place by surface tension. A miniaturized version of the float zone process, known as laser-heated pedestal growth (LHPG), uses a CO_2 laser to form the molten zone. In all cases, the sample is moved so that the molten zone travels through the entire sample. This molten zone can be seeded to produce oriented or single crystals and has been used for the growth of metal crystals

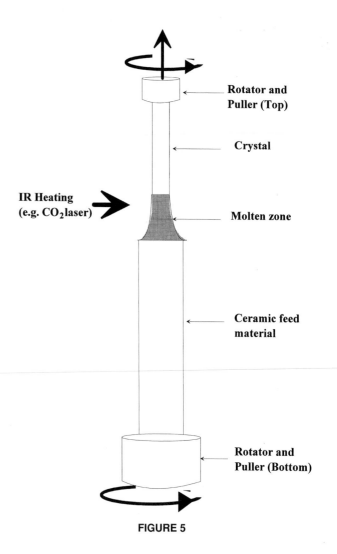

FIGURE 5

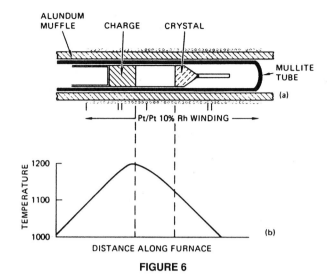

FIGURE 6

as well as a variety of complex oxides, but it is not useful if a material contains a volatile component. A related technique, the traveling solvent floating zone technique (TSFZ), passes a flux composition along the length of a sample to produce crystalline product material. Crystals of the cuprate superconductor, $La_{1.85}Sr_{.15}CuO_4$, were grown by TSFZ.

6. Vapor-Phase Transport or Deposition

All of the previous crystal growth approaches involve the use of melts for crystal growth. Bulk crystals can also be grown from a vapor phase. In the simplest situation, condensation of a suitable vapor produces crystals such as the growth of ice crystals from water vapor. Crystals of the II–VI semiconductors (e.g., CdS or ZnS) grow as single crystals from their constituent gases, cadmium or zinc and sulfur. This technique is called *physical vapor deposition* (PVD; see Fig. 6). In a related technique, chemical vapor transport, a reagent, often a halide, is introduced into the system which reacts at an elevated temperature to form exclusively gaseous products. The growth system, usually a sealed tube, is in a temperature gradient. The gaseous species travel to a second point in the system and the temperature is such that they decompose to reform the original material, which grows as crystals.

IV. THIN-FILM GROWTH TECHNIQUES AND PARAMETERS

Crystalline thin films have achieved an importance in modern technology by providing properties that are difficult to achieve in bulk materials and allowing integration of materials having different functionalities—for instance, by fabricating multiple layers of different materials—to produce a desired performance. Thin films are used in a wide variety of applications, including optics, as decorative coatings or memory disks, electrical uses such as semiconductor devices, and thermal layers such as heat sinks or barrier layers. All thin-film deposition processes involve source materials for growth and a mechanism for transporting those species to the substrate surface where deposition takes place. Amorphous, polycrystalline, or epitaxial single-crystal films can be grown, depending on the match of crystalline structure of the material being grown to the substrate lattice, as well as experimental control over the rates, substrate temperature, and other deposition parameters.

Growth processes for epitaxial thin film are essentially the same as those relevant to bulk crystals except for the influence of the substrate at early stages of growth. Epitaxial growth is obtained when the deposited layer follows the structural arrangement of the substrate, whether the substrate is the same material (homoepitaxy) or a different chemical substance (heteroepitaxy). Major factors in matching a desired thin film to a particular substrate are the similarity in lattice parameters and coefficients of thermal expansion.

Many thin-film growth techniques involve deposition from the vapor phase. Such approaches are generally applicable to the growth of any desired material and allow a wide range of control over the temperature of the substrate during growth. Also, the growing surface can be analyzed *in situ*, which has been key to the growth of high-quality epitaxial thin films.

Source material for the thin film to be grown may be solid, liquid, or gas. In physical vapor deposition (PVD) methods, the solid source is vaporized by heating, by evaporation using an energetic beam of electrons, by sputtering via energetic ions, or by ablation using a focused laser beam. Solid sources may also be chemically converted to gas-phase species (for example, Ga to $GaCl_3$) which are then used in deposition processes categorized as chemical vapor deposition (CVD).

A. Molecular Beam Epitaxy

Molecular beam epitaxy (MBE) is a carefully controlled form of vacuum evaporation. Effusion cells for evaporation or sublimation create beams of atoms of individual elements in an ultrahigh vacuum (UHV) system for delivery to the substrate surface where solidification occurs. These molecular beams can be started or stopped in less than the time it takes to grow a single atomic or molecular layer. A schematic diagram for an MBE system for the growth of GaAs/GaAlAs films is shown in Fig. 7. Stoichiometry of the film is controlled both by the substrate temperature and the rates at which the various component

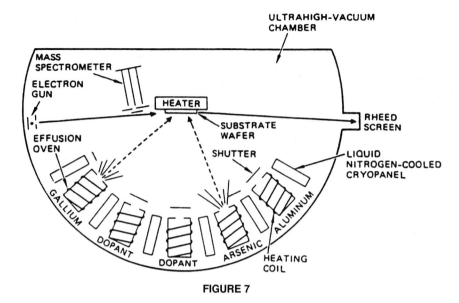

FIGURE 7

species are delivered to the growing film. In order to provide sources of species that cannot be evaporated, metal–organic (MO) sources are sometimes used, a technique called MOMBE. A special case of MBE is atomic layer epitaxy (ALE). Here, a monolayer of each constituent of the desired phase is deposited sequentially to form the desired compound. For example, deposition of sequential monolayers of Zn and S will form ZnS. Because films are grown under UHV conditions, physical and chemical properties of the film can be measured *in situ* using reflection high-energy electron diffraction (RHEED) and Auger electron spectroscopy (AES) studies which improves experimental control over the structure of the resulting film. MBE-like technology has been used to grow crystalline films of a wide range of materials including Si, Si–Ge, II–VI semiconductors, III–V semiconductors, cuprate superconductors, and metals.

B. Metal Organic Chemical Vapor Deposition

Thin-film growth using organometallic (OM) sources for chemical vapor deposition or vapor phase epitaxy (MOCVD or OMVPE) has wide commercial application, particularly for III–V semiconductor growth and silicon device microfabrication. The MOCVD method provides conformal coverage of complex shapes and suitable deposition rates for a commercial process. The impact of this approach has included the efficient fabrication of photonic devices, such as semiconductor lasers, detectors, solar cells, etc., based on quantum-well, superlattice, and other two-dimensional structures. The OMVPE process can be used to produce virtually atomically abrupt interfaces for changes in the chemical composition or doping levels of the layers.

A simplified schematic diagram for an MOCVD reactor utilizing three constituents is shown in Fig. 8. In "cold wall" techniques, source gases are chosen that do not decompose or react until they come in contact with the hot substrate, which is usually heated on a susceptor (e.g., a carbon block) with rf induction heating. Heating may also be provided using internal or external lamps. Reactor designs may be vertical or horizontal and are chosen with a particular substrate size and desired reaction taken into consideration. In "hot wall" techniques, the reactor is contained within a furnace and the entire reactor is heated to maintain the reactant elements or compounds in the vapor state until they react, and deposition on the substrate, and often elsewhere in the system as well, takes place. In both cases, residual gases and gaseous products are passed out of the system by flow-through of inert gases or via pumping.

C. Laser Ablation

Laser ablation or pulsed laser deposition (PLD) is another important technique for the growth of epitaxial thin films, particularly of multicomponent oxides. A conceptually straightforward technique, PLD uses an eximer laser operating in the vacuum UV, 193 or 248 nm, which is impacted on a target surface to produce a plume of gas-phase species from the target surface. The material ejected from the target is deposited on a substrate.

The PLD process is highly versatile since targets and substrates can easily be changed without fear of cross-contamination. Many of the issues in PLD deposition are similar to those encountered in other vapor-phase deposition processes. A unique feature of PLD is its high deposition rate. Drawbacks to the technique include the

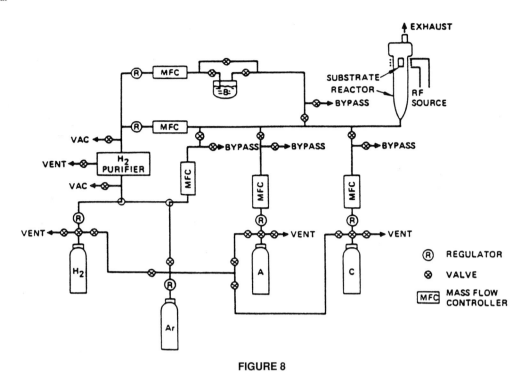

FIGURE 8

formation of particles during ablation that settle on the film and the highly focused nature of the plume which makes scale-up of the technique difficult. Still, PLD has proved to be a particularly useful approach for the growth of thin films of high-temperature superconducting cuprates. More recently, it has been employed for the growth of films of various ferroelectrics, ferrites, and other oxides.

D. Sputtering

Sputtering, which involves the creation of vapor-phase species by the erosion of a target through bombardment by ions in a plasma, is an important research and commercial film growth technique. Effects of the plasma, including scattering of particles by the plasma gas, negative ion ejection from the target, resputtering effects, and ion bombardment of the growing film, have received much study. These processes are important for controlled growth of films of desired composition and reproducible structures.

Sputtering targets can be metals, glasses, or ceramics. Films of various metals can be grown provided a high-purity sputtering atmosphere is maintained. In reactive sputtering, which uses a metal or composite target in an atmosphere containing a reactive gas such as oxygen, the composition of the grown film is shifted from that of the target. Reactive sputtering is often used for the growth of oxides or nitrides using metal targets. For example, TiO_2 films can be produced by sputtering a metallic titanium target in an oxygen-containing plasma. The use of reac-

tive sputtering can lower the costs of targets, result in the growth of higher purity films, and provide higher film-growth rates. Sputtering is readily scaled so that highly uniform films can be grown even on large substrates.

V. SELECTION OF CRYSTAL GROWTH APPROACH AND CONDITIONS

Choosing the appropriate crystal growth approach and source materials for the growth of a desired material is based on a number of different considerations. The form in which the material will be used, such as thin film or bulk part, is highly relevant. Various growth techniques can be considered if only tiny crystals are needed (for example, for X- ray structural studies) while a more limited number of growth techniques are likely to yield large bulk crystals suitable for neutron studies or bulk optics applications. Knowledge of the physical and chemical properties of the material to be grown, including melting point and melting behavior, component vapor pressures, and reactivities, can be very helpful in this selection.

A particularly valuable form of data for the crystal grower is the equilibrium phase diagram. While bulk crystal growth is a steady-state rather than an equilibrium process, much of the information needed to engineer a crystal growth process appears in the phase diagram.

The temperature–composition (T–X) relationship for a simple binary phase diagram is shown in Fig. 4. The pure components A and B both melt congruently, or without decomposition, at temperatures T_A and T_B, respectively, and thus could individually be grown using melt growth techniques—for example, using Czochralski growth. In addition, the diagram indicates a simple eutectic point, E, that is a minimum melting mixture of the constituents with a melting temperature below either T_A or T_B. This indicates that either compound can be grown indirectly within the phase field which is in equilibrium (as a solid) with a liquid in a temperature range below its melting temperature but above T_E. Below T_E, a solid mixture of A and B exists. To grow a crystal of one compound from solution, the crystal must be separated from the remaining melt before T_E is reached.

In general, the selection of crystal growth parameters and the allowed variations of parameters (degrees of freedom) are determined from the phase diagram and are governed by the Gibbs phase rule. The Gibbs phase rule can be written:

$$P + F = C + 2$$

where P is the number of phases present at equilibrium; F, the degrees of freedom of the system (temperature, pressure, and composition,); and C, the number of components of the system. In equilibrium phase diagrams, the rule allows the crystal grower to determine the number of degrees of freedom remaining. Those degrees of freedom represent the control parameters for crystal growth.

Unfortunately, phase diagrams are known for only a small number of chemical systems. Thus, an empirical approach to the growth of crystals typically first involves the exploration of the growth method preferred by the experimenter or for which equipment is available.

VI. DISTRIBUTION COEFFICIENT, MASS TRANSPORT, AND HEAT TRANSFER

The growth of very high-quality and/or large-size single crystals requires detailed refinements to any growth approach. Crystal growth processes tend to be very complex and involve a large number of growth parameters. In general, the resulting crystal reflects the high sensitivity of growth to tiny fluctuations in some of the growth parameters, and absolute reproducibility can be difficult to obtain. To optimize growth, a detailed knowledge of factors influencing growth in the particular chemical system of interest is needed. Several of those factors are briefly discussed here.

Impurities are often deliberately doped into a crystal—for example, to control the luminescent properties of a laser crystal or the electrical conductivity of a semiconductor crystal. The selective incorporation (or rejection) of those components at the growth interface is a parameter known as the *distribution coefficient* (or *segregation coefficient*). If the distribution coefficient, k, defined as the concentration of constituent A in the solid over the concentration of that constituent in the liquid, $[A]_s/[A]_l$, is less than unity ($k < 1$), then the component will be rejected from the solid and will tend to concentrate close to the growth interface. The buildup in concentration of the rejected dopant can lead to a morphological stability problem. On the other hand, if $k > 1$, there will be a depletion of that component close to the interface compared with the bulk composition. Diffusion processes responding to concentration and thermal gradients are required to homogenize the composition at the growing surface and can limit the growth rates possible for some crystals.

Macroscopic mass and heat transport play a central role in crystal growth. Molecules, clusters, or atoms must be transported macroscopic distances to the growth interface, where they may be incorporated into the growing crystal. This transport can be fast or slow compared with the growth kinetics. The rate at which a crystal grows can be limited by interfacial kinetics or by the transport process (diffusion limited). The temperature gradient (ΔT) is a significant factor that the crystal grower must control. Large gradients can lead to large convective processes, which, when extreme, cause unstable growth interfaces.

VII. CRYSTAL GROWTH: THEORY AND EXPERIMENT

Computer-sided analyses of the interactions of natural convection and the shape of the crystal–melt interface have established the influence of these parameters in crystal growth from a congruent melt. New algorithms are being developed to model the interface shape, velocity, and pressure fields in the melts as well as the temperature distribution in the crystal and the melt. Such studies have shown that, for moderate convection levels, the boundary layer model is an oversimplification of the interactions between complex flow patterns and the dopant field. This approach has revealed that the concentration gradient next to the crystal is not radially uniform, but that as much as a 60% variation in radial segregation is calculated. Unfortunately, accurate comparisons of calculated models with actual systems requires detailed determination of the variation of thermophysical properties with temperature, data generally unavailable. However, where such data exist, excellent agreement between theoretical models and experimental data has been shown.

SEE ALSO THE FOLLOWING ARTICLES

CHEMICAL THERMODYNAMICS • CRYSTALLOGRAPHY •
ELECTROLYTE SOLUTIONS, THERMODYNAMICS • ELEC-
TROLYTE SOLUTIONS, TRANSPORT PROPERTIES • KINET-
ICS (CHEMISTRY) • METAL ORGANIC CHEMICAL VAPOR
DEPOSITION (MOCVD) • MOLECULAR BEAM EPITAXY
• NONELECTROLYTE SOLUTIONS, THERMODYNAMICS •
SINTERING • SOLID-STATE CHEMISTRY

BIBLIOGRAPHY

Dryburgh, P. M. (1987). "Advanced Crystal Growth," Prentice Hall,
Engelwood Cliffs, NJ.

Elwell, D., and Scheel, H. J. (1975). "Crystal Growth from High Temperature Solutions," Academic Press, New York.

Herman, M. A., and Sitter, H. (1997). "Molecular Beam Epitaxy: Fundamentals and Current Status," Springer Series in Materials Science,
Vol. 7. Springer-Verlag, New York.

Holden, A., and Morrison, P. (1993). "Crystals and Crystal Growing,"
MIT Press, Cambridge, MA.

Hurle, D. T. J. (1994). "Handbook of Crystal Growth," Vols. 1–3. Elsevier, Amsterdam.

Laudise, R. A. (1970). "The Growth of Single Crystals," Prentice Hall,
Engelwood Cliffs, NJ.

Nishinaga, T., Nishioka, K., Harada, J., and Sasaki, A. (1997). "Advances in the Understanding of Crystal Growth Mechanisms,"
Elsevier, Amsterdam.

Pimpinelli, A., and Villain, J. (1999). "Physics of Crystal Growth,"
Cambridge Univ. Press, London.

Ritter, G., Matthai, C., Takai, O., Rocher, A. M., Cullis, A. G.,
Ranganathan, S., and Kuroda, K. (1998). "Recent Developments in
Thin Film Research: Epitaxial Growth and Nanostructures," Elsevier,
Amsterdam.

Smith, D. L. (1995). "Thin-Film Deposition: Principles and Practice,"
McGraw-Hill, New York.

Vossen, J. L., and Kern, W. (1991). "Thin Film Processes II," Academic
Press, New York.

Crystallization Processes

Ronald W. Rousseau
Georgia Institute of Technology

GLOSSARY

Crystallizer The vessel or process unit in which crystallization occurs.

Growth The increase in crystal size due to deposition of solute on crystal surfaces.

Magma The mixture of crystals and mother liquor in the crystallizer.

Mode of crystallization The means by which a thermodynamic driving force for crystallization is created.

Mother liquor The liquid solution from which crystals are formed.

MSMPR crystallizer A vessel operating in a continuous manner in which crystallization occurs and whose contents are perfectly mixed. As a result of perfect mixing, all variables descriptive of the mother liquor and crystals are constant throughout the vessel and are identical to corresponding variables in the product stream leaving the vessel.

Nucleation The formation of new crystals.

Primary nucleation The formation of crystals by mechanisms that do not involve existing crystals of the crystallizing species; includes both homogeneous and heterogeneous nucleation mechanisms.

Secondary nucleation The formation of crystals through mechanisms involving existing crystals of the crystallizing species.

Solubility The equilibrium solute concentration. The dimensions in which solubility is expressed include, but are not limited to, mass or mole fraction, mass or mole ratio of solute to solvent, and mass or moles of solute per unit volume of solvent or solution.

Supersaturation The difference between existing and equilibrium conditions; the quantity represents the driving force for crystal nucleation and growth.

CRYSTALLIZATION PROCESSES addressed in this discussion are used in the chemical, petrochemical, pharmaceutical, food, metals, agricultural, electronics, and other industries. Moreover, the principles of crystallization are important in all circumstances in which a solid crystalline phase is produced from a fluid, even when the solid is not a product of the process. Much has been done in recent years to improve the understanding of crystallization, and a large portion of the research on the topic has dealt with mechanisms of nucleation and growth. Especially important has been elucidation of the effects of process variables on the rates at which these phenomena occur. Additionally, extensive progress has been achieved in modeling both steady-state and dynamic behavior of crystallization systems of industrial importance. The primary elements of the discussion that follows are the principles that influence yield, morphology, and size distribution of crystalline products.

I. OBJECTIVES OF CRYSTALLIZATION PROCESSES

Several examples of objectives that may be satisfied in crystallization processes are given in the following discussion. Soda ash (sodium carbonate) is recovered from brine by contacting the brine with carbon dioxide that reacts with sodium carbonate to form sodium bicarbonate. Sodium bicarbonate, which has a lower solubility than sodium carbonate, crystallizes as it is formed. The primary objective of the crystallizers used in this process is separation of a high percentage of sodium bicarbonate from the brine in a form that facilitates segregation of the crystals from the mother liquor. The economics of crystal separation from the mother liquor are affected primarily by the variables that control the flow of liquid through the cake of crystals formed on a filter or in a centrifuge. For example, the flow rate of liquid through a filter cake depends on the available pressure drop across a filter, liquid viscosity, and the size distribution of crystals collected on the filter. With a fixed available pressure drop and defined liquid properties, the crystal size distribution controls filter throughput and, concomitantly, the production rate from the process.

Crystallization can be used to remove solvent from a liquid solution. For example, concentration of fruit juice requires the separation of solvent (water) from the natural juice. The common procedure is evaporation, but the derived juices may lose flavor components or undergo thermal degradation during the evaporative process. In freeze concentration, the solvent is crystallized (frozen) in relatively pure form to leave behind a solution with a higher solute concentration than the original mixture. Significant advantages in product taste have been observed in the application of this process to concentrations of various types of fruit juice.

The elimination of small amounts of an impurity from a product species may be an objective of crystallization. In such instances, a multistep crystallization–redissolution–recrystallization process may be required to produce a product that meets purity specifications. For example, in the manufacture of the amino acid L-isoleucine, the product is first recovered in acid form, redissolved, neutralized, and then recrystallized in order to exclude the impurity L-leucine and other amino acids from the product.

A simple change in physical properties also can be achieved by crystallization. In the process of making soda ash, referred to earlier, the sodium bicarbonate crystals are subjected to heat that causes the release of carbon dioxide and produces low-density sodium carbonate crystals. The density of these crystals is incompatible with their use in glass manufacture, but a more acceptable crystal can be obtained by contacting the sodium carbonate crystals with water to form crystalline sodium carbonate monohydrate. Drying the resulting crystals removes the water of hydration and produces a dense product that is acceptable for glass manufacture.

Separation of a chemical species from a mixture of similar compounds can be achieved by crystallization. The mode of crystallization may fall in the realm of what is known as melt crystallization. In such processes, the mother liquor largely is comprised of the melt of the crystallizing species, and, subsequent to its crystallization, crystals formed from the mother liquor are remelted to produce the product from the crystallizer. *Para(p)*-xylene can be crystallized from a mixture that includes *ortho* and *meta* isomers in a vertical column that causes crystals and mother liquor to move countercurrently. Heat is added at the bottom of the column to melt the *p*-xylene crystals; a portion of the melt is removed from the crystallizer as product and the remainder flows up the column to contact the downward-flowing crystals. Effluent mother liquor, consisting almost entirely of the *ortho* and *meta* isomers of xylene, is removed from the top of the column.

Production of a consumer product in a form suitable for use and acceptable to the consumer also may be an objective of a crystallization process. For example, sucrose (sugar) can be crystallized in various forms. However, different cultures are accustomed to using sugar that has a particular appearance and, unless the commodity has that appearance, the consumer may consider the sugar to be unacceptable.

In all these processes, there are commmon needs: to form crystals, to cause them to grow, and to separate the crystals from the residual liquid. While conceptually simple, the operation of a process that utilizes crystallization can be very complex. The reasons for such complexity involve the interaction of the common needs and process requirements on product yield, purity, and, uniquely, crystal morphology and size distribution. In the following discussion, the interactions will be discussed and general principles affecting crystallizer operation will be outlined. More extensive discussion of the subject matter can be found in the bibliography at the end of the chapter.

II. EQUILIBRIUM AND MASS AND ENERGY BALANCES

A. Solid–Liquid Equilibrium

The solubility of a chemical species in a solvent refers to the amount of solute that can be dissolved at constant temperature, pressure, and solvent composition (including the presence of other solutes). In other words, it is the concentration of the solute in the solvent at equilibrium.

As with all multiphase systems, the Gibbs phase rule provides a useful tool for determining the number of intensive variables (ones that do not depend on system mass) that can be fixed independently:

$$N_{DF} = N_c - N_p + 2 \qquad (1)$$

N_{DF} is the number of degrees of freedom, N_c is the number of components, and N_p is the number of phases in the system. The number of degrees of freedom represents the number of independent variables that must be specified in order to fix the condition of the system. For example, the Gibbs phase rule specifies that a two-component, two-phase system has two degrees of freedom. If temperature and pressure are selected as the specified variables, then all other intensive variables—in particular, the composition of each of the two phases—are fixed, and solubility diagrams of the type shown for a hypothetical mixture of R and S in Fig. 1 can be constructed.

Several features of the hypothetical system described in Fig. 1 illustrate the selection of crystallizer operating

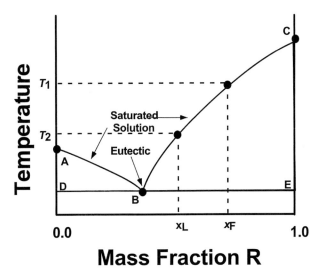

FIGURE 1 Hypothetical solubility diagram of eutectic-forming system.

conditions and the limitations placed on the operation by the system properties. The curves AB and BC represent solution compositions that are in equilibrium with solids whose compositions are given by the lines AD and CE, respectively. If AD and CE are vertical and coincident with the left and right extremes, the crystals are pure S and R, respectively. Crystallization from any solution whose equilibrium composition is to the left of a vertical line through point B will produce crystals of pure S, while solutions with an equilibrium composition to the right of the line will produce crystals of pure R. A solution whose composition falls on the line through B will produce a mixture of crystals of R and S.

Now suppose a saturated solution at temperature T_1 is fed to a crystallizer operating at temperature T_2. Since it is saturated, the feed has a mole fraction of R equal to x_F. The maximum production rate of crystals occurs when the solution leaving the crystallizer is saturated, meaning that the crystal production rate, m_{prod}, depends on the value of T_2:

$$m_{prod} = m_F x_F - m_L x_L \qquad (2)$$

where m_F is the feed rate to the crystallizer and m_L is the solution flow rate leaving the crystallizer. Note that the lower limit on T_2 is given by the eutectic point, and that attempts to operate the crystallizer at a temperature other than the eutectic value will result in a mixture of crystals of R and S.

When certain solutes crystallize from aqueous solutions, the crystals are hydrated salts, which means that the crystals contain water and solute in a specific stoichiometric ratio. The water in such instances is referred to as *water of hydration*, and the number of water molecules associated with each solute molecule may vary with the crystallization temperature.

Potassium sulfate provides an example of such behavior. When it crystallizes from an aqueous solution above $40°C$, the crystals are anhydrous K_2SO_4, while below $40°C$ each molecule of K_2SO_4 that crystallizes has 10 molecules of water associated with it. The hydrated salt, $K_2SO_4 \cdot 10H_2O(s)$, is called potassium sulfate decahydrate. Another solute that forms hydrated salts is magnesium sulfate, which can incorporate differing amounts of water depending upon the temperature at which crystallization occurs (see Table I).

The solubility diagrams of several species are shown in Fig. 2, and these illustrate the importance of solubility behavior in the selection of the mode of crystallization. For example, consider the differences between potassium nitrate and sodium chloride: The solubility of potassium nitrate is strongly influenced by the system temperature, whereas the opposite is true for sodium chloride. As a consequence, (1) a high yield of potassium nitrate crystals can be obtained by cooling a saturated feed solution,

TABLE I Water of Hydration for MgSO$_4$

Form	Name	wt% MgSO$_4$	Conditions
MgSO$_4$	Anhydrous magnesium sulfate	0.0	>100°C
MgSO$_4 \cdot$H$_2$O	Magnesium sulfate monohydrate	87.0	67 to 100°C
MgSO$_4 \cdot$6 H$_2$O	Magnesium sulfate hexahydrate	52.7	48 to 67°C
MgSO$_4 \cdot$7 H$_2$O	Magnesium sulfate heptahydrate	48.8	2 to 48°C
MgSO$_4 \cdot$12 H$_2$O	Magnesium sulfate dodecahydrate	35.8	−4 to 2°C

but (2) cooling a saturated sodium chloride solution accomplishes little crystallization, and vaporization of water is required to increase the yield.

The effect of water of hydration on solubility can be seen in Fig. 2. Note, for example, that sodium sulfate has two forms in the temperature range of the solubility diagram: sodium sulfate decahydrate (Na$_2$SO$_4 \cdot$10H$_2$O), which is known as Glauber's salt, and anhydrous sodium sulfate. Since a transition from Glauber's salt to anhydrous sodium sulfate occurs at approximately 34°C, crystals recovered from a crystallizer operating above about 34°C will be anhydrous, but those from a crystallizer operating below this temperature will contain 10 waters of hydration. Also observe the effect of water of hydration on solubility characteristics; clearly, cooling crystallization could be used to recover significant yields of Glauber's salt but evaporative crystallization would be required to obtain high yields of the anhydrous salt.

Mixtures of multiple solutes in a single solvent are encountered in a number of processes—for example, in the recovery of various chemicals from ores or brines. Expres-

sion of the complex solubility behavior in such systems by graphical means usually is limited to systems of two solutes. The interaction of added solutes on solubility is illustrated by the plot of equilibrium behavior for potassium nitrate–sodium nitrate–water in Fig. 3. As before, the curves in the diagram trace solution compositions that are in equilibrium with solid solutes. Points A, D, G, and J are based on the solubilities of pure potassium nitrate, while points C, F, I, and L are based on solubilities of pure sodium nitrate. Curves AB, DE, GH, and JK represent compositions of solutions in equilibrium with solid potassium nitrate at 30, 50, 70, and 100°C, respectively. Curves BC, EF, HI, and KL represent compositions of solutions in equilibrium with solid sodium nitrate. Should the solution condition, including temperature, correspond to points B, E, H, K or any condition on the curve connecting these points, crystals of both solutes would be formed by cooling.

A second type of solubility behavior is exhibited by mixtures that form solid solutions. Consider, for example, a hypothetical system containing R and S whose

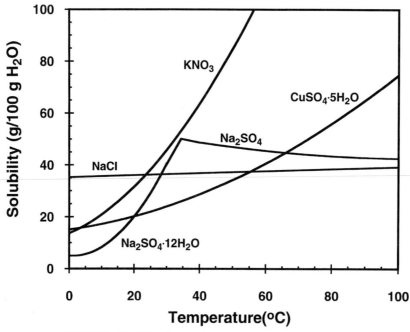

FIGURE 2 Solubility diagram for several common substances.

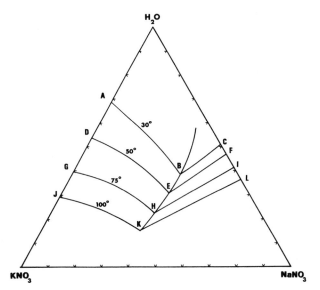

FIGURE 3 Solubility diagram of KNO₃ and NaNO₃ mixtures in water.

equilibrium behavior is described in Fig. 4. The phase envelope is drawn based on the compositions of coexisting liquid and solid phases at equilibrium. The pure component R has a melting point at pressure P equal to T_2 while the melting point of pure S is T_1. The system behavior can best be described by the following example: Consider a mixture of R and S at temperature T_A and having a mass fraction of R equal to z_M. From the phase diagram, the mixture is a liquid. As the liquid is cooled, a solid phase forms when the temperature reaches T_B and the system is allowed to come to equilibrium; the solid-phase composition corresponds to a mass fraction of R equal to x_B. On cooling the liquid further, the ratio of solid to liquid in-

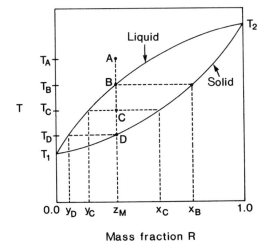

FIGURE 4 Hypothetical solubility diagram of mixture without a eutectic at constant pressure: x, solid; y, liquid; z, combined.

creases and at T_C the mass fraction of R in the liquid is y_C and in the solid it is x_C. At T_D the liquid phase disappears, leaving a solid with a mass fraction of R equal to z_M.

Systems that exhibit behavior of the type illustrated in Fig. 4 cannot be purified in a single crystallization stage. They represent situations in which multiple stages or continuous-contacting devices may be useful. The principles of such operations are analogous to those of other countercurrent contacting operations—for example, distillation, absorption, and extraction.

Variables other than temperature and the presence of other solutes can influence solubility. For example, the effect of a nonsolvent on solubility sometimes is used to bring about recovery of a solute. Figure 5 shows the solubility of L-serine in aqueous solutions containing varying amounts of methanol. Note that increasing methanol content reduces the solubility by more than an order of magnitude, and this characteristic can be used to obtain a high yield in the recovery of L-serine.

There is increasing interest in the crystallization of solutes from supercritical-fluid solvents. In such instances, solubilities often are correlated by an equation of state. Such concepts are beyond the scope of the current discussion but are presented elsewhere in the encyclopedia.

Although this discussion provides insight to the types of solubility behavior that can be exhibited by various systems, it is by no means a complete survey of the topic. Extensive solubility data and descriptions of more complex equilibrium behavior can be found in the literature. Published data usually consist of the influence of temperature on the solubility of a pure solute in a pure solvent; seldom are effects of other solutes, co-solvents, or pH considered. As a consequence, solubility data on a system of interest should be measured experimentally, and the solutions used in the experiments should be as similar as possible to those expected in the process. Even if a crystallizer has been designed and the process is operational, obtaining solubility data using mother liquor drawn from the crystallizer or a product stream would be wise. Moreover, the solubility should be checked periodically to see if it has changed due to changes in the upstream operations or raw materials.

There have been advances in the techniques by which solid–liquid equilibria can be correlated and, in some cases, predicted. These are described in references on phase-equilibrium thermodynamics.

B. Mass and Energy Balances

Illustrating the formulation of mass and energy balances is simplified by restricting the analysis to systems whose crystal growth kinetics are sufficiently fast to utilize essentially all of the supersaturation provided by the crystallizer; in other words, the product solution

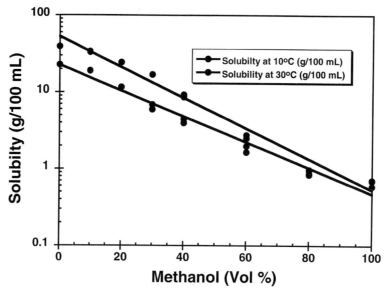

FIGURE 5 Effect of methanol on solubility of L-serine.

is assumed to be saturated. Under such conditions (referred to in the crystallization literature as Class II or fast-growth behavior), the solute concentration in the mother liquor can be assigned a value corresponding to saturation. Should the supersaturation in the mother liquor be so great as to affect the solute balance, the operation is said to follow Class I or slow-growth behavior. In Class I behavior, the operating conditions affect the rate at which solute is crystallized, and an expression coupling the rate of growth to a solute balance must be used to describe the system. Such treatment will be considered beyond the scope of this discussion.

The solution of mass and energy balances requires solubility and enthalpy data on the system of interest. Various methods of presenting solubility data were given earlier, and the use of solubilities to estimate crystal production rates from a cooling crystallizer was demonstrated by the discussion of Eq. (2). Subsequent to determining the yield, the rate at which heat must be removed from the crystallizer can be calculated from an energy balance:

$$m_C \hat{H}_C + m_L \hat{H}_L - m_F \hat{H}_F = Q \qquad (3)$$

where m_F, m_C, and m_L are feed rate, crystal production rate, and mother liquor flow rate, respectively; $\hat{H}$ is specific enthalpy of the stream corresponding to the subscript; and Q is the required rate of heat transfer to the crystallizer. As m_F, m_C, and m_L are known or can be calculated from a simple mass balance, determination of Q requires estimation of specific enthalpies. These are most conveniently obtained from enthalpy-composition diagrams, which are available in the general literature for a number of substances.

If specific enthalpies are unavailable, they can be estimated based on defined reference states for both solute and solvent. Often the most convenient reference states are crystalline solute and pure solvent at an arbitrarily chosen reference temperature. The reference temperature selected usually corresponds to that at which the heat of crystallization, $\Delta \hat{H}_c$, of the solute is known. (The heat of crystallization is approximately equal to the negative of the heat of solution.) For example, if the heat of crystallization is known at T_{ref}, then reasonable reference conditions would be the solute as a solid and the solvent as a liquid, both at T_{ref}. The specific enthalpies could be estimated then as:

$$\hat{H}_F = x_F \Delta \hat{H}_c + C_{p_F}(T - T_{ref}) \qquad (4)$$

$$\hat{H}_C = C_{p_C}(T - T_{ref}) \qquad (5)$$

$$\hat{H}_L = x_L \Delta \hat{H}_c + C_{p_L}(T - T_{ref}) \qquad (6)$$

where x_F and x_L are the mass fractions of solute in the feed and mother liquor, respectively. All that is required now to determine the required rate of heat transfer is the indicated heat capacities, which can be estimated based on system composition or measured experimentally.

Now suppose some of the solvent is evaporated in the crystallizer. Independent balances can be written on total and solute masses:

$$m_F = m_V + m_L + m_C \qquad (7)$$

$$x_F m_F = x_L m_L + x_C m_C \qquad (8)$$

Assuming that the streams leaving the crystallizer are in equilibrium, there is a relationship between the temperature (or pressure) at which the operation is conducted and x_L and x_C. In addition, an energy balance must be satisfied:

$$m_F \hat{H}_F + Q = m_V \hat{H}_V + m_L \hat{H}_L + m_C \hat{H}_C \qquad (9)$$

The specific enthalpies in the above equation can be determined as described earlier, provided the temperatures of the product streams are known. Evaporative cooling crystallizers (described more completely in Section V) operate at reduced pressure and may be considered adiabatic. In such circumstances, Eq. (9) is modified by setting $Q = 0$. As with many problems involving equilibrium relationships and mass and energy balances, trial-and-error computations are often involved in solving Eqs. (7) through (9).

III. NUCLEATION AND GROWTH KINETICS

The kinetics of crystallization have constituent phenomena in crystal nucleation and growth. The rates at which these occur are dependent on driving forces (usually expressed as supersaturation), physical properties, and process variables, but relationships between these quantities and crystallization kinetics often are difficult to express quantitatively. As a result, empirical or qualitative links between a process variable and crystallization kinetics are useful in providing guidance in crystallizer design and operation and in developing strategies for altering the properties of crystalline products.

Nucleation and growth can occur simultaneously in a supersaturated environment, and the relative rates at which these occur are primary determinants of the characteristics of the crystal size distribution; one way of influencing product size distributions is through the control of variables such as supersaturation, temperature, and mixing characteristics. Obviously, those factors that increase nucleation rates relative to growth rates lead to a crystal size distribution consisting of smaller crystals. In the discussion that follows, an emphasis will be given to the general effects of process variables on nucleation and growth, but the present understanding of these phenomena does not allow quantitative *a priori* prediction of the rates at which they occur.

A. Supersaturation

Supersaturation is the thermodynamic driving force for both crystal nucleation and growth; and therefore, it is the key variable in setting the mechanisms and rates by which these processes occur. It is defined rigorously as the deviation of the system from thermodynamic equilibrium and is quantified in terms of chemical potential,

$$\Delta \mu_i = \mu_i - \mu_i^* = RT \ln \frac{a_i}{a_i^*} \qquad (10)$$

where μ_i is the chemical potential of solute i at the existing conditions of the system, μ_i^* is the chemical potential of the solute equilibrated at the system conditions, and a_i and a_i^* are activities of the solute at the system conditions and at equilibrium, respectively. Less abstract definitions involving measurable system quantities are often used to approximate supersaturation; these involve either temperature or concentration (mass or moles of solute per unit volume or mass of solution or solvent) or mass or mole fraction of solute. Recommendations have been made that it is best to express concentration in terms of moles of solute per unit mass of solvent. For systems that form hydrates, the solute should include the water of hydration, and that water should be deducted from the mass of solvent.

Consider, for example, a system at temperature T with a solute concentration C, and define the equilibrium temperature of a solution having a concentration C as T^* and the equilibrium concentration of a solution at T as C^*. These quantities may be used to define the following approximate expressions of supersaturation:

1. The difference between the solute concentration and the concentration at equilibrium, $\Delta C_i = C_i - C_i^*$
2. For a solute whose solubility in a solvent increases with temperature, the difference between the temperature at equilibrium and the system temperature, $\Delta T = T^* - T$
3. the supersaturation ratio, which is the ratio of the solute concentration and the equilibrium concentration, $S_i = C_i / C_i^*$
4. The ratio of the difference between the solute concentration and the equilibrium concentration to the equilibrium concentration, $\sigma_i = (C_i - C_i^*)/C_i^* = S_i - 1$, which is known as relative supersaturation.

Any of the above definitions of supersaturation can be used over a moderate range of system conditions, but as outlined in the following paragraph, the only rigorous expression is given by Eq. (10).

The definitions of supersaturation ratio and relative supersaturation can be extended to any of the other variables used in the definition of supersaturation. For example, defining $S_{a_i} = a_i / a_i^*$ gives:

$$\frac{\Delta \mu_i}{RT} = \ln S_{a_i} = \ln \frac{\gamma_i C_i}{\gamma_i^* C_i^*} \qquad (11)$$

Therefore, for ideal solutions or for $\gamma_i \approx \gamma_i^*$,

$$\frac{\Delta \mu_i}{RT} \approx \ln \frac{C_i}{C_i^*} = \ln S_i \qquad (12)$$

Furthermore, for low supersaturations (say, $S_i < 1.1$),

$$\frac{\Delta \mu_i}{RT} \approx S_i - 1 = \sigma_i \qquad (13)$$

The simplicity of Eq. (13) results in the use of relative supersaturation in most empirical expressions for nucleation and growth kinetics. While beguilingly simple, and correct in limiting cases, great care should be taken in extending such expressions beyond conditions for which the correlations were developed.

For ionic solutes, $a_i = a_\pm^\nu$, which leads to $S_{a_i} = (a_\pm/a_\pm^*)^\nu$ and

$$\frac{\Delta \mu_i}{RT} = \nu \ln S_{a_i} = \nu \ln \frac{\gamma_{i\pm} C_i}{\gamma_{i\pm}^* C_i^*} \tag{14}$$

Again, for $\gamma_{i\pm} \approx \gamma_{i\pm}^*$,

$$\frac{\Delta \mu_i}{RT} \approx \nu \ln \frac{C_i}{C_i^*} = \nu \ln S_i \tag{15}$$

B. Primary Nucleation

The term *primary nucleation* is used to describe both homogeneous and heterogeneous nucleation mechanisms in which solute crystals play no role in the formation of new crystals. Primary nucleation mechanisms involve the formation of crystals through a process in which constituent crystal units are stochastically combined. Both homogeneous and heterogeneous nucleation require relatively high supersaturations, and they exhibit a high-order dependence on supersaturation. As will be shown shortly, the high-order dependence has a profound influence on the character of crystallization processes in which primary nucleation is the dominant means of crystal formation.

The classical theoretical treatment of primary nucleation that produces a spherical nucleus results in the expression:

$$B^\circ = A \exp\left(-\frac{16\pi \epsilon_{surf}^3 v^2}{3k^3 T^3 [\ln(\sigma + 1)]^2}\right)$$

$$\overset{\sigma < 0.1}{\approx} A \exp\left(-\frac{16\pi \epsilon_{surf}^3 v^2}{3k^3 T^3 \sigma^2}\right) \tag{16}$$

where k is the Boltzmann constant, ϵ_{surf} is the interfacial surface energy per unit area, v is molar volume of the crystallized solute, and A is a constant.

The theory shows that the most important variables affecting the rates at which primary nucleation occur are interfacial energy ϵ_{surf}, temperature T, and supersaturation σ. The high-order dependence of nucleation rate on these three variables, especially supersaturation, is important because, as shown by an examination of Eq. (16), a small change in any of the three variables could produce an enormous change in nucleation rate. Such behavior gives rise to the often observed phenomenon of having a clear liquor transformed to a slurry of very fine crystals with only a slight increase in supersaturation, for example by decreasing the solution temperature.

The effect of exogenous solid matter (as in heterogeneous nucleation) in the supersaturated solution is equivalent to that of a catalyst in a reactive mixture. Namely, it is to reduce the energy barrier to the formation of a new phase. In effect, the solid matter reduces the interfacial energy ϵ_{surf} by what may amount to several orders of magnitude.

The classical nucleation theory embodied in Eq. (16) has a number of assumptions and physical properties that cannot be estimated accurately. Accordingly, empirical power-law relationships involving the concept of a metastable limit have been used to model primary nucleation kinetics:

$$B^\circ = k_N \sigma_{max}^n \tag{17}$$

where k_N and n are parameters fit to data and σ_{max} is the supersaturation at which nuclei are observed when the system is subjected to a specific protocol. Although Eq. (17) is based on empiricism, it is consistent with the more fundamental Eq. (16).

C. Secondary Nucleation

Secondary nucleation is the formation of new crystals through mechanisms involving existing solute crystals; in other words, crystals of the solute *must* be present for secondary nucleation to occur. Several features of secondary nucleation make it important in the operation of industrial crystallizers: First, continuous crystallizers and seeded batch crystallizers have crystals in the magma that can participate in secondary nucleation mechanisms. Second, the requirements for the mechanisms of secondary nucleation to be operative are fulfilled easily in most industrial crystallizers. Finally, many crystallizers are operated in a low supersaturation regime so as to maximize yield, and at such supersaturations the growth of crystals is more likely to produce desired morphologies and high purity; these low supersaturations can support secondary nucleation but not primary nucleation.

1. Mechanisms

Secondary nucleation can occur through several mechanisms, including initial breeding, contact nucleation (also known as collision breeding), and shear breeding. Although a universal expression for the kinetics of secondary nucleation does not exist, a working relationship often can be obtained by correlating operating data from a crystallizer with a semi-empirical expression. Guidance as to the form of the expression and the variables that it should include can be obtained by understanding the various mechanisms of secondary nucleation.

Initial breeding results from immersion of seed crystals in a supersaturated solution, and it is thought to be caused by dislodging extremely small crystals that were formed on the surface of larger crystals during drying. Although this mechanism is unimportant in continuous and unseeded batch crystallization, it can have a significant impact on the operation of seeded batch crystallizers. The number of crystals formed by initial breeding, has been found to be proportional to the surface area of crystals used to seed a batch crystallizer. Characteristics of the resulting distribution are affected strongly by the growth kinetics of nuclei resulting from initial breeding, and the phenomenon of growth-rate dispersion (which will be discussed later) can lead to erroneous conclusions regarding the nucleation kinetics.

Shear breeding results when supersaturated solution flows by a crystal surface and carries with it crystal precursors believed formed in the region of the growing crystal surface. High supersaturation is required for shear breeding to produce significant numbers of nuclei.

Contact nucleation in industrial processes results from collisions of crystals with the impeller used for circulation of the magma or with other crystallizer internals such as baffles, pipe and crystallizer walls, and even other crystals. Careful experimental studies have shown that the number of crystals produced by collisions between crystals and these objects depends upon the collision energy, supersaturation at impact, supersaturation at which crystals mature, material of the impacting object, area and angle of impact, and system temperature. The collision energy for contact nucleation is small and does not necessarily result in the macroscopic degradation or attrition of the contacted crystal.

Nucleation from collisions between crystals in the circulating magma and the rotor in a circulation pump or an agitator usually dominate nucleation resulting from other collisions. The operating variables in systems of this type can be manipulated to some extent, thereby modifying nucleation rates and the concomitant crystal size distribution. For example, internal classification can be used to keep larger crystals away from energetic collisions with an impeller, but doing so may create other problems with stability of the crystal size distribution. The rotational speed of an impeller can be changed if there are appropriate controls on the pump or agitator. Caution must be exercised, however, for a reduction in circulation velocity can reduce heat-transfer coefficients and increase fouling (encrustation) on heat-transfer surfaces. Moreover, the crystals in the magma must be kept suspended or crystal morphology and growth rates could be affected adversely. Impact

energy may have a high-order dependence on rotational speed and, if that is the case, modest changes in this variable could alter nucleation rates substantially. The fraction of the impact energy transmitted from an impeller to the crystal can be manipulated by changing the material of construction of the impeller. The influence of using soft materials to coat impellers or crystallizer internals may vary from one crystalline system to another; those systems in which the crystal face is soft may be more susceptible to nucleation rate changes than those crystalline systems where the face is hard.

Supersaturation has been observed to affect contact nucleation, but the mechanism by which this occurs is not clear. There are data that infer a direct relationship between contact nucleation and crystal growth; these data showed that the number of nuclei produced by an impact was proportional to the linear growth rate of the impacted face. This could indicate that the effect of supersaturation is to alter growth rates and, concomitantly, the characteristics of the impacted crystal faces; alternatively, what appears to be a mechanistic relationship actually could be a result of both nucleation and growth depending upon supersaturation.

Another theory that could account for the effect of supersaturation on contact nucleation is based on the view that nuclei formed cover a range of sizes that includes the critical nucleus. Since only the nuclei larger than the critical nucleus are stable, the relationship of the size of the critical nucleus to supersaturation reflects the dependence of contact nucleation on supersaturation. This concept, which has been referred to as a survival theory, seems to have been refuted by measurements of the sizes of crystals formed by collisions. These sizes are much larger than the critical nucleus, and the survival theory would have little influence on the number of nuclei that survive.

Evidence of the formation of polymolecular clusters in supersaturated solutions may provide a mechanistic interpretation of the effect of supersaturation on contact nucleation kinetics. These clusters may participate in nucleation, although the mechanism by which this would occur is not clear. One model that has been proposed, however, calls for the formation of a semi-ordered region consisting of molecular clusters awaiting incorporation into the crystal lattice. Collisions or fluid shear of the region containing high cluster concentrations could then result in these clusters serving as secondary nuclei. In such a model, the variables that influence formation and diffusion of the clusters also influence crystal growth rates and nucleation.

2. Kinetic Expressions

Irrespective of the actual mechanisms by which contact nucleation occurs, empirical power-law expressions provide a useful means of correlating nucleation kinetics and using the resulting correlations in process analysis and control. The correlations generally take the form:

$$B° = k_N \sigma^i M_T^j N^k \qquad (18)$$

where k_N, i, j, k are positive parameters obtained from data correlation, M_T is the magma density (mass of solids per unit volume of slurry or solvent in the magma), and N is the rotational velocity of the impeller or pump rotor. For convenience, either crystal growth rate or mean residence time, both of which are directly related to supersaturation, may be substituted for σ in Eq. (18).

If primary nucleation dominates the process, i tends to larger values (say greater than 3), j and k approach zero, and Eq. (18) approaches Eq. (17). Should crystal–impeller and/or crystal–crystallizer impacts dominate, j approaches 1; on the other hand, if crystal–crystal contacts dominate, j approaches 2.

The ease with which nuclei can be produced by contact nucleation is a clear indication that this mechanism is dominant in many industrial crystallization operations. Research on this nucleation mechanism is continuing with the objective of building an understanding of the phenomenon that will allow its successful inclusion in models describing commercial systems.

D. Fundamentals of Crystal Growth

Crystal growth rates may be expressed as (1) the linear advance rate of an individual crystal face, (2) the change in a characteristic dimension of a crystal, or (3) the rate of change in mass of a crystal or population of crystals. These different expressions are related through crystal geometry; it is often convenient to use the method of measurement as the basis of the growth rate expression or, in certain instances, the method used to analyze a crystallization process will require that growth rate be defined in a specific way. For example, the use of a population balance to describe crystal size distribution requires that growth rate be defined as the rate of change of a characteristic dimension.

Single-crystal growth kinetics involve the advance rate of an individual crystal face normal to itself or the rate of change in crystal size associated with exposure to a supersaturated solution. The advance rate of a single crystal face can be quantified by observation of the face through a calibrated eyepiece of an optical microscope, which allows examination of the structure of the advancing crystal face and isolation of surface-reaction kinetics from mass-transfer kinetics (these phenomena will be discussed later). An additional advantage of single-crystal systems is that it is possible to examine crystal growth kinetics without interference from competing processes such as nucleation.

Multicrystal-magma studies usually involve examination of the rate of change of a characteristic crystal dimension or the rate of increase in the mass of crystals in a magma. The characteristic dimension in such analyses depends upon the method used in the determination of crystal size; for example, the second largest dimension is measured by sieve analyses, while an equivalent spherical diameter is determined by both electronic zone sensing and laser light scattering instruments. A relationship between these two measured dimensions and between the measured quantities and the actual crystal dimensions can be derived from appropriate shape factors. Volume and area shape factors are defined by the equations:

$$v_{crys} = k_{vol} L^3 \qquad \text{and} \qquad a_{crys} = k_{area} L^2 \qquad (19)$$

where v_{crys} and a_{crys} are volume and area of a crystal, k_{vol} and k_{area} are volume and area shape factors, and L is the characteristic dimension of the crystal. Suppose an equivalent spherical diameter L_{sphere} is obtained from an electronic zone-sensing instrument, and the actual dimensions of the crystal are to be calculated. Assume for the sake of this example that the crystals have a cubic shape. Let L_{cube} be the edge length of the crystal and k_{vol}^{sphere} and k_{vol}^{cube} be the volume shape factors for a sphere and a cube, respectively. Since the volume of the crystal is the same, regardless of the arbitrarily defined characteristic dimension,

$$v_{crys} = k_{vol}^{sphere} L_{sphere}^3 = k_{vol}^{cube} L_{cube}^3 \qquad (20)$$

Since k_{vol}^{sphere} is $\pi/6$ and k_{vol}^{cube} is 1.0, the numerical relationship between L_{cube} and L_{sphere} is given by:

$$L_{cube} = \left(\frac{k_{vol}^{sphere}}{k_{vol}^{cube}} \right)^{1/3} L_{sphere} = \left(\frac{\pi}{6} \right)^{1/3} L_{sphere} \qquad (21)$$

The rate of change of a crystal mass dm_{crys}/dt can be related to the rate of change in the crystal characteristic dimension ($dL/dt = G$) by the equation:

$$\frac{dm_{crys}}{dt} = \frac{d(\rho k_{vol} L^3)}{dt} = 3\rho k_{vol} L^2 \left(\frac{dL}{dt} \right) \qquad (22)$$

where ρ is crystal density. Since $k_{area} = a_{crys}/L^2$,

$$\frac{dm_{crys}}{dt} = 3\rho (k_{vol}/k_{area}) a_{crys} G \qquad (23)$$

At least two resistances contribute to the kinetics of crystal growth. These resistances apply to (1) integration of the crystalline unit (e.g., solute molecules) into the crystal surface (i.e., lattice), and (2) molecular diffusion or bulk transport of the unit from the surrounding solution to the crystal surface. As aspects of molecular diffusion and mass transfer are covered elsewhere, the current discussion will focus only on surface incorporation.

1. Mechanisms

Among the many models that have been proposed to describe surface-reaction kinetics are those that assume crystals grow by layers and others that consider growth to occur by the movement of a continuous step. Each physical model results in a specific relationship between growth rate and supersaturation and, although none can predict growth kinetics *a priori*, insights regarding the effects of process variables on growth can be obtained. Because of the extensive literature on the subject, only the key aspects of the physical models and (in one case) the resulting relationship between growth and supersaturation predicted by each theory will be discussed here.

The model used to describe the growth of crystals by layers is based on a two-step, birth-and-spread mechanism. In one of the steps (birth) a two-dimensional nucleus is formed on the crystal surface, and in the second step (spread) the two-dimensional nucleus grows to cover the crystal surface. When one or the other of the steps is controlling growth rates, simplifications of the more complicated dependence of growth rate on supersaturation can be developed to give what are known as the mononuclear two-dimensional nucleation theory and the polynuclear two-dimensional nucleation theory. In the mononuclear two-dimensional nucleation theory, surface nucleation occurs at a finite rate while the spreading across the surface occurs at an infinite rate. The reverse is true for the polynuclear two-dimensional nucleation theory. Theoretical relationships have been derived between growth rate and supersaturation for each of these conditions but are considered beyond the scope of this discussion.

The screw-dislocation theory (sometimes referred to as the BCF theory because of its development by Burton, Cabrera, and Frank) is based on a mechanism of continuous movement in a spiral or screw of a step or ledge on the crystal surface. The theory shows that the dependence of growth rate on supersaturation can vary from a parabolic relationship at low supersaturations to a linear relationship at high supersaturations. In the BCF theory, growth rate is given by:

$$G = k_G \left(\frac{\epsilon \sigma^2}{b} \right) \tanh \left(\frac{b}{\epsilon \sigma} \right) \qquad (24)$$

where ϵ is screw dislocation activity and b is a system-dependent quantity that is inversely proportional to temperature. It can be shown that the dependence of growth rate on supersaturation is linear if the ratio $b/\epsilon\sigma$ is large, but the dependence becomes parabolic as the ratio becomes small. It is possible, then, to observe variations in the dependence of growth rate on supersaturation for a given crystal-solvent system.

An empirical approach also can be used to relate growth kinetics to supersaturation by simply fitting growth-rate data with a power-law function of the form:

$$G = k_G \sigma^g \qquad (25)$$

where k_G and g are system-dependent constants. Such an approach is valid over modest ranges of supersaturation, and the power-law function approximates the fundamental expressions derived from the above models.

2. Impurities

The presence of impurities can alter growth rates substantially, usually by decreasing them. Furthermore, as described in Section IV.B, impurities can alter crystal morphology through their effects on the growth rates of crystal faces. Mechanisms include: (1) adsorption of an impurity on the crystal surface or at specific growth sites such as kinks, thereby blocking access to the site by a growth unit; (2) formation of complexes between an impurity and a growth unit; and (3) incorporation of an impurity into a growing crystal and creating defects or repelling the addition of a growth unit to the subsequent crystal layer. Few of these mechanistic views result in predictive capabilities, and it is usual to rely on experimental data that are often correlated empirically.

Because impurities most often result in reduced crystal growth rate, feedstocks to laboratory and bench-scale units should be as similar as possible to that expected in the full-scale unit. The generation of impurities in upstream process units can depend on the way those units are operated, and protocols of such units should follow a consistent practice. It is equally important to monitor the composition of recycle streams so as to detect any accumulation of impurities that might lead to a reduction in growth rates.

The solvent from which a material is crystallized influences crystal morphology and growth rate. These effects have been attributed to two sets of factors. One has to do with the effects of solvent on viscosity, density, and diffusivity and, therefore, mass transfer. The second factor is concerned with the structure of the interface between crystal and solvent; a solute–solvent system that has a high solubility is likely to produce a rough interface and, concomitantly, large crystal growth rates.

E. Crystal Growth in Mixed Crystallizers

Population balances on crystals in a crystallizer require a definition of growth rates in terms of the rate of change of a characteristic dimension:

$$G = \frac{dL}{dt} \qquad (26)$$

Furthermore, the solution of a differential population balance requires that the relationship between growth rate and size of the growing crystals be known. When all crystals in the magma grow at a constant and identical rate, the crystal–solvent system is said to follow the McCabe ΔL law, while systems that do not are said to exhibit anomalous growth.

Two theories have been used to explain growth-rate anomalies: size-dependent growth and growth-rate dispersion. As with systems that follow the ΔL law, anomalous growth by crystals in a multicrystal magma produces crystal populations with characteristic forms. Unfortunately, it is difficult to determine the growth mechanism from an analysis of these forms. This means that either size-dependent growth or growth-rate dispersion may be used to correlate population density data without a certainty that the correct source of anomalous growth has been identified. Determining the actual source of anomalous growth is not trivial, but it may be worthwhile since alignment between a mathematical model and system behavior enhances the utility of the model.

Size-dependent crystal growth results when the rate of growth depends on the size of the growing crystal. Certainly, this may be the case if bulk transport is the controlling resistance to crystal growth, and the literature abounds with expressions for the appropriate mass-transfer coefficients. In the more common situation in which surface integration controls growth rate, there are no mechanistic relationships between growth rate and crystal size, and simple empirical expressions are called upon for that purpose.

Growth-rate dispersion is the term used to describe the behavior of similar sized crystals in the same population exhibiting different growth rates or growth rates that vary with time. The consequences of growth-rate dispersion are illustrated in Fig. 6, which shows the growth of a crystal population that has been immersed in a supersaturated solution. The spread of the distribution increases as the crystal population grows; the slower growing crystals form the tail of the advancing distribution while the faster growing ones form the leading edge. If all crystals in the population grew at the same rate, the distribution would advance uniformly along the size axis. Two causes of growth-rate dispersion have been observed. In one, the growth rate of each crystal in a population is nearly constant, but crystals in the population may grow at a different rate; in the other, the growth rate of an individual crystal fluctuates about a mean value.

The consequences of anomalous growth depends upon the process involved, and this will be pointed out in the discussion on population balances.

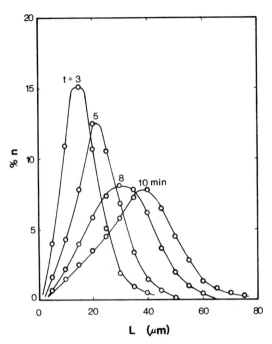

FIGURE 6 Transient population density plot showing growth-rate dispersion.

IV. PURITY, MORPHOLOGY, AND SIZE DISTRIBUTIONS

Crystal properties can be divided into two categories: those based on the individual crystal and those involving all crystals of a given population. The three characteristics of the section title compose what is often called *crystal quality*. They represent, along with yield, the most important criteria in the operation of a crystallizer. In the following discussion, some of the factors that influence purity and morphology are described and an introduction is given to methods of quantifying crystal size distributions.

A. Crystal Purity

The purity of a crystalline product depends on the nature of the other species in the mother liquor from which the crystals are produced, the physical properties of the mother liquor, and the processing that occurs between crystallization and the final product (downstream processing). Impurities can find their way into the final product through a number of mechanisms: the formation of occlusions, trapping of mother liquor in physical imperfections of the crystals or agglomerates, adsorption of species onto crystal surfaces, as part of chemical complexes (hydrates or solvates), or through lattice substitution.

Occlusions find their way into the crystal structure when the supersaturation in close proximity to the crystal surface is high enough to lead to an unstable surface. Such instability leads to the creation of dendrites, which then join to trap mother liquor in pools of liquid within the crystal. Occlusions are often visible and can be avoided through careful control of the supersaturation in the crystallizer.

Mother liquor can be flushed from a cake of crystals on a filter or centrifuge by washing with a liquid that also may dissolve a small portion of the cake mass. To be effective, the wash liquid must be spread uniformly over the cake and flow through the porous material without significant channeling. Such washing is hindered when the crystals themselves have significant cracks, crevices, or other manifestations of breakage or the mother liquor has a viscosity that is significantly greater than the wash liquid. In the latter event, significant channeling (also called *fingering*) may reduce the effectiveness of the wash process.

Lattice substitution requires that the incorporated impurity be of similar size and function to the primary crystallizing species. In other words, the impurity must fit into the lattice without causing significant dislocations. An example of such a system is found in the crystallization of L-isoleucine in the presence of trace quantities of L-leucine. The two species have similar molecular structures, differing only by one carbon atom in the position of a methyl side group. In this system, the incorporation of L-leucine in L-isoleucine crystals is proportional to the concentration of L-leucine in the mother liquor. Moreover, the shape of the recovered crystals changes as the content of L-leucine in recovered crystal increases.

B. Crystal Morphology

Both molecular and macroscopic concepts are important in crystal morphology. Molecular structures (i.e., the arrangements of molecules in specific lattices) can greatly influence the properties of a crystalline species and variations from a single structure lead to the prospect of polymorphic systems. In such systems, the molecular species of the crystal can occupy different locations depending on the conditions at which the crystal is formed, and both microscopic and macroscopic properties of the crystal can vary depending on the polymorph formed. There is, in general, a single stable polymorph for prevailing conditions, but that polymorph may not have been formed during the crystallization process. In such cases, system thermodynamics will tend to force transformation from the unstable polymorph to the stable one at rates that may vary from being nearly instantaneous to infinitely slow. Additional discussion of the molecular structures of crystalline materials has been provided elsewhere.

The characteristic macroscopic shape of a crystal results, in large measure, from the internal lattice structure; surfaces are parallel to planes formed by the constituent units of the crystal. Moreover, although the Law of Constant Interfacial Angles is a recognition that angles between corresponding faces of all crystals of a given substance are constant, the faces of individual crystals of that substance may exhibit varying degrees of development. As a result, the general shape or habit of a crystal may vary considerably.

Crystal morphology (i.e., both form and shape) affects crystal appearance; solid–liquid separations such as filtration and centrifugation; product-handling characteristics such as dust formation, agglomeration, breakage, and washing; and product properties such as bulk density, dissolution kinetics, catalytic activity, dispersability, and caking.

The shape of a crystal can vary because the relative rates of growth of crystal faces can change with system conditions; faster growing faces become smaller than faces that grow more slowly and in the extreme may disappear from the crystal altogether. For illustration, consider the two-dimensional crystal shown in Fig. 7a and the process variables that would cause the habit to be modified to the forms shown in Figs. 7b and c. The shape of the crystal depends on the ratio of the growth rate of the horizontal faces, G_h, to the growth rate of the vertical faces, G_v. For the shapes shown in Fig. 7,

$$\left(\frac{G_h}{G_v}\right)_b < \left(\frac{G_h}{G_v}\right)_a < \left(\frac{G_h}{G_v}\right)_c \qquad (27)$$

Growth rates depend on the presence of impurities, system temperature, solvent, mixing, and supersaturation, and the importance of each may vary from one crystal face to another. Consequently, an alteration in any or all of these variables can result in a change of the crystal shape.

Modeling intermolecular and intramolecular interactions through molecular mechanics calculations has advanced significantly in the past decade, and it has provided the basis for prediction of the equilibrium shape of

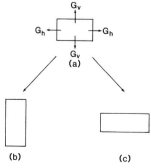

FIGURE 7 Effect of facial growth rates on crystal shape.

a known crystalline species. While not yet uniformly successful on a quantitative basis, the definition and modeling of crystal lattice potential energy equations has provided an understanding of crystal growth and morphology on the molecular level. Derivation of external crystal morphology from internal lattice structures via simulation has been proven possible for several organic compounds. Numerical minimization techniques, coupled with the appropriate valence and nonbonded energy expressions, have enabled accurate determination of favorable molecular arrangements within a wide variety of molecular crystals.

The shape of crystals obtained as a result of following a specific crystallization protocol may be unsatisfactory and, as a result, methods for modifying the habit of considerable interest. The predictive capabilities cited in the preceding paragraph are of great utility in such an instance as they may be used to determine factors leading to the unsatisfactory shape and guide subsequent experiments in which a more desirable shape is sought. Inevitably, such a search involves extensive laboratory or bench-scale experiments to determine processing variations that will lead to a desired crystal shape.

As an example of the variations in shape that can be exhibited by a single crystalline material, consider the forms of potassium sulfate shown in Fig. 8. Clearly, the processing characteristics and particulate properties of the differently shaped potassium sulfate crystals will vary.

The mechanisms and variables affecting crystal shape can be categorized as follows:

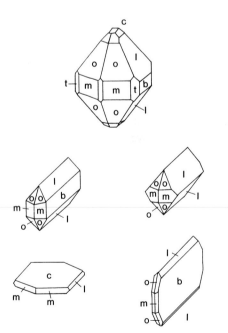

FIGURE 8 Shapes of K_2SO_4 crystals. [From Mullin, J. W. (1993). "Crystallization," 3rd ed. Butterworth-Heinemann, London. With permission.]

1. Intrinsic growth rates
   a. *Temperature*: The growth rates of individual crystal faces depend on temperature, typically following an Arrhenius rate law:
   $$G = G_0 \exp\left(-\frac{\Delta E_G}{RT}\right) \qquad (28)$$
   If different crystal faces have different activation energies, variation of the temperature at which crystallization takes place modifies individual growth rates to varying degrees and results in a modified crystal shape.
   b. *Mixing*: The intensity of mixing may determine the degree to which bulk mass transfer is involved in growth kinetics, and this can influence the resulting crystal shape.
   c. *Supersaturation*: The dependence of growth kinetics on supersaturation may vary from one crystal face to another. Accordingly, different prevailing supersaturations can lead to different crystal shapes.

2. Interfacial behavior
   a. *Solvent*: Different solvents exhibit different interactions with crystal faces and can alter crystal shape. A change in solvent also can alter the stoichiometry of the crystal (e.g., from a hydrated to an anhydrate stoichiometry), which can produce crystals with quite different morphology.
   b. *Surfactants*: Addition of a surfactant to a crystallizing system can influence the crystal shape in a manner illustrated schematically in Fig. 9. Here, surfactant molecules are shown being attracted to crystal faces in varying ways; the hydrophilic head groups favor the horizontal faces, while the hydrophobic tail groups are preferentially attracted to the vertical faces. A growth unit must displace the surfactant to approach a growing crystal face. As hydrophilic interactions are typically much stronger than hydrophobic ones, the growth unit preferentially enters the vertical faces and growth in the horizontal direction is favored.

3. Access to growth site
   a. *Blockage by species attracted to growth site*: Impurities may preferentially locate at a kink or other favored growth site and block growth at that site. A difference in the character of the kink or growth site from one face to another could result in modification of the crystal shape.
   b. *Species partially fitting into crystal lattice*: In these instances, an impurity molecule is comprised of two parts, one that fits into the crystal lattice

Crystal

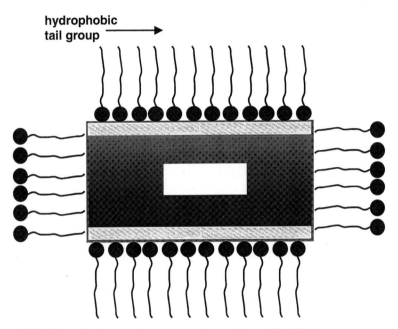

FIGURE 9 Attachment of surfactant molecules to crystal surfaces.

and a second that does not. The part that does not fit repulses incoming growth units or causes significant interatomic stress because of its position in the crystal lattice. If such species are purposely added to the crystallizing system to modify crystal morphology, they are referred to as tailor-made additives.

C. Crystal Size Distributions

Most crystallization processes produce particles whose sizes cover a range of varying breadth. If the particles consist of single crystals, the resulting distribution is a crystal size distribution (CSD); on the other hand, if the particles consist of agglomerates or other combination of multiple crystals, the distribution is a particle size distribution. In either case, the distribution is expressed in terms of either population (number) or mass. The population distribution relates the number of crystals at each size to the size, while the mass distribution expresses how mass is distributed over the size range. In the following paragraphs, methods for describing and using distribution functions will be outlined.

Size distribution is a major determinant of the properties of crystalline products, especially appearance, and to downstream processing and handling of crystalline materials. Solid–liquid separation by filtration or centrifugation can be straightforward with a desired CSD, but it can be disastrous when an inappropriate one increases resistance to liquor flow through a filter or centrifuge cake. Likewise, CSD affects other downstream processing such as the removal of impurities and mother liquor by washing, dissolution or reaction of the crystals, and transporting or storing crystals.

Crystal size distributions may be expressed by: (1) histograms, which are the amount or fraction of mass or

TABLE II Sieved KNO₃ Crystals from Hypothetical 1-Liter Sample

Sieve no., i	L (μm)	ΔM_i (g/L)	$M(L)$ (g/L)	$F(L)$ (frac)	$\bar{L}_i$ (μm)	ΔN_i (no./L)	$N(L)$ (frac)	m (g/μm·L)	n (no./μm·L)
1	707	0	32.974	1.000		1611			
2	500	7.296	25.678	0.779	603.5	16	1595	0.0352	0.076
3	354	11.512	14.166	0.430	427.0	70	1525	0.0789	0.480
4	240	9.011	5.154	0.156	297.0	163	1362	0.0790	1.430
5	177	3.145	2.009	0.061	208.5	164	1198	0.0499	2.610
6	125	1.322	0.687	0.021	151.0	182	1016	0.0254	3.500
7	88	0.462	0.225	0.007	106.5	181	834	0.0125	4.900
8	63	0.159	0.066	0.002	75.5	175	659	0.0064	7.000
9	44	0.055	0.011	0.000	53.5	171	488	0.0029	9.000
10	0	0.011	0.000	0.000	22.0	488	0	0.0002	11.100
Total		32.974				1611			

number over each increment in size; (2) cumulative distributions, which are the total or fraction of mass or number below (or above) a given size; and (3) density functions, which are the derivatives (with respect to size) of cumulative distributions. These definitions will be illustrated by considering a hypothetical potassium nitrate system from which a 1-liter slurry sample has been withdrawn, filtered, washed, dried, and sieved to give the results shown in Table II.

The first three and the sixth columns give the sieve data and should be read as follows: All of the sieved matter passed through the 707-μm sieve, and 7.296 g remained on the 500-μm sieve and had an arithmetic average size of 603.5 μm. Similar descriptions can be given for crystals that remained on the other sieves and pan ($L = 0$). The total crystal mass recovered was 39.974 g. A histogram of the mass distribution from these data is shown in Fig. 10.

The method by which crystals are sized gives either number or mass of crystals in a given size range. The sieve analysis in the above example gives mass distributions, so that the histogram is constructed in terms of crystal mass, and a cumulative mass distribution, $M(L)$, can be defined as the mass of crystals in the sample passing through the sieve of size L. In other words,

$$M(L) = \sum_{L=0}^{L(i)} \Delta M_i \qquad (29)$$

Such calculations give the mass of crystals below size L, and the results are shown in column 4 of Table II. Column 5 gives the cumulative mass fraction distribution:

$$F(L) = \frac{M(L)}{M_{\text{total}}} \qquad (30)$$

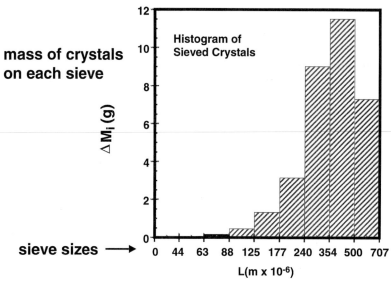

mass of crystals on each sieve

sieve sizes ⟶

FIGURE 10 Histogram of size distribution from example.

Transforming a mass distribution to a number distribution, or vice versa, requires a relationship between the measured and desired quantities. The mass of a single crystal, m_{crys}, is related to crystal size by the volume shape factor, k_{vol} (see Eq. (19)):

$$m_{\text{crys}} = \rho k_{\text{vol}} L^3 \qquad (31)$$

Consequently, the number of crystals on a sieve in the example, ΔN_i, can be estimated by dividing the total mass on sieve i by the mass of an average crystal on that sieve. If the crystals on that sieve are assumed to have a size equal to the average of the sieve through which they have passed and the one on which they are held, $\bar{L}_i = (L_{i-1} + L_i)/2$, then:

$$\Delta N_i = \frac{\Delta M_i}{\rho k_{\text{vol}} \bar{L}_i^3} \qquad (32)$$

Potassium nitrate crystals have a density of 2.11×10^{-12} g/μm^3, which allows for the determination of the estimated crystal numbers on each sieve in Table II. A cumulative number distribution, $N(L)$, and a cumulative number fraction distribution, $F(L)$, can be calculated using methods similar to those for calculating $M(L)$ and $W(L)$.

Mass and population densities are estimated from the respective cumulative number and cumulative mass distributions:

$$m(\bar{L}) = \frac{\Delta M_i}{\Delta L_i} \qquad (33)$$

$$n(\bar{L}) = \frac{\Delta N_i}{\Delta L_i} \qquad (34)$$

So that if $\Delta L_i \to 0$,

$$m(L) = \frac{dM(L)}{dL} \Rightarrow M(L) = \int_0^\infty m \, dL \qquad (35)$$

and

$$n(L) = \frac{dN(L)}{dL} \Rightarrow N(L) = \int_0^\infty n \, dL \qquad (36)$$

Equations (33) and (34) are used to obtain the last two columns of Table II.

In the example, all of the results are for the given sample size of 1 liter and the quantities estimated have units reflecting that basis. This basis volume is arbitrary, but use of the calculated quantities requires care in defining this basis consistently in corresponding mass and population balances. The volume of clear liquor in the sample is an alternative, and sometimes more convenient, basis.

Moments of a distribution provide information that can be used to characterize particulate matter. The jth moment of the population density function $n(L)$ is defined as:

$$m_j = \int_0^\infty L^j n(L) \, dL \qquad (37)$$

From Eq. (37), it can be demonstrated that the total number of crystals, the total length, the total area, and the total volume of crystals, all in a unit of sample volume, can be evaluated from the zeroth, first, second, and third moments of the population density function. Moments of the population density function also can be used to estimate number-weighted, length-weighted, area-weighted, and volume- or mass-weighted quantities. These averages are calculated from the general expression:

$$\bar{L}_{j+1,j} = \frac{m_{j+1}}{m_j} \qquad (38)$$

where $j = 0$ for a number-weighted average, 1 for a length-weighted average, 2 for an area-weighted average, and 3 for a volume- or mass-weighted average.

Crystal size distributions may be characterized usefully (though only partially) by a single crystal size and the spread of the distribution about that size. For example, the dominant crystal size represents the size about which the mass in the distribution is clustered. It is defined as the size, L_D, at which a unimodal mass density function is a maximum, as shown in Fig. 11; in other words, the dominant crystal size L_D is found where dm/dL is zero. (The data used to construct Fig. 11 are from Table II.) As the mass density is related to the population density by:

$$m = \rho k_v n L^3 \qquad (39)$$

the dominant crystal size can be evaluated from the population density by:

$$\frac{d(nL^3)}{dL} = 0 \qquad \text{at} \qquad L = L_D \qquad (40)$$

The spread of the mass-density function about the dominant size is the coefficient of variation (c.v.) of the CSD.

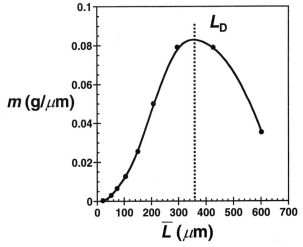

FIGURE 11 Mass–density function with single mode showing dominant size.

It is defined by:

$$\text{c.v.} = \frac{\sigma}{L_D} \qquad (41)$$

and estimated from the moments of the distribution:

$$\text{c.v.} = \left[\frac{m_3 m_5}{m_4^2} - 1 \right]^{1/2} \qquad (42)$$

This is especially useful for systems that cannot be described by an analytical distribution function.

V. CRYSTALLIZER CONFIGURATION AND OPERATION

Crystallization equipment can vary in sophistication from a simple stirred tank to a complicated multiphase column, and the protocol can range in complexity from simply allowing a vat of liquor to cool to the careful manipulation required of batch cyclic operations. In principle, the objectives of these systems are the same: to produce a product meeting specifications on quality at an economical yield. This section will examine some of the considerations that go into the selection of a crystallizer so as to meet these objectives.

One of the first decisions that must be made is whether the crystallizer operation is to be batch or continuous. In general, the advantages of each type of operation should be weighed in choosing one over the other, but more often the decision rests on whether the other parts of the process are batch or continuous. If they are batch, then it is likely that the crystallizer also should be batch.

The equipment required for batch crystallization can be very simple. For example, some crystalline materials are produced by simply allowing a charge of hot liquor to cool. After the crystals have formed, the magma is discharged through a filter or the liquor may be decanted and the settled slurry filtered. Very large crystals can be obtained by allowing encrustations formed on the walls of these crystallizers to grow undisturbed; after the system has come to equilibrium, the liquor is drained and the crystals are removed by scraping them from the surface.

Batch crystallizers can be used in a campaign to produce a particular product and in a second campaign to produce another product. Generally, it is not possible to operate continuous processes in this way. Batch crystallizers can handle viscous or toxic systems more easily than can continuous systems, and interruption of batch operations for periodic maintenence is less difficult than dealing with interruptions in continuous processes. The latter factor may be especially important in biological processes that require frequent sterilization of equipment. Batch crystallizers can produce a narrow crystal size distribution, whereas special processing features are required to narrow the distribution obtained from a continuous crystallizer. The effects of operating variables on crystal size distributions will be discussed in Section VI.

The throughput per unit crystallizer volume is greater for a continuous system. Batch units have several operating steps in a cycle—charging, heating or cooling, crystallizing, discharging, and cleaning—and the unit production rate is based on the total cycle time, even though the formation of crystals may occur only during a small portion of the cycle.

It may be easier to operate a continuous system so that it reproduces a particular crystal size distribution than it is do reproduce crystal characteristics from a batch unit. Moreover, the coupling of several transient variables and nucleation make it difficult to model and control the operation of a batch crystallizer.

A. Relationship of Solubility to Mode of Operation

The driving force for crystal formation can be generated through a variety of means, including cooling or heating to reduce or increase the system temperature, evaporating solvent, evaporative (flash) cooling, inducing a chemical reaction (when the reaction product is sparingly soluble, the process is called *precipitation*), adjusting pH, salting out through the addition of a nonsolvent, direct-contact cooling with a refrigerant, or some other means. All of these modes of operation can be implemented in either a batch or a continuous process. In addition, two or more of the modes may be combined to enhance the product yield.

As discussed in an earlier section, solubility is intrinsic to the solute-solvent system, and the relationship of solubility to temperature often determines the mode by which a crystallizer is operated. Recall for example that the solubility of NaCl (see Fig. 2) is essentially independent of temperature, while $Na_2SO_4 \cdot 10H_2O$ has a solubility that exhibits a strong dependence on temperature. Consequently, cooling a sodium chloride solution cannot generate significant product yield; solvent evaporation is the primary mode of NaCl production. On the other hand, reducing the temperature of a saturated solution of sodium sulfate generates substantial product and may be used alone or in combination with evaporation.

Cooling crystallizers utilize a heat sink to remove both the sensible heat from the feed stream and the heat of crystallization released or, in some cases absorbed, as crystals are formed. The heat sink may be no more than the ambient surroundings of a batch crystallizer, or (as is more likely) it may be cooling water or another process stream.

Evaporative crystallizers generate supersaturation by removing solvent from the mixture, thereby increasing the solute concentration. They may be operated under vacuum, and in those circumstances it is necessary to have a vacuum pump or ejector as a part of the unit. If the boiling point elevation—the increase in boiling temperature due to the presence of the solute—is low, mechanical recompression of the vapor obtained from solvent evaporation may be used in some cases to produce a heat source to drive the operation.

Evaporative-cooling crystallizers are fed with a liquor whose temperature is such that solvent flashes upon feed entry to the crystallizer. They typically are operated under vacuum, and flashing of solvent increases the solute concentration in the remaining liquor while simultaneously reducing the temperature of the magma. The mode of this operation can be reduced to that of a simple cooling crystallizer by returning condensed solvent to the crystallizer body.

Salting-out crystallization operates through the addition of a nonsolvent to the magma in a crystallizer. The selection of the nonsolvent is based on the effect of the solvent on solubility, cost, properties that affect handling, interaction with product requirements, and ease of recovery. Adding a nonsolvent to the system increases the complexity of the process; it increases the volume required for a given residence time and produces a highly nonideal mixture of solvent, nonsolvent, and solute.

Melt crystallization operates with heat as a separating agent, but a crystalline product is not generated in the process. Instead, crystals formed during the operation are remelted and the melt is removed as the product. Such operations are often used to perform the final purification of products after prior separation units; for example, the purity of an acrylic acid feed may be increased from 99.5 to 99.9%. Melt crystallizers do not require solids handling units nor do they utilize solid–liquid separation equipment. Finally, in some instances the use of melt crystallization can eliminate the use of solvents, thereby reducing the environmental impact of the process.

B. Crystallizers

The basic requirements of a crystallization system are (1) a vessel to provide sufficient residence time for crystals to grow to a desired size, (2) mixing to provide a uniform environment for crystal growth, and (3) a means of generating supersaturation. Crystallization equipment is manufactured and sold by several vendors, but some chemical companies design their own crystallizers based on expertise developed within their organizations. Rather than attempt to describe the variety of special crystallizers that can be found in the marketplace, this section will provide a brief general survey of types of crystallizers that utilize the modes outlined above.

The forced-circulation crystallizer is a simple unit designed to provide high heat-transfer coefficients in either an evaporative or a cooling mode. Figure 12 shows a

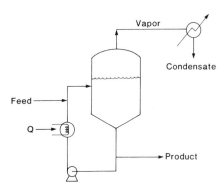

FIGURE 12 Schematic diagram of a forced-circulation evaporative crystallizer.

schematic diagram of an evaporative forced-circulation crystallizer that withdraws a slurry from the crystallizer body and pumps it through a heat exchanger. Heat transferred to the circulating magma causes evaporation of solvent as the magma is returned to the crystallizer. This type of unit is used to control circulation rates and velocities past the heat transfer surfaces, and the configuration shown is especially useful in applications requiring high rates of evaporation. A calandria that provides heat transfer through natural convection is an alternative to forced-circulation systems.

Scale formation on the heat exchanger surfaces or at the vapor–liquid surface in the crystallizer can cause operational problems with evaporative crystallizers. These can be overcome by avoiding vaporization or excessive temperatures within the heat exchanger and by properly introducing the circulating magma into the crystallizer. For example, introducing the circulating magma a sufficient distance below the surface of the magma in the crystallizer prevents vaporization upon re-entry and forces it to occur at a well-mixed zone above the point of re-entry. Alternatively, the magma may be introduced so as to induce a swirling motion that dislodges encrustations from the wall of the crystallizer at the vapor-liquid interface.

Figure 13 shows a schematic diagram illustrating the configuration of a surface cooling (indirect heat transfer) crystallizer. Heat can be transferred to a coolant in an external heat exchanger, as shown, or in coils or a jacket

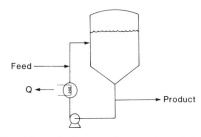

FIGURE 13 Schematic diagram of a forced-circulation, surface-cooling crystallizer.

within the crystallizer. An external cooling surface requires the use of a circulation pump, but this expense may be mitigated by obtaining a higher heat-transfer coefficient than would result with the use of coils or a jacketed vessel. The rate of heat transfer Q from the circulation loop of a cooling crystallizer must be sufficient to reduce the temperature of the feed and to remove the heat of crystallization of the solute. Assuming that no substantial crystallization occurs in the heat exchanger and limiting the difference between entering and leaving temperatures of the circulating magma $(T_{in} - T_{out})$, so as to minimize formation of encrustations, the required magma circulation rate $\dot{m}_{circ}$ can be determined from the equation:

$$\dot{m}_{circ} = \frac{Q}{[C_P(T_{in} - T_{out})]_{circ}} \qquad (43)$$

where C_P is the heat capacity of the circulating magma. The methods by which Q can be evaluated were discussed in Section II. It is not uncommon to limit the decrease in magma temperature to about 3 to 5°C; therefore, both the circulation rate and heat-transfer surface must be large.

The feed to cooling crystallizers should be rapidly mixed with the magma so as to minimize the occurrence of regions of high supersaturation. Such regions lead to excessive nucleation, which is detrimental to the crystal size distribution. The type of pump used in the circulation loop also can lead to degradation of the crystal size distribution; an inappropriate pump causes crystal attrition through abrasion, fracture, or shear, and most commercial systems use specially designed axial-flow pumps that provide high flow rates and low shear.

Direct-contact refrigeration can be used if either the operating temperature of the crystallizer is low in comparison to the temperature of available cooling water or there are severe problems with encrustations. In such an operation, a refrigerant is mixed with the crystallizer contents and vaporized at the magma surface. On vaporizing, the refrigerant removes sufficient heat from the magma to cool the feed and to remove the heat of crystallization. The refrigerant vapor must be compressed, condensed, and recycled for the process to be economical. Moreover, the refrigerant must be insoluble in the liquor to minimize losses and product contamination.

Special devices for classification of crystals may be used in some applications. Figure 14 shows a draft-tube-baffle (DTB) crystallizer that is designed to provide preferential removal of both fines and classified product. As shown, feed is introduced to the fines circulation line so that any nuclei formed upon introduction of the feed can be dissolved as the stream flows through the fines-dissolution heat exchanger. The contents of the crystallizer are mixed by the impeller, which forces the slurry to flow in the indicated direction. A quiescent zone is formed between the

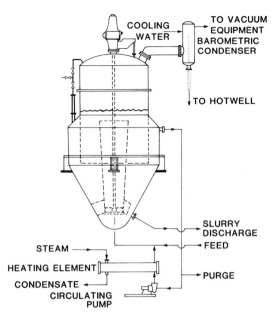

FIGURE 14 Draft-tube-baffle crystallizer. [Courtesy of Signal Swenson Division.]

baffle extending into the chamber and the outside wall of the crystallizer. Flow through the quiescent zone can be adjusted so that crystals below a certain size (determined by settling velocity) are removed in the fines-dissolution circuit. In the elutriation leg, crystals below a certain size are preferentially swept back into the crystallizer by the flow of recycled mother liquor; accordingly, larger crystals, which have a higher settling velocity, are removed preferentially from the system.

A second major type of crystallizer with special channeling devices is comprised of those having configurations like the Oslo crystallizer shown in Fig. 15. The objective of this unit is to form a supersaturated solution by evaporation in the upper chamber and to have crystal growth in the lower (growth) chamber. The use of the downflow pipe in the crystallizer provides good mixing in the growth chamber. As shown, the lower chamber has a varying diameter, which can provide some internal classification of crystals. The lowest portion of the chamber has the smallest diameter and can be considered perfectly mixed; as the chamber diameter increases, the upward velocity of the slurry decreases and larger crystals tend to settle. In principle, only small crystals are supposed to leave the chamber in the circulating slurry, to flow through the circulation pump, and to enter the upper chamber. As the probability of a crystal colliding with the impeller decreases with decreasing crystal size, the internal classification provided by the Oslo crystallizer could provide some control of contact nucleation.

Melt crystallizers can be operated in a variety of ways. In one, feed enters the crystallizer and contacts a slurry

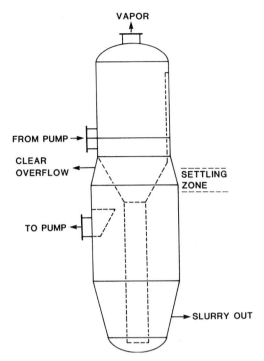

FIGURE 15 Oslo crystallizer.

of mother liquor and crystals of the desired product. The crystals are forced to move in a specific direction by gravity or rotating blades. As they flow towards the appropriate end of the crystallizer, the crystals encounter a heated region and are melted. A portion of the melt is removed as product, while the remainder flows countercurrently to the crystals, thereby providing some refining and removing impure adhering liquid.

In a second method of operation, the feed material is circulated through a bank of tubes, each of which has a diameter of up to about 8 cm. The walls of the tubes are cooled, and material crystallizes on them throughout a fixed operating period. At the end of that period, the remaining liquid is sent to a holding tank for further processing, and then the tubes are heated slowly to cause partial melting of the adhering solids. This step is known as "sweating," and the impure "sweated" liquid produced is removed from the crystallizer and held for further processing. Finally, the product is obtained by adding additional heat to the tubes and melting the remaining adhering solids. The actual sequencing of these steps and the reprocessing of residual and sweated liquids may be quite complicated.

VI. POPULATION BALANCES AND CRYSTAL SIZE DISTRIBUTIONS

A balance on the population of crystals in a crystallizer can be used to relate process variables to the crystal size

distribution of the product or intermediate material. Such balances are not independent of those on mass and energy, and their solution requires an independent expression for nucleation kinetics.

In formulating a population balance, crystals are assumed sufficiently numerous for the population distribution to be treated as a continuous function. One of the key assumptions in the development of a simple population balance is that all crystal properties, including mass (or volume), surface area, and so forth are defined in terms of a single crystal dimension referred to as the characteristic length. For example, Eq. (19) relates the surface area and volume of a single crystal to a characteristic length L. In the simple treatment provided here, shape factors are taken to be constants. These can be determined by simple measurements or estimated if the crystal shape is simple and known—for example, for a cube $k_{area} = 6$ and $k_{vol} = 1$.

The beginning point for any balance is the following statement:

$$\text{input} + \text{generation} - \text{output} - \text{consumption}$$

$$= \text{accumulation} \qquad (44)$$

where each of the terms may be expressed as a rate or an amount. In a population balance, the number of entities (such as crystals) is the balanced quantity and each of the terms has dimensions of number of crystals per unit time for a differential balance or number of crystals for an integral balance. The principles involved in formulating a balance are outlined in the following sections, and they provide guidance in developing corresponding balances for systems whose configurations do not conform to those described here.

A. Perfectly Mixed, Continuous Crystallizers

The balance equation must be constructed for a control volume, which for a perfectly mixed crystallizer may be assumed to be the total volume of the crystallizer V_T. Then, a balance on the number of crystals in any size range (say, L_1 to $L_2 = L_1 + \Delta L$) must account for crystals that enter and leave that size range by: (1) convective flow, (2) crystal growth, (3) crystal agglomeration, and (4) crystal breakage. Agglomeration and breakage can be detected through careful inspection of product particles, and they can be quite significant in some processes. For simplicity, however, they will be assumed negligible in the present analysis. The rate of crystal growth G will be defined as in Eq. (26); i.e., the rate of change of the characteristic crystal dimension L:

$$G = \frac{dL}{dt}$$

Then,

growth rate into the size range $= V_T(Gn)_{L_1}$ (45)

growth rate out of the size range $= V_T(Gn)_{L_2}$ (46)

removal rate of crystals in the size range $= V_{out}\displaystyle\int_{L_1}^{L_2} n\, dL$ (47)

feed rate of crystals in the size range $= V_{in}\displaystyle\int_{L_1}^{L_2} n_{in}\, dL$ (48)

accumulation rate in the crystallizer $= \dfrac{\partial}{\partial t}\displaystyle\int_{L_1}^{L_2} nV_T\, dL$ (49)

Substituting the terms from Eqs. (46) through (49) into Eq. (44) gives:

$$V_T(Gn)_{L_1} + V_{in}\int_{L_1}^{L_2} n_{in}\, dL$$

$$= V_T(nG)_{L_2} + V_{out}\int_{L_1}^{L_2} n\, dL + \frac{\partial}{\partial t}\int_{L_1}^{L_2} nV_T\, dL \quad (50)$$

Manipulation of this equation leads to

$$\frac{\partial(nG)}{\partial L} + \frac{V_{out}n}{V_T} - \frac{V_{in}n_{in}}{V_T} = -\frac{\partial n}{\partial t} \quad (51)$$

Equation (51) may be used as a starting point for the analysis of any crystallizer that has a well-mixed active volume and for which crystal breakage and agglomeration can be ignored. As an illustration of how the equation can be simplified to fit specific system behavior, suppose the feed to the crystallizer is free of crystals and that it is operating at steady state. Then, $n_{in} = 0$ and $\partial n/\partial t = 0$. Now suppose that the crystal growth is invariant with size and time; in other words, assume the system follows the McCabe ΔL law and therefore exhibits neither size-dependent growth nor growth-rate dispersion. Then,

$$\frac{\partial(nG)}{\partial L} = G\frac{\partial n}{\partial L} \quad (52)$$

Defining a mean residence time $\tau = V_T/V_{out}$ and applying the aforementioned restrictions leads to

$$G\frac{dn}{dL} + \frac{n}{\tau} = 0 \quad (53)$$

(τ is often referred to as the drawdown time to reflect the fact that it is the time required to empty the contents from the crystallizer.) Integrating Eq. (53) with the boundary condition $n = n^\circ$ at $L = 0$:

$$n = n^\circ \exp\left(-\frac{L}{G\tau}\right) \quad (54)$$

If the crystallizer has a clear feed, growth is invariant, but if the magma volume V_T is allowed to vary, the population balance gives:

$$\frac{\partial n}{\partial t} + \frac{\partial(nG)}{\partial L} + n\frac{\partial(\ln V_T)}{\partial t} + \frac{V_{out}n}{V_T} = 0 \quad (55)$$

The system model that led to the development of the last two equations is referred to as the mixed-suspension, mixed-product removal (MSMPR) crystallizer.

Under steady-state conditions, the rate at which crystals are produced by nucleation must be equal to the difference in rates at which crystals leave and enter the crystallizer. Accordingly, for a clear feed,

$$V_T B^\circ = V_{out}\int_0^\infty n\, dL \Rightarrow B^\circ = \frac{1}{\tau}\int_0^\infty n\, dL \quad (56)$$

For crystallizers following the constraints given above,

$$B^\circ = n^\circ G \quad (57)$$

For a given set of crystallizer operating conditions, nucleation and growth rates can be determined by measuring the population density of crystals in a sample taken from either the well-mixed zone of a crystallizer or the product stream flowing from that zone. Sample analyses are correlated with Eqs. (54) and (57), and nucleation and growth rates are determined from those correlations. The sample must be representative of the crystal population in the crystallizer (or leaving the well-mixed unit), and experience with such measurements is invaluable in performing this analysis properly. Figure 16 shows a plot of

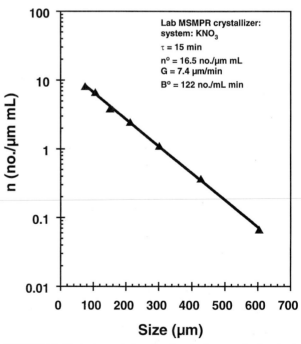

FIGURE 16 Typical population density plot from perfectly mixed, continuous crystallizer.

typical population density data obtained from a crystallizer meeting the stated assumptions. The slope of the plot of such data may be used to obtain the growth rate while the intercept can be used to estimate nucleation rate.

Many industrial crystallizers operate in a well-mixed or near well-mixed manner, and the equations derived above can be used to describe their performance. Also, the simplicity of the equations describing an MSMPR crystallizer make experimental equipment configured so as to meet the assumptions leading to Eq. (54) useful in determining nucleation and growth kinetics. From a series of runs at different operating conditions, correlations of nucleation and growth kinetics with appropriate process variables can be obtained (see, for example, the discussions of Eqs. (18) and (25)). The resulting correlations can then be used to guide either crystallizer scale-up or the development of an operating strategy for an existing crystallizer.

It is often very difficult to measure supersaturation, especially in systems that have high growth rates. Even though the supersaturation in such systems is so small that it can be neglected in writing a solute mass balance, it is important in setting nucleation and growth rates. In such instances it is convenient to substitute growth rate for supersaturation by combining Eqs. (18) and (25). This gives:

$$B^\circ = k_{nuc} G^i M_T^j N^k \tag{58}$$

The constant k_{nuc} depends on process variables other than supersaturation, magma density, and intensity of mixing; these include temperature and presence of impurities. If sufficient data are available, these variables may be separated from the constant by adding more terms in a power-law correlation. k_{nuc} is specific to the operating equipment and not transferable from one equipment scale to another. The system-specific constants i and j are obtainable from experimental data and may be used in scale-up, although j may vary considerably with mixing conditions.

As shown by Eq. (54), growth rate G can be obtained from the slope of a plot of the log of population density against crystal size; nucleation rate B° can be obtained from the same data by using the relationship given by Eq. (57), with n° being the intercept of the population density plot. Nucleation rates obtained by these procedures should be checked by comparison with values obtained from a mass balance (see the later discussion of Eq. (66)).

The perfectly mixed crystallizer of the type described in the preceding discussion is highly constrained. Alteration of the characteristics of crystal size distributions produced by such systems can be accomplished only by modifications of the nucleation and growth kinetics of the system being crystallized. Indeed, examination of Eq. (54) shows that once nucleation and growth kinetics

are fixed, the crystal size distribution is determined in its entirety. In addition, such distributions have the following characteristics:

- Mass density function (from Eq. (39)):

$$m = \rho k_{vol} n^\circ L^3 \exp\left(-\frac{L}{G\tau}\right) \tag{59}$$

- Dominant crystal size (from Eq. (40)):

$$L_D = 3G\tau \tag{60}$$

- Moments of n (from Eq. (37)):

$$m_i = i! n^\circ (G\tau)^{i+1} \tag{61}$$

- Total number of crystals per unit volume:

$$N_T = \int_0^\infty n\, dL = m_0 = n^\circ G\tau \tag{62}$$

- Total length of crystals per unit volume:

$$L_T = \int_0^\infty nL\, dL = m_1 = n^\circ (G\tau)^2 \tag{63}$$

- Total surface area of crystals per unit volume:

$$A_T = k_{area} \int_0^\infty nL^2\, dL = k_{area} m_2 = 2k_{area} n^\circ (G\tau)^3 \tag{64}$$

- Total solids volume per unit volume:

$$V_{TS} = k_{vol} \int_0^\infty nL^3\, dL = k_{vol} m_3 = 6k_{vol} n^\circ (G\tau)^4 \tag{65}$$

- The coefficient of variation of the mass density function (from Eq. (42)) is 50%.
- The magma density M_T (mass of crystals per unit volume of slurry or liquor) is the product of the crystal density, the volumetric shape factor, and the third moment of the population density function:

$$M_T = 6\rho k_{vol} n^\circ (G\tau)^4 \tag{66}$$

System conditions often allow for the measurement of magma density, and in such cases is should be used as a constraint in evaluating nucleation and growth kinetics from measured population densities. This approach is especially useful in instances of uncertainty in the determination of population densities from sieving or other particle sizing techniques.

B. Preferential Removal of Crystals

As indicated above, crystal size distributions produced in a perfectly mixed crystallizer are constrained by the nature of the system. This is because both liquor and solids have the same residence time distributions, and it is the crystal residence time distribution that gives the population density function the characteristic exponential form in Eq. (54). Nucleation and growth kinetics can influence the population density function, but they cannot alter the form of the functional dependence of n on L.

Crystallizers are made more flexible by the introduction of selective removal devices that alter the residence time distributions of materials flowing from the crystallizer. Three removal functions—clear-liquor advance, classified-fines removal, and classified-product removal—and their idealized removal devices will be used here to illustrate how design and operating variables can be manipulated to alter crystal size distributions. Idealized representations of the three classification devices are illustrated in Fig. 17.

Clear-liquor advance from what is called a *double draw-off crystallizer* is simply the removal of mother liquor without simultaneous removal of crystals. The primary action in classified-fines removal is preferential withdrawal from the crystallizer of crystals of a size below some specified value; this may be coupled with the dissolution of the crystals removed as fines and the return of the resulting solution to the crystallizer. Classified-product removal is carried out to remove preferentially those crystals of a size larger than some specified value. In the following discussion, the effects of each of these selective removal functions on crystal size distributions will be described in terms of the population density function n. Only the ideal solid–liquid classification devices will be examined. It is convenient in the analyses to define flow rates in terms of clear liquor. Necessarily, then, the population density function is defined on a clear-liquor basis.

Clear-liquor advance is used for two purposes: (1) to reduce the quantity of liquor that must be processed by the solid–liquid separation equipment (e.g., filter or centrifuge) that follows the crystallizer, and (2) to separate the residence time distributions of crystals and liquor. The reduction in liquor flow through the separation equipment can allow the use of smaller equipment for a fixed production rate or increased production through fixed equipment. Separating the residence time distributions of crystals and liquor means that crystals will have an average residence time longer than that of the liquor. This should, in principle, lead to the production of larger crystals, but because the crystallizer is otherwise well mixed, the crystal population density will have the same form as that for the MSMPR crystallizer (Eq. (54)).

The analysis goes as follows: Let V_{in}, V_{CL}, and V_{out} represent volumetric flow rates of clear liquor fed to the crystallizer, of clear-liquor advance, and of output slurry respectively. The population density function is given by the expression:

$$n = n^\circ \exp\left(-\frac{L}{G\tau_{prod}}\right) \qquad (67)$$

where $\tau_{prod} = V_T/V_{out}$. Increasing V_{CL} decreases V_{out} and thereby increases the residence time of the crystals in the crystallizer. Unless the increase in magma density results in significant increases in nucleation, the utilization of clear-liquor advance will produce an increase in the dominant crystal size. Often the increase is much greater than that predicted from theory, and it is suspected that this is because the stream being removed as clear liquor actually contains varying amounts of fines. If this is the case,

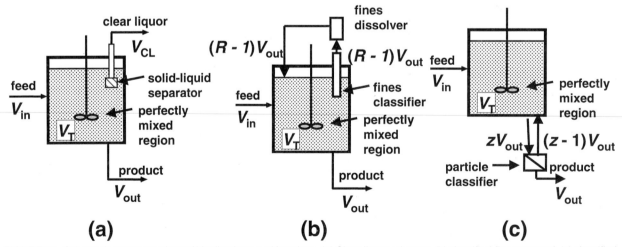

FIGURE 17 Schematic representations of idealized removal functions. (a) Clear-liquor advance, (b) classified-fines removal, (c) classified-product removal.

both clear-liquor advance and fines-removal are enhancing crystal size.

As an idealization of classified-fines removal, assume that two streams are withdrawn from the crystallizer, one corresponding to the product stream and the other a fines-removal stream. Designate the flow rate of the clear solution in the product stream to be V_{out} and set the flow rate of the clear solution in the fines-removal stream to be $(R-1)V_{out}$. Also, assume that the device used to separate fines from the larger crystals functions so that only crystals below an arbitrary size L_F are in the fines-removal stream and that all crystals below size L_F have an equal probability of being withdrawn as fines. Under these conditions, the crystal size distribution is characterized by two mean residence times, one for the fines and the other for crystals larger than L_F, that are related by the equations:

$$\tau = \frac{V_T}{V_{out}} \qquad \text{(for } L > L_F) \qquad (68)$$

$$\tau_F = \frac{V_T}{R V_{out}} = \frac{\tau}{R} \qquad \text{(for } L \leq L_F) \qquad (69)$$

where V_T is the total volume of clear solution in the crystallizer.

For systems following invariant growth, the crystal population density in each size range will decay exponentially with the inverse of the product of growth rate and residence time. For a continuous distribution, the population densities of the classified fines and the product crystals must be the same at $L = L_F$. Accordingly, the population density for a crystallizer operating with classified-fines removal is given by:

$$n = n^\circ \exp\left[-\frac{RL}{G\tau}\right] \qquad \text{(for } L \leq L_F) \qquad (70)$$

$$n = n^\circ \exp\left[-\frac{(R-1)L_F}{G\tau}\right] \exp\left[-\frac{L}{G\tau}\right] \quad \text{(for } L > L_F) \qquad (71)$$

Figure 18 shows how the population density function changes with the addition of classified-fines removal. The lines drawn are for a hypothetical system, but they illustrate qualitatively what can be demonstrated analytically; that is, fines removal increases the dominant crystal size, but it also increases the spread of the distribution.

A simple method for implementation of classified-fines removal is to remove slurry from a settling zone in the crystallizer. The settling zone can be created by constructing a baffle that separates the zone from the well-mixed portion of the vessel—recall, for example, the draft-tube-baffle crystallizer described in Section V—or, in small-

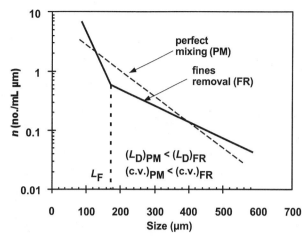

FIGURE 18 Population density plot for product from crystallizer with idealized classified-fines removal.

scale systems, by simply inserting a length of pipe or tubing of appropriate diameter into the well-mixed crystallizer chamber. The separation of crystals in the settling zone is based on the dependence of settling velocity on crystal size. Crystals entering the settling zone and having a settling velocity greater than the upward velocity of the slurry remain in the crystallizer. As the cross-sectional area of a settling zone is invariant, the flow rate of slurry through the zone determines the cut-size L_F, and it also determines the parameter R used in Eqs. (69) through (71).

In a crystallizer equipped with classified-product removal, crystals above some coarse size L_C are removed at a rate Z times the removal rate of smaller crystals. This can be accomplished by using an elutriation leg, a hydrocyclone, or a screen to separate larger crystals for removal from the system. Using the analysis of classified-fines removal as a guide, it can be shown that the crystal population density is given by the equations:

$$n = n^\circ \exp\left[-\frac{L}{G\tau}\right] \qquad \text{(for } L \leq L_C) \qquad (72)$$

$$n = n^\circ \exp\left[\frac{(Z-1)L_C}{G\tau}\right] \exp\left[-\frac{ZL}{G\tau}\right] \quad \text{(for } L > L_C) \qquad (73)$$

where τ is defined as the residence time V_T/V_{out}. Figure 19 shows the effects of classified-product removal on crystal size distribution; the dominant crystal size is reduced and the spread of the distribution becomes narrower. Note that it is impossible for crystals smaller than L_C to leave the idealized classified-product crystallizer illustrated in Fig. 17c. Accordingly, the population densities shown on Fig. 19 for the classified-product crystallizer represent conditions *inside* the perfectly mixed region of the unit.

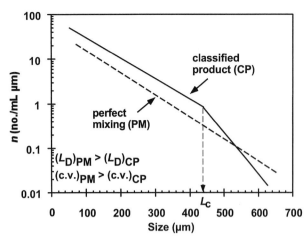

FIGURE 19 Population density plot for crystals in crystallizer with idealized classified-product removal.

If both fines and product are removed on a classified basis, the population density will be given by the equations:

$$n = n° \exp\left[-\frac{RL}{G\tau}\right] \qquad \text{(for } L \leq L_F) \qquad (74)$$

$$n = n° \exp\left[-\frac{(R-1)L_F}{G\tau}\right] \exp\left[-\frac{L}{G\tau}\right]$$
$$\text{(for } L_F < L < L_C) \qquad (75)$$

$$n = n° \exp\left[-\frac{(R-1)L_F}{G\tau}\right] \exp\left[\frac{(Z-1)L_C}{G\tau}\right]$$
$$\times \exp\left[-\frac{ZL}{G\tau}\right] \qquad \text{(for } L \geq L_C) \qquad (76)$$

Selection of a crystallizer that has both classified-fines and classified-product removal is done to combine the best features of each: increased dominant size and narrower distribution. Figure 20 illustrates the effects of both removal functions on population density. Note that this plot of population density results from sampling the magma within a crystallizer, not from sampling the product stream, which for the ideal classification devices considered here can only have crystals larger than L_C. As discussed earlier for the classified-product crystallizer, the population densities shown in Fig. 20 represent those found *in* the crystallizer.

The model of the crystallizer and selective removal devices that led to Eqs. (74) through (76) is referred to as the R-Z crystallizer. It is an obvious idealization of actual crystallizers because of the perfect cuts assumed at L_F and L_C. However, it is a useful approximation to many systems and it allows qualitative analyses of complex operations.

Although many commercial crystallizers operate with some form of selective crystal removal, such devices can be difficult to operate because of fouling of heat-exchanger surfaces or blinding of screens. In addition, classified-product removal can lead to cycling of the crystal size distribution. Often such behavior can be minimized or even eliminated by increasing the fines-removal rate.

Moments of the population density function given by Eqs. (74) through (76) can be evaluated in piecewise fashion:

$$m_i = \int_0^{L_F} L^i n\, dL + \int_{L_F}^{L_C} L^i n\, dL + \int_{L_C}^{\infty} L^i n\, dL \quad (77)$$

Equation (77) is used to estimate the moments of the population density function within the crystallizer, not of the product distribution. (Recall that moments of the distribution within the crystallizer are often required for kinetic equations.) Assuming perfect classification, moments of the product distribution can be obtained from the expression:

$$m_{i,\text{prod}} = \int_{L_C}^{\infty} L^i n\, dL \qquad (78)$$

Moments can be used to characterize the material produced from or contained in a crystallizer with classified-fines or classified-product removal or to evaluate the effect of these selective removal functions on product characteristics. All that is required is the use of the equations derived earlier to relate special properties, such as coefficient of variation to the operational parameters R and Z.

C. Batch Crystallization

As with continuous crystallizers, the mode by which supersaturation is generated affects the crystal yield and size distribution; however, it is the *rate* at which such supersaturation is generated that is most important in determining product characteristics. Furthermore, there are infinite

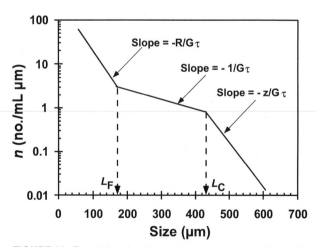

FIGURE 20 Population density plot for crystals in crystallizer with idealized classifiedfines and classified-product removal.

possibilities in selecting cooling profiles, $T(t)$, or vapor generation profiles, $V(t)$, or time dependencies of precipitant or nonsolvent addition rates.

For illustrative purposes, consider that the protocol for a cooling crystallizer can involve either natural cooling—cooling resulting from exposure of the crystallizer contents to a heat sink without intervention of a control system—or manipulation of cooling to reduce the system temperature in a specific manner. In both cases, the instantaneous heat-transfer rate is given by:

$$Q = UA(T - T_{sink}) \qquad (79)$$

where U is a heat-transfer coefficient, A is the area available for heat transfer, T is the temperature of the magma, and T_{sink} is the temperature of the cooling fluid. If T_{sink} is a constant, the maximum heat-transfer rate and, therefore, the highest rate at which supersaturation is generated are at the beginning of the process. This protocol can lead to excessive primary nucleation and the formation of encrustations on the heat-transfer surfaces.

The objective of programmed cooling is to control the rate at which the magma temperature is reduced so that supersaturation remains constant at some prescribed value, usually below the metastable limit associated with primary nucleation. Typically the batch is cooled slowly at the beginning of the cycle and more rapidly at the end. An analysis that supports this approach is presented later. In size-optimal cooling, the objective is to vary the cooling rate so that the supersaturation in the crystallizer is adjusted to produce an optimal crystal size distribution.

Protocols similar to those described above for cooling crystallizers exist for crystallization modes involving evaporation of solvent and the rate at which a non solvent or a reactant is added to a crystallizer.

A population balance can be used to follow the development of a crystal size distribution in batch crystallizer, but both the mathematics and physical phenomena being modeled are more complex than for continuous systems at steady state. The balance often utilizes the population density defined in terms of the total crystallizer volume, rather than on a specific basis: $\bar{n} = nV_T$. Accordingly, the general population balance given by Eq. (51) can be modified for a batch crystallizer to give:

$$\frac{\partial(nV_T)}{\partial t} + \frac{\partial(GnV_T)}{\partial L} = \frac{\partial \bar{n}}{\partial t} + \frac{\partial(G\bar{n})}{\partial L} = 0 \qquad (80)$$

The solution to this equation requires both an initial condition ($\bar{n}$ at $t = 0$) and a boundary condition (usually obtained by assuming that crystals are formed at zero size):

$$\bar{n}(0, t) = \bar{n}^\circ(t) = \frac{B^\circ(t)}{G(0, t)} \qquad (81)$$

The identification of an initial condition associated with the crystal size distribution is very difficult. If the system is seeded, the initial condition becomes:

$$\bar{n}(L, 0) = \bar{n}_{seed}(L) \qquad (82)$$

where $\bar{n}_{seed}$ is the population density function of the seed crystals. If the system is unseeded, the nuclei often are assumed to form at size zero.

The rate of cooling, or evaporation, or addition of diluent required to maintain specified conditions in a batch crystallizer often can be determined from a population-balance model. Moments of the population density function are used in the development of equations relating the control variable to time. As defined earlier, the moments are

$$m_i = \int_0^\infty L^i \bar{n} \, dL \qquad (83)$$

Recognizing that the zeroth moment is the total number of crystals in the system, it can be shown that:

$$\frac{dm_0}{dt} = \bar{n}^\circ G = B^\circ = \frac{dN_T}{dt} \qquad (84)$$

Moment transformation of Eq. (80) leads to the following relationship:

$$\frac{\partial m_j}{\partial t} = jGm_{j-1} \qquad (85)$$

Combining Eq. (85) with the relationships of moments to distribution properties developed in Section VI.A for $j = 1, 2, 3$ gives:

$$\frac{dm_1}{dt} = Gm_0 \xrightarrow{m_0=N_T} \frac{dL_T}{dt} = GN_T \qquad (86)$$

$$\frac{dm_2}{dt} = 2Gm_1 \xrightarrow{m_1=L_T} \frac{dA_T}{dt} = 2Gk_{area}L_T \qquad (87)$$

$$\frac{dm_3}{dt} = 3Gm_2 \xrightarrow{k_{area}m_2=A_T} \frac{dM_T}{dt} = 3G\rho\left(\frac{k_{vol}}{k_{area}}\right)A_T \qquad (88)$$

where N_T is the total number of crystals, L_T is total crystal length, A_T is total surface area of the crystals, and M_T is the total mass of crystals in the crystallizer. In addition to a population balance, a solute balance must also be satisfied:

$$\frac{d(V_T C)}{dt} + \frac{dM_T}{dt} = 0 \qquad (89)$$

where V_T is the total volume of the system, and C is solute concentration in the solution.

The above equations can be applied to any batch crystallization process, regardless of the mode by which supersaturation is generated. For example, suppose a model is needed to guide the operation of a seeded batch crystallizer so that solvent is evaporated at a rate that gives a constant crystal growth rate G and no nucleation; in other words, supersaturation is to be held constant and only those crystals added at the beginning of the run are in the crystallizer. Model development proceeds as follows: combining the solute balance, Eq. (89), with Eq. (88),

$$\frac{d(V_T C)}{dt} + \frac{3\rho A_T k_{vol} G}{k_{area}} = 0 \qquad (90)$$

Recognizing that the process specification requires C to be a constant and taking the derivative of Eq. (90):

$$C \frac{d^2 V_T}{dt^2} + 3\rho \left(\frac{k_{vol}}{k_{area}}\right) G \frac{dA_T}{dt} = 0 \qquad (91)$$

$$\Downarrow \text{Eq. (87)}$$

$$C \frac{d^2 V_T}{dt^2} + 6\rho k_{vol} G^2 L_T = 0 \qquad (92)$$

Taking the derivative of the last equation:

$$C \frac{d^3 V_T}{dt^3} + 6\rho k_{vol} G^2 \frac{dL_T}{dt} = 0 \qquad (93)$$

$$\Downarrow \text{Eq. (86)}$$

$$C \frac{d^3 V_T}{dt^3} + 6\rho k_{vol} G^3 N_T = 0 \qquad (94)$$

Suppose that the batch crystallizer is seeded with a mass of crystals with a uniform size of $\bar{L}_{seed}$. The number of seed crystals is N_{seed}, and, as the operation is to be free from nucleation, the number of crystals in the system remains the same as the number of seed crystals. The initial values of total crystal length, total crystal surface area, total crystal mass, and system volume are

$$L_T(0) = N_{seed} \bar{L}_{seed} \qquad (95)$$

$$A_T(0) = k_{area} N_{seed} \bar{L}_{seed}^2 \qquad (96)$$

$$M_T(0) = \rho k_{vol} N_{seed} \bar{L}_{seed}^3 \qquad (97)$$

$$V_T(0) = V_{T0} \qquad (98)$$

On integrating Eq. (94), the following dependence of system volume on time can be obtained:

$$C(V_{T0} - V_T) = k_{vol} \rho N_{seed} \big[(Gt)^3 + 3(Gt)^2 \bar{L}_{seed} + 3(Gt) \bar{L}_{seed}^2 \big] \qquad (99)$$

Therefore, for the specified conditions, the evaporation rate $(-dV_T/dt)$ is a parabolic (second-order) function of time, and the rate of heat input to the crystallizer must be controlled to match the conditions called for by Eq. (99).

If a cooling mode is used to generate supersaturation, an analysis similar to that given above can be used to derive an appropriate dependence of system temperature on time. The result depends upon the relationship of solubility to temperature. If that relationship is linear, the cooling rate varies with time in a parabolic manner; i.e.,

$$-\frac{dT}{dt} = C_1 t^2 + C_2 t + C_3 \qquad (100)$$

An approximation to the temperature–time relationship that serves as a good starting point for establishing a fixed protocol is given by:

$$T = T_0 - (T_0 - T_{final}) \left(\frac{t}{\tau}\right)^3 \qquad (101)$$

where τ is the overall batch run time.

It is clear that stringent control of batch crystallizers is critical to obtaining a desired crystal size distribution. It is also obvious that the development of a strategy for generating supersaturation can be aided by the types of modeling illustrated above. However, the initial conditions in the models were based on properties of seed crystals added to the crystallizer. In operations without seeding, initial conditions are determined from a model of primary nucleation.

D. Effects of Anomalous Growth

Throughout this section, crystals have been assumed to grow according to the McCabe ΔL law. This has simplified the analyses of both continuous and batch crystallizers and, indeed, crystal growth often follows the ΔL law. However, as outlined in Section III, size-dependent growth and growth-rate dispersion contribute to deviations from the models developed here. Both of these phenomena lead to similar results: In continuous, perfectly mixed crystallizers, the simple expression for population density given by Eq. (54) is no longer valid. Both size-dependent growth and growth-rate dispersion due to the existence of a random distribution of growth rates among crystals in a magma lead to curvature in plots of $\ln n$ vs. L. Models for both causes of this behavior exist but are considered beyond the scope of the present discussion. In batch crystallization, the effects of anomalous growth lead to a broadening of the distribution, as was illustrated in Fig. 6.

E. Summary

The discussion presented here has focused on the principles associated with formulating a population balance and applying simplifying conditions associated with specific crystallizer configurations. The continuous and batch systems used as examples were idealized so that the principles could be illustrated, but the concepts can be applied to more complicated configurations. Additionally, there has been a growing body of work on aspects of population balance formulation that greatly extends the ability to describe complex systems. Such work has involved anomalous crystal growth, crystal agglomeration, and crystal breakage and necessarily results in substantially more complex models.

SEE ALSO THE FOLLOWING ARTICLES

CRYSTAL GROWTH • CRYSTALLOGRAPHY • PRECIPITATION REACTIONS • SEPARATION AND PURIFICATION OF BIOCHEMICALS • SOLID-STATE CHEMISTRY • X-RAY ANALYSIS

BIBLIOGRAPHY

Moyers, G. C., and Rousseau, R. W. (1986). *In* "Handbook of Separation Process Technology" (R. W. Rousseau, ed.), Wiley, New York.

Mullin, J. W. (1993). "Industrial Crystallization," 3rd ed. Butterworth-Heinemann, London.

Myerson, A. S. (1993). "Handbook of Industrial Crystallization," Butterworth-Heinemann, London.

Randolph, A. D., and Larson, M. A. (1988). "Theory of Particulate Processes," 2nd ed. Academic Press, San Diego, CA.

Rousseau, R. W. (1993). *In* "Kirk-Othmer Encyclopedia of Chemical Technology," Vol. 7, 4th ed., pp. 683–730, John Wiley & Sons, New York.

Rousseau, R. W. (1997). *In* "Encyclopedia of Separation Technology," Vol. 1 (D. M. Ruthven, ed.), pp. 393–439, Wiley Interscience, New York.

Tavare, N. S. (1995). "Industrial Crystallization: Process Simulation, Analysis and Design," Plenum, New York.

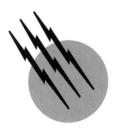

Crystallography

Jeffrey R. Deschamps
Judith L. Flippen-Anderson

Laboratory for the Structure of Matter, Naval Research Laboratory

GLOSSARY

Absorption edge Sharp discontinuity in the variation of the linear absorption coefficient with the wavelength of the incident radiation. The discontinuity occurs when the energy of the incident radiation, $E = h\nu$, matches the excitation energy of an electron in an atom of the sample.

Anomalous dispersion A phenomenon that influences the intensities of X-ray reflections and causes a difference in the intensity of equivalent reflections. The effect is particularly important in studies of single crystals in polar space groups and is used in some experiments to determine phase information.

Bragg reflection When X-rays strike a crystal they are diffracted only when the Bragg equation, $n\lambda = 2d \sin\theta$ (where n is an integer and d is the spacing of a set of lattice planes), is satisfied. The diffracted beam is considered a *reflection*.

Bravais lattice One of the 14 possible arrays of points repeated periodically in three-dimensional space such that the arrangement of points about any one point is identical in every respect to the arrangement of points about any other point in the lattice.

Centrosymmetric A structure or space group containing an inversion center is centrosymmetric, if there is no inversion center it is noncentrosymmetric.

Diffractometer An instrument used to measure the position (i.e., Bragg angle) and relative intensity of the diffraction pattern produced by a crystalline material.

Lattice Any repetitive pattern can be described by noting the motif (the unit of pattern that repeats by translation) and the translation interval. In the case of a three-dimensional pattern such as a crystal, the lattice describes translations in three dimensions. It is an imaginary, mathematical construct characterized by three translations, **a**, **b**, **c**, and three angles, α, β, γ.

Miller indices A set of integers with no common factors, inversely proportional to the intercepts with the crystal axes of a lattice plane.

Orientation matrix A matrix relating the crystal axes to the instrument axes such that one can predict the values of the instrument angles (2θ, ω, χ, and Φ) for a given reflection of the crystal.

Patterson function A Fourier summation that uses the squares of the structure factor magnitudes as coefficients. The peaks in this map correspond to vectors between atoms. The peak height is related to the scattering powers of the atoms at the two ends of the vector. The region around the origin gives information about bonded distances.

Phase problem A central problem of crystallography. The intensities of the different reflections allow derivation of the amplitude of the structure factors but not their phases. The phases are required in order to calculate the electron density, which is a "map" showing the position of atoms in the unit cell.

Point group A group of symmetry operations that leave unmoved at least one point within the object to which they apply.

Polar space group Space group in which the origin is not fixed by symmetry and hence must be defined (e.g., the space group $P2_1$).

Reciprocal lattice A set of imaginary points constructed in such a way that the direction of a vector from one point to another coincides with the direction of a normal to the real space planes within the crystal. The separation of those points (absolute value of the vector) is equal to the reciprocal of the real inter-planar distance.

Space group Identical atom groups are usually symmetrically arranged within the crystal lattice. The symmetry relating the groups may be due to rotations, inversions, mirror planes, or some other relational operation. The space group constitutes a mathematical shorthand description of the symmetry operations required to produce the unit cell.

Special position A point left invariant by at least two symmetry operations of the space group.

Structure factor F_{hkl}, complex quantity corresponding to the amplitude and phase of the diffraction maximum associated with the reciprocal lattice point hkl:

$$F_{hkl} = \sum_{j=1}^{N} f_j \exp[2\pi i(hx_j + ky_j + lz_j)]$$

where N is the number of atoms in the unit cell and x_j, y_j, z_j are the fractional coordinates of the jth atom.

Torsion angle If a group of four atoms (ABCD) is projected onto a plane normal to the bond between B and C, the angle between bonds connecting A and B, and C and D is the torsion angle.

Unit cell Parallelepiped bounded by three noncoplanar vectors **a**, **b**, **c** with angles α, β, γ that repeats by translation. If this unit is the smallest volume that meets these criteria it is referred to as the *primitive unit cell*.

X-ray Electromagnetic radiation with wavelengths in the range 0.01 to 1.0 nm (0.1 to 10 Å). The wavelength range most commonly used in diffraction experiments is between 0.71 and 1.54 Å. Shorter wavelengths require longer path-lengths (distance from crystal to detector) in order to resolve adjacent peaks in the diffraction pattern. Both the sample and the air along the beam path can significantly attenuate longer wavelengths.

CRYSTALLOGRAPHY is a broadly encompassing discipline that involves a variety of fields of study. The primary concern of modern crystallography is the three-dimensional arrangement of atoms in matter. Although the term most often refers to studies of crystalline solids (either single crystals or crystalline powders) using X-ray or neutron diffraction, it encompasses a much broader range of methodologies.

I. INTRODUCTION

In the late 1660s, crystallography began as the study of the macroscopic geometry of crystals. Crystals were grouped into systems on the basis of the symmetry of their external shapes. Based on these observations, seven crystal systems were identified: triclinic, monoclinic, trigonal, tetragonal, hexagonal, orthorhombic, and cubic. It was theorized that the observed crystal shapes could be built up by stacking minute balls. This idea was refined, and in the late 18th century it was thought that crystals were composed of elementary building blocks (now referred to as the *unit cell*). This was supported by the orderly cleavage angles of calcite, which suggested a regular stacking of these elementary blocks. Further studies in the 1800s led to derivation of the 32 different *point groups*, the *Bravais lattices*, and the 230 *space groups*. All of these advances were made before any direct observations of the arrangement of atoms within a crystal were possible.

In 1912, Von Laue reasoned that the arrangement of atoms in crystals could help him measure the wavelength of X-rays. Based on experiments with copper sulfate, he demonstrated the ability of crystals to act as three-dimensional diffraction gratings. Crystallography had entered a new era: the analysis of the arrangement of atoms in a crystal by careful analysis of the diffraction pattern of that crystal. Crystal structures were now viewed as being built up from repeating units of an atomic pattern rather than the regular stacking of solid shapes.

As new types of scattering were discovered, they became incorporated into the discipline, and crystallography came to include structural studies of all classes of substances. Absorption, diffraction, or other scattering methods are used to study crystals, powders, amorphous materials, surfaces, liquids, and gases. The physics and methodologies associated with these techniques are also part of the science of crystallography.

Crystallographic studies play a vital role in materials science, chemistry, pharmacology, mineralogy, polymer science, and molecular biology. Accurate knowledge of molecular structures is a prerequisite for rational drug design and structure-based functional studies. Crystallography is the only method for determining the "absolute" configuration of a molecule. Absolute configuration is a critical property in biological systems, as changes in this may alter the response of the biologic system.

A requirement for the high accuracy of crystallographic structures is that a good crystal must be found, and this is often the rate-limiting step. Additionally, only limited information about the dynamic behavior of the molecule is available from a single diffraction experiment.

In the past three decades, new developments in detectors, increases in computer power, and powerful graphics capabilities have contributed to a dramatic increase in the number of materials characterized by crystallography. Synchrotron sources offer the possibility of time-resolved studies of physical, chemical, and biochemical processes in the millisecond to nanosecond range; the ability to study the nearest neighbors of cations present at parts per million concentrations; and the possibility of recording small-angle scattering data and powder data in seconds. Charge-density studies have been made on numerous light atom structures and are beginning to provide new insights into bonding of transition metals. Rietveld refinement is revolutionizing the study of powders and is being extended to fibers. Direct methods of structure solution are being applied successfully to structures of over 1000 atoms. The Human Genome Project has created many opportunities for crystallographic studies of biological macromolecules and resulted in intense activity in the areas of structural genomics and proteomics. Polarized neutrons are being used to determine the spin structure of magnetic materials and to probe the surface structure of such materials. The review that follows attempts to describe the fundamentals of crystallography and capture the excitement and diversity within the discipline.

II. EVOLUTION OF CRYSTALLOGRAPHY

It was not very long ago that X-ray diffraction data were collected on photographic film, intensities of spots on the film (corresponding to data points) were "measured" by eye, and Fourier transforms were performed with Beevers–Lipson or Patterson–Tunell strips and summed by hand. Three-dimensional electron-density maps were plotted by hand, one section at a time, traced onto glass sheets, and stacked in frames for interpretation. As long as all calculations were done by hand, small, flat molecules were the ones most amenable to study. Early computers were difficult to program and had limited storage capacity, but they made it possible to solve and refine crystal structures of molecules of moderate size in 6 months to a year.

The rapid advances in crystallography owe much to the development of computer-controlled diffractometers for data collection, high-speed computers for data analyses, and, most recently, powerful graphics devices for displaying structures with the ability to perform real-time manipulations. Tasks that required months of effort can now be accomplished in minutes with the aid of a computer.

A. The Early Years

The first experiments in X-ray diffraction were recorded on film. By 1913, W. H. Bragg had constructed the first "X-ray spectrometer" to allow a more careful study of X-rays. This instrument also proved useful in studies on crystals. Using measurements made with this X-ray spectrometer, Bragg's son determined the structures of fluorspar, cuprite, zincblende, iron pyrites, sodium nitrate, and the calcite group of minerals.

The first crystal structures reported were of substances crystallizing in cubic space groups. The structure of diamond was determined in 1913 by the Braggs, from symmetry considerations using the observed intensities to discriminate among possible structures. Their model established the carbon–carbon single bond distance of 1.52 Å and confirmed that bonds to carbon are directed tetrahedrally. The younger Bragg combined symmetry arguments with the notion that the scattering power of atoms is related to their atomic weight to explain the structures of the alkyl halides. These concepts were extended to ZnS, CaF_2, and FeS_2. The alkyl halide structures show anions surrounded by cations and cations surrounded by anions, demonstrating conclusively that discrete "molecules" of the type NaCl do not exist in crystals of ionic materials. Pauling published the first intermetallic structure, Mg_2Sn, in 1923. In that same year, Dickerson and Raymond showed that hexamethylene tetramine consisted of discrete molecules, each having the same structure, with C–N distance of 1.44 Å in a body-centered cubic lattice of edge 7.02 Å.

1. Heavy-Atom Methods

The elder Bragg realized that the periodic pattern in the electron-density distribution could be represented by a

Fourier summation. The coefficients of this summation became known as the *structure factors*. This allowed the solution of structures where positions of atoms were not restricted to special positions. The determination of the structure of diopside, $CaMg(SiO_3)_2$, in 1928 was the first example of the use of this method.

In 1934, Patterson showed that a Fourier series with $|F|^2$ as coefficients could be summed without knowledge of the phases and would reveal interatomic vectors. Since the weight of a peak in the Patterson function is proportional to the product $Z_i Z_j$ of the atomic numbers of the atoms at the ends of the vector, vectors involving heavy atoms stand out among light atom–light atom vectors. In 1936, Harker showed that symmetry properties of the crystal caused vector density to accumulate on certain planes and lines (later known as the Harker section) in the Patterson function. These two papers were the foundation of the heavy-atom method for crystal structure solution. The method assumes that phases calculated from the heavy-atom positions will be sufficiently accurate that a Fourier synthesis (using $|F|$ as opposed to $|F|^2$) will reveal the positions of more atoms, thus allowing solution of the structure. Phase information from the new atoms could then be added to the Fourier synthesis to locate more new atoms and so on, until the full structure was revealed.

The method of isomorphous replacement was used first to solve the structure of the alums by Lipson and Beevers in 1935. If a centrosymmetric light-atom structure and its heavy-atom derivative differ only in the presence of the heavy atom, then the differences in the intensity of equivalent reflections can be used to determine the signs of the structure factors. Robertson applied this method to the phthalocyanines in 1936. These molecules form an isomorphous series and crystallize in $P2_1/a$. By comparing intensities of nickel phthalocyanine and the unsubstituted molecule, Robertson was able to assign phases to all but a few of the measured reflections from the $h0l$ projection. Harker generalized the method to the noncentrosymmetric case in 1956. Isomorphous replacement can be combined with anomalous dispersion to obtain phase information for large molecules.

2. Absolute Configuration

Friedel's law states that the scattering from the front and back sides of a plane, hkl, are the same. This means that the measured intensities of the 111 and $\overline{111}$ reflections (and all other "Friedel" pairs) should be equivalent. However, in a 1930 study on ZnS by Coster, Knol, and Prins, it was noted that the 111 reflection was not equivalent to its Friedel mate. In the case of the ZnS crystal, the {111} faces are prominent but, even by visual examination, do not appear identical. One face is shiny and the other is dull.

The crystal structure can be regarded as alternating layers of Zn and S atoms perpendicular to the {111} direction. Looking at one layer of Zn atoms, we find it lies closer to one of the two adjacent layers of S atoms. If we assume that the short Zn–S spacing is a "bonding" interaction and the long Zn . . . S gap is a van der Waals contact, we do not expect cleavage between bonded layers. This implies that one of the {111} faces corresponds to a layer of S atoms and the other to a layer of Zn atoms.

It was fortuitous that the intensities in the Coster, Knol, and Prins study were measured using $AuL\alpha$ radiation ($\lambda = 1.276$ Å). The K-absorption edge of Zn is 1.283 Å. Thus, the measured intensities of Friedel pairs were not the same. This was the first example of anomalous scattering. The difference due to anomalous scattering is greatest when data are collected near an absorption edge of a heavy atom in the structure. It was nearly 20 years after the ZnS experiment that Bijvoet realized this principle could be used to determine the absolute configuration of the sodium rubidium salt of (+)-tartaric acid.

B. Modern Crystallography

The early years can be characterized as a period in which the size of a problem amenable to analysis was computationally and instrumentally limited. Data collection and analysis were manual processes. This began to change in the 1950s.

1. Data Collection

Progress in crystallographic data collection can be charted by examining early issues of *Acta Crystallographica*, published by the International Union of Crystallography (IUCr). Although a manual "diffractometer" was available in 1913, the primary method for collecting intensity data for crystallographic studies relied on X-ray cameras and film. In 1948, about two-thirds of the crystal structures reported used a camera and film to collect the intensity data, and less than one-quarter used a combination of film and a diffractometer. Despite the existence of automated single-crystal diffractometers, the situation was little changed by 1962.

Use of the Cambridge Structure Database (CSD) allows a more systematic study of the evolution of data collection methods (Fig. 1) with the caveat that this database references only organic and organometallic structures. Although automated diffractometers were first available in about 1955 it was not until the mid-1960s that this new technology made much of an impact. At that time, only a few hundred structures were being added to the CSD each year. By the mid-1970s, over 1000 structures were added to the CSD each year and data for over 80% were

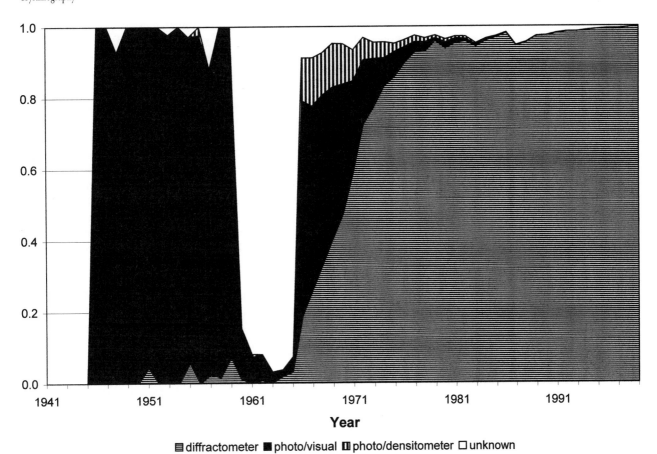

FIGURE 1 Changes in data-collection methods with time. Using structures entered in the Cambridge Structure Database, changes in data collection strategies were determined by examining the fraction of structures reported in one year for various data collection methods (photograph with visual estimation of intensities, photograph with densitometer, diffractometer, or unknown). The large fraction listed as unknown between 1960 and 1965 is likely due to changes occurring at that time in how data were collected and how data-collection methods were reported.

collected on a diffractometer. In 1997, over 15,000 structures were added to the CSD and virtually all were collected on a diffractometer. The explosive growth of this database cannot be attributed solely to improvements in data collection, but certainly the routine use of automated instrumentation had a significant impact.

2. Structure Solution

Data collection was not the only beneficiary of the post-World War II progress in crystallography. No general method existed for solving unknown structures without heavy atoms until the advent of direct methods—a means of determining the values of phases from relationships among the structure factor magnitudes associated with those phases. The earliest structure solved by such methods was decaborane by use of some inequalities derived by Harker and Kasper. However, this technique was limited to centrosymmetric structures. The major effort in the 1950s

concerned the development of the mathematical aspects of crystal structure analysis. The first general procedures for solving both centrosymmetric and noncentrosymmetric structures was developed in the early part of the 1960s. Use of this method grew with the power of computers and computer programs. It is now the most widely used method of solving crystal structures of moderate size. Efforts are currently being made to apply direct methods to very large structures such as proteins. For a more detail discussion of structure solution methods, see Section IV.H.

3. Charge-Density Distribution

The first papers to explore the difference between the results of X-ray and neutron diffraction experiments appeared in the early 1970s. Systematic differences between the positional and thermal parameters determined by the two techniques were reported. These differences were attributed to the difference in how neutrons and X-rays interact with atoms. Neutrons are diffracted primarily from the nucleus; hence, neutron diffraction produces information about nuclear position. X-rays are diffracted by electrons and therefore yield information about the distribution of electrons in the molecule. It was logical to extend these ideas and attempt to map the redistribution of electron density that occurs on bonding.

A number of different kinds of mapping have been done. Subtracting core-electron densities from experimental electron densities (i.e., $p(\mathbf{r})$) should reveal details of the redistribution of valence electrons on bonding. The term "valence density" is used to describe the difference function:

$$p_{\text{valence}}(\mathbf{r}) = p(\mathbf{r}) - \sum_{\text{atoms}} p_{i,core}(\mathbf{r})$$

To define either of these functions the positions and thermal parameters must be known. One way to approach this is to use neutrons to determine the positional and thermal parameters and to use these parameters in conjunction with X-ray, spherical-atom scattering factors to calculate structure factors for the "promolecule" using:

$$F_{\text{calc},N} = \sum_{\text{atoms}} f_i \exp(2\pi i \mathbf{H} \cdot \mathbf{r}_i) T_i$$

$F_{\text{calc},N}$ is the X-ray structure factor calculated from neutron positions. The deformation density or $X-N$ map then corresponds to:

$$p_{\text{deformation}}^{X-N}(\mathbf{r}) = \frac{1}{V} \times \sum_{H} (F_{\text{obs},X} - F_{\text{calc},N}) \exp(-2\pi i \mathbf{H} \cdot \mathbf{r})$$

Since both F_{obs} and F_{calc} contain the effects of thermal motion, this deformation density map is thermally smeared. Its resolution is limited by the maximum $(\sin\theta)/\lambda$ value of the data.

Given that there are few neutron-diffraction facilities in the world and that it is difficult to correct adequately for systematic effects in the two experiments (i.e., absorption, extinction, thermal diffuse scattering, multiple reflections), it would be desirable to study bonding effects using exclusively X-ray data. There are several approaches to this problem.

The $X-X$ formalism is similar to the $X-N$ formalism described above, except that the calculated values for positional and thermal parameters are derived from refinement of high-angle X-ray data [$(\sin\theta)/\lambda > 0.70\,\text{Å}^{-1}$]. For deformation density maps, neutral spherical atoms are subtracted from the observed density; in valence density maps, Hartree–Fock core-electron densities are used to evaluate F_{calc}. Comparison of $X-X$ and $X-N$ maps shows that they do indeed yield the same qualitative information. Bonding density shows in the middle of bonds, and lone-pair density is in the correct location. However, $X-X$ maps

systematically underestimate lone-pair peak heights and place hydrogen atoms too close to the atoms to which they are bonded. The experiments must be conducted at low temperature (i.e., $-75°C$ or preferably less) for this method to succeed.

Other approaches include refinement of separate parameters for core and valence electrons or the direct refinement of a deformation model. The major advantage of the direct refinement methods is that they make no assumptions about the $(\sin\theta)/\lambda$ dependence of bonding features.

Errors in the deformation density maps arise from errors in both the model and the data. Demands on diffraction methodology and interpretation are many times more severe than those relevant to an average stereochemical investigation. The following considerations are extremely important:

1. The X-ray data set must be complete (all symmetry-related reflections measured) to a $(\sin\theta)/\lambda$ limit of about $1.3\,\text{Å}^{-1}$. This implies the use of short-wavelength radiation such as $MoK\alpha$ ($\lambda = 0.7107\,\text{Å}$) or $AgK\alpha$ ($\lambda = 0.5612\,\text{Å}$). The maximum value of $(\sin\theta)/\lambda$ is 0.65 for $CuK\alpha$ ($\lambda = 1.5418\,\text{Å}$).

2. The data must be corrected for absorption, extinction, and thermal diffuse scattering. Multiple reflections must be either avoided or eliminated. If these conditions are met, internal agreements should be of the order of 2%; i.e.,

$$\left[\sum_{H} F^2(\mathbf{H}) - \langle F^2(\mathbf{H}) \rangle \right] \bigg/ \sum_{H} F^2(\mathbf{H}) = 0.02$$

4. Rietveld Analysis

There has been a renaissance in powder diffraction in recent years because Rietveld refinement allows determinations of positional and thermal parameters from powder data, even when the diffraction peaks are not well separated in the recorded pattern. Rietveld analysis is not a method of structure solution and can only be applied when cell dimensions and space group are known and when a reasonable model exists for the structure.

In a polycrystalline sample, information may be lost as a result of the random orientation of the crystallites. A more serious loss of information can result from the overlap of independent diffraction peaks in the powder pattern. Using the total integrated intensities of the separate groups of overlapping peaks in the least-squares refinement of a structure leads to the loss of all the information contained in the often-detailed profile of these composite peaks. Rietveld developed a refinement method that uses the profile intensities of the composite peaks instead of the integrated quantities. This is a pattern-fitting method of structure refinement and allows extraction of the maximum amount of information contained in the powder pattern.

A powder pattern is recorded in a step-scan mode with a step width of 0.02 to 0.03° 2θ. No attempt is made to allocate observed intensity to individual reflections or to resolve overlapping reflections. Instead, the intensity of the powder diffraction pattern is calculated as a stepwise function of the angle, 2θ. Refinement allows calculation of the shifts in the parameters that will improve the fit of the calculated powder pattern to the observed one. The quantity minimized is

$$\sum_i w_i [y_i(\text{obs}) - y_i(\text{calc})]^2$$

where $y_i(\text{obs})$ is the observed intensity of the ith step of the profile, and $y_i(\text{calc})$ includes the usual structural parameters (i.e., positional parameters x_i, y_i, z_i; thermal parameter B_{ij}; and site-occupancy parameters p_j). However, the model must also include instrumental and sample parameters: $2\theta_0$ (overall scale), overall temperature factor, profile breadth ($\mathbf{H}^2 = U \tan 2\theta + V \tan \theta + W$), profile asymmetry, background, preferred orientation, lattice parameters, and wavelength. The agreement factors most often quoted are

$$R_{\text{weighted pattern}} = \left[\frac{\sum w_i [y_i(\text{obs}) - (1/c) y_i(\text{calc})]^2}{\sum w_i [y_i(\text{obs})]^2} \right]^{1/2}$$

and

$$R_{\text{Bragg}} = \frac{\sum |I_k(\text{"obs"}) - I_k(\text{calc})|}{\sum I_k(\text{"obs"})}$$

In R_{Bragg}, "obs" has quotation marks because $I_k(\text{"obs"})$ is calculated by partitioning the intensity.

To date, most of the papers published using this method have been neutron-diffraction studies from reactor sources. Advantages of neutron data include minimal preferred orientation, no polarization, and neutron absorption cross-sections smaller than X-ray values by a factor of 10^4, scattering independent of θ, and for fixed-wavelength experiments the peak shape is simple.

In X-ray experiments using radiation from conventional sources, peak shape is complicated by both α_1–α_2 splitting (at high angles) and by the fact that the peak shape is neither Lorentzian nor Gaussian, but is better described by a convolution of these two functions called the *Voigt function*. The pseudo-Voigt function used in many programs is an approximation of the Voigt function that can be evaluated much more quickly.

Spallation neutron sources, time-of-flight experiments with reactor sources, and synchrotron sources all have special problems in defining peak shapes, and this limits the precision of the resulting parameters. However, precision comparable to single-crystal X-ray diffraction can now be obtained from neutron diffraction at a fixed wavelength with Rietveld refinement. For X-rays from conventional sources, the precision of positional parameters is comparable to the single-crystal case, but thermal parameters are less reliable by a factor of two or three. Considerable effort is being expended to improve profile functions for the various X-ray and neutron sources.

5. Small-Angle Scattering

Scattering at small angles is derived from large structural units—units whose dimension D is much larger than the wavelength of the radiation used in the experiment. The acronyms SAXS and SANS refer to small-angle X-ray scattering and neutron scattering, respectively.

Different sorts of small-angle experiments are typical of the kind of material studied and yield a characteristic type of pattern. Low-angle data from ordered or semi-ordered systems give Bragg peaks at specific values of scattering vector. Examples include aligned structures with long-range periodicity, such as two-dimensional biological structures or samples such as opal that present long-range order.

Scattering from polymers in dilute solution or from biological materials yields patterns that look rather like the Debye–Scherrer rings observed in wide-angle data from powders. When scattering arises from the spherical particles in a mono-disperse system, the pattern is a Bessel function consisting of a succession of peaks of diminishing magnitude that are broad relative to Bragg peaks. Analysis of the pattern yields the radius of the particles.

In some circumstances, multiple Bragg reflections give rise to scattering in the small-angle region. One of the advantages of the "tunable" sources of X-rays and neutrons is that multiple Bragg scattering can be avoided by choosing wavelengths larger than the lattice spacing.

Small-angle scattering can be applied to a wide variety of materials. In polymer science, it has been used to investigate chain conformation in amorphous polymers, the state of mixing in polymer blends, the compatibility ranges of polymer blends, and the measurement of domain structure and molecular conformation within those domains. In biological materials, examples include measurement of the radius of gyration of proteins in solution; aggregation of chlorophyll into micelles; diffraction patterns of semi-ordered materials such as muscle, collagen, etc.; and studies of the shapes and constitution of viruses. Separation processes, such as those used in refining of metals or extraction of tar sand, frequently involve micelle formation. Small-angle scattering can give information regarding the shape, size, and degree of polymerization of the aggregates. Studies of materials such as cements, zeolites, and catalysts involve the measurement of size and distribution of pores and measurements of specific surface.

6. Extended X-Ray Absorption Fine Structure

Extended X-ray absorption fine structure (EXAFS) is a technique for studying the local environment of a specific atomic species in a complex matrix. Because the interaction of X-rays with the material under study is absorption rather than diffraction, the technique can be applied to gases, liquids, and amorphous solids as well as to crystals.

In the experiment, the X-ray absorption coefficient is measured from slightly below to about 1000 eV above the absorption edge for the atomic species whose environment is to be studied. Through analysis of the fine structure above the edge it is possible to determine the coordination of the atom. The actual position of the edge gives information about the oxidation state of the absorbing atom, while structure at energies at or just below the edge provides information about bound states associated with the absorbing atom and hence about the symmetry of the environment. The technique can be applied to any atom with $Z > 15$ to yield a determination of radial distance to a precision of ± 0.02 to 0.01 Å. For elements with high values of Z, it may be preferable to use the L-absorption edge rather than the K edge because higher fluxes are available; however, theory for the L edge is not yet well developed.

Extended X-ray absorption fine structure has applications in many branches of science. In molecular biology, it has been used to study Ca^{+2} transport in membranes, binding of oxygen in hemoglobin, and other coordination problems. It is an invaluable tool for the study of amorphous substances such as glass, since the manufacture of glasses with particular mechanical and thermal properties depends on structure. Catalysis has become extremely important in energy development, resource utilization, pollution abatement, refining of metals, etc. The chemical state and atomic environment of an atomic species in a catalyst *in situ* while reduced with hydrogen, chemisorbed with oxygen, heated, quenched, etc., can be determined with EXAFS. This allows the design of heterogeneous catalysts that are tailored from precise knowledge of electronic and structural parameters.

Synchrotron sources have revolutionized EXAFS studies. The intensity of the source and high collimation make it possible to collect the relevant data in about 20 minutes in a sample such as Cu metal with a resolution of 1 eV at 8.8 keV. To collect similar data with a rotating anode source would take about 2 weeks, and the precision would be reduced by a factor of about 100.

7. Implications of New X-Ray and Neutron Sources

Synchrotron sources present a unique combination of properties that are very attractive for X-ray scattering, absorption, and diffraction experiments. The radiation produced has extreme brightness over a broad spectral range. Synchrotron sources are five to six orders of magnitude brighter than the bremsstrahlung (that part of the X-ray spectrum caused by the slowing of electrons on impact with the target, also referred to as *white radiation*) available from a conventional rotating anode source. Monochromators currently available produce resolution of 0.1 eV at 8 keV. The radiation is naturally collimated with a divergence of the order of 2×10^{-4} radians, is plane polarized, and has a precise time structure (subnanosecond pulses repeated every 0.5 to 1 μsec). These properties allow experiments that simply cannot be done with conventional sources: EXAFS on dilute samples (parts per million range), measurement of the magnitude and angular dependence of the real and imaginary components of anomalous dispersion, and determination of the structure of a protein using only one derivative and three wavelengths. (In this context, the use of anomalous dispersion is formally equivalent to multiple isomorphous replacement with the added feature that the isomorphism is exact.) Determinations of cation-site distribution in minerals and diffraction from monolayers on surfaces have many applications in such areas as catalysis or materials science. Perhaps the most exciting application is the ability to do time-resolved studies of physical, chemical, and biological processes using small-angle scattering, powder diffraction, and other scattering techniques.

One early example was the study by Larson of temperature and temperature gradients in silicon during pulsed-laser annealing. In this example, the duration of the laser pulse was 15 nsec and that of the synchrotron X-ray pulse was 0.15 nsec. The laser bursts were synchronized so that the probing X-rays arrived at 20, 55, and 155 nsec after the laser pulse. The experiments showed that the lattice temperature of silicon reaches the melting point during the 15-nsec pulse and remains at the melting point during the high reflectivity phase, after which time the temperature rapidly subsides. Temperature gradients at the liquid–solid interface were measured for the first time and were found to be in the range of 10^7 °K/cm. Larson received the Warren Award for Diffraction Physics for this pioneering work in nanosecond, time-resolved X-ray diffraction.

The new pulsed spallation sources (such as the European Spallation Neutron Source at Forschungszentrum Jülich GmbH, or the Spallation Neutron Source at Oak Ridge National Laboratory) provide spectra substantially richer in short-wavelength neutrons than those available from the reactor sources. Pulse duration and repetition are source parameters, but the time structure can be exploited by a variety of techniques. Essentially the same types of experiments are done at both the reactor and spallation sources. The higher neutron fluxes available with the new sources allow experiments to be done on smaller samples and/or in shorter times than was previously possible.

8. Contribution of Diffraction to Molecular Biology

Molecular biologists seek to unravel the mysteries of the cell by mapping gene location, function, and control. By understanding all of these we gain insight into how and when genes are turned on and how these might be used to perform useful tasks. One goal of such studies is to alter (i.e., reengineer) a natural biologic process to perform some other function. Examples of this include attempts to modify metal-binding proteins such that a protein that was originally selective for calcium is selective for zinc or copper. This reengineered metal-binding protein could then be used to construct a sensor for zinc or copper. Other examples include modifying enzymes for use in industrial processes rather than biologic processes. In order to effect these changes, the relationship between structure and function must be understood.

The structure of a macromolecule can be "known" at a number of levels. Primary structure is the linear sequence of building blocks (i.e., amino acids) from which the protein is built. For example, the β chain of human hemoglobin contains 144 amino acids, of which the first five have the sequence valine, histamine, leucine, threonine, and proline. The term "secondary structure" refers to local interactions that determine the conformation of the polypeptide chain and the interchain hydrogen bonding scheme. "Extended chain," "α-helix," and "β-sheet" are terms used to describe secondary structure. Tertiary structure is the three-dimensional arrangement of atoms within the macromolecule, while quaternary structure describes the arrangement and interaction of aggregates of the macromolecules themselves. Diffraction techniques are widely used to study the secondary, tertiary, and sometimes quaternary structures of macromolecules.

Single crystals of macromolecules may be studied by X-ray diffraction. A wide variety of techniques are now available for macromolecular structure solution. In the past, heavy-atom multiple isomorphous replacement (MIR) was the most common method of structure solution; however, anomalous dispersion is becoming more common as a method of structure solution. The relatively small protein, crambin, was an early example of using anomalous dispersion to effect structure solution of a macromolecule. In a few cases, neutron-diffraction studies have been carried out on single crystals of proteins. In myo-globin, for example, at 1.8 Å resolution, the negative density of the protons (which form about half of the scattering material in the cell) makes the polypeptide chain stand out clearly in the Fourier maps. Soaking the same crystal in D_2O has allowed identification of exchangeable protons.

Neutron- and X-ray-scattering experiments using non-Bragg scattering can be used to study the size, shape, and aggregation of micelles. Contrast variation can be used to study the internal structure of viruses.

The average scattering densities from the protein coat and the RNA interior of viruses are different. Each can be "matched" by a different H_2O/D_2O ratio. Thus, for virus particles in solution, the matched phase becomes "invisible" to the neutron beam and allows the radial distribution of scattering of the other component to be recorded. For spherical viruses, this allows measurement of the thickness of the protein coat and the degree of interpenetration of the two phases. Enzymes and other proteins bind some substrates so well that detailed, atomic-level analyses of the structure of the native protein can be compared with those of protein–substrate complexes, protein–inhibitor complexes, and proteins with catalytic groups bound to allosteric sites. Detailed comparisons between such structures have greatly enhanced our understanding of the mechanisms of biological catalysis.

Molecular graphics programs are now available that can display the full three-dimensional structure of a macromolecule and zoom in on any portion of it. It is possible to examine the active site and to attempt to fit known substrates and/or inhibitors into that site. The availability of coordinates for macromolecules in the Protein Data Bank (see Section VI.A.8) has allowed many fruitful applications of this type.

III. STRUCTURE OF A CRYSTAL

A. Choice of Unit Cell

A crystal is a multifaceted solid, similar in appearance to an unpolished gemstone. Internally it consists of a basic pattern, known as the *repeating unit*, of molecules that repeats itself by translation, in three dimensions, to the edges of the crystal. By choosing one corner of the repeating unit to be the origin, one can use three translational vectors, having both length and direction, to construct a parallelepiped that contains the entire basic pattern. This parallelepiped is defined as the *unit cell* and one need only determine the contents of the unit cell, the basic pattern, to know the structure of the entire crystal. The general symbols for the unit-cell vectors are **a**, **b**, **c** and for their magnitudes a, b, c. The coordinate axes, or directions of the sides of the unit cell, are referred to in general as the

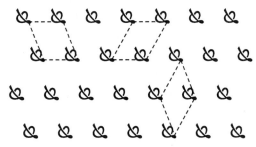

FIGURE 2 Choice of unit cell. In the absence of symmetry, the unit cell may be chosen in a variety of ways. Each cell contains one unit of pattern. All such cells have the same volume.

x, y, and z axes. The interaxial angles are denoted by α, β, and γ. The unit cell can be defined in a variety of ways, but for a given pattern, the volumes of each cell will be equal (Fig. 2).

If the repeating unit itself has no internal symmetry, then the choice of unit cells is infinite. However, if the basic repeating unit contains additional symmetry, this influences the choice of the unit cell. If there are planes or axes of symmetry, the cell edges are generally chosen to be parallel or perpendicular to these directions. This places restrictions on how the unit cell may be defined and gives rise to seven crystal systems: triclinic, monoclinic, orthorhombic, tetragonal, hexagonal, trigonal, and cubic (see Table I).

B. Diffraction Pattern

Friedrich, Knipping, and Laue first demonstrated the fact that crystals could act as three-dimensional diffraction gratings for X-rays in 1912. This work not only established the wave nature of X-rays but also established the relationship between a crystal and its diffraction pattern. The size and shape of the repeating unit in the crystal determine the position of the diffraction spots they recorded on film. It was some time later before it was realized that the intensity of the spots is related to the distribution of atoms within the unit cell.

TABLE I Seven Crystal Systems and Their Unit Cell Constraints

Crystal System	Conditions imposed on cell geometry
Triclinic	$a \neq b \neq c$; $\alpha \neq \beta \neq \gamma$
Monoclinic	$a \neq b \neq c$; $\alpha = \gamma = 90°$
Orthorhombic	$a \neq b \neq c$; $\alpha = \beta = \gamma = 90°$
Tetragonal	$a = b$; $\alpha = \beta = \gamma = 90°$
Trigonal	$a = b$; $\alpha = \beta = 90° \gamma = 120°$ (hexagonal axes)
	$a = b = c$; $\alpha = \beta = \gamma$ (trigonal axes)
Hexagonal	$a = b$; $\alpha = \beta = 90° \gamma = 120°$
Cubic	$a = b = c$; $\alpha = \beta = \gamma = 90°$

Bragg's contribution was to recognize the similarity between diffraction in a crystal and reflection in a mirror plane (Fig. 3). Consider a set of parallel planes with spacing d and an incoming beam of monochromatic X-rays at a glancing angle, θ. The condition for constructive interference is that the path difference between waves "reflected" from successive planes must be an integral number of wavelengths (i.e., $AB + BC = n\lambda$). However, $AB = BC = d \sin\theta$; thus, $n\lambda = 2d \sin\theta$, which is Bragg's law. Note that the angle between the incident beam direction and the reflected beam is 2θ.

The smaller the interplanar spacing d, the higher the angle at which the diffraction maximum or "reflection" is observed. This implies that large unit cells will give diffraction patterns with small spacing between the spots, while small unit cells will give patterns with wide spacing. Thus, lattices can describe both the crystal and its diffraction pattern. Since the lattice of the diffraction pattern is inversely proportional to the crystal lattice, it is defined as the reciprocal lattice.

In crystal space we can define a set of parallel planes of spacing d, and note that the first of these planes (for which the distance from the origin is d) has intercepts with the edges of the unit cell of $\mathbf{a}/h$, $\mathbf{b}/k$, $\mathbf{c}/l$. The Miller's index of that set of planes is then hkl, where h, k, and l are small integers with no common factor. In Fig. 4 we have a plane with intercepts $\mathbf{a}/3$, $\mathbf{b}/4$, $\mathbf{c}/2$. The Miller's index is 342. The set of parallel planes of index 342 will give rise to a spot in the diffraction pattern with index 342. The reflection 684 can be regarded either as the second-order reflection from the planes 342 or as the first-order reflection from a set of parallel planes with spacing $d/2$. The number of diffraction planes possible for a given structure is directly related to the lengths of $\mathbf{a}$, $\mathbf{b}$, and $\mathbf{c}$.

C. Basic Formulas of Crystallography

The crystal structure is a pattern that repeats in three dimensions, and Fourier series can represent repetitive patterns.

Let the position of the jth atom in the unit have fractional coordinates x_j, y_j, z_j (in this notation, x_j means $x/\mathbf{a}$). Then, the vector from the origin to the jth atom would be $\mathbf{r}_j = \mathbf{a}x_j + \mathbf{b}y_j + \mathbf{c}z_j$. The vector representing the diffracted beam direction is $\mathbf{H} = h\mathbf{a}^* + k\mathbf{b}^* + l\mathbf{c}^*$ where hkl are indices of the reflecting plane and $\mathbf{a}^*$, $\mathbf{b}^*$, $\mathbf{c}^*$ are the base vectors of the reciprocal lattice. The direction of the diffracted beam is given in terms of the indices hkl. The set of planes hkl cuts $\mathbf{a}$ into h divisions, $\mathbf{b}$ into k divisions, and $\mathbf{c}$ into l divisions. The phase difference for unit translation along $\mathbf{a}$ is $2\pi h$. Thus, if $h = 3$, a ray scattered by an electron at $\mathbf{a} = 1$ would be $2\pi 3$ or three wavelengths out of phase with one scattered by an electron at the origin.

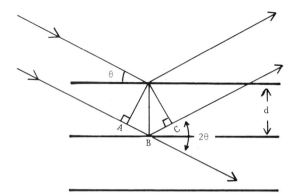

FIGURE 3 Bragg's law. The distance AB = BC = $d \sin\theta$. Constructive interference occurs if the path difference is a whole number of wavelengths. Thus, $n\lambda = 2d\sin\theta$.

The amplitude of the wave scattered by the plane hkl is

$$F_{hkl} = V \int_0^1 \int_0^1 \int_0^1 \rho(xyz)$$

$$\times \exp[2\pi i(hx + ky + lz)]\, dx\, dy\, dz$$

where $\rho(xyz)$ is the electron density at the point x, y, z in the unit cell. The quantity $(hx + ky + lz)$ is the vector product $(\mathbf{H} \cdot \mathbf{r})$.

By the properties of Fourier series, F_{hkl} is a Fourier coefficient of $\rho(xyz)$ so that:

$$\rho(xyz) = \frac{1}{V} \sum_h \sum_k \sum_l F_{hkl} \exp[-2\pi i(\mathbf{H} \cdot \mathbf{r})]$$

where the summations in h, k, and l each run from $-\infty$ to $+\infty$. Note the change in sign of the exponent between the two expressions. F_{hkl} is the Fourier transform of the electron density in the cell. The electron density is the inverse Fourier transform of structure factors.

If the electron density is a superposition of N atomic densities, then the structure factor expression can be rewritten as:

$$F_{hkl} = \sum_{j=1}^{N} f_j \exp[2\pi i(hx_j + ky_j + lz_j)]$$

The summation is over all atoms in the cell and the scattering factor of the jth atom is

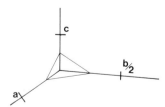

FIGURE 4 Miller's index. The plane has intercepts **a**/3, **b**/4, **c**/2. The Miller's index is 342.

$$f_j = V \int_{-\infty}^{+\infty} \rho_j(uvw)$$

$$\times \exp[2\pi i(hu + kv + lw)]\, du\, dv\, dw$$

where $\rho_j(uvw)$ is the electron density of the jth atom referred to x_j, y_j, z_j as origin. Thus, if we know the position of an atom, then we can calculate the phases. Conversely, if we know the phases, then we can calculate the electron density and hence the positions of the atoms. The central problem in crystallography is that the phases are not observed in the diffraction experiment. Solving a structure consists of finding positions for the atoms or of finding phases for the structure amplitudes. Details of the major methods for structure solution are found in Appendix II.

IV. STEPS IN CRYSTAL STRUCTURE ANALYSIS

The steps outlined here would be typical of those used for a moderately complex structure. Notes are also included on alternative techniques that may be more applicable to macromolecular crystallography.

A. Growing Crystals

All strategies for the growth of crystals for diffraction experiments are aimed at bringing a concentrated solution of a homogeneous population of molecules very slowly toward a state of minimum solubility. The goal is to achieve a limited degree of supersaturation, from which the system can relax by formation of a crystalline precipitate. Many techniques developed for achieving these ends have been described.

Crystallization techniques used in routine synthetic methods tend not to produce crystals of the quality required for structural work. Suitable crystals must be grown slowly at near-equilibrium conditions. This implies low supersaturation ratios and small gradients. Generally, supersaturation is achieved by changing the composition of a solution containing the sample to be analyzed (and possibly other additives) or by altering the temperature. In either case, the concentration of the sample is driven beyond its saturation limit, whereby the sample is forced out of solution and crystal formation may result.

When crystals are grown from solution, changing the solvent may have a pronounced effect on their habit and size. Properties that may influence crystal growth, such as density, viscosity, dielectric constant, and solubility, may be varied over a wide range by mixing two or more solvents.

A small surface-to-volume ratio is useful for slow evaporation of solvent. For small samples, an NMR (nuclear magnetic resonance) tube may be a suitable crystallization vessel.

Very insoluble compounds may be crystallized using a method known as *reactant diffusion*. In this method, reactants A and B are allowed to mix by diffusion; the very insoluble product C will crystallize in the zone of mixing. Sparingly soluble compounds can sometimes be crystallized from boiling solvent in a soxhlet extractor. Crude product is placed in the thimble and the reservoir is seeded.

Sublimation is effective for some classes of compounds. Vacuum sublimation reduces the temperature required and so increases the range of compounds for which it is suitable. If the temperature gradient is too large, only microcrystals will be formed. Large crystals grow at the expense of small ones only when the process is carried out slowly.

Vapor diffusion is a method that works well with milligram quantities. The solute is dissolved in a solvent in which it is relatively soluble. A small container of this solution is placed inside a closed beaker with a second solvent in which the solute is only sparingly soluble. The two solvents must be miscible in one another and the second solvent should be the more volatile. Suitable solvent pairs include ethanol/ether, benzene/ligroin, and water/ethanol. Diffusion of one liquid into another is also effective. The solute is dissolved in the solvent in which it is more soluble. Crystals form at the interface between the two solvents.

A form of vapor diffusion is commonly used for the growth of protein crystals. For proteins (and other macromolecules) the solution properties modified to achieve supersaturation include increasing the concentration of an additive (e.g., a precipitant), decreasing total solution volume, changing the solution pH, and/or changing the temperature. Although a wide variety of experimental setups have been used in protein crystallization, the most common technique is "hanging-drop microvapor diffusion" (HDMVD). In HDMVD, a droplet of 4 to 20 μl containing protein and precipitant is suspended from a glass coverslip which is sealed above a reservoir of a solution at a higher precipitant concentration. Since the droplet is at a lower precipitant concentration than the reservoir, the net migration of water vapor occurs from the droplet to the reservoir, resulting in a decrease in drop volume. The decrease in drop volume results in increased precipitant and protein concentrations, which should drive the protein out of solution. Crystals form if conditions are favorable. More often, protein precipitates as an amorphous solid along the bottom of the droplet.

Once initial conditions are found where crystals have formed, additional experiments often must be performed to perfect crystal growth in order to produce X-ray-diffraction-quality crystals. The crystals must be single, with no satellite growths or twinning, and on the order of 0.1 mm on each edge. Fine-tuning conditions, adding detergents or counterions to the precipitant, and seeding droplets with microcrystals are all techniques used to prepare large, single crystals from initial successful experiments.

B. Microscopic Examination

The crystal chosen for analysis must have a uniform internal structure and be of an appropriate size and shape. The first criterion implies that the substance is pure—the crystal contains no voids or inclusions and is not bent, cracked, distorted, or composed of crystallites. It must be a single crystal, and ideally it should not be twinned (a twinned crystal has two or more different orientations of the lattice growing together).

The size of crystal required is determined by the conditions of the experiment. For X-ray diffraction, approximately 0.1 mm is preferred; for neutrons, an order of magnitude larger is appropriate. A crystal with roughly equal dimensions and well-defined edges is ideal; however, many crystals grow as plates or needles. Although a crystal with a highly asymmetric shape is far from ideal, useful structural information can often be obtained from such a crystal. In some cases, a more uniformly shaped fragment may be cut from a larger crystal with a sharp razor blade.

Examination with a binocular microscope allows a rapid screening of crystals. A few that appear suitable should then be examined more closely through crossed polarizers. As the crystals are rotated, they should either appear uniformly dark in all orientations or they should be bright and extinguish (appear uniformly dark) every 90° of rotation. An unsuitable crystal may show dark and light regions simultaneously, or regions that do not extinguish, or different regions that display different colors.

C. Mounting a Crystal

The usual way of mounting a crystal for X-ray diffraction is to glue it to the end of a glass fiber that is mounted in a brass or aluminum pin. For photographic work, it is necessary to align a real lattice vector (Weissenberg technique) or a reciprocal lattice vector (precession technique) so that it is perpendicular to the X-ray beam. On a diffractometer, it is necessary for the crystal to be well centered in the X-ray beam, but alignment is neither necessary nor advisable since the aligned position is the one that often leads to overlapping reflections.

Materials that are air or moisture sensitive may be stabilized by covering the crystal with a light coating of oil or the mounting glue (provided that the crystal is not soluble

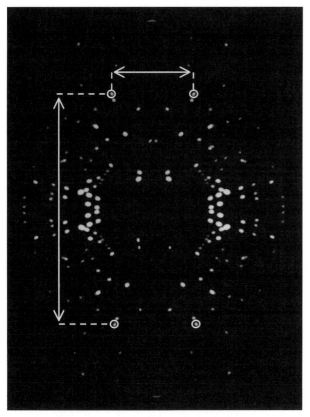

FIGURE 5 Polaroid rotation photograph.

in the solvent of the glue). Very sensitive substances may have to be mounted in capillary tubes under an inert atmosphere, with all mounting operations carried out in a dry box.

D. Preliminary Photographs

Most diffractometers are equipped with a Polaroid camera for a quick check of crystal quality. Alternatively a rotation photograph can be acquired with an area detector. The spots on the photograph should have similar shapes, without tails or streaks (see Fig. 5). A more sensitive check of crystal quality is provided by omega scans of a number of reflections measured with different orientations of the crystal. If the diffraction pattern falls off rapidly with angle, low-temperature data collection is advisable; however, some crystals crack when cooled. If suitable crystals are scarce, it may be advisable to collect a dataset at room temperature and then cool the sample to improve the resolution of the experiment

E. Establishing the Orientation Matrix

The orientation matrix and the cell dimensions are established by determining the setting angles of about eight to ten reflections. These may be found using information available from the preliminary photograph or by allowing the computer-controlled diffractometer to search for them. The indices of these reflections are then determined (usually with a computer program). The cell dimensions are then refined by least squares. High-angle reflections are most sensitive, but fairly strong reflections are required. For this reason, it is common to use relatively low-angle data to establish a preliminary unit cell and matrix, and to recalculate the matrix when the intensities of the high-angle data have been established.

F. Data Collection

It is always advisable to collect symmetry-equivalent data (reflection intensities). The degree of agreement between equivalent reflections allows assessment of crystal quality, absolute configuration, stability of the counting chain, suitability of absorption correction, and other systematic effects. For charge density studies, a complete dataset should be collected to a resolution of at least $\sin\theta/\lambda = 1.3$ Å^{-1} with an internal agreement of about 2%. For determination of absolute configuration, effects of decomposition, absorption, and extinction errors are minimized if hkl and $\overline{hkl}$ are measured consecutively at $+2\theta$ and -2θ. This may be done for a few dozen of the most sensitive reflections or for the whole dataset. For large datasets, it is common practice to collect data in shells.

G. Data Reduction

The process of deriving structure amplitudes $|F_{hkl}|$ from the observed intensities I_{hkl} is known as *data reduction*. A number of geometrical factors influence the intensities observed in a diffraction experiment. The most important of these are Lorentz, polarization, absorption, and extinction corrections.

The first three corrections are normally applied to the observed intensities in the process of calculating the structure amplitudes, $|F_{hkl}|$. The Lorentz factor corrects for the relative speeds with which different reflections pass through the reflecting position. Since X-rays are polarized on reflection, and the degree of polarization depends on experimental conditions, a correction must be applied to account for how the polarization affects the observed intensities. Frequently, Lorentz and polarization corrections are combined. Absorption depends on the average path-length of both the incident and the reflected beam and hence may be very different for symmetry-related reflections.

Extinction is an interference process. An extinction correction is generally made during least-squares refinement, making it a correction applied to the model rather than to the observations. A detailed discussion of the absorption, extinction, and other factors affecting diffraction intensity can be found in Appendix I.

H. Solving the Phase Problem

The central problem in crystallography arises because the experimental data yield only the modulus of the structure factor, $|F_{hkl}|$ and not the phase. The phase is required in order to evaluate the electron density in the unit cell, but it cannot be measured directly.

Several methods have been developed to determine the phases of the complex structure factors, $|F_{hkl}|$, with no prior knowledge of atomic positions. These methods include multiple isomorphous replacement, single-isomorphous replacement with anomalous dispersion, multiple-wavelength and single-wavelength anomalous dispersion, heavy-atom, and direct methods. With the exception of direct methods, all of these methods take advantage of the scattering properties of "heavy" atoms (transition, actinides, and lanthanides). For very large molecules, such as proteins, neither direct nor heavy-atom methods are generally used. Molecular replacement (MR) also provides a powerful phasing method for the structure analysis. A model structure similar to the structure being analyzed is required for MR.

If the positions of the atoms are known, both the magnitude and phase of the structure factor can be calculated. The heavy-atom method of structure solution depends on this. If at least one atom in the structure is heavy enough to be located in a Patterson function, that position can be used to calculate phase angles. A Fourier summation using observed structure amplitudes and these calculated phases would reveal the heavy atom and some others. The additional atoms are included in the structure factor calculation, providing a better estimate of the phase angles. This process can be facilitated by use of the tangent formula to extend and refine the new phases and is repeated until all of the atoms are located.

Multiple isomorphous replacement (MIR) is a powerful method for determination of phases. MIR depends upon the phasing power of heavy-metal atoms bound to a compound in such a fashion that the positions of other atoms in the crystal are minimally perturbed (i.e., isomorphous derivatives). The MIR method requires that the crystallographer prepare more than one derivative of the parent crystal. These often turn out to be unstable, not isomorphous, or to have the metal bound with too low an occupancy to be useful. Even when successful, this method requires the collection and reduction of numerous datasets from multiple crystals. Anomalous dispersion methods can be combined with isomorphous replacement to circumvent some of the disadvantages.

Advances in the phasing of macromolecular data have been made by the use of a phenomenon called *anomalous dispersion* or *anomalous scattering*. Single isomorphous replacement with anomalous scattering (SIRAS) takes advantage of both the phasing power and anomalous scattering properties of certain heavy atoms. An advantage of the SIRAS technique is that data, in some cases, can be collected using a conventional Cu-Kα radiation source. The SIRAS approach to phasing obviates the need for crystallization and data collection from multiple samples. The technique does require careful data collection but produces all of the information needed to determine phases from a single dataset collected from only one crystal.

A crystal containing an anomalous scattering atom may be used to collect data at multiple wavelengths and allow the phases to be determined using multiple-wavelength anomalous dispersion (MAD). A single crystal can be analyzed at multiple wavelengths, generating a variation in scattering factors that allows direct determination of crystal structures. With careful measurements, even weak signals from a single crystal can provide the necessary phasing information. Data must be collected using "tunable" X-ray radiation available only at synchrotron facilities.

A paper by J. Karle (see Bibliography) offers the possibility of nearly direct phasing for protein crystals. The potential for analysis of the data using this single-wavelength anomalous dispersion (SAD) technique is yet to be explored. Given the collection of sufficiently accurate anomalous dispersion data, one dataset at one wavelength may provide all the information required to determine the phases.

Direct methods of phase determination do not depend on any *a priori* structural information. Phases are determined from statistical relationships among the intensities. Powerful computer programs have facilitated the use of these methods for centrosymmetric and noncentrosymmetric structures of moderate size. Considerable effort has been made to extend these methods to protein structures, and structures containing over 1000 non-hydrogen atoms have been solved by direct methods.

In general, the conformationally rigid portions of the molecule are most easily located. Once the major features of the molecule have been recognized, the difference map provides a powerful means both for checking the accuracy of the partial model and for completing it. A Fourier summation using $\Delta F = |F_{obs} - F_{calc}|$ as coefficients and phases calculated from the known portion of the structure yields a map of the discrepancies between the crystal and the model. An atom present in the crystal but not included in the model will appear as a peak. Difference maps are also useful for locating light atoms in the presence of heavy ones, such as the location of hydrogen atoms in compounds of first-row elements or carbon atoms in a tungsten compound.

I. Refinement

Once approximate locations have been determined for all, or almost all, of the atoms in the structure, it must be refined or made more precise. The standard method for refinement of structures, known at *atomic resolution* (not for macromolecules), is full-matrix least squares. For structures that are not known at atomic resolution, or for very large structures, sparse-matrix or simulated annealing is used for refinement. Since the structure factors are not linear functions of the parameters, the process is an iterative one. Hydrogen atoms are not generally included in the early stages of refinement but are included in the final cycles of refinement.

The function minimized is $\sum \omega \Delta^2$, where ω is the weight assigned to a particular observation and Δ is the difference between the observed and calculated values of $\boldsymbol{F}_{hkl}$ (for a refinement based on $|\boldsymbol{F}|$) or $\boldsymbol{F}_{hkl}^2$ (for a refinement based on $|\boldsymbol{F}|^2$). A convenient parameter, referred to as the R-factor, for monitoring the progress of the refinement is

$$R = \sum (||F_o| - |F_c||) \Big/ \sum |F_c|$$

Correct structures generally have R-values under 0.10, and those that are well behaved are frequently under 0.05.

J. Determination of Absolute Configuration

If a particular torsional angle has a positive sign in the right-handed enantiomer, it will have a negative sign in the left-handed molecule. Thus, determination of the absolute configuration in a chiral molecule can be regarded as the determination of the correct signs for the torsional angles. A torsional angle ABCD is positive if a clockwise rotation will cause the bond AB to eclipse the bond CD.

Determination of absolute configuration by X-ray crystallography requires a structural study in the presence of dispersive scatterers. Thus, if at least one atom in the structure is an anomalous (i.e., dispersive) scatterer, Friedel's law breaks down and reflections from two sides of the same plane are no longer equal. The differences in intensity are generally small, so careful measurement is required. Coster et al. demonstrated that a noncentrosymmetric crystal structure could be distinguished from its inverted image using these differences. Later, Bijvoet realized this principle was more general and used it to determine the absolute configuration of the sodium rubidium salt of (+)-tartaric acid. In 1983, Flack developed a method for distinguishing a noncentrosymmetric structure from its inverse. Using the method of Flack, any noncen-

trosymmetric crystal is treated as a twin by inversion and the contribution of the two components evaluated during refinement as the Flack parameter. In the case where the arrangement of atoms in the model and crystal are in agreement, the contribution of the Flack parameter is zero. If the model and crystal are inverted with respect to each other, the Flack parameter is one and the model needs to be inverted.

K. Derived Parameters

The parameters produced directly by the least-squares refinement are the positions of the atoms and their thermal parameters. Bond lengths, bond angles, and torsional angles are derived from these positions. An examination of short inter- and intramolecular contacts may provide information about hydrogen bonding, van der Waals forces, packing forces, etc.

Room-temperature studies of organic compounds generally show appreciable thermal motion with U_{ij} values of the order of 0.04 Å^2. This corresponds to a root-mean-square vibration amplitude of 0.2 Å and is a reminder that some caution is required in comparing bond lengths from diffraction experiments with those determined by spectroscopic and theoretical work. Bond lengths can sometimes be "corrected" for thermal motion, but it is generally preferable to reduce thermal motion by cooling the crystal. Many investigators routinely collect data at temperatures in the range of 100 to 200 °K.

A single crystal structure determination provides valuable information on chemical connectivity, relative conformation, and, under the proper experimental conditions, absolute configuration. However, an understanding of structure–function relationships requires correlating features from a number of different structures. The existence of computer-searchable databases of structural data greatly enhances the possibilities for such comparisons (see Section VI).

V. COMPARISON OF X-RAY AND NEUTRON DIFFRACTION

For a structure with N atoms, each with atomic scattering amplitude f_i and position $\mathbf{r}_i$ in the unit cell, the structure factor for the Bragg reflection of index $\mathbf{h}$ is

$$F(\mathbf{h}) = \sum_{i=1}^{N} f_i(\mathbf{h}) T_i(\mathbf{h}) \exp(2\pi i \mathbf{h} \cdot \mathbf{r}_i)$$

where T_i is the temperature factor of the ith atom. Differences between neutron diffraction and X-ray diffraction lie in the scattering amplitudes $f_i(\mathbf{h})$ and in the temperature parameters $T_i(\mathbf{h})$. In X-ray diffraction, X-rays are scattered by the electrons of an atom, and the scattering factors $f_i(\mathbf{h})$ are strongly dependent on scattering angle. At a scattering angle of $0°$, f_i is proportional to Z, the atomic number of the scattering atom. As the scattering angle increases, the scattering factor $f_i(\mathbf{h})$ decreases. Thermal motion causes the scattering to fall off even more strongly. These factors limit the resolution available in an experiment and make it difficult to determine the positions of light atoms accurately in the presence of heavy ones or to distinguish among heavy atoms that have similar atomic numbers.

In neutron diffraction, the scattering is primarily from the nucleus. Since the diameter of the nucleus is small relative to the wavelength of thermal neutrons, the scattering factor is a constant, characteristic of the particular nucleus and independent of scattering angle. There is no simple relationship between the scattering amplitudes and the nuclear mass or charge. Nuclei with similar atomic number can have significantly different scattering amplitudes. Hence, neutron diffraction can distinguish among near neighbors in the periodic table and so is useful in the study of alloys.

Table II shows the relative scattering lengths of a number of atoms for X-rays and neutrons. If the average contribution to the intensity of a structure factor is

$$P(i) = f_i^2 \bigg/ \sum_{i=1}^{n} f_i^2$$

Then, for a compound such as benzene with equal numbers of hydrogen and carbon atoms, carbon will contribute 97% and hydrogen will contribute 3%. Thus, in the X-ray ex-

periment, positions of hydrogen atoms will be more poorly determined than positions of carbon atoms. In the neutron experiment on the same compound, hydrogen would contribute 24% and carbon 76%. (Because of the negative sign on the scattering length of hydrogen, a Fourier summation will show "holes" rather than peaks at hydrogen positions.) For deuterobenzene, the contributions from C and D (in the neutron-scattering experiment) are virtually identical, with each contributing 50% of the scattering. Thus, neutron diffraction can locate hydrogen and deuterium atoms with the same precision as carbon, nitrogen, and oxygen, while X-rays cannot. For this reason, studies of hydrogen bonding in biologically significant compounds such as amino acids and sugars were among the early experiments using neutron diffraction.

As the precision of the experiments and the sophistication of the refinements improved, it became obvious that there were systematic differences in the positional and thermal parameters from the two experiments that were much larger than expected from the estimated standard deviations. The differences are very pronounced for hydrogen atoms. Even in the most precise, low-temperature studies, electron density maxima for hydrogen atoms were displaced by as much as 0.2 Å from the positions of protons determined from neutron diffraction. Since the C–H and O–H bond lengths in the neutron experiment agree with spectroscopic measurements, it was recognized that the apparent shortening of the bonds to hydrogen observed in X-ray experiments is a bonding effect. The position of maximum electron density does not coincide with the position of the nucleus because the formation of the covalent bond perturbs the electron-density distribution in the atom.

Similar, but smaller, effects are observed for first-row atoms, C, N, O, etc. Typically, the discrepancy is of the order of 0.01 Å, but it depends critically on the range of Bragg angles included in the X-ray refinement. If the refinement is based on very-high-order data $[(\sin\theta)/\lambda > 1.00 \text{ Å}^{-1}]$, the discrepancies will be much reduced; however, few organic compounds scatter to such high angles even at liquid-nitrogen temperature.

Systematic differences in temperature factors also reflect the very real differences in the scattering processes in the two experiments. In aromatic molecules, thermal vibration parameters are greater for X-rays than for neutrons in the plane of the ring and smaller perpendicular to the ring, implying that electron density is smeared in the plane of the ring by covalent bonding and contracted in the perpendicular direction.

Atoms with lone pairs of electrons show significant differences in both positional and thermal parameters in the two experiments because the lone-pair density is not centered on the nucleus. These very real differences suggest strongly that neutron diffraction is the preferred tool

TABLE II Scattering Amplitudes for Selected Elements $(10^{-12}$ cm$)^a$

Element	Atomic number	Scattering amplitude	
		X-ray	Neutron
^{1}H	1	0.28	−0.37
^{2}D	1	0.28	0.67
^{12}C	6	1.69	0.66
^{14}N	7	1.97	0.94
^{16}O	8	2.26	0.58
^{32}S	16	4.51	0.28
Cl^b	17	4.79	0.96
Br^b	35	9.87	0.68
^{127}I	53	14.95	0.52
^{238}U	92	25.92	0.84

^{a} For X-rays, the scattering amplitude at $0°$ is given by $(e^2/mc^2)f_0$ or $((0.282 \times 10^{-12}$ cm$) \times$ atomic number). Values for neutrons are taken from "International Tables for X-Ray Crystallography," Vol. IV, 1974, pp. 270–271.

^{b} These values are for the elements in their natural isotopic abundances.

for determining atomic parameters, whereas the X-rays measure the electron density in the solid. The combination of the two techniques provides a means of studying bonding effects (this topic was covered in greater detail in Section II.B.3).

VI. RESULTS

The determination of a single crystal structure may answer a question about the connectivity of a molecule or some detail of its conformation. The direct results of structure analysis are the positional, thermal, and occupancy parameters of atoms in the asymmetric unit. Bond lengths and angles, torsion angles, and intermolecular associations (such as hydrogen bonding) are all derived from these basic structural parameters. While the final R-value that a structure refines to is often a good indicator of the quality of a structural determination, a plot showing the thermal ellipsoids (Fig. 6) can also give an indication of the quality of a structural determination. Errors in the structure determination and thermal motion in the molecule can distort these ellipsoids.

Understanding a complex process (such as the mechanism of a reaction, the biological activity of a class of drugs, the phenomenon of one-dimensional conduction, or any other structure–property relationship) requires the detailed analysis of a large number of structures and correlation with results from other disciplines. Databases provide the kind of information required for structure–function studies and are very valuable to the scientific community. The topics presented here are intended to be illustrative of the many ways that the databases have been used. There are many other uses not documented here.

A. Crystallographic Databases

At present there are eight major databases for crystallographic results continuously maintained and updated in different laboratories in Europe and North America. In each case the data are available in machine-readable form, and considerable effort has been expended to develop efficient computational algorithms for searching the files and correlating the data. Perhaps the most important feature of the databases is that the data have been checked and errors are corrected when possible or flagged if uncorrected. Some of these databases have been incorporated into commercial packages. Some are available in both printed and computer-readable form. In the brief summary that follows, they are listed in alphabetical order.

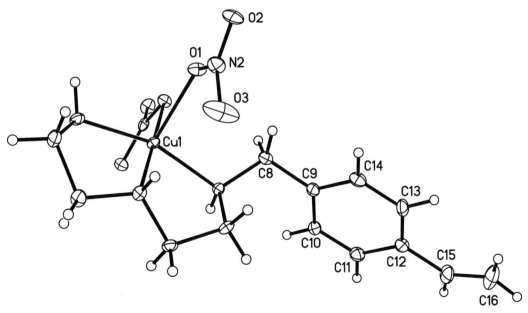

FIGURE 6 Thermal ellipsoid plot from a crystallographic structure determination. Hydrogen atoms are plotted as small balls with an arbitrary radius; non-hydrogen atoms are plotted as ellipsoids, with the axes corresponding to the thermal parameters for the atom. Notice that the ellipsoids gradually increase in size and become more asymmetric moving out the chain from C8 to C16. The ellipsoids can become large and more asymmetric for noncovalently bounds groups such as the coordinated nitro group (N_2, O_1, O_2, and O_3). In both of these cases, the axes of the ellipsoids are aligned with expected vibrational modes and do not indicate any major problems with the structure. If the axes were poorly aligned and/or the ellipsoids more asymmetric, this could be an indication of problems with the structural determination.

1. Cambridge Structural Database

(http://www.ccdc.cam.ac.uk/prods/csd/csd.html; Cambridge Crystallographic Data Center (CCDC), University Chemical Laboratory, Lensfield Road, Cambridge, CB2 1EW, U.K.) The Cambridge Structure Database (CSD) is the largest database of experimentally determined organic and metallo-organic crystal structures in the world; inorganic carbon compounds (such as carbonates and cyanides) are excluded. The CCDC also provides a suite of graphical search, retrieval, data manipulation, and visualization software for use with the database. The CSD contains bibliographic data, tables of connectivity, atomic positions, cell dimensions, and quality indicators for virtually all three-dimensional structures of organic compounds published since 1935. The CSD currently contains over 210,000 entries and is growing at a rate of approximately 15,000 entries per year. The CSD is available to scientists throughout the world.

2. Crystal Data

(http://www.nist.gov/srd/nist3.htm; Crystallographic Section, National Bureau of Standards, Washington, D.C. 20234.) This database contains lattice parameters for all crystals whose dimensions have been reported by X-ray, neutron, or electron diffraction on single crystals or fully indexed powders. Data include name, formula, cell dimensions, space groups, number of molecules in the cell, density (measured and calculated), bibliographical data, crystal habit, melting point, etc. Crystal Data accepts data from the other databases.

3. Electron Density Data Base

(Prof. H. Burzlaff, Lehrstuhl fur Kristallographie, Institut fur Angewandte Physik der Universitat, Bismarckstrasse 10, D-91054 Erlangen, Germany.) This database contains accurate structure factors for crystal structures whose electron densities have been carefully determined. This is the type of data required for studying bonding effects, covalency in organometallics, and other details of electron distribution.

4. Inorganic Crystal Structure Database

(http://crystal.fiz-karlsruhe.de/portal/cryst/ab_icsd.html.) The Inorganic Crystal Structure Database (ICSD) was initiated in 1978 at the Institute for Inorganic Chemistry at the University of Bonn. Today the database is produced by FIZ Karlsruhe (P.O. Box 2465, D-76012 Karlsruhe, Germany) in cooperation with NIST. "Inorganic" is defined to exclude metals, alloys, and compounds with C–H and C–C bonds (with the exception of graphites).

The data stored include chemical name, chemical formula, density, lattice parameters, space groups, atomic coordinates, oxidation state, temperature factors, remarks regarding conditions of measurement, R-values, and bibliographical references. Online access to the file is available. It is also possible to lease the entire database. The database now contains over 53,000 entries and is updated twice a year.

5. Metals Data File

(Manager, CAN/SND, Canada Institute for Scientific and Technical Information, National Research Council (NRC) of Canada, Ottawa, Canada K1A OS2.) The Metals file contains structural data for metal and alloy structures determined since 1913 based on either powder or single-crystal diffraction. Under an exclusive license from the NRC, Toth Information Systems has maintained and updated the database. If available, the following information is included: formula, cell dimensions, structure type, Pearson symbol, atomic coordinate, temperature parameters, occupancy factors, R-values, method of refinement, instrument used, radiation, and bibliographical information. New software for manipulating the file has been developed by Toth Information Systems.

6. Nucleic Acid Database

(http://ndb-mirror-2.rutgers.edu/NDB/ndb.html; Dr. H. M. Berman, Department of Chemistry, Rutgers University, 610 Taylor Road, Piscataway, NJ 08854-8087.) The goal of the Nucleic Acid Database (NDB) project is to assemble and distribute structural information about nucleic acids. Online access to the NDB is freely available. A variety of tools have been developed in conjunction with the NDB to provide a robust user interface.

7. Powder Diffraction File

(http://www.icdd.com/_productsold/pdf2.htm; International Center for Diffraction Data (ICDD), 1601 Park Lane, Swarthmore, PA 19081.) The Powder Diffraction File (PDF) is a compilation of powder diffraction patterns produced and published by the Joint Committee on Powder Diffraction Standards of the ICDD. The PDF is the world's largest and most complete collection of X-ray powder diffraction patterns. The 1999 release of the PDF contains about 115,000 patterns, of which 20,000 are organic and 95,000 inorganic. Included in the database are calculated powder patterns for almost 40,000 compounds. The PDF is used for identification of crystalline materials by matching d-spacings and diffraction intensity measurements. Each pattern includes a table of interplanar d-spacings, relative intensities, and Miller indices, as well as additional helpful information such as chemical formula, compound name, mineral name, structural formula, crystal system, physical data, experimental parameters, and references.

8. Protein Data Bank

(http://www.rcsb.org/pdb/; Dr. H.M. Berman, Department of Chemistry, Rutgers University, 610 Taylor Road, Piscataway, NJ 08854-8087.) The Protein Data Bank (PDB) is the single international repository for the processing and distribution of three-dimensional macromolecular structure data (primarily determined experimentally by X-ray crystallography and NMR). In June of 1999, Brookhaven National Laboratory ceased its operation of the PDB when the Research Collaboratory for Structural Bioinformatics (RCSB) took over. The RCSB operates under a contract from the U.S. National Science Foundation with additional support from the Department of Energy and two units of the National Institutes of Health. Contents of the PDB are in the public domain, but the original work as well as the PDB should be properly cited whenever referred to. Affiliated centers in Australia, England, and Japan undertake distribution of data in their respective areas.

B. Structure Correlation

The working hypothesis behind structure-correlation studies is that changes observed in a structural fragment or subunit in a number of different environments occur along a potential-energy valley in the parameter space of that fragment. Each observed structure is a sample point. An example of this type of study examines the role of iodine in the binding of thyroid hormone. Since iodine-containing thyroid hormones are protein bound during most of their metabolic lifetime, attractive interactions with nucleophiles could play an important role in that binding. Short contacts between iodine bound to carbon and nucleophiles (such as O, N, and S) were studied in an effort to better understand these interactions. The shortest contacts are essentially linear with the lone pair of the nucleophile directed towards the C–I vector. Such contacts are similar to hydrogen bonds and have been estimated to contribute attractive energy of about 3 kcal/iodine atom. The importance of such interactions is further supported by the observation of a short (2.96 Å) I–O contact in the crystal structure of pre-albumin with bound thyroxine.

C. Reaction Coordinate

A chemical reaction can be represented by a plot of energy as a function of reaction coordinate. Figure 7 shows

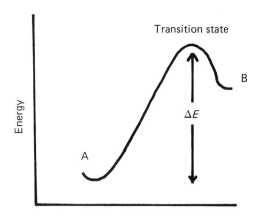

FIGURE 7 Reaction coordinate. A plot of energy as a function of reaction coordinate for molecule A yielding product B.

the starting material A going to product B through a transition state with activation energy of ΔE, which can be obtained from kinetic data. It is fairly obvious that we can determine the structures of the compounds A and B. Less obvious is that the databases provide a means of looking at the reaction path in some detail. Each individual structure provides a snapshot at one point along the reaction coordinate, but a whole family of structures can plot out a curve related to the potential-energy surface of the reaction.

An early example of this approach was provided by the study of the interactions between amino and carbonyl groups in nucleophilic addition reactions. By examining the data from six crystal structures, Burgi, Dunitz, and Schefter were able to show that interaction with the nucleophile causes the carbon of the carbonyl group to be displaced from the plane of its three substituents toward the approaching nucleophile. The direction of approach of the lone pair on the nucleophile is at an angle roughly 109° to the carbonyl bond (not perpendicular to the plane of the carbonyl). In addition, the displacement of the carbonyl carbon out of the plane of its substituents yields a smooth curve when plotted as a function of the observed C–N separation.

As further confirmation of the validity of this approach, the general conclusions have been reinforced by comparison with SCF-LCGO calculations on the system $CH_2O + H^- \rightarrow CH_3O^-$. The calculated reaction path for the nucleophilic attack of hydride ion on formaldehyde shows very close resemblance to the one predicted by the method of structure correlation.

A different type of correlation is provided in a set of papers published in 1984 by Kirby and colleagues on the length of the C–O bond. This investigation was prompted by the observation that C–O bond lengths for acetals showed an unusually broad range of values. To determine whether this was a general phenomenon or one peculiar to acetals, the geometry of nearly 2400 ethers and esters was investigated. The results show clearly that there is substantial, systematic variation in C–O bond lengths. The fragment of interest is defined as R_1–O–R_2. Data are divided into four categories, depending on whether R_1 is methyl, primary, secondary, or tertiary, and each of those categories is then divided according to the effective electronegativity of the group R_2. The four categories of R_2 are alkyl, aryl, enol/ether, and acyl/carboxylic ester, giving 16 categories in all. The shortest C–O bonds (1.418 Å) are found in compounds where R_1 is methyl and R_2 is alkyl. Within each of the four categories of R_1, the length of the R_1–O bond increases with increasing electronegativity of the R_2 group. The effective electronegativity of a group can be estimated from the pK_a of its conjugate acid. In the subset of 2-(aryloxy)-tetrahydropyrans, there is a linear relationship between the length of the exocyclic C–O bond and the pK_a of the leaving group. This implies a linear relationship between the bond length and the free energy of activation for the hydrolysis reaction or any other reaction in which the C–O bond is cleaved. The consequences of these generalizations to the chemistry and reactivities of acetals and glycosides are fully explored in these papers. In the third paper in the series, the authors plot the reaction coordinate for six aryl tetrahydropyranyl acetals. Both relative free energy and pK_a are plotted as a function of the bond length. The Morse function is also plotted, as is a reaction coordinate–energy contour diagram. This work by Kirby et al. is one of the most comprehensive attempts at deriving structure–reactivity relationships so far available.

D. Drug Design

The coupling of the resources of databases with molecular graphics devices and QSAR (quantitative structure–activity relationships) techniques raises a tantalizing possibility that therapeutically useful new structures might be predictable. Detailed comparison of structures of known agonists and antagonists is the major strategy currently in use. Similarities in the three-dimensional structure of portions of the molecules allow identification of the features required for binding. Differences in other portions of the structure may account for the agonist/antagonist response after binding. Once the pharmacophore has been identified, binding studies can be combined with structural comparisons to map the receptor site.

A sum or superposition of active molecules can be used to define the available volume within a receptor. Inactive molecules then define excluded volumes—volumes occupied either by the receptor itself or by a cofactor. Any new drug must be designed to present the correct pharmacophore and to occupy only the available volume.

Examples of this include mapping of the methionine binding site in the enzyme S-adenosyl ATP transferase by Marshall and colleagues, and the postulating of a common site of action for gamma-butyrolactone analogs and picrotoxinin. On the basis of this model, it has been proposed that the convulsant activity of these compounds is related to their ability to block the passage of chloride ions through channels. The model would appear to be applicable to a number of convulsant and anticonvulsant drugs.

E. Crystallography and Molecular Mechanics

Molecular mechanics is an empirical method for calculation of properties of molecules such as molecular geometry, heat of formation, strain energy, dipole moment, and vibrational frequencies. Different programs use different parameter sets and reproduce these physical properties with different degrees of fidelity. Parameters are assumed to be transferable from one type of molecule to another.

The geometrical parameters used in molecular-mechanics programs are frequently derived from crystallographic data. However, the predictive value of the method is limited by the datasets used to derive the parameters. For instance, a parameter set derived from uncrowded hydrocarbons is not likely to predict structures of crowded hydrocarbons with satisfactory accuracy. Transition states, small rings, unusual states of hybridization, and other electronic effects may require special treatment. These caveats notwithstanding, the method has enjoyed considerable success.

It is not uncommon for molecular-mechanics calculations to be used to provide a starting point for the refinement of electron diffraction data. A recent example is provided by the study of bicyclo[3.2.0]heptane. Many other examples are available in the literature.

Molecular mechanics can also be useful in interpretation of crystal structures, particularly in differentiating electronic and steric effects or in estimating the effects of packing forces. In the structure of lepidoptrene, a C–C bond was observed to have a length of 1.64 Å in the X-ray experiment, while molecular mechanics predicted a length of 1.57 Å. Mislow and colleagues demonstrated that the additional lengthening of the bond beyond the amount expected from steric strain was caused by "through-bond" coupling of adjacent π systems.

Although molecular-mechanics calculations can be useful in assessing which aspects of a structure control conformation, it should be pointed out that conformational parameters from molecular mechanics tend to be less reliable then bond lengths or bond angles. Also, conformational parameters in crystal structures are most influenced by packing forces. Molecular-mechanics calculations have been used to suggest alternative conformations (having energy similar to that observed in the crystal) that might exist in the gas phase or in solution. Additionally, molecular-mechanics has been used to optimize the geometry of a pharmacophore in model studies of drug–receptor binding, to evaluate the interaction energies between dinucleoside monophosphates and cationic intercalators such as ethidium bromide, and to interpret conformational polymorphism. Although some caution is obviously necessary, molecular mechanics and crystallography can provide complementary information in a variety of cases.

APPENDIX I: FACTORS AFFECTING INTENSITIES

A number of geometrical factors influence the intensities observed in a diffraction experiment. The most important of these are Lorentz, polarization, absorption, and extinction corrections. The first three corrections are normally applied to the observed intensities in the process of calculating the structure amplitudes $|F_{hkl}|$. Many researchers feel that the absence of this correction is the single largest source of systematic error in crystal structures in the current literature. Extinction is a correction made during least-squares refinement and so is a correction applied to the model rather than to the observations. However, the correct formulation for extinction depends on both the polarization and the path-length (calculated during absorption correction) and so is discussed with these other correction terms.

A. Absorption

The absorption of X-rays by crystals obeys the relation $I = I_0 \exp(-\mu t)$, where μ is the linear absorption coefficient, in units of (length)$^{-1}$, and t is the thickness. The general effect is to reduce the intensity of reflections at low $\sin \theta$. If the crystal is centrosymmetric in cross-section, neglect of this correction will have little effect on positional parameters, although scale and thermal parameters may be very strongly affected. If the crystal cross-section is not centrosymmetric, all structural parameters will be systematically wrong. Neglect of absorption is probably the largest single source of systematic error in published structures.

Absorption corrections can be done by collecting φ-scans and calculating an empirical correction or by indexing and measuring crystal faces and calculating a face-indexed absorption correction. For a face-indexed correction, the size and shape of the crystal must be precisely determined. Once the crystal shape is established, a number of techniques are available for calculating the integrated path-length for incident and diffracted beams for all reflections.

For crystals with no reentrant angles, the most common methods are the analytical method and the method of Gaussian quadrature. In both cases, the objective is to evaluate the integral,

$$A = \frac{1}{V} \int_{\text{crystal}} \exp(-\mu T)\, dv$$

This can be done analytically if the crystal is divided into a number of polyhedra in each of which the path-length is a linear function of the coordinates. De Meulenaer and Tompa first programmed this method in 1965. Calculation time is independent of the severity of absorption. The factor-limiting accuracy in cases of severe absorption is the precision of measuring the crystal, particularly in its shortest direction.

Gaussian quadrature is a numerical integration method that evaluates an integral by summing an appropriate polynomial. It uses a nonisometric grid in which the interval is subdivided symmetrically about the midpoint, with large spacings near the middle and smaller ones toward the edges. This tends to put the maximum number of grid points near the surface of the crystal where the change of absorption with path-length is largest. The number of grid points determines the precision of the calculation. Thus, if a $4 \times 4 \times 8$ grid gives a precision of 2% in calculated transmission for $\mu = 2.5$ cm^{-1}, a grid of $8 \times 8 \times 16$ is required to produce the same precision if $\mu = 5.0$ cm^{-1}. By choosing sufficient points, the Gaussian method can reproduce the analytical result to any desired precision; however, for strongly absorbing crystals, the analytical method is the method of choice.

When a crystal has reentrant angles, neither of these methods can be used, and one possible resort is numerical integration with an isometric grid; however, the grid must be very fine in order to achieve reasonable precision. For crystals mounted in capillaries, crystals with irregular shapes, and other pathological cases, absorption may be dealt with by measurement of a transmission surface as proposed by Huber and Kopfmann in 1969. However, these problems can be dealt with by the program DIFABS, developed by Walker and Stuart in 1983, which models the absorption surface by a Fourier series in polar coordinates. The coefficients are obtained by minimizing the sum of squares of residuals between observed and absorption-modified values of the structure factors. The chief virtue of the method is that it can be used even when the crystal is no longer available

B. Lorentz Factor

The Lorentz factor corrects for the relative speeds with which different reflections pass through the reflecting position. The intensity of a reflection produced by a moving crystal depends on the time taken for the corresponding reciprocal lattice point to pass through the sphere of reflection. Using $1/(\omega S \cos\theta)$ as the angular velocity, the definition $S = 1/d$, and Bragg's law, $\lambda = 2d \sin\theta$, the correction takes the form:

$$d/(\omega \cos\theta) = \lambda/(2\omega \sin\theta \cos\theta) = \lambda/(\omega \sin 2\theta)$$

The factor $1/(\sin 2\theta)$ is the Lorentz factor.

C. Polarization

X-rays are polarized on reflection, and the degree of polarization depends on experimental conditions. Neutrons are not polarized on reflection from ordinary crystals. X-rays produced from an X-ray tube or a rotating-anode generator are unpolarized; that is, all directions of the electric vector normal to the direction of propagation are equally represented. Thus, the beam can be regarded as composed of two components, one polarized parallel to the reflection plane and one perpendicular. The relative intensities of the two components are

$$I_{\parallel} = I_{\perp} = \tfrac{1}{2} I_0$$

For an ideally imperfect crystal, the intensity of radiation scattered in a particular direction is proportional to $\sin^2\phi$, where ϕ is the angle between the electric vector and the direction of observation. It follows that the parallel component will be attenuated by reflection but the perpendicular component will not. Thus, the relative intensities of the two components after reflection at an angle θ will be $I_{\parallel} = \tfrac{1}{2} I_0 \sin^2(90 - 2\theta) = \tfrac{1}{2} I_0 \cos^2 2\theta$ and $I_{\perp} = \tfrac{1}{2} I_0 \sin^2 90° = \tfrac{1}{2} I_0$. Thus, the beam is partially polarized at all angles and completely polarized at $2\theta = 90°$.

For data monochromated by means of a β filter, the polarization correction is

$$p = \tfrac{1}{2}(1 + \cos^2 2\theta)$$

Frequently, Lorentz and polarization corrections are combined to give:

$$Lp = (1 + \cos^2 2\theta)/2 \sin 2\theta$$

These corrections are applied to the observed intensity to derive structure amplitudes:

$$|F_{hkl}| = \sqrt{I_{hkl}/Lp}$$

In many modern diffractometers, monochromatization of the primary beam is achieved by Bragg reflection from a suitable crystal. Three of the most commonly used monochromator crystals are quartz ($d = 3.35$ Å), LiF ($d = 2.01$ Å), and highly oriented graphite ($d = 3.35$ Å). The monochromator may be installed in the incident beam or the diffracted beam and may be mounted with its axis

parallel or perpendicular to the equatorial plane of the diffractometer. If the monochromator is in the incident beam and mounted so that its axis is perpendicular to the equatorial plane, the component that was attenuated by reflection from the monochromator is attenuated again by the sample, so that the polarization correction for the twice-reflected beam becomes:

$$p = \left(1 + \cos^2 2\theta_m \cos^2 2\theta\right)\big/\left(1 + \cos^2 2\theta_m\right)$$

For a monochromator mounted with its axis parallel to the equatorial plane, reflection from the monochromator attenuates the beam normal to the equatorial plane, while reflection from the sample attenuates the parallel component, so that

$$p = \left(\cos^2 2\theta_m \cos^2 2\theta\right)\big/\left(1 + \cos^2 2\theta_m\right)$$

These formulas assume that the monochromator crystal is an ideally mosaic crystal. For a perfect or non-mosaic crystal, the factor $\cos^2 2\theta_m$ should be replaced by $|\cos\theta_m|$.

In practice, the polarization ratio seldom corresponds to either of these ideal values and may not even lie between them. This has led some investigators to recast the equations in the form:

$$p = (1 + K \cos^2 2\theta)/(1 + K)$$

for the monochromator axis perpendicular to the equatorial plane, and

$$p = (K + \cos^2 2\theta)/(1 + K)$$

for the parallel orientation, where K is the actual measured value of the polarization ratio for the monochromator in question. The value of K is different for different wavelengths. For routine structural work with MoKα radiation, the error in assuming that the monochromator is an ideal mosaic will generally be small. However, maximum error occurs at $\theta = 45°$ and so is important in the case of very precise studies that rely on high-angle data. Vincent and Flack have developed a method for determining K, the polarization ratio, without special equipment.

D. Extinction

As early as 1922, Darwin realized that absorption is not the only effect that attenuates the X-ray beam as it passes through the crystal. He described two phenomena, which he designated as primary and secondary extinction, and showed how they could be treated mathematically.

Primary extinction is an interference process. If a set of planes is in a position to reflect, the reflected rays may also be reflected a second time. Since there is a phase change of $\pi/2$ on reflection, a beam that has been reflected n times will be exactly out of phase with one that has reflected $(n-2)$ times. This causes the reflected intensity to be proportional to |F| rather than |F|^2. A crystal for which this is true is called an *ideally perfect* crystal. Such crystals are rare. It is much more common to encounter crystals where $I \propto |F|^n$ where I < n < 2 but nearer to 2.

Darwin modeled the phenomena of extinction by assuming that crystals were made up of mosaic blocks, slightly misaligned with respect to one another. In perfect crystals, the blocks are assumed to be large and the misalignments small. Secondary extinction occurs because the planes first encountered by the incident beam reflect so strongly that deeper planes receive less radiation and so reflect with less power than they otherwise would have done. This effect is pronounced for strong reflections with |F/V| of the order of 0.1×10^{-24} cm^{-3}.

According to the mosaic model, the effect is expected in crystals where the mosaic blocks are small with respect to the size of the crystal. A mosaic crystal in which the blocks are sufficiently misaligned that secondary extinction is negligible is called an *ideally imperfect* crystal. Since such crystals are seldom encountered in experiments, it is convenient to correct for extinction in least-squares refinement. In most current programs, the correction is based on Zachariasen's 1967 formalism for isotropic extinction, in which the mosaic blocks are assumed spherical:

$$F_c^* = k|F_c|\left(1 + 2r^* Q_0 \bar{T} p_2 / p_1\right)^{-1/4}$$

where k is the scale factor, F_c is the calculated value of the structure factor, $r^* = \beta[1 + (\beta/g)^2]^{1/2}$ where $\beta = 2\bar{t}/3\lambda$, $\bar{t}$ is the mean path length in a single domain, and g is related to the mosaic spread distribution and is frequently assumed to be Gaussian. For X-rays,

$$Q_0 = \left(\frac{e^2 FK}{mc^2 V}\right)\frac{\lambda^3}{\sin 2\theta}$$

where e^2/mc^2 is the classical radius of an electron, and K is the polarization ratio. For neutrons,

$$Q_0 = \left(\frac{F}{V}\right)^2 \frac{\lambda^2}{\sin 2\theta}$$

The term $\bar{T}$ is the mean path-length in the crystal and represents an integration of incident and diffracted beams over all diffraction paths in the crystal. This is normally evaluated during the calculation of absorption correction. If absorption is small in the crystal, $\bar{T}$ may be arbitrarily set to some value such as 0.03 cm. Finally, the term p_n is a polarization term generally assumed to have the form $1 + \cos^{2n} 2\theta$ appropriate to filtered radiation. If a monochromator is used, the appropriate form of the polarization factor should be incorporated into the extinction calculation. For very precise work, the actual polarization ratio of the monochromator should be determined experimentally. For extinction correction, the term refined is r^*.

It contains both a breadth parameter and a misalignment parameter. If $\beta \gg g$, then the broadening of the diffraction peak is dominated by mosaic spread, and we have a type I extinction. Type II extinction, when $\beta \ll g$, is less commonly encountered and corresponds to the situation where the misalignment is small and the breadth of the diffraction peaks is controlled by small domain size. Separation of the terms is possible if two determinations are made on the same crystal with different wavelengths. This is rarely done.

Coppens and Hamilton first extended the treatment of extinction to an anisotropic model in 1970. The crystal is modeled as if it were composed of ellipsoidal particles whose misorientations follow a Gaussian probability distribution. Since there is no need for the distribution of mosaic blocks to obey any symmetry in the crystal, symmetry-equivalent data are not averaged in this treatment.

E. Anomalous Dispersion

When the wavelength of the incident X-ray beam is close to the absorption edge of a scattering atom, the atomic scattering factor for that atom becomes complex:

$$f = f_0 + \Delta f' + i\Delta f''$$

where f_0 is the normal scattering factor for wavelengths far from the absorption edge, and $\Delta f'$ and $i\Delta f''$ are correction terms. The quantity $\Delta f'$ is usually negative, and $i\Delta f''$ is always $\pi/2$ radians ahead of the real part in phase. For structural work, the corrections are assumed to be independent of scattering angle. In addition, $\Delta f' = 0$ for wavelengths longer than the absorption edge.

Four aspects of anomalous dispersion important in normal structural work in crystallography are

1. Determination of absolute configuration
2. Solution of the phase problem (structure solution)
3. Distinguishing among atoms of similar scattering power
4. Avoiding systematic errors in structures with polar space groups

Knowledge of the absolute configuration is extremely important in physiologically active materials, since biological systems discriminate strongly between enantiomorphous forms of their substrates. Anomalous scattering with phase change causes a breakdown in Friedel's law, and $I_{hkl} \neq I_{\bar{h}\bar{k}\bar{l}}$. This effect was first exploited for determination of the absolute configuration of the sodium rubidium salt of (+)-tartaric acid using Zr radiation ($\lambda = 6.07$ Å), whose wavelength is slightly shorter than that of the absorption edge of Rb ($\lambda = 6.86$ Å). The experiment coupled with the known relationship between the stereochemistry of (+)-tartaric acid and (+)-glyceraldehyde showed that Fischer's arbitrary choice had been correct. With modern data collection techniques, the determination of absolute configuration is relatively uncomplicated. In favorable cases, the method can be applied to a compound with no atom heavier than oxygen when the incident radiation is CuKα ($\lambda = 1.54$ Å). Rabinovich and Hope have determined the absolute sign of the torsional angles in the achiral compound 4, 4'-dimethylchalcone, $C_{17}H_{16}O$. These authors believe that the determination of absolute configuration may be possible with hydrocarbons.

In structure solution, anomalous scattering without phase change is formally equivalent to isomorphous replacement. An anomalous-difference Patterson function is analogous to an isomorphous-difference Patterson and so contains peaks only for vectors between anomalously scattering atoms and vectors between anomalous scatterers and normal scatterers. Vectors between normal scatterers do not appear.

Anomalous dispersion with phase change ($\Delta f'' \neq 0$), can be used to determine the phase angles from noncentrosymmetric crystals. In the case where the position of the anomalous scatterer is known, the procedure requires that differences in intensity between be measurable for a significant number of intensities.

Anomalous dispersion provides an elegant means to distinguish among near neighbors in the periodic table that would otherwise have similar scattering power. Alloys such as β-brass (Cu–Zn) and Cu_2MnAl were early examples of structures determined by this technique. The tunability of the synchrotron source allows choice of a wavelength close to the absorption edge of an element in the sample to maximize the anomalous component of scattering.

Neglecting to correct for anomalous dispersion will obviously introduce a small error in the magnitudes and in the phases of the calculated structure factors. It was long assumed that such errors would have little effect on atomic positions. However, Cruickshank and McDonald have pointed out that neglect of the correction will always cause errors in thermal parameters, and in the case of polar space groups very serious errors in coordinates can arise. The size of the error varies directly with $\Delta f''$ and inversely with the resolution of the data. For a moderately heavy atom such as Co ($Z = 27$), the error in coordinates can be as large as 0.06 Å in the experiment done with Cu radiation. Neglect of anomalous dispersion will cause errors of the order of 0.005 Å for structures with atoms no heavier than oxygen with Cu radiation or sulfur with Mo radiation. The error caused by including $\Delta f''$ and choosing the wrong enantiomer is twice as large.

F. Scaling of Data

Virtually all methods of solving structures require a reasonable estimate of the relative scale factor between the observed and calculated structure factors. Most computer programs that calculate scale and temperature factors are based on Wilson's 1942 equation:

$$\langle I_{hkl} \rangle = \sum_i f_i^2$$

which says that the local average value of the intensity is equal to the sum of the squares of the scattering factors. It is assumed that the average is taken over a sufficiently narrow range of $(\sin\theta)/\lambda$ so that the f values can be treated as constants:

$$I_{hkl} = k \sum_i f_i^2 \exp[-2B(\sin^2\theta/\lambda^2)]$$

$$I_{hkl} \bigg/ \sum_i f_i^2 = k \exp[-2B(\sin^2\theta/\lambda^2)]$$

$$\log_e \left[I_{hkl} \bigg/ \sum_i f_i^2 \right] = \log_e k - 2B(\sin^2\theta/\lambda^2)$$

Thus, a plot of $\log_e[I_{hkl}/\sum_i f_i^2]$ versus $(\sin^2\theta)/\lambda^2$ will give a straight line of slope $2B$ and intercept k.

A number of conditions must hold so that the values of B and k are reasonable. The sampling interval should be small, so 40 to 50 intervals of $(\sin\theta)/\lambda$ are required for three-dimensional datasets. Weak reflections must be included. Elimination of all reflections with $I < 3\sigma(I)$ will cause the average intensity to be too high in the high ranges of $(\sin\theta)/\lambda$, and that will cause an underestimate of the temperature factor and a corresponding overestimate of the scale factor. Including weak reflections at half of the local minimum observed intensity is better than leaving them out, but a Baysian fill is probably a better strategy. Large excursions on both sides of the best-fit line are quite common and reflect the facts that Wilson's formula was derived for a random distribution of atoms and real structures contain many repetitions of certain bonded and nonbonded distances. The inflection points (the regions in the graph where the experimental points cross the straight line) are relatively constant for different types of structures. Hall and Subramanian recommend an "inflection point least squares" in which the least-squares line is fitted to 15 points: the five lowest angle points, five points nearest to $(\sin^2\theta)/\lambda^2 = 0.15$, and five points nearest $(\sin^2\theta)/\lambda^2 = 0.26$.

G. Thermal Diffuse Scattering

Thermal diffuse scattering (TDS) arises mainly from low-frequency acoustic modes in the crystal. First- and second-order TDS cause the scattering density to peak under Bragg peaks, with the degree of peaking related to the velocity of sound in the crystal. The effect is not removed in normal data-reduction techniques and is different in different directions in the crystal. In normal structure determination, TDS is ignored. The result is a systematic decrease in apparent thermal parameters. Since TDS increases with increased $(\sin \theta)/\lambda$, it enhances the apparent intensity of high-order diffraction data.

In very precise work, such as the determination of charge-density distribution, it is extremely important that the effect be eliminated or accounted for. It should be noted that the amount of TDS included in a diffraction profile will generally be different for X-ray and neutron experiments, since it depends on such experimental conditions as primary beam divergence, wavelength spread crystal dimensions, and counter aperture. Extensive calculations are required to correct for TDS, and most formulations demand that the elastic constants of the crystal be known. However, cooling the crystal can reduce the effect. If α is defined as $I(\text{TDS})/I(\text{Bragg})$, cooling from room temperature to liquid-nitrogen temperature will reduce α by a factor of 5. Cooling to liquid-helium temperature will reduce α by another factor of five. Facilities for X-ray diffraction experiments down to liquid-nitrogen temperature are fairly common. Helium cryostats are rare.

APPENDIX II: METHODS OF STRUCTURE SOLUTION

A. Trial-and-Error Methods

The earliest structures that were determined by X-ray diffraction were mineral structures with relatively high symmetry. Intensities were measured as strong, medium, and weak, and most of the atoms sat on special positions in the cell. From knowledge of the density of the material, its chemical formula, and the space group, one could postulate trial structures and see whether the pattern of intensities matched. The method cannot generally be applied to molecular structures in low-symmetry space groups.

B. Transform Methods

A crystal can be regarded as convolution of the lattice with the unit-cell contents. By the convolution theorem, the transform of a convolution is the product of the transforms. The diffraction pattern is the transform of the crystal structure and so must be the product of the delta function representing the reciprocal lattice and the transform of the unit cell contents.

From the properties of a delta function we know that the product $f(x)\,\delta(x - x_0)$ has values only at x_0 since $\delta(x - x_0) = 0$ if $x \neq x_0$. Thus, the diffraction pattern of a crystal can be regarded as the transform of the unit-cell contents sampled at the points of the reciprocal lattice. This implies that a direct plot of the weighted reciprocal lattice can give some information about the structure. The method is used frequently to solve the structures of polynuclear aromatic hydrocarbons. In these compounds the dominant features of the diffraction pattern are the benzene transform and the fringe function showing the separation of the molecules. The molecules generally crystallize in centrosymmetric space groups with one short axis roughly perpendicular to the plane of the molecule.

Figure 8a shows the calculated transform of a regular hexagon, the transform of a benzene ring. It is characterized by positive and negative regions with strong positive peaks at a distance $0.8\ \text{Å}^{-1}$ from the origin. If the hexagon is tilted along an axis, the distance perpendicular to the axis of tilt will be foreshortened in crystal space. The corresponding distances in the transform will be elongated. Simple geometry will allow calculation of the angle of tilt.

Determination of the separation is relatively simple. Consider two centrosymmetric molecules related by a center of symmetry with a separation of molecular centers

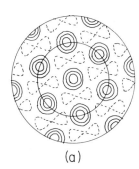
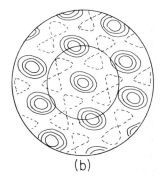
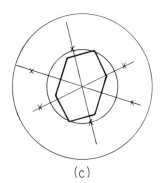

(a)   (b)   (c)

FIGURE 8 Transform of a hexagon. The circle has a radius of $0.8\ \text{Å}^{-1}$. (a) Regular hexagon; (b) tilted hexagon; (c) geometric construction to determine tilt of hexagon in crystal space.

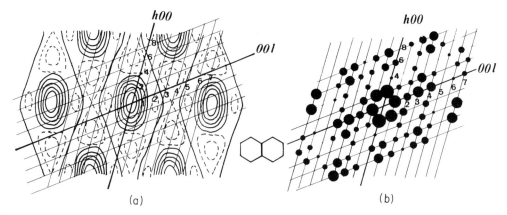

FIGURE 9 (a) Transform of naphthalene with correctly oriented reciprocal lattice; inset shows naphthalene orientation used to calculate the transform. (b) Weighted reciprocal lattice of naphthalene for comparison with transform. Reflections such as 202 and 801, which lie on regions of the transform where density is changing, are much more sensitive to orientation than are those such as 203 that lie in the middle of strong areas.

of 5 Å. Since the molecules are identical in shape and orientation, the combined transform will be that of a single molecule crossed by straight fringes. Regions in the combined transform are strong only if the corresponding region in the single transform is strong, but weak regions arise either from weak regions in the single transform or from the zeros of the fringe system. A line perpendicular to the fringes is the direction of the line joining the centers of molecules, and the separation of the molecules is reciprocal to the spacing of the fringes.

If the transform and the reciprocal lattice are drawn on the same scale, the correct relative orientation of one to the other can be established by matching strong areas of the diffraction pattern with strong areas in the transform. For fine adjustment of the orientation, attention must be paid to those reflections most sensitive to orientation effects, those lying on rising or falling regions of the transform. See, for example, reflections 202 and 801 in Fig. 9. By contrast, the 203 reflection lies well within a strong area of the transform and its value will not be affected by even fairly large changes in orientation.

C. Heavy-Atom Methods

In X-ray diffraction, the scattering power of an atom is proportional to the square of the atomic number, Z^2. If a molecule contains a heavy atom (high Z) and that atom can be located, then a set of phase angles can be calculated for the dataset that are approximations to the true phases. A Fourier synthesis calculated with observed structure amplitudes and phases appropriate to the heavy atom will give a map that contains the heavy atom, some light atoms, and some noise. Phases based on the known atom positions are better estimates of the true phase than the heavy atom alone. The iterative procedure is repeated until all atoms are located.

The location of the heavy atom can be determined from the Patterson function:

$$P(uvw) = \frac{2}{V} \sum_h \sum_k \sum_l (F_{hkl})^2$$
$$\times [\cos 2\pi(hu + kv + lw)]$$
$$= \int_{\text{volume of cell}} p(xyz)$$
$$\times p(x + u, y + v, z + w)\,dv$$

The Patterson function can be calculated directly from the intensities with no previous knowledge of the phases. It is the self-convolution of the electron density. This means that a peak at uvw represents a vector between two atoms whose separation is equal to the vector distance from the origin to the point uvw. The weight of that peak is proportional to the product of the atomic numbers of the atoms at each end of the vector. Vectors between heavy atoms tend to dominate these maps and hence allow the position of the heavy atom to be determined.

Figure 10 shows a hypothetical molecule with one heavy atom and its Patterson function. Note that the Patterson function is always centrosymmetric. If the heavy atom were iodine and the light atoms were carbons, I–I vectors would have weight 2809; I–C vectors, 318; and C–C vectors, 36.

If two molecules are related by a center of symmetry as in Fig. 11, then in addition to the intramolecular vectors near the origin of the vector maps, there are intermolecular vectors. The I–C and C–C vectors are double-weight vectors because both molecules give the same pattern of vectors. However, the I–I vectors, which represent vectors between iodine atoms across the center of symmetry, are single-weight peaks.

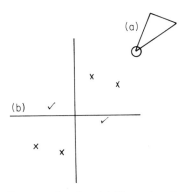

FIGURE 10 Patterson function. (a) Three-atom "molecule" and (b) its vector set. If there are N peaks in the Fourier, there are N^2 peaks in the Patterson function. Of these, N are superimposed at the origin and $N(N-1)$ are distributed through the cell. The atom marked O is heavy; X indicates a heavy-atom–light-atom vector; ✔ indicates a light-atom–light-atom vector.

If the reader were to make a copy of Fig. 11 on transparent paper, place the origin of the transparent map on an I–I vector, and mark the places where the two maps overlap with a mark corresponding to the lower intensity in the overlapping functions, the original structure would be recovered. This is the basis for the "difference function," one of the methods for recovering the electron density from the Patterson function. If the heavy-atom peak had been a double-weight peak, two superpositions would be necessary to recover the original function.

There are constraints on the relative size of the scattering contribution of the heavy atoms. If the heavy atom

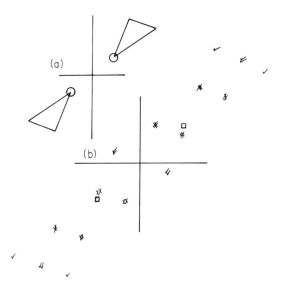

FIGURE 11 Symmetry in the Patterson function. (a) Two molecules related by a center of symmetry, and (b) the corresponding vector set. The symbol □ indicates a heavy-atom–heavy-atom vector. Note that peaks corresponding to a vector distance between an atom and its symmetry mate are single-weight peaks. All other peaks are double weight.

is too light, phases calculated from its position are poor estimates of the true phases and it may be very difficult to find correct atom positions in a very noisy Fourier map. If it is too heavy, the scattering of the heavy atom will dominate to such an extent that the precision of the light-atom parameters may be seriously affected.

The rule of thumb is that the ratio $\sum Z_{heavy}^2 / \sum Z_{light}^2$ should be approximately 1; however, the method will tolerate large deviations in either direction. For instance, the structure of vitamin B_{12} ($C_{63}H_{88}N_{14}O_{14}PCo \cdot H_2O$) was solved using the phases from the cobalt atom as a starting point. The Z^2 ratio is about 0.17!

A Sim-weighted Fourier is a Fourier series phased by the known portion of the structure with coefficients weighted according to the probability that the phase is correct. This is a very useful technique for improving the signal-to-noise ratio in poorly phased Fourier maps.

D. Isomorphous Replacement

Two compounds are perfectly isomorphous if the only difference in their electron-density maps corresponds to the site of a replaceable atom. The method requires two isomorphous derivatives in the centrosymmetric case and three or more isomorphous derivatives in the noncentrosymmetric case. As direct methods have improved, the use of this method for general organic structures has declined; however, many all-protein structures have been solved by this method or by the combination of isomorphous replacement and anomalous scattering.

The centrosymmetric case is straightforward. Consider two derivatives, A and B, for which the light-atom portions are identical but the replaceable atoms are different. Then,

$$F_A = F_L + F_{AR}$$

$$F_B = F_L + F_{BR}$$

$$\Delta F_{AB} = F_A - F_B = F_{AR} - F_{BR}$$

The magnitudes $|F_A|$ and $|F_B|$ are available from the data collection; the magnitudes and signs of the replaceable components are available once the positions of the replaceable atoms are known. Thus, the four possible sign combinations corresponding to the left-hand side of the third equation are calculated, and the combination that gives best agreement with the right-hand side is accepted. Reflections whose phases are not well determined are omitted from the Fourier synthesis.

In the noncentrosymmetric case, the use of two isomorphous derivatives leads to a twofold ambiguity in phase and it is necessary to have a third derivative with a heavy atom in a different position in order to resolve the ambiguity. This is illustrated in Fig. 12. Again, let

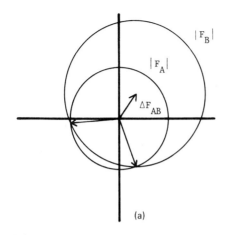

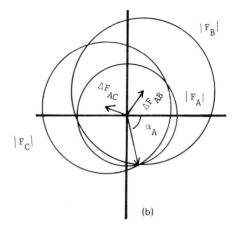

FIGURE 12 Isomorphous replacement. (a) The difference ΔF_{AB} between the structure factors for the two isomorphs A and B provides the center for $|F_B|$. (b) The ambiguity in phase can be resolved by a third derivative.

$$F_A = F_L + F_{AR}$$

$$F_B = F_L + F_{BR}$$

$$\Delta F_{AB} = F_{AR} + F_{BR}$$

The circles of radius $|F_A|$ represents all possible values for the phase of F_A. The magnitude and phase of ΔF are indicated by the vector and a circle of radius F_B that is drawn using the tip of ΔF as center. The two points of intersection correspond to the phase combinations that satisfy the third equation. A third derivative introduces a new equation,

$$\Delta F_{AC} = F_{AR} - F_{CR}$$

Drawing a third circle with radius $|F_C|$ and center at ΔF_{AC} resolves the ambiguity.

E. Anomalous Scattering

When the energy of the radiation used in the experiment lies near an absorption edge for one (or more) of the atoms in the crystal, the scattering factor for that atom becomes complex:

$$f = f_0 + f' + if''$$

In this circumstance, Friedel's law breaks down and $|F|^2_{hkl} \neq |F|^2_{\overline{hkl}}$. For centrosymmetric crystals, the data are measured twice: once with a radiation for which the heavy atom scatters normally, and once for scattering anomalously. The two sets of data must be scaled carefully because the differences tend to be small. If the imaginary component is small, then anomalous scattering is exactly analogous to isomorphous replacement.

The results can be shown in a Harker diagram. In the centrosymmetric case with only one type of anomalous scatterer (Fig. 13a), the first circle (i.e., F_N) corresponds to the amplitude of the structure factor with no anomalous

contribution. The second circle (F_{NH}) corresponds to the amplitude of the structure factor with an anomalous contribution. $F_{H'}$ and $F_{H''}$ represent the real and imaginary components of contribution of the anomalous scatterers. Since there is only one anomalous scatterer, the vectors $F_{H'}$ and $F_{H''}$ are perpendicular. The vector sum of $F_{H'}$ and $F_{H''}$ becomes the center for the F_{NH} structure amplitude. The circles intersect in two places but there is no ambiguity, since the phase must lie on the real axis.

In the noncentrosymmetric case (Fig. 13b), the same procedure again leaves us with a phase ambiguity. In structures of moderate size, it may be sufficient to choose the solution near the phase of the heavy atom. This will not always be the correct choice but frequently leads to the correct solution.

In protein structures, the heavy atom is generally far too light to use to discriminate between the two choices. The ambiguity can be resolved by using the other member of the Bivoet pair (Fig. 13c) or by using data from an isomorphous derivative (not shown). In any case, three datasets are required.

An interesting variation on the method is possible using synchrotron radiation. If the protein contains one anomalous scatterer, the tunability of the source can be exploited to collect datasets at different wavelengths. Some protein structures contain a large number of atoms whose anomalous dispersion corrections are so weak they may be neglected. If they also contain only a few anomalously scattering atoms all of one type, a simple special case of the general system of equations results:

$$|F_{\lambda h}|^2 = \left|F_{1,\mathbf{h}}^n\right|^2 + [1 + Q(Q + 2\cos\delta_{\lambda 2})]\left|F_{2,\mathbf{h}}^n\right|^2$$
$$+ 2(1 + Q\cos\delta_{\lambda 2})\left|F_{1,\mathbf{h}}^n\right|\left|F_{2,\mathbf{h}}^n\right|\cos\left(\phi_{1,\mathbf{h}}^n - \phi_{2,h}^n\right)$$
$$+ 2Q\sin\delta_{\lambda 2}\left|F_{1,\mathbf{h}}^n\right|\left|F_{2,\mathbf{h}}^n\right|\sin\left(\phi_{1,h}^n - \phi_{2,\mathbf{h}}^n\right)$$

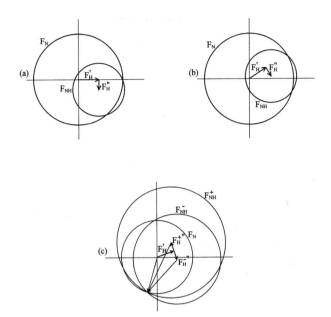

FIGURE 13 Anomalous dispersion. The first circle is F_N. The vectors $F_{H'}$ and $F_{H''}$ are drawn and their vector sum becomes the center for F_{NH}. This gives a twofold ambiguity, which can be resolved in the (a) centrosymmetric case by noting that the phase must lie on the real axis. In the (b) noncentrosymmetric case, the ambiguity can be resolved by using the negative anomalous component as shown in (c) or an isomorphous derivative (not shown). F_{NH}^- and F_{NH}^+ refer to F_{hkl} and $F_{\overline{hkl}}$, respectively.

where Q is the ratio $f_{\lambda 2}^a / f_{2,\mathbf{h}}^n$, $|F_{1,\mathbf{h}}^n|$ is the magnitude of the structure factor for the normally scattering atoms, $|F_{2,\mathbf{h}}^n|$ is the normal part of the structure factor for the anomalously scattering atoms, and $\phi_{1,\mathbf{h}}^n$ and $\phi_{2,\mathbf{h}}^n$ are the associated phase angles. There is a second equation for the Friedel mate, and it is the same as the previous one except for a minus sign before the last term. Thus, there will be two independent equations for each wavelength at which data are collected, plus a third equation resulting for the trigonometric identity $\sin^2 x + \cos^2 x = 1$. This set of equations can be solved if data are collected at two or more wavelengths. Generally, one dataset is collected at energies below the absorption edge so that $\Delta f'' = 0$; a second set is collected above the absorption edge, with $\Delta f'' \neq 0$ and $\Delta f'$ having the same value as for the first experiment. The actual values for the real and imaginary parts of the scattering factor are determined experimentally before the choice of wavelengths is made. The obvious advantage is that the isomorphism is exact. In principle, the data can be collected on one crystal; in practice, radiation damage will require the use of several different crystals. A generalization of the theory can take into account any number of types of anomalous scatterers and any number of anomalous scatterers within each type.

F. Direct Methods

Methods of structure solution that attempt to evaluate the phases of the structure factors without recourse to struc-

tural information are known as direct methods. It is not necessary to determine all of the phases. In general, about $10\times$ the number of non-hydrogen atoms in the cell is sufficient.

Virtually all of the direct-method programs currently available make use of normalized structure factors, $E(\mathbf{h})$, which "correct" the structure factor $F(h)$ for fall-off with angle caused by the temperature factor and the scattering factors of the atoms,

$$E(\mathbf{h}) = F(\mathbf{h}) \Big/ \left(\varepsilon_{\mathbf{h}}^{-1/2} \sum f_i^2 \right)^{1/2}$$

If the shapes of the scattering curves are similar, the values of $E(\mathbf{h})$ can be calculated from the relation:

$$E(\mathbf{h}) = \left[\varepsilon_{\mathbf{h}}^{-1/2} \sum_{j=1}^{N} Z_j \Big/ (\sigma_2)^{1/2} \right] \exp(2\pi i \mathbf{h} \cdot \mathbf{r}_j)$$

where N is the number of atoms in the cell, and $\varepsilon_{\mathbf{h}}$ is the average intensity multiple of the $\mathbf{h}$th reflection. The average value of $|E|^2(\mathbf{h}) = 1$. The most frequently used relationship in direct methods for centrosymmetric structures is the $\sum_2$ or triplet relationship:

$$S(E_{\mathbf{h}}) \approx S \sum_k E_{\mathbf{k}} E_{\mathbf{h}-\mathbf{k}}$$

where $S(E_{\mathbf{h}})$ is the sign of the reflection hkl, and $\approx$ means "probably equal to." The probability associated with this relationship is

$$P_+(E_{\mathbf{h}}) = \tfrac{1}{2} + \tfrac{1}{2}\tanh\left(\sigma_3(\sigma_2)^{-3/2}|E_{\mathbf{h}}|\sum_k E_{\mathbf{k}}E_{\mathbf{h-k}}\right)$$

where $\sigma_n = \sum_{i=1}^{N} Z_i^n$ and Z is the atomic number. For example,

$$\mathbf{h} = 63\bar{3} \quad \text{and} \quad \mathbf{k} = 790$$
$$\mathbf{h-k} = \overline{163}$$

In this example from a centrosymmetric structure, if all of these reflections are strong, it is probable that the sign of $F(63\bar{3})$ is positive. The probability increases as the magnitude of the normalized structure factors increase. The steps in solving a centrosymmetric structure (where the phases can have values of 0 or π) are

1. Evaluate E terms. This includes calculation of the normalized structure factors and sorting of the reflections among eight subgroups defined by the parity of the h, k, and l indices.
2. Form $\sum_2$ relationships with the strongest reflections—those that have E values greater than some arbitrary value such as 1.2 or 1.5.
3. Determine phase. Historically, phases were determined by the symbolic addition method. In this method, origin-determining reflections are given signs and a few others, chosen from the strongest $|E|$ values with the most $\sum_2$ interactions, are given symbols (a, b, c, etc.). Signs can be used because, in the case of a centrosymmetric structure, the phases can only be 0 or π. Thus, $S_{\mathbf{h}}$ means the sign of the structure factor for the $\mathbf{h}$ reflection. The value for $S_{\mathbf{h}}$ could arise in one of several ways. One way is its assignment as described above. Alternatively, $S_{\mathbf{h}}$ could acquire a value through the triplet relationship as follows: If $S_{\mathbf{k}}$ is know to be positive, and $S_{\mathbf{h-k}}$ is known to be a, then:

$$S_{\mathbf{h}} = S_{\mathbf{k}}S_{\mathbf{h-k}} = a$$

During the course of the analysis, relationships appear among the symbols, such as $ac = e$. Manipulation of these relationships usually allows the number of unknowns to be reduced at the end of phase determination.
4. Calculate the E map—a Fourier summation using E values as coefficients and phases determined in step 3.

In the noncentrosymmetric case, the solution is more difficult since the phase can take any value between 0 and 2π. Hence, a different set of relationships was developed for this case:

$$\varphi_{\mathbf{h}} = \langle \varphi_{\mathbf{k}} + \varphi_{\mathbf{h-k}} \rangle_{\mathbf{k_r}}$$
$$\varphi_{\mathbf{h}} = \frac{\sum_{\mathbf{k_r}} |E_{\mathbf{k}}E_{\mathbf{h-k}}|(\varphi_{\mathbf{k}} + \varphi_{\mathbf{h-k}})}{\sum_{\mathbf{k_r}} |E_{\mathbf{k}}E_{\mathbf{h-k}}|}$$

The symbol $\mathbf{k_r}$ implies that $\mathbf{k}$ ranges only over those vectors associated with large $|E|$ values.

The process of choosing initial origin-determining phases is similar to the centrosymmetric case, but an additional enantiomorph-determining reflection must also be specified. The symbols are assigned in the same way as before and result in assignments such as $\phi_{\mathbf{h}} = 2a - b$. These are converted to numerical values, and each set is then expanded and refined using the tangent formula. The first computer programs for structure solution used this method. The symbolic addition procedure generates only a few alternatives for the values of the phases which must be considered because the number of resulting unknown symbols is usually no more than three for four.

In multisolution (i.e., multitrial) methods, a small number of phases are assigned arbitrarily to fix the origin and, in the case of noncentrosymmetric space groups, the enantiomorph. Additional reflections are each assigned many different starting values in the hope that one (or more) of these sets of starting conditions will lead to a solution. Some programs use random-number generators to set starting values for some 20 to 200 phases, which are then extended and refined by the tangent formula:

$$\tan\phi_{\mathbf{h}} = \frac{\sum_{\mathbf{k}} w_{\mathbf{k}}|E_{\mathbf{k}}E_{\mathbf{h-k}}|\sin(\phi_{\mathbf{k}} + \phi_{\mathbf{h-k}})}{\sum_{\mathbf{k}} w_{\mathbf{k}}|E_{\mathbf{k}}E_{\mathbf{h-k}}|\cos(\phi_{\mathbf{k}} + \phi_{\mathbf{h-k}})}$$

The weighting function in the tangent formula is useful in some approaches. A new phase $\phi_{\mathbf{h}}$ is assigned a weight that is the minimum of $0.2\alpha_{\mathbf{h}}$ and 1.0:

$$w_0 = \min(0.2\alpha_{\mathbf{h}}, 1.0)$$

Although this allows rapid development of a phase set for each trial, it tends to lead to an incorrect centrosymmetric solution in the case of polar space groups. For Hall–Irwin weights, $w = w_0 f(\alpha/\alpha_{\text{est}})$, where $f = 1.0$ for $\alpha < \alpha_{\text{est}}$ and decreases for $\alpha > \alpha_{\text{est}}$. This weighting tends to conserve the enantiomorph and has been incorporated into more recent versions of the multisolution programs. The major difference in treatments for centric and acentric datasets lies in the values assigned to the extra reflections in the starting set. In the centric case, the possible values are 0 and π; in the acentric case, general reflections have four possible values, $\pm\pi/2$ and $\pm3\pi/2$.

A recent modification to the multisolution approach consists of phase refinement alternated with the imposition of constraints by peak picking in real space. This approach, referred to as a *dual-space method*, differs from other multisolution methods in that the phases are initially assigned values by computing structure factors for a randomly positioned set of atoms. The occurrence of two Fourier transformations per cycle results in an algorithm that is computationally intensive. However, the new approach also extends the limits of these programs to much larger structures. It has proven to be capable of solving complete structures containing as many as 2000 independent non-hydrogen atoms (provided that accurate diffraction data have been measured to a resolution of 1.2 Å or better and some heavy atoms are present in the structure).

There has been progress in the experimental evaluation of triplet-phase invariants. The phenomenon of simultaneous diffraction has long been considered a nuisance in single-crystal work. If three reflections lie on the sphere of reflection simultaneously, there is a power transfer that tends to enhance such weak reflections at the expense of strong ones. In 1977, it was discovered that the shape of the simultaneous diffraction profile is sensitive to the phase of the triplet that gave rise to it. The sense of the asymmetry is opposite for positive and negative triplets. This effect has been observed with ordinary mosaic crystals of relatively heavy scatterers ($ZnWO_4$) using CuKα radiation from a fine-focus tube. This technique could be enhanced by the use of synchrotron radiation, since it is desirable that the beam be monochromatic, intense, and highly collimated. It is envisioned that experimental phase determination could be used to establish a starting set of 50 or so triplets and that the tangent formula or some similar technique could then expand these relationships.

APPENDIX III: METHODS OF REFINEMENT

Once a trial structure has been proposed, improvements in the values of the parameters are sought so that the model corresponds as closely as possible to reality. Exact agreement between observed and calculated structure factors would yield a difference Fourier synthesis that was flat and an R-value of zero. The methods of refinement discussed here are in the context of small molecules where the ratio of reflections to parameters is commonly of the order of 10:1. Such a degree of overdetermination does not exist in protein structures. The modification of these techniques that would be required to handle proteins is not discussed in this review.

A. Difference Fourier

A Fourier synthesis with coefficients $\Delta F = |F_0| - |F_c|$ reflects differences between the crystal and the model. The difference map is routinely used during structure solution to check the integrity of the model as it is being developed. Large peaks in the map ($Z/3$ to $Z/2$) correspond to atoms not yet included in the model. Smaller peaks may indicate a slightly misplaced atom or wrong scattering type

or incorrect thermal parameters. Holes indicate electron density in the model where none exists in the crystal (see Fig. 14).

It is common to use difference maps to find hydrogen atoms once the positions of the heavier atoms have been refined isotropically. In centrosymmetric structures, the phases are either positive or negative and are usually correct for all but the smallest structure factors by this stage. Hydrogen atoms will have electron densities in the range 0.6 to 0.9 $\Delta e\,\text{Å}^{-3}$. Phase errors in the noncentrosymmetric case cause difference maps to be less well defined. At the end of refinement, a difference map should show no significant features.

Since it is much faster to calculate a Fourier synthesis than to carry out least squares, refinement by difference synthesis was popular before the advent of modern

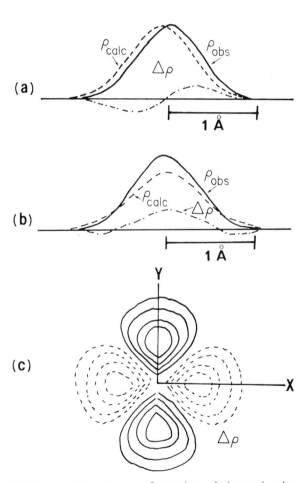

FIGURE 14 Difference map. Comparison of observed, calculated, and difference electron density for (a) an atom displaced by 0.1 Å from its correct position, and (b) a temperature factor overestimated. (c) Difference density observed when an isotropic temperature factor is used in the model and the thermal vibration of the crystal is actually anisotropic. Dashed contours are negative.

computers. For small- and medium-sized molecules (<400 atoms), least squares is the preferred method of refinement, and difference maps are used mainly for error detection.

B. Least-Squares Analysis

The set of parameters that minimizes the sum of the squares of the difference between the observed and calculated values of the structure factors is the most satisfactory set. The function minimized is $\sum_{hkl} \omega \Delta^2$ where $\Delta = |F_0| - |F_c|$, and ω is the weight. The minimization condition is

$$\sum_{hkl} \omega \Delta \frac{\partial F_c}{\Delta p_j} = 0$$

But, the relationship between Δ and the parameters is nonlinear. For a set of parameters close to the true values, Δ may be expanded as a function of the parameters by a Taylor series truncated to first order:

$$\Delta(p + \varepsilon) = \Delta(p) - \sum_{i=1}^{n} \varepsilon_i \frac{\partial |F_c|}{\partial p_i}$$

where p stands for the whole set of n parameters, ε_i is a small change in the ith parameter, and the minus sign reminds us that $\Delta = |F_0| - |F_c|$, and it is F_c that must be changed.

The *normal equations* have the form:

$$\sum \omega \frac{\partial |F|^2}{\partial p_i} \varepsilon_1 + \sum \omega \frac{\partial |F|}{\partial p_1} \frac{\partial |F|}{\partial p_2} \varepsilon_2 + \cdots$$

$$= \sum \omega \Delta \frac{\partial |F|}{\partial p_1}$$

$$\sum \omega \frac{\partial |F|}{\partial p_1} \frac{\partial |F|}{\partial p_2} \varepsilon_1 + \sum \omega \frac{\partial |F|^2}{\partial p_2} \varepsilon_2 + \cdots$$

$$= \sum \omega \Delta \frac{\partial |F|}{\partial p_2}$$

The left-hand side yields an $n \times n$ matrix, symmetrical about its diagonal. Computer programs store only the upper diagonal half of this matrix. For a 30-atom structure there are 90 positional parameters, 180 temperature parameters, and a scale factor. Thus, $n = 271$. Evaluation of the left-hand side involves evaluating $271 \times 270/2 = 36{,}585$ terms, with the sum over all reflections formed for each term. The matrix is then inverted and solved for the shifts in the parameters. These shifts are applied and the process continues in an iterative manner.

C. Constrained Refinement

There are a number of reasons for wishing to add constraints to a refinement, such as when the resolution is limited, the structure is disordered, or light atoms are being refined in the presence of heavy ones. Examples of where this type of refinement has been used are inorganic coordination compounds with triphenylphosphine as a ligand. The benzene ring can be constrained to the shape of a regular hexagon with attached hydrogen atoms in their correct positions. Only six positional parameters need be refined: three for the center of mass and three for the orientation angles. The model for thermal motion may allow individual thermal parameters for each atom of the group, or a single temperature parameter for the whole group.

The advantage of using rigid-body constraints is that the number of parameters is reduced. In addition, calculation of the coefficients of the derivatives of the individual atom coordinates with respect to the rigid-body coordinates is performed only once (not for each reflection). Thus, the refinement will require less computer storage and less computer time. As an added bonus, the refinement tends to converge more quickly than an unconstrained refinement. However, it must be kept in mind that a group constrained to an incorrect geometry will result in a structure that is systematically wrong. If the number and quality of data will allow, the constraints may be released at the end of the refinement. Hamilton's R-value ratio test may help to determine whether the change in R-value on release of constraints is significant.

The "hard" constraints described above reduce the number of parameters needed to solve the structure and hence reduce the size of the matrix. "Soft" constraints do not change the size of the matrix but increase the number of observational equations. Bond length and angles are permitted to deviate from standard values by a fixed amount, or a ring may be constrained to be planar. This technique can be very useful in dealing with highly correlated parameters.

In general, the addition of constraints at the end of refinement will cause the R-value to rise slightly. Strict enforcement of an inappropriate constraint will increase the effect. It is unusual for constraints to be in place during the whole refinement; if they are, it is important that they be weighted appropriately.

D. Thermal Motion Analysis

Most modern crystal-structure analyses refine values for the anisotropic vibrational parameters U_{ij} for individual atoms in the structure. This model assumes that the vibrations are harmonic and that vibrations of neighboring atoms are uncorrelated. If systematic effects such as absorption and extinction have been ignored, if the quality of the crystals is poor, or if the weighting scheme is inappropriate, these parameters may have little physical meaning.

Hirshfeld has proposed a convenient test for reasonableness of a set of vibration parameters. Since bond-stretching vibrations can be assumed to have much higher energy than other intramolecular modes such as torsion or angle bending, bonds can be considered "rigid" to a first approximation. If two atoms A and B are bonded, it is expected that the component of vibration along the bond will be the same for both atoms. For carbon atoms, agreement within 0.001 Å^2 is expected for low-temperature structure determinations where bonding effects have been accounted for. In the spherical atom approximation, the discrepancy is more likely to be of the order of 0.005 Å^2. In molecular crystals, vibrations of neighboring atoms are correlated to some degree. For relatively rigid molecules, translational and vibrational oscillations are the major contributors to the internal modes of vibration. If the rigid body model is a good approximation, it is possible to "correct" bond lengths for thermal vibration.

The reason for the correction can be seen in Fig. 15. If a molecule is undergoing librational motion, with a root-mean-square amplitude of libration ω about an axis through 0, then an atom at a distance l from the center of libration sweeps out an arc. However, the X-ray analysis will place the center of the atom at C, the centroid of the electron distribution. Thus, bond distances are systematically shortened by an amount $PC = l - l \cos \omega \approx l\omega^2/2$. In three dimensions, the corrections are additive.

The rigid-body models require 12 parameters for centrosymmetric molecules and 21 for noncentrosymmetric. In the former, both the translational and librational axes pass through the center of gravity of the molecule. In the noncentrosymmetric case, the libration axes do not necessarily coincide, and an additional tenser, S, is required to account for the correlation between the librational and translational motions.

Two indicators are used to test the goodness of fit between the refined values of the U_{ij} parameters and those calculated from the rigid-body model. The root-mean-square difference between observed and calculated values, given by $\langle (\Delta U_{ij})^2 \rangle^{1/2}$, would be considered excellent in the region of 2×10^{-4} Å^2 but only fair at 2×10^{-3} Å^2. $R = \sum |\Delta U_{ij}| / \sum U_{ij}$ would be considered excellent at 0.04.

Obviously, the interatomic vectors in a crystal structure can be interpreted as bond distances only if thermal vibrations are negligible. To correct for the effects of thermal vibration, a model for that vibration is required. Often it is better to reduce the effects of thermal motion by working at the lowest available temperature than to attempt an *a posteriori* correction.

SEE ALSO THE FOLLOWING ARTICLES

CRYSTAL GROWTH • CRYSTALLIZATION PROCESSES • FERROMAGNETISM • MICROSCOPY • PHASE TRANSFORMATIONS, CRYSTALLOGRAPHIC ASPECTS • X-RAY SMALL-ANGLE SCATTERING • X-RAY, SYNCHROTRON AND NEUTRON DIFFRACTION

BIBLIOGRAPHY

Carter, C. W., Jr., and Sweet, R. M., Eds. (1997). "Methods in Enzymology," Vol. 277, "Macromolecular Crystallography." Academic Press, New York.

Dunitz, J. D. (1979). "X-Ray Analysis and Structure of Organic Molecules," Cornell University Press, Ithaca, NY.

Ewald, P. P., Ed. (1962). "Fifty Years of X-Ray Diffraction," N.V.A. Oosthoek's Uitgeversmaatschappij, Utrecht.

Journal of Physics D, Applied Physics. (1991). Special issue on structural aspects of crystal growth, **24**(2), February 14.

Karle, J. (1994). *J. Chem. Inf. Comput. Sci.* **34**, 381–390.

Ladd, M. C. F., and Palmer, R. A., Eds. (1980). "Theory and Practice of Direct Methods in Crystallography," Plenum, New York.

McPherson, A. (1982). "Preparation and Analysis of Protein Crystals," Wiley, New York.

Schmidt, P. W., Ed. (1983). "Proc. symp. small-angle scattering." *Trans. Am. Crystallogr. Assoc.* **19**.

Schoenborn, B. P., Ed. (1985). "Proc. symp. structure determination with synchrotron radiation." *Trans. Am. Crysrallogr. Assoc.* **21**.

Wyckoff, H. W., Hirs, C. H. W., and Timasheff, S. N., Eds. (1985). "Methods in Enzymology," Vol. 114, "Diffraction Methods for Biological Macromolecules." Academic Press, New York.

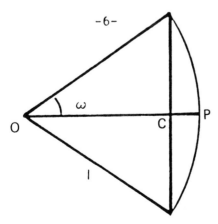

FIGURE 15 Rigid-body motion. An atom at a distance *l* from the center of libration sweeps out in an arc. Least-squares analysis places the atom at *C* rather than *P*.

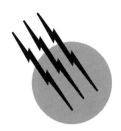

Cybernetics and Second-Order Cybernetics

Francis Heylighen

Free University of Brussels

Cliff Joslyn

Los Alamos National Laboratory

GLOSSARY

Variety A measure of the number of possible states or actions.
Entropy A probabilistic measure of variety.
Self-organization The spontaneous reduction of entropy in a dynamic system.
Control Maintenance of a goal by active compensation of perturbations.
Model A representation of processes in the world that allows predictions.
Constructivism The philosophy that models are not passive reflections of reality, but active constructions by the subject.

CYBERNETICS is the science that studies the abstract principles of organization in complex systems. It is concerned not so much with what systems consist of, but how they function. Cybernetics focuses on how systems use information, models, and control actions to steer toward and maintain their goals, while counteracting various disturbances. Being inherently transdisciplinary, cybernetic reasoning can be applied to understand model and design systems of any kind: physical, technological, biological, ecological, psychological, social, or any combination of those. Second-order cybernetics in particular studies the role of the (human) observer in the construction of models of systems and other observers.

I. HISTORICAL DEVELOPMENT OF CYBERNETICS

A. Origins

Derived from the Greek *kybernetes*, or "steersman," the term "cybernetics" first appears in Antiquity with Plato and in the 19th century with Ampère, who both saw it as the science of effective government. The concept was revived and elaborated by the mathematician Norbert Wiener in his seminal 1948 book, whose title defined it as "Cybernetics, or the study of control and communication in the animal and the machine." Inspired by wartime and prewar results in mechanical control systems such as servomechanisms and artillery targeting systems, and the contemporaneous development of a mathematical theory of communication (or information) by Claude Shannon, Wiener set out to develop a general theory of organizational and control relations in systems.

Information Theory, Control Theory, and Control Systems Engineering have since developed into independent disciplines. What distinguishes cybernetics is its emphasis on control and communication not only in engineered, artificial systems, but also in evolved, natural systems such as organisms and societies, which set their own goals, rather than being controlled by their creators.

Cybernetics as a specific field grew out of a series of interdisciplinary meetings held from 1944 to 1953 that brought together a number of noted postwar intellectuals, including Wiener, John von Neumann, Warren McCulloch, Claude Shannon, Heinz von Foerster, W. Ross Ashby, Gregory Bateson, and Margaret Mead. Hosted by the Josiah Macy Jr. Foundation, these became known as the Macy Conferences on Cybernetics. From its original focus on machines and animals, cybernetics quickly broadened to encompass minds (e.g., in the work of Bateson and Ashby) and social systems (e.g., Stafford Beer's management cybernetics), thus recovering Plato's original focus on the control relations in society.

Through the 1950s, cybernetic thinkers came to cohere with the school of General Systems Theory (GST), founded at about the same time by Ludwig von Bertalanffy, as an attempt to build a unified science by uncovering the common principles that govern open, evolving systems. GST studies systems at all levels of generality, whereas Cybernetics focuses more specifically on goal-directed, functional systems which have some form of control relation. While there remain arguments over the relative scope of these domains, each can be seen as part of an overall attempt to forge a transdisciplinary "Systems Science."

Perhaps the most fundamental contribution of cybernetics is its explanation of purposiveness, or goal-directed behavior, an essential characteristic of mind and life, in terms of control and information. Negative feedback control loops which try to achieve and maintain goal states were seen as basic models for the autonomy characteristic of organisms: their behavior, while purposeful, is not strictly determined by either environmental influences or internal dynamical processes. They are in some sense "independent actors" with a "free will." Thus cybernetics foresaw much current work in robotics and autonomous agents. Indeed, in the popular mind, "cyborgs" and "cybernetics" are just fancy terms for "robots" and "robotics." Given the technological advances of the postwar period, early cyberneticians were eager to explore the similarities between technological and biological systems. Armed with a theory of information, early digital circuits, and Boolean logic, it was unavoidable that they would hypothesize digital systems as models of brains, and information as the "mind" to the machine's "body."

More generally, cybernetics had a crucial influence on the birth of various modern sciences: control theory, computer science, information theory, automata theory, artificial intelligence and artificial neural networks, cognitive science, computer modeling and simulation science, dynamical systems, and artificial life. Many concepts central to these fields, such as complexity, self-organization, self-reproduction, autonomy, networks, connectionism, and adaptation, were first explored by cyberneticians during the 1940s and 1950s. Examples include von Neumann's computer architectures, game theory, and cellular automata; Ashby's and von Foerster's analysis of self-organization; Braitenberg's autonomous robots; and McCulloch's artificial neural nets, perceptrons, and classifiers.

B. Second-Order Cybernetics

Cybernetics had from the beginning been interested in the similarities between autonomous, living systems and machines. In this postwar era, the fascination with the new control and computer technologies tended to focus attention on the engineering approach, where it is the system designer who determines what the system will do. However, after the control engineering and computer science disciplines had become fully independent, the remaining cyberneticists felt the need to clearly distinguish themselves from these more mechanistic approaches, by emphasizing autonomy, self-organization, cognition, and the role of the observer in modeling a system. In the early 1970s this movement became known as *second-order cybernetics*.

They began with the recognition that all our knowledge of systems is mediated by our simplified representations— or *models*—of them, which necessarily ignore those aspects of the system which are irrelevant to the purposes for which the model is constructed. Thus the properties of the systems themselves must be distinguished from those of their models, which depend on us as their creators. An engineer working with a mechanical system, on the other hand, almost always know its internal structure and behavior to a high degree of accuracy, and therefore tends to de-emphasize the system/model distinction, acting as if the model *is* the system.

Moreover, such an engineer, scientist, or "first-order" cyberneticist, will study a system as if it were a passive, objectively given "thing," that can be freely observed, manipulated, and taken apart. A second-order cyberneticist working with an organism or social system, on the other hand, recognizes that system as an agent in its own right, interacting with another agent, the observer. As quantum mechanics has taught us, observer and observed cannot be separated, and the result of observations will depend on their interaction. The observer too is a cybernetic system, trying to construct a model of another cybernetic system. To understand this process, we need a "cybernetics of cybernetics," i.e., a "meta" or "second-order" cybernetics.

These cyberneticians' emphasis on such epistemological, psychological, and social issues was a welcome complement to the reductionist climate which followed on the great progress in science and engineering of the day. However, it may have led them to overemphasize the novelty of their "second-order" approach. First, it must be noted that most founding fathers of cybernetics, such as Ashby, McCulloch, and Bateson, explicitly or implicitly agreed with the importance of autonomy, self-organization, and the subjectivity of modeling. Therefore, they can hardly be portrayed as "first-order" reductionists. Second, the intellectual standard bearers of the second-order approach during the 1970s, such as von Foerster, Pask, and Maturana, were themselves directly involved in the development of "first-order" cybernetics in the 1950s and 1960s. In fact, if we look more closely at the history of the field, we see a continuous development toward a stronger focus on autonomy and the role of the observer, rather than a clean break between generations or approaches. Finally, the second-order perspective is now firmly ingrained in the foundations of cybernetics overall. For those reasons, the present article will discuss the basic concepts and principles of cybernetics as a whole, without explicitly distinguishing between "first-order" and "second-order" ideas, and introduce cybernetic concepts through a series of models of classes of systems.

It must further be noted that the sometimes ideological fervor driving the second-order movement may have led a bridge too far. The emphasis on the irreducible complexity of the various system-observer interactions and on the subjectivity of modeling has led many to abandon formal approaches and mathematical modeling altogether, limiting themselves to philosophical or literary discourses. It is ironic that one of the most elegant computer simulations of the second-order idea that models affect the very system they are supposed to model was not created by a cyberneticist, but by the economist Brian Arthur. Moreover, some people feel that the second-order fascination with self-reference and observers observing observers observing themselves has fostered a potentially dangerous detachment from concrete phenomena.

C. Cybernetics Today

In spite of its important historical role, cybernetics has not really become established as an autonomous discipline. Its practitioners are relatively few, and not very well organized. There are few research departments devoted to the domain, and even fewer academic programs. There are many reasons for this, including the intrinsic complexity and abstractness of the subject domain, the lack of up-to-date textbooks, the ebb and flow of scientific fashions, and the apparent overreaching of the second-order movement. But the fact that the Systems Sciences (including General Systems Theory) are in a similar position indicates that the most important cause is the difficulty of maintaining the coherence of a broad, interdisciplinary field in the wake of the rapid growth of its more specialized and application-oriented "spin-off" disciplines, such as computer science, artificial intelligence, neural networks, and control engineering, which tended to sap away enthusiasm, funding, and practitioners from the more theoretical mother field.

Many of the core ideas of cybernetics have been assimilated by other disciplines, where they continue to influence scientific developments. Other important cybernetic principles seem to have been forgotten, though, only to be periodically rediscovered or reinvented in different domains. Some examples are the rebirth of neural networks, first invented by cyberneticists in the 1940s, in the late 1960s and again in the late 1980s; the rediscovery of the importance of autonomous interaction by robotics and AI in the 1990s; and the significance of positive feedback effects in complex systems, rediscovered by economists in the 1990s. Perhaps the most significant recent development is the growth of the *complex adaptive systems* movement, which, in the work of authors such as John Holland, Stuart Kauffman, and Brian Arthur and the subfield of *artificial life*, has used the power of modern computers to simulate and thus experiment with and develop many of the ideas of cybernetics. It thus seems to have taken over the cybernetics banner in its mathematical modeling of complex systems across disciplinary boundaries, however, while largely ignoring the issues of goal-directedness and control.

More generally, as reflected by the ubiquitous prefix "cyber," the broad cybernetic philosophy that systems are defined by their abstract relations, functions, and information flows, rather than by their concrete material or components, is starting to pervade popular culture, albeit it in a still shallow manner, driven more by fashion than by deep understanding. This has been motivated primarily by the explosive growth of information-based technologies including automation, computers, the Internet, virtual reality, software agents, and robots. It seems likely that as the applications of these technologies become increasingly complex, far-reaching, and abstract, the need will again be felt for an encompassing conceptual framework, such as cybernetics, that can help users and designers alike to understand the meaning of these developments.

Cybernetics as a theoretical framework remains a subject of study for a few committed groups, such as the *Principia Cybernetica Project*, which tries to integrate cybernetics with evolutionary theory, and the *American Society for Cybernetics*, which further develops the second-order approach. The *sociocybernetics* movement actively pursues a cybernetic understanding of social systems. The cybernetics-related programs on autopoiesis, systems dynamics, and control theory also continue, with applications in management science and even psychological therapy. Scattered research centers, particularly in Central and Eastern Europe, are still devoted to specific technical applications, such as biological cybernetics, medical cybernetics, and engineering cybernetics, although they tend to keep closer contact with their field of application than with the broad theoretical development of cybernetics. General Information Theory has grown as the search for formal representations which are not based strictly on classical probability theory.

There has also been significant progress in building a semiotic theory of information, where issues of the semantics and meaning of signals are at last being seriously considered. Finally, a number of authors are seriously questioning the limits of mechanism and formalism for interdisciplinary modeling in particular, and science in general. The issues here thus become what the ultimate limits on knowledge might be, especially as expressed in mathematical and computer-based models. What's at stake is whether it is possible, in principle, to construct models, whether formal or not, which will help us understand the full complexity of the world around us.

II. RELATIONAL CONCEPTS

A. Distinctions and Relations

In essence, cybernetics is concerned with those properties of systems that are independent of their concrete material or components. This allows it to describe physically very different systems, such as electronic circuits, brains, and organizations, with the same concepts, and to look for isomorphisms between them. The only way to abstract a system's physical aspects or components while still preserving its essential structure and functions is to consider relations: How do the components differ from or connect to each other? How does the one transform into the other?

To approach these questions, cyberneticians use high level concepts such as *order, organization, complexity, hierarchy, structure, information,* and *control,* investigating how these are manifested in systems of different types. These concepts are *relational,* in that they allow us to analyze and formally model different abstract properties of systems and their dynamics, for example, allowing us to

ask such questions as whether complexity tends to increase with time.

Fundamental to all of these relational concepts is that of *difference* or *distinction.* In general, cyberneticians are not interested in a phenomenon in itself, but only in the difference between its presence and absence, and how that relates to other differences corresponding to other phenomena. This philosophy extends back to Leibniz, and is expressed most succinctly by Bateson's famous definition of information as "a difference that makes a difference." Any observer necessarily starts by conceptually separating or distinguishing the object of study, the *system,* from the rest of the universe, the *environment.* A more detailed study will go further to distinguish between the presence and absence of various properties (also called dimensions or attributes) of the systems. For example, a system such as billiard ball can have properties, such as a particular color, weight, position, or momentum. The presence or absence of each such property can be represented as a binary, Boolean variable, with two values: "yes," the system has the property, or "no," it does not. G. Spencer Brown, in his book "Laws of Form," has developed a detailed calculus and algebra of distinctions, and shown that this algebra, although starting from much simpler axioms, is isomorphic to the more traditional Boolean algebra.

B. Variety and Constraint

This binary approach can be generalized to a property having multiple discrete or continuous values, for example, which color or what position or momentum. The conjunction of all the values of all the properties that a system at a particular moment has or lacks determines its *state.* For example, a billiard ball can have color *red,* position *x* and momentum *p.* The set of all possible states that a system can be in defines its *state space.* An essential component of cybernetic modeling is a quantitative measure for the size of the state space, or the number of distinct states. This measure is called *variety.* Variety represents the freedom the system has in choosing a particular state, and thus the uncertainty we have about which state the system occupies. Variety V is defined as the number of elements in the state space S, or, more commonly, as the logarithm to the basis two of that number: $V = \log_2(|S|)$.

The unit of variety in the logarithmic form is the *bit.* A variety of one bit, $V = 1$, means that the system has two possible states, that is, one difference. In the simplest case of n binary variables, $V = \log_2(2^n) = n$ is therefore equal to the minimal number of independent dimensions. But in general, the variables used to describe a system are neither binary nor independent. For example, if a particular type of berry can, depending on its degree of ripeness, be either small and green or large and red (recognizing only two states of size and color), then the variables "color" and "size" are completely dependent on each other, and the total variety is 1 bit rather than the 2 you would get if you would count the variables separately.

More generally, if the actual variety of states that the system can exhibit is smaller than the variety of states we can potentially conceive, then the system is said to be constrained. *Constraint C* can be defined as the difference between maximal and actual variety: $C = V_{max} - V$. Constraint is what reduces our uncertainty about the system's state, and thus allows us to make nontrivial predictions. For example, in the previously cited example if we detect that a berry is small, we can predict that it will also be green. Constraint also allows us to formally model relations, dependencies, or couplings between different systems, or aspects of systems. If you model different systems or different aspects, or dimensions of one system together, then the joint state space is the Cartesian product of the individual state spaces: $S = S_1 \times S_2 \times \cdots S_n$. Constraint on this product space can thus represent the mutual dependency between the states of the subspaces, like in the berry example, where the state in the color space determines the state in the size space, and vice versa.

C. Entropy and Information

Variety and constraint can be measured in a more general form by introducing probabilities. Assume that we do not know the precise state s of a system, but only the probability distribution $P(s)$ that the system would be in state s. Variety can then be expressed through a formula equivalent to *entropy*, as defined by Boltzmann for statistical mechanics:

$$H(P) = -\sum_{s \in S} P(s) \cdot \log P(s). \qquad (1)$$

H reaches its maximum value if all states are equiprobable, that is, if we have no indication whatsoever to assume that one state is more probable than another state. Thus it is natural that in this case entropy H reduces to variety V. Again, H expresses our uncertainty or ignorance about the system's state. It is clear that $H = 0$, if and only if the probability of a certain state is 1 (and of all other states 0). In that case we have maximal certainty or complete information about what state the system is in.

We defined constraint as that which reduces uncertainty, that is, the difference between maximal and actual uncertainty. This difference can also be interpreted in a different way, as *information*, and historically H was introduced by Shannon as a measure of the capacity for information transmission of a communication channel. Indeed, if we get some information about the state of the system (e.g., through observation), then this will reduce our uncertainty about the system's state, by excluding—or reducing the probability of—a number of states. The information I we receive from an observation is equal to the degree to which uncertainty is reduced: $I = H(\text{before}) - H(\text{after})$. If the observation completely determines the state of the system ($H(\text{after}) = 0$), then information I reduces to the initial entropy or uncertainty H.

Although Shannon came to disavow the use of the term "information" to describe this measure, because it is purely syntactic and ignores the *meaning* of the signal, his theory came to be known as Information Theory nonetheless. H has been vigorously pursued as a measure for a number of higher-order relational concepts, including complexity and organization. Entropies, correlates to entropies, and correlates to such important results as Shannon's 10th Theorem and the Second Law of Thermodynamics have been sought in biology, ecology, psychology, sociology, and economics.

We also note that there are other methods of weighting the state of a system which do not adhere to probability theory's additivity condition that the sum of the probabilities must be 1. These methods, involving concepts from fuzzy systems theory and possibility theory, lead to alternative information theories. Together with probability theory these are called Generalized Information Theory (GIT). While GIT methods are under development, the probabilistic approach to information theory still dominates applications.

D. Modeling Dynamics

Given these static descriptions of systems, we can now model their dynamics and interactions. Any process or change in a system's state can be represented as a transformation: $T: S \to S: s(t) \to s(t+1)$. The function T by definition is *one-to-one* or *many-to-one*, meaning that an initial state $s(t)$ is always mapped onto a single state $s(t+1)$. Change can be represented more generally as a relation $R \subset S \times S$, thus allowing the modelling of *one-to-many* or *many-to-many* transformations, where the same initial state can lead to different final states. Switching from states s to probability distributions $P(s)$ allows us to again represent such indeterministic processes by a function: $M: P(s, t) \to P(s, t+1)$. M is a stochastic process, or more precisely, a *Markov chain*, which can be represented by a matrix of transition probabilities: $P(s_j(t+1) \mid s_i(t)) = M_{ij} \in [0, 1]$.

Given these process representations, we can now study the dynamics of variety, which is a central theme of cybernetics. It is obvious that a one-to-one transformation will conserve all distinctions between states and therefore the variety, uncertainty or information. Similarly, a many-to-one mapping will erase distinctions, and thus reduce variety, while an indeterministic, one-to-many mapping will increase variety and thus uncertainty. With a general many-to-many mapping, as represented by a Markov process, variety can increase or decrease, depending on the initial probability distribution and the structure of the transition matrix. For example, a distribution with variety 0 cannot decrease in variety and will in general increase, while a distribution with maximum variety will in general decrease. In the following sections we will discuss some special cases of this most general of transformations.

With some small extensions, this dynamical representation can be used to model the interactions between systems. A system A affects a system B if the state of B at time $t + 1$ is dependent on the state of A at time t. This can be represented as a transformation $T: S_A \times S_B \rightarrow S_B: (s_A(t), s_B(t)) \rightarrow s_B(t + 1)$. s_A here plays the role of the *input* of B. In general, B will not only be affected by an outside system A, but in turn affect another (or the same) system C. This can be represented by another a transformation $T': S_A \times S_B \rightarrow S_C: (s_A(t), s_B(t)) \rightarrow s_C(t + 1)$. s_C here plays the role of the *output* of B. For the outside observer, B is a process that transforms input into output. If the observer does not know the states of B, and therefore the precise transformations T and T', then B acts as a *black box*. By experimenting with the sequence of inputs $s_A(t)$, $s_A(t + 1)$, $s_A(t + 2)$, ..., and observing the corresponding sequence of outputs $s_C(t + 1)$, $s_C(t + 2)$, $s_C(t + 3)$, ..., the observer may try to reconstruct the dynamics of B. In many cases, the observer can determine a state space S_B so that both transformations become deterministic, without being able to directly observe the properties or components of B.

This approach is easily extended to become a full theory of automata and computing machines, and is the foundation of most of modern computer science. This again illustrates how cybernetic modeling can produce useful predictions by only looking at *relations* between variables, while ignoring the physical components of the system.

III. CIRCULAR PROCESSES

In classical, Newtonian science, causes are followed by effects, in a simple, linear sequence. Cybernetics, on the other hand, is interested in processes where an effect feeds back into its very cause. Such circularity has always been difficult to handle in science, leading to deep conceptual problems such as the logical paradoxes of self-reference. Cybernetics discovered that circularity, if modelled adequately, can help us to understand fundamental phenomena, such as self-organization, goal-directedness, identity, and life, in a way that had escaped Newtonian science. For example, von Neumann's analysis of reproduction as the circular process of self-construction anticipated the discovery of the genetic code. Moreover, circular processes are in fact ubiquitous in complex, networked sys-

tems such as organisms, ecologies, economies, and other social structures.

A. Self-Application

In simple mathematical terms, circularity can be represented by an equation representing how some phenomenon or variable y is mapped, by a transformation or process f, onto itself:

$$y = f(y). \tag{2}$$

Depending on what y and f stand for, we can distinguish different types of circularities. As a concrete illustration, y might stand for an image, and f for the process whereby a video camera is pointed at the image, the image is registered and transmitted to a TV screen or monitor. The circular relation $y = f(y)$ would then represent the situation where the camera points at the image shown on its own monitor. Paradoxically, the image y in this situation is both cause and effect of the process, it is both object and representation of the object. In practice, such a video loop will produce a variety of abstract visual patterns, often with complex symmetries.

In discrete form, Eq. (2) becomes $y_{t+1} = f(y_t)$. Such equations have been extensively studied as iterated maps, and are the basis of *chaotic dynamics* and *fractal geometry*. Another variation is the equation, well known from quantum mechanics and linear algebra:

$$ky = f(y). \tag{3}$$

The real or complex number k is an *eigenvalue* of f, and y is an *eigenstate*. Eq. (3) reduces to the basic Eq. (2) if $k = 1$ or if y is only defined up to a constant factor. If $k = \exp(2\pi i \, m/n)$, then $f^n(y)$ is again y. Thus, imaginary eigenvalues can be used to model periodic processes, where a system returns to the same state after passing through n intermediate states.

An example of such periodicity is the self-referential statement (equivalent to the liar's paradox): "this statement is false." If we start by assuming that the statement is true, then we must conclude that it is false. If we assume it is false, then we must conclude it is true. Thus, the truth value can be seen to oscillate between true and false, and can perhaps be best conceived as having the equivalent of an imaginary value. Using Spencer Brown's calculus of distinctions, Varela has proposed a similar solution to the problem of self-referential statements.

B. Self-Organization

The most direct application of circularity is where $y \in S$ stands for a system's state in a state space S, and f for a dynamic transformation or process. Equation (2) then states that y is a fixpoint of the function f, or an *equilibrium* or *absorbing* state of the dynamic system: if the system reaches state y, it will stop changing. This can be generalized to the situation where y stands for a subset of the state space, $y \subset S$. Then, every state of this subset is sent to another state of this subspace: $\forall x \in y: f(x) \in y$. Assuming y has no smaller subset with the same property, this means that y is an *attractor* of the dynamics. The field of dynamical systems studies attractors in general, which can have any type of shape or dimension, including zero-dimensional (the equilibrium state discussed above), one-dimensional (a limit cycle, where the system repeatedly goes through the same sequence of states), and fractal (a so-called "strange" attractor).

An attractor y is in general surrounded by a *basin* $B(y)$: a set of states outside y whose evolution necessarily ends up inside: $\forall s \in B(y), s \notin y, \exists n$ such that $f^n(s) \in y$. In a deterministic system, every state either belongs to an attractor or to a basin. In a stochastic system there is a third category of states that can end up in either of several attractors. Once a system has entered an attractor, it can no longer reach states outside the attractor. This means that our uncertainty (or statistical entropy) H about the system's state has decreased: we now know for sure that it is not in any state that is not part of the attractor. This spontaneous reduction of entropy or, equivalently, increase in order or constraint, can be viewed as a most general model of *self-organization*.

Every dynamical system that has attractors will eventually end up in one of these attractors, losing its freedom to visit any other state. This is what Ashby called the *principle of self-organization*. He also noted that if the system consists of different subsystems, then the constraint created by self-organization implies that the subsystems have become mutually dependent, or mutually adapted. A simple example is magnetization, where an assembly of magnetic spins that initially point in random directions (maximum entropy), end up all being aligned in the same direction (minimum entropy, or mutual adaptation). Von Foerster added that self-organization can be enhanced by random perturbations ("noise") of the system's state, which speed up the descent of the system through the basin, and makes it leave shallow attractors so that it can reach deeper ones. This is the *order from noise* principle.

C. Closure

The "attractor" case can be extended to the case where y stands for a complete state space. Equation (2) then represents the situation where every state of y is mapped onto another state of y by f. More generally, f might stand for a group of transformations, rather than a single transformation. If f represents the possible dynamics of the system with state space y, under different values of external parameters, then we can say that the system is organizationally closed: it is invariant under any possible dynamical transformation. This requirement of *closure* is implicit in traditional mathematical models of systems. Cybernetics, on the other hand, studies closure explicitly, with the view that systems may be open and closed simultaneously for different kinds of properties f_1 and f_2. Such closures give systems an unambiguous *identity*, explicitly distinguishing what is inside from what is outside the system.

One way to achieve closure is self-organization, leaving the system in an attractor subspace. Another way is to expand the state space y into a larger set y^* so that y^* recursively encompasses all images through f of elements of y: $\forall x \in y: x \in y^*; \forall x' \in y^*: f(x') \in y^*$. This is the traditional definition of a set y^* through *recursion*, which is frequently used in computer programming to generate the elements of a set y^* by iteratively applying transformations to all elements of a starting set y.

A more complex example of closure is *autopoiesis* ("self-production"), the process by which a system recursively produces its own network of physical components, thus continuously regenerating its essential organization in the face of wear and tear. Note that such "organizational" closure is not the same as thermodynamic closure: the autopoietic system is open to the exchange of matter and energy with its environment, but it is autonomously responsible for the way these resources are organized. Maturana and Varela have postulated autopoiesis to be the defining characteristic of living systems. Another fundamental feature of life, *self-reproduction*, can be seen as a special case of autopoiesis, where the self-produced components are not used to rebuild the system, but to assemble a copy of it. Both reproduction and autopoiesis are likely to have evolved from an autocatalytic cycle, an organizationally closed cycle of chemical processes such that the production of every molecule participating in the cycle is catalysed by another molecule in the cycle.

D. Feedback Cycles

In addition to looking directly at a state y, we may focus on the *deviation* $\Delta y = (y - y_0)$ of y from some given (e.g., equilibrium) state y_0, and at the "feedback" relations through which this deviation depends on itself. In the simplest case, we could represent this as $\Delta y(t + \Delta t) = k \Delta y(t)$. According to the sign of the dependency k, two special cases can be distinguished.

If a positive deviation at time t (increase with respect to y_0) leads to a negative deviation (decrease with respect to y_0) at the following time step, the feedback is *negative*

($k < 0$). For example, more rabbits eat more grass, and therefore less grass will be left to feed further rabbits. Thus, an increase in the number of rabbits above the equilibrium value will lead, via a decrease in the supply of grass, at the next time step to a decrease in the number of rabbits. Complementarily, a decrease in rabbits leads to an increase in grass, and thus again to an increase in rabbits. In such cases, any deviation from y_0 will be suppressed, and the system will spontaneously return to equilibrium. The equilibrium state y_0 is *stable*, resistant to perturbations. Negative feedback is ubiquitous as a control mechanism in machines of all sorts, in organisms (for example, in homeostasis and the insulin cycle), in ecosystems, and in the supply/demand balance in economics.

The opposite situation, where an increase in the deviation produces further increases, is called *positive* feedback. For example, more people infected with the cold virus will lead to more viruses being spread in the air by sneezing, which will in turn lead to more infections. An equilibrium state surrounded by positive feedback is necessarily unstable. For example, the state where no one is infected is an unstable equilibrium, since it suffices that one person become infected for the epidemic to spread. Positive feedbacks produce a runaway, explosive growth, which will only come to a halt when the necessary resources have been completely exhausted. For example, the virus epidemic will only stop spreading after all people that could be infected have been infected. Other examples of such positive feedbacks are arms races, snowball effects, increasing returns in economics, and the chain reactions leading to nuclear explosions. While negative feedback is the essential condition for stability, positive feedbacks are responsible for growth, self-organization, and the amplification of weak signals. In complex, hierarchical systems, higher-level negative feedbacks typically constrain the growth of lower-level positive feedbacks.

The positive and negative feedback concepts are easily generalized to networks of multiple causal relations. A causal link between two variables, $A \rightarrow B$ (e.g., *infected people $\rightarrow$ viruses*), is positive if an increase (decrease) in A produces an increase (decrease) in B. It is negative, if an increase produces a decrease, and vice versa. Each loop in a causal network can be given an overall sign by multiplying the signs ($+$ or $-$) of each of its links. This gives us a simple way to determine whether this loop will produce stabilization (negative feedback) or a runaway process (positive feedback). In addition to the sign of a causal connection, we also need to take into account the *delay* or *lag* between cause and effect, e.g., the rabbit population will only start to increase several weeks after the grass supply has increased. Such delays may lead to an oscillation, or limit cycle, around the equilibrium value.

Such networks of interlocking positive and negative feedback loops with lags are studied in the mathematical field of System Dynamics, a broad program modelling complex biological, social, economic and psychological systems. System Dynamics' most well-known application is probably the "Limits to Growth" program popularized by the Club of Rome, which continued the pioneering computer simulation work of Jay Forrester. System dynamics has since been popularized in the Stella software application and computer games such as SimCity.

IV. GOAL-DIRECTEDNESS AND CONTROL

A. Goal-Directedness

Probably the most important innovation of cybernetics is its explanation of goal-directedness or purpose. An autonomous system, such as an organism, or a person, can be characterized by the fact that it pursues its own goals, resisting obstructions from the environment that would make it deviate from its preferred state of affairs. Thus, goal-directedness implies regulation of—or control over—perturbations. A room in which the temperature is controlled by a thermostat is the classic simple example. The setting of the thermostat determines the preferred temperature or goal state. Perturbations may be caused by changes in the outside temperature, drafts, opening of windows or doors, etc. The task of the thermostat is to minimize the effects of such perturbations, and thus to keep the temperature as much as possible constant with respect to the target temperature.

On the most fundamental level, the goal of an autonomous or autopoietic system is survival, that is, maintenance of its essential organization. This goal has been built into all living systems by natural selection: those that were not focused on survival have simply been eliminated. In addition to this primary goal, the system will have various subsidiary goals, such as keeping warm or finding food, that indirectly contribute to its survival. Artificial systems, such as thermostats and automatic pilots, are not autonomous: their primary goals are constructed in them by their designers. They are *allopoietic*: their function is to produce something other ("allo") than themselves.

Goal-directedness can be understood most simply as suppression of deviations from an invariant goal state. In that respect, a goal is similar to a stable equilibrium, to which the system returns after any perturbation. Both goal-directedness and stability are characterized by *equifinality*: different initial states lead to the same final state, implying the destruction of variety. What distinguishes them is that a stable system automatically returns to its equilibrium state, without performing any work or effort. But a goal-directed system must actively intervene to achieve and maintain its goal, which would not be an equilibrium otherwise.

Control may appear essentially conservative, resisting all departures from a preferred state. But the net effect can be very dynamic or progressive, depending on the complexity of the goal. For example, if the goal is defined as the distance relative to a moving target, or the rate of increase of some quantity, then suppressing deviation from the goal implies constant change. A simple example is a heat-seeking missile attempting to reach a fast moving enemy plane.

A system's "goal" can also be a subset of acceptable states, similar to an attractor. The dimensions defining these states are called the *essential variables*, and they must be kept within a limited range compatible with the survival of the system. For example, a person's body temperature must be kept within a range of approximately 35–40°C. Even more generally, the goal can be seen as a gradient, or "fitness" function, defined on state space, which defines the degree of "value" or "preference" of one state relative to another one. In the latter case, the problem of control becomes one of on-going optimization or maximization of fitness.

B. Mechanisms of Control

While the perturbations resisted in a control relation can originate either inside (e.g., functioning errors or quantum fluctuations) or outside of the system (e.g., attack by a predator or changes in the weather), functionally we can treat them as if they all come from the same, external source. To achieve its goal in spite of such perturbations, the system must have a way to block their effect on its essential variables. There are three fundamental methods to achieve such regulation: buffering, feedback and feedforward (see Fig. 1).

Buffering is the passive absorption or damping of perturbations. For example, the wall of the thermostatically controlled room is a buffer: the thicker or the better insulated it is, the less effect fluctuations in outside temperature will have on the inside temperature. Other examples are the shock absorbers in a car, and a reservoir, which provides a regular water supply in spite of variations in rain

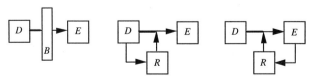

FIGURE 1 Basic mechanisms of regulation, from left to right: buffering, feedforward and feedback. In each case, the effect of disturbances *D* on the essential variables *E* is reduced, either by a passive buffer *B*, or by an active regulator *R*.

fall. The mechanism of buffering is similar to that of a stable equilibrium: dissipating perturbations without active intervention. The disadvantage is that it can only dampen the effects of uncoordinated fluctuations; it cannot systematically drive the system to a non-equilibrium state, or even keep it there. For example, however well-insulated, a wall alone cannot maintain the room at a temperature higher than the average outside temperature.

Feedback and feedforward both require *action* on the part of the system to suppress or compensate the effect of the fluctuation. For example, the thermostat will counteract a drop in temperature by switching on the heating. *Feedforward* control will suppress the disturbance *before* it has had the chance to affect the system's essential variables. This requires the capacity to *anticipate* the effect of perturbations on the system's goal. Otherwise the system would not know which external fluctuations to consider as perturbations, or how to effectively compensate their influence before it affects the system. This requires that the control system be able to gather early information about these fluctuations.

For example, feedforward control might be applied to the thermostatically controlled room by installing a temperature sensor outside of the room, which would warn the thermostat about a drop in the outside temperature, so that it could start heating before this would affect the inside temperature. In many cases, such advance warning is difficult to implement, or simply unreliable. For example, the thermostat might start heating the room, anticipating the effect of outside cooling, without being aware that at the same time someone in the room switched on the oven, producing more than enough heat to offset the drop in outside temperature. No sensor or anticipation can ever provide complete information about the future effects of an infinite variety of possible perturbations, and therefore feedforward control is bound to make mistakes. With a good control system, the resulting errors may be few, but the problem is that they will accumulate in the long run, eventually destroying the system.

The only way to avoid this accumulation is to use feedback, that is, compensate an error or deviation from the goal *after* it has happened. Thus feedback control is also called *error-controlled regulation*, since the error is used to determine the control action, as with the thermostat which samples the temperature inside the room, switching on the heating whenever that temperature reading drops lower than a certain reference point from the goal temperature. The disadvantage of feedback control is that it first must allow a deviation or error to appear before it can take action, since otherwise it would not know which action to take. Therefore, feedback control is by definition imperfect, whereas feedforward could in principle, but not in practice, be made error-free.

The reason feedback control can still be very effective is continuity: deviations from the goal usually do not appear at once, they tend to increase slowly, giving the controller the chance to intervene at an early stage when the deviation is still small. For example, a sensitive thermostat may start heating as soon as the temperature has dropped one tenth of a degree below the goal temperature. As soon as the temperature has again reached the goal, the thermostat switches off the heating, thus keeping the temperature within a very limited range. This very precise adaptation explains why thermostats in general do not need outside sensors, and can work purely in feedback mode. Feedforward is still necessary in those cases where perturbations are either discontinuous, or develop so quickly that any feedback reaction would come too late. For example, if you see someone pointing a gun in your direction, you would better move out of the line of fire immediately, instead of waiting until you feel the bullet making contact with your skin.

C. The Law of Requisite Variety

Control or regulation is most fundamentally formulated as a reduction of variety: perturbations with high variety affect the system's internal state, which should be kept as close as possible to the goal state, and therefore exhibit a low variety. So in a sense control prevents the transmission of variety from environment to system. This is the opposite of information transmission, where the purpose is to maximally conserve variety.

In active (feedforward and/or feedback) regulation, each disturbance from D will have to be compensated by an appropriate counteraction from the regulator R (Fig. 1). If R would react in the same way to two different disturbances, then the result would be two different values for the essential variables, and thus imperfect regulation. This means that if we wish to completely block the effect of D, the regulator must be able to produce at least as many counteractions as there are disturbances in D. Therefore, the variety of R must be at least as great as the variety of D. If we moreover take into account the constant reduction of variety K due to buffering, the principle can be stated more precisely as:

$$V(E) \geq V(D) - V(R) - K \qquad (4)$$

Ashby has called this principle the *law of requisite variety*: in active regulation *only variety can destroy variety*. It leads to the somewhat counterintuitive observation that the regulator must have a sufficiently *large* variety of actions in order to ensure a sufficiently *small* variety of outcomes in the essential variables E. This principle has important implications for practical situations: since the variety of perturbations a system can potentially be confronted with

is unlimited, we should always try maximize its internal variety (or diversity), so as to be optimally prepared for any foreseeable or unforeseeable contigency.

D. Components of a Control System

Now that we have examined control in the most general way, we can look at the more concrete components and processes that constitute a control system, such as a simple thermostat or a complex organism or organization (Fig. 2). As is usual in cybernetics, these components are recognized as *functional*, and may or may not correspond to *structural* units.

The overall scheme is a feedback cycle with two inputs: the goal, which stands for the preferred values of the system's essential variables, and the disturbances, which stand for all the processes in the environment that the system does not have under control but that can affect these variables. The system starts by observing or sensing the variables that it wishes to control because they affect its preferred state. This step of *perception* creates an internal representation of the outside situation. The information in this representation then must be processed in order to determine: (1) in what way it may affect the goal and (2) what is the best reaction to safeguard that goal.

Based on this interpretation, the system then decides on an appropriate action. This action affects some part of the environment, which in turn affects other parts of the environment through the normal causal processes or dynamics of that environment. These dynamics are influenced by the set of unknown variables which we have called the disturbances. This dynamical interaction affects among others the variables that the system keeps under observation. The change in these variables is again perceived by the system, and this again triggers interpretation, decision and action, thus closing the control loop.

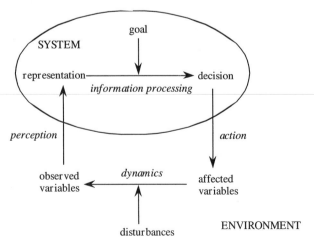

FIGURE 2 Basic components of a control system.

This general scheme of control may include buffering, feedforward and feedback regulation. Buffering is implicit in the dynamics, which determines to what degree the disturbances affect the observed variables. The observed variables must include the essential variables that the system wants to keep under control (feedback or error-controlled regulation) in order to avoid error accumulation. However, they will in general also include various nonessential variables, to function as early warning signals for anticipated disturbances. This implements feedforward regulation.

The components of this scheme can be as simple or as complex as needed. In the thermostat, perception is simply a sensing of the one-dimensional variable room temperature; the goal is the set-point temperature that the thermostat tries to achieve; information processing is the trivial process of deciding whether the perceived temperature is lower than the desired temperature or not; and action consists of either heating, if the temperature is lower, or doing nothing. The affected variable is the amount of heat in the room. The disturbance is the amount of heat exchanged with the outside. The dynamics is the process by which inside heating and heath exchange with the outside determine inside temperature.

For a more complex example, we may consider a board of directors whose goal is to maximize the long-term revenue of their company. Their actions consist of various initiatives, such as publicity campaigns, hiring managers, starting up production lines, saving on administrative costs, etc. This affects the general functioning of the company. But this functioning is also affected by factors that the board does not control such as the economic climate, the activities of competitors, the demands of the clients, etc. Together these disturbances and the initiatives of the board determine the success of the company, which is indicated by variables such as amount of orders, working costs, production backlog, company reputation, etc. The board, as a control system, will interpret each of these variables with reference to their goal of maximizing profits, and decide about actions to correct any deviation from the preferred course.

Note that the control loop is completely symmetric: if the scheme in Fig. 2 is rotated over $180°$, environment becomes system while disturbance becomes goal, and vice versa. Therefore, the scheme could also be interpreted as two interacting systems, each of which tries to impose its goal on the other one. If the two goals are incompatible, this is a model of conflict or competition; otherwise, the interaction may settle into a mutually satisfactory equilibrium, providing a model of compromise or cooperation.

But in control we generally mean to imply that one system is more powerful than the other one, capable of suppressing any attempt by the other system to impose *its* preferences. To achieve this, an *asymmetry* must be built into the control loop: the actions of the system (controller) must have more effect on the state of the environment (controlled) than the other way around. This can also be viewed as an *amplification* of the signal travelling through the control system: weak perceptual signals, carrying information but almost no energy, lead to powerful actions, carrying plenty of energy. This asymmetry can be achieved by weakening the influence of the environment, e.g., by buffering its actions, and by strengthening the actions of the system, e.g., by providing it with a powerful energy source. Both cases are illustrated by the thermostat: the walls provide the necessary insulation from outside perturbations, and the fuel supply provides the capacity to generate enough heat. No thermostatic control would be possible in a room without walls or without energy supply. The same requirements applied to the first living cells, which needed a protective membrane to buffer disturbances, and a food supply for energy.

E. Control Hierarchies

In complex control systems, such as organisms or organizations, goals are typically arranged in a hierarchy, where the higher-level goals control the settings for the subsidiary goals. For example, your primary goal of survival entails the lower-order goal of maintaining sufficient hydration, which may activate the goal of drinking a glass of water. This will in turn activate the goal of bringing the glass to your lips. At the lowest level, this entails the goal of keeping your hand steady without spilling water.

Such hierarchical control can be represented in terms of the control scheme of Fig. 2 by adding another layer, as in Fig. 3. The original goal 1 has now itself become the result of an action, taken to achieve the higher level goal 2. For example, the thermostat's goal of keeping the temperature at its set-point can be subordinated to the higher order goal of keeping the temperature pleasant to the people present without wasting fuel. This can be implemented by adding an infrared sensor that perceives whether there are people present in the room, and if so, then setting the thermostat at a higher temperature T_1, otherwise setting it to the lower temperature T_2. Such control layers can be added arbitrarily by making the goal at level n dependent on the action at level $n + 1$.

A control loop will reduce the variety of perturbations, but it will in general not be able to eliminate all variation. Adding a control loop on top of the original loop may eliminate the residual variety, but if that is not sufficient, another hierarchical level may be needed. The required number of levels therefore depends on the regulatory ability of the individual control loops: the weaker that ability, the more hierarchy is needed. This is Aulin's *law of*

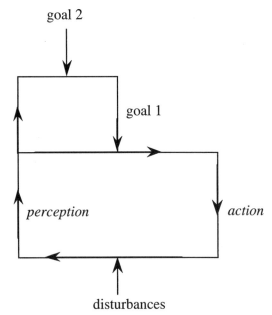

FIGURE 3 A hierarchical control system with two control levels.

requisite hierarchy. On the other hand, increasing the number of levels has a negative effect on the overall regulatory ability, since the more levels the perception and action signals have to pass through, the more they are likely to suffer from noise, corruption, or delays. Therefore, if possible, it is best to maximize the regulatory ability of a single layer, and thus minimize the number of requisite layers. This principle has important applications for social organizations, which have a tendency to multiply the number of bureaucratic levels. The present trend toward the flattening of hierarchies can be explained by the increasing regulatory abilities of individuals and organizations, due to better education, management and technological support.

Still, when the variety becomes really too great for one regulator, a higher control level must appear to allow further progress. Valentin Turchin has called this process a *metasystem transition*, and proposed it as a basic unit, or "quantum," of the evolution of cybernetic systems. It is responsible for the increasing functional complexity which characterizes such fundamental developments as the origins of life, multicellular organisms, the nervous system, learning, and human culture.

V. COGNITION

A. Requisite Knowledge

Control is not only dependent on a requisite variety of actions in the regulator: the regulator must also *know* which action to select in response to a given perturbation. In the simplest case, such knowledge can be represented as a one-to-one mapping from the set D of perceived disturbances to the set R of regulatory actions: $f: D \rightarrow R$, which maps each disturbance to the appropriate action that will suppress it. For example, the thermostat will map the perception "temperature too low" to the action "heat," and the perception "temperature high enough" to the action "do not heat." Such knowledge can also be expressed as a set of production rules of the form "if *condition* (perceived disturbance), then *action*." This "knowledge" is embodied in different systems in different ways, for example, through the specific ways designers have connected the components in artificial systems, or in organisms through evolved structures such as genes or learned connections between neurons as in the brain.

In the absence of such knowledge, the system would have to try out actions blindly, until one would by chance eliminate the perturbation. The larger the variety of disturbances (and therefore of requisite actions), the smaller the likelihood that a randomly selected action would achieve the goal, and thus ensure the survival of the system. Therefore, increasing the variety of actions must be accompanied by increasing the constraint or selectivity in choosing the appropriate action, that is, increasing knowledge. This requirement may be called the *law of requisite knowledge*. Since all living organisms are also control systems, life therefore implies knowledge, as in Maturana's often quoted statement that "to live is to cognize."

In practice, for complex control systems control actions will be neither blind nor completely determined, but more like "educated guesses" that have a reasonable probability of being correct, but without a guarantee of success. Feedback may help the system to correct the errors it thus makes before it is destroyed. Thus, goal-seeking activity becomes equivalent to heuristic problem-solving.

Such incomplete or "heuristic" knowledge can be quantified as the conditional uncertainty of an action from R, given a disturbance in D: $H(R \mid D)$. (This uncertainty is calculated as in Eq. (1), but using conditional probabilities $P(r \mid d)$). $H(R \mid D) = 0$ represents the case of no uncertainty or complete knowledge, where the action is completely determined by the disturbance. $H(R \mid D) = H(R)$ represents complete ignorance. Aulin has shown that the law of requisite variety (4) can be extended to include knowledge or ignorance by simply adding this conditional uncertainty term (which remained implicit in Ashby's non-probabilistic formulation of the law):

$$H(E) \geq H(D) + H(R \mid D) - H(R) - K \qquad (5)$$

This says that the variety in the essential variables E can be reduced by (1) increasing buffering K, (2) increasing variety of action $H(R)$, or (3) decreasing the uncertainty

$H(R \mid D)$ about which action to choose for a given disturbance, that is, increasing knowledge.

B. The Modeling Relation

In the above view of knowledge, the goal is implicit in the condition–action relation, since a different goal would require a different action under the same condition. When we think about "scientific" or "objective" knowledge, though, we conceive of rules that are independent of any particular goal. In higher order control systems that vary their lower order goals, knowledge performs the more general function of making *predictions*: "what will happen if this condition appears and/or that action is performed?" Depending on the answer to that question, the control system can then choose the best action to achieve its present goal.

We can formalize this understanding of knowledge by returning to our concept of a *model*. We now introduce *endo-models*, or models *within* systems, as opposed to our previous usage of *exo-models*, or models *of* systems. Figure 4 shows a model (an endo-model), which can be viewed as a magnification of the feedforward part of the general control system of Fig. 2, ignoring the goal, disturbances and actions.

A model starts with a system to be modeled, which we here call the "world," with state space $W = \{w_i\}$ and dynamics $F_a: W \rightarrow W$. The dynamics represents the temporal evolution of the world, like in Fig. 2, possibly under the influence of an action a by the system. The model itself consists of internal model states, or representations $R = \{r_j\}$ and a modeling function, or set of prediction rules, $M_a: R \rightarrow R$. The two are coupled by a perception function $P: W \rightarrow R$, which maps states of the world onto their representations in the model. The prediction M_a succeeds if it manages to anticipate what will happen to the representation R under the influence of action a. This means that the predicted state of the model $r_2 = M_a(r_1) = M_a(P(w_1)$ must be equal to the state of the model created by perceiving the actual state of the world

w_2 after the process $F_a: r_2 = P(w_2) = P(F_a(w_1))$. Therefore, $P(F_a) = M_a(P)$.

We say that the mappings P, M_a and F_a must *commute* for M to be a good model which can predict the behavior of the world W. The overall system can be viewed as a *homomorphic* mapping from states of the world to states of the model, such that their dynamical evolution is preserved. In a sense, even the more primitive "condition-action" rules discussed above can be interpreted as a kind of homomorphic mapping from the events ("disturbances") in the world to the actions taken by the control system. This observation was developed formally by Conant and Ashby in their classic paper "Every good regulator of a system must be a model of that system." Our understanding of "model" here, however, is more refined, assuming that the control system can consider various predicted states $M_a(r_1)$, without actually executing any action a. This recovers the sense of a model as a representation, as used in Section I.B and in science in general, in which observations are used merely to confirm or falsify predictions, while as much as possible avoiding interventions in the phenomenon that is modelled.

An important note must be made about the *epistemology*, or philosophy of knowledge, implied by this understanding. At first sight, defining a model as a homomorphic mapping of the world would seem to imply that there is an objective correspondence between objects in the world and their symbolic representations in the model. This leads back to "naive realism" which sees true knowledge as a perfect reflection of outside reality, independent of the observer. The homomorphism here, however, does not conserve any objective structure of the world, only the type and order of phenomena as perceived by the system. A cybernetic system only perceives what points to potential disturbances of its own goals. It is in that sense intrinsically subjective. It does not care about, nor has it access to, what "objectively" exists in the outside world. The only influence this outside world has on the system's model is in pointing out which models make inaccurate predictions. Since an inaccurate prediction entails poor control, this is a signal for the system to build a better model.

C. Learning and Model-Building

Cybernetic epistemology is in essence *constructivist*: knowledge cannot be passively absorbed from the environment, it must be actively constructed by the system itself. The environment does not instruct, or "in-form," the system, it merely weeds out models that are inadequate, by killing or punishing the system that uses them. At the most basic level, model-building takes place by variation-and-selection or trial-and-error.

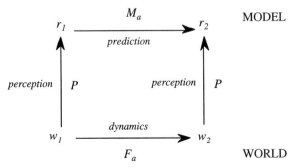

FIGURE 4 The modeling relation.

Let us illustrate this by considering a primitive aquatic organism whose control structure is a slightly more sophisticated version of the thermostat. To survive, this organism must remain in the right temperature zone, by moving up to warmer water layers or down to colder ones when needed. Its perception is a single temperature variable with 3 states $X = \{too\ hot,\ too\ cold,\ just\ right\}$. Its variety of action consists of the 3 states $Y = \{go\ up,\ go\ down,\ do\ nothing\}$. The organism's control knowledge consists of a set of perception-action pairs, or a function $f: X \rightarrow Y$. There are $3^3 = 27$ possible such functions, but the only optimal one consists of the rules *too hot* $\rightarrow$ *go down*, *too cold* $\rightarrow$ *go up*, and *just right* $\rightarrow$ *do nothing*. The last rule could possibly be replaced by either *just right* $\rightarrow$ *go up* or *just right* $\rightarrow$ *go down*. This would result in a little more expenditure of energy, but in combination with the previous rules would still keep the organism in a negative feedback loop around the ideal temperature. All 24 other possible combinations of rules would disrupt this stabilizing feedback, resulting in a runaway behavior that will eventually kill the organism.

Imagine that different possible rules are coded in the organism's genes, and that these genes evolve through random mutations each time the organism produces offspring. Every mutation that generates one of the 24 combinations with positive feedback will be eliminated by natural selection. The three negative feedback combinations will initially all remain, but because of competition, the most energy efficient combination will eventually take over. Thus internal variation of the control rules, together with natural selection by the environment eventually results in a workable model.

Note that the environment did not *instruct* the organism how to build the model: the organism had to find out for itself. This may still appear simple in our model with 27 possible architectures, but it suffices to observe that for more complex organisms there are typically millions of possible perceptions and thousands of possible actions to conclude that the space of possible models or control architectures is absolutely astronomical. The information received from the environment, specifying that a particular action or prediction is either successful or not, is far too limited to select the right model out of all these potential models. Therefore, the burden of developing an adequate model is largely on the system itself, which will need to rely on various internal heuristics, combinations of pre-existing components, and subjective selection criteria to efficiently construct models that are likely to work.

Natural selection of organisms is obviously a quite wasteful method to develop knowledge, although it is responsible for most knowledge that living systems have evolved in their genes. Higher organisms have developed a more efficient way to construct models: *learning*. In learning, different rules compete with each other within the same organism's control structure. Depending on their success in predicting or controlling disturbances, rules are differentially rewarded or reinforced. The ones that receive most reinforcement eventually come to dominate the less successful ones. This can be seen as an application of control at the metalevel, or a metasystem transition, where now the goal is to minimize the perceived difference between prediction and observation, and the actions consist in varying the components of the model.

Different formalisms have been proposed to model this learning process, beginning with Ashby's *homeostat*, which for a given disturbance searched not a space of possible actions, but a space of possible sets of disturbance $\rightarrow$ action rules. More recent methods include neural networks and genetic algorithms. In genetic algorithms, rules vary randomly and discontinuously, through operators such as mutation and recombination. In neural networks, rules are represented by continuously varying connections between nodes corresponding to sensors, effectors and intermediate cognitive structures. Although such models of learning and adaptation originated in cybernetics, they have now grown into independent specialisms, using labels such as "machine learning" and "knowledge discovery."

D. Constructivist Epistemology

The broad view espoused by cybernetics is that living systems are complex, adaptive control systems engaged in circular relations with their environments. As cyberneticians consider such deep problems as the nature of life, mind, and society, it is natural that they be driven to questions of philosophy, and in particular epistemology.

As we noted, since the system has no access to how the world "really" is, models are subjective constructions, not objective reflections of outside reality. As far as they can know, for knowing systems these models effectively *are* their environments. As von Foerster and Maturana note, in the nervous system there is no *a priori* distinction between a perception and a hallucination: both are merely patterns of neural activation. An extreme interpretation of this view might lead to solipsism, or the inability to distinguish self-generated ideas (dreams, imagination) from perceptions induced by the external environment. This danger of complete relativism, in which any model is considered to be as good as any other, can be avoided by the requirements for *coherence* and *invariance*.

First, although no observation can prove the truth of a model, different observations and models can mutually confirm or support each other, thus increasing their joint reliability. Thus, the more coherent a piece of knowledge is with all other available information, the more reliable it is. Second, percepts appear more "real" as they vary less between observations. For example, an *object* can be defined as that aspect of a perception that remains invariant when the point of view of the observer is changed. In the formulation of von Foerster, an object is an eigenstate of a cognitive transformation. There is moreover invariance over observers: if different observers agree about a percept or concept, then this phenomenon may be considered "real" by *consensus*. This process of reaching consensus over shared concepts has been called "the social construction of reality." Gordon Pask's Conversation Theory provides a sophisticated formal model of such a "conversational" interaction that ends in an agreement over shared meanings.

Another implication of constructivism is that since all models are constructed by some observer, this observer must be included in the model for it to be complete. This applies in particular to those cases where the process of model-building affects the phenomenon being modelled. The simplest case is where the process of observation itself perturbs the phenomenon, as in quantum measurement, or the "observer effect" in social science. Another case is where the predictions from the model can perturb the phenomenon. Examples are self-fulfilling prophecies, or models of social systems whose application in steering the system changes that very system and thus invalidates the model. As a practical illustration of this principle, the complexity theorist Brian Arthur has simulated the seemingly chaotic behavior of stock exchange-like systems by programming agents that are continuously trying to model the future behavior of the system to which they belong, and use these predictions as the basis for their own actions. The conclusion is that the different predictive strategies cancel each other out, so that the long-term behavior of the system becomes intrinsically unpredictable.

The most logical way to minimize these indeterminacies appears to be the construction of a *metamodel*, which represents various possible models and their relations with observers and the phenomena they represent. For example, as suggested by Stuart Umpleby, one of the dimensions of a metamodel might be the degree to which an observation affects the phenomenon being observed, with classical, observer-independent observations at one extreme, and quantum observation closer to the other extreme.

However, since a metamodel is a still a model, built by an observer, it must represent itself. This is a basic form of self-reference. Generalizing from fundamental epistemological restrictions such as the theorem of Gödel and the Heisenberg indeterminacy principle, Lars Löfgren has formulated a *principle of linguistic complementarity*, which implies that all such self-reference must be partial: languages or models cannot include a complete representation of the process by which their representations are connected to the phenomena they are supposed to describe. Although this means that no model or metamodel can ever be complete, a metamodel still proposes a much richer and more flexible method to arrive at predictions or to solve problems than any specific object model. Cybernetics as a whole could be defined as an attempt to build a universal metamodel, that would help us to build concrete object models for any specific system or situation.

SEE ALSO THE FOLLOWING ARTICLES

ARTIFICIAL INTELLIGENCE • ARTIFICIAL NEURAL NETWORKS • CELLULAR AUTOMATA • COGNITIVE SCIENCES • DISCRETE SYSTEMS MODELING • HUMAN-COMPUTER INTERACTION • HUMANOID ROBOTS • MATHEMATICAL MODELING • ROBOTICS, COMPUTER SIMULATIONS FOR • SELF-ORGANIZING SYSTEMS • SYSTEM THEORY

BIBLIOGRAPHY

Ashby, W. R. (1956–1999). "Introduction to Cybernetics," Methuen, London (electronically republished at http://pcp.vub.ac.be/books/IntroCyb.pdf).

François, C. (ed.) (1997). "International Encyclopedia of Systems and Cybernetics," Saur, Munich.

Geyer, F. (1995). "The challenge of sociocybernetics," *Kybernetes* **24**(4), 6–32.

Heims, S. J. (1991)."Constructing a Social Science for Postwar America: The Cybernetics Group," MIT Press, Cambridge, MA.

Heylighen, F., Joslyn, C., and Turchin, V. (eds.) (1991–2001). "Principia Cybernetica Web" (http://pcp.vub.ac.be).

Holland, J. (1995). "Hidden Order: How Adaptation Builds Complexity," Addison-Wesley, Reading, MA.

Klir, G. (1991). "Facets of Systems Science," Plenum, New York.

Maturana, H., and Varela, F. (1998). "The Tree of Knowledge," (revised edition), Shambhala Press, Boston.

Richardson, G. P. (1991). "Feedback Thought in Social Science and Systems Theory," University of Pennsylvania Press, Philadelphia.

Meystel, A. (1996). "Intelligent systems: A semiotic perspective," *Int. J. Intell. Control Systems* **1**, 31–57

Umpleby, S., and Dent, E. (1999). "The origins and purposes of several traditions in systems theory and cybernetics," *Cybernetics and Systems* **30**, 79–103.

von Foerster, H. (1995). "Cybernetics of Cybernetics" (2nd ed.), FutureSystems Inc., Minneapolis, MN.

Dams, Dikes, and Levees

Robert B. Jansen

Consulting Civil Engineer

I. Types of Dams
II. History of Dams
III. Appurtenant or Related Works
IV. Design of Dams
V. Construction of Dams
VI. Safety of Dams
VII. Weighing Benefits and Detriments of Dams

GLOSSARY

Dam A barrier constructed for storage of water for any of various purposes, including urban water supply, irrigation, flood control, power production, navigation, soil conservation, debris retention, and recreation. A multiple-purpose project may provide several of these benefits.

Dike A barrier, usually an embankment of earth or stone, for retaining a body of water or a stream. It may be an auxiliary embankment related to a main dam.

Levee An embankment (sometimes considered to be the same as a dike) designed to control a stream in a defined course or to prevent a body of water from intruding into areas to be protected. River walls and sea walls also serve such purposes.

DAMS HAVE BEEN keys to development in many lands, providing sustenance and protection for human survival. In the United States, systems of dams, locks, powerhouses, and boating facilities on the Ohio, Tennessee, and Mississippi Rivers are outstanding examples of projects serving multiple purposes. Of historical interest in other regions are storages or diversions for hydraulic mining, log ponds, saw mills, and canal transportation. Dams enable diversion of water for agriculture, as well as regulated discharges to natural streams for preservation of fish and wildlife. There are more than 45,000 large dams (over 15 meters high) in the world, of various types consisting of embankments and concrete and masonry structures. Public safety is a primary consideration in their design, construction, and operation.

By general definition, dams include dikes and levees. Dikes have been used for land reclamation, such developments being famous in Europe, particularly exemplified by the large-scale projects built in the Ijsselmeer (Zuider Zee) in the Netherlands. In ancient Egypt, seasonal diversions of excess flows of the Nile were made into extensive basins enclosed by dikes. In some regions, dikes used for dividing wet lowlands have been called levees. River levees are usually comparatively low earthfills, sometimes offset at a distance from the river to contain overbank flows in a floodplain. Some of the earliest river levees were those along the Nile in Egypt. Among the other important levee systems developed in major watersheds are those on the

Mississippi River, those in California on the American and Sacramento Rivers, and those that enclose islands in the San Joaquin delta.

I. TYPES OF DAMS

Dams may be classified by their materials and by their design to carry the imposed loads.

A. Embankments

Embankments may be earthfills or rockfills or combinations thereof. Their impervious elements may be interior cores of earthfill or walls composed of conventional or asphaltic concrete or of masonry (in older designs). In other cases, a waterproof membrane has been established on the upstream face of the embankment by constructing a concrete slab, which is a common design for rockfills but is seldom seen on newer earthfills. An example of a concrete-faced rockfill is the 187-meter-high (614-foot-high) Aguamilpa Dam in Mexico, completed in 1994. Other materials have been used for facings, including asphalt, timber planks, steel plates, and synthetic sheets. The local materials for the embankment itself may come from quarries and borrow areas or from excavations for structures and conduits.

Earthfills are the most common type of dam. An early earth dam was generally composed of the same material throughout, constituting a "homogeneous" section. As designs evolved, strength and seepage control were enhanced by cross sections with zones of various gradations. Most modern earthfill dams are zoned, careful attention being given to the transitions between zones. Even those embankments that are predominantly single zones are provided with drainage, such as by horizontal and inclined layers of coarse materials. A typical zoned earthfill dam has an interior core of compacted fine-grained soil of low permeability bolstered by a succession of outer pervious zones grading from sand to gravel, crushed rock, cobbles, or rockfill. An eminent earthfill is Tajikistan's Nurek Dam, which heads the list of the world's high dams.

An earthfill that is now less common is the hydraulic fill dam, which contains soils that have been conveyed to the site in a liquid mix, typically from a dredged deposit. This involved continuous ponding. At some projects on alluvial foundations, seepage underflow was controlled by dumping fine-grained soils into water-filled trenches. Semihydraulic fills were constructed of soils hauled and dumped at the site and then sluiced into place. Generally the sluicing of either type was done with the objective of depositing fine-grained earth in the center and coarse materials in the outer parts of the dam section. However, this was a relatively crude process that created marginally stable embankments. Hydraulic fills have lost favor to earthfills placed and compacted by heavy equipment, which give more reliable structural performance, particularly under seismic loading.

B. Gravity Dams

A gravity dam is a structure designed to withstand loads by its own weight and by its resistance to sliding and overturning on its foundation. Newer dams of this type are typically composed of unreinforced concrete monoliths with seals at the joints. Foremost among the world's gravity dams is the 285-meter-high (935-foot-high) Grande Dixence in Switzerland, completed in 1961. In past centuries, many gravity dams were constructed of stone masonry. New masonry dams, of both gravity and arch designs, are being built in India and China and in other lands where the cost of labor is low. Also of historic interest is the filled crib, the stability of which depends upon the weight of earth or rockfill enclosed in cells formed by a framework of timbers or concrete beams. Such dams, common a hundred years ago, were usually faced with timber planking.

C. Arch Dams

An arch dam is a concrete or masonry structure that carries load to a large extent by lateral thrust into its abutments and therefore can be thinner than a gravity dam. The modern arch is usually curved in both plan and section. Some of the early arch dams were conservatively proportioned so that much of their resistance was due to weight. A notable example of this arched gravity type is Hoover Dam. Some true arch dams of record-breaking heights are the Soviet-engineered Inguri (272 m/892 ft) and Chirkei (233 m/764 ft), Vaiont Dam in Italy (262 m/860 ft), Ertan Dam in China (240 m/787 ft), Mauvoisin Dam in Switzerland (250 m/820 ft), El Cajón Dam in Honduras (234 m/768 ft), and the Swiss Contra Dam (220 m/722 ft).

D. Buttress Dams

Buttress dams are concrete or masonry structures usually comprising inclined panels or arches supported by buttresses. In the Ambursen type, the upstream face is a sloping plane formed by flat slabs bearing directly, or on corbels, at the upstream edges of the buttresses. A special type is the massive-head dam, in which the face is formed by enlarging the upstream ends of the buttresses so that they join to eliminate the need for a spanning deck slab. Figure 1 shows the concrete section of Pueblo Dam, consisting of 23 massive-head blocks. This structure, completed in 1975, was bolstered in 1999 against potential

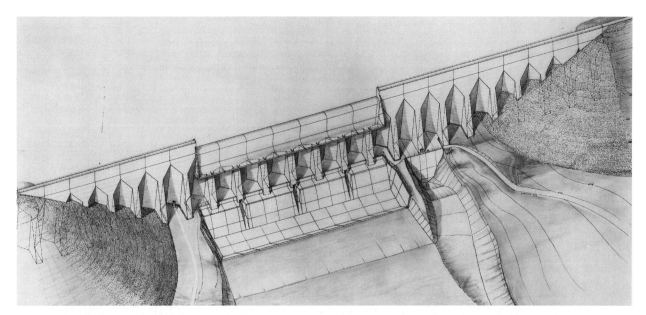

FIGURE 1 Massive-head, buttress-type concrete section of Pueblo Dam on the Arkansas River in Colorado, 533 m (1750 ft) long and 76 m (250 ft) high above the lowest foundation. The overflow spillway in the middle has a crest length of 168 m (550 ft). The dam, which was completed in 1975, extends as earth embankment at each end of the concrete section.

foundation sliding by adding a large roller-compacted concrete plug in the spillway plunge pool and a rock-bolted block at the toe of the dam. Discharges into the modified stilling basin were begun in the spring of 2000.

A multiple-arch dam has a series of arches joined at the buttresses. A unique extension of this concept is seen at Coolidge Dam in Arizona, composed of three large domes supported by massive buttresses. Engineering of multiple-arch dams was advanced notably in the first third of the 20th century at sites where they had particular economic advantages, such as at Florence Lake Dam in the Sierra Nevada of California (Fig. 2). Fewer have been built more recently; however, one of the most remarkable structures of this type, completed in 1968, is Daniel Johnson (Manicouagan No. 5) Dam in Québec, Canada (Fig. 3). It is 214 m (703 ft) high and 1314 m (4311 ft) long, comprising 13 arches, the largest spanning 161.5 m (530 ft).

E. Roller-Compacted Concrete Dams

An innovation that has won worldwide acceptance in recent times is roller-compacted concrete (RCC, or rollcrete), which has been adopted for construction of new gravity and arch dams, as well as rehabilitation of conventional dams. A rollcrete mix typically may consist of sand, coarse aggregate such as gravel or crushed rock, portland cement plus pozzolan (commonly fly ash), and water. Only limited use has been made of admixtures for water reduction and set retardation, or for air entrainment in special

FIGURE 2 Florence Lake Dam in California, 46 m (150 ft) high and 962 m (3156 ft) long, with 58 arches. This dam was completed in 1926.

FIGURE 3 Daniel Johnson Dam in Québec, Canada, a multiple-arch structure 214 m (703 ft) high, with the highest arch spanning 161.5 m (530 ft).

applications. The mix is hauled, placed, and compacted by heavy equipment. RCC has the advantages of rapid construction, low cement content and consequent reduced heat generation, less exacting materials requirements, and usually lower cost than in conventional concrete construction. The first RCC dams were gravity structures. Their achieved heights increased rapidly in the 1990s, exemplified by the Japanese Miyagase Dam (155 m/509 ft) in 1996. Yet higher RCC dams are being planned, including Longtan Dam in China, with an ultimate height of 216 m (709 ft). As methods have advanced, arches of this type have become feasible. An outstanding example is Shapai Dam, 128 m (420 ft) high, completed in 1999 on an upper reach of the Chaopohe River in Sichuan Province, China. At that project, use was made of grout-enriched roller-compacted concrete (GERCC) adjacent to forms, a new technique that involves pumping cement grout onto the uncompacted RCC and applying a large internal vibrator to secure a mix of high quality. This and other innovations in the design and construction of RCC dams, including the Chinese-developed sloped-layer placement, have improved bonding and control of seepage, which were problems in the early dams of this type.

F. Cofferdams

Cofferdams are water barriers, usually temporary, for dewatering sites for construction of dams and other waterworks. Steel sheetpile cellular cofferdams typically are filled with sand, gravel, or broken rock. Because of the limited space that they occupy, they have particular advantages where river navigation must be continued during construction of the main facilities. Embankment cofferdams are used commonly where the river is diverted around the worksite. In many projects, the upstream cofferdam has been incorporated into the permanent embankment dam. Retention of the downstream cofferdam is less common, and in such cases may require modifications to ensure internal drainage of the final dam.

G. Saddle Dams

Saddle dam is the term given to a water barrier built in a topographic depression or gap on the rim of a reservoir. It is typically an auxiliary feature to increase the water storage capability. Most saddle dams are relatively small embankments, but some are of substantial size. A current example is the saddle dam of the Eastside Reservoir Project in Riverside County, CA, completed in 1999. It is an earth-core rockfill 130 ft (40 m) high and 2300 ft (701 m) long, with volume of 2,500,000 yd^3. The foundation is a rock (phyllite and schist) ridge that required grouting to depths of as much as 100 ft (30 m).

H. Diversion Dams

A diversion dam is a structure constructed across a stream to turn it into another channel or conduit. It may be a temporary facility, as a cofferdam, for passage of flow during construction of principal works, or it may be permanent for diversion and conveyance of water for various continuing purposes.

I. Determination of Dam Type

The most suitable type of dam may depend upon many factors, including environment, topography, geology, weather, site accessibility, sources of materials, stream diversion requirements, and the potential for floods and earthquakes.

Earthfills are adaptable to most sites but may be less favored at narrow canyons where they might be subjected to large deformation and where river diversion and flood control would be difficult. They have advantages on alluvial or glacial foundations, in comparison with concrete dams, which are most suited for construction on rock. However, placement and compaction of earthfills may be hampered by excessive rainfall or frost. Soils used in embankments generally have lower shear strength (friction plus cohesion) than does typical rockfill, which is characterized by high frictional resistance to sliding.

Rockfills may be favored where rock sources are plentiful but earth materials are scarce. Dams of this type with earth impervious zones can be susceptible to differential settlement in narrow gorges. Rockfills typically have high shearing strength and therefore can be built with steep faces, with attendant benefits of economic volume and accommodation in topographically limited sites. Because neither rockfills nor earthfills are resistant to sustained overtopping, they usually require separate spillways that may be expensive.

Concrete dams have distinct advantages in their ability to stand overpour. They usually require sound rock foundations, which may preclude some laminated shales and sandstones. Nonuniform rock formations, including those with faults and shear zones, pose special challenges in concrete dam design. For arches, such conditions might be decisive in weighing their comparative merits, or even their feasibility. Gravity dams are well suited for construction on most rock sites of any width. Their accommodation of spillways and outlets may give them an advantage over embankments. Some low gravity dams have been built on soil or gravel foundations, but in most cases the potential for detrimental deformation or erosion would rule against such construction.

Where site conditions have not been amenable to incorporating a spillway in an arch, short tunnels have been economical alternatives for passing floods. Arches and buttress dams may have favorably economic proportions, even though the unit costs of their construction are relatively high. Buttress dams can be built at nearly any site

that would accommodate a gravity dam. Their comparatively reduced concrete volumes have given them an advantage in regions where sources of materials were limited and costs of labor were not excessive. However, offsetting factors are the details of the slender structural members, which may necessitate high-strength concrete and steel reinforcement and sometimes precautions to counter the effects of weathering and ambient temperature variations. Because of their smaller bases, buttress and arch dams usually pose less concern about uplift by foundation water pressures than do gravity dams. However, on some foundations with low bearing capacity, buttresses of modest size have been built on concrete base slabs to spread the load. In such a case, uplift may require special attention.

Over the years, advances in technology have altered preferences for the various dam types. Rockfills have become more attractive as improvements have been made in quarrying, zoning, placement, and compaction, as well as in details of face slabs. Roller-compacted concrete has won wider acceptance as measures for seepage control have been developed and as use of RCC for armoring or bolstering embankments has gained favor.

II. HISTORY OF DAMS

Dams were among the earliest structures, having been vital in the development of water resources in Asia and the Middle East. Historically, dams have been associated closely with the rise and fall of civilizations. Many of the ancient water projects survived for centuries and then deteriorated. Without water, the societies that it sustained withered away. The causes of decline were many, including the changes in peoples from generation to generation. Some works were wasted by neglect. Others suffered from inherent problems, such as sedimentation of the reservoirs.

Table I is a list of some notable historical dams. The historian Herodotus wrote of a dam on the River Nile attributed to Menes, the king of Egypt, about 3000 BC. Some translations of the ancient records suggest that the dam was at Kosheish, about 25 miles upstream from Cairo, and was constructed of stone masonry. The remains of other masonry dams are found in Egypt, including the Sadd-el-Kafara Dam in the Wadi Garawi about 20 miles south of Cairo, the origin of which has been dated to between about 2900 BC and 2700 BC. Measurement of the remnants has suggested that this structure was as high as 14 m (46 ft) and consisted of a central zone of gravel, rubble, and fines, bolstered by rockfill sections faced with stepped cut limestone blocks. It evidently was destroyed by flood.

In Babylonia and Assyria, there were diversions from the Tigris and Euphrates Rivers and their tributaries as early as 2100 BC, accomplished by earth and stone dams.

TABLE I Some Notable Historical Dams

Year built[a]	Dam name	Country	Stream	Dam type	Height(m)
3000 BC ±	Kosheish	Egypt	Nile	Masonry	15 (?)
2700 BC ±	Sadd-el-Kafara	Egypt	Wadi el-Garawi	Earth/masonry	14
2100 BC ±	Adheim	Iraq	Adheim	Masonry	17
2000 BC ±	Marduk	Iraq	Tigris	Earth	12
750 BC ±	Sadd-el-Arim	Yemen	Wadi Dhana	Earth/masonry	14
591 BC ±	Shaopi	China	—	Earth	10
240 BC ±	—	China	Gukow	Stone-crib	30
150 BC ±	Alcantarilla	Spain	Guajaraz	Masonry/earth	20
110 AD ±	Proserpina	Spain	Pardillas	Masonry/earth	19
162 AD	Kaerumataike	Japan	Yodo	Earth	17
190 AD ±	Cornalbo	Spain	Albarregas	Earth	24
960 AD	Band-I-Amir	Iran	Kur	Masonry	9
1050 AD ±	Bhojpur	India	Kaliasot	Earth	27
1300 AD ±	Kebar	Iran	Kebar	Masonry arch	26
1500 AD	Mudduk Masur	India	—	Earth	33
1594 AD	Alicante (Tibi)	Spain	Monegre	Arched gravity	41
1671 AD	St. Ferreol	France	Laudot	Earth/masonry	36
1791 AD	Valdeinfierno	Spain	Luchena	Arched gravity	36
1800 AD ±	Mir Alam	India	From Esee River	Multiple arch	12
1854 AD	Zola	France	—	Masonry arch	42
1866 AD	Gouffre d'Enfer	France	Furan	Arched gravity	60
1897 AD	Periyar	India	Periyar	Concrete gravity	54
1904 AD	Cheesman	U.S.	South Platte	Arched gravity	72
1905 AD	New Croton	U.S.	Croton	Masonry gravity	90.5
1910 AD	Buffalo Bill	U.S.	Shoshone	Concrete arch	99
1915 AD	Arrowrock	U.S.	Boise	Arched gravity	107
1931 AD	Salt Springs	U.S.	N. Fk. Mokelumne	Rockfill	100
1932 AD	Owyhee	U.S.	Owyhee	Arched gravity	127
1936 AD	Hoover (Boulder)	U.S.	Colorado	Arched gravity	221

[a]The ± sign indicates approximate date of construction.

In the second and first millennia BC, diversion dams were built in southern Arabia, the largest evidently being the 14-meter-high (46-ft-high) Sadd-el-Arim Dam (also referred to as Sadd Ma'rib or Sudd-al-Arim) near Ma'rib, capital of Sheba (Saba), in the region now known as Yemen. It has been described as an earth embankment with an upstream face protected by a layer of stones. Its construction has been dated to about 750 BC. This is believed to be the structure in the Wadi Dhana (also identified as Wadi Denne or Wadi Sadd or Wadi Saba), where remnants of canal intakes and a spillway have been found.

Other notable dams were built in Sri Lanka (formerly Ceylon), India, Baluchistan (now part of Pakistan), and China in the millennia BC. A Chinese dam of special interest is the 10-meter-high (33-ft-high) earthfill forming Shaopi Reservoir, which was completed in 591 BC and is still operating for supply of irrigation water in Anhui Province.

Important water projects were developed in Spain by the Romans. The Alcantarilla Dam near Toledo, believed to have been constructed in the second century BC, was composed of masonry and concrete buttressed by earthfill. Later Spanish waterworks built by the Romans include Proserpina and Cornalbo Dams, near Mérida. In these and many other dams in the Roman empire, effective use was made of stone masonry. Records dating to about 560 AD tell of a curved dam on a tributary of the Khabur River at the Turkish-Syrian border near Daras. Some historians regard it as the first known arch dam. It disappeared long ago.

Significant advances were made in the engineering of dams in the Middle Ages. An important, long-surviving arch dam of the Mongol period in Persia was built on the Kebar River near Qum, southwest of Teheran in about 1300 AD. About 26 m (85 ft) high and 55 m (180 ft) long, it was composed of cemented rubble masonry with mortared stone block facing. An arch-gravity dam of similar

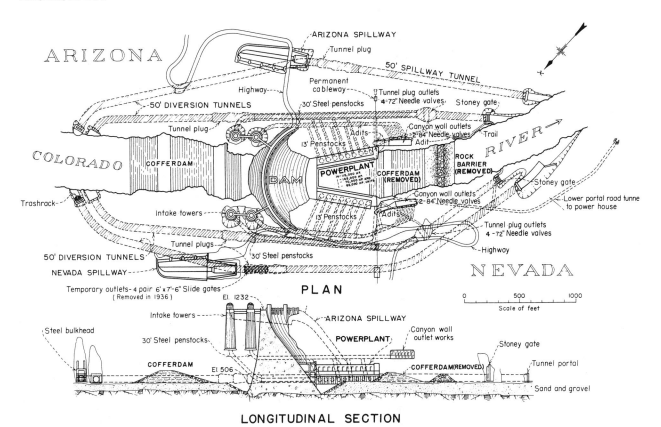

FIGURE 4 Hoover Dam on the Colorado River.

composition was constructed in 1384 AD near Almansa in Spain. After successive enlargements, the latest in 1921 AD, the height of the Almansa Dam reached 25 m (82 ft). This was surpassed by another curved masonry dam in Spain, the Alicante (Tibi) Dam, completed in 1594 AD. For nearly three centuries, this was the highest dam in the world, towering 41 m (135 ft) above its base. Its height was increased to 46 m (151 ft) in 1943.

The earliest true multiple-arch dam in the records is Mir Alam Dam, constructed in about 1800 AD near Hyderabad, India. This mortared masonry structure, about 12 m (40 ft) high, had 21 vertical arches. In the 19th century, significant progress was made in the engineering of dams. Methods of analysis of concrete and masonry structures were advanced in England and France. Engineers in the United States built notable dams, including Old Croton Dam in 1842 for New York City; Mill River Dam in 1862 for New Haven, CN; Lake Cochituate Dam in 1863 for Boston; and Druid Lake Dam in 1871 for Baltimore. Many earth dams, as high as 38 m (125 ft), were constructed for irrigation projects in the western states. Dams made of stone-filled log cribs were developed for storage of water for hydraulic mining in the California gold fields. Larger reservoirs were created by dumped rockfills with timber plank facing.

An American precedent was established in the construction in 1904 of Colorado's Cheesman Dam, 72 m (236 ft) high and arched even though it was a full gravity section. New Croton Dam, a gravity structure 90.5 m (297 ft) high completed in 1905, was one of the last large American dams made of masonry. It was the highest dam in the world, but was surpassed in 1910 by Buffalo Bill (Shoshone) Dam in Wyoming at 99 m (325 ft). This is a true arch, as is Pathfinder Dam, another Wyoming dam, completed in 1909.

Many advances were made in the engineering of concrete dams in the 20th century as design and construction methods evolved rapidly. Hoover (Boulder) Dam, completed in 1936 on the Colorado River, is an arched gravity concrete structure 221 m (726 ft) high (Fig. 4). On the same river is Glen Canyon Dam, an arch dam completed in 1966. It is 216 m (710 ft) high. Dworshak Dam, a straight concrete gravity structure completed in 1973 on the North Fork of the Clearwater River in Idaho, is 219 m (717 ft) high (Fig. 5). Shasta Dam (1945) in California is of the arched gravity type, 183 m (600 ft) high (Fig. 6). Mossyrock Dam (1968) in the State of Washington is an arch 185 m (607 ft) high (Fig. 7). Other monumental achievements in concrete construction have been Vaiont Dam (1961) in Italy, and three Swiss Dams: Mauvoisin

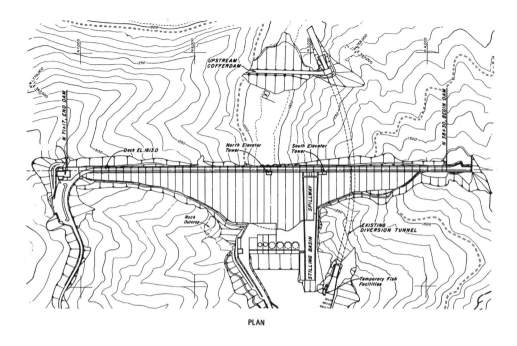

PLAN

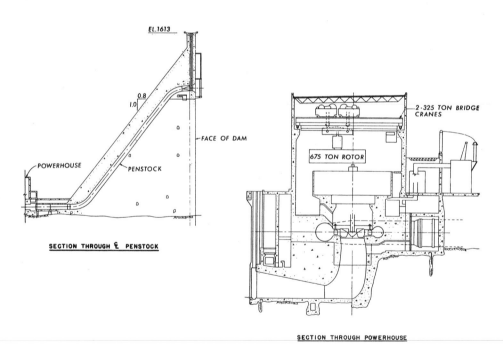

FIGURE 5 Dworshak Dam on the North Fork of the Clearwater River in Idaho.

(completed originally in 1957 and raised in 1988), Grande Dixence (1961), and Contra (1965) (see Fig. 8). These were followed by Inguri, Sayano-Shushensk, and Chirkei, all shown in Table II.

Major advances have also been made in the engineering of embankment dams. Outstanding examples are

California's Oroville Dam (1968), an earthfill 230 m (755 ft) high (Fig. 9); Mica Dam (1973), an earthfill-rockfill 242 m (794 ft) high in western Canada; La Esmeralda (Chivor) (1976), an earthfill-rockfill 237 m (777 ft) high in Colombia (Fig. 10); Manuel M. Torres (Chicoasén) (1980), an earthfill-rockfill 261 m (856 ft) high in

FIGURE 6 Shasta Dam on the Sacramento River in California.

Mexico; Alberto Lleras Camargo (Guavio) (1989), an earthfill-rockfill 243 m (797 ft) high in Colombia; Tehri (under construction), an earthfill-rockfill 261 m (856 ft) high in India; and Nurek (1980), an earthfill 300 m (984 ft) high in Tajikistan. A notable example of a rockfill with a concrete face is the 160-meter-high Foz do Areia Dam, completed in 1980 in Brazil. Figure 11 is a section of this dam, showing the geometry of its zones and the varied thicknesses of the compacted rockfill layers.

As the 21st century dawned, building of large dams in China was continuing at a vigorous pace, including Ertan (1999), an arch 240 m (787 ft) high, as well as Hongjiadu, a concrete-faced rockfill 182 m (597 ft) high, and Three Gorges, a 181-m (594-ft) gravity dam, both under construction. The Chinese have adopted and advanced new concepts in the design and construction of roller-compacted concrete dams and concrete-faced rockfills.

Sharing of knowledge about dams throughout the world has been facilitated by the work of the International Commission on Large Dams (ICOLD), founded in 1928. Its Congress in Beijing in 2000 listed 83 member countries. This cooperation has enabled significant advances in dam engineering with consequent enhancement of national economies and public safety.

III. APPURTENANT OR RELATED WORKS

A. Spillways

Spillways are facilities for conveying excess streamflows past a dam. In important projects on major rivers, spillways are sized to discharge extreme floods. At a concrete gravity dam, the spillway is typically made part of the main structure, providing an overflow that is either uncontrolled or regulated by gates (Fig. 12). At an embankment dam, the spillway is usually a channel (chute) or conduit in a separate location in or on an abutment or at a farther site on the reservoir rim (Fig. 13). The types of spillways include overflow with an ogee-crest shape and shaft and tunnel with a glory hole (morning glory), funnel-shaped entrance. Another type that has been used under special conditions is the side channel, in which the crest is approximately parallel to a discharge channel consisting of a trough leading to a tunnel or chute. Hoover Dam has two spillways of this kind (Fig. 4). At some arch dams, the spillway design entails free fall from the crest to a plunge pool created by a smaller dam immediately downstream. At other types of dams, where the streambed would be subject to erosion by high-velocity flow, the potentially destructive energy of the discharging water is dissipated in a stilling basin. Terminal structures also include chutes and aprons with baffles, and flip (deflector) buckets, sometimes referred to as ski-jump spillways, which loft a trajectory so it will impinge at a safe distance.

At many spillways, gates and other devices have been used to control discharges. In the past, flashboards often were placed on the crest to raise the reservoir level. A few of these are still seen at older projects. Some of them were simple, temporary barriers fabricated of planks or timber panels designed to fall and be washed away at predetermined flood levels. Others were, in effect, permanent automatic gates that lowered in response to rising water pressure. A typical flashboard of this type is hinged and counterweighted so that it rotates and lies down to conform to the fixed crest shape during flood passage and then rises to resume its retention of the normal water surcharge. In modern controlled spillways, conventional designs include hinged steel gates of the radial (taintor) type lifted by electric hoists on decks overhead. To safeguard against power outages during storms, provision often is made for standby generators using diesel or gasoline fuel.

B. Outlets

Outlets are facilities for water discharges to serve normal project requirements, as well as to augment flood control capacity in some cases. They are usually classified

TABLE II Some High Dams of the World[a]

Dam name	Country	Stream	Dam type	Height (m)	Year completed
Nurek	Tajikistan	Vakhsh	Earth	300	1980
Grande Dixence	Switzerland	Dixence	Gravity	285	1961
Inguri	Georgia	Inguri	Arch	272	1980
Yusufeli[b]	Turkey	Coruh	Earth/rock	270	Underway
Vaiont[c]	Italy	Vaiont	Arch	262	1961
Manuel M. Torres (Chicoasén)	Mexico	Grijalva	Earth/rock	261	1980
Tehri	India	Bhagirathi	Earth/rock	261	Underway
Mauvoisin	Switzerland	Drance de Bagnes	Arch	250	1957/1988
Deriner	Turkey	Coruh	Arch	247	Underway
Alberto Lleras C. (Guavio)	Colombia	Guavio	Earth/rock	243	1989
Mica	Canada	Columbia	Earth/rock	242	1973
Sayano-Shushensk	Russia	Yenisey	Gravity/arch	242	1987
Ertan	China	Yalongjiang	Arch	240	1999
La Esmeralda (Chivor)	Colombia	Batá	Earth/rock	237	1976
El Cajón	Honduras	Humuya	Arch	234	1985
Chirkei	Russia	Sulak	Arch	233	1978
Oroville	U.S.	Feather	Earth	230	1968
Bhakra	India	Sutlej	Gravity	226	1963
Luzzone	Switzerland	Brenno di Luzzone	Arch	225	1963/1998
Hoover	U.S.	Colorado	Arch/gravity	221	1936
Contra	Switzerland	Verzasca	Arch	220	1965
Mratinje	Montenegro	Piva	Arch	220	1976
Dworshak	U.S.	N. Fork Clearwater	Gravity	219	1973
Glen Canyon	U.S.	Colorado	Arch	216	1966
Toktogul	Kyrgyzstan	Naryn	Gravity	215	1978
Daniel Johnson	Canada	Manicouagan	Multiple arch	214	1968
Keban	Turkey	Firat (Euphrates)	Earth/rock/gravity	208	1974
Zimapan	Mexico	Moctezuma	Arch	207	1994
Karun III	Iran	Karun	Arch	205	Underway
Lakhwar	India	Yamuna	Earth	204	Underway
Dez	Iran	Dez	Arch	203	1962
Almendra	Spain	Tormes	Arch	202	1970
Berke	Turkey	Ceyhan	Arch	201	Underway
Khudoni	Georgia	Inguri	Arch	201	Underway
Karun I	Iran	Karun	Arch	200	1976
Kölnbrein	Austria	Malta	Arch	200	1977
San Roque	Philippines	Agno	Earth/rock	200	Underway
Itaipú	Brazil/Paraguay	Paraná	Earth/rock/gravity	196	1983
Altinkaya	Turkey	Kizil Irmak	Earth/rock	195	1988
New Bullards Bar	U.S.	North Yuba	Arch	194	1970
Seven Oaks	U.S.	Santa Ana	Earth/rock	193	1999
New Melones	U.S.	Stanislaus	Earth/rock	191	1979
Miel I	Colombia	Miel	RCC	188	Underway
Aguamilpa	Mexico	Santiago	Rock	187	1994
Kurobe	Japan	Kurobe	Arch	186	1964
Swift	U.S.	Lewis	Earth	186	1958
Zillergründl	Austria	Ziller	Arch	186	1987
Mossyrock	U.S.	Cowlitz	Arch	185	1968
Oymapinar	Turkey	Manavgat	Arch	185	1984
Atatürk	Turkey	Firat (Euphrates)	Earth/rock	184	1990
W.A.C. Bennett	Canada	Peace	Earth	183	1967
Shasta	U.S.	Sacramento	Arch/gravity	183	1945

[a] Dams 183 m (600 ft) or higher.

[b] Yusufeli Dam design was complete in 1999; financing and construction contracts were set in 2000.

[c] Slide of a canyon wall in 1963 made Vaiont Reservoir inoperable.

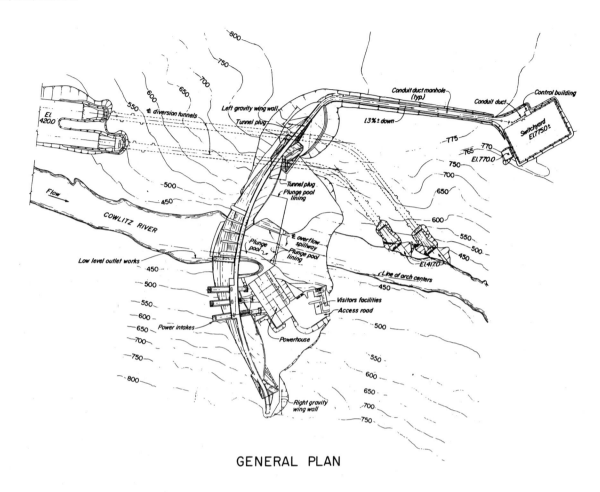

GENERAL PLAN

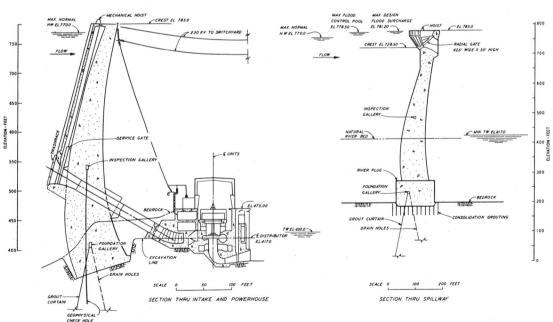

FIGURE 7 Mossyrock Dam on the Cowlitz River in Washington.

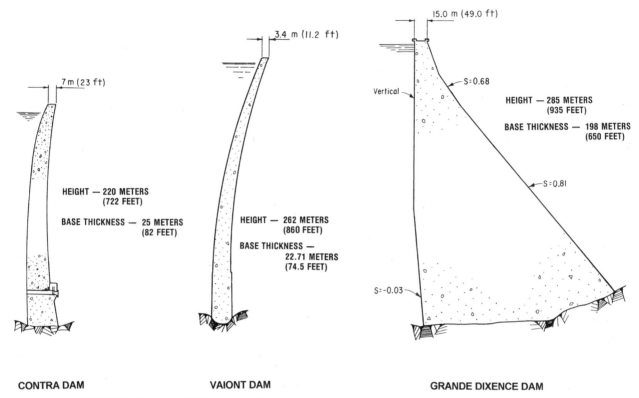

FIGURE 8 Sections of high concrete dams, showing the contrast in volume of arch and gravity structures.

by their purposes, including irrigation, municipal or industrial water supply, power, flood, and river release. River outlets at concrete dams are commonly called *sluices*. Outlets may be of high value in releasing stored water to lower the reservoir level in case of emergency. Where dissipation of the energy of the discharges is required, it is conventionally accomplished by devices and structures similar in concept to those used for this purpose at spillways. The flows through outlets are regulated by gates or valves, which may be of a variety of types. Depending upon the purposes to be served, the outlets may discharge to the open air or they may convey the flows into penstocks for power installations or into open or closed conduits for other water supply needs. To satisfy water quality requirements at some projects, the outlet works have been designed with multiple-level intakes that enable withdrawals at optimum temperatures and chemical content. At Oroville Dam, the inclined intake structure for the penstocks leading to the underground Edward Hyatt Powerplant was provided with water-level control shutters so that selection could be made of cool water for salmon and steelhead in the Feather River or of warmer water for rice cultivation in the region, as the season demands. (At some other projects, downstream temperature requirements have been met by warming ponds.) At Shasta Dam,

also in California, a multilevel temperature control facility was added in 1997, at a cost of $80 million, to enable selective withdrawal for protection of endangered salmon species.

C. Other Works

Facilities that may be closely related to dams are powerhouses, navigation locks, and fish ladders. Where such works adjoin or are integral parts of the water barrier, they must meet the same structural demands as the dam itself.

IV. DESIGN OF DAMS

Methods of design of dams and other water barriers have progressed substantially in recent decades, drawing fully from the interdisciplinary work of engineers, geologists, hydrologists, and seismologists, and from research that has expanded the horizons in this field. Increasingly effective use has been made of physical and mathematical models in hydraulic and structural analysis. Observations of the performances of dams during earthquakes and the examination of histories of accidents and failures have shown the way to safer dams.

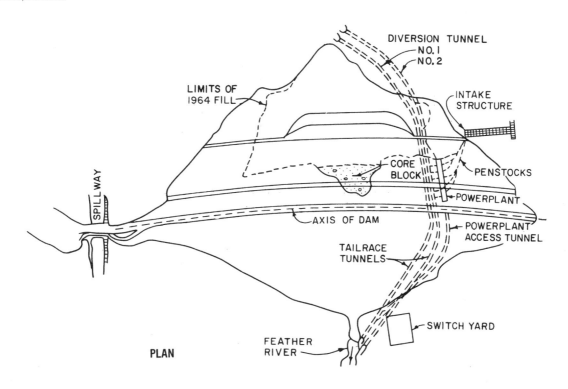

ZONED EARTHFILL

HEIGHT — 230 METERS
(755 FEET)

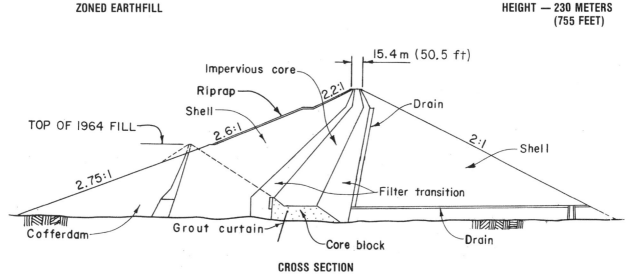

FIGURE 9 Oroville Dam on the Feather River in California.

A. Earthfills

Design of earthfills has evolved as the science of soil mechanics has advanced and become an integral part of the broad field of geotechnical engineering. One of the important fundamentals is the relationship between the optimum degree of compaction and the moisture of the soil being placed. Rudimentary methods of slope stability analysis, while still useful, have been joined with more sophisticated procedures for assessing the response to loading, particularly to the effects of seismic vibration and consequent variation in internal water pressures. The value of the new techniques has been measured by instrumental monitoring of real structural behavior. Control of seepage through earth zones is ensured by various design elements, which may include foundation cutoffs, impervious earth

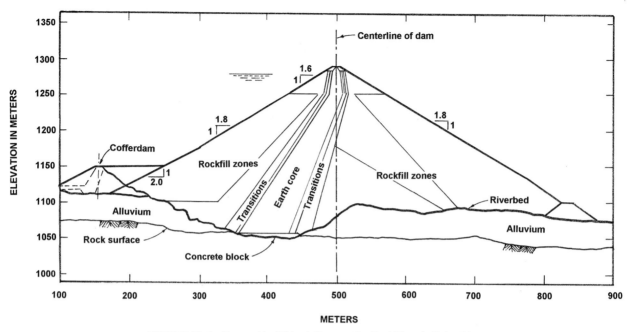

FIGURE 10 La Esmeralda (Chivor) Dam on the Batá River in Colombia.

blankets on the reservoir floor immediately upstream from the dam, internal filter and transition zones, drains, or pressure relief wells downstream from the embankment.

B. Rockfills

Design of rockfills also has benefited from these advances, although the proportioning of this type of embankment is still comparatively simple, drawing amply from successful precedent. A fundamental consideration in the design of rockfill embankments is the gradation of the constituent materials, which influences stress–strain characteristics and permeability. The actual condition of the *in situ* fill is determined not only by the sizes and durability of the particles, but also by the thickness of the placed layers and the degree of compaction, a basic objective being to

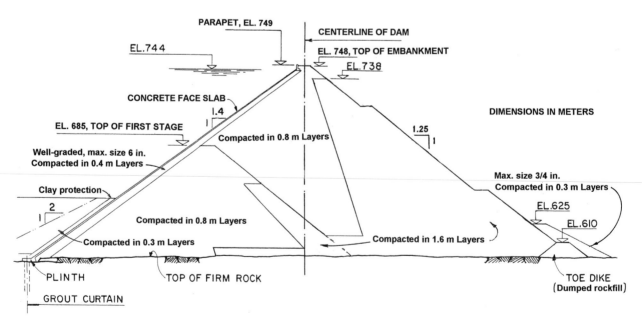

FIGURE 11 Foz do Areia Dam, a rockfill on the Iguaçu River in Brazil.

FIGURE 12 Piedra del Aguila Dam on the Limay River in Argentina, a concrete gravity structure 174 m (571 ft) high. The project was completed in 1992. The spillway is in the right of the view.

ensure that settlement will be minimal so that the impervious element will not be broken.

The design of concrete-faced rockfill dams has been greatly improved in recent times and this type has won acceptance at many new projects. Primary advances have been made in the required degree of compaction, in the details of the plinth and the face slab, and in the gradations of the rockfill zones, especially under the face slab and at its edges, where finer rock sizes are specified. Careful attention is given to the reinforcement and anchoring of the plinth and to treatment of the foundation under and immediately downstream from it. To limit deformation at the perimeter joint of the face slab, typical specifications call for the underlying fill to be fine rock filter that is specially compacted. Rockfill stability is analyzed by procedures that have been applied to earthfills, assessing the potential for sliding on assumed internal planar or curved surfaces,

FIGURE 13 Alicura Dam on the Limay River in Argentina, an earthfill 130 m (427 ft) high, completed in 1984. The dam is on the left, the powerhouse is in the center, and the spillway is on the right.

using friction factors determined by experimentation or experience with comparable materials at other projects.

C. Gravity Dams

A typical gravity dam comprises spillway and nonoverflow blocks, some of them including outlet works and intake facilities for control of water releases. Low-level outlets for river discharges are located typically in the spillway section. Intakes for penstocks leading to a powerhouse are commonly incorporated into a nonoverflow section. A long-standing objective has been to avoid tension in the unreinforced concrete mass. To establish an adequate base width to ensure stability, the downstream faces of the blocks are usually inclined no steeper than about 55 degrees with the horizontal, which is expressed as a 0.7:1 slope. The actual geometry in each project is determined by analysis of the forces that might be imposed, including the loads from extreme floods and earthquake, as well as from possible silt and ice forces on the upstream face. The overflow crest in the spillway section is conventionally shaped to conform to the trajectory of a discharging stream. The mass concrete of the common dam of this type does not have steel reinforcement, but it is provided in parts subject to tensile stresses, as in walls and piers and around galleries and interior valve and gate chambers. A gravity dam typically is founded on rock, which may be grouted to reduce its permeability. The foundation may also be provided with drains to control uplift pressures. Both of these treatments are common in modern dams.

D. Arch Dams

Design of arch dams has evolved as techniques for determining their optimal geometry have been refined. Structural behavior is analyzed by complex computer programs and verified by instrumental measurement of actual response to loading. These advances have enabled a significant change from the early simple designs to present-day, thin, high-strength concrete sections with varied curvature to optimize stress distributions. For the analysis of arches, as well as other kinds of dams, widespread use is being made of the finite element method, which entails mathematically modeling the structure as a composite of individual elements joined at nodes to form a mesh. Nodal deflections and element stresses are computed by solving a series of simultaneous equations.

E. Buttress Dams

The stability of a buttress dam depends primarily on the slope of the upstream face, which typically has been set between about 45 and 50 degrees with the horizontal. The buttresses of flat-slab and multiple-arch dams must be designed to carry the loads transmitted by the slabs or arches as well as to resist lateral forces, including importantly the impact from seismic load. They may be solid, single walls, or double-walled structures. Whatever the case, both the upstream watertight members and the buttresses are provided with steel reinforcement. For lateral stability, the buttresses of many dams of this type have counterforts or stiffeners incorporated as flanges or pilasters. Some of them have also been braced by struts in the spans between buttresses.

V. CONSTRUCTION OF DAMS

A. Stream Diversion

Although some dams are built offstream, as in pumped storage projects and for other special purposes, most are located where construction requires isolation of the work areas from natural streamflows. This is commonly accomplished by cofferdams that enable temporary redirection of discharges through a tunnel in an abutment or over a restricted part of the streambed. In the latter alternative, the construction may be done in stages so the configuration of the cofferdams changes as the work progresses. In the construction of concrete dams, diversion may be facilitated by passing the river over low blocks or through sluices.

B. Earthfill Dams

Construction of earthfill dams has evolved with the development of high-capacity hauling and compacting equipment. The fill is compacted in layers, commonly using sheepsfoot rollers for finer grained soils, rubber-tired rollers for impervious materials well-graded from gravel to sand to silt and clay, and vibratory rollers for pervious zones. Because of the inherent susceptibility of some soils to erosion, strict attention must be given especially to the gradation and compaction of material at the interfaces between zones and between the embankment and its foundation and adjoining structures. At the boundaries, protection is ensured by measures that prevent the migration of fine particles, including intermediate filter zones composed of compatibly graded coarser materials, concrete membranes, and grouting of the underlying rock formation. Because the natural conditions at each project are unique and may not be known adequately until construction is well under way, the final details of an earthfill may have to be determined as the work at the damsite proceeds. Designers and allied specialists, including geologists, therefore must continue to be involved.

Oroville Dam, the highest dam in the United States, was completed in 1968 and is a landmark structure that introduced important innovations and set the stage for construction of very high embankments. It has a concrete core block 39 m (128 ft) high with a volume of 193,000 m^3 (252,000 yd^3) which sealed a gorge and provided a base for construction of the dam's earth core. The block served as a key element in the first year of construction by retaining fill in a zone immediately upstream, affording flood protection while the works for diversion of the Feather River were under way. The outer zones (shells) of the dam, as seen in Fig. 9, are composed of gravels from dredger tailings left by gold mining in the area. A 668-ton, self-propelled wheel excavator removed these materials from the pits at a rate of 5700 tons per hour. Its rotating wheel, 30 ft in diameter, had eight large buckets. The materials were transported to hoppers at loading stations in the borrow area by a system of conveyor belts and by 100-ton, bottom-dump trucks loaded by draglines. Forty-car railroad trains were loaded and traveled 19 km (12 miles) to the damsite, where they were unloaded two cars at a time by a car dumper with swivel couplings that enabled rotating the cars and dumping without separating the cars from the rest of the train. Special laboratory facilities were constructed for testing the large gravel and cobble sizes. This involved the largest testing machine of its type in the United States which could accommodate a specimen of earth and rock 36 inches in diameter and 7.5 ft high.

C. Rockfill Dams

Close coordination of design and construction engineers is also required throughout the development of a rockfill project. The ultimate foundation treatment and the optimum use of the available materials may have to be decided as the site conditions are exposed. The foundations for rockfills are usually sound rock, but at some projects dense glacial or alluvial deposits have served as a base for pervious embankment zones. The properties of the rock sources influence construction procedures. Since about 1960, the increasingly common use of vibratory rollers for compaction of rockfill in relatively thin layers has greatly improved the quality of this type of dam. Before that, the established practice was to dump and sluice the rockfill in high lifts without mechanical compaction, which resulted in embankments that were susceptible to excessive deformation and consequent cracking of the impervious elements (concrete face slab or earth core). In current projects, the common transport of rockfill to the damsite from the quarry or other source is by heavy dump trucks or bottom-dump wagons. Final placement and leveling of the rockfill layers are generally by bulldozer for the large sizes and often by grader for the finer material. The amount of water applied to the fill to enhance the compactive effort varies with the quality of the rockfill, the hardest and least friable rock requiring a minimum of sprinkling.

Illustrative of current U.S. practice in construction of large embankment dams is the Eastside Reservoir Project of the Metropolitan Water District of Southern California, completed in 1999. It comprises three earth-core rockfill dams with total volume of 84 million m³ (110 million yd³), 70% of which is rockfill. The project involved the largest available heavy equipment and machinery and set new records for handling earth and rock materials. Loading of the rock required five hydraulic shovels, each with 17 m³ (22 yd³) capacity, plus large front-end loaders. Dump trucks weighing up to 350 tons when full hauled the rock to the embankment, where it was spread by 100-ton bulldozers. The rockfill was compacted by smooth-drum vibratory rollers. The materials for the earth cores of the dams were excavated and moved by belt loaders at the source and hauled to the fill by 120-yd³ belly-dump trailers. The three dams required a total of about 10 million yd³ of processed rock for filters and drains. The larger of two rock-crushing plants for this purpose had a maximum design capacity of 3200 tons per hour.

D. Concrete Dams

Construction of concrete dams starts with site preparation that entails foundation excavation to suitable rock, placement of concrete for sealing and shaping the excavated surfaces, deep grouting to establish a seepage curtain under the structure, sometimes shallow grouting to consolidate the rock, and drilling of holes for drainage. For construction of the dam itself, careful attention is given to the aggregates of the concrete mix. Mass concrete especially requires control of aggregate temperatures to help in limiting cracking. Cooling the placed blocks of gravity dams has been done by passing cold water through embedded piping and by installation of surface insulation.

VI. SAFETY OF DAMS

A. Elements of Safe Dams

Populations in many countries live in close proximity to dams and depend upon their benefits and their safety for survival. These vital structures require a high level of defensive engineering to anticipate and cope with potentially adverse conditions. They must incorporate multiple safeguards to control the hazards that may be inherent at each site. There are reliable ways to design and maintain safe dams, drawing from advanced technology and from the experiences at reservoirs that have failed. The characteristics of the types of dams and their foundations provide guidance. Gravity dams founded on sound rock are intrinsically stable, as are arches with unyielding abutments. Clean, hard, well-compacted rockfill is highly resistant to deformation and internal erosion. Earthfills properly zoned for seepage control and protected at structural and foundation boundaries can endure under a full range of loadings. Buttress dams composed of high-quality, well-reinforced concrete, constructed on stable rock and braced laterally, have served for many years.

B. Instrumentation and Surveillance

Public confidence in the safety of dams is sustained by long records of trouble-free performance. Risks have been limited by continuously improved engineering and by strict surveillance programs, including overview by governmental agencies. These efforts are focused primarily on control of those usually small amounts of water that might escape from the reservoir through the dam or through its foundation. Safeguards must also be provided against extreme floods and seismic events, although the probabilities of their occurrence are generally very low.

Dams must be watched to ensure early detection of abnormal conditions or deviations in their performance. This surveillance relies heavily upon regular visual inspection by experienced observers, but also includes analysis of data from instrumentation. In all cases, the types and numbers of installed instruments should be determined so that problems can be pinpointed and the volume of data to be processed is not so excessive that there may be delays in giving an alert.

An earthfill dam needs to be monitored to detect deformation, cracking, leakage, and erosion. Variations in conditions must be watched to understand their causes and the possible adverse indications. Particular attention must be directed to foundation and structural interfaces where seepage may be conducive to erosion. This requires that regular measurements be made of the water passing through the dam and its foundation and that individual paths of leakage be traced. Monitoring is done commonly at weirs at or near the downstream toe of the embankment, giving attention also to any indications of eroded materials. Internal instruments for measuring water pressures (piezometers), earth pressures, deformation, and temperatures are also useful in assessing the dam's behavior under the changing conditions that are imposed. Diagnostic devices include borehole cameras, hydrophones, and current meters for examining the inside of the embankment. Emphasis is placed on quick analysis of surveillance readings to highlight any deviations from established norms. This has been facilitated by electronic data entry, telemetering, and computer processing. However, most problems at embankments have been found by inspectors on the ground.

Water pressures at the bases of concrete dams and in the foundations are measured by uplift cells (pore pressure cells) and by piezometers in holes drilled in the rock. Structural movement is monitored by surveys, as well as by joint meters, and pendula (plumblines) in interior shafts. Other instrumentation may include strain meters, stress meters, thermometers, and weirs or other devices for measuring leakage.

In seismically active regions, accelerometers may be installed on dams and in their environs to measure patterns of response to tremors. For embankments, significant data may also be obtained by gauging variations in internal water pressures in earthfill zones during earthquakes.

C. Notable Events

Only a few dams have suffered much damage from major earthquakes, even where there were many reservoirs in the zones of high intensity. A notable exception was Sheffield Dam, an earthfill dam 7.6 m (25 ft) high that failed in the Santa Barbara, CA, quake in June 1925, with a Richter magnitude of 6.3 and an epicenter about 11 km (7 mi) from the damsite. Another example was Lower Van Norman (San Fernando), an earthfill dam 43 m (142 ft) high that was damaged, but survived without reservoir loss, in the 1971 San Fernando earthquake (magnitude 6.6). In both of these cases, the problems were caused by liquefaction of soils, a phenomenon that has been studied extensively in more recent times, with the result that the necessary safeguards have been developed and are widely accepted.

Some of the most significant problems at concrete dams have been related to conditions in the natural sites rather than to the body of the dam itself. Notable cases include St. Francis, Malpasset, and Vaiont Dams.

St. Francis (Fig. 14) was an arched concrete gravity structure 62 m (205 ft) high in San Francisquito Canyon about 72 km (45 mi) north of Los Angeles, CA. It collapsed just before midnight on March 12, 1928. Large blocks of the dam were washed thousands of feet down the canyon. A single monolith was left standing. The basic weaknesses that led to the failure were in the rock foundation, consisting of schist susceptible to sliding on one abutment

FIGURE 14 The remains of St. Francis Dam after failure in 1928.

and water-softened conglomerate on the other (west) abutment, the two formations joining at a geologic fault.

Malpasset Dam was a concrete arch 61 m (200 ft) high on Le Reyan River in southern France. It broke on December 2, 1959. The break occurred within the foundation, as an abutment slid along a weak seam in the rock and the arch rotated about its other end. After the failure, most of the arch on the sliding side had disappeared along with much of the underlying rock.

Vaiont Dam is a concrete arch 262 m (860 ft) high on Vaiont River near Belluno, Italy. On October 9, 1963, there was a massive slide of a canyon wall into the reservoir, causing overtopping of the dam to a height of about 100 m (330 ft) above the crest. Tremors from the slide were recorded on seismographs in a large part of western and central Europe. Despite the tremendous overload, the arch remained intact, attesting to the inherent strength of this kind of dam. However, the flood wave downstream destroyed the town of Longarone at the junction of the Vaiont and Piave Rivers, where 2600 people died. The reservoir was made inoperable.

Rogun Dam in Tajikistan was expected to be the world's highest dam (335 m). Its site is upstream from Nurek Reservoir on the Vakhsh River. The Rogun design was similar to that of Nurek Dam, with earth core and gravel shells and rockfill facing zones. Construction was progressing when, upon independence from the Soviet Union in 1991, political turmoil and civil war in Tajikistan led to suspension of the project. A large flood in 1993 severely damaged the incomplete works. The future of Rogun Dam is uncertain.

VII. WEIGHING BENEFITS AND DETRIMENTS OF DAMS

The benefits of dams are so interwoven in the fabric of society that some of them may be unrecognized. Others may be known but taken for granted, as are the light switch and the water faucet. Dams are essential to meeting the world's rapidly growing needs for food and energy. The primary benefits of drinking water, irrigation of crops, groundwater recharge, river flow maintenance (including dilution to improve water quality), power generation, flood control, navigation, fire protection, sewage transport, recreation (swimming, boating, fishing, and river rafting), wildlife enhancement, and protection from salinity intrusion engender secondary and tertiary effects including growth of industry, commerce, employment, and local and federal tax bases supporting schools; medical services; and other community needs. In arid regions, reservoirs have improved temperature ranges, increased humidity, and enhanced the environment for migratory birds and other crea-

tures. The advantages of water power as a clean, perpetual resource are evident in comparison with nuclear and fossil-fuel generation. The contributions of hydroelectric powerplants to pollution control can be demonstrated by calculating the volumes of toxic particulates and gases that would be discharged to the atmosphere by alternative coal-fired powerplants of equal capacity. Energy generation from the renewable water source is of global importance in reducing emissions of harmful ozone-depleting chemicals to the atmosphere.

Objective analysis recognizes that dams have both positive and negative effects on fish. The challenge is to mitigate the harm while preserving the advantages. Ideal solutions are elusive. Benefits include the creation of lake fish habitat and the maintenance of downriver flows to sustain migrating fish. Streams without dams also present hazards to fish, including dangers from debris carried by uncontrolled floods and from low water leaving them and their eggs vulnerable to predators. Extensive study has been given to the hardships encountered by salmon in their long journeys, including commercial fishing, the harsh ocean environment, and the degradation of spawning habitat in watersheds by pollution and erosion. Only a very small percentage of the migrants survive the round trip to return to their birthplaces. The decline in the numbers of salmon, as well as the reduced numbers of other species of wildlife, may be attributable to many factors. Herring, an important source of food for salmon, are no longer numerous. Possible causes are disease, predators, shifts in ocean conditions, and increases in industrial pollution. On the Columbia River, where scientists have conducted long-term, exhaustive studies of salmon losses, a major factor has been found to be predation by the many thousands of terns nesting on islands of alluvium dredged from the navigation channels.

The interruption of fish migration by dams is an age-old problem, which engineers have tried to solve by various means, including fish hatcheries, fish ladders, screens, locks, elevators, bypasses, barging, trucking, and pumping. These have been effective at some projects, but there is a continuing need for improved designs. Hatcheries have provided benefits, but the fish produced do not generally survive as well as wild species. There may be substantial delays in passage of fish at dams, whether going upstream or downstream, during which exposure to disease and predators is prolonged. Migrants traveling downstream may be injured severely in passing through spillways and powerplant turbines. At some projects, deflectors have been installed on the faces of spillways to prevent deep plunging of the overflow so that supersaturation of nitrogen and other gases and consequent injury to fish are reduced. Other beneficial measures include scheduling of spillway discharges during downstream migrant passage and placing screens on penstock intakes, as well as transporting fish around the dams. Some designs provide for interception of fish upstream from reservoirs and directing them through bypass channels to holding facilities to await transport. In the system of the Snake and Columbia Rivers in Washington, this has involved barging over hundreds of miles. Fish detoured around dams may not only escape physical injury but are exposed to less risk from nitrogen supersaturation. The possible effect of transport on the spawning instinct is under continuing study.

Some activists advocate removal of dams on certain rivers to allow fish passage. This option can simultaneously eliminate existing benefits and detriments. Consideration must be given to the infrastructures in the regions dependent upon the dams and their reservoirs. Experience has shown the destruction of a dam to be difficult and costly and not without its own negative effects, including reservoir sediment release. However, in a few unusual cases, demolition has been the selected alternative. In 1999, under a ruling by the Federal Energy Regulatory Commission, the Edwards Dam, a small 162-year-old timber and concrete structure on the Kennebec River in Maine, was breached to allow return of fish from the Atlantic Ocean. In the year 2000, plans were under way to remove dams on the Elwha River in Washington and on forks of Battle Creek in California. Elwha Dam and Glines Canyon Dam on the Elwha, completed in 1913 and 1927, respectively, were not provided with facilities for salmon protection and thus had a major impact on the economy of the local Native American tribe. The cost of removing the dams and restoring the river to its natural condition is projected to exceed $100 million. The several dams to be removed in the Battle Creek watershed are low diversion structures. The restoration plan also includes installation of fish ladders and screens at certain dams that remain, as well as increased stream releases to improve the environment for fish and wildlife. The cost of this plan is expected to be about $50 million. Today, in the development and operation of water projects, full attention must be given to achieving an optimum balance of environmental and economic factors. This demands that dams and other works meet strict standards for preservation of the river and enhancement of conditions in the region.

SEE ALSO THE FOLLOWING ARTICLES

EARTHQUAKE ENGINEERING • GEOENVIRONMENTAL ENGINEERING • HYDROELECTRIC POWER STATIONS • HYDROGEOLOGY • STREAMFLOW • WATER RESOURCE SYSTEMS

BIBLIOGRAPHY

Jansen, R. B. (1983). "Dams and Public Safety," U.S. Govt. Printing Office, Denver, Co.

Jansen, R. B., ed. (1988). "Advanced Dam Engineering for Design, Construction, and Rehabilitation," Van Nostrand-Reinhold, New York.

Jansen, R. B. (1988). "Dam safety in America," *Hydro Rev.*, 7(3), pp. 10–20.

Kollgaard, E. B., and Chadwick, W. L., eds. (1988). "Development of Dam Engineering in the United States," Pergamon Press, New York.

Leonards, G. A., ed. (1987). "Dam Failures," Elsevier, Amsterdam.

National Research Council (1983). "Safety of Existing Dams: Evaluation and Improvement," National Academy Press, Washington, D.C.

National Research Council (1985). "Safety of Dams: Flood and Earthquake Criteria," National Academy Press, Washington, D.C.

Dark Matter in the Universe

Steven N. Shore
Indiana University South Bend

Virginia Trimble
University of California, Irvine, and University of Maryland, College Park

GLOSSARY

Baryons Protons and neutrons that are the basic constituents of luminous objects and which take part in nuclear reactions. These are strongly interacting particles, which feel the electromagnetic and nuclear, or strong, forces.

Cosmic background radiation (CBR) Relic radiation from the Big Bang, currently having a temperature of 2.74 K. It separated out from the matter at the epoch at which the opacity to scattering of the universe, because of cooling and expansion, fell to small enough values that the probability of interaction between the matter and the primordial photons became small compared with unity. It is essentially an isotropic background and its temperature (intensity) fluctuations serve to place limits on the scale of the density fluctuations in the universe at a redshift of about 1000. The redshift is defined as $\lambda/\lambda_0 = 1 + z$, where λ is the currently observed wavelength of a photon and λ_0 is the wavelength at which it was emitted from a distant object.

Deceleration parameter Constant that measures both the density of the universe, compared with the critical density necessary for a flat space–time metric (called $\Omega = \rho/\rho_c$), and the departure of the velocity–redshift relation from linearity (q_0).

Faber–Jackson relation Relation between the bolometric luminosity of a galaxy and its core velocity dispersion. This permits the determination of the intrinsic luminosity of elliptical galaxies without the need to assume a population model for the galaxy. It is used in the measurement of distances to galaxies independent of their Hubble types.

Hubble constant (H_0) Constant of proportionality for the rate of velocity of recession of galaxies as a function of redshift. Its current value is 71 ± 8 km sec^{-1} Mpc^{-1}. Often, in order to scale results to this empirically determined constant, it is quoted as $h = H_0/(100$ km sec^{-1} Mpc^{-1}). The inverse of the Hubble constant is a measure of the age of the universe, called the *Hubble time*, and is about 1.5×10^{10} years.

Leptons Lightest particles, especially the electron, muon, and tau and their associated neutrinos. These particles interact via the weak and electromagnetic forces (electroweak).

Tully–Fisher relation Relation between the maximum orbital velocity observed for a galaxy's 21-cm neutral hydrogen (the line width) and the bolometric luminosity of the galaxy.

Units Solar mass ($M_\odot$), 2×10^{33} g; solar luminosity ($L_\odot$), 4×10^{33} erg sec^{-1}; parsec (pc), 3×10^{18} cm (3.26 light years).

DARK MATTER is the subluminous matter currently required to explain the mass defect observed from visible material on scales ranging from galaxies to superclusters of galaxies. The evidence shows that the need for some form of invisible but gravitationally influential matter is present on length scales larger than the distance between stars in a galaxy. This article examines the methods used for determining galaxies and clusters of galaxies and discusses the cosmological implications of various explanations of dark matter.

I. HISTORICAL PROLOGUE

The existence of a species of matter that can not be seen but merely serves some mechanical function in the universe can be traced to Aristotelian physics and was clerly supported throughout the development of physical models prior to relativity. The concept of the *aether*, at first something apart from normal material in being imponderable eventually metamorphosed into that of a fluid medium capable of supporting gravitation and electromagnetic radiation. However, since it was the medium that was responsible both for generating and transmitting such forces, its nature was intimately tied to the overall structure of the universe and could not be separated from it. It had to be assumed and could not easily be studied. Nineteenth century

attempts, using high-precision measurements, to observe the anisotropy of the propagation of light because of the motion of the earth, the Michelson–Morley experiment being the best known, showed that the aether's mechanical properties could not conform to those normally associated with terrestrial fluids.

A version of the search for some cosmic form of dark matter began in the late 17th century with Halley's question of the origin of the darkness of the night sky. Later enunciated in the 19th century as *Olber's paradox,* it required an understanding of the flux law for luminous matter and was argued as follows: in an infinite universe, with an infinite number of luminous bodies, the night sky should be at least as bright as the surface of a star, if not infinitely bright. Much as in current work on dark matter, the argument was based on the assumption of the premise of a cosmological model and of the dynamical (for in this case phenomenological), character of a particular observable.

In order to circumvent the ramifications of the paradox without questioning its basic premises, F. Struve, in the 1840s, argued for the existence of a new kind of matter in the cosmos, one capable of extinguishing the light of the stars as a function of their distance, hence path length through the medium. The discovery of dark nebulae by Barnard and Wolf at the turn of the century, of stellar extinction by Kapteyn about a decade later, and of the reddening of distant globular clusters by Trumpler in the 1930s served to support the contention that this new class of matter was an effective solution to the *missing light* problem. However, as J Hershel realized very soon after Struve's original suggestion of interstellar dust, this cannot be a viable solution for the paradox. In an infinite universe, the initially cold, dark matter would eventually come into thermal equilibrium. The resulting glow would eventually reach, if not the intensity of the light of a stellar surface, an intensity still substantially above the levels required to reproduce the darkness of the sky between the stars.

An additional thread in this history is the explanation of the distribution of the nebulae, those objects we now know to be extragalactic. In detailed statistical studies. Charlier, Lundmark, and later Hubble, among others, noted a *zone of avoidance,* which was located within some 20° of the galactic plane. Several new forces were postulated to explain this behavior, all of which were removed by the discovery of the particulate nature of the dust of the interstellar medium and the expansion of the universe. However, the explanatory power of the concepts of dark matter and unknown forces of nature serves as a prototype for many, of the current questions, although the circumstances and nature of the evidence have dramatically altered in time.

The origin of the current problem of dark matter (hereafter called *DM*) really starts in the late 1930s with the discovery of galaxy clusters by Shapley. The fact that the universe should be filled with galaxies that are riding on the expanding space–time was not seen as a serious problem for early cosmology, but the fact that these objects cluster seems quite difficult to understand. Their distribution appears to be very anisotropic, and was seen at the time to be the evidence that the initial conditions of the expansion may be drastically altered by later gravitational developments of the universe.

About the same time, Zwicky, analyzing the distribution of velocities in clusters of galaxies, argued that the structures on scales of megaparsecs (Mpc) cannot be bound without the assumption of substantial amounts of nonluminous material being bound along with the galaxies. His was also the first attempt to apply the virial theorem to the analysis of the mass of such structures. Oort, in his analysis of the gravitational acceleration perpendicular to the galactic plane, also concluded that only about half of the total mass was in the form of visible stars.

With the advent of large-scale surveys of the velocities of galaxies in distant clusters, the problem of DM has become central to cosmology. It is the purpose of this article to examine some of the issues connected with current cosmological requirements for some form of nonluminous, gravitationally interacting matter and the evidence of its presence on scales from galactic to supercluster.

II. KINEMATIC STUDIES OF THE GALAXY

The Milky Way Galaxy (the Galaxy) is a spiral, consisting of a central spheroid and a disk of stars extending to several tens of kiloparsecs. A halo is also present and appears to envelope the entire system extending to distances of over 50 kpc from the center. It is this halo that is assumed to be the seat of most of the DM. Since it is essentially spherical, it contributes to the mass that lies interior to any radius and also to the vertical acceleration. The stars of which it is composed have high-velocity dispersions, of order 150 km sec^{-1}, and are therefore distributed over a larger volume while still bound to the Milky Way.

Galactic studies of dark matter come in two varieties: local, meaning the solar neighborhood (a region of several hundred parsecs radius), and large-scale, on distances of many kiloparsecs. The study of the kinematics of stars in the Galaxy produced the first evidence for "missing mass," and there is more detailed information available for the Milky Way than for any other system. Further, many of the methods that have been applied, or will be applied in future years, to extragalactic systems have been developed for the Galaxy. So that the bases of the evidence can be better understood, this section will be a more detailed discussion of some of the methods used for the galactic mass determination.

A. Local Evidence

The local evidence for DM in the Galaxy comes from the study of the vertical acceleration relative to the galactic plane, the so-called "Oort criterion." This uses the fact that the gravitational acceleration perpendicular to the plane of the Galaxy structures the stellar distribution in that direction and is dependent on the surface density of material in the plane. The basic assumption is that the stars, which are collisionless, form a sort of "gas" whose vertical extent is dependent on the velocity dispersion and the gravitational acceleration. For instance, taking the pressure of the stars to be $\rho\sigma^2$, where ρ is the stellar density and σ is their velocity dispersion in the z direction, then vertical hydrostatic equilibrium in the presence of an acceleration K_z gives the scale height of an "isothermal" distribution:

$$z_0 \approx \sigma^2 / K_z. \tag{1}$$

This represents the mean height that stars, in their oscillations through the plane, will reach. It is sometimes called the *scale height* of the stellar distribution.

The gravitational potential is determined by the Poisson equation:

$$\nabla^2 \Phi = -4\pi G\rho = \nabla \cdot \mathbf{K}, \tag{2}$$

where G is the gravitational constant and $\mathbf{K}$ is the total gravitational acceleration. Separation of this equation into radial and vertical components, assuming that the system is planar and axisymmetric, permits the determination of the vertical acceleration. It can be assumed that the stars in the plane are orbiting the central portions of the galaxy, and that

$$K_r = \frac{\Theta^2(r)}{r}. \tag{3}$$

Here, Θ is the circular velocity at distance r from the galactic center. Under this assumption, the Poisson equation becomes

$$\frac{\partial}{\partial z} K_z = -4\pi G\rho + \frac{2\Theta}{r}\frac{d\Theta}{dr}, \tag{4}$$

where the radial gradient of the circular velocity is determined from observation. From the observation of the radial gradient in the rotation curve, it is possible to determine the vertical acceleration from dynamics of stars above the plane and thus determine the space mass density in the solar neighborhood.

The argument proceeds using the fact that the vertical motion of a star through the plane is that of a harmonic oscillator. The maximum z velocity occurs when the star passes through the plane. From the vertical gradient in the z component of the motion, V_z, one obtains the acceleration, and the mass density follows from this. Assuming that stars start from free fall through the plane, their total energy is given by

$$\Phi_0 + \tfrac{1}{2} V_{z,0}^2 = \tfrac{1}{2} V_z^2, \tag{5}$$

so that assuming that $V_{z,0} = 0$, in other words, that the star falls from a large distance having started at rest, it follows that the maximum velocity through the plane of a galaxy is determined by the potential, thus the acceleration, at the extremum of the orbit (apogalactic distance). The vertical acceleration is a function of the surface mass density. Thus, star counts can be used to constrain, from luminous material, the mass in the solar neighborhood. Note that $K_z \sim -4\pi G\Sigma$, where Σ is the surface density, given by

$$\Sigma = \int_0^z \rho(r, z')\, dz'. \tag{6}$$

It should therefore be possible to determine the gravitational acceleration from observations of the vertical distribution of stars, obtain the required surface density, and compare this with the observations of stars in the plane.

The mass required to explain K_z observed in the solar neighborhood, about $0.15 M_\odot\ \mathrm{pc}^{-3}$, exceeds the observed mass by a factor of about two. This is not likely made up by the presence of numerous black holes or more exotic objects in the plane since, for instance, the stability of the solar system is such as to preclude the numerous encounters with such gravitational objects in its lifetime. It appears that there must be a considerable amount of uniformly distributed, low-mass objects or some other explanation for the mass.

Recent observations with the FUSE satellite in the ultraviolet and with ISO in the infrared reveal significant abundances of H_2 in the interstellar medium. ISO data for the edge-on galaxy NGC 891 show enough in this cold and shocked gas to account for the *disk* (only) component of the missing mass. It thus appears plausible that for the relatively low mass component of disk galaxies— the disk—there is no evidence for a low-velocity exotic source of dark matter. The same, however, is not true for the halo, as we will soon discuss.

The velocity dispersion of stars is given by the gravitational potential in the absence of collisions. Thus, from the determination of the mass density in the solar neighborhood with distance off the plane, the vertical component of the gravitational acceleration can be found, leading to a comparison with the velocity dispersion of stars in the z direction. The resulting comparison of the dynamic and static distributions determines the amount of mass required for the acceleration versus the amount of mass observed.

Most determinations of K_z find that only one-third to one-half of the mass required for the dynamics is observed in the form of luminous matter. In spite of several attempts to account for this mass in extremely low luminosity M dwarf stars, the lowest end of the mass distribution of stellar masses, the problem remains. The mass of the observed halo is not sufficient to explain the vertical acceleration, and the mass function does not appear to extend to sufficiently low masses to solve this problem.

Another local determination of the mass of the Galaxy, this time in the plane, can be made from the observation of the maximum velocity, relative to the so-called local standard of rest, of stars orbiting the galactic center. The argument is best illustrated for a point mass. In the case of a circular orbit about a point, the orbital velocity, Θ, is given by centrifugal acceleration being balanced by the gravitational attraction of the central mass,

$$\Theta(r) = \left(\frac{GM}{r}\right)^{1/2}, \tag{7}$$

and it is easily seen that the escape velocity is $v_{\mathrm{esc}}(r) = \sqrt{2}\,\Theta(r)$ for all radii. The coefficient is slightly lower in the case of an extended mass distribution, but the argument follows the same basic form. The local standard of rest (LSR) is found from the velocity of the sun relative to the most distant objects, galaxies and quasi-stellar objects (QSOs), and the globular cluster system of our galaxy. The maximum orbital velocity observed in the solar neighborhood should then be about $0.4 v_{\mathrm{LSR}}$, or in the case of our system about $70\ \mathrm{km\ sec}^{-1}$ in the direction of solar motion. For escape from the system, at a distance of 8.5 kpc with a $\Theta_0 = 250\ \mathrm{km\ sec}^{-1}$, the escape velocity from the solar circle should be only about $300\ \mathrm{km\ sec}^{-1}$. The mass of the Galaxy can be determined from the knowledge of the distance of the sun from the galactic center (determined from mass distributions and from the brightness of variable stars in globular clusters) and from the shape of the rotation curve as a function of distance from the galactic center.

B. Large-Scale Determinations

From the stellar orbits, one knows the rate of galactic differential rotation in the vicinity of the sun, a region about 3 kpc in radius and located about 8.5 kpc from the galactic center. The measurement of galactic rotational motion can be greatly extended through the use of millimeter molecular line observations, such as OH masers, and the 21-cm line of neutral hydrogen. Most of the disk is accessible, but at the expense of a new assumption. It is assumed that the gas motions are good tracers of the gravitational field and that random motions are small and unimportant in comparison with the rotational motions. It is also assumed that the gas is coplanar with the stellar distribution and that the motion is largely circular, with no large-scale, noncircular flows being superimposed. These assumptions are not obviously violated in the Galaxy, but may be problematic in many extragalactic systems, especially barred spiral galaxies. The maximum radial velocity along lines of sight interior to the solar radius occurs at the *tangent points* to the orbits. This method of tangent points presumes that the differential rotation of the Galaxy produces a gradient in the radial velocity along any line of sight through the disk for gas and stars viewed from the sun and that the maximum velocity occurs at a point at which the line of sight is tangent to some orbit. The distance can be determined from knowledge of the distance of the sun from the galactic center, about 8 kpc. Using Θ_{HI} gives a measure of the mass interior to the point r and thus the cumulative mass of the Galaxy. Recent work has extended this to include large molecular clouds, which also orbit the galactic center.

The determination of the orbits and distances of stars and gas clouds outside of the solar orbit is difficult since the tangent point method cannot be applied for extrasolar orbits, but from the study of molecular clouds and 21-cm absorption and stellar velocities of standard stars it appears that the rotation curve outside of the solar circle is still flat, or perhaps rising. The argument that this implies a considerable amount of extended, dark mass follows from an extension of the argument for the orbital velocity about a point mass, $\Theta^2(r) \sim M(r)/r$. Since the observations support $\Theta = const$, it appears that $M(r) \sim r$. The scale length for the luminous matter is small, about 5 kpc, and this is substantially smaller than the radial distances where the rotation curve has been observed to still be flat (of order 15 kpc). The same behavior has been observed in external galaxies, as we will discuss shortly. Thus, there is a substantial need for the Galaxy to have a large amount of dark matter in order to account for dynamics on scales larger than 10 kpc.

As mentioned, mass measurements of the Galaxy from stellar and gas rotation curves beyond the distance of the sun from the center are subject to several serious problems, which serve as warnings for extragalactic studies. One is that there appear to be substantial departures of the distribution from planarity. That is, the outer parts of the disk appear to be warped. This means that much of the motion may not be circular and there may be sizable vertical motions, which means that the motions are not strictly indicative of the mass. For the inner galactic plane, there is some evidence of noncircular motion perhaps caused by barlike structures in the mass distribution, thus affecting the mass studies.

An extension of the Oort method for the vertical acceleration, now taken to the halo stars, can be used to measure the total mass of the Galaxy. One looks at progressively more distant objects. These give handles on several quantities. For instance, the maximum velocity of stars falling through the plane can be compared with the maximum distances to which stars are observed to be bound to the

galactic system. Halo stars of known intrinsic luminosity, such as RR Lyrae variable stars, can be studied with some certainty in their intrinsic brightness. From the observed brightness, the distance can be calculated. Thus, the distance within the halo can be found. From an observation of the vertical component of the velocity, one can obtain the total mass of the galactic system lying interior to the star's orbit. The same can be done for the globular clusters, the most distant of which should be nearly radially infalling toward the galactic center. These methods give a wide range of masses, from about $4 \times 10^{11} M_\odot$ to as high as $10^{12} M_\odot$. to distances of 50–100 kpc.

The phenomenological solution to the problem of subluminous mass increasing in fraction of the galactic population with increasing length scale measured assumes that the mass-to-light ratio, M/L, is a function of distance from the galactic center. For low-mass stars, this number is of order 1–5 in solar units ($M_\odot/L_\odot$), but for the required velocity distribution in the galaxy, this must be as high as 100. There are few objects, except Jupiter-sized or smaller masses, which have this property. The reason is simple—nuclear reactions that are responsible for the luminosity of massive objects, greater than about $0.08 M_\odot$, cannot be so inefficient as to produce this high value. Even the fact that the flux distribution can be redistributed in wavelength because of the effects of opacity in the atmospheres of stars of very low mass will still leave them bright enough to observe in the infrared, if not the visible, and their total (bolometric) luminosities are still high enough to provide M/L ratios that are less than about 10. Black holes, neutron stars, or cold white dwarfs would appear good candidates within the Galaxy. But limits can be set on the space density of these objects from X-ray surveys since they would accerete material from the interstellar medium and emit X rays as a result of the infall. The observed X-ray background and the known rate of production of such relics of stellar evolution are both too low to allow these objects to serve as the sole solution to the disk DM problem. Some other explanation is required.

III. LARGE-SCALE STRUCTURE OF THE GALAXY

A. Multiwavelength Observations

1. Radio and Far-Infrared (FIR) Observations

Neutral and molecular hydrogen (H I and H$_2$), form only a small component of the total mass of the galactic system. Both 21-cm and CO (millimeter) observations can only account for about one-third of the total galactic mass being in the form of diffuse or cloud matter. The IRAS satellite, which performed an all-sky survey between 12 and

100 μm, did not find a significant population of optically unseen but FIR bright point sources in the galaxy. Instead, it showed that the emission from dust in the diffuse interstellar medium is consistent with the amount of neutral gas present in the plane and that there is not a very sizable component of the galaxy at high-galactic latitude. This severely constrains conventional explanations for the DM since any object would come into equilibrium at temperatures of the same order as an interstellar grain and would likely be seen by the IRAS satellite observations. As we mentioned, however, spectra obtained with ISO show diffuse molecular gas in unexpectedly large abundance in spirals.

2. X-Ray Observations

These show the interstellar medium to possess a very hot, supernova-heated phase with kinetic temperatures of order 10^6–10^7 K. But here again it is only a small fraction of the total mass of luminous matter. The spatial extent of this gas is greater than that revealed by the IR and radio searches but is still consistent with the galactic potential derived from the optical studies. As mentioned, X-ray observations also constrain the space density of collapsed but subluminous objects, such as black holes, through limits on the background produced by such objects accreting interstellar gas and dust.

3. Ultraviolet Observations

The stars that produce the UV radiation are the most massive or most evolved members of the disk, being either young stars or the central stars of planetary nebulae. They have very low M/L ratios, are easily seen in the UV, and would not be likely candidates for DM. There is no compelling evidence from other wavelength regions that the missing matter is explained by relatively ordinary matter. Baryonic matter at temperatures from 10^7 K to below the microwave background of 3 K can be ruled out by the currently available observations.

B. The Components of the Galactic Disk and Halo

Star counts as a function of magnitude in specific directions (that is, as a function of l_{II} and b_{II}, the galactic longitude and latitude, respectively) provide the best information on the mass of the visible matter and the structure of the galaxy. This has been discussed most completely by Bahcall and Soniera (1981). It depends on the luminosity function for stars, their distribution in spectral type, and their evolution. However, in the end, it provides more of a constraint on the evolution of the stellar population than

on the question of whether the system is supported by a substantial fraction of dark matter.

Although such studies concentrate on the luminous matter, the reckoning of scale heights for the different stellar populations provides an estimating method for the vertical and radial components of the acceleration—direct mass modeling. Such studies are also sensitive to the details of reddening from dust both in and off of the galactic plane, metallicity effects as a function of distance from the galactic center, and evolutionary effects connected with the processes of star formation in different parts of the galaxy. Supplementing the space density studies with space motions (proper motion, tangential to the line of sight, can be found for some classes of high-velocity stars and can be added to the radial velocities to give an overall picture of the stellar kinematics) permits both a kinematic and photometric determination of the galactic mass. These studies have extended the determination of the vertical acceleration to distances of more than 50 kpc into the halo.

Another constraint on the mass of the Galaxy comes from the tidal radii of the Magellanic Clouds and from the masses and mass distributions of globular clusters. The Magellanic Clouds (the Clouds) are fairly large compared with their distances from the galactic center. They therefore feel a differential gravitational acceleration that counteracts their intrinsic self-gravity and tends to rip them apart as they move around the Galaxy. The maximum distance that a star can be from the Clouds before it is more bound to the Galaxy than the Clouds, or conversely the minimum distance to which the Clouds can approach the Galaxy without suffering tidal disruption, provides a measure of the total mass of the Galaxy. The distribution of the globular clusters does not provide any strong support for a DM halo. In addition, the clusters appear to be bound, even though they are the highest mass separate components of the galactic system, without involking any nonluminous mass. In order to account for the dynamics and to be consistent with the known ages of the clusters, the typical M/L ratio is about 3, characteristic of low-mass stars. So it appears that objects of order 10^6–$10^8 M_\odot$, and with sizes of up to 100 pc, are *not* composed of large quantities of dark mass.

C. Theoretical Studies

The spiral structure of disk galaxies has been a longstanding problem since the discovery of the extragalactic nature of these objects. The first suggestion that the patterns seen in spirals might be due to some form of intrinsic collective mode of a disk gravitational instability was from Lindblad and was elaborated by Lin and Shu (1965) as the "quasistationary density wave" model. The picture requires a collective mode in which the stars in the disk behave like a self-gravitating but collisionless fluid (there is no real viscous dissipation), which clumps in a spatially periodic wave. This wave of density feeds back into the gravitational potential, which then serves as the means for supporting the disk wave structure.

It was soon realized that these waves are not stationary. Further, because of angular momentum transport by tidal coupling of the inner and outer parts of the disk, the waves will wind up and dissipate in the absence of a continuing forcing. Halos have been shown to help stabilize the disk, suggesting that perhaps the very existence of spiral galaxies is an indication of some DM being distributed over a large volume of the Galaxy. Ostriker and Peebles (1973), by analysis of the stability of disks against collapse into barlike configurations, came up with an independent argument supporting the need for an extensive halo having at least two-thirds the total mass of a disk galaxy. Their criterion was established by the stability analysis of a self-gravitating ellipsoid, showing that a disk is unstable to the formation of a bar if the ratio of the kinetic energy to the gravitational potential is large, that is, if $T/W > 0.14$. Here, T is the kinetic energy of the stars and W is the gravitational potential energy. This was perhaps the first paper to argue for the existence of a hot halo on the basis of the stability of a simple model system. Further modeling supports this conclusion, which now forms one of the cornerstones of galactic structure models: if the halo is not massive enough to stabilize the disk of a spiral galaxy, the system will collapse to a bar embedded in a more extended spheroid of stars.

The implication is that, since there are many disks that do not possess large-scale, barlike structures, there may be a halo associated with many of these systems. In addition, the fact that a disk is observed is indication enough that there should be a considerable quantity of mass associated with it. Thus, the search was stimulated for rotation curves that remain flat or non-Keplerian (not point-mass-like) at large distances from the galactic center.

IV. DARK MATTER IN THE LOCAL GROUP

A. Magellanic Clouds

The total mass of the Galaxy can be determined from the stability of the Magellanic Clouds. These two Irr galaxies orbit the Milky Way, being slowly tidally disrupted by the interaction with the disk of the Galaxy. The tidal radius of the Clouds and the fact that they have been stable and bound to each other for far longer than a single orbit (these satellites of the Galaxy appear as old as the Milky Way) provide an upper limit to the mass of the Galaxy and Clouds to a distance of about 60 kpc. This is

not far from that obtained from the study of halo stars and clusters, and provides an upper limit of about $10^{12} M_\odot$. The required M/L is thus about 30. The total mass of the Magellanic Clouds may be individually underestimated, but this appears to be a good limit on the total mass of the system and is in qualitative agreement with the mass required to explain the dynamics of the Local Group (the galaxy cluster to which the Galaxy belongs) as well.

B. M 31 and the Rotation Curves of External Galaxies

The study of the rotation curve of the Andromeda Galaxy, M 31, the nearest spiral galaxy to ours and one which forms a sort of binary system with the Milky Way, shows that the rotation curve is flat to the distance at which the stars are too faint to determine the rotation curve. The M/L ratio is close to 30, about the same as that found for the Galaxy. In order to place this result in context, it is necessary to describe the process whereby the rotation curves are determined for extragalactic systems.

First, one can assume that the disk is radially supported by the revolution of collisionless stars about the mass lying interior to their radial distance from the galactic center. This assumption makes the mass within a distance r, $M = M(r)$, a function of radius, and then allows the determination of the orbital velocity assuming that the mean eccentricity of the orbits is small or zero. A slit is placed along, and at various angles to, the symmetry axes of the galaxy, and the radial velocity of the stars is determined. The disk is assumed to be circular, so the inclination of the galaxy can be determined from the ellipticity of the projected disk in the line of sight. There are several methods for fitting the mass model to this rotation curve, most of which assume a power-law form for the rotation curve with distance and fit the coefficients of the density distribution required to produce the observed velocities by the solution of the dynamical equations in the radial direction. This density profile is then integrated to give the mass interior to a given radius.

Several assumptions go into this method, most of which appear theoretically well justified. The first is that the system is collisionless. There is little evidence that the stars in a galaxy undergo close encounters—they are simply too small in comparison with the distances that separate them. However, there are very massive objects present in the disk, the Giant Molecular Clouds, which may have masses as great as $10^7 M_\odot$ and which appear to control the velocity dispersion of the constituent stars. These clouds are also the sites for star formation, and as such they form the basis for the interaction of the gas and star components of the Galaxy. They may also transfer angular momentum through the Galaxy, making the disk appear viscous. Such interactions act much like encounters to change the velocity distribution to something capable of supporting a flat rotation curve. However, this remains speculation.

The observation of the rotation curve using neutral hydrogen line emission is a better way of determining the mass of a galaxy at a large distance. First, it appears that the H I distribution is considerably more extended than that of the bright stars; H I maps extend, at times, to four or five times the optical radius of the disk. Even at these large distances, studies show that the rotation curves remain flat! This is a very difficult observation to understand if the sole means of support for the rotation is the mechanism just described. Rather, it would seem that some form of extended mass distribution of very high M/L objects is required. The radial extent of such masses may be as great as 100 kpc, where the optical radii of these galaxies is smaller than 30 kpc.

The measurement of the rotation curve for the Local Group galaxies is complicated by the fact that the systems are quite close and therefore require low-resolution observations to determine the global rotation curves. This is also an aid in that the small-scale structure can be studied in detail in order to determine the effects of departures from circular orbits on the final mass determination.

The study of radial variations of velocity curves is essentially the same for external galaxies as for ours. One assumes that there is a density distribution of the form $\rho_*(r, z)$ of the stars and that this is the only contributor to the rotation curve for the Galaxy. One then obtains $\Theta(r)$, from which one can solve the equation of motion for radial support alone:

$$\Theta^2 = 4\pi (1 - e^2)^{1/2} \int_0^r \frac{\rho(\xi)\xi^2 \, d\xi}{(r^2 - \xi^2 e^2)^{1/2}}, \qquad (8)$$

where e is the eccentricity of the spheroid assumed for the mass distribution. One can then expand ρ in a power series of the form

$$\rho(r) = \frac{a_{-2}}{r^2} + \frac{a_{-1}}{r} + a_0 + \cdots \qquad (9)$$

and perform a similar expansion for the velocity at a given distance. The solution is then mapped term by term. Once finished, one can then integrate the entire mass of the disk by assuming that the spheroid is homogeneous, so that

$$M(r) = 4\pi (1 - e^2)^{1/2} \int_0^r \rho(\xi)\xi^2 \, d\xi. \qquad (10)$$

Such models are, of course, extremely simplified (the method was introduced by Schmidt in the 1950s and is still useful for exposition), but they can illustrate the problem of the various assumptions one must make in obtaining quantitative estimates of the mass from the rotation curve. Since the rotation curves remain flat, one can also assume that the M/L ratio is a function of distance from the galac-

tic center. This is folded into the final result after the mass to a given radius has been determined.

Unlike our system, where one cannot be sure what is or is not a halo star, one can determine with some ease the velocity dispersions for face-on galaxies perpendicular to the galactic plane with the distance from the center of the system. In a procedure much like the Oort method described for the Galaxy, it is possible to place constraints on the amount of missing mass in the disk by using the determination of surface densities from the dynamical information. External galaxies appear to have values similar to our system for the dark matter in the plane, about one-half of the matter being observed in stars.

Many disk systems show warps in their peripheries; M 31, the Galaxy, and other systems show distortions of both their stellar and H I distributions in a bending of the plane on opposite parts of the disk. This argues that the halos of these galaxies cannot be too extensive, beyond the observable light, or one would not be able to observe such warps. Calculations show that these will damp, from dynamical friction (dissipation) and also viscous effects, if the halo is too extensive. However, in the case of some of the current models for the distribution of DM in clusters, it is not clear whether this can also constrain a more evenly distributed DM component (one less tied to the galaxies but more evenly distributed in the clusters than the halos).

V. BINARY GALAXIES

Binary galaxies may also be used to determine the M/L of the DM. The process is much like that employed for a binary star, with the exception that the galaxies do not actually execute an orbit within the time scale or our observations. Instead, one must perform the determination statistically. For a range of separations and velocities, one can determine a statistical distribution of the M/L values, which can then be compared with the mean separations and masses obtained from the rotation curves of the galaxies involved. Again, there is strong evidence for a large amount of unseen matter, often reaching M/L values greater than 100. The method has also been extended to small galactic groups, with similar results.

Using large samples of binary galaxies, so that the effects of orbital eccentricity, inclination, and mass transfer can be accounted for, recent studies have shown that about a factor of a three times the observed mass is required to explain binary galaxy dynamics. Summarized, the estimates of M/L range from about 10 to 50. The wide range reflects uncertainties in both the Hubble constant, which is required for the specification of the separation in physical units, and the projection of the velocities of the galaxies in the line of sight. Unlike clusters, binary galaxies are rather sensitive to the projection factors, since one is dealing only with two objects at a time.

Measurement of the masses of the component galaxies can also add information to this picture, and there have been efforts to obtain the rotation curves for the galaxies in binaries and small groups. The interpretation of the velocity differences, the quantity which, along with the projected separation, gives the mass of the system, requires some idea of the potential in which the galaxies move. If there is very extended DM around each of the component galaxies, the use of point-mass or simple spheroidal models will fail to give meaningful masses. Also, since some of the binary galaxies may be barred spirals and others may show subtle and not yet observed tidal effects, it is likely that the mass estimates from these systems should be taken with considerable caution.

Small groups of galaxies, the so-called Hickson groups, can be used to provide an example of dynamical interaction which lies between the clusters and the binaries. Here, however, the picture is also complex. Interactions between the constituent galaxies can alter their mass distributions and affect their orbits in a manner that vitiates the determination of the orbital properties for the member galaxies. If these groups are bound and stable, the M/L must be of order 100. However, Rubin provided evidence, from the study of individual galaxies, that the rotation curves do not show the flat behavior of isolated galaxies. It is possible that the halos of the galaxies have been altered in the groups. Deep photographs of the groups provide some handle on this. Complicated mass distributions of the luminous matter are observed, indicating that the galaxies have been strongly interacting with one another. Mergers, collisions, and tidal interactions at a distance are all indicated in the images. Clearly, more detailed study is required of these objects before they can be used as demonstrations of the need for DM.

VI. THE LOCAL SUPERCLUSTER

A. Virgo and the Virgocentric Flow

The Local Group sits on the periphery of the Virgo supercluster, containing several hundred bright galaxies and the Virgo cluster centered on the giant elliptical galaxy M 87, which contains as much as $10^{14} M_\odot$. Because of its mass and proximity, its composition and dynamics have been well studied. Virial analyses indicate that the total cluster is bound by DM, which amounts to about 10 times the observed mass (a figure that is typical of clusters and which will be discussed in more detail later). Because of its closeness, however, the Virgo supercluster has an even more dramatic effect on motion of the Local Group, which can be used to independently estimate its total mass.

From the motion of the Local Group with respect to the Virgo cluster, it is possible to estimate the total mass of the cluster. The argument is a bit like that of Newton's apple analogy. In the free expansion of the universe, local regions may be massive enough to retard the flow of the galaxies away from each other. This is responsible for the stability of groups of galaxies in the first place. They are locally self-gravitating. In the case of the motion of the Galaxy toward Virgo, there are several ways of measuring this velocity.

One of the most elegant is the dipole anisotropy of the cosmic background radiation (CBR). The idea of using the CBR as a fixed light source relative to which the motion of the observer can be obtained is sometimes called the "new aether drift." The motion of the observer produces a variation in the effective temperature of the background that is directly proportional to the redshift of the observer relative to the local Hubble flow. For motions of the order of 300 km sec^{-1}, for example, the variation in the temperature of the CBR should be about 1 mK, with a dipole (that is, consine) dependence on angle. The CBR shows a dipole temperature distribution, about one-third of which can be explained by the motion of the Local Group toward Virgo, about 250–300 km sec^{-1}, called the *virgocentric velocity*. Since the Local Group is part of the supercluster system that contains Virgo, it is not surprising that we are at least partially bound to the cluster. The total mass implied by this motion is consistent with the virial mass determined for the cluster, which implies a considerable amount of dark matter (about 10 times the observable mass). The dipole anisotropy, however, is only partially, explained by the virgocentric flow. There must be some other larger scale motion, about a factor of two larger, responsible for most of the variation in the CBR. This seems to result from the bulk motion of the local supercluster toward the Hydra-Centaurus supercluster ($l_{\rm II} \approx 270°$, $b_{\rm II} \approx 30°$). This brings up the question of large-scale deviations, on the scale of superclusters, of the expansion from the Hubble law.

B. The Large-Scale Streaming in the Local Galaxies: Dark Matter and the Hubble Flow

Distance determination can be a problem for galaxies outside of the Local Group and Virgo cluster. This has to do with our inability to resolve individual stars and H II regions. Several methods have evolved that allow for determination of the intrinsic brightness of galaxies independent of their Hubble flow velocities. The *Tully–Fisher relation* correlates the absolute magnitude of a galaxy, either in the blue or in the infrared, with the width of the 21-cm line. Although this depends on the inclination of the galaxy, this can be taken into account from optical imaging, and it provides a calibration for the intrinsic luminosity of the galaxy from which, using the apparent magnitude, the distance can be obtained directly, without cosmological assumptions. Another calibrator, especially useful for elliptical and gas-poor galaxies, is the *Faber–Jackson relation*. This uses the velocity dispersion observed for the core of elliptical galaxies and spheroids to determine the intrinsic luminosity of the parent galaxy. It is representative of a wide class of objects, indicative of dissipative formation of the systems, and is

$$L \sim \sigma^n, \tag{11}$$

where σ is the velocity dispersion of the nucleus and $n \approx 4$ from most studies.

Using these methods, it is possible to obtain the distance to a galaxy independent of the Hubble velocity; one can then look for systematic deviations from the isotropy of the expansion of the galaxies in the vicinity of the Local Group. Observations show that there is a large-scale deviation of galaxy motion in the vicinity of the Galaxy. Deviations at large displacement to the virgocentric velocity of order 600–1000 km sec^{-1} show that there is a large departure from the Hubble law relative to the uniform expansion that is usually assumed. The characteristic scale associated with this deviation is about 50–100 Mpc, much larger than the size usually associated with clusters of galaxies but on a scale of superclusters. No mass can clearly be identified with this gravitationally perturbing concentration, but it has been argued that it may be a group of galaxies of order $10^{14} M_\odot$, about the size associated with a very large cluster or small supercluster of galaxies. The mystery is its low luminosity, but it is located near the galactic plane, which would account for the difficulty in observing its constituent galaxies. This may be indicative of other large-scale deviations on a larger scale of gigaparsecs.

VII. CLUSTERS OF GALAXIES

From the first recognition of galaxies as stellar systems like the Milky Way, it has been clear that their distribution is markedly inhomogeneous. Early studies by Shapley indicated large (factor of two) density fluctuations in their distribution, while later work has expanded the complexity of this distribution to larger density contrasts. A variety of large-scale structures is observed in the galactic clustering, which gives rise to the idea of a hierarchy. First, there are clusters, megaparsec scale concentrations of luminous objects with total masses typically of $10^{13} - 10^{14} M_\odot$. These contrast with voids, the most famous of which is the Bootes Void, in which the density of galaxies is about 1/10 that of the averages. Clusters of galaxies cluster themselves into

structures called superclusters, and it appears that even these may have some clustering characteristics, although they are sufficiently rare that the statistical information is shaky at best. Clusters can be dynamically and morphologically separated from the background of galaxies and these can be studied more or less in isolation from one another.

Several catalogs are available for galaxy clusters, generally identified with the names Abell and Zwicky. In addition, large-scale sky counts of galaxy frequency per square degree have been made by Shane and Wirtanen (the Lick survey), Zwicky, and the Jagellonian surveys. These do not contain any dynamical information; they are number counts only. But from information about the spatial correlation on the two-dimensional surface of the sky, the three-dimensional properties of galaxies can be crudely determined.

This situation greatly improved with the Center for Astrophysics redshift survey, a complete radial velocity study for galaxies within 15,000 km sec^{-1} of the Local Group (covering scales $< 0.1c$). With velocity information available, it is possible to place the galaxies in question in three-dimensional space and to look at the detailed distribution of both the galaxies and clusters on the scale of 100 Mpc. This has given the first evidence for the nature of the large-scale structure of the universe and the need to consider dark matter in the context of both clusters and superclusters of galaxies. First, we examine the basis for the determination of mass within the clusters and the arguments on the reality of clustering. Then, we discuss the ways in which this information can be used to address the distribution of, and need for, the DM on scales comparable with the size of the visible universe.

The problem of cold versus hot dark matter has been addressed by a number of techniques. Large-scale structures have been detailed in the neighborhood of the Local Group. Two big structures, dubbed the Great Attractor (Lynden-Bell *et al.*, 1988) and the Great Wall (Geller and Huchra, 1990), have been discovered on size scales of order 100 Mpc. The presence of a large-scale deviation from the velocities of the Hubble flow was the signature of the Attractor, a comparatively local region possibly associated with a supercluster.

The Harvard Center for Astrophysics redshift survey of about 15,000 galaxies in the northern hemisphere has revealed several superstructures and, in general, a distinctly inhomogeneous distribution for local luminous matter. Its implication for dark matter is that, at least on the scale of clusters and individual galaxies, there is considerable power in the small-scale length for which CDM is likely responsible. A statistically more distant sample has, however, been analyzed using the IRAS database. These infrared-detected galaxies provide a sample with very different selection criteria than the optically based, and optically based, surveys previously available. The procedure for analyzing such datasets is to follow up on the IR detections with random samples of redshifts of a comparatively large number of galaxies (of order several thousand) from which distances are obtained. The results, reported by Saunders *et al.* (1991), find a significant excess in the galaxy correlation function on scales in excess of 20 Mpc, a size large in comparison with that expected for ordinary CDM. Precisely how this will stand up in extended surveys based on deeper cuts of the existing catalogs, and how this compares with the source counts from COBE and other IR missions, remains to be seen.

A. The Virial Theorem

One of the first suggestions that there is a compelling need to include unseen matter in the universe came from the original determination of velocity dispersions in clusters of galaxies. The argument is quite similar to that used for a gas confined in a gravitational field. Galaxies in a cluster may collide, a problem to which we will later return, but for the moment we will assume that the galaxies are in independent orbits about the center of mass of the clusters. The velocity dispersion is the result of the distribution of orbits of the galaxies about the cluster center, and the radial extent of the galaxies in space results from the fact that they are bound to the cluster. A dynamical system in hydrostatic equilibrium obeys the virial theorem, which states that

$$2T + W = 0, \tag{12}$$

where T is the kinetic energy and W is the gravitational potential energy. The total energy of the system must be *at most zero*, since the system is assumed to be bound. Therefore, the total energy is given by $E = T + W = \frac{1}{2}W < 0$, and the velocity dispersion is given approximately by

$$\sigma^2 = \frac{G M_{\text{virial}}}{\langle R \rangle}. \tag{13}$$

The mass determined from the statistical distribution of orbits depends on the size of the system, and this requires some estimate of the density profile of the cluster. The rms value of the group radius usually suffices for the component of the radial velocity. It is assumed that the orbits are randomly distributed and that they have not been seriously altered since the formation of the cluster. Also, an assumption is embedded in this equation for the dispersion: the mass distribution has remained unaltered on a time scale comparable with the orbital time scale for the galaxies about the center of mass.

The argument that clusters must be bound and stable comes from the distribution of galaxy velocities within the clusters. The dynamical time scale, also called the crossing time, is the period of a galaxy orbit through the center of the cluster potential. The characteristic time scales are of order 10^9 years, much shorter than the Hubble time (the expansion time scale for the universe, taken as the inverse of the Hubble constant).

There have been numerous simulations of the evolution of galaxies in clusters, most of which show evidence for encounters of the cluster members. There is also strong evidence for interaction in the numerous disturbed galaxies in these clusters and the presence of cD galaxies, giant ellipticals with extended halos, in the cluster centers. The latter are assumed to grow by some form of accretion, also called *cannibalism*, of the neighboring galactic systems.

B. Interactions in Groups of Galaxies

Galaxies in clusters collide. This simple fact is responsible for many of the problems in interpreting the virial theorem because it assumes that the dynamical system is intrinsically dissipationless. Collisions redistribute angular momentum and mass and perturb orbits in stochastic and time-dependent ways. None of these effects can be included in the virial theorem formulation, but instead require a full Fokker–Planck treatment of the dynamics.

Observations of small groups of galaxies, the Hickson groups, reveal a complex array of interactions—shells, bridges, and tails are found for many of the members of these small groups. The extent to which this alters the mass distribution of the groups is not known nor is it known how this affects the determination of the M/L ratio for the group.

For clusters of galaxies, the situation is even more complicated. Many cD galaxies have been found to be embedded in shell systems, which are believed to arise from accretion of low-mass disk galaxies by the giant ellipticals. The shells have been used to argue for the mass distribution of the cDs, although the detailed mass distribution cannot be determined from the use of the shells. Instead, they serve as a warning about using simple collisionless arguments for the evolution of the cluster galaxy distribution.

C. The Distribution of Clusters and Superclustering

The two-point correlation function, $\xi(r)$, is the measure of the probability that there will be two galaxies within a fixed distance r of each other. The probability of finding a galaxy in volume element dV_1 within a distance r of another in volume dV_2 is given by

$$dP(1, 2) = n^2[1 + \xi(r_{1,2})]\, dV_1\, dV_2, \tag{14}$$

where n is the mean number density of galaxies. Defined in this way, it is a good measurement of the trend of galaxies to cluster. If there is a clustering, the correlation is asymptotically vanishing at large distance, very positive at small distance, and negative for some interval. It is only a requirement that the cumulant be positive definite (normalizable). The two-point function is not, however, the sole discriminant of clustering. The problem with the equations of motion is that, in an expanding background and under the influence of gravity, the mass points evolve in a highly nonlinear way. The equations of motion do not close at any order, and there is a hierarchy to the distribution of the correlations of different orders.

The correlation function seems to have a nearly universal dependence on separation, $\xi(r) \sim r^{-1.8}$, whether for galaxy–galaxy or cluster–cluster correlations. The coefficient is different, which indicates that the hierarchy is not quite exact and that there may be further differences lurking on the scale of superclusters. It is not likely, however, that this will be easily studied, considering the size of the samples and the extreme difficulty in determining the extent to which these largest scale structures can be separated from the clusters.

D. The Cosmic Microwave Background

The November 1989 launch of the Cosmic Background Explorer, COBE, revolutionized the study of large-scale structure. Designed to operate in the far-infrared and millimeter portions of the spectrum, COBE performed an all-sky survey of the diffuse cosmic background radiation (CBR).

The CBR is truly Planckian; that is, it is a uniform-temperature blackbody spectrum. The temperature is 2.735 K, with an error of less than 60 mK for the wavelength region between 400 μm and 1 cm. The dipole anisotropy, caused by motion of the local standard of rest toward the Virgo cluster, is easily detected in the data with an amplitude of 3.3 mK (with less than a 10% error) in the direction $\alpha = 11^h.1$ and $\delta = -6^o.3$. The lack of a strong quadrupole signature in the background is also very important. This places limits on the distribution of large mass concentrations at large redshift and also on strong mass concentrations at distances comparable to the Virgo cluster (about 10–100 Mpc).

An important limit provided by the shape of the spectrum is the variation of temperature over the sky due to scattering of the CBR by hot gas. Called the Sunyaev–Zeldovich effect, relativistic or near-relativistic electrons with temperature T_e. scatter the background photons, boosting their energies to $\nu \approx k_B T_e / h$, where h is Planck's constant and k_B is the Boltzmann constant. This scattering removes the photons from the long-wavelength part of the CBR, resulting in a change in the intensity with position. The limit provided by the COBE results is that y, the dimensionless Sunyaev–Zeldovich parameter, is $\leq 10^{-3}$, ruling out a substantial contribution by a hot intracluster medium to the temperature of the CBR.

The major result from COBE was the detection of large-scale ($>10°$) temperature fluctuations, $\Delta T / T \approx (29 \pm 1) - (35 \pm 2)$ μK (the smaller amplitude corresponds to the larger angular scales). These are the signature of the fluctuations in the matter distribution imposed during inflation that cross the horizon at the decoupling epoch and produce local perturbations in the gravitational field. Even without the further constraint of galaxy and cluster formation, these temperature variations require a large dark matter component. First, they are very low amplitude. There is little time for them to grow large enough to drive clustering on scales of tens of megaparsecs unless there is something massive but unseen. Second, they do make it out of the earliest stages of the expansion without damping. And most significantly, they are very large scale.

VIII. COSMOLOGICAL CONSTRAINTS AND EXPLANATIONS OF THE DARK MATTER

Several candidates have been suggested for the particular constituents of DM. Since this is a field of enormous variety and unusual richness of ideas, many of which change on short life times, only a generic summary will be provided. This should be sufficient to direct the interested reader to the literature.

A. Cosmological Preliminaries

In attempting to find a likely candidate for the DM, one must first look at some of the constraints placed on models by the cosmological expansion. First, the particles must survive the process of annihilation and creation that dominates their statistical population fluctuations during the early phases of the expansion. The cosmology is given by the Friedmann–Robertson–Walker metric for a homogenous, isotropically expanding universe with radial coordinate r and angular measures θ and ϕ:

$$ds^2 = dt^2 - \frac{R^2(t)\, dr^2}{1 - kr^2} - R^2(t)(d\theta^2 + \sin^2\theta\, d\phi^2), \quad (15)$$

where $R(t)$ is the scale factor, the rate of the expansion being given by

$$\frac{d}{dt}(\rho V) + p \frac{dV}{dt} = 0, \quad (16)$$

which is the entropy equation, with the volume $V \sim R^3$ and ρ and p being the energy density and pressure, respectively, and

$$H^2(t) = \left(\frac{\dot{R}}{R}\right)^2 = \frac{8\pi G\rho}{3} - \frac{k}{R^2}, \qquad (17)$$

which is essentially the Hubble law. The value of k is critical here to the argument, since it is the factor that determines the curvature of the universe. The critical solution, with $\Omega = 1$, has a deceleration parameter $q_0 = 0$. Here, $\Omega = \rho/\rho_c$, where ρ_c is the density needed to give a flat space–time, $\rho_c = 3H_0^2/8\pi G \approx 2 \times 10^{-29}h^2$ g cm^{-3}, where H_0 is the present value of the Hubble constant and $h = H_0/(100$ km sec^{-1} Mpc$^{-1})$. This solution is favored by inflationary cosmologies. During the radiation-dominated era, the energy density varies like R^{-4} and the pressure is given by $\frac{1}{3}\rho$, so that during the earliest epoch, $\rho \sim t^{-2} \sim T^{-2}$.

The horizon is given by

$$l_H = \int_{t_0}^{t} \frac{dt'}{R(t')}. \qquad (18)$$

Perturbations in the expanding universe have a chance of becoming important when they grow to the scale of the horizon. Others will damp because they cannot be causally connected and will simply locally evolve and die away. The expansion then produces fluctuations in the background radiation on the scale of the horizon at the epoch at which regions begin to coalesce.

The entropy of the expanding universe is constant, so that the number density of particles at the time of their formation is related to the temperature of the background, $n_i T^{-3} =$ const, from which we can define the ratio of the photon to baryon number density as a measure of the entropy, $S = n_\gamma/n_b = 10^9$, where $n_\gamma \sim T^3$. For particle creation, the particles will freeze out of equilibrium when the energy in the background is small compared with their rest mass energy, or when $T \leq m_i c^2/k_B$, where k_B is the Boltzmann constant. Thus, for particles that are very massive, during the radiation-dominated phase, the temperature can be very high indeed. Put another way, the equilibrium temperature is defined as $T_{eq} = 10^{11}m_i(\text{GeV})$ K. Since the temperature during the radiation-dominated era varies like R^{-1}, this implies that $R/R_0 \sim 10^{-11}m_i$, where R_0 is the present radius. The point is that the particles which may constitute nonbaryonic explanations for the DM are created in a very early phase of the universe, one which presented a very different physical environment than anything we currently know directly. The cosmic background radiation has a temperature of 2.735 K at the present epoch.

B. The Cosmological Constant

The cosmological constant, Λ, was first introduced by Einstein in an attempt to maintain Mach's principle that local inertia should be determined by the distribution of matter on the largest scale. Although ignored whenever possible, the constant is easily included in the field equations through an added term $\Lambda/3$ on the right-hand side of Eq. (17). This behaves like a negative pressure, driving the expansion, and is presumed to arise from vacuum energy. Observations had, until recently, been consistent with a vanishingly small Λ. But recent data for type Ia supernovae strongly point to a value of $\Omega_\Lambda \approx 0.7$ (although still with $\Omega = \Omega_{\text{matter}} + \Omega_\Lambda = 1$), which in light of the nucleosynthesis limits would make this term the dominant contributor to the energy density. The data show an additional acceleration is required to explain why supernovae of this type, which have been shown to be standard candles with intrinsically small luminosity dispersions, are fainter and more distant for $z > 0.5$ than would be predicted from standard cosmological models with $\Omega = 1$.

C. Global Constraints on the Presence and Nature of Dark Matter

1. Nucleosynthesis

There are few direct lines of evidence for the total mass density of the universe. One of the most direct comes from the analysis of the abundance of the light elements. Specifically, in the early moments of the Big Bang, within the first few minutes, the isotopes of hydrogen and helium froze out. While ^{4}He can be generated in stars by nuclear reactions, deuterium cannot. In the Big Bang, however, it can be produced by several different reactions,

$$\text{p} + \text{n} \rightarrow {}^2\text{D} + \gamma, \qquad \text{p} + \text{p} + \text{e} \rightarrow {}^2\text{D} + \gamma,$$
$$^3\text{He} + \gamma \rightarrow {}^2\text{D} + \text{n},$$

and the deuterium can be destroyed very efficiently by

$$^2\text{D} + \text{n} \rightarrow {}^3\text{H}, \qquad {}^2\text{D} + \text{p} \rightarrow {}^3\text{He} + \gamma.$$

The critical feature of all of these reactions is that they depend on the rate of expansion of the universe during the nucleosynthetic epoch. The drop in both the density and temperature serves to throttle the reactions, producing several useful indicators of the density.

The argument continues: if the rate of expansion is very rapid, that is, if the density is much lower than closure, then the deuterium can be rapidly produced but not effectively destroyed because of the drop in the reaction rate for the subsequent parts of the nuclear network. However, if the density is high, the rate of expansion will be slow enough for the chain of reactions to go to equilibrium abundances of ^{4}He because of the consumption of the ^{2}D. Thus, the D/H ratio, the mass fraction of ^{4}He, and the ^{3}He/^{4}He ratio can be used to constrain the value of $\Omega = \rho/\rho_c$, where ρ_c is the critical density required for a flat universe ($2 \times 10^{-29} h^2$ g cm^{-3}). The larger is Ω, the smaller is the D/H ratio. Current observations of the primordial abundances of helium isotopes and of deuterium provide $\Omega_{matter} \approx 0.1-0.3$.

This appears to provide a number significantly lower than that obtained from the virial masses for clusters and from clustering arguments. Since the baryons in the early universe determined the abundances of the elements that emerged from the Big Bang, the abundances of the light isotopes provide a strong constraint on the fraction of DM that can be in baryonic form, luminous or not.

2. Isotropy of the CBR

Recent balloon observations, especially the BOOMERANG experiment, have probed the intermediate angular scales of the CBR fluctuations. These data show a strong peaking on a scale of about 1°, which corresponds to an angular wave number of $l \approx 200$. This is presumably revealing the acoustic (pressure) fluctuations that were generated during the inflationary epoch and survived by stretching through the expansion to $z \approx 1000$ and indicate the size of the horizon. For angular scales smaller than about 7 arcmin $\times \Omega^{1/2}$, the perturbations are damped because they are lower than the Silk mass, see below, Eq. (23). The spectrum is best reproduced by an open ΛCDM model (CDM with a cosmological constant). Further improvements in the angular correlation tests will be achieved after MAP and PLANCK are launched in this decade.

3. The Formation of Galaxies and Clusters of Galaxies

Recent work on the distribution of galaxies has centered on the idea that the visible matter does not represent the distribution of mass overall. The picture that is developing along these lines, called "biased galaxy formation," takes as its starting point the assumption that galaxies are unusual phenomena in the mass spectrum. One usually assumes that the galaxies are the result of perturbations in the expanding universe at some epoch during the radiation-dominated period and therefore are representative of the matter distribution. However, biasing argues that the galaxies are the product of unusually large density fluctuations and that the subsequent development of these perturbations is dominated by the smoother background distribution of dark matter—that which never coalesces into galactic mass objects. Several mechanisms have been suggested, all of which may be reasonable in some part

of the evolution of the early universe. One picture uses explosions, or bubbles, formed from the first generation of stars and protogalaxies to redistribute mass, alter the structure of the microwave background, and erase the initial perturbation spectrum.

Gravitational clustering alone is insufficient to explain the dramatic density contrasts observed between clusters and voids, unless one invokes large perturbations at the epoch of decoupling of matter and radiation. The current density contrast is of order 2–10, which implies that at the recombination era the perturbations in the density must be of order 10^{-3}. Either the fluctuations in the density at this period are isothermal, in which case the density variation is not reflected in the temperature fluctuations in the cosmic background radiation, or there must be some other mechanism for the formation of the galaxies. In an adiabatic variation, $\delta T/T = (1/3) \times \delta\rho/\rho$, which is clearly ruled out by observations on scales from arcseconds to degrees to a level of 10^{-5}. If the density fluctuations are smaller than this value, there is not enough time for them to grow to the sizes implied by the large-scale structure by the present epoch. Further, quasars also appear to show clustering and to be members of clusters, at $z = 1-4$, so this pushes the growth of nonlinear perturbations to even earlier periods in the history of the universe. If the density of matter is really the cosmological critical value, that is, if $\Omega = 1$, then there cannot be a simple explanation of the distribution of luminous matter simply by the effects of primordial fluctuations.

Biasing mechanisms can be combined with the DM in assuming that the massive particles are the ones which show the large-scale perturbations. Here, the difference between hot and cold DM scenarios shows up most clearly. If the matter is formed hot and stays hot, it will damp out all of the small-scale perturbations. If formed cold, these will be the ones that will grow most rapidly, with gravitational effects later sorting the smallest fragments of the Big Bang into the larger scale hierarchical structures of clusters and superclusters. The topologies of the resultant universes differ between these two extremes, with the cold matter showing the larger contrast between voids and clusters and an apperance that is best characterized by filaments rather than lumps. It should be added that biasing can act on both scenarios and that the final appearance of the simulations can depend on the statistical method chosen to determine the biasing.

4. Cold versus Hot Dark Matter: New Results

The choice between cold versus hot dark matter scenarios depends on the smoothness of the mass distribution but particularly on the concentration of galaxies within the large-scale filaments that link clusters. Too hot a DM component and all the substructure washes out, too cold and it grows too quickly. As with Goldilocks, there is always a third option—warm DM—but the choice of particle is less clear. The distinction between the different scenarios rests on whether the particles are relativistic and/or massless, or thermal and/or baryonic, in which case the former support the largest physical perturbations but prevent the coalescence of substructure, while the latter exacerbate its growth.

D. The Particles from the Beginning of Time

The initial scale at which all of the particle theories start is with the Planck mass, the scale at which quantum perturbations are on the same scale for gravity as the event horizon. This gives

$$m_{\text{Planck}} = \left(\frac{hc}{2\pi G} \right)^{1/2} \approx 1.7 \times 10^{19} \text{ GeV.} \qquad (19)$$

This is the scale at which the ultimate unification of forces is achieved, independent of the particle theories. It depends only on the gravitational coupling constant. It is then a question of where the next *grand unification* scale occurs (the GUT transition). This is usually placed at $10^{12}-10^{15}$ GeV, considerably later and cooler for the background. The particles of which the background is assumed to be composed are usually assumed to arise subsequent to this stage in the expansion, and will therefore have masses well above those of the baryons, in general, but considerably below that of the Planck scale.

1. Inflationary Universe and Soon After

The most recent interest by theorists in the need for DM comes from the *inflationary universe* model. In this picture, originated by Guth and Linde, the universal expansion begins in a vacuum state that undergoes a rapid expansion leading to an initially flat space–time, which then suffers a phase transition at the time when the temperature falls below the Planck mass. At this point, the universe reinflates, now containing matter and photons, and it is this state which characterizes the current universe. One of the compelling questions that this scenario addresses is why the universe is so nearly flat. That is, the universe, were it significantly different from a critical solution at the initial instant, would depart by large factors from this state by the present epoch. The same is true for the isotropy. The universe at the time of appearance of matter must have been completely isotropized, something which cannot be understood within the framework of noninflationary pictures of the Big Bang.

The fact that, in these models, $\Omega = 1$ is a requirement, while all other determinations of the density parameter

yield far smaller values (open universes), fuels the search for some form of nonbaryonic DM. The ratio $\Omega / \Omega_{\text{observed}}$ is approximately 10–100, of the same order as that required from clusters of galaxies.

During the expansion, particles can be created and annihilated through interactions; thus,

$$\frac{dn}{dt} = -D(T)n^2 - 3\frac{\dot{R}}{R}n + C(t), \qquad (20)$$

where the first term is the annihilation, which is temperature dependent; the second is the dilution of the number density because of the expansion; and the third is the creation term, which depends on time and (implicitly) temperature. Let us examine what happens if the rate of creation depends on a particle whose rate of creation is in equilibrium with its destruction. Then, if a two-body interaction is responsible for the creation of new particles and they have a mass m, the Saha equation provides the number in equilibrium,

$$n_{\text{eq}} \sim (mT)^{3/2} \exp(-m/T), \qquad (21)$$

and the rate of particle production becomes a function of the particle mass. Thus, the more massive particles can be destroyed effectively in the early universe but will survive subsequently because of the rapid expansion of the background and dilution of the number density. Any DM candidate will therefore have a sensitive dependence on its mass for the epoch at which the separation from radiation occurs as well as the strength of the particle's interaction with matter and radiation.

Following their freeze-out phase, the particles will be coupled to the expanding radiation and matter until their path length becomes comparable with the horizon scale. This phase, called *decoupling*, also depends on the strength of the interaction with the radiation and matter. After this epoch, the particles can freely move in the universe, which is now "optically thin" to them. It is this free-streaming that determines the scale over which the particles can allow for perturbations to grow. For instance, in the case of the massive cold particles, those created with very low velocity dispersions, the scale over which they can freely move without interacting gravitationally with the baryonic matter is determined by the mass of the baryon clumps. For cold DM, this means that the slow particles, those with energies of less than 10 eV, can be trapped within galactic potentials. Those with energies of 100 eV would still effectively be trapped by clusters. If the particles are hot, they cannot be trapped by these clumps and can, in fact, erase the mass perturbations that would lead to the formation of these smaller structures.

As mentioned in discussing galaxy formation in general, simulations of the DM show that for the hot particles, the mass scale that survives the primordial perturbations and damping is the size of superclusters, about $10^{15}-10^{16} M_\odot$. The superclusters thus form first, then these fragment and separate from the background and form clusters and their constituent galaxies from the "top down." If the particles are cold, they cannot erase the smallest scales of the perturbations, on the order of galaxy size ($10^{12}-10^{13} M_\odot$), and clusters and superclusters form gravitationally from the "bottom up." The problem is that there is little observational evidence to distinguish which of these is the most likely explanation. In effect, it is the simulations of the galactic interactions, placed in an expanding cosmology and with very restrictive assumptions often being made about the interactions between galaxies within the clusters thus formed, which are used as argumentation for one or the other scenario. In fact, it is for this reason that the models must be called *scenarios* since there cannot be detailed analytic solution of the complex problem and the solutions are only sufficient within the limitations of the precise assumptions that have been applied in the calculations.

The critical mass for gravitational instability and galaxy formation is the *Jeans mass*, the length (and therefore mass) scale at which material perturbations become self-gravitating and separate out in the expanding background. For species in equilibrium with radiation, it depends on both the number density of the dominant particle and its mass as well as on the temperature of the background radiation. In the case of the expanding universe, since during the radiation-dominated era the mass density and temperature are intimately and inseparably linked, the Jeans mass depends only on the mass of the particle that is assumed to be the dominant matter species:

$$M_J \sim \rho_i \left(\frac{a_s^2}{G\rho_i} \right)^{3/2} = \left(\frac{T^3}{Gm_i^2 n_i} \right)^{1/2}$$
$$\approx 3 \times 10^{-9} m_{i,\text{GeV}}^{-1} M_\odot, \tag{22}$$

where a_s is the sound speed and n_i is the number density of the dominant species. The smaller the mass of the particle, the larger the initially unstable mass. This collapse must compete against the fact that viscosity from the interaction with background photons tends to damp out the fluctuations in the expanding medium. The critical mass for stability against dissipation within the radiation background is given by the *Silk mass*, below which the photons, by scattering and the effective viscosity that is represents, damp perturbations. This gives a mass of

$$M_s = 3 \times 10^{13} (\Omega h^2)^{-5/2} M_\odot. \tag{23}$$

Below this mass, which is of the same order as a galactic mass, perturbations are damped out during the early stages of the expansion. This provides a minimum scale on which galaxies can be conceived to be formed. Other particles,

however, can also serve to damp out the perturbations at an even earlier epoch if they are hot enough.

2. The Particle Zoo

a. Baryons. Baryons, the protons and neutrons, are the basic building blocks of ordinary matter. They interact via the strong and electromagnetic forces and constitute the material out of which all luminous material is composed.

The basic form of baryonic matter, hot or cold gas, can be ruled out on several counts. One has been discussed in the section on nucleosynthesis in the Big Bang; the baryonic density is constrained by isotopic ratios to be small. Another constraint is provided by the absorption lines formed from cold neutral gas through the intergalactic medium in lines of sight to distant quasars. There are not sufficiently high column densities observed along any of the lines of sight to explain the missing mass. The absorption from Lyman α, the ground-state transition of neutral hydrogen, is sufficiently strong and well enough observed that were the gas in neutral form, it would certainly show this transition viewed against distant light sources. While such narrow absorption lines are seen in QSO spectra, they are not sufficiently strong to account alone for the dark matter.

In addition, hot gas is observed in clusters of galaxies in the form of diffuse X-ray halos surrounding the galaxies and cospatial with the extent of their distribution. Here again, the densities required to explain the observations are not sufficient to bind the clusters or to account for the larger scale missing mass since the X-ray emission seems to be confined only to the clusters and not to pervade the universe at large scale.

It is possible that the matter could be in the form of black holes formed from ordinary material in a collapsed state, but the required number, along with the difficulties associated with their formation, makes this alternative tantalizing but not very useful at present. Of course, football- to planet-sized objects could be present in very large numbers, forming the lowest mass end of the mass spectrum that characterizes stars. Aside from the usual problem of not being able theoretically to form these objects in the required numbers, there are also the constraints of the isotropy of the microwave background and of the upper limit to the contribution of such masses to the infrared background. Since these objects would behave like ordinary matter, they should come into thermal equilibrium with the background radiation and hence be observable as fluctuations in the background radiation on the scale of clusters of galaxies and smaller. Such a variation in the temperature of the CBR is not observed. In addition, it would have been necessary for the baryons to have participated in the nucleosynthetic events of the first few minutes, and this also constrains the rate of expansion to rule out these as the primary components of the universe. If inflation did occur and the universe is really an $\Omega = 1$ world, then the critical missing mass cannot be made up by baryons without doing significant damage to the understanding of the nuclear processing of the baryonic matter.

b. Neutrinos. The electron neutrino (ν_e) is the lightest lepton, and in fact the lightest fermion, making it an interesting particle for explaining DM. It cannot decay into lower mass species if it is only available in one helicity, and thus would survive from the epoch at which it decoupled from matter and radiation during the radiation-dominated era of the expansion.

It is well known that neutrino processes, being moderated by the weak force, permit the particles to escape detection by many of the classical tests. That these are weak particles means that they decouple from the expansion sooner than the photons and can freely stream on scales larger than those of the photons, also remaining hotter than the photon gas and therefore more extended in their distribution if they accrete onto galaxies or in clusters of galaxies. Should the neutrino have a large enough mass, about 30 eV, the predicted abundance of the three known species is sufficient to close the universe and possibly account for the gross features of the missing mass. Limits place this maximum mass for the electron neutrino species as ≤ 10 eV. Other problems with this explanation include the rates of nucleosynthesis in the early universe, particularly the neutron to proton ratio, which is fixed by the abundance of light leptons and the number of neutrino flavors.

The argument that the ν_e mass must be nonvanishing is important. If the neutrino is massive, it may be nonrelativistic. This implies that it decoupled from the microwave background at some time before the baryons, but later than the massless particles. Since the particle decoupling depends on the temperature of the radiation, this determines the mean free-streaming velocity. The capture efficiency of baryons to this background of particles is dependent on their mass and velocity. The escape velocity from a galaxy is several hundred km sec^{-1}, while for galaxy clusters it is much higher. Thus, for galactic halo DM to be explained by ν_e, it is necessary that the particles be cold, that is, they must have free-streaming velocities less than 100 km sec^{-1}. The restriction of the particle mass to less than 10 eV by the observations of the width of the burst of neutrinos from Supernova 1987 A in the Large Magellanic Cloud likely adds fuel to the argument that the neutrino is an unlikely candidate for DM.

If ν_e is the dominant component of DM, an analog of the Silk mass is possible, below which fluctuations are damped because of the weak interaction of matter with the relic neutrinos. The density parameter is now given by $\Omega = 0.01 h^{-2} m_\nu$ (eV) and

$$M_\nu = 5 \times 10^{14} (\Omega_\nu h^2)^{-2} M_\odot, \qquad (24)$$

which is closer to the scale of clusters than galaxies. With current limits, it looks as if ν_e can neither provide $\Omega_\nu = 1$ nor seriously alter the mass scale for galaxy formation, but may nonetheless be important in nucleosynthesis.

c. Alternative particles. After exhausting classical explanations of DM, one must appeal to far more exotic candidates. It is this feature of cosmology that has become so important for particle theorists. Denied laboratory access to the states of matter that were present in the early stages of the Big Bang, they have attempted to use the universe as a probe of the conditions of energy and length that dominate the smallest scales of particle interactions at a fundamental level. Above the scale of energy represented by proton decay experiments, the only probable site for the study of these extreme states of matter is in the cosmos.

Many grand unified theories, or GUTs, contain broken symmetries that must be mediated by the addition of new fields. In order to carry these interactions, particles have to be added to the known spectrum from electroweak and quantum chromodynamics (QCD) theory. One such particle, the axion, is a moderator of the charge-parity (CP) violation observed for hadrons. It is a light particle, but not massless, which is weakly interacting with matter and should therefore decouple early in its history from the background radiation. Being of light mass and stable, it will survive to the epoch of galaxy formation and may provide the particles needed for the formation of the gravitational potentials that collect matter to form the galaxy-scale perturbations.

It is perhaps easiest to state that the axion has the attractive feature that its properties can largely be inferred from models of DM behavior rather than the other way around, and that the formation of these particles, or something much like them, is a requirement in most GUTs. Simulations of galaxy clustering make use of these and other particles with very generalized properties in order to limit the classes of possible explanations for the DM of galaxies and clusters of galaxies.

Supersymmetry, often called SUSY, is the ultimate exploitation of particle symmetries—the unification of fermions and bosons. Since there appears to be a characteristic scale of mass at which this can be effected and since supersymmetry assigns to each particle of one type a partner of the opposite type but of different mass, it is possible that the different particles decouple from the primordial radiation at different times (different temperature scales at which they are no longer in thermal equilibrium) and subsequently freely move through the universe at large. This provides a natural explanation for the mass scales observed in the expanding hierarchy of the universe.

The photon is the lightest boson, massless and with integer spin. It is a stable particle. Its supersymmetric (SUSY) partner is the *photino*, the lowest mass of the SUSY zoo. The photino freezes out of the background at very high temperatures, of order $T_{\text{photino}} \approx 4(m_{\text{SUSY,f}}/100 \, (\text{GeV})^{4/3}$ MeV, where $m_{\text{SUSY,f}}$ is the mass of the SUSY particle (a fermion) by which the photino interacts with the background fermions. If the mass is very small, less than 4 MeV, the decoupling occurs early in the expansion, leading to relativistic particles and free streaming on a large scale; above this, the particle can come into equilibrium with the background through interactions that will reduce its streaming velocity, and annihilation processes, like those of monopoles, will dominate its equilibrium abundance. The final abundance of the photino is dominated by the strength of the SUSY interactions at temperatures higher than T_{photino}.

If the gravitino, the partner of the graviton, exists, then it is possible for the photino to decay into this SUSY fermion and a photon. Limits on the half-life can be as short as seconds or longer than days, suggesting that massive photinos may be responsible for mediating some of the nucleosynthesis in the late epoch of the expansion. If the photinos decay into photons, this raises the temperature of the background at precisely the time at which it may alter the formation of the light isotopes, photodisintegrating ^{4}He and changing the final abundances from nucleosynthesis. Recent attempts to place limits on the processes have shown that the photinos, in order to be viable candidates for DM, must be stable, or have a large mass (greater than a few GeV) if they are nonrelativistic, or have a very low mass (<100 eV). There are problems with large numbers of low-mass particles for the same reason there are problems with ν_e, since these would tend to wash out much of the smallscale structure while not being well bound to galaxy halos.

Strings are massive objects that are the product of GUTs. They are linear, one-dimensional objects that behave like particles. Arguments about their structure and behavior have shown that, as gravitating objects, they can serve as sites for promoting galaxy formation, and may be responsible for the biasing of the DM to form potential wells that accrete the baryons out of which galaxies are composed. There is no experimental support for their existence, but there is considerable interest in them from the point of view of particle theories. Strings have the property that they can either come in infinite linear or closed forms. The closed forms serve as the best candidates for inclusion in the DM scenarios.

IX. FUTURE PROSPECTS

The current status of DM is very confusing, especially in light of the wealth of possible models for its explanation. None of the particles presently known can explain the behavior of matter on scales above the galactic, while there are a number of hypothetical particles that can do so for larger scales. The masses of galaxies are such that, in order to explain their halos, the particles of which they are composed must be nonrelativistic, that is, they must have velocities less than a few hundred km sec^{-1}. Thus, cold DM appears the best candidate for the constituents of galactic halos. On the scale of clusters and superclusters, however, it is still not clear whether this is a firm limit. Biased galaxy formation seems a viable explanation of the distribution of luminous matter, but here again there is a wealth of explanation and not much data with which to test it. A summary of the evidence for DM is given in Table I, and a list of candidates is given in Table II.

The best candidate for testing some of the models is space observation. The Hubble Space Telescope is a high-resolution optical and ultraviolet spectroscopic and imaging instrument capable of reaching 30th magnitude and of performing high-resolution observations of galaxies to at least $z = 4$. Surveys of redshift distributions in clusters of galaxies with very high velocity resolution should aid greatly in the details of virial mass calculations, while the imaging should delineate the extent to which interactions between cluster galaxies have played a role in the formation of the observed galactic distributions. The Cosmic Background Explorer (COBE), launched in November 1989, was designed to look at the isotropy of the microwave background, near the peak of the CBR spectrum. COBE has placed strict limits on the scales on which the background shows temperature variations, thereby delimiting the scale of adiabatic perturbations in the expanding universe at the recombination epoch.

TABLE I Summary of Evidence for Dark Matter in Different Environments

Evidence	$\langle \text{M/L} \rangle$	Ω
Galactic stars and clusters	1	0.001
Vertical galactic gravitational acceleration	2–3	0.002
Disk galaxy dynamics	10	0.01
Irregular and dwarf galaxies	10–100	0.101–0.1
Binary galaxies and small groups	10–100	0.01–0.1
Rich clusters and superclusters	100–300	0.2 ± 0.1
Large-scale structure	1000	0.5–1.0
Baryosynthesis	10–100	≤0.1
Inflationary universe	1000h	1.0

TABLE II Nonbaryonic Dark-Matter Candidates and Their Properties[a]

Candidate/particle	Approximate mass	Predicted by	Astrophysical effects
G(R)	—	Non-Newtonian gravitation	Mimics DM on large scales
Λ (cosmological constant)	—	General relativity	Provides $\Omega = 1$ without DM
Axion, majoron, Goldstone boson	10^{-5} eV	QCD; PQ symmetry breaking	Cold DM
Ordinary neutrino	10–100 eV	GUTs	Hot DM
Light higgsino, photino, gravitino, axino, sneutrino[b]	10–100 eV	SUSY/SUGR	Hot DM
Para-Photon	20–400 eV	Modified QED	Hot/warm DM
Right-handed neutrino	500 eV	Superweak interaction	Warm DM
Gravitino, etc.[b]	500 eV	SUSY/SUGR	Warm DM
Photino, gravitino, axino, mirror-particle, simpson neutrino[b]	keV	SUSY/SUGR	Warm /cold DM
Photino, sneutrino, higgsino, gluino, heavy neutrino[b]	MeV	SUSY/SUGR	Cold DM
Shadow matter	MeV	SUSY/SUGR	Hot/cold (like baryons)
Preon	20–200 TeV	Composite models	Cold DM
Monopoles	10^{16} GeV	GUTs	Cold DM
Pyrgon, maximon, Perry Pole, newtorites, Schwarzschild	10^{19} GeV	Higher dimension theories	Cold DM
Supersymmetric strings	10^{19} GeV	SUSY/SUGR	Cold DM
Quark nuggets, nuclearites	10^{15} g	QCD, GUTs	cold DM
Primordial (mini) black holes	10^{15-30} g	General relativity	Cold DM
Cosmic strings, domain walls	10^{8-10} M	GUTs	Promote galaxy formation, though small contributor to Ω

[a] QCD, quantum chromodynamics; PQ, Peccei–Quinn; GUTs, grand unified theories; SUSY, supersymmetry; SUGR, supergravity; QED, quantum electrodynamics.

[b] Of these various supersymmertric partners predicted by assorted versions of SUSY/SUGR, only one, the lightest, can be stable and contribute to Ω, but the theories do not at present tells us which one it will be or the mass to be expected.

Laboratory tests of particle theories should also contribute to the DM problem. The plans for the Superconducting Supercollider show that it should be able to detect the Higgs boson, responsible for the mass of particles in grand unified theories, and this should also feed the modeling of supersymmetric interactions. As the knowledge of the behavior of the electroweak bosons increases, we should also have a clearer picture of the role played by SUSY in the early universe and in the generation of the particles which are responsible for (at least) some of the DM.

On a final note, it may be said, possibly, we have this all wrong. The study of matter that cannot be seen but only felt on very large scales is obviously one driven by models and calculations. These have assumptions built into them that may or may not be justified in the context of the particular studies. Therefore, it may be the case that DM is more an expression of our ignorance of the details of the universe at large distances and on the cosmological scale than we currently believe. However, the need for invoking DM is very widespread in astrophysics: it is required in many explanations, models for it come in many varieties, its presence is indicated by different methods of mass determination, and it is a phenomenon that involves only classical dynamics. Most simple alternatives proposed so far have been very specifically tailored to the individual problems and are often too *ad hoc* to serve as fundamental explanations of all of the phenomena that require the presence of some form of DM. This is a field rich in speculation, but it is also a field rich in quantitative results. As more data are accumulated on the dynamics of galaxies and clusters of galaxies, we will clearly be able to distinguish between the "everything we know is wrong" school and that which detects the fingerprint of the early universe and its processes in the current world.

APPENDIX: THE DISCRETE VIRIAL THEOREM

Consider the motion of a particle about the center of mass of a cluster. The gravitational potential is the result of the interaction with all particles j, not the same as i, for the ith particle:

$$\Phi \equiv -G \sum_{i<j} \frac{m_j}{R_{ij}}, \qquad (25)$$

where the masses are allowed to be different. The equations of motion for this particle is

$$m_i \mathbf{v}_i = -m_i \nabla \Phi. \tag{26}$$

Taking the product of this equation with the position of the ith particle (the scalar product) gives

$$\sum_i \mathbf{r_i} m_i \ddot{\mathbf{r}} = -2T + \frac{1}{2}\ddot{\mathbf{I}}, \tag{27}$$

where I is the moment of inertia of the cluster and T is the kinetic energy of the particles. For discrete particles, this is easily calculated. Now assume that R_{ij} is the scalar distance between the ith and jth particles, Then the equation of motion yields the discrete vitrial theorem:

$$\ddot{\mathbf{I}} = \sum_i m_i v_i^2 - G \sum_i \sum_{i<j} \frac{m_i m_j}{R_{ij}} = 2T + W. \tag{28}$$

Here W is the gravitational energy of the cluster, summed over all masses. This is the usual form of the virial theorem, with the additional term for the secular variation of the moment of inertia included. It is usually assumed that the system has been started in equilibrium, so that the geometry of the configuration is constant. This implies that one can ignore the variation in I. Now, assume that all particles have the same mass, m. This gives

$$-G \sum_{i<j} m_i^2 \langle R_{ij} \rangle^{-1} + \sum_i m_i \langle v_i^2 \rangle = 0. \tag{29}$$

We obtain only a two-dimensional spatial picture and a one-dimensional velocity picture of a three-dimensional distribution of galaxies in a cluster. The virial theorem applies to the fullphase space of the constituent masses and is three dimensional, of course, in spite of the fact that we can only see the line-of-sight velocities. Therefore, in order to obtain mass estimates from the virial constraint, we must make some assumption about the isotropy of the velocity distribution. The simplest and conventional assumption is that the radial velocity is related to the mean square velocity by $\langle v_i^2 \rangle = 3\langle V_{\text{rad},i}^2 \rangle$ when averaged over the various orientations of the cluster. The *virial mass estimator* is then given by

$$M_{\text{VT}} = \frac{3\pi N}{2G} \left(\frac{\sum_i V_{\text{rad},i}^2}{\sum_{i<j} \rho_{ij}^{-1}} \right), \tag{30}$$

where now N is the number of (identical) particles and ρ_{ij} is the projected separation of the objects i and j in the sky. The coefficient results from assuming that these are randomly oriented and that $\langle R_{ij}^{-1} \rangle = (2/\pi)\rho_{ij}^{-1}$.

Several assumptions have been built into this derivation, among which are the assumptions of isotropy of the orbits of stars about the center of the cluster and the stability of the shape of the cluster. Notice that there will be

an additional term in the equation of motion if there is an oscillation of the cluster with time and that this will reduce the estimated virial mass (although it will in general be a small term since it is inversely proportional to the square of the Hubble time). An additional potential is contributed by the DM, but it may not enter into the virial argument in the same way as this discrete particle picture. In the preceding discussion, the constituent moving galaxies were assumed to be the only cause of the gravitational field of the clusters. Now if there is a large, rigid mass distribution that forms the potential in which the galaxies move, the mass estimates from the virial theorem are incorrect. There will always be an additional term, W_{DM}, which adds to the gravitational energy of the visible galaxies but does not contribute to the mass, and this can alter the determination of the mass responsible for the binding of the visible tracers of the dynamics. One should exercise caution before accepting uncritically the statements of the virial arguments by carefully examining whether they are based on valid, perhaps case-by-case, assumptions.

SEE ALSO THE FOLLOWING ARTICLES

COSMIC INFLATION • COSMOLOGY • DENSE MATTER PHYSICS • GALACTIC STRUCTURE AND EVOLUTION • NEUTRINO ASTRONOMY • STELLAR STRUCTURE AND EVOLUTION

BIBLIOGRAPHY

Aaronson, M., Bothun, G., Mould, J., Huchra, J., Schommer, R. A., and Cornell, M. E. (1986). "A distance scale from the infrared magnitude/HI velocity–width relation. V. Distance moduli to 10 galaxy clusters, and positive detection of bulk supercluster motion toward the microwave anisotropy," *Astrophys. J.* **302**, 536.

Athanassoula, E., Bosma, A., and Papaioannou, S. (1987). "Halo parameters of spiral galaxies," *Astron. Astrophys.* **179**, 23.

Bahcall, J. N., and Soniera, R. M. (1981). *Astrophys. J. Suppl.* **47**, 357.

Bahcall, N. A., Ostriker, J. P., Perlmutter, S., and Steinhardt, P. J. (1999). "The cosmic triangle: Revealing the structure of the universe," *Science* **284**, 1481.

Bergström, L. (1999) "Non-baryonic dark matter," *Nucl. Phys. B (Proc. Suppl.)* **70**, 31.

Blumenthal, G. R., Faber, S. M., Primack, J. R., and Rees, M. J. (1984). "Formation of galaxies and large-scale structure with cold dark matter," *Nature* **311**, 517.

Boesgaard, A. M., and Steigman, G. (1985). "Big bang nucleosynthesis: Theories and observations," *Annu. Rev. Astron. Astrophys.* **23**, 319.

Dekel, A., and Rees, M. J. (1987). "Physical mechanisms for biased galaxy formation," *Nature* **326**, 455.

Dressler, A. (1984). "The evolution of galaxies in clusters," *Annu. Rev. Astron. Astrophys.* **22**, 185.

Faber, S. M., and Gallagher, J. S. (1979). "Masses and mass-to-light ratios of galaxies," *Annu. Rev. Astron. Astrophys.* **17**, 135.

Geller, M., and Huchra, J. (1990). *Science* **246**, 897.

Holt, S. S., Bennett, C. L., and Trimple, V. (eds.). (1991). "After the First Three Minutes," American Institute of Physics Press, New York.

Kashlinsky, A., and Jones, B. J. T. (1991). "Large-scale structure of the universe," *Nature* **349,** 753.

Lynden-Bell, D., *et al.* (1988). *Ap. J.* **326,** 19.

Lin, C. C., and Shu, F. H. (1965). *Astrophys. J.* **140,** 646.

Ostriker, J. P., and Peebles, P. J. E. (1973). *Astrophys. J.* **186,** 467.

Partridge, B. (2000), "The universe as a laboratory for gravity," *Class. Quantum Grav.* **17,** 2411.

Peebles, J. (1980). "Large-Scale Structure of the Universe," Princeton University Press, Princeton, NJ.

Peebles, P. J. E. (1993). "Principles of Physical Cosmology," Princeton University Press, Princeton, NJ.

Rubin, V. (1983). "Dark matter in spiral galaxies," *Sci. Am.* **248** (6), 96.

Saunders, W., *et al.* (1991). "The density field of the local universe," *Nature* **349,** 32.

Silk, J. (2000). "Cosmology and structure formation," *Nucl. Phys. B (Proc. Suppl.)* **81,** 2.

Sato, K., (ed.). (1999). "Cosmological Parameters and the Evolution of the Universe," Kluwer, Dordrecht, The Netherlands.

Spiro, M., Aubourg, E., and Palanque-Delabroille, N. (1999), "Baryonic dark matter," *Nucl. Phys. B (Proc. Suppl.)* **70,** 14.

Trimble, V. (1988). "Dark matter in the universe: Where, what, why?" *Contemp. Phys.* **29,** 373.

van Moorsel, G. A. (1987). "Dark matter associated with binary galaxies," *Astron. Astrophys.* **176,** 13.

Vilenkin, A. (1985). "Cosmic strings and domain walls," *Phys. Rep.* **121,** 264.

White, S. D. M., Frenk, C. S., Davis, M., and Efstathiou, G. (1987). "Clusters, filaments, and voids in a universe dominated by cold dark matter," *Astrophys. J.* **313,** 505.

Zeld'ovich, Ya. B. (1984). "Structure of the universe," *Sov. Sci. Rev. E Astrophys. Space Phys.* **3,** 1.

Databases

Alvaro A. A. Fernandes
Norman W. Paton
University of Manchester

GLOSSARY

Application A topic or subject for which information systems are needed, e.g., genomics.

Conceptual model A data model that describes application concepts at an abstract level. A conceptual model for an application may be amenable to implementation using different database management systems.

Concurrency control Mechanisms that ensure that each individual accessing the database can interact with the database as if they were the only user of the database, with guarantees as to the behavior of the system when many users seek to read from or write to the same data item at the same time.

Database A collection of data managed by a *database management system*.

Database management system A collection of services that together give comprehensive support to applications requiring storage of large amounts of data that are to be shared by many users.

DBMS *See* **database management system**.

Data model A collection of data types which are made available to application developers by a DBMS.

Data type A set of instances of an application concept with an associated set of legal operations in which instances of the concept can participate. Some data types are *primitive* (e.g., integer, string) insofar as they are supported by the DBMS directly (and, hence, are not application specific). Some are *application specific* (e.g., gene). Some data types are *atomic* (e.g., integer) and some are *complex* (e.g., address, comprising street name, state, etc.). Some data types are *scalar* (e.g., integer) and some are *bulk*, or *collection* (e.g., sets, lists).

Persistence The long-term storage of an item of data in the form supported by a data model.

Referential integrity The requirement that when a reference is made to a data item, that data item must exist.

Secondary storage management The ability to store data on magnetic disks or other long-term storage devices, so that data outlive the program run in which they were created.

Security Mechanisms that guarantee that no users see or modify information that they are not entitled to see or modify.

A DATABASE is a collection of data managed by a *database management system* (DBMS). A DBMS provides facilities for describing the data that are to be stored, and is engineered to support the long-term, reliable storage of large amounts of data (Atzeni *et al.*, 1999). DBMSs also provide query language and/or programing language interfaces for retrieving and manipulating database data. Many organizations are dependent in a significant way upon the reliability and efficiency of the DBMSs they deploy.

I. DATABASE MANAGEMENT SYSTEMS

A **database management system** (DBMS) provides a collection of services that together give comprehensive support to applications requiring storage of large amounts of data that are to be shared by many users. Among the most important services provided are:

1. Secondary storage management. The ability to store data on magnetic disks or other long-term storage devices, so that data outlive the program run in which they were created. Sophisticated secondary storage management systems are required to support effective storage and access to large amounts of data. Storage managers typically include facilities for providing rapid access to specific data items using *indexes*, minimize the number of distinct accesses required to the secondary store by *clustering* related data items together, and seek wherever possible to make effective use of *buffers* that retain recently accessed data items in the memory of the computer on which the database system or the application is running.

2. Concurrency control. Large databases are valuable to the organizations that hold them, and often have to be accessed or updated by multiple users at the same time. Uncontrolled updates to a shared store can lead to the store being corrupted as a result of programs interfering with each other in unanticipated ways. As a result, DBMSs incorporate mechanisms that ensure that each individual accessing the database can interact with the database as if they were the only user of the database, with guarantees as to the behavior of the system when many users seek to read from or write to the same data item at the same time.

3. Recovery. Many database applications are active 24 hr a day, often with many different programs making use of the database at the same time, where these programs are often running on different computers. As a result, it is inevitable that programs accessing a database will fail while using the database, and even that the computer running the

DBMS can go down while it is in use. The DBMS must be as resilient as possible in the face of software or hardware failures, with a view to minimizing the chances of a database being corrupted or information lost. As a result, DBMSs include recovery facilities that often involve some temporary *replication* of data so that enough information is available to allow automatic recovery from different kinds of failure.

4. Security. Most large database systems contain data that should not be visible to or updateable by at least some of the users of the database. As a result, comprehensive mechanisms are required to guarantee that no users see information that they are not entitled to see, and more important still, that the users who are able to modify the contents of the database are carefully controlled—staff should not be able to modify their own salaries, etc. As a result, database systems provide mechanisms that allow the *administrator* of a database to control who can do what with the database.

DBMSs can thus be seen as having many responsibilities. Like a bank, it is good if a DBMS provides a wide range of facilities in an efficient way. However, in the same way as it is of overriding importance that a bank does not lose your money, it is crucial that a DBMS must keep track of the data that are entrusted to it, minimizing errors or failures. Thus the provision of a database management facility is essentially a serious matter, and substantial database failures would be very damaging to many organizations. However, the details of how the aforementioned services are supported are quite technical, and the remainder of this chapter focuses on the facilities provided to users for describing, accessing, and modifying database data. Details on how databases are implemented can be found in Elmasri and Navathe (2000) and Garcia-Molina *et al.* (2000).

II. DATA MODELS

In the context of database technology, by **data model** is meant a collection of data types which are made available to application developers by a DBMS. By **application** is meant a topic or subject for which information systems are needed, e.g., genomics. The purpose of a data model is to provide the formal framework for modeling application requirements regarding the persistent storage, and later retrieval and maintenance, of data about the entities in the conceptual model of the application and the relationships in which entities participate.

Each data type made available by the data model becomes a building block with which developers model application requirements for persistent data. The permissible operations associated with each data type become the building blocks with which developers model application requirements for interactions with persistent data. Data models, therefore, differ primarily in the collection of data types that they make available. The differences can relate to structural or behavioral characteristics. Structural characteristics determine what states the database may be in (or, equivalently, are legal under the data model). Behavioral characteriztics determine how database states can be scrutinized and what transitions between database states are possible under the data model.

A DBMS is said to implement a data model in the sense that:

1. It ensures that every state of every database managed by it only contains instances a data type that is well-defined under the implemented data model.
2. It ensures that every retrieval of data and every state transition in every database managed by it are the result of applying an operation that is (and only involves types that are) well defined under the implemented data model.

From the point of view of application developers then, the different structural or behavioral characteriztics associated with different data models give rise to the pragmatic requirement of ensuring that the data model chosen to model application requirements for persistent data does not unreasonably distort the conceptual view of the application that occurs most naturally to users. The main data models in widespread use at the time of this writing meet this pragmatic requirement in different degrees for different kinds of application.

A. The Relational Data Model

The relational data model makes available one single data type, referred to as a *relation*. Informally, a relation can be thought of as a table, with each row corresponding to an instance of the application concept modeled by that relation and each column corresponding to the values of a property that describes the instances. Rows are referred to as *tuples* and column headers as *attributes*. More formal definitions now follow.

Let a **domain** be a set of atomic values, where a value is said to be *atomic* if no further structure need be discerned in it. For example, integer values are atomic, and

so, quite often, are strings. In practice, domains are specified by choosing one data type, from the range of primitive data types available in the software and hardware environment, whose values become elements of the domain being specified.

A **relation schema** $R(A_1, \ldots, A_n)$ describes a relation R of **degree** n by enumerating the *n-attributes* $A_1, \ldots, A_n$ that characterize R. An **attribute** names a role played by some domain D in describing the entity modeled by R. The domain D associated with an attribute A is called the **domain** of A and is denoted by $dom(A) = D$. A **relational database schema** is a set of relation schemas plus the following **relational integrity constraints**:

1. *Domain Constraint*: Every value of every attribute is atomic.
2. *Key Constraint*: Every relation has a designated attribute (or a concatenation thereof) which acts as the **primary key** for the relation in the sense that the value of its primary key identifies the tuple uniquely.
3. *Entity Integrity Constraint*: No primary key value can be the distinguished NULL value.
4. *Referential Integrity Constraint*: If a relation has a designated attribute (or a concatenation thereof) which acts as a **foreign key** with which the relation refers to another, then the value of the foreign key in the referring relation must be a primary key value in the referred relation.

For example, a very simple relational database storing nucleotide sequences might draw on the primitive data type string to specify all its domains. Two relation names might be DNA-sequence and organism. Their schemas can be represented as in Fig. 1. In this simple example, a DNA_sequence has an identifying attribute sequence_id, one or more accession nos by which the DNA_sequence is identified in other databases, the identifying attribute protein_id which the DNA_sequence codes for and the organism_ids of the organisms where the DNA_sequence has been identified. An organism has an identifying attribute organism_id, perhaps a common_name and the up_node of the organism in the adopted taxonomy.

A **relation instance** $r(R)$ of a relation schema $R(A_1, \ldots, A_n)$ is a subset of the Cartesian product of the domains of the attributes that characterize R. Thus, a relation instance of degree n is a set, each element of which is an *n*-tuple of the form $\langle v_1, \ldots, v_n \rangle$ such that

DNA_sequence			
sequence_id	accession_no	protein_id	organism_id

organism		
organism_id	common_name	up_node

FIGURE 1 Relation schemas for DNA_sequence and organism.

DNA_sequence				organism		
sequence_id	accession_no	protein_id	organism_id	organism_id	common_name	up_node
TRBG361	X56734	AA40058	Trif. repens	Bras. napus	rape	Brassicaceae
BNBGL	X82577	CAA57913	Bras. napus	Trif. repens	white clover	Papilionoideae
...	...	...	...	...	...	...

FIGURE 2 Relation instances under the schemas in Fig. 1.

either $v_i \in dom(A_i)$, $1 \le i \le n$, or else v_i is the distinguished NULL value. A **relational database state** is a set of relation instances. Legal relation instances for DNA_sequence and organism from Fig. 1 might be as depicted in Fig. 2.

Notice, in Fig. 2, how each tuple in DNA_sequence is an element of the Cartesian product:

$$dom(\text{sequence\_id}) \times dom(\text{accession\_no})$$
$$\times dom(\text{protein\_id}) \times dom(\text{organism\_id})$$

and correspondingly in the case of organism. Notice also, in Figs. 1 and 2, that all the relational integrity constraints are satisfied, as follows. The domain constraint is satisfied due to the fact that the domain of every attribute is the primitive data type string, whose instances have been assumed to be atomic. The designated primary keys for DNA_sequence and organism are, respectively, sequence_id and organism_id, and since they are intended as unique identifiers of the entities modeled, the key constraint is satisfied. The entity integrity constraint is satisfied assuming that tuples that are not shown also do not contain NULL as values for the primary keys. There is a designated foreign key, organism_id, from DNA_sequence to organism, and since the values of organism_id appear as primary key values in organism, the referential integrity constraint is satisfied.

The relational data model has several distinctive features as a consequence of its simple mathematical characterization:

- Since a relation instance is a set of tuples, no order can be assumed in which tuples occur when retrieved or otherwise are operated on.
- Since there are no duplicate elements in a set, each tuple in a relation instance is unique. In other words, the concatenation of all the attributes is always a candidate key for the relation.
- Since values are atomic, the relational model cannot ascribe to an attribute a domain whose elements are collections (e.g., sets, or lists). For example, in Fig. 2, if a DNA_sequence has more than one accession_no, this cannot be captured with sets of strings as attribute values.

The languages with which data in relation instances can be retrieved and manipulated (thereby effecting state transitions) are discussed in Section III.A.

The relational model was proposed in a form very close to that described in this section in Codd (1970). An excellent formal treatment can be found in Abiteboul *et al.* (1995).

B. Object-Oriented Data Models

Rather than a single data type, as is the case with the relational data model, an object-oriented data model makes available to application developers a collection of type constructors with which complex, application-specific types can be built. The immediate effect of this approach, in contrast with the relational case, is that more complex application data can be modeled more directly in the database. Instead of having to break down one application concept into many relations, the complex structure needed to model the concept is often directly represented by a combination of type constructors provided by the data model.

The most basic notion in object-oriented data models is that of an *object*, whose purpose is to model application entities. An **object** is characterized by having an *identity*, by being in a certain *state* (which conforms to a prescribed *structure*) and by being capable of interaction via a collection of *methods* (whose specifications characterize the *behavior* of the object).

The **identity** of an object is unique. It is assigned to the object by the DBMS when the object is created, and deleted when the object is destroyed. In contrast with a relational primary key, between these two events the identity of the object cannot be changed, neither by the DBMS nor by application systems. Another contrast is that the identity of an object is distinct and independent from both the state of the object and the methods available for interacting with it.

Objects that are structurally and behaviorally similar are organized into *classes*. A **class** has an associated extent which is a set of the identifiers of objects whose structural and behavioral similarity the class reflects. An **object-oriented schema** is a collection of classes plus integrity constraints.

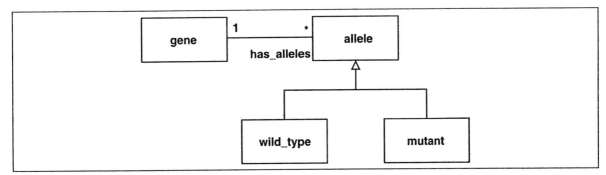

FIGURE 3 A fragment of an object-oriented schema diagram.

In contrast to the relational case, there is no widespread agreement on which integrity constraints are an intrinsic part of an object-oriented model, but one of the most often used stipulates how many objects of each of the classes related by some relationship type can participate in relationship instances. For binary relationships (which are the most commonly used), there are three possibilities, viz., *one-to-one*, *one-to-many*, and *many-to-many*.

Figure 3 depicts in diagrammatic form a small fragment of an object-oriented database schema with four classes, viz., gene, allele, wild_type, and mutant. The latter two are subclasses of allele as indicated by the arrow head and explained later. Note also the relationship type from gene to allele named has_alleles. The annotations 1 and * specify a one-to-many constraint to the effect that an object of class gene may be related to many objects of class allele, but one allele is only related to one gene.

The **structure** of an object is characterized by a collection of *properties* and *relationships*. An object is in a certain **state** when all the properties and relationships defined by its structure are assigned legal values. Both

properties and **relationships** resemble attributes in the relational model, with the following main differences:

1. The domain of a property need not be atomic insofar as the type of its values can be structures or collections by the application of the available **type constructors** (typically, tuple_of, set_of, list_of, and bag_of).
2. The domain of a relationship is a set of object identities (or power set thereof).

Figure 4 shows how the structure of the gene class from Fig. 3 might be specified in an object-oriented data model. Notice the use of type constructors in assigning domains to the property citation and the relationship has_alleles. Notice the nonatomic domain codon_structure. Notice also that the identity of individual gene objects is not dependent on any of the specifications in Fig. 4: It is the DBMS that assigns identifiers, manages their use (e.g., in establishing a relationship between gene objects and allele objects), and annuls them if objects are destroyed. Notice, finally, that

```
tuple_of codon_structure{
        initiation_codon        :   string;
        codon_sequence          :   list_of(string);
        termination_codon       :   string; }
class gene{
        properties
                standard_name   :   string;
                sequence        :   codon_structure;
                function        :   string;
                citation        :   list_of(string);
                protein_id      :   protein;
                organism_id     :   organism;
        relationships
                has_alleles     :   set_of(allele);
        methods
                ... }
```

FIGURE 4 A structure for the gene class from Fig. 3.

```
class gene{
        properties
            ...
        relationships
            ...
        methods
            specification
                    compute_G+C_density    ()           -> float;
                    retrieve_region_mRNA   (int, int)    -> list_of(mRNA);
                    ...
            definition
                    ...
    ... }
```

FIGURE 5   (Partial) behavior of the gene class from Fig. 4.

while an object-oriented database allows relationships to be expressed as simple properties (e.g., protein_id and organism_id establish relationships between gene and protein and organism, respectively), it is only by explicitly expressing them as relationships that an object-oriented DBMS would have the information needed to enforce the kind of referential integrity that relational systems enforce for attributes designated as foreign keys.

The **behavior** of an object is characterized by a collection of *method specifications*. A **method specification** is an application-defined declaration of the name and parameter types for a method that can be invoked on the object. By **method invocation** on an object o with identity i is meant, roughly, that the method call refers to i (thereby identifying o as the object to which the method is addressed) and is meant to be executed in the context of the current state of o using o's assigned behavior (where that execution may itself invoke methods defined on objects other than o). A **method definition** is an executable specification of a function or procedure.

Typically, **accessor methods** (i.e., those that return the current value of a property or relationship) and **update methods** (i.e., those that modify to a given value the current value of a property or relationship) are automatically generated at schema compilation time. A popular approach to naming such methods is to prefix the property or relationship name by get_ and set_, respectively. If type constructors are involved then more methods are made available, accordingly. For example, with reference to Fig. 4, methods would be generated that implement (implicit) specifications such as get_citation() -> list_of(string), or insert_citation(string). The former invoked on a gene object o would return a list of strings currently assigned as the value of the citation property of o, while the latter would cause a state transition in o by adding the string passed as argument to the list of strings currently assigned as the value of the citation property of o.

Figure 5 shows how behavior for the gene class from Fig. 4 might be specified (but method definitions are omitted). Thus, a gene object will respond to a request for its density of guanine and cytosine pairs with a float value. Also, given a region inside it (as defined by a start and an end position in its nucleotide sequence) a gene object will respond to a request for mRNA found inside the region with a list of identifiers of mRNA objects (assuming a class mRNA to have been declared in the schema).

The requirement that one must always invoke methods to interact with objects gives rise to a distinctive feature of object-oriented approaches, viz., **encapsulation**, by which is meant that the state of an object is protected from direct manipulation or scrutiny. This is beneficial in the sense that it allows application developers to ignore the structural detail of the state of objects: all they need to know is the behavior of the object, i.e., which methods can be invoked on it. This, in turn, allows the detailed structure of objects to change without necessarily requiring changes in the application programs that use them.

The approach of organizing objects into classes gives rise to another distinctive feature of object-oriented approaches, viz., **inheritance**, by which is meant a subsumption relationship between the extents of two classes C and C'. Thus, a class C is said to **extend** (or **specialize**, or **inherit from**) a class C' if all objects in the extent of C have the same properties, relationships, and compatible method specifications as (but possibly more than) the objects in the extent of C'. In this case, C is said to be a **subclass** of C'. Thus, inheritance in object-oriented data models is both structural and behavioral (but the compatibility constraints for the latter are beyond the scope of this discussion). There are many motivations for inheritance mechanisms as described, including a stimulus to reuse, a more direct correspondence with how applications are conceptualized, etc.

In the presence of inheritance, not only can a method name be **overloaded** (i.e., occur in more than one method specification), but it can also be **overridden** (i.e., occur in more than one method definition). In this case, method invocations can be attended to by more than one method definition and the decision as to which one to use usually depends on the type of the object that the method was invoked on. This selection process which binds a method definition to a method invocation can only be carried out at run time, therefore this distinctive object-oriented feature is known as **late** (or **dynamic**) **binding**.

To summarize, an object-oriented data model has the following main distinctive features:

- Objects have not only structure but also behavior.
- Objects have a unique identity that is independent of their state and behavior.
- Values can be drawn from complex domains resulting from the application of type constructors such as `tuple_of`, `set_of`, `list_of`, and `bag_of`.
- The state of objects is encapsulated.
- Objects are organized into classes by structural and behavioral similarity.
- Structural and behavioral inheritance is supported.
- Overloading, overriding, and late binding are supported.

Because an object-oriented model brings together structure and behavior, much of the functionality required by applications to interact with application objects can be modeled inside the database itself. Nevertheless, most object-oriented database systems provide application program interfaces that allow a programming language to retrieve and manipulate objects.

Unlike the relational model, object-oriented data models cannot all be traced back to a single major proposal. The closest thing to this notion is a manifesto (found, e.g., in Atkinson *et al.* (1990)) signed by prominent members of the research community whose recommendations are covered by and large in the treatment given in this section. The ODMG industry consortium of object database vendors has proposed an object-oriented data model that is widely seen as the *de facto* standard (Cattell *et al.*, 2000).

C. Object-Relational Data Models

While object-oriented data models typically relegate relations to the status of one among many types that can be built using type constructors (in this case, e.g., `tuple_of`), object-relational data models retain the central role relations have in the relational data model.

However, they relax the domain constraint of the relational data model (thereby allowing attribute values to be drawn from complex domains), they incorporate some of the distinctive features of the object-oriented model such as inheritance and assigned behavior with encapsulated states, and they allow for database-wide functionality to be specified by means of *rules* (commonly referred to as *triggers*, as described in Section IV.B) that react to specific interactions with application entities by carrying out some appropriate action.

As a consequence of the central role that relations retain in object-relational data models, one crucial difference with respect to the object-oriented case is that the role played by object identity is relaxed to an optional, rather than mandatory, feature. Thus, an object-relational DBMS stands in an evolutionary path regarding relational ones, whereas object-oriented ones represent a complete break with the relational approach. In this context, notice that while a `tuple_of` type constructor may allow a relation type to be supported, each tuple will have an identity, and attribute names will be explicitly needed to retrieve and interact with values.

Such differences at the data model level lead to pragmatic consequences of some significance at the level of the languages used to interact with application entities. In particular, while object-oriented data models naturally induce a navigational approach to accessing values, this leads to chains of reference of indefinite length that need to be traversed, or navigated.

In contrast, object-relational data models retain the associative approach introduced by the relational data model. Roughly speaking, this approach is based on viewing the sharing of values between attributes in different relations as inherently establishing associations between relation instances. These associations can then be exploited to access values across different relations without specifically and explicitly choosing one particular chain of references. These issues are exemplified in Sections III.B and III.C

Since object-relational models are basically hybrids, their notions of schema and instance combine features of both relational and object-oriented schemas and instances. The object-relational approach to modeling the *gene* entity type is the same as the object-oriented one depicted in Figs. 4 and 5 but for a few differences, e.g., relationships are usually not explicitly supported as such by object-relational data models.

As in the object-oriented case, object-relational data models were first advocated in a concerted manner by a manifesto (Stonebraker *et al.*, 1990) signed by prominent members of the research community. A book-length treatment is available in Stonebraker and Brown (1999). Any undergraduate textbook on Database Systems (e.g., Atzeni *et al.*, 1999; Elmasri and Navathe, 2000) can be used to complement the treatment of this and the section that follows.

```
CREATE SCHEMA genomics AUTHORIZATION DBA;
CREATE TABLE  genomics.DNA_sequence(
        sequence_id        VARCHAR(10)    NOT NULL
        accession_no       VARCHAR(10)    NOT NULL,
        protein_id         VARCHAR(10)    NOT NULL,
        organism_id        VARCHAR(10)    NOT NULL
        PRIMARY KEY (sequence_id),
        FOREIGN KEY (organism_id)
            REFERENCES genomics.organism(organism_id)
            ON DELETE CASCADE   ON UPDATE CASCADE );
CREATE TABLE  genomics.organism(
        organism_id        VARCHAR(10)    NOT NULL,
        common_name        VARCHAR(50),
        up_node            VARCHAR(30)
        PRIMARY KEY (organism_id) );
```

FIGURE 6 SQL declarations for the schemas in Fig. 1.

III. DATABASE LANGUAGES

A DBMS must support one or more languages with which application developers can set up, populate, scrutinize, and maintain data. In principle, different purposes are best served by different database languages, but in practice, DBMSs tend to use one language to cater for more than one purpose.

The kinds of **database languages** that must in principle be supported in DBMSs are:

1. **Data definition languages**, which are used to declare schemas (perhaps including application-specific integrity constraints)
2. **Data manipulation languages**, which are used to retrieve and manipulate the stored data

Data manipulation languages can be further categorized as follows:

1. **Query languages**, which most often are high-level, declarative, and computationally incomplete (i.e., not capable of expressing certain computations, typically due to the lack of support for updates or for recursion or iteration)
2. **Procedural languages**, which most often are low-level, imperative, and computationally complete

Besides these, most DBMSs provide interfacing mechanisms with which developers can implement applications in a general-purpose language (referred to in this context as a **host language**) and use the latter to invoke whatever operations are supported over the stored data (because they are part of the behavior of either the data types made available by the data model or the application types declared in the application schema).

A. Relational Database Languages

SQL is the ISO/ANSI standard for a relational database language. SQL is both a data definition and a data manipulation language. It is also both a query language and capable of expressing updates. However, SQL is not computationally complete, since it offers no support for either recursion or iteration. As a consequence, when it comes to the development of applications, SQL is often embedded in a host language, either one that is specific to the DBMS being used or a general purpose language for which a query language interface is provided.

Figure 6 shows how the schema illustrated in Fig. 1 can be declared in SQL. Notice that SQL includes constructs to declare integrity constraints (e.g., a referential integrity on from DNA_sequence to organism) and even an action to be performed if it is violated (e.g., cascading deletions, by deleting every referring tuple when a referenced tuple is deleted).

SQL can also express insertions, deletions and updates, as indicated in Fig. 7. Note in Fig. 7 that on insertion it is possible to omit null values by listing only the attributes for which values are being supplied. Note also that the order of insertion in Fig. 7 matters, since referential integrity constraints would otherwise be violated. Finally, note that, because of cascading deletions, the final statement will also delete all tuples in DNA_sequence that refer to the primary key of the tuple being explicitly deleted in organism.

After the operations in Fig. 7 the state depicted in Fig. 2 will have changed to that shown in Fig. 8.

As a relational query language, SQL always returns results that are themselves relation instances. Thus, the basic constructs in SQL cooperate in specifying aspects of the schema as well as the instantiation of a query result. Roughly, the SELECT clause specifies the names of attributes to appear in the result, the FROM clause specifies

```
INSERT INTO genomics.organism(organism_id, up_node)
VALUES      ('Poly. tinctorium', 'Polygonaceae');

INSERT INTO genomics.DNA_sequence
VALUES      ('AB003089', 'AB003089', 'BAA78708', 'Poly. tinctorium');

DELETE FROM genomics.organism
WHERE       organism_id = 'Bras. napus';
```

FIGURE 7 Using SQL to effect state transitions in the relation instances from Fig. 2.

the names of relations contributing data to the result, and the WHERE clause specifies, in terms of the relations (and their attributes) mentioned in the FROM clause, the conditions which each tuple in the result must satisfy. SQL queries tend to be structured around this combination of SELECT, FROM and WHERE clauses. Figure 9 shows example SQL queries.

Query RQ1 in Fig. 9 returns a unary table, each tuple of which records the organism_id of organisms that share the common_name of "white_clover." Query RQ2 returns a binary table relating each common_name found in the organism table with the protein_ids produced by their identified genes. Figure 10 shows the relations instances returned by RQ1 and RQ2.

SQL also supports aggregations (e.g., COUNT and AVG, which, respectively, count the number of tuples in a result and compute the average value of a numeric attribute in the result), groupings, and sorting.

Embedding SQL into a host language is another approach to retrieving and manipulating data from relational databases. Vendors typically provide a host language for SQL of which the fragment in Fig. 11 is an illustration. The fragment in Fig. 11 uses a CURSOR construct to scan the organism relation instance for organisms with no common name. When one is found, rather than leaving the value unassigned, the program uses the UPDATE construct to set the common_name attribute to the string None.

SQL is legally defined by ISO/ANSI standards which are available from those organizations. For comprehensive treatment, a good source is Melton and Simon, 1993. A detailed treatment of the relational algebra and calculi which underpin SQL can be found in Abiteboul *et al.* (1995).

B. Object-Oriented Database Languages

In the object-oriented case, the separation between the languages used for data definition, querying and procedural manipulation is more explicit than in the relational case. This is because in object-oriented databases, the syntactic style of SQL is circumscribed largely to the query part of the data manipulation language.

Also, rather than make use of vendor-specific host languages, object-oriented DBMSs either provide interfaces to general-purpose languages or else the DBMS itself supports a persistent programming language strategy to application development (i.e., one in which a distinction between the data space of the program and the database is deliberately not made, which leads to applications that need not explicitly intervene to transfer data from persistent to transient store and back again).

The *de facto* standard for object-oriented databases is the proposal by the ODMG consortium of vendors. The ODMG standard languages are ODL, for definition, and OQL (which extends the query part of SQL), for querying. The standard also defines interfaces for a few widely used general-purpose languages. Figure 4 could be declared in ODL as shown in Fig. 12.

Note how ODL allows inverse relationships to be named, as a consequence of which referential integrity is enforced in both directions.

Two OQL queries over the gene class in Fig. 12 are given in Fig. 13. Query OQ1 returns a set of complex values, i.e., name-cited pairs, where the first element is the standard_name of an instance of the gene class and the second is the list of strings stored as the value of the citation attribute for that instance. Query OQ2 returns the common_name of organisms associated with genes that have alleles. Note the use of the COUNT aggregation function over a collection value. Note, finally, the denotation g.organism_id.common_name (known as a *path expression*). Path expressions allow a navigational style of access.

For ODMG-compliant DBMSs, the authoritative reference on ODL and OQL is (Cattell *et al.*, 2000). A more

DNA_sequence					organism		
sequence_id	accession_no	protein_id	organism_id		organism_id	common_name	up_node
TRBG361	X56734	AA40058	Trif. repens		Trif. repens	white clover	Papilionoideae
AB003089	AB003089	BAA78708	Poly. tinctorium		Poly. tinctorium	NULL	Polygonaceae
...	...	...	...		...	...	...

FIGURE 8 Updated relation instances after the commands in Fig. 7.

```
RQ1:    SELECT  organism_id
        FROM    organism
        WHERE   common_name = 'white clover';

RQ2:    SELECT  o.common_name, s.protein_id
        FROM    organism AS o, DNA_sequence AS s
        WHERE   o.organism_id = s.organism_id;
```

FIGURE 9 Using SQL to query the database state in Fig. 9.

formal treatment of some of the issues arising in object-oriented languages can be found in Abiteboul *et al.* (1995).

C. Object-Relational Database Languages

The proposed standard for object-relational database languages is SQL-99. Figure 14 shows how Fig. 4 could be specified in SQL-99. Note the use of ROW TYPE to specify a complex domain, the use of REF to denote tuple identifiers and the use of type constructors such as SET and LIST. Note also that, unlike ODL (cf. Fig. 12), in SQL-99 inverse relationships are not declared. Note, finally, how gene is modeled as including operations, as indicated by the keyword FUNCTION introducing the behavioral part of the specification of gene.

Two SQL-99 queries over the gene type in Fig. 14 are given in Fig. 15. Query ORQ1 returns a binary table relating the standard_name of each gene with the common_name of organisms where the gene is found. Query ORQ2 returns the common_name of organisms associated with genes that have alleles. Note that in SQL-99 path expressions use the symbol -> to dereference identifiers and (not shown in Fig. 15) the symbol '..' to denote attributes in row types.

SQL-99 is legally defined by ISO/ANSI standards which are available from those organizations.

IV. ADVANCED MODELS AND LANGUAGES

The previous sections have described mainstream database models and languages, presenting facilities that are in widespread use today. This section introduces several extensions to the mainstream data models and languages with facilities that allow the database to capture certain application requirements more directly.

RQ1	RQ2	
organism_id	common_name	protein_id
Trif. repens	white clover	AA40058
	rape	CAA57913
	...	...

FIGURE 10 Results of the SQL queries in Fig. 9.

A. Deductive Databases

Deductive database systems can be seen as bringing together mainstream data models with logic programming languages for querying and analyzing database data. Although there has been research on the use of deductive databases with object-oriented data models, this section illustrates deductive databases in the relational setting, in particular making use of Datalog as a straightforward deductive database language (Ceri *et al.*, 1990).

A **deductive (relational) database** is a Datalog program. A **Datalog program** consists of an *extensional database* and an *intensional database*. An **extensional database** contains a set of facts of the form:

$$p(c_1, \ldots, c_m)$$

where p is a predicate symbol and $c_1, \ldots, c_m$ are constants. Each predicate symbol with a given arity in the extensional database can be seen as analogous to a relational table. For example, the table organism in Fig. 2 can be represented in Datalog as:

```
organism('Bras. napus', 'rape',
    'Brassicaceae',).
organism ('Trif. repens',
    'white clover', 'Papilionoideae',).
```

Traditionally, constant symbols start with a lower case letter, although quotes can be used to delimit other constants.

An **intensional database** contains a set of rules of the form:

$$p(x_1, \ldots, x_m) :- q_1(x_{11}, \ldots, x_{1k}), \ldots, q_j(x_{j1}, \ldots, x_{jp})$$

where p and q_i are predicate symbols, and all argument positions are occupied by variables or constants.

Some additional terminology is required before examples can be given of Datalog queries and rules. A **term** is either a constant or a variable—variables are traditionally written with an initial capital letter. An atom $p(t_1, \ldots, t_m)$ consists of a predicate symbol and a list of arguments, each of which is a term. A **literal** is an atom or a negated atom $\neg p(t_1, \ldots, t_m)$. A **Datalog query** is a conjunction of literals.

```
             DECLARE  o_id              CHAR(10);
                      o_common_name     CHAR(50);
                      CURSOR organism_cursor
                      IS      SELECT organism_id, common_name
                              FROM organism;
             BEGIN    OPEN    organism_cursor;
                      LOOP    FETCH   organism_cursor
                              INTO    o_id, o_common_name;
                              EXIT WHEN organism_cursor%NOTFOUND;
                              IF      o_common_name IS NULL
                              THEN    UPDATE  organism
                                      SET     common_name = 'None'
                                      WHERE   organism_id = o_id;
                      END LOOP;
                      CLOSE   organism_cursor;
             END;
```

FIGURE 11 Embedding SQL in a host language to effect state transitions.

```
    class gene{
        (extent        Genes)
        attribute      string          standard_name ;
        attribute      struct          codon_structure {
                                       string initiation_codon,
                                       list<string> codon_sequence,
                                       string termination_codon } sequence;
        attribute      string          function;
        attribute      list<string>    citation;
        attribute      protein         protein_id;
        attribute      organism        organism_id;
        relationship   set<alleles>    has_alleles
                       inverse         allele ::  is_allele_of;
            ... }
```

FIGURE 12 ODL to specify the gene class in Fig. 4.

```
    OQ1:    SELECT   struct(name   :  g.standard_name
                             cited :  g.citation)
            FROM     g IN gene;

    OQ2:    SELECT   g.organism_id.common_name
            FROM     g AS gene
            WHERE    COUNT(g.has_alleles) > 0;
```

FIGURE 13 Using OQL to query the gene class in Fig. 12.

```
CREATE ROW TYPE  codon_structure(
                 initiation_codon      CHAR(3),
                 codon_sequence        LIST(CHAR(3)),
                 termination_codon     CHAR(3));
CREATE TABLE     codon_sequence        OF TYPE        codon_structure;

CREATE TYPE      gene(  standard_name  VARCHAR(50),
                        sequence       REF (codon_sequence),
                        function       VARCHAR(30),
                        citation       LIST(VARCHAR(30)),
                        protein_id     REF(protein_tuple) REFERENCES (protein),
                        organism_id    REF(organism_tuple) REFERENCES (organism),
                        has_alleles    SET(REF(allele_tuple)) REFERENCES (allele),
                        FUNCTION ...
                        ... );
```

FIGURE 14 SQL-99 to specify the gene entity in Fig. 3.

```
ORQ1:   SELECT   standard_name, SET(organism_id -> common_name)
        FROM     gene;

ORQ2:   SELECT   g.organism_id -> common_name
        FROM     gene AS g
        WHERE    (SELECT COUNT(a) FROM a IN g.has_alleles)) > 0;
```

FIGURE 15 Using SQL-99 to query the gene type in Fig. 14.

Figure 16 shows some queries and rules expressed in Datalog. The queries DQ1 and DQ2 are expressed using a set comprehension notation, in which the values to the left of the | are the result of the Datalog query to the right of the |.

Query DQ1 retrieves the organism identifier of 'white clover'. Informally, the query is evaluated by unifying the query literal with the facts stored in the extensional database. Query DQ2 retrieves the identifiers of the proteins in 'white clover'. However, the predicate species_protein representing the relationship between the common name of a species and the identifiers of its proteins is not part of the extensional database, but rather is a rule defined in the intensional database. This rule, DR1 in Fig. 16, is essentially equivalent to the SQL query RQ2 in Fig. 9, in that it retrieves pairs of species common names and protein identifiers that satisfy the requirement that the proteins are found in the species. In DR1, the requirement that a dna_Sequence be found in an organism is represented by the atoms for dna_Sequence and organism sharing the variable OID, i.e., the organism identifier of the organism must be the same as the organism identifier of the dna_Sequence.

Deductive databases can be seen as providing an alternative paradigm for querying and programming database applications. A significant amount of research has been conducted on the efficient evaluation of recursive queries in deductive languages (Ceri *et al.*, (1990), an area in which deductive languages have tended to be more powerful than other declarative query languages. Recursive queries are particularly useful for searching through tree or graph structures, for example in a database representing a road network or a circuit design. A further feature of deductive languages is that they are intrinsically declara-

tive, so programs are amenable to evaluation in different ways—responsibility for finding efficient ways of evaluating deductive programs rests as much with the query optimizer of the DBMS as with the programmer.

Although there have been several commercial deductive database systems, these have yet to have a significant commercial impact. However, some of the features of deductive languages, for example, relating to recursive query processing, are beginning to be adopted by relational vendors, and are part of the SQL-99 standard.

B. Active Databases

Traditional DBMSs are **passive** in the sense that commands are executed by the database (e.g., query, update, delete) as and when requested by the user or application program. This is fine for many tasks, but where a DBMS simply executes commands without paying heed to their consequences, this places responsibilities on individual programmers that they may struggle to meet (Paton and Diaz, 1990).

For example, imagine that it is the policy of an organization managing a repository of DNA sequence data that whenever a new sequence is provided with a given sequence_id, the previous sequence is recorded, along with the date when the change takes place and the name of the user making the update. This straightforward policy may, however, be difficult to enforce in practice, as there may be many different ways in which a new replacement can be provided—for example, through a program that accepts new information from a WWW interface, or through an interactive SQL interface. The policy should be enforced in a consistent manner, no matter how the sequence has come to be updated. In an **active database**, the database system is able to respond automatically to

```
DQ1:    { OID | organism(OID, 'white clover', _) }

DR1:    species_protein(CommonName, ProteinId) :-
            organism(OID, CommonName, _),
            dna_Sequence(_, _, ProteinId, OID).

DQ2:    { ProteinId | species_protein('white clover', ProteinId) }
```

FIGURE 16 Using datalog to query the database state in Fig. 8.

events of relevance to the database, such as the updating of a table, and thus the database can provide centralized support for policies such as those described earlier.

Active behavior is described using **event-condition-action** or **ECA** rules. These rules have three components:

1. **Event:** a description of the happening to which the rule is to respond. In the example above, this could be the updating of the relation storing the sequence data.
2. **Condition:** a check on the context within which the event has taken place. The condition is either a check on the parameters of the event or a query over the database.
3. **Action:** a program block that indicates the reaction that the system should take to the event in a context where the condition is true.

Active rules will now be illustrated using the example described above, based around a table `sequence` with attributes `sequence_id` and `sequence`. A record will be kept of modified sequences in the table `modified_sequence`, with attributes `sequence_id`, `old_sequence`, `date_of_change`, `user` and `update_type`.

The active rule in Fig. 17 is written using the SQL-99 syntax. In the SQL standard, active rules are known as **triggers**. The event the trigger is monitoring is `after update of sequence on sequence`, which is raised after any change to the `sequence` attribute of the table `sequence`. The tuple prior to the update taking place is referred to within the trigger as `oldseq`. The example is of a `row` trigger, which means that the trigger responds to updates to individual tuples, so if a single update statement modifies 10 rows in the `sequence` table, the trigger will be executed 10 times. In this example there is no condition—the action is executed regardless of the state of the database, etc.

Writing correct triggers is not altogether straightforward. For the example application, additional triggers

would be required, for example, to log the deletion of an existing sequence entry, so that a modification of a sequence as a delete followed by an insert is not omitted from the log. Considerable skill is required in the development and maintenance of large rule bases. However, it is clear that active rules can be used for many different purposes—supporting business rules, enforcing integrity constraints, providing fine tuned authorization mechanisms, etc. As a result, the principal relational database products support active mechanisms, and the SQL-99 standard should encourage consistency across vendors in due course.

V. DISTRIBUTION

A conventional DBMS generally involves a single **server** communicating with a number of user **clients**, as illustrated in Fig. 18. In such a model, the services provided by the database are principally supported by the central server, which includes a secondary storage manager, concurrency control facilities, etc., as described in Section I.

In relational database systems, clients generally communicate with the database by sending SQL query or update statements to the server as strings. The server then compiles and optimizes the SQL, and returns any result or error reports to the client.

Clearly there is a sense in which a client-server database is distributed, in that application programs run on many clients that are located in different parts of the network. However, in a client-server context there is a single server managing a single database described using a single data model. The remainder of this section introduces some of the issues raised by the relaxation of some of these restrictions.

A. Distributed Databases

In a distributed database, data is stored in more than one place, as well as being accessed from multiple places, as

```
create trigger logUpdate
after update of sequence on sequence
referencing old as oldseq
for each row
begin
        insert into modified_sequence values (
                oldseq.sequence_id,
                oldseq.sequence,
                SYSDATE,
                USER,
                'delete');
end
```

FIGURE 17 Using an active rule to log changes to the `sequence` table.

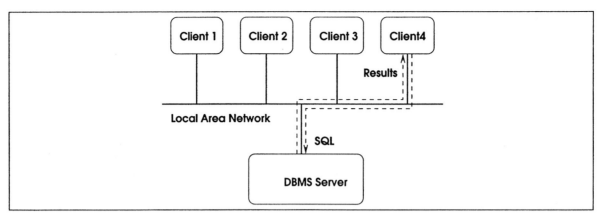

FIGURE 18 A client-server database architecture.

in the client-server model. A typical distributed DBMS architecture is illustrated in Fig. 19, in which there are two servers, each of which is accessed by multiple clients.

Distributed databases can come into being for different reasons. For example, a multinational organization may have distinct databases for each of its national subsidiaries, but some tasks may require access to more than one of these databases (e.g., an item that is out of stock in one country may be available in another). Another situation in which multiple databases may have to be used together is where organizations merge, and it becomes necessary to support some measure of interoperation across their independently developed information systems.

The context within which a distributed database comes into being significantly influences the complexity of the resulting system. The nature of a distributed database can be characterized by the following features:

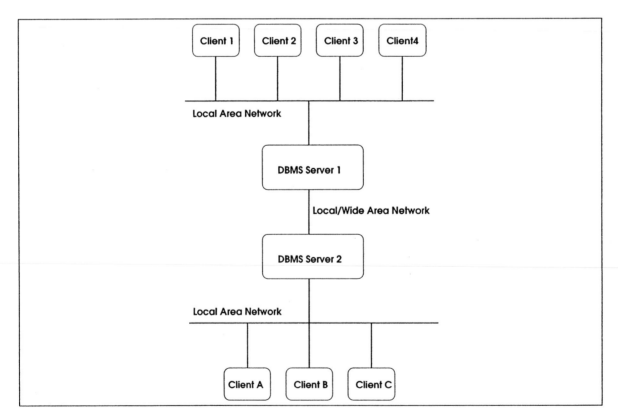

FIGURE 19 A distributed database architecture.

1. **Fragmentation.** The fragmentation of data in a distributed database describes how individual data items are partitioned across the database servers. For example, in the genomic database of Fig. 1, one form of fragmentation could lead to all the sequences from an `organism` being stored on the same server. This might happen, for example, if the sequencing of an organism is being carried out at a single site. This form of fragmentation is known as **horizontal fragmentation**, as the tuples in the tables `organism` and `DNA_sequence` are partitioned on the basis of their `organism_id` values. In **vertical fragmentation**, some of the attributes of a table are placed on one server, and other attributes on another.

   In many applications, it is difficult to arrange a straightforward fragmentation policy—for example, if distributed databases are being used where two organizations have recently merged, it is very unlikely that the existing databases of the merging organizations will have made use of similar fragmentation schemes.

2. **Homogeneity.** In a **homogeneous** distributed database system, all the servers support the same kind of database, e.g., where all the servers are relational, and in particular where all the servers have been provided by the same vendor. It is common, however, for more than one database paradigm to be represented in a distributed environment, giving rise to a **heterogeneous** distributed database.

3. **Transparency.** In Fig. 19, a user at `Client 1` can issue a request that must access or update data from either of the servers in the distributed environment.

The level of transparency supported indicates the extent to which users must be aware of the distributed environment in which they are operating (Atzeni *et al.*, 1999). Where there is **fragmentation transparency**, requests made by the user need not take account of where the data is located. Where there is **replication transparency**, requests made by the user need not take account of where data has been replicated in different servers, for example, to improve efficiency. Where there is **language transparency**, the user need not be concerned if different servers support different query languages. The greater the level of transparency supported, the more infrastructure is required to support transparent access. Distributed query processing and concurrency control present particular challenges to database systems developers (Garcia-Molina *et al.*, 2000).

B. Data Warehouses

Traditional database systems have tended to be designed with an emphasis on **on line transaction processing** (OLTP), in which there are large numbers of short transactions running over the database. In our biological example, a typical transaction might add a new sequence entry into the database. Such transactions typically read quite modest amounts of data from the database, and update one or a few tuples.

However, the data stored in an organization's databases are often important for management tasks, such as decision support, for which typical transactions are very different from those used in OLTP. The phrase **on line**

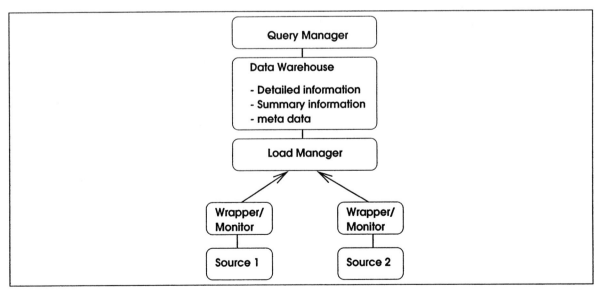

FIGURE 20 A data warehouse architecture.

analytical processing (OLAP) has been used to characterize this different kind of access to databases, in which typical transactions read rather than update the database, and may perform complex *aggregation* operations on the database. As the data required for OLAP often turn out to be stored in several OLTP databases, it often proves to be necessary to replicate data in a system designed specifically to support OLAP—such a replicated store is known as a **data warehouse**.

A data warehouse architecture is illustrated in Fig. 20 (Anahory and Murray, 1997). A data warehouse contains three principal categories of information: detailed information (for example, in the genomic context, this may be the raw sequence data or other data on the function of individual protein); summary information (for example, in the genome context, this may record the results of aggregations such as the numbers of genes associated with a given function in each species); and meta data (which describes the information in the warehouse and where it came from).

An important prerequisite for the conducting of effective analyses is that the warehouse contains appropriate data, of adequate quality, which is sufficiently up-to-date. This means that substantial effort has to be put into loading the data warehouse in the first place, and then keeping the warehouse up-to-date as the sources change. In Fig. 20, data essentially flow from the bottom of the figure to the top. Sources of data for the warehouse, which are often databases themselves, are commonly **wrapped** so that they are syntactically consistent, and **monitored** for changes that may be relevant to the warehouse (e.g., using active rules). The **load manager** must then merge together information from different sources, and discard information that doesn't satisfy relevant integrity constraints. The **query manager** is then responsible for providing comprehensive interactive analysis and presentation facilities for the data in the warehouse.

VI. CONCLUSIONS

This chapter has provided an introduction to databases, and in particular to the models, languages, and systems that allow large and complex data-intensive applications to be developed in a systematic manner. DBMSs are now quite a mature technology, and almost every organization of any size uses at least one such system. The DBMS will probably be among the most sophisticated software systems deployed by an organization.

This is not to say, however, that research in database management is slowing down. Many data-intensive applications do not use DBMSs, and developments continue,

with a view to making database systems amenable for use in more and more different kinds of application. For example, spatial, temporal, multimedia, and semistructured data are difficult to represent effectively in many database systems, and many issues relating to distributed information management continue to challenge researchers.

SEE ALSO THE FOLLOWING ARTICLES

DATA MINING AND KNOWLEDGE DISCOVERY • DATA STRUCTURES • DESIGNS AND ERROR-CORRECTING CODES • INFORMATION THEORY • OPERATING SYSTEMS • PROJECT MANAGEMENT SOFTWARE • REQUIREMENTS ENGINEERING (SOFTWARE) • SOFTWARE ENGINEERING • SOFTWARE MAINTENANCE AND EVOLUTION • WWW (WORLD-WIDE WEB)

BIBLIOGRAPHY

Abiteboul, S., Hull, R., and Vianu, V. (1995). *Foundations of Databases*, Addison-Wesley. ISBN 0-201-53771-0.

Anahory, S., and Murray, D. (1997). *Data Warehousing in the Real World*, Addison-Wesley. ISBN 0-201-17519-3.

Atkinson, M., Bancilhon, F., DeWitt, D., Dittrich, K., Maier, D., and Zdonik, S. B. (1990). "The Object-Oriented Database System Manifesto," *In* [(Kim *et al.*, 1990)], pp. 223–240.

Atzeni, P., Ceri, S., Paraboschi, S., and Torlone, R. (1999). *Database Systems: Concepts, Languages and Architectures*, McGraw-Hill.

Cattell, R. G. G., Barry, D. K., Berler, M., Eastman, J., Jordan, D., Russell, C., Schadow, O., Stanienda, T., and Velez, F. (2000). *The Object Data Standard: ODMG 3.0*, Morgan Kaufman, ISBN 1-55860-647-5.

Ceri, S., Gottlob, G., and Tanca, L. (1990). *Logic Programming and Databases*, Springer-Verlag, Berlin. ISBN 0-387-51728-6.

Codd, E. F. "A Relational Model of Data for Large Shared Data Banks," *Communications of the ACM* **13**(6): 377–387, June 1970; Also in CACM **26**(1) January 1983 pp. 64–69.

Elmasri, R., and Navathe, S. B. (2000). *Fundamentals of Database Systems*, Addison-Wesley, Reading, MA, USA, 3rd. edition, ISBN 0-201-54263-3.

Garcia-Molina, H., Ullman, J., and Widom, J. (2000). *Database System Implementation*, Prentice Hall. ISBN 0-13-040264-8.

Kim, W., Nicolas, J.-M., and Nishio, S. (eds.). (1990). *Deductive and Object-Oriented Databases (First International Conference DOOD'89, Kyoto)*, Amsterdam, The Netherlands, Elsevier Science Press (North-Holland), ISBN 0-444-88433-5.

Melton, J., and Simon, A. R. (1993). *Understanding the New SQL: A Complete Guide*, Morgan Kaufman, ISBN 1-55860-245-3.

Paton, N., and Diaz, O. (1999). "Active Database Systems," *ACM Computing Surveys* **1**(31), 63–103.

Stonebraker, M., and Brown, P. (1999). *Object-Relational DBMS: Tracking the Next Great Wave*, Morgan Kaufman, ISBN 1-55860-452-9.

Stonebraker, M., Rowe, L. A., Lindsay, B. G., Gray, J., Carey, M. J., Brodie, M. L., Bernstein, P. A., and Beech, D. "Third-Generation Database System Manifesto—The Committee for Advanced DBMS Function," *SIGMOD Record* **19**(3): 31–44, September 1990.

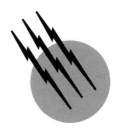

Data Mining and Knowledge Discovery

Sally I. McClean

University of Ulster

GLOSSARY

Association rules link the values of a group of attributes, or variables, with the value of a particular attribute of interest which is not included in the group.

Data mining process takes place in four main stages: Data Pre-processing, Exploratory Data Analysis, Data Selection, and Knowledge Discovery.

Data mining tools are software products; a growing number of such products are becoming commercially available. They may use just one approach (single paradigm), or they may employ a variety of different methods (multi-paradigm).

Deviation detection is carried out in order to discover Interestingness in the data. Deviations may be detected either for categorical or numerical data.

Interestingness is central to Data Mining where we are looking for new knowledge which is nontrivial. It allows the separation of novel and useful patterns from the mass of dull and trivial ones.

Knowledge discovery in databases (KDD) is the main objective in Data Mining. The two terms are often used synonymously, although some authors define Knowledge Discovery as being carried out at a higher level than Data Mining.

DATA MINING is the process by which computer programs are used to repeatedly search huge amounts of data, usually stored in a Database, looking for useful new patterns. The main developments that have led to the emergence of Data Mining have been in the increased volume of data now being collected and stored electronically, and an accompanying maturing of Database Technology. Such developments have meant that traditional Statistical Methods and Machine Learning Technologies have had to be extended to incorporate increased demands for fast and scaleable algorithms.

In recent years, Database Technology has developed increasingly more efficient methods for data processing and data access. Simultaneously there has been a convergence between Machine Learning Methods and Database Technology to create value-added databases with an increased capability for intelligence. There has also been a convergence between Statistics and Database Technology.

I. DATA MINING AND KNOWLEDGE DISCOVERY

A. Background

The main developments that have led to the emergence of Data Mining as a promising new area for the discovery of knowledge have been in the increased amount of data now available, with an accompanying maturing of Database Technology. In recent years Database Technology has developed efficient methods for data processing and data access such as parallel and distributed computing, improved middleware tools, and Open Database Connectivity (ODBC) to facilitate access to multi-databases.

Various Data Mining products have now been developed and a growing number of such products are becoming commercially available. Increasingly, Data Mining systems are coming onto the market. Such systems ideally should provide an integrated environment for carrying out the whole Data Mining process thereby facilitating end-user Mining, carried out automatically, with an interactive user interface.

B. The Disciplines

Data Mining brings together the three disciplines of Machine Learning, Statistics, and Database Technology. In the Machine Learning field, many complex problems are now being tackled by the development of intelligent systems. These systems may combine Neural Networks, Genetic Algorithms, Fuzzy Logic systems, Case-Based Reasoning, and Expert Systems. Statistical Techniques have become well established as the basis for the study of Uncertainty. Statistics embraces a vast array of methods used to gather, process, and interpret quantitative data. Statistical Techniques may be employed to identify the key features of the data in order to explain phenomena, and to identify subsets of the data that are interesting by virtue of being significantly different from the rest. Statistics can also assist with prediction, by building a model from which some attribute values can be reliably predicted from others in the presence of uncertainty. Probability Theory is concerned with measuring the likelihood of events under uncertainty, and underpins much of Statistics. It may also be applied in new areas such as Bayesian Belief Networks, Evidence Theory, Fuzzy Logic systems and Rough Sets.

Database manipulation and access techniques are essential to efficient Data Mining; these include Data Vi-

sualization and Slice and Dice facilities. It is often the case that it is necessary to carry out a very large number of data manipulations of various types. This involves the use of a structured query language (SQL) to perform basic operations such as selecting, updating, deleting, and inserting data items. Data selection frequently involves complex conditions containing Boolean operators and statistical functions, which thus require to be supported by SQL. Also the ability to join two or more databases is a powerful feature that can provide opportunities for Knowledge Discovery.

C. Data Mining Objectives and Outcomes

Data Mining is concerned with the search for new knowledge in data. Such knowledge is usually obtained in the form of rules which were previously unknown to the user and may well prove useful in the future. These rules might take the form of specific rules induced by means of a rule induction algorithm or may be more general statistical rules such as those found in predictive modeling. The derivation of such rules is specified in terms of Data Mining tasks where typical tasks might involve classifying or clustering the data.

A highly desirable feature of Data Mining is that there be some high-level user interface that allows the end-user to specify problems and obtain results in as friendly as matter as possible. Although it is possible, and in fact common, for Data Mining to be carried out by an expert and the results then explained to the user, it is also highly desirable that the user be empowered to carry out his own Data Mining and draw his own conclusions from the new knowledge. An appropriate user interface is therefore of great importance.

Another secondary objective is the use of efficient data access and data processing methods. Since Data Mining is increasingly being applied to large and complex databases, we are rapidly approaching the situation where efficient methods become a *sine qua non*. Such methods include Distributed and Parallel Processing, the employment of Data Warehousing and accompanying technologies, and the use of Open Database Connectivity (ODBC) to facilitate access to multi-databases.

D. The Data Mining Process

The Data Mining process may be regarded as taking place in four main stages (Fig. 1):

- Data Pre-processing
- Exploratory Data analysis
- Data Selection
- Knowledge Discovery

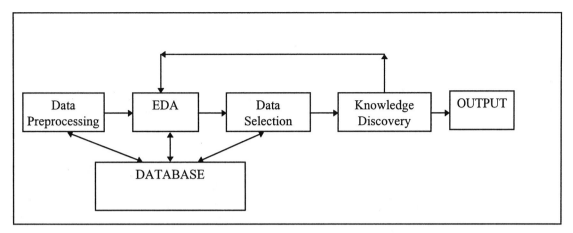

FIGURE 1 The Data Mining Process.

Data Pre-processing is concerned with data cleansing and reformatting, so that the data are now held in a form that is appropriate to the Mining algorithms and facilitates the use of efficient access methods. Reformatting typically involves employing missing value handling and presenting the data in multidimensional views suitable for the multidimensional servers used in Data Warehousing.

In Exploratory Data Analysis (EDA), the miner has a preliminary look at the data to determine which attributes and which technologies should be utilized. Typically, Summarization and Visualization Methods are used at this stage.

For Data Selection, we may choose to focus on certain attributes or groups of attributes since using all attributes at once is likely to be too complex and time consuming. Alternatively, for large amounts of data, we may choose to sample certain tuples, usually chosen at random. We may then carry out Knowledge Discovery using the sample, rather than the complete data, thus speeding up the process enormously. Variable reduction techniques or new variable definition are alternative methods for circumventing the problems caused by such large data sets.

Knowledge Discovery is the main objective in Data Mining and many different technologies have been employed in this context. In the Data Mining Process we frequently need to iterate round the EDA, Data Selection, Knowledge Discovery part of the process, as once we discover some new knowledge, we often then want to go back to the data and look for new or more detailed patterns.

Once new knowledge has been mined from the database, it is then reported to the user either in verbal, tabular or graphical format. Indeed the output from the Mining process might be an Expert System. Whatever form the output takes, it is frequently the case that such information is really the specification for a new system that will use the knowledge gained to best advantage for the user and domain in question. New knowledge may feed into the business process which in turn feeds back into the Data Mining process.

E. Data Mining Tasks

1. Rule Induction

Rule induction uses a number of specific beliefs in the form of database tuples as evidence to support a general belief that is consistent with these specific beliefs. A collection of tuples in the database may form a relation that is defined by the values of particular attributes, and relations in the database form the basis of rules. Evidence from within the database in support of a rule is thus used to induce a rule which may be generally applied.

Rules tend to be based on sets of attribute values, partitioned into an antecedent and a consequent. A typical "if then" rule, of the form "if antecedent = true, then consequent = true," is given by "if a male employee is aged over 50 and is in a management position, then he will hold an additional pension plan." Support for such a rule is based on the proportion of tuples in the database that have the specified attribute values in both the antecedent and the consequent. The degree of confidence in a rule is the proportion of those tuples that have the specified attribute values in the antecedent, which also have the specified attribute values in the consequent.

Rule induction must then be combined with rule selection in terms of interestingness if it is to be of real value in Data Mining. Rule-finding and evaluation typically require only standard database functionality, and they may be carried out using embedded SQL. Often, if a database is very large, it is possible to induce a very large number of rules. Some may merely correspond to well-known domain knowledge, whilst others may simply be of little interest to the user. Data Mining tools must therefore support the selection of **interesting** rules.

2. Classification

A commonly occurring task in Data Mining is that of classifying cases from a dataset into one of a number of well-defined categories. The categories are defined by sets of attribute values, and cases are allocated to categories according to the attribute values that they possess. The selected combinations of attribute values that define the classes represent **features** within the particular context of the classification problem. In the simplest cases, classification could be on a single binary-valued attribute, and the dataset is partitioned into two groups, namely, those cases with a particular property, and those without it. In general it may only be possible to say which class the case is "closest to," or to say how likely it is that the case is in a particular category.

Classification is often carried out by **supervised Machine Learning**, in which a number of training examples (tuples whose classification is known) are presented to the system. The system "learns" from these how to classify other cases in the database which are not in the training set. Such classification may be probabilistic in the sense that it is possible to provide the probability that a case is any one of the predefined categories. **Neural Networks** are one of the main Machine Learning technologies used to carry out classification. A probabilistic approach to classification may be adopted by the use of **discriminant functions**.

3. Clustering

In the previous section, the classification problem was considered to be essentially that of learning how to make decisions about assigning cases to known classes. There are, however, different forms of classification problem, which may be tackled by **unsupervised learning**, or clustering. Unsupervised classification is appropriate when the definitions of the classes, and perhaps even the number of classes, are not known in advance, e.g., market segmentation of customers into similar groups who can then be targeted separately.

One approach to the task of defining the classes is to identify clusters of cases. In general terms, **clusters** are groups of cases which are in some way similar to each other according to some measure of **similarity**. Clustering algorithms are usually iterative in nature, with an initial classification being modified progressively in terms of the class definitions. In this way, some class definitions are discarded, whilst new ones are formed, and others are modified, all with the objective of achieving an overall goal of separating the database tuples into a set of cohesive categories. As these categories are not predetermined, it is clear that clustering has much to offer in the process of Data Mining in terms of discovering **concepts**, possibly within a concept hierarchy.

4. Summarization

Summarization aims to present concise measures of the data both to assist in user comprehension of the underlying structures in the data and to provide the necessary inputs to further analysis. Summarization may take the form of the production of graphical representations such as bar charts, histograms, and plots, all of which facilitate a visual overview of the data, from which sufficient insight might be derived to both inspire and focus appropriate Data Mining activity. As well as assisting the analyst to focus on those areas in a large database that are worthy of detailed analysis, such visualization can be used to help with the analysis itself. Visualization can provide a "drill-down" and "drill-up" capability for repeated transition between summary data levels and detailed data exploration.

5. Pattern Recognition

Pattern recognition aims to classify objects of interest into one of a number of categories or classes. The objects of interest are referred to as **patterns**, and may range from printed characters and shapes in images to electronic waveforms and digital signals, in accordance with the data under consideration. Pattern recognition algorithms are designed to provide automatic identification of patterns, without the need for human intervention. Pattern recognition may be **supervised**, or **unsupervised**.

The relationships between the observations that describe a pattern and the classification of the pattern are used to design **decision rules** to assist the recognition process. The observations are often combined to form **features**, with the aim that the features, which are smaller in number than the observations, will be more reliable than the observations in forming the decision rules. Such **feature extraction** processes may be application dependent, or they may be general and mathematically based.

6. Discovery of Interestingness

The idea of interestingness is central to Data Mining where we are looking for new knowledge that is nontrivial. Since, typically, we may be dealing with very large amounts of data, the potential is enormous but so too is the capacity to be swamped with so many patterns and rules that it is impossible to make any sense out of them. It is the concept of interestingness that provides a framework for separating out the novel and useful patterns from the myriad of dull and trivial ones.

Interestingness may be defined as deviations from the norm for either categorical or numerical data. However, the initial thinking in this area was concerned with categorical data where we are essentially comparing the deviation between the proportion of our target group with a particular property and the proportion of the whole population with the property. Association rules then determine where particular characteristics are related.

An alternative way of computing interestingness for such data comes from statistical considerations, where we say that a pattern is interesting if there is a statistically significant association between variables. In this case the measure of interestingness in the relationship between two variables A and B is computed as:

$$\text{Probability of } (A \text{ and } B) \text{-Probability of}$$

$$(A)^* \text{Probability of } (B).$$

Interestingness for continuous attributes is determined in much the same way, by looking at the deviation between summaries.

7. Predictive Modeling

In Predictive Modeling, we are concerning with using some attributes or patterns in the database to predict other attributes or extract rules. Often our concern is with trying to predict behavior at a future time point. Thus, for business applications, for example, we may seek to predict future sales from past experience.

Predictive Modeling is carried out using a variety of technologies, principally Neural Networks, Case-Based Reasoning, Rule Induction, and Statistical Modeling, usually via Regression Analysis. The two main types of predictive modeling are **transparent** (explanatory) and **opaque** (black box). A transparent model can give information to the user about why a particular prediction is being made, while an opaque model cannot explain itself in terms of the relevant attributes. Thus, for example, if we are making predictions using Case-Based Reasoning, we can explain a particular prediction in terms of similar behavior commonly occurring in the past. Similarly, if we are using a statistical model to predict, the forecast is obtained as a combination of known values which have been previously found to be highly relevant to the attribute being predicted. A Neural Network, on the other hand, often produces an opaque prediction which gives an answer to the user but no explanation as to why this value should be an accurate forecast. However, a Neural Network can give extremely accurate predictions and, where it may lack in explanatory power, it more than makes up for this deficit in terms of predictive power.

8. Visualization

Visualization Methods aim to present large and complex data sets using pictorial and other graphical representations. State-of-the-art Visualization techniques can thus assist in achieving Data Mining objectives by simplifying the presentation of information. Such approaches are often concerned with summarizing data in such a way as to facilitate comprehension and interpretation. It is important to have the facility to handle the commonly occurring situation in which it is the case that too much information is available for presentation for any sense to be made of it—the "haystack" view. The information extracted from Visualization may be an end in itself or, as is often the case, may be a precursor to using some of the other technologies commonly forming part of the Data Mining process.

Visual Data Mining allows users to interactively explore data using graphs, charts, or a variety of other interfaces. Proximity charts are now often used for browsing and selecting material; in such a chart, similar topics or related items are displayed as objects close together, so that a user can traverse a topic landscape when browsing, or searching for information. These interfaces use colors, filters, and animation, and they allow a user to view data at different levels of detail. The data representations, the levels of detail and the magnification, are controlled by using mouse-clicks and slider-bars.

Recent developments involve the use of "virtual reality," where, for example, statistical objects or cases within a database may be represented by graphical objects on the screen. These objects may be designed to represent people, or products in a store, etc., and by clicking on them the user can find further information relating to that object.

9. Dependency Detection

The idea of dependency is closely related to interestingness and a relationship between two attributes may be thought to be interesting if they can be regarded as dependent, in some sense. Such patterns may take the form of statistical dependency or may manifest themselves as **functional dependency** in the database. With functional dependency, all values of one variable may be determined from another variable. However, statistical dependency is all we can expect from data which is essentially random in nature.

Another type of dependency is that which results from some sort of causal mechanism. Such causality is often represented in Data Mining by using Bayesian Belief Networks which discover and describe. Such causal models allow us to predict consequences, even when circumstances change. If a rule just describes an association, then we cannot be sure how robust or generalizable it will be in the face of changing circumstances.

10. Uncertainty Handling

Since real-world data are often subject to uncertainty of various kinds, we need ways of handling this uncertainty. The most well-known and commonly used way of handling uncertainty is to use classical, or Bayesian, probability. This allows us to establish the probabilities, or support, for different rules and to rank them accordingly. One well-known example of the use of Bayesian probability is provided by the Bayesian Classifier which uses Bayes' Theorem as the basis of a classification method. The various approaches to handling uncertainty have different strengths and weaknesses that may make them particularly appropriate for particular Mining tasks and particular data sets.

11. Sequence Processing

Sequences of data, which measure values of the same attribute at a sequence of different points, occur commonly. The best-known form of such data arises when we collect information on an attribute at a sequence of time points, e.g., daily, quarterly, annually. However, we may instead have data that are collected at a sequence of different points in space, or at different depths or heights. Statistical data that are collected at a sequence of different points in time are known as time series.

In general, we are concerned with finding ways of describing the important features of a time series, thus allowing Predictive Modeling to be carried out over future time periods. There has also been a substantial amount of work done on describing the relationship between one time series and another with a view to determining if two time series co-vary or if one has a causal effect on the other. Such patterns are common in economic time series, where such variables are referred to as leading indicators. The determination of such leading indicators can provide new knowledge and, as such, is a fertile area for Data Mining.

The methods used for Predictive Modeling for the purpose of sequence processing are similar to those used for any other kind of Predictive Modeling, typically Rule Induction and Statistical Regression. However, there may be particular features of sequences, such as seasonality, which must be incorporated into the model if prediction is to be accurate.

F. Data Mining Approaches

As has already been stated, Data Mining is a multidisciplinary subject with major input from the disciplines of Machine Learning, Database Technology and Statistics but also involving substantial contributions from many other areas, including Information Theory, Pattern Recognition, and Signal Processing. This has led to many different approaches and a myriad of terminology where different communities have developed substantially different terms for essentially the same concepts. Nonetheless, there is much to gain from such an interdisciplinary approach and the synergy that is emerging from recent developments in the subject is one of its major strengths.

G. Advantages of Data Mining

Wherever techniques based on data acquisition, processing, analysis and reporting are of use, there is potential for Data Mining. The collection of consumer data is becoming increasingly automatic at the point of transaction. Automatically collected retail data provide an ideal arena for Data Mining. Highly refined customer profiling becomes possible as an integral part of the retail system, eschewing the need for costly human intervention or supervision. This approach holds the potential for discovering interesting or unusual patterns and trends in consumer behavior, with obvious implications for marketing strategies such as product placement, customized advertising, and rewarding customer loyalty. The banking and insurance industries have also well-developed and specialized data analysis techniques for customer profiling for the purpose of assessing credit worthiness and other risks associated with loans and investments. These include using Data Mining methods to adopt an integrated approach to mining criminal and financial data for fraud detection.

Science, technology, and medicine are all fields that offer exciting possibilities for Data Mining. Increasingly it is the vast arrays of automatically recorded experimental data that provide the material from which may be formed new scientific knowledge and theory. Data Mining can facilitate otherwise impossible Knowledge Discovery, where the amount of data required to be assimilated for the observation of a single significant anomaly would be overwhelming for manual analysis.

Both modern medical diagnosis and industrial process control are based on data provided by automated monitoring systems. In each case, there are potential benefits for efficiency, costs, quality, and consistency. In the Health Care environment, these may lead to enhanced patient care, while application to industrial processes and project management can provide a vital competitive advantage.

Overall, however, the major application area for Data Mining is still Business. For example a recent survey of Data Mining software tools (Fig. 2) showed that over three-quarters (80%) are used in business applications, primarily in areas such as finance, insurance, marketing and market segmentation. Around half of the vendor tools surveyed were suited to Data Mining in medicine and industrial applications, whilst a significant number are most useful in scientific and engineering fields.

II. THE TECHNOLOGIES

A. Machine Learning Technologies

1. Inferencing Rules

Machine Learning, in which the development of Inferencing Rules plays a major part, can be readily applied

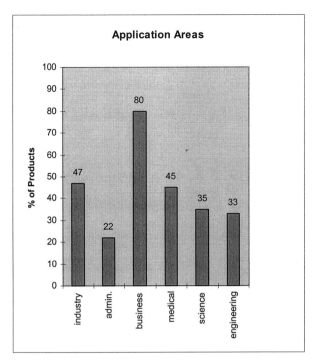

FIGURE 2 Application areas for Data Mining Tools.

to Knowledge Discovery in Databases. Records (tuples) in a database may be regarded as training instances that attribute-value learning systems may then use to discover patterns in a file of database records. Efficient techniques exist which can handle very large training sets, and include ways of dealing with incomplete and noisy data.

Logical reasoning may be automated by the use of a logic programming system, which contains a language for representing knowledge, and an **inference engine** for automated reasoning. The induction of logical definitions of relations has been named **Inductive Logic Programming (ILP)**, and may also be used to compress existing relations into their logical definitions. Inductive learning systems that use ILP construct logical definitions of target relations from examples and background knowledge. These are typically in the form of if–then rules, which are then transformed into clauses within the logic programming system. ILP systems have applications in a wide variety of domains, where Knowledge Discovery is achieved via the learning of relationships. Inference rules may be implemented as **demons**, which are processes running within a program while the program continues with its primary task.

2. Decision Trees

A Decision Tree provides a way of expressing knowledge used for classification. A Decision Tree is constructed by using a **training set** of cases that are described in terms of a collection of attributes. A sequence of tests is carried out

on the attributes in order to partition the training set into ever-smaller subsets. The process terminates when each subset contains only cases belonging to a single class. Nodes in the Decision Tree correspond to these tests, and the leaves of the tree represent each subset. New cases (which are not in the training set) may be classified by tracing through the Decision Tree starting at its root and ending up at one of the leaves.

The choice of test at each stage in "growing" the tree is crucial to the tree's predictive capability. It is usual to use a selection criterion based on the gain in classification information and the information yielded by the test. In practice, when "growing" a Decision Tree, a small working set of cases is used initially to construct a tree. This tree is then used to classify the remaining cases in the training set: if all are correctly classified, then the tree is satisfactory. If there are misclassified cases, these are added to the working set, and a new tree constructed using this augmented working set. This process is used iteratively until a satisfactory Decision Tree is obtained. Overfitting of the data and an associated loss of predictive capability may be remedied by **pruning** the tree, a process that involves replacing sub-trees by leaves.

3. Neural Networks

Neural Networks are designed for pattern recognition, and they thus provide a useful class of Data Mining tools. They are primarily used for classification tasks. The Neural Network is first trained before it is used to attempt to identify classes in the data. Hence from the initial dataset a proportion of the data are partitioned into a **training set** which is kept separate from the remainder of the data. A further proportion of the dataset may also be separated off into a **validation set** that is used to test performance during training along with criteria for determining when the training should terminate, thus preventing **overtraining**.

A Neural Network is perhaps the simplest form of parallel computer, consisting of a (usually large) set of simple processing elements called **neurons**. The neurons are connected to one another in a chosen configuration to form a network. The types of connectivities or network architectures available can vary widely, depending on the application for which the Neural Network is to be used. The most straightforward arrangements consist of neurons set out in layers as in the **feedforward network**. Activity feeds from one layer of neurons to the next, starting at an initial input layer.

The **Universal Approximation Theorem** states that a single layer net, with a suitably large number of hidden nodes, can well approximate any suitably smooth function. Hence for a given input, the network output may be compared with the required output. The total mean square error function is then used to measure how close the actual output is to the required output; this error is reduced by a technique called **back-error propagation**. This approach is a **supervised** method in which the network learns the connection weights as it is taught more examples. A different approach is that of **unsupervised learning**, where the network attempts to learn by finding statistical features of the input training data.

4. Case-Based Reasoning

Case-Based Reasoning (CBR) is used to solve problems by finding similar, past cases and adapting their solutions. By not requiring specialists to encapsulate their expertise in logical rules, CBR is well suited to domain experts who attempt to solve problems by recalling approaches which have been taken to similar situations in the past. This is most appropriate in domains which are not well understood, or where any rules that may have been devised have frequent exceptions. CBR thus offers a useful approach to building applications that support decisions based on past experience.

The quality of performance achieved by a case-based reasoner depends on a number of issues, including its experiences, and its capabilities to adapt, evaluate, and repair situations. First, partially matched cases must be retrieved to facilitate reasoning. The retrieval process consists of two steps: recalling previous cases, and selecting a best subset of them. The problem of retrieving applicable cases is referred to as the **indexing problem**. This comprises the matching or similarity-assessment problem, of recognizing that two cases are similar.

5. Genetic Algorithms

Genetic Algorithms (GA's) are loosely based on the biological principles of genetic variation and natural selection. They mimic the basic ideas of the evolution of life forms as they adapt to their local environments over many generations. Genetic Algorithms are a type of **evolutionary algorithm**, of which other types include Evolutionary Programming and Evolutionary Strategies.

After a new generation is produced, it may be combined with the population that spawned it to yield the new current population. The size of the new population may be curtailed by selection from this combination, or alternatively, the new generation may form the new population. The genetic operators used in the process of generating offspring may be examined by considering the contents of the population as a **gene pool**. Typically an individual may then be thought of in terms of **a binary string** of fixed length, often referred to as a **chromosome**. The genetic operators that define the offspring production process are usually a combination of **crossover** and **mutation** opera-

tors. Essentially these operators involve swapping part of the binary string of one parent with the corresponding part for the other parent, with variations depending on the particular part swapped and the position and order in which it is inserted into the remaining binary string of the other parent. Within each child, mutation then takes place.

6. Dynamic Time-Warping

Much of the data from which knowledge is discovered are of a temporal nature. Detecting patterns in sequences of time-dependent data, or time series, is an important aspect of Data Mining, and has applications in areas as diverse as financial analysis and astronomy. Dynamic Time-Warping (DTW) is a technique established in the recognition of patterns in speech, but may be more widely applied to other types of data.

DTW is based on a dynamic programming approach to aligning a selected template with a data time series so that a chosen measure of the distance between them, or error, is minimized. The measure of how well a template matches the time series may be obtained from the table of cumulative distances. A **warping path** is computed through the grid from one boundary point to another, tracing back in time through adjacent points with minimal cumulative distance. This warping path defines how well the template matches the time series, and a measure of the fit can be obtained from the cumulative distances along the path.

B. Statistical and other Uncertainty-Based Methods

1. Statistical Techniques

Statistics is a collection of methods of enquiry used to gather, process, or interpret quantitative data. The two main functions of Statistics are to describe and summarize data and to make inferences about a larger population of which the data are representative. These two areas are referred to as Descriptive and Inferential Statistics, respectively; both areas have an important part to play in Data Mining. Descriptive Statistics provides a toolkit of methods for data summarization while Inferential Statistics is more concerned with data analysis.

Much of Statistics is concerned with statistical analysis that is mainly founded on statistical inference or hypothesis testing. This involves having a Null Hypothesis (H_0): which is a statement of null effect, and an Alternative Hypothesis (H_1): which is a statement of effect. A test of significance allows us to decide which of the two hypotheses (H_0 or H_1) we should accept. We say that a result is significant at the 5% level if the probability that the discrepancy between the actual data and what is expected assuming the null hypothesis is true has probability less that 0.05 of occurring. The significance level therefore tells us where to **threshold** in order to decide if there is an effect or not.

Predictive Modeling is another Data Mining task that is addressed by Statistical methods. The most common type of predictive model used in Statistics is **linear regression**, where we describe one variable as a linear combination of other known variables. A number of other tasks that involve analysis of several variables for various purposes are categorized by statisticians under the umbrella term multivariate analysis.

Sequences of data, which measure values of the same attribute under a sequence of different circumstances, also occur commonly. The best-known form of such data arises when we collect information on an attribute at a sequence of time points, e.g., daily, quarterly, annually. For such time-series data the trend is modeled by fitting a regression line while fluctuations are described by mathematical functions. Irregular variations are difficult to model but worth trying to identify, as they may turn out to be of most interest.

Signal processing is used when there is a continuous measurement of some sort—the signal—usually distorted by noise. The more noise there is, the harder it is to extract the signal. However, by using methods such as filtering which remove all distortions, we may manage to recover much of the original data. Such filtering is often carried out by using **Fourier transforms** to modify the data accordingly. In practice, we may use **Fast Fourier Transforms** to achieve high-performance signal processing. An alternative method is provided by **Wavelets**.

All of Statistics is underpinned by classical or Bayesian probability. Bayesian Methods often form the basis of techniques for the automatic discovery of *classes* in data, known as **clustering** or **unsupervised learning**. In such situations Bayesian Methods may be used in computational techniques to determine the optimal set of classes from a given collection of unclassified instances. The aim is to find the most likely set of classes given the data and a set of prior expectations. A balance must be struck between data fitting and the potential for class membership prediction.

2. Bayesian Belief Networks

Bayesian Belief Networks are graphical models that communicate causal information and provide a framework for describing and evaluating probabilities when we have a network of interrelated variables. We can then use the graphical models to evaluate information about external interventions and hence predict the effect of such interventions. By exploiting the dependencies and interdependencies in the graph we can develop efficient algorithms that calculate probabilities of variables in graphs which

are often very large and complex. Such a facility makes this technique suitable for Data Mining, where we are often trying to sift through large amounts of data looking for previously undiscovered relationships.

A key feature of Bayesian Belief Networks is that they discover and describe causality rather than merely identifying associations as is the case in standard Statistics and Database Technology. Such causal relationships are represented by means of **DAGs (Directed Acyclic Graphs)** that are also used to describe conditional independence assumptions. Such conditional independence occurs when two variables are independent, conditional on another variable.

3. Evidence Theory

Evidence Theory, of which Dempster–Shafer theory is a major constituent, is a generalization of traditional probability which allows us to better quantify uncertainty. The framework provides a means of representing data in the form of a **mass function** that quantifies our degree of belief in various propositions. One of the major advantages of Evidence Theory over conventional probability is that it provides a straightforward way of quantifying ignorance and is therefore a suitable framework for handling missing values.

We may use this Dempster–Shafer definition of mass functions to provide a lower and upper bound for the probability we assign to a particular proposition. These bounds are called the **belief** and **plausibility**, respectively. Such an interval representation of probability is thought to be a more intuitive and flexible way of expressing probability, since we may not be able to assign an exact value to it but instead give lower and upper bounds.

The Dempster–Shafer theory also allows us to transform the data by changing to a higher or lower granularity and reallocating the masses. If a rule can be generalized to a higher level of aggregation then it becomes a more powerful statement of how the domain behaves while, on the other hand, the rule may hold only at a lower level of granularity.

Another important advantage of Evidence Theory is that the Dempster–Shafer law of combination (the orthogonal sum) allows us to combine data from different independent sources. Thus, if we have the same frame of discernment for two mass functions which have been derived independently from different data, we may obtain a unified mass assignment.

4. Fuzzy Logic

Fuzzy logic maintains that all things are a matter of degree and challenges traditional two-valued logic which holds that a proposition is either true or it is not. Fuzzy Logic is defined via a **membership function** that measures the degree to which a particular element is a member of a set. The membership function can take any value between 0 and 1 inclusive.

In common with a number of other Artificial Intelligence methods, fuzzy methods aim to simulate human decision making in uncertain and imprecise environments. We may thus use Fuzzy Logic to express expert opinions that are best described in such an imprecise manner. Fuzzy systems may therefore be specified using natural language which allows the expert to use vague and imprecise terminology. Fuzzy Logic has also seen a wide application to control theory in the last two decades.

An important use of fuzzy methods for Data Mining is for classification. Associations between inputs and outputs are known in fuzzy systems as **fuzzy associative memories** or **FAM**s. A FAM system encodes a collection of compound rules that associate multiple input statements with multiple output statements We combine such multiple statements using logical operators such as conjunction, disjunction and negation.

5. Rough Sets

Rough Sets were introduced by Pawlak in 1982 as a means of investigating structural relationships in data. The technique, which, unlike classical statistical methods, does not make probability assumptions, can provide new insights into data and is particularly suited to situations where we want to reason from qualitative or imprecise information. Rough Sets allow the development of **similarity measures** that take account of semantic as well as syntactic distance. Rough Set theory allows us to eliminate redundant or irrelevant attributes. The theory of Rough Sets has been successfully applied to knowledge acquisition, process control, medical diagnosis, expert systems and Data Mining. The first step in applying the method is to generalize the attributes using domain knowledge to identify the concept hierarchy. After generalization, the next step is to use reduction to generate a minimal subset of all the generalized attributes, called a **reduct**. A set of general rules may then be generated from the reduct that includes all the important patterns in the data. When more than one reduct is obtained, we may select the best according to some criteria. For example, we may choose the reduct that contains the smallest number of attributes.

6. Information Theory

The most important concept in Information Theory is **Shannon's Entropy**, which measures the amount of information held in data. Entropy quantifies to what extent the data are spread out over its possible values. Thus high entropy means that the data are spread out as much as possible while low entropy means that the data are nearly all concentrated on one value. If the entropy is low, therefore, we have high information content and are most likely to come up with a strong rule.

Information Theory has also been used as a measure of interestingness which allows us to take into account how often a rule occurs and how successful it is. This is carried out by using the **J-measure**, which measures the amount of information in a rule using Shannon Entropy and multiplies this by the probability of the rule coming into play. We may therefore rank the rules and only present the most interesting to the user.

C. Database Methods

1. Association Rules

An Association Rule associates the values of a given set of attributes with the value of another attribute from outside that set. In addition, the rule may contain information about the frequency with which the attribute values are associated with each other. For example, such a rule might say that "75% of men, between 50 and 55 years old, in management positions, take out additional pension plans."

Along with the Association Rule we have a **confidence threshold** and a **support threshold**. Confidence measures the ratio of the number of entities in the database with the designated values of the attributes in both A and B to the number with the designated values of the attributes in A. The support for the Association Rule is simply the proportion of entities within the whole database that take the designated values of the attributes in A and B.

Finding Association Rules can be computationally intensive, and essentially involves finding all of the covering attribute sets, A, and then testing whether the rule "A implies B," for some attribute set B separate from A, holds with sufficient confidence. Efficiency gains can be made by a combinatorial analysis of information gained from previous passes to eliminate unnecessary rules from the list of candidate rules. Another highly successful approach is to use **sampling** of the database to estimate whether or not an attribute set is covering. In a large data set it may be necessary to consider which rules are interesting to the user. An approach to this is to use **templates**, to describe the form of interesting rules.

2. Data Manipulation Techniques

For Data Mining purposes it is often necessary to use a large number of data manipulations of various types. When searching for Association Rules, for example, tuples with certain attribute values are grouped together, a task that may require a sequence of conditional data selection operations. This task is followed by counting operations to determine the cardinality of the selected groups. The nature of the rules themselves may require further data manipulation operations such as summing or averaging of data values if the rules involve comparison of numerical attributes. Frequently knowledge is discovered by combining data from more than one source—knowledge which was not available from any one of the sources alone.

3. Slice and Dice

Slice and Dice refers to techniques specifically designed for examining cross sections of the data. Perhaps the most important aspect of Slice and Dice techniques is the facility to view cross sections of data that are not physically visible. Data may be sliced and diced to provide views in orthogonal planes, or at arbitrarily chosen viewing angles. Such techniques can be vital in facilitating the discovery of knowledge for medical diagnosis without the requirement for invasive surgery.

4. Access Methods

For Data Mining purposes it is often necessary to retrieve very large amounts of data from their stores. It is therefore important that access to data can be achieved rapidly and efficiently, which effectively means with a minimum number of input/output operations (I/Os) involving physical storage devices. Databases are stored on direct access media, referred to generally as **disks**. As disk access times are very much slower than main storage access times, acceptable database performance is achieved by adopting techniques whose objective is to arrange data on the disk in ways which permit stored records to be located in as few I/Os as possible.

It is valuable to identify tuples that are logically related, as these are likely to be frequently requested together. By locating two logically related tuples on the same page, they may both be accessed by a single physical I/O. Locating logically related tuples physically close together is referred to as clustering. Intra-file clustering may be appropriate if sequential access is frequently required to a set of tuples within a file; inter-file clustering may be used if sets of tuples from more than one file are frequently requested together.

D. Enabling Technologies

1. Data Cleansing Techniques

Before commencing Data Mining proper, we must first consider all data that is erroneous, irrelevant or atypical, which Statisticians term **outliers**.

Different types of outliers need to be treated in different ways. Outliers that have occurred as a result of human error may be detected by consistency checks (or integrity constraints). If outliers are a result of human ignorance, this may be handled by including information on changing definitions as **metadata**, which should be consulted when outlier tests are being carried out. Outliers of distribution are usually detected by outlier tests which are based on the deviation between the candidate observation and the average of all the data values.

2. Missing Value Handling

When we carry out Data Mining, we are often working with large, possibly heterogeneous data. It therefore frequently happens that some of the data values are missing because data were not recorded in that case or perhaps was represented in a way that is not compatible with the remainder of the data. Nonetheless, we need to be able to carry out the Data Mining process as best we can. A number of techniques have been developed which can be used in such circumstances, as follows:

- All tuples containing missing data are eliminated from the analysis.
- All missing values are eliminated from the analysis.
- A typical data value is selected at random and imputed to replace the missing value.

3. Advanced Database Technology

The latest breed of databases combines high performance with multidimensional data views and fast, optimized query execution. Traditional databases may be adapted to provide query optimization by utilizing Parallel Processing capabilities. Such **parallel databases** may be implemented on parallel hardware to produce a system that is both powerful and scaleable. Such postrelational Database Management Systems represent data through nested multidimensional tables that allow a more general view of the data of which the relational model is a special case. **Distributed Databases** allow the contributing heterogeneous databases to maintain local autonomy while being managed by a global data manager that presents a single data view to the user. **Multidimensional servers** support the multidimensional data view that represents multidimensional data through nested data. In the three-dimensional case, the data are stored in the form of a **data cube**, or in the case of many dimensions we use the general term **data hypercube**.

The Data Warehousing process involves assembling data from heterogeneous sources systematically by using **middleware** to provide connectivity. The data are then cleansed to remove inaccuracies and inconsistencies and transformed to give a consistent view. **Metadata** that maintain information concerning the source data are also stored in the warehouse. Data within the warehouse is generally stored in a distributed manner so as to increase efficiency and, in fact, parts of the warehouse may be replicated at local sites, in **data marts**, to provide a facility for departmental decision-making.

4. Visualization Methods

Visualization Methods aim to present complex and voluminous data sets in pictorial and other graphical representations that facilitate understanding and provide insight into the underlying structures in the data. The subject is essentially interdisciplinary, encompassing statistical graphics, computer graphics, image processing, computer vision, interface design and cognitive psychology.

For exploratory Data Mining purposes, we require flexible and interactive visualization tools which allow us to look at the data in different ways and investigate different subsets of the data. We can highlight key features of the display by using color coding to represent particular data values. Charts that show relationships between individuals or objects within the dataset may be color-coded, and thus reveal interesting information about the structure and volume of the relationships. Animation may provide a useful way of exploring sequential data or time series by drawing attention to the changes between time points. **Linked windows**, which present the data in various ways and allow us to trace particular parts of the data from one window to another, may be particularly useful in tracking down interesting or unusual data.

5. Intelligent Agents

The potential of Intelligent Agents is increasingly having an impact on the marketplace. Such agents have the capability to form their own goals, to initiate action without instructions from the user and to offer assistance to the user without being asked. Such software has been likened to an intelligent personal assistant who works out what is needed by the boss and then does it. Intelligent Agents are essentially software tools that interoperate with other software to exchange information and services. They act as an intelligent layer between the user and the data, and facilitate tasks that serve to promote the user's overall goals. Communication with other software is achieved by exchanging messages in an **agent communication language**. Agents may be organized into a federation or agency where a number of agents interact to carry out different specialized tasks.

6. OLAP

The term OLAP (On-line Analytical Processing) originated in 1993 when Dr. E. F. Codd and colleagues developed the idea as a way of extending the relational database paradigm to support business modeling. This development took the form of a number of rules that were designed to facilitate fast and easy access to the relevant data for purposes of management information and decision support. An OLAP Database generally takes the form of a multidimensional server database that makes management information available interactively to the user. Such multidimensional views of the data are ideally suited to an analysis engine since they give maximum flexibility for such database operations as Slice and Dice or drill down which are essential for analytical processing.

7. Parallel Processing

High-performance parallel database systems are displacing traditional systems in very large databases that have complex and time-consuming querying and processing requirements. Relational queries are ideally suited to parallel execution since they often require processing of a number of different relations. In addition to parallelizing the data retrieval required for Data Mining, we may also parallelize the data processing that must be carried out to implement the various algorithms used to achieve the Mining tasks. Such processors may be designed to (1) share memory, (2) share disks, or (3) share nothing. Parallel Processing may be carried out using shared address space, which provides hardware support for efficient communication. The most scaleable paradigm, however, is to share nothing, since this reduces the overheads. In Data Mining, the implicitly parallel nature of most of the Mining tasks allows us to utilize processors which need only interact occasionally, with resulting efficiency in both speed-up and scalability.

8. Distributed Processing

Distributed databases allow local users to manage and access the data in the local databases while providing some sort of global data management which provides global users with a global view of the data. Such global views allow us to combine data from the different sources which may not previously have been integrated, thus providing the potential for new knowledge to be discovered. The constituent local databases may either be homogeneous and form part of a design which seeks to distribute data storage and processing to achieve greater efficiency, or they may be heterogeneous and form part of a legacy system where the original databases might have been developed using different data models.

E. Relating the Technologies to the Tasks

Data Mining embraces a wealth of methods that are used in parts of the overall process of Knowledge Discovery in Databases. The particular Data Mining methods employed need to be matched to the user's requirements for the overall KDD process.

The tools for the efficient storage of and access to large datasets are provided by the Database Technologies. Recent advances in technologies for data storage have resulted in the availability of inexpensive high-capacity storage devices with very fast access. Other developments have yielded improved database management systems and Data Warehousing technologies. To facilitate all of the Data Mining Tasks, fast access methods can be combined with sophisticated data manipulation and Slice and Dice techniques for analysis of Data Warehouses through OLAP to achieve the intelligent extraction and management of information.

The general tasks of Data Mining are those of description and prediction. Descriptions of the data often require Summarization to provide concise accounts of some parts of the dataset that are of interest. Prediction involves using values of some attributes in the database to predict unknown values of other attributes of interest. Classification, Clustering, and Pattern Recognition are all Data Mining Tasks that can be carried out for the purpose of description, and together with Predictive Modeling and Sequence Processing can be used for the purpose of prediction. All of these descriptive and predictive tasks can be addressed by both Machine Learning Technologies such as Inferencing Rules, Neural Networks, Case-Based Reasoning, and Genetic Algorithms, or by a variety of Uncertainty Methods.

Data Mining methods are used to extract both patterns and models from the data. This involves modeling dependencies in the data. The model must specify both the structure of the dependencies (i.e., which attributes are inter-dependent) and their strengths. The tasks of Dependency Detection and modeling may involve the discovery of empirical laws and the inference of causal models from the data, as well as the use of Database Technologies such as Association Rules. These tasks can be addressed by Machine Learning Technologies such as Inferencing Rules, Neural Networks and Genetic Algorithms, or by Uncertainty Methods, including Statistical Techniques, Bayesian Belief Networks, Evidence Theory, Fuzzy Logic and Rough Sets.

The tasks of Visualization and Summarization play a central role in the successful discovery and analysis of patterns in the data. Both of these are essentially based on

Statistical Techniques associated with Exploratory Data Analysis. The overall KDD process also encompasses the task of Uncertainty Handling. Real-world data are often subject to uncertainty of various kinds, including missing values, and a whole range of Uncertainty Methods may be used in different approaches to reasoning under uncertainty.

III. DATA MINING FOR DIFFERENT DATA TYPES

A. Web Mining and Personalization

Developments in Web Mining have been inexorably linked to developments in e-commerce. Such developments have accelerated as the Internet has become more efficient and more widely used. Mining of click streams and session log analysis allows a web server owner to extract new knowledge about users of the service, thus, in the case of e-commerce, facilitating more targeted marketing. Similarly, personalization of web pages as a result of Data Mining can lead to the provision of a more efficient service.

B. Distributed Data Mining

Recent developments have produced a convergence between computation and communication. Organizations that are geographically distributed need a decentralized approach to data storage and decision support. Thus the issues concerning modern organizations are not just the size of the database to be mined, but also its distributed nature. Such developments hold an obvious promise not only for what have become traditional Data Mining areas such as Database Marketing but also for newer areas such as e-Commerce and e-Business.

Trends in DDM are inevitably led by trends in Network Technology. The next generation Internet will connect sites at speeds of the order of 100 times faster than current connectivity. Such powerful connectivity to some extent accommodates the use of current algorithms and techniques. However, in addition, new algorithms, and languages are being developed to facilitate distributed data mining using current and next generation networks.

Rapidly evolving network technology, in conjunction with burgeoning services and information availability on the Internet is rapidly progressing to a situation where a large number of people will have fast, pervasive access to a huge amount of information that is widely accessible. Trends in Network Technology such as bandwidth developments, mobile devices, mobile users, intranets, information overload, and personalization leads to the conclusion that mobile code, and mobile agents, will, in the near future, be a critical part of Internet services. Such developments must be incorporated into Data Mining technology.

C. Text Mining

Text may be considered as sequential data, similar in this respect to data collected by observation systems. It is therefore appropriate for Data Mining techniques that have been developed for use specifically with sequential data to be also applied to the task of Text Mining. Traditionally, text has been analyzed using a variety of information retrieval methods, including natural language processing. Large collections of electronically stored text documents are becoming increasingly available to a variety of end-users, particularly via the World Wide Web. There is great diversity in the requirements of users: some need an overall view of a document collection to see what types of documents are present, what topics the documents are concerned with, and how the documents are related to one another. Other users require specific information or may be interested in the linguistic structures contained in the documents. In very many applications users are initially unsure of exactly what they are seeking, and may engage in browsing and searching activities.

General Data Mining methods are applicable to the tasks required for text analysis. Starting with textual data, the Knowledge Discovery Process provides information on commonly occurring phenomena in the text. For example, we may discover combinations of words or phrases that commonly appear together. Information of this type is presented using **episodes**, which contain such things as the base form of a word, grammatical features, and the position of a word in a sequence. We may measure, for example, the support for an episode by counting the number of occurrences of the episode within a given text sequence.

For Text Mining, a significant amount of pre-processing of the textual data may be required, dependent on the domain and the user's requirements. Some natural language analysis may be used to augment or replace some words by their parts of speech or by their base forms. Post-processing of the results of Text Mining is usually also necessary.

D. Temporal Data Mining

Temporal Data Mining often involves processing time series, typically sequences of data, which measure values of the same attribute at a sequence of different time points. Pattern matching using such data, where we are searching for particular patterns of interest, has attracted considerable interest in recent years. In addition to traditional statistical methods for time series analysis, more recent work on sequence processing has used association rules developed by the database community. In addition Temporal Data Mining may involve exploitation of efficient

methods of data storage, fast processing and fast retrieval methods that have been developed for temporal databases.

E. Spatial Data Mining

Spatial Data Mining is inexorably linked to developments in Geographical Information Systems. Such systems store spatially referenced data. They allow the user to extract information on contiguous regions and investigate spatial patterns. Data Mining of such data must take account of spatial variables such as distance and direction. Although methods have been developed for Spatial Statistics, the area of Spatial Data Mining per se is still in its infancy. There is an urgent need for new methods that take spatial dependencies into account and exploit the vast spatial data sources that are accumulating. An example of such data is provided by remotely sensed data of images of the earth collected by satellites.

F. Multimedia Data Mining

Multimedia Data Mining involves processing of data from a variety of sources, principally text, images, sound, and video. Much effort has been devoted to the problems of indexing and retrieving data from such sources, since typically they are voluminous. A major activity in extracting knowledge from time-indexed multimedia data, e.g., sound and video, is the identification of episodes that represent particular types of activity; these may be identified in advance by the domain expert. Likewise domain knowledge in the form of metadata may be used to identify and extract relevant knowledge. Since multimedia contains data of different types, e.g., images along with sound, ways of combining such data must be developed. Such problems of Data Mining from multimedia data are, generally speaking, very difficult and, although some progress has been made, the area is still in its infancy.

G. Security and Privacy Aspects of Data Mining

As we have seen, Data Mining offers much as a means of providing a wealth of new knowledge for a huge range of applications. The knowledge thus obtained from databases may be far in excess of the use to which the data owners originally envisaged for the database. However, such data may include sensitive information about individuals or might involve company confidential information. Care must therefore be taken to ensure that only authorized personnel are permitted to access such databases. However, it may be possible to get around this problem of preserving the security of individual level data by using anonymization techniques and possibly only providing a sample of the data for Mining purposes. In addition, it is often the case that, for purposes of Data Mining, we do not need to use individual level data but instead can utilize aggregates.

For Database Technology, intrusion detection models must be developed which protect the database against security breaches for the purpose of Data Mining. Such methods look for evidence of users running huge numbers of queries against the database, large volumes of data being downloaded by the user, or users running their own imported software on portions of the database.

H. Metadata Aspects of Data Mining

Currently, most data mining algorithms require bringing all together data to be mined in a single, centralized data warehouse. A fundamental challenge is to develop distributed versions of data mining algorithms so that data mining can be done while leaving some of the data in place. In addition, appropriate protocols, languages, and network services are required for mining distributed data to handle the mappings required for mining distributed data. Such functionality is typically provided via metadata.

XML (eXtensible Markup Language) is fast emerging as a standard for representing data on the World Wide Web. Traditional Database Engines may be used to process semistructured XML documents conforming to Data Type Definitions (DTDs). The XML files may be used to store metadata in a representation to facilitate the mining of multiple heterogeneous databases. **PMML** (predictive Mark-up Language) has been developed by the Data Mining community for the exchange of models between different data sites; typically these will be distributed over the Internet. Such tools support interoperability between heterogeneous databases thus facilitating Distributed Data Mining.

IV. KEY APPLICATION AREAS

A. Industry

Industrial users of databases are increasingly beginning to focus on the potential for embedded artificial intelligence within their development and manufacturing processes. Most industrial processes are now subject to technological control and monitoring, during which vast quantities of manufacturing data are generated. Data Mining techniques are also used extensively in process analysis in order to discover improvements which may be made to the process in terms of time scale and costs.

Classification techniques and rule induction methods are used directly for quality control in manufacturing. Parameter settings for the machinery may be monitored and

evaluated so that decisions for automatic correction or intervention can be taken if necessary. Machine Learning technologies also provide the facility for failure diagnosis in the maintenance of industrial machinery.

Industrial safety applications are another area benefiting from the adoption of Data Mining technology. Materials and processes may need to be classified in terms of their industrial and environmental safety. This approach, as opposed to experimentation, is designed to reduce the cost and time scale of safe product development.

B. Administration

There is undoubtedly much scope for using Data Mining to find new knowledge in administrative systems that often contain large amounts of data. However, perhaps because the primary function of administrative systems is routine reporting, there has been less uptake of Data Mining to provide support and new knowledge for administrative purposes that in some other application areas.

Administrative systems that have received attention tend to be those in which new knowledge can be directly translated into saving money. An application of this type is provided by the Inland Revenue that collects vast amounts of data and may potentially save a lot of money by devising ways of discovering tax dodges, similarly for welfare frauds. Another successful application of Data Mining has been to the health care system where again new discoveries about expensive health care options can lead to huge savings. Data Mining is also likely to become an extremely useful tool in criminal investigations, searching for possible links with particular crimes or criminals.

C. Business

As we might expect, the major application area of Data Mining, so far, has been Business, particularly the areas of Marketing, Risk Assessment, and Fraud Detection.

In Marketing, perhaps the best known use of Data Mining is for customer profiling, both in terms of discovering what types of goods customers tend to purchase in the same transaction and groups of customers who all behave in a similar way and may be targeted as a group. Where customers tend to buy (unexpected) items together then goods may be placed on nearby shelves in the supermarket or beside each other in a catalogue. Where customers may be classified into groups, then they may be singled out for customized advertising, mail shots, etc. This is known as micro marketing. There are also cases where customers of one type of supplier unexpectedly turn out to be also customers of another type of supplier and advantage may be gained by pooling resources in some sense. This is known as cross marketing.

Another use of Data Mining for Business has been for Risk Assessment. Such assessment of credit worthiness of potential customers is an important aspect of this use which has found particular application to banking institutions where lending money to potentially risky customers is an important part of the day-to-day business. A related application has been to litigation assessment where a firm may wish to assess how likely and to what extent a bad debtor will pay up and if it is worth their while getting involved in unnecessary legal fees.

Case Study I. A supermarket chain with a large number of stores holds data on the shopping transactions and demographic profile of each customer's transactions in each store. Corporate management wants to use the customer databases to look for global and local shopping patterns.

D. Database Marketing

Database Marketing refers to the use of Data Mining techniques for the purposes of gaining business advantage. These include improving a company's knowledge of its customers in terms of their characteristics and purchasing habits and using this information to classify customers; predicting which products may be most usefully offered to a particular group of customers at a particular time; identifying which customers are most likely to respond to a mail shot about a particular product; identifying customer loyalty and disloyalty and thus improving the effectiveness of intervention to avoid customers moving to a competitor; identifying the product specifications that customers really want in order to improve the match between this and the products actually offered; identifying which products from different domains tend to be bought together in order to improve cross-marketing strategies; and detecting fraudulent activity by customers.

One of the major tasks of Data Mining in a commercial arena is that of market segmentation. Clustering techniques are used in order to partition a customer database into homogeneous segments characterized by customer needs, preferences, and expenditure. Once market segments have been established, classification techniques are used to assign customers and potential customers to particular classes. Based on these, prediction methods may be employed to forecast buying patterns for new customers.

E. Medicine

Potential applications of Data Mining to Medicine provide one of the most exciting developments and hold much promise for the future. The principal medical areas which have been subjected to a Data Mining approach, so far, may be categorized as: diagnosis, treatment, monitoring, and research.

The first step in treating a medical complaint is diagnosis, which usually involves carrying out various tests and observing signs and symptoms that relate to the possible diseases that the patient may be suffering from. This may involve clinical data, data concerning biochemical indicators, radiological data, sociodemographic data including family medical history, and so on. In addition, some of these data may be measured at a sequence of time-points, e.g., temperature, lipid levels. The basic problem of diagnosis may be regarded as one of classification of the patient into one, or more, possible disease classes.

Data Mining has tremendous potential as a tool for assessing various treatment regimes in an environment where there are a large number of attributes which measure the state of health of the patient, allied to many attributes and time sequences of attributes, representing particular treatment regimes. These are so complex and interrelated, e.g., the interactions between various drugs, that it is difficult for an individual to assess the various components particularly when the patient may be presenting with a variety of complaints (multi-pathology) and the treatment for one complaint might mitigate against another.

Perhaps the most exciting possibility for the application of Data Mining to medicine is in the area of medical research. Epidemiological studies often involve large numbers of subjects which have been followed-up over a considerable period of time. The relationship between variables is of considerable interest as a means of investigating possible causes of diseases and general health inequalities in the population.

Case Study II. A drug manufacturing company is studying the risk factors for heart disease. It has data on the results of blood analyses, socioeconomic data, and dietary patterns. The company wants to find out the relationship between the heart disease markers in the blood and the other relevant attributes.

F. Science

In many areas of science, automatic sensing and recording devices are responsible for gathering vast quantities of data. In the case of data collected by remote sensing from satellites in disciplines such as astronomy and geology the amount of data are so great that Data Mining techniques offer the only viable way forward for scientific analysis.

One of the principal application areas of Data Mining is that of space exploration and research. Satellites provide immense quantities of data on a continuous basis via remote sensing devices, for which intelligent, trainable image-analysis tools are being developed. In previous large-scale studies of the sky, only relatively small amount of the data collected have actually been used in manual attempts to classify objects and produce galaxy catalogs.

Not only has the sheer amount of data been overwhelming for human consideration, but also the amount of data required to be assimilated for the observation of a single significant anomaly is a major barrier to purely manual analysis. Thus Machine Learning techniques are essential for the classification of features from satellite pictures, and they have already been used in studies for the discovery of quasars. Other applications include the classification of landscape features, such as the identification of volcanoes on the surface of Venus from radar images. Pattern recognition and rule discovery also have important applications in the chemical and biomedical sciences. Finding patterns in molecular structures can facilitate the development of new compounds, and help to predict their chemical properties. There are currently major projects engaged in collecting data on the human gene pool, and rule-learning has many applications in the biomedical sciences. These include finding rules relating drug structure to activity for diseases such as Alzheimer's disease, learning rules for predicting protein structures, and discovering rules for the use of enzymes in cancer research.

Case Study III. An astronomy cataloguer wants to process telescope images, identify stellar objects of interest and place their descriptions into a database for future use.

G. Engineering

Machine Learning has an increasing role in a number of areas of engineering, ranging from engineering design to project planning. The modern engineering design process is heavily dependent on computer-aided methodologies. Engineering structures are extensively tested during the development stage using computational models to provide information on stress fields, displacement, load-bearing capacity, etc. One of the principal analysis techniques employed by a variety of engineers is the finite element method, and Machine Learning can play an important role in learning rules for finite element mesh design for enhancing both the efficiency and quality of the computed solutions.

Other engineering design applications of Machine Learning occur in the development of systems, such as traffic density forecasting in traffic and highway engineering. Data Mining technologies also have a range of other engineering applications, including fault diagnosis (for example, in aircraft engines or in on-board electronics in intelligent military vehicles), object classification (in oil exploration), and machine or sensor calibration. Classification may, indeed, form part of the mechanism for fault diagnosis.

As well as in the design field, Machine Learning methodologies such as Neural Networks and Case-Based Reasoning are increasingly being used for engineering

project management in an arena in which large scale international projects require vast amounts of planning to stay within time scale and budget.

H. Fraud Detection and Compliance

Techniques which are designed to register abnormal transactions or data usage patterns in databases can provide an early alert, and thus protect database owners from fraudulent activity by both a company's own employees and by outside agencies. An approach that promises much for the future is the development of adaptive techniques that can identify particular fraud types, but also be adaptive to variations of the fraud. With the ever-increasing complexity of networks and the proliferation of services available over them, software agent technology may be employed in the future to support interagent communication and message passing for carrying out surveillance on distributed networks.

Both the telecommunications industry and the retail businesses have been quick to realize the advantages of Data Mining for both fraud detection and discovering failures in compliance with company procedures. The illegal use of telephone networks through the abuse of special services and tariffs is a highly organized area of international crime. Data Mining tools, particularly featuring Classification, Clustering, and Visualization techniques have been successfully used to identify patterns in fraudulent behavior among particular groups of telephone service users.

V. FUTURE DEVELOPMENTS

Data Mining, as currently practiced, has emerged as a subarea of Computer Science. This means that initial developments were strongly influenced by ideas from the Machine Learning community with a sound underpinning from Database Technology. However, the statistical community, particularly Bayesians, was quick to realize that they had a lot to contribute to such developments. Data Mining has therefore rapidly grown into the interdisciplinary subject that it is today.

Research in Data Mining has been led by the KDD (Knowledge Discovery in Databases) annual conferences, several of which have led to books on the subject (e.g., Fayyad *et al.*, 1996). These conferences have grown in 10 years from being a small workshop to a large independent conference with, in Boston in 2000, nearly 1000 participants. The proceedings of these conferences are still the major outlet for new developments in Data Mining.

Major research trends in recent years have been:

- The development of scalable algorithms that can operate efficiently using data stored outside main memory

- The development of algorithms that look for local patterns in the data—data partitioning methods have proved to be a promising approach
- The development of Data Mining methods for different types of data such as multimedia and text data
- The developments of methods for different application areas

Much has been achieved in the last 10 years. However, there is still huge potential for Data Mining to develop as computer technology improves in capability and new applications become available.

SEE ALSO THE FOLLOWING ARTICLES

ARTIFICAL NEURAL NETWORKS • COMPUTER ALGO-RITHMS • DATABASES • FOURIER SERIES • FUNCTIONAL ANALYSIS • FUZZY SETS, FUZZY LOGIC, AND FUZZY SYS-TEMS • STATISTICS, BAYESIAN • WAVELETS

BIBLIOGRAPHY

Adriaans, P., and Zantinge, D. (1996). "Data Mining," Addison-Wesley, MA.

Berry, M., and Linoff, G. (1997). "Data Mining Techniques for Marketing, Sales and Customer Support," Wiley, New York.

Berson, A., and Smith, S. J. (1997). "Data Warehousing, Data Mining, and Olap," McGraw-Hill, New York.

Bigus, J. (1996). "Data Mining With Neural Networks," McGraw-Hill, New York.

Breiman, L., Friedman, J. H., Olshen, R. A., and Stone, C. J. (1984). "Classification and Regression Trees," Wadsworth, Belmont.

Cabena, P., Hadjinian, P., Stadler, R., Verhees, J., and Zanasi, A. (1997). "Discovering Data Mining from Concept to Implementation," Prentice-Hall, Upper Saddle River, NJ.

Fayyad, U. M., Piatetsky-Shapiro, G., Smyth, P., and Uthurusamy, R. (1996). "Advances in Knowledge Discovery and Data Mining," AAAI Press/The MIT Press, Menlo Park, CA.

Freitas, A. A., and Lavington, S. H. (1998). "Mining Very Large Databases With Parallel Processing," Kluwer, New York.

Groth, R. (1997). "Data Mining: A Hands on Approach to Information Discovery," Prentice-Hall, Englewood Cliffs, NJ.

Inmon, W. (1996). "Using the Data Warehouse," Wiley, New York.

Kennedy, R. L., Lee, Y., Van Roy, B., and Reed, C. D. (1997). "Solving Data Mining Problems Through Pattern Recognition," Prentice-Hall, Upper Saddle River, NJ.

Lavrac, N., Keravnou, E. T., and Zupan, B. (eds.). (1997). "Intelligent Data Analysis in Medicine and Pharmacology," Kluwer, Boston.

Mattison, R. M. (1997). "Data Warehousing and Data Mining for Telecommunications," Artech House, MA.

Mitchell, T. (1997). "Machine Learning," McGraw-Hill, New York.

Ripley, B. (1995). "Pattern Recognition and Neural Networks," Cambridge University Press, Cambridge.

Stolorz, P., and Musick, R. (eds.). (1997). "Scalable High Performance Computing for Knowledge Discovery and Data Mining," Kluwer, New York.

Weiss, S. M., and Indurkhya, N. (1997). "Predictive Data Mining: A Practical Guide" (with Software), Morgan Kaufmann, San Francisco, CA.

Wu, X. (1995). "Knowledge Acquisition from Databases," Ablex, Greenwich, CT.

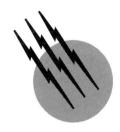

Data Mining, Statistics

David L. Banks
U.S. Department of Transportation

Stephen E. Fienberg
Carnegie Mellon University

GLOSSARY

Bagging A method that averages predictions from several models to achieve better predictive accuracy.

Boosting A method that uses outcome-based weighting to achieve improved predictive accuracy in a classification rule.

Classification The statistical problem of assigning new cases into categories based on observed values of explanatory variables, using previous information on a sample of cases for which the categories are known.

Clustering The operation of finding groups within multivariate data. One wants the cases within a group to be similar and cases in different groups to be clearly distinct.

Curse of dimensionality The unfortunate fact that statistical inference becomes increasingly difficult as the number of explanatory variables becomes large.

Nonparametric regression Methods that attempt to find a functional relationship between a response variable and one or more explanatory variables without making strong assumptions (such as linearity) about the form of the relationship.

Principle of parsimony This asserts that a simpler model usually has better predictive accuracy than a complex model, even if the complex model fits the original sample of data somewhat better.

Projection pursuit An exploratory technique in which one looks for low-dimensional projections of high-dimensional data that reveal interesting structure or define useful new explanatory variables.

Recursive partitioning An approach to regression or classification in which one fits different models in different regions of the space of explanatory variables and the data are used to identify the regions that require distinct models.

Smoothing Any operation that does local averaging or local fitting of the response values in order to estimate a functional relationship.

DATA MINING is an emerging analytical area of research activity that stands at the intellectual intersection of statistics, computer science, and database management. It deals with very large datasets, tries to make fewer theoretical assumptions than has traditionally been done in statistics, and typically focuses on problems of classification, clustering, and regression. In such domains, data mining often uses decision trees or neural networks as models and frequently fits them using some combination of techniques such as bagging, boosting/arcing, and racing. These domains and techniques are the primary focus of the present article. Other activities in data mining focus on issues such as causation in large-scale systems (e.g., Spirtes *et al.* 2001; Pearl, 2000), and this effort often involves elaborate statistical models and, quite frequently, Bayesian methodology and related computational techniques (e.g., Cowell *et al.*, 1999, Jordan, 1998). For an introductory discussion of this dimension of data mining see Glymour *et al.* (1997).

The subject area of data mining began to coalesce in 1990, as researchers from the three parent fields discovered common problems and complementary strengths. The first KDD workshop (on Knowledge Discovery and Data Mining) was held in 1989. Subsequent workshops were held in 1991, 1993, and 1994, and these were then reorganized as an annual international conference in 1995. In 1997 the *Journal of Data Mining and Knowledge Discovery* was established under the editorship of Usama Fayyad. It has published special issues on electronic commerce, scalable and parallel computing, inductive logic programming, and applications to atmospheric sciences. Interest in data mining continues to grow, especially among businesses and federal statistical agencies; both of these communities gather large amounts of complex data and want to learn as much from their collections as is possible.

The needs of the business and federal communities have helped to direct the growth of data mining research. For example, typical applications in data mining include the following:

- Use of historical financial records on bank customers to predict good and bad credit risks.
- Use of sales records and customer demographics to perform market segmentation.
- Use of previous income tax returns and current returns to predict the amount of tax a person owes.

These three examples imply large datasets without explicit model structure, and the analyses are driven more by management needs than by scientific theory. The first example is a classification problem, the second requires clustering, and the third would employ a kind of regression analysis.

In these examples the raw data are numerical or categorical values, but sometimes data mining treats more exotic situations. For example, the data might be satellite photographs, and the investigator would use data mining techniques to study climate change over time. Or they can be other forms of images such as those arising from functional magnetic resonance imaging in medicine, or robot sensors. Or the data might be continuous functions, such as spectra from astronomical studies of stars. This review focuses upon the most typical applications, but the reader can obtain a fuller sense of the scope by examining the repository of benchmark data maintained at the University of California at Irvine (Bay, 1999). This archive is resource for the entire machine learning community, who use it to test and tune new algorithms. It contains, among other datasets, all those used in the annual KDD Cup contest, a competitive comparison of data mining algorithms run by the organizers of the KDD conference.

It is unclear whether data mining will continue to gain intellectual credibility among academic researchers as a separate discipline or whether it will be subsumed as a branch of statistics or computer science. Its commercial success has created a certain sense of distance from traditional research universities, in part because so much of the work is being done under corporate auspices. But there is broad agreement that the content of the field has true scientific importance, and it seems certain that under one label or another, the body of theory and heuristics that constitutes the core of data mining will persist and grow.

The remainder of this article describes the two practical problems whose tension defines the ambit of data mining, then reviews the three basic kinds of data mining applications: classification, clustering, and regression. In parallel, we describe three relatively recent ideas that are characteristic of the intellectual cross-fertilization that occurs in data mining, although they represent just a small part of the work going on in this domain. Boosting is discussed in the context of classification, racing is discussed in the context of clustering, and bagging is discussed in the context of regression. Racing and bagging have broad applicability, but boosting is (so far) limited in application to classification problems. Other techniques for these problems of widespread interest not discussed here include support vector machines and kernel smoothing methods (e.g., Vapnik, 2000).

I. TWO PROBLEMS

Data mining exists because modern methods of data collection and database management have created sample sizes large enough to overcome (or partially overcome) the limitations imposed by the curse of dimensionality. However, this freedom comes at a price—the size of the datasets severely restricts the complexity of the calculations that can be made. These two issues are described in the following subsections.

A. The Curse of Dimensionality

The curse of dimensionality (COD) was first described by Richard Bellman, a mathematician, in the context of approximation theory. In data analysis, the term refers to the difficulty of finding hidden structure when the number of variables is large.

For all problems in data mining, one can show that as the number of explanatory variables increases, the problem of structure discovery becomes harder. This is closely related to the problem of variable selection in model fitting. Classical statistical methods, such as discriminant analysis and multiple linear regression, avoided this problem by making the strong model assumption that the mathematical relationship between the response and explanatory variables was linear. But data miners prefer alternative analytic approaches, since the linearity assumption is usually wrong for the kinds of applications they encounter.

For specificity, consider the case of regression analysis; here one looks for structure that predicts the value of the response variable Y from explanatory variables $X \in \mathbb{R}^p$. Thus one might want to predict the true amount of tax that an individual owes from information reported in the previous year, information declared on the current form, and ancillary demographic or official information. The COD arises when p is large.

There are three essentially equivalent descriptions of the COD, but each gives a usefully different perspective on the problem:

1. For large p nearly all datasets are too sparse.
2. The number of regression functions to consider grows quickly (faster than exponentially) with p, the dimension of the space of explanatory variables.
3. For large p, nearly all datasets are multicollinear (or concurve, the nonparametric generalization of multicollinearity).

These problems are minimized if data can be collected using an appropriate statistical design, such as Latin hypercube sampling (Stein, 1987). However, aside from simulation experiments, this is difficult to achieve.

The first version of the curse of dimensionality is most easily understood. If one has five points at random in the unit interval, they tend to be close together, but five random points in the unit square and the unit cube tend to be increasingly dispersed. As the dimension increases, a sample of fixed size provides less information on local structure in the data.

To quantify this heuristic about increasing data sparsity, one can calculate the side length of a p-dimensional subcube that is expected to contain half of the (uniformly random) data in the p-dimensional unit cube. This side-length is $(0.5)^{1/p}$, which increases to 1 as p gets large. Therefore the expected number of observations in a fixed volume in $\mathbb{R}^p$ goes to zero, which implies that the data can provide little information on the local relationship between X and Y. Thus it requires very large sample sizes to find local structure in high dimensions.

The second version of the COD is related to complexity theory. When p is large there are many possible models that one might fit, making it difficult for a finite sample properly to choose among the alternative models. For example, in predicting the amount of tax owed, competing models would include a simple one based just on the previous year's payment, a more complicated model that modified the previous year's payment for wage inflation, and a truly complex model that took additional account of profession, location, age, and so forth.

To illustrate the explosion in the number of possible models, suppose one decides to use only a polynomial model of degree 2 or less. When $p = 1$ (i.e., a single explanatory variable), there are seven possible regression models:

$$Y = \beta_0 + \epsilon,$$
$$Y = \beta_0 + \beta_1 X_1 + \epsilon,$$
$$Y = \beta_0 + \beta_1 X_1 + \beta_2 X_1^2 + \epsilon,$$
$$Y = \beta_1 X_1 + \epsilon,$$
$$Y = \beta_0 + \beta_2 X_1^2 + \epsilon,$$
$$Y = \beta_1 X_1 + \beta_2 X_1^2 + \epsilon,$$
$$Y = \beta_0 + \beta_2 X_1^2 + \epsilon,$$

where ϵ denotes noise in the observation of Y. For $p = 2$ there are 63 models to consider (including interaction terms of the form $X_1 X_2$), and simple combinatorics shows that, for general p, there are $2^{1+p+p(p+1)/2} - 1$ models of degree 2 or less. Since real-world applications usually explore much more complicated functional relationships than low-degree polynomials, data miners need vast quantities of data to discriminate among the many possibilities.

Of course, it is tempting to say that one should include all the explanatory variables, and in the early days of data mining many naive computer scientists did so. But statisticians knew that this violated the principle of parsimony and led to inaccurate prediction. If one does not severely limit the number of variables (and the number of transformations of the variables, and the number of interaction terms between variables) in the final model, then one ends up *overfitting* the data. Overfit happens when the chosen model describes the accidental noise as well as the true signal. For example, in the income tax problem it might by chance happen that people whose social security numbers end in 2338 tend to have higher incomes. If the model fitting process allowed such frivolous terms, then this chance relationship would be used in predicting tax obligation, and such predictions would be less accurate than those obtained from a more parsimonious model that excluded misinformation. It does no good to hope that one's dataset lacks spurious structure; when p is large, it becomes mathematically certain that chance patterns exist.

The third formulation of the COD is more subtle. In standard multiple regression, multicollinearity arises when two or more of the explanatory variables are highly correlated, so that the the data lie mostly inside an affine subspace of $\mathbb{R}^p$ (e.g., close to a line or a plane within the p-dimensional volume). If this happens, there are an uncountable number of models that fit the data about equally well, but these models are dramatically different with respect to predictions for future responses whose explanatory values lie outside the subspace. As p gets large, the number of possible subspaces increases rapidly, and just by chance a finite dataset will tend to concentrate in one of them.

The problem of multicollinearity is aggravated in nonparametric regression, which allows nonlinear relationships in the model. This is the kind of regression most frequently used in data mining. Here the analogue of multicollinearity arises when predictors concentrate on a smooth manifold within $\mathbb{R}^p$, such as a curved line or sheet inside the p-volume. Since there are many more manifolds than affine subspaces, the problem of concurvity in nonparametric regression distorts prediction even more than does multicollinearity in linear regression.

When one is interested only in prediction, the COD is less of a problem for future data whose explanatory variables have values close to those observed in the past. But an unobvious consequence of large p is that nearly all new observation vectors tend to be far from those previously seen. Furthermore, if one needs to go beyond simple prediction and develop interpretable models, then the COD can be an insurmountable obstacle. Usually the most one can achieve is local interpretability, and that happens only where data are locally dense. For more detailed discussions of the COD, the reader should consult Hastie and Tibshirani (1990) and Scott and Wand (1991).

The classical statistical or psychometrical approach to dimensionality reduction typically involves some form of principal component analysis or multidimensional scaling. Roweis and Saul (2000) and Tenenbaum *et al.* (2000) suggest some novel ways to approach the problem of nonlinear dimensionality reduction in very high dimensional problems such as those involving images.

B. Massive Datasets

Massive datasets pose special problems in analysis. To provide some benchmarks of difficulty, consider the following taxonomy by dataset size proposed by Huber (1995):

Type	Size (bytes)
Tiny	10^2
Small	10^4
Medium	10^6
Large	10^8
Huge	10^{10}
Monster	10^{12}

Huber argues that the category steps, which are factors of 100, correspond to reasonable divisions at which the quantitative increase in sample size compels a qualitative change in the kind of analysis.

For tiny datasets, one can view all the data on a single page. Computational time and storage are not an issue. This is the realm in which classical statistics began, and most problems could be resolved by intelligent inspection of the data tables.

Small datasets could well defeat tabular examination by humans, but they invite graphical techniques. Scatterplots, histograms, and other visualization techniques are quite capable of structure discovery in this realm, and modern computers permit almost any level of analytic complexity that is wanted. It is usually possible for the analyst to proceed adaptively, looking at output from one trial in order to plan the next modeling effort.

Medium datasets begin to require serious thought. They will contain many outliers, and the analyst must develop automatic rules for detecting and handling them. Visualization is still a viable way to do structure discovery, but when the data are multivariate there will typically be more scatterplots and histograms than one has time to examine individually. In some sense, this is the first level at which the entire analytical strategy must be automated, which limits the flexibility of the study. This is the lowest order in the taxonomy in which data mining applications are common.

Large datasets are difficult to visualize; for example, it is possible that the point density is so great that a scatterplot is completely black. Many common statistical procedures, such as cluster analysis or regression and classification based on smoothing, become resource-intensive or impossible.

Huge and monster datasets put data processing issues at the fore, and analytical methods become secondary. Simple operations such as averaging require only that data be read once, and these are feasible at even the largest size, but analyses that require more than $\mathcal{O}(n)$ or perhaps $\mathcal{O}(n \log(n))$ operations, for n the sample size, are impossible.

The last three taxa are a mixed blessing for data mining. The good news is that the large sample sizes mitigate the curse of dimensionality, and thus it is possible, in principle, to discover complex and interesting structure. The bad news is that many of the methods and much of the theory developed during a century of explosive growth in statistical research are impractical, given current data processing limitations.

From a computational standpoint, the three major issues in data mining are processor speed, memory size, and algorithmic complexity. Speed is proportional to the number of floating point operations (flops) that must be performed. Using PCs available in 2000, one can undertake analyses requiring about about 10^{13} flops, and perhaps up to 10^{16} flops on a supercomputer. Regarding memory, a single-processor machine needs sufficient memory for about four copies of the largest array required in the analysis, and the backup storage (disk) should be able to hold an amount equal to about 10 copies of the raw data (otherwise one has trouble storing derived calculations and exploring alternative analyses). The algorithmic complexity of the analysis is harder to quantify; at a high level it depends on the number of logical branches that the analyst wants to explore when planning the study, and at a lower level it depends on the specific numerical calculations employed by particular model fitting procedures.

Datasets based on extracting information from the World Wide Web or involving all transactions from banks or grocery stores, or collections of fMRI images can rapidly fill up terabytes of disk storage, and all of these issues become relevant. The last issue, regarding algorithmic complexity, is especially important added on top of the sheer size of some datasets. It implies that one cannot search too widely in the space of possible models, and that one cannot redirect the analysis in midstream to respond to an insight found at an intermediate step.

For more information on the issues involved in preparing superlarge datasets for analysis, see Banks and Parmigiani (1992). For additional discussion of the special problems proposed in the analysis of superlarge datasets, see the workshop proceedings on massive datasets produced by the National Research Council (1997).

II. CLASSIFICATION

Classification problems arise when one has a training sample of cases in known categories and their corresponding explanatory variables. One wants to build a decision rule that uses the explanatory variables to predict category membership for future observations. For example, a common data mining application is to use the historical records on loan applicants to classify new applicants as good or bad credit risks.

A. Methods

Classical classification began in 1935, when Sir Ronald Fisher set his children to gather iris specimens. These specimens were then classified into three species by a botanist, and numerical measurements were made on each flower. Fisher (1936) then derived mathematical formulas which used the numerical measurements to find hyperplanes that best separated the three different species.

In modern classification, the basic problem is the same as Fisher faced. One has a training sample, in which the correct classification is known (or assumed to be accurate with high probability). For each case in the training sample, one has additional information, either numerical or categorical, that may be used to predict the unknown classifications of future cases. The goal is use the information in the learning sample to build a decision function that can reliably classify future cases.

Most applications in data mining use either logistic regression, neural nets, or recursive partitioning to build the decision functions. The next three subsections describe these approaches. For information on the more traditional discriminant analysis techniques, see Press (1982).

1. Logistic Regression

Logistic regression is useful when the response variable is binary but the explanatory variables are continuous. This would be the case if one were predicting whether or not an customer is a good credit risk, using information on their income, years of employment, age, education, and other continuous variables.

In such applications one uses the model

$$P[Y = 1] = \frac{\exp(X^{\mathrm{T}}\theta)}{1 + \exp(X^{\mathrm{T}}\theta)}, \tag{1}$$

where $Y = 1$ if the customer is a good risk, X is the vector of explanatory variables for that customer, and θ are the unknown parameters to be estimated from the data. This model is advantageous because, under the transformation

$$p = \ln \frac{P[Y = 1]}{1 - P[Y = 1]}$$

one obtains the linear model $p = X^{\mathrm{T}}\theta$. Thus all the usual machinery of multiple linear regression will apply.

Logistic regression can be modified to handle categorical explanatory variables through definition of dummy variables, but this becomes impractical if there are many categories. Similarly, one can extend the approach to cases in which the response variable is polytomous (i.e., takes more than two categorical values). Also, logistic regression can incorporate product interactions by defining new explanatory variables from the original set, but this, too, becomes impractical if there are many potential interactions. Logistic regression is relatively fast to implement, which is attractive in data mining applications that have large datasets. Perhaps the chief value of logistic regression is that it provides an important theoretical window on the behavior of more complex classification methodologies (Friedman *et al.*, 2000).

2. Neural Nets

Neural nets are a classification strategy that employs an algorithm whose architecture is intended to mimic that of a brain, a strategy that was casually proposed by von Neumann. Usually, the calculations are distributed across multiple nodes whose behavior is analogous to neurons, their outputs are fed forward to other nodes, and these results are eventually accumulated into a final prediction of category membership.

To make this concrete, consider a simple perceptron model in which one is attempting to use a training set of historical data to teach the network to identify customers who are poor credit risks. Figure 1 shows a hypothetical situation; the inputs are the values of the explanatory variables and the output is a prediction of either 1 (for a good risk) or −1 (for a bad risk). The weights in the nodes are developed as the net is trained on the historical sample,

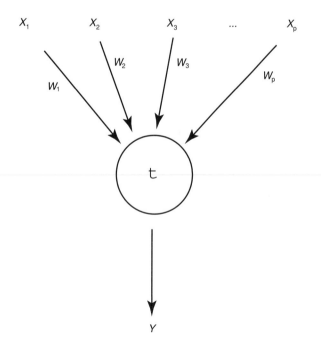

FIGURE 1 Simple perceptron.

and may be thought of as estimated values of model parameters. As diagrammed in Fig. 1, the simple perceptron fits the model

$$y = \text{signum}\left\{ \sum_{i=1}^{p} w_i x_i + \tau \right\},$$

where the weights w_i and the threshhold parameter τ are estimated from the training sample.

Unlike logistic regression, it is easy for the simple perceptron to include categorical explanatory variables such as profession or whether the applicant has previously declared bankruptcy, and there is much less technical difficulty in extending the perceptron to the prediction of polytomous outcomes. There is no natural way, however, to include product interaction terms automatically; these still require hand-tuning.

The simple perceptron has serious flaws, and these have been addressed in a number of ways. The result is that the field of neural networks has become complex and diverse. The method shown in Fig. 1 is too primitive to be commonly used in modern data mining, but it serves to illustrate the basic ideas.

As a strategy, neural networks go back to the pioneering work of McCulloch and Pitts (1943). The computational obstacles in training a net went beyond the technology then available. There was a resurgence of attention when Rosenblatt (1958) introduced the perceptron, an early neural net whose properties were widely studied in the 1960s; fundamental flaws with the perceptron design were pointed out by Minsky and Papert (1969). The area languished until the early 1980s, when the Hopfield net (Hopfield, 1982) and the discovery of the backpropagation algorithm [Rumelhart *et al.* (1985) were among several independent inventors] led to networks that could be used in practical applications. Additional information on the history and development of these ideas can be found in Ripley (1996).

The three major drawbacks in using neural nets for data mining are as follows:

1. Neural nets require such computer-intensive that fitting even Huber's large datasets can be infeasible. In practice, analysts who are intent on applying neural nets to their problems typically train the net on only a small fraction of the learning sample and then use the remainder to estimate the accuracy of their classifications.
2. They are difficult to interpret. It is hard to examine a trained network and discover which variables are most influential and what are the functional relationships between the variables. This impedes the kind of scientific insight that is especially important to data miners.

3. Neural nets do not automatically provide statements of uncertainty. At the price of greatly increased computation one can use statistical techniques such as cross-validation, bootstrapping, or jackknifing to get approximate misclassification rates or standard errors or confidence intervals.

Nonetheless, neural nets are one of the most popular tools in the data mining community, in part because of their deep roots in computer science.

Subsection IV.A.4 revisits neural nets in the context of regression rather than classification. Instead of describing the neural net methodology in terms of the nodes and connections which mimic the brain, it develops an equivalent but alternative representation of neural nets as a procedure for fitting a mathematical model of a particular form. This latter perspective is the viewpoint embraced by most current researchers in the field.

3. Recursive Partitioning

Data miners use recursive partitioning to produce decision trees. This is one of the most popular and versatile of the modern classification methodologies. In such applications, the method employs the training sample to recursively partition the set of possible explanatory measurements. The resulting classification rule can displayed as a decision tree, and this is generally viewed as an attractive and interpretable rule for inference.

Formally, recursive partitioning splits the training sample into increasingly homogeneous groups, thus inducing a partition on the space of explanatory variables. At each step, the algorithm considers three possible kinds of splits using the vector of explanatory values X:

1. Is $X_i \leq t$ (univariate split)?
2. Is $\sum_{i=1}^{p} w_i x_i \leq t$ (linear combination split)?
3. Does $x_i \in S$ (categorical split, used if x_i is a categorical variable)?

The algorithm searches over all possible values of t, all coefficients $\{w_i\}$, and all possible subsets S of the category values to find the split that best separates the cases in the training sample into two groups with maximum increase in overall homogeneity.

Different partitioning algorithms use different methods for assessing improvement in homogeneity. Some seek to minimize Gini's index of diversity, others use a "twoing rule," and hybrid methods can switch criteria as they move down the decision tree. Similarly, some methods seek to find the greatest improvement on both sides of the split, whereas other methods choose the split that achieves maximum homogeneity on one side or the other. Some meth-

ods grow elaborate trees and then prune back to improve predictive accuracy outside the training sample (this is a partial response to the kinds of overfit concerns that arise from the curse of dimensionality).

However it is done, the result is a decision tree. Figure 2 shows a hypothetical decision tree that might be built from credit applicant data. The first split is on income, a continuous variable. To the left-hand side, corresponding to applicants with incomes less than $25,000, the next split is categorical, and divides according to whether the applicant has previously declared bankruptcy. Going back the the top of the tree, the right-hand side splits on a linear combination; the applicant is considered a good risk if a linear combination of their age and income exceeds a threshold. This is a simplistic example, but it illustrates the interpretability of such trees, the fact that the same variable may be used more than once, and the different kinds of splits that can be made.

Recursive partitioning methods were first proposed by Morgan and Sonquist (1963). The method became widely popular with the advent of CART, a statistically sophisticated implementation and theoretical evaluation developed by Breiman et al. (1984). Computer scientists have also contributed to this area; prominent implementations of decision trees include ID3 and C4.5 (Quinlan, 1992). A treatment of the topic from a statistical perspective is given by Zhang and Singer (1999). The methodology extends to regression problems, and this is described in Section IV.A.5 from a model-fitting perspective.

B. Boosting

Boosting is a method invented by computer scientists to improve weak classification rules. The idea is that if one has a classification procedure that does slightly better than chance at predicting the true categories, then one can apply this procedure to the portions of the training sample that are misclassified to produce new rules and then weight all the rules together to achieve better predictive accuracy. Essentially, each rule has a weighted vote on the final classification of a case.

The procedure was proposed by Schapire (1990) and improved by Freund and Schapire (1996) under the name AdaBoost. There have been many refinements since, but the core algorithm for binary classification assumes one has a weak rule $g_1(\mathbf{X})$ that takes values in the set $\{1, -1\}$ according to the category. Then AdaBoost starts by putting equal weight $w_i = n^{-1}$ on each of the n cases in the training sample. Next, the algorithm repeats the following steps K times:

1. Apply the procedure g_k to the training sample with weights $w_1, \ldots, w_n$.

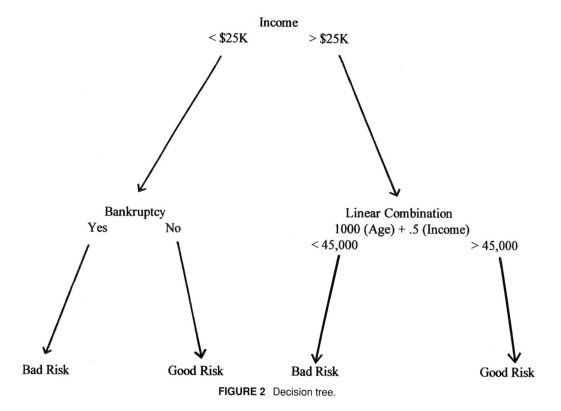

FIGURE 2 Decision tree.

2. Find the empirical probability p_w of misclassification under these weightings.
3. Calculate $c_k = \ln[(1 - p_w)/p_w]$.
4. If case i is misclassified, replace w_i by $w_i \exp c_k$.
   Then renormalize so that $\sum_i w_i = 1$ and go to step 1.

The final inference is the sign of $\sum_{k=1}^{K} c_k g_k(X)$, which is a weighted sum of the determinations made by each of the K rules formed from the original rule g_1.

This rule has several remarkable properties. Besides provably improving classification, it is also resistant to overfit, which arises when K is large. The procedure allows quick computation, and thus can be made practical even for huge datasets, and it can be generalized to handle more than two categories. Boosting therefore provides an automatic and effective way to increase the capability of almost any classification technique. As a new method, it is the object of active research; Friedman *et al.* (2000) describe the current thinking in this area, linking it to the formal role of statistical models such as logistic regression and generalized additive models.

III. CLUSTER ANALYSIS

Cluster analysis is the term that describes a collection of data mining methods which take observations and form groups that are usefully similar. One common application is in market segmentation, where a merchant has data on the purchases made by a large number of customers, together with demographic information on the customers. The merchant would like to identify clusters of customers who make similar purchases, so as to better target advertising or forecast the effect of changes in product lines.

A. Clustering Strategies

The classical method for grouping observations is hierarchical agglomerative clustering. This produces a cluster tree; the top is a list of all the observations, and these are then joined to form subclusters as one moves down the tree until all cases are merged in a single large cluster. For most applications a single large cluster is not informative, however, and so data miners require a rule to stop the agglomeration algorithm before complete merging occurs. The algorithm also requires a rule to identify which subcluster should be merged next for each stage of the tree-building process.

Statisticians have not found a universally reliable rule to determine when to stop a clustering algorithm, but many have been suggested. Milligan and Cooper (1985) described a large simulation study that included a range of realistic situations. They found that no rule dominated all the others, but that the cubic clustering criterion was rarely bad and often quite good. However, in practice, most analysts create the entire tree and then inspect it to find a point at which further linkages do not seem to further their purpose. For the market segmentation example, a merchant might be pleased to find clusters that can be interpreted as families with children, senior citizens, yuppies, and so forth, and would want to stop the clustering algorithm when further linkage would conflate these descriptive categories.

Similar diversity exists when choosing a subcluster merging rule. For example, if the point clouds associated with the final interpretable clusters appear ellipsoidal with similar shape and orientation, then one should probably have used a joining rule that connects the two subclusters whose centers have minimum Mahalanobis (1936) distance. Alternatively, if the final interpretable clusters are nonconvex point clouds, then one will have had to discover those by using some kind of nearest neighbor joining rule. Statisticians have devised many such rules and can show that no single approach can solve all clustering problems (Van Ness, 1999).

In data mining applications, most of the joining rules developed by statisticians require infeasible amounts of computation and make unreasonable assumptions about homogeneous patterns in the data. Specifically, large, complex data sets do not generally have cluster structure that is well described by sets of similar ellipsoids. Instead, data miners expect to find structures that look like sheets, ellipsoids, strings, and so forth (rather as astronomers see when looking at the large-scale structure of the universe).

Therefore, among agglomerative clustering schemes, data miners almost always use nearest neighbor clustering. This is one of the fastest clustering algorithms available, and is basically equivalent to finding a minimum spanning tree on the data. Using the Prim (1957) algorithm for spanning trees, the computation takes $\mathcal{O}(n^2)$ comparisons (where n is the number of observations). This is feasible for medium datasets in Huber's taxonomy. Furthermore, nearest neighbor methods are fairly robust at finding the diverse kinds of structure that one anticipates.

As an alternative to hierarchical agglomerative clustering, some data miners use k-means cluster analysis, which depends upon a strategy pioneered by MacQueen (1967). Starting with the assumption that the data contain a prespecified number k of clusters, this method iteratively finds k cluster centers that maximize between-cluster distances and minimize within-cluster distances, where the distance metric is chosen by the user (e.g., Euclidean, Mahalanobis, sup norm, etc.). The method is useful when one has prior beliefs about the likely number of clusters in the data. It also can be a useful exploratory tool. In the computer science community, k-means clustering is known as the quantization problem and is closely related to Voronoi tesselation.

The primary drawback to using k-means clustering in data mining applications is that exact solution requires extensive computation. Approximate solutions can be obtained much more rapidly, and this is the direction in which many researchers have gone. However, the quality of the approximation can be problematic.

Both k-means and agglomerative cluster analysis are strongly susceptible to the curse of dimensionality. For the market segmentation example, it is easy to see that customers could form tight clusters based upon their first names, or where they went to elementary school, or other misleading features that would not provide commercial insight. Therefore it is useful to do variable selection, so that clustering is done only upon features that lead to useful divisions. However, this requires input from the user on which clusters are interpretable and which are not, and encoding such information is usually impractical. An alternative is to use robust clustering methods that attempt to find a small number of variables which produce well-separated clusters. Kaufman and Rousseeuw (1990) review many ideas in this area and pay useful attention to computational issues.

For medium datasets in Huber's taxonomy, it is possible to use visualization to obtain insight into the kinds of cluster structure that may exist. This can provide guidance on the clustering algorithms one should employ. Swayne *et al.* (1997) describe software that enables data miners to see and navigate across three-dimensional projections of high-dimensional datasets. Often one can see groups of points or outliers that are important in the application.

B. Racing

One of the advances that came from the data mining synergy between statisticians and computer scientists is a technique called 'racing' (Maron and Moore, 1993). This enables analysts to do much larger searches of the space of models than was previously possible.

In the context of cluster analysis, suppose one wanted to do variable selection to discover which set of demographic features led to, say, 10 consumer clusters that had small intracluster variation and large intercluster variation. One approach would be to consider each possible subset of the demographic features, run the clustering algorithm, and then decide which set of results had the best cluster separation. Obviously, this would entail much computation.

An alternative approach is to consider many feature subsets simultaneously and run the clustering algorithm for perhaps 1% of the data on each subset. Then one compares the results to see if some subsets lead to less well-defined clusters than others. If so, those subsets are eliminated and the remaining subsets are then tested against each other on a larger fraction of the data. In this way one can weed out poor feature subset choices with minimal computation and reserve resources for the evaluation of the very best candidates.

Racing can be done with respect to a fixed fraction of the data or a fixed amount of runtime. Obviously, this admits the possibility of errors. By chance a good method might appear poor when tested with only a fraction of the data or for only a fixed amount of computer time, but the probability of such errors can be controlled statistically, and the benefit far outweighs the risk. If one is using racing on data fractions, the problem of chance error makes it important that the data be presented to the algorithm in random order, otherwise the outcome of the race might depend upon subsamples that are not representative of the entire dataset.

Racing has much broader application than cluster analysis. For classification problems, competing classifiers can be tested against each other, and those with high misclassification rates are quickly eliminated. Similarly, for regression problems, competing regression models are raced to quickly eliminate those that show poor fit. In both applications one can easily obtain a 100-fold increase in the size of the model space that is searched, and this leads to the discovery of better classifiers and better regression functions.

IV. NONPARAMETRIC REGRESSION

Regression is a key problem area in data mining and has attracted a substantial amount of research attention. Among dozens of new techniques for nonparametric regression that have been invented over the last 15 years, we detail seven that are widely used. Section IV.A describes the additive model (AM), alternating conditional expectation (ACE), projection pursuit regression (PPR), neural nets (NN), recursive partitioning regression (RPR), multivariate adaptive regression splines (MARS), and locally weighted regression (LOESS). Section IV.B compares these techniques in terms of performance and computation, and Section IV.C describes how bagging can be used to improve predictive accuracy.

A. Seven Methods

The following seven methods may seem very different, but they employ at most two distinct strategies for addressing the curse of dimensionality. One strategy fits purely local models; this is done by RPR and LOESS. The other strategy uses low-dimensional smoothing to achieve flexibility in fitting specific model forms; this is done by AM, ACE, NN, and PPR. MARS combines both strategies.

1. The Additive Model (AM)

Many researchers have developed the AM; Buja *et al.* (1989) describe the early development in this area. The

simplest AM says that the expected value of an observation Y_i can be written

$$\mathrm{E}[Y_i] = \theta_0 + \sum_{j=1}^{p} f_j(X_{ij}), \qquad (2)$$

where the functions f_j are unspecified but have mean zero.

Since the functions f_j are estimated from the data, the AM avoids the conventional statistical assumption of linearity in the explanatory variables; however, the effects of the explanatory variables are still additive. Thus response is modeled as a sum of arbitrary smooth univariate functions of explanatory variables. One needs about 100 data points to estimate each f_j, but under the model given in (1), the requirement for data grows only linearly in p.

The backfitting algorithm is the essential tool used in estimating an additive model. This algorithm requires some smoothing operation (e.g., kernel smoothing or nearest neighbor averages; Hastie and Tibshirani, 1990) which we denote by $Sm(\cdot|\cdot)$. For a large classes of smoothing operations, the backfitting algorithm converges uniquely.

The backfitting algorithm works as follows:

1. At initialization, define functions $f_j^{(0)} \equiv 0$ and set $\theta_0 = \bar{Y}$.
2. At the ith iteration, estimate $f_j^{(i+1)}$ by

$$f_j^{(i+1)} = Sm\left(Y - \theta_0 - \sum_{k \neq j} f_k^i \,\middle|\, X_{1j}, \ldots, X_{nj}\right)$$

for $j = 1, \ldots, p$.
3. Check whether $|f_j^{(i+1)} - f_j^{(i)}| < \delta$ for all $j = 1, \ldots, p$, for δ the prespecified convergence tolerance. If not, go back to step 2; otherwise, take the current $f_j^{(i)}$ as the additive function estimate of f_j in the model.

This algorithm is easy to code. Its speed is chiefly determined by the complexity of the smoothing function.

One can generalize the AM by permitting it to add a few multivariate functions that depend on prespecified explanatory variables. Fitting these would require bivariate smoothing operations. For example, if one felt that prediction of tax owed in the AM would improved by including a function that depended upon both a person's previous year's declaration and their current marital status, then this bivariate smoother could be used in the second step of the backfitting algorithm.

2. Alternating Conditional Expectations (ACE)

A generalization of the AM allows a smoothing transformation of the response variable as well the smoothing of the p explanatory variables. This uses the ACE algorithm, as developed by Breiman and Friedman (1985), and it fits the model

$$E[g(Y_i)] = \theta_0 + \sum_{j=1}^{p} f_j(X_{ij}). \qquad (3)$$

Here all conditions are as stated before for (1), except that g is an arbitrary smooth function scaled to ensure the technically necessary requirement that $\text{var}[g(Y)] = 1$ (if not for this constraint, one could get a perfect fit by setting all functions to be identically zero).

Given data Y_i and X_i, one wants to find g, θ_0, and $f_1, \ldots, f_p$ such that $E[g(Y_i) \mid X_i] - \theta_0 - \sum_{j=1}^{p} f_j(X_{ij})$ is well described as independent error. Thus one solves

$$(\hat{g}, \hat{f}_1, \ldots, \hat{f}_p) = \operatorname*{argmin}_{(g, f_1, \ldots, f_p)}$$

$$\times \left\{ \sum_{i=1}^{n} \left[g(Y_i) - \sum_{j=1}^{p} f_j(X_{ij}) \right]^2 \right\},$$

where $\hat{g}$ is constrained to satisfy the unit-variance requirement. The algorithm for achieving this is described by Breiman and Friedman (1985); they modify the backfitting algorithm to provide a step that smoothes the left-hand side while maintaining the variance constraint.

ACE analysis returns sets of functions that maximize the linear correlation between the sum of the smoothed explanatory variables and the smoothed response variable. Therefore ACE is more similar in spirit to the multiple correlation coefficient than to multiple regression. Because ACE does not directly attempt a regression analysis, it has certain undesirable features; for example, small changes in the data can lead to very different solutions (Buja and Kass, 1985), it need not reproduce model transformations, and, unlike regression, it treats the explanatory and response variables symmetrically.

To redress some of the drawbacks in ACE, Tibshirani (1988) devised a modification called AVAS, which uses a variance-stabilizing transformation in the backfitting loop when fitting the explanatory variables. This modification is somewhat technical, but in theory it leads to improved properties when treating regression applications.

3. Projection Pursuit Regression (PPR)

The AM uses sums of functions whose arguments are the natural coordinates for the space $\mathbb{R}^p$ of explanatory variables. But when the true regression function is additive with respect to pseudovariables that are linear combinations of the explanatory variables, then the AM is inappropriate. PPR was developed by Friedman and Stuetzle (1981) to address such situations.

Heuristically, imagine there are two explanatory variables and suppose the regression surface is shaped like a sheet of corrugated aluminum. If that sheet is oriented to make the corrugations parallel to the axis of the first explanatory variable (and thus perpendicular to the second), then AM works well. When the aluminum sheet is rotated slightly so that the corrugations do not parallel a natural axis, however, AM fails because the true function is a nonadditive function of the explanatory variables. PPR would succeed, however, because the true function can be written as an additive model whose functions have arguments that are linear combinations of the explanatory variables.

PPR combines the backfitting algorithm with a numerical search routine, such as Gauss–Newton, to fit models of the form

$$E[Y_i] = \sum_{k=1}^{r} f_k(\alpha_k' X_i). \qquad (4)$$

Here the $\alpha_1, \ldots, \alpha_r$ are unit vectors that define a set of r linear combinations of explanatory variables. The linear combinations are similar to those used for principal components analysis (Flury, 1988). These vectors need not be orthogonal, and are chosen to maximize predictive accuracy in the model as estimated by cross-validation.

Operationally, PPR alternates calls to two routines. The first routine conditions on a set of pseudovariables given by linear combinations of original variables; these are fed into the backfitting algorithm to obtain an AM in the pseudovariables. The other routine conditions on the estimated AM functions from the previous step and then searches to find linear combinations of the original variables which maximize the fit of those functions. By alternating iterations of these routines, the result converges to a unique solution.

PPR is often hard to interpret for $r > 1$ in (3). When r is allowed to increase without bound, PPR is consistent, meaning that as the sample size grows, the estimated regression function converges to the true function.

Another improvement that PPR offers over AM is this it is invariant to affine transformations of the data; this is often desirable when the explanatory variables are measured in the same units and have similar scientific justifications. For example, PPR might be sensibly used to predict tax that is owed when the explanatory variables are shares of stock owned in various companies. Here it makes sense that linear combinations of shares across commercial sectors would provide better prediction of portfolio appreciation than could be easily obtained from the raw explanatory variables.

4. Neural Nets (NN)

Many neural net techniques exist, but from a statistical regression standpoint (Barron and Barron, 1988), nearly all variants fit models that are weighted sums of sigmoidal functions whose arguments involve linear combinations of the data. A typical feedforward network uses a model of the form

$$E[Y] = \beta_0 + \sum_{i=1}^{m} \beta_i f\left(\alpha_i^T x + \gamma_{i0}\right),$$

where $f(\cdot)$ is a logistic function and the β_0, γ_{i0}, and α_i are estimated from the data. Formally, this approach is similar to that in PPR. The choice of m determines the number of hidden nodes in the network and affects the smoothness of the fit; in most cases the user determines this parameter, but it is also possible to use statistical techniques, such as cross-validation to assess model fit, that allow m to be estimated from the data.

Neural nets are widely used, although their performance properties, compared to alternative regression methods, have not been thoroughly studied. Ripley (1996) describes one assessment which finds that neural net methods are not generally competitive. Schwarzer *et al.* (1986) review the use of neural nets for prognostic and diagnostic classification in clinical medicine and reach similar conclusions. Another difficulty with neural nets is that the resulting model is hard to interpret. The Bayesian formulation of neural net methods by Neal (1996) provides a some remedy for this difficulty.

PPR is very similar to neural net methods. The primary difference is that neural net techniques usually assume that the functions f_k are sigmoidal, whereas PPR allows more flexibility. Zhao and Atkeson (1992) show that PPR has similar asymptotic properties to standard neural net techniques.

5. Recursive Partitioning Regression (RPR)

RPR has become popular since the release of CART (Classification and Regression Tree) software developed by Breiman *et al.* (1984). This technique has already been described in the context of classification, so this subsection focuses upon its application to regression. The RPR algorithm fits the model

$$E[Y_i] = \sum_{j=1}^{M} \theta_j I_{R_j}(X_i),$$

where the $R_1, \ldots, R_M$ are rectangular regions which partition $\mathbb{R}^p$, and $I_{R_j}(X_i)$ denotes an indicator function that takes the value 1 if and only if $X_i \in R_j$ and is otherwise zero. Here θ_j is the estimated numerical value of all responses with explanatory variables in R_j.

RPR was intended to be good at discovering local low-dimensional structure in functions with high-dimensional global dependence. RPR is consistent; also, it has an attractive graphic representation as a decision tree, as il-

lustrated for classification in Fig. 2. Many common functions are difficult for RPR, however; for example, it approximates a straight line by a stairstep function. In high dimensions it can be difficult to discover when the RPR piecewise constant model closely approximates a simple smooth function.

To be concrete, suppose one used RPR to predict tax obligation. The algorithm would first search all possible splits in the training sample observations and perhaps divide on whether or not the declared income is greater than $25,000. For people with lower incomes, the next search might split the data on marital status. For those with higher incomes, subsequent split might depend on whether the declared income exceeds $75,000. Further splits might depend on the number of children, the age and profession of the declarer, and so forth. The search process repeats in every subset of the training data defined by previous divisions, and eventually there is no potential split that sufficiently reduces variability to justify further partitions. At this point RPR fits the averages of all the training cases within the most refined subsets as the estimates θ_j and shows the sequence of chosen divisions in a decision tree. (Note: RPR algorithms can be more complex; e.g., CART "prunes back" the final tree by removing splits to achieve better balance of observed fit in the training sample with future predictive error.)

6. Multivariate Adaptive Regression Splines (MARS)

Friedman (1991) proposed a data mining method that combines PPR with RPR through use of multivariate adaptive regression splines. It fits a model formed as a weighted sum of multivariate spline basis functions (tensor-spline basis functions) and can be written as

$$E[Y_i] = \sum_{k=0}^{q} a_k B_k(X_i),$$

where the coefficients a_k are estimated by (generalized) cross-validation fitting. The constant term is obtained by setting $B_0(X_1, \ldots, X_n) \equiv 1$, and the multivariate spline terms are products of univariate spline basis functions:

$$B_k(x_1, \ldots, x_n) = \prod_{s=1}^{r_k} b\left(x_{i(s,k)} \mid t_{s,k}\right), \qquad 1 \le k \le r.$$

The subscript $i(s, k)$ identifies a particular explanatory variable, and the basis spline for that variable puts a knot at $t_{s,k}$. The values of q, the $r_1, \ldots, r_q$, the knot locations, and the explanatory variables selected for inclusion are determined from the data adaptively.

MARS output can be represented and interpreted in a decomposition similar to that given by analysis of variance. It is constructed to work well if the true function has local dependence on only a few variables. MARS can automatically accommodate interactions between variables, and it handles variable selection in a natural way.

7. Locally Weighted Regression (LOESS)

Cleveland (1979) developed a technique based on locally weighted regression. Instead of simply taking a local average, LOESS fits the model $E[Y] = \theta(x)'x$, where

$$\hat{\theta}(x) = \underset{\theta \in \mathbb{R}^p}{\operatorname{argmin}} \sum_{i=1}^{n} w_i(x)(Y_i - \theta'X_i)^2 \qquad (5)$$

and w_i is a weight function that governs the influence of the ith datum according to the direction and distance of X_i from x.

LOESS is a consistent estimator but may be inefficient at finding relatively simple structures in the data. Although not originally intended for high-dimensional regression, LOESS uses local information with advantageous flexibility. Cleveland and Devlin (1988) generalized LOESS to perform polynomial regression rather than the linear regression $\theta'X_i$ in (4), but from a data mining perspective, this increases the cost of computation with little improvement in overall predictive accuracy.

B. Comparisons

Nonparametric regression is an important area for testing the performance of data mining methods because statisticians have developed a rich theoretical understanding of the issues and obstacles. Some key sources of comparative information are as follows:

- Donoho and Johnstone (1989) make asymptotic comparisons among the more mathematically tractable techniques. They indicate that projection-based methods (PPR, MARS) perform better for radial functions but kernel-based methods (LOESS) are superior for harmonic functions. (Radial functions are constant on hyperspheres centered at $\mathbf{0}$, while harmonic functions vary sinusoidally on such hyperspheres.)
- Friedman (1991) describes simulation studies of MARS, and related work is cited by his discussants. Friedman focuses on several criteria; these include scaled versions of mean integrated squared error (MISE) and predictive-squared error (PSE). From the standpoint of data mining practice, the most useful conclusions are (1) for data that are pure noise in 5 and 10 dimensions, and sample sizes of 50, 100, and 200, AM and MARS are comparable and both are unlikely to find false structure, and (2) if test data are generated from the following additive function of five variables

$$Y = 0.1 \exp(4X_1) + \frac{4}{1 + \exp(-20X_2 + 10)}$$
$$+ 3X_3 + 2X_4 + X_5,$$

with five additional noise variables and sample sizes of 50, 100, and 200, MARS had a slight tendency to overfit, particularly for the smallest sample sizes.

All of these simulations, except for the pure noise condition, have large signal-to-noise ratios.

- Barron (1991, 1993) proved that in a narrow sense, the MISE for neural net estimates in a certain class of functions has order $\mathcal{O}(1/m) + \mathcal{O}(mp/n) \ln n$, where m is the number of nodes, p is the dimension, and n is the sample size. Since this is linear in dimension, it evades the COD; similar results have been obtained by Zhao and Atkeson (1992) for PPR, and it is probable that a similar result holds for MARS. These findings have limited practical value; Barron's class of functions consists of those whose Fourier transform $\tilde{g}$ satisfies $\int |\omega| |\tilde{g}(\omega)| \, d\omega < c$ for some fixed c. This excludes such simple cases as hyperflats, and the class becomes smoother as dimension increases.
- De Veaux *et al.* (1993) tested MARS and a neural net on two functions; they found that MARS was faster and had better MISE.
- Zhang and Singer (1999) applied CART, MARS, multiple logistic regression, and other methods in several case studies and found that no method dominates the others. CART was often most interpretable, but logistic regression led more directly to estimation of uncertainty.

The general conclusions are (1) parsimonious modeling is increasingly important in high dimensions, (2) hierarchical models using sums of piecewise-linear functions are relatively good, and (3) for any method, there are datasets on which it succeeds and datasets on which it fails.

In practice, since it is hard to know what methods works best in a given situation, data miners usually hold out a part of the data, apply various regression techniques to the remaining data, and then use the models that are built to estimate the hold-out sample. The method that achieves minimum prediction error is likely to be the best for that application.

Most of the software detailed in this section is available from the StatLib archive at `http://lib.stat.cmu.edu`; this includes AM, ACE, LOESS, and PPR. Both MARS and CART are commercially available from Salford Systems, Inc., at `http://www.salford-systems.com`. Splus includes versions of RPR, the (generalized) AM, ACE, PPR, and LOESS; this is commercially available from Mathsoft, Inc., at `http://www.splus.mathsoft.com`.

C. Bagging

Bagging is a strategy for improving predictive accuracy by model averaging. It was proposed by Breiman (1996), but has a natural pedigree in Bayesian work on variable selection, in which one often puts weights on different possible models and then lets the data update those weights.

Concretely, suppose one has a training sample and a nonparametric regression technique that takes the explanatory variables and produces an estimated response value. Then the simplest form of bagging proceeds by drawing K random samples (with replacement) from the training sample and applying the regression technique to each random sample to produce regression rules $T_1(X), \ldots, T_K(X)$. For a new observation, say X^*, the bagging predictor of the response Y^* is $K^{-1} \sum_{k=1}^{K} T_k(X^*)$.

The idea behind bagging is that model fitting strategies usually have high variance but low bias. This means that small changes in the data can produce very different models but that there is no systematic tendency to produce models which err in particular directions. Under these circumstances, averaging the results of many models can reduce the error in the prediction that is associated with model instability while preserving low bias.

Model averaging strategies are moving beyond simple bagging. Some employ many different kinds of regression techniques rather than just a single method. Others modify the bagging algorithm in fairly complex ways, such as arcing (Brieman, 1998). A nice comparison of some of the recent ideas in this area is given by Dietterich (1998), and Hoeting *et al.* (1999) give an excellent tutorial on more systematic Bayesian methods for model averaging. Model averaging removes the analyst's ability to interpret parameters in the models used and can only be justified in terms of predictive properties.

SEE ALSO THE FOLLOWING ARTICLES

ARTIFICIAL NEURAL NETWORKS ● DATABASES ● DATA STRUCTURES ● INFORMATION THEORY ● STATISTICS, BAYESIAN ● STATISTICS, MULTIVARIATE

BIBLIOGRAPHY

Banks, D. L., and Parmigiani, G. (1992). "Preanalysis of superlarge data sets," *J. Quality Technol.* **24,** 930–945.

Barron, A. R. (1991). "Complexity regularization with aapplications to artificial neural networks." *In* "Nonparametric Functional Estimation" (G. Roussas, ed.), pp. 561–576, Kluwer, Dordrecht.

Barron, A. R. (1993). "Universal approximation bounds for superposi-tions of a sigmoidal function," *IEEE Trans. Information Theory* **39,** 930–945.

Barron, A. R., and Barron, R. L. (1988). "Statistical learning networks: A unifying view," *Comput. Sci. Stat.*, **20,** 192–203.

Bay, S. D. (1999). "The UCI KDD Archive," http://kdd.ics.uci.edu. Department of Information and Computer Science, University of California, Irvine, CA.

Breiman, L. (1996). "Bagging predictors," *Machine Learning* **26,** 123–140.

Breiman, L. (1998). "Arcing classifiers," *Ann. Stat.* **26,** 801–824.

Breiman, L., and Friedman, J. (1985). "Estimating optimal transforma-tions for multiple regression and correlation," *J. Am. Stat. Assoc.* **80,** 580–619.

Breiman, L., Friedman, J., Olshen, R.A., and Stone, C. (1984). "Classi-fication and Regression Trees," Wadsworth, Belmont, CA.

Buja, A., and Kass, R. (1985). "Discussion of 'Estimating optimal trans-formations for multiple regression and correlation,' by Breiman and Friedman," *J. Am. Stat. Assoc.* **80,** 602–607.

Buja, A., Hastie, T. J., and Tibshirani, R. (1989). "Linear smoothers and additive models," *Ann. Stat.* **17,** 453–555.

Cleveland, W. (1979). "Robust locally weighted regression and smooth-ing scatterplots," *J. Am. Stat. Assoc.* **74,** 829–836.

Cleveland, W., and Devlin, S. (1988). "Locally weighted regression: An approach to regression analysis by local fitting," *J. Am. Stat. Assoc.* **83,** 596–610.

Cowell, R. G., Dawid, A. P., Lauritzen, S. L., and Spiegelhalter, D. G. (1999). "Probabilistic Networks and Expert Systems," Springer-Verlag, New York.

Deitterich, T. (1998). "An experimental comparison of three methods for constructing ensembles of decision trees: Bagging, boosting, and randomization," *Machine Learning* **28,** 1–22.

De Veaux, R. D., Psichogios, D. C., and Ungar, L. H. (1993). "A com-parison of two nonparametric estimation schemes: MARS and neural networks," *Computers Chem. Eng.* **17,** 819–837.

Donoho, D. L., and Johnstone, I. (1989). "Projection based approxima-tion and a duality with kernel methods," *Ann. Stat.* **17,** 58–106.

Fisher, R. A. (1936). "The use of multiple measurements in taxonomic problems," *Ann. Eugen.* **7,** 179–188.

Flury, B. (1988). "Common Principal Components and Related Multi-variate Models," Wiley, New York.

Freund, Y., and Schapire, R. E. (1996). "Experiments with a new boost-ing algorithm." *In* "Machine Learning: Proceedings of the Thirteenth International Conference," pp. 148–156, Morgan Kaufmann, San Mateo, CA.

Friedman, J. H. (1991). "Multivariate additive regression splines," *Ann. Stat.* **19,** 1–66.

Friedman, J. H., and Stuetzle, W. (1981). "Projection pursuit regression," *J. Am. Stat. Assoc.* **76,** 817–23.

Friedman, J. H., Hastie, T., and Tibshirani, R. (2000). "Additive logistic regression: A statistical view," *Ann. Stat.* **28,** 337–373.

Glymour, C., Madigan, D., Pregibon, D., and Smyth, P. (1997). "Statis-tical themes and lessons for data mining," *Data Mining Knowledge Discovery* **1,** 11–28.

Hastie, T. J., and Tibshirani, R. J. (1990). "Generalized Additive Models," Chapman and Hall, New York.

Hoeting, J. A., Madigan, D., Raftery, A. E., and Volinsky, C. T. (1999). "Bayesian model averaging: A tutorial," *Stat. Sci.* **14,** 382–417.

Hopfield, J. J. (1982). "Neural networks and physical systems with emer-gent collective computational abilities," *Proc. Natl. Acad. Sci. USA* **79,** 2554–2558.

Huber, P. J. (1994). "Huge data sets." *In* "Proceedings of the 1994 COMP-STAT Meeting," (R. Dutter and W. Grossmann, eds.), pp. 221–239, Physica-Verlag, Heidelberg.

Jordan, M. I., (ed.). (1998). "Learning in Graphical Models," MIT Press, Cambridge, MA.

Kaufman, L., and Rousseeuw, P. J. (1990). "Finding Groups in Data: An Introduction to Cluster Analysis," Wiley, New York.

MacQueen, J. (1967). "Some methods for classification and analysis of multivariate observations." In "Proceedings of the Fifth Berkeley Symposium on Mathematical Statistics and Probability," pp. 281–297, University of California Press, Berkeley, CA.

Mahalanobis, P. C. (1936). "On the generalized distance in statistics," Proc. Natl. Inst. Sci. India 12, 49–55.

Maron, O., and Moore, A. W. (1993). "Hoeffding races: Accelerating model selection search for classification and function approximation." In "Advances in Neural Information Processing Systems 6," pp. 38–53, Morgan Kaufmann, San Mateo, CA.

McCulloch, W. S., and Pitts, W. (1943). "A Logical calculus of the ideas immanent in nervous activity," Bull. Math. Biophys. 5, 115–133.

Milligan, G. W., and Cooper, M. C. (1985). "An examination of procedures for determining the number of clusters in a dataset," Psychometrika 50, 159–179.

Minsky, M., and Papert, S. A. (1969). "Perceptrons: An Introduction to Computational Geometry," MIT Press, Cambridge, MA.

Morgan, J. N., and Sonquist, J. A. (1963). "Problems in the analysis of survey data and a proposal," J. Am. Stat. Assoc. 58, 415–434.

National Research Council. (1997). "Massive Data Sets: Proceedings of a Workshop," National Academy Press, Washington, DC.

Neal, R. (1996). "Bayesian Learning for neural networks," Springer-Verlag, New York.

Pearl J. (1982). "Causality," Cambridge University Press, Cambridge.

Pearl, J. (2000). "Causality: Models, Reasoning and Inference," Cambridge University Press, Cambridge.

Press, S. J. (1982). "Applied Multivariate Analysis: Using Bayesian and Frequentist Methods of Inference," 2nd ed., Krieger, Huntington, NY.

Prim, R. C. (1957). "Shortest connection networks and some generalizations." Bell Syst. Tech. J. 36, 1389–1401.

Quinlan, J. R. (1992). "C4.5 : Programs for Machine Learning," Morgan Kaufmann, San Mateo, CA.

Ripley, B. D. (1996). "Pattern Recognition and Neural Networks," Cambridge University Press, Cambridge.

Rosenblatt, F. (1958). "The perceptron: A probabilistic model for information storage and organization in the brain," Psychol. Rev. 65, 386–408.

Roweis, S. T., and Saul, L. K. (2000). "Nonlinear dimensionality reduction by local linear embedding," Science 290, 2323–2326.

Rumelhart, D., Hinton, G. E., and Williams, R. J. (1986). "Learning representations by back-propagating errors," Nature 323, 533–536.

Schapire, R. E. (1990). "The strength of weak learnability," Machine Learning 5, 197–227.

Schwarzer, G., Vach, W., and Schumacher, M. (1986). "On misuses of artificial neural networks for prognostic and diagnostic classification in oncology," Stat. Med. 19, 541–561.

Scott, D. W., and Wand, M. P. (1991). "Feasibility of multivariate density estimates," Biometrika 78, 197–206.

Spirtes, P., Glymour, C., and Scheines, R. (2001). "Causation, Prediction, and Search," 2nd ed., MIT Press, Cambridge, MA.

Stein, M. L. (1987). "Large sample properties of simulations using latin hypercube sampling," Technometrics 29, 143–151.

Swayne, D. F., Cook, D., and Buja, A. (1997). "XGobi: Interactive dynamic graphics in the X window system," J. Comput. Graphical Stat. 7, 113–130.

Tennenbaum, J. B., de Silva, V., and Langford, J. C. (2000). "A global geometric framework for nonlinear dimensionality reduction," Science 290, 2319–2323.

Tibshirani, R. (1988). "Estimating optimal transformations for regression via additivity and variance stabilization," J. Am. Stat. Assoc. 83, 394–405.

Van Ness, J. W. (1999). "Recent results in clustering admissibility." In "Applied Stochastic Models and Data Analysis," (H. Bacelar-Nicolau, F. Costa Nicolau, and J. Janssen, eds.), pp. 19–29, Instituto Nacional de Estatistica, Lisbon, Portugal.

Vapnik, V. N. (2000). "The Nature of Statistical Learning," 2nd ed., Springer-Verlag, New York.

Zhang, H., and Singer, B. (1999). "Recursive Partitioning in the Health Sciences," Springer-Verlag, New York.

Zhao, Y., and Atkeson, C. G. (1992). "Some approximation properties of projection pursuit networks." In "Advances in Neural Information Processing Systems 4" (J. Moody, S. J. Hanson, and R. P. Lippmann, eds.), pp. 936–943, Morgan Kaufmann, San Mateo, CA.

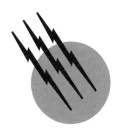

Data Structures

Allen Klinger

University of California, Los Angeles

GLOSSARY

Algorithm Regular procedure (like a recipe) that terminates and yields a result when it is presented with input data.

Binary search tree Data structure used in a search.

Binary tree Tree in which each entry has no, one, or two successors; a data structure that is used to store many kinds of other data structures.

Bit Binary digit.

Circular linkage Pointers permitting more than one complete traversal of a data structure.

Data structure An abstract idea concerning the organization of records in computer memory; a way to arrange data in a computer to facilitate computations.

Deque Double-ended queue (inputs and outputs may be at both ends in the most general deques).

Double linkage Pointers in both directions within a data structure.

Field Set of adjoining bits in a memory word; grouped bits treated as an entity.

Graph Set of nodes and links.

Hash (hashing) Process of storing data records in a disorderly manner; the hash function calculates a key and finds an approximately random table location.

Heap Size-ordered tree; all successor nodes are either (consistently) smaller or larger than the start node.

Key Index kept with a record; the variable used in a sort.

Linear list One-dimensional data set in which relative order is important. In mathematical terms, linear list elements are totally ordered.

Link Group of bits that store an address in (primary, fast) memory.

Linked allocation Method for noncontiguous assignment of memory to a data structure; locations or address for the next element stored in a part of the current word.

List Treelike structure useful in representing recursion.

Minimal spanning tree Least sum-of-link path that connects all nodes in a graph with no cycles (closed circuits).

Node Element of a data structure consisting of one or more memory words. Also, an entity in a graph connected to others of its kind by arcs or links.

Pointer See link.

Port Location where data enter or exit a data structure.

Priority queue Queue where each item has a key that governs output; replacing physical position of records in the data structure by their key values.

Quad-tree (also quadtree) A tree where every node has four or fewer successor elements; abbreviated form of quaternary tree.

Queue Linear list with two ports, one for inputs; the other for outputs; data structure supporting algorithms that operate in a first-in–first-out manner.

Recursion Repeated use of a procedure by itself.

Ring Circularly linked data structure.

Search Operation of locating a record or determining its absence from a data structure.

Sequential allocation Assignment of memory by contiguous words.

Sort Permutation of a set of records to put them into a desired order.

Stack One-port linear list that operates in a last-in–first-out manner. This structure is heavily used to implement recursion.

Tag Bit used to signal whether a pointer is as originally intended; if the alternative, the pointer is a thread.

Thread Alternate use for a pointer in a data structure.

Tree Hierarchical data structure.

Trie Tree data structure employing (1) repeated subscripting, and (2) different numbers of node successors. A data structure that is particularly useful for multiway search; structure from information re*trie*val; useful means for search of linguistic data.

DATA STRUCTURES are conceptual tools for planning computer solution of a problem. For a working definition of the term use *the way information is organized in computer memory*. The memory-organization approach taken is closely intertwined with the algorithms that can be designed. That makes *algorithms* (or *analysis of algorithms*) closely related to *data structures*. In practice, computer programs implement algorithms. Data structures have an impact on computer operation. They determine available memory space for other programs and fix the running time of their own program and the ability of other routines to function.

Data structures are places to put useful things inside a computer. The choice of one kind rather than another is a process of arranging key information to enable its use for some purpose. Another way to define data structures is as practical ways to implement computation on digital computers, taking account of characteristics of the hardware, operating system, and programming language.

Data structures are ways of describing relationships among entities. Recognizing a relationship and its properties helps in writing computer programs. Many ways to arrange data are related to, or are like, well-known physical relationships between real objects. Examples include people in business situations or families, books on a shelf, newspaper pages, and cars on a highway. Sometimes data structures are dealt with in combination with all their implementation details. This combination is called an *abstract data type*.

There is a close connection between kinds of data structures and algorithms, one so close that in some cases it is equivalence. An algorithm is *a regular procedure to perform calculations*. This is seen in examples such as a search for a name in a list, as in using phone directories. A directory involves data-pairs (name, telephone number)— usually alphabetically organized by initial letter of the last name. There is a practical usefulness of alphabetization. As with any data structuring, this idea and its implementation enable use of a simpler algorithm. A central data structuring notion is arrangement, particularly in the sense *systematic* or *ordered*. When data are disordered, the algorithm would need to read every record until the one wanted was found. As in this example, data structure selection is an essential choice governing how computers are to be used in practice. Data structures are fundamental to all kinds of computing situations.

Descriptions of the simplest data structure entities and explanations of their nature follow in succeeding sections. Basic data structures are *stack, queue*, and other *linear lists; multiple-dimension arrays; (recursive) lists*; and *trees* (including *forests* and *binary trees*). *Pointer or link* simply means computer data constituting a memory location. *Level* indicates position in a structure that is hierarchical. Link, level, and the elementary structures are almost intuitive concepts. They are fairly easily understood by reference to their names or to real-life situations to which they relate. Evolving computer practice has had two effects. First, the impact of the World Wide Web and Internet browsers has acquainted many computer users with two basic ideas: *link (pointer)* and *level*. Second, computer specialists have increased their use of *advanced data structures*. These may be understandable from their names or descriptive properties. Some of these terms are *tries, quad-trees (quadtrees, quaternary trees), leftist-trees, 2–3 trees, binary search trees*, and *heap*. While they are less common data structures and unlikely to be part of a first course in the field, they enable algorithmic procedures in applications such as image transmission, geographic data, and library search.

The basic data structure choice for any computing task involves use of either (1) *reserved contiguous memory*, or (2) *links, pointers*—locators for related information elements. *Data structuring* concerns deciding among methods that possess value for *some* information arrangements. Often the decision comes down to a mathematical analysis that addresses how frequently one or another kind of data is present in a given problem. The data structure may need to be chosen to deal with any of several issues, for example: (1) things to be computed, (2) characteristics of the programming language, (3) aspects of the operating system, (4) available digital hardware, and (5) ways that results may be used. Thus, *data structures* is a technical term that covers: (1) practical aspects of the computing domain needed to effectively and efficiently use computers, and (2) theoretical aspects involving mathematical analysis issues.

I. INTRODUCTION

Data structures make it easier for a programmer to decide and state how a computing machine will perform a given task. To actually execute even such a simple algorithm as multiplication of two numbers with whole and fractional parts (as $2\frac{1}{4}$ times $3\frac{1}{3}$) using a digital device, a suitable data structure must be chosen since the numeric information must be placed in the machine somehow. How this is done depends on the representation used in the data structure—for example, the number of fields (distinct information entities in a memory word) and their size in binary digits or bits. Finally, the actual program must be written as a step-by-step procedure to be followed involving the actual operations on the machine-represented data. (These operations are similar to those a human would do as a data processor performing the same kind of algorithm.)

The topic of data structures fits into the task of programming a computer as mathematical reasoning leads from word descriptions of problems to algebraic statements. (Think of "*rate* times *time* equals *distance*" problems.) In other words, data structure comparisons and choices resemble variable definitions in algebra. Both must be undertaken in the early stages of solving a problem. Data structures are a computer-oriented analogy of "let *x* be the unknown" in algebra; their choice leads to the final design of a program system, as the assignment of *x* leads to the equation representing a problem.

The first three sections of this article introduce basic data structure tools and their manipulation, including the concepts of sequential and linked allocation, nodes and fields, hierarchies or trees, and ordered and nonordered storage in memory. The section on elementary and advanced structures presents stacks, queues, and arrays,

as well as inverted organization and multilists. The final sections deal with using data structures in algorithm design. Additional data structure concepts, including binary search trees, parent trees, and heaps, are found in these sections, as is a survey of sorting and searching algorithms.

II. MEMORY ALLOCATION AND ALGORITHMS

A rough data structure definition is a manner for recording information that permits a program to execute a definite method of movement from item to item. An analysis is needed before programming can be done to decide which data structure would be the best choice for the situation (data input, desired result).

Examples of tasks we could want the computer to perform might include:

1. Selection of individuals qualified for a task (typist, chemist, manager)
2. Output of individuals with a common set of attributes (potential carpool members for certain regions, bridge players vested in the company retirement plan and over 50 years in age, widget-twisters with less than two years of seniority)
3. Output of individuals with their organizational relationships (division manager, staff to division manager, department director, branch leader, section chief, group head, member of technical staff)

In data structure terms, all these outputs are lists, but the last item is more. In ordinary terms, it is an *organization chart*. That is a special kind of data structure possessing *hierarchy*. As a data structure, the relationship possesses a hierarchical quality so it is not just a list but also a *tree*. Trees can be represented by a linear list data structure through the use of a special place marker that signifies *change hierarchy level*. In text representation, either a period or a parenthesis is used for that purpose. Whether a data structure is a list or is represented by a list is another of the unusual aspects of this field of knowledge. Nevertheless, these are just two aspects of the problem being dealt with, specifically:

1. The nature of the physical data relationship
2. The way the data may be presented to the computer

Two kinds of storage structures are familiar from everyday life: *stack* and *queue*. Real examples of queues are waiting lines (banks, post office), freeway traffic jams, and so forth. Physical stacks include piles of playing cards in solitaire and rummy games and unread material on desks.

Data structure methods help programs get more done using less space. Storage space is valuable—there is only a limited amount of *primary computer storage (random access, fast,* or *core* memory; RAM stands for *random access memory*). Programs have been rewritten many times, often to more efficiently use fast memory. Many techniques evolved from memory management programming lore, some from tricks helpful in speeding the execution of particular algorithms. Peripheral memory devices are commonly referred to as *external or secondary storage.* They include compact-disc read-only memory (*CD-ROM, disks, drums,* and *magnetic tape*). Such devices enable retaining and accessing vast volumes of data. The use of secondary stores is an aspect of advanced data structuring methodology often dealt with under *file management* or *database management.* This article is mainly concerned with fast memory (primary storage).

Improvements in storage use or in execution speed occur when program code is closely matched to machine capabilities. Data structure ideas do this by using special computer hardware and software features. This can involve more full use of *bits* in computer words, *fields* of words in primary storage, and *segments* of files in secondary storage.

Data structure ideas can improve algorithm performance in both the static (space) and dynamic (time) aspects of memory utilization. The static aspects involve the actual memory allocation. The dynamic aspects involve the evolution of this space over time (measured in central processor unit, or CPU, cycles) as programs execute. Note that algorithms act on the data structure. Program steps may simply access or read out a data value or rearrange, modify, or delete stored information. But, they also may combine or eliminate partially allotted sets of memory words. Before-and-after diagrams help in planning dynamic aspects of data structures. Ladder diagrams help in planning static allocations of fast memory. Memory allocation to implement a data structure in available computer storage begins with the fundamental decision whether to use linked or sequential memory. Briefly, linked allocation stores wherever space is available but requires advanced reservation of space in each memory word for pointers to show where to go to get the next datum in the structure. Memory that is sequentially allocated, by contrast, requires reservation of a fixed-sized block of words, even though many of these words may remain unused through much of the time an algorithm is being followed as its program executes.

We now present an example that illustrates several other data structure concepts; the notions of *arrays* (tables) and *graphs,* which we will describe below, are used in handling scheduling problems.

1. Example (Precedence)

A room has several doors. Individuals arrive at the room at different times. This information can be arranged in several ways. Listing individuals in arrival order is one possibility. Another is listing individuals with their arrival times—that is, in a table. A graph is yet another way; see Fig. 1.

A graph can represent the precedence relations via its physical structure (its nodes and its edges), as in Fig. 1. The arrows indicate "successor-of-a-node." J(ones) is no node's successor (was first); Sh(arp) has no successors (was last). G(reen) and B(rown) arrived simultaneously after S(mith) and before Sh(arp).

Computer storage of precedence information via a table leads to a linear list data structure. The alternative, the graph, could be stored in numeric form as node pairs. The arrows in the graph become locators (links, pointers) to data words in computer memory. Ordinal information (first, second, etc.) also could be stored in the data words and could facilitate certain search algorithms. Thus, alternatives exist for presenting facts about time or order of each individual's entry. Decisions to be made about those facts are derived from the data structure contents. One possible data structure is a table giving entry times. Another is a graph showing entry order. Since order can be derived from time data, an algorithm that operates on the data in the table (remove two entries, subtract their time values, etc.) can give information that is explicit in the other data structure via the graph arrows (or locator in computer memory). This is a general property of data structures. That is, data represented can shorten computations. It is also true that computing can reduce storage needs.

Priority of arrival is analogous to scheduling such things as jobs to build a house or courses to complete a degree. Computer algorithms that take advantage of data structure properties are useful tools for planning complex schedules. Efficient scheduling algorithms combine elementary data structures with the fact that graph elements without predecessors can be removed while preserving priorities.

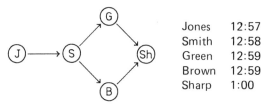

Jones	12:57
Smith	12:58
Green	12:59
Brown	12:59
Sharp	1:00

FIGURE 1 Precedence graph and table of arrival times.

A *node* is the smallest elementary data structure unit. Physical nodes are contiguous words in fast memory. It is useful to think of the realization of a data structure as a network of nodes embedded in memory, possibly connected by links. Parts of nodes and portions of memory words referred to as entities by programs are called *fields*. A one-bit field is called a *flag*, or sometimes a *tag*.

Simple sequential assignment of memory can implement *list structures*. Three elementary data structures are illustrated as follows by list examples. A pile of papers on a desk is a stack. A group of cars in a one-lane traffic bottleneck is a queue. A collection of candies can be a bag. Actually, in computer terms the data objects (papers, cars, and candies) are stored in the named data structure, and all three are generally called *linear lists*. The data type (e.g., real, integer, complex) is another attribute that must be noted in planning the data structure. It can change the space needed by each node (complex values require more bits).

An array is a multidimensional (not linear) data structure. An appointment schedule for the business week, hour by hour and day by day, is an example of an array. A mathematical function can generate data for an array structure. For example, the four-dimensional array shown in Table I (with entries in exponential notation) was obtained by substituting numeric values for x, y, w, and z in the expression:

$$w = x^{3.5} + 2y^z$$

The exponential notation for numerical items displayed here is also called *floating point*. The notation is useful in this table because there is a wide range of values for the w variable for similar x, y, and z.

In sequential allocation of storage, data are grouped together one memory word after another within a given structure. In linked allocation, a field or node part, usually a section of one memory word, is used to store the next location. Reserving more memory than is currently required to hold the data on hand is necessary in sequential allocation since the number of elements stored in a structure may grow during execution of programs. In contrast, with sequential methods, linked allocation does not require reserving extra memory words; instead, it adds bits needed to store pointers as overhead in every node.

Although less memory is generally needed with linked allocation than with sequential allocation, the latter facilitates more rapid access to an arbitrary node. This is so since an arbitrary node is located through what may be a lengthy chain of pointers in the former case. Sequential allocation access to the contents of the data structure involves both the start value (a pointer to the abstract structure) and an index to the items it holds.

III. HIERARCHICAL DATA STRUCTURES

Tree data structures enable representation of multidirectional or hierarchic relationships among data elements. There are several related data structures (for example, *trees*, *forests*), and the most general hierarchic structure, the *list*, is usually indicated as a special structure written with capitalized first letter, *List*. Capital L lists represent recursion. A List differs from either a tree or a forest by permitting nodes to point to their ancestors. All these structures are usually represented in memory by yet another one that also is hierarchic, the *binary tree*. This is so because there are straightforward ways to implement *binary tree traversal*.

A *binary tree* is *a hierarchic structure with every element having exactly no, one, or two immediate successors*. When another structure (*tree*, *forest*, *List*) is given a binary tree representation, some of the "linked" memory allocation needed to store the data it holds becomes available in a form that is more sequential. This is related to a descriptive term about binary trees, the notion of *completeness*. A binary tree is said to be *complete* when all the nodes that are present at a level are in sequence, beginning at the left side of the structure, with no gaps.

There are many definite ways to move through all the items in a binary tree data structure (several of them correspond to analogous procedures for general trees, forests, and Lists). Some of the most useful are *pre-order* (where a root or ancestor is visited before its descendants; after it is evaluated, first the left successor is visited in preorder, then the right), *in-order* (also called *symmetric order*), and *post-order*. In post-order, the ancestor is seen only after the post-order traversal of the left subtree and then the right. In-order has the in-order traversal of the left subtree preceding the evaluation of the ancestor. Traversal of the two parts occurs before visiting the right sub-tree in the in-order fashion.

Trees may be ordered; that is, all the "children" of any node in the structure are indexed (first, second, etc.). An ordered tree is the most common kind used in computer programs because it is easily represented by a binary tree; within each sibling set is a linear-list-like structure, which is useful for data retrieval; a nonordered tree is much like a

TABLE I Four-Dimension Array: $w = x^{3.5} + 2y^z$

x	y	z	w
1.25	50	0.25	7.50
3.75	3	3.75	8.82×10^5
2.35	32	3.75	8.82×10^5
64	21	1.75	2.10×10^6
...			

bag or set. A node in either kind of tree has degree equal to the number of its children, that is, the length of its sublist. Another basic tree term is *level*. This signifies the number of links between the first element in the hierarchy, called the *root*, and the datum. As a result, we say that the root is at level zero. A tree node with no successors is called a *tip* or *leaf*.

Strings, linear lists with space markers added, are often used to give text depictions of trees. There are several ways to visually depict the hierarchical relationship expressed by tree structures: nested parentheses, bar indentation, set inclusion, and decimal points—essentially the same device used when library contents are organized, for example, by the Dewey decimal system. All of these methods can be useful. The kind of information that can be stored by a tree data structure is both varied and often needed in practical situations. For example, a tree can describe the chapter outline of a book (Fig. 2a). On personal computers or web browsers, the starting location is called the *root* or *home*. It lists highest level information.

A binary tree is particularly useful because it is easy to represent in computer memory. To begin with, the data words (the contents stored at each node in the structure) can be distinguished from successor links by a one-bit flag. Another one-bit flag indicates a "left/right" characteristic of the two pointers. Thus, at a cost of very little dedicated storage, basic information that is to be kept can be stored with the same memory word field assignments. The binary tree version of Fig. 2a is Fig. 2b.

Binary trees always have tip nodes with unused pointers (link fields). Hence, at the small cost of a tag bit to indicate whether the link field is in use, this space can be used for traversal. The idea is that any pointer field not used to locate other data in the binary tree structure can be used to describe where to go in a traversal (via pre-order, post-order, etc.). Hence, these tip fields can be used to speed algorithm execution. These pointer memory location values may be useful in indicating where the algorithm should go next. If a tag bit indicates that the purpose is data structure traversal, the locator is called a *thread*.

IV. ORDER: SIMPLE, MULTIPLE, AND PRIORITY

A. Linear and Indexed Structures

The simplest data structures are lists (also called data strings). Even the ordinary list data structure can be set up to have one or another of several possible input and output properties. There are ways to use such input and output properties to give advantages in different situations. Hence, careful choice of the simple structure can contribute to making a computer program fast and space-efficient.

Two kinds of simple lists are stacks and queues. A *stack* is a data store where both input and output occur at the same location: the general name for such a place is a *port*. Stacks are characterized by last-in–first-out data handling. A *queue* is a two-port store characterized by first-in–first-out data handling. A queue is a more versatile data structure than a stack. A method for storing a queue in a fixed fast memory space uses circular wrapping of the data (modulo storage). There is another kind of linear list—a *deque* or *double-ended queue*, which has two ports. In the most general deque, input and output can take place at both ports.

Linear lists are data structures with entries related only through their one-dimensional relative positions. Because of this, from size and starting location of the structure, both first and last nodes can be located easily.

Data stored in a stack are accessible to the program in the reverse of the input order, a property that is useful in situations involving recursion. Evaluating an algebraic formula involving nested expressions is a recursive process that is facilitated by use of a stack to store left parentheses and symbols (operands) while the program interprets operators.

Stacks, *queues*, and *deques* all make heavy use of linear order among the data elements. However, for some computing purposes, a data structure that pays no attention to order is very useful. One elementary structure that has this nature is called a *bag*. Bags are structures where elements may be repeatedly included. An example of a bag is

DIRECTORY-BAG
M. JONES, 432-6101;
M. JONES, 432-6101; M. JONES, 432-6101;
A. GREEN, 835-7228; A. GREEN, 835-7228;

One use of bags is in looping as a supplementary and more complex index.

Up to this point we have considered only one type of elementary structure, in the sense that all the ones we have dealt with have been one dimensional. In mathematics, a one-dimensional entity is a kind of array known as a *vector*. A vector can be thought of as a set of pairs: an index and a data item. The convention is that data items are all the same type and indices are integers. With the same convention (both indices and data present) allowing an index to be any ordered string of integers yields an elementary data structure called an *array*. If the ordered string of integers has n-many elements, the array is associated with that number of dimensions. In algebra, mathematics applies two-dimensional arrays to simultaneous linear equations, but the data structure idea is of a more general concept. It allows any finite number of dimensions.

A *dense array* is one where data items are uniformly present throughout the structure. In a block array, data

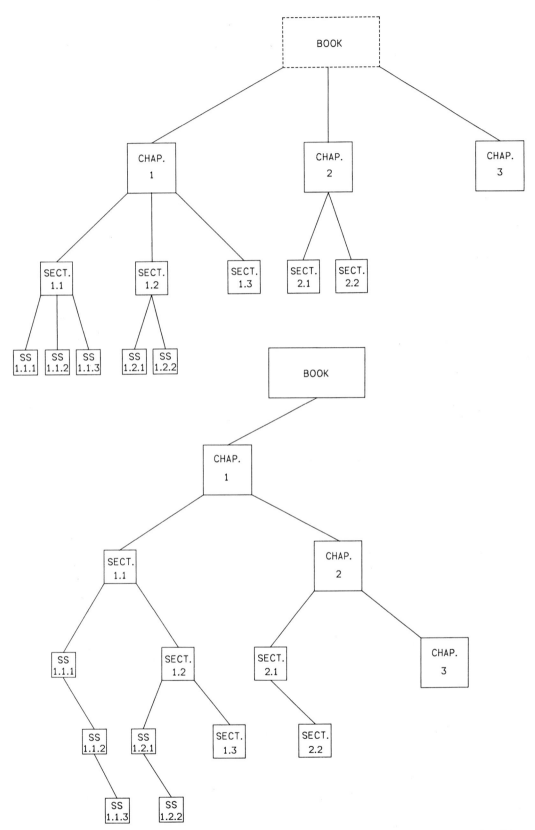

FIGURE 2 Trees and a book chapter outline. (a) General tree representation; (b) binary tree representation. Chap., chapter; Sect., section; SS, subsection.

items may be present only in certain intervals of coordinate directions. Both dense and block arrays are good candidates for sequential storage, in contrast with sparse arrays. A *sparse array* characteristically has a number of data items present much smaller than the product of the maximum coordinate dimensions, and data that are arbitrarily distributed within it. Linked allocation is preferred for sparse arrays. Sparse arrays occur in routing applications. A table using a binary variable to show plants where goods are manufactured and destinations where they are sold would have many *zeros* signifying *plant-not-connected-to-market*. Use of a linked representation for storing this sparse array requires much less storage than by sequential means.

Compressed sequential allocation is another array storage method. It uses the concept of a base location and stores only nonzero data items. It has the advantage that the nonzero data items are stored in sequential order, so access is faster than with linked representation. For r nonzero elements, where r is not very small compared to the product of the array dimensions mn, the storage required for compressed sequential allocation is $2r$. The base locations are essentially list heads (see below), and searching a list of r of them takes, on the average, of the order of $log_2 r$ steps. Arrays are also called *orthogonal lists*, the idea being that the several coordinate directions are perpendicular; rows and columns in a two-dimensional matrix are examples. In an $m \times n$ matrix A with elements $a[i, j]$, a typical element $A[k, l]$ belongs to two orthogonal linear lists (row list $A[k, *]$, column list $A[*, l]$, where the asterisk represents all the values, i.e., $1, \ldots, m$ or $l, \ldots, n$).

B. Linkage

The method of double linkage of all nodes has advantages in speeding algorithm execution. This technique facilitates insertion and deletion in linear lists. Another advanced method has pointers that link the beginning and end of a data structure. The resulting data structures enable simplified traversal algorithms. This technique is usually called *circular linkage* or sometimes simply *rings*; it is a valuable and useful method when a structure will be traversed several times and is entered each time at a starting point other than the starting datum ("top" in a stack, "front" in a queue).

Whenever frequent reversal of traversal direction or insertion or deletion of data elements occurs, a doubly linked structure should be considered. Two situations where this often occurs are in discrete simulation and in algebra. Knuth described a simulation where individuals waiting for an elevator may tire and disappear. Likewise, the case of polynomial addition requires double linkage because sums of like terms with different signs necessitate deletions.

A special type of queue called a *priority queue* replaces the physical order of the data elements by the numerical value of a field in the nodes stored. Data are output from a priority queue in the order given by these numerical values. In order to do this, either the queue must be scanned in its entirety each time a datum is to be output, or the insertion operation must place the new node in its proper location (based on its field value). Whatever the approach, this structure introduces the need for several terms discussed below in greater detail. A *key* is a special field used for ordering a set of records or nodes. The term *key* often signifies "data stored in secondary storage as an entity." The process of locating the exact place to put an item with a particular key is called a *search*. The process of ordering a set of items according to their key values is called a *sort*.

Even though at some time during program execution a data structure may become empty, it could be necessary to locate it. A special memory address (pointer, locator) for the data structure makes it possible for that to be done. This address is called a *list head*. (List heads are traversal aides.)

Multilists are a generalization of the idea of orthogonal lists in arrays. Here, several lists exist simultaneously and use the same data elements. Each list corresponds to some logical association of the data elements. That is, if items are in a list they possess a common attribute.

Two other advanced structures, *inverted lists* and *doubly linked trees*, enable more efficiency in executing certain kinds of algorithms by using more pointers. Inverted organization takes one table, such as Table I, into many. The added pointers make data structure elements correspond to the subtables.

V. SEARCHING AND SORTING TECHNIQUES

The two topics, searching and sorting, are related in the sense that it is simpler to locate a record in a list that has been sorted. Both are common operations that take place many times in executing algorithms. Thus, they constitute part of the basic knowledge of the field of data structures. There are many ways to define special data structures to support searching and sorting. There is also a direct connection between searching itself, as an algorithmic process, and tree data structures.

Sorting tasks appear in so many applications that the efficiency of sort algorithms can have a large effect on overall computer system performance. Consequently, choice of a sort method is a fundamental data structure decision. The ultimate decision depends on the characteristics of the data sets to be sorted and the uses to which the sorted data will be put. The requests for a sorted version of a list, their nature, and the rate of re-sort on given lists must all be taken into account. Several measures of merit are used to make quantitative comparisons between sorting techniques; memory requirement is one such measure. In essence, a sort yields a permutation of the entries in an input list. However, this operation takes place not on the information contained in the individual entries, but rather on a simpler and often numerical entity known as a *key*.

A. Sorting Algorithms

There are very many different sorting algorithms. Some of them have descriptive names, including *insertion sort*, *distribution sorting*, and *exchange sorting*. Another kind, *bubble sort*, is based on a simple idea. It involves a small key rising through a list of all others. When the list is sorted, that key will be above all larger values. Some sorting methods rely on special data structures. One such case is *heap sort*.

A heap is a size-ordered complete binary tree. The root of the tree is thus either the largest of the key values or the least, depending on the convention adopted. When a heap is built, a new key is inserted at the first free node of the bottom level (just to the right of the last filled node), then exchanges take place (bubbling) until the new value is in the place where it belongs.

Insertion sort places each record in the proper position relative to records already sorted.

Distribution sort (also called *radix sort*) is based on the idea of partitioning the key space into successively finer sets. When the entire set of keys has been examined, all relative positions in the list have been completely determined. (Alphabetizing a set is an example of a radix sort.)

When a large sorted list is out of order in a relatively small area, *exchange sorts* can be useful. This is a kind of strategy for restoring order. The process simply exchanges positions of record pairs found out of order. The list is sorted when no exchanges can take place.

Another sorting strategy takes the most extreme record from an unsorted list, ends a sorted list to it, then continues the process until the unsorted list is empty. This approach is called *sorting by selection*.

Counting sort algorithms determine the position of a particular key in a sorted list by finding how many keys are greater (or less) than that chosen. Once the number is determined, no further relative movement of the key position is found.

Merging two sorted lists requires only one traversal of each list—the key idea in *merg sort*. To sort a list by merging, one begins with many short sorted lists. Often those "runs" of elements in a random list that are already in order form one of them. The process merges them two at a time. The result is a set of fewer long lists. The procedure repeats until a single list remains.

B. Searching: Algorithms and Data Structures

Searching can be thought of as relatively simple in contrast with the many kinds of sorting algorithms. The simple sequential search proceeds stepwise through the data structure. If the item sought is present, it will be found. If it is present, the worst case is when it is last; otherwise (that is, when it is not in the list) the search is both unsuccessful and costly in terms of the number of items examined.

A much better way to search, on the average, in terms of the number of comparisons required either to find an item or to conclude that it is not in the data structure being examined is called a *binary search*. The approach is based on sorting the records into order before beginning the search and then successively comparing the entry sought first with one in the data structure middle. The process continues successively examining midpoints of subsets. At each stage, approximately half the data structure can be discarded. The result is on the average order of magnitude of $log_2 r$ search steps for a file of size r.

Any algorithm for searching an ordered table of length N by comparisons can be represented by a binary tree data structure. That structure is called a *binary search tree* when the nodes are labeled sequentially (left to right by 1 to N in the N-node case). This fact can also be stated as follows: *There is an equivalent algorithm for searching an ordered table of length N and a binary search tree* (i.e., a data structure). Focusing on the recursive definition of a binary search tree, first, all are binary trees; second, binary search trees have their node information fields consists of keys; finally, each key satisfies the conditions that:

1. All keys in the left subtree are less.
2. All keys in the right subtree are greater.
3. Left and right subtrees are binary search trees.

The many special cases of such recursively defined binary search trees (data structure) each correspond to a search algorithm. Movement through a binary search tree is like going to a place in the file. That place has a key, which is to be examined and compared to the one sought.

Two special cases of binary search trees (and, hence, search algorithms) are *binary* and *Fibonaccian*. Each of these algorithms is relatively easy to implement in terms of computations used to create the underlying search tree data structure. By contrast, *interpolation search* is a valuable method but is more complex in terms of computations.

To obtain the search tree that corresponds to binary search of 16 records, assume that the numbering $1, \ldots,$ 16 gives the keys. If we entered the file at the midpoint and entered each subfile at its midpoint, we would have a binary search tree with key value 8 at the root, left successor 4, and right successor 12. Fibonacci search replaces the division-by-two operations necessary in a binary search by simpler addition and subtraction steps. The Fibonacci series consists of numbers that are each the sum of the immediately preceding two in that sequence. A Fibonacci series begins with two ones, and has values:

$$1, 1, 2, 3, 5, 8, 13, 21, 34, 55, \ldots$$

In a Fibonaccian search, the elements of the binary search tree are either Fibonacci numbers or derived from them; the root is a Fibonacci number, as are all nodes reached by only left links. Right links lead to nodes whose values are the ancestor plus the difference between it and its left successor. That is, the difference between the ancestor and left successor is added to the ancestor to get the right successor value. Fibonaccian binary search trees have a total number of elements one less than a Fibonacci number.

VI. TREE APPLICATIONS

A. Parent Trees and Equivalence Representation

In a *parent tree* data structure, each successor points to its ancestor. Hence, such a structure can be stored in memory as a sequential list of (node, parent-link) pairs, as illustrated by Fig. 3. The parent tree representation facilitates "bottom-up" operations, such as finding the (1) root, (2) depth in the tree, and (3) ancestors (i.e., all nodes in the chain from the selected one to the root). Another advantage is in savings in link overhead: Only one link per node is required, compared to two per node in the conventional (downward-pointer binary tree) representation. The disadvantage of the parent representation is that it is inefficient for problems requiring either enumeration of all nodes in a tree or top-down exploration of tree sections. Further, it is valid only for nonordered trees. Trees where sibling order is not represented are less versatile data structures. For example, search trees cannot be represented as parent trees, since the search is guided by the order on keys stored in the data structure information fields.

Equivalence classes are partitions of a set into subsets whose members can be treated alike. Binary relations can be given explicitly by a list of the objects so paired with the symbol ":" indicating that objects on either side stand in some relation to one another. Sometimes ":" is a symbol for the equivalence relation; another form is "==". These are read "can be replaced by" and "is identically equal to," respectively.

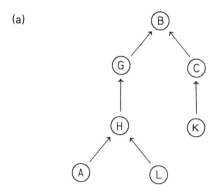

(a)

(b)

NODE	MEMORY ADDRESS
A	1018
B	1182
H	1221
K	1050
G	1129
L	1281
C	1084

(c)

	NODE NAME	NODE LOCATION	PARENT NAME	PARENT-LINK LOCATES
206	A	1018	H	1221
207	B	1182	–	—
208	H	1221	G	1129
209	K	1050	C	1084
210	G	1129	B	1182
211	L	1281	H	1221
212	C	1084	B	1182

FIGURE 3 A parent tree and its memory representation. (a) Parent tree; (b) locations of data found in the parent tree; (c) a sequential table storing the parent tree information.

An equivalence relation of a set of objects is defined by: (1) the list of equivalent pairs, or (2) a rule that permits generation of equivalent pairs (possibly only a subset).

1. Example (Pictures)

The parts of a picture that together compose objects can be members of equivalence classes. A possible rule for generating related pairs is "Two black squares are equivalent if they share an edge." Then, each equivalence class represents a closed pattern.

An algorithm that solves the general equivalence class problem scans a list of input data pairs. The process isolates equivalent pairs and creates parent trees. The idea is that the algorithm creates a node for each of the two elements in a given pairing and one (e.g., always the one on the left) is made the root. If a new equivalent pair also has an element already in a parent tree, the other item is adjoined to the existing tree (under the relation of this program). Whether two nodes are equivalent can be determined by finding and comparing the roots of their corresponding equivalence trees to determine whether they are in the same partition.

Parent trees are advantageous for algorithms for deciding equivalence, since the most time-consuming step, retrieval of the root of a tested node, is speeded by this data structure. Thus, to organize a set of nodes into equivalence classes, use an algorithm to put them into parent trees.

B. Spanning Trees and Precedence

Spanning trees are illustrated in Fig. 4(a) for a set of four nodes. As the term implies, a spanning tree is a tree that contains all the nodes. Suppose the tree is viewed as a graph so that links are no longer solely pointers (memory addresses of the next datum) but edges with weights. This enables the concept of a *minimal spanning tree*. This idea is simply the spanning tree of least total edge values. The concept is illustrated by the following example and by Fig. 4(b).

1. Example (Minimal Spanning Tree)

. We have four nodes, a, b, c, and d, and seek to build a minimal spanning tree given the edge weights:

$$(a, b) = 6, \quad (b, d) = 12$$
$$(a, c) = 3, \quad (c, d) = 16$$

The tree with the form $(a\,(b\,(d),\,c)$ has total weight 21, which is less than that for the tree with d linked to c instead of b—i.e., $a\,(b,\,c\,(d))$. Both trees just described appear in Fig. 4 (second row at right).

The minimal spanning tree algorithm uses the following idea. Initially, all the nodes are members of different sets. There are as many sets as nodes. Each node has itself as the only member of its set. As the algorithm proceeds, at each stage it groups more nodes together, just as in equivalence methods. The algorithm stops when all nodes are in one set. The parent tree data structure is the best one to use to implement the minimal spanning tree algorithm.

Yet another algorithm concept and its implications for data structure selection arise from the precedence situation introduced earlier. To create an efficient scheduling procedure, observe that any task that does not require completion of others before it is begun can be started at any time. In data structure terms, this is equivalent to an algorithm using the following fact: *Removal of a graph element without a predecessor does not change the order of priorities stored in the graph.*

VII. RANDOMNESS, ORDER, AND SELECTIVITY

The issue of order is central to data structures, but it is also the opposite of a basic mathematical notion, randomness. Currently, data encryption is a valued use of computation. The basic coding method involves creating a form of disorder approaching randomness to render files unintelligible

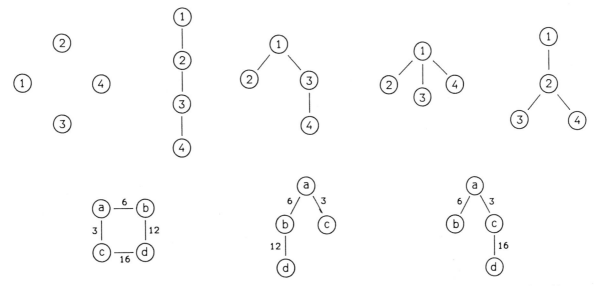

FIGURE 4 Spanning trees. (a) Four nodes with four spanning trees, all with node one at the root; (b) four nodes with edge weights and two different spanning trees.

to an intruder. Factoring a very large integer into a product of prime numbers, finding the greatest common divisor of two integers, and devising new algorithms to calculate fundamental quantities are all central issues in restructuring data that relate to encryption algorithms. This topic is beyond the scope of this article, but is addressed in the second of the three data structure volumes by Knuth.

Regarding disorder versus order, π, the ratio of the circumference of a circle to its diameter, has many known and some esoteric qualities. For example,

1. It is irrational.
2. It is not known whether its successive digits are truly random.
3. Successive digits of π exhibit local nonrandomness (e.g., there are seven successive 3's occurring starting at the 710,100th decimal digit, as noted by Gardner).
4. π is well approximated by several easily computed functions: 22/7, 355/113, $[2143/22]^{1/4}$, this last noted by Ramanujan.

Selecting where to search in the plane is aided by both *quad-trees* and *2–3 trees*. Quad-trees are like binary trees but regulary involve *four or fewer success to every tree node*. As a data structure, quad-trees enable algorithms to process spatial issues, including hidden-line elimination in graphic display. The three-dimensional analog is *octress* (eight or fewer successor nodes). The somewhat similar 2–3 trees require all internal nodes (i.e., nonterminals) to have either two or three successors.

A *trie* is a data structure that uses repeated subscripting and is used to enable multiway search. The term comes from the word re*trie*val. It originated in the field of information retrieval and has developed in part because of machine translation. The benefit of using the trie approach is that it helps reduce the number of ways to search. It does this by eliminating impossible cases. (For example, in English words, a "q" is never followed by anything but a space [Iraq] or the letter "u". That involves a considerable reduction from 26 or 27 other letters or space symbol.) Entries in the nodes of a trie are instructions that describe how to proceed further in a symbol-by-symbol search.

Hashing is a very useful computer technique for locating (storing) data records. The hashing method deliberately introduces disorder. In other words, hashing uses randomness in storing. The rationale is that *on the average* this achieves a significant speed improvement in locating a data record. The index of the item to be stored is called a *key*, and the function that takes the key into a table location is called a *hash function*. Usually, this is done in a way that chops up the numeric key value, introducing a somewhat random element into the actual

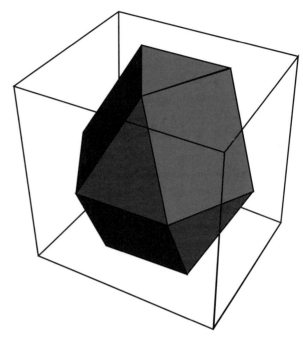

FIGURE 5 Cuboctahedron.

pattern of data storage. (This is what lead to the term *hashing*.)

Both fractions and whole numbers led to ideas that are essentially data structures. For fractions, see the universal resource locators in the bibliography. For whole numbers, addition of a long list of large numbers vertically arranged is a much easier matter than all in a row, if they are written in Arabic numerals.

A geometric realization of the value of lists and data structures in general is shown in Fig. 5, a view of a 14-faced solid generally called a *cuboctahedron*. The image was generated by the Mathematica utility from the data structures below. Coordinates of the vertices were calculated by analytic geometry. Once each triple, x, y, z, was known, the graphic routine needed to sequence vertices, generate faces, and then color the result. Two of the lists that were involved in that process follow. (Notice that there are quadrilateral and triangular faces; likewise, each list consists of sublists.) Braces, "{"and"}", initiate and conclude, respectively, each list/sublist. The integers in the "Faces" list range from one to 12. Each one is a specific three-dimensional point or corner of the solid object. The actual coordinates of these points appear in the 1 through 12 order in the "Vertices" list.

1. Faces = {{10, 11, 12}, {7, 8, 9}, {4, 5, 9, 8}, {8, 3, 4}, {2, 3, 8, 7}, {7, 1, 2}, {1, 6, 9, 7}, {9, 5, 6}, {10, 2, 3}, {1, 6, 12}, {11, 4, 5}, {3, 4, 11, 10}, {1, 2, 10, 12}, {11, 12, 6, 5}}

2. Vertices = {{1, 0, 0}, {1/2, $(3^{1/2})/2$, 0}, {−1/2, $(3^{1/2})/2$, 0}, {−1, 0, 0}, {−1/2, $−(3^{1/2})/2$, 0}, {1/2, $−(3^{1/2})/2$, 0}, {1/2, $(3^{1/2})/4$, 1}, {−1/2, $(3^{1/2})/4$, 1}, {0, $−(3^{1/2})/4$, 1}, 0, $(3^{1/2})/4$, −1, {−1/2, $−(3^{1/2})/4$, −1}, {1/2, $−(3^{1/2})/4$, −1}}

VIII. CONCLUSION

The subject of data structures clearly contains a myriad of technical terms. Each of the topics discussed has been briefly mentioned in this article. A basic text on data structures would devote many pages to the fine points of use. Yet the pattern of the subject is now clear. Before programming can begin, planning the algorithms and the data they will operate on must take place. To have efficient computing, a wide range of decisions regarding the organization of the data must be made. Many of those decisions will be based on ideas of how the algorithms should proceed (e.g., "put equivalent nodes into the same tree"). Others will be based on a detailed analysis of alternatives. One example is taking into account the likely range of values and their number. This determines possible size of a data table. Both table size and retrieval of elements within it impact key choices. Two computer data structure considerations are always *memory needed* and *processing time* or *algorithm execution speed.*

Many aspects of data structures and algorithms involve data stored where access is on a secondary device. When that is the situation, procedures deal with search and sorting. Sorting occupies a substantial portion of all the computing time used and contains numerous alternate algorithms. Alternative means exist because data sometimes make them advantageous.

As in the case of sorting, it is always true that actual choice of how an algorithmic task should be implemented can and should be based on planning, analysis, and tailoring of a problem that is to be solved. The data structure also needs to take into account the computer hardware characteristics and operating system.

Explosive development of computer networks, the World Wide Web, and Internet browsers means that many technical terms discussed in this article will join links, pointers, and hierarchy in becoming common terms, not solely the province of computer experts using data structures.

Universal Resource Locators

Mathematics and computer data structures—http://www.cs.ucla.edu/~klinger/inprogress.html
Egyptian fractions—http://www.cs.ucla.edu/~klinger/efractions.html
Baseball arithmetic—http://www.cs.ucla.edu/~klinger/bfractions.html
Thinking about number—http://www.cs.ucla.edu/~klinger/5fractions.html
Cuboctahedron image—http://www.cs.ucla.edu/~klinger/tet1.jpg

SEE ALSO THE FOLLOWING ARTICLES

C AND C++ PROGRAMMING LANGUAGE • COMPUTER ALGORITHMS • COMPUTER ARCHITECTURE • DATABASES • DATA MINING • EVOLUTIONARY ALGORITHMS AND METAHEURISTICS • QUEUEING THEORY

BIBLIOGRAPHY

Aho, A., Hopcroft, J., and Ullman, J. (1983). "Data Structures and Algorithms," Addison-Wesley, Reading, MA.

Dehne, F., Tamassia, R., and Sack, J., eds. (1999). "Algorithms and Data Structures," (Lecture Notes in Computer Science), Proc. 6th Int. Workshop, Vancouver, Canada, August 11–14, Springer-Verlag, New York.

Graham, R., Knuth, D., and Patashnik, O. (1988). "Concrete Mathematics," Addison-Wesley, Reading, MA.

Knuth, D. (1968). "The Art of Computer Programming," Vol. I, "Fundamental Algorithms," Addison-Wesley, Reading, MA.

Knuth, D. (1973). "The Art of Computer Programming," Vol. III, "Sorting and Searching," Addison-Wesley, Reading, MA.

Knuth, D. (1981). "The Art of Computer Programming," Vol. II, "Seminumerical Algorithms," Addison-Wesely, Reading, MA.

Meinel, C. (1998), "Algorithmen und Datenstrukturen im VLSI-Design [Algorithms and Data Structures in VLSI Design]," Springer-Verlag, New York.

Sedgewick, R. (1990). "Algorithms in C," Addison-Wesley, Reading, MA.

Sedgewick, R. (1999). "Algorithms in C++: Fundamentals, Data Structures, Sorting, Searching," 3rd ed., Addison-Wesley, Reading, MA.

Waite, M., and Lafore, R. (1998). "Data Structures and Algorithms in Java," Waite Group.

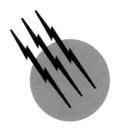

Data Transmission Media

John S. Sobolewski

University of New Mexico

GLOSSARY

Bit Short for binary digit, the smallest unit of information in a binary system. A bit represents the choice between a one (mark) or a zero (space) condition.

Carrier Continuous frequency or periodic pulses capable of being modulated or changed by an information signal.

Channel Path for transmission of information between two or more points using some form of transmission medium.

Channel capacity Number of information bits per second that can be transmitted over a channel.

Demodulation Process of retrieving information from a modulated carrier wave. The opposite of modulation.

Medium, bounded Transmission medium that constrains and guides or conducts information-carrying signals, also called a guided transmission medium. Examples include wires and optical fibers.

Medium, transmission Transmission medium that provides the physical channel or path used for transmission of information between two or more points.

Medium, unbounded Transmission medium that permits signals to be transmitted but does not guide or constrain them; also called an unguided transmission medium. Examples include radiowave transmission using space or air as the transmission medium.

Modulation Process of varying some characteristic of a carrier wave in accordance with the instantaneous value or sample of the information to be transmitted.

Multiplexing Division of a higher bandwidth communication channel into two or more lower bandwidth channels.

Repeater Device whose function is to retime and restore received signals back to their original form and strength and to retransmit them further down the channel.

Signal Aggregate of waves propagated along a transmission channel toward a receiver at the destination.

ELECTRONIC COMMUNICATION systems depend upon transmission media for sending voice, data, video, or other signals from one point to another. Among the physical media are twisted pairs, coaxial cable, free space, and optical fibers. The distance of transmission can be very small, as between computer memory and the central processing unit, or very large, as between a corporate headquarters and an overseas subsidiary. Without the transmission media, communication systems could not exist.

In recent years, there has been a virtual explosion in consumer demand for bandwidth due to the traffic generated by the Internet (300% growth per year), electronic mail, faxes, cellular telephones, mobile computers, teleconferencing, as well as conventional data and video transmission. This demand for new services and higher bandwidth has resulted in the deployment of new communication infrastructure and new technologies for the more effective use of the transmission media already in place. Every indication is that these trends will continue in the foreseeable future.

I. BASIC SIGNAL TYPES

Information is transmitted over communication channels either as analog signals or digital signals. Analog signals are in the form of a continuous time-varying physical quantity, such as voltage magnitude or frequency, that reflects the variations of the information or signal source with time. Speech, radio, and conventional television signals are examples of analog signals. Digital signals consist of a stream of discrete pulses with well-defined states. Data generated by digital computers and their peripherals or data terminals are digital in nature and, in general, consist of two discrete levels corresponding to digital ones and zeros, as illustrated in Fig. 1.

It is important to note that a communication system can be designed to carry either analog or digital signals by using appropriate signal conversion devices. This applies to systems using any type of transmission medium—wire,

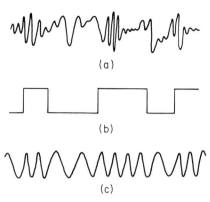

FIGURE 1 Any information can be transmitted in (a) analog or (b) digital form. Conversion from one to the other is available; (c) shows a digital signal converted to an analog signal using frequency modulation.

coaxial cable, optical fibers, or free space. Furthermore, since analog signals can be converted to digital signals and vice versa, any type of information can be transmitted either in digital or analog form that is compatible with the transmission medium over which it will travel with no significant loss of information.

In the past, most communication channels were designed for transmission of analog signals within a certain range of frequencies. If we need to transmit digital computer data for long distances over such channels, we can use a special device that converts the digital data into an analog signal consisting of a continuous range of frequencies, as illustrated in Fig. 1. We can thus use existing analog channels for transmission of digital data. Conversely, it is possible to convert analog voice or television signals into digital signals and transmit them over channels designed for digital signals. The capability of converting analog to digital signals and digital to analog signals is important, since the need for transmission of digital data is growing rapidly while many of the communication channels in existence are analog in nature. However, digital transmission has advantages over analog transmission, and consequently digital technology is evolving rapidly. Over the past two decades, millions of miles of digital transmission channels have been installed by communication carriers, mostly to carry data and digitized speech. This trend will continue, and communication channels installed in the future will be almost exclusively digital in nature, not only for transmission of data but also because digital channels are more cost effective for virtually all communication needs.

II. CHARACTERISTICS OF SOME COMMON SIGNALS

Before describing the concept of a communication channel and the various transmission media in greater detail, it is necessary to describe the characteristics of some of the most common types of signals that are routinely transmitted over existing communication channels. Even though significant economic payoff can result from designing communication channels that can carry all types of signals, some of the signal characteristics are so different that different communication architectures are needed. The sections that follow describe the characteristics of program and speech, which are analog signals; data, which is a digital signal; and video, which can be considered a hybrid between analog and digital.

A. Program Signals

Program signals include radio broadcasts, the audio portion of television programs, and high-fidelity music programs that are distributed over cable systems to subscribing customers. Such signals are characterized by a greater average volume, dynamic range in volume, duration, and bandwidth (which results in greater fidelity) than typical telephone speech signals. A high dynamic range of volume is needed to reproduce faithfully the variety of program material transmitted, which may include speech, music, and special sound effects. The bandwidth required may be as high as 30 to 15,000 Hz for high-quality music for FM stereo or the audio portion of television programs, even though only a fraction of this bandwidth is used for speech during, say, a newscast.

B. Speech Signals

Speech signals are usually transmitted over telephone channels. The telephone set transforms the voice signals into electrical analog signals which are transmitted to the local telephone exchange and then through the wide-area telephone network to the receiving party. Although speech typically covers frequencies from 30 to 10,000 Hz, most of the energy is in the range from 200 to 3500 Hz. Since the human ear is not very sensitive to small changes in frequencies and since humans can correct for things such as missing syllables or words, we do not need to reproduce speech signals precisely to achieve acceptable quality of transmission. Consequently, most telephone communications are bandlimited to between 200 and 3500 Hz to save transmission costs. This savings comes about since costs are directly related to the bandwidth (i.e., the range of frequencies transmitted) and (as will be described later) bandlimiting allows a greater number of voice channels to be multiplexed over a high-bandwidth channel. It must be pointed out, however, that the nominal bandwidth of a voice channel is defined as 4000 Hz, with the additional bandwidth allowing a guard band on either side of the speech signal to reduce interference between channels.

The time required to set up a telephone connection can vary from a few seconds to tens of seconds, the communication is almost always two-way, and the two parties continuously transmit (talk), listen (receive), or pause until the call is terminated. A wide dynamic range in volume is needed, although not as large as for program signals. Conversations require immediate delivery of the signal (i.e., delays are not tolerated), but communication is relatively tolerant of noise on the channel.

C. Data Signals

Servers, workstations, or personal computers usually produce digital signals consisting of streams of binary digits (bits) representing ones or zeros. Unlike the speech signal, which contains much redundant information and is relatively tolerant of noise or errors on the channel, the digital data signal may have little or no redundancy; therefore, errors cannot be tolerated. In fact, bits (parity or cyclic redundancy bits) are usually added to the basic digital signal to detect errors and correct them by retransmission. By addition of even more bits and use of error-correcting codes, errors can be detected and corrected without retransmission of the data.

Other important characteristics are shown in Table I, which summarizes the differences between digital and

TABLE I   Differences in Characteristics between Data and Telephone Voice Signals

Data signal	Telephone voice signal
Desirable to set up a connection in one second or less	One second to tens of seconds to set up a connection
One- or two-way transmission	Two-way transmission in most cases
Error-free received data	Tolerant of noise and some errors on the channel
Little or no redundancy	Much inherently redundant information
Transmission usually in bursts	Transmit or receive continuously until call is disconnected
Data can usually be stored and transmitted later when convenient	Not tolerant of transmission delays
Transmission has high peaks, peak-to-average ratio as high as 1000 to 1	Transmission rate relatively constant
Connection may be required 24 hours/day, 7 days/week (e.g., bank cash machine)	Duration of connection usually several minutes
May require a wide range of bandwidths, from thousands to tens of millions of bits per second	Requires fixed bandwidth of about 4000 Hz

telephone voice signals. It is important to recognize that those differences are significant and that telephone and computer users have fundamentally different requirements. With the large increase in data transmission since the beginning of the computer era, and the even greater increases projected for the future, there will be continued rapid progress in the development of new network architectures designed specifically for data transmission.

D. Video Signals

Most video signals carried over communication channels today are color television signals. A video transmission system must deal with four important factors when transmitting images of moving objects, specifically:

1. Perception of the distribution of luminance (degree of shade between light and dark)
2. Perception of a three-dimensional image (width, height, and depth)
3. Perception of motion related to the two factors above
4. Perception of color (hues and tints)

Monochrome video transmission deals with the first three factors. In color video transmission, color is combined with the monochrome (black and white) picture. At every moment, the video signal must integrate luminance and color from a scene in three dimensions, as a function of time, and combine them into a complex electrical signal for transmission. What makes this more complex is that time itself is variable since the scene is changing all the time. The integration of visual information is carried out by a process called *scanning*.

The scanning process consists of taking a horizontal strip across the image and scanning it from left to right beginning at the top. When the right-hand end of a strip is reached, the next lower horizontal strip is scanned. Luminance values are translated on each scanning interval into voltage or current variations and are transmitted. In the U.S., a total of 525 horizontal scan lines make a whole picture frame. Scanning a whole frame is repeated at a rate of 30 frames per second. This rate takes advantage of the persistence of vision of the human eye, thus eliminating flicker and giving the perception of motion, as in motion pictures. Because of the amount of information that must be transmitted to produce flicker-free images with acceptable horizontal and vertical resolution, video signals require a bandwidth of 6 MHz in the U.S., as compared with a 4-kHz bandwidth required for acceptable transmission of voice signals.

To permit decoding of the video signal at the receiver, it is necessary to transmit vertical and horizontal synchronizing pulses interleaved with the picture information. These pulses synchronize the transmitter and receiver by identifying the start of every horizontal scan line that corresponds to the top of the picture and the subsequent horizontal scan lines that make up the rest of the picture. Since these synchronizing pulses are essentially digital signals while the picture information in each horizontal scan line is in analog form, video signals may be thought of as hybrid signals.

III. COMMUNICATION CHANNELS

A block diagram of an electronic communication system is shown in Fig. 2. It consists of an input signal, a transmitter, a communication channel, a receiver, and an output signal from the receiver. The transmitter modifies the input message signal into a form suitable for transmission over the channel, which is the transmission path for providing communication between transmitter and receiver. The purpose of the receiver is to recreate the original message signal at the output. For example, when digital data signals must be sent over a communication channel designed primarily for analog signals, the transmitter has to convert the digital signals to analog signals by a process known as *modulation*. The receiver then demodulates the analog signal back to digital form and passes it on to its ultimate destination.

The communication channel can consist of various media such as wire, coaxial cable, optical fibers, or free space, in which case the signal is radiated as an electromagnetic wave as in conventional television or radio broadcasting.

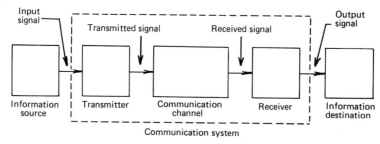

FIGURE 2 Model of a communication system.

Communication channels have practical limitations, such as bandwidth, and suffer from various impairments, such as nonlinearities as well as noise and interference, which may be introduced from within or from outside the channel. Transmitters and receivers must be carefully designed to match the signals to be transmitted to the physical properties of the communication channel and to minimize the effects of channel impairments on the quality of reception.

We can classify communication channels as analog or digital depending on the basic signals they transmit. Analog channels can be further classified as voice, program, or video channels depending on their intended use and the kind of signals they carry. They may be characterized by their bandwidth, which is the range of frequencies they transmit. Thus, a voiceband channel may be called a 4-kHz channel while broadband channels may be 48- or 240-kHz channels. Digital channels are usually characterized by their bit rate. Channels with a bit rate of 1540 kbps (kilobits per second), for example, are used with increasing frequency and can be used to transmit simultaneously several lower bit rate channels through multiplexing, as explained in Section VII.

Analog and digital channels can also be classified as simplex, half-duplex, or full-duplex. Simplex channels transmit in one direction only and are used for radio broadcast but seldom for data communication, which usually requires two-way transmission. Half-duplex channels can send in both directions but only in one direction at any given time. In other words, they provide nonsimultaneous two-way communication. Full-duplex channels, on the other hand, allow simultaneous two-way communication.

IV. CHANNEL CAPACITY

In practice, a communication channel represents a financial investment, hence the goal of communication engineers is to maximize the return on that investment. In the case of digital transmission, this is done by maximizing the channel capacity, which may be defined as the maximum rate at which information can be sent over the channel with an arbitrarily small probability of error.

In 1928, Nyquist showed that for binary transmission (i.e., the transmitted signal has one of two possible values at any one time) $2W$ bits/sec can be transmitted over a channel of bandwidth W in the absence of noise. In the general case, if we use M-ary ($M > 2$) rather than pure binary transmission by sending one of M different and distinguishable signal values at any one time, then in the absence of noise we have

$$C = 2W \log_2 M \qquad (1)$$

where C is the channel capacity in bits/sec and W the channel bandwidth in Hz. Thus, if we have a channel bandwidth

W of 3300 Hz, then in the absence of noise we can transmit up to 6600 bits/sec using pure binary transmission ($M = 2$). However, if we use a transmission scheme that makes it possible to send one of four distinguishable values at any given instant ($M = 4$), then we increase the channel capacity to 13,200 bits/sec. Similarly, if we can send one of eight distinguishable values at any instant, then each level in effect represents $\log_2 8 = 3$ bits, and the channel capacity becomes 19,800 bits/sec using this transmission scheme.

For a channel of bandwidth W, it is highly desirable to increase M in the above equation to maximize the channel capacity. The question arises, however, as to the number of different signaling values that can be transmitted and be separately distinguished at the receiver in the presence of noise, distortion, limits on signal power, and other channel impairments that occur in practice. In 1948, Shannon proved that if signals are sent with signal power S over a channel with Gaussian noise (amplitude of noise signal follows a Gaussian distribution) of power N, then the capacity C of the channel in bits per second is

$$C = W \log_2(1 + S/N) \qquad (2)$$

where W is the bandwidth of the channel. This is one of the fundamental laws of communication and gives the maximum signaling rate over a communication channel in terms of three parameters that are known or measurable. We can design elaborate coding schemes or modulation techniques, but we will never be able to increase the capacity unless we increase either the available bandwidth or the signal-to-noise ratio S/N.

If we take the previous example of the channel with a bandwidth of 3300 Hz and if we assume a signal-to-noise ratio of 63, then the maximum possible rate at which data could be transmitted over this channel is $3300 \log_2(1 + 63) = 19{,}800$ bits/sec, no matter how many different signaling levels our transmission scheme may have at any one time. In fact, for this particular channel, if we tried to use a transmission scheme with one of the 16 distinguishable signal values at any instant to increase the channel capacity, then in the absence of noise, relation (1) gives a capacity of 26,400 bits/sec. However, with the given signal-to-noise ratio of 63, the capacity is limited to the 19,800 bits/sec, implying that 16 different values would not be distinguishable on this channel unless the signal-to-noise ratio was increased, which could be done by increasing the signal power of the transmitted signal.

Unfortunately, as explained in Section VIII, practical communication channels include impairments other than Gaussian noise, and the limit given by relation (2) is very difficult or costly to achieve in practice. However, it clearly identifies the two resources that designers can use to increase the channel capacity, namely, channel bandwidth and transmitted signal power. Sometimes these resources are limited, in which case the channels may be classified as bandlimited, as in the case of telephone voice channels, or power-limited, as in the case of, say, a satellite channel.

V. BASEBAND TRANSMISSION

Signals can be transmitted over a communication channel using either baseband or broadband transmission. In baseband transmission, the information signal is sent over the channel directly without modification. In broadband transmission, the information signal is modified by superimposing it on a higher-frequency signal, called the *carrier*, which "carries" the information over the channel. Broadband transmission, or carrier transmission as it is sometimes called, uses modulation techniques that are described in the next section.

Because there is no modification of the information signal, baseband transmission systems can be used for digital or analog signals. In practice, they require an electrical or light (e.g., optical fiber) conductor to carry the baseband signal and are used primarily for transmission over relatively short distances. Separate transmitters and receivers are seldom required (e.g., the source and destination act as transmitter and receiver, respectively), and in those cases in which they are required they are relatively simple devices. The simplicity of baseband transmission makes it very common. In fact, it is used for short-distance signal transmission amongst most components of electronic devices or systems. Transmission of signals between components within a computer, a computer and its local peripheral devices, and an audio amplifier and the speakers are all examples of baseband transmission. Others include transmission over local area networks in a building or factory and over telephone subscriber loops between a telephone set and the local exchange. Longer telephone links, such as those between exchanges, do not usually carry baseband signals but use higher carrier frequencies to carry many signals simultaneously.

Because of the relatively short distances involved, the bandwidth of baseband channels can be relatively high. The major limitations of such channels include stray capacitance, inductance, and resistance which cause distortion, especially of the leading and trailing edges of digital signals. These limitations will be described in greater detail in Sections IX and XII.

VI. BROADBAND TRANSMISSION

In broadband transmission, information signals are processed and superimposed on a carrier that is more suitable for transmission over a particular channel. This is known as *modulation*, which is defined as the process by which some characteristic (e.g., amplitude, frequency, or phase) of a carrier wave is varied or modulated in accordance with the instantaneous value or samples of the information signal to be transmitted. The information signal, which can be analog or digital, is called the *modulating wave*, while the result of the modulation process is called the *modulated wave*. Demodulation restores the modulated signal to its original form at the receiving end of the communication channel.

At first, modulation techniques were used primarily for radio broadcasts where a relatively low-bandwidth speech or music program signal was used to modulate a much higher frequency carrier, resulting in a modulated signal whose frequency spectrum was suited for radio transmission. Subsequently these techniques were used for telephone transmission, because it was realized that telephone lines had a bandwidth greater than that needed for speech. By using several voice signals to modulate carriers of different frequencies, several voice signals can be transmitted (or multiplexed) over a single telephone line, thus reducing the cost per voice channel.

Modulation is used for two other important reasons. First, it is used to convert digital signals for transmission over analog channels that would otherwise destroy the digital signal. Second, it is used to convert voice or other analog signals to digital form to achieve lower noise or less distortion. Modulation, detection, and demodulation are complex topics, and only the most basic principles can be described here. A more complete treatment is given in Schwartz (1980) and Proakis (1983).

A. Analog Modulation

In analog modulation, a sine-wave carrier is modulated by an analog signal such as voice of a program signal. The sine-wave carrier $c(t)$ may be represented by:

$$c(t) = A_c \sin(2\pi f_c t + \theta_c) \qquad (3)$$

where A_c is the carrier amplitude, f_c is the carrier frequency, and θ_c is the phase. The value of A_c, f_c, and θ_c can be varied by the message or modulating signal $m(t)$ to form the modulated wave $s(t)$. In all cases, the carrier frequency f_c must be much greater than the highest frequency component of the modulating signal $m(t)$.

In amplitude modulation (AM), the amplitude of the carrier is varied in accordance with the message signal $m(t)$. This is done by adding the product of $c(t)$ and a fraction k of $m(t)$ to the carrier $c(t)$, yielding:

$$\begin{aligned} s(t) &= c(t) + km(t)c(t) \\ &= A_c(1 + km(t))\sin(2\pi f_c + \theta_c) \qquad (4) \end{aligned}$$

The absolute value of $s(t)$ has the same shape as $m(t)$ provided that the absolute value of $km(t)$ is always less than 1.

By taking the Fourier transform of $s(t)$, it may be shown that it consists of two symmetrical sidebands, each of bandwidth m, on either side of the carrier frequency f_c, where m is the bandwidth of the message signal $m(t)$. Consequently, one of the sidebands is sometimes removed through filtering to conserve bandwidth, and the carrier is suppressed to reduce transmission power. This variation of AM is known as *single sideband suppressed carrier* (SSB/SC) modulation and was commonly used for voice and data transmission over telephone circuits.

In frequency modulation (FM), the amplitude of $s(t)$ is constant, but the instantaneous frequency $f_i(t)$ of the modulated wave $s(t)$ changes linearly with the amplitude of the message signal $m(t)$ and is given by:

$$f_i(t) = f_c + k_f m(t) \qquad (5)$$

which yields

$$s(t) = A_c \sin(2\pi(f_c + k_f m(t))t + \theta_c) \qquad (6)$$

where k_f is a constant and determines the maximum frequency deviation that can occur. In phase modulation (PM), the amplitude and frequency are maintained constant, but the instantaneous phase of the modulated wave $s(t)$ changes linearly with $m(t)$, yielding:

$$s(t) = A_c \sin(2\pi f_c t + k_\theta m(t)) \qquad (7)$$

where k_θ is a constant and determines the maximum phase shift that can occur.

Amplitude and frequency modulation are used widely for radio program signals. An important feature of frequency and phase modulation is that they are more immune to noise and interference than amplitude modulation. This advantage, however, is achieved at the expense of a more complex receiver and greater bandwidth requirement. In fact, frequency and phase modulation provide a convenient means for obtaining better noise performance at the expense of bandwidth. Figure 3 illustrates the three modulation techniques in the case where the modulating signal is a digital rather than an analog signal.

B. Digital–Analog Modulation

The three analog modulation techniques just described can also be used for digital signals. In the case of binary digital signals, the modulation process involves keying (switching) the amplitude, frequency, or phase of the carrier between two different values according to the message signal. This is a form of digital–analog modulation and results in three distinct techniques known as *amplitude shift keying* (ASK), *frequency shift keying* (FSK), and *phase shift*

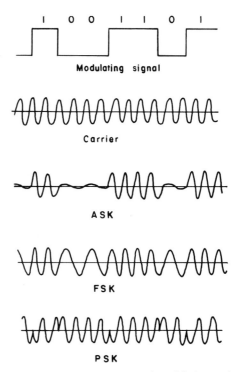

FIGURE 3 Three basic techniques of modulating a sinusoidal carrier using a digital modulating signal: ASK, FSK, and PSK.

keying (PSK). They are illustrated in Fig. 3 and are extensively used in older modems (modulator–demodulators) which allow digital signals to be transmitted over analog channels. Figure 4 illustrates a simple demodulation (or detection) scheme for amplitude shift keying. The

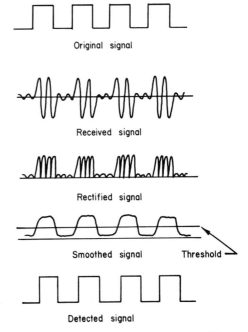

FIGURE 4 Demodulation used with ASK.

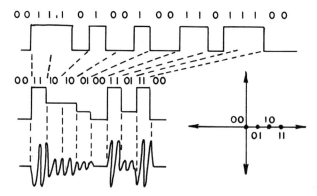

FIGURE 5 Amplitude modulation with four states and the corresponding signal-space diagram showing how the four amplitudes are coded in terms of magnitude and phase.

received signal is rectified and smoothed using a low-pass filter, and then the message signal is reproduced by passing the smoothed signal through a trigger circuit with the thresholds indicated.

C. Multiple Level Modulation

The preceding discussion assumed binary signals. With multiple level (M-level) modulation, it is possible to transmit more than two amplitudes, frequencies, or phases. If four levels of amplitude are used, as shown in Fig. 5, each distinguishable level can represent a dibit (two bits) of information. This doubles the effective bandwidth but only at the expense of greater susceptibility to noise. If eight levels were used, a tribit (three bits) could be sent at any one time, but the noise susceptibility would be even greater.

The noise susceptibility of M-level amplitude modulation can be reduced by using M-level phase modulation. Thus, dibits 00, 01, 10, and 11 could be represented by phase shifts of 90°, 0°, 180°, and 270°, respectively. M-level amplitude and phase modulation can be combined, and the result is sometimes known as *quadrature amplitude modulation* (QAM). Figure 6a shows a signal-space diagram (also known as the *signal constellation*) for a QAM system in which a combination of two levels of amplitude and four levels of phase shift (eight possible combinations) are used to transmit a tribit of information in one of eight possible (amplitude, phase) states. Figure 6b shows a more complicated modulation technique in which a quadbit (four bits) of information can be transmitted in one of 16 possible combinations of phase and amplitude modulation. Note that this scheme is much more complex than the others described previously. A synchronizing signal first establishes the absolute phase. The first bit, Q_1, of each quadbit determines the signal element amplitude to be transmitted. The next three bits, Q_2, Q_3, and Q_4, determine the phase shift or change relative to the absolute phase of the preceding element.

M-level modulation techniques such as QAM are finding increasing use in modern data-transmission systems. A 9600-bit/sec QAM modem, for example, maps a quadbit of information into one of 16 (2^4) possible combinations of amplitude and phase, as shown in the space-signal diagram of Fig. 6b and transmits 2400 quadbits/sec.

A related modulation technique is trellis code modulation (TCM), in which redundant code bits are added and transmitted with the data. A 14,400-bits/sec TCM modem, for example, assembles six data bits and generates a redundant seventh bit using two of the data bits and a binary convolutional encoding scheme. The resulting seven bits are mapped into one of 128 (2^7) possible combinations of amplitude and phase, much like QAM, and 2400 of these seven-bit symbols are transmitted each second to obtain the 14,400-bits/sec effective data transmission rate. The 14,400 comes about because only six of the seven bits represent actual data.

At the receiving end, the three encoded bits (the redundant bit and the two data bits used to generate it) select one of the eight (2^3) subsets, each consisting of 16 amplitude–phase combinations, while the remaining four bits select one of the 16 combinations from the selected subset. The redundant seventh bit ensures that only certain sequences

Tribit	Relative Amplitude	Phase Shift (deg)
000	1	0
001	1	90
010	1	270
011	1	180
100	2	0
101	2	90
110	2	270
111	2	180

(a)

Quadbit Q_1	Relative Amplitude	Absolute Phase (deg)
0	3	0 , 90 , 180 , 270
1	5	0 , 90 , 180 , 270
0	$\sqrt{2}$	45 , 135 , 225 , 315
1	$3\sqrt{2}$	45 , 135 , 225 , 315

Quadbits $Q_2Q_3Q_4$	Phase Shift (deg)
000	45
001	0
010	90
011	135
100	270
101	315
110	225
111	180

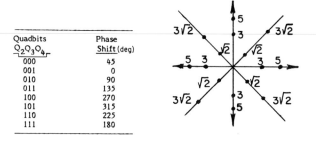

(b)

FIGURE 6 M-ary modulation using a combination of amplitude and phase modulation. (a) Two levels of amplitude and four levels of phase shift; (b) a more complicated technique.

of amplitude–phase combinations are valid and that any two valid sequences of combinations are far apart on the signal-space diagram to help in the decoding.

The major differences between QAM and TCM are the encoder that generates the redundant bit and the resulting signal-space diagram which, for TCM, has twice as many amplitude–phase combinations because of the introduction of the redundant bit. By careful choice of the signal-space diagram and the convolution code to generate the redundant bit, TCM can have a noise tolerance of twice that of QAM (see Payton, 1985) which translates to higher transmission rates over bandwidth-limited channels. Typical modems using QAM are limited to 9600 bits/sec while newer TCM modems can operate reliably at speeds up to 33,600 bits/sec over telephone channels. Newer so-called 56,000 bits/sec modems are asymmetric in that they can typically receive data at up to 56,000 bits/sec but transmit at only up to 33,600 bits/sec using different modulation techniques for the transmit and receive directions. Because such modems can use data compression, they can achieve effective throughput rates two or more times the above rates on compressible files.

D. Pulse Modulation

All the modulation techniques described previously use a continuous sinusoidal carrier wave that is modulated by analog or digital signals. Because of the continuous carrier wave, these are sometimes referred to as continuous-wave (CW) modulation techniques. In pulse modulation, the carrier is not a continuous wave but a periodic pulse train whose amplitude, duration, or position is varied in accordance with the message. Pulse amplitude (PAM), pulse duration (PDM), and pulse position (PPM) modulation are illustrated in Fig. 7. Note that PPM consists of equal-width pulses derived from the trailing edge of PDM pulses. PPM has an advantage over PDM since the latter can require significant transmitter power for transmitting pulses of long duration.

E. Pulse Code Moudulation

In pulse amplitude modulation, the amplitude of the pulse can assume any value between zero and some maximum value. Pulse code modulation (PCM) is derived from PAM but is distinguished from the latter by two additional signal-processing steps, called *quantizing* and *encoding*, that take place before the signal is transmitted. Quantizing replaces the exact amplitude of the samples with the nearest value from a limited set of specific amplitudes. The sample amplitude is then encoded, and the codes are transmitted typically as binary codes. This means that, unlike other modulation techniques described so far, in PCM

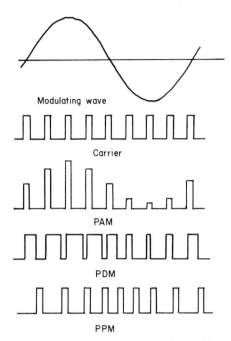

FIGURE 7 PAM, PDM, and PPM using a sinusoidal modulating wave and a pulse carrier.

both the sampling time and the amplitude are in discrete form.

Representing an exact sample amplitude by one of 2^n predetermined and discrete amplitudes introduces an error called *quantization noise* that can be made negligible by using a sufficiently large number of quantizing levels. Studies have shown that using 8 bits per sample to represent one of 256 quantizing levels provides a satisfactory signal-to-noise ratio for speech signals (see Rey, 1983). The sampling rate is usually determined from the sampling theorem, which states that a baseband (information) signal of finite energy with no frequency components higher than W Hz is completely specified by the amplitudes of its samples taken at a rate of $2W$/sec. The corollary of the sampling theorem states that a baseband analog channel can be used to transmit a train of independent pulses at a maximum rate that is twice the channel bandwidth W. These results are important in determining appropriate sample rates and bandwidths for conversion between analog and digital signals.

Applying the sampling theorem to speech signals that are limited to 4000 Hz, we find that they need to be sampled 8000 times/sec to be completely specified. Using PCM with 8 bits to represent one of 256 discrete amplitude samples, 8×8000 or 64,000 bits/sec are required to transmit the 4000-Hz voice signal. If we now use the corollary to the sampling theorem, we find that a channel with a bandwidth of 32,000 Hz is required to transmit the 64,000 bits/sec needed to specify the 4000-Hz voice signal. Although it is true that PCM requires more bandwidth than the baseband analog signal (32,000 Hz bandwidth for the 4000-Hz voice signal in the above example), this is more than offset by the following:

1. PCM has very high immunity to noise.
2. PCM repeater design is relatively simple.
3. The PCM signal can be completely reconstructed at each repeater location by a process called *regeneration*.
4. PCM provides a uniform modulation technique suitable for other signals on many different types of media including wire, coaxial cable, free space, and optical fibers.
5. PCM is compatible with time division multiplexing.

F. Comparison of Modulation Techniques

The various modulation techniques that have been described have distinct advantages and disadvantages in terms of cost, immunity to noise, and other impairments commonly encountered in communication channels. Amplitude modulation (AM) is widely used for radio programs. It is easy to maintain and relatively low in cost, but it is susceptible to noise. Frequency modulation is more effective in terms of noise tolerance and more suited for data transmission than AM. Phase modulation is more complex and costly but is relatively immune to noise and theoretically makes the best use of bandwidth for a given transmission rate. Various forms of phase and hybrids of phase and amplitude modulation (QAM and TCM) are increasingly used for data communication over analog channels at rates up to 56,000 bits/sec. Although it requires higher bandwidth, PCM has the advantage that the signal is regenerative and has greatest immunity to noise. Optical fibers, with their very high bandwidth, are well suited for PCM. Consequently, fibers and PCM are rapidly becoming the two leading technologies for transmission of data and digitized analog signals.

VII. MULTIPLEXING

Multiplexing is a technique that allows a number of lower bandwidth communication channels to be combined and transmitted simultaneously over one higher bandwidth channel. At the receiving end, demultiplexing recovers the original lower bandwidth channels. The main purpose of multiplexing is to make efficient use of the full bandwidth of a communication channel and achieve a lower per channel cost. The three basic multiplexing methods in use are space-division multiplexing, frequency-division multiplexing, and time-division multiplexing.

A. Space-Division Multiplexing

Space-division multiplexing refers to the physical grouping together of many individual channels or transmission paths to save physical space. A large number of wire pairs, coaxial cables, and/or optical fibers are usually grouped together to form a larger cable, such as the one illustrated in Fig. 11a. Each wire pair, fiber, or coaxial cable in the main cable is a communication channel giving a high total aggregate bandwidth. With each such individual channel in the cable capable of being frequency- or time-division multiplexed, such cables have enough bandwidth to carry more than 100,000 two-way voice channels in a cable diameter of under 3 inches.

B. Frequency-Division Multiplexing

As illustrated in Fig. 8, frequency-division multiplexing (FDM) divides the frequency spectrum of a higher bandwidth channel into many individual smaller bandwidth communication channels. Signals on these channels are transmitted at the same time but at different carrier frequencies. Guard bands are needed between the frequency channels to reduce interchannel interference.

Perhaps the most familiar example of FDM is radio broadcasting. Stations broadcast continuously but are assigned and occupy a different frequency in the radio spectrum. Program signals use amplitude or frequency modulation to modulate a carrier whose frequency is allocated to that station. Tuning circuits in the radio receiver select a given frequency and allow the signal from one station to be separated from the others. The signals are in the form of modulated electromagnetic waves using the atmosphere as the communication channel.

Frequency-division multiplexing for voice signals over telephone channels is very similar. At one end, a number of modulators with carrier frequencies differing by 4000 Hz modulate the various channels onto the higher bandwidth channel. At the receiving end, an equal number of demodulators tuned to the same frequency bands as the modulators receive and demodulate the multiplexed signal into the corresponding channels. This is

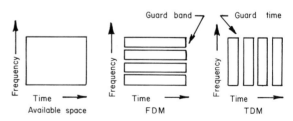

FIGURE 8 Relationship between frequency and time in FDM and TDM.

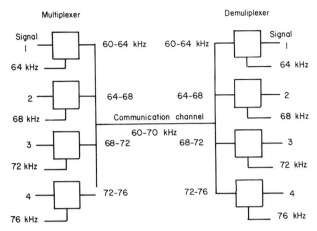

FIGURE 9 Frequency-division multiplexing

illustrated in Fig. 9 for one-way transmission, for the sake of simplicity.

C. Time-Division Multiplexing

Time-division multiplexing (TDM) operates by giving the entire channel bandwidth to a stream of characters or bits from each of the low-speed channels for a small segment of time, as illustrated in Fig. 8. This allows many low-bandwidth channels to be accommodated on a high-bandwidth channel by interleaving them in the time domain. Guard times are used to separate time slices, and the transmitting and receiving ends must be synchronized. A familiar example of a particular type of TDM might be the input–output bus of a computer servicing many peripherals for short periods of time, one at a time.

D. Multiplex Hierarchies

As networks grow and get more complex, hierarchies of multiplexing are required whereby low-bandwidth channels are multiplexed onto high-bandwidth channels, which in turn are multiplexed onto even higher bandwidth channels, and so on. In the FDM hierarchy, multiplex levels correspond to increasingly higher frequency bands.

In the TDM hierarchy, multiplex levels correspond to increasingly higher pulse rates. For example, using PCM, each voice channel requires 64,000 bps. Twenty-four voice channels can be time-division multiplexed onto a T1 carrier operating at 1.544 Mbps, while four T1 carriers can be multiplexed onto a T2 carrier operating at 6.312 Mbps. A T4M carrier operates at 274.176 Mbps and can carry traffic from 42 T2 carriers or up to 4032 voice channels.

When hierarchies of multiplexing are used, it is important to have accurate carrier frequencies (pilot tones) for FDM and stable timing signals for TDM to synchronize transmitting and receiving stations.

VIII. TYPES OF DATA TRANSMISSION MEDIA

Data transmission media provide the physical communication channel to interconnect the transmitting and receiving ends. Media that constrain and guide or conduct communication signals are called *bounded* or *guided media*. Media that permit signals to be transmitted but do not guide them are called *unbounded media*. Bounded media include open wires or other conductors, twisted wire pairs, coaxial cables, waveguides, and optical fibers which are conductors of light. The atmosphere and outer space are examples of unbounded or unguided media for transmitting broadcast radio, television, terrestrial microwave radio, and satellite communications. Table II summarizes the various media, the main applications, the typical frequency bands or bit rates used, and the relative frequency of use.

It should be noted that the multiplexing hierarchies described in Section VII.D imply hierarchies in the transmission media. A single intercontinental voice or data message, for example, may use the following hierarchy of transmission media:

1. Twisted pair between source and the local telephone exchange

TABLE II Commonly Used Transmission Media

Medium used	Main application	Typical frequency range	Relative use
Open conductor	Very short distances	Varies with distance	Very high
Twisted wire pair	Up to 50 miles	Varies with distance	Very high
Coaxial cable	Short to long haul	3–500 MHz	High
Waveguide	Short distances	3–12 GHz	Low
Optical fiber	Short or long haul	Up to 16 Gbits/sec	High
Atmosphere/microwave radio	Short haul	10.7–11.7 GHz	Medium
	Long haul	3.7–6.425 GHz	High
Space/satellite	Long haul	5.925–6.425 GHz up; 3.7–4.2 GHz down	Medium
		14.0–14.5 GHz up; 11.7–12.2 GHz down	High
Atmosphere/broadcast radio	Short to long distances	Varies with distance	Very high

2. Microwave transmission between the local exchange and the central office which switches lines onto higher capacity trunks and vice versa
3. Optical fiber cable between the central office and the toll office which switches the continental trunk to an overseas trunk
4. Satellite transmission or optical fiber between toll offices on the two continents
5. Optical fiber from the intercontinental toll office to the overseas local exchange
6. Twisted pair between the overseas local exchange and the final destination

IX. TRANSMISSION IMPAIRMENTS

In practice, communication signals are degraded by physical limitations of the transmission media such as bandwidth, impairments arising from within the channel such as echo, and impairments introduced from outside the channel such as impulse noise. The impact of such impairments depends on the type of signal transmitted and whether the channel is designed for analog or digital transmission. They can attenuate, amplify, or severely distort the signal as it passes through the communication channel and introduce errors. Impulse noise, which can be caused by electrical storms, for example, has little effect on the reception of analog speech signals because of amplitude limiters in the channel and the relative tolerance of the human ear to occasional errors. However, such noise can obliterate digital data signals.

The planning, design, installation, operation, and maintenance of communication channels for transmission of analog or digital signals depend upon an understanding of the impairments to which these channels are susceptible and the effect of these impairments on the signal. Some commonly encountered impairments are briefly described below. Other impairments related to specific transmission media will be described in later sections.

A. White or Gaussian Noise

White noise has a normal (Gaussian) amplitude distribution and causes the background hiss occasionally encountered over telephone or radio channels.

B. Impulse Noise

Impulse noise usually consists of a short-duration (less than 100 to 200 milliseconds), high-amplitude burst of noise energy that is much greater than normal peaks of message circuit noise. The short duration does not unduly impair analog speech signals but can obliterate digital data

signals. For this reason, data signals are usually transmitted in blocks containing error-detecting codes so that when errors are detected, the data can be retransmitted. Because of its relative short duration, impulse noise results in greater error rates at higher data rates since at higher rates it becomes increasingly difficult to distinguish between a data and noise pulse. Impulse noise is usually caused by lightning storms and voltage transients by electromechanical switching systems. The objective for telephone lines is to have no more than 15 impulse counts in 15 minutes for 50% of all telephone communications.

C. Cross-Talk Noise

Cross-talk is usually caused by capacitive or inductive coupling between adjacent channels. It can be a problem in high-speed digital circuits. If it is intelligible on analog channels (i.e., a conversation on an adjacent circuit can be overheard), it becomes particularly objectionable, not only because it impairs the conversation but also because it creates a loss of privacy. Methods to control cross-talk include shielding conductors, separating tightly coupled circuits, impedance matching of lines, and suppressing nonlinearities. Telephone companies have a goal of limiting the cross-talk index, which is the actual percentage probability of receiving an audible, intelligible speech signal on a call, to under 0.5% on most of their channels.

D. Quantizing Noise

Quantizing noise is caused by errors introduced when quantizing an analog signal into one of 2^n discrete amplitudes. It can be controlled by using a code with a suitably large value of n (e.g., 8 for voice signals). A codec is a coder-decoder that is used in PCM systems for analog-to-digital and digital-to-analog conversion.

E. Impedance Mismatch

If long wires, wire pairs, and coaxial cables are not terminated in their characteristic impedance, reflections can take place, which can cause high standing-wave ratios that appear as noise in the system. This can be overcome by terminating long lines with a load impedance that is equal to or very close to the characteristic impedance of the line, as explained in Section XII.

F. Attenuation Distortion

Attenuation distortion (sometimes called *amplitude/frequency distortion*) occurs when the relative magnitude of different frequency components of a signal are altered during transmission over a channel. A transmission medium that is ideal has a flat frequency response for the range of frequencies over which the spectrum of the information signal is nonzero. The distortion is caused by capacitive and inductive reactances in series and in parallel with the electrical conductors comprising the channel. Attenuation distortion is much less a problem in voice transmission than in data transmission because of the redundant nature of speech signals as opposed to the small amount of redundancy in data signals, where the loss or alteration of one bit usually alters the meaning of the code word in which the bit is contained.

G. Envelope-Delay Distortion

A signal that carries information has a phase component in addition to amplitude and frequency. To obtain an undistorted signal, it is required that the transmission medium have not only a flat amplitude-versus-frequency characteristic but also a linear phase-versus-frequency characteristic. As in the case of attenuation distortion, typical transmission media do not exhibit ideal phase response because of distributed stray inductance, capacitance, resistance, and conductance which affect the time or phase relationship between various frequency components of a transmitted signal. This is known as *envelope delay*, or *group distortion*, and causes various frequency components of a complex signal to propagate through the transmission medium at different velocities, causing the composite signal at the receiver to be distorted. If the distortion is sufficiently large, late-arriving energy from one pulse can interfere with the start of the next pulse, resulting in a phenomenon known as *intersymbol interference*. Equalizers are devices that attempt to equalize the envelope delay and compensate for the different components that make up amplitude distortion.

H. Frequency Translation

Frequency translation is a phenomenon that results in a frequency shift whereby all frequency components in the modulated signal are shifted by a constant amount. It is generally due to oscillator drift or frequency offset in the carrier wave equipment. Frequency translation can be especially troublesome in systems that transmit FDM signals. This type of system relies heavily on channel filters in the modulation–demodulation process. Since frequency translation causes a shift in the energy spectrum for which the system was designed, the spectrum shift causes some of the desired information-signal energy to encounter the undesirable amplitude and phase-distortion characteristics that are usually found at the band edges of filters. If the frequency error is sufficiently large, the system is seriously degraded, especially for high-rate data transmission. This

is because many modems use carriers that are synchronized between the receiver and transmitter, and frequency errors in the media appear as errors in the modems. On telephone channels, the objective is to hold the frequency shift to ± 2 Hz for acceptable performance.

I. Phase Jitter

Phase jitter refers to small changes in the phase of the received signal and appears as a slight frequency modulation on the carrier or baseband signal. It often takes the form of multiples or submultiples of ac power frequencies. It is usually caused by reactive coupling between equipment associated with power lines or jitters in signals used for timing (clocks). Phase jitter is of little consequence in voice transmission since the human ear is not sensitive to small changes in phase or frequency. However, phase jitter can seriously deteriorate data transfer since it causes the data pulses to jitter, and, if large variations occur, then one data pulse may try to occupy the time slot of another and cause an error, especially in TDM systems. In addition, phase jitter may make synchronization of the transmitter and receiver more difficult.

J. Harmonic Distortion

Harmonic distortion is usually caused by clipping or limiting the transmitted signal, which causes higher harmonics of the transmitted signal at the receiver. In most cases, this is not a serious problem.

K. Echo

Talker echo is a telephone line impairment that is caused by reflections when long lines are not terminated in their characteristic impedance. It causes part of the speaker signal to be reflected back and can seriously interfere with the talker's speaking process. If the elapsed time (i.e., the delay) is long and the echo path loss is inadequate (i.e., there is not enough loss to attenuate the echo sufficiently), the interference may be so great that speaking is nearly impossible. Impedance mismatches are usually caused by hybrids in the signal path. Hybrids are circuits used for converting a two-wire circuit to a four-wire circuit and vice versa. Echoes can be controlled using echo suppressors, which detect echoes and automatically insert a loss over the listener's speaking path to prevent the echo from reaching the talker. When both speak at the same time, the suppressor inserts a loss in both directions, resulting in undesirable clipping of speech signals. An echo canceler, rather than inserting a loss in the return path, uses the transmitted speech signal to generate a replica of the echo and subtracts this signal from the echo, thus canceling it.

Using echo suppressors and cancellers reduces the echo to tolerable levels on 99% of all telephone connections in which this problem is encountered (Rey, 1983).

L. Inductive Interference

Because power and telephone companies tend to serve the same customers and share the same right of way, telephone wires are exposed to electromagnetic fields created by power distribution systems. If the telephone lines are not perfectly balanced, a voltage difference between the two conductors of a cable pair may develop in the presence of a strong electromagnetic field that can result in an audible noise level. This type of inductive interference is sometimes known as *electromagnetic interference*, or EMI for short. Telephone sets and transmission media are designed to have high tolerance of 60-Hz hum, but higher harmonics can cause problems. This is minimized by balancing lines, grounding cable shields, and separating power and telephone media where necessary. As a last resort, magnetic core devices are used to minimize this type of noise.

M. Message Circuit Noise

Message circuit noise is a signal composed of noise from a number of sources such as power hum, inductive interference, Gaussian noise, central office noise, and impulse noise, which have been previously discussed. It causes impairments similar to those caused by its various component parts.

N. Ground Noise

Ground noise is caused by voltage differences between the receiving and transmitting ends due to current flow through the ground return path. Transmission media consisting of a single wire over a ground, as shown in Fig. 10a, are especially susceptible to this type of noise.

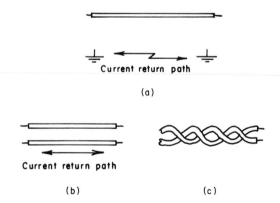

FIGURE 10 Single and paired cables. (a) Single cable over ground; (b) open pair; (c) twisted pair.

X. BOUNDED TRANSMISSION MEDIA

Bounded transmission media constrain and guide communication signals. They include single-wire (or single conductor) lines, paired cable, coaxial cable, optical fibers, and waveguides. The first three media are electrical conductors, optical fibers are conductors of light, and waveguides constrain and guide electromagnetic waves.

A. Single-Wire Lines

Single-wire lines use a single wire or some other electrical conductor to provide a path for an electric current with an electrical ground providing the return path, as shown in Fig. 10a. Since this is the medium used almost exclusively to interconnect components in all types of electronic systems over short distances (most connections between components on a circuit board or even on a silicon chip are examples of a single-wire-over-ground transmission medium), a substantial portion of Section XII will be devoted to this medium, with emphasis on its proper use in circuits where high bandwidths are involved.

B. Paired Cable

Because of ground noise problems associated with a single conductor when transmitting signals over longer distances, two-conductor paired cable is used, with the second cable providing the return path for the signal current. The two cables can be parallel to each other (open wire pair) or twisted (twisted pair), as shown in Figs. 10b and c. Open wires are susceptible to cross-talk and electromagnetic interference from power lines. As explained in Section XII, twisted pairs reduce the distance between the conductors and reduce the inductive cross-talk by reducing the area of the current loop. Consequently, twisted pairs have replaced open pairs for telephone transmission except in some rural areas. To further reduce susceptibility to noise and electromagnetic interference, cable pairs can be shielded. In many cases, space-division multiplexing is used by combining many twisted pairs into a larger cable. Twisted pairs are stranded into a ropelike form called a *binder group*, and several binder groups are twisted together around a common axis to form the cable core. A protective sheath is wrapped around the core, resulting in a cable like that in Fig. 11a. Paired cable is made in a number of standard sizes and may contain from two to several thousand cable pairs.

Paired cable is the main type of medium for local telephone and data transmission. It can be installed easily using commonly available tools. Today, it is typically used for relatively short distances (normally less than 10 miles). It has bandwidth limitations and is susceptible to noise.

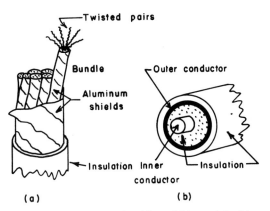

FIGURE 11 (a) Multipair cable and (b) coaxial cable.

As frequency or bit rates of transmission increase, the repeaters needed to amplify or regenerate the signals must be placed closer together to overcome the increased effect of the transmission impairments. With proper design, twisted pair cables can carry about 12 voice channels per pair with repeaters every 2 to 4 miles.

Table III briefly summarizes the various categories of unshielded twisted-pair cabling systems commonly used today and their main applications. Most new building cabling today consists of category 5 cable because it can be used for both telephone service and local area networking at up to 100 Mbits/sec. New standards are currently being discussed for higher category cables to extend local area network support up to 1000-Mbits/sec or gigabit ethernet.

C. Coaxial Cable

Coaxial and twinaxial (two coaxial cables in a single jacket) cables are used for analog or digital transmission requiring high bandwidth. As shown in Fig. 11b, a coaxial cable consists of an inner conductor completely surrounded by an outer conductor, with high-quality insulation between the two. The outer conductor is surrounded by a protective sheath. As with twisted pairs, coaxial cables can be bundled together with twisted pairs to form a larger cable with very high total bandwidth.

Coaxial cables have various characteristic impedances, with 50, 75, and 93 ohms being the most common. They have the advantage of operating at very high frequencies, which allows them to carry a very large number of analog or digital channels using FDM. Since the outer conductor is grounded and provides an effective shield that improves with increasing frequencies, the cross-talk between adjacent coaxial cables decreases with frequency, rather than increasing as is the case with unshielded cable pairs.

In the telephone network, coaxial cable was used primarily on intercity routes for the long-haul network where heavy traffic existed. The Bell L5 coaxial cable, for example, had a capacity of 10,800 voice circuits per carrier channel with repeaters every mile. In these applications, the installed cost per voice circuit is lower than for paired cable. Old intercontinental submarine cables also used coaxial cables, but in this environment the design, operational, and reliability requirements are different from cables used on land. More recently, coaxial cables were being used for local area networks within or between buildings. Such networks interconnect a large number of workstations and computers using baseband transmission with TDM or broadband transmission with FDM. They have a low incidence of errors, support high data rates and can be tapped into easily, and devices such as taps, controllers, splitters, couplers, multiplexers, and repeaters are readily available.

Coaxial cables are also used extensively in community antenna television (CATV) or cable TV systems for distribution of television and music programs. However, coaxial cables for voice and data are being rapidly replaced by other media. Specifically, modern local area networks tend to use category 5 (or better) twisted pairs for intrabuilding cabling and optical fibers for interbuilding cabling. Similarly, coaxial cables carrying voice signals have or are being replaced by optical fibers, since the latter are much more cost effective because of their high bandwidth capability.

D. Waveguides

A waveguide is a rectangular or circular pipe, usually made of copper, that confines and guides very high-frequency electromagnetic waves between two locations. Compared with coaxial cable, it has a very low attenuation at microwave frequencies, which is its main advantage. Its main disadvantages are that it must be manufactured with extreme uniformity to achieve the low attenuation and great care must be taken during installations to minimize sharp bends, which also increase attenuation. These disadvantages have limited the application of waveguides primarily to carrying 3- to 15-GHz-range signals from the base of microwave radio towers to the dish antennas at the top.

TABLE III Categories of Twisted Pair Cabling Systems

Category	Maximum data rate/(Mbits/sec)	Typical application
1	<1	Old telephone service
2	4	IBM token-ring networks
3	16	10 Mbits/sec ethernet
4	20	IBM 16 Mbits/sec token ring
5	155	100 Mbits/sec ethernet, 155 Mbits/sec ATM

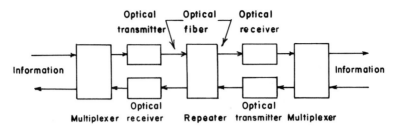

FIGURE 12 Typical optical fiber communication system.

E. Optical Fibers

Optical fibers have become the preferred medium for terrestrial communication because they can carry gigabits of information per second over short or long distances. The basis of fiberoptic, or lightguide, transmission systems is an optical fiber, which is a thin, flexible glass or plastic fiber through which light is transmitted. An optical fiber is actually a waveguide that guides the propagation of optical frequency waves directly or through total internal reflection at the fiber boundaries. A fiberoptic transmission system is shown in Fig. 12 and is similar to a conventional transmission system, except that the transmitter uses a light emitting diode (LED) or a laser diode (LD) to change electrical signals into light signals while the receiver uses a photodiode or similar device to convert the light signals back into electrical signals.

Compared with wire, twisted pairs, or coaxial cable, optical fibers offer advantages so great in terms of information-carrying capacity and signal protection that they have replaced many of the older media used in the last decade. These advantages include the following:

1. *Large bandwidth*. Optical fibers with suitable optical transceivers have bandwidths of tens of GHz compared with an upper limit of 500 MHz for coaxial cable. These bandwidths are achievable because the information-carrying capacity of a transmission medium increases with the carrier frequency and for optical fibers the carrier is light, which has frequencies several orders of magnitude higher than the highest radio frequencies. With currently available systems, a single fiber cable can accommodate 150,000 voice-frequency circuits, and this will increase rapidly with the introduction of new optical fiber technologies.

2. *Low losses*. As shown in Fig. 13, the attenuation of an optical fiber is essentially independent of the modulation frequency, whereas copper wires and coaxial cables exhibit increasing attenuation with increased frequency. The advantage of the flat attenuation response of low-loss optical fibers is that repeaters need not be placed so close together. While a 44.7-Mbit/sec T3 carrier in a digital telephone system has a standard repeater spacing of 1.1 km for coaxial cable and 6.4 km for older optical fibers, optical systems with repeater spacing of more than 100 km are being planned using new, low-loss fibers. This means that interoffice trunks will require no repeaters since the distances involved are usually

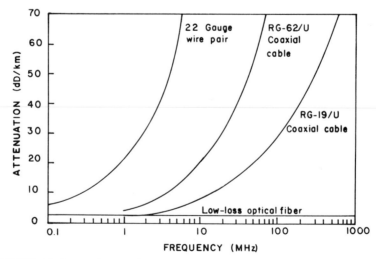

FIGURE 13 Attenuation as a function of frequency for some commonly used media.

shorter than the above repeater spacing for new fibers.

3. *Electromagnetic interference (EMI) immunity.* Because optical fibers are made of insulating materials such as glass and plastic, they are not affected by ordinary electromagnetic fields. With conductors such as copper wire, shielding is required in many cases to reduce the effects of stray magnetic fields. The EMI immunity of optical fibers also offers the potential for signal protection through low error rates of 1 per 10^{10} bits or better. This has the potential for reducing the cost and overhead associated with the complex error checking and error-correction mechanisms that have to be used with other transmission media.

4. *Small size and light weight.* A fiber with a 0.062-mm diameter and a 3.5-mm protective jacket has far more information-carrying capacity and can easily replace a cable consisting of 900 twisted copper wire pairs with an 8-cm outer diameter. The smaller size for a given capacity and the fact that glass and plastic weigh less than copper mean that fibers are lighter by a factor of 10 to 100 than copper cables of equivalent capacity. Both size and weight are important factors when considering overcrowded conduits running under city streets.

5. *Security.* Unlike electrical conductors, optical fibers do not radiate energy. Consequently, eavesdropping techniques that can be used on the former are useless with fibers. Furthermore, it is very difficult to tap an optical fiber since, in general, this affects the light transmission enough to be detected.

6. *Safety and electrical isolation.* Because fiber is an insulator, it provides electrical isolation between data source and destination. This avoids ground noise often encountered when systems are interconnected by electrical conductors. Furthermore, fibers present no electrical spark hazards and can be used in applications where electrical codes and common sense prohibit the use of electrical conductors.

Despite their many important advantages, optical fibers have several disadvantages. They are a relatively new technology and not yet understood well by enough designers, who are comfortable designing with electrical conductors and who base their solutions on what they know best. This situation is improving rapidly as industry-wide standards related to this new technology continue to emerge; the costs of light sources, detectors, fibers, and other related hardware continue to fall with increased use; and more technicians are trained to handle, install, and maintain optical fiber systems. Since it has been established that this technology can outperform cable as a transmission medium in many situations, the next decade should see a great increase in its use.

1. Types of Fibers

Optical fibers are made of plastic, glass, or silica. Plastic fibers are the least efficient but tend to be larger, more economical, and more rugged. Glass or silica fibers are much smaller, and their lower attenuation loss makes them more suited to high-capacity channels. A third type of fiber, with size and performance intermediate to the other two types, has a glass or silica core with a plastic cladding.

The basic optical fiber consists of two concentric layers—the inner core and the outer cladding, which has a refractive index lower than the core. The construction is such that light injected into the core always strikes the core-to-cladding interface at an angle greater than the critical angle and is therefore reflected back into the core. Since the angles of incidence and reflection are equal, the light wave continues to zigzag down the length of the core by total internal reflection, as shown in Fig. 14a. In effect, the light is trapped in the core, and the cladding not only provides protection to the core but also may be thought of as the "insulation" that prevents the light from escaping. The exact characteristics of light propagation depend on fiber size, construction, and composition, as well as on the nature of the light source. Although fiber performance can be reasonably approximated by considering light as rays, more exact analysis must deal with field theory and solutions to Maxwell's equations, which show that light does not travel randomly through a fiber but is channeled into modes that represent allowed solutions to the field equations. Consequently, it is more informative to classify fibers by refractive index profiles and the number of transmission modes supported.

The two main types of refractive index profiles are step and graded. In a step-index fiber, the core has a uniform refractive index n_1 with a distinct change or step to a lower

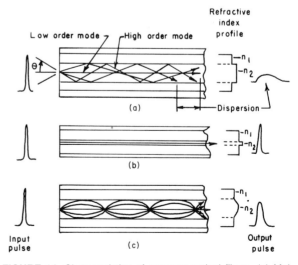

FIGURE 14 Characteristics of common optical fibers. (a) Multimode step index; (b) single-mode step index; (c) multimode graded index.

index n_2 for the cladding. In a graded index fiber, the refractive index of the core is not uniform but is highest at the center and decreases until it matches that of the cladding. The index profiles and the light-wave propagation supported by these different fibers are illustrated in Fig. 14.

The simplest type of fiber is the multimode step-index fiber, which usually has a core diameter from 0.05 to 1.0 mm. The relatively large core allows many modes of light propagation, and since some rays follow longer paths than others, their original relationship is not preserved. The result is that a narrow pulse of light has a tendency to spread as it travels down the fiber, as shown in Fig. 14a. This spreading of light pulses is known as *modal dispersion* and is 15 to 40 nsec/km for typical multimode step fibers. Consequently, such fibers tend to be used for transmission over short to medium distances.

Modal dispersion can be reduced or eliminated by using a fiber with a core diameter small enough that the fiber propagates efficiently only the lowest order mode along its axis, as shown in Fig. 14b. Single-mode, step-index fibers typically have core diameters between 0.002 and 0.01 mm and are very efficient for very high-speed, long-distance transmission. Their small size, however, makes them relatively difficult to work with, especially when fibers must be linked.

Modal dispersion can also be reduced by using graded-index fibers, as shown in Fig. 14c. The core is a series of concentric rings, each with a lower refractive index moving from the center of the core to the cladding boundary. Since light travels faster in a medium of lower refractive index, light farther from the fiber axis travels faster. This means that in a graded-index fiber, high-order modes have a faster average velocity than lower order modes; therefore, all modes tend to arrive at any point at nearly the same time. Light is refracted successively by the different layers of the core, with the result that the path of travel appears to be nearly sinusoidal. In such fibers, modal dispersion is typically well under 10 nsec/km and sometimes even less than 1 nsec.

As in the case of electrical conductors, optical fibers are protected by enclosing many fibers in a protective sheath, which is usually made of some tough plastic material. Because the fibers are so thin, the cable is usually strengthened by adding steel wire or Kevlar aramid yarn with a high Young's modulus to give the cable assembly flexibility and tensile strength. Cables are also available containing both optical fibers and electrical conductors, with the latter usually being used to provide power for remotely located repeaters.

2. Signal Degradation

Signal degradation in optical fiber systems is caused by one or more of the following:

1. *Attenuation.* Attenuation, or transmission loss (dimming of light intensity), is caused by absorption and scattering. Absorption is the optical equivalent to electrical conductor resistance and is usually caused by fiber impurities that absorb light energy and turn it into heat. The amount of absorption by these impurities depends on their concentration and the light wavelength used. High-absorption regions, such as that occurring in the 950-nm wavelength range due to hydroxyl ion (OH^-) impurities, should be avoided, for example. Scattering results from imperfections in fibers. Rayleigh scattering comes from the atomic and molecular structure of the core and from density and composition variations that are introduced during manufacturing. Unintentional variations in core diameter, microscopic bends in the fiber, and small discontinuities in the core-to-cladding interface also cause scattering loss. For commercially available fibers, attenuation ranges from about 1 db/km (decibel per kilometer) for premium small-core fibers to over 2000 db/km for large-core plastic fibers. Since attenuation losses vary with the light wavelength used, it is important for optical performance to match carefully the characteristics of the light-source transmitter and the fiber.

2. *Dispersion.* Dispersion, a measure of the widening of a light pulse as it travels along the fiber, is usually expressed in nsec/km. Dispersion limits the bandwidth or information-carrying capacity of a fiber, since input pulses must be separated enough in time that dispersion does not cause adjacent pulses to overlap at the destination to ensure that the receiver can distinguish them. Dispersion can be considered as the optical analog of envelope delay distortion in electrical conductors. Dispersion can be of two types. Modal dispersion arises from the different length of paths traveled by the different modes. Material dispersion is due to different velocities of different wavelengths. In single-mode fibers, which exhibit no modal dispersion, material dispersion is the sole frequency-limiting mechanism. High-quality optical fibers have a dispersion of about 1 nsec/km or even less.

3. *Other losses.* Interconnection is a critical part of optical fiber communication links. Fibers must be connected or spliced to provide low-loss coupling through the junction. Precise alignment results in low loss, but the small size of fiber cores together with dimensional variations in core diameter and alignment mechanisms make this a formidable task. Factors affecting these losses include:

- Differences in core diameters of fibers to be joined
- Differences in the numerical aperture of fibers to be joined; the numerical aperture (NA) is a measure of the light-gathering capability of a fiber and is equal to $\sin \theta$ where θ is the half-angle of the cone within which all incident light is totally internally reflected by the fiber core, as shown in Fig. 14a.

- Core-to-cladding eccentricity, which can cause misalignment
- Core ellipticity, which can reduce the contact area
- Lateral misalignment of the axes of the two cores to be joined
- Angular misalignment of the two cores
- Fiber-end separation
- Tilting angle between the two cores if the fiber is not cut on an axis that is perfectly perpendicular to the core axis

A splice, which refers to a permanent interconnection, has a typical loss of about 0.1 db. A connection refers to detachable or temporary interconnections, which in practice have less precise alignments, and results in a loss of up to 4 db using inexpensive connectors for large fibers.

3. Light Sources and Detectors

In an optical fiber communication system, the light source must efficiently convert electrical energy (current and voltage) into optical energy in the form of light. A good source must be

1. Small and bright, to permit the maximum transfer of light into the core of the fiber
2. Fast, to respond to rapidly changing modulating signals encountered in high-bandwidth data systems
3. Monochromatic (i.e., produces light within a narrow band of wavelengths), to limit dispersion within the fiber
4. Reliable, with a lifetime in the tens of thousands or preferably in the hundreds of thousands of hours of operation

The most commonly used light sources are gallium arsenide light emitting diodes (LEDs) and laser diodes (LDs). These devices are small, with sizes compatible with the cores of fibers, and emit light wavelengths in the range of 800–900, 1300, and 1530–1560 nm. In general, LEDs are not as monochromatic as LDs; consequently, they tend to be used for applications that require bit-rate capacities of several hundred Mbits/sec or less. They have a typical lifetimes of 100,000 hours and a light-output power characteristic that is almost linear over a large range of driving current and hence can be used for transmitting analog signals using amplitude modulation. LDs produce light with a much narrower spectrum than LEDs and can transfer more power into the fiber. This combination of characteristics allows them to be used in systems working at Gbit/sec rates, and even at hundreds of Gbits/sec using cavity lasers and dense wavelength-division multiplexing (DWDM).

Optical detectors convert optical energy into electrical energy. Devices most commonly used for this purpose include PIN and avalanche photodiodes (APDs). A PIN diode is a specially made diode with intrinsic material between the P and N materials to give a faster response time to incident light energy. The photodiodes are usually made of silicon because of their sensitivity in the 750- to 950-nm wavelength region. Silicon PIN diodes convert light power input to electrical current output with a quantum efficiency of over 90% and a response time of 0.5 μsec or less. APDs have much faster response times than PIN diodes and can be used at higher bit rates.

4. Modulation in Optical Fiber Systems

Although the light output power of LEDs and LDs is relatively linear over a wide range of drive currents indicating they are suitable for continuous-wave modulation (e.g., amplitude modulation), today's optical fiber communication systems are more suitable to digital rather than analog operation. In the digital mode, the light source is switched on and off, which greatly simplifies the detection process at the destination.

The selection of line-signaling format is an important consideration in optical communication systems. It is desirable to use a format that is self-clocking at the receiver and conserves output power at the source, especially when using LD sources, since they should be driven to high power output for only short intervals to improve their lifetimes. Some of the popular digital signaling formats are illustrated in Fig. 15, including:

1. Nonreturn to zero (NRZ), in which a change of state occurs only if there is a 1-to-0 or a 0-to-1 transition. A string of ones is a continuous "on" condition, while a string of zeros is a continuous "off" condition.

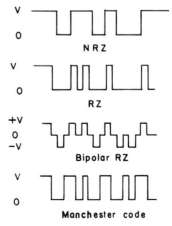

FIGURE 15 Examples of binary signal formats for optical fibers.

2. Return to zero (RZ) format, in which there is a complete pulse transmitted for each logic one condition. Note that the pulse width must be less than the bit interval to permit the return to zero condition.

3. Bipolar RZ format, in which a pulse is transmitted for each logical one and zero condition. Again, the pulse width must be less than the bit interval to permit the return to zero condition.

4. Manchester code format, in which, by convention, a logic zero is defined as a positive-going transition, while a logic one is defined as a negative-going transition occurring at times other than bit-time boundaries.

For optical systems, the RZ and Manchester code formats are good candidates because they are self-clocking and because the relatively short pulses, even for a continuous string of logical ones, conserve the life of LDs. Note, however, that these formats require at least twice the bandwidth of the NRZ format for a given bit rate. Pulse modulation techniques such as PDM and PPM are also sometimes used.

5. Wavelength-Division Multiplexing

In older conventional fiberoptic systems, a single light frequency or wavelength carries information along the fiber. Wavelength-division multiplexing (WDM), increases the information-carrying capacity of a fiber by assigning multiple incoming optical signals to specific light frequencies (or wavelengths) within a designated frequency band and multiplexing them onto a single fiber. This is the optical equivalent of conventional frequency-division multiplexing described in Section VII.B. The term *dense wavelength division multiplexing* (DWDM) is usually reserved for optical systems that use more than eight different optical wavelengths to simultaneously carry information over a single fiber.

The technologies that support WDM and DWDM include new types of laser diodes and optical amplifiers. The diodes can produce multiplicity of closely spaced wavelengths in the 1530- to 1560-nm region, with a spacing between wavelengths of 0.8 nm. The optical amplifiers consist of optical fibers doped with the rare element erbium (EDFAs, or erbium-doped fiber amplifiers) which can amplify all the wavelengths as a group when pumped by an external light source. This allows the optical signals to be boosted directly without converting them to electrical form. These technologies can support 32 channels (wavelengths) each capable of carrying 10 Gbits/sec, or a total of 320 Gbits/sec per fiber, and even faster systems are being planned.

A significant advantage of WDM is that each individual wavelength or channel can carry an independent stream of data at the same or at different bit rates as the other channels. Besides greatly improving the capacity of optical fibers, WDM therefore provides a mechanism for a unified optical communication infrastructure capable of meeting the new service demands of tomorrow.

F. Outlook for Bounded Media

Conducting wires, such as conductors over grounds, strip lines, and twisted pairs, provide a convenient mechanism for interconnecting electronic components to make up more complex systems. They are also widely used for local area networks within buildings and for connecting telephone sets to the local central office. Of all transmission media, they are the most pervasive and are likely to remain so in the near future since they are by far the least costly. Compared to other transmission media, they are also by far the most limited in terms of bandwidth, distances spanned, and susceptibility to interference. Because they radiate electromagnetic energy, they can interfere with other equipment and should not be used in an environment where data security is important. Although they will be increasingly replaced by other media for high transmission rates and distances greater than a few kilometers, they will remain the most pervasive medium for use over short distances because of their simplicity and low cost.

Coaxial cables have made inroads in high-frequency system interconnects, cable television, high-bandwidth telephone transmission systems, and local area networks. They will continue to be used for the former two applications but are rapidly being replaced by other media for the latter two applications. Specifically, most intrapremises local area networks now use category 5 (or better) twisted pairs, while optical fiber systems are rapidly replacing coaxial cable for wide area telephony and data transmission.

Major improvements in light sources, detectors, splicing technology, dense wavelength-division multiplexing equipment, and saliton transmission systems, together with the emergence of standards and reduction in component costs, make optical fiber systems the transmission medium of choice for future high-bandwidth, long-distance, terrestrial communication systems. The dominant application areas will continue to be telephone, video and data trunk lines, local area and local distribution networks where the distance and bandwidth requirements exceed the capabilities of category 5 wiring, and high-interference environments where electrical conductors would be too costly to shield. In the more distant future, optical and electro-optical technologies may be used to solve interconnection problems in high-speed digital systems, even for very short distances (Goodman et al., 1984). Today, electrical conductors are used for this purpose almost exclusively.

XI. UNBOUNDED TRANSMISSION

Bounded media require a physical connection between two points to conduct or guide current, electromagnetic waves, or light. In unbounded media, no such physical connection is required. Space or air is the transmission medium for electromagnetic waves, which are therefore unbounded by the medium. In the past, this medium was used almost exclusively for some forms of television or radio transmission, which includes broadcast, microwave, and satellite transmission. More recently, the use of this transmission medium has been increasing rapidly for voice and data communication using a variety of cell telephone and mobile computing technologies. The major advantages of using space or air as the transmission medium include:

1. No physical path is needed between the source and destination. This is extremely important since providing a voice line or cable between two points may be economically infeasible or physically impossible, as in the case of communication with a mobile computer.
2. The source (transmitter) and destination (receiver) can both be mobile.
3. Transmission can be point to point or point to many points simultaneously (multi-access) as in broadcast radio.
4. A broad spectrum from low to high bandwidth is available.
5. The method can be quickly implemented and no right of way is required for the transmission path.

Radio transmission also has some disadvantages. It is subject to interference and a number of propagation anomalies that may require considerable engineering effort to overcome. It also requires space in the crowded radio transmission spectrum and therefore may require licensing and design approval from regulatory agencies.

A. Radio Frequency Spectrum

In radio transmission, electromagnetic power from a radio transmitter is coupled by the transmitter antenna into air or free space. In radio reception, electromagnetic radio waves are intercepted by a receiving antenna and coupled into a receiver for detection. The antennas can be omnidirectional as in broadcast radio or highly directional as in microwave radio transmission. With an omnidirectional antenna, a receiver can receive signals (as well as noise and interference) from other transmitters located anywhere within receiving distance. With directional transmitting and receiving antennas, signals are transmitted in a very narrow angular sector (e.g., 1° for microwave transmission), thus a receiver can only detect signals emanating from a narrow sector. If the transmitting and receiving antennas are aimed at each other, good communication can be established with relatively low transmitted power and little interference to other radio links, a key point in conservation of the limited radio frequency spectrum shown in Table IV. Such point-to-point radio communication means that frequencies can be reused in the same general vicinity without interference. This should be contrasted with onmidirectional broadcast systems where frequencies can be reused only if transmitters using the same frequencies are located sufficiently far apart so as not to cause mutual interference at local receivers. Table IV shows some of the most common radio frequency bands in current use and their typical applications, characteristics, and principal modulation methods.

B. Signal Degradation

Radio waves, like light waves, are subject to reflection and refraction. They are also subject to attenuation losses due to atmospheric and natural phenomena such as rain, snow, and fog. These result in three major types of signal degradation: multipath interference, fading, and attenuation losses.

Reflections from the earth's surface, the ionosphere, natural or manmade objects, and atmospheric refraction can create multiple paths between the transmitting and receiving antennas. Depending on the relative path distance, the reflected wave is shifted in phase with respect to the original wave, which can cause interference at the receiver called *multipath interference*. Since the amount of phase shift is frequency dependent, the combined received signal is also frequency dependent, which can lead to serious problems in wideband transmission.

Fading is caused by abnormal changes in the refractive index of the atmosphere. Normally, the atmosphere refracts or bends radio waves back toward the surface of the earth. However, abnormal distribution of temperature, humidity, and heavy ground fog can cause radio waves to be bent toward the surface much more than normal so that they never reach the receiving antenna, causing changes in received signal strength or even complete loss. Variation or, specifically, reduction of received signal strength at different periods in time is called *fading*.

As transmission frequencies increase, path attenuation losses due to the atmosphere also increase. More serious losses are caused by fog, snow, and especially rain, which becomes very significant at frequencies above 4 GHz. The effects of these losses are fading and increased error rates. They are usually allowed for during the design process by using published meteorological data of the region in which a radio link is located.

TABLE IV   Radio Frequency Spectrum

Frequency	Name	Characteristics	Principal modulation methods[a]		Typical uses
			Analog	Digital	
3–30 KHz	Very low frequency (VLF)	Very noisy, very limited bandwidth	Seldom used	ASK, FSK, PSK	Sonar, long-range navigation
30–300 KHz	Low frequency (LF)	As for VLF	Seldom used	ASK, FSK, PSK	Navigational aids, radio beacons
300–3000 KHz	Medium frequency (MF)	Noisy, limited bandwidth, large transmitting antenna	AM	ASK, FSK, PSK	Commercial AM radio, maritime radiotelephone, distress calls
3–30 MHz	High frequency (HF)	Noisy, long distance but subject to fading, subject to interference, moderate antenna size	AM	ASK, FSK, PSK	Shortwave radio, CB radio, ship to coast, point to point
30–300 MHz	Very high frequency (VHF)	Line-of-sight range, moderate noise and interference, small antenna size	AM, FM	FSK, PSK, others	Television, FM radar, land and air mobile traffic, police
300–3000 MHz	Ultrahigh frequency (UHF)	Line-of-sight, high bandwidth, low noise, high congestion in some areas	FM	FSK, PSK, others	UHF television, radar, space telemetry, microwave links
3–30 GHz	Superhigh frequency (SHF)	Line-of-sight, bandwidth to 500 MHz, low noise, narrow antenna beams	FM	FSK, PSK, others	Microwave links, radar, satellite communication
30–30 GHz	Extremely high frequency (EHF)	Line-of-sight, bandwidth to 1 GHz, low noise, high attenuation, very small antenna	FM	FSK, PSK, others	Radar landing systems, satellite communication

[a] ASK = amplitude shift keying; FSK = frequency shift keying; PSK = phase shift keying.

C. Terrestrial Microwave Radio

Terrestrial microwave transmission uses highly directional antennas for line-of-sight propagation paths using frequencies in the 4- to 12-GHz range. The antennas are usually parabolic with diameters ranging from 12 inches to several feet, depending upon their spacing. The spacing between repeater stations depends upon the geographical terrain over a given route, the technology used in the terminal equipment, and the transmitter power permitted by the local regulatory agency. For long-distance transmission, typical repeater spacings are 20 to 30 miles, but longer spacings are possible in areas where atmospheric conditions result in little fading.

Microwave links can carry several thousand voice channels using frequency-division multiplexing. They are an important source of competition with coaxial systems and have the advantages that no physical facility is required to guide the microwave energy between separate stations and that these stations can normally be 20 to 30 miles apart as opposed to about 1 mile for coaxial systems.

Using the earth's atmosphere as the transmission medium, however, results in multipath interference and fading. This can be minimized using frequency diversity,

which means the transmission of the same radio signal over different microwave frequencies at the same time. Since, for a given set of atmospheric conditions, radio signals at different frequencies experience different degrees of multipath interference and fading, the strongest received signal is selected for retransmission or detection. Frequency diversity has the disadvantage of using more of the available radio frequency spectrum; consequently, space diversity is sometimes used as an alternative approach, especially for paths that experience severe fading. Space diversity uses two receiving antennas, usually mounted on the same tower and separated vertically by several wavelengths. By switching from the regular to the diversity antenna whenever the signal level drops, the received signal can be maintained nearly constant. Neither frequency nor space diversification is effective in countering attenuation caused by rain. The only remedy for this is the use of greater transmission power or shorter repeater spacing.

D. Satellite Transmission

Satellite transmission consists of a line-of-sight propagation path from a ground station to a communications satellite (up link) and back to an earth station (down link). The satellite is usually placed in a geosynchronous orbit about 22,300 miles above the earth so that it appears stationary from any point from which it is visible and acts like a repeater in the sky. The ground station includes the antennas, buildings, and electronics necessary to transmit, receive, multiplex, and demultiplex signals. The frequency spectrum used is similar to that used for terrestrial microwave radio. The ground station antenna is usually highly directional, while the satellite antenna has a larger beam width to cover a larger portion of the earth's surface and to be able to communicate with many widely separated earth stations simultaneously.

The capability of a satellite to act as a repeater for many different earth stations is called *multiple access* (MA). Three main methods are currently in use to accomplish this: frequency-division multiple access (FDMA), time-division multiple access (TDMA), and code-division multiple access (CDMA). In FDMA, circuits between different earth stations are assigned different frequency bands within the allowable bandwidth. In TDMA, the entire bandwidth is allocated to each earth station for a short time, just like in time-division multiplexing. CDMA (also called *spread spectrum multiple access*) uses a pseudorandom code and operates in both time and frequency domains. It is effective against jamming techniques and is used primarily in military satellite communications.

Satellites have had a dramatic impact on communication topologies and pricing. They can be used for transmission of program video, voice, or data signals almost anywhere on earth, no matter how remote the location or whether it is fixed or mobile like a ship. They provide multichannel capabilities, wide bandwidths, and high data rates. The transmission cost is independent of the distance between the source and destination. Furthermore, only one satellite repeater is required for most transmissions. This characteristic of satellite transmission may make it superior to terrestrial systems such as microwave or coaxial cable, since the latter systems require many repeaters in tandem to cover long distances, and amplification of the signal by each repeater tends to increase the effects of distortion and noise.

Because of the greater distance between the earth and the satellite repeater, attenuation and transmission delays can cause problems. Attenuation can be overcome by using high gain, narrow beams, and path elevation angles greater than 20° for the ground antennas. The high path elevation reduces the distance the signals travel through the atmosphere to reduce attenuation and fading. However, rain attenuation can still be a problem, especially at the higher carrier frequencies, but it can be minimized by a form of space diversity since rain is unlikely to occur at two widely separated ground stations. The total transmis-

sion delay is approximately 0.5 sec. which is much higher than for terrestrial transmission media and is due to the much longer distance the signal must travel (a minimum of $2 \times 22,300$, or 44,600 miles). This delay can impair the quality of voice communication but has the greatest detrimental effect on data transmission unless communication protocols are designed to match the characteristics of this transmission medium. Low-orbit satellites (250 to 1000 miles above the earth) can overcome this delay problem, and such systems are beginning to be deployed even though the number of satellites required is greater and their liftimes are shorter than for geosynchronous satellites.

E. Cellular Telephones

Cellular telephones use a combination of low-powered radio transmitters and receivers (transceivers) to provide voice and data telephone service to mobile users. They include small antennas and work by transmitting information to cellular towers which cover a given area or cell. The towers within these cells are connected to a central switching station (usually by fiberoptic cable), which in turn is connected to the rest of the telephone system. Cellular calls are therefore received by the towers using the atmosphere as the transmission medium and are relayed to the telephone network. The cells overlap so that when the caller moves from one cell to another the cell towers automatically transfer the call so that the communication is uninterrupted.

Cellular phone networks cover most metropolitan areas and their coverage is rapidly expanding into rural areas. While simple in theory, reliable production systems are very complex. Older systems used the 824- to 893-MHz band, but newer systems operate at much higher frequencies (around 1900 MHz) and use PCM rather than analog modulation techniques. A good description of cellular telephone technologies can be found in Macario (1993).

F. Other Unbounded Media Transmission Systems

Communications satellites and terrestrial microwave radio are by far the most frequent systems using unbounded media for transmission of data. However, many other systems using unbounded media are available and are commonly used, especially for transmission of voice and program signals. A partial listing and a brief description of these systems is given below. A more detailed description may be found in Freeman (1981) and Pooch et al. (1983).

1. *Radio and television broadcast.* These systems use omnidirectional antennas to broadcast program signals to receivers in the local area. AM radio uses the band from

535 to 1605 KHz, while FM radio uses 88 to 108 MHz. Television uses the very high and ultrahigh frequency bands to provide the bandwidth required for acceptable transmission of video program signals. A receiver can select one of many local stations by tuning to the frequency allocated to the desired station, which is a form of FDM.

2. *High-frequency radio.* As little as 40 years ago, high-frequency (HF) radio carried almost all transatlantic telephone traffic. Today it is used primarily for short-wave radio and for telephone communications to ships at sea and to countries not connected by cable and with no satellite antenna. Long-distance communication is possible with HF radio since the frequencies employed allow the radio waves to be reflected by the ionosphere back to earth until they reach their final destination in one or more hops. For this reason it is sometimes called *ionospheric radio transmission.* Because of the movement and changes of the ionosphere, HF radio is subject to fading, interference, distortion, and periodic blackouts. Thus, it is rarely used for data transmission unless extreme precautions are taken for detection and correction of errors.

3. *Tropospheric-scatter radio transmission.* In tropospheric-scatter radio transmission, the troposphere is used to scatter and reflect a fraction of the incident radio waves back to earth. With two highly directional antennas pointed at the troposphere (which is about 6 miles up, as opposed to 30 miles for the ionosphere), enough radiowave energy is reflected back to result in very reliable over-the-horizon communication systems with effective ranges of 100 to 600 miles. Such systems are used where it is not possible or economical to use land lines or even microwave radio. Island chains or very rugged mountain ranges are good examples of terrain where such systems are used. The carrier frequencies are in the 900- to 5000-MHz range and can carry several hundred voice circuits over short distances of 100 miles, although 72 circuits is more typical over longer distances. It is subject to fading, which is usually overcome using space diversity with two receivers and transmitters at each end of the link with the antennas separated by at least 30 wavelengths to achieve reliable transmission.

4. *Packet radio.* Packet radio can be used to allow mobile computers to communicate with a server that is connected to an omnidirectional home base transceiver. Data to be transmitted to a computer are formatted into fixed-length packets, which include the address of the destination at the beginning and cyclic redundancy check characters at the end of the packet. Each computer receives all the transmitted packets and discards those that are not addressed to it. Packets with detected errors are not acknowledged by the computer, which causes the home base to retransmit the message after a certain time. When computers transmit to the home base, it is possible that two or more computers may transmit simultaneously, in which case the messages are not acknowledged. If a computer does not receive an acknowledgment within a predetermined time, it reschedules the packet for retransmission at a random time. Eventually, all packets will be received correctly. Different frequencies can be used for the server-to-computer and computer-to-server links to allow concurrent reception and transmission. A variation of packet radio can allow any workstation or computer to communicate with any other without the master–slave relationship described above.

5. *Light transmission systems.* Light-transmission systems use a light-emitting or laser diode as a transmitter and a photodiode or phototransistor as a receiver. As in microwave radio, the transmission path must be line-of-sight, and air is the transmission medium. The problem with such systems is that extreme care must be taken to align the light source and detector. Even then, problems can occur in heavy rain, fog, or snow, which attenuate the transmitted light signal and impair error-free reception. Bounded communication using optical fibers overcomes these problems.

6. *Wireless ethernet networks.* New technologies for 10- and 100-Mbits/sec ethernet are emerging to provide more effective point-to-point wireless connectivity and to support mobile computer users. A typical point-to-point wireless ethernet network may include several remote locations, each with a unidirectional antenna, connected to a central site equipped with an omnidirectional antenna that is connected to the enterprise ethernet. Similar technologies allow mobile computers with omnidirectional antennas to communicate with a local ethernet network much like cellular phones can communicate with the terrestrial telephone system. Every indication is that the use of such systems will grow very rapidly in the near future.

G. Outlook for Unbounded Transmission Media

The ease of establishing communication links will remain the primary attraction of links using unbounded transmission media. The multiple access capability of some systems using these media is also an important advantage. New developments in systems such as packet radio and wireless ethernet are of particular importance since they promise inexpensive communication between fixed or mobile workstations and computers. With satellite communications significantly reducing the cost for long-distance transmission, the main competition to systems using unbounded transmission media are optical fibers. Perhaps the greatest obstacle to the growth of use of unbounded media is the limited frequency spectrum, but with improvements in antenna design, integrated circuits operating at high frequencies, and advances in modulation techniques to use bandwidth even more efficiently, systems using unbounded media will always play a major role in communications. An important reason for the move toward digital television is that the latter can use bandwidth more effectively and will free up much of the current television frequency spectrum to other applications.

XII. INTERCONNECTION CONSIDERATIONS FOR ELECTRICAL CONDUCTORS

It has been mentioned in Section X that electrical conductors are and will continue to be the predominant transmission medium for interconnecting components within a system. As bandwidths and bit rates increase, it becomes increasingly important to take into account the stray inductance and capacitance that influence the performance of these media. These stray reactances are strongly dependent on the geometry, conductor length, and proximity to the ground plane. By carefully positioning the conductor we can reduce one stray component, but only at the expense of increasing the other. Because of the relatively slow rate of change of stray inductance and capacitance with the distance from the ground plane, the conductor length is the major factor determining the stray reactances.

The current or voltage propagation delay T_d in a conductor is given by:

$$Td = l\sqrt{LC} \qquad (8)$$

where l is the length of the conductor in meters, L is its distributed inductance, and C is its distributed capacitance per unit length. The conductor is said to be "short" if its propagation delay T_d is much less than the rise and fall times of the pulses it is expected to transmit. If its length is such that its T_d is comparable to or longer than the rise and fall times, or greater than $0.35/f$ where f is the greatest frequency per second for which it will be used, the conductor is termed as being "long," in which case transmission line theory should be used to predict its characteristics.

The transmission line parameters relevant to our discussion are the characteristic impedance and the reflection coefficient. The characteristic impedance Z_0 is the impedance seen at one end of an infinitely long line and is approximately by:

$$Z_0 = \sqrt{LC} \qquad (9)$$

The reflection coefficient r is the ratio of the reflected voltage v_r to the incident voltage v_i if a conductor is not terminated in its characteristic impedance and is given by:

$$r = \frac{v_r}{v_i} = \frac{Z_L - Z_0}{Z_L + Z_0} \qquad (10)$$

where Z_L is the terminating or load impedance. If a long conductor is not terminated or matched in its characteristic impedance, reflections will be present and ringing will occur if pulses with sharp leading and trailing edges are transmitted over the line. Such reflections and ringing increase the inherent electrical noise of the system in which they occur. The impedance ranges of transmission lines that are usually encountered in practice are given below. Note that a strip line is a rectangular conductor over a ground with the width of the conductor begin much greater than its thickness. This type of conductor is encountered in printed circuits, for example.

Types of transmission line	Characteristic impedance (Ω)
Wire over ground	80–400
Twisted pair	80–200
Coaxial pair	40–120
Strip line	20–140

The above relations are useful in determining the length and placement of lines. In high-impedance circuits, it is important to minimize capacitance by making the line as short as possible and placing it as far as possible from the ground. In low-impedance circuits, it is important to minimize the inductance by making the line as short as possible and placing it as close to ground as possible. For long lines, the lines should be terminated in their characteristic impedance whenever possible to minimize reflections and hence circuit noise.

The presence of stray electromagnetic fields in a system causes cross-talk voltage or current to be induced in neighboring conductors. This cross-talk may be predominantly capacitive or inductive. Capacitive cross-talk is due to stray capacitance between conductors and is predominant in circuits with large voltage swings and small currents (i.e., high-impedance circuits). The capacitive cross-talk voltage v_c is given by:

$$v_c = k_c\varepsilon(\text{voltage swing} \times \text{length of line}$$
$$\times \text{function of spacing})/\text{rise time of voltage} \quad (11)$$

where ε is the dielectric constant of the medium between the conductors and k_c is a constant. In general, the closer the lines the greater the v_c. Spacing the lines farther apart and away from the ground reduces v_c but has the effect of increasing inductive cross-talk. When the voltage swings are small and the currents are large (i.e., low-impedance circuits), inductive cross-talk usually predominates. It arises chiefly because of mutual inductance coupling between conductors and is generally given by:

$$i_i = k_i(\text{current swing} \times \text{length of line}$$
$$\times \text{function of spacing})/\text{rise time of current} \quad (12)$$

where i_i is the cross-talk current, which can be reduced if the lines are farther from each other and as close as possible to the ground.

If more than one line is driven at the same time, the inductive currents in the passive line are additive. This is not the case with capacitive cross-talk. An upper bound is reached for v_c if the passive line is surrounded by active lines. Both types of cross-talk may be reduced by using special lines. Twisted pairs, coaxial lines, and multiple shielded lines all have zero mutual impedance in theory and therefore have zero inductive cross-talk. In practice, the following numbers are useful:

Type of wiring	Mutual inductance (μH/m)
"Neat" in bundles	1–7
Point-to-point over ground	0.6
Twisted pairs	0.06
Coaxial cable	0.006
Multiple shielded lines	As small as necessary

In summary, it is impossible to eliminate all the stray inductance and capacitance of conductors. It is possible only to trade one against the other by varying the characteristic impedance. In interconnecting components or systems, the goal should be to use point-to-point connections over a ground plane. All lines should be coaxial or twisted pairs and matched at least at the receiving end. Practical considerations may not permit this, but the more we approach this type of interconnections, the less trouble will be experienced with system noise and cross-talk and therefore with getting a high-speed system to operate successfully.

XIII. STANDARDS

Modern communication systems represent complex and rapidly changing technologies. Effective communications on a local, country, and worldwide basis require that communications equipment and transmission media work together in harmony. Thus, there is need for widely accepted standards. Such standards are established by a number of organizations, including the International Telegraph and Telephone Consultative Committee (CCITT), the International Organization for Standardization (ISO), the American National Standards Institute (ANSI), the Electronic Industries Association (EIA), the European Computer Manufacturers Association (ECMA), the National Communication System (Federal Standards–Telecommunications), and the National Bureau of Standards (NBS). Standards most relevant to data communication are compiled and summarized in Folts (1982).

XIV. SUMMARY

Advances in computer and communication technologies are transforming our society. They have created entire new industries and are changing the way we learn, do research, conduct business, communicate with each other, and even entertain ourselves. These changes have led to an exponential increase in communication traffic generated by the Internet, cellular telephones, faxes, modems, teleconferencing, and data and video services. Ubiquitous connectivity is removing the constraints of space and time to access services from anywhere at any time. This has led to enormous increases in the demand for additional communication bandwidth capacity, and the marketplace has responded with a variety of new technologies that make more effective use of all traditional transmission media including optical fibers, free space, and the atmosphere, as well as the older coaxial and paired cables previously used for voice communication. These technologies have created, and will continue to create, a vast global network for the rapid movement of information in the form of voice, video, and data.

SEE ALSO THE FOLLOWING ARTICLES

DIGITAL SPEECH PROCESSING • NETWORKS FOR DATA COMMUNICATION • OPTICAL FIBER COMMUNICATIONS • OPTICAL INFORMATION PROCESSING • RADIO SPECTRUM UTILIZATION • SIGNAL PROCESSING, ACOUSTIC • SIGNAL PROCESSING, ANALOG • SIGNAL PROCESSING, DIGITAL • SIGNAL PROCESSING, GENERAL • VOICEBAND DATA COMMUNICATIONS

BIBLIOGRAPHY

Folts, H. C., ed. (1982). "McGraw-Hill's Compilation of Data Communications Standards," McGraw-Hill, New York.

Freeman, R. L. (1981). "Telecommunication Transmission Handbook," John Wiley & Sons, New York.

Goodman, J. W., Leonberger, F. I., Kung, S. Y., and Athale, R. A. (1984). *Proc. IEEE Special Issue on Optical Computing* **72** (7), 850–866.

Macario, M. (1993). "Cellular Radio: Principles and Design," McGraw-Hill, New York.

Payton, J., and Qurèshi, S. (1985). "Trellis encoding: what it is and how it affects data transmission," *Data Commun*. May.

Pooch, U. W., Greene, W. G., and Moss, G. G. (1983). "Telecommunications and Networking," Little, Brown & Co., Boston, MA.

Proakis, J. G. (1983). "Digital Communications," McGraw-Hill, New York.

Rey, R. F., ed. (1983). "Engineering and Operations in the Bell System," 2nd ed. AT&T Bell Laboratories, Murray Hill, NJ.

Schwartz, M. (1980). "Information Transmission, Modulation and Noise," 3rd ed. McGraw-Hill, New York.

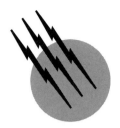

Dense Matter Physics

George Y. C. Leung
Southeastern Massachusetts University

GLOSSARY

Adiabat Equation of state of matter that relates the pressure to the density of the system under a constant entropy.

Baryons Elementary particles belonging to a type of fermions that includes the nucleons, hyperons, delta particles, and others. Each baryon is associated with a baryon number of one, which is a quantity conserved in all nuclear reactions.

Bosons Elementary particles are divided into two classes called *bosons* and *fermions*. The bosons include the photons, phonons, and mesons. At thermal equilibrium, the energy distribution of identical bosons follows the Bose–Einstein distribution.

Degenerate electrons System of electrons that occupy the lowest allowable momentum states of the system, thus constituting the absolute ground state of such a system.

Fermions Class of elementary particles that includes the electrons, neutrions, nucleons, and other baryons. Identical fermions obey Pauli's exclusion principle and follow the Fermi–Dirac distribution at thermal equilibrium.

Isotherm Equation of state of matter that relates the pressure to the density of the system at constant temperature.

Neutrinos Neutral, massless fermions that interact with matter through the weak interaction. Neutrinos are produced, for example, in the decay of the neutrons.

Neutronization Form of nuclear reaction in which the neutron content of the reaction product is always higher than that of the reaction ingredient. It occurs in dense matter as its density increases from 10^7 to 10^{12} g/cm^3.

Nuclear matter Matter substance forming the interior of a nucleus. Its density is approximately 2.8×10^{12} g/cm^3, which is relatively independent of the nuclear species. It is composed of nearly half neutrons and half protons.

Phonons Lattice vibrations of a solid may be decomposed into a set of vibrational modes of definite frequencies. Each frequency mode is composed of an integral number of quanta of definite energy and momentum.

These quanta are called *phonons*. They are classified as bosons.

Photons Particle–wave duality is an important concept of quantum theory. In quantum theory, electromagnetic radiation may be treated as a system of photons endowed with particle properties such as energy and momentum. A photon is a massless boson of unit spin.

Quarks Subparticle units that form the elementary particles. There are several species of quarks, each of which possesses, in addition to mass and electric charge, other fundamental attributes such as c-charge (color) and f-charge (flavor).

Superconductivity Electrical resistance of a superconductor disappears completely when it is cooled below the critical temperature. The phenomenon is explained by the fact that due to the presence of an energy gap in the charge carriers' (electrons or protons) energy spectrum, the carriers cannot be scattered very easily, and the absence of scattering leads to superconductivity.

Superfluidity Superfluidity is the complete absence of viscosity. The conditions leading to superconductivity also lead to superfluidity in the proton or electron components of the substance. In the case of neutron matter, the neutron component may turn superfluid due to the absence of scattering. The critical temperatures for the proton and neutron components in neutron matter need not be the same.

DENSE MATTER PHYSICS is the study of the physical properties of material substance compressed to high density. The density range begins with hundreds of grams per cubic centimeter and extends to values 10 to 15 orders of magnitudes higher. Although such dense matter does not occur terrestrially, it exists inside stellar objects such as the white dwarf stars, neutron stars, and black holes and possibly existed during the early phase of the universe. Dense matter physics therefore provides the scientific basis for the investigation of these objects.

I. BACKGROUND AND SCOPE

Matter is the substance of which all physical objects are composed. The density of matter is the ratio of its mass to volume and is a measure of the composition of matter and the compactness of the constituent entities in it. In units of grams per cubic centimeter (g/cm^3) the density of water is 1.0 g/cm^3, and the densities of all macroscopic objects on earth do not exceed roughly 20 g/cm^3. However, some stellar objects are believed to be formed of matter with much higher densities. In the 1920s, a star called Sirius B, a binary companion of the star Sirius, was found to be a highly compact object having the mass of our sun but the size of a planet, and thus must be composed of matter of very high density, estimated to reach millions of g/cm^3. Sirius B is now known to belong to a class of stellar objects called the *white dwarf stars*. Dense matter physics began as an effort to understand the structure of the white dwarf stars. It matured into a branch of science devoted to the study of the physical properties of dense matter of all types that may be of interest to astrophysical and cosmological investigations.

Since the type of matter under study cannot be found terrestrially, it is impossible to subject it to direct laboratory examination. Hence, the study of dense matter physics is mainly theoretical in nature. In the 1920s the emergence of quantum mechanics was making a strong impact on physics, and a theory of dense matter based on the quantum mechanical behavior of electrons at high density was constructed. It marked the dawn of dense matter physics, and this theory remains valid today for the study of white dwarf stars. The subsequent identification of other compact stellar objects such as the neutron stars and black holes greatly intensified the study of dense matter physics. We survey here what can be expected theoretically from dense matter and the implications of current theories on the structure of these compact stellar objects.

On the experimental side, the study is benefited by the fact that if the concept of matter density is extended to include microscopic bodies such as the atomic nuclei, then a substance called *nuclear matter*, which possesses extremely high density, may be identified. Through nuclear physics study, it is then possible to subject matter with such high density to laboratory examinations. Such experimental information provides an invaluable guide to the study of matter forming the neutron stars.

Compact stellar objects are mainly the remains of stars whose nuclear fuels have been exhausted and are drained of the nuclear energy needed to resist the pull of the gravitational force. As the gravitational force contracts the stellar body, it also grows in strength. This unstable situation is described as gravitational collapse, which continues until a new source of reaction strong enough to oppose the gravitational force becomes available. The search for the physical properties of dense matter responsible for resistances to gravitational collapse is an important aspect in dense matter physics since the results have important astrophysical implications.

The structure and stability of a compact stellar object depend on its composition and the equation of state of the form of matter that it is composed of. The equation of state expresses the pressure generated by the matter substance as a function of its density and temperature. The determination of the composition and the equation of state of dense matter is a prime objective in dense matter studies. These topics are discussed in Sections III and IV after a brief introduction of the basic theoretical method involved is presented in Section II.

Compact stellar objects perform rotations, pulsations, and emissions, and to understand these processes we would need to know, in addition to the equation of state, the properties of dense matter under nonequilibrium conditions. These are called the *transport properties*, which include the electrical and thermal conductivity and viscosity. These intrinsic properties of dense matter are discussed in Section V. The effects of a strong magnetic field on the transport properties, however, are not included in this writing. Properties related to radiative transfer, such as emissivity and opacity, are discussed in Section VI. However, radiative transfer by photons in dense matter is completely superceded by conductive transfer, and since it does not play an important role, the photon emissivity and opacity in dense matter will not be discussed. Instead, Section VI concentrates on the much more interesting topic of neutrino emissivity and opacity in dense matter.

The properties of dense matter will be discussed in several separate density domains, each of which is characterized by typical physical properties. In the first density domain, from 10^2 to 10^7 gm/cm^3, the physical properties are determined to a large extent by the electrons among the constituent atoms. The electrons obey an important quantum mechanical principle called *Pauli's exclusion principle* which forbids two electrons to occupy the same quantum state in a system. All electrons must take up quantum states that differ in energy, and as the electron density is increased, more and more of the electrons are forced to take on new high-energy quantum states. Consequently, the total energy of the electrons represents by far the largest share of energy in the matter system. It is also responsible for the generation of an internal pressure in the system. All white dwarf stars are believed to be composed of matter with densities falling in this domain, which is known to sustain stable stellar configurations. The physical mechanism mentioned here for electrons is central to establishing the physical properties of dense matter at all density domains, and for this reason it is first introduced in Section II.

The second density domain ranges from 10^7 to 10^{12} g/cm^3, where nuclear physics plays a key role. Above 10^7 g/cm^3 the constituent atomic nuclei of the dense matter experiences nuclear transmutations. In general, an increase in density above this point leads to the appearance of nuclei that are richer in neutron content than those occurring before. This process, called *neutronization*, continues throughout the entire density domain. The process also suppresses the increase in electron number with increase in matter density and thus deprives the matter system of its major source of energy. Matter with densities belonging

to this density domain experiences a gradual reduction in compressibility with increasing density and is no longer able to sustain stable stellar configurations after its density exceeds 10^8 g/cm^3.

As matter density approaches 10^{12} g/cm^3, some nuclei become so rich in neutrons that they cease to bind the excess neutrons; nuclei now appear to be immersed in a sea of neutrons. The onset of such a phenomenon is called *neutron drip*, a term suggesting that neutrons are dripping out of the nuclei. This leads to the third density domain ranging from 10^{12} to 10^{15} g/cm^3. A rapid increase in neutron density accompanying an increasing matter density leads to the production of energetic neutrons, since neutrons (like electrons) obey Pauli's exclusion principle. Hence, the same quantum mechanical mechanism characterizing the first density domain becomes operative here. As soon as neutrons were discovered experimentally in the 1930s, this mechanism was invoked to suggest the possible existence of stable neutron stars, long before neutron stars were actually identified in astronomical observations. Unlike the electrons, however, neutrons interact among themselves with nuclear forces that are comparatively strong and must be handled with great care. The average density of atomic nuclei, or nuclear matter density, is of the order of 10^{14} g/cm^3. Much of the needed physics in understanding matter with density in this density domain must come from nuclear physics.

Our understanding of matter with densities above 10^{15} g/cm^3 is very tentative; for this reason we shall, for discussion, assign matter with densities above 10^{15} g/cm^3 into the fourth and last density domain. In this area we shall discuss the physical basis for topics such as hyperonic matter, pion condensation, and quark matter.

Since the study of dense matter is highly theoretical in nature, we begin our discussion with an introduction to the basic theoretical method needed for such an investigation in establishing the composition of dense matter and its equation of state.

II. BASIC THEORETICAL METHOD

Matter may first be considered as a homogeneous system of atoms without any particular structure. Such a system may be a finite portion in an infinite body of such substance, so that boundary effects on the system are minimized. The system in the chosen volume possesses a fixed number of atoms. At densities of interest all atoms in it are crushed, and the substance in the system is best described as a plasma of atomic nuclei and electrons. We begin by studying pure substances, each formed by nuclei belonging to a single nuclear species, or nuclide. The admixture of other nuclides in a pure substance can be accounted for and will be considered after the general method of investigation is introduced.

The physical properties of the system do not depend on the size of the volume, since it is chosen arbitrarily, but depend on parameters such as density, which is obtained by dividing the total mass of the system by its volume. The concept of density will be enlarged by introducing a host of densitylike parameters, all of which are obtained by dividing the total quantities of these items in the system by its volume. The term "density" will be qualified by calling it *mass density* whenever necessary. In addition, parameters such as the electron number density and the nuclei number density will be introduced.

Each nuclide is specified by its atomic number Z and mass number A. Z is also the number of protons in the nucleus and A the total number of protons and neutrons there. Protons and neutrons have very nearly the same mass and also very similar nuclear properties; they are often referred to collectively as *nucleons*. Each nuclide will be designated by placing its mass number as a left-hand superscript to its chemical symbol, such as ^{4}He and ^{56}Fe.

Let the system consist of N atoms in a volume V. The nuclei number density is

$$n_A = N/V \qquad (1)$$

and the electron number density is

$$n_e = NZ/V \qquad (2)$$

which is the same as the proton number density, because the system is electrically neutral. The mass of a nuclide is given in nuclear mass tables to high accuracy. To determine the mass density of a matter system we usually do not need to know the nuclear mass to such accuracy and may simply approximate it by the quantity Am_p, where $m_p = 1.67 \times 10^{-24}$ g is the proton mass. The actual mass of a nucleus should be slightly less than that because some of the nuclear mass, less than 1%, is converted into the binding energy of the nucleus. The mass density ρ of the matter system is given simply by:

$$\rho = NAm_p/V \qquad (3)$$

For example, for a system composed of electrons and helium nuclei, for which $Z = 2$ and $A = 4$, a mass density of 100 g/cm^3 corresponds to a nuclei number density of $n_A = 4 \times 10^{24}$ cm^{-3} and an electron number density $n_e = 2n_A$.

Like all material substances, dense matter possesses an internal pressure that resists compression, and there is a definite relation between the density of a substance and the pressure it generates. The functional relationship between pressure and density is the equation of state of the substance and is a very important physical property for astrophysical study of stellar objects. The structure of a stellar object is determined mainly by the equation of state of the stellar substance. In this section, the method for the derivation of the equation of state will be illustrated.

Unlike a body of low-density gas, whose pressure is due to the thermal motions of its constituent atoms and is therefore directly related to the temperature of the gas, dense matter derives its pressure from the electrons in the system. When matter density is high and its electron number density is correspondingly high, an important physical phenomenon is brought into play which determines many physical properties of the system. We shall illustrate the application of this phenomenon to the study of dense matter with densities lying in the first density domain, 10^2–10^7 g/cm^3.

A. Pauli's Exclusion Principle

It is a fundamental physical principle that all elementary particles, such as electrons, protons, neutrons, photons, and mesons, can be classified into two major categories called *fermions* and *bosons*. The fermions include electrons, protons, and neutrons, while the bosons include photons and mesons. Atomic nuclei are regarded as composite systems and need not fall into these classes. In this study, one fermionic property, Pauli's exclusion principle, plays a particularly important role. The principle states that *no two identical fermions should occupy the same quantum state in a system*. Let us first explain the meaning of identical fermions. Two fermions are identical if they are of the same type and have the same spin orientation. The first requirement is clear, but the second deserves further elaboration. A fermion possesses intrinsic angular momentum called *spin* with magnitude equal to $\frac{1}{2}, \frac{3}{2}, \frac{5}{2}, \ldots$ basic units of the angular momentum (each basic unit equals Planck's constant divided by 2π). The electron spin is equal to $\frac{1}{2}$ of the basic unit, and electrons are also referred to as *spin-half particles*. Spin is a vector quantity and is associated with a direction. In the case of electron spin, its direction may be specified by declaring whether it is oriented parallel to or antiparallel to a chosen direction. The fact that there are no other orientations besides these two is a result of nature's quantal manifestation, the recognition of which laid the foundation of modern atomic physics. These two spin orientations are simply referred to as *spin-up* and *spin-down*. All spin-up electrons in a system are identical to each other, as are all spin-down electrons. The term identical electrons is thus defined. Normally, these two types of electrons are evenly distributed, since the system does not prefer one type of orientation over the other.

We come now to the meaning of a quantum state. Each electron in a dynamical system is assigned a quantum state. The quantum state occupied by an electron may be specified by the electron momentum, denoted by **p**, a vector that is expressable in Cartesian coordinates as p_x, p_y, p_z. When an electron's momentum is changed to $\mathbf{p}'$ in a dynamical process, it moves to a new quantum state. The momentum associated with the new quantum state $\mathbf{p}'$ must, however, differ from $\mathbf{p}$ by a finite amount. For a system of uniformly distributed electrons in a volume V, this amount may be specified in the following way. If each of the three momentum components is compared, such as p_x and p'_x, they must differ by an amount equal to h/L, where h is Planck's constant and $L = V^{1/3}$. In other words, in the space of all possible momenta for the electrons, each quantum state occupies a size of h^3/V, which is the product of the momentum differences in all three components.

B. Fermi Gas Method

According to Pauli's exclusion principle, all electrons of the same spin orientation in a system must possess momenta that differ from each other by the amount specified above. The momenta of all electrons are completely fixed if it is further required that the system exist in its ground state, which is the lowest possible energy state for the system because there is just one way that this can be accomplished, which is for the electrons to occupy all the low-momentum quantum states before they occupy any of the high-momentum quantum states. In other words, there exists a fixed momentum magnitude p_F, and, in the ground-state configuration, all states with momentum magnitudes below p_F are occupied while the others with momentum magnitudes above p_F are unoccupied. Graphically, the momenta of the occupied states plotted in a three-dimensional momentum space appear like a solid sphere centered about the point of zero momentum. The summation of all occupied quantum states is equivalent to finding the volume of the sphere, since all states are equally spaced from each other inside the sphere. Let the radius of the sphere be given by p_F, called the *Fermi momentum*, then the integral for the volume of the sphere with the momentum variable expressed in the spherical polar coordinates is given by:

$$\int_0^{p_F} 4\pi p^2 \, dp = \left(\frac{4\pi}{3}\right) p_F^3 \qquad (4)$$

where p is the magnitude of $\mathbf{p}$, $p = |\mathbf{p}|$. The number of electrons accommodated in this situation is obtained by dividing the above momentum volume by h^3/V and then multiplying by 2, which is to account for electrons with two different spin orientations. The result is proportional to the volume V which appears because the total number of electrons in the system is sought. The volume V is divided out if the electron number density is evaluated, which is given by:

$$n_e = (8\pi/3h^3)p_F^3 \qquad (5)$$

Equation (5) may be viewed as a relation connecting the electron number density n_e to the Fermi momentum p_F. Henceforth, p_F will be employed as an independent variable in establishing the physical properties of the system.

The total kinetic energy density (energy per unit volume) of the electrons can be found by summing the kinetic energies of all occupied states and then dividing by the volume. Each state of momentum $\mathbf{p}$ possesses a kinetic energy of $p^2/2m_e$, where m_e is the electron mass, and the expression for the electron kinetic energy density is

$$\varepsilon_e = \frac{2}{h^3} \int_0^{p_F} 4\pi p^2 \, dp \frac{p^2}{2m_e} = \frac{2\pi}{5h^3 m_e} p_F^5 \qquad (6)$$

The average kinetic energy per electron is obtained by dividing ε_e by n_e:

$$\varepsilon_e/n_e = 0.6\left(p_F^2/2m_e\right) \qquad (7)$$

where $p_F^2/2m_e$ is the kinetic energy associated with the most energetic electrons in the system, and here we see that the average kinetic energy of all electrons in the system is just six-tenths of it. This is an important result to bear in mind. It tells us that p_F is a representative parameter for the system and may be used for order of magnitude estimates of many of the energy-related quantities. The method described here for finding the energy of the system of electrons may be applied to other fermions such as protons and neutrons and is referred to generally as the *Fermi gas method*.

It is now instructive to see what would be the average electron kinetic energy in a matter system that is composed of ^{4}He nuclei at a density of 100 g/cm^3. From the electron number density that we have computed before, $n_e = 8 \times 10^{24}$ cm^{-3}, we find that

$$\frac{\varepsilon_e}{n_e} = \frac{0.3h^2c^2}{m_e c^2}\left(\frac{3}{4\pi}n_e\right)^{2/3} \approx 300 \text{ eV} \qquad (8)$$

where eV stands for electron volts. In evaluating expressions such as Eq. (8), we shall follow a scheme that reduces all units needed in the problem to two by inserting factors of h or c at appropriate places. These two units are picked to be electron volts (eV) for energy and centimeters (cm) for length. Thus, instead of expressing the electron mass in grams it is converted into energy units by multiplying by a factor of c^2, and $m_e c^2 = 0.511 \times 10^6$ eV. The units for h can also be simplified if they are combined with a factor of c, since $hc = 1.24 \times 10^{-4}$ eV cm. This scheme is employed in the evaluation of Eq. (8) as the insertions of the c factors are explicitly displayed.

The average kinetic energy per electron given by Eq. (8) is already quite formidable, and it increases at a rate proportional to the two-thirds power of the matter density. For comparison, we estimate the average thermal energy per particle in a system, which may be approximated by the expression $k_B T$, where k_B is Boltzmann's constant and T the temperature of the system in degrees Kelvin (K). The average thermal energy per particle at a temperature of 10^6 K is about 100 eV. Consequently, in the study of compact stellar objects whose core densities may reach millions of g/cm^3 while the temperature is only 10^7 K, the thermal energy of the particles may justifiably be ignored.

Electrons belonging to the ground state of a system are called *degenerate electrons*, a term that signifies the fact that all electron states whose momenta have magnitudes below the Fermi momentum p_F are totally occupied. Thus, whenever electrons are excited dynamically or thermally to states with momenta above p_F, leaving states with momenta below p_F empty, such a distribution of occupied states is called *partially degenerate*. Partially degenerate electrons will be discussed later in connection with systems at high temperatures.

C. Pressure

The internal pressure P of a system of particles is given by the thermodynamic expression,

$$P = -\frac{dE}{dV}\bigg|_N \qquad (9)$$

where E is the total energy of the system and N its particle number in the volume V. For example, in the case of the electron gas, $E = \varepsilon_e V$ and $N = n_e V$. The evaluation of the derivative in Eq. (9) is best carried out by using p_F as an independent variable and converting the derivative into a ratio of the partial derivatives of E and V with respect to p_F, while keeping N fixed. For a noninteracting degenerate electron system, the pressure can also be derived from simple kinematical considerations, and the result is expressed as:

$$P_e = \frac{2}{h^3} \int_0^{p_F} 4\pi p^2 \, dp \frac{pv}{3} \qquad (10)$$

where v is the velocity of the electron, $v = p/m_e$. By either method, the electron pressure is evaluated to be

$$P_e = \frac{8\pi}{15 m_e h^3} p_F^5 \qquad (11)$$

Pressure due to a degenerate Fermi system is called degenerate pressure.

Pressure generated by the nuclei may be added directly to the electron pressure, treating both as partial pressures. They contribute additively to the total pressure of the system. Since nuclei are not fermions (with the sole exception of the hydrogen nuclei, which are protons), they do not possess degenerate pressure but only thermal pressure,

which is nonexistent at zero temperature. In the case of the hydrogen nuclei, the degenerate pressure they generate may also be neglected when compared with the electron pressure, since the degenerate pressure is inversely proportional to the particle mass; the proton mass being 2000 times larger than the electron mass makes the proton pressure 2000 times smaller than the elctron pressure (bearing in mind also that the proton number density is the same as the electron number density). Thus, the pressure from the system P is due entirely to the electron pressure:

$$P = P_e \qquad (12)$$

D. Relativistic Electrons

Because the electrons have very small mass, they turn relativistic at fairly low kinetic energies. Relativistic kinematics must be employed for the electrons when the kinetic energy approaches the electron rest energy $m_e c^2$. For a noninteracting degenerate electron system, its most energetic electrons should reach this energy threshold when the electron number density is $n_e = 2 \times 10^{30}$ cm^{-3}, which translates into a helium matter density of $\rho = 3 \times 10^7$ g/cm^3.

In the relativistic formalism, the evaluation of the electron number density of the system remains unchanged since it depends only on p_F. The evaluation of the electron energy density must, however, be modified by replacing the individual electron kinetic energy from the expression $p^2/2m_e$ to the relativistic expression,

$$e_k = \left(p^2 c^3 + m_e^2 c^4\right)^{1/2} - m_e c^2 \qquad (13)$$

so that

$$\varepsilon_e = \frac{2}{h^3} \int_0^{p_F} 4\pi p^2 \, dp \, e_k \qquad (14)$$

The electron pressure may also be found from Eq. (9) without modification, but if Eq. (10) is used, the electron velocity must be modified to the relativistic form, which is

$$v = \frac{p}{\left(p^2 c^2 + m_e^2 c^4\right)^{1/2}} \qquad (15)$$

In summary, the physical quantities of a noninteracting degenerate electron system are

$$n_e = \frac{8\pi}{3\lambda_e^3} t^3 \qquad (16a)$$

$$\varepsilon_e = \frac{\pi m_e c^2}{\lambda_e^3} \{t(2t^2 + 1)(t^2 + 1)^{1/2} - \ln[t + (t^2 + 1)^{1/2}]\} \qquad (16b)$$

$$P_e = \frac{\pi m_e c^2}{\lambda_e^3} \left\{ \frac{t}{3}(2t^2 - 3)(t^2 + 1)^{1/2} + \ln[t + (t^2 + 1)^{1/2}] \right\} \qquad (16c)$$

where $\lambda_e = h/m_e c$ is the electron Compton wavelength having the dimension of a length and $t = p_F/m_e c$, which is dimensionless.

E. Equation of State

The mass density of dense matter given by Eq. (3) may be expressed in terms of the electron number density as:

$$\rho = m_p n_e / Y_e \tag{17}$$

where Y_e, called the *electron fraction*, is the ratio of the number of electrons to nucleons in the system. For a pure substance, it is just $Y_e = Z/A$, but for a mixed substance, where a variety of nuclides are present, it is given by:

$$Y_e = \sum_i x_i \left(\frac{Z_i}{A_i} \right) \tag{18}$$

where the subscript i designates the nuclide type and x_i denotes the relative abundance of that nuclide in the system. For example, $Y_e = 1.0$ for a pure hydrogen system, $Y_e = 0.5$ for a pure ^{4}He system, and $Y_e = \frac{26}{56} = 0.464$ for a pure ^{56}Fe system. All pure substances that are composed of nuclides lying between He and Fe in the periodic table have electron fractions bounded by the values 0.464 and 0.5. Since pressure depends on n_e of a system while its mass density depends on n_e/Y_e, it is clear that matter systems with the same Y_e would have the same equation of state.

In Fig. 1, typical equations of state of dense matter at zero temperature are plotted for the cases of $Y_e = 1.0$ and $Y_c = 0.5$. The pressure is due entirely to that of a non-interacting degenerate electron system and is computed from Eq. (16c). Pressure is expressed in units of dynes per square centimeter (dyn/cm^2). From the approximate linearity of these curves on a log–log plot, it is apparent that the pressure and density are related very nearly by a power law. It is therefore both convenient and useful to express the equation of state in the form:

$$P \propto \rho^\Gamma \tag{19}$$

where

$$\Gamma = \frac{d(\ln P)}{d(\ln \rho)} \tag{20}$$

is called the *adiabatic index* of the equation of state. Γ should be approximately constant in the density range belonging to the first density domain. We find $\Gamma = \frac{5}{3}$ at the low-density end of the density domain, and it decreases slowly to values approaching $\Gamma = \frac{4}{3}$ at the high-density end of the regime. The magnitude of Γ is a measure of the "stiffness" of the equation of state, in the language of astrophysics. A high Γ means the equation of state is stiff (the substance is hard to compress). Such information

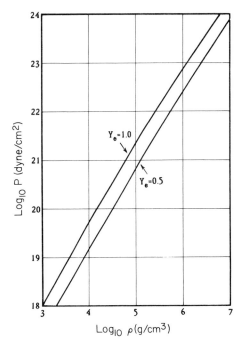

FIGURE 1 Two equations of state at zero temperature, one for matter composed of helium nuclei ($Y_e = 0.5$) and the other for matter composed of hydrogen nuclei ($Y_e = 1.0$), computed by the Fermi gas method.

is of course of utmost importance to the study of stellar structure.

The numerical values of the equation of state of dense matter composed of helium nuclei at zero temperature are listed in Table I. In spite of the fact that this equation of state is obtained by neglecting electrostatic interaction, it should still be applicable to the study of white dwarf stars which are expected to be composed predominantly

TABLE I Equation of State of Matter Composed of Helium Nuclei at Zero Temperature

t	ρ (g/cm^3)	P (dyn/cm^2)	$\mathrm{Log}_{10}\,\rho$	$\mathrm{Log}_{10}\,P$	Γ
0.037	1.00×10^2	6.70×10^{15}	2.0	15.826	$\frac{5}{3}$
0.054	3.16×10^2	4.58×10^{16}	2.5	16.661	$\frac{5}{3}$
0.080	1.00×10^3	3.12×10^{17}	3.0	17.494	$\frac{5}{3}$
0.117	3.16×10^3	2.11×10^{18}	3.5	18.324	1.66
0.172	1.00×10^4	1.43×10^{19}	4.0	19.156	1.66
0.252	3.16×10^4	9.63×10^{19}	4.5	19.984	1.65
0.371	1.00×10^5	6.41×10^{20}	5.0	20.807	1.64
0.544	3.16×10^5	4.16×10^{21}	5.5	21.619	1.61
0.798	1.00×10^6	2.59×10^{22}	6.0	22.414	1.57
1.172	3.16×10^6	1.52×10^{23}	6.5	23.182	1.51
1.720	1.00×10^7	8.36×10^{23}	7.0	23.922	1.45
2.524	3.16×10^7	4.31×10^{24}	7.5	24.634	1.40
3.706	1.00×10^8	2.13×10^{25}	8.0	25.328	1.37

of dense helium matter. As we shall see below, the electrostatic interaction is not going to be important at high densities. The exact composition of the star is also not important in establishing a representative equation of state for the substance forming the white dwarf star, since all nuclides from He to Fe yield very similar Y_e factors; therefore, any mixture of these nuclides would give very nearly the same equation of state. The maximum core density of the largest white dwarf star should not exceed 10^7 g/cm^3.

F. Electrostatic Interaction

In a plasma of electrons and nuclei, the most important form of interaction is the electrostatic interaction among the charged particles, which could modify the total energy of the system. Several estimates of the electrostatic interaction energy will be given here beginning with the crudest.

Each electron in the plasma experiences attraction from the surrounding nuclei and repulsion by the other electrons. Its electrostatic interaction does not depend on the overall size of the system chosen for the investigation, since it is electrically neutral and should depend only on the local distribution of the electric charges. The positive charges carried by the protons are packed together in units of Z in the nucleus, while the negative charges carried by the electrons are distributed in the surrounding space. To investigate the electrostatic energy due to such a charge distribution, let us imagine that the system can be isolated into units, each occupied by a single nucleus of charge Z and its accompanying Z electrons. The volume of the unit is given by Z/n_e, since $1/n_e$ is the volume occupied by each electron and there are Z electrons in the unit. The shape of the unit may be approximated by a spherical cell with the nucleus residing at its center. In each spherical cell, the electrons are distributed in a spherically symmetric way about the center. In other words, if a set of spherical polar coordinates were introduced to describe this cell, with the coordinate origin at the nucleus, the electron distribution can only be a function of the radial variable r and not of the polar angle variables. With such a charge distribution, the cell appears electrically neutral to charges outside the cell, and all electrostatic energy of such a unit must be due to interactions within the cell. The radius of the cell r_s is given by the relation $(4\pi/3)r_s^3 = Z/n_e$. A more accurate treatment would call for the replacement of the spherical cell by polygonal cells of definite shapes by assuming that the nuclei in the system are organized into a lattice structure, but the energy corrections due to such considerations belong to higher orders and may be ignored for the time being.

The determination of the radial distribution of the electrons within the cell is complicated by the fact that it cannot be established by classical electrostatic methods, because in this case with cell size in atomic dimensions, Pauli's exclusion principle plays a major role. As a first estimate we may assume the electrons are distributed uniformly within the cell. The interaction energy due to electron–nucleus interaction is

$$E_{N-e} = -3/2(Ze)^2/r_s \qquad (21)$$

and that due to electron–electron interaction is

$$E_{e-e} = 3/5(Ze)^2/r_s \qquad (22)$$

giving a total electrostatic energy per cell of

$$E_s = E_{N-e} + E_{e-e} = -9/10(Ze)^2/r_s \qquad (23)$$

In these expressions, the electric charge is expressed in the Gaussian (c.g.s.) units, so the atomic fine-structure constant is given by $\alpha = 2\pi e^2/hc = (137.04)^{-1}$. Dividing Eq. (23) by Z gives us the average electrostatic energy per electron, which, if expressed as a function of the electron number density of the system, is

$$\varepsilon_s = E_s/Z = -(9e^2/10)(4\pi Z^2 n_e/3)^{1/3} \qquad (24)$$

Note that the interaction energy is negative and thus tends to reduce the electron pressure in the system. Quantitatively, for an electron number density of $n_e = 8 \times 10^{24}$ cm^{-3} and a corresponding helium matter density of 100 g/cm^3, we find $\varepsilon_s = 65$ eV, which is about 20% of the average electron kinetic energy. At higher densities, the relative importance of the electrostatic interaction energy to the total energy of the system is actually reduced.

G. Thomas–Fermi Method

The assumption of uniform electron distribution in a cell, though expedient, is certainly unjustified, since the electrons are attracted to the nucleus, thus the electron distribution in the neighborhood of the nucleus should be higher. The exact determination of the electron distribution in a cell is a quantum mechanical problem of high degree of difficulty. The best-known results on this problem are derived from the Thomas–Fermi method, which is an approximate method for the solution of the underlying quantum mechanical problem. The method assumes the following:

1. It is meaningful to introduce a radially dependent Fermi momentum $p_F(r)$ which determines the electron number density, as in Eq. (5), so that

$$n_e(r) = 8\pi/3h^3[p_F(r)]^3 \qquad (25)$$

2. The maximum electron kinetic energy, given by $p_F^2/2m_e$, now varies radially from point to point, but, for

a system in equilibrium, electrons at all points must reach the same maximum total energy, kinetic and potential, thus

$$[p_F(r)]^2/2m_e - e\,\Phi(r) = \mu \qquad (26)$$

where $\Phi(r)$ is the electrostatic potential energy, and μ the chemical potential, a constant independent of r (μ replaces p_F in establishing the electron number density in the cell).

3. The electrostatic potential obeys Gauss' law:

$$\nabla^2\Phi = -4\pi[Ze\,\delta(r) - en_e(r)] \qquad (27)$$

where $\delta(r)$ is the mathematical delta function. From these three relations the electron distribution may be solved. Since the first relation is based on Eq. (5), which is derived from a system of uniformly distributed electron gas, its accuracy depends on the actual degree of variation in electron distribution in the cell. The second relation can be improved by adding to the electrostatic potential a term called the *exchange energy*, which is a quantum mechanical result. The improved method which includes the exchange energy is called the Thomas–Fermi–Dirac method.

The solution of these three relations for $n_e(r)$ is quite laborious, but it has been done for most of the stable nuclides. After it is found, the electron pressure in the system can be evaluated by employing Eq. (16c), setting p_F in the expression equal to $p_F(r_s)$, which is the Fermi momentum of the electrons at the cell boundary. There is no net force acting at the cell boundary, and the pressure there is due entirely to the kinetic energy of the electrons. Thus, the electron pressure can be evaluated by Eq. (16c). Note that this condition would not be fulfilled by an interior point in the cell. Denoting the degenerate electron pressure evaluated for noninteracting electrons by p_e, the pressure P of the matter system corrected for electrostatic interaction by means of the Thomas–Fermi–Dirac method may be expressed as:

$$P = F(\xi)P_e \qquad (28)$$

where ξ is a dimensionless parameter related to r_s and Z:

$$\xi = (0.62)Z^{-1/3}(a_0/r_s) \qquad (29)$$

where $a_0 = (h/2\pi)^2/e^2m_e = 5.292 \times 10^{-9}$ cm is the Bohr radius and $F(\xi)$ a rather complicated function that is displayed graphically in Fig. 2. The dashed lines represent the correction factor derived from the Thomas–Fermi method, and it holds for all nuclides. The correction factor derived from the Thomas–Fermi–Dirac method depends on Z, and curves for two extreme cases are shown. Curves for nuclides of intermediate Z should fall between these two curves. The correction curve based on the assumption of uniform electron distribution is shown by the dotted curve for comparison. It is computed for the case of Fe mat-

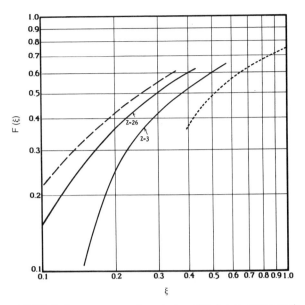

FIGURE 2 Pressure corrections factor $F(\xi)$ due to electrostatic interaction determined from the Thomas–Fermi method (dashed line), the Thomas–Fermi–Dirac method (solid lines), and by the assumption of uniform electron distribution (dotted line).

ter. All curves tend to converge for large values of ξ. The relation between matter density and ξ is given by:

$$\rho \approx 12AZ\xi^3 \text{ g/cm}^3 \qquad (30)$$

The values of $F(\xi)$ for all the curves shown in Fig. 2 are less than unity, and the corrected pressure is always less than that given by the noninteracting degenerate electron system. As ξ increases, $F(\xi)$ tends toward unity, implying that the electrostatic correction on the pressure becomes less and less significant with increasing density. At $\xi = 1$, the pressure of the system is about 20% less than that given by the electron degenerate pressure. For that value of ξ, the density of He matter is about 10^2 g/cm^3, while the density of Fe matter is already 10^4 g/cm^3. Thus, the proper treatment of electrostatic interaction is more relevant to the high-Z nuclei than to the low-Z nuclei. None of the curves shown in Fig. 2 is applicable to dense He matter, whose ξ values would be larger than unity. For He matter, electrostatic correction based on uniform electron distribution should be applicable. At a He matter density of 10^4 g/cm^3, the electrostatic correction accounts for a 5% reduction in pressure. Electrostatic correction based on uniform electron distribution may be applied to all matter system whenever the ξ values exceed unity.

Even though the Thomas–Fermi–Dirac method works quite satisfactorily for dense matter systems, the reader should, however, be reminded that it is not suitable for studying common metals whose densities are rather low. In these cases the cells are so large that the electrons fall into orbital motions around the nucleus and must be handled differently.

H. High Temperatures

At high temperatures, of the order of millions of degrees Kelvin, the thermal energy of the particles in the system becomes comparable to the average kinetic energy of the electrons, in which case the zero-temperature equation of state must be corrected for thermal effects. We use the term "finite temperature" to denote a situation where temperature shows a perceptible effect on the energy state of the system. The proper treatment of finite temperature will first be discussed by ignoring the electrostatic interaction, which will be discussed later.

Without electrostatic interaction, the nuclei behave like a classical ideal gas, and the pressure P_A they contribute to the system is given by the classical ideal gas law:

$$P_A = \rho/Am_p k_B T \qquad (31)$$

The electrons, on the other hand, must be handled by the Fermi gas method. At finite temperatures electrons no longer occupy all the low-energy states; instead, some electrons are thermally excited to high-energy states. Through quantum statistics it is found that the probability for occupation of a quantum state of momentum **p** is given by the function,

$$f(\mathbf{p}) = \{1 + \exp[(\varepsilon(\mathbf{p}) - \mu)/k_B T]\}^{-1} \qquad (32)$$

where $\varepsilon(\mathbf{p}) = p^2/2m_e$ and μ is the chemical potential, which in the present noninteracting case is given by the Fermi energy ε_F:

$$\mu = \varepsilon_F = p_F^2/2m_e \qquad (33)$$

It accounts for the energy needed to introduce an additional particle into the system. The Fermi–Dirac distribution $f(\mathbf{p})$ is the distribution of occupied quantum states in an identical fermion system in thermal equilibrium. Since the distribution that we shall employ depends on only the magnitude of the momentum and not its direction, it shall henceforth be written as $f(p)$. The finite temperature parameters of a noninteracting electron system are to be evaluated from:

$$n_e = \frac{2}{h^3} \int 4\pi p^2 \, dp \; f(p) \qquad (34a)$$

$$\varepsilon_e = \frac{2}{h^3} \int 4\pi p^2 \, dp \; f(p) \varepsilon(p) \qquad (34b)$$

$$P_e = \frac{2}{h^3} \int 4\pi p^2 \, dp \; f(p) \frac{pv}{3} \qquad (34c)$$

where the range of the p integration extends from zero to infinity. Note that, for relativistic electrons, $\varepsilon(p)$ in

Eq. (34b) should be replaced by e_k of Eq. (13) and v in Eq. (34c) should be changed from $v = p/m_e$ to that given in Eq. (15). The integrals in these equations cannot be evaluated analytically, and tabulated results of these integrals are available.

It is also useful to evaluate an entropy density (entropy per unit volume) for the system. The entropy density due to the electrons may be expressed in terms of quantities already evaluated as

$$s_e = (P_e + \varepsilon_e - n_e \mu)/T \qquad (35)$$

The entropy density due to the nuclei is

$$s_A = (P_A + \varepsilon_A)/T = (5/3)P_A/T \qquad (36)$$

where ε_A is the energy density of the nuclei. The entropy density has the same dimensions as Boltznman's constant k_B.

The equation of state relates the pressure of the system to its density and temperature. Since the pressure is determined by two independent parameters, it is hard to display the result. Special cases are obtained by holding either the temperature constant or the entropy of the system constant as the density is varied. An equation of state that describes a thermodynamic process in which the temperature remains constant is called an *isotherm*, and an equation of state that describes an adiabatic process for which the entropy of the system remains constant is called an *adiabat*. In the case where the total number of particles in the system remains unchanged as its volume is varied, constant entropy means the entropy per particle remains a constant, which is obtained by dividing the entropy density given by either Eq. (35) or Eq. (36) by the particle number density. In Fig. 3, typical isotherms and adiabats for a system of (noninteracting) neutrons at high densities are shown. Such a system of neutrons is called *neutron matter*, which is an important form of dense matter and will be discussed in detail later. Neutrons are fermions, and the method of Fermi gas employed for the study of an electron system may be applied in the same manner to the neutron system. In Fig. 3 it is evident that the adiabats have much steeper rises than the isotherms, and this feature holds for all matter systems in general.

The inclusion of electrostatic interaction to a finite-temperature system can be dealt with by extending the Thomas–Fermi–Dirac method described before. At a finite temperature the electron distribution in the system's quantum states must be modified from that of degenerate electrons to that of partially degenerate electrons as determined by the Fermi–Dirac distribution of Eq. (32). Results of this type have been obtained, but they are too elaborate to be summarized here. Interested readers are referred to

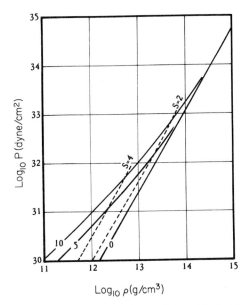

FIGURE 3 Isotherms (solid lines) and adiabats (dashed lines) for a dense system of neutrons computed by the Fermi gas method with complete neglect of nuclear interaction. Isotherms shown are computed for temperatures equivalent to $k_B T = 0, 5,$ and 10 MeV, and the adiabats for entropy per neutron equal to $S/k_B = 2$ and 4.

references listed in the bibliography, where reference to the original articles on this topic can be found.

This section concentrates mainly on simple matter systems with densities lying in the first density domain, but the concepts discussed are applicable to the study of dense matter at all densities. At higher densities, the electrostatic interaction is replaced by nuclear interaction as the dominant form of interaction. The discussion of nuclear interaction will be deferred to the next section.

III. COMPOSITION OF DENSE MATTER

The composition of dense matter will be discussed separately for the four density domains delineated in Section I. In the first density domain where density is 10^2–10^7 g/cm^3, dense matter can in principle exist in any composition, especially at low temperatures. The constituent nuclei of the matter system may belong to any nuclide or combination of nuclides listed in the nuclear table. A sufficient number of electrons must be present to render the system electrically neutral. Stated in such general terms, not much more needs to be added. However, since dense matter is normally found in the interior of stellar objects, its composition naturally depends on nucleosynthesis related to stellar evolution. A review of the stellar evolution process will suggest the prevalent compositions of dense matter in this density domain and the conditions under which they exist.

A star derives its energy through the fusion of hydrogen nuclei to form ^{4}He. It is called *hydrogen burning*, and its ignition temperature is about 10^7 K. The resulting helium nuclei are collected at the core of the star. As the helium core becomes hot enough and dense enough (estimated to be 1.5×10^8 K and 5×10^4 g/cm^3), nuclear reactions based on ^{4}He occur, initiating the helium flash. Helium burning generates the following sequence of relatively stable products: ^{12}C $\rightarrow$ ^{16}O $\rightarrow$ ^{20}Ne $\rightarrow$ ^{24}Mg. These ^{4}He-related processes account for the fact that after ^{1}H and ^{4}He, which are respectively the most and second most abundant nuclear species in stellar systems, ^{16}O is the third most abundant, ^{12}C the fourth, and ^{20}Ne the fifth. Both hydrogen burning and helium burning proceed under high temperatures and high densities. The reason is nuclei are electrically charged and do not normally come close enough for reactions to take place unless their thermal energy is large enough to overcome the repulsive electrical force. The repulsive force increases with the electric charge Z of the nuclei, which is why reactions of this type begin with low-Z nuclei and advance to high-Z nuclei as core temperature rises and density increases. These are exothermic reactions in which energy is generated.

As helium burning proceeds, the helium-exhausted core contracts sufficiently to initiate ^{12}C $+$ ^{12}C and ^{16}O $+$ ^{16}O reactions, called *carbon burning* and *oxygen burning*, respectively. The final products of these processes are ^{24}Mg, ^{28}Si, and ^{32}S. Upon conclusion of oxygen burning, the remanent nuclei may not wait for the further contraction of the core to bring about reactions such as ^{24}Mg $+$ ^{24}Mg, since the electrical repulsion between them is very large and the temperature and density needed to initiate such reactions are correspondingly high. Instead, these nuclei react with photons to transform themselves in successive steps to the most tightly bound nuclide in the nuclear table, which is ^{56}Fe, in a process called *photodisintegration rearrangement*. Therefore, the most likely composition of matter with a density reaching the high end of the first density domain is of electrons and ^{56}Fe nuclei.

A. Neutronization

As matter density advances into the second density domain, 10^7–10^{12} g/cm^3, the ground-state composition of the dense matter system changes with density. For example, if a dense matter system composed of ^{56}Fe nuclei at 10^7 g/cm^3 is compressed, then as density increases the constituent nuclei transmute in a sequence such as Fe $\rightarrow$ Ni $\rightarrow$ Se $\rightarrow$ Ge and so on. Transmutations occur because the availability of high-energy electrons in a dense system is making such reactions possible, and transmutations result if the transmuted nuclei lower the total energy of the matter system. This process is called *neutronization*, because the resulting nuclide from such a reaction is always richer in neutron content than that entering into the reaction.

The physical principle involved can be seen from the simplest of such reactions, in which an electron, e, is captured by a proton, p, to produce a neutron, n, and a neutrino, v:

$$e + p \rightarrow n + v \qquad (37)$$

This is an endothermic reaction. For it to proceed, the electron must carry with it a substantial amount of kinetic energy, the reason being that there is a mass difference between the neutron and proton that, expressed in energy units, is $(m_n - m_p)c^2 = 1.294$ MeV, where 1 MeV $= 10^6$ eV. However, when matter density exceeds 10^7 g/cm^3, electrons with kinetic energies comparable to the mass difference become plentiful, and it is indeed energetically favorable for the system to have some of the high-energy electrons captured by protons to form neutrons. The neutrinos from the reactions in general escape from the system since they interact very weakly with matter at such densities. Reactions based on the same principle as Eq. (37) are mainly responsible for neutronization of the nuclei in the matter system.

Let us denote the mass of a nucleus belonging to a certain nuclide of atomic number Z and mass number A by $M(Z, A)$ and compare the energy contents of different pure substances, each of which is composed of nuclei of a single species. Each system of such a pure substance is characterized by a nuclei number density n_A. Since n_A is different for different systems, it is best to introduce instead a nucleon number density n_B. All systems with the same density have the same n_B, where the subscript B refers to baryons, a generic term including nucleons and other nucleonlike particles. Each baryon is assigned a unit baryon number, which is a quantity conserved in all particle reactions. Thus, n_B will not be changed by the reaction shown in Eq. (37) or the neutronization process in general, which makes it a useful parameter. It is related to the nuclei number density by $n_B = A n_A$ and to the electron number density by $n_e = n_B Z/A$.

At zero temperature, the energy density of such a system is given by:

$$\varepsilon = (n_B/A)M(Z, A)c^2 + \varepsilon_s + \varepsilon_e \qquad (38)$$

where ε_s is the electrostatic interaction energy density and ε_e the degenerate electron energy density, which is to be calculated from Eq. (16b) for relativistic electrons. The masses of most stable nuclides and their isotopes have been determined quite accurately and are listed in the nuclear table. One may therefore compute ε according to Eq. (38) for all candidate nuclides using listed masses from the nuclear table. As a rule, only even–even nuclides, which are nuclides possessing even numbers of protons

and even numbers of neutrons, are needed for consideration. These nuclides are particularly stable and are capable of bringing the dense matter system to a low energy state.

The electrostatic interaction energy density ε_s may be estimated from methods discussed in Section II. Even though electrostatic interaction energy plays a minor role in determining the pressure of the system once the density is high, it has an effect on the composition of the system, which depends on a relatively small amount of energy difference. The ε_s may be estimated by evaluating the lattice energy due to the electrostatic interaction energy of a system of nuclei that form into a lattice, while the electrons are assumed to be distributed uniformly in space. The lattice giving the maximum binding is found to be the body-centered-cubic lattice, and the corresponding lattice energy is

$$\varepsilon_L = -0.89593(Ze)^2/r_s \qquad (39)$$

where r_s is the cell radius defined for Eq. (21) and is related to n_B by:

$$r_s = (4\pi n_B/3A)^{-1/3}$$

For each nuclide, the energy density of the system evaluated from Eq. (38) is a function of n_B. Next, plot the quantity ε/n_B versus $1/n_B$ as shown in Fig. 4, where the curves corresponding to nuclides ^{62}Fe, ^{62}Ni, and ^{64}Ni are drawn. From the curves indicated by ^{62}Fe and ^{62}Ni, it is obvious that a matter system composed of ^{62}Ni nuclei would have an energy lower than that composed of ^{62}Fe nuclei at the densities shown. The ground-state composition of

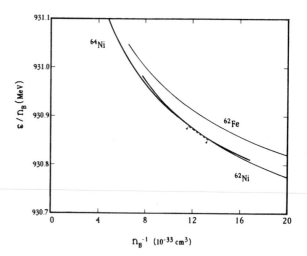

FIGURE 4 Schematic energy curves for matter composed of nuclides ^{62}Fe, ^{62}Ni, and ^{64}Ni illustrating the tangent construction method for establishing the domain of a first-order phase transition in matter. The short dashed line is a line tangent to the ^{62}Ni and ^{64}Ni curves. The triangular pointers indicate the points of tangency. [After Leung, Y. C. (1984). "Physics of Dense Matter," Science Press, Beijing, China. By permission of Science Press.]

dense matter is determined by nuclides whose curves form the envelope to the left of all the curves, as illustrated, for example, by the curves for ^{62}Ni and ^{64}Ni in Fig. 4. Two of these curves on the envelope cross at a certain density. This means that a change in composition occurs in the neighborhood of that density. Such a change, called a *first-order phase transition*, begins at a density below the density where the curves cross, when some of the ^{62}Ni nuclei are being rearranged into ^{64}Ni, and ends at a density above that intersection when all of the ^{62}Ni nuclei are changed into ^{64}Ni. Throughout a first-order phase transition, the pressure of the system remains constant, which is a requirement of thermal equilibrium. The exact density at which a phase transition begins and the density at which it ends can be found from the tangent construction method shown in Fig. 4. By this method one draws a straight-line tangent to the convex curves on the envelope as shown by the dotted lines. The values of n_B at the points of tangency correspond to the onset and termination of phase transition. All points on the tangent exhibit the same pressure since the expression for pressure, Eq. (9), may be converted to read:

$$P = -\frac{\partial(\varepsilon/n_B)}{\partial(1/n_B)} \quad (40)$$

Within the range of phase transition, the tangent now replaces the envelope in representing the energy of the matter system.

The sequence of nuclides constituting the zero-temperature ground state of dense matter in the density range 10^7–10^{11} g/cm^3 obtained by this method with ε_s replaced by ε_L is shown in Table II. As indicated, neutronization begins at a density of 8.1×10^6 g/cm^3, which is quite close to 10^7 g/cm^3, and that is why we choose 10^7 g/cm^3 to mark the begining of the second density domain. Note that the sequence of nuclides in the table shows a relative increase in neutron content with density, which is indicated by the diminishing Z/A ratios.

The lattice structure can be destroyed by thermal agitation. At each density, there is a corresponding melting temperature T_m that can be estimated by comparing the lattice energy with its thermal energy. It is usually taken to be

$$k_B T_m = -c_m^{-1}(Ze)^2/r_s \quad (41)$$

where $c_m \approx 50$–100. Thus, for Fe matter at $\rho = 10^8$ g/cm^3, the melting temperature is about $T_m \approx 2 \times 10^8$ K.

B. Nuclear Semiempirical Mass Formula

As matter density exceeds 4×10^{11} g/cm^3 or so, the ground-state composition of the matter system may pre-

TABLE II Ground-State Composition of Dense Matter in the Second Density Domain[a]

Nuclide	Z	Z/A	ρ_{max} (g/cm^3)
^{56}Fe	26	0.4643	8.1×10^6
^{62}Ni	28	0.4516	2.7×10^8
^{64}Ni	28	0.4375	1.2×10^9
^{84}Se	34	0.4048	8.2×10^9
^{82}Ge	32	0.3902	2.2×10^{10}
^{80}Zn	84	0.3750	4.8×10^{10}
^{78}Ni	28	0.3590	1.6×10^{11}
^{76}Fe	26	0.3421	1.8×10^{11}
^{124}Mo	42	0.3387	1.9×10^{11}
^{122}Zr	40	0.3279	2.7×10^{11}
^{120}Sr	38	0.3166	3.7×10^{11}
^{118}Kr	36	0.3051	4.3×10^{11}

[a] ρ_{max} is the maximum density at which the nuclide is present. [From Baym, G., Pethick, C., and Sutherland, P. (1971). *Astrophysical J.* **170**, 299. Reprinted with permission of *The Astrophysical Journal*, published by the University of Chicago Press; © 1971 The American Astronomical Society.]

fer nuclides that are so rich in neutrons that these nuclides do not exist under normal laboratory conditions, and their masses would not be listed in nuclear tables. To understand these nuclides, theoretical models of the nucleus must be constructed to deduce their masses and other properties. This is a difficult task since nuclear forces are complicated, and the problem is involved. A preliminary investigation of this problem should rely on as many empirical facts about the nucleus as possible. For the present purpose, the nuclear semiempirical mass formula, which interpolates all known nuclear masses into a single expression, becomes a useful tool for suggesting the possible masses of these nuclides.

The nuclear mass is given by the expression:

$$M(Z, A) = (m_p + m_e)Z + m_n(A - Z) - B(Z, A) \quad (42)$$

where m_p, m_e, and m_n denote the proton, electron, and neutron mass, espectively, and $B(Z, A)$ the binding energy of the nucleus. The nuclear semiempirical mass formula for even–even nuclei expresses the binding energy in the following form:

$$B(Z, A) = a_V A - a_S A^{2/3} - a_C Z^2 A^{-1/3}$$
$$- \frac{a_A(A - 2Z)^2}{A} + a_P A^{-3/4} \quad (43)$$

where the coefficients have been determined to be $a_V = 15.75$ MeV, $a_S = 17.8$ MeV, $a_C = 0.710$ MeV, $a_A = 23.7$ MeV, and $a_P = 34$ MeV. This mass formula is not just a best-fit formula since it has incorporated many theoretical elements in its construction. This feature, hopefully, will make it suitable for extension to cover unusual nuclides.

Applying Eqs. (42) and (43) to the computation of the matter energy density given by Eq. (38), one finds matter composition with density approaching 4×10^{11} g/cm^3. The general result is that as matter density increases, the constituent nuclei of the system become more and more massive and at the same time become more and more neutron rich. In general, nuclei become massive so as to minimize the surface energy which is given by the term proportional to a_S in Eq. (43), and they turn neutron rich so as to minimize the electrostatic interaction energy within the nuclei, given by the term proportional to a_C in Eq. (43). In general, the nuclear semiempirical mass formula is not believed to be sufficiently accurate for application to nuclides whose Z/A ratios are much below $Z/A \approx 0.3$. For these nuclides, we must turn to more elaborate theoretical models of the nucleus.

C. Neutron Drip

When the nuclei are getting very large and very rich in neutron content, some of the neutrons become very loosely bound to the nuclei; as density is further increased, unbound neutrons begin to escape from the nuclei. The nuclei appear to be immersed in a sea of neutrons. This situation is called *neutron drip*, a term suggesting that neutrons are dripping out of the giant nuclei to form a surrounding neutron sea. At zero temperature, this occurs at a density of about 4×10^{11} g/cm^3. When it occurs, matter becomes a two-phase system, with one phase consisting of the nuclei and the other consisting of neutrons (with possibly a relatively small admixture of protons). The nuclei may be visualized as liquid drops suspended in a gas consisting of neutrons. These two phases coexist in phase equilibrium. The energy densities of these two phases are to be investigated separately. If the surface effects around the nuclei may be neglected, then these two phases are assumed to be uniform systems, each of which exhibits a pressure and whose neutron and proton components reach certain Fermi energies [cf. Eq. (26)]. At phase equilibrium, the pressures of these two phases must be identical, and the Fermi energies of the respective neutron and proton components of these two phases must also be identical. The composition of matter in the density domain of neutron drip is established by the system that is in phase equilibrium and at the same time reaching the lowest possible energy state.

The evaluation of the Fermi energies for the neutron and proton components depends on the nature of the nuclear interaction. Since the nuclear interaction is not as well known as the electrostatic interaction, the results derived for post-neutron drip densities are not as well established

as those found for the first density domain. At the present time, several forms of effective nuclear interactions constructed specifically to explain nuclear phenomenology are quite helpful for this purpose. Effective nuclear interaction is to be distinguished from realistic nuclear interaction, which is derived from nuclear scattering data and is regarded as a more fundamental form of nuclear interaction.

D. Liquid Drop Model

A nucleus in many respects resembles a liquid drop. It possesses a relatively constant density over its entire volume except near the surface, and the average interior density is the same for nuclei of all sizes. A model of the nucleus that takes advantage of these features is the liquid drop model. It considers the total energy of the nucleus to be the additive sum of its bulk energy, surface energy, electrostatic energy, and translational energy. The bulk energy is given by multiplying the volume of the nucleus by the energy density of a uniform nuclear matter system with nuclear interaction included. The surface energy is usually taken to be a semiempirical quantity based on calculations for nuclei of finite sizes and on experimental results on laboratory nuclei. In a two-phase situation the surface energy must be corrected for nucleon concentration outside the nucleus. The electrostatic energy involves the interaction among all charged particles inside and outside the nucleus. The position of the nuclei may again be assumed to form a body-centered-cubic lattice. Both electrostatic lattice energy and electrostatic exchange energy contribute to it. The translational energy is due to the motion of the nucleus.

At high temperatures, the bulk energy must be computed according to a system of partially degenerate nucleons. This is usually done by employing effective nuclear interaction, in which case the nucleons are assumed to be uncorrelated, and the Fermi–Dirac distribution of Eq. (32) may be directly applied to the nucleons. The problem is much more complicated if the realistic nuclear interaction is employed, in which case particle correlations must be included, and for this reason the computation is much more elaborate. The surface energy shows a reduction with temperature. This result can be extracted from finite nuclei calculations. The lattice energy also shows temperature modification, since nuclei at the lattice points agitate with thermal motions. The same thermal motion also contributes to the translational energy of the nuclei, which may be assumed to possess thermal velocities given by the Boltzmann distribution.

The results of a study based on the liquid drop model of the nucleus is depicted in Fig. 5, where the variation of the size of a nucleus (its A number) with matter

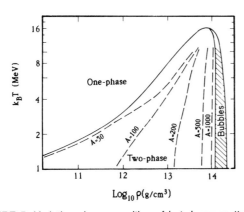

FIGURE 5 Variations in composition of hot dense matter with $Y_e = 0.25$. A two-phase system exists in the region under the uppermost solid line and a one-phase system above it. The dashed lines indicate the types of equilibrium nuclides participating in the two-phase system. [After Lamb, D. Q., Lattimer, J. M., Pethick, C. M., and Ravendall, D. G. (1978). *Phys. Rev.* **41**, 1623. By permission of the authors.]

density and temperature is shown. The study is done with an effective nuclear interaction called the *Skyrme interaction* and for a matter system having an overall electron fraction $Y_e = 0.25$. A wide range of temperatures, expressed in terms of $k_B T$ in units of MeV, is included. Note that the temperature corresponding to $k_B T = 1$ Mev is $T = 1.16 \times 10^{10}$ K. The dashed lines indicate the A numbers of the nuclei at various densities and temperatures. The solid line forms the boundary separating a one-phase system from a two-phase system. The conditions for a two-phase system are included in the plot under the solid line. The two-phase condition disappears completely when matter density exceeds 2×10^{14} g/cm^3. Just before that density, there exists a density range where the nuclei in the system merge and trap the surrounding neutrons into a form of bubbles. This range is indicated by the cross-hatched area in the plot and is labeled "Bubbles."

E. Neutron Matter

In general, when matter density exceeds 2×10^{14} g/cm^3, all nuclei are dissolved into a homogeneous system, and for a range of densities above that, matter is composed almost entirely of neutrons. The admixture of protons is negligibly small (about 1%), because all protons must be accompanied by an equal number of electrons which, being very light, contribute a large amount of kinetic energy to the system. Hence, all ground-state systems tend to avoid the presence of electrons. Such a nearly pure neutron system is called *neutron matter*.

The average density of the atomic nucleus, quite independent of its species, is approximately 2.8×10^{14} g/cm^3; therefore, many of the methods employed for the study of

the nucleus may be applied to the study of neutron matter. As we have mentioned before, the distribution of mass density of a nucleus has been found to be quite uniform over its entire volume, and that uniform mass density is very nearly the same from one nuclide to the other. For a heavy nucleus, its volume is extended so as to maintain a mass density common to nuclei of all sizes. This remarkable fact is described as nuclear saturation. Nuclear substance possessing this mass density is called *nuclear matter*, and its mass density, called *nuclear density*, is found to be 2.8×10^{14} g/cm^3.

Neutron matter, however, is not nuclear matter. Neutron matter consists nearly entirely of neutrons, whereas nuclear matter consists of neutrons and protons in roughly equal fractions. While nuclear matter forms bound units or nuclei, neutron matter is an unbound system. The ground-state composition of an extended system is not given by that of nuclear matter but neutron matter. Neutron matter does not exist terrestrially; its existence is only inferred theoretically. Neutron matter may be studied in analogy to nuclear matter. To accomplish this there are methods based on realistic nuclear interaction expressed in the form of nuclear potentials, the best known of which are the Brueckner–Bethe–Goldstone method and the constrained variational method. There are also methods based on the use of effective nuclear interaction. These methods combine the Hartree–Fock method with some form of effective nuclear potential. Though they are considered less fundamental than the two methods mentioned before, they usually yield more direct and accurate results at the nuclear density where they are designed to perform. Since neutron matter exists for a range of densities, the choice of an applicable method depends also on how well these methods may be extended to cover such a wide range of densities. In this respect, a phenomenological model called the *relativistic mean field model* seems very attractive. We shall return to these methods in the next section when the equation of state of neutron matter is discussed.

It has been seriously proposed that known nuclear interaction forces would make neutron matter superfluid and the proton component in it superconductive when its temperature is below a critical temperature T_c which is estimated to be as high as 10^9–10^{10} K. These phenomena would have profound effects on the transport properties of neutron matter and will be discussed in more detail in Sections V and VI.

Since neutrons, like electrons, are fermions and obey Pauli's exclusion principle, many features of a dense degenerate electron system are exhibited by neutron matter. In spite of having a strong nuclear interaction, these features still play dominant roles in the neutron system. Thus, results based on a dense degenerate neutron system serve as useful guides to judge the properties of neutron matter over a wide range of densities. Such guidance is particularly needed when hyperons and other massive particles begin to make their appearance in dense matter, because the precise nature of their interactions is still poorly known.

F. Baryonic Matter

As matter density advances beyond 10^{15} g/cm^3 and into the fourth and last density domain, many new and unexpected possibilities may arise. It is fairly sure that light hyperons appear first. Hyperons are like nucleons in every aspect except that they are slightly more massive and carry a nonzero attribute called *hypercharge* or *strangeness*. They are $\Lambda(1115)$ and $\Sigma(1190)$, where the numbers in the parentheses give their approximate masses, expressed as mc^2 in units of MeV. Λ is electrically neutral, but there are three species of Σ with charges equal to $+1, 0$, and -1 of the proton charge. More exact values of their masses may be found in a table of elementary particles. These particles are produced in reactions such as:

$$p + e \rightarrow \Lambda + \nu \tag{44a}$$

$$p + e \rightarrow \Sigma^0 + \nu \tag{44b}$$

$$n + e \rightarrow \Sigma^- + \nu \tag{44c}$$

The hyperons appear as soon as the kinetic energies of the neutrons exceed the mass difference between neutron and these particles, in a manner similar to the neutronization process.

In order of ascending mass, the next group of nucleon-like particles to appear are the $\Delta(1232)$, which come in four species with electric charges equal to $+2, +1, 0$, and -1 of the proton charge. Δ are produced as excited states of the nucleons. The nucleons, hyperons, and Δ fall into the general classification of baryons. All baryons are assigned the baryon number $+1$ and antibaryons the number of -1. The baryon number is conserved additively in all particle reactions. Matter systems composed of nucleons, hyperons, Δ, and possibly other baryons are called *baryonic matter*.

G. Pion Condensation

Pions, $\pi(140)$, fall into a class of elementary particles called *mesons* which are quite different from the baryons. They participate in nuclear reactions with the baryons, but they are bosons and not restricted by Pauli's exclusion principle. They are produced in a large variety of particle interactions. They do not normally appear in a matter system, because energy is needed to create them, and their presence would mean an increase in the total energy of the matter system. They do appear, however, if their interaction with the nucleons creates an effective mass for the

nucleon that is lower than its actual mass by an amount comparable to the mass of the pion. This occurs when the baryonic matter reaches some critical density, estimated to be of the order of 10^{15} g/cm^3. This result is not definite since the elementary particle interaction is still far from being understood. Different estimates yield rather different results. When the pions do appear, they may possess very similar momenta as required by the interaction, which is momentum dependent. Such a state of a boson system is called a *condensate*. The appearance of a pion condensate in dense matter is called *pion condensation*.

H. Quark Matter

With the advent of the quark theory some of the traditional notions about elementary particles must be revised. The baryons and mesons, which have been traditionally called *elementary particles*, must be viewed as composite states of quarks. Current quark models have achieved such success that it is hardly in doubt that quark theory must play a key role in the understanding of the subparticle world. Quarks are spin-half fermions possessing fractional electric charges such as $\frac{2}{3}$, $\frac{1}{3}$ and $-\frac{1}{3}$ of the proton charge. Besides mass, spin, and electric charge, they possess additional attributes such as c-charge (also called *color*) and f-charge (also called *flavor*). The electric charge will be called *q-charge* in the present terminology. In a quark model, the baryons mentioned before are the bound states of three quarks, and the mesons are the bound states of quark and antiquark paris. Multiquark states involving more than three quarks are also possible. The interactions of the quarks are governed by quantum chromodynamics (QCD) through their c-charges, quantum flavodynamics (QFD) through their f-charges, and quantum electrodynamics (QED) through their q-charges. The bound-state configurations of the quarks are described by QCD, which is a highly nonlinear theory, and for this reason the transition from one form of manifestation to another can occur abruptly.

Such an abrupt transition is believed to be responsible for the distinct boundary of a nucleon, for example. Inside the nucleon boundary, where three quarks are in close proximity to each other, the effective quark interaction is very weak, but once any one of them reaches the boundary, the interaction turns strong so rapidly that the quark does not have enough energy to penetrate the boundary, thus, in effect, the quarks are confined inside the boundary. Each nucleon, or baryon, therefore occupies a volume.

What if the baryon density inside a matter system is so high that the volumes occupied by them are crushed? Naturally, the boundaries will be gone, and all quarks merge to form a uniform system. Such a state of dense matter is called *quark matter*. The transition to quark matter has been estimated to occur at densities as low as 2 to 4×10^{15} g/cm^3 at zero temperature and at even lower densities at high temperatures. Current estimates of the average quark-confining energy is about 200 MeV per quark; therefore, if the temperature is above 2×10^{12} K or a corresponding thermal energy of $k_B T = 200$ MeV, the nucleons are vaporized and the quarks set free, forming a sort of quark gas. These results, we hasten to add, are only tentative.

There is very little confidence in postulating anything beyond the quark matter, and for this reason we shall end our discussion at this point.

IV. EQUATION OF STATE OF DENSE MATTER

The equation of state of a substance is a functional relation between the pressure generated by the substance and its density and temperature. It reflects the composition and internal structure of the substance. The equation of state of dense matter is of prime importance to astrophysical study of compact stellar objects such as white dwarf stars and neutron stars. It will be discussed in detail for the first three density domains. Our knowledge about dense matter in the fourth density domain is insufficient to provide accurate quantitative evaluation of its properties at this time.

A. The First Density Domain

For the first density domain (10^2–10^7 g/cm^3), we shall concentrate our discussion on a matter system composed of Fe^{56} nuclei, which represents the most stable form of matter system in that domain and shall be referred to as Fe matter. The equation of state of Fe matter at zero temperature in this density range is determined basically by the degenerate electrons in the system. Electrostatic interaction among the charged particles plays a relatively minor role, which has been demonstrated in Section II. Nevertheless, the degenerate electron pressure must be corrected for electrostatic effects. For Fe matter with densities below 10^4 g/cm^3, electrostatic interaction must be computed on the basis of the Thomas–Fermi–Dirac method which is discussed in Section II; for Fe matter with densities above 10^4 g/cm^3, the electrostatic interaction may be computed on the assumption that electrons are distributed uniformly around the nuclei.

Figure 6 shows the ground-state equation of the state of dense matter at zero temperature in a log–log plot over the first three density domains. The portion below 10^7 g/cm^3 belongs to the equation of state for Fe matter in the first density domain. The numerical values of this equation of state are given in Table III under the heading BPS. The

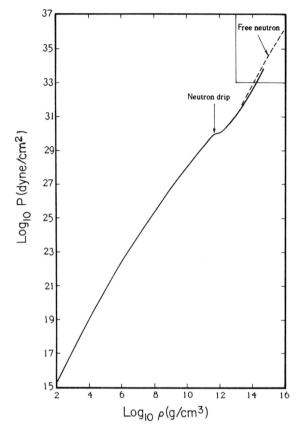

FIGURE 6 The ground-state equation of state of dense matter. The numerical values of the solid line are listed in Table III under the heading BPS. The onset of neutron drip is indicated. The different versions of the equation of state in the high-density region (outlined by rectangle) are given in Fig. 7. For correlation purposes, the equation of state of a noninteracting neutron system, indicated by "free neutron," is drawn here.

adiabatic indices are listed under Γ. The equation of state in this density domain should be quite accurate since its composition is well established.

At high temperatures, degenerate electrons become partically degenerate in a way described by Eq. (32). Finite temperature equations of state will not be shown here. General behaviors of the isotherms and adiabats should be similar to those shown in Fig. 3. At densities below 10^4 g/cm^3, where the energy correction due to electrostatic interaction is important, the Thomas–Fermi–Dirac method must be extended to include thermal effects. Interested readers are referred to references listed in the Bibliography for further details.

B. The Second Density Domain

The ground-state composition of matter in the second density domain (10^7–10^{12} g/cm^3) is described in Section III. The composition varies with density due to neutronization

TABLE III   Ground-State Equation of State of Dense Matter

		BPS		
ρ (g/cm^3)	P (dyn/cm^2)	Z	A	Γ
2.12×10^2	5.82×10^{15}	26	56	—
1.044×10^4	9.744×10^{18}	26	56	1.796
2.622×10^4	4.968×10^{19}	26	56	1.744
1.654×10^5	1.151×10^{21}	26	56	1.670
4.156×10^5	5.266×10^{21}	26	56	1.631
1.044×10^6	2.318×10^{22}	26	56	1.586
2.622×10^6	9.755×10^{22}	26	56	1.534
1.655×10^7	1.435×10^{24}	28	62	1.437
3.302×10^7	3.833×10^{24}	28	62	1.408
1.315×10^8	2.604×10^{25}	28	62	1.369
3.304×10^8	8.738×10^{25}	28	64	1.355
1.045×10^9	4.129×10^{26}	28	64	1.344
2.626×10^9	1.272×10^{27}	34	84	1.340
1.046×10^{10}	7.702×10^{27}	32	82	1.336
2.631×10^{10}	2.503×10^{28}	30	80	1.335
6.617×10^{10}	8.089×10^{28}	28	78	1.334
1.049×10^{11}	1.495×10^{29}	28	78	1.334
2.096×10^{11}	3.290×10^{29}	40	122	1.334
4.188×10^{11}	7.538×10^{29}	36	118	1.334
4.460×10^{11}	7.890×10^{29}	40	126	0.40
6.610×10^{11}	9.098×10^{29}	40	130	0.40
1.196×10^{12}	1.218×10^{30}	42	137	0.63
2.202×10^{12}	1.950×10^{30}	43	146	0.93
6.248×10^{12}	6.481×10^{30}	48	170	1.31
1.246×10^{13}	1.695×10^{31}	52	200	1.43
2.210×10^{13}	3.931×10^{31}	58	241	1.47
6.193×10^{13}	1.882×10^{32}	79	435	1.54
1.262×10^{14}	5.861×10^{32}	117	947	1.65
2.761×10^{14}	2.242×10^{33}	—	—	1.82
5.094×10^{14}	7.391×10^{33}	—	—	2.05

FP[b]		SKM	
ρ (g/cm^3)	P (dyn/cm^2)	ρ (g/cm^3)	P (dyn/cm^2)
4.11×10^{13}	1.01×10^{32}	4.11×10^{13}	8.00×10^{31}
5.64×10^{13}	1.52×10^{32}	5.64×10^{13}	1.42×10^{32}
9.75×10^{13}	3.36×10^{32}	9.75×10^{13}	4.34×10^{32}
1.55×10^{14}	8.60×10^{32}	1.55×10^{14}	1.23×10^{33}
2.31×10^{14}	2.30×10^{33}	2.31×10^{14}	3.21×10^{33}
3.21×10^{14}	5.71×10^{33}	3.29×10^{14}	7.64×10^{33}
4.97×10^{14}	1.82×10^{34}	4.51×10^{14}	1.67×10^{34}
7.72×10^{14}	6.22×10^{34}		
1.23×10^{14}	2.21×10^{35}		

RMF	
ρ (g/cm^3)	P (dyn/cm^2)
3.36×10^{14}	9.69×10^{33}
4.60×10^{14}	3.39×10^{34}
6.25×10^{14}	9.04×10^{34}
8.46×10^{14}	1.99×10^{35}
1.14×10^{15}	3.73×10^{35}
2.25×10^{15}	1.12×10^{36}
7.42×10^{15}	5.03×10^{36}

[a] From Baym, G., Pethick, C., and Sutherland, P. (1971). *Astrophysical Journal* **170**, 199. Reprinted with permission of *The Astrophysical Journal*, published by the University of Chicago Press; ©1971 The American Astronomical Society.

[b] From Friedman, B., and Pandharipande, V. R. (1981). *Nuclear Physics* **A361**, 502. Reprinted with permission of *Nuclear Physics*; ©1981 North-Holland Publishing Co., Amsterdam.

of the constituent nuclei. As the composition changes from one form to another, a first-order phase transition is involved and, over the density range where phase transition occurs, the pressure remains constant. Therefore, in a detailed plot of the equation of state in this density domain, pressure rises with density except at regions of phase transition, where it remains constant. The equation of state appears to rise in steps with increasing density, but since the steps are quite narrow, the equation of state in this density domain may be approximated by a smooth curve.

The pressure of the matter system is due entirely to the degenerate electrons. In establishing the ground-state composition of the matter system, electrostatic interaction energy in the form of lattice energy has been included, but it is quite negligible as far as the pressure of the system is concerned. The pressure in this density domain does not rise as rapidly with density as it does in the first density domain because the ground-state composition of the system shows a gradual decline in the Z/A values with density. The composition is quite well established up to a density of 4×10^{11} g/cm^3, which marks the onset of neutron drip. The composition consists of nuclei found under normal laboratory conditions or their nearby isotopes, and their masses can be either measured or extrapolated from known masses with reasonable certainty. The ground-state equation of state at zero temperature in this density range is shown in Fig. 6. The numerical values of the equation of state before neutron drip, together with the atomic number Z and mass number A of the constituent nuclei at these densities, are listed in Table III under the heading BPS.

C. The Third Density Domain

The third density domain (10^{12}–10^{15} g/cm^3) begins properly with the onset of neutron drip, which occurs at a density of 4×10^{11} g/cm^3. With the onset of neutron drip, matter is composed of giant nuclei immersed in a sea of neutrons. Nucleons inside the nuclei coexist with nucleons outside the nuclei forming a two-phase system. Theoretical studies of such a system rely heavily on the nuclear liquid drop model which gives a proper account of the different energy components in a nucleus. The nuclear liquid drop model is described in Section III. At each density, matter is composed of a particular species of nuclei characterized by certain Z and A numbers, which exist in phase equilibrium with the neutron sea outside which has a much lower Z/A ratio. Such a two-phase system constitutes the ground-state composition of matter in this density range.

Once the ground-state composition is established, the equation of state can be found as before by establishing the electron fraction Y_e of the system and proceeding to evaluate the electron pressure. The electron pressure remains the main pressure component of the system until neutron pressure takes over at higher densities. Electron number density, however, is kept fairly constant throughout the neutron drip region. It is the neutron number density and not the electron number density that is rising with increasing matter density. The suppression in electron number density with increasing density has kept the system pressure relatively constant in the density range between 4×10^{11}–10^{12} g/cm^3. This is shown in Fig. 6. Eventually, enough neutrons are produced, and the system pressure is taken over entirely by the neutron pressure. Then the equation of state shows a rapid rise subsequent to 10^{12} g/cm^3.

The numerical values of this portion of the equation of state are listed in Table III under the heading BPS. They are computed by means of the so-called Reid soft-core potential, which is determined phenomenologically by fitting nucleon–nucleon scattering data at energies below 300 MeV, as well as the properties of the deuteron. It is considered one of the best realistic nuclear potentials applicable to nuclear problems. The computation is done in the elaborate pair approximation which includes correlation effects between pairs of nucleons. In practice, this usually means solving the Brueckner–Bethe–Goldstone equations for the nucleon energy of each quantum state occupied by the nucleons. The summation of the nucleon energies for all occupied states yields the energy density of the system. Computations with the realistic nuclear potential are very involved, and several additional corrections are needed to achieve agreement with empirical results. Extension of the method to include finite temperature calculations has not been attempted.

A second form of approach is described as the independent particle approximation, in which case all particles are uncorrelated and move in the system without experiencing the presence of the others except through an overall nuclear potential. The success of the method depends on the adequacy of the effective potential that is prescribed for interaction between each pair of nucleons. A large variety of nuclear effective potentials has been devised. Potentials are usually expressed in functional forms depending on the separation between the interacting pair of nucleons. These potentials depend not only on the particle distance but also on their spin orientations. Some even prescribe dependence on the relative velocity between the nucleons. One effective nuclear potential deserving special mention is the Skyrme potential which, like the Reid soft-core potential, belongs to the class of velocity-dependent potentials. Computations based on effective nuclear potentials seem to constitute the only viable method in dealing with the neutron drip problem at finite temperatures. Some results of finite temperature equations of state in the neutron

drip region have been obtained by means of the Skyrme potential. The difficulty in working with a neutron drip system lies in the treatment of phase equilibrium, which must be handled with delicate care. The nuclear liquid drop model serves to reduce much of that work to detailed algebraic manipulations. Still, quantitative results of finite-temperature equations of state in the neutron drip region are scarce.

D. Neutron Matter Region

As matter density increases towards 10^{12} g/cm^3, the constituent nuclei become so large and their Z/A ratio so low that they become merged with the neutron sea at a density of approximately 2×10^{14} g/cm^3, where the phenomenon of neutron drip terminates, and the ground state of the matter system is represented by nearly pure neutron matter. Since neutron matter is so similar to nuclear matter, all successful theories that describe nuclear matter properties have been applied to predict the properties of neutron matter. The ground-state equation of neutron matter that is determined by pair approximation is listed in Table III under BPS. It terminates at a density of 5×10^{14} g/cm^3, which is the upper limit of its applicability. There are also results obtained by the constrained variational method and the independent particle method with the Skyrme potential. These differential equations of state are compared in Fig. 7. It gives us some idea as to how unsettled the issue remains at the present time.

The numerical results of the ground-state equation of state obtained by the constrained variational method are listed in Table III under the heading FP. They are evaluated from a form of realistic nuclear potential by a method that solves the nuclear many-body problem by means of a variational technique. The finite temperature equation of state evaluated by this method is also available.

The numerical results of the ground-state equation of state evaluated from the Skyrme potential are listed in Table III under the heading SKM. Being an effective potential, the Skyrme potential contains adjustable parameters that are established by fitting nuclear properties. As the potential is tried out by different investigators, new sets of potential parameters are being proposed. The results given here are based on a recent set of parameters designated as SKM in the literature.

There are also attempts to formulate the nuclear interaction problem in a relativistic formalism. This seems to be quite necessary if the method is to be applicable to matter density in the 10^{15} g/cm^3 region. A simplified nuclear interaction model based on the relativistic quantum field theory seems quite attractive. It is called the *relativistic mean field model*. In this model, nuclear interaction is described by the exchange of mesons. In its simplest form, only two types of mesons, a scalar meson and a vector meson, both electrically neutral, are called upon to describe nuclear interaction at short range. The scalar meson is called the σ-meson and the vector meson the ω-meson. Employed on a phenomenological level, the model requires only two adjustable parameters to complete its description, and when it is applied to study the nuclear matter problem it successfully predicts nuclear saturation. These two parameters may then be adjusted so that saturation occurs at the right density and with the correct binding energy. The model is then completely prescribed and may then be extended to study the neutron matter problem. The equation of state thus obtained for neutron matter at zero temperature is plotted in Fig. 7, together with others for comparison. The numerical results are listed in Table III under the heading RMF.

The relativistic mean field model shows great promise as a viable model for dense matter study. In the present form it has some defects. For example, it over-predicts the value of the compression modulus of nuclear matter. Also, the inclusion of a few other types of mesons may be needed to improve on the description of nuclear interaction. In particular, a charge vector meson that plays no role in the nuclear matter system contributes nevertheless to the neutron matter system and should be included. Since empirical results on dense matter are limited, it is difficult to fix the added parameters due to the new mesons in a phenomenological way. Future improvements of this model may have to be based on new results coming from experimental heavy ion collision results.

The diagonal line labeled "causality limit" in Figure 7 represents an equation of state given by $P = c\rho$. Any substance whose equation of state extends above this line would give rise to a sound speed exceeding the speed of light, which is not allowed because it would violate causality, thus the causality limit defines an upper limit to all equations of state. It is drawn here to guide the eye.

The pressure of neutron matter rises rapidly with increasing density. Such an equation of state is called *stiff*. This feature is already exhibited by a free neutron system that is devoid of interaction. The equation of state of a free neutron system is plotted alongside the others in Fig. 7 for comparison and is labeled "free neutron." The stiffness of the free neutron equation of state had led scientists in the 1930s to suggest the possible existence of neutron stars in the stellar systems almost as soon as neutrons were discovered. Neutron stars were detected a quarter of century later.

E. Pion Condensation Region

The stiffness of the equation of state of neutron matter could be greatly modified should pions be found to form a condensate in neutron matter. A pion condensate occurs in a matter system only if the interaction between the nucleons and pions lowers the interaction energy sufficiently to account for the added mass and kinetic energy of the pions. The possibility of pion condensation has been suspected for a long time since pion–nucleon scattering experiments have revealed a strong attractive interaction between pions and nucleons, and this attractive mode could produce the pion condensation phenomenon. However, a quantitatively reliable solution of this problem is yet to be established. Nuclear interactions are difficult to deal with, and most theoretical results are not considered trustworthy unless they can be collaborated by some empirical facts. In the pion condensation problem, no empirical verification is available, and many of the estimates about pion condensation, though plausible, cannot be easily accepted.

The effect of pion condensation on the equation of state of neutron matter is to lower the pressure of the system, because it adds mass to the system without contributing the comparable amount of kinetic energy. The softening of the equation of state is a reflection of the presence of an attractive interaction that brings about pion condensation

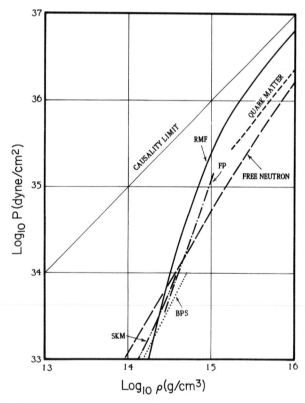

FIGURE 7 Different versions of the equation of state for ground-state neutron matter. Numerical values of curves indicated by BPS, FP, SKM, and RMF are listed in Table III under the same headings. See text for discussion of free neutron, quark matter, and causality limit.

and serves to lower the pressure. Some estimates have suggested that pion condensation should occur at a neutron matter density of around 5×10^{15} g/cm^3. No quantitative results, however, will be presented here for the pion condensation situation.

F. Baryonic Matter

As the density of neutron matter increases, the Fermi energy of the neutron system increases accordingly. If we neglect interactions among the particles, then as soon as the neutron Fermi energy reaches the energy corresponding to the mass difference of the neutron and Λ, the ground state of the matter system favors the replacement of the high-energy neutrons by Λ. Λ is the least massive baryon heavier than the nucleons. Based on an estimate in which interactions are neglected, Λ should appear in neutron matter at a density of 1.6×10^{15} g/cm^3. On a scale of ascending mass, the next group of baryons after the nucleons and Λ is the Σ, which consists of three charged species, and after that is the Δ, which consists of four charged species. Similar estimates put the appearance of Σ at 2.9×10^{15} g/cm^3 and the appearance of Δ at 3.7×10^{15} g/cm^3.

The above estimates emphasize the mass of the baryons. It turns out that the electric charge of the baryon is also important. For example, when a Σ^- appears, not only does it replace an energetic neutron but, because of its charge, it replaces an energetic electron as well. Again, based on estimates in which interactions are neglected, Σ^- appears at a density of 10^{15} g/cm^3 at Δ^- at 1.3×10^{15} g/cm^3, which are lower values than those given above.

To study baryonic matter properly, particle interactions must be included. Unfortunately, our knowledge of baryon interaction is still quite limited, and the interaction strengths among the different species of baryons are not yet established with sufficient accuracy. Therefore, much of our understanding of the baryonic matter is based on theoretical conjectures of the nature of baryon interaction.

One of the best methods appropriate to the study of baryonic matter is the relativistic mean field model mentioned before. The model considers all particle interactions to the mediated by two mesons, the scalar σ-meson and the vector ω-meson, both electrically neutral. Particle interaction is established once the coupling constants of these mesons with different baryons are known. There are theoretical justifications to believe that the ω-meson interacts with all baryons at equal strength, in the same manner that the electric field interacts with all electric charges equally no matter which particles carry them. In analogy with the electric charge, each baryon is associated with a unit of baryonic charge and its antiparticle with a negative unit of baryonic charge. The baryonic force mediated by the ω-meson behaves very much like the electric force in the

sense that like charges repel and unlike charges attract. It is repulsive between baryons but attractive between baryons and antibaryons. If this is the case, the coupling constants of the ω-meson with all baryons in the relativistic mean field model may be taken to be the same as that adopted for the nucleons. One possible fallacy in this reasoning is that the coupling constants in the relativistic mean field model represent effective coupling constants that incorporate modifications due to the presence of neighboring particles. Therefore, even though the concept of a baryonic charge to which the ω-meson couples is correct, the effective ω-meson coupling constants with different baryons in the relativistic mean field model need not be the same.

For the σ-meson, the best estimate of its coupling constants is probably by means of the quark model which assigns definite quark contents to the baryons and σ-meson. If the quarks are assumed to interact equally among themselves, and given the fact that particle attributes such as electric charge and hypercharge must be conserved in the interaction process, it is possible to deduce that the σ-meson couples equally between nucleons and Δ, and also equally between Λ and Σ, but its coupling with Λ and Σ is only two-thirds of the coupling with the nucleon and Δ.

Incorporating these couplings in the relativistic mean field model, we can deduce the equation of state for baryonic matter. It shows a slight decrease in pressure compared with that due to neutron matter when matter density exceeds 10^{15} g/cm^3. The effect of admixing other baryons in a neutron matter system is relatively minor on the system's equation of state. On the other hand, the effect of particle interaction on the equation of state is quite significant. Therefore, a proper understanding of the equation of state of matter in the 10^{15} g/cm^3 range awaits better knowledge of particle interactions as well as methods in dealing with a many-body system.

G. Quark Matter

In a quark model of the elementary particles, the nucleon is viewed as the bound state of three quarks, which interact via the exchange of gluons as determined by quantum chromodynamics. The size of the nucleon is therefore determined by the confining radius of the quark interaction. When matter density reaches the point that the average separation of the nucleons is less than its confining radius, the individual identity of the nucleon is lost, and the matter system turns into a uniform system of quarks forming quark matter.

Quarks are fermions and like electrons and neutrons obey Pauli's exclusion principle. There are different species of quarks, and the known quarks are given the names of up quark (u), down quark (d), strange quark (s), and charm quark (c). There may be others. Inside a nucleon, their effective masses are estimated to be (expressed in mc^2) under 100 MeV for the u and d quarks, approximately 100–200 MeV for the s quark, and approximately 2000 MeV for the c quark. If quark interaction is neglected, the equation of state of quark matter may be deduced in the same manner as it is for a degenerate electron system. The treatment of quark interaction turns out to be not as difficult as it is for neutrons. In a quark matter, the effective quark interaction is relatively weak, and perturbative treatment of the interaction is possible. This interesting feature of quantum chromodynamics prescribes that the effective interaction of the quarks should decrease in strength as the quarks interact in close proximity to each other, and this has been verified to be true in experiments. The equation of state of quark matter at zero temperature computed with quark interaction has been obtained. One version of it is plotted in Fig. 7 for illustration.

The transition from baryonic matter to quark matter is a first-order phase transition that may be established by a tangent construction method on the energy density of the baryonic matter and that of the quark matter in a manner similar to that discussed in Section III for the neutronization process and illustrated in Fig. 4. The onset of transition from neutron matter to quark matter has been estimated to begin at densities around 1.5 to 4×10^{15} g/cm^3.

In Table III, only the basic ground-state equation of state of dense matter is presented. It serves to suggest the possible behavior of matter compressed to high densities. It would be too elaborate to detail all aspects of the equation of state at finite temperatures. At the present time, the study of dense matter is being actively pursued.

V. TRANSPORT PROPERTIES OF DENSE MATTER

Properties of dense matter under nonequilibrium conditions, such as during the transfer of mass and conduction of heat and electricity, are of much physical interest. These are the transport properties, and knowledge of them is important in understanding stellar structure and stellar evolution. We discuss here the following transport properties of dense matter: electrical conductivity, heat conductivity, and shear viscosity.

A. Electrical Conductivity

The electrical conductivity of a substance is given by the ratio of the induced electric current density to the applied electric field. In the case of a metal, the electric current is due to the flow of conduction electrons. If all conduction electrons are given an average drift velocity v_d, then the current density j_e is given by:

$$j_e = nev_d \tag{45}$$

where n is the number density of the conduction electrons and e the electric charge of the electrons. The electrons acquire the drift velocity as they are accelerated by the electric field. When an electron collides with either the lattice or another electron, its drift velocity is redirected randomly, and the subsequent velocity is averaged to zero statistically. Thus, after each collision the electron may be assumed to be restored to thermal equilibrium and begins anew with zero drift velocity. The average drift velocity is therefore determined by the mean time between collisions τ, during which an electron is accelerated to its drift velocity by the electric field $\mathcal{E}$:

$$v_d = (e\mathcal{E}/m_e)\tau \tag{46}$$

τ is also called the relaxation time. Putting these two expressions together gives the electrical conductivity:

$$\sigma = (ne^2/m_e)\tau \tag{47}$$

The electrical conductivity is in units of inverse seconds, or sec^{-1}, in the cgs system of units and is related to the mks system of units by sec$^{-1} = (9 \times 10^9 \, \Omega \, \text{meter})^{-1}$. When the electrons in the system are degenerate, the electron mass in Eq. (47) is replaced by p_F/c, where p_F is the Fermi momentum of the degenerate electron system. The relaxation time is the most crucial parameter in this investigation and must be related to the electron density, the average electron speed, the number and types of scatterers in the system, and the scattering cross sections of the electrons with different types of scatterers.

When the temperature is below the melting temperature T_m given by Eq. (41), matter in the density range of 10^2–10^{14} g/cm^3 is in a solid state possessing a crystalline structure. The constituent nuclei are organized into a lattice while the electrons are distributed more or less uniformly in the space between. In the neutron drip region, 10^{12}–10^{15} g/cm^3, neutrons are also outside the nuclei. The relaxation time of the electrons is determined by the frequency of scatterings with the lattice nuclei and with the other electrons and neutrons. When there are several scattering mechanisms present, the relaxation times due to different mechanisms are found by the inverse sum of their reciprocals:

$$\tau^{-1} = \tau_1^{-1} + \tau_2^{-1} + \cdots \tag{48}$$

Upon electron scattering, the lattice vibrates and the vibration propagates collectively like sound waves through the lattice. In quantized form the sound waves behave like particles, called *phonons*. Electron scattering from the lattice is usually described as electron–phonon scattering. Phonons increase rapidly in number with temperature. We therefore expect the electric conductivity due to electron–phonon scattering to decrease with increasing temperature.

Electron–electron scatterings have minor effects on the electric conductivity, because in dense matter all the electrons are not bound to the nuclei. Electrons lose energy only if they are scattered from bound electrons. Elastic scattering of electrons does not alter the current being transported. Hence, this form of scattering does not affect electric conductivity and may be ignored.

Let us imagine that the dense matter in consideration is formed from a dynamical process, as in the formation of a neutron star, in which case the material substance is adjusting to reach the proper density while it is being cooled. If the solidification rate is faster than the nuclear equilibrium rate, there will be large admixtures of other nuclei with the equilibrium nuclei. The nonequilibrium nuclei act as impurities in the system. Also, there may be defects in the system due to rapid rates of cooling and rotation. Electrons and phonons are scattered by these impurities and defects which also limit the transport of electric charge by electrons.

There are estimates of the electric conductivity in dense matter that consider all the above discussed features. Typical results for a wide range of densities at a temperature of 10^8 K are shown in Fig. 8. Electric conductivities due to different scattering mechanisms are shown separately. The total conductivity should be determined by the lowest lying portions of the curves, since the total conductivity, like the relaxation time, is found by taking the inverse sum of the reciprocals of the conductivities due to different scattering mechanisms.

Electron–phonon scattering should be the dominant mechanism. The conductivity due to electron–phonon scattering has a temperature dependence of approximately T^{-1}. Electron–impurities scattering depends on the impurity concentration and the square of the charge difference between the impurity charge and the equilibrium nuclide charge. Let us define a concentration factor,

$$x_{\text{imp}} = \sum \left(\frac{n_i}{n_A} \right)(Z_i - Z)^2 \qquad (49)$$

where the summation is over all impurity species designated by subscript i; n_i and Z_i are, respectively, the number density and charge of each impurity species. The conductivity curve due to impurity scattering σ_{imp} is drawn in Fig. 8 with an assumed $x_{\text{imp}} = 0.01$. It is independent of T.

The electron–phonon curve would not be correct at the low-density end where the melting temperature is below 10^8 K. Above melting temperature, the lattice structure would not be there, and electrons would not be impeded by scattering from phonons but instead from the nuclei. Electric conductivity due to electron–nuclei scattering is shown in the possible melting region.

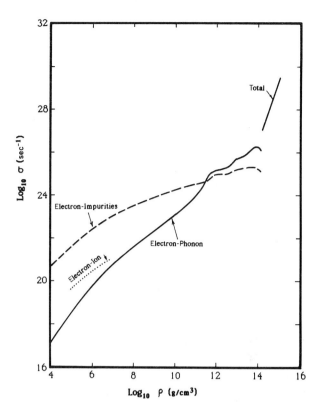

FIGURE 8 The electrical conductivity of dense matter at a temperature of $T = 10^8$ K. For densities below 2×10^{12} g/cm^3, the solid line corresponds to the electrical conductivity limited by the mechanism of electron–phonon scattering while the system is in a solid state. Possible melting occurs in the low-density region at this temperature, and electron–phonon scattering is replaced by electron-ion scattering. The electrical conductivity due to electron–ion scattering is drawn as a dotted line. The electrical conductivity due to electron–impurities scattering is drawn as a dashed line, taking an arbitrary impurity concentration factor of $x_{\text{imp}} = 0.01$. For densities above 2×10^{12} g/cm^3, the electrical conductivity is limited mainly by the mechanism of electron–proton scattering, and it is drawn as a solid line and labeled "total." [From Elliott, F., and Itoh, N., *Astrophys. J.* **206**, 218, 1976; **230**, 847, 1979. Reprinted with permission of *The Astrophysical Journal*, published by the University of Chicago Press; ©1976 and 1979. The American Astronomical Society.]

In the neutron matter region, 10^{14}–10^{15} g/cm^3, the system is composed basically of three types of particles—neutrons, protons, and electrons—of which both protons and electrons act as carriers for electrical conduction. The main difficulty in dealing with this situation is to take into full account the interactions among the particles. With the presence of a strong attractive interaction among the nucleons, it is quite likely that the protons will be paired to turn the system into a superconducting state when the temperature falls below the critical temperature, in which case the electric conductivity becomes infinite. The critical temperature T_c is estimated to be as high as 10^9–10^{10} K. Such a temperature, though it appears high, corresponds actually to a thermal energy $k_B T$ that is small compared with the Fermi energies of the particles present.

If the system is not in a superconducting state, or in other words it is in a normal state, its electrical conductivity is due mainly to electrons as carriers. The electron relaxation time is determined by scattering by protons and has been evaluated. The electrical conductivity of neutron matter in the normal state is plotted in Fig. 8 in the density range 10^{14}–10^{15} g/cm^3.

B. Thermal Conductivity

The thermal conductivity of a substance is given by the ratio of the amount of heat transferred per unit area per unit time to the temperature gradient. Kinetic theory of dilute gas yields the following expression for thermal conductivity:

$$k = (1/3)n c_s \bar{v}^2 \tau \qquad (50)$$

where n is the carrier number density, c_s its specific heat, $\bar{v}$ the average thermal velocity, and τ the relaxation time of the carriers. The thermal conductivity is expressed in units of erg/cm K sec. When there are several types of carriers participating in the transport of heat, the final thermal conductivity is the sum of individual conductivities due to different types of carriers. In the case of a solid, the important carriers are the electrons and phonons, but in general the phonon contribution to thermal conduction is negligible compared with the electron contribution.

In thermal conduction by electrons, no electric current is generated while energy is being transported. The type of electron–phonon scattering important for thermal conduction is different from that of electrical conduction. At each point in the system, the electron number density obeys the Fermi–Dirac distribution of Eq. (32), which assigns a higher probability of occupation of high-energy states when the temperature is high than when the temperature is low. When a thermal gradient exists in the system, neighboring points have different electron distributions. Electrons moving from a high-temperature point to a low-temperature point must lose some of their energy to satisfy the distribution requirement. If this can be accomplished over a short distance, the thermal resistivity of the substance is high, or its thermal conductivity is low. The most important mechanism of energy loss is through inelastic scatterings of electrons by phonons at small angles. Such scatterings constitute the major source of thermal resistivity. On the other hand, elastic scattering of electrons do not lead to energy loss and aid in thermal conduction. The frequency of inelastic scattering to that of elastic scattering depends on the thermal distribution of phonons. The temperature dependence of thermal conductivity due to electron–phonon scatterings is therefore complicated. For matter density below 10^8 g/cm^3, it is relatively indepen-

dent of temperature, but above that it shows a decrease with increasing temperature.

Although electron–electron scatterings do not contribute to the electric conductivity, they contribute to the thermal conductivity by redistributing the electron energies. The thermal conductivity due to impurity scattering k_{imp} is directly related to the electric conductivity due to impurity scattering σ_{imp} by the Wiedemann–Franz rule:

$$k_{imp} = \pi^2 T / 3 (k_B/e)^2 \sigma_{imp} \qquad (51)$$

The thermal conductivities due to different mechanisms at 10^8 K are shown in Fig. 9. The thermal conductivity due

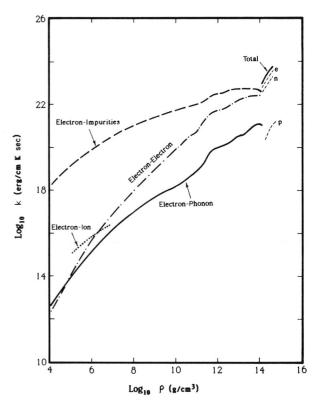

FIGURE 9 The thermal conductivity of dense matter at a temperature of $t = 10^8$ K. For densities below 2×10^{12} g/cm^3, the solid line corresponds to the thermal conductivity limited by the mechanism of electron–phonon scattering. If the lattice is melted, the solid line should be replaced by the dotted line, which is the thermal conductivity due to electron–ion scattering. The thermal conductivity due to electron–electron scattering is given by the dot-dashed line, and that due to electron–impurities scattering is given by the dashed line, where the impurity concentration factor is assumed to be $x_{imp} = 0.01$. For densities above 2×10^{12} g/cm^3, the thermal conductivity received contributions from the electrons, neutrons, and protons, which are drawn in thin dashed lines and marked by e, n, and p, respectively. The total thermal conductivity from these carriers is drawn in a solid line and labeled "total." [From Elliott, F., and Itoh, N., *Astrophys. J.* **206**, 218, 1976; **230**, 847, 1979. Reprinted with permission of *The Astrophysical Journal*, published by the University of Chicago Press; © 1976 and 1979. The American Astronomical Society.]

to impurity scattering is drawn with $x_{imp} = 0.01$, and it has a temperature dependence linear in T. The thermal conductivity due to electron–electron scattering is expected to have a temperature dependence of T^{-1}. The thermal conductivity due to electron–ion scattering is shown for the region where 10^8 K is expected to be above the melting temperature. The total thermal conductivity of dense matter is determined by the mechanism that yields the lowest thermal conductivity at that density, since the total thermal conductivity, like the relaxation time, is found from the inverse sum of the reciprocals of the conductivities due to different mechanisms.

In the neutron matter region, 10^{14}–10^{15} g/cm^3, all three types of particles—neutrons, protons, and electrons—contribute to the thermal conductivity

$$k = k_e + k_n + k_p$$

where the subscripts e, n, and p denote contributions to the thermal conductivity by electrons, neutrons, and protons, respectively. When the particles are in a normal stare (i.e., not in a superfluid or superconducting state), the thermal conductivity is determined primarily by the highly mobile electrons, whose motion is impeded largely by scatterings with the protons and other electrons and much less by scatterings with the neutrons. The neutron contribution to the thermal conductivity is substantial because of its high number density. Neutrons encounter neutron–proton scattering and neutron–neutron scattering in the process. The proton contribution to the thermal conductivity is small because the proton number density is low, but otherwise the protons contribute in a manner similar to the neutrons. The thermal conductivities due to these three components for a system in the normal state are shown in short dashed lines in Fig. 9. The total conductivity is drawn as a solid line there.

The system may also become superfluid when its temperature falls below the critical temperature of $T_c \approx 10^9$–10^{10} K. The critical temperature of the protons is in general different from that of the neutrons, and therefore it is possible that while one turns superfluid, the other remains normal. Also, when the temperature falls below the critical temperature of a certain type of particles (say, the neutrons), there remains a normal component of neutrons in the system. This situation is usually described by a two-fluid model that consists of both the superfluid and normal fluid components. In general, scattering of particles off the superfluid component is negligible for transport purposes.

If only the protons turn superfluid (and superconducting) while the neutrons remain normal, the thermal conductivity due to the superfluid protons vanishes. The thermal conductivity found for the system in the normal state is basically unaltered, because the protons give a very

small contribution to the thermal conductivity, as shown in Fig. 9.

If the neutrons turn superfluid while the protons remain normal, the superfluid component of the neutrons gives vanishing thermal conductivity, and the contribution by the normal component of the neutrons to the thermal conductivity diminishes rapidly with decreasing temperature below the critical temperature, because the number density of the normal neutrons decreases rapidly with decreasing temperature. The thermal conductivity in this case is therefore determined entirely by the electron contribution to the thermal conductivity, which is modified slightly from the normal matter case due to the absence of scattering by superfluid. The net result is that the thermal conductivity is only slightly reduced from the normal case shown in Fig. 9. However, if both the neutrons and protons turn superfluid, the thermal conductivity is due entirely to electron–electron scattering, and the general result is indicated by the extension of the electron–electron curve for densities below 10^{14} g/cm^3.

C. Shear Viscosity

When a velocity gradient exists in a fluid, a shearing stress is developed between two layers of fluid with differential velocities. The shear viscosity is given by the ratio of the shearing stress to the transverse velocity gradient. Elementary kinetic theory suggests that the shear viscosity of a dilute gas is given by:

$$\eta = \frac{1}{3}nm\bar{v}^2\tau \tag{52}$$

where n is the molecular density, m the mass of each molecule, $\bar{v}$ the average thermal velocity of the molecules, and τ the relaxation time. Viscosity is expressed in units of g/cm/sec, which is also called *poise*. In appearance it is similar to the expression for thermal conductivity with the exception that the specific heat per particle in the thermal conductivity is being replaced by the particle mass. Consequently, we may suspect that the electron component of the viscosity should behave similarly to the thermal conductivity. This, however, is not true due to the fact that the relaxation time involved relates to different aspects of the scattering mechanism. Also, there is an additional component to the viscosity. The solid lattice can make a great contribution to the total viscosity. Unfortunately, the determination of the lattice viscosity is very difficult, and no adequate work has been performed to determine its properties at the present time. The following discussion relates only to the electron viscosity.

The electron relaxation time is determined by electron scatterings by phonons, impurities, electrons, and nuclei. Shearing stress is developed when electrons belonging to fluid layers of different velocities are exchanged. Thus, viscosity is related to mass transfer or the transfer of electrons. This is similar to electric conduction where the transfer of electrons gives rise to charge transfer and is different from heat conduction which involves the adjustment of electron energy distributions. The evaluation of the relaxation times for the viscosity due to different scattering mechanisms is similar to that for the electrical conductivity.

The viscosity of dense matter at 10^8 K due to different scattering mechanisms is shown in the Fig. 10. The temperature dependence of the viscosity due to electron–phonon scattering is approximately T^{-1}, as in the case

of the electrical conductivity. This is also true of the viscosity due to electron-impurities scattering, which is independent of temperature as in the case of the electrical conductivity. While electron–electron scatterings do not contribute to the electrical conductivity, they play a role in viscosity giving rise to a temperature dependence of T^{-2}.

In the neutron matter region, 10^{14}–10^{15} g/cm³, all three types of particles—neutrons, protons, and electrons— contribute to the viscosity. The contributions from neutrons, protons, and electrons to the viscosity are shown separately in Fig. 10 in this density range by dashed lines. The total viscosity is drawn as a solid line. They all have a temperature dependence of T^{-2}.

When the temperature drops below the critical temperature T_c, superfluid proton and/or neutron components appear. The behavior of the viscosity in the superfluid state is very similar to the thermal conductivity. If the protons turn superfluid, the viscosity is basically unaltered from the normal viscosity, because the proton contribution is small. If the neutrons turn superfluid, the superfluid component of the neutrons has vanishing viscosity. The viscosity is dominated by the electron contribution which is determined by the electron–electron scattering and electron-proton scattering mechanisms. When both protons and neutrons turn superfluid, then the viscosity is determined entirely by electron–electron scattering, and the general result is indicated by the extension of the electron–electron curve for densities below 10^{14} g/cm³.

VI. NEUTRINO EMISSIVITY AND OPACITY

In a nonequilibrium situation where a temperature gradient exists in a substance, energy is transported not only by means of thermal conduction, as discussed in the last section, but also by radiation. Parameters of radiative transfer intrinsic to the matter system are its emissivity and opacity. There are two major forms of radiation. In one form, the radiative energy is transmitted by photons and in the other by neutrinos.

In a dense matter system whose constituent electrons are highly degenerate (i.e., when the electron Fermi energy is high compared with the thermal energy $k_B T$), the degenerate electrons cannot avail themselves as effective scatterers for the passage of energy carriers created by the thermal gradient, and the thermal conductivity is correspondingly high. Therefore, energy transport is far more effective through heat conduction than it would be for radiative transfer. For this reason, the problem of photon emissivity and opacity in dense matter receives very little attention. Most astrophysical studies of photon emissivity and opacity are performed for relatively low-density

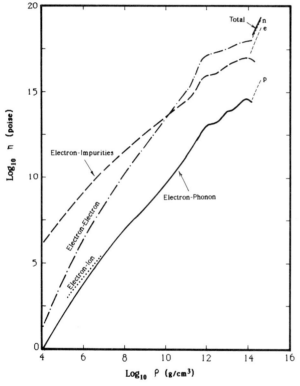

FIGURE 10 The viscosity of dense matter at a temperature of $T = 10^8$ K. For densities below 2×10^{12} g/cm³, the viscosity due to electron–phonon scattering is drawn as a solid line. If the lattice is melted, the viscosity is due to electron–ion scattering, which is drawn as a dotted line. The viscosity due to electron–electron scattering is drawn as a dot-dashed line, and the viscosity due to electron-impurities scattering is drawn as a dashed line, where the impurity concentration factor is taken to be $x_{imp} = 0.01$. For densities above 2×10^{12} g/cm³, the viscosity receives contributions from neutrons, electrons, and protons, and they are shown as thin dashed lines and labeled n, e, and p, respectively. The total viscosity from these three components is drawn as a solid line and labeled "total." [From Elliott, F., and Itoh, N., *Astrophys. J.* **206**, 218, 1976; **230**, 847, 1979. Reprinted with permission of *The Astrophysical Journal*, published by the University of Chicago Press; ©1976 and 1979 The American Astronomical Society.]

substances, such as those forming the interiors of luminous stars. Since our subject is dense matter, these parameters will not be discussed here. Interested readers are referred to astrophysical texts suggested in the bibliography. More relevant to dense matter physics is the topic of neutrino emissivity and opacity.

A. Neutrino Emissivity

The reaction described by Eq. (37) is a sample reaction in which a neutrino is produced. Indeed, neutrinos are produced throughout the second density domain whenever neutronization occurs, and in these processes neutrinos are produced even at zero temperature. At high temperature there are other reactions that are more effective in neutrino production. Reactions involving neutrinos belong to a class of interaction called the *weak interaction*. The coupling constant for the weak interaction is much smaller than that of the electromagnetic interaction. Neutrino interaction rates are in general slower than comparable photon interaction rates by a factor of at least 10^{20}. Neutrinos can pass through thick layers of substance and experience no interaction. For example, the neutrino mean free path through matter with density similar to our sun is about a billion solar radii. Hence, once they are produced in a star they are lost into space, and they serve as an efficient cooling mechanism for hot and dense stellar objects.

During a supernova process, the collapse of the stellar core raises the core temperature to as high as 10^{11} K. It quickly cools to a temperature of 10^9 K through neutrino emission as the core stabilizes into a neutron star. Neutrino emission dominates photon emission until the temperature drops to 10^8 K. It is estimated that neutrino cooling dominates for at least the first few thousand years of a neutron star after its formation; therefore, the neutron cooling mechanisms deserve attention. We are primarily interested in those neutrino emission mechanisms that supersede photon emission under similar conditions. They determine therefore the cooling rate of neutron stars in the early period, and they also play a major role in the dynamics of a supernova process.

An important neutrino production process is the modified Urca reactions which involve neutrons, protons, and electrons:

$$n + n \rightarrow n + p + e + \bar{\nu} \tag{53a}$$

$$n + p + e \rightarrow n + n + \nu \tag{53b}$$

where ν and $\bar{\nu}$ denote neutrino and antineutrino, respectively. In the following we shall not distinguish neutrinos from antineutrinos and address them collectively as neutrinos. These two reactions are very nearly the inverse of each other, and when they are occurring at equal rates, the number of neutrons, protons, and electrons in the system

is unaltered, while neutrinos are being produced continuously. Urca is the name of a casino in Rio de Janeiro. Early pioneers of neutrino physics saw a parallel between nature's way of extracting energy from the stellar systems and the casino's way of extracting money from its customers, so they named the reactions after the casino. Reactions (53a,b) are modifications of the original Urca reactions by adding an extra neutron to the reaction. This increases the energy range over which neutrinos may be produced and thus improves the production rate.

Our current understanding of the weak interaction theory is provided by the Weinberg–Salam–Glashow theory. Even though most of the neutrino reactions to be discussed in this section have never been verified under laboratory conditions, they are nevertheless believed to be correct, and quantitative estimates of their reaction rates are reliable. When reactions (53a) occurs, a transition is made from a state of two neutrons to a state of a neutron, proton, electron, and neutrino. The theory evaluates the transition probability from an initial state of two neutrons that occupy quantum states of definite momenta to a final state of four particles of definite momenta. Whenever a transition is made, a neutrino of a specific energy is produced. The total neutrino energy emitted from the system per unit time is found by summing the neutrino energies of all allowed transitions multiplied by their respective transition probabilities. The neutrino emissivity is the total neutrino energy emitted per unit time per unit volume of the substance.

Consider a neutron matter system in the density range of 10^{14}–10^{15} g/cm^3 that is composed mainly of neutrons with a small admixture (about 1%) of protons and electrons. Quantum states occupied by the particles are given by the Fermi–Dirac distribution f_j of Eq. (32). Note that the temperature dependence of the final result comes from the temperature factor in the distribution. Designating the initial-state neutrons by subscripts 1 and 2, the final-state neutron by 3, and the proton, electron, and neutrino by p, e, and ν, respectively, we find that the emissivity of the neutron matter system is given by the following equation:

$$\begin{aligned}
E_{\text{Urca(a)}} = \int & \left[d^3 n_1 f_1 \right] \left[d^3 n_2 f_2 \right] \left[d^3 n_3 (1 - f_3) \right] \\
& \times \left[d^3 n_p (1 - f_p) \right] \left[d^3 n_e (1 - f_e) \right] \\
& \times \left[d^3 n (1 - f_\nu) \right] W(i \rightarrow f) e_\nu \tag{54}
\end{aligned}$$

where the summations over quantum states are represented by integrations over $d^3 n_j = V(d^3 p_j / h^3)$, $e_\nu = c|\mathbf{p}_\nu|$ is the neutrino energy, and $W(i \rightarrow f)$ is the transition probability from initial state i to final state f per unit time and per unit volume of the system. It has the following structure:

$$W(\text{i} \to \text{f}) = (2\pi)^4 \, \delta(\mathbf{p}_\text{i} - \mathbf{p}_\text{f}) \, \delta(E_\text{i} - E_\text{f})$$
$$\times \sum |H(\text{i} \to \text{f})|^2 \quad (55)$$

where $\mathbf{p}_\text{i}$ and $\mathbf{p}_\text{f}$ denote the total momenta of the initial and final states, respectively; E_i and E_f, the total energies of the initial and final states, respectively; and $H(\text{i} \to \text{f})$, the matrix element of the Hamiltonian describing the interaction: The summation symbol $\sum$ indicates the summations over all spin orientations of the particles. The mathematical delta functions are here to ensure that only energy–momentum-conserving initial and final states are included in this evaluation. In Eq. (54), the initial state particles are assigned distributions f_j while the final state particles are assigned distributions $(1 - f_j)$ because the particles produced in the reaction must be excluded from states that are already occupied; therefore, they must take up states that are not occupied, which are expressed by $(1 - f_j)$. The chemical potentials in the distributions f_j determine the particle numbers in the system. They are related by $\mu_1 = \mu_2 = \mu_3$ and $\mu_\text{p} = \mu_\text{e}$, while in most cases $\mu_\nu = 0$. The quantity μ_ν will be different from zero only in circumstance where the neutrino opacity is so high that the neutrinos are trapped momentarily and become partially degenerate.

If the neutrons and protons in the system are assumed to be noninteracting, the emissivity due to be noninteracting, the emissivity due to the reaction (53a) can be easily evaluated, and the result may be expressed conveniently as:

$$E_{\text{Urca(a)}} = (6.1 \times 10^{19} \text{ erg/cm}^3 \text{ sec})(\rho/\rho_0)^{2/3}(T_9)^8 \quad (56)$$

where $\rho_0 = 2.8 \times 10^{14} \text{ g/cm}^3$ is the density of nuclear matter, and the symbol T_9 stands for temperature in units of 10^9 K. Here, the emissivity (in units of ergs per cubic centimeter per second) is evaluated for a dense system of neutron matter whose proton and electron contents are determined by the ground-state requirement under reaction Eq. (37). It is expressed in this form because neutron matter exists only with densities comparable to nuclear matter density, and T_9 is a typical temperature scale for neutron star and supernova problems.

An interesting point to note is that this emissivity is given by eight powers of temperature. This comes about for the following reason. For the range of temperature considered, the thermal energy is still small compared with the degenerate or Fermi energy ε_F of the fermions in the system (except neutrinos), and most of the easily accessible quantum states are already occupied, leaving only a small fraction of quantum states in each species contributing to the reaction (of the order of $k_\text{B} T/\varepsilon_\text{F}$ per species). Since there are two fermion species in the initial state and three (not counting neutrinos) in the final state, a factor of T^5 is introduced. The allowed neutrino states are restricted only

by energy conservation, and their number is given by the integration over $d^3 n_\nu \delta(E_\text{i} - E_\text{f})$, which is proportional to the square of the neutrino energy e_ν^2. This, together with the neutrino energy term in Eq. (54), gives e_ν^3. Since e_ν must be related to $k_\text{B} T$, which is the only energy variable in the problem, all together they give the T^8 dependence to the expression. The above deduction has general applicability, and if there were one less fermion in both the initial and final states, then the emissivity from such a process should be proportional to T^6.

The emissivity due to reaction (53b), $E_{\text{Urca(b)}}$, can be shown to be of comparable magnitude to that evaluated above, and the total emissivity due to the modified Urca process is simply twice that of Eq. (56):

$$E_{\text{Urca}} = (1.2 \times 10^{20} \text{ erg/cm}^3 \text{ sec})(\rho/\rho_0)^{2/3}(T_9)^8 \quad (57)$$

At high density, muons would appear alongside electrons, and neutrino production reactions similar to those of Eqs. (53a) and (53b) become operable, but with the electrons there replaced by muons. Similar results are obtained, but muons do not appear in dense matter until its density exceeds $8 \times 10^{14} \text{ g/cm}^3$, and the emissivity due to the additional muon processes is only a minor correction.

A more significant consideration is the inclusion of nuclear interaction, which has been neglected in the above evaluation. Nuclear interaction appears in the evaluation of the matrix element $H(\text{i} \to \text{f})$. When nuclear interaction is included, the total emissivity due to modified Urca process is changed to

$$E_{\text{Urca}} = (7.4 \times 10^{20} \text{ erg/cm}^3 \text{ sec})(\rho/\rho_0)^{2/3}(T_9)^8 \quad (58)$$

which is a factor of six higher than that evaluated without the inclusion of nuclear interaction.

Other important neutrino production mechanisms in neutron matter are

$$\text{n} + \text{n} \to \text{n} + \text{n} + \nu + \bar{\nu} \quad (59a)$$

$$\text{n} + \text{p} \to \text{n} + \text{p} + \nu + \bar{\nu} \quad (59b)$$

in each of which a pair of neutrinos is produced as the nucleons scatter from each other. Neutrino emissivities for these processes are evaluated to be

$$E_{\text{nn}} = (1.8 \times 10^{19} \text{ erg/cm}^3 \text{ sec}) (\rho/\rho_0)^{1/3}(T_9)^8 \quad (60a)$$

$$E_{\text{np}} = (2.0 \times 10^{19} \text{ erg/cm}^3 \text{ sec}) (\rho/\rho_0)^{2/3}(T_9)^8 \quad (60b)$$

where the subscript nn denotes reaction (59a) and np denotes (59b). These emissivities are evaluated with nuclear interaction taken into consideration. The processes are, however, not as effective as the modified Urca processes in neutrino production.

If neutron matter turns superfluid after its temperature falls below the critical temperature T_c, then the neutrino production rates evaluated above must be reduced; T_c

should be in the range 10^9–10^{10} K. Superfluidity is explained by the fact that an energy gap appears in the energy spectrum of the particle, just above its Fermi energy. What that means is that a group of quantum states whose energies fall in the energy gap are excluded from the system. Normally, in an inelastic scattering process, the neutrons or protons are scattered into these states, but since the states are absent they must be excited into much higher energy states, and scattering becomes difficult and less likely to occur. The consequence is that neutrons and protons may move about relatively freely without being impeded by scatterings. This is the explanation of superfluidity. By the same token, the above neutrino production mechanisms that depend on the scattering of neutrons and protons are similarly reduced. Qualitatively, if superfluidity occurs in both the neutron and proton components, E_{Urca} and E_{np} are reduced by a factor of $\exp[-(\Delta_n + \Delta_p)/k_B T]$, where Δ_n and Δ_p are the width of the energy gaps for neutron and proton superfluidity, respectively. For E_{nn}, the reduction is $\exp[-2\Delta_n/k_B T]$. If superfluidity occurs in just one component, the reductions are obtained by setting the energy gap of the normal component in the above expressions to zero.

In the case of pion condensation in neutron matter, interactions involving pions and nucleons also modify the neutrino production rate. Some estimates have shown that the neutrino production rates thus modified can be significant. However, due to the uncertainty in our present understanding of the pion condensation problem, no emissivity for this situation will be quoted here.

There are also neutrino production mechanisms not involving the direct interaction of two nucleons, such as the following:

1. Pair annihilation

$$e + e^+ \rightarrow \nu + \bar{\nu} \tag{61a}$$

2. Plasmon decay

$$\text{plasmon} \rightarrow \nu + \bar{\nu} \tag{61b}$$

3. Photoannihilation

$$e + \gamma \rightarrow e + \nu + \bar{\nu} \tag{61c}$$

4. Bremsstrahlung

$$e + (Z, A) \rightarrow (Z, A) + e + \nu + \bar{\nu} \tag{61d}$$

5. Neutronization

$$e + (Z, A) \rightarrow (Z - 1, A) + \nu \tag{61e}$$

where e^+ denotes a positron, plasmon a photon propagating inside a plasma, and (Z, A) a nucleus of atomic number Z and mass number A. A photon in free space

cannot decay into a neutrino pair, because energy and momentum cannot be conserved simultaneously in the process. However, when a photon propagates inside a plasma, its relation between energy and momentum is changed in such a way that the decay becomes possible. Quanta of the electromagnetic wave in a plasma are called *plasmons*. The neutrino production rates for these processes have been found to be relatively insignificant at typical neutron star densities and temperatures and will not be listed here.

B. Neutrino Opacity

When a radiation beam of intensity $I(\text{erg/cm}^2/\text{sec})$ is incident on a substance of density $\rho(\text{g/cm}^2)$, the amount of energy absorbed from the beam per unit volume per unit time $E(\text{erg/cm}^3/\text{sec})$ is proportional to the opacity $\kappa = E/\rho I$. The opacity is expressed in units of cm^2/g. Each neutrino emission process has an inverse process corresponding to absorption. In addition to absorption, scattering can also impede the passage of neutrinos through the medium. Both absorption and scattering contribute to the opacity. Some of the more important processes are listed below.

1. Scattering by neutrons

$$\nu + n \rightarrow \nu + n \tag{62a}$$

2. Scattering by protons

$$\nu + p \rightarrow \nu + p \tag{62b}$$

3. Scattering by electrons

$$\nu + e \rightarrow \nu + e \tag{62c}$$

4. Scattering by nuclei

$$\nu + (Z, A) \rightarrow \nu + (Z, A) \tag{62d}$$

5. Absorption by nucleons

$$\nu + n \rightarrow p + e \tag{62e}$$

6. Absorption by nuclei

$$\nu + (Z, A) \rightarrow (Z + 1, A) + e \tag{62f}$$

Similar processes occur for antineutrinos, which shall not be displayed here.

For each of these reactions, a reaction cross section is evaluated from the Weinberg–Salam–Glashow theory of weak interactions. The cross section represents an area effective in obstructing the incident beam of neutrinos. The opacity may be expressed in terms of the reaction cross sections as follows:

$$\kappa = \rho^{-1} \sum n_j \sigma_j \tag{63}$$

where n_j denotes the number density of the particles that react with the neutrino and σ_j their reaction cross sections, and the summation $\sum$ is over all the reactions listed above. Often, neutrino opacity is expressed by a neutrino mean free path, which is defined as:

$$\lambda = (\rho\kappa)^{-1} \tag{64}$$

Among these reactions, the contribution of reaction (62f) to the opacity of dense matter is quite negligible because it entails the production of electrons; since electrons in the system are already highly degenerate, it is difficult to accommodate the newly produced electrons.

The most important reaction in this regard is reaction (62d) where the neutrino is scattered by the nuclei. This is the result of coherent scattering, in which all nucleons in a nucleus participate as a single entity in the process. The cross section of a coherent process involving A nucleons is proportional to A^2 times the cross section of scattering from a single nucleon. This reaction therefore dominates all others when the matter system is composed of giant nuclei, which is the case when matter density is below nuclear density, that is, before nuclei dissolve into neutron matter.

For neutron matter the important reactions are scatterings by neutrons, protons, and electrons as indicated by reactions (62a), (62b), and (62c). Neutrinos are scattered elastically by the nucleons and nuclei since the scatterers are massive. The neutrinos may change directions after scattering but do not lose their energies to the scatterers. They lose energy only if they are scattered by electrons. Electron scattering is therefore an important process in lowering the energies of the high-energy neutrinos, bringing them into thermal equilibrium with all neutrinos should the neutrinos be trapped in the system for a duration long enough for this to happen. Even though neutrinos interact very weakly and are therefore very difficult to confine, neutrino trapping is in fact believed to occur at the moment when the collapsing stellar core reaches the point of rebound initiating the explosive supernova process. Therefore, a great deal of attention has been given to the problem of neutrino opacity and the issue of neutrino trapping.

The cross sections for these processes in the reference frame of the matter system are evaluated to be as follows:

1. Neutrino–electron scattering

$$\sigma_e \approx (1/4)\sigma_0\left(e_\nu/m_ec^2\right)(\varepsilon_f/mc),$$
$$e_\nu \ll \varepsilon_F \tag{65}$$

2. Neutrino–nucleon scattering

$$\sigma_N \approx (1/4)\sigma_0\left(e_\nu/m_ec^2\right)^2,$$
$$e_\nu \ll m_nc^2 \tag{66}$$

3. Neutrino–nucleus scattering

$$\sigma_A \approx (1/16)\sigma_0\left(e_\nu/m_ec^2\right)^2 A^2[1 - (Z/A)],$$
$$e_\nu \ll 300A^{-1/3}\,\text{MeV} \tag{67}$$

where $\sigma_0 = 1.76 \times 10^{-44}$ cm^2 is a typical weak interaction cross section, e_ν the neutrino energy, and ε_F the Fermi enegy of the electrons. The total neutron opacity of the substance is given by:

$$\kappa = \rho^{-1}[n_e\sigma_e + n_N\sigma_N + n_A\sigma_A] \tag{68}$$

According to the Weinberg–Salam–Glashow theory, neutrinos also interact with quarks, and dense quark matter also emits and absorbs neutrinos. However, since our understanding of quark matter is still far from complete, no results related to neutrino emissivity and opacity in quark matter will be quoted at this time.

SEE ALSO THE FOLLOWING ARTICLES

ATOMIC PHYSICS • BINARY STARS • DARK MATTER IN THE UNIVERSE • NEUTRINOS • NEUTRON STARS • NUCLEAR PHYSICS • PARTICLE PHYSICS, ELEMENTARY • PLASMA SCIENCE AND ENGINEERING • QUANTUM THEORY • STELLAR STRUCTURE AND EVOLUTION • SUPERCONDUCTIVITY

BIBLIOGRAPHY

Chandrasekhar, S. (1939). "An Introduction to the Study of Stellar Structure," Dover, New York.
Clayton, D. D. (1968). "Principles of Stellar Evolution and Nucleosynthesis," McGraw-Hill, New York.
Gasiorowicz, S. (1979). "The Structure of Matter: A Survey of Modern Physics," Addison-Wesley, Reading, MA.
Leung, Y. C. (1984). "Physics of Dense Matter," Science Press, Beijing, China.
Schwarzschild, M. (1958). "Structure and Evolution of the Stars," Dover, New York.
Shapiro, S. L., and Teukolsky, S. A. "Black Holes, White Dwarfs, and Neutron Stars, The Physics of Compact Objects," Wiley, New York.
Zeldovich, Ya. B., and Novikov, I. D. (1971). "Relativistic Astrophysics," Chicago University Press, Chicago, IL.

Designs and Error-Correcting Codes

K. T. Phelps
C. A. Rodger
Auburn University

GLOSSARY

Binary sum, $u + v$ Componentwise addition of u and v with $1 + 1 = 0$ (exclusive or).
Binary word A string (or vector or sequence) of 0's and 1's
Code A set of (usually binary) words, often all of the same length.
Combinatorial design A collection of subsets of a set satisfying additional regularity properties.
Decoding Finding the most likely codeword (or message) transmitted.
Distance, $d(u, v)$ The number of coordinates in which u and v differ.
Encoding The assignment of codewords to messages.
Information rate The fraction of information per transmitted bit.
Weight, $wt(u)$ The number of nonzero bits in the word u.

I. INTRODUCTION

Both error-correcting codes and combinatorial designs are areas of discrete (not continuous) mathematics that began in response to applied problems, the first in making the electronic transmission of information reliable and the second in the design of experiments with results being statistically analyzed. It turns out that there is substantial overlap between these two areas, mainly because both are looking for uniformly distributed subsets within certain finite sets. In this article we provide a brief introduction to both areas and give some indication of their interaction.

A. Error-Correcting Codes

Error-correcting codes is a branch of discrete mathematics, electrical engineering, and computer science that has developed over the past 50 years, largely in response to the dramatic growth of electronic transfer and storage of information. Coding Theory began in the late 1940s and early 1950s with the seminal work of Shannon (1948), Hamming (1950), and Golay (1949). Error-correcting codes' first significant application was in NASA's deep space satellite communications. Other important applications since then have been in storage devices (e.g., compact discs), wireless telephone channels, and geopositioning systems. They are now routinely used in all satellite communications and mobile wireless communications systems.

Since no communication system is ideal, information can be altered, corrupted, or even destroyed by *noise*. Any communication system needs to be able to recognize or *detect* such errors and have some scheme for recovering the information or *correcting* the error. In order to protect against the more likely errors and thus improve the reliability, redundancy must be incorporated into the message. As a crude example, one could simply transmit the message several times in the expectation that the majority will appear correctly at their destination, but this would greatly increase the cost in terms of time or the rate of transmission (or space in storage devices).

The most basic error control scheme involves simply detecting errors and requesting retransmission. For many communications systems, requests for retransmission are impractical or impose unacceptable costs on the communication system's performance. In deep space communications, a request for retransmission would take too long. In speech communications, noticeable delays are unacceptable. In broadcast systems, it is impractical given the multitude of receivers. The problem of correcting errors and recovering information becomes of paramount importance in such constrained situations.

Messages can be thought of as words over some alphabet, but for all practical purposes, all messages are simply strings of 0's and 1's, or *binary words*. Information can be partitioned or blocked up into a sequence of binary words or messages of fixed length k. A (*block*) *code*, C, is a set of binary words of fixed length n, each element of which is called a *codeword*. Mathematically, codewords can be considered to be vectors of length n with elements being chosen from a finite field, normally of order 2, but in some cases from the field $GF(2^r)$. [So, for example, the binary codeword 01101 could also be represented as the vector $(0, 1, 1, 0, 1)$.] There are also convolutional codes where the codewords do not have fixed length (or have infinite length), but these will be discussed later. *Encoding* refers to the method of assigning messages to codewords which are then transmitted. Clearly, the number of codewords has to be at least as large as the number of k-bit messages. The *rate* of the code is k/n, since k bits of information result in n bits being transmitted.

The key to being able to detect or correct errors that occur during transmission is to have a code, C, such that no two codewords are close. The *distance* between any two words u and v, denoted by $d(u, v)$, is simply the number of coordinates in which the two words differ. The *weight* of u, $wt(u)$, is just the number of nonzero coordinates in u. Using the binary sum (so $1 + 1 = 0$), we have $d(u, v) = wt(u + v)$.

Under the assumptions that bits are more likely to be transmitted correctly than incorrectly (a natural assumption) and that messages are equally likely to be transmitted (this condition can be substantially relaxed), it is easy to show that for any received word w, the most likely codeword originally transmitted is the codeword $c \in C$ for which $d(c, w)$ is least; that is, the most likely codeword sent is the one closest to the received word. This leads to the definition of the *minimum distance* $d(C) = \min_{c_1, c_2 \in C} \{d(c_1, c_2)\}$ of a code C, the distance between the two closest codewords in C. Clearly, if a word w is received such that $d(c, w) \leq t = \lfloor (d(C) - 1)/2 \rfloor$ for some $c \in C$, then the unique closest codeword to w is c; therefore, c is the most likely codeword sent and we *decode* w to c. (Notice that if c was the codeword that was originally transmitted, then at most t bits were altered in c during transmission to result in w being received.) So decoding each received word to the closest codeword (known as *maximum likelihood decoding*, or MLD) will always result in correct decoding provided at most t errors occur during transmission. Furthermore, if c_1 and c_2 are the two closest codewords [so $d(c_1, c_2) = d(C)$], then it is clearly possible to change $t + 1$ bits in c_1 so that the resulting word w satisfies $d(c_1, w) > d(c_2, w)$. Therefore, if these $t + 1$ bits are altered during the transmission of c_1, then, using MLD, we would incorrectly decode the received word w to c_2. Since MLD results in correct decoding for C no matter which codeword is transmitted and no matter which set of up to t bits are altered during transmission, and since this is not true if we replace t with $t + 1$, C is known as a t-*error-correcting code*, or as an $(n, |C|, d)$ code where $d = d(C) \geq 2t + 1$.

The *construction problem* is to find an $(n, |C|, d)$ code, C, such that the minimum distance d (and thus t) is large—this improves the error-correction ability and thus the reliability of transmission; and where $|C|$ is large, so the rate of transmission ($\log_2 |C|/n = k/n$) is closer to 1. Since messages are usually blocked up into k-bit words, one usually has $\log_2 |C| = k$. Clearly, these aims compete against each other. The more codewords one packs together in C, the harder it is to keep them far apart. In practice one needs to make decisions about the reliability of the channel and the need to get the message transmitted correctly and then weigh that against the cost of decreasing the rate of transmission (or increasing the amount of data to be stored). It is possible to obtain bounds on one of the three parameters n, d, and $|C|$ in terms of the other two, and in some cases families of codes have been constructed which meet these bounds. Two of these families are discussed in this article: the *perfect codes* are codes that meet the Hamming Bound and are described in Section II; the *maximum distance separable* (or MDS) codes are codes that meet the Singleton Bound and are described in Section IV.

The second problem associated with error-correcting codes is the *encoding problem*. Each message m is to be assigned a unique codeword c, but this must be done efficiently. One class of codes that have an efficient encoding algorithm are *linear* codes; that is, the *binary sum* of any pair of codewords in the code C is also a codeword in C (here, *binary sum* means componentwise binary addition of the binary digits or the *exclusive or* of the two codewords). This means that C is a vector space and thus has a basis which we can use to form the rows of a *generating* matrix G. Then each message m is encoded to the codeword $c = mG$. One might also require that C have the additional property of being *cyclic*; that is, the cyclic shift $c' = x_n x_1 x_2 \ldots x_{n-1}$ of any codeword $c = x_1 x_2 \ldots x_n$ (where $x_i \in \{0, 1\}$) is also a codeword for every codeword c in C. If C is cyclic and linear, then encoding can easily be completed using a shift register design. The representation of a code is critical in decoding. For example, Hamming codes (see Section II) possess a cyclic representation but also have equivalent representations that are not cyclic.

The final main problem associated with error-correcting codes is the *decoding problem*. It is all well and good to know from the design of the code that all sets of up to t errors occurring during transmission result in a received word, w, that is closer to the codeword c that was sent than it is to any other codeword; but given w, how do you efficiently find c and recover m? Obviously, one could test w against each possible codeword and perhaps eventually decode which is closest, but some codes are very big. Not only that, but it can also be imperative that decoding be done extremely quickly, as the following example demonstrates.

The introduction of the compact disc (CD) by Phillips in 1979 revolutionized the recording industry. This may not have been possible without the heavy use of error-correcting codes in each CD. (Errors can occur, for example, from incorrect cutting of the CD.) Each codeword on each CD represents less than 0.0001 sec of music, is represented by a binary word of length 588 bits, and is initially selected from a Reed-Solomon code (see Section III) that contains 2^{192} codewords. Clearly, this is an application where all decoding must take place with no delay, as nobody will buy a CD that stops the music while the closest codeword is being found! It turns out that not only are the Reed-Solomon codes excellent in that they meet the Singleton Bound (see Section III), but they also have

an extremely efficient decoding algorithm which finds the closest codeword without having to compare the received word to all 2^{192} codewords.

Again, the class of linear codes also has a relatively efficient decoding algorithm. Associated with each linear code C is the dual code $C^\perp$ consisting of all vectors (codewords) such that the dot product with any codeword in C is 0 (again using *xor* or *binary* arithmetic). This is useful because of the fact that if H is a generating matrix for $C^\perp$, then $Hw^T = 0$ if and only if w is a codeword. H is also known as the *parity check* matrix for C. The word $s = Hw^T$ is known as the *syndrome* of w. Syndromes are even more useful because it turns out that for each possible syndrome s, there exists a word e_s with the property that a closest codeword to *any* received word w with syndrome s is $w + e_s$. This observation is taken even further to obtain a very efficient decoding algorithm for the Reed-Solomon codes that can deal with the 2^{196} codewords in real time; it incorporates the fact that these codes are not only linear but also cyclic.

Another family of codes that NASA uses is the *convolutional codes*. Theoretically, these codes are infinite in length, so a completely different decoding algorithm is required in this case (see Section V).

In the following sections, we focus primarily on the construction of some of the best codes, putting aside discussion of the more technical problem of describing decoding algorithms for all except the convolutional codes in Section V. This allows the interaction between codes and designs to be highlighted.

B. Combinatorial Designs

Although (combinatorial) designs were studied earlier by such people as Euler, Steiner, Kirkman, it was Yates (1936) who gave the subject a shot in the arm in 1935 by pointing out their use in statistics in the design of experiments. In particular, he defined what has become known as an (n, k, λ) *balanced incomplete block design* (BIBD) to be a set V of n elements and a set B of subsets (called *blocks*) of V such that

1. Each block has size $k < n$.
2. Each pair of elements in V occur as a subset of exactly λ blocks in B.

Fisher and Yates (1938) went on to find a table of small designs, and Bose (1939) soon after began a systematic study of the existence of such designs. Bose made use of finite geometries and finite fields in many of his constructions.

A natural generalization of BIBD is to replace (2) with

2′. Each $(t + 1)$-element subset of V occurs as a subset of exactly λ blocks of B.

Such designs are known as $(t + 1)$ *designs*, which can be briefly denoted by $S_\lambda(t + 1, k, n)$; in the particular case when $\lambda = 1$, they are known as *Steiner* $(t + 1)$ *designs* which are denoted by $S(t + 1, k, n)$. By elementary counting techniques, one can show that if $s < t$, then an $S_\lambda(t + 1, k, n)$ design is also an $S_\mu(s, k, u)$ design where $\mu = \lambda \binom{n-s}{t+1-s} / \binom{k-s}{t+1-s}$. Since μ must be an integer, this provides several necessary conditions for the existence of a $(t + 1)$ design.

For many values of n, k, and t, an $S(t + 1, k, n)$ design cannot exist. A *partial* $S(t + 1, k, n)$ design then is a set of k subsets of an n set where any $(t + 1)$ subset is contained in at most one block. This is equivalent to saying that any two k subsets intersect in at most t elements. Partial designs are also referred to as *packings*, and much research has focused on finding maximum packings for various parameters.

There are very few results proving the existence of s designs once $s \geq 3$. Hanani found exactly when there exists an $S_\lambda(3, 4, v)$ [also called Steiner Quadruple Systems; see Lindner and Rodger (1975) for a simple proof and Hartman and Phelps (1992) for a survey], and Teirlinck (1980) proved that there exists an $S_\lambda(s, s + 1, v)$ whenever $\lambda = ((s + 1)!)^{2s+1}$, $v \geq s + 1$, and $v \equiv s$ (modulo $((s + 1)!)^{2s+1}$). Otherwise, just a few s designs are known [see Colbourn and Dinitz (1996)]. Much is known about their existence when $s = 2$. In particular, over 1000 papers have been written [see Colbourn and Rosa (1999)] about $S_\lambda(2, 3, v)$ designs (known as *triple systems*, and as *Steiner triple systems* if $\lambda = 1$). We only need to consider designs with $\lambda = 1$ in this article.

Certainly, $(t + 1)$ designs and maximum packings are of interest in their own right, but they also play a role in the construction of good codes. To see how, suppose we have a $(t + 1)$ design (or packing) (V, B). For each block b in B, we form its *characteristic vector* c_b of length n, indexed by the elements in V, by placing a 1 in position i if $i \in B$ and placing a 0 in position i if $i \notin B$. Let C be the code $\{c_b \mid b \in B\}$. Then C is a code of length n in which all codewords have exactly k 1's (we say each codeword has *weight* k). The fact that (V, B) is a $t + 1$ design (or packing) also says something about the minimum distance of C: since each pair of blocks intersect in at most t elements, each pair of codewords have at most t positions where both are 1, so each pair of codewords disagree in at least $2k - 2t$ positions, so $d(C) \geq 2k - 2t$. This connection is considered in some detail in Section III. We also show in Section II that the codewords of weight d in perfect codes together form the characteristic vectors of a $(t + 1)$ design.

There is much literature on the topic of combinatorial designs [see Colbourn and Dinitz (1996) for an encyclopedia of designs and Lindner and Rodger (1997) for an intro-

ductory text], and this topic is also considered elsewhere in this encyclopedia, so here we restrict our attention to designs that arise in connection with codes.

II. PERFECT CODES

Let C be a code of length n with minimum distance d. Let $t = \lfloor (d - 1)/2 \rfloor$. Then, as described in Section I, for each codeword c in C, and for each binary word w of length n with $d(w, c) \leq t$, the *unique* closest codeword to w is c. Since we can choose any i of the n positions to change in c in order to form a word of length n distance exactly i from c, the number of words distance i from c is $\binom{n}{i} = n!/(n - i)!i!$. So the total number of words of length n that are distance at most t from c is $\binom{n}{0} + \binom{n}{1} + \cdots + \binom{n}{t}$, one of which is c, thus the number of words of length n distance at most t from some codeword in C is $|C| \sum_{i=0}^{t} \binom{n}{i}$ (by the definition of t, no codeword is within distance t of two codewords). Of course, the total number of binary words of length n is 2^n. Therefore, it must be the case that

$$|C| \leq 2^n \Big/ \sum_{i=0}^{t} \binom{n}{i}.$$

This bound is known as the *Hamming Bound* or the *sphere packing bound*. Any code that satisfies equality in the Hamming Bound is known as a *perfect code*, in which case $d = 2t + 1$.

From the argument above, it is clear that for any perfect code, each word of length n must be within distance t of a unique codeword (if a code is not perfect, then there exist words for which the distance to any closest codeword is more than t). In particular, if C is a perfect code with $d = 2t + 1$, then the codewords of minimum weight d in C are the characteristic vectors of the blocks of an $S(t + 1, 2t + 1, n)$ design. To see this, note that each word w of weight $t + 1$ is within distance t of a unique codeword c, where clearly c must have weight $d = 2t + 1$. Equivalently, each $(t + 1)$ subset is contained in a unique d subset, the characteristic vector of which is a codeword. In fact, for any codeword $c \in C$, one can define the neighborhood packing,

$$NS(c) = \{x + c \mid x \in C \text{ and } d(x, c) = d\}.$$

Then, the code C will be perfect *if and only if* every neighborhood packing is, in fact, the characteristic vectors of an $S(t + 1, 2t + 1, n)$ design. To see the converse, suppose C is a code with every $NS(c)$ an $S(t + 1, 2t + 1, n)$ design. If w is any word, let $c \in C$ be the closest codeword and assume $d(c, w) \geq t + 1$. Choose any $t + 1$ coordinates where c and w disagree. Since $NS(c)$ is an $S(t + 1, 2t + 1, n)$ design, these coordinates uniquely determine a block of size

$2t + 1$ in the design and, hence, a codeword c' such that $d(c, c') = 2t + 1$. So c' disagrees with c in the same $t + 1$ coordinates as does w, and thus, c' agrees with w in these $t + 1$ coordinates. Thus,

$$d(c', w) \le d(c, c') - (t + 1) + d(c, w) - (t + 1)$$
$$\le d(c, w) - 1,$$

which contradicts the assumption that c was the closest codeword and thus $d(c, w) \le t$.

It turns out that perfect binary codes are quite rare. The Hamming codes are an infinite family of perfect codes of length $2^r - 1$, for any $r \ge 2$, in which the distance is 3 and the number of codewords is $2^{2^r - 1 - r}$. A linear Hamming code can always be formed by defining its parity check matrix, H, to be the $r \times n$ matrix in which the columns form all nonzero binary words of length r. Notice also that, in view of the comments earlier in this section, the codewords of weight 3 in any Hamming code form a Steiner triple system, $S(2, 3, n)$.

The other perfect binary code is the Golay code, which has length 23, distance 7, and 2^{12} codewords.

Tietäväinen (1973), based on work by van Lint, showed that these are the only perfect binary codes and, in fact, generalized this result to codes over finite fields [see also Zinov'ev and Leont'ev (1973)]. For a survey of results on perfect codes, see van Lint (1975).

III. CONSTANT WEIGHT CODES

A *constant weight code* (CW) with parameters n, d, w is a set C of binary words of length n all having weight w such that the distance between any two codewords is at least d. All nontrivial (n, d, w) CW codes have $d \le 2w$. Let $A(n, d, w)$ be the largest number of codewords in any CW code with these parameters. The classic problem then is to detemine this number or find the best upper and lower bounds on $A(n, d, w)$.

Binary CW codes have found application in synchronization problems, in areas such as optical codedivision multiple-acces (CDMA) communications systems, frequency-hopping spread-spectrum communications, modile radio, radar and sonar signal design, and the construction of protocol sequences for multiuser collision channel without feedback. Constant weight codes over other alphabets have received some attention, but so far there have been few applications. We will only discuss the binary CW codes.

Constant weight codes have been extensively studied, and a good reference is MacWilliams and Sloane (1977). Eric Raines and Neil Sloane maintain a table of the best known lower bounds on $A(n, d, w)$ on the web

site: http://www.research.att.com/njas/codes/Andw/. We will present an overview of this topic with an emphasis on the connections with designs.

Since the sum of any two binary words of the same weight always has even weight, we have $A(n, 2\delta - 1, w) = A(n, 2\delta, w)$. We will assume from now on that the distance d is even. We also have $A(n, d, w) = A(n, d, n - w)$, since whenever two words are distance d apart, so are their complements. This means one only needs to consider the case $w \le n/2$.

The connection between CW codes and designs is immediate. In terms of sets, a CW code is just a collection of w subsets of an n set where the intersection of any two w subsets contains at most $t = w - \frac{d}{2}$ elements. Equivalently, a CW code is a partial $S(w - \frac{d}{2} + 1, w, n)$ Steiner system. We then have

$$A(n, d, w) \le \frac{n(n - 1) \cdots (n - w + d/2)}{w(w - 1) \cdots (d/2)}$$

with equality if and only if a Steiner system $S(w - \frac{d}{2} + 1, w, n)$ exists.

The interest in CW codes also comes from the problem of finding linear (or nonlinear) codes (n, M, d) of maximum size M. Obviously, $A(n, d, w)$ is an upper bound on the number of words of a given weight in such a maximum code. Conversely, such codes (or their cosets) can give lower bounds for $A(n, d, w)$. In particular, the stronger version of the Hamming Bound (given in the section on perfect codes) was originally proved using $A(n, 2t + 2, 2t + 1)$.

$A(n, 2t + 2, 2t + 1)$ is just the number of blocks in a maximum partial $S(t + 1, 2t + 1, n)$ design or packing. If C is a t-error-correcting code, then for any $c \in C$, the number of blocks in a neighborhood packing $|NS(c)| \le A(n, 2t + 2, 2t + 1)$. The number of words that are distance $t + 1$ from c but not distance t from any other codeword is

$$\binom{n}{t + 1} - \binom{2t + 1}{t + 1} |NS(c)| \ge \binom{n}{t + 1}$$
$$- \binom{2t + 1}{t + 1} A(n, 2t + 2, 2t + 1).$$

Each such word is distance $t + 1$ from at most $\lfloor n/t + 1 \rfloor$ other codewords. Thus, summing over all $c \in C$, each such word is counted at most this many times. This gives a stronger version of the Hamming bound:

$$|C| \left(\left(\sum_{i=0}^{t} \binom{n}{i} \right) + \frac{\binom{n}{t+1} - \binom{2t+1}{t+1} A(n, 2t + 2, 2t + 1)}{\lfloor n/(t + 1) \rfloor} \right)$$
$$\le 2^n.$$

Constant weight codes cannot be linear, since this would mean the zero vector was in the code, but one can have a code with all nonzero words having the same weight. These codes are sometime referred to as linear *equidistant codes*. The dual of the Hamming code (also called the *simplex code*) is an example of such a code. In fact, it has been proved that the only such codes are formed by taking several copies of a simplex code. The proofs that all such codes are generalized simplex codes come explicitly from coding theory (Bonisoli, 1983) and also implicitly from results on designs and set systems (Teirlinck, 1980). There is a close connection between linear equidistant codes and finite geometries. The words of a simplex code correspond to the hyperplanes of projective space [over $GF(2)$] just as the words of weight 3 in the Hamming code correspond to lines in this projective space. [For connections between codes and finite geometries, see Black and Mullin (1976).]

Another variation on CW codes are optical orthogonal codes (OOC) which were motivated by an application to optical CDMA communication systems. Briefly, an (n, w, t_a, t_b) OOC is a CW code, C, of length n and weight w such that for any $c = (c_0, c_1, \ldots, c_{n-1}) \in C$, and each $y \in C$, $c \neq y$ and each $i \not\equiv 0 \pmod{n}$,

$$\sum_{j=0}^{n-1} c_j c_{j+i} \leq t_a, \tag{1}$$

and

$$\sum_{j=0}^{n-1} c_j y_{j+i} \leq t_c. \tag{2}$$

Equation (1) is the autocorrelation property, and Eq. (2) is the cross-correlation property. Most research has focused on the case where $t_a = t_c = t$, in which case we refer to an (n, w, t) OOC. Again, one can reformulate these properties in terms of (partial) designs or packings. In this case, an OOC is a collection of w subsets of the integers $(\bmod\, n)$, such that for subsets $c, b \in C$,

$$|(c + i) \cap (c + j)| \leq t_a \; (i \neq j), \tag{3}$$

and

$$|(c + i) \cap (b + j)| \leq t_c. \tag{4}$$

Here, $c + i = \{x + i \pmod{n} \mid x \in c\}$.

An OOC code is equivalent to a cyclic design or packing. A code or packing is said to be cyclic if every cyclic shift of a codeword (or block) is another codeword. The set of all cyclic shifts of a codeword is said to be an *orbit*. A representative from that orbit is often called a base block. An (n, w, t) OOC is a set of base blocks for a cyclic (partial) $S(t + 1, w, n)$ design or packing (assuming $t < w$). Conversely, given such a cyclic partial $S(t + 1, w, n)$ design or packing, one can form an (n, w, t) OOC by taking one representative block or codeword from each orbit.

IV. MAXIMUM DISTANCE SEPARABLE CODES

For any linear code C, recall that the minimum distance equals the minimum weight of any nonzero codeword. Also, if C has dimension k, then $C^\perp$ has dimension $n - k$ and any parity check matrix H of C has rank $n - k$. If $c \in C$ is a codeword of minimum weight, $wt(c) = d$, then $Hc^\top = 0$ implies that d columns of H are dependent, but no $d - 1$ columns are dependent. Since H has rank $n - k$, every $n - k + 1$ columns of H are dependent. Thus,

$$d \leq n - k + 1.$$

This is known as the *Singleton Bound*, and any code meeting equality in this bound is known as a *maximum distance separable* code (or MDS code).

There are no interesting binary MDS codes, but there are such codes over other alphabets, for example, the Reed-Solomon codes used in CD encoding of order 256 (Reed-Solomon codes are described below). Even though such codes are treated mathematically as codes with 256 different "digits," each still has an implementation as a binary code, since each of the digits in the finite field $GF(2^8)$ can be represented by a binary word of length 8; that is, by one byte. So the first step in encoding the binary string representing all the music onto a CD is to divide it into bytes and to regard each such byte, as a field element in $GF(2^8)$.

We now consider a code $C \subseteq F^n$ as a set of codewords over the alphabet F, where F is typically the elements of a finite field. The are several equivalent definitions for a linear code C of length n and dimension k to be an MDS code:

1. C has minimum distance $d = n - k + 1$,
2. Every k column of G, the generating matrix for C, is linearly independent.
3. Every $n - k$ column of H, the parity check matrix for C, is linearly independent.

Note, that from item (3) above C is MDS if and only if $C^\perp$ is MDS.

If one arranges the codewords of C in an $|C| \times n$ array, then from item (2) this array will have the property that for any choice of k columns (or coordinates) and any word of length k, $w \in F^k$, there will be exactly one row of this array that has w in these coordinates. An *orthogonal array* is defined to be a $q^k \times n$ array with entries from a set F, $|F| = q$, with precisely this property: restricting the array to any k columns, every word $w \in F^k$ occurs exactly once in this $q^k \times k$ subarray. Two rows of an orthogonal array can agree in at most $k - 1$ coordinates, which means that they must disagree in at least $n - (k - 1)$ coordinates. Thus, the distance between any two rows of an orthogonal array is $d = n - k + 1$. Obviously, the row vectors of an orthogonal array are also codewords of an MDS code, except that orthogonal arrays (and MDS codes) do not need to be linear and exist over arbitrary alphabets.

Orthogonal arrays were also introduced in the design of experiments in statistical studies and are closely related to designs and finite geometries. In fact, the construction for Reed-Solomon codes was first published by Bush as a construction of orthogonal arrays (see Bush, 1952; Colbourn and Dinitz, 1996).

There are various representations of the Reed-Solomon codes. We present the most perspicuous one. Again, let F denote a finite field and let $F_k[x]$ denote the space of all polynomials of degree less than k with coefficients from F. Choose $n > k$ different (nonzero) field elements α_1, $\alpha_2, \ldots, \alpha_n \in F$. For each polynomial $f(x) \in F_k[x]$, form the *valuation vector* $c_f = (f(\alpha_1), f(\alpha_2), \ldots, f(\alpha_n))$. Define

$$C = \{c_f \mid f(x) \in F_k[x]\}.$$

First, we note that C is a linear code, since $c_f + c_g = c_{f+g}$. Second, for any two different polynomials $f(x)$, $g(x) \in F_k[x]$, we have $c_f \neq c_g$; if c_f and c_g were equal, then the polynomial $f(x) - g(x)$ would have n roots but degree $< k (< n)$. This means that $|F_k[x]| = |F^k| = q^k = |C|$, and thus, C has dimension k and length n. Finally, since $f(\alpha_i) = 0$ if and only if α_i is a root of $f(x)$, and moreover any polynomial of degree $\leq k - 1$ has at most $k - 1$ roots, then c_f has at most $k - 1$ zeros and at least $n - k + 1$ nonzero entries. Therefore, the minimum distance for C is $n - k + 1$ and C is an MDS code.

Reed-Solomon codes also have a representation as a cyclic code and a relatively efficient decoding algorithm (see Hoffman *et al.*, 1991, for example).

V. CONVOLUTIONAL CODES

Convolutional codes are practical codes, having been adopted for use by both NASA and the European Space Agency. In fact, they encode messages twice: first using a Reed-Solomon code, then the resulting codeword is encoded using a convolutional code.

Convolutional codes are infinite length codes that are both linear and cyclic. The messages to be considered are strung together into a stream of bits which form a single message m that is encoded by feeding m into a shift register (see Fig. 1). Initially, μ codewords are formed: for $1 \leq i \leq \mu$, and for each *tick* $t \geq 0$, the contents of certain registers are added together to form the tth bit in the output of the codeword $c_i = c_{i,0}, c_{i,1}, c_{i,2}, \ldots$. The single codeword c to which m is encoded is then formed by $c = c_{1,0} c_{2,0}, \ldots, c_{\mu,0} c_{1,1} c_{2,1}, \ldots, c_{\mu,1}, \ldots$.

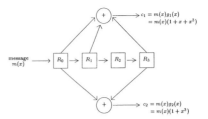

FIGURE 1 Convolutional code encoding.

Algebraically, one can represent the registers whose contents are added to form c_i by the polynomial $g_i = g_{i,0} x^0 + g_{i,1} x + \ldots g_{i,\ell} x^\ell$, where $g_{i,j} = 1$ if the contents in register R_j are used when forming c_i, and is 0 otherwise. Then by writing the message $m = m_0 m_1 m_2, \ldots$ in polynomial form $m(x) = m_0 + m_1 x + m_2 x^2 + \cdots$, it turns out that $c_i(x) = m(x) g_i(x)$. This representation of convolutional code encoding makes it clear why the code is cyclic and linear.

Convolutional codes can also be represented graphically. Assuming one message bit is moved into the shift register at each tick, the vertices of the directed graph are the 2^ℓ binary words $(r_0, r_1, \ldots, r_{\ell-1})$ that can make up the contents of the first ℓ registers $R_0, R_1, \ldots, R_{\ell-1}$, and there is a directed edge from $r = (r_0, r_1, \ldots, r_{\ell-1})$ to $s = (s_0, s_1, \ldots, s_{\ell-1})$ if and only if $s_i = r_{i-1}$ for $1 \leq i \leq \ell - 1$. So there is a directed edge from r to s if in one tick the contents of the first ℓ registers can change from r into s. Furthermore, the directed edge from $(m_{t+\ell}, m_{t+\ell-1}, \ldots, m_t)$ to $(m_{t+\ell+1}, m_{t+\ell}, \ldots, m_{t+1})$ is labeled with the tth bits of $c_1, c_2, \ldots, c_\mu$; so this label is the contribution to c made at the tth tick. This directed graph is essentially the transition diagram of the finite state machine formed by the shift register.

Representing a convolutional code with this directed graph D is helpful in understanding both encoding and decoding. Codewords simply correspond to walks in D with the codeword being the concatenation of the labels on the edges in the walk. If message bits are moved into the shift register one at a time, then there are exactly two arcs directed out of each vertex (corresponding to a message bit of a 0 or a 1 being moved into R_0), so the walk can head off in one of two possible directions. It is this observation that makes it clear what is involved in decoding. At each tick, decoding one message bit requires deciding which of the two directions to take from the current vertex (state). This decision cannot be based on knowing the entire received word, since it has arbitrarily long length. Instead, one gathers the next τ ticks worth of the received word, which together form, say, w', then finds which walk W emanating from the current state most closely matches w', then finally makes a decoding decision to take one step along W. This process can be efficiently implemented by using the Viterbi Decoder. Deciding on how large to make τ can affect the codeword to which w is decoded (see Hoffman *et al.*, 1991, for example).

In deciding which convolutional code to use, choices have to made about $g_1(x), \ldots, g_\mu(x)$ and the number k of message symbols to move into the shift register at each tick, usually chosen to be 1. The rate of the code is then k/μ.

SEE ALSO THE FOLLOWING ARTICLES

COMMUNICATION SATELLITE SYSTEMS • DATABASES • DISCRETE MATHEMATICS AND COMBINATORICS • WIRELESS COMMUNICATIONS

BIBLIOGRAPHY

Beth, Th., Jungnickel, D., and Lenz, H. (1999, 2000). "Design Theory," Vols. 1 and 2, Cambridge Univ. Press, Cambridge, UK.

Blake, I. F., and Mullin, R. C. (1976). "The Mathematical Theory of Coding Theory," Academic Press, New York.

Bonisoli, A. (1983) "Every equidistant linear code is a sequence of dual Hamming codes," *Ars Combinatoria* **18**, 181–186.

Bose, R. C. (1939) "On the construction of balanced incomplete block designs," *Ann. Eugen.* **9**, 353–399.

Bush, K. A. (1952) "Orthogonal arrays of index unity," *Ann. Math. Stat.* **23**, 426–434.

Colbourn C. J., and Dinitz, J. H., eds. (1996). "The CRC Handbook of Combinatorial Designs," CRC Press, Boca Raton, FL.

Colbourn, C. J., and Rosa, A. (1999) "Triple Systems," Oxford Univ. Pess, Oxford.

Fisher, R. A., and Yates, F. (1938). "Statistical Tables for Biological, Agricultural and Medical Research," Oliver & Boyd, Edinburgh.

Golay, M. J. E. (1949). "Notes on digital coding," *Proc. IEEE* **37**, 657.

Hamming, R. S. (1950) "Error-detecting and error-correcting codes," *Bell Syst. Tech. J.* **29**, 147–160.

Hanani, H. (1960) "On quadruple systems," *Canad. J. Math.*, **12**, 145–157.

Hartman, A., and Phelps, K. T. (1992) "Steiner Quadruple Systems, Contemporary Design Theory" (J. H. Dinitz and D. R. Stinson, eds.), Wiley, New York.

Hoffman, D. G., Leonard, D. A., Lindner, C. C., Phelps, K. T., Rodger, C. A., and Wall, J. R. (1991). "Coding Theory: The Essentials," Dekker, New York.

Lindner, C. C., and Rodger, C. A. (1997) "Design Theory," CRC Press, Boca Raton, FL.

van Lint, J. H. (1975) "A survey of perfect codes," *Rocky Mount. J. Math.* **5**, 199–224.

MacWilliams, F. J., and Sloane, N. J. A. (1977). "The Theory of Error-Correcting Codes," North-Holland, Amsterdam.

Shannon, C. E. (1948) "A mathematical theory of communication," *Bell Syst. Tech. J.* **27**, 379–423 and 623–656.

Teirlinck, L. (1980). "On projective and affine hyperplanes," *J. Combinatorial Theory, Ser. A* **28**, 290–306.

Tietäväinen, A. (1973) "On the nonexistence of perfect codes over finite fields," *SIAM J. Appl. Math.* **24**, 88–96.

Yates, F. (1939) "Complex experiments," *J. R. Stat. Soc.* **2**, 181–247.

Yates, F. (1936) "Incomplete randomized blocks," *Ann. Eugen.* **7**, 121–140.

Zinov'ev, V. A., and Leont'ev, V. K. (1973). "The nonexistence of perfect codes over Galois fields," *Probl. Control Inf. Theory* **2**(2), 123–132.

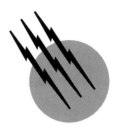

Diamond Films, Electrical Properties

Friedel Sellschop
Simon Connell
Charles Levitt
Elias Sideras-Haddad
*Schonland Research Centre for Nuclear Sciences
and Department of Physics, University of the
Witwatersrand*

I. Natural Diamond
II. Synthetic Diamond
III. Diamond in Electronics: Present and Potential
IV. Conclusions

GLOSSARY

Diamond types Type Ia contains nitrogen, typically in amounts that range from hundreds of parts per million to thousands of parts per million. In the subset Type IaA, the nitrogen is believed to occur as two nitrogen atoms in nearest neighbor substitutional sites. It can also appear in the Type IaB form which is believed to occur as four substitutional nitrogen atoms in association with a vacancy. Type Ib diamond has nitrogen, generally at lower concentration than in Type Ia, in single substitutional sites. Type II diamond has very low concentrations of nitrogen, with the subsets Types IIa and IIb, where Type IIb contains small amounts of boron which makes it a p-type semiconductor.

Defects The diamond crystal lattice is typically imperfect either as a result of actual breakdown of the perfect crystal ordering over limited distances, or as a result of the inclusion during growth of a mineral species.

Radiation damage The perfection of the crystal lattice of diamond can be disrupted as a consequence of the interaction of energetic particles (electrons, positrons, protons, alphas, and heavy ions) and of photons. This transfer of energy to the lattice carbon atoms can lead to their removal from the lattice sites and the consequent production of interstitials and vacancies. At higher concentrations, these may aggregate into larger structures.

HPHT diamond Man-made diamond produced at high pressure and high temperature permitting diamond growth in the diamond stable region of the carbon phase diagram.

CVD diamond Man-made diamond, generally in thin films, produced by chemical vapor deposition, in metastable conditions on a chosen substrate. In the event that this substrate is single crystal diamond, homoepitaxial growth may be achieved producing single crystal diamond. Deposition on substrates other than single crystal diamond generally leads to the production of polycrystalline diamond. Growth under these conditions is believed to be mediated by surface hydrogen chemistry.

p-Type semiconductor The description of a semiconductor with acceptor characteristics, such as boron doped diamond.

n-Type semiconductor The description of a semiconductor with donor characteristics, such as believed would be the case for lithium doped diamond.

THAT DIAMOND is defect-rich, with the consequent influence on its physical properties, is not surprising. With a bandgap of 5.4 eV, the fundamental absorption edge is expected to be at this energy, but only a small minority of diamonds reflect this. For some 98% of natural diamonds, the absorption edge is at significantly lower energy values (higher wavelengths). This displacement of more than 1 eV is a stark indication of the role that must be attributed to defects.

In fact this observation has been used for some considerable time for the classification of natural diamonds: those few with fundamental absorption edge at the theoretically predicted value are termed Type II diamonds; those with displaced fundamental absorption edge are classified as Type I diamonds. Subsequently this classification has been further refined in light of observations, so that the Type I category is subdivided into two groups: the most populous one is termed Type Ia, and the rare class Type Ib. The Type Ib natural diamond is characterized by an electron spin resonance signal. The Type II class has also undergone a subdivision into the rare Type IIb which is semiconducting, leaving the Type IIa diamonds as those Type IIs that are electrical insulators.

The infrared spectrum of diamond also reflects this type division. Figures 1 and 2 show the distinction that has already been drawn using the ultraviolet end of the spectrum. The infrared spectra, which are shown alongside, also reveal substantial differences. Both Types I and II diamonds show an absorption band in the range 1500 to 2700 cm^{-1}. This is the two-phonon absorption band intrinsic to diamond. But the Type Ia shows in addition an absorption band from 900 to 1500 cm^{-1}, corresponding to a one-phonon absorption regime. For pure diamond, one-phonon absorption is not allowed on symmetry arguments, hence the symmetry must be broken by some defect. Both absorption bands have additional finer structure. This leads to some further subdivision of the most populous class Type Ia diamonds: in the resolved finer structure of the two-phonon absorption band if the peak of maximum absorption is at 1282 cm^{-1} the diamond is tagged as being of Type IaA. If, however, the maximum absorption peaks at 1185 cm^{-1} it is classified as being Type IaB. It is rare to find a diamond that is purely Type IaA or Type IaB, a mixture of the two most commonly occurs. In Type Ib diamonds the maximum in the one-phonon absorption band occurs at 1130 cm^{-1}. Hence the evident distinction of different defects in all the defined diamond "types" is clear.

Defects in crystalline materials such as diamond, can be of basically two kinds: inclusions of "foreign" materials such as specific minerals, and structural anomalies, including foreign atoms, in the otherwise well-ordered lattice. In natural diamonds we may anticipate that in its genesis out of the magma in the upper mantle of the Earth, it is likely to embrace during its crystallization such cogenetic minerals as are present. Similarly, it is to be expected that the surface of diamond will saturate the dangling carbon bonds with ambient elements.

Man-made diamond is of two types: diamond grown at high temperature and high pressure (HTHP diamond) in simulation of the natural conditions of diamond growth, and diamond grown in metastable conditions (such as by chemical vapor deposition: "CVD diamond"). The growth environments of these two are quite different and will thus be differently reflected in the inclusion chemistry of bulk and surface.

The second form of defect, namely, that of structure anomalies, consists of those defects that are inherent and those that are induced, in particular by radiation both natural and man-made, and annealing.

In this review, the structure that will be followed, is to consider first of all natural diamond, then the HPHT, and finally the CVD diamond types.

I. NATURAL DIAMOND

A. Foreign Matter Inclusions

Man has recognized diamond as something special for many centuries. The earliest written records are biblical (e.g., Exodus XXVIII, verses 15 et seq, ca. 1200 BC). Early scientific examination revealed that natural diamond was rich in defects. In a series of papers in the period 1820 to 1862, Sir David Brewster using microscopic observations with polarization measurements, could confidently assert that

.... it seems, indeed, to be a general truth, that there are comparatively few diamonds without cavities and flaws, and that this mineral is a fouler stone than any other used in jewellery.

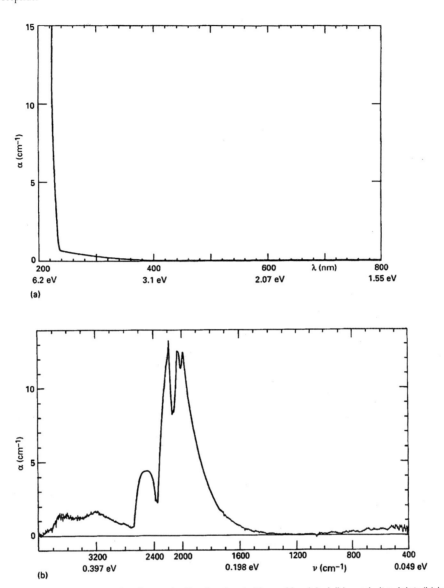

FIGURE 1 Absorption spectra of a diamond with a low level of impurities (a) visible and ultraviolet; (b) infrared.

In 1904, Bauer and Spencer inferred the presence of liquid or gaseous inclusions of H_2O and CO_2 in diamonds. The work of Kaiser and Bond (1959) in using a vacuum fusion method to release volatiles in diamond and identify the molecular components, heralded new insights, revealing nitrogen as a major impurity in natural diamond , together with some CO and H_2.

Subsequently Melton and Giardini (1974) analyzed the gases released on crushing diamond and could conclude that CO, H_2, H_2O, CO_2, N_2, CH_4, and Ar were present.

Fesq *et al.* (1975) tackled the analysis of some 1000 diamonds using an ultrasensitive multielement technique namely that of instrumental neutron activation analysis (INAA), which by virtue of the penetrating power of the incident neutrons and of the emitted photons, analyzes the whole bulk of the sample. This technique is in general sensitive to all elements of A-value greater than 23 (Na). These workers were careful to separate diamonds with (visible, under $50\times$ magnification, with crossed polaroids) inclusions, from those that in terms of these criteria could be defined as inclusion-free. The expectation was that the inclusion-free group would show very little impurities, and the with-inclusions group would reflect in its trace elemental impurity picture the chemistry of the inclusions observed in diamond and in the kimberlite matrix. The six most common minerals cogenetic with diamond were analyzed by the same INAA technique. To characterize the chemistry of these six mineral types, statistically significant positive and negative interelement correlations

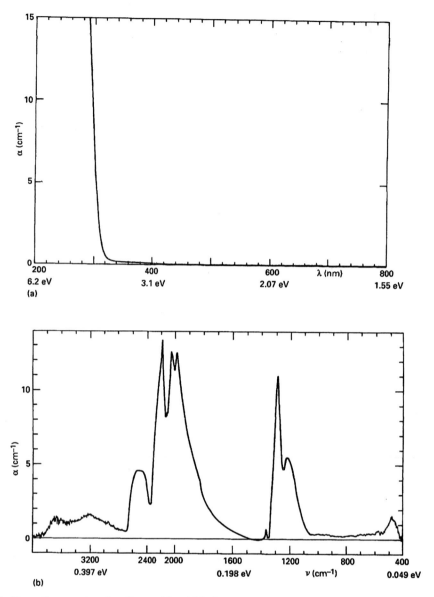

FIGURE 2 Absorption spectra of a diamond in which A centers are the predominant impurities (a) visible and ultraviolet; (b) infrared.

were established. These signatures for the six main mineral types are presented in Table I.

The INAA produced a plethora of elemental information, for both the with-inclusions (as expected) and the without-inclusions (not expected) diamonds. The high degree of sensitivity of this technique necessarily leads one to pose the question as to whether one is sampling elemental "noise." This is not the case: interelement correlations on the one hand substantiate the existence, albeit on the trace level, of a real chemistry, and on the other hand give the surprising result that the correlation functions, although perhaps weaker for the with-inclusions diamonds, reveal the same features for diamonds with and without

inclusions. An example is given in Table II for the predominantly lithophile elements.

This identical composition of the trace element chemistry of all natural diamonds (with and without visible inclusions) suggests the ubiquituous presence of small (submicroscopic) inclusions, the chemistry of which dominates the observed impurity composition even in the presence of visible mineral inclusions. We may compare this composition with that of the commonly observed inclusions: an example of an interelement correlation is shown in Fig. 3, and is compared with the correlation fields that characterize some known mineral assemblages relevant to diamond. It is evident that the submicroscopic inclusion

TABLE I Statistically Significant Element Correlations Used for the Identification of Cogenetic Mineral Inclusions in Diamonds

Mineral	High (positive) correlation between	Low (negative) correlation between
1. Olivine, enstatite	O, Mg	Al, Ca, Na, Sc
2. Garnet	O, Al, Sc, Mn, Cr, Mg, and heavy REE (rare earth elements)	Na, light REE
3. Diopside	O, Ca, Na, Mg, Al and light REE	Sc, Mn and heavy REE
4. Rutile, ilmenite	O, Ti, Fe, Ta	Ca, Al, Sc, Mn
5. Chrome spinel	O, Cr, Mg, Fe	Ca, Sc, Na
6. Sulfides	S, Fe, Ni, Cu, Co	O, Mg

chemistry for all diamonds corresponds to the garnet and diopside composition field.

It is tempting—but reasonable—to infer that these submicroscopic inclusions are simply droplets of quenched or temperature re-equilibrated magma which are trapped during diamond growth—the lithophile elements which are found to intercorrelate so significantly in all diamonds, are characteristic constituents of basic or ultrabasic magmas from which diamonds are believed to have crystallized.

Natural diamonds are therefore all characterized by the common defect of the presence of droplets of the parental magma, submicroscopic in size. It is in addition not uncommon to have visible (identifiable) mineral inclusions as a defect.

The colorful concept of 'magma droplets' should not be taken too literally. It is probably possible to attribute in a specific diamond sample the magma droplets to recognized mineral species.

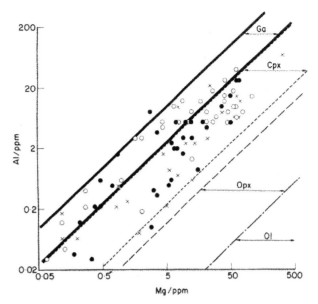

FIGURE 3 The relationship between aluminum and magnesium in diamonds from Premier (○), Finsch (●) and Jagersfontein (×) kimberlites, South Africa, Superimposed on this diagram are lines representing the extreme compositions for garnets (Ga-solid lines) and clinopyroxenes (Cpx-small dashed lines) from both garnet peridotite and eclogite assemblages, as well as the most aluminous orthopyroxene (opx) and olivine (01).

This conclusion of the ubiquitous presence of magma droplets was reached on the evidence of elemental data from A ~ 23 and upwards. The lighter elements are inaccessible to INAA, but amenable to charged particle analysis (Sellschop, 1992), which confirms the historic earlier work that hydrogen and oxygen share with nitrogen the feature of being minor or even major impurities. It is likely that the hydrogen and oxygen are present in the form of

TABLE II Interelement Correlation of Predominantly Lithophile Elements in Diamonds from South Africa[a]

	O	Na	Mg	Al	K	Ca	Sc	Ti	V	Cr	Mn	Fe
O	1	0.75	0.80	0.76	0.76	0.77	0.74	0.84	0.77	0.74	0.78	0.77
Na	*0.57*	1	0.86	0.90	0.86	0.92	0.87	0.87	0.90	0.77	0.87	0.86
Mg	*0.67*	*0.69*	1	0.89	0.92	0.88	0.89	0.87	0.90	0.87	0.94	0.92
Al	*0.61*	*0.83*	*0.81*	1	0.85	0.92	0.97	0.87	0.93	0.85	0.93	0.89
K	*0.62*	*0.75*	*0.88*	*0.77*	1	0.85	0.83	0.86	0.85	0.82	0.85	0.85
Ca	*0.59*	*0.88*	*0.78*	*0.94*	*0.82*	1	0.90	0.90	0.92	0.82	0.91	0.88
Sc	*0.62*	*0.74*	*0.86*	*0.96*	*0.80*	*0.91*	1	0.84	0.94	0.87	0.94	0.89
Ti	*0.60*	*0.79*	*0.76*	*0.86*	*0.77*	*0.87*	*0.82*	1	0.88	0.78	0.92	0.86
V	*0.64*	*0.83*	*0.83*	*0.91*	*0.79*	*0.93*	*0.91*	*0.82*	1	0.90	0.94	0.92
Cr	*0.60*	*0.52*	*0.81*	*0.75*	*0.73*	*0.73*	*0.83*	*0.69*	*0.84*	1	0.86	0.84
Mn	*0.63*	*0.77*	*0.92*	*0.94*	*0.81*	*0.89*	*0.95*	*0.84*	*0.91*	*0.80*	1	0.94
Fe	*0.69*	*0.78*	*0.91*	*0.92*	*0.79*	*0.86*	*0.91*	*0.82*	*0.88*	*0.74*	*0.95*	1

[a] Roman print represents the interelement correlation in all diamonds analyzed in this study; italics represent the interelement correlations in diamonds without observable mineral inclusions only.

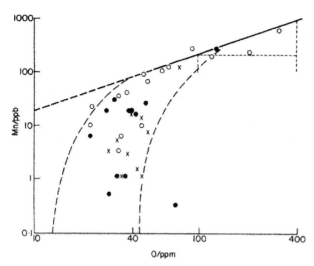

FIGURE 4 The relationship between oxygen and manganese in diamonds without observable mineral inclusions from Premier (○), Finsch (●) and Jagersfontein (×) kimberlites, South Africa. The field between dashed lines extending to the X-axis, indicates the excess oxygen present in an inferred H_2O–CO_2 rich fluid phase. The solid line represents the line of best fit.

water in the magma droplets, which should also contain gases such as CO and CO_2. Independent evidence has been obtained of the strong presence of oxygen in natural diamond from fast neutron activation analysis (Bibby *et al.*, 1974): good interelement correlations are obtained with elemental information derived from the INAA analyses, but an excess of oxygen is always found, an example of which is shown in Fig. 4. This excess of about 35 ppm is believed to be that associated with the water and gas richness of the magma droplets.

A foreign inclusion of particular interest is boron. This is found in a rare class of natural diamonds that is electrically semiconducting (Type IIb). The acceptor responsible for this property has been shown to be boron (Lightowers and Collins, 1976; Sellschop *et al.*, 1974). It is believed to be in the substitutional sites in the diamond lattice.

We have been considering a common defect in the bulk of diamonds. A brief comment is warranted on the surface of natural diamond. Ion channeling confirms the expectation that there is some small measure of crystal disorder in the outermost few monolayers. Charged particle analysis reveals that hydrogen and oxygen are typically found on the surface. The relative population of these two species can be controlled by the environment in which the diamond surface is prepared, as for example in polishing. In an oxygen-poor environment hydrogen saturates the surface dangling bonds of the diamond. It is established that the final monolayer of hydrogen is chemically bound to the surface carbon atoms (Jans *et al.*, 1994).

B. Structural Defects

Optical interrogation of natural diamond reveals a plethora of spectral features, many of which are due to point defects, as, for example, vacancies, interstitials, and substitutional impurities, in the lattice of the diamond. In Type Ia diamonds, in the infrared absorption spectrum the two- and three-phonon absorptions are prominent, but evident also is the forbidden one-phonon absorption between about 900 and 1500 cm^{-1}, which must therefore be defect-activated. As pointed out earlier, from the observation of numbers of diamonds, it became evident that the one-phonon region could be deconvoluted into two characteristic distributions, coded as "A" and "B". The structure believed to be responsible for the Type IaA centre is two nearest-neighbor nitrogen atoms in substitutional sites, as shown in cartoon form in Fig. 5. Also shown in this figure are the IaB center which is believed to consist of four nitrogen atoms in substitutional sites surrounding a vacancy, and the single isolated substitutional nitrogen center that characterizes Type Ib diamond, of which the IR spectral component is often coded as "C". Sometimes linear features that line up with the {100} planar directions are seen and attributed with some caution to platelets of nitrogen: the presence of these platelets correlates with a peak that is observed in absorption at 7.3 μm (1370 cm^{-1}).

Further defect structures can readily be conceived. In summary, the situation found experimentally is as shown in Table III. This extensive inventory of data reflects optical features that are due to the creation of defects as a consequence of damage caused by irradiation, as, for example, by energetic electrons or by neutrons, often followed by thermal annealing, as also of features that are ascribed to the role of foreign atoms, in particular nitrogen and boron. Let us consider some of these detailed features.

1. Unmodified Natural Diamonds

It would appear that Type IIa diamonds are the form of natural diamond that is most free from foreign atoms. This

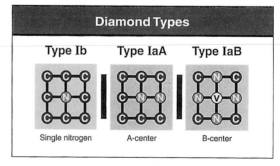

FIGURE 5 Schematic diagrams illustrating the occurrence of substitutional impurities in different diamond types, classified into such types on the basis of distinctive optical features.

does not mean that they are the most perfect diamonds since they may still have dislocations and mosaic structure, which can be seen in X-ray topographs and by measurement of the X-ray rocking curves (Sellschop *et al.*, 2000). Although the concentration is low, nevertheless nitrogen is measured as still present (Sellschop, 1992). The one-phonon absorption features are absent. The Type IIb diamonds are unique in being electrically conductive and are usually a delicate blue in colour. Analysis has shown (Lightowers and Collins, 1976; Sellschop *et al.*, 1977) that these properties are due to the presence of boron at very low levels. As shown in Table III, the boron gives rise to absorption lines at 0.305 and 0.347 eV. The boron absorption arises by purely electronic transitions. Type IIb diamond is thus an acceptor. Nitrogen is present and being a donor might compensate to some extent the acceptor boron. However, the nitrogen levels lie at least 2 eV below the conduction band, which is unusually deep. Hence at room temperature the thermal energy is insufficient to enable the electrons from these levels to reach the conduction-band.

Type I diamonds are characterized by the presence of nitrogen. Optical absorption associated with nitrogen is due to one-phonon processes involving the lattice vibrations of the carbon atoms. A dominant example is the "A" feature which is associated with the lines observed at 1282, 1205, and 1099 cm^{-1}, which are interpreted as being the three C–N stretching vibrations. The A feature is believed to be due to two nitrogen atoms in adjacent substitutional lattice sites (see Fig. 5). The only C–N bending vibration believed to have been observed so far, is the line at 481 cm^{-1}.

Also prominent is the "B" feature, with lines observed at 1429, 1370, 1333, 1176, 1000, 775, and 328 cm^{-1}. (The 1370-cm^{-1} line is the strongest of these). There is less certainty as to the structure of the B center, but it is speculated that it consists of four adjacent substitutional nitrogen atoms, surrounding a vacancy (see Fig. 5).

Type Ib diamond is identified by its electron-spin resonance signal; it is paramagnetic. The associated optical features are lines at 1130 cm^{-1} and a subsidiary peak at 1344 cm^{-1}. The most simple interpretation is that it is due to isolated single substitutional nitrogen atoms.

We have thusfar associated optical features with clusters of one, two, and four nitrogen substitutions. In naturally occurring diamond, three adjacent substitutional nitrogen atoms, surrounding a vacancy, the so-called N3 center seen at in the UV at 415 nm, occur quite commonly. This gives rise to a paramagnetic signal.

In the one-phonon absorption region, the peak at 1370 cm^{-1} can be prominent. This correlates with the observation of "platelets" of nanometer or even few-micrometer dimensions consisting of one or two extra atomic layers that appear in the (100) planar orientation

as squeezed into the diamond lattice. At one time it was assumed that these were composed of nitrogen, but this is by no means assured.

There are many further optical centers, shown in Table III. Interesting, for example, in naturally occurring diamond, are the sharp lines at 1405 and 3107 cm^{-1}. They are ascribed to the C–H bond vibrations. Absolute measurements of hydrogen concentration in natural diamond suggest that in light of no apparent correlation with the strength of these two lines, only a minor part of the hydrogen actually present is in this form. It is also reported that these two lines can be produced by electron irradiation and subsequent annealing. It is possible that this procedure mobilizes part of the already present hydrogen creating some C–H bonds. It is possible that much of the hydrogen is present in the form of water in the magmatic inclusions that we concluded earlier were ubiquitous in all natural diamond.

2. Natural Diamonds, Modified

A great deal of engineering of the diamond lattice and its inclusions can be accomplished by damage induced by irradiation, with and without subsequent annealing. As irradiation sources, electrons and neutrons have most generally been used. A common result of irradiation is the formation of an optical center at 1.673 eV known as the GR1 center. In fact, it is one of the members of a typical vibronic band that stretches through GR2, GR3, ... to GR8 which appears at 3.01 eV. The most conceptually simple of irradiation effects is the creation of a vacancy in the diamond lattice, with a consequent interstitial elsewhere in the crystal. The GR band is interpreted as reflecting the neutral vacancy. A second vibronic system, with zero phonon line at 3.149 eV, is known as the ND1 center, and is interpreted as the negative vacancy. Annealing experiments suggest that this simple vacancy defect is stable up to at least about 900 K.

Subjecting a Type Ib diamond to irradiation and annealing conditions above 800 K causes a reduction of the GR and ND1 systems produced initially by the irradiation, accompanied by the growth of a vibrational band with zero phonon line at 1.945 eV. This center is interpreted as a structure consisting of a single substitutional nitrogen atom plus a neutral vacancy. A second center is observed at 2.156 eV corresponding to the structure of a single nitrogen atom plus a negatively charged vacancy (Fig. 6).

For the case of Type Ia diamond, under the same irradiation and annealing conditions, the GR and ND1 centers initially formed are substantially reduced in intensity, accompanied, however, by the formation of two vibronic bands with zero phonon lines at 2.463 and 2.498 eV, respectively. These are labeled as the H3 and H4

TABLE III Defect Centers in Natural Diamond

Structure	Code	Energy eV	Comment		Reference	Alternate (cm⁻¹)	Units (μm)	Add. info.			
		0.041	In some la			328	30.5	Band B	Bend modes in platelets?		
		0.06				481	20.8	Band A	Only bend vibr obs.		
		0.096			*	775	12.9	Band B			
		0.124			*	1000	10.0	Band B	Narrow, in coat		
		0.125	>Irrad		Clark and Davey (1984)*	1008	9.92	B			
		0.136			*	1099	9.1	Band A	C–N str vibr		
		0.139	>n irrad.		Clark and Davey (1984)*	1121	8.92				
		0.14	Type lb		*	1130	8.85	Strong yellow	Paramag. P1?		^{15}N shift
		0.146	1-Phonon abs	("B")	*	1176	8.5	Band B			
		0.147	B		*	1185	8.44	Even # N's			
		0.148	>n irrad.		Clark and Davey (1984)*	1193	8.38				
		0.149			*	1205	8.30	Band A	C–N str vibr		
		0.159	1-Phonon abs	("A")	*	1282	7.8	Band A	C–N str vibr		
		0.165			**	1333	7.5	Band B	Raman freq.		
		0.167	Type lb		**	1344	7.44	no shift w. ^{15}N	Correl. 1130		
		0.17	Platelet	B[1]	**	1370	7.3	Band B	Strongest mem.		
		0.174	C–H bond		**	1405	7.2	sp² vibr.	N–H bond ?	Also prod e irrad + anneal.	
		0.178			**	1429	7.0	Band B			
	H 1a	0.18	In la & 1 lB	>Irr + anneal	**	1450	6.90	1 N interst.	Diff. ch. st. to	1502 cm⁻¹	^{15}N shift
		0.186	In lb only	>Irr + anneal	**	1502	6.66	1 N interst.	Diff. ch. st.to	1450 cm⁻¹	^{15}N shift
		0.19	>irrad	Ann. 400°C	**	1531	6.53	C inntersts.	As 1531 decr.	1570 incr	
		0.195	Platelets	>Irr + ann.	**	1570	6.37	C inntersts.	1570 inc. as	Ann. 1531	
		0.212	>hi. N irrad.	+Ann. @ 600°C	**	1706	5.86	1 N involved			^{15}N shift
		0.23	la > irrad. +	Ann. @ ~500°C	**	1856	5.39	di-N center	IaA?		
		0.305	Boron		**	2460	4.065	Blue			
		0.312	>n irr + ann. 300°C		**	2520	3.97	Weak line			
		0.345			***	2786	3.59	Weak line	Correl. 3107, 1405		
		0.347	Boron		***	2790	3.584	Blue			
		0.347(3)	Uncompen. Boron		***	2802	3.57	zpl due to X-itions betw. Bound hole states			
		0.385	C–H bond	Stretch mode	***	3107	3.219	N–H? sp²	FWHM 4 cm⁻¹	Also prod e irrad + ann. 680°C	
N–V⁻–N	H 2	1.257	Type I	Irrad + anneal		10142	0.986				
		1.424				11494	0.870	C-C vibr			
		1.65	>Irrad + ann. 800°C			13299	0.752				
V⁰	GR 1	1.665	Immobile <1120 K			13423	0.745	>Irrad	Tetra hedral center		
V⁰	GR 1	1.673	>Irrad		Clark et al. (1979)	13495	0.741	>Irrad	Tetra hedral center blue	Fast lumines.	

continues

TABLE III *(continued)*

Structure	Code	Energy eV	Comment		Reference	Alternate (cm⁻¹)	Units (μm)	Add. info.	
N–V⁰		1.945	lb	>Irr + anneal		15674	0.638	Fast luminesce	N–V vibronic center@Nsub.
		2.086	>Irrad	>Irr + anneal	800°C Clark et al. (1979)	16807	0.595	Type I only	corr. W. 0.425 μm
		2.156		>Irr + anneal		17391	0.575	1 N in ⟨001⟩ split interst. + V along ⟨001⟩ axis	
		2.367	>Irrad	Strong photochromic		19084	0.524	1 or more interststs. + 1 N	
	3 H	2.46				19841	0.504		
N–V⁰–N	H 3	2.463	"A" + V	>Irr + anneal	800°C Clark et al. (1979)	19881	0.503	Fast luminescence	Green PL emiss
N₄ + V₂⁰	H 4	2.498	"B" + V	> Irr + anneal 800°C		20161	0.496	Fast luminescence	
		2.535		>Irrad.		20450	0.489	1 or more interts. + 1 N	
	N 2	2.596				20921	0.478	Correl. nat occur. w N 3	
		2.721				21930	0.456	Weak yellow lum under UV	
		2.807	>Irr + anneal			22624	0.442	1 N + interst.	
		2.85				22989	0.435	Blue emiss due decor disloc	
	GR 2	2.88				23202	0.431		
		2.916	>Irr + ann 800°C			23529	0.425	Corr w. 0.595 μm	
N3 + V⁰	N 3	2.985	In laB	Need str. A or B	Clark et al. (1979)	24096	0.415	Fast luminesc. paramag. P2	Pale yellow Nat occur.
	GR 8	3.01				24272	0.412		
V⁻	ND 1	3.149	lb	>Irrad.	Clark et al. (1979)	25381	0.394	Decays to V⁰	ann.out@1120K
		3.15					0.393	Fast luminescence	
		3.188		>Irr + anneal			0.389	1 N + interst.	
	N 4	3.603					0.344	Correl. W. N 3	
	N 4	3.61	In laB		Clark et al. (1979)		0.343		
	5RL	4.582	lb	>Irrad + anneal			0.271	Fast luminescence	C interst.
	N 9	5.26	In laB		Clark et al. (1979)		0.236		
	Fund. Abs. edge	5.4				43478	0.230		

* = 1 phonon
** = 2 phonon
*** = 3 phonon

jpfs 010127

centers, respectively. These centers are identified as having structures corresponding to A + V⁰ (=H3) (=2N + V⁰) and B + V⁰ (H4) (=4N + 2V⁰), respectively (Fig. 7). Some of the diffusing vacancies are evidently trapped at

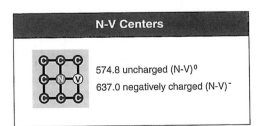

N-V Centers

574.8 uncharged (N-V)⁰
637.0 negatively charged (N-V)⁻

FIGURE 6 The two types of nitrogen-vacancy (N-V) centers (uncharged and negatively charged), where a single nitrogen atom is attached to a single vacancy.

the A and the B centers. Lines are also produced in this procedure at 2.086 and 2.916 eV, the strength of these two correlating with one another.

Irradiation effects are also observed in the IR spectrum: so, for example, following neutron irradiation absorption peaks are observed at 0.139 and 0.148 eV. Other irradiation and annealing features in the IR are, in both Types Ia and Ib, a line at 0.18 eV ("H1a") and for Type Ib only a line at 0.186 eV, both due perhaps to a single nitrogen atom. Similarly, lines believed due to carbon interstitials are found at 0.19 eV (annealing at 400°C) and 0.195 eV. During annealing, as the 0.19 eV line decreases, the 0.195-eV line increases in strength. For irradiation of Type Ia diamonds, followed by annealing at ∼500°C, a line at 0.23 eV is produced. Following high-dose neutron

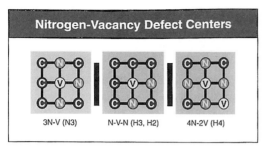

FIGURE 7 Diamonds may have defect centers that have 2, 3 and 4 substitutional nitrogen impurities in combination with a vacancy or vacancies. These include the N3 system (three nitrogen atoms surrounding a vacancy), the H3 and H2 systems (an "A" aggregate associated with a vacancy [uncharged and negatively charged respectively]), and the H4 system (a "B" aggregate bound to an additional vacancy).

irradiation, and subsequent annealing at 600°C, a line is produced at 0.212 eV. A weak line is produced following neutron irradiation and annealing at 300°C, at 0.312 eV. Structures corresponding to all these IR lines are still largely speculative.

II. SYNTHETIC DIAMOND

A. High-Temperature High-Pressure Growth (HTHP)

Diamond has been grown, consistent with the phase diagram of carbon (Fig. 8), by the HTHP process for many decades now (Sellschop *et al.*, 2000). Direct conversion from graphite to diamond is rarely used, the common approach being to include a catalyst in the growth cell. The most common catalysts used are Ni, Co, and Fe.

Not surprisingly, the catalyst features in the defect chemistry of the synthetic diamond produced. So if Ni is employed as catalyst (Fig. 9), it can be found present in

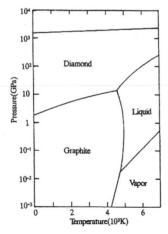

FIGURE 8 Pressure–temperature phase diagram of carbon.

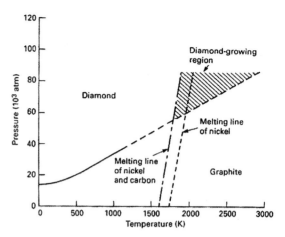

FIGURE 9 The diamond–graphite equilibrium line and the melting line of nickel and the nickel-carbon eutectic (From Bovenkerk *et al.* (1959). *Nature* **184**, 1094–1098.)

substitutional positions with a negative charge state. This is observed in nitrogen containing crystals grown from a nickel based alloy (Kanda and Sekine, 1994). Nickel is found in interstitial positions with a positive charge state, as observed in nitrogen-free crystals grown from nickel to which nitrogen getters (Ti, Zr) and/or boron have been added (Kanda and Sekine, 1994).

Unless special attention is given to gettering out the nitrogen in the growth process, nitrogen is found in all HPHT synthetic diamonds, and specifically in single substitutional sites, in other words as Type Ib. Unlike natural Type Ib diamond where analysis has shown (Sellschop, 1992) that only a small part of the total nitrogen present is in the single substitutional form, all the nitrogen in synthetic diamond appears to be in Type Ib form. Even this has to be qualified, however, if the synthesis takes place at higher temperatures, adjacent substitutional pairs of nitrogen atoms can be produced, in other words an "A" center as in Type IaA natural diamond. As in natural diamond, we find in the absorption spectrum the two-phonon band intrinsic to diamond as well as the nitrogen-mediated one-phonon band.

We may expect for synthetic diamond to produce the defect structures consequent on irradiation and subsequent annealing as was the case for natural diamond of equivalent type. In this way the H2 line at 1.257 eV appears (Clark *et al.*, 1992), as does the H3 center at 2.463 eV. From more detailed study it was concluded that the H2 center has the structure $N-V^{-}-N$, while the H3 center has the structure $N-V^{0}-N$. The line observed at 1.424 eV, is shown to be associated with a localized vibrational mode involving a carbon–carbon vibration.

If boron is included in the growth capsule, it is readily incorporated in the diamond, with characteristic blue hue, and electrical properties as for natural Type IIb diamonds.

A prominent feature of synthetic diamond is the inhomogeneous distribution of impurity concentration for the different growth sectors. Even within a specific growth sector, it is not homogeneous. For nitrogen, the direction dependence of the concentration, at room temperature, goes as

$$[111] > [100] > [113] > [110]$$

while at low temperature one finds

$$[100] > [111].$$

For boron, it appears that

$$[111] > [110] > [100] = [113] > [115].$$

Nickel is limited to [111] sectors.

B. Chemical Vapor Deposition (CVD)

Fairly recently it has been discovered that it is possible to produce diamond films by a chemical vapor deposition (CVD) technique, despite the fact that this takes place under conditions far removed from those of the diamond stable region in the phase diagram. Typically as feed gas, a hydrocarbon gas (such as methane) in an admixture with hydrogen is used. These films are generally polycrystalline unless they are grown on a single-crystal diamond as a substrate, when homoepitaxial growth can be achieved (Butler and Woodin, 1993).

A wide variety of different technologies are used which have been thoroughly chronicled. Raman spectroscopy appears to be a commonly used diagnostic technique. The two- and three-phonon bands are found to be present, and also the one-phonon band when nitrogen is incorporated. Many other defects are reported, embracing ones familiar to us from the study of natural diamond, but new defects are observed that were not identified in natural diamond. So, for example, for combustion-produced CVD films (Morrison and Glass, 1994), the defects in the IR at 1130, 1175 and 1371 cm^{-1} are observed, consistent with Table III. In addition defects are found at 1235 and 1755 cm^{-1} and ascribed to the influence of boron which was added to the growth mix. In particular a number of lines are found in the 2800- to 3000-cm^{-1} region (e.g., at 2820, 2833, 2850, 2880, 2920, 2960, 2980, 3025, 3081 cm^{-1}), all of which are believed to be associated with the C–H stretch vibrations of the C–H bond.

It is interesting to note that the prime candidates for the C–H bond in natural diamond, the 3107- and 1405-cm^{-1} lines, apparently do not appear in CVD diamond.

In UV, defects are recognized from the evidence in natural diamond (Table III) at, for example, 3.188, 2.985, 2.88, 2.463, 2.156, 1.945, and 1.673 eV. Additional defect lines are reported at, for example, 2.45, 2.330, 2.326, 2.282, 2.234, and 2.21 eV.

It is interesting to note that some of the defect features that in natural diamond had to be produced by irradiation followed by annealing, appear as intrinsic in some CVD diamonds.

III. DIAMOND IN ELECTRONICS: PRESENT AND POTENTIAL

The anticipation that diamond has a significant role to play in electronic and optoelectronic applications has long been current. The earliest evidence for this is in the use of natural diamond as a detector of ionizing particles. This was successfully demonstrated but did not lead to a general use of diamond in this role. This was in all likelihood due to the perception that diamond was an expensive material, on the one hand, and that *natural* diamond was variable in its defect characteristics over which one had no control, on the other hand.

The advent of synthetic diamond, produced in simulation of the genesis of diamond in nature, by a high-temperature high-pressure (HTHP) process, did not alter this situation. This was no doubt due to the reality that the HTHP process was almost exclusively deployed in the production of very small diamonds, grits in fact, which found ready application in industry for cutting, grinding, sawing, and drilling. With the more recent development of chemical vapor deposition (CVD) diamond the prospect of having a diamond source that was entirely controllable and thus reproducible, was enthusiastically embraced. Even though the CVD product was generally polycrystalline and available only in thin films, enthusiasm was evident that at last the development of diamond devices could be contemplated. Considerable resource, both human and material, has been dedicated to the development of detectors, for example, in the complex large detector systems that now are an essential part of large accelerator experiments.

However, it should be noted that at the same time, the HTHP process has increasingly been extended to produce large single crystals of high quality. Hence the availability of single-crystal diamond of reproducible quality is becoming a reality which is bound to be a stimulus to the application of diamond in these fields.

The physical properties of diamond such as the large band gap (5.45 eV), the high carrier mobility, high breakdown voltage, the excellent thermal conductivity, *inter alia*, are suggestive of superior performance. In a well-known assessment by Johnson (1965), the cut-off frequency of diamond as a transistor was reckoned to be much higher than that for Ge, Si, and GaAs. More specifically, if (admittedly selected) parameters of diamond, including the high electron and hole mobilities, high breakdown field, high saturated electron velocity, high thermal conductivity, and low permittivity are used, then the Johnson figure of merit is more than 8000 times higher for diamond than for silicon. The hypothetical frequency performance of MESFETS for a range of wide band gap semiconductors was calculated by Trew *et al.* (1991) to be much better in the case of diamond than for GaAs or SiC. A generally optimistic view is prevalent in regard to the role of diamond in the electronics field. In stark contrast, Collins (1992) has warned against over-optimistic projections.

A rare form of natural diamond (the so-called Type IIb) is a p-type semiconductor. It has been shown that this is due to the presence of boron (Lightowers and Collins, 1976; Sellschop *et al.*, 1974). It is thus not surprising that early efforts at producing diamond devices should have been based on the use of Type IIb natural diamond. Davidson (1995) has given a good summary of these early endeavors. For example, lithium and arsenic have been implanted into p-type natural diamond, and following a high-temperature anneal, it has been demonstrated that p–n junctions have been formed. Arsenic implantation into natural Type IIb diamond has been successfully used to produce a npn transistor. Geis *et al.* (1988) reported on the fabrication of a permeable base transistor in natural p-type diamond with the aid of ion beam assisted etching.

Synthetic boron-doped diamond has also been used. Geis and Angus (1992) have reported forming point contact transistors in such materials. Similarly boron-doped CVD synthetic diamond films have been deployed for device purposes. Gildenblatt *et al.* (1991) have produced a thin-film field-effect transistor using a boron-doped homoepitaxially grown diamond. Using similar CVD material, Grot *et al.* (1990) have fabricated mesa-isolated recessed-gate field-effect transistors.

These diamond devices tend to suffer from excessive series resistance at room temperature and from a strong temperature dependence of the channel resistance.

Nevertheless these early demonstrations serve to indicate the potential for achieving the active electronic functions expected of a successful semiconductor material. At present this would seem to be most evident for diamond of natural, HTHP synthetic, and homoepitaxial CVD origin. Clearly, it would be advantageous if one could successfully produce heteroepitaxial CVD diamond, an ambition which has yet to be realized. In the absence of this source of material, increasing attention has been devoted to polycrystalline CVD diamond.

Polycrystalline diamond (PCD) has been extensively explored as a radiation detector, particularly to exploit its radiation hardness as, for example, in large detector assemblies (Adam *et al.*, 2000) where the inhospitable environment does not permit conventional materials to survive

for the periods desired. In medical applications the biological equivalence of (polycrystalline) diamond is seen as an advantage (Keddy *et al.*, 1987). Various devices have been reported in PCD. Okano *et al.* (1991) have produced a pn junction diode in PCD–CVD. A PCD–CVD diode has been made by Miyata *et al.* (1991). Tessmer *et al.* (1992) have reported on the fabrication of a PCD–CVD field effect transistor. Hence, it appears that some active electronic devices can be realized in PCD–CVD. While useful workable performance can be achieved, it must be appreciated that theoretical performance for mobility, lifetime, and other carrier properties will be thwarted by grain boundaries, dislocations, and other defects associated with polycrystalline material.

In fact, most of the devices associated with conventional semiconductor materials have been demonstrated in diamond. This is true not only for the standard items which include resistors, capacitors, and transistors but also in regard to sensors. Davidson (1995) has reported on a sensitive high-temperature thermistor, a light-sensitive switch, an optical radiation detector, an optoelectronic switch, and a fast room temperature IR detector in diamond. A novel (static) application has been pioneered by Kalbitzer (2000) on the use of diamond as a high-density memory storage device.

Very interesting is the attention Davidson (1995) draws to the fact that boron-doped diamond films have piezoresistance. This effect should be very useful for high-temperature sensors since it appears that the effect is reserved for high temperatures, which is not the case for the equivalent current silicon technology. One can picture the application of this effect in diamond micro-accelerometers. Recognizing the unsurpassed stiffness of diamond, it should be very effective also as a pressure sensor.

Despite the uniqueness of the physical parameters of diamond which so seductively suggest its superior performance as a semiconductor, despite the fact as indicated earlier by the many examples cited, that the potential of diamond as a device material has been demonstrated, the unmitigated widespread success still has to happen. This remains an urgent challenge to both the device engineer and the diamond growth engineer since a large part of the inhibiting difficulty has its origins, as surveyed in considerable detail in the early part of this paper, in the defect richness of all forms of diamond. Methods of doping, such as ion implantation, add to the defect concentration, since annealing of diamond is not as simple as that for silicon.

IV. CONCLUSIONS

Diamond, irrespective of its mode of origin, is evidently a defective material. The most important of these defects has been summarized in what is intended to be an easily accessible form. Where the structure of the defect is at least speculated, this has been indicated.

Nevertheless, the physical properties of diamond are so special, representing an extreme in a remarkable number of phenomena, that it remains an attractive material for many actual and perceived applications. These physical features include *inter alia*, its extreme hardness. It is the strongest material known (highest bulk modulus: 1.2×10^{12} N/m^2; lowest compressibility: 8.3×10^{-13} m^2/N), it has the highest known value of thermal conductivity at room temperature (2×10^3 W/m/K), it has a very low thermal expansion coefficient (at 293K: 0.8×10^{-6}), broad optical transparency from the deep UV to the far IR, it is a good electrical insulator (room temperature resistivity is $\sim 10^{16}$ Ω cm), it can be doped to change its resistivity over the range 10 to 10^{16} Ω cm, so becoming a semiconductor with a wide band gap of 5.4 eV, it is resistant to chemical attack, and is biologically compatible.

From a consideration of its wide band gap, the low thermal coefficient of expansion, the high thermal conductivity, the high breakdown field, the saturated electron velocity, and the hole mobility in natural semiconducting diamond, the conviction is still fervently held that excellent high-power, high-frequency, high-temperature semiconducting devices can and will be fabricated.

One may thus quite confidently predict that progress will be deliberate in the immediate future in both improving the control of synthetic diamond production and independently thereof in new and innovative methods of annealing, so as to routinely produce material of lower defect concentration (Sellschop *et al.*, 2000) and to move rapidly from the historic applications in cutting, polishing, and grinding, to the much more sophisticated optoelectronic and electronic applications that take full benefit of the extreme physical properties of this remarkable material.

ACKNOWLEDGMENTS

The authors wish to record their appreciation for permission to use published figures as follows: Figures 1, 2 and 9 from *Properties and Applications of Diamond* by Wilks and Wilks, reprinted by permission of Butterworth Heinemann. Figures 3 and 4 from *The Properties of Diamond*, chapter by Sellschop, reprinted by permission of Academic Press. Figures 5, 6 and 7 reprinted with permission from *C. P. Smith et al.* "GE POL Diamonds Before and After" *Gems and Gemology* **36** (2000) 192–215. Copyright GIA. Figure 8 reprinted with permission from the Institution of Electrical Engineers, UK, *Properties and Growth of Diamond*, ed. G. Davies, EMIS Datarevie series No. 9, page 404 (Hisao Kanda). The authors acknowledge with appreciation the support of the National Research Foundation of South Africa and, in particular for the provision of samples, Messrs De Beers Industrial Diamonds (Pty) Ltd. The encouragement of Peter Rose is recorded with appreciation.

SEE ALSO THE FOLLOWING ARTICLES

Boron Hydrides • Chemical Vapor Deposition • Electron Probe Analysis of Minerals • Mineralogy and Instrumentation • Thin-Film Transistors

BIBLIOGRAPHY

Adam, W. *et al.* (2000). *Nuclear Instruments and Methods in Physics Research* **A453**, 141–148.

Adam, W. *et al.* (2000). *Nuclear Instruments and Methods in Physics Research* **A447**, 244–250.

Bauer, M., and Spencer, L. I. (1904). "Precious Stones," C. Griffin and Co., London.

Bibby, D. M., Fesq, H. W., and Sellschop, J. P. F. (1974). *SAJ. Sci.* **70**, 377–378.

Bovenkerk, H. P., Bundy, F. P., and Hall, H. T. *et al.* (1959). *Nature* **184**, 1094–1098.

Brewster, D. (1820). *Edinburgh Philosophical J.* **11**, 334.

Butler, J. E., and Woodin, R. L. (1993). *Phil. Trans. R. Soc. Lond. A* **342**, 209.

Clark, C. D., Collins, A. T., and Woods, G. S. (1992). *In* "The Properties of Natural and Synthetic Diamond" (J. Field, ed.), pp. 35–79, Academic Press, London.

Clark, C. D., and Davey, S. T. (1984). *J. Phys. C. Solid State Phys.* **17**, 1127–1140.

Clark, C. D., Mitchell, E. W. J., and Parsons, B. J. (1979). *In* "The Properties of Diamond" (J. Field, ed.), pp. 23–77, Academic Press.

Collins, A. T. (1992). *Materials Sci. Eng.* **B11**, 257.

Davidson, J. L. (1995). *In* "Wide Band Gap Electronic Materials" (M. A. Prelas, P. Gielisse, G. Popovici, B. V. Spitsyn, and T. Stacy, eds.), p. 143, Kluwer Academic Publishers, New York.

Fesq, H. W., Bibby, D. M., Erasmus, C. S., Kable, E. J. D., and Sellschop, J. P. F. (1975). *Phys. Chem. Earth* **17**, 195.

Geis, M. W., and Angus, J. C. (1992). *In Scientific American* (October), 84.

Geis, M. W., Efremow, N. N., and Rathman, D. D. (1988). *J. Vac. Sci. Technol.* **A6**, 1953.

Gildenblatt, G. S., Grot, S. A., and Badzian, A. R. (1991). *Proc. IEEE* **79**, 647.

Grot, S. A., Gildenblatt, G. S., Hatfield, C. W., Wronski, C. R., Badzian, A. R., Badzian, T., and Messier, R. (1990). *IEEE Electron Device Lett.* **11**, 100.

Jans, S., Kalbitzer S., Oberschachtsiek, P., and Sellschop, J. P. F. (1994). *Nucl. Instr. Meth. Phys. Res.* **B85**, 321.

Johnson, A. (1965). *RCA Rev.* **26**, 163.

Kanda, H., and Sekine, T. (1994). *In* "Properties and Growth of Diamond" (G. Davies, ed.), pp. 415–426, EMIS Datareviews Series No. 9, 1994 INSPEC, IEE, London.

Kaiser, W., and Bond, W. L. (1959). *Phys. Rev.* **115**, 857.

Kalbitzer, S., Klatt, Ch., Sellschop, J. P. F., Connell, S. H., Nilen, R. W. N., Brink, D. J., Müller, G., Voglmeier, L., Möller, H., Prinsloo, L. C., Friedland, E., Hauser, T., and Demanet, C. M. (2001). *Applied Physics* **A72**, 1–32.

Keddy, R. J., Nam, T. L., and Burns, R. C. (1987). *Phys. Med. Biol. (UK)* **32**, 751.

Lightowlers, E. C., and Collins, A. T. (1976). "Diamond research 1976" (Suppl. *Ind. Diam. Rev.*), 14–21.

Trew, R. J., Yan, J., and Mock, P. M. (1991). *Proc. IEEE* **79**, 602.

Melton, C. E., and Giardini, A. A. (1974). *Am. Min.* **59**, 775.

Miyata, K., Dreifus, D. L., Das, K., Glass, J. T., and Kobashi, K. (1991). *In* "Second International Symposium on Diamond Materials," The Electrochemical Society Meeting, Washington DC.

Morrison, P. W., and Glass, J. T. (1994). *In* "Properties and Growth of Diamond" (G. Davies, ed.), pp. 380–390, EMIS Datareview Series No. 9, INSPEC, IEE, UK. 1994.

Okano, K., Kiyota, H., Iwasaki, T., and Nakamura, Y. (1991). *Solid State Electr.* **34,** 139.

Sellschop, J. P. F. (1992). *In* "The Properties of Natural and Synthetic Diamond" (J. Field, ed.), pp. 81–179, Academic Press.

Sellschop, J. P. F., Bibby, D. M., Keddy, R. J., Mingay, D. W., and Renan, M. J. (1974). "Abstracts Diamond Conference," unpublished.

Sellschop, J. P. F., Connell, S. H., Nilen, R. W. N., Freund, A. K., Hoszowska, J., Detlefs, C., Hustache, R., Burns, R. C., Rebak, M., Hansen, J. O., Welch, D. L., and Hall, C. E. (2000). *New Diamond and Frontier Carbon Technology* **10,** 253–282.

Sellschop, J. P. F., Renan, M. J., Keddy, R. J., Mingay, D. W., and Bibby, D. M. (1977). *Int. J App. Rad. Isotopes* **28,** 277–279.

Tessmer, A. J., Plano, L. S., and Dreifus, D. L. (1992). *In* "Proceedings of the Fiftieth Annual Device Research Conference," MIT.

Dielectric Gases

L. G. Christophorou
S. J. Dale

Oak Ridge National Laboratory

GLOSSARY

Breakdown voltage Critical voltage under which a gas makes the transition from an insulator to a conductor.

Dielectric gas Relatively poor conductor or nonconductor of electricity to high applied electrical stress; gas with high breakdown voltage.

Electrical breakdown (or electrical discharge or spark) of a gas Dramatic event whereby the electrical conductivity of an electrically stressed gas increases by many orders of magnitude in times ranging from nanoseconds to milliseconds.

Electronegative gas Electron-attaching gas in which stable negative ions (parent and/or fragment) are produced.

Synergism of gas mixtures Property of gas mixtures whereby they have dielectric strengths exceeding the partial-pressure-weighted dielectric strengths of the individual components making up the gas mixture.

Tailoring gas dielectrics Combination, on the basis of fundamental physicochemical knowledge, of two or more gases to optimize their overall dielectric properties.

A DIELECTRIC GAS IS a gaseous medium consisting of one or more components with a high breakdown voltage (i.e., a gaseous medium that is a relatively poor conductor or a nonconductor of electricity to high applied electrical strength). This article outlines the basic physical processes that determine the dielectric properties of gases and discusses the breakdown strengths of dielectric gases and the main uses of gaseous dielectrics.

I. INTRODUCTION

A gas in its normal state is a perfect insulator. If, however, the gas is electrically stressed, the free electrons present in it gain energy from the applied electric field; when the level of the applied field is such that an appreciable number of these free electrons possess kinetic energies high enough to ionize the gas atoms or molecules, the gas "breaks down" (i.e., it makes the transition from an insulator to a conductor). The minimum critical voltage under which the electrical conductivity of the gas increases by many orders of magnitude is known as the breakdown voltage and the phenomenon itself as electrical breakdown, electrical discharge, or spark.

As mentioned earlier, a dielectric gas is a relatively poor conductor or a nonconductor of electricity to high applied electrical strength (i.e., a gas with a high breakdown voltage). As such, it is used to insulate electrically various types of high-voltage equipment (see Section IV). As we shall see in Section III, the magnitude of the breakdown voltage depends not only on the nature, number density, and temperature of the gas, but also on many other factors such as the type of applied voltage and the geometry, material, and surface condition of the electrodes. The breakdown voltage varies considerably from one gaseous medium to another, and it can be—for certain electronegative gases, for example—over six times larger than the breakdown voltage of atmospheric air, which is the "traditional" gas dielectric (see Section III).

Depending on the form of the applied voltage and the nature and density of the gas, the transition of a gaseous medium from an insulator to a conductor occurs in times ranging from nanoseconds to milliseconds. The transition is critically determined by the behavior of electrons, ions, and photons in the gas, especially by those processes that produce or deplete free electrons. While a multiplicity of physical processes and species, both neutral and charged, play a role in determining the dielectric properties of a gas, it seems that the electron is the key particle, and its interactions with the gas molecules are the critical processes. Knowledge of these processes often allows a prediction of the dielectric properties of the gas and a choice of the appropriate gaseous medium for specific uses.

II. BASIC PHYSICAL PROCESSES AND PROPERTIES

A. Basic Physical Processes

The basic physical processes that determine the properties of dielectric gases involve excited and unexcited atoms and molecules, electrons, positive and negative ions, and photon interactions with the gas and the electrodes. We shall focus on those physical processes that are associated with the gas itself (not with the electrodes), and in Table I we list the principal ones. Basically, all these processes affect

TABLE I   Principal Physical Processes in Electrically Stressed Gas Dielectrics

Process number	Process representation	Process description
Group A.	*Electron–molecule interactions*	
1	$e + AB \rightarrow AB + e$	Elastic electron scattering (direct)
2	$e + AB \rightarrow AB^* + e$	Inelastic electron scattering (dirct)
3a	$e + AB \Rightarrow A + B + e$	Dissociation by electron impact
3b	$\Rightarrow A + B^* + e$	Dissociative excitation by electron impact
4a	$e + AB \rightarrow AB^+ + 2e$	Ionization by electron impact
4b	$e + AB \Rightarrow A + B^+ + 2e$	Dissociative ionization by electron impact
5a	$e + AB \rightarrow AB^{-*} \rightarrow AB^-$	Parent negative-ion formation
5b	$\Rightarrow A + B^-$	Dissociative attachment
5c	$\rightarrow AB (AB^*) + e$	Elastic (inelastic) electron scattering (indirect)
6	$e + AB \rightarrow A^+ + B^- + e$	Ion-pair formation
Group B.	*Photon–molecule interactions*	
7	$h\nu + AB \rightarrow AB^*$	Photonabsorption
8a	$h\nu + AB \rightarrow AB^+ + e$	Photoionization
8b	$\Rightarrow A + B^+ + e$	Dissociative photoionization
9	$h\nu + AB \rightarrow A + B$	Photodissociation
10	$h\nu + AB^- (B^-) \rightarrow AB (B) + e$	Photodetachment
11a	$AB^* + C \rightarrow AB + C^+ + e$	Penning ionization
11b	$B^* + C \rightarrow B + C^+ + e$	Penning ionization involving highly excited atoms (e.g., Rydberg states)
Group C.	*"Secondary" interactions*	
12	$e + AB^* \rightarrow AB (AB^*)$	Electron–positive ion recombination
13	$B^+ + A^- \rightarrow AB (AB^*)$	Positive ion–negative ion recombination
14	$AB^- + C \rightarrow AB + C + e$	Collisional detachment
15	$AB^- + C \rightarrow ABC + e$	Associative detachment
16	$AB^- + C \rightarrow AB + C^-$	Electron transfer
17	$AB^- + nC \rightarrow AB^- C_n \quad n \geq 1$	Cluster formation

the dielectric behavior of the gas by their effect(s) on the number density and energies of the free electrons present in the electrically stressed gas. Both the numbers and the energies of the free electrons depend on the gas itself and the density-reduced electric field E/N (E is the applied electric field and N the gas number density). Let us briefly look at the processes in Table I and their expected effect(s) on the dielectric properties of the gas. For convenience, we distinguish three groups of interactions: (1) electron–molecule, (2) photon–molecule, and (3) "secondary." In Table I, AB represents an unexcited and AB* an excited diatomic or polyatomic molecule, and the double arrows indicate that the reaction can produce a multiplicity of products.

1. Electron–Molecule Interactions

Processes 1 and 2 in Table I are direct elastic and inelastic electron scattering, respectively. Along with process 5c (indirect elastic and inelastic electron scattering, whereby the colliding electron is temporarily captured by the molecule and then released), they crucially determine the energies of the free electrons present in the stressed gas. Their cross sections depend on the electron energy itself and the details of the molecular (atomic) electronic structure. The inelastic electron scattering processes involve excitation of rotational, vibrational, and electronic states, while the elastic scattering does not change the internal energy of the molecule (atom). Although direct electron scattering is nonresonant (i.e., it occurs over a wide range of electron energies), indirect electron scattering is resonant; it usually is very efficient at low ($\lesssim 20$ eV) energies and rather significant in establishing the "electron slowing-down" properties of the dielectric gas. Obviously, polyatomic molecules are, as a rule, more efficient in slowing down the electrons than are small molecules or atoms.

Processes 3a and 3b represent the dissociation of molecules by electrons. They can proceed via a multiplicity of channels and thus produce a variety of neutral fragments, some of which, such as B* in 3b, may be excited and posses sufficient internal energy to ionize an impurity species C present in the dielectric (process 11b) and in this way to increase electron production (B* can also eject electrons when it collides with a surface). Processes 3a and 3b slow down the electrons present in the gas, as do processes 1, 2, and 5c, but in addition they produce free radicals which can change the number density and the composition of the dielectric gas.

Processes 4a and 4b are the principal ways by which new electrons are generated by electron–molecule collisions (and by which existing ones are slowed down); through process 4b a multiplicity of positive ions can be produced.

Processes 5a and 5b are the main reactions which deplete the electrons present in the dielectric, producing parent 5a and fragment 5b negative ions. In this way the electrons are prevented from causing ionization of the gas. To this end, besides the electron-attachment cross section or the electron-attachment rate constant as a function of electron energy, the binding energy of the attached electron (otherwise known as the electron affinity) must be large to prevent electron detachment (i.e., release of the attached electron). Process 5 has been studied under "isolated conditions" (i.e., very low pressures) and under "multiple collision conditions" (i.e., high pressures) in which the effect of the medium can often be significant. It is a resonant process occurring in the energy range from 0 to ~ 20 eV, depending on AB. It crucially affects the breakdown voltage and other properties of the gas dielectric because of its dominant role on the number density of the electrons. Gases with large electron-attachment cross sections are called electronegative or electron-attaching; the cross sections for the electron-attachment resonances decrease with increasing energy, and hence low-energy free electrons can be more efficiently removed from the dielectric (via this process) than can higher energy electrons. Not all gases are electronegative, however, but the good gas dielectrics are, or else contain electronegative additives.

The last process in group A in Table I, process 6, occurs at higher energies than 5, and although it produces negative ions, it does not deplete electrons, it slows them down. In the energy range of interest, its cross section is not appreciable, and its effect on the gas dielectric properties is thought not to be significant.

2. Photon–Molecule Interactions

In the second group of reactions are those between dielectric gas molecules and photons, produced by deexcitation of excited species at high E/N (such as AB* and B* in Table I and more likely, excited species produced by recombination processes). Here, three types of processes can increase the number of free electrons in the gas dielectric: photoionization of AB (processes 8a and 8b), photoionization of the negative ions (photodetachment, process 10), and Penning ionization (process 11a or 11b, due to excited species produced via 7 or 3b, respectively). Of course, photons can collide with the electrodes and inject new electrons into the gas dielectric in this way.

3. "Secondary" Interactions

The number density of electrons and ions, and thus the associated space charge effects in nonuniform fields, can be further affected by what can be termed secondary reactions. These can deplete electrons and positive ions (reaction 12) or positive and negative ions (reaction 13); can collisionally detach electrons from negative ions (reactions 14 and 15) or convert one ionic species to another (reaction 16) (and thus change its stability); or can change the ion's size (and thus its mobility) by clustering (reaction 17). While the role of these processes may be less obvious than the roles of groups A and B, it can, depending on the prevailing conditions, be most significant. For example, the gas dielectric behavior under steep-fronted voltage pulses is affected by the availability of "initiating" electrons produced by reactions 14 and 15. Similarly, corona stabilization (Section III.B) is influenced by the electron–ion (12) and ion–ion (13) recombination processes.

Understanding of the phenomena preceding the transition of the gas from an insulator to a conductor (prebreakdown phenomena) and the mechanisms involved in discharge initiation and development invariably requires basic knowledge on at least a fraction of the processes in Table I. This knowledge comes from two sources: low-pressure beam experiments and high-pressure swarm experiments. In high-pressure swarm experiments, as in electrically stressed gas dielectrics, the free electrons attain an equilibrium energy distribution $f(\varepsilon, E/N)$, and the measured electron transport coefficients (the electron drift velocity w and the ratio D_T/μ of the transverse diffusion coefficient D_T to the electron mobility μ) are related to the cross sections for the microscopic electron–molecule interactions through $f(\varepsilon, E/N)$. In principle, from a measurement of $w(E/N)$ and $D_T/\mu(E/N)$ and a knowledge of the electron scattering cross section, $f(\varepsilon, E/N)$ can be calculated through the Boltzmann transport equation or by Monte Carlo methods. If, then, for a given gas dielectric the various cross sections are known, they can be integrated over $f(\varepsilon, E/N)$ and used along with the appropriate charge conservation equations to determine the current growth in the gas and predict its breakdown voltage. In practice this is difficult because neither the cross sections nor $f(\varepsilon, E/N)$ is known for the majority of the dielectric gases or gas mixtures, so one resorts to the more easily accessible swarm coefficients to predict the discharge development and behavior.

From high-pressure swarm studies, the coefficients for excitation, detachment, and ion–molecule reactions are obtained as functions of E/N, as well as the primary ionization coefficient α and the effective electron attachment coefficient η. Most of the data are on the latter two coefficients. The coefficients α and η are most significant and are related to the respective ionization, $\sigma_i(\varepsilon)$, and attachment, $\sigma_a(\varepsilon)$, cross sections and $f(\varepsilon, E/N)$ by

$$\frac{\alpha}{N}\left(\frac{E}{N}\right) = \left(\frac{2}{m}\right)^{1/2} w^{-1} \int_I^\infty f\left(\varepsilon, \frac{E}{N}\right) \varepsilon^{1/2} \sigma_i(\varepsilon)\, d\varepsilon,$$

(1)

$$\frac{\eta}{N_a}\left(\frac{E}{N}\right) = \left(\frac{2}{m}\right)^{1/2} w^{-1} \int_0^\infty f\left(\varepsilon, \frac{E}{N}\right) \varepsilon^{1/2} \sigma_a(\varepsilon)\, d\varepsilon,$$

(2)

where I is the ionization onset energy, N the total gas number density, and N_a the attaching gas number density; for a unary electronegative gas dielectric $N = N_a$, but for mixtures containing electronegative and nonelectronegative components $N_a < N$.

B. Dielectric Properties

Having elaborated briefly on the basic physical processes occurring in electrically stressed dielectric gases, we can appropriately ask the question: How can the dielectric properties of a gaseous medium be optimized based on knowledge of such processes? To illustrate the type of answer one can get to this question let us see how knowledge of the electron-attaching, electron slowing-down, and electron impact-ionization properties of gases allows one to choose and to tailor gaseous dielectrics. This can be seen by referring to Fig. 1. When the value of E/N is low (e.g., 1.24×10^{-16} V cm^2 in Fig. 1 for N$_2$), $f(\varepsilon, E/N)$ lies at low energies, and the number of electrons capable of ionizing the gas is negligible (i.e., the gas is an insulator). As the voltage is increased, however, $f(\varepsilon, E/N)$ shifts to higher energies, and for sufficiently high E/N values the number of electrons capable of ionizing the gas is such that the gas makes the transition from an insulator to a conductor. In Fig. 1, $f(\varepsilon, E/N)$ is shown for N$_2$ at the limiting value of E/N, $(E/N)_{\text{lim}}$ ($\simeq 1.3 \times 10^{-15}$ V cm^2) (i.e., the value of E/N at which breakdown occurs). Even at this high E/N value only a small fraction of electrons possess sufficient energy to induce ionization, which, nonetheless, for a non-electron-attaching gas such as N$_2$, is sufficient to promote gas breakdown. This is designated in Fig. 1 by the shaded area α, which is a measure of the ionization coefficient α/N [Eq. (1)].

For a non-electron-attaching gas and a uniform field, knowledge of α provides a measure of $(E/N)_{\text{lim}}$ through the so-called Townsend breakdown criterion,

$$\gamma(e^{\alpha d} - 1) = 1.$$

(3)

In Eq. (3), αd is the number of electrons generated by an electron leaving the cathode and arriving as an electron avalanche at the anode (at a distance d from the cathode), and γ is the so-called secondary ionization coefficient, defined as the number of secondary electrons produced per primary ionization. These secondary-electron processes include (1) electron emission from the cathode as it is struck and photons, positive ions, and metastable molecules and (2) gas processes such as photoionization of the gas. Physically, Eq. (3) states that when each initial

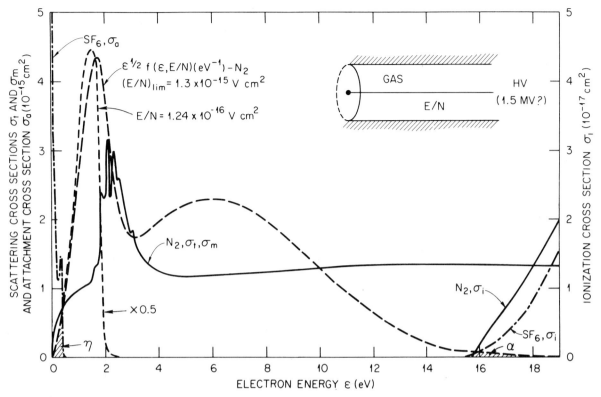

FIGURE 1 Ionization cross section $\sigma_i(\varepsilon)$ for N_2 and SF_6 close to the ionization onset I. Electron-scattering cross section as a function of ε for N_2 and electron-attachment cross section $\sigma_a(\varepsilon)$ for SF_6. Normalized electron energy distribution function $\varepsilon^{1/2}f(\varepsilon, E/N)$ as a function of ε for N_2 at two values of E/N. [From Christophorou, L. G., et al. (1984). *IEEE Trans, Elect. Insul.* **EI-19**, 550–566.]

electron produces a successor, the current maintains itself and becomes independent of γ.

For an electronegative gas the situation is different because in this case the free electrons can be effectively prevented from initiating breakdown if they attach themselves to gas molecules and form stable negative ions. The total attachment cross section $\sigma_a(\varepsilon)$ for SF_6 is shown in Fig. 1. In general, $\sigma_a(\varepsilon)$ is large at very low energies, and thus only electrons with energies at the extreme low-energy range can be removed efficiently by electron attachment. The shaded area in Fig. 1 that is designated by η is a measure of the effective electron attachment coefficient η/N_a [Eq. (2)]. Knowledge of α/N and η/N_a allows one to predict $(E/N)_{\lim}$, which for static uniform fields and a unary gas is defined as the value of E/N at which $\alpha = \eta$, namely,

$$\gamma[\exp\{(\alpha - \eta)d\} - 1] = 1 - \eta/\alpha. \qquad (4)$$

Both α/N and η/N_a are functions of E/N (see Figs. 2 and 3). Most often, one measures the so-called effective ionization coefficient $(\alpha - \eta)/N$ (see Fig. 2), and the static uniform field breakdown strength $(E/N)_{\lim}$ is defined as the value of E/N for which

$$(\alpha - \eta)/N \equiv \bar{\alpha}/N = 0. \qquad (5)$$

Thus (see Fig. 2) for pure SF_6, $(E/N)_{\lim} = 3.61 \times 10^{-15}$ V cm^2, and it decreases as N_2 is added in the binary mixture.

For nonuniform fields, $\bar{\alpha}$ is a function of the position between the electrodes, and the breakdown voltage V_s can be calculated from the so-called streamer criterion,

$$\int_0^{x_0} \bar{\alpha}(x)\, dx = k, \qquad (6)$$

where x_0 is the length at which the electron avalanche reaches the critical number of electrons ($\sim 10^8$) in the avalanche tip for causing streamer formation. The k is a constant characteristic of the gas. It is generally accepted that photoionization processes play a role in the propagation of the streamer.

For the gas dielectric strength to be optimized, not only must $\sigma_i(\varepsilon)$ be small and $\sigma_a(\varepsilon)$ be large [and extend to high energies to maximize the overlap of $\sigma_a(\varepsilon)$ and $f(\varepsilon, E/N)$, as in Eq. (2)], but the electron energies must be as low as possible because a low-lying $f(\varepsilon, E/N)$ minimizes α and maximizes η. Many studies, however, clearly show that high breakdown strengths require large $\sigma_a(\varepsilon)$. This can be seen from the examples in Fig. 3. As the attachment-rate

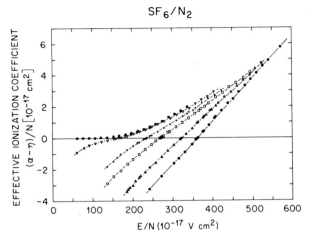

FIGURE 2 Effective ionization coefficient $(\alpha - \eta)/N$ as a function of the density-reduced electric field E/N. Curves are for the following gases and $(E/N)_{lim}(10^{-17} \text{ V cm}^2)$: ■ N_2, 130; ▽ 1% SF_6, 160; + 10% SF_6, 235; □ 20% SF_6, 269; △ 50% SF_6, 323; and ● 100% SF_6, 361. [Data from Aschwanden, T. (1984). *In* "Gaseous Dielectrics IV" (L. G. Christophorou and M. O. Pace, eds.), p. 30, Pergamon Press, New York.]

constant, as a function of the mean electron energy $\langle \varepsilon \rangle$, $k_a(\langle \varepsilon \rangle)$, increases, so does the breakdown voltage V_s^R relative to SF_6. Knowledge of $k_a(\langle \varepsilon \rangle)$ or $\sigma_a(\varepsilon)$ led to the identification of many excellent unary gas dielectrics such as the perfluorocarbons (see Table II, which is discussed in Section III).

It is thus apparent that the dielectric properties of gases can be optimized by a combination of two or more gases (i.e., by tailoring multicomponent gas mixtures) designed, for example, to provide the best effective combination of electron-attaching and electron slowing-down components. Basic knowledge on the processes in Table I offers several ways to the systematic development of dielectric gas mixtures. Thus knowledge on the electron-attachment cross section guided the choice of unary gas dielectrics or electronegative components for dielectric gas mixtures, and knowledge on electron scattering at low energies guided the choice of buffer gases for mixtures containing electronegative additives. Of practical significance are mixtures of the strongly electron-attaching gases in Table II with abundant, inert, and inexpensive buffer gases (e.g., N_2), with which they act synergistically: the buffer gas(es) scatter electrons into the energy range in which the electronegative gas(es) capture electrons most efficiently.

Examples of the various types of observed uniform field behavior of the breakdown voltage $(V_s)_{mix}$ of binary gas mixtures with respect to those $(V_s)_{A,B}$ of the individual components A, B as a function of gas composition are shown in Fig. 4. Figure 4a shows the behavior of $(V_s)_{mix}$ for binary mixtures of electronegative gases whose $k_a(\langle \varepsilon \rangle)$ is independent of gas number density N. The $(V_s)_{mix}$ is nearly equal to the sum of the partial-pressure-weighted

V_s of the constituent component gases. In Fig. 4b, examples are given of binary mixtures of a buffer gas that slows down electrons efficiently (CO, N_2, or CO_2) and the electronegative gas SF_6. At all gas compositions, the measured $(V_s)_{mix}$ exceeds the partial-pressure-weighted values of the individual components. This has been referred to as synergism. Similar results are shown in Fig. 4c for mixtures of the strongly electronegative gas c-C_4F_8 and the nonpolar weakly electron-attaching buffer gas CF_4 or the polar buffer gases CHF_3 or 1,1,1-CH_3CF_3, which slow down electrons efficiently via dipole scattering. The $(V_s)_{mix}$ for the polar gas-containing mixtures far exceeds the partial-pressure-weighted V_s, especially for small percentages of the electronegative gases.

In Fig. 4c the behavior of curve 1 is interesting in that for certain gas compositions the $(V_s)_{mix}$ exceeds the V_s of either component. This has been observed for other binary mixtures (e.g., 1-C_3F_6/c-C_4F_8; 1-C_3F_6/SO_2; SO_2/SF_6; C_3F_8/SF_6; and OCS/SF_6) for which the electron-attachment properties of one or both of the constituent gases depend on the total gas pressure and the mixture composition. This is clearly seen by the data in Fig. 4d, which show the variation of $(V_s)_{mix}$ for SF_6/1-C_3F_6 with relative composition and total gas number density.

Many studies on gas mixtures identified binary gas mixtures [e.g., SF_6/N_2, perfluorocarbon/SF_6, and perfluorocarbon/N_2 (or CHF_3)] that can be useful for applications.

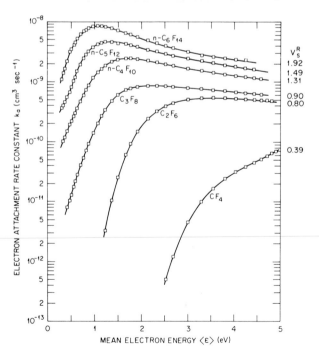

FIGURE 3 Total electron-attachment rate constant k_a as a function of the mean electron energy $\langle \varepsilon \rangle$ for the perfluoroalkanes C_NF_{2N+2} ($N = 1$–6) and their dc uniform field breakdown voltages relative to SF_6. [From Christophorou, L. G., *et al.* (1984), *IEEE Trans. Elect. Insul.* **EI-19**, 550–566.]

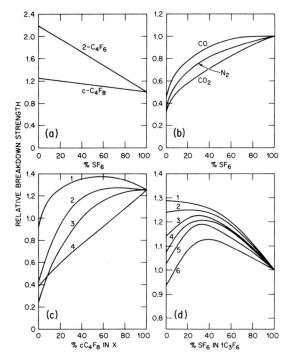

FIGURE 4 Relative breakdown strength of various gas mixtures. In (c), curves 1–4 are for x = 1-C_3F_6, 1,1,1-CH_3CF_3, CHF_3, and CF_4, respectively. In (d), curves 1–6 are for N_{total} (10^{19} cm^{-3}) = 10.00, 7.52, 5.02, 3.76, 2.51, and 1.67 respectively. [Based on data from Christophorou, L. G. (ed.), (1982). "Gaseous Dielectrics III," Pergamon Press, New York; Christophorou, L. G. (ed.). (1984). "Electron–Molecule Interactions and Their Applications," Academic Press, New York; Christophorou, L. G., *et al.* (1984). *IEEE Trans. Elect. Insul.* **EI-19,** 550–566; and Electric Power Research Institute. (1982). "Gases superior to SF_6 for insulation and interruption," **EPRI·EL-2620,** Electric Power Research Institute, Palo Alto, CA [prepared by Westinghouse Electric Corporation].

The SF_6/N_2 mixtures are particularly attractive. Their relatively high dielectric strength (for SF_6 concentrations $\gtrsim 40\%$) have made the SF_6/N_2 mixtures candidates for high-voltage gas-insulated equipment, especially in low-temperature environments ($< -40°C$) where pure SF_6 would condense at the normal operating pressures (~ 4.5 atm); a 50/50 mixture of N_2/SF_6 has a dielectric strength $\sim 85\%$ that of pure SF_6 in uniform fields.

Ternary gaseous mixtures have also been developed to optimize as many of the desirable dielectric properties and characteristics as possible. Such gaseous dielectrics, composed of N_2, SF_6, and small amounts of strongly electronegative perfluorocarbons, show promise for applications.

III. BREAKDOWN STRENGTH

A. Uniform Fields

The range of breakdown potentials for gases is considerable. This can be seen from the selected data on the dc uniform field breakdown strengths V_s^R in Table II, which were measured at approximately atmospheric pressures: The V_s of air is ~ 50 times higher than that of Ne, and the V_s of SF_6 is ~ 3.3 times higher than that of air. The highest known V_s (~ 2.5 times higher than SF_6) are exhibited by strongly electron-attaching polyatomic gases such as the perfluorocarbons in Table II and other polyhalogenated molecules. Weakly electron-attaching or non-electron-attaching gases (Table II) have low V_s values. Nonelectronegative molecular gases with large electron-scattering cross sections have reasonably high V_s values compared, for instance, with the rare gases, in which low-energy electron scattering is totally elastic.

The V_s of a gaseous medium is expected, in accordance with Paschen's law, to be a function only of Nd_s (the product of the gas number density N and the electrode separation d_s); thus, for sufficiently high values of N to the right of the Paschen minimum, $V_s/Nd_s = (E/N)_{lim}$ should be independent of N. This relationship holds for

TABLE II Relative dc Uniform Field Breakdown Strengths V_s^R of Some Dielectric Gases[a]

Gas	$V_s^{R b,c}$	Comments
SF_6	1	Most common dielectric gas besides air
C_3F_8	0.90	Strongly and very strongly electron-attaching gases, especially at low energies
n-C_4F_{10}	1.31	
c-C_4F_8	~ 1.35	
1,3-C_4F_6	~ 1.50	
c-C_4F_6	~ 1.70	
2-C_4F_8	~ 1.75	
2-C_4F_6	~ 2.3	
c-C_6F_{12}	~ 2.4	
CF_3H	0.27	Weakly electron-attaching; some (CO, N_2O) are effective in electron slowing-down
CO_2	0.30	
CF_4	0.39	
CO	0.40	
N_2O	0.44	
H_2	0.18	Virtually non-electron-attaching
Air	~ 0.30	
N_2	0.36	Non-electron-attaching but efficient in electron slowing-down
Ne	0.006	Non-electron-attaching and not efficient in electron slowing-down
Ar	0.07	

[a] Based on data in Christophorou (1984), Meek and Craggs (1978), and Christophorou *et al.* (1984).

[b] Some of the values given are for quasi-uniform fields and may thus be lower than their uniform field values.

[c] The relative values can be put on an absolute scale by multiplying by 3.61×10^{-15} V cm^2, the $(E/N)_{lim}$ of SF_6.

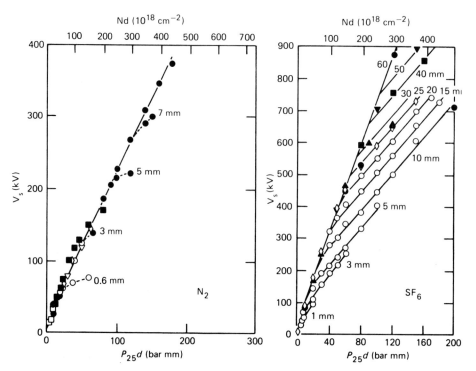

FIGURE 5 Breakdown voltages of N_2 and SF_6 at high Pd ($T = 298$ K). [From Meek, J. M., and Craggs, J. D. (eds.). (1978). "Electrical Breakdown of Gases," Wiley, New York.]

most gases at low field strengths and moderate pressures (see Fig. 5). At high field strengths, the V_s varies less than linearly with increasing N for a fixed d_s. This is illustrated by the uniform field data for N_2 and SF_6 shown in Fig. 5 for a number of N and d_s values. It is clear from these data that the deviations are larger for the electronegative gas SF_6 than for the nonelectronegative gas N_2. This pertains to the cause of these deviations, which are generally attributed to the effects of electrode geometry and surface. Changes in field uniformity by surface roughness affects (increases) the value of $\bar{\alpha}/N$ (more so than α/N) principally by decreasing the value of η/N due to the shift of $f(\varepsilon, E/N)$ to higher energies as the field increases at imperfections, scratches, and surface projections.

It has recently been found that Paschen's law can be violated, in a way opposite to that just described [i.e., $(E/N)_{lim}$ increasing rather than decreasing with increasing N], in cases in which it would normally be expected to hold. An example of this type of behavior is shown in Fig. 6, where the $(E/N)_{lim}$ of 1-C_3F_6 (perfluoropropylene) is plotted versus N (or compressibility-corrected pressure). It is seen that $(E/N)_{lim}$ increases with N over a given range, contrary to that of SF_6, which is independent of N. The increase of $(E/N)_{lim}$ with N relates to the decrease with N of $\bar{\alpha}/N$ due to the increase with N of η/N for 1-C_3F_6 (no such increase occurs for SF_6). The

saturation in the dependence of $(E/N)_{lim}$ on P is related to the saturation of the increase in $k_a(\langle\varepsilon\rangle)$ with increasing P. A similar violation of Paschen's law has been observed for other gases (e.g., OCS, N_2O, SO_2, and C_3F_8), all of which have electron-attaching properties that are functions of N.

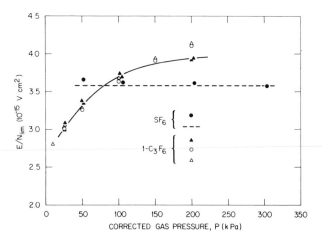

FIGURE 6 $(E/N)_{lim}$ versus pressure P, corrected for compressibility, for SF_6 and 1-C_3F_6; the various symbols refer to measurements of different authors. [From Christophorou, L. G. (ed.). (1982). "Gaseous Dielectrics III," Pergamon Press, New York; Christophorou, L. G., and pace, M. O. (eds.). (1984). "Gaseous Dielectrics IV," Pergamon Press, New York.]

B. Nonuniform Fields

A nonuniform field is often characterized by the so-called field utilization factor n, defined as the ratio of the average field E_a to the maximum field E_m, and is a function of the electrode geometry. When the maximum electrical stress in a gas gap is large enough, localized ionization occurs near the high-field electrode without breakdown. This phenomenon is generally known as corona. The charges generated by corona are separated by the electric field and cause field distortion. The space charge produced by corona profoundly affects the breakdown characteristics of the gas. This is illustrated by the breakdown voltage/pressure characteristics of electronegative gases exemplified in Fig. 7 for SF_6. Up to a pressure P_1 the breakdown voltage V_s increases almost linearly with P and considerably exceeds the corona inception voltage V_i. Subsequently, V_s goes through a maximum and then falls, up to pressure P_c' that may coincide with the critical pressure P_c, defined as the pressure for which $V_s = V_i$ (i.e., the pressure at which breakdown occurs directly at the inception voltage without any preceding corona). The pressure range over which $V_s > V_i$ is called the corona stabilization region. In this region, homopolar space charge generated by corona reduces the field near the high-stress electrode. The size and shape of this region and the value of P_c are functions of the field utilization factor n, the applied voltage, and the nature of the gas. Studies have shown that the smaller the field utilization factor (i.e., the more nonuni-

form the field), the higher the value of P_c (i.e., corona stabilization occurs over a wider range of pressures).

Under impulse conditions, the behavior of V_s at $P > P_1$ depends on the impulse risetime and on the availability of initiating electrons (i.e., the behavior is strongly dependent on the statistics of discharge initiation). For wavefronts $\gtrsim 100$ μsec, there is time for corona shielding to develop, and the V_s versus P curve is similar to that for dc. For short risetimes (a few microseconds or less), the characteristic is relatively flat between P_i and P_c (see Fig. 7).

In view of the possible significance of the corona stabilization region in the reliability of gas-insulated equipment, systematic studies have been undertaken to tailor dielectric gas mixtures to have improved stabilization. Most such efforts have focused on increasing corona-controlled breakdown of compressed electronegative mixtures containing SF_6 or $SF_6 + N_2$ for both polarities by increasing the negative and positive space charge, respectively, with small amounts of additives that are more electronegative than SF_6 (such as certain perfluorocarbons) for negative polarity and with additives whose ionization threshold energy is much lower than that of SF_6 for positive polarity. Often as little as 1–5% of a suitable electronegative perfluorocarbon additive can effect a substantial increase in the corona stabilization. However, for such mixtures the positive polarity is usually lowered compared with SF_6 or SF_6/N_2. An improvement in the corona stabilization for both positive and negative polarity can be achieved by an additive that is electronegative and has a low ionization potential.

C. Effect of Particles

Another form of nonuniform field breakdown in a gaseous dielectric is that from free conducting particles. Free conducting particles subjected to an electric field in a gaseous medium between the electrodes become levitated when the force exerted on them by the electric field exceeds the gravitational force. With an ac voltage, the particles will bounce on the lower electrode, and the bounce height will increase with the applied voltage. With a dc voltage, the particles can cross the gap between the electrodes as soon as they are levitated. Tests with free conducting particles have shown that the worst type of particles are long, metallic needles or wires. Figure 8 shows the effect of free wire particles in a coaxial electrode system in SF_6 on the ac breakdown voltage as a function of gas pressure and wire length. It is seen that the breakdown voltage is virtually independent of gas pressure for the longer particles and that particle-initiated breakdown occurs at fields considerably lower than those for the uncontaminated system. The effect is usually larger for positive polarity voltages.

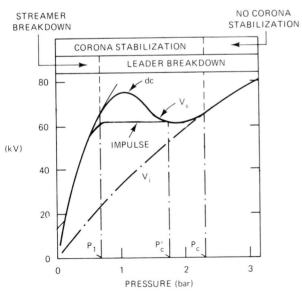

FIGURE 7 Positive dc corona in a point-plane SF_6 gap (radius of point tip, $r_0 = 2$ mm; electrode gap $d = 20$ mm). [From Farish, O. (1983). In "Proceedings XVI International Conference on Phenomena in Ionized Gases" (W. Bötticher et al., (eds.). p. 187, Düsseldorf, Germany.)

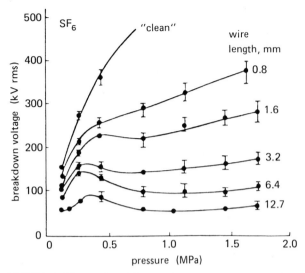

FIGURE 8 The ac breakdown voltage versus pressure in a 150 mm/250 mm coaxial geometry containing free copper wires with 0.4 mm diameter and lengths ranging from 0.8 to 12.7 mm. [From Cookson, A. H., *et al.* (1971). *IEEE Trans. Power App. Syst.* **PAS-90,** 871.]

Although the free conducting particles cause reductions in the V_s of quasi-uniform field electrode geometries, the V_s is higher when the particles are fixed to an electrode. The fixed-particle breakdown voltage/pressure characteristic exhibits the familiar corona stabilization region discussed in the previous section, where $V_s > V_i$ (see Fig. 7). It also has been found that the V_s for free particles with ac voltage corresponds closely to the impulse breakdown voltage with fixed particles. This indicates that the free-particle breakdown mechanism is similar to that of impulse breakdown. For impulse voltages, the corona stabilization process for the fixed particle fails and the impulse breakdown voltage is considerably lower than for ac over the pressure range where the corona stabilization mechanism is dominant. Observations of particle breakdown have shown that as the particle approaches an electrode, discharges occur between the particle and the electrode as a result of their different potentials. The result is a sudden change in the electric field at the particle tip that faces the main gap, which is equivalent to an impulse voltage applied to the particle.

Particle-initiated breakdown is one of the most severe imperfections in gas-insulated apparatus, seriously reducing the dielectric strength of gases and the reliability of gas-insulated apparatus. Obviously, the best way to alleviate the effect of conducting particles is to remove them from the equipment. Various techniques to remove them and to promote particle motion or scavenging into low-field particle traps have been studied and are in use in gas-insulated equipment.

D. Voltage Waveform Effects and Time to Breakdown

When a voltage of sufficient magnitude ($\geq V_s$ for dc) is suddenly applied to a gas-insulated electrode gap, or a gas-insulated conductor, breakdown does not occur instantaneously, but after a finite time $t = t_s + t_f$. The t_s is called the statistical time lag and is the time that elapses between the application of the voltage V ($\geq V_s$) and the occurrence of a free electron in the stressed gas volume which initiates the breakdown process. The t_f is called the formative time lag and is the time interval between the occurrence of the free electron and the collapse of the voltage (i.e., breakdown).

The statistical time lag t_s can vary from nanoseconds to milliseconds (or longer) depending on the time the initiatory electron becomes available when $V \geq V_s$. Initiatory electrons can be produced by cosmic radiation, natural radioactivity from materials, field emission from the cathode surface, or collisional detachment from negative ions in the case of electronegative gases. In experimental apparatus they can be produced intentionally by, for instance, an ultraviolet source, in which case t_s is reduced considerably.

The formative time lag usually varies from nanoseconds to microseconds and is influenced by the overvoltage [$\Delta(V - V_s)$], the field distribution, and, for unsymmetrical fields, the polarity. This is apparent from the data in Fig. 9 on N_2 and SF_6, which were obtained with a square impulse generator having ~2 nsec rise time and ~150 kV amplitude. It is evident that the higher the overvoltage (voltage in excess of V_s), the shorter the t_f at which breakdown occurs. It is also seen that t_f is significantly higher for negative polarity and that it varies considerably with field uniformity, especially for electronegative gases (Fig. 9). The voltage–time characteristics of dielectric gases are of practical importance for the insulation coordination and overvoltage protection of gas-insulated equipment.

E. Gases with Insulator/Conductor Properties

In a number of technologies a gas is needed which is both a good conductor and a good insulator. For example, in pulsed power technologies the key element is a fast repetitive switch. Among a number of switching devices, the diffuse gas discharge switch is promising for a system such as inductive energy storage. The operation of a diffuse discharge switch for inductive storage is characterized by two distinct stages: (1) the conducting (storing) stage, when E/N is small (~3×10^{-17} V cm²), and (2) the transferring stage (when the stored energy in the inductor is transferred to the load), when E/N is large

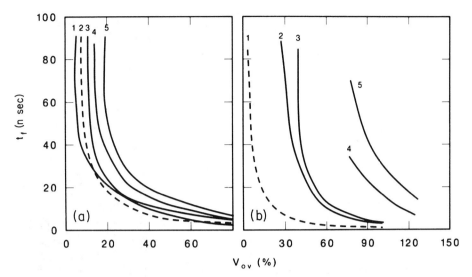

FIGURE 9 Formative time lag versus percentage of overvoltage in (a) N_2 and (b) SF_6 for different field distributions $n \equiv E_a/E_{max}$ (n values of 0.84, 0.54, and 0.32 correspond, respectively, to approximately homogenous, weakly inhomogenous, and inhomogenous fields), $d = 4$ mm, and $P = 1$ bar. In (a), for N_2, curves 1–5 are (respectively) for $n = 0.84$, polarity $+$; $n = 0.54$, polarity $+$; $n = 0.54$, polarity $-$; $n = 0.32$, polarity $+$; and $n = 0.32$, polarity $-$. In (b), for SF_6, curves 1–5 are (respectively) for $n = 0.84$, polarity $+$; $n = 0.54$, polarity $+$; $n = 0.54$, polarity $-$; $n = 0.32$, polarity $+$; and $n = 0.32$, polarity $-$. [Data from Peiffer, W. (1984). *In* "Gaseous Dielectrics IV" (L. G. Christophorou and M. O. Pace, eds.), pp. 329, 331, Pergamon Press, New York.]

($\gtrsim 150 \times 10^{-17}$ V cm^2). The successful operation of such switching devices depends on the availability of a gas that is a good conductor in the conducting stage and a good insulator in the transferring stage. To optimize conduction under the low-E/N conditions of the conducting stage, the electrons produced by the external source (e-beam or laser) must remain free and must have as large a drift velocity w as possible. To optimize the insulating properties under the high-E/N conditions of the transferring stage, the gas must effectively remove electrons by attachment (have a large attachment rate constant at high E/N). These requirements are schematically illustrated in Fig. 10.

Gas mixtures with such desirable characteristics have been reported by various authors. In particular, it has been shown that binary gas mixtures of buffer gases such as Ar and CH_4, whose electron-scattering cross sections have a Ramsauer–Townsend minimum at low energies (~ 0.5 eV), and electron-attaching gases such as C_2F_6, and C_3F_8, which attach electrons efficiently at high E/N and have much-reduced electron-attachment rate constants at low E/N, are most appropriate for diffuse discharge switching applications. Such mixtures have distinct maxima in the $w(E/N)$ at E/N values appropriate for the conducting stage of the switch, w values in excess of 10^7 cm sec^{-1}, and breakdown strength $\gtrsim 150 \times 10^{-17}$ V cm^2 for mixtures containing $\gtrsim 10\%$ of the attaching gas.

IV. USES OF DIELECTRIC GASES

The most abundant "traditional" dielectric gas is atmospheric air. It naturally insulates overhead transmission lines that crisscross the countryside. Overhead transmission lines up to 800 kV are presently in service, and

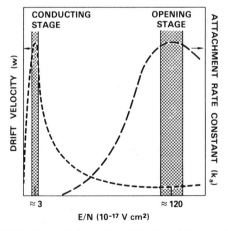

FIGURE 10 Schematic illustration of the desirable characteristics of the $w(E/N)$ and $k_a(E/N)$ functions of the gaseous medium in an externally (e-beam) sustained diffuse discharge switch. Indicated are rough estimates of the E/N values for the conducting and the opening stages of the switch. [From Christophorou, L. G., *et al.* (1983). *In* "Proceedings 4th IEEE International Pulsed Power 1983, Albuquerque, New Mexico" (T. H. Martin and M. F. Rose, eds.), p. 702, Texas Tech University Press, Lubbock, TX.]

FIGURE 11 A view of Oak Ridge National Laboratory's electrostatic tandemtype heavy ion accelerator, which is insulated with SF_6. (Courtesy of Oak Ridge National Laboratory.)

lines for 1200 kV and higher voltages are under consideration for the future. Early uses of gas dielectrics included the use of N_2 and CO_2 (at pressures up to ~20 atm) in high-voltage standard capacitors. The distinct advantages of gaseous dielectrics (e.g., low weight, excellent recovery characteristics, low dielectric losses, compactness of equipment, and environmental advantages) resulted in their deployment for electrical insulation and arc interruption purposes in a variety of electrical equipment. In this section we briefly mention some of their main uses.

A. Research Equipment

High-voltage power supplies and Van de Graaff-type accelerators are examples of laboratory equipment that employ gas dielectrics for electrical insulation. Most such equipment uses SF_6. Other gases or gas mixtures (e.g., N_2/CO_2 and N_2/SF_6) have occasionally been used.

One of the largest Van de Graaff-type accelerators is the Holifield Heavy Ion Accelerator at Oak Ridge National Laboratory in Tennessee. It uses 0.7 MPa of SF_6 as an insulating gas. With this machine a dc voltage of 34 MV was achieved. A view of this electrostatic tandem-type heavy ion accelerator is shown in Fig. 11.

B. Circuit Breakers

Prior to the 1950s, high-voltage power circuit breakers mainly relied on oil and compressed air for insulation and current interruption. In 1956 the first high-voltage power circuit breaker was put into service at 115 kV, using SF_6 as the insulating and interrupting medium. Since then, circuit breakers with SF_6 gas have been put into service in all distribution and transmission voltage classes from 34.5 through 800 kV with interrupting capabilities in excess of 80 kA. Virtually all new circuit breakers today use a single-pressure chamber (usually at ~6 atm) in which the gas is further compressed with a piston attached to the moving contact and provides an axial gas blast for the are interruption (see Fig. 12). Earlier versions of SF_6 interrupters were of the two-pressure design, where the gas blast for the arc interruption was provided to the arc interruption chamber from a high-pressure (~14.5 atm) reservoir.

A basic requirement for an arc-interrupting medium, besides a high dielectric strength, is a high rate of recovery of its dielectric strength. The ability of SF_6 to recover its dielectric strength quickly after arc interruption along with its rapid thermal recovery and high degree of molecular reconstitution after arcing make it the most attractive arc-interruption gaseous medium.

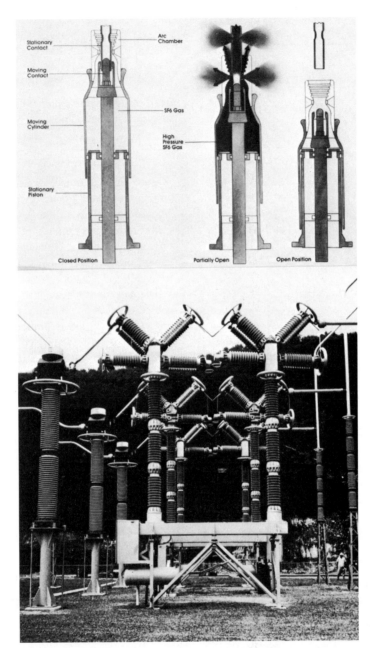

FIGURE 12 Circuit breakers using SF_6 gas. (Courtesy of Westinghouse Electric Corporation.)

A search for other gases/mixtures for use in circuit breakers showed that while no other gaseous medium outperforms SF_6 in circuit breakers, some gases/mixtures can be considered. The performance ratings (based on the critical interruption current I_c for a given surge impedance Z_0) relative to SF_6 of some such media are listed in Table III.

C. Compressed-Gas Insulated Substations

The excellent dielectric strength and arc extinguishing properties of SF_6 facilitated the development of circuit breakers with metal casings, generally called metal clad or dead-tank circuit breakers because the casing or tank is at ground potential. This naturally led to the development of other substation components such as disconnect switches, arresters, interconnecting links (buses), and cables for getaways using the metalclad technology.

The totally enclosed metalclad substations using SF_6 gas (Fig. 13) are significantly more compact than the open-air substations and switchyards. Often, the land required is reduced by as much as 40 times, and the resultant savings in the cost of land are thus significant. Furthermore,

TABLE III Arc Interruption Capabilities of Gases and Gas Mixtures[a,b]

Gas/mixture	$Z_0 = 450 \ \Omega$		$Z_0 = 225 \ \Omega$	
	I_c (kA)	$SF_6 = 100$	I_c (kA)	$SF_6 = 100$
SF_6	21.0	100	26.3	100
SF_6/N_2 (75/25)	17.8	85	20.4	78
$CH_4/CClF_2CF_3$ (50/50)	17.8	85	20.2	77
CF_2CFCF_3/SF_6 (75/25)	17.0	81	20.0	76
CF_3SO_2F/SF_6 (50/50)	16.5	79	18.3	70
SF_6/He (75/25)	15.4	73	20.4	78
SF_6/N_2 (50/50)	14.9	71	17.2	65
CF_2CFCF_3	14.8	70	17.8	68
SF_6/He (50/50)	14.7	70	19.7	75
$CClF_2CF_3/SF_6$ (75/30)	14.0	67	17.6	67
$CHClF_2/SF_6$ (75/25)	13.8	66	14.7	56
$CBrF_3/SF_6$ (75/25)	11.6	55	14.5	55
CF_3SO_2F/SF_6 (75/25)	11.4	54	13.8	52
CF_4	11.1	53	14.6	56
$CBrF_3$	11.1	53	16.8	64
$CClF_2CF_3$	10.8	51	15.4	59

[a] Total pressure = 0.62 Mpa.
[b] Data from Electric Power Research Institute. (1982). "Gases superior to SF_6 for insulation and interruption," EPRI · EL-2620, Electric Power Research Institute, Palo Alto, CA [prepared for Westinghouse Electric Corporation.]

metalclad substations are free from industrial and coastal pollution problems and are esthetically more pleasing. The compactness of gas-insulated substations also offers the option of locating them in the basements of buildings, thus allowing the transmission of power at high voltages straight into urban centers for distribution.

D. Compressed-Gas Insulated Cables

Transmission voltage cables using a compressed gas as the electrical insulation medium have been in commercial use since 1971. The voltage rating ranges from 138 to 800 kV. Such cables are used as getaways from substations to overhead transmission lines, in crossings of two or more transmission lines, in tunnels from hydroelectric and pumped storage generator plants, in river crossings, and so on. They are especially attractive when high-voltage and high-power ratings are required in limited space, when metalclad substations are connected directly to gas-insulated cables, or when a reduction in fire hazard is required, as in tunnels.

The most common design of compressed-gas insulated cables consists of an aluminum center conductor and a concentrically located outer aluminum enclosure. The conductor is supported by epoxy insulators in the enclosure. SF_6 is the only dielectric gas used for gas-insulated substations and cables. Besides being an excellent dielectric, SF_6 also has good heat transfer characteristics for removing the heat generated by the resistive losses of the conductor to the enclosure. Usually the SF_6 pressure ranges from 3 to 5 bar. A schematic of a typical design of a gas-insulated cable is shown in Fig. 14, and a typical installation is shown in Fig. 13. Some designs include built-in particle traps to improve the reliability of the system.

FIGURE 13 Totally enclosed metalclad substation using SF_6 gas as the insulating medium. (Courtesy of Westinghouse Electric Corporation.)

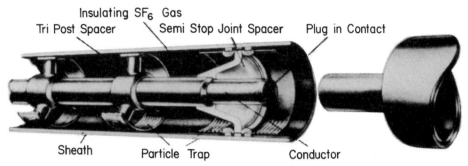

Insulating SF₆ Gas
Tri Post Spacer   Semi Stop Joint Spacer   Plug in Contact

Sheath   Particle Trap   Conductor

FIGURE 14 Schematic of a compressed-gas insulated (CGI) cable. (Courtesy of Westinghouse Electric Corporation.)

By far the largest use of gas-insulated cables has been in installations that are 100–200 m in length. The longest cable is 700 m (420 kV/1386 MVA). Research and development have been conducted to develop gas-insulated cables using SF_6 for 1200-kV transmission. Three-conductor cables (three-in-one), where the three-phase conductors are arranged in a single enclosure, are also used up to 362 kV transmission voltage. Gas-insulated technology is being further developed by considering flexible enclosure designs for easier installation.

Compressed-gas insulated cables have about two to three times the load-carrying capability of comparable oil-filled cables. Typically, a gas-insulated transmission cable at a 242 kV system voltage can carry 3000 A, equivalent to 1250 MVA, while at 1200 kV, 10,000 A, equivalent to 16 GVA, has been demonstrated.

V. CONCLUDING REMARKS

Recent advances in basic research, especially in electron- and ion-collision physics, have resulted in improved understanding of the dielectric properties of gases. This understanding, in turn, has aided efforts to identify, improve, and tailor new dielectric gases for a variety of electrical insulation needs.

Basic research is still needed to provide better understanding and a sounder scientific basis for the expected expansion in the uses of dielectric gases as insulants in high-voltage transmission and distribution and other high-voltage electrical equipment. The full potential of and benefit from the use of gas dielectrics is yet to be realized. To optimize this benefit, a well-balanced growth in basic and applied research and industrial development in this area is necessary.

SEE ALSO THE FOLLOWING ARTICLES

ATOMIC AND MOLECULAR COLLISIONS • MOLECULAR ELECTRONICS • NANOELECTRONICS • POWER TRANSMISSION, HIGH VOLTAGE

BIBLIOGRAPHY

Christophorou, L. G. (ed.). (1980). "Gaseous Dielectrics II," Pergamon Press, Oxford.

Christophorou, L. G. (ed.). (1982). "Gaseous Dielectrics III," Pergamon Press, Oxford.

Christophorou, L. G. (ed.). (1984). "Electron–Molecule Interactions and Their Applications," Vols. 1 and 2, Academic Press, New York.

Christophorou, L. G., and Pace, M. O. (eds.). (1984). "Gaseous Dielectrics IV," Pergamon Press, Oxford.

Christophorou, L. G., and Sauers, I. (eds.). (1991). "Gaseous Dielectrics VI," Plenum Press, New York.

Christophorou, L. G., Sauers, I., James, D. R., Rodrigo, H., Pace, M. O., Carter, J. G., and Hunter, S. R. (1984). *IEEE Trans. Elect. Insul.* **EI-19,** 550–566.

Kunhardt, E. E., and Luessen, L. H. (eds.). (1983). "Electrical Breakdown and Discharges in Gases," Plenum Press, New York.

Meek, J. M., and Craggs, J. D. (eds.). (1978). "Electrical Breakdown of Gases," Wiley, New York.

Nasser, E. (1971). "Fundamentals of Gaseous Ionization and Plasma Electronics," Wiley, New York.

Special Issue. (1990). *IEEE Trans. Elect. Insul.* **25** (1, February).

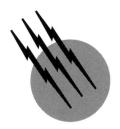

Differential Equations, Ordinary

Anthony N. Michel

University of Notre Dame

GLOSSARY

Equilibrium or rest position For the system of equations $x' = f(t, x)$, any point x_e such that $f(t, x_e) = 0$ for all t is an equilibrium point or a rest position.

Fundamental matrix If $\{\phi_1, \ldots, \phi_n\}$ denotes a set of linearly independent solutions for the equation $x' = A(t)x$, then the matrix $\Phi = [\phi_1, \ldots, \phi_n]$ is a fundamental matrix of $x' = A(t)x$.

Initial-value problem The system of ordinary differential equations $x' = f(t, x)$ along with the initial data $x(\tau) = \xi$ is an initial value problem, where τ denotes initial time and ξ denotes the initial condition or the initial state.

nth-Order ordinary differential equations Equation of the form $y^{(n)} = h(t, y^{(1)}, \ldots, y^{(n-1)})$ is an nth-order ordinary differential equation, where $y^{(i)}$ denotes the ith derivative of y with respect to t.

Qualitative theory of ordinary differential equations Study of families of solutions of ordinary differential equations, such as, for example, the (stability) properties of solutions near an equilibrium point.

Solution of an initial-value problem n-Vector-valued function ϕ is a solution of an initial-value problem if ϕ satisfies the equation $x' = f(t, x)$ and if $\phi(\tau) = \xi$.

State transition matrix For the initial value problem $x' = A(t)x, x(\tau) = \xi$, the matrix given by $\Phi(t, \tau) \overset{\Delta}{=} \Phi(t)\Phi^{-1}(\tau)$ is the state transition matrix, where Φ denotes a fundamental matrix for $x' = A(t)x$ and Φ^{-1} denotes the inverse of Φ; the unique solution of the initial value problem is then $\phi(t, \tau, \xi) = \Phi(t, \tau)\xi$.

System of linear homogeneous differential equations System of equations given by $x' = A(t)x$ where x is an n vector and $A(t)$ denotes an $n \times n$ (time-varying) matrix is a linear homogeneous system of ordinary differential equations.

System of ordinary differential equations The system of equations given by $x' = f(t, x)$ where x is an n vector, t is real (time), f is a function, and x' denotes differentiation of x with respect to t is a system of n ordinary differential equations of the first order.

EQUATIONS containing the derivatives or differentials of one or more dependent variables, with respect to one or more independent variables, are called *differential equations*. If such equations contain only ordinary derivatives of one or more dependent variables, with respect to a single independent variable, then one speaks of *ordinary differential equations*. Equations involving the partial derivatives of one or more dependent variables of two or more independent variables are called *partial differential equations*.

I. INTRODUCTION

In what follows, we will concern ourselves only with ordinary differential equations. The study of such equations may be divided into qualitative theory and quantitative theory (e.g., the numerical solution of differential equations). We will concern ourselves almost exclusively only with a qualitative theory for such equations.

Ordinary differential equations, which were first considddered in the seventeenth century by Leibnitz and Newton, arise in nearly all disciplines of science (physics, chemistry, biology, and the like) and engineering (aerospace, chemical, civil, electrical, mechanical, nuclear, and so forth) as well as in economics and societal systems. It is not an overstatement to say that a very great deal of applied mathematics involves in some way the study of differential equations.

The study of ordinary differential equations can be from fairly elementary and down-to-earth to rather advanced and abstract levels. In the current treatment we will follow a path between these two extremes. The study of ordinary differential equations at a very basic level requires as a prerequisite some knowledge of elementary calculus. At an intermediate level, the study of such equations demands some background in real variables and linear algebra, while at the advanced level, the study of differential equations may involve facts from measure and integration theory as well as functional analysis.

II. INITIAL-VALUE PROBLEMS

A. First-Order Ordinary Differential Equations

We let D denote a domain (i.e., an open, non-empty, and connected set) in the R^{n+1} space. We call R^{n+1} the (t, x) space, and we denote elements of R^{n+1} by $(t, x_1, \ldots, x_n) = (t, x)$ and elements of R^n by $(x_1, \ldots, x_n) = x$. Next we consider n functions $f_i, i = 1, \ldots, n$, which map D into the real numbers R. To express this, we write $f_i: D \to R$. We assume that each f_i is continuous at all points in D and we express this by writing $f_i \in C(D)$. Finally, we let $x_i^{(n)}$ denote the nth derivative of x_i with respect to t (provided that it exists) (i.e., $d^n x_i / dt^n = x_i^{(n)}$). In particular, when $n = 1$, we frequently write

$$x_i^{(1)} \triangleq x_i' = dx_i/dt$$

We call the system of equations given by:

$$x_i' = f_i(t, x_1, \ldots, x_n), \qquad i = 1, \ldots, n \qquad \text{(E}_i\text{)}$$

a **system of n ordinary differential equations of the first order.** By a **solution** of the system of ordinary differential equations (E$_i$) we shall mean n continuously differentiable functions $\phi_1, \ldots, \phi_n$ defined on an interval $J = (a, b)$ (recall that (a, b) is the set of all t in R with the property $a < t < b$) such that $(t, \phi_1(t), \ldots, \phi_n(t)) \in D$ for all $t \in J$ and such that

$$\phi_i'(t) = f_i(t, \phi_1(t), \ldots, \phi_n(t))$$
$$i = 1, \ldots, n$$

for all $t \in J$.

Next, we let $(\tau, \xi_1, \ldots, \xi_n) \in D$. Then the **initial-value problem** associated with (E$_i$) is given by:

$$x_i' = f_i(t, x_1, \ldots, x_n), \qquad i = 1, \ldots, n$$
$$x_i(\tau) = \xi_i, \qquad i = 1, \ldots, n \qquad \text{(I}_i\text{)}$$

A set of functions $(\phi_1, \ldots, \phi_n)$ is a **solution** of (I$_i$) if $(\phi_1, \ldots, \phi_n)$ is a solution of (E$_i$) on some interval J containing τ and if $(\phi_1(\tau), \ldots, \phi_n(\tau)) = (\xi_1 \ldots, \xi_n)$.

In Fig. 1, the solution of a hypothetical initial-value problem is given when $n = 1$. Note that at (τ, ξ), $\phi'(\tau) = f(\tau, \phi(\tau)) = m$ is the slope of line L in the figure.

In dealing with systems of equations, we find it convenient to use vector notation, such as:

$$x = \begin{bmatrix} x_1 \\ \vdots \\ x_n \end{bmatrix}, \qquad \xi = \begin{bmatrix} \xi_1 \\ \vdots \\ \xi_n \end{bmatrix}, \qquad \phi = \begin{bmatrix} \phi_1 \\ \vdots \\ \phi_n \end{bmatrix}$$

$$f(t, x) = \begin{bmatrix} f_1(t, x_1, \ldots, x_n) \\ \vdots \\ f_n(t, x_1, \ldots, x_n) \end{bmatrix}$$

$$= \begin{bmatrix} f_1(t, x) \\ \vdots \\ f_n(t, x) \end{bmatrix}$$

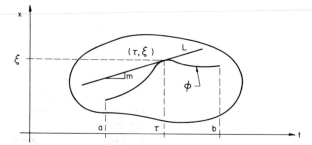

FIGURE 1 Solution of an initial-value problem; t interval $J = (a, b)$, m(slope of line L) = $f(\tau, \phi(\tau))$.

$$x' = \begin{bmatrix} x'_1 \\ \vdots \\ x'_n \end{bmatrix}, \qquad \begin{bmatrix} \int_\tau^t f_1(s, \phi(s))\, ds \\ \vdots \\ \int_\tau^t f_n(s, \phi(s))\, ds \end{bmatrix}$$

$$= \int_\tau^t f(s, \phi(s))\, ds$$

With this notation we can express the initial-value problem (I_i) by:

$$x' = f(t, x), \qquad x(\tau) = \xi \qquad \text{(I)}$$

It is an easy matter to verify that the initial-value problem (I) can equivalently be expressed by an integral equation of the form:

$$\phi(t) = \xi + \int_\tau^t f(s, \phi(s))\, ds \qquad \text{(V)}$$

where ϕ denotes the solution of (I).

B. Classification of Systems of First-Order Differential Equations

We are now ready to classify systems of first-order differential equations in a variety of ways:

1. If in (I), $f(t, x) \equiv f(x)$ for all (t, x) in D, then

$$x' = f(x) \qquad \text{(A)}$$

and we call (A) an **autonomous system** of first-order ordinary differential equations.

2. If $(t + T, x) \in D$ when $(t, x) \in D$ and if $f(t, x) = f(t + T, x)$ for all $(t, x) \in D$ then (I) assumes the form

$$x' = f(t, x) = f(t + T, x) \qquad \text{(P)}$$

Such a system is called a **periodic system** of first-order differential equations of period T. The smallest $T > 0$ for which (P) is true is the **least period** of this system of equations.

3. When in (I), $f(t, x) = A(t)x$, where $A(t) = [a_{ij}(t)]$ is a real $n \times n$ matrix with elements $a_{ij}(t)$ which are defined and at least piecewise continuous on a t interval J, then we have

$$x' = A(t)x \qquad \text{(LH)}$$

and we speak of a **linear homogeneous system** of ordinary differential equations.

4. If for (LH) $A(t)$ is defined for all real t and if there is a $T > 0$ such that $A(t) = A(t + T)$ for all t, then we have

$$x' = A(t)x = A(t + T)x \qquad \text{(LP)}$$

This system is called a **linear periodic system** of ordinary differential equations.

5. If in (I), $f(t, x) = A(t)x + g(t)$, where $g(t)^{\mathrm{T}} = [g_1(t), \dots, g_n(t)]$ and where $g_i : J \to R$, then we have

$$x' = A(t)x + g(t) \qquad \text{(LN)}$$

In this case we speak of a **linear nonhomogeneous system of ordinary differential equations.**

6. If in (I), $f(t, x) = Ax$, where $A = [a_{ij}]$ is a real $n \times n$ matrix with constant coefficients, then we have

$$x' = Ax \qquad \text{(L)}$$

This type of system is called a **linear, autonomous, homogeneous** system of ordinary differential equations.

C. nth-Order Ordinary Differential Equations

Thus far we have concerned ourselves with systems of first-order ordinary differential equations. It is also possible to characterize initial value problems by means of nth-order ordinary differential equations. To this end, we let h be a real function which is defined and continuous on a domain D of the real $(t, y_1, \dots, y_n)$ space. Then

$$y^{(n)} = h\left(t, y^{(1)}, \dots, y^{(n-1)}\right) \qquad (\text{E}_n)$$

is an nth-**order ordinary differential equation.** A **solution** of (E_n) is a real function ϕ which is defined on a t interval $J = (a, b)$ which has n continuous derivatives on J and satisfies $(t, \phi(t), \dots, \phi^{(n-1)}(t)) \in D$ for all $t \in J$ and

$$\phi^{(n)}(t) = h\left(t, \phi(t), \dots, \phi^{(n-1)}(t)\right)$$

for all $t \in J$.

Now for a given $(\tau, \xi_1, \dots, \xi_n) \in D$, the initial value problem for (E_n) *is*

$$y^{(n)} = h\left(t, y, y^{(1)}, \dots, y^{(n-1)}\right)$$
$$y^{(\tau)} = \xi_1, \dots, y^{(n-1)}(\tau) = \xi_n \qquad (\text{I}_n)$$

A function ϕ is a solution of (I_n) if ϕ is a solution of Eq. (E_n) on some interval containing τ and if $\phi(\tau) = \xi_1, \dots, \phi^{(n-1)}(\tau) = \xi_n$.

As in the case of systems of first-order equations, we single out several special cases.

1. Consider equations of the form

$$y^{(n)} + a_{n-1}(t)y^{(n-1)} + \cdots + a_1(t)y^{(1)} + a_0(t)y = g(t)$$
$$\text{(1)}$$

where $a_{n-1}(t), \dots, a_0(t)$ are real continuous functions defined on the interval J. We refer to Eq. (1) as a **linear homogeneous ordinary differential equation of order** n.

2. If in (1) we let $g(t) \equiv 0$, then

$$y^{(n)} + a_{n-1}(t)y^{(n-1)} + \cdots + a_1(t)y^{(1)} + a_0(t)y = 0 \quad \text{(2)}$$

We call Eq. (2) a **linear homogeneous ordinary differential equation of order** n.

3. If in (2) we have $a_i(t) \equiv a_i$, $i = 0, 1, \ldots, n-1$, then

$$y^{(n)} + a_{n-1}y^{(n-1)} + \cdots + a_1 y^{(1)} + a_0 y = 0 \qquad (3)$$

and we call Eq. (3) a **linear, autonomous, homogeneous ordinary differential equation of order** n.

We can, of course, also define **periodic** and **linear periodic ordinary differential equations of order** n in the obvious way.

It turns out that the theory of nth-order ordinary differential equations reduces to the theory of a system of n first-order ordinary differential equations. To this end, we let $y = x_1$, $y^{(1)} = x_2, \ldots, y^{(n-1)} = x_n$ in Eq. (I_n). Then we have the system of first-order ordinary differential equations:

$$
\begin{aligned}
x_1' &= x_2 \\
x_2' &= x_3 \\
&\;\;\vdots \\
x_n' &= h(t, x_1, \ldots, x_n)
\end{aligned}
\qquad (4)
$$

which is defined for all $(t, x_1, \ldots, x_n) \in D$. Assume that the vector $\phi = (\phi_1, \ldots, \phi_n)^{\mathrm{T}}$ is a solution of (4) on an interval J. Since $\phi_2 = \phi_1'$, $\phi_3 = \phi_2', \ldots, \phi_n = \phi_1^{(n-1)}$, and since

$$
\begin{aligned}
&h(t, \phi_1(t), \ldots, \phi_n(t)) \\
&\quad = h\big(t, \phi_1(t), \ldots, \phi_1^{(n-1)}(t)\big) = \phi_1^{(n)}(t)
\end{aligned}
$$

it follows that the first component ϕ_1 of the vector ϕ is a solution of Eq. (E_n) on the interval J. Conversely, if ϕ_1 is a solution of (E_n) on J, then the vector $(\phi, \phi^{(1)}, \ldots, \phi^{(n-1)})^{\mathrm{T}}$ is clearly a solution of Eq. (4). Moreover, if $\phi_1(\tau) = \xi_1, \ldots, \phi_1^{(n-1)}(\tau) = \xi_n$, then the vector ϕ satisfies $\phi(\tau) = \xi = (\xi_1, \ldots, \xi_n)^{\mathrm{T}}$. The converse is also true.

D. Examples of Initial-Value Problems

We conclude the present section with several representative examples.

1. Consider the second-order ordinary differential equation given by:

$$m\, d^2 x/dt^2 + g(x) = 0 \qquad (5)$$

where $m > 0$ is a constant and $g: R \to R$ is continuous. This equation, along with $x(0) = \xi_1$ and $x'(0) = \xi_2$ speci-

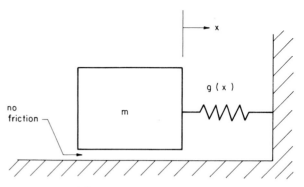

FIGURE 2 Mass–spring system.

fied, constitutes an initial-value problem. If we let $x_1 = x$ and $x_2 = x'$, then Eq. (5) can equivalently be represented by a system of two first-order ordinary differential equations given by:

$$x_1' = x_2, \qquad x_2' = -(1/m)g(x_1) \qquad (6)$$

with $x_1(0) = \xi_1$ and $x_2(0) = \xi_2$.

Equation (5) can be used to describe a variety of physical phenomena. Consider, for example, the mass–spring system depicted in Fig. 2. When the system is in the rest position, then $x = 0$; otherwise, x is positive in the direction of the arrow or negative otherwise. The function $g(x)$ denotes the restoring force of the spring while the mass is expressed by m (in a consistent set of units).

There are several well-known special cases for Eq. (5):

(a) If $g(x) = kx$, where $k > 0$ is known as Hooke's constant, then Eq. (5) is a linear ordinary differential equation called the *harmonic oscillator*.

(b) If $g(x) = k(1 + a^2 x^2)x$, where $k > 0$ and $a^2 > 0$ are parameters, then Eq. (5) is a nonlinear ordinary differential equation and one refers to the resulting system as a "mass and a hard spring."

(c) If $g(x) = k(1 - a^2 x^2)x$, where $k > 0$ and $a^2 > 0$ are parameters, then Eq. (5) is a nonlinear ordinary differential equation and one refers to the resulting system as a "mass and a soft spring."

Alternatively, Eq. (5) can be used to describe the behavior of the pendulum shown in Fig. 3 with $\theta = x$. In this case, the restoring force $g(x)$ is specified by:

$$g(x) = m(g/l)\sin x$$

2. Using Kirchhoff's voltage and current laws, the circuit of Fig. 4 can be modeled by the linear system of equations:

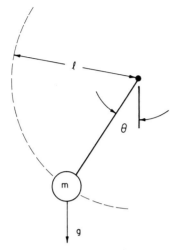

FIGURE 3 Simple pendulum.

$$v_1' = -\frac{1}{C_1}\left(\frac{1}{R_1}+\frac{1}{R_2}\right)v_1 + \frac{1}{R_2C_2}v_2 + \frac{v}{R_1C_1}$$

$$v_2' = -\frac{1}{C_1}\left(\frac{1}{R_1}+\frac{1}{R_2}\right)v_1 - \left(\frac{R_2}{L}-\frac{1}{R_2C_1}\right)v_2$$

$$+\frac{R_2}{L}v_3 + \frac{v}{R_1C_1}$$ (7)

$$v_3' = \frac{1}{R_2C_2}v_1 - \frac{1}{R_2C_2}v_2$$

In order to complete the description of this circuit, we need to specify the initial conditions $v_1(0)$, $v_2(0)$, and $v_3(0)$.

III. FUNDAMENTAL THEORY

In this section, we address the following questions:

1. Under what conditions has the initial-value problem (I) *at least one solution* for a given set of initial data (τ, ξ)?
2. Under what conditions has (I) *exactly one solution* for given (τ, ξ)?

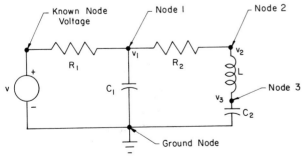

FIGURE 4 An example of an electric circuit.

3. What is the extent of the time interval over which one or more solutions exist for (I)?
4. How do solutions for (I) behave when the initial data (τ, ξ) (or some other parameters for the differential equation) are varied?

The significance of the preceding questions is brought further to light when the following examples are considered.

1. For the initial-value problem,

$$x' = -\operatorname{sgn} x, \qquad x(0) = 0, \quad t \geq 0 \qquad (8)$$

where

$$\operatorname{sgn} x = \begin{cases} 1, & x > 0 \\ 0, & x = 0 \\ -1, & x < 0 \end{cases}$$

no continuously differentiable function ϕ exists which satisfies Eq. (8). Hence, *no solution* (as defined in Section II) exists for the present initial-value problem.

2. The initial-value problem,

$$x' = 1/(2x), \qquad x(0) = 0, \quad t \geq 0 \qquad (9)$$

has *two solutions* given by $\phi(t) = \pm t^{1/2}$ which exist for all $t \geq 0$.

3. The initial-value problem,

$$x' = 1 + x^2, \qquad x(0) = 0, \quad t \geq 0 \qquad (10)$$

has the *unique solution* given by $\phi(t) = \tan t$. This solution exists only when $0 \leq t < \pi/2$, since ϕ is not continuously differentiable at $t = \pi/2$. In this case, we say that this solution has *finite escape time*.

4. The initial-value problem given by:

$$x' = ax, \qquad x(\tau) = \xi \qquad (11)$$

where a is a fixed parameter, has a unique solution given by $\phi(t) = \phi(t, \tau, \xi) = \xi e^{a(t-\tau)}$ which exists for all real t. Note that the solution ϕ is continuous with respect to the parameters, a, τ, and ξ.

A. Existence of Solutions

In order to simplify our presentation, we will consider in the next few results one-dimensional initial value problems (i.e., we will assume that for (I), $n = 1$). Later in this section we will show how these results are modified for higher dimensional systems. Thus, we have a domain $D \in R^2$, we are given $(\tau, \xi) \in D$ and $f \in C(D)$, and *we seek a solution for the initial-value problem*,

$$x' = f(t, k), \qquad x(\tau) = \xi \qquad (I')$$

In doing so, it suffices to find a solution of the integral equation:

$$\phi(t) = \xi + \int_\tau^t f(s, \phi(s)) \, ds \qquad (V')$$

One way of solving the above problem is by considering approximations to a solution first; an ε-**approximate solution** for (I') on an interval J containing τ is a real-valued function ϕ which is piecewise continuously differentiable on J and satisfies $\phi(\tau) = \xi$, $(t, \phi(t)) \in D$ for all $t \in J$, and which satisfies:

$$|\phi'(t) - f(t, \phi(t))| < \varepsilon$$

at all points t of J where $\phi'(t)$ exists.

Let us now consider a subset S of D defined by:

$$S = \{(t, x) \in D : |t - \tau| \le a, |x - \xi| \le b\} \subset D$$

Since $f \in C(D)$, there is an $M \ge 0$ such that $|f(t, x)| \le M$ for all $(t, x) \in S$. Now define $c \overset{\Delta}{=} \min\{a, b/M\}$ and, depending on the size of M relative to a, define one of the triangular regions shown either in Fig. 5a or 5b.

It is now not too difficult to prove the following results:

If $f \in C(D)$ and if c is as defined above, then for any $\varepsilon > 0$ there is an ε-approximate solution of (I') on the interval $|t - \tau| \le c$.

Indeed, for a given $\varepsilon > 0$, such a solution will be of the form:

$$\phi(t) = \phi(t_j) + f(t_j, \phi(t_j))(t - t_j)$$
$$t_j \le t \le t_{j+1}, \qquad j = 0, 1, 2, \ldots, m-1 \qquad (12)$$

when $t \ge \tau = t_0$ (the case in which $t \le \tau$ is modified in the obvious way). In Eq. (12), the choice of m and of $\max|t_{j+1} - t_j|$ will depend on ε but not on t, but in any

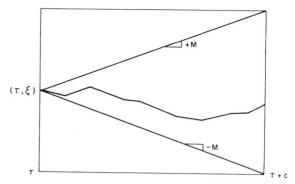

FIGURE 6 Typical ε-approximation solution.

case, we have $\sum_{j=0}^{m-1} |t_{j+1} - t_j| = c$. In Fig. 6, a typical ε-approximate solution is shown.

Next, let us consider a monotone decreasing sequence of real numbers with limit zero, and let us denote this sequence by $\{\varepsilon_m\}$. An example of such a sequence would be the case when $\varepsilon_m = 1/m$, where m denotes all positive integers greater or equal to one. Corresponding to each ε_m, let us consider now an ε_m-approximate solution which we denote by ϕ_m. Next, let us consider the family of ε_m-approximate solutions $\{\phi_m\}$, $m = 1, 2, \ldots$. This family $\{\phi_m\}$ is an example of an *equicontinuous* family of functions. Now, according to *Ascoli's lemma*, an equicontinuous family $\{\phi_m\}$, as constructed above, will contain a subsequence of functions, which we denote by $\{\phi_{m_k}\}$, which converges uniformly on the interval $[\tau - c, \tau + c]$ to a continuous function ϕ; that is,

$$\lim_{k \to \infty} \phi_{m_k}(t) = \phi(t), \qquad \text{uniformly in } t \qquad (13)$$

Now it turns out that ϕ is continuously differentiable on the interval $(\tau - c, \tau + c)$ and that it satisfies the integral equation (V') and, hence, the initial-value problem (I'). In other words, ϕ is a solution of (I').

The preceding discussion gives rise to the **Cauchy–Peano existence theorem:**

If $f \in C(D)$ and $(\tau, \xi) \in D$, then the initial-value problem (I') has a solution defined on $|t - \tau| \le c$ where c is as defined in Fig. 5.

We mention in passing that a special case of Eq. (12),

$$\phi(t_{j+1}) = \phi(t_j) + f(t_j, \phi(t_j))(t_{j+1} - t_j) \qquad (14)$$

is known as **Euler's method** of solving ordinary differential equations.

It should be noted that the above result yields only a *sufficient condition*. In other words, when f is not continuous on the domain D, then the initial-value problem (I') may or may not possess a solution in the sense defined in Section II.

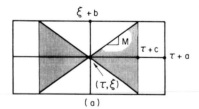

(a)

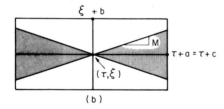

(b)

FIGURE 5 (a) Case $c = b/M$. (b) Case $c = a$.

The above result asserts the existence of a solution (I′) "locally," that is, only on a sufficiently short time interval (determined by $|t - \tau| \le c$). In general, this assertion cannot be changed to existence of a solution for all $t \ge \tau$ (or for all $t \le \tau$) as the following example shows:

The initial-value problem,

$$x' = x^2, \qquad x(\tau) = \xi$$

has a solution $\phi(t) = \xi[1 - \xi(t - \tau)]^{-1}$ which exists forward in time for $\xi > 0$ only until $t = \tau + \xi^{-1}$.

B. Continuation of Solutions

Our next task is to determine if it is possible to extend a solution ϕ to a larger interval than was indicated above ($|t - \tau| \le c$). The answer to this is affirmative. To see this, suppose that $f \in C(D)$ and suppose also that f is bounded on D. Suppose also that by some procedure, as above, it was possible to show that ϕ is a solution of the scalar differential equation,

$$x' = f(t, x) \qquad (E')$$

on an interval $J = (a, b)$. Using expression (V′) for ϕ it is an easy matter to show that the limit of $\phi(t)$ as t approaches a from the right exists and that the limit of $\phi(t)$ as t approaches b from the left exists; that is,

$$\lim_{t \to a^+} \phi(t) = \phi(a^+)$$

and

$$\lim_{t \to b^-} \phi(t) = \phi(b^-)$$

Now clearly, if the point $(a, \phi(a^+)) \in D$ (resp., if $(b, \phi(b^-)) \in D$), then by repeating the procedure given in the above results (ε-approximate solution result and Peano–Cauchy theorem), the solution ϕ can be continued to the left past the point $t = a$ (resp., to the right past the point $t = b$). Indeed, it should be clear that repeated applications of these procedures will make it possible to continue the solution ϕ to the boundary of D. This is depicted in Fig. 7. It is worthwhile to note that the solution ϕ in this figure exists over the interval J' and not over the interval $\tilde{J}$.

We summarize the preceding discussion in the following **continuation result**:

If $f \in C(D)$ and if f is bounded on D, then all solutions of (E′) can be extended to the boundary of D. These solutions are then **noncontinuable.**

C. Uniqueness of Solutions

Next, we address the question of uniqueness of solutions. To accomplish this, we require the following concept:

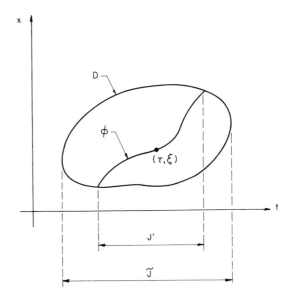

FIGURE 7 Continuation of a solution ϕ to ∂D.

$f \in C(D)$ is said to satisfy a **Lipschitz condition** on D (with respect to x) with **Lipschitz constant** L if

$$|f(t, \bar{\bar{x}}) - f(t, \bar{x})| \le L|\bar{\bar{x}} - \bar{x}| \qquad (15)$$

for all $(t, \bar{x})(t, \bar{\bar{x}})$ in D. The function f is said to be **Lipschitz continuous** in x on D in this case.

For example, it can be shown that if $\partial f(t, x)/\partial x$ exists and is continuous on D, then f will be Lipschitz continuous on any compact and convex subsect D_0 of D.

In order to establish a uniqueness result for solutions of the initial value problem (I′), we will also require a result known as the **Gronwall inequality**: Let r and k be continuous nonnegative real functions defined on an interval $[a, b]$ and let $\delta \ge 0$ be a constant. If

$$r(t) \le \delta + \int_a^t k(s)r(s)\, ds \qquad (16)$$

then

$$r(t) \le \delta \exp\left(\int_a^t k(s)\, ds \right) \qquad (17)$$

Now suppose that for (I′) the Cauchy–Peano theorem holds and suppose that for *one given* $(\tau, \xi) \in D$, *two* solutions ϕ_1 and ϕ_2 exist over some interval $|t - \tau| \le d, d > 0$. On the interval $\tau \le t \le \tau + d$ we now have, using (V′) to express ϕ_1 and ϕ_2,

$$\phi_1(t) - \phi_2(t) = \int_\tau^t [f(s, \phi_1(s)) - f(s, \phi_2(s))]\, ds \quad (18)$$

Now if, in addition, f is Lipschitz continuous in x, then Eq. (18) yields:

$$|\phi_1(t) - \phi_2(t)| \le \int_\tau^t L|\phi_1(s) - \phi_2(s)|\, ds$$

Letting $r(t) = |\phi_1(t) - \phi_2(t)|$, $\delta = 0$, and $k(t) \equiv L$, and applying the Gronwall inequality, we now obtain:

$$|\phi_1(t) - \phi_2(t)| \leq 0 \quad \text{for all} \quad \tau \leq t \leq \tau + d$$

Hence, it must be true that $\phi_1(t) = \phi_2(t)$ on $\tau \leq t \leq \tau + d$. A similar argument will also work for the interval $\tau - d \leq t \leq \tau$.

Summarizing, we have the following **uniqueness result**:

If $f \in C(D)$ and if f satisfies a Lipschitz condition on D with Lipschitz constant L, then the initial-value problem (I$'$) has at most one solution on any interval $|t - \tau| \leq d$, $d > 0$.

If the solution ϕ of (I$'$) is unique, then the ε-approximate solutions constructed before will tend to ϕ as $\varepsilon \to 0^+$ and this is the basis for justifying Euler's method—a numerical method of constructing approximations to ϕ. Now, if we assume that f satisfies a Lipschitz condition, an alternative classical method of approximation is the **method of successive approximations**. Specifically, let $f \in C(D)$ and let S be the rectangle in D centered at (τ, ξ) shown in Fig. 5 and let c be defined as in Fig. 5. Successive approximations for (I$'$), or equivalently for (V$'$), are defined as:

$$\phi_0(t) = \xi$$

$$\phi_{m+1}(t) = \xi + \int_\tau^t f(s, \phi_m(s)) \, ds, \quad (19)$$

$$m = 0, 1, 2, \ldots$$

for $|t - \tau| \leq c$.

The following result is the basis for justifying the method of successive approximations:

If $f \in C(D)$ and if f is Lipschitz continuous on S with constant L, then the successive approximations ϕ_m, $m = 0, 1, 2, \ldots$, given in Eq. (19) exist on $|t - \tau| \leq c$, are continuous there, and converge uniformly, as $m \to \infty$, to the unique solution of (I$'$).

D. Continuity of Solutions with Respect to Parameters

Our next objective is to study the dependence of solutions ϕ of (I$'$) on initial data (τ, ξ). In this connection, we find it advantageous to highlight this dependence by writing $\phi(t) = \phi(t, \tau, \xi)$.

Now suppose that $f \in C(D)$ and suppose that f satisfies a Lipschitz condition on D with Lipschitz constant L. Furthermore, suppose that ϕ and ψ solve:

$$x' = f(t, x) \qquad \text{(E}')$$

on an interval $|t - \tau| \leq d$ with $\psi(\tau) = \xi_0$ and $\phi(\tau) = \xi$. Then, by using (V$'$), we obtain:

$$|\phi(t, \tau, \xi) - \psi(t, \tau, \xi_0)| \leq |\xi - \xi_0|$$

$$+ \int_\tau^t L|\phi(s, \tau, \xi) - \psi(s, \tau, \xi_0)| \, ds$$

and by using the Gronwall inequality with $\delta = |\xi - \xi_0|$, $L \equiv k(s)$, and $r(t) = |\phi(t) - \psi(t)|$, we obtain the estimate:

$$|\phi(t, \tau, \xi) - \psi(t, \tau, \xi_0)| \leq |\xi - \xi_0| \exp(L|t - \tau|)$$

$$t \in |t - \tau| \leq d \quad (20)$$

If, in particular, we consider a sequence of initial conditions $\{\xi_m\}$ having the property that $\xi_m \to \xi_0$ as $m \to \infty$, then it follows from Eq. (20) that $\phi(t, \tau, \xi_m) \to \phi(t, \tau, \xi_0)$, uniformly in t on $|t - \tau| \leq d$.

Summarizing, we have the following **continuous dependence result**:

Let $f \in C(D)$ and assume that f satisfies a Lipschitz condition on D. Then, the unique solution $\phi(t, \tau, \xi)$ of (I$'$), existing on some bounded interval containing τ, depends continuously on ξ, uniformly in t.

This means that if $\xi_m \to \xi_0$ then $\phi(t, \tau, \xi_m) \to \phi(t, \tau, \xi_0)$, uniformly in t on $|t - \tau| \leq d$ for some $d > 0$.

In a similar manner we can show that $\phi(t, \tau, \xi)$ will depend continuously on the initial time τ. Furthermore, if the differential equation (E$'$) depends on a parameter, say μ, then the solutions of the corresponding initial value problem may also depend in a continuous manner on μ, provided that certain safeguards are present. We consider a specific case in the following.

Consider the initial-value problem,

$$x' = f(t, x, \mu), \qquad x(\tau) = \xi_\mu = \mu + 1 \qquad (\text{I}'_\mu)$$

where μ is a scalar parameter. Let f satisfy Lipschitz conditions with respect to x and μ for $(t, x) \in D$ and for $|\mu - \mu_0| < c$. Using an argument similar to the one employed in connection with Eq. (20), we can show that the solution $\phi(t, \tau, \xi_\mu, \mu)$ of (I$'_\mu$), where ξ_μ depends continuously on μ, is a continuous function of μ (i.e., as $\mu \to \mu_0$, $\xi_\mu \to \xi_{\mu 0}$, and $\phi(t, \tau, \xi_\mu, \mu) \to \phi(t, \tau, \xi_{\mu 0}, \mu_0)$).

As an example, consider the initial-value problem

$$x' = x + \mu t, \qquad x(\tau) = \xi_\mu = \mu + 1 \qquad (21)$$

The right-hand side of Eq. (21) has a Lipschitz constant with respect to x equal to one and with respect to μ equal to $|a - b|$, where $J = (a, b)$ is assumed to be a bounded t-interval. The solution of (I$'_\mu$) is

$$\phi(t, \tau, \xi_\mu, \mu) = [\mu(\tau + 2) + 1]e^{(t-\tau)} - \mu(t + 1) \quad (22)$$

At $t = \tau$, we have $\phi(\tau, \tau, \xi_\mu, \mu) = \mu + 1 = x(\tau)$. Now what happens when $\mu \to 0$? In this case, Eq. (21) becomes:

$$x' = x, \qquad x(\tau) = 1 \qquad (23)$$

while the solution of Eq. (23) is

$$\phi(t) = e^{t-\tau} \tag{24}$$

If we let $\mu \to 0$ in Eq. (22), then we also obtain:

$$\phi(t, \tau, \xi_0, 0) = e^{t-\tau} \tag{25}$$

as expected.

E. Systems of Equations

In the interests of simplicity, we have considered thus far in this section the one-dimensional initial-value problem (I'). It turns out that the preceding results can be restated and proved for initial-value problems (I) involving systems of equations (E). In doing so, one must replace absolute values of scalars by norms of vectors or matrices, convergence of scalars by convergence of vectors and matrices, and so forth.

Rather than go through the task of restating the preceding results for systems of equations, we present as an illustration an additional result. Specifically, consider the linear system of equations

$$x' = A(t)x + g(t) = f(t, x) \tag{LN}$$

where $A(t) = [a_{ij}(t)]$ is an $n \times n$ matrix and $g(t)$ is an n-vector. We assume that $a_{ij}(t)$, $i, j = 1, \ldots, n$, and $g_i(t)$, $i = 1, \ldots, n$, are real and continuous functions defined on an interval J.

By making use of the taxicab norm given by

$$|x| = \sum_{i=1}^{n} |x_i|$$

it is an easy matter to show that for t on any compact subinterval J_0 of J there exists $L_0 \geq 0$ such that

$$|f(t, \bar{x}) - f(t, \bar{\bar{x}})| \leq \left(\sum_{i=1}^{n} \max_{1 \leq j \leq n} |a_{ij}(t)| \right) |\bar{x} - \bar{\bar{x}}|$$
$$\leq L_0 |\bar{x} - \bar{\bar{x}}| \tag{26}$$

If we now invoke the results of the present section (rephrased for systems of equations), we obtain the following:

Suppose that A(t) and g(t) in (LN) are defined and continuous on an interval J. Then for any τ in J and any ξ in R^n, Eq. (LN) has a unique solution satisfying $x(\tau) = \xi$. This solution exists on the entire interval J and is continuous in (t, τ, ξ). If A and g depend continuously on parameters $\lambda \in R^l$, then the solution will also vary continuously with λ.

It is emphasized that the above result is a *global* result, since the solution ϕ exists over the *entire* interval J. On the other hand, as noted before, our earlier results in this section will in general be of a *local* nature.

F. Differentiability of Solutions with Respect to Parameters

In some of the preceding results we investigated the continuity of solutions with respect to parameters. Next, we address the question of differentiability of solutions with respect to parameters for the initial-value problem,

$$x' = f(t, x), \qquad x(\tau) = \xi \tag{I}$$

Again we assume that $f \in C(D)$, $(\tau, \xi) \in D \subset R^{n+1}$, where D is a domain. In addition, we assume that $f(t, x)$ is differentiable with respect to $x_1, \ldots, x_n$ and we form the Jacobian matrix $f_x(t, x)$ given by:

$$f_x(t, x) = \frac{\partial f}{\partial x}(t, x)$$

$$= \begin{bmatrix} \frac{\partial f_1}{\partial x_1}(t, x) & \frac{\partial f_1}{\partial x_2}(t, x) & \cdots & \frac{\partial f_1}{\partial x_n}(t, x) \\ \vdots & \vdots & & \vdots \\ \frac{\partial f_n}{\partial x_1}(t, x) & \frac{\partial f_n}{\partial x_2}(t, x) & \cdots & \frac{\partial f_n}{\partial x_n}(t, x) \end{bmatrix} \tag{27}$$

Under the above conditions, it can be shown that when f_x exists and is continuous, then the solution ϕ of (I) depends smoothly on the parameters of the problem. More specifically,

Let $f \in C(D)$, let f_x exist, and let $f_x \in C(D)$. If $\phi(t, \tau, \xi)$ is the solution of (E) such that $\phi(\tau, \tau, \xi) = \xi$, then ϕ is continuously differentiable in (t, τ, ξ). Each vector-valued function $\partial \phi / \partial \xi_i$ or $\partial \phi / \partial \tau$ will solve:

$$y' = f_x(t, \phi(t, \tau, \xi))y$$

as a function of t while

$$\frac{\partial \phi}{\partial \tau}(\tau, \tau, \xi) = -f(\tau, \xi)$$

and

$$\frac{\partial \phi}{\partial \xi}(\tau, \tau, \xi) = E_n$$

where E_n denotes the $n \times n$ identity matrix.

G. Comparison Theory

We conclude this section by touching on the **comparison theory** for differential equations. This theory is useful in continuation of solutions, in establishing estimates on bounds of solutions, and, as we will see later, in stability theory. Before we can present some of the main results of this theory we need to introduce a few concepts.

Once again we consider the scalar initial-value problem,

$$x' = f(t, x), \qquad x(\tau) = \xi \qquad (\text{I}')$$

where $f \in C(D)$ and $(\tau, \xi) \in D$. For (I') we define the **maximal solution** ϕ_{M} as that noncontinuable solution of (I') having the property that if ϕ is any other solution of (I'), then $\phi_{\text{M}}(t) \geq \phi(t)$ as long as both solutions are defined. The **minimal solution** ϕ_{m} of (I') is defined in a similar manner. It is not too difficult to prove that ϕ_{M} and ϕ_{m} actually exist.

In what follows, we also need the concept of **upper right Dini derivative** $D^+ x$. Given $x : (\alpha, \beta) \to R$ and $x \in C(\alpha, \beta)$ (i.e., x is a continuous real-valued function defined on the interval (α, β)), we define:

$$D^+ x(t) = \lim_{h \to 0^+} \sup[x(t + h) - x(t)]/h$$

$$\overset{\Delta}{=} \overline{\lim}[x(t + h) - x(t)]/h$$

where lim sup denotes the limit supremum. The **lower right Dini derivative** $D^- x$ is defined similarly, replacing the lim sup by the lim inf.

We will also require the concept of differential inequalities, such as, for example, the inequality,

$$D^+ x(t) \leq f(t, x(t)) \qquad \text{on} \quad D \qquad (28)$$

Any function ϕ satisfying Eq. (28) is called a *solution* for (28). Differential inequalities involving $D^- x$ are defined similarly.

Our first comparison result is now as follows:

Suppose that the maximal solution ϕ_{M} of (I') stays in D for all $t \in [\tau, T]$. If a continuous function $\psi(t)$ with $\psi(\tau) = \xi$ satisfies

$$\psi'(t) \overset{\Delta}{=} D^+ \psi(t) \leq f(t, \psi(t)) \qquad \text{on} \quad D$$

then it is true that

$$\psi(t) \leq \phi_{\text{M}}(t) \qquad \text{for all} \quad t \in [\tau, T]$$

A similar result involving minimal solutions can also be established.

The above result can now be applied to systems of equations to obtain estimates for the norms of solutions. We have the following:

Let $f \in C(D)$, $D \subset R^{n+1}$, and let ϕ be a solution of

$$x' = f(t, x), \qquad x(\tau) = \xi \qquad (\text{I})$$

Let $F(t, v)$ be a scalar-valued continuous function such that

$$|f(t, x)| \leq F(t, |x|) \qquad \text{for all} \quad (t, x) \in D$$

where $|f(t, x)|$ denotes any one of the equivalent norms of $f(t, x)$ on R^n. If $\eta \leq |\phi(\tau)|$ and if v_{M} denotes the maximal solution of the scalar comparison equation given by:

$$v' = F(t, v), \qquad u(\tau) = \eta \qquad (29)$$

then

$$|\phi(t)| \leq v_{\text{M}}(t)$$

for as long as both functions exist. (Here, $|\phi(t)|$ denotes the norm of $\phi(t)$.)

As an application of this result, suppose that $f(t, x)$ in (E) is such that

$$|f(t, x)| \leq A|x| + B$$

for all $t \in J$ and for all $x \in R^n$, where $J = [t_0, T]$, and where $A > 0$, $B > 0$ are parameters. Then Eq. (29) assumes the form:

$$v' = Au + B, \qquad v(\tau) = \eta$$

According to the above result, we now have

$$v_{\text{M}}(t) = e^{A(t - \tau)}(\eta - B/A) + B/A$$

Since $v_{\text{M}}(t)$ exists for all $t \in J$, then so do the solutions of (E). Also, if ϕ is any solution of (E) with $|\phi(\tau)| \leq \eta$, then the estimate,

$$|\phi(t)| \leq e^{A(t - \tau)}(\eta - (\beta/A)) + (\beta/A)$$

is true.

IV. LINEAR SYSTEMS

Both in the theory of differential equations and in their applications, linear systems of ordinary differential equations are extremely important. In this section we first present the general properties of linear systems. We then turn our attention to the special cases of linear systems of ordinary differential equations with constant coefficients and linear systems of ordinary differential equations with periodic coefficients. We also address some of the properties of nth-order linear ordinary differential equations.

A. Linear Homogeneous and Nonhomogeneous Systems

We first consider linear homogeneous systems,

$$x' = A(t)x \qquad (\text{LH})$$

As noted in Section III, this system possesses unique solutions for every $(\tau, \xi) \in D$ where $x(\tau) = \xi$,

$$D = \{(t, x) : t \in J = (a, b), \ x \in R^n (\text{or } x \in C^n)\}$$

when each element $a_{ij}(t)$ of matrix $A(t)$ is continuous over J. These solutions exist over the entire interval $J = (a, b)$

and they depend continuously on the initial conditions. In applications it is typical that $j = (-\infty, \infty)$. We note that $\phi(t) \equiv 0$, for all $t \in J$, is a solution of (LH), with $\phi(\tau) = 0$. This is called the **trivial solution** of (LH).

In this section we consider matrices and vectors which will be either real or complex valued. In the former case, the field of scalars for the x space is the field of real numbers ($F = R$) and in the latter case, the field for the x space is the field of complex numbers ($F = C$).

Now let V denote the set of all solutions of (LH) on J; let α_1, α_2 be scalars (i.e., $\alpha_1, \alpha_2 \in F$); and let ϕ_1, ϕ_2 be solutions of (LH) (i.e., $\phi_1, \phi_2 \in V$). Then it is easily verified that $\alpha_1 \phi_1 + \alpha_2 \phi_2$ will also be a solution of (LH) (i.e., $\alpha_1 \phi_1 + \alpha_2 \phi_2 \in V$). We have thus shown that V is a vector space. Now if we choose n linearly independent vectors $\xi_1, \ldots, \xi_n$ in the n-dimensional x-space, then there exist n solutions $\phi_1, \ldots, \phi_n$ of (LH) such that $\phi_1(\tau) = \xi_1, \ldots, \phi_n(\tau) = \xi_n$. It is an easy matter to verify that this set of solutions $\{\phi_1, \ldots, \phi_n\}$ is linearly independent and that it spans V. Thus, $\{\phi_1, \ldots, \phi_n\}$ is a basis of V and any solution ϕ can be expressed as a linear combination of the vectors $\phi_1, \ldots, \phi_n$.

Summarizing, we have

The set of solutions of (LH) *on the interval J forms an n-dimensional vector space.*

In view of the above result it now makes sense to define a **fundamental set of solutions** for (LH) as a set of n linearly independent solutions of (LH) on J. If $\{\phi_1, \ldots, \phi_n\}$ is such a set, then we can form the matrix:

$$\Phi = [\phi_1, \ldots, \phi_n] \qquad (30)$$

which is called a **fundamental matrix** of (LH).

In the following, we enumerate some of the important properties of fundamental matrices. All of these properties are direct consequences of definition (30) and of the properties of solutions of (LH). We have

1. *A fundamental matrix Φ of* (LH) *satisfies the matrix equation:*

$$X' = A(t)X \qquad (31)$$

where $X = [x_{ij}]$ denotes an $n \times n$ matrix. (Observe that Eq. (31) consists of a system of n^2 first-order ordinary differential equations.)

2. *If Φ is a solution of the matrix equation* (31) *on an interval J and if τ is any point of J, then*

$$\det \Phi(t) = \det \Phi(\tau) \exp\left(\int_\tau^t \text{tr } A(s) \, ds \right)$$

for every $t \in J$

(Here, $\det \Phi$ denotes the determinant of Φ and $\text{tr } A$ denotes the trace of the matrix A.) This result is known as **Abel's formula**.

3. *A solution Φ of matrix equation* (31) *is a fundamental matrix of* (LH) *if and only if its determinant is nonzero for all $t \in J$. (This result is a direct consequence of Abel's formula.)*

4. *If Φ is a fundamental matrix of* (LH) *and if C is any nonsingular constant $n \times n$ matrix, then ΦC is also a fundamental matrix of* (LH). *Moreover, if Ψ is any other fundamental matrix of* (LH), *then there exists a constant $n \times n$ nonsingular matrix P such that $\Psi = \Phi P$.*

In the following, we let $\{e_1, e_2, \ldots, e_n\}$ denote the set of vectors $e_1^T = (1, 0, \ldots, 0)$, $e_2^T = (0, 1, 0, \ldots, 0), \ldots, e_n^T = (0, \ldots, 0, 1)$. We call a fundamental matrix Φ of (LH) whose columns are determined by the linearly independent solutions $\phi_1, \ldots, \phi_n$ with

$$\phi_1(\tau) = e_1, \ldots, \phi_n(\tau) e_n, \qquad \tau \in J$$

the **state transition matrix** Φ for (LH). Equivalently, if Ψ is *any* fundamental matrix of (LH), then the matrix Φ determined by:

$$\Phi(t, \tau) \triangleq \Psi(t)\Psi^{-1}(\tau) \qquad \text{for all} \quad t, \tau \in J$$

is said to be the **state transition matrix** of (LH).

We now enumerate several properties of state transition matrices. All of these are direct consequences of the definition of state transition matrix and of the properties of fundamental matrices. In the following we let $\tau \in J$, we let $\phi(\tau) = \xi$, and we let $\Phi(t, \tau)$ denote the state transition matrix for (LH) for all $t \in J$. Then,

1. *$\Phi(t, \tau)$ is the unique solution of the matrix equation:*

$$\frac{\partial}{\partial t} \Phi(t, \tau) \triangleq \Phi'(t, \tau) = A(t)\Phi(t, \tau)$$

with $\Phi(\tau, \tau) = E$, the $n \times n$ identity matrix.

2. *$\phi(t, \tau)$ is nonsingular for all $t \in J$.*

3. *For any $t, \sigma, \tau \in J$, we have*

$$\Phi(t, \tau) = \Phi(t, \sigma)\Phi(\sigma, \tau)$$

4. *$[\Phi(t, \tau)]^{-1} \triangleq \Phi^{-1}(t, \tau) = \Phi(\tau, t)$ for all $t, \tau \in J$.*

5. *The unique solution $\phi(t, \tau, \xi)$ of* (LH) *with $\phi(\tau, \tau, \xi) = \xi$ specified, is given by:*

$$\phi(t, \tau, \xi) = \Phi(t, \tau)\xi \qquad \text{for all} \quad t \in J \qquad (32)$$

In engineering and physics applications, $\phi(t)$ is interpreted as representing the "state" of a (dynamical) system represented by (LH) at time t and $\phi(\tau) = \xi$ is interpreted as representing the "state" at time τ. In Eq. (32), $\Phi(t, \tau)$

relates the "states" of (LH) at t and τ. This motivated the name "state transition matrix."

Let us now consider a couple of specific examples.

1. For the system of equations

$$x_1' = 5x_1 - 2x_2$$
$$x_2' = 4x_1 - x_2 \tag{33}$$

we have

$$A(t) \equiv A = \begin{bmatrix} 5 & -2 \\ 4 & -1 \end{bmatrix} \qquad \text{for all} \quad t \in (-\infty, \infty)$$

Two linearly independent solutions for Eq. (33) are

$$\phi_1(t) = \begin{bmatrix} e^{3t} \\ e^{3t} \end{bmatrix}, \qquad \phi_2(t) = \begin{bmatrix} e^t \\ 2e^t \end{bmatrix}$$

The matrix

$$\Phi(t) = \begin{bmatrix} e^{3t} & e^t \\ e^{3t} & 2e^t \end{bmatrix}$$

satisfies the equation $\Phi' = A\Phi$ and

$$\det \Phi(t) = e^{4t} \neq 0 \qquad \text{for all} \quad t \in (-\infty, \infty)$$

Thus, Φ is a fundamental matrix for Eq. (33). Also, in view of Abel's formula, we obtain:

$$\det \Phi(t) = \det \Phi(\tau) \exp\left[\int_\tau^t \operatorname{tr} A(s)\, ds\right]$$

$$= e^{4\tau} \exp\left[\int_\tau^t 4\, ds\right] = e^{4t}$$

$$\text{for all } t \in (-\infty, \infty)$$

as expected. Finally, since

$$\Phi^{-1}(t) = \begin{bmatrix} 2e^{-3t} & -e^{-3t} \\ -e^{-t} & -e^{-t} \end{bmatrix}$$

we obtain for the transition matrix of Eq. (33),

$$\Psi(t)\Psi^{-1}(\tau) = \begin{bmatrix} 2e^{3(t-\tau)} - e^{t-\tau} & -e^{3(t-\tau)} + e^{t-\tau} \\ 2e^{3(t-\tau)} - 2e^{t-\tau} & -e^{3(t-\tau)} + 2e^{t-\tau} \end{bmatrix}$$

2. For the system,

$$x_1' = x_2, \qquad x_2' = tx_2 \tag{34}$$

we have

$$A(t) \begin{bmatrix} 0 & 1 \\ 0 & t \end{bmatrix} \qquad \text{for all} \quad t \in (-\infty, \infty)$$

Two linearly independent solutions of Eq. (34) are

$$\phi_1(t) = \begin{bmatrix} 0 \\ 1 \end{bmatrix}, \qquad \phi_2(t) = \begin{bmatrix} \int_\tau^t e^{\eta^2/2}\, d\eta \\ e^{t^2/2} \end{bmatrix}$$

The matrix

$$\Phi(t) = \begin{bmatrix} 1 & \int_\tau^t e^{\eta^2/2} \\ 0 & e^{t^2/2} \end{bmatrix}$$

satisfies the matrix equation $\Phi' = A(t)\Phi$ and

$$\det \Phi(t) = e^{t^2/2} \qquad \text{for all} \quad t \in (-\infty, \infty)$$

Therefore, Φ is a fundamental matrix for Eq. (34). Also, in view of Abel's formula, we have

$$\det \Phi(t) = \det \Phi(\tau) \exp\left[\int_\tau^t \operatorname{tr} A(s)\, ds\right]$$

$$= e^{\tau^2/2} \exp\left[\int_\tau^t \eta\, d\eta\right] = e^{-t^2/2}$$

$$\text{for all} \quad t \in (-\infty, \infty)$$

as expected. Also, since

$$\Phi^{-1}(t) = \begin{bmatrix} 1 & -e^{-t^2/2} \int_\tau^t e^{\eta^2/2}\, d\eta \\ 0 & -e^{-t^2/2} \end{bmatrix}$$

we obtain for the state transition of Eq. (34),

$$\Phi(t)\Phi^{-1}(\tau) = \begin{bmatrix} 1 & e^{-\tau^2/2} \int_\tau^t e^{\eta^2/2}\, d\eta \\ 0 & e^{(t^2-\tau^2)/2} \end{bmatrix}$$

Finally, suppose that $\phi(\tau) = \xi = [1, 1]^T$. Then

$$\phi(t, \tau, \xi) = \Phi(t, \tau)\xi = \begin{bmatrix} 1 + e^{-\tau^2/2} \int_\tau^t e^{\eta^2/2}\, d\eta \\ e^{(t^2-\tau^2)/2} \end{bmatrix}$$

Next, we consider linear nonhomogeneous systems,

$$x' = A(t)x + g(t) \tag{LN}$$

We assume that $A(t)$ and $g(t)$ are defined and continuous over $R = (-\infty, \infty)$ (i.e., each component $a_{ij}(t)$ of $A(t)$ and each component $g_k(t)$ of $g(t)$ is defined and continuous on R). As noted in Section III, system (LN) has for any $\tau \in R$ and any $\xi \in R^n$ a unique solution satisfying $x(\tau) = \xi$. This solution exists on the *entire* real line R and is continuous in (t, τ, ξ). Furthermore, if A and g depend continuously on parameters $\lambda \in R^l$, then the solution will also vary continuously with λ. Indeed, if we differentiate the function

$$\phi(t, \tau, \xi) = \Phi(t, \tau)\xi + \int_\tau^t \Phi(t, \eta)g(\eta)\, d\eta \tag{35}$$

with respect to t to obtain $\phi'(t, \tau, \xi)$, and if we substitute ϕ and ϕ' into (LN) (for x), then it is an easy matter to verify that Eq. (35) is in fact the unique solution of (LN) with $\phi(t, \tau, \xi) = \xi$.

We note that when $\xi = 0$, then Eq. (35) reduces to

$$\phi_p(t) = \int_\tau^t \Phi(t, \eta)g(\eta)\,d\eta \tag{36}$$

and when $\xi \neq 0$ but $g(t) \equiv 0$, then Eq. (35) reduces to

$$\phi_h(t) = \Phi(t, \tau)\xi \tag{37}$$

Thus, the solution of (LN) may be viewed as consisting of a component due to the "forcing term" $g(t)$ and another component due to the initial data ξ. This type of separation is in general possible only in linear systems of differential equations. We call ϕ_p the **particular solution** and ϕ_h the **homogeneous solution** of (LN).

Before proceeding to linear systems with constant coefficients, we introduce adjoint equations. Let Φ be a fundamental matrix for the linear homogeneous system (LH). Then,

$$(\Phi^{-1})' = -\Phi^{-1}\Phi'\Phi^{-1} = -\Phi^{-1}A(t)$$

Taking the conjugate transpose of both sides, we obtain:

$$(\Phi^{*-1})' = -A^*(t)\Phi^{*-1}$$

This implies that Φ^{*-1} is a fundamental matrix for the system:

$$y' = -A^*(t)y, \qquad t \in J \tag{38}$$

We call Eq. (38) the **adjoint** to (LH), and we call the matrix equation,

$$Y' = -A^*(t)Y, \qquad t \in J$$

the **adjoint** to matrix equation (31).

One of the principal properties of adjoint systems is summarized in the following result:

If Φ is a fundamental matrix for (LH), then Ψ is a fundamental matrix for its adjoint (38) if and only if

$$\Psi^*\Phi = C$$

where C is some constant nonsingular matrix.

B. Linear Systems with Constant Coefficients

We now turn our attention to linear systems with constant coefficients. For purposes of motivation, we first consider the scalar initial-value problem:

$$x' = ax, \qquad x(\tau) = \xi \tag{39}$$

It is easily verified that Eq. (39) has the solution:

$$\phi(t) = e^{a(t-\tau)}\xi \tag{40}$$

It turns out that a similar result holds for the system of linear equations with constant coefficients,

$$x' = Ax \tag{L}$$

By making use of the Weierstrass M test, it is not difficult to verify the following result:

Let A be a constant $n \times n$ matrix which may be real or complex and let $S_N(t)$ denote the partial sum of matrices defined by the formula,

$$S_N(t) = E + \sum_{k=1}^N \frac{t^k}{k!}A^k \tag{41}$$

where E denotes the $n \times n$ identity matrix and k! stands for k factorial. Then each element of the matrix $S_N(t)$ converges absolutely and uniformly on any finite t interval $(-a, a)$, $a > 0$, as $N \to \infty$.

This result enables us to define the matrix,

$$e^{At} = E + \sum_{k=1}^\infty \frac{t^k}{k!}A^k \tag{42}$$

for any $-\infty < t < \infty$.

It should be clear that when $A(t) \equiv A$, system (LH) reduces to system (L). Consequently, the results we established above for (LH) are also applicable to (L). Now by making use of these results, the definition of e^{At} in Eq. (42), and the convergence properties of $S_N(t)$ in Eq. (42), it is not difficult to establish several important properties of e^{At} and of (L). To this end we let $J = (-\infty, \infty)$ and $\tau \in j$, and we let A be a given constant $n \times n$ matrix for (L). Then the following is true:

1. *$\Phi(t) \triangleq e^{At}$ is a fundamental matrix for (L) for $t \in J$.*
2. *The state transition matrix for (L) is given by $\Phi(t, \tau) = e^{A(t-\tau)} \triangleq \Phi(t - \tau), t \in J$.*
3. *$e^{At_1}e^{At_2} = e^{A(t_1+t_2)}$ for all $t_1, t_2 \in J$.*
4. *$Ae^{At} = e^{At}A$ for all $t \in J$.*
5. *$(e^{At})^{-1} = e^{-At}$ for all $t \in J$.*
6. *The unique solution ϕ of (L) with $\phi(\tau) = \xi$ is given by:*

$$\phi(t, \tau, \xi) = e^{A(t-\tau)}\xi \tag{43}$$

Notice that solution (43) of (L) such that $\phi(\tau) = \xi$ depends on t and τ only via the difference $t - \tau$. This is the typical situation for *general autonomous* systems that satisfy uniqueness conditions. Indeed, if $\phi(t)$ is a solution of

$$x' = F(x), \qquad x(0) = \xi$$

then clearly $\phi(t - \tau)$ will be a solution of

$$x' = F(x), \qquad x(\tau) = \xi$$

Next, we consider the "forced" system of equations,

$$x' = Ax + g(t) \qquad (44)$$

where $g: J \to R^n$ is continuous. Clearly, Eq. (44) is a special case of (LN). In view of Eq. (35) we thus have

$$\phi(t) = e^{A(t-\tau)}\xi + e^{At} \int_\tau^t e^{-A\eta} g(\eta)\, d\eta \qquad (45)$$

for the solution of Eq. (44).

Next, we address the problem of evaluating the state transition matrix. While there is no general procedure for evaluating such a matrix for a time-varying matrix $A(t)$, there are several such procedures for determining e^{At} when $A(t) \equiv A$. In the following, we consider two such methods.

We begin by recalling the Laplace transform. To this end, we consider a vector $f(t) = [f_1(t), \ldots, f_n(t)]^T$, where $f_i: [0, \infty) \to R$, $i = 1, \ldots, n$. Letting s denote a complex variable, we define the *Laplace transform* of f_i as:

$$\hat{f}_i(s) = \mathscr{L}[f_i(t)] \triangleq \int_0^\infty f_i(t)e^{-st}\, dt \qquad (46)$$

provided, of course, that the integral in Eq. (46) exists. (In this case f_i is said to be Laplace transformable.) Also, we define the Laplace transform of the vector $f(t)$ by:

$$\hat{f}(s) = [\hat{f}_1(s), \ldots, \hat{f}_n(s)]^T,$$

and we define the Laplace transform of a matrix $C(t) = [c_{ij}(t)]$ similarly. Thus, if $c_{ij}: [0, \infty) \to R$ and if each c_{ij} is Laplace transformable, then the Laplace transform of $C(t)$ is defined by:

$$\hat{C}(s) = \mathscr{L}[c_{ij}(t)] = [\mathscr{L}c_{ij}(t)] = [\hat{c}_{ij}(s)]$$

Now consider the initial value problem,

$$x' = Ax, \qquad x(0) = \xi \qquad (47)$$

Taking the Laplace transform of both sides of Eq. (47), we obtain:

$$sx(s) - \xi = Ax(s)$$

or

$$(sE - A)x(s) = \xi$$

or

$$x(s) = (sE - A)^{-1}\xi \qquad (48)$$

where E denotes the $n \times n$ identity matrix. It can be shown by analytic continuation that $(sE - A)^{-1}$ exists for all s, except at the eigenvalues of A (i.e., except at those values of s where the equation $\det(sE - A) = 0$ is satisfied). Taking the inverse Laplace transform of Eq. (48) (i.e., by reversing the procedure and obtaining, for example,

in Eq. (46) $f_i(t)$ from $f_i(s)$ we obtain for the solution of Eq. (47),

$$\phi(t) = \mathscr{L}^{-1}[(sE - A)^{-1}]\xi = \Phi(t, 0)\xi = e^{At}\xi \qquad (49)$$

where $\mathscr{L}^{-1}[\hat{f}(s)] = f(t)$ denotes the inverse Laplace transform of $\hat{f}(s)$. It follows from Eqs. (49) and (48) that

$$\hat{\Phi}(s) = (sE - A)^{-1}$$

and

$$\Phi(t, 0) \triangleq \Phi(t) = \mathscr{L}^{-1}[(sE - A)^{-1}] = e^{At} \qquad (50)$$

Finally, note that when the initial time $\tau \neq 0$, we can immediately compute $\Phi(t, \tau) = \Phi(t - \tau) = e^{A(t-\tau)}$.

Next, let us consider a "forced" system of the form:

$$x' = Ax + g(t), \qquad x(0) = \xi \qquad (51)$$

and let us assume that the Laplace transform of g exists. Taking the Laplace transform of both sides of Eq. (51) yields:

$$s\hat{x}(s) - \xi = A\hat{x}(s) + \hat{g}(s)$$

or

$$(sE - A)\hat{x}(s) = \xi + \hat{g}(s)$$

or

$$\hat{x}(s) = (sE - A)^{-1}\xi + (sE - A)^{-1}\hat{g}(s)$$
$$= \hat{\Phi}(s)\xi + \hat{\Phi}(s)\hat{g}(s) \triangleq \hat{\phi}_h(s) + \hat{\phi}_p(s) \qquad (52)$$

Taking the inverse Laplace transform of both sides of Eq. (52) and using Eq. (45), we obtain:

$$\phi(t) = \phi_h(t) + \phi_p(t)$$
$$= \mathscr{L}^{-1}[(sE - A)^{-1}]\xi + \mathscr{L}^{-1}[(sE - A)^{-1}\hat{g}(s)]$$
$$= \Phi(t, 0)\xi + \int_0^t \Phi(t - \eta)g(\eta)\, d\eta \qquad (53)$$

Therefore,

$$\phi_p(t) = \int_0^t \Phi(t - \eta)g(\eta)\, d\eta \qquad (54)$$

as expected. We call the expression in Eq. (54) the **convolution** of Φ and g. Clearly, convolution of Φ and g in the time domain corresponds to multiplication of Φ and g in the s domain.

Let us now consider the specific initial-value problem,

$$\begin{aligned} x_1' &= -x_1 + x_2, & x_1(0) &= -1 \\ x_2' &= -2x_2 + u(t), & x_2(0) &= 0 \end{aligned} \qquad (55)$$

where

$$u(t) = \begin{cases} 1 & \text{for } t > 0 \\ 0 & \text{elsewhere} \end{cases}$$

We have, in this case,

$$(sE - A) = \begin{bmatrix} s+1 & -1 \\ 0 & s+2 \end{bmatrix}$$

$$\hat{\Phi}(s) = (sE - A)^{-1}$$

$$= \begin{bmatrix} \dfrac{1}{s+1} & \left(\dfrac{1}{s+1} - \dfrac{1}{s+2} \right) \\ 0 & \dfrac{1}{s+2} \end{bmatrix}$$

and

$$\Phi(t) = e^{At} = \mathcal{L}^{-1}[\hat{\Phi}(s)]$$

$$= \begin{bmatrix} e^{-t} & (e^{-t} - e^{-2t}) \\ 0 & e^{-2t} \end{bmatrix}$$

It now follows that

$$\phi_{\mathrm{h}}(t) = \begin{bmatrix} e^{-t} & (e^{-t} - e^{-2t}) \\ 0 & e^{-2t} \end{bmatrix} \begin{bmatrix} -1 \\ 0 \end{bmatrix}$$

$$= \begin{bmatrix} -e^{-t} \\ 0 \end{bmatrix}$$

Also, since $\hat{u}(s) = 1/s$, we have

$$\hat{\phi}_{\mathrm{p}}(s) = \begin{bmatrix} \dfrac{1}{s+1} & \left(\dfrac{1}{s+1} - \dfrac{1}{s+2} \right) \\ 0 & \dfrac{1}{s+2} \end{bmatrix} \begin{bmatrix} 0 \\ \dfrac{1}{s} \end{bmatrix}$$

$$= \begin{bmatrix} \dfrac{1}{2} \left(\dfrac{1}{s} \right) + \dfrac{1}{2} \left(\dfrac{1}{s+2} \right) - \dfrac{1}{s+1} \\ \dfrac{1}{2} \left(\dfrac{1}{s} \right) - \dfrac{1}{2} \left(\dfrac{1}{s+2} \right) \end{bmatrix}$$

and

$$\phi_{\mathrm{p}}(t) = \begin{bmatrix} \frac{1}{2} + \frac{1}{2}e^{-2t} - e^{-t} \\ \frac{1}{2} - \frac{1}{2}e^{-2t} \end{bmatrix}$$

Therefore, the solution of the initial value problem (55) is

$$\phi(t) = \phi_{\mathrm{p}}(t) + \phi_{\mathrm{h}}(t) = \begin{bmatrix} \frac{1}{2} - 2e^{-t} + \frac{1}{2}e^{-2t} \\ \frac{1}{2} - \frac{1}{2}e^{-2t} \end{bmatrix}$$

A second method of evaluating e^{At} and of solving initial value problems for (L) and Eq. (44) involves the transformation of A into a Jordan canonical form. Specifically, it is shown in linear algebra that for every complex $n \times n$ matrix A there exists a nonsingular $n \times n$ matrix P (i.e., $\det P \neq 0$) such that the matrix,

$$J = P^{-1}AP \tag{56}$$

is in the canonical form,

$$J = \begin{bmatrix} J_0 & & & \\ & J_1 & 0 & \\ & 0 & \ddots & \\ & & & J_s \end{bmatrix} \tag{57}$$

where J_0 is a diagonal matrix with diagonal elements $\lambda_1, \ldots, \lambda_k$ (not necessarily distinct); that is,

$$J_0 = \begin{bmatrix} \lambda_1 & & & \\ & \ddots & & 0 \\ & 0 & & \\ & & & \lambda_k \end{bmatrix} \tag{58}$$

and each J_p is an $n_p \times n_p$ matrix of the form ($p = 1, \ldots, s$):

$$J_q = \begin{bmatrix} \lambda_{k+p} & 1 & 0 & \cdots & 0 \\ 0 & \lambda_{k+p} & 1 & & \vdots \\ \vdots & \vdots & \vdots & & \vdots \\ & & & \cdots & 1 \\ 0 & 0 & 0 & \cdots & \lambda_{k+p} \end{bmatrix} \tag{59}$$

where λ_{k+p} need not be different from λ_{k+q} if $p \neq q$ and $k + n_1 + \cdots + n_s = n$. The numbers λ_i, $i = 1, \ldots, k+s$, are the eigenvalues of A (i.e., the roots of the equation $\det(\lambda E - A) = 0$). If λ_i is a simple eigenvalue of A (i.e., it is not a repeated root of $\det(\lambda E - A) = 0$), then it appears in the block J_0. The blocks $J_0, J_1, \ldots, J_s$ are called **Jordan blocks** and J is called the **Jordan canonical form** of A.

Returning to the subject at hand, we consider once more the initial value problem (47) and let P be a real $n \times n$ nonsingular matrix which transforms A into a Jordan canonical form J. Consider the transformation $x = Py$ or, equivalently, $y = P^{-1}x$. Differentiating both sides with respect to t, we obtain:

$$y' = P^{-1}x' = P^{-1}APy = Jy$$
$$y(\tau) = P^{-1}\xi \tag{60}$$

The solution of Eq. (60) is given as

$$\psi(t) = e^{J(t-\tau)} P^{-1} \xi \tag{61}$$

Using Eq. (61) and $x = Py$, we obtain for the solution of Eq. (47),

$$\phi(t) = P e^{J(t-\tau)} P^{-1} \xi \tag{62}$$

In the case in which A has n *distinct* eigenvalues $\lambda_1, \ldots, \lambda_n$, we can choose $P = [p_1, p_2, \ldots, p_n]$ in such a way that p_i is an eigenvector corresponding to the eigenvalue λ_i, $i = 1, \ldots, n$ (i.e., $p_i \neq 0$ satisfies the equation

$\lambda_i p_i = A p_i$). Then the Jordan matrix $J = P^{-1}AP$ assumes the form:

$$J = \begin{bmatrix} \lambda_1 & & & 0 \\ & \ddots & & \\ 0 & & & \\ & & & \lambda_n \end{bmatrix}$$

Using the power series representation, Eq. (42), we immediately obtain the expression:

$$e^{Jt} = \begin{bmatrix} e^{\lambda_1 t} & & & 0 \\ & \ddots & & \\ 0 & & & \\ & & & e^{\lambda_n t} \end{bmatrix}$$

In this case, we have the expression for the solution of Eq. (47):

$$\phi(t) = P \begin{bmatrix} e^{\lambda_1(t-\tau)} & & & 0 \\ & \ddots & & \\ 0 & & & \\ & & & e^{\lambda_n(t-\tau)} \end{bmatrix} P^{-1}\xi$$

In the general case when A has repeated eigenvalues, we can no longer diagonalize A and we have to be content with the Jordan form given by Eq. (57). In this case, $P = [v_1, \ldots, v_n]$, where the v_i denote generalized eigenvectors. Using the power series representation, Eq. (42) and the very special nature of the Jordan blocks (58) and (59), it is not difficult to show that in the case of repeated eigenvalues we have

$$e^{Jt} = \begin{bmatrix} e^{J_0 t} & & & 0 \\ & e^{J_1 t} & & \\ & & \ddots & \\ 0 & & & e^{J_s t} \end{bmatrix} \quad -\infty < t < \infty$$

where

$$e^{J_0 t} = \begin{bmatrix} e^{\lambda_1 t} & & & 0 \\ & \ddots & & \\ 0 & & & \\ & & & e^{\lambda_k t} \end{bmatrix}$$

and

$$e^{t J_i} = e^{\lambda_{k+i} t} \begin{bmatrix} 1 & t & \cdots & t^{n_i-1}/(n_i-1)! \\ 0 & 1 & \cdots & t^{n_i-2}/(n_i-2)! \\ \vdots & \vdots & & \vdots \\ 0 & 0 & & 1 \end{bmatrix}$$

$$i = 1, \ldots, s$$

From Eq. (62), it now follows that the solution of Eq. (47) is given by:

$$\phi(t) = P \begin{bmatrix} e^{J_0(t-\tau)} & & & \\ & e^{J_1(t-\tau)} & & 0 \\ & & \ddots & \\ 0 & & & e^{J_s(t-\tau)} \end{bmatrix} P^{-1}\xi$$

As a specific example of the above procedure of determining the state transition matrix, consider the initial-value problem:

$$x_1' = -x_1 + x_2, \qquad x_1(0) = 1$$
$$x_2' = -2x_2, \qquad x_2(0) = 2$$

In this case, we have

$$A = \begin{bmatrix} -1 & 1 \\ 0 & -2 \end{bmatrix}$$

with eigenvalues $\lambda_1 = -1$ and $\lambda_2 = -2$ and with corresponding eigenvectors,

$$P_1 = \begin{bmatrix} 1 \\ 0 \end{bmatrix}, \qquad P_2 = \begin{bmatrix} -1 \\ 1 \end{bmatrix}$$

We thus have

$$P = [p_1, p_2] = \begin{bmatrix} 1 & -1 \\ 0 & 1 \end{bmatrix}, \qquad P^{-1} = \begin{bmatrix} 1 & 1 \\ 0 & 1 \end{bmatrix}$$

and

$$J = \begin{bmatrix} \lambda_1 & 0 \\ 0 & \lambda_2 \end{bmatrix} = \begin{bmatrix} -1 & 0 \\ 0 & -2 \end{bmatrix}$$

Furthermore, we obtain:

$$\begin{bmatrix} \phi_1(t) \\ \phi_2(t) \end{bmatrix} = P e^{Jt} P^{-1}\xi$$

$$= \begin{bmatrix} 1 & -1 \\ 0 & 1 \end{bmatrix} \begin{bmatrix} e^{-t} & 0 \\ 0 & e^{-2t} \end{bmatrix} \begin{bmatrix} 1 & 1 \\ 0 & 1 \end{bmatrix} \begin{bmatrix} 1 \\ 2 \end{bmatrix}$$

$$= \begin{bmatrix} 3e^{-t} - 2e^{-2t} \\ 2e^{-2t} \end{bmatrix}$$

C. Linear Systems With Periodic Coefficients

Next, we consider linear homogeneous periodic systems of the form:

$$x' = A(t)x, \qquad -\infty < t < \infty \qquad \text{(P)}$$

where the elements of A are continuous functions on R and where

$$A(t) = A(t + T) \qquad \text{(63)}$$

for some $T > 0$ which is a period of A.

The principal result for (P) which we shall present here (called **Floquet theory**) involves the logarithm of a matrix, the existence of which is not too difficult to establish. We have the following:

Let B be a nonsingular n × n matrix. Then there exists an n × n matrix F, called the logarithm of B, such that

$$e^F = B$$

Using properties of the fundamental matrix as well as the concept of the logarithm of a matrix, we are in a position to prove the following fundamental result for (P):

Let Eq. (63) be true and let A(t) be continuous on $(-\infty, \infty)$. If $\Phi(t)$ is a fundamental matrix for (P), then so is $\Phi(t + T)$, $-\infty < t < \infty$. Moreover, corresponding to every Φ, there exists a nonsingular matrix P which is also periodic with period T and a constant matrix R, such that

$$\Phi(t) = P(t)e^{tR} \tag{64}$$

It is not difficult to show, using Eq. (64), that if the fundamental matrix Φ for (P) is known over *any* interval of length T, then it is automatically known for all $-\infty < t < \infty$. For example, if $\Phi(t)$ is known for all t over the interval $[t_0, t_0 + T]$, then we can show that

$$\Phi(t) = P(t)e^{tT^{-1}\log C} \tag{65}$$

where $C = \Phi(t_0)^{-1}\Phi(t_0 + T)$. Thus, since $P(t)$ is periodic, $\Phi(t)$ as given in Eq. (65) will be known for all t over $(-\infty, \infty)$.

It can also be shown that even though the fundamental matrix Φ does not determine R uniquely in Eq. (64), the set of all fundamental matrices of (P), and hence of $A(t)$, determines uniquely all quantities associated with e^{TR} which are invariant under a similarity transformation. Specifically, the set of all fundamental matrices of $A(t)$ determine a unique set of eigenvalues of the matrix e^{TR}, $\lambda_1, \ldots, \lambda_n$, which are called the **Floquet multipliers** associated with $A(t)$. None of these vanishes since $\Pi \lambda_i = \det e^{TR} \neq 0$. Also, the eigenvalues of R, $\rho_1, \ldots, \rho_n$, are called the **characteristic exponents** of $A(t)$.

Now let us suppose that all ρ_i are such that Re $\rho_i < 0$ (i.e., the real part of each ρ_i is negative) and let $\alpha = \min_i |\text{Re } \rho_i|$. Now, if we arrange things so that R in Eq. (64) is in Jordan canonical form, then it is a simple matter to show that there exists a $k > 0$ such that for each component $\phi_i(t)$ of the solution $\phi(t)$,

$$|\phi_i(t)| \leq ke^{\alpha t} \quad \text{for all} \quad t \geq 0 \tag{66}$$

and $|\phi_i(t)| \to 0$ as $t \to \infty$. In other words, if the eigenvalues ρ_i, $i = 1, \ldots, n$, of R have negative real parts, then the norm of any solution of (P) tends to zero as $t \to \infty$ at an exponential rate.

Finally, by using Eq. (64) we can write:

$$P(t) = \Phi(t)e^{-tR} \tag{67}$$

which in turn can be used to see that $AP - P' = PR$. Thus, for the transformation,

$$x = P(t)y \tag{68}$$

we compute

$$x' = A(t)x = A(t)P(t)y$$
$$= P'(t)y + P(t)y' = (P(t)y)'$$

or

$$y' = P^{-1}(t)[A(t)P(t) - P'(t)]y$$
$$= P^{-1}(t)(P(t)R)y = Ry$$

This shows that *transformation (68) reduces the linear, homogeneous, periodic system* (P) *to*

$$y' = Ry \tag{69}$$

a linear homogeneous system with constant coefficients. Since $P(t)$ is nonsingular, we are thus able to deduce the properties of the solutions of (P) from those of Eq. (69), provided of course that we can determine the matrix $P(t)$.

D. Linear *n* th-Order Ordinary Differential Equations

We conclude this section by considering some of the more important aspects of linear *n*th-order ordinary differential equations. We shall consider equations of the form,

$$y^{(n)} + a_{n-1}(t)y^{(n-1)} + \cdots + a_1(t)y^{(1)} + a_0(t)y = b(t) \tag{70}$$

$$y^{(n)} + a_{n-1}(t)y^{(n-1)} + \cdots + a_1(t)y^{(1)} + a_0(t)y = 0 \tag{71}$$

and

$$y^{(n)} + a_{n-1}y^{(n-1)} + \cdots + a_1y^{(1)} + a_0y = 0 \tag{72}$$

In Eqs. (70) and (71) the functions $a_k(t)$ and $b(t)$, $k = 1, \ldots, n - 1$, are continuous on some appropriate time interval J. If we define the differential operator L_n by:

$$L_n = \frac{d^n}{dt^n} + a_{n-1}(t)\frac{d^{n-1}}{dt^{n-1}} + \cdots + a_1(t)\frac{d}{dt} + a_0(t) \tag{73}$$

then we can rewrite Eqs. (70) and (71) more compactly as

$$L_n y = b(t) \tag{74}$$

and

$$L_n y = 0 \qquad (75)$$

respectively. We can rewrite Eq. (72) similarly by defining a differential operator L in the obvious way.

Following the procedure in Section II, we can reduce the study of Eq. (71) to the study of the system of n first-order ordinary differential equations,

$$x' = A(t)x \qquad \text{(LH)}$$

where $A(t)$ is the **companion matrix** given by:

$$A(t) = \begin{bmatrix} 0 & 1 & 0 & \cdots & 0 \\ 0 & 0 & 1 & \cdots & 0 \\ \vdots & \vdots & \vdots & & \\ 0 & 0 & 0 & \cdots & 1 \\ -a_0(t) & -a_1(t) & -a_2(t) & \cdots & -a_{n-1}(t) \end{bmatrix}$$

$$(76)$$

Since $A(t)$ is continuous on J, we know from Section III that there exists a unique solution $\phi(t)$, for all $t \in J$, to the initial-value problem,

$$x' = A(t)x, \qquad x(\tau) = \xi, \qquad \tau \in J$$

where $\xi = (\xi_1, \dots, \xi_n)^T \in R^n$. The first component of this solution is a solution of $L_n y = 0$ satisfying $y(\tau) = \xi_1, y'(\tau) = \xi_2, \dots, y^{(n-1)}(\tau) = \xi_n$.

Now let $\phi_1, \dots, \phi_n$ be n solutions of Eq. (75). Then we can easily show that the matrix,

$$\Phi(t) = \begin{bmatrix} \phi_1 & \phi_2 & \cdots & \phi_n \\ \phi_1' & \phi_2' & \cdots & \phi_n' \\ \vdots & \vdots & & \vdots \\ \phi_1^{(n-1)} & \phi_2^{(n-1)} & \cdots & \phi_n^{(n-1)} \end{bmatrix}$$

is a solution of the matrix equation,

$$X' = A(t)X \qquad (77)$$

where $A(t)$ is defined by Eq. (76). We call the determinant of Φ the **Wronskian** for Eq. (75) with respect to the solutions $\phi_1, \dots, \phi_n$ and we denote it by:

$$W(\phi_1, \dots, \phi_n) = \det \Phi(t)$$

Note that $W(\phi_1, \dots, \phi_n)(t)$ depends on $t \in J$. Since Φ is a solution of matrix equation (77), then by Abel's formula it follows that for any $\tau \in J$ and for any $t \in J$,

$$W(\phi_1, \dots, \phi_n)(t) = \det \Phi(\tau) \exp\left[\int_\tau^t \operatorname{tr} A(s)\, ds \right]$$

$$= W(\phi_1, \dots, \phi_n)(\tau)$$

$$\times \exp\left\{ \int_\tau^t -a_{n-1}(s)\, ds \right\} \quad (78)$$

As an example, consider the second-order differential equation

$$t^2 y'' + t y' - y = 0, \qquad 0 < t < \infty$$

which can be written equivalently as:

$$y'' + (1/t)y' - (1/t^2)y = 0, \qquad 0 < t < \infty \quad (79)$$

The functions $\phi_1(t = t$ and $\phi_2(t) = 1/t$ are clearly solutions of Eq. (79). We now form the matrix,

$$\Phi(t) = \begin{bmatrix} \phi_1 & \phi_2 \\ \phi_1' & \phi_2' \end{bmatrix} = \begin{bmatrix} t & 1/t \\ 1 & 1/t^2 \end{bmatrix}$$

which yields the Wronskian

$$W(\phi_1, \phi_2)(t) = \det \Phi(t) = -2/t, \qquad t > 0$$

In the notation of Eq. (76) we have $a_1(t) = 1/t$, $a_0(t) = -1/t^2$, and thus $a_1(s) = 1/s$. In view of Eq. (78), we have for any $\tau > 0$,

$$W(\phi_1, \phi_2)(t) = \det \Phi(t)$$

$$= W(\phi_1, \phi_2)(\tau) \exp\left\{ \int_\tau^t -a_1(s)\, ds \right\}$$

$$= -(2/\tau)e^{\ln(\tau/t)} = -2/t, \qquad t > 0$$

as expected.

Similarly as in the case of systems of equations, we can prove the following result for nth-order differential equations:

A set of n solutions of Eq. (75), $\phi_1, \dots, \phi_n$, is linearly independent on J if and only if $W(\phi_1, \dots, \phi_n)(t) \neq 0$ for all $t \in J$. Moreover, every solution of Eq. (75) is a linear combination of any set of n linearly independent solutions.

The above result enables us to make the following definition:

*A set of n linearly independent solutions of Eq. (75) on j, $\phi_1, \dots, \phi_n$, is called a **fundamental set** of solutions for Eq. (75).*

Next, we turn our attention to nonhomogeneous linear nth-order ordinary differential equations of the form (70). As shown in Section II, the study of Eq. (70) reduces to the study of the system of n first-order ordinary differential equations,

$$x' = A(t)x + g(t) \qquad (80)$$

where $A(t)$ is given by Eq. (76) and $g(t) = [0, \dots, 0, b(t)]^T$. Recall that for given $\tau \in J$ and given $x(\tau) = \xi \in R^n$, Eq. (80) has a unique solution given by $\phi = \phi_h + \phi_p$, where $\phi_h(t) = \Phi(t, \tau)\xi$ is a solution of (LH),

$\phi(t, \tau)$ denotes the state transition matrix of $A(t)$, and ϕ_p is a particular solution of Eq. (80), given by:

$$\phi_p(t) = \int_\tau^t \Phi(t, s)g(s)\,ds$$

$$= \Phi(t) \int_\tau^t \Phi^{-1}(s)g(s)\,ds$$

We now specialize this result from the n-dimensional system (80) to the corresponding nth-order equation (70) to obtain the following result:

If $\{\phi_1, \dots, \phi_n\}$ is a fundamental set for the equation $L_n y = 0$, then the unique solution ψ of the equation $L_n y = b(t)$ satisfying $\psi(\tau) = \xi_1, \dots, \psi^{(n-1)}(\tau) = \xi_n$ is given by:

$$\psi(t) = \psi_h(t) + \psi_P(t) = \psi_h(t) + \sum_{k=1}^n \phi_k(t)$$

$$\times \int_\tau^t \frac{W_k(\phi_1, \dots, \phi_n)(s)}{W(\phi_1, \dots, \phi_n)(s)} b(s)\,ds \tag{81}$$

Here, ψ_h is the solution of $L_n y = 0$ such that $\psi(\tau) = \xi_1$, $\psi'(\tau) = \xi_2, \dots, \psi^{(n-1)}(\tau) = \xi_n$, and $W_k(\phi_1, \dots, \phi_n)(t)$ is obtained from $W(\phi_1, \dots, \phi_n)(t)$ by replacing the kth column in $W(\phi_1, \dots, \phi_n)(t)$ by $(0, \dots, 0, 1)^T$.

We apply the above example to the second-order differential equation,

$$y'' + (1/t)y' - (y/t^2) = b(t), \qquad 0 < t < \infty$$

where $b(t)$ is a real continuous function for all $t > 0$. From the example involving Eq. (79) we have $\phi_1(t) = t$, $\phi_2(t) = 1/t$, and $W(\phi_1, \phi_2)(t) = -2/t$, $t > 0$. Also,

$$W_1(\phi_1, \phi_2)(t) = \begin{vmatrix} 0 & 1/t \\ 1 & -1/t^2 \end{vmatrix} = -\frac{1}{t},$$

$$W_1(\phi_1, \phi_2)(t) = \begin{vmatrix} t & 0 \\ 1 & 1 \end{vmatrix} = t$$

From Eq. (81) we now have

$$\psi(t) = \psi_h(t) + \psi_p(t)$$

$$= \psi_h(t) + \frac{t}{2} \int_\tau^t b(s)\,ds - \frac{1}{2t} \int_\tau^t s^2 b(s)\,ds$$

Next, we consider nth-order ordinary differential equations with constant coefficients given by Eq. (72) which can equivalently be written as $L_n y = 0$, where

$$L_n = \frac{d^n}{dt^n} + a_{n-1}\frac{d^{n-1}}{dt^{n-1}} + \cdots + a_1\frac{d}{dt} + a_0$$

We assume that $J = (-\infty, \infty)$, we call

$$p(\lambda) = \lambda^n + a_{n-1}\lambda^{n-1} + \cdots + a_1\lambda + a_0 \tag{82}$$

the **characteristic polynomial** of the differential equation (72), and we call

$$p(\lambda) = 0 \tag{83}$$

the **characteristic equation** of Eq. (72). The roots of $p(\lambda)$ are called the **characteristic roots** of Eq. (72).

We see that the study of Eq. (72) reduces to the study of the system of first-order ordinary differential equations with constant coefficients given by $x' = Ax$, where

$$A = \begin{bmatrix} 0 & 1 & 0 & 0 & \cdots & 0 \\ 0 & 0 & 1 & 0 & \cdots & 0 \\ \vdots & \vdots & \vdots & \vdots & & \vdots \\ 0 & 0 & 0 & 0 & \cdots & 1 \\ -a_0 & -a_1 & -a_2 & -a_3 & \cdots & -a_{n-1} \end{bmatrix} \tag{84}$$

The following result, which is proved in a straightforward manner, connects Eq. (72) and $x' = Ax$ with A given by Eq. (84):

The characteristic polynomial of A in Eq. (83) is precisely the characteristic polynomial $p(\lambda)$ given by Eq. (82), that is,

$$p(\lambda) = \det(\lambda E_n - A)$$

The next result enumerates a fundamental set for Eq. (72):

Let $\lambda_1, \dots, \lambda_s$ be the distinct roots of the characteristic equation (83) and suppose that λ_i has multiplicity m_i, $i = 1, \dots, s$, with $\sum_{i=1}^s m_i = n$. Then the following set of functions is a fundamental set for Eq. (72):

$$t^k e^{\lambda_i t}, \qquad k = 0, 1, \dots, m_i - 1,$$
$$i = 1, \dots, s \tag{85}$$

As a specific example, consider:

$$p(\lambda) = (\lambda - 2)(\lambda + 3)^2(\lambda + i)(\lambda - i)(\lambda - 4)^4 \tag{86}$$

Then, $n = 9$, and $\{e^{2t}, e^{-3t}, te^{-3t}, e^{-it}, e^{+it}, e^{4t}, te^{4t}, t^2 e^{4t}, t^3 e^{4t}\}$ is a fundamental set for the differential equation corresponding to the characteristic equation (86).

We conclude this section by considering adjoint equations. Corresponding to the operator L_n given in Eq. (73), we define a second linear operator L_n^+ of order n, which we call the **adjoint** of L_n, as follows. The domain of L_n^+ is the set of all continuous functions defined on J such that $[\bar{a}_j(t)y(t)]$ has j continuous derivatives on J. (Here, $\bar{a}_j(t)$ denotes the complex conjugate of $a_j(t)$.) For each function y, define:

$$L_n^+ y = (-1)^n y^{(n)} + (-1)^{n-1}(\bar{a}_{n-1}y)^{n-1}$$

$$+ \cdots + (-1)(\bar{a}_1 y)' + \bar{a}_0 y$$

The equation,

$$L_n^+ y = 0, \qquad t \in J$$

is called the **adjoint equation** to $L_n y = 0$.

When Eq. (75) is written in companion form (LH) with $A(t)$ given by Eq. (76), then the adjoint system is $z' = -A^*(t)z$, where

$$A^*(t) = \begin{bmatrix} 0 & 0 & \cdots & 0 & -\bar{a}_0(t) \\ 1 & 0 & \cdots & 0 & -\bar{a}_1(t) \\ 0 & 1 & \cdots & 0 & -\bar{a}_2(t) \\ \vdots & \vdots & & \vdots & \vdots \\ 0 & 0 & \cdots & 1 & -\bar{a}_{n-1}(t) \end{bmatrix}$$

This adjoint system can be written in component form as:

$$z_1' = \bar{a}_0(t)z_n$$
$$z_j' = -z_{j-1} + \bar{a}_{j-1}(t)z_n, \qquad 2 \le j \le n \quad (87)$$

If $\psi = [\psi_1, \psi_2, \ldots, \psi_n]^T$ is a solution of Eq. (87) and if $a_j \psi_n$ has j derivatives, then

$$\psi_n' - (\bar{a}_{n-1}\psi_n) = -\psi_{n-1}$$

and

$$\psi_n'' - (\bar{a}_{n-1}\psi_n)' = -\psi_{n-1}' = \psi_{n-1} - (\bar{a}_{n-2}\psi_n)$$

or

$$\psi_n'' - (\bar{a}_{n-1}\psi_n)' + (\bar{a}_{n-2}\psi_n) = \psi_{n-2}$$

Continuing in this manner, we see that ψ_n solves $L_n^+ \psi = 0$.

V. STABILITY

Since there are no general rules for determining explicit formulas for the solutions of systems of ordinary differential equations (E), the analysis of initial-value problems (I) is accomplished along two lines: (1) a quantitative approach is used which usually involves the numerical solution of such problems by means of simulations on a digital computer, and (2) a qualitative approach is used which is usually concerned with the behavior of families of solutions of a given differential equation and which usually does not seek specific explicit solutions. As mentioned in Section I, we will concern ourselves primarily with qualitative aspects of ordinary differential equations.

The principal results of the qualitative approach include stability properties of an equilibrium point (rest position) and the boundedness of solutions of ordinary differential equations. We shall consider these topics in the present section.

A. The Concept of an Equilibrium Point

We shall concern ourselves with systems of equations,

$$x' = f(t, x) \qquad \text{(E)}$$

where $x \in R^n$. When discussing *global results*, such as global asymptotic stability, we shall always assume that $f: R^+ \times R^n \to R^n$. On the other hand, when considering *local results*, we shall usually assume that $f: R^+ \times B(h) \to R^n$ for some $h > 0$, where $R^+ = [0, \infty)$, $B(h) = \{x \in R^n: |x| < h\}$ and $|\cdot|$ is any one of the equivalent norms on R^n. On some occasions we assume that $t \in R = (-\infty, \infty)$ rather than $t \in R^+$. Unless otherwise stated, we assume that for every (t_0, ξ), $t_0 \in R^+$, the initial-value problem,

$$x' = f(t, x), \qquad x(t_0) = \xi \qquad \text{(I)}$$

possesses a unique solution $\phi(t, t_0, \xi)$ which is defined for all $t \ge t_0$ and which depends continuously on the initial data (t_0, ξ). Since it is natural in this section to think of t as representing time, we shall use the symbol t_0 in (I) to represent the initial time (rather than τ as was done earlier). Furthermore, we shall frequently use the symbol x_0 in place of ξ to represent the initial state.

A point $x_e \in R^n$ is called an **equilibrium point** of (E) (at time $t^* \in R^+$) if $f(t, x_e) = 0$ for all $t \ge t^*$. Other terms for equilibrium point include **stationary point, singular point, critical point**, and **rest position**.

We note that if x_e is an equilibrium point of (E) at t^*, then it is an equilibrium point at all $\tau \ge t^*$. Note also that in the case of autonomous systems,

$$x' = f(x) \qquad \text{(A)}$$

and in case of T-periodic systems,

$$x' = f(t, x), \qquad f(t, x) = f(t + T, x) \qquad \text{(P)}$$

a point $x_e \in R^n$ is an equilibrium at some t^* if and only if it is an equilibrium point at all times. Also note that if x_e is an equilibrium (at t^*) of (E), then the transformation $s = t - t^*$ reduces (E) to $dx/ds = f(s + t^*, x)$ and x_e is an equilibrium (at $s = 0$) of this system. For this reason, we shall henceforth assume that $t^* = 0$ in the above definition and we shall not mention t^* further. Note also that if x_e is an equilibrium point of (E), then for any $t_0 \ge 0$, $\phi(t, t_0, x_e) = x_e$ for all $t \ge t_0$ (i.e., x_e is a unique solution of (E) with initial data given by $\phi(t_0, t_0, x_e) = x_e$).

As a specific example, consider the simple pendulum introduced in Section II, which is described by equations of the form,

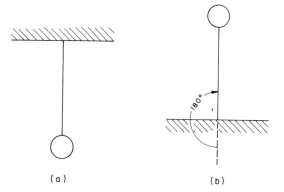

FIGURE 8 (a) Stable and (b) unstable equilibria of the simple pendulum.

$$x_1' = x_2$$
$$x_2' = k \sin x_1, \qquad k > 0 \tag{88}$$

Physically, the pendulum has two equilibrium points. One of these is located as shown in Fig. 8a and the second point is located as shown in Fig. 8b. However, the *model* of this pendulum, described by Eq. (88), has countably infinitely many equilibrium points located in R^2 at the points $(\pi n, 0)$, $n = 0, \pm 1, \pm 2, \ldots$.

An equilibrium point x_e of (E) is called an **isolated equilibrium point** if there is an $r > 0$ such that $B(x_e, r) \subset R^n$ contains no equilibrium points of (E) other than x_e itself. (Here, $B(x_e, r) = \{x \in R^n : |x - x_e| < r\}$.)

All equilibrium points of Eq. (88) are isolated equilibria in R^2. On the other hand, for the system,

$$x_1' = -ax_1 + bx_1 x_2$$
$$x_2' = -bx_1 x_2 \tag{89}$$

where $a > 0$, $b > 0$ are constants, every point on the positive x_2 axis is an equilibrium point for Eq. (89).

It should be noted that there are systems with *no equilibrium points* at all, as is the case, for example, in the system,

$$x_1' = 2 + \sin(x_1 + x_2) + x_1$$
$$x_2' = 2 + \sin(x_1 + x_2) - x_1 \tag{90}$$

Many important classes of systems possess only *one equilibrium point*. For example, the linear homogeneous system,

$$x' = A(t)x \tag{LH}$$

has a unique equilibrium at the origin if $A(t_0)$ is nonsingular for all $t_0 \geq 0$. Also, the system,

$$x' = f(x) \tag{A}$$

where f is assumed to be continuously differentiable with respect to all of its arguments and where

$$J(x_e) = \frac{\partial f}{\partial x}(x)\bigg|_{x = x_e}$$

denotes the $n \times n$ Jacobian matrix defined by $\partial f / \partial x = [\partial f_i / \partial x_j]$ has an isolated equilibrium at x_e if $f(x_e) = 0$ and $J(x_e)$ is nonsingular.

Unless otherwise stated, we shall assume throughout this section that a given equilibrium point is an isolated equilibrium. Also, we shall assume, unless otherwise stated, that in a given discussion, the equilibrium of interest is located at the origin of R^n. This assumption can be made without any loss of generality. To see this, assume that $x_e \neq 0$ is an equilibrium point of (E) (i.e., $f(t, x_e) = 0$ for all $t \geq 0$). Let $w = x - x_e$. Then $w = 0$ is an equilibrium of the transformed system,

$$w' = F(t, w) \tag{91}$$

where

$$F(t, w) = f(t, w + x_e) \tag{92}$$

Since Eq. (92) establishes a one-to-one correspondence between the solutions of (E) and Eq. (91), we may assume henceforth that (E) possesses the equilibrium of interest located at the origin. The equilibrium $x = 0$ will sometimes be referred to as the **trivial solution** of (E).

B. Definitions of Stability and Boundedness

We now state several definitions of stability of an equilibrium point, in the sense of Lyapunov.

The equilibrium $x = 0$ of (E) is **stable** if for every $\varepsilon > 0$ and any $t_0 \in R^+$ there exists a $\delta(\varepsilon, t_0) > 0$ such that:

$$|\phi(t, t_0, \xi)| < \varepsilon \qquad \text{for all} \quad t \geq t_0 \tag{93}$$

whenever

$$|\xi| < \delta(\varepsilon, t_0) \tag{94}$$

It is an easy matter to show that if the equilibrium $x = 0$ satisfies Eq. (93) for a single t_0 when Eq. (94) is true, then it also satisfies this condition at every initial time $t_0' > t_0$. Hence, in the preceding definition it suffices to take the single value $t = t_0$ in Eqs. (93) and (94).

Suppose that the initial-value problem (I) has a unique solution ϕ defined for t on an interval J containing t_0. By the **motion** through $(t_0, \xi = x(t_0))$ we mean the set $\{t, \phi(t) : t \in J\}$. This is, of course, the graph of the function ϕ. By the **trajectory** or **orbit** through $(t_0, \xi = x(t_0))$ we mean the set $C(x(t_0)) = \{\phi(t) : t \in J\}$. The **positive semitrajectory** (or **positive semiorbit**) is defined as $C^+(x(t_0)) = \{\phi(t) : t \in J$ and $t \geq t_0\}$. Also, the **negative trajectory** (or **negative semiorbit**) is defined as $C^-(x(t_0)) = \{\phi(t) : t \in J$ and $t \leq \tau\}$.

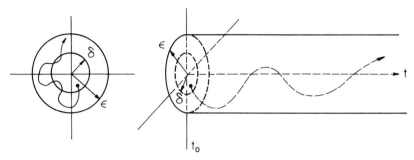

FIGURE 9 Stability of an equilibrium point.

Now, in Fig. 9 we depict the behavior of the trajectories in the vicinity of a stable equilibrium for the case $x \in R^2$.

When $x_e = 0$ is stable, by choosing the initial points in a sufficiently small spherical neighborhood, we can force the graph of the solution for $t \geq t_0$ to lie entirely inside a given cylinder.

In the above definition of stability, δ depends on ε and t_0 (i.e., $\delta = \delta(\varepsilon, t_0)$). If δ is independent of t_0 (i.e., $\delta = \delta(\varepsilon)$). then the equilibrium $x = 0$ of (E) is said to be **uniformly stable**.

The equilibrium $x = 0$ of (E) is said to be **asymptotically stable** if (1) it is stable, and (2) for every $t_0 \geq 0$ there exists an $\eta(t_0) > 0$ such that $\lim_{t \to \infty} \phi(t, t_0, \xi) = 0$ whenever $|\xi| < \eta$. Furthermore, the set of all $\xi \in R^n$ such that $\phi(t, t_0, \xi) \to 0$ as $t \to \infty$ for some $t_0 \geq 0$ is called the **domain of attraction** of the equilibrium $x = 0$ of (E). Also, if for (E) condition (2) is true, then the equilibrium $x = 0$ is said to be **attractive**.

The equilibrium $x = 0$ of (E) is said to be **uniformly asymptotically stable** if (1) it is uniformly stable, and (2) there is a $\delta_0 > 0$ such that for every $\varepsilon > 0$ and for any $t_0 \in R^+$ there exists a $T(\varepsilon) > 0$, independent of t_0, such that $|\phi(t, t_0, \xi)| < \varepsilon$ for all $t \geq t_0 + T(\varepsilon)$ whenever $|\xi| < \delta_0$.

In Fig. 10 we depict property (2), for uniform asymptotic stability, pictorially. *By choosing the initial points in a sufficiently small spherical neighborhood at $t = t_0$, we can force the graph of the solution to lie inside a given cylinder for all $t > t_0 + T(\varepsilon)$.* Condition (2) can be rephrased by saying that there exists a $\delta_0 > 0$ such that

$\lim_{t \to \infty} \phi(t + t_0, t_0, \xi) = 0$ uniformly in (t_0, ξ) for $t_0 \geq 0$ and for $|\xi| \leq \delta_0$.

In applications, we are frequently interested in the following special case of uniform asymptotic stability: the equilibrium $x = 0$ of (E) is **exponentially stable** if there exists an $\alpha > 0$, and for every $\varepsilon > 0$ there exists a $\delta(\varepsilon) > 0$, such that $|\phi(t, t_0, \xi)| \leq \varepsilon e^{\alpha(t-t_0)}$ for all $t \geq t_0$ whenever $|\xi| < \delta(\varepsilon)$ and $t \geq 0$.

In Fig. 11, the behavior of a solution in the vicinity of an exponentially stable equilibrium $x = 0$ is shown.

The equilibrium $x = 0$ of (E) is said to be **unstable** if it is not stable. In this case, there exists a $t_0 \geq 0$, $\varepsilon > 0$, a sequence $\xi_m \to 0$ of initial points, and a sequence $\{t_m\}$ such that $|\phi(t_0 + t_m, t_0, \xi_m)| \geq \varepsilon$ for all m, $t_m \geq 0$.

If $x = 0$ is an unstable equilibrium of (E), it still can happen that all the solutions tend to zero with increasing t. Thus, instability and attractivity are compatible concepts. Note that the equilibrium $x = 0$ is necessarily unstable if every neighborhood of the origin contains initial points corresponding to unbounded solutions (i.e., solutions whose norm $|\phi(t, t_0, \xi)|$ grows to infinity on a sequence $t_m \to \infty$). However, it can happen that a system (E) with unstable equilibrium $x = 0$ may have only bounded solutions.

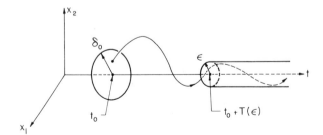

FIGURE 10 Attractivity of an equilibrium point.

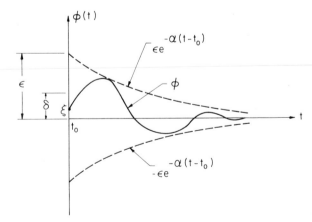

FIGURE 11 Exponential stability of an equilibrium point.

The above concepts pertain to *local properties of an equilibrium*. In the following definitions, we consider some *global characterizations of an equilibrium*.

A solution $\phi(t, t_0, \xi)$ of (E) is **bounded** if there exists a $\beta > 0$ such that $|\phi(t, t_0, \xi)| < \beta$ for all $t \geq t_0$, where β may depend on each solution. System (E) is said to possess **Lagrange stability** if for each $t_0 \geq 0$ and ξ the solution $\phi(t, t_0, \xi)$ is bounded.

The solutions of (E) are **uniformly bounded** if for any $\alpha > 0$ and $t_0 \in R^+$ there exists a $\beta = \beta(\alpha) > 0$ (independent of t_0) such that if $|\xi| < \alpha$, then $|\phi(t, t_0, \xi)| < \beta$ for all $t \geq t_0$.

The solutions of (E) are **uniformly ultimately bounded** (with bound B) if there exists a $B > 0$ and if corresponding to any $\alpha > 0$ and $t_0 \in R^+$ there exists a $T = T(\alpha)$ (independent of t_0) such that $|\xi| < \alpha$ implies that $|\phi(t, t_0, \xi)| < \beta$ for all $t \geq t_0 + T$.

In contrast to the boundedness properties given in the preceding three paragraphs, the concepts introduced earlier as well as those stated in the following are usually referred to as stability (respectively, instability) **in the sense of Lyapunov**.

The equilibrium $x = 0$ of (E) is **asymptotically stable in the large** if it is stable and if every solution of (E) tends to zero as $t \to \infty$. In this case, the domain of attraction of the equilibrium $x = 0$ of (E) is all of R^n. Note that in this case, $x = 0$ is the *only* equilibrium of (E).

The equilibrium $x = 0$ of (E) is **uniformly asymptotically stable in the large** if (1) it is uniformly stable, and (2) for any $\alpha > 0$ and any $\varepsilon > 0$, and $t_0 \in R^+$, there exists $T(\varepsilon, \alpha) > 0$, independent of t_0 such that if $|\xi| < \alpha$, then $|\phi(t, t_0, \xi)| < \varepsilon$ for all $t \geq t_0 + T(\varepsilon, \alpha)$.

Finally, the equilibrium $x = 0$ of (E) is **exponentially stable in the large** if there exists $\alpha > 0$ and for any $\beta > 0$, there exists $k(\beta) > 0$ such that $|\phi(t, t_0, \xi)| \leq k(\beta)|\xi|e^{-\alpha(t-t_0)}$ for all $t \geq t_0$ whenever $|\xi| < \beta$.

At this point it may be worthwhile to consider some specific examples:

1. The scalar equation,

$$x' = 0 \tag{95}$$

has for any initial condition $x(0) = c$ the solution $\phi(t, 0, c) = c$. All solutions are equilibria of Eq. (95). The trivial solution is *stable*; in fact, it is *uniformly stable*. However, it is not asymptotically stable.

2. The scalar equation,

$$x' = ax, \qquad a > 0 \tag{96}$$

has for every $x(0) = c$ the solution $\phi(t, 0, c) = ce^{at}$ and $x = 0$ is the only equilibrium of Eq. (96). This equilibrium is *unstable*.

3. The scalar equation,

$$x' = -ax, \qquad a > 0 \tag{97}$$

has for every $x(0) = c$ the solution $\phi(t, 0, c) = ce^{-at}$ and $x = 0$ is the only equilibrium of Eq. (97). This equilibrium is *exponentially stable in the large*.

4. The scalar equation,

$$x' = [-1/(t + 1)]x \tag{98}$$

has for every $x(t_0) = c, t_0 \geq 0$, a unique solution of the form $\phi(t, t_0, c) = (1 + t_0)c/(t + 1)$ and $x = 0$ is the only equilibrium of Eq. (98). This equilibrium is *uniformly stable* and *asymptotically stable in the large*, but it is not uniformly asymptotically stable.

C. Some Basic Properties of Autonomous and Periodic Systems

Making use of the properties of solutions and using the definitions of stability in the sense of Lyapunov it is not difficult to establish the following general stability results for systems described by:

$$x' = f(x) \tag{A}$$

and

$$x' = f(t, x), \qquad f(t, x) = f(t + T, x) \tag{P}$$

1. *If the equilibrium $x = 0$ of (P) (or of (A)) is stable, then it is in fact uniformly stable.*

2. *If the equilibrium $x = 0$ of (P) (or of (A)) is asymptotically stable, then it is uniformly asymptotically stable.*

D. Linear Systems

Next, by making use of the general properties of the solutions of linear autonomous homogeneous systems,

$$x' = Ax, \qquad t \geq 0 \tag{L}$$

and of linear homogeneous systems (with $A(t)$ continuous),

$$x' = A(t)x, \qquad t \geq t_0, \quad t_0 \geq 0 \tag{LH}$$

the following results are easily verified:

1. *The equilibrium $x = 0$ of (LH) is stable if and only if the solutions of (LH) are bounded. Equivalently, the equilibrium $x = 0$ of (LH) is stable if and only if*:

$$\sup_{t \geq t_0} |\Phi(t, t_0)| \overset{\Delta}{=} c(t_0) < \infty$$

where $|\Phi(t, t_0)|$ denotes the matrix norm induced by the vector norm used on R^n and sup denotes supremum.

2. *The equilibrium $x = 0$ of (LH) is uniformly stable if and only if:*

$$\sup_{t_0 \geq 0} c(t_0) \stackrel{\Delta}{=} \sup_{t_0 \geq 0} \left(\sup_{t \geq t_0} |\Phi(t, t_0)| \right) \stackrel{\Delta}{=} c_0 < \infty$$

3. *The following statements are equivalent:* (a) *The equilibrium $x = 0$ of (LH) is asymptotically stable.* (b) *The equilibrium $x = 0$ of (LH) is asymptotically stable in the large.* (c) $\lim_{t \to \infty} = 0$.

4. *The equilibrium $x = 0$ of (LH) is uniformly asymptotically stable if and only if it is exponentially stable.*

5. (a) *The equilibrium $x = 0$ of (L) is stable if all eigenvalues of A have nonpositive real parts and every eigenvalue of A that has a zero real part is a simple zero of the characteristic polynomial of A.* (b) *The equilibrium $x = 0$ of (L) is asymptotically stable if and only if all eigenvalues of A have negative real parts. In this case, there exist constants $k > 0$, $\sigma > 0$ such that $|\Phi(t, t_0)| \leq k \exp[-\sigma(t - t_0)]$, $t_0 \leq t \leq \infty$, where $\Phi(t, t_0)$ denotes the state transition matrix of (L).*

We shall find it convenient to use the following convention, which has become standard in the literature: A real $n \times n$ matrix A is called **stable** or a **Hurwitz matrix** if all of its eigenvalues have negative real parts. If at least one of the eigenvalues has a positive real part, then A is called **unstable**. A matrix A which is neither stable nor unstable is called **critical** and the eigenvalues of A with zero real parts are called **critical eigenvalues**.

Thus, the equilibrium $x = 0$ of (L) is asymptotically stable if and only if A is stable. If A is unstable, then $x = 0$ is unstable. If A is critical, then the equilibrium is stable if the eigenvalues with zero real parts correspond to a simple zero of the characteristic polynomial of A; otherwise, the equilibrium may be unstable.

Next, we consider the stability properties of linear periodic systems,

$$x' = A(t)x, \qquad A(t) = A(t + T) \qquad \text{(PL)}$$

where $A(t)$ is a continuous matrix for all $t \in R$. For such systems, the following results follow directly from Floquet theory:

1. *The equilibrium $x = 0$ of (PL) is uniformly stable if all eigenvalues of R in Eq. (64) have nonpositive real parts and any eigenvalue of R having zero real part is a simple zero of the characteristic polynomial of R.*

2. *The equilibrium $x = 0$ of (PL) is uniformly asymptotically stable if and only if all eigenvalues of R have negative real parts.*

E. Two-Dimensional Linear Systems

Before we present the principal results of the Lyapunov theory for general systems (E), we consider in detail the behavior of trajectories near an equilibrium point $x = 0$ of two-dimensional linear systems of the form

$$
\begin{aligned}
x_1' &= a_{11}x_1 + a_{12}x_2 \\
x_2' &= a_{21}x_1 + a_{22}x_2
\end{aligned}
\qquad (99)
$$

We can rewrite Eq. (99) equivalently as

$$x' = Ax \qquad (100)$$

where

$$A = \begin{bmatrix} a_{11} & a_{12} \\ a_{21} & a_{22} \end{bmatrix} \qquad (101)$$

When $\det A \neq 0$ system (99) will have one and only one equilibrium point, namely $x = 0$. We shall classify this equilibrium point (and, hence, system (99)) according to the following cases which the eigenvalues λ_1, λ_2 of A can assume:

1. λ_1, λ_2 are real and $\lambda_1 < 0$, $\lambda_2 < 0$: $x = 0$ is an asymptotically stable equilibrium point called a **stable node**.

2. λ_1, λ_2 are real and $\lambda > 0$, $\lambda_2 > 0$: $x = 0$ is an unstable equilibrium point called an **unstable node**.

3. λ_1, λ_2 are real and $\lambda_1 \lambda_2 < 0$: $x = 0$ is an unstable equilibrium point called a **saddle**.

4. λ_1, λ_2 are complex conjugates and Re $\lambda_1 =$ Re $\lambda_2 < 0$: $x = 0$ is an asymptotically stable equilibrium point called a **stable focus**.

5. λ_1, λ_2 are complex conjugates and Re $\lambda_1 =$ Re $\lambda_2 > 0$: $x = 0$ is an unstable equilibrium point called an **unstable focus**.

6. λ_1, λ_2 are complex conjugates and Re $\lambda_1 =$ Re $\lambda_2 = 0$: $x = 0$ is a stable equilibrium called a **center**.

Using the results of Section IV, it is possible to solve Eq. (99) explicitly and verify that the qualitative behavior of the trajectories near the equilibrium $x = 0$ is as shown in Figs. 12–14 for the cases of a stable node, unstable node,

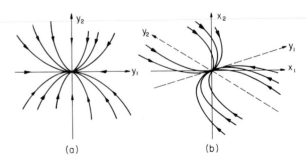

FIGURE 12 Trajectories near a stable node.

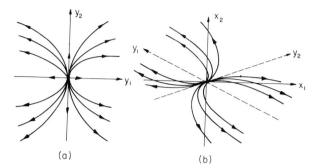

FIGURE 13 Trajectories near an unstable node.

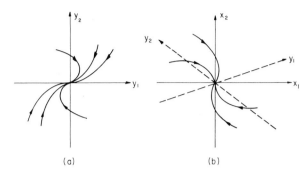

FIGURE 15 Trajectories near a stable node (repeated eigenvalue case).

and saddle, respectively. (The arrows on the trajectories point into the direction of increasing time.) In these cases, the figures labeled (a) correspond to systems in which A is in Jordan canonical form while the figures labeled (b) correspond to systems in which A is in some arbitrary form. In a similar manner, it is possible to verify that the qualitative behavior of the trajectories near the equilibrium $x = 0$ is as shown in Figs. 15–18 for the cases of a stable node (repeated eigenvalue case), an unstable focus, a stable focus, and a center, respectively. (For purposes of convention, we assumed in these figures that when λ_1, λ_2 are real and not equal, then $\lambda_1 > \lambda_2$.)

F. Lyapunov Functions

Next, we present general stability results for the equilibrium $x = 0$ of a system described by (E). Such results involve the existence of realvalued functions $v: D \to R$. In the case of local results (e.g., stability, instability, asymptotic stability, and exponential stability results), we shall usually only require that $D = B(h) \subset R^n$ for some $H > 0$, or $D = R^+ \times B(h)$. (Recall that $R^+ = (0, \infty)$ and $B(h) = \{x \in R^n: |x| < h\}$ where $|x|$ denotes any one of the equivalent norms of x on R^n.) On the other hand, in the case of global results (e.g., asymptotic stability in the large, exponential stability in the large, and uniform boundedness of solutions), we have to assume that $D = R^n$ or

$D = R^+ \times R^n$. Unless stated otherwise, we shall always assume that $v(t, 0) = 0$ for all $t \in R^+$ (resp., $v(0) = 0$).

Now let ϕ be an arbitrary solution of (E) and consider the function $t \mapsto v(t, \phi(t))$. If v is continuously differentiable with respect to all of its arguments, then we obtain (by the chain rule) the derivative of v with respect to t along the solutions of (E), $v'_{(E)}$, as:

$$v'_{(E)}(t, \phi(t)) = \frac{\partial v}{\partial t}(t, \phi(t)) + \nabla v(t, \phi(t))^{\mathrm{T}} f(t, \phi(t))$$

Here, ∇v denotes the gradient vector of v with respect to x. For a solution $\phi(t, t_0, \xi)$ of (E), we have

$$v(t, \phi(t)) = v(t_0, \xi) + \int_{t_0}^{t} v'_{(E)}(\tau, \phi(\tau, t_0, \xi))\, d\tau$$

The above observations motivate the following: let $v: R^+ \times R^n \to R$ (resp., $v: R^+ \times B(h) \to R$) be continuously differentiable with respect to all of its arguments and let ∇v denote the gradient of v with respect to x. Then $v'_{(E)}: R^+ \times R^n \to R$ (resp., $v'_{(E)}: R^+ \times B(h) \to R$) is defined by:

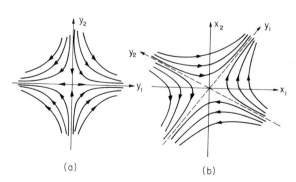

FIGURE 14 Trajectories near a saddle.

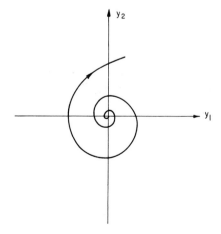

FIGURE 16 Trajectory near an unstable focus.

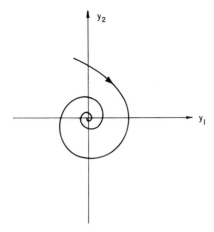

FIGURE 17 Trajectory near a stable focus.

$$v'_{(E)}(t, x) = \frac{\partial v}{\partial t}(t, x) + \sum_{i=1}^{n} \frac{\partial v}{\partial x_i}(t, x) f_i(t, x)$$

$$= \frac{\partial v}{\partial t}(t, x) + \nabla v(t, x)^{\mathrm{T}} f(t, x) \qquad (102)$$

We call $v'_{(E)}$ the **derivative of v (with respect to t) along the solutions of** (E).

It is important to note that in Eq. (102) the derivative of v with respect to t, along the solutions of (E), is evaluated *without having to solve* (E). The significance of this will become clear later. We also note that when $v: R^n \to R$ (resp., $v: B(h) \to R$), then Eq. (102) reduces to $v'_{(E)}(t, x) = \nabla v(x)^{\mathrm{T}} f(t, x)$. Also, in the case of autonomous systems (A), if $v: R^n \to R$ (resp., $v: B(h) \to R$), we have

$$v'_{(A)}(x) = \nabla v(x)^{\mathrm{T}} f(x) \qquad (103)$$

Occasionally, we shall require only that v be continuous on its domain of definition and that it satisfy locally a

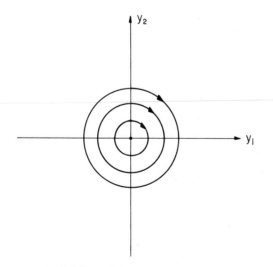

FIGURE 18 Trajectories near a center.

Lipschitz condition with respect to x. In such cases we define the **upper right-hand derivative of v with respect to t along the solutions of** (E) by:

$$v'_{(E)}(t, x) = \lim_{\theta \to 0^+} \sup(1/\theta)\{v(t + \theta,$$

$$\phi(t + \theta, t, x)) - v(t, x)\}$$

$$= \lim_{\theta \to 0^+} \sup(1/\theta)\{v(t + \theta,$$

$$x + \theta \cdot f(t, x)) - v(t, x)\} \qquad (104)$$

When v is continuously differentiable, then Eq. (104) reduces to Eq. (102).

We now give several important properties that v functions may possess. In doing so, we employ **Kamke comparison functions** defined as follows: a continuous function $\psi: [0, r_1] \to R^+$ (resp., $\psi: (0, \infty) \to R^+$) is said to belong to the **class K** (i.e., $\psi \in K$), if $\psi(0) = 0$ and if ψ is strictly increasing on $[0, r_1]$ (resp., on $[0, \infty)$). If $\psi: R^+ \to R^+$, if $\psi \in K$ and if $\lim_{r \to \infty} \psi(r) = \infty$, then ψ is said to belong to **class KR**.

We are now in a position to characterize v-functions in several ways. In the following, we assume that $v: R^+ \times R^n \to R$ (resp., $v: R^+ \times B(h) \to R$), that $v(0, t) = 0$ for all $t \in R^+$, and that v is continuous.

1. v is **positive definite** if, for some $r > 0$, there exists a $\psi \in K$ such that $v(t, x) \geq \psi(|x|)$ for all $t \geq 0$ and for all $x \in B(r)$.

2. v is **decrescent** if there exists a $\psi \in K$ such that $|v(t, x)| \leq \psi(|x|)$ for all $t \geq 0$ and for all $x \in B(r)$ for some $r > 0$.

3. $v: R^+ \times R^n \to R$ is **radially unbounded** if there exists a $\psi \in KR$ such that $v(t, x) \geq \psi(|x|)$ for all $t \geq 0$ and for all $x \in R^n$.

4. v is **negative definite** if $-v$ is positive definite.

5. v is **positive semidefinite** if $v(t, x) \geq 0$ for all $x \in B(r)$ for some $r > 0$ and for all $t \geq 0$.

6. v is **negative semidefinite** if $-v$ is positive semidefinite.

The definitions involving the above concepts, when $v: R^n \to R$ or $v: B(h) \to R$ (where $B(h) \subset R^n$ for some $h > 0$) involve obvious modifications. We now consider several specific cases:

1. The function $v: R^3 \to R$ given by $v(x) = x^{\mathrm{T}}x = x_1^2 + x_2^2 + x_3^2$ is positive definite and radially unbounded. (Here, x^{T} denotes the transpose of x.)

2. The function $v: R^3 \to R$ given by $v(x) = x_1^2 + (x_2 + x_3)^2$ is positive semidefinite (but not positive definite).

3. The function $v: R^2 \to R$ given by $v(x) = x_1^2 + x_2^2 - (x_1^2 + x_2^2)^3$ is positive definite but not radially unbounded.

4. The function $v: R^3 \to R$ given by $v(x) = x_1^2 + x_2^2$ is positive semidefinite (but not positive definite).

5. The function $v: R^2 \to R$ given by $v(x) = x_1^4/(1 + x_1^4) + x_2^4$ is positive definite but not radially unbounded.

6. The function $v: R^+ \times R^2 \to R$ given by $v(t, x) = (1 + \cos^2 t)x_1^2 + 2x_2^2$ is positive definite, decrescent, and radially unbounded.

7. The function $v: R^+ \times R^2 \to R$ given by $v(t, x) = (x_1^2 + x_2^2)\cos^2 t$ is positive semidefinite and decrescent.

8. The function $v: R^+ \times R^2 \to R$ given by $v(t, x) = (1 + t)(x_1^2 + x_2^2)$ is positive definite and radially unbounded but not decrescent.

9. The function $v: R^+ \times R^2 \to R$ given by $v(t, x) = x_1^2/(1 + t) + x_2^2$ is decrescent and positive semidefinite but not positive definite.

10. The function $v: R^+ \times R^2 \to R$ given by $v(t, x) = (x_2 - x_1)^2(1 + t)$ is positive semidefinite but not positive definite or decrescent.

Of special interest are functions $v: R^n \to R$ that are **quadratic forms** given by:

$$v(x) = x^T B x = \sum_{i,k=1}^{n} b_{ik} x_i x_k \qquad (105)$$

where $B = [b_{ij}]$ is a real symmetric $n \times n$ matrix (i.e., $B^T = B$). Since B is symmetric, it is diagonizable and all of its eigenvalues are real. For Eq. (105) one can prove the following:

1. v is positive definite (and radially unbounded) if and only if all principal minors of B are positive; that is, if and only if

$$\det \begin{bmatrix} b_{11} & \cdots & b_{1k} \\ \vdots & & \vdots \\ b_{k1} & \cdots & b_{kk} \end{bmatrix} > 0, \qquad k = 1, \ldots, n$$

2. v is negative definite if and only if

$$(-1)^k \det \begin{bmatrix} b_{11} & \cdots & b_{1k} \\ \vdots & & \vdots \\ b_{k1} & \cdots & b_{kk} \end{bmatrix} > 0, \qquad k = 1, \ldots, n$$

3. v is **definite** (i.e., either positive definite or negative definite) if and only if all eigenvalues are nonzero and have the same sign.

4. v is **semidefinite** (i.e., either positive semidefinite or negative semidefinite) if and only if the nonzero eigenvalues of B have the same sign.

5. If λ_m and λ_M denote the smallest and largest eigenvalues of B and if $|x|$ denotes the Euclidean norm of x, then $\lambda_m |x|^2 \leq v(x) \leq \lambda_M |x|^2$ for all $x \in R^n$. (The Euclidean norm of x is defined as $(x^T x)^{1/2} = (\sum_{i=1}^{n} x_i^2)^{1/2}$.)

6. v is **indefinite** (i.e., in every neighborhood of the origin $x = 0$, v assumes positive and negative values) if and only if B possesses both positive and negative eigenvalues.

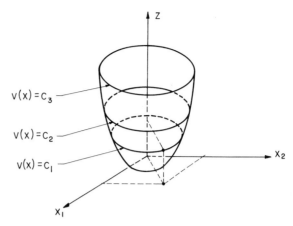

FIGURE 19 Surface described by a quadratic form.

Quadratic forms (105) have some interesting geometric properties. To see this, let $n = 2$, and assume that both eigenvalues of B are positive so that v is positive definite and radially unbounded. In R^3, let us now consider the surface determined by:

$$z = v(x) = x^T B x \qquad (106)$$

This equation describes a cup-shaped surface as depicted in Fig. 19. Note that corresponding to every point on this cup-shaped surface there exists one and only one point in the $x_1 x_2$ plane. Note also that the loci defined by $C_i = \{x \in R^2 : v(x) = c_i \geq 0\}$, $c_i = $ const, determine closed curves in the $x_1 x_2$ plane as shown in Fig. 20. We call these curves **level curves**. Note that $C_0 = \{0\}$ corresponds to the case in which $z = c_0 = 0$. Note also that this function v can be used to *cover the entire* R^2 plane with closed curves by selecting for z all values in R^+.

In the case when $v = x^T B x$ is a positive definite quadratic form with $x \in R^n$, the preceding comments are still true; however, in this case, the closed curves C_i must be replaced by closed hypersurfaces in R^n and a simple geometric visualization as in Figs. 19 and 20 is no longer possible.

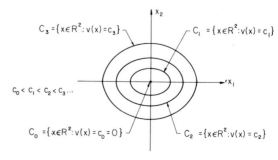

FIGURE 20 Level curves determined by a quadratic form.

G. Lyapunov Stability and Instability Results: Motivation

Before we summarize the principal Lyapunov-type of stability and instability results, we give a geometric interpretation of some of these results in R^2. To this end, we consider the system of equations,

$$x_1' = f_1(x_1, x_2), \qquad x_2' = f_2(x_1, x_2) \qquad (107)$$

and we assume that f_1 and f_2 are such that for every $(t_0, x_0), t_0 \geqslant 0$, Eq. (107) has a unique solution $\phi(t, t_0, x_0)$ with $\phi(t_0, t_0, x_0) = x_0$. We also assume that $(x_1, x_2)^{\mathrm{T}} = (0, 0)^{\mathrm{T}}$ is the only equilibrium in $B(h)$ for some $h > 0$.

Next, let v be a positive definite, continuously differentiable function with nonvanishing gradient ∇v on $0 < |x| \leqslant h$. Then, $v(x) = c, c \geqslant 0$, defines for sufficiently small constants $c > 0$ a family of closed curves C_i which cover the neighborhood $B(h)$ as shown in Fig. 21. Note that the origin $x = 0$ is located in the interior of each such curve and in fact, $C_0 = \{0\}$.

Now suppose that all trajectories of Eq. (107) originating from points on the circular disk $|x| \leq r_1 < h$ cross the curves $v(x) = c$ from the exterior toward the interior when we proceed along these trajectories in the direction of increasing values of t. Then, we can conclude that these trajectories approach the origin as t increases (i.e., the equilibrium $x = 0$ is in this case asymptotically stable).

In terms of the given v function, we have the following interpretation. For a given solution $\phi(t, t_0, x_0)$ to cross the curve $v(x) = r, r = v(x_0)$, the angle between the outward normal vector $\nabla v(x_0)$ and the derivative of $\phi(t, t_0, x_0)$ at $t = t_0$ must be greater than $\pi/2$; that is,

$$v_{(107)}'(x_0) = \nabla v(x_0) f(x_0) < 0$$

For this to happen at all points, we must have $v_{(107)}'(x) < 0$ for $0 < |x| \leq r_1$. The same results can be arrived at from an analytic point of view. The function $V(t) = v(\phi(t, t_0, x_0))$ decreases monotonically as t increases. This implies that the derivative $v'(\phi(t, t_0, x_0))$ along the solution $(\phi(t, t_0, x_0))$ must be negative definite in $B(r)$ for $r > 0$ sufficiently small.

Next, let us assume that, Eq. (107) has only one equilibrium (at $x = 0$) and that v is positive definite and radially unbounded. It turns out that in this case, the relation $v(x) = c, c \in R^+$, can be used to cover all of R^2 by closed curves of the type shown in Fig. 21. If for arbitrary (t_0, x_0), the corresponding solution of Eq. (107), $\phi(t, t_0, x_0)$, behaves as already discussed, then it follows that the derivative of v along this solution, $v'(\phi(t, t_0, x_0))$, will be negative definite in R^2.

Since the foregoing discussion was given in terms of an arbitrary solution of Eq. (107), we may suspect that the following results are true:

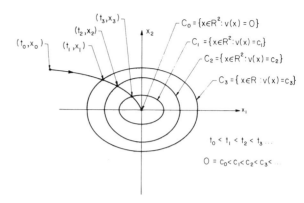

FIGURE 21 Trajectory near an asymptotically stable equilibrium point.

1. *If there exists a positive definite function v such that $v_{(107)}'$ is negative definite, then the equilibrium $x = 0$ of Eq. (107) is asymptotically stable.*

2. *If there exists a positive definite and radially unbounded function v such that $v_{(107)}'$ is negative definite for all $x \in R^2$, then the equilibrium $x = 0$ of Eq. (107) is asymptotically stable in the large.*

Continuing our discussion by making reference to Fig. 22, let us assume that we can find for Eq. (107) a continuously differentiable function $v: R^2 \to R$ which is indefinite and has the properties discussed below. Since v is indefinite, there exist in each neighborhood of the origin points for which $v > 0$, $v < 0$, and $v(0) = 0$. Confining our attention to $B(k)$, where $k > 0$ is sufficiently small, we let $D = \{x \in B(k): v(x) < 0\}$. ($D$ may consist of several subdomains.) The boundary of D, ∂D, as shown in Fig. 22, consists of points in $\partial B(k)$ and of points determined by $v(x) = 0$. Assume that in the interior of D, v is bounded. Suppose $v_{(107)}'(x)$ is negative definite in D

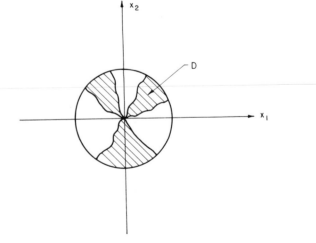

FIGURE 22 Instability of an equilibrium point.

and that $x(t)$ is a trajectory of Eq. (107) which originates somewhere on the boundary of D $(x(t_0) \in \partial D)$ with $v(x(t_0)) = 0$. Then, this trajectory will penetrate the boundary of D at points where $v = 0$ as t increases and it can never again reach a point where $v = 0$. In fact, as t increases, this trajectory will penetrate the set of points determined by $|x| = k$ (since; by assumption, $v'_{(107)} < 0$ along this trajectory and $v < 0$ in D). But, this indicates that the equilibrium $x = 0$ of Eq. (107) is unstable.

We are once more led to a conjecture:

3. *Let a function $v: R^2 \rightarrow R$ be given which is continuously differentiable and which has the following properties*: (a) *There exist points x arbitrarily close to the origin such that $v(x) < 0$; they form the domain D bounded by the set of points determined by $v = 0$ and the disk $|x| = k$.* (b) *In the interior of D, v is bounded.* (c) *In the interior of D, $v'_{(107)}$ is negative. Then, the equilibrium $x = 0$ of Eq. (107) is unstable.*

H. Principal Lyapunov Stability and Instability Theorems

It turns out that results of the type given above for Eq. (107) are true for general systems (E). These results, which are proved by standard δ–ε arguments, comprise the **direct method of Lyapunov**, which is also sometimes called the **second method of Lyapunov**. The reason for this nomenclature is clear: Results of the kind presented here allow us to make qualitative statements about whole families of solutions of (E), without actually solving this equation.

In the following, we enumerate some of the more important results of the direct method. We shall assume that $v: R^+ \times B(h) \rightarrow R$ (resp., $v: R^+ \times R^n \rightarrow R$).

1. *If there exists a continuously differentiable positive definite function v with a negative semidefinite (or identically zero) derivative $v'_{(E)}$, then the equilibrium $x = 0$ of (E) is stable.*

As an example, consider the system given by:

$$x'_1 = x_2, \qquad x'_2 = -x_2 - e^{-t}x_1 \qquad (108)$$

which has an equilibrium at $(x_1, x_2)^T = (0, 0)^T$. For Eq. (108), choose the positive definite function $v(t, x_1, x_2) = x_1^2 + e^t x_2^2$. We obtain $v'_{(108)}(t, x_1, x_2) = -e^t x_2^2$ which is negative semidefinite. We conclude that the equilibrium $x = 0$ of Eq. (108) is *stable*.

2. *If there exists a continuously differentiable, positive definite, decrescent function v with negative semidefinite derivative $v'_{(E)}$, then the equilibrium $x = 0$ of (E) is uniformly stable.*

As an example, consider the simple pendulum,

$$x'_1 = x_2, \qquad x'_2 = -k \sin x_1 \qquad (109)$$

where $k > 0$ is a constant. As noted earlier, Eq. (109) has an isolated equilibrium at $x = 0$. Choose $v(x_1, x_2) = \frac{1}{2}x_2^2 + k \int_0^{x_1} \sin \eta \, d\eta$, which is continuously differentiable and positive definite. Also, since v does not depend on t, it will automatically be decrescent. Furthermore, $v'_{(109)}(x_1, x_2) = (k \sin x_1)x'_1 + x_2 x'_2 = (k \sin x_1)x_2 + x_2(-k \sin x_1) = 0$. Therefore, the equilibrium $x = 0$ of Eq. (109) is *uniformly stable*.

3. *If there exists a continuously differentiable, positive definite, decrescent function v with a negative definite derivative $v'_{(E)}$, then the equilibrium $x = 0$ of (E) is uniformly asymptotically stable.*

For an example, the system,

$$\begin{aligned}
x'_1 &= (x_1 - c_2 x_2)(x_1^2 + x_2^2 - 1) \\
x'_2 &= (c_1 x_1 + x_2)(x_1^2 + x_2^2 - 1)
\end{aligned} \qquad (110)$$

has an isolated equilibrium at the origin $x = 0$. Choosing $v(x) = c_1 x_1^2 + c_2 x_2^2$, we obtain $v'_{(110)}(x) = 2(c_1 x_1^2 + c_2 x_2^2)(x_1^2 + x_2^2 - 1)$. If $c_1 > 0, c_2 > 0$, then v is positive definite (and decrescent) and $v'_{(110)}$ is negative definite in the domain $x_1^2 + x_2^2 < 1$. Therefore, the equilibrium $x = 0$ of Eq. (110) *is uniformly asymptotically stable*.

4. *If there exists a continuously differentiable, positive definite, decrescent, and radially unbounded function v such that $v'_{(E)}$ is negative definite for all $(t, x) \in R^+ \times R^n$, then the equilibrium $x = 0$ of (E) is uniformly asymptotically stable in the large.*

As an example, consider the system,

$$\begin{aligned}
x'_1 &= x_2 + c x_1(x_1^2 + x_2^2) \\
x'_2 &= -x_1 + c x_2(x_1^2 + x_2^2)
\end{aligned} \qquad (111)$$

where c is a real constant. Note that $x = 0$ is the only equilibrium. Choosing the positive definite, decrescent, and radially unbounded function $v(x) = x_1^2 + x_2^2$, we obtain $v'_{(111)}(x) = 2c(x_1^2 + x_2^2)^2$. We conclude that if $c = 0$, then $x = 0$ of Eq. (111) is *uniformly stable* and if $c < 0$, then $x = 0$ of Eq. (111) is *uniformly asymptotically stable in the large*.

5. *If there exists a continuously differentiable function v and three positive constants c_1, c_2, and c_3 such that*

$$c_1 |x|^2 \leq v(t, x) \leq c_2 |x|^2$$

$$v'_{(E)}(t, x) \leq -c_3 |x|^2$$

for all $t \in R^+$ and for all $x \in B(r)$ for some $r > 0$, then the equilibrium $x = 0$ of (E) is exponentially stable.

6. *If there exist a continuously differentiable function v and three positive constants c_1, c_2, and c_3 such that*

$$c_1|x|^2 \leq v(t, x) \leq c_2|x|^2$$

$$v'_{(E)}(t, x) \leq -c_3|x|^2$$

for all $t \in R^+$ and for all $x \in R^n$, then the equilibrium $x = 0$ of (E) is exponentially stable in the large.

As an example, consider the system,

$$x'_1 = -a(t)x_1 - bx_2$$
$$x'_2 = bx_1 - c(t)x_2 \tag{112}$$

where b is a real constant and where a and c are real and continuous functions defined for $t \geq 0$ satisfying $a(t) \geq \delta > 0$ and $c(t) \geq \delta > 0$ for all $t \geq 0$. We assume that $x = 0$ is the only equilibrium for Eq. (112). If we choose $v(x) = \frac{1}{2}(x_1^2 + x_2^2)$, then $v'_{(112)}(t, x) = -a(t)x_1^2 - c(t)x_2^2 \leq -\delta(x_1^2 + x_2^2)$. Hence, the equilibrium $x = 0$ of Eq. (112) is exponentially stable in the large.

7. If there exists a continuously differentiable function v defined on $|x| \geq R$ (where R may be large) and $0 \leq t \leq \infty$, and if there exist $\psi_1, \psi_2 \in KR$ such that $\psi_1(|x|) \leq v(t, x) \leq \psi_2(|x|)$, $v'_{(E)}(t, x) \leq 0$ for all $|x| \geq R$ and for all $0 \leq t < \infty$, then the solutions of (E) are uniformly bounded.

8. If there exists a continuously differentiable function v defined on $|x| \geq R$ (where R may be large) and $0 \leq t < \infty$, and if there exist $\psi_1, \psi_2 \in KR$ and $\psi_3 \in K$ such that $\psi_1(|x|) \leq v(t, x) \leq \psi_2(|x|)$, $v'_{(E)}(t, x) \leq -\psi_3(|x|)$ for all $|x| \geq R$ and $0 \leq t < \infty$, then the solutions of (E) are uniformly ultimately bounded.

As an example, consider the system,

$$x' = -x - \sigma, \qquad \sigma' = -\sigma - f(\sigma) + x \tag{113}$$

where $f(\sigma) = \sigma(\sigma^2 - 6)$. There are isolated equilibrium points at $x = \sigma = 0$, $x = -\sigma = 2$, and $x = -\sigma = -2$. Choosing the radially unbounded and decrescent function $v(x, \sigma) = \frac{1}{2}(x^2 + \sigma^2)$, we obtain $v'_{(113)} = (x, \sigma) = -x^2 - \sigma^2(\sigma^2 - 5) \leq -x^2 - (\sigma^2 - \frac{5}{2})^2 + \frac{25}{4}$. Also $v'_{(113)}$ is negative for all (x, σ) such that $x^2 + \sigma^2 > R^2$, where, for example, $R = 10$ will do. Therefore, all solutions of Eq. (113) are uniformly bounded and, in fact, uniformly ultimately bounded.

9. The equilibrium $x = 0$ of (E) is unstable (at $t = t_0 \geq 0$) if there exists a continuously differentiable, decrescent function v such that $v'_{(E)}$ is positive definite (negative definite) and if in every neighborhood of the origin there are points x such that $v(t_0, x) > 0(v(t_0, x) < 0)$.

Reconsider system (111), this time assuming that $c > 0$. If we choose $v(x) = x_1^2 + x_2^2$, then $v'_{(111)}(x) = 2c(x_1^2 + x_2^2)^2$ and we can conclude from the above result that the equilibrium $x = 0$ of (E) is unstable.

10. Let there exist a bounded and continuously differentiable function $v: D \to R$, $D = \{(t, x) \geq t_0, x \in B(h)\}$, with the following properties: (a) $v'_{(E)}(t, x) = \lambda v(t, x) + w(t, x)$, where $\lambda > 0$ is a constant and $w(t, x)$ is either identically zero or positive semidefinite; (b) in the set $D_1 = \{(t, x): t = t_1, x \in B(h_1)\}$ for fixed $t_1 \geq t_0$ and with arbitrarily small h_1, there exist values x such that $v(t_1, x) > 0$. Then the equilibrium $x = 0$ of (E) is unstable.

As a specific example, consider:

$$x'_1 = x_1 + x_2 + x_1 x_2^4$$
$$x'_2 = x_1 + x_2 - x_1^2 x_2 \tag{114}$$

which has an isolated equilibrium $x = 0$. Choosing $v(x) = (x_1^2 - x_2^2)/2$, we obtain $v'_{(114)}(x) = \lambda v(x) + w(x)$, where $w(x) = x_1^2 x_2^4 + x_1^2 x_2^2$ and $\lambda = 2$. It follows from the above result that the equilibrium $x = 0$ of Eq. (114) is unstable.

11. Let there exist a continuously differentiable function v having the following properties: (a) For every $\varepsilon > 0$ and for every $t \geq 0$, there exist points $\bar{x} \in B(\varepsilon)$ such that $v(t, \bar{x}) < 0$. We call the set of all points (t, x) such that $x \in B(h)$ and such that $v(t, x) < 0$ the "domain $v < 0$." It is bounded by the hypersurfaces which are determined by $|x| = h$ and $v(t, x) = 0$ and it may consist of several component domains. (b) In at least one of the component domains D of the domain $v < 0$, v is bounded from below and $0 \in \partial D$ for all $t \geq 0$. (c) In the domain D, $v'_{(E)} \leq -\Psi(|v|)$, where $\psi \in K$. Then, the equilibrium $x = 0$ of (E) is unstable.

As an example, consider the system,

$$x'_1 = x_1 + x_2$$
$$x'_2 = x_1 - x_2 + x_1 x_2 \tag{115}$$

which has an isolated equilibrium at the origin $x = 0$. Choosing $v(x) = -x_1 x_2$, we obtain $v'_{(115)}(x) = -x_1^2 - x_2^2 - x_1^2 x_2$. Let $D = \{x \in R^2: x_1 > 0, x_2 > 0, \text{ and } x_1^2 + x_2^2 < 1\}$. Then, for all $x \in D$, $v < 0$ and $c'_{(115)} < 2v$. We see that the above result is applicable and conclude that the equilibrium $x = 0$ of Eq. (115) is unstable.

The results given in items 1–11 are also true when v is continuous (rather than continuously differentiable). In this case, $v'_{(E)}$ must be interpreted in the sense of Eq. (104).

For the case of systems (A),

$$x' = f(x) \tag{A}$$

it is sometimes possible to relax the conditions on $v'_{(A)}$ when investigating the asymptotic stability of the equilibrium $x = 0$, by insisting that $v'_{(A)}$ be only negative semidefinite. In doing so, we require the following concept: A set Γ of points in R^n is **invariant (with respect to** (A)) if every solution of (A) starting in Γ remains in Γ for all time.

We are now in a position to state our next result which is part of **invariance theory** for ordinary differential equations.

12. *Assume that there exists a continuously differentiable, positive definite, and radially unbounded function* $v: R^n \rightarrow R$ *such that* (a) $v'_{(A)}(x) \leq 0$ *for all* $x \in R^n$, *and* (b) *the origin* $x = 0$ *is the only invariant subset of the set* $E = \{x \in R^n : v'_{(A)}(x) = 0\}$. *Then, the equilibrium* $x = 0$ *of* (A) *is asymptotically stable in the large.*

As a specific example, consider the *Liénard equation* given by:

$$x'_1 = x_2, \qquad x'_2 = -f(x_1)x_2 - g(x_1) \qquad (116)$$

where it is assumed that f and g are continuously differentiable for all $x_1 \in R$, $g(x_1) = 0$ if and only if $x_1 = 0$, $x_1 g(x_1) > 0$ for all $x_1 \neq 0$ and $x_1 \in R$,

$$\lim_{|x_1| \to \infty} \int_0^{x_1} g(\eta)\, d\eta = \infty$$

and $f(x_1) > 0$ for all $x_1 \in R$. Then $(x_1, x_2) = (0, 0)$ is the only equilibrium of Eq. (116). Let us choose the v function,

$$v(x_1, x_2) = \frac{1}{2}x_2^2 + \int_0^{x_1} g(\eta)\, d\eta$$

(for Eq. (116)) which is positive definite and radially unbounded. Along the solutions of Eq. (116) we have $v'_{(116)}(x_1, x_2) = -x_2^2 f(x_1) \leq 0$ for all $(x_1, x_2) \in R^2$. It is easily verified that the set E in the above theorem is in our case the x_1 axis. Furthermore, a moment's reflection shows that the largest invariant subset (with respect to Eq. (116)) of the x_1 axis is the set $\{(0, 0)^T\}$. Thus, by the above result, the origin $x = 0$ of Eq. (116) is *asymptotically stable in the large.*

I. Linear Systems Revisited

One of the great drawbacks of the Lyapunov theory, as developed above, is that there exist no general rules of choosing v functions, which are called **Lyapunov functions**. However, for the case of linear systems,

$$x' = Ax \qquad (L)$$

it is possible to construct Lyapunov functions, as shown by the following result:

Assume that the matrix A has no eigenvalues on the imaginary axis. Then, there exists a Lyapunov function v of the form:

$$v(x) = x^T B x, \qquad B = B^T \qquad (117)$$

whose derivative $v'_{(L)}$, given by:

$$v'_{(L)}(x) = -x^T C x$$

where

$$-C = A^T B + B A, \qquad C = C^T \qquad (118)$$

is definite (i.e., negative definite or positive definite).

The above result shows that if A is a stable matrix, then for (L), our earlier Lyapunov result for asymptotic stability in the large constitutes also necessary conditions for asymptotic stability. Also, if A is an unstable matrix with no eigenvalues on the imaginary axis, then according to the above result, our earlier instability result given in item 9 above yields also necessary conditions for instability.

In view of the above result, the v function in Eq. (117) is easily constructed by assuming a definite matrix C (either positive definite or negative definite) and by solving the **Lyapunov matrix equation** (118) for the $n(n + 1)/2$ unknown elements of the symmetric matrix B.

J. Domain of Attraction

Next, we should address briefly the problem of estimating the domain of attraction of an equilibrium $x = 0$. Such questions are important when $x = 0$ is not the only equilibrium of a system or when $x = 0$ is not asymptotically stable in the large.

For purposes of discussion, we consider a system,

$$x' = f(x) \qquad (A)$$

and we assume that for (A) there exists a Lyapunov function v which is *positive definite* and *radially unbounded*. Also, we assume that over some domain $D \subset R^n$ containing the origin, $v'_{(A)}(x)$ is *negative*, except at the origin, where $v'_{(A)} = 0$. Now let C_i denote $C_i = \{x \in R^n : v(x) \leq c_i\}$, $c_i > 0$. Using similar reasoning as was done in connection with Eq. (107), we can show that as long as $C_i \subset D$, C_i will be a subset of the domain of attraction of $x = 0$. Thus, if $c_i > 0$ is the largest number for which this is true, then it follows that C_i will be contained in the domain of attraction of $x = 0$. The set C_i will be the best estimate that we can obtain for our particular choice of v function.

K. Converse Theorems

Above we showed that for system (L) there exist actually converse Lyapunov (stability and instability) theorems. It turns out that for virtually every result which we gave above (in items 1–11) there exists a converse. Unfortunately, these **Lyapunov converse theorems** are not much help in constructing v functions in specific cases. For purposes of illustration, we cite here an example of such a converse theorem:

If f and $f_x = \partial f / \partial x$ are continuous on the set $(R^+ \times B(r))$ for some $r > 0$, and if the equilibrium $x = 0$ of (E) is uniformly asymptotically stable, then there exists a Lyapunov function v which is continuously differentiable on $(R^+ \times B(r_1))$ for some $r_1 > 0$ such that v is positive definite and decrescent and such that $v'_{(E)}$ is negative definite.

L. Comparison Theorems

Next, we consider once more some comparison results for (E), as was done in Section III. We shall assume that $f: R^+ \times B(r) \to R^n$ for some $r_1 > 0$ and that f is continuous there. We begin by considering a scalar ordinary differential equation of the form,

$$y' = G(t, y) \qquad (\tilde{C})$$

where $y \in R$, $t \in R^+$, and $F: R^+ \times [0, r) \to R$ for some $r > 0$. Assume that G is continuous on $R^+ \times [0, r)$ and that $G(t, 0) = 0$ for all t. Also, assume that $y = 0$ is an isolated equilibrium for $(\tilde{C})$.

The following results are the basis of the **comparison principle** in the stability analysis of the isolated equilibrium $x = 0$ of (E):

Let f and G be continuous on their respective domains of definition. Let $v: R^+ \times B(r) \to R$ be a continuously differentiable, positive-definite function such that

$$v'_{(E)}(t, x) \leq G(t, v(t, x)) \qquad (119)$$

Then, the following statements are true: (1) If the trivial solution of $(\tilde{C})$ is stable, then the trivial solution of system (E) is stable. (2) If v is decrescent and if the trivial solution of $(\tilde{C})$ is uniformly stable, then the trivial solution of (E) is uniformly stable. (3) If v is decrescent and if the trivial solution of $(\tilde{C})$ is uniformly asymptotically stable, then the trivial solution of (E) is uniformly asymptotically stable. (4) If there are constants $a > 0$ and $b > 0$ such that $a|x|^b \leq v(t, x)$, if v is decrescent, and if the trivial solution of $(\tilde{C})$ is exponentially stable, then the trivial solution of (E) is exponentially stable. (5) If $f: R^+ \times R^n \to R^n$, $G: R^+ \times R \to R$, $v: R^+ \times R^n \to R$ is decrescent and radially unbounded, if Eq. (119) holds for all $t \in R^+$, $x \in R^n$, and if the solutions of $(\tilde{C})$ are uniformly bounded (uniformly ultimately bounded), then the solutions of (E) are also uniformly bounded (uniformly ultimately bounded).

The above results enable us to analyze the stability and boundedness properties of an n-dimensional system (E), which may be complex, in terms of the corresponding properties of a one-dimensional comparison system $(\tilde{C})$, which may be quite a bit simpler. The generality and effectiveness of the above results can be improved and extended by considering *vector-valued comparison equations and vector Lyapunov functions.*

M. Lyapunov's First Method

We close this section by answering the following question: Under what conditions does it make sense to linearize a nonlinear system about an equilibrium $x = 0$ and then deduce the properties of $x = 0$ from the corresponding linear system? This is known as **Lyapunov's first method** or **Lyapunov's indirect method.**

We consider systems of n real nonlinear first-order ordinary differential equations of the form:

$$x' = Ax + F(t, x) \qquad (PE)$$

where $F: R^+ \times B(h) \to R^n$ for some $h > 0$ and A is a real $n \times n$ matrix. Here, we assume that Ax constitutes the **linear part** of the right-hand side of (PE) and $F(t, x)$ represents the remaining terms which are of order higher than one in the various components of x. Such systems may arise in the process of linearizing nonlinear equations of the form:

$$x' = g(t, x) \qquad (G)$$

or they may arise in some other fashion during the modeling process of a physical system.

To be more specific, let $g: R \times D \to R^n$ where D is some domain in R^n. If g is continuously differentiable on $R \times D$ and if ϕ is a given solution of (E) defined for all $t \geq t_0 \geq 0$, then we can **linearize** (G) **about** ϕ in the following manner. Define $y = x - \phi(t)$ so that

$$
\begin{aligned}
y' &= g(t, x) - g(t, \phi(t)) \\
&= g(t, y + \phi(t)) - g(t, \phi(t)) \\
&= (\partial g / \partial t)(t, \phi(t))y + G(t, y)
\end{aligned}
$$

Here,

$$G(t, y) = [g(t, y + \phi(t)) - g(t, \phi(t))] - (\partial g / \partial x)(t, \phi(t))y$$

is $\mathfrak{o}(|y|)$ as $|y| \to 0$ uniformly in t on compact subsets of $[t_0, \infty)$.

Of special interest is the case when g is independent of t (i.e., when $g(t, x) \equiv g(x)$) and $\phi(t) = \xi_0$ is a constant (equilibrium point). Under these conditions we have

$$y' = Ay + G(y)$$

where $A = (\partial g / \partial x)(x)|_{x = \xi_0}$, where $(\partial g / \partial x)(x)$ denotes the Jacobian of $g(x)$.

By making use of the result for the Lyapunov function (117), we can readily prove the following results:

1. *Let A be a real, constant, and stable $n \times n$ matrix and let $F: R^+ \times B(h) \to R^n$ be continuous in (t, x) and satisfy $F(t, x) = \mathfrak{o}(|x|)$ as $|x| \to 0$, uniformly in $t \in R^+$.*

Then, the trivial solution of (PE) *is uniformly asymptotically stable.*

As a specific example, consider the *Liénard equation* given by:

$$x'' + f(x)x' + x = 0 \qquad (120)$$

where $f : R \to R$ is a continuous function with $f(0) > 0$. We can rewrite Eq. (120) as:

$$x_1' = x_2, \qquad x_2' = -x_1 - f(0)x_2 + [f(0) - f(x_1)]x_2$$

and we can apply the above result with $x = (x_1, x_2)^{\mathrm{T}}$,

$$A = \begin{bmatrix} 0 & 1 \\ -1 & -f(0) \end{bmatrix}$$

$$F(t, x) = \begin{bmatrix} 0 \\ [f(0) - f(x_1)]x_2 \end{bmatrix}$$

Noting that A is a stable matrix and that $F(t, x) = o(|x|)$ as $|x| \to 0$, uniformly in $t \in R^+$, we conclude that the trivial solution $(x, x') = (0, 0)$ of Eq. (120) is uniformly asymptotically stable.

2. *Assume that A is a real $n \times n$ matrix with no eigenvalues on the imaginary axis and that at least one eigenvalue of A has positive real part. If $F : R^+ \times B(h) \to R^n$ is continuous and satisfies $F(t, x) = o(|x|)$ as $|x| \to 0$, uniformly in $t \in R^+$, then the trivial solution of* (PE) *is unstable.*

As a specific example, consider the simple pendulum,

$$x'' + k \sin x = 0, \qquad k > 0 \qquad (121)$$

Note that $x_e = \pi$, $x_e' = 0$ is an equilibrium of Eq. (121). Let $y = x - x_e$ so that

$$y'' + a \sin(y + \pi) = y'' - ay + a(\sin(y + \pi) + y) = 0$$

This equation can be put into the form (PE) with

$$A = \begin{bmatrix} 0 & 1 \\ a & 0 \end{bmatrix}$$

$$F(t, x) = \begin{bmatrix} 0 \\ a(\sin(y + \pi) + y) \end{bmatrix}$$

Applying the above result we conclude that the equilibrium point $(\pi, 0)$ is *unstable*.

SEE ALSO THE FOLLOWING ARTICLES

ARTIFICIAL NEURAL NETWORKS • CALCULUS • COMPLEX ANALYSIS • DIFFERENTIAL EQUATIONS, PARTIAL • FUNCTIONAL ANALYSIS • MEASURE AND INTEGRATION

BIBLIOGRAPHY

Antsaklis, P. J., and Michel, A. N. (1997). "Linear Systems," McGraw-Hill, New York.

Boyce, W. E., and DiPrima, R. C. (1997). "Elementary Differential Equations and Boundary Value Problems," John Wiley & Sons, New York.

Brauer, F., and Nohel, J. A. (1969). "Qualitative Theory of Ordinary Differential Equations," Benjamin, New York.

Carpenter, G. A., Cohen, M., and Grossberg, S. (1987). "Computing with neural networks," *Science* **235**, 1226–1227.

Coddington, E. A., and Levinson, N. (1955). "Theory of Ordinary Differential Equations," McGraw-Hill, New York.

Hale, J. K. (1969). "Ordinary Differential Equations," Wiley, New York.

Halmos, P. R. (1958). "Finite Dimensional Vector Spaces," Van Nostrand, Princeton, NJ.

Hille, E. (1969). "Lectures on Ordinary Differential Equations," Addison-Wesley, Reading, MA.

Hoffman, K., and Kunze, R. (1971). "Linear Algebra," Prentice-Hall, Englewood Cliffs, NJ.

Hopfield, J. J. (1984). "Neurons with graded response have collective computational properties like those of two-state neurons," *Proc. Nat. Acad. Sci. U.S.A.* **81**, 3088–3092.

Kantorovich, L. V., and Akilov, G. P. (1964). "Functional Analysis in Normed Spaces," Macmillan, New York.

Michel, A. N. (1983). "On the status of stability of interconnected systems," *IEEE Trans. Automat. Control* **28**(6), 639–653.

Michel, A. N., and Herget, C. J. (1993). "Applied Algebra and Functional Analysis," Dover, New York.

Michel, A. N., and Miller, R. K. (1977). "Qualitative Analysis of Large-Scale Dynamical Systems," Academic Press, New York.

Michel, A. N., and Wang, K. (1995). "Qualitative Theory of Dynamical Systems," Dekker, New York.

Michel, A. N., Farrell, J. A., and Porod, W. (1989). "Qualitative analysis of neural networks," *IEEE Trans. Circuits Syst.* **36**(2), 229–243.

Miller, R. K., and Michel, A. N. (1982). "Ordinary Differential Equations," Academic Press, New York.

Naylor, A. W., and Sell, G. R. (1971). "Linear Operator Theory in Engineering and Science," Holt, Rinehart & Winston, New York.

Royden, H. L. (1963). "Real Analysis," Macmillan, New York.

Rudin, W. (1953). "Principles of Mathematical Analysis," McGraw-Hill, New York.

Simmons, G. F. (1972). "Differential Equations," McGraw-Hill, New York.

Differential Equations, Partial

Martin Schechter
University of California, Irvine

GLOSSARY

Boundary Set of points in the closure of a region not contained in its interiors.
Bounded region Region that is contained in a sphere of finite radius.
Eigenvalue Scalar λ for which the equation $Au = \lambda u$ has a nonzero solution u.
Euclidean *n* dimensional space $\mathbb{R}^n$ Set of vectors $x = (x_1, \ldots, x_n)$ where each component x_j is a real number.
Partial derivative Derivative of a function of more than one variable with respect to one of the variables keeping the other variables fixed.

A PARTIAL DIFFERENTIAL EQUATION is an equation in which a partial derivative of an unknown function appears. The order of the equation is the highest order of the partial derivatives (of an unknown function) appearing in the equation. If there is only one unknown function $u(x_1, \ldots, x_n)$, then a partial differential equation for u is of the form:

$$F\left(x_1, \ldots, x_n, u, \frac{\partial u}{\partial x_1}, \ldots, \frac{\partial u}{\partial x_n}, \frac{\partial^2 u}{\partial x_1^2}, \ldots, \frac{\partial^k u}{\partial x_1^k}, \ldots\right) = 0$$

One can have more than one unknown function and more than one equation involving some or all of the unknown functions. One then has a system of j partial differential equations in k unknown functions. The number of equations may be more or less than the number of unknown functions. Usually it is the same.

I. IMPORTANCE

One finds partial differential equations in practically every branch of physics, chemistry, and engineering. They are also found in other branches of the physical sciences and in the social sciences, economics, business, etc. Many parts of theoretical physics are formulated in terms of partial differential equations. In some cases, the axioms require that the states of physical systems be given by solutions of partial differential equations. In other cases, partial differential equations arise when one applies the axioms to specific situations.

II. HOW THEY ARISE

Partial differential equations arise in several branches of mathematics. For instance, the Cauchy–Riemann equations,

$$\frac{\partial u(x, y)}{\partial x} = \frac{\partial v(x, y)}{\partial y}, \qquad \frac{\partial u(x, y)}{\partial y} = -\frac{\partial v(x, y)}{\partial x}$$

must be satisfied if

$$f(z) = u(x, y) + i v(x, y)$$

is to be an analytic function of the complex variable $z = x + iy$. Thus, the rich and beautiful branch of mathematics known as analytic function theory is merely the study of solutions of a particular system of partial differential equations.

As a simple example of a partial differential equation arising in the physical sciences, we consider the case of a vibrating string. We assume that the string is a long, very slender body of elastic material that is flexible because of its extreme thinness and is tightly stretched between the points $x = 0$ and $x = L$ on the x axis of the x, y plane. Let x be any point on the string, and let $y(x, t)$ be the displacement of that point from the x axis at time t. We assume that the displacements of the string occur in the x, y plane. Consider the part of the string between two close points x_1 and x_2. The tension T in the string acts in the direction of the tangent to the curve formed by the string. The net force on the segment $[x_1, x_2]$ in the y direction is

$$T \sin \varphi_2 - T \sin \varphi_1$$

where φ_i is the angle between the tangent to the curve and the x axis at x_i. According to Newton's second law, this force must equal mass times acceleration. This is

$$\int_{x_1}^{x_2} \rho \, \partial^2 y / \partial t^2 \, dx$$

where ρ is the density (mass per unit length) of the string. Thus, in the limit

$$T \frac{\partial}{\partial x} \sin \varphi = \rho \frac{\partial^2 y}{\partial t^2}$$

We note that $\tan \varphi = \partial y / \partial x$. If we make the simplifying assumption (justified or otherwise) that

$$\cos \varphi \approx 1, \qquad \frac{\partial}{\partial x} \cos \varphi \approx 0$$

we finally obtain:

$$T \partial^2 y / \partial x^2 = \rho \, \partial^2 y / \partial t^2$$

which is the well-known equation of the vibrating string.

The derivation of partial differential equations from physical laws usually brings about simplifying assumptions that are difficult to justify completely. Most of the time they are merely plausibility arguments. For this reason, some branches of science have accepted partial differential equations as axioms. The success of these axioms is judged by how well their conclusions describe past observations and predict new ones.

III. SOME WELL-KNOWN EQUATIONS

Now we list several equations that arise in various branches of science. Interestingly, the same equation can arise in diverse and unrelated areas.

A. Laplace's Equation

In n dimensions this equation is given by:

$$\Delta u = 0$$

where

$$\Delta = \frac{\partial^2}{\partial x_1^2} + \cdots + \frac{\partial^2}{\partial x_n^2}$$

It arises in the study of electromagnetic phenomena (e.g., electrostatics, dielectrics, steady currents, magnetostatics), hydrodynamics (e.g., irrotational flow of a perfect fluid, surface waves), heat flow, gravitation, and many other branches of science. Solutions of Laplace's equation are called *harmonic functions*.

B. Poisson's Equation

$$\Delta u \equiv f(x), \qquad x = (x_1, \ldots, x_n)$$

Here, the function $f(x)$ is given. This equation is found in many of the situations in which Laplace's equation appears, since the latter is a special case.

C. Helmholtz's Equation

$$\Delta u \pm \alpha^2 u = 0$$

This equation appears in the study of elastic waves, vibrating strings, bars and membranes, sound and acoustics, electromagnetic waves, and the operation of nuclear reactors.

D. The Heat (Diffusion) Equation

This equation is of the form:

$$u_t = a^2 \Delta u$$

where $u(x_1, \ldots, x_n, t)$ depends on the variable t (time) as well. It describes heat conduction or diffusion processes.

E. The Wave Equation

$$\Box u \equiv (1/c^2) - \Delta u = 0$$

This describes the propagation of a wave with velocity c. This equation governs most cases of wave propagation.

F. The Telegraph Equation

$$\Box u + \sigma u_t = 0$$

This applies to some types of wave propagation.

G. The Scalar Potential Equation

$$\Box u = f(x, t)$$

H. The Klein–Gordon Equation

$$\Box u + \mu^2 u = 0$$

I. Maxwell's Equations

$$\nabla \times \mathbf{H} = \sigma \mathbf{E} + \varepsilon \, \partial \mathbf{E}/\partial t$$

$$\nabla \times \mathbf{E} = -\mu \, \partial \mathbf{H}/\partial t$$

Here $\mathbf{E}$ and $\mathbf{H}$ are three-dimensional vector functions of position and time representing the electric and magnetic fields, respectively. This system of equations is used in electrodynamics.

J. The Cauchy–Riemann Equations

$$\frac{\partial u}{\partial x} = \frac{\partial v}{\partial y}, \qquad \frac{\partial u}{\partial y} = -\frac{\partial v}{\partial x}$$

These equations describe the real and imaginary parts of an analytic function of a complex variable.

K. The Schrödinger Equation

$$-\frac{h^2}{2m} \Delta \psi + V(x)\psi = i h \frac{\partial \psi}{\partial t}$$

This equation describes the motion of a quantum mechanical particle as it moves through a potential field. The function $V(x)$ represents the potential energy, while the unknown function $\psi(x)$ is allowed to have complex values.

L. Minimal Surfaces

In three dimensions a surface $z = u(x, y)$ having the least area for a given contour satisfies the equation:

$$\left(1 + u_y^2\right)u_{xx} - 2u_x u_y u_{xy} + \left(1 + u_x^2\right)u_{yy} = 0$$

where $u_x = \partial u/\partial x$, etc.

M. The Navier–Stokes Equations

$$\frac{\partial u_j}{\partial t} + \sum_k \frac{\partial u_j}{\partial x_k} u_k + \frac{1}{\rho} \frac{\partial p}{\partial x_j} = \gamma \Delta u_j$$

$$\sum_k \frac{\partial u_k}{\partial x_k} = 0$$

This system describes viscous flow of an incompressible liquid with velocity components u_k and pressure p.

N. The Korteweg–Devries Equation

$$u_t + cuu_x + u_{xxx} = 0$$

This equation is used in the study of water waves.

IV. TYPES OF EQUATIONS

In describing partial differential equations, the following notations are helpful. Let $\mathbb{R}^n$ denote Euclidean n dimensional space, and let $\mathbf{x} = (x_1, \ldots, x_n)$ denote a point in $\mathbb{R}^n$. One can consider partial differential equations on various types of manifolds, but we shall restrict ourselves to $\mathbb{R}^n$. For a real- or complex-valued function $u(x)$ we shall use the following notation:

$$D_k u = \partial u / i \, \partial x_k, \qquad 1 \le k \le n$$

If $\mu = (\mu_1, \ldots, \mu_n)$ is a multi-index of nonnegative integers, we write:

$$x^\mu = x_1^{\mu_1} \ldots x_n^{\mu_n}, \qquad |\mu| = \mu_1 + \cdots + \mu_n$$
$$D^\mu = D_1^{\mu_1} \cdots D_n^{\mu_n}$$

Thus, D^μ is a partial derivative of order $|\mu|$.

A. Linear Equations

The most general linear partial differential equation of order m is

$$Au \equiv \sum_{|\mu| \le m} a_\mu(x) D^\mu u = f(x) \tag{1}$$

It is called linear because the operator A is linear, that is, satisfies:

$$A(\alpha u + \beta v) = \alpha A u + \beta A v$$

for all function u, v and all constant scalars α, β. If the equation cannot be put in this form, it is nonlinear. In Section III, the examples in Sections A–K are linear.

B. Nonlinear Equations

In general, nonlinear partial differential equations are more difficult to solve than linear equations. There may be no solutions possible, as is the case for the equation:

$$|\partial u/\partial x| + 1 = 0$$

There is no general method of attack, and only special types of equations have been solved.

1. Quasilinear Equation

A partial differential equation is called *quasilinear* if it is linear with respect to its derivatives of highest order. This means that if one replaces the unknown function $u(x)$ and all its derivatives of order lower than the highest by known functions, the equation becomes linear. Thus, a quasilinear equation of order m is of the form:

$$\sum_{|\mu|=m} a_\mu(x, u, Du, \ldots, D^{m-1}u)D^\mu u$$

$$= f(x, u, Du, \ldots, D^{m-1}u) \qquad (2)$$

where the coefficients a_μ depend only on x, u, and derivatives of u up to order $m - 1$. Quasilinear equations are important in applications. In Section III, the examples in Sections L–N are quasilinear.

2. Semilinear Equation

A quasilinear equation is called *semilinear* if the coefficients of the highest-order derivatives depend only on x. Thus, a semilinear equation of order m is of the form:

$$Au \equiv \sum_{|\mu|=m} a_\mu(x)D^\mu u = f(x, u, Du, \ldots, D^{m-1}u) \quad (3)$$

where A is linear. Semilinear equations arise frequently in practice.

C. Elliptic Equations

The quasilinear equation (2) is called *elliptic* in a region $\Omega \subset \mathbb{R}^n$ if for every function $v(x)$ the only real vector $\xi = (\xi_1, \ldots, \xi_n) \, \varepsilon \mathbb{R}^n$ that satisfies:

$$\sum_{|\mu|=m} a_\mu(x, v, Dv, \ldots, D^{m-1}v)\xi^\mu = 0$$

is $\xi = 0$. It is called *uniformly elliptic* in Ω if there is a constant $c_0 > 0$ independent of x and v such that:

$$c_0|\xi|^m \leqq \left| \sum_{|\mu|=m} a_\mu(x, v, Dv, \ldots, D^{m-1}v)\xi^\mu \right|$$

where $|\xi|^2 = \xi_1^2 + \cdots \xi_n^2$. The equations in Sections A–C and J–L are elliptic equations or systems.

D. Parabolic Equations

When $m = 2$, the quasilinear equation (2) becomes:

$$\sum_{j,k} a_{jk}(x, u, Du)\frac{\partial^2 u}{\partial x_j \partial x_k} = f(x, u, Du) \qquad (4)$$

We shall say that Eq. (4) is *parabolic* in a region $\Omega \subset \mathbb{R}^n$ if for every choice of the function $v(x)$, the matrix $\Lambda = (a_{jk}(x, v, Dv))$ has a vanishing determinant for each $x \in \Omega$. The equation in Section III.D is a parabolic equation.

E. HYPERBOLIC EQUATIONS

Equation (4) will be called *ultrahyperbolic* in Ω if for each $v(x)$ the matrix $\Lambda = (a_{jk}(x, v, Dv))$ has some positive, some negative, and no vanishing eigenvalues for each $x \in \Omega$. It will be called *hyperbolic* if all but one of the eigenvalues of Λ have the same sign and none of them vanish in Ω. The equations in Sections III.E–H are hyperbolic equations.

The only time Eq. (4) is neither parabolic nor ultrahyperbolic is when all the eigenvalues of Λ have the same sign with none vanishing. In this case, Eq. (4) is elliptic as described earlier.

F. Equations of Mixed Type

If the coefficients of Eq. (1) are variable, it is possible that the equation will be of one type in one region and of another type in a different region. A simple example is

$$\partial^2 u/\partial x^2 - x\partial^2 u/\partial y^2 = 0$$

in two dimensions. In the region $x > 0$ it is hyperbolic, while in the region $x < 0$ it is elliptic. (It becomes parabolic on the line $x = 0$.)

G. Other Types

The type of an equation is very important in determining what problems can be solved for it and what kind of solutions it will have. As we saw, some equations can be of different types in different regions. One can define all three types for higher order equations, but most higher order equations will not fall into any of the three categories.

V. PROBLEMS ASSOCIATED WITH PARTIAL DIFFERENTIAL EQUATIONS

In practice one is rarely able to determine the most general solution of a partial differential equation. Usually one looks for a solution satisfying additional conditions. One may wish to prescribe the unknown function and/or some of its derivatives on part or all of the boundary of the region in question. We call this a *boundary-value problem*. If the equation involves the variable t (time) and the additional conditions are prescribed at some time $t = t_0$, we usually refer to it as an *initial-value problem*.

A problem associated with a partial differential equation is called *well posed* if:

1. A solution exists for all possible values of the given data.
2. The solution is unique.
3. The solution depends in a continuous way on the given data.

The reason for the last requirement is that in most cases the given data come from various measurements. It is important that a small error in measurement of the given data should not produce a large error in the solution. The method of measuring the size of an error in the given data and in the solution is a basic question for each problem. There is no standard method; it varies from problem to problem.

The kinds of problems that are well posed for an equation depend on the type of the equation. Problems that are suitable for elliptic equations are not suitable for hyperbolic or parabolic equations. The same holds true for each of the types. We illustrate this with several examples.

A. Dirichlet's Problem

For a region $\Omega \subset \mathbb{R}^n$, Dirichlet's problem for Eq. (2) is to prescribe u and all its derivatives up to order $\frac{1}{2}m - 1$ on the boundary $\partial\Omega$ of Ω. This problem is well posed only for elliptic equations. If $m = 2$, only the function u is prescribed on the boundary. If Ω is unbounded, one may have to add a condition at infinity.

It is possible that the Dirichlet problem is not well posed for a linear elliptic equation of the form (1) because 0 is an eigenvalue of the operator A. This means that there is a function $w(x) \not\equiv 0$ that vanishes together with all derivatives up to order $\frac{1}{2}m - 1$ on $\partial\Omega$ and satisfies $Aw = 0$ in Ω. Thus, any solution of the Dirichlet problem for Eq. (1) is not unique, for we can always add a multiple of w to it to obtain another solution. Moreover, it is easily checked that one can solve the Dirichlet problem for Eq. (1) only if:

$$\int_\Omega f(x)w(x)\,dx$$

is a constant depending on w and the given data. Thus, we cannot solve the Dirichlet problem for all values of the given data. When Ω is bounded, one can usually remedy the situation by considering the equation,

$$Au + \varepsilon u = f \qquad (5)$$

in place of Eq. (1) for ε sufficiently small.

B. The Neumann Problem

As in the case of the Dirichlet problem, the Neumann problem is well posed only for elliptic operators. The Neumann problem for Eq. (2) in a region Ω is to prescribe on $\partial\Omega$ the normal derivatives of u from order $\frac{1}{2}m$ to $m - 1$. (In both the Dirichlet and Neumann problems, exactly $\frac{1}{2}m$ normal derivatives are prescribed on $\partial\Omega$. In the Dirichlet problem, the first $\frac{1}{2}m$ are prescribed [starting from the zeroth-order derivative—u itself]. In the Neumann problem the next $\frac{1}{2}m$ normal derivatives are prescribed.)

As in the case of the Dirichlet problem, the Neumann problem for a linear equation of the form (1) can fail to be well posed because 0 is an eigenvalue of A. Again, this can usually be corrected by considering Eq. (5) in place of Eq. (1) for ε sufficiently small.

When $m = 2$, the Neumann problem consists of prescribing the normal derivative $\partial u / \partial n$ of u on $\partial\Omega$. In the case of Laplace's or Poisson's equation, it is easily seen that 0 is indeed an eigenvalue if Ω is bounded, for then any constant function is a solution of:

$$\Delta w = 0 \quad \text{in} \quad \Omega, \qquad \partial w / \partial n = 0 \quad \text{on} \quad \partial\Omega \qquad (6)$$

Thus, adding any constant to a solution gives another solution. Moreover, we can solve the Neumann problem:

$$\Delta u = f \quad \text{in} \quad \Omega, \qquad \frac{\partial u}{\partial n} = g \quad \text{on} \quad \partial\Omega \qquad (7)$$

only if

$$\int_\Omega f(x)\,dx = \int_{\partial\Omega} g\,ds$$

Thus, we cannot solve Eq. (7) for all f and g. If Ω is unbounded, one usually requires that the solution of the Neumann problem vanish at infinity. This removes 0 as an eigenvalue, and the problem is well posed.

C. The Robin Problem

When $m = 2$, the Robin problem for Eq. (2) consists of prescribing:

$$Bu = \alpha\,\partial u / \partial n + \beta u \qquad (8)$$

on $\partial\Omega$, where $\alpha(x), \beta(x)$ are functions and $\alpha(x) \neq 0$ on $\partial\Omega$. If $\beta(x) \equiv 0$, this reduces to the Neumann problem. Again this problem is well posed only for elliptic equations.

D. Mixed Problems

Let B be defined by Eq. (8), and assume that $\alpha(x)^2 + \beta(x)^2 \neq 0$ on $\partial\Omega$. Consider the boundary value problem consisting of finding a solution of Eq. (4) in Ω and prescribing Bu on $\partial\Omega$. On those parts of $\partial\Omega$ where $\alpha(x) = 0$, we are prescribing Dirichlet data. On those parts of $\partial\Omega$ where $\beta(x) = 0$, we are prescribing Neumann data. On the remaining sections of $\partial\Omega$ we are prescribing Robin data. This is an example of a mixed boundary-value problem in which one prescribes different types of data on different parts of the boundary. Other examples are provided by parabolic and hyperbolic equations, to be discussed later.

E. General Boundary-Value Problems

For an elliptic equation of the form (2) one can consider general boundary conditions of the form:

$$B_j u \equiv \sum_{|\mu| \leq m_j} b_{j\mu} D^\mu u$$

$$= g_j \quad \text{on} \quad \partial\Omega, \qquad 1 \leq j \leq m/2 \qquad (9)$$

Such boundary-value problems can be well posed provided the operators B_j are independent in a suitable sense and do not "contradict" each other or Eq. (2).

F. The Cauchy Problem

For Eq. (2), the Cauchy problem consists of prescribing all derivatives of u up to order $m - 1$ on a smooth surface S and solving Eq. (2) for u in a neighborhood of S. An important requirement for the Cauchy problem to have a solution is that the boundary conditions not "contradict" the equation on S. This means that the coefficient of $\partial^m u/\partial n^m$ in Eq. (2) should not vanish on S. Otherwise, Eq. (2) and the Cauchy boundary conditions,

$$\partial^k u/\partial n^k = g_k \quad \text{on} \quad S, k = 0, \ldots, n - 1 \qquad (10)$$

involve only the function g_k on S. This is sure to cause a contradiction unless f is severely restricted. When this happens we say that the surface S is *characteristic* for Eq. (2). Thus, for the Cauchy problem to have a solution without restricting f, it is necessary that the surface S be noncharacteristic. In the quasilinear case, the coefficient of $\partial^m u/\partial n^m$ in Eq. (2) depends on the g_k. Thus, the Cauchy data (10) will play a role in determining whether or not the surface S is characteristic for Eq. (2). The Cauchy–Kowalewski theorem states that for a noncharacteristic analytic surface S, real analytic Cauchy data g_k, and real analytic coefficients a_μ, f in Eq. (2), the Cauchy problem (2) and (10) has a unique real analytic solution in the neighborhood of S. This is true irrespective of the type of the equation. However, this does not mean that the Cauchy problem is well posed for all types of equations. In fact, the hypotheses of the Cauchy–Kowalewski theorem are satisfied for the Cauchy problem,

$$u_{xx} + u_{yy} = 0, \qquad y > 0$$

$$u(x, 0) = 0, \qquad u_y(x, 0) = n^{-1} \sin nx$$

and, indeed, it has a unique analytic solution:

$$u(x, y) = n^{-2} \sinh ny \sin nx$$

The function $n^{-1} \sin nx$ tends uniformly to 0 as $n \to \infty$, but the solution does not become small as $n \to \infty$ for $y \neq 0$. It can be shown that the Cauchy problem is well posed only for hyperbolic equations.

VI. METHODS OF SOLUTION

There is no general approach for finding solutions of partial differential equations. Indeed, there exist linear partial differential equations with smooth coefficients having no solutions in the neighborhood of a point. For instance, the equation:

$$\frac{\partial u}{\partial x_1} + i\frac{\partial u}{\partial x_2} + 2i(x_1 + ix_2)\frac{\partial u}{\partial x_3} = f(x_3) \qquad (11)$$

has no solution in the neighborhood of the origin unless $f(x_3)$ is a real analytic function of x_3. Thus, if f is infinitely differentiable but not analytic, Eq. (11) has no solution in the neighborhood of the origin. Even when partial differential equations have solutions, we cannot find the "general" solution. We must content ourselves with solving a particular problem for the equation in question. Even then we are rarely able to write down the solution in closed form. We are lucky if we can derive a formula that will enable us to calculate the solution in some way, such as a convergent series or iteration scheme. In many cases, even this is unattainable. Then, one must be satisfied with an abstract theorem stating that the problem is well posed. Sometimes the existence theorem does provide a method of calculating the solution; more often it does not.

Now we describe some of the methods that can be used to obtain solutions in specific situations.

A. Separation of Variables

Consider a vibrating string stretched along the x axis from $x = 0$ and $x = \pi$ and fixed at its end points. We can assign the initial displacement and velocity. Thus, we are interested in solving the mixed initial and boundary value problem for the displacement $u(x, t)$:

$$\Box u = 0, \qquad\qquad 0 < x < \pi, \quad t > 0 \quad (12)$$

$$u(x, 0) = f(x),$$

$$u_t(x, 0) = g(x), \qquad 0 \leq x \leq \pi \qquad (13)$$

$$u(0, t) = u(\pi, t) = 0, \qquad t \geq 0 \qquad (14)$$

We begin by looking for a solution of Eq. (12) of the form:

$$u(x, t) = X(x)T(t)$$

Such a function will be a solution of Eq. (12) only if:

$$\frac{T''}{T} = c^2 \frac{X''}{X} \tag{15}$$

Since the left-hand side of Eq. (15′) is a function of t only and the right-hand side is a function of x only, both sides are constant. Thus, there is a constant K such that $X''(x) = KX(x)$. If $K = \lambda^2 > 0$, X must be of the form:

$$X = Ae^{-\lambda x} + Be^{\lambda x}$$

For Eq. (14) to be satisfied, we must have

$$X(0) = X(\pi) = 0 \tag{15'}$$

This can happen only if $A = B = 0$. If $K = 0$, then X must be of the form:

$$x = A + Bx$$

Again, this can happen only if $A = B = 0$. If $K = -\lambda^2 < 0$, then X is of the form:

$$X = A \cos \lambda x + B \sin \lambda x$$

This can satisfy Eq. (15) only if $A = 0$ and $B \sin \lambda \pi = 0$. Thus, the only way that X should not vanish identically is if λ is an integer n. Moreover, T satisfies:

$$T'' + n^2 c^2 T = 0$$

The general solution for this is

$$T = A \cos nct + B \sin nct$$

Thus,

$$u(x, t) = \sin nx (A_n \cos nct + B_n \sin nct)$$

is a solution of Eqs. (12) and (14) for each integer n. However, it will not satisfy Eq. (13) unless $f(x)$, $g(x)$ are of a special form. The linearity of the operator $\square$ allows one to add solutions of Eqs. (12) and (14). Thus,

$$u(x, t) = \sum_{n=1}^{\infty} \sin nx (A_n \cos nct + B_n \sin nct) \tag{16}$$

will be a solution provided the series converges. Moreover, it will satisfy Eq. (13) if:

$$f(x) = \sum_{n=1}^{\infty} A_n \sin nx,$$

$$g(x) = \sum_{n=1}^{\infty} nc B_n \sin nx$$

This will be true if $f(x)$, $g(x)$ are expandable in a Fourier sine series. If they are, then the coefficients A_n, B_n are given by:

$$A_n = \frac{2}{\pi} \int_0^\pi f(x) \sin nx \, dx,$$

$$B_n = \frac{2}{nc\pi} \int_0^\pi g(x) \sin nx \, dx$$

With these values, the series (16) converges and gives a solution of Eqs. (12)–(14).

B. Fourier Transforms

If we desire to determine the temperature $u(x, t)$ of a system in $\mathbb{R}^n$ with no heat added or removed and initial temperature given, we must solve:

$$u_t = a^2 \Delta u, \qquad x \varepsilon \mathbb{R}^n, \quad t > 0 \tag{17}$$

$$u(x, 0) = \varphi(x), \qquad x \varepsilon \mathbb{R}^n \tag{18}$$

If we apply the Fourier transform,

$$\hat{f}(\xi) = (2\pi)^{-n/2} \int e^{-i\xi x} f(x) \, dx \tag{19}$$

where $\xi x = \xi_1 x_1 + \cdots + \xi_n x_n$, we obtain:

$$\hat{u}_t(\xi, t) + a^2 |\xi|^2 \hat{u}(\xi, t) = 0$$

The solution satisfying Eq. (18) is

$$\hat{u}(\xi, t) = e^{-a^2 |\xi|^2 t} \hat{\varphi}(\xi)$$

If we now make use of the inverse Fourier transform,

$$f(x) = (2\pi)^{-n/2} \int e^{i\xi x} \hat{f}(\xi) \, d\xi$$

we have

$$u(x, t) = \int K(x - y, t) \varphi(y) \, dy$$

where

$$K(x, t) = (2\pi)^{-n} \int e^{ix\xi - a^2 |\xi|^2 t} \, d\xi$$

If we introduce the new variable,

$$\eta = at^{1/2} \xi - \tfrac{1}{2} i a^{-1} t^{-1/2} x$$

this becomes

$$(2\pi)^{-n} a^{-n} t^{-n/2} e^{-|x|^2/4a^2 t} \int e^{-|\eta|^2} \, d\eta$$

$$= (4\pi a^2 t)^{-n/2} e^{-|x|^2/4a^2 t}$$

This suggests that a solution of Eqs. (17) and (18) is given by:

$$u(x, t) = (4\pi a^2 t)^{-n/2} \int e^{-|x-y|^2/4a^2 t} \varphi(y) \, dy \tag{20}$$

It is easily checked that this is indeed the case if φ is continuous and bounded. However, the solution is not unique unless one places more restriction on the solution.

C. Fundamental Solutions, Green's Function

Let

$$K(x, y) = \frac{|x - y|^{2-n}}{(2 - n)\omega_n} + h(x), \qquad n > 2,$$

$$K(x, y) = \frac{\log 4}{2\pi} + h(x), \qquad n = 2$$

where $\omega_n = 2\pi^{n/2}/\Gamma(\frac{1}{2}n)$ is the surface area of the unit sphere in $\mathbb{R}^n$ and $h(x)$ is a harmonic function in a bounded domain $\Omega \subset \mathbb{R}^n$ (i.e., $h(x)$ is a solution of $\Delta h = 0$ in Ω). If the boundary $\partial\Omega$ of Ω is sufficiently regular and $h \in C^2(\bar{\Omega})$, then Green's theorem implies for $y \in \Omega$:

$$u(y) = \int_\Omega K(x, y)\Delta u(x)\, dx$$

$$+ \int_{\partial\Omega} \left(u(x)\frac{\partial K(x, y)}{\partial n} - K(x, y)\frac{\partial u}{\partial n}\right) dS_x \quad (21)$$

for all $u \in C^2(\bar{\Omega})$. The function $K(x, y)$ is called a *fundamental solution* of the operator Δ. If, in addition, $K(x, y)$ vanishes for $x \in \partial\Omega$, it is called a *Green's function*, and we denote it by $G(x, y)$. In this case,

$$u(y) = \int_{\partial\Omega} u(x)\frac{\partial G(x, y)}{\partial n}\, dS_x, \qquad y \in \Omega \quad (22)$$

for all $u \in C^2(\bar{\Omega})$ that are harmonic in Ω. Conversely, this formula can be used to solve the Dirichlet problem for Laplace's equation if we know the Green's function for Ω, since the righthand side of Eq. (22) is harmonic in Ω and involves only the values of $u(x)$ on $\partial\Omega$. It can be shown that if the prescribed boundary values are continuous, then indeed Eq. (22) does give a solution to the Dirichlet problem for Laplace's equation.

It is usually very difficult to find the Green's function for an arbitrary domain. It can be computed for geometrically symmetric regions. In the case of a ball of radius R and center 0, it is given by:

$$G(x, y) = K(x, y) - (|y|/R)^{2-n} K(x, R^2|y|^{-2}y)$$

D. Hilbert Space Methods

Let

$$P(D) = \sum_{|\mu|=m} a_\mu D^\mu$$

be a positive, real, constant coefficient, homogeneous, elliptic partial differential operator of order $m = 2r$. This means that $P(D)$ has only terms of order m, and

$$c_0|\xi|^m \le P(\xi) \le C_0|\xi|^m, \qquad \xi \in \mathbb{R}^n \quad (23)$$

holds for positive constants c_0, C_0. We introduce the norm,

$$|v|_r = \left(\int |\xi|^m |\hat{v}(\xi)|^2\, d\xi\right)^{1/2}$$

for function v in $C_0^\infty(\Omega)$, the set of infinitely differentiable functions that vanish outside Ω. Here, Ω is a bounded domain in $\mathbb{R}^n$ with smooth boundary, and $\hat{v}(\xi)$ denotes the Fourier transform given by Eq. (19). By Eq. (23) we see that $(P(D)v, v)$ is equivalent to $|v|_r^2$ on $C_0^\infty(\Omega)$, where

$$(u, v) = \int_\Omega u(x)\overline{v(x)}\, dx$$

Let

$$a(u, v) = (u, P(D)v), \qquad u, v \in C_0^\infty(\Omega) \quad (24)$$

If $u \in C^m(\bar{\Omega})$ is a solution of the Dirichlet problem,

$$P(D)u = f \quad \text{in} \quad \Omega \quad (25)$$

$$D^\mu u = 0 \quad \text{on} \quad \partial\Omega, \qquad |\mu| < r \quad (26)$$

then it satisfies

$$a(u, v) = (f, v), \qquad v \in C_0^\infty(\Omega) \quad (27)$$

Conversely, if $\partial\Omega$ is sufficiently smooth and $u \in C^m(\bar{\Omega})$ satisfies Eq. (27), then it is a solution of the Dirichlet problem, Eqs. (25) and (26). This is readily shown by integration by parts. Thus, one can solve Eqs. (25) and (26) by finding a function $u \in C^m(\bar{\Omega})$ satisfying Eq. (27). Since $a(u, v)$ is a scalar product, it would be helpful if we had a theorem stating that the expression (f, v) can be represented by the expression $a(u, v)$ for some u. Such a theorem exists (the Riesz representation theorem), provided $a(u, v)$ is the scalar product of a Hilbert space and

$$|(f, v)| \le Ca(v, v)^{1/2} \quad (28)$$

We can fit our situation to the theorem by completing $C_0^\infty(\Omega)$ with respect to the $|v|_r$ norm and making use of the fact that $a(v, v)^{1/2}$ and $|v|_r$ are equivalent on $C_0^\infty(\Omega)$ and consequently on the completion $H_0^r(\Omega)$. Moreover, inequality (28) follows from the Poincare inequality,

$$\|v\| \le M^r |v|_r, \qquad v \in C_0^\infty(\Omega) \quad (29)$$

which holds if Ω is contained in a cube of side length M. Thus, by Schwarz's inequality,

$$|(f, v)| \le \|f\|\, \|v\| \le \|f\| M^r |v|_r \le ca(v, v)^{1/2}$$

The Riesz representation theorem now tells us that there is a $u \in H_0^r(\Omega)$ such that Eq. (27) holds. If we can show that u is in $C^m(\bar{\Omega})$, it will follow that u is indeed a solution of the Dirichlet problem, Eqs. (25) and (26). As it stands now, u is only a *weak solution* of Eqs. (25) and (26). However, it can be shown that if $\partial\Omega$ and f are sufficiently smooth, then u will be in $C^m(\bar\Omega)$ and will be a solution of Eqs. (25) and (26).

The proof of the Poincare inequality (29) can be given as follows. It sufficies to prove it for $r = 1$ and Ω contained in the slab $0 < x_1 < M$. Since $v \in C_0^\infty(\Omega)$,

$$v(x_1, \ldots, x_n)^2 = \left(\int_0^{x_1} v_{x1}(t, x_2, \ldots, x_n)\, dt \right)^2$$

$$\leq x_1 \int_0^{x_1} v_{x1}(t, x_2, \ldots, x_n)^2\, dt$$

$$\leq M \int_0^M v_{x1}(t, x_2, \ldots, x_n)^2\, dt$$

Thus,

$$\int_0^M v(x_1, \ldots, x_n)^2\, dx_1 \leq M^2 \int_0^M v_{x1}(t, x_2, \ldots, x_n)^2\, dt$$

If we now integrate over $x_2, \ldots, x_n$, we obtain:

$$\|v\| \leq M \|v_{x1}\|$$

But, by Parseval's identity,

$$\int |v_{x1}|^2\, dx = \int |\hat v_{x1}|^2\, d\xi$$

$$= \int \xi_1^2 |\hat v|^2\, d\xi \leq \int |\xi|^2 |\hat v|^2\, d\xi = |v|_1^2$$

E. Iterations

An important method of solving both linear and nonlinear problems is that of *successive approximations*. We illustrate this method for the following Dirichlet problem:

$$\Delta u = f(x, u), x \in \Omega \tag{30}$$

$$u = 0 \quad \text{on} \quad \partial\Omega \tag{31}$$

We assume that the boundary of Ω is smooth and that $f(x, t) = f(x_1 \ldots, x_n, t)$ is differentiable with respect to all arguments. Also we assume that:

$$|f(x, t)| \leq N, \qquad x \in \Omega, \quad -\infty < t < \infty \tag{32}$$

$$|\partial f(x, t)/\partial t| \leq \psi(t), \quad x \in \Omega \tag{33}$$

where $\psi(t)$ is a continuous function.

First, we note that for every compact subset G of Ω there is a constant C such that:

$$\max_G |\nabla v| \leq C(\sup_\Omega |\Delta v| + \sup_\Omega |v|) \tag{34}$$

for all $v \in C^2(\Omega)$. Assume this for the moment, and let $w(x)$ be the solution of the Dirichlet problem,

$$\Delta w = -N \quad \text{in} \quad \Omega, \qquad w = 0 \quad \text{on} \quad \partial\Omega \tag{35}$$

It is clear that $w(x) \geq 0$ in Ω. For otherwise it would have a negative interior mininum in Ω. At such a point, one has $\partial^2 w/\partial x_k^2 \geq = 0$ for each k, and consequently. $\Delta w \geq 0$ contradicting Eq. (35). Since $w \in C(\bar\Omega)$, there is a constant C_t such that:

$$0 \leq w(x) \leq C_1, \qquad x \in \Omega$$

Let

$$K = \max_{|t| \leq C_1} \psi(t)$$

Then, by Eq. (33):

$$|\partial f(x, t)/\partial t| \leq K, \qquad |t| \leq C_1 \tag{36}$$

Consequently,

$$f(x, t) - f(x, s) \leq K(t - s) \tag{37}$$

when $-C_1 \leq s \leq t \leq C_1$. We define a sequence $\{u_k\}$ of functions as follows. We take $u_0 = w$ and once u_{k-1} has been defined, we let u_k be the solution of the Dirichlet problem:

$$Lu_k \equiv \Delta u_k - Ku_k = f(x, u_{k-1}) - Ku_{k-1} \tag{38}$$

in Ω with $u_k = 0$ on $\partial\Omega$. The solution exists by the theory of linear elliptic equations. We show by induction that:

$$-w \leq u_k \leq u_{k-1} \leq w \tag{39}$$

To see this for $k = 1$, note that:

$$L(u_1 - w) = f(x, w) - Kw - \Delta w + Kw \geq 0$$

From this we see that $u_1 \leq w$ in Ω. If $u_1 - w$ had an interior positive maximum in Ω, we would have

$$L(u_1 - w) = \Delta(u_1 - w) - K(u_1 - w) < 0$$

at such a point. Thus, $u_1 \leq w$ in Ω. Also we note:

$$\Delta(u_1 + w) = f(x, w) + K(u_1 - w) + \Delta w \leq K(u_1 - w)$$

This shows that $u_1 + w$ cannot have a negative minimum inside Ω. Hence, $u_1 + w \geq 0$ in Ω, and Eq. (39) is verified for $k = 1$. Once we know it is verified for k, we note that:

$$L[u_{k+1} - u_k] = f(x, u_k) - f(x, u_{k-1})$$
$$- K(u_k - u_{k-1}) \geq 0$$

by Eq. (37). Thus, $u_{k-1} \leq u_k$ in Ω. Hence,

$$\Delta(u_{k+1} + w) = f(x, u_k) - Ku_k + Ku_{k+1}$$
$$+ \Delta w \leq K(u_{k+1} - u_k)$$

Again, we deduce from this that $u_{k+1} + w \geq 0$ in Ω. Hence, Eq. (39) holds for $k + 1$ and consequently for all k. In particular, we see that the u_k are uniformly bounded in Ω, and by Eq. (38) the same is true of the functions Δu_k.

Hence, by Eq. (34), the first derivatives of the u_k are uniformly bounded on compact subsets of Ω. If we differentiate Eq. (38), we see that the sequence $\Delta(\partial u_k/\partial x_j)$ is uniformly bounded on compact subsets of Ω (here we make use of the continuous differentiability of f). If we now make use of Eq. (34) again, we see that the second derivatives of the u_k are uniformly bounded on compact subsets of Ω. Hence, by the Ascoli–Arzela theorem, there is a subsequence that converges together with its first derivatives uniformly on compact subsets of Ω. Since the sequence u_k is monotone, the whole sequence must converge to a continuous function u that satisfies $|u(x)| \leq w(x)$. Hence, u vanishes on $\partial\Omega$. By Eq. (38), the functions Δu_k must converge uniformly on compact subsets, and by Eq. (34), the same must be true of the first derivatives of the u_k. From the differentiated Eq. (38) we see that the $\Delta(\partial u_k/\partial x_j)$ converge uniformly on bounded subsets and consequently the same is true of the second derivatives of the u_k by Eq. (34). Since the u_k converge uniformly to u in Ω and their second derivatives converge uniformly on bounded subsets, we see that $u \in C^2(\Omega)$ and $\Delta u_k \to \Delta u$. Hence,

$$\Delta u = \lim \Delta u_k$$

$$= \lim[f(x, u_{k-1}) + K(u_k - u_{k-1})] = f(x, u)$$

and u is the desired solution.

It remains to prove Eq. (34). For this purpose we let $\varphi(x)$ be a function in $C_0^\infty(\Omega)$ which equals one on G. Then we have, by Eq. (21),

$$\varphi(y)v(y) = \int_\Omega K(x, y)\Delta(\varphi(x)v(x))\,dx$$

(the boundary integrals vanish because φ is 0 near $\partial\Omega$). Thus, if $y \in G$,

$$v(y) = \int_\Omega K\{\varphi\Delta v + 2\nabla\varphi \cdot \nabla v + v\Delta\varphi\}\,dx$$

$$= \int_\Omega \{\varphi K\Delta v - 2v\nabla K \cdot \nabla\varphi - vK\Delta\varphi\}\,dx$$

by integration by parts. We note that $\nabla\varphi$ vanishes near the singularity of K. Thus, we may differentiate under integral sign to obtain:

$$\frac{\partial v(y)}{\partial y_j} = \int_\Omega \left\{\varphi\frac{\partial K}{\partial y_j}\Delta v - 2v\frac{\partial\nabla k}{\partial y_j} \cdot \nabla\varphi - v\frac{\partial k}{\partial y_j}\Delta\varphi\right\}\,dx$$

Consequently,

$$\left|\frac{\partial v}{\partial y_j}\right| \leq \sup_\Omega |\Delta v| \int_\Omega \left|\frac{\partial K}{\partial y_j}\right|\,dx$$

$$+ \sup_\Omega |v| \int_\Omega \left\{|\nabla\varphi|\left|\frac{\partial K}{\partial y_j}\right| + |\Delta\varphi|\left|\frac{\partial K}{\partial y_j}\right|K\right\}\,dx$$

and all of the integrals are finite. This gives Eq. (34), and the proof is complete.

F. Variational Methods

In many situations, methods of the calculus of variations are useful in solving problems for partial differential equations, both linear and nonlinear. We illustrate this with a simple example. Suppose we wish to solve the problem,

$$-\sum \frac{\partial}{\partial x_k}\left[p_k(x)\frac{\partial u(x)}{\partial x_k}\right] + q(x)u(x) = 0 \quad \text{in} \quad \Omega \quad (40)$$

$$u(x) = g(x) \quad \text{on} \quad \partial\Omega \quad (41)$$

Assume that $p_k(x) \geq c_0, q(x) \geq c_0, c_0 > 0$ for $x \in \Omega$, Ω bounded, $\partial\Omega$ smooth, and that g is in $C^1(\partial\Omega)$. We consider the expression,

$$a(u, v) = \int_\Omega \left\{\frac{1}{2}\sum p_k(x)\frac{\partial u(x)}{\partial x_k}\frac{\partial v(x)}{\partial x_k}\right.$$

$$\left. + q(x)u(x)v(x)\right\}\,dx$$

and put $a(u) = a(u, u)$. If $u \in C^2(\Omega) \cap C^0(\bar{\Omega})$, satisfies, Eq. (41), and

$$a(u) \leq a(w) \quad (42)$$

for all w satisfying Eq. (41), then it is readily seen that u is a solution of Eq. (40). Let v be a smooth function which vanishes on $\partial\Omega$. Then, for any scalar β,

$$a(u) \leq a(u + \beta v) = a(u) + 2\beta a(u, v) + \beta^2 a(v)$$

and, consequently,

$$2\beta a(u, v) + \beta^2 a(v)^2 \geq 0$$

for all β. This implies $a(u, v) = 0$. Integration by parts now yields Eq. (40). The problem now is to find a u satisfying Eq. (42). We do this as follows. Let H denote the Hilbert space obtained by completing $C^2(\bar{\Omega})$ with respect to the norm $a(w)$, and let H_0 be the subspace of those functions in H that vanish on $\partial\Omega$. Under the hypotheses given it can be shown that H_0 is a closed subspace of H. Let w be an element of H satisfying Eq. (41), and take a sequence $\{v_k\}$ of functions in H_0 such that:

$$a(w - v_k) \to d = \inf_{v \in H_0} a(w - v)$$

The parallelogram law for Hilbert space tells us that:

$$a(2w - v_j - v_k) + a(v_j - v_k)$$

$$= 2a(w - v_j) + 2a(w - v_k)$$

But,

$$a(2w - v_j - v_k) = 4a\left(w - \frac{1}{2}(v_j + v_k)\right) \geq 4d$$

Hence,

$$4d + a(v_j - v_k) \leq 2a(w - v_j) + 2a(w - v_k) \to 4d$$

Thus,

$$a(v_j - v_k) \to 0$$

and $\{v_k\}$ is a Cauchy sequence in H_0. Since H_0 is complete, there is a $v \in H_0$ such that $a(v_k - v) \to 0$. Put $u = w - v$. Then clearly u satisfies Eqs. (41) and (42), hence $a(u, v) = 0$ for all $v \in H_0$. If u is in $C^2(\Omega)$, then it will satisfy Eq. (40) as well. The rub is that functions in H need not be in $C^2(\Omega)$. However, the day is saved if the $p_k(x)$ and $q(x)$ are sufficiently smooth. For then one can show that functions $u \in H$ which satisfy $a(u, v) = 0$ for all $v \in H_0$ are indeed in $C^2(\Omega)$.

G. Critical Point Theory

Many partial differential equations and systems are the Euler–Lagrange equations corresponding to a real valued functional G defined on some Hilbert space E. Such a functional usually represents the energy or related quantity of the physical system governed by the equations. In such a case, solutions of the partial differential equations correspond to critical points of the functional. For instance, if the equation to be solved is

$$-\Delta u(x) = f(x, u), x \in \Omega \subset \mathbb{R}^n \qquad (43)$$

then solutions of Eq. (43) are the critical points of the functional,

$$G(u) = \|\nabla u\|^2 - 2\int_\Omega F(x, u)\, dx$$

where the norm is that of $L^2(\Omega)$ and

$$F(x, t) = \int_0^t f(x, s)\, ds$$

The history of this approach can be traced back to the calculus of variations in which equations of the form:

$$G'(u) = 0 \qquad (44)$$

are the Euler–Lagrange equations of the functional G. The original method was to find maxima or minima of G by solving Eq. (44) and then show that some of the solutions are extrema. This approach worked well for one-dimensional problems. In this case, it is easier to solve Eq. (44) than it is to find a maximum or minimum of G. However, in higher dimensions it was realized quite early that it is easier to find maxima and minima of G than it is to solve Eq. (44). Consequently, the tables were turned, and critical point theory was devoted to finding extrema of G. This approach is called the *direct method* in the calculus of variations. If an extremum point of G can be identified, it will automatically be a solution of Eq. (44).

The simplest extrema to find are global maxima and minima. For such points to exist one needs G to be semi-bounded. However, this in itself does not guarantee that an extremum exists. All that can be derived from the semiboundedness is the existence of a sequence $\{u_k\} \subset E$ such that:

$$G(u_k) \to c, \, G'(u_k) \to 0 \qquad (45)$$

where c is either the supremum or infimum of G. The existence of such a sequence does not guarantee that an extremum exists. However, it does so if one can show that the sequence has a convergent subsequence. This condition is known as the *Palais–Smale condition*. A sequence satisfying Eq. (45) is called a *Palais–Smale sequence*.

Recently, researchers have discovered several sets of sufficient conditions on G which will guarantee the existence of a Palais–Smale sequence. If, in addition, the Palais–Smale condition holds, then one obtains a critical point. These conditions can apply to functionals which are not semibounded. Points found by this method are known as *mountain pass points* due to the geometrical interpretation that can be given. The situation can be described as follows. Suppose Q is an open set in E and there are two points e_0, e_1 such that $e_0 \in Q, e_1 \notin \bar{Q}$, and

$$G(e_0), G(e_1) \le b_0 = \inf_{\partial Q} G \qquad (46)$$

Then there exists a Palais–Smale sequence, Eq. (45), with c satisfying:

$$b_0 \le c < \infty \qquad (47)$$

When G satisfies Eq. (46) we say that it exhibits mountain pass geometry.

It was then discovered that other geometries (i.e., configurations) produce Palais–Smale sequences, as well. Consider the following situation. Assume that

$$E = M \oplus N$$

is a decomposition of E into the direct sum of closed subspaces with

$$\dim N < \infty$$

Suppose there is an $R > 0$ such that

$$\sup_{N \cap \partial B_R} G \le b_0 = \inf_M G$$

Then, again, there is a sequence satisfying Eqs. (45) and (47). Here, B_R denotes the ball of radius R in E and ∂B_R denotes its boundary.

In applying this method to solve Eq. (43), one discovers that the asymptotic behavior of $f(x, t)$ at ∞ plays an important role. One can consider several possibilities. When

$$\limsup_{|t| \to \infty} |f(x, t)/t| = \infty \qquad (48)$$

we say that problem (43) is *superlinear*. Otherwise, we call it *sublinear*. If

$$f(x,t)/t \to b_\pm(x) \ as \ t \to \pm\infty \qquad (49)$$

and the $b_\pm$ are different, we say that it has a nonlinearity at ∞. An interesting special case of Eq. (49) is when the $b_\pm(x)$ are constants. If $b_-(x) \equiv a$ and $b_+(x) \equiv b$, we say that (a, b) is in the Fučík spectrum $\sum$ if:

$$-\Delta u = bu^+ - au^-$$

has a nontrivial solution, where $u^\pm = \max\{\pm u, 0\}$. Because of its importance in solving Eq. (43), we describe this spectrum. It has been shown that emanating from each eigenvalue λ_ℓ of $-\Delta$, there are curves $\mu_\ell(a)$, $\nu_{\ell-1}(a)$ (which may coincide) which are strictly decreasing at least in the square $S = [\lambda_{\ell-1}, \lambda_{\ell+1}]^2$ and such that $(a, \mu_\ell(a))$ and $(a, \nu_{\ell-1}(a))$ are in $\sum$ in the square S. Moreover, the regions $b > \mu_\ell(a)$ and $b < \nu_{\ell-1}(a)$ are free of $\sum$ in S. These curves are known exactly only in the one-dimensional case (ordinary differential equations). In higher dimensions it is not known in general how many curves of $\sum$ emanate from each eigenvalue. It is known that there is at least one (when $\mu_\ell(a)$ and $\nu_{\ell-1}(a)$ coincide). If there are two or more curves emanating from an eigenvalue, the status of the region between them is unknown in general. If the eigenvalue is simple, there are at most two curves of $\sum$ emanating from it, and any region between them is not in $\sum$. On the other hand, examples are known in which many curves of $\sum$ emanate from a multiple eigenvalue. In the higher dimensional case, the curves have not been traced asymptotically.

If

$$f(x,t)/t \to \lambda_\ell \ as \ |t| \to \infty \qquad (50)$$

where λ_ℓ is one of the eigenvalues of $-\Delta$, we say that Eq. (43) has asymptotic resonance. One can distinguish several types. One can have the situation:

$$f(x,t) = \lambda_\ell t + p(x,t) \qquad (51)$$

where $p(x,t) = o(|t|^\beta)$ as $|t| \to \infty$ for some $\beta < 1$. Another possibility is when $p(x,t)$ satisfies:

$$|p(x,t)| \le V(x) \in L^2(\Omega) \qquad (52)$$

and

$$p(x,t) \to p_\pm(x) \ a.e. \ as \ t \to \pm\infty$$

A stronger form occurs when

$$p(x,t) \to 0 \ as \ |t| \to \infty \qquad (53)$$

and

$$|P(x,t)| \le W(x) \in L^1(\Omega) \qquad (54)$$

where

$$P(x,t) := \int_0^t p(x,s)\,ds$$

This type of problem is more difficult to solve; it is called *strong resonance*. Possible situations include:

$$P(x,t) \to P_\pm(x) \ as \ t \to \pm\infty \qquad (55)$$

and

$$P(x,t) \to P_0(x) \ as \ |t| \to \infty \qquad (56)$$

What is rather surprising is that the stronger the resonance, the more difficult it is to solve Eq. (43). There is an interesting connection between Eq. (43) and nonlinear eigenvalue problems of the form:

$$G'(u) = \beta u \qquad (57)$$

for functionals. This translates into the problem

$$-\Delta u = \lambda f(x, u) \qquad (58)$$

for partial differential equations. It can be shown that there is an intimate relationship between Eqs. (44) and (57) (other than the fact that the former is a special case of the latter). In fact, the absence of a certain number of solutions of Eq. (44) implies the existence of a rich family of solutions of Eq. (57) on all spheres of sufficiently large radius, and vice versa. The same holds true for Eqs. (43) and (58).

H. Periodic Solutions

Many problems in partial differential equations are more easily understood if one considers periodic functions. Let

$$Q = \left\{ x \in \mathbb{R}^n : 0 \le x_j \le 2\pi \right\}$$

be a cube in $\mathbb{R}^n$. By this we mean that Q consists of those points $x = (x_1, \ldots, x_n) \in \mathbb{R}^n$ such that each component x_j satisfies $0 \le x_j \le 2\pi$. Consider n-tuples of integers $\mu = (\mu_1, \ldots, \mu_n) \in \mathbb{Z}^n$, where each $\mu_j \ge 0$, and write $\mu x = \mu_1 x_1 + \cdots + \mu_n x_n$. For $t \in \mathbb{R}$ we let H_t be the set of all series of the form:

$$u = \sum a_\mu e^{i\mu x} \qquad (59)$$

where the a_μ are complex numbers satisfying:

$$\alpha_{-\mu} = \bar{\alpha}_\mu$$

(this is done to keep the functions u real valued), and

$$\|u\|_t^2 = (2\pi)^n \sum (1 + \mu^2)^t |\alpha_\mu|^2 < \infty \qquad (60)$$

where $\mu^2 = \mu_1^2 + \cdots + \mu_n^2$. It is not required that the series (59) converge in anyway, but only that Eq. (60) hold. If

$$u = \sum a_\mu e^{i\mu x}, \qquad v = \sum \beta_\mu e^{i\mu x}$$

are members of H_t, we can introduce the scalar product:

$$(u, v)_t = (2\pi)^n \sum (1 + \mu^2)^t \alpha_\mu \beta_{-\mu} \qquad (61)$$

With this scalar product, H_t becomes a Hilbert space. If

$$f(x) = \sum \gamma_\mu e^{i\mu x} \qquad (62)$$

we wish to solve

$$-\Delta u = f \qquad (63)$$

In other words, we wish to solve:

$$\sum \mu^2 \alpha_\mu e^{i\mu x} = \sum \gamma_\mu e^{i\mu x}$$

This requires:

$$\mu^2 \alpha_\mu = \gamma_\mu \quad \forall \mu$$

In order to solve for α_μ, we must have

$$\gamma_0 = 0 \qquad (64)$$

Hence, we cannot solve for all f. However, if Eq. (64) holds, we can solve Eq. (63) by taking:

$$\alpha_\mu = \gamma_\mu / \mu^2 \quad when \quad \mu \neq 0 \qquad (65)$$

On the other hand, we can take α_0 to be any number we like, and u will be a solution of Eq. (63) as long as it satisfies Eq. (65). Thus, we have Theorem 0.1:

Theorem 0.1. *If f, given by Eq. (62), is in H_t and satisfies Eq. (64), then Eq. (63) has a solution $u \in H_{t+2}$. An arbitrary constant can be added to the solution.*

If we wish to solve:

$$-(\Delta + \lambda)u = f \qquad (66)$$

where $\lambda \in \mathbb{R}$ is any constant, we want $(\mu^2 - \lambda)\alpha_\mu = \gamma_\mu$, or

$$\alpha_\mu = \gamma_\mu / (\mu^2 - \lambda) \quad when \quad \mu^2 \neq \lambda \qquad (67)$$

Therefore, we must have

$$\gamma_\mu = 0 \quad when \quad \mu^2 = \lambda \qquad (68)$$

Not every λ can equal some μ^2. This is certainly true if $\lambda < 0$ or if $\lambda \notin \mathbb{Z}$. But even if λ is a positive integer, there may be no μ such that:

$$\lambda = \mu^2 \qquad (69)$$

For instance, if $n = 3$ and $\lambda = 7$ or 15, there is no μ satisfying Eq. (69). Thus, there is a subset:

$$0 = \lambda_0 < \lambda_1 < \cdots < \lambda_k < \cdots$$

of the positive integers for which there are n-tuples μ satisfying Eq. (69). For $\lambda = \lambda_k$, we can solve Eq. (66) only if Eq. (68) holds. In that case, we can solve by taking α_μ to be given by Eq. (67) when Eq. (69) does not hold, and

to be given arbitrarily when Eq. (69) does hold. On the other hand, if λ is not equal to any λ_k, then, Eq. (69) never holds, and we can solve Eq. (66) by taking the α_μ to satisfy Eq. (67). Thus, we have Theorem 0.2:

Theorem 0.2. *There is a sequence $\{\lambda_k\}$ of nonnegative integers tending to $+\infty$ with the following properties. If $f \in H_t$ and $\lambda \neq \lambda_k$ for every k, then there is a unique solution $u \in H_{t+2}$ of Eq. (66). If $\lambda = \lambda_k$ for some k, then one can solve Eq. (66) only if f satisfies Eq. (68). The solution is not unique; there is a finite number of linearly independent periodic solutions of:*

$$(\Delta + \lambda_k)u = 0 \qquad (70)$$

which can be added to the solution.

The values λ_k for which Eq. (70) has a nontrivial solution (i.e., a solution which is not $\equiv 0$) are called *eigenvalues*, and the corresponding nontrivial solutions are called *eigen functions*.

To analyze the situation a bit further, suppose $\lambda = \lambda_k$ for some k, and $f \in H_t$ is given by Eq. (62) and satisfies Eq. (68). If $v \in H_t$ is given by:

$$v = \sum \beta_\mu e^{i\mu x} \qquad (71)$$

then

$$(f, v)_t = (2\pi)^n \sum (1 + \mu^2)^t \gamma_\mu \beta_{-\mu}$$

by Eq. (61). Hence, we have

$$(f, v)_t = 0$$

for all $v \in H_t$ satisfying Eq. (71) and

$$\beta_\mu = 0 \quad when \quad \mu^2 \neq \lambda_k \qquad (72)$$

On the other hand,

$$(\Delta + \lambda_k)v = \sum (\lambda_k - \mu^2)\beta_\mu e^{i\mu x} \qquad (73)$$

Thus, v is a solution of Eq. (70) if and only if it satisfies Eq. (72). Conversely, if $(f, v)_t = 0$ for all v satisfying Eqs. (71) and (72), then f satisfies Eq. (68). Combining these, we have Theorem 0.3:

Theorem 0.3. If $\lambda = \lambda_k$ for some k, *then there is a solution $u \in H_{t+2}$ of Eq. (66) if and only if:*

$$(f, v)_t = 0 \qquad (74)$$

for all $v \in H_t$ satisfying:

$$(\Delta + \lambda_k)v = 0 \qquad (75)$$

Moreover, any solution of Eq. (75) can be added to the solution of Eq. (66).

What are the solutions of Eq. (75)? By Eq. (73), they are of the form:

$$v = \sum_{\mu^2 = \lambda_k} \beta_\mu e^{i\mu x} = \sum_{\mu^2 = \lambda_k} [a_\mu \cos \mu x + b_\mu \sin \mu x]$$

where we took $\beta_\mu = \alpha_\mu - i b_\mu$. Thus, Eq. (74) becomes:

$$(f, \cos \mu x)_t = 0, \ (f, \sin \mu x)_t = 0 \ \ when \ \ \mu^2 = \lambda_k$$

The results obtained for periodic solutions are indicative of almost all regular boundary-value problems.

I. Sobolev's Inequalities

Certain inequalities are very useful in solving problems in partial differential equations. The following are known as the Sobolev inequalities.

Theorem 0.4. *For each $p \geq 1$, $q \geq 1$ satisfying:*

$$\frac{1}{p} \leq \frac{1}{q} + \frac{1}{n} \tag{76}$$

there is a constant C_{pq} such that:

$$|u|_q \leq C_{pq}(|\nabla u|_p + |u|_p), \qquad u \in C^1(Q) \tag{77}$$

where

$$|u|_q = \left(\int_Q |u|^q \, dx \right)^{1/q} \tag{78}$$

and

$$|\nabla u| = \left(\sum_{k=1}^{n} \left| \frac{\partial u}{\partial x_k} \right|^2 \right)^{1/2} \tag{79}$$

As a corollary, we have Corollary 0.1:

Corollary 0.1. $|u|_q \leq C_q \|u\|_1$, $u \in H_1$, *where* $1 \leq q \leq 2^* := 2n/(n-2)$.

SEE ALSO THE FOLLOWING ARTICLES

CALCULUS • DIFFERENTIAL EQUATIONS, ORDINARY • GREEN'S FUNCTIONS

BIBLIOGRAPHY

Bers, L., John, F., and Schechter, M. (1979). "Partial Differential Equations," American Mathematical Society, Providence, RI.

Courant, R., and Hilbert, D. (1953, 1962). "Methods of Mathematical Physics: I, II," Wiley–Interscience, New York.

Gilbarg, D., and Trudinger, N. S. (1983). "Elliptic Partial Differential Equations of Second Order," Springer-Verlag, New York.

Gustofsen, K. E. (1987). "Partial Differential Equations and Hilbert Space Methods," Wiley, New York.

John, F. (1978). "Partial Differential Equations," Springer-Verlag, New York.

Lions, J. L., and Magenes, E. (1972). "Non-Homogeneous Boundary Value Problems and Applications," Springer-Verlag, New York.

Powers, D. L. (1987). "Boundary Value Problems," Harcourt Brace Jovanovich, San Diego, CA.

Schechter, M. (1977). "Modern Methods in Partial Differential Equations: An Introduction," McGraw–Hill, New York.

Schechter, M. (1986). "Spectra of Partial Differential Operators," North-Holland, Amsterdam.

Treves, F. (1975). "Basic Linear Partial Differential Equations," Academic Press, New York.

Zauderer, E. (1983). "Partial Differential Equations of Applied Mathematics," Wiley, New York.

Diffractive Optical Components

R. Magnusson
University of Connecticut

D. Shin
University of Texas at Arlington

GLOSSARY

Bragg condition A condition for constructive wave addition in a periodic lattice.

Diffraction efficiency A measure of the power density (intensity) of a diffracted wave normalized to that of the incident wave.

Diffraction grating A structure consisting of a periodic spatial variation of the dielectric constant, conductivity, or surface profile.

Evanescent wave An exponentially decaying, nonpropagating wave.

Grating equation An expression that provides the directions θ_i of the diffracted waves produced by a plane wave with wavelength λ_0 incident at θ_{in} on a periodic structure with period Λ. For a grating in air it is $\sin\theta_i = \sin\theta_{in} - i\lambda_0/\Lambda$ where i is an integer labeling the diffraction orders.

Grating fill factor Fraction of the grating period occupied by the high-refractive-index material, n_H.

Grating vector A vector, analogous to the wave vector $\mathbf{k}$ normal to the grating fringes with magnitude $K = 2\pi/\Lambda$, where Λ is the grating period.

Monochromatic plane wave A plane wave with single frequency and wavelength.

Planar grating A spatially modulated periodic structure with the modulated region confined between parallel planes.

Plane of incidence A plane containing the wave vectors of the incident and diffracted waves.

Plane wave An electromagnetic wave with infinite transverse dimensions. It is an idealized model of a light beam where the finite transverse extent of a real beam is ignored for mathematical expediency; this works well if the beam diameter is large on the scale of the wavelength. The planes of constant phase are parallel everywhere.

Reflection grating A grating where the diffracted wave of interest emerges in the space of the cover material.

Surface-relief grating A periodic corrugated surface.

TE (TM) polarized optical wave An electromagnetic wave with its electric (magnetic) vector normal to the plane of incidence.

Transmission grating A grating where the diffracted wave of interest emerges in the space of the substrate material.

Wave vector A vector normal to the planes of constant phase associated with a plane wave. Its magnitude is $k_0 = 2\pi/\lambda_0$, where λ_0 is the wavelength.

DIFFRACTIVE OPTICAL COMPONENTS consist of fine spatial patterns arranged to control propagation of light. Lithographic patterning of dielectric surfaces, layers, or volume regions yields low-loss structures that affect the spatial distribution, spectral content, energy content, polarization state, and propagation direction of an optical wave. Applications include spectral filters, diffractive lenses, antireflection surfaces, beam splitters, beam steering elements, laser mirrors, polarization devices, beam-shaping elements, couplers, and switches. These components are widely used in lasers, fiber-optic communication systems, spectroscopy, integrated optics, imaging, and other optical systems.

I. INTRODUCTION

A. Overview

Application of fabrication methods similar to those used in the semiconductor industry enables optical components with features on the order of and smaller than the wavelength of light. Diffractive optical components possessing multiple phase levels in glass, for example, require sequential high-precision mask alignment and etching steps. Such microscopic surface-relief structures can effectively transform the phasefront of an incident optical beam to redirect or focus it. These components are compact, lightweight, and replicable in many cases. Diffractive optical elements can be used independently or in conjunction with conventional refractive elements for design flexibility and enhanced functionality. Progress

in microfabrication methods, including nanotechnology, will undoubtedly further advance this field. For example, complex three-dimensional diffractive components called photonic crystals are under rapid development. Additionally, progress in modeling and design methodologies (for example, finite-difference time domain analysis and genetic algorithm optimization methods) is occurring in parallel.

This article addresses key aspects of diffractive optics. Common analytical models are described and their main results summarized. Exact numerical methods are applied when precise characterization of the periodic component is required, whereas approximate models provide analytical results that are valuable for preliminary design and improved physical insight. Numerous examples of the applications of diffractive optical components are presented. These are optical interconnects, diffractive lenses, and subwavelength elements including antireflection surfaces, polarization devices, distributed-index components, and resonant filters. Finally, recording of gratings by laser interference is presented and an example fabrication process summarized.

B. The Generic Problem

The diffraction grating is the fundamental building block of diffractive optical components. These may be planar or surface-relief gratings made of dielectrics or metals. Figure 1 illustrates a single-layer diffraction grating and key descriptive parameters including the grating period, Λ, the grating fill factor, f, the grating refractive indices (n_H, n_L), thickness, d, and refractive indices of the cover and substrate media (n_C, n_S). The grating shown has a rectangular profile; if $f = 0.5$ the profile is said to be square. An incident plane wave at angle θ_{in} is dispersed into several diffracted waves (diffraction orders labeled by the integer i as shown) propagating both forwards and backwards.

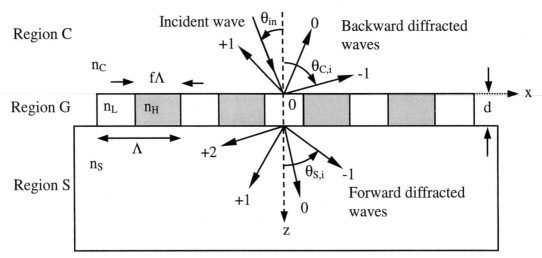

FIGURE 1 Geometry of diffraction by a rectangular grating.

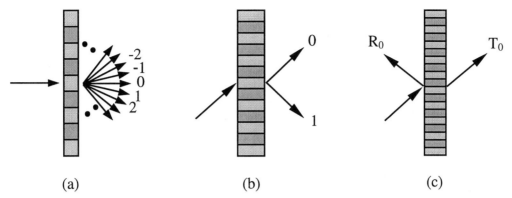

FIGURE 2 Operating regimes of transmissive diffraction gratings: (a) multiwave regime, (b) two-wave regime, and (c) zero-order regime.

The propagation angle of each diffracted wave may be obtained from the grating equation given by

$$n_P \sin \theta_{P,i} = n_C \sin \theta_{in} - i\lambda_0/\Lambda \qquad (1)$$

where λ_0 is the free-space wavelength, i is an integer, and P is either C or S depending on the region under consideration.

Exact electromagnetic analysis can be applied to find the intensity of each diffracted wave shown in Fig. 1. In such calculations, which provide purely numerical results due to the inherent complexity of the full problem, all the waves indicated as well as evanescent waves (not shown) must be included. Simplifying assumptions yield approximate results that are exceedingly useful in the understanding and design of diffractive optical components. Design procedures may additionally include numerical optimization techniques such as simulated annealing or genetic algorithms as well as iterative Fourier transform methods.

C. Operating Regimes

Figure 2 defines the operating regimes of transmissive diffractive elements. Figure 2(a) illustrates the multiwave regime where multiple output diffracted waves are gen-

erated by a grating with a large period on the scale of the wavelength. Note that reflections at refractive-index discontinuities as well as waves propagating in the $-z$ direction (generated by diffraction) are ignored in this figure. In Fig. 2(b), under Bragg conditions, only two main waves are assumed to be dominant with reflected waves neglected. Figure 2(c) defines the zero-order regime in which the grating period is smaller than the wavelength such that all higher diffracted orders are cut off (evanescent). Each of these cases can be modeled approximately resulting in analytical solutions for the diffraction efficiencies, η, associated with the various orders.

Figure 3 provides examples of reflection-type diffractive elements. Typically, a single dominant Bragg-diffracted wave prevails as shown. The waveguide reflection grating is an important element in integrated optical devices and in fiber-optical communication systems for wavelength division multiplexing and other applications. In a waveguide geometry, the incident and diffracted waves are treated as guided modes (not infinite plane waves). Figure 4 illustrates the filtering property of a bulk reflection grating (distributed Bragg mirror such as used for cavity reflection in vertical-cavity lasers); note the efficient reflection within a specific wavelength range.

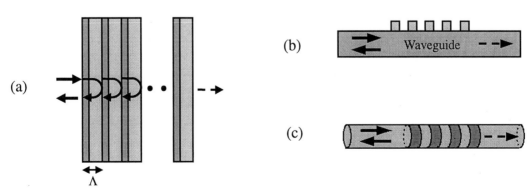

FIGURE 3 Examples of reflection gratings: (a) bulk grating, (b) waveguide grating, and (c) fiber grating.

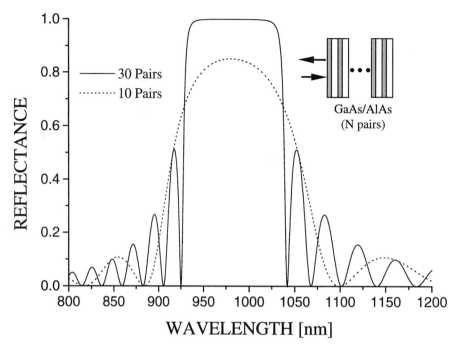

FIGURE 4 Spectral reflectance of a distributed Bragg reflector consisting of 10 and 30 pairs of alternating GaAs/AlAs layers with quarter-wavelength thicknesses at $\lambda_0 = 980$ nm. The cover region material is $Al_{0.7}Ga_{0.3}As$ and the substrate is GaAs. Refractive indices used in the simulation are $n_{GaAs} = 3.512$, $n_{AlGaAs} = 3.151$, and $n_{AlAs} = 3.007$.

D. Application Example: Monochromator

Diffraction gratings are commonly used in spectroscopy. For example, a typical monocromator is schematically illustrated in Fig. 5. For polychromatic light incident on a reflection grating as shown, each component wavelength of a specific diffracted order is spatially (angularly) sepa-

rated according to the grating equation. The angular spread (dispersion) per unit spectral interval is given, differentiating the grating equation for a constant angle of incidence, by

$$\frac{d\theta}{d\lambda} = \left| \frac{i}{\Lambda \cos \theta_{C,i}} \right| \tag{2}$$

The exit slit controls the spectral content of the output light by blocking the undesirable wavelength components around the central one. A similar instrument is the spectrograph where the slit is replaced by an array of detectors. The output spectrum is picked up by the array such that a prescribed spectral band can be associated with a given detector element.

II. GRATING DIFFRACTION MODELS

A. Rigorous Coupled-Wave Analysis

Stimulated by advances in digital computers, various grating-diffraction analysis techniques, which solve the electromagnetic Maxwell's equations numerically either in differential or integral form, have been developed. Rigorous coupled-wave analysis (RCWA) is widely applied; this method is briefly summarized herein using the diffraction geometry given in Fig. 1.

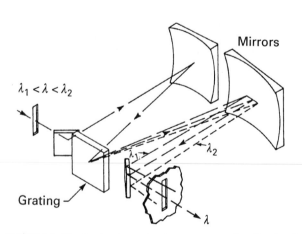

FIGURE 5 The basic elements of a monocromator. The input spectrum contains wavelengths in the range $\lambda_1 < \lambda < \lambda_2$. The grating spreads the light within the diffraction order (say $i = 1$) in which the instrument is operated. No other diffracted orders are shown in the drawing for clarity. The width of the slit defines the spectral contents of the output light. [From Oriel Corporation (1989). "Oriel 1989 Catalog," Volume II.]

An incident TE-polarized plane wave may be expressed as

$$E_{y,in}(x, z) = \exp[-jk_0 n_C (x \sin\theta_{in} + z \cos\theta_{in})] \quad (3)$$

where $k_0 = 2\pi/\lambda_0$ and $j = \sqrt{-1}$. In the homogeneous regions (regions C and S) adjacent to the grating layer, the diffracted fields may be expressed as superpositions of plane waves (so-called Rayleigh expansion) and, thus, the total electric fields in each region are given by

$$E_y(x, z < 0) = E_{y,in}(x, z)$$
$$+ \sum_{i=-\infty}^{\infty} r_i \exp[-j(k_{x,i}x - k_{z,C,i}z)] \quad (4)$$

$$E_y(x, z > d) = \sum_{i=-\infty}^{\infty} t_i \exp[-j\{k_{x,i}x + k_{z,S,i}(z - d)\}], \quad (5)$$

where r_i and t_i are amplitudes of i-th order backward- and forward-diffracted waves, respectively. The x component of the diffracted wave vector, $k_{x,i}$, is given, from the Floquet condition, by

$$k_{x,i} = k_0(n_C \sin\theta_{in} - i\lambda_0/\Lambda) \quad (6)$$

and the z component $k_{z,P,i}$ is

$$k_{z,P,i} = \begin{cases} \left(k_0^2 n_P^2 - k_{x,i}^2\right)^{1/2}, & k_0^2 n_P^2 > k_{x,i}^2 \\ -j\left(k_{x,i}^2 - k_0^2 n_P^2\right)^{1/2}, & k_0^2 n_P^2 < k_{x,i}^2 \end{cases},$$
$$\text{where } P = C \text{ or } S, \quad (7)$$

where the real values of $k_{z,P,i}$ correspond to propagating waves and the imaginary values correspond to evanescent waves.

Inside the grating region ($0 \le z \le d$), the total electric field may be expressed as a coupled-wave expansion

$$E_y(x, 0 \le z \le d) = \sum_{i=-\infty}^{\infty} S_i(z) \exp(-jk_{xi}x), \quad (8)$$

where $S_i(z)$ is the complex amplitude of i-th diffracted wave. The total field satisfies the Helmholtz equation

$$\frac{d^2 E_y(x, z)}{dx^2} + \frac{d^2 E_y(x, z)}{dz^2} + k_0^2 \varepsilon(x) E_y(x, z) = 0, \quad (9)$$

where $\varepsilon(x)$ is the periodic relative permittivity (dielectric constant) modulation given by

$$\varepsilon(x) = \begin{cases} n_L^2, & 0 \le x \le f\Lambda \\ n_H^2, & \text{otherwise} \end{cases}. \quad (10)$$

Due to the periodicity, $\varepsilon(x)$ can be expressed as a Fourier series

$$\varepsilon(x) = \sum_{h=-\infty}^{\infty} \varepsilon_h \exp(jhKx), \quad (11)$$

where ε_h is the h-th Fourier harmonic coefficient and $K = 2\pi/\Lambda$ is the amplitude of grating vector. Inserting Eqs. (8) and (11) into the Helmholtz Equation (9) results in an infinite set of coupled-wave equations given by

$$\frac{d^2 S_i(z)}{dz^2} = k_{xi}^2 S_i(z) - k_0^2 \sum_{h'=-\infty}^{\infty} \varepsilon_{i-h'} S_{h'}(z) \quad (12)$$

where h' is an integer. The coupled-wave Eq. (12) can be cast into an eigenvalue problem as $d^2 S_i(z)/dz^2 = q^2 S_i(z)$. Numerical solution involves a truncated $(N \times N)$ matrix equation $[\ddot{S}_i(z)] = [A][S_i(z)]$ with solutions of the form

$$S_i(z) = \sum_{m=1}^{N} W_{i,m} \left[C_m^+ \exp(-q_m z) + C_m^- \exp\{q_m(z - d)\} \right], \quad (13)$$

where $W_{i,m}$ and q_m are the eigenvectors and positive square roots of the eigenvalues of matrix $[A]$, respectively. The C_m^+ and C_m^- are unknown constants to be determined by the boundary conditions. From the total electric field Eqs. (4), (5), and (8) with (13), the tangential magnetic fields (x component) can be obtained by Maxwell's curl equation

$$H_x(x, z) = \frac{1}{j\omega\mu_0} \frac{dE_y(x, z)}{dz}, \quad (14)$$

where ω is the frequency and μ_0 is the permeability of free space. Finally, the unknown quantities of interest r_i and t_i as well as C_m^+ and C_m^- can be obtained by solving the boundary-condition equations obtained by matching the tangential electric and magnetic fields at boundaries $z = 0$ and $z = d$. The diffraction efficiencies of backward (C) and forward (S) diffracted waves are defined as

$$\eta_{r,i} = Re\left(\frac{k_{z,C,i}}{k_{z,C,0}}\right) |r_i|^2 \quad (15)$$

$$\eta_{t,i} = Re\left(\frac{k_{z,S,i}}{k_{z,C,0}}\right) |t_i|^2. \quad (16)$$

For lossless gratings, energy conservation yields $\sum_{i=-\infty}^{\infty} (\eta_{r,i} + \eta_{t,i}) = 1$.

The accuracy of the RCWA method depends on the number of space harmonics (N) retained in the truncation of matrices. An arbitrary grating shape with a dielectric constant modulation $\varepsilon(x, z)$ may be implemented by replacing the grating layer with a stack of digitized rectangular grating layers with $\varepsilon(x)$ only. Rigorous coupled-wave analysis is a noniterative, stable numerical technique that is relatively straightforward in implementation. It has been successfully applied to planar and surface-relief gratings (with arbitrary grating profiles), dielectric (lossless and lossy) and metallic gratings, transmission and reflection gratings, isotropic and anisotropic gratings, multiplexed gratings, phase-shifted gratings, and photonic crystals. Extensions to include transversely finite beams (such as Gaussian laser beams) have also been implemented as found in the literature.

The calculated results presented in Figs. 4, 14, 16, and 19 were generated with RCWA.

B. Transmittance Diffraction Theory

The diffraction efficiencies of transmission gratings with low spatial frequency ($\Lambda \gg \lambda$) may be estimated using formulation based on the transmittance function. In this region, sometimes called the Raman–Nath diffraction regime, multiple diffracted output waves are typically observed and polarization effects are weak. The pertinent model is illustrated in Fig. 6. A unit-amplitude plane wave is incident upon a low-spatial-frequency grating and is diffracted into multiple forward-propagating waves. Backward propagating diffracted waves are ignored. The transmittance of the grating may be expressed as the convolution

$$T(x) = \tau(x) * \sum_{h=-\infty}^{\infty} \delta(x - h\Lambda), \tag{17}$$

where $\tau(x)$ is the transmittance of each period of the grating and h is an integer. Since $T(x)$ is a periodic function, it may be expanded into a Fourier series as

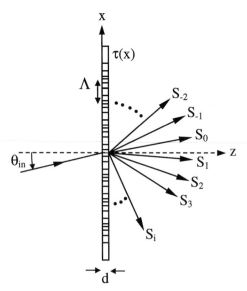

FIGURE 6 Transmittance theory model.

$$T(x) = \sum_{i=-\infty}^{\infty} S_i \exp(jiKx). \tag{18}$$

In the transmittance approach, the emerging diffracted field is well approximated by the product of the input field and the transmittance of the structure. The Fourier coefficients, S_i, are thus identified as the amplitudes of the diffracted waves, which can be calculated by

$$S_i = \frac{1}{\Lambda} \int_{-\Lambda/2}^{\Lambda/2} \tau(x) \exp(-jiKx) \, dx. \tag{19}$$

The diffraction efficiency of i-th diffracted order is defined as

$$\eta_i = |S_i|^2 \tag{20}$$

with the corresponding power conservation law being $\sum_{i=-\infty}^{\infty} \eta_i = 1$ for lossless structures.

Representative results including analytical expressions of diffraction efficiencies are given in Tables Ia and Ib for common planar and surface-relief grating shapes. Absorption and amplitude transmittance gratings exhibit much lower efficiencies than phase gratings. Diffraction efficiency depends strongly on the shape of grating modulation. In particular, the phase gratings with a sawtooth (or blazed) profile (both planar and surface-relief type) can transfer the input energy completely to a particular diffracted order.

The array of dielectric cylinders shown in the inset of Fig. 7 is an example of a low-spatial-frequency grating. In the figure, diffraction-efficiency measurements are plotted for structures with two different cylinder diameters (equal to the grating period) 20 and 50 μm under a TE-polarized normally incident HeNe laser ($\lambda_0 = 632.8$ nm) illumination. This structure generates multiple forward-propagating diffracted orders with similar diffraction efficiencies, functioning as a fan-out element in optical interconnections or as a multiple imaging device. In this experiment, 63 and 159 forward-propagating orders were observed for these gratings with $\Lambda = 20$ and 50 μm, respectively, in agreement with the number of propagating orders (N), predicted by the grating equation, $N = 2\Lambda/\lambda_0 + 1$. The calculated results in Fig. 7 found by

$$S_i = \int_0^1 \exp[-j2g(1 - y^2)^{1/2}] \cos(i\pi y) \, dy, \tag{21}$$

where $y = x/\Lambda$ and $g = \pi \Delta nd/\lambda_0$ show reasonably good agreement with experimental results. The agreement improves as λ_0/Λ becomes smaller. Results experimentally obtained with TM-polarized incident waves show a small but observable difference ($\sim$8% higher efficiency in the region near $i = 0$ for $\Lambda = 50$ μm) with respect to results for TE-polarized light.

TABLE Ia   Summary of Diffraction Efficiency Results from Transmittance Theory (Planar Gratings)

Grating type		Phase gratings	Absorption gratings	Transmittance gratings
$\tau(x)$		$\exp[-jk_0\Delta n(x)d]$	$\exp[-jk_0\{\alpha_0+\Delta\alpha(x)\}d]$	$\tau_0+\Delta\tau(x)$
Sinusoidal		$J_i^2(2\gamma)$ (33.8%)	$\exp[-2\psi_0]\mathrm{I}_i^2(\psi_1)$ (4.80%)	$\begin{cases}\tau_0^2, & i=0 \\ (\tau_1/2)^2, & i=\pm1 \\ 0, & i\neq\pm1\end{cases}$ (6.25%)
Square		$\begin{cases}\cos^2(2\gamma), & i=0 \\ 0, & i=\text{even} \\ (2/i\pi)^2\sin^2(2\gamma), & i=\text{odd}\end{cases}$ (40.5%)	$\begin{cases}\exp[-2\psi_0]\cosh^2(\psi_1), & i=0 \\ 0, & i=\text{even} \\ \exp[-2\psi_0](2/i\pi)^2\sinh^2(\psi_1), & i=\text{odd}\end{cases}$ (10.13%)	$\begin{cases}\tau_0^2, & i=0 \\ 0, & i=\text{even} \\ (2\tau_1/\pi i)^2, & i=\text{odd}\end{cases}$ (10.13%)
Sawtooth		$[2\gamma+i\pi]^{-2}\sin^2(2\gamma)$ (100%)	$\exp[-2\psi_0]\left\{(i\pi)^2+(\psi_1^2)\right\}^{-1}\sinh^2(\psi_1)$ (1.86%)	$\begin{cases}\tau_0^2, & i=0 \\ (\tau_1/\pi i)^2, & i\neq0\end{cases}$ (2.53%)
Definitions	$A_0=n_0,\alpha_0,\tau_0$ $A_1=n_1,\alpha_1,\tau_1$	$\gamma=\pi n_1 d/\lambda_0$	$\psi_0=\alpha_0 d,\ \psi_1=\alpha_1 d$	

Note: Maximum achievable efficiency ($\eta_{\max}$) is given in the parenthesis. $\theta_{\text{in}}=0$.

C. Two-Wave Coupled-Wave Analysis

Two-wave coupled-wave theory (Kogelnik's theory) treats diffraction from a bulk-modulated, planar structure operating in a Bragg regime. Such gratings can be realized by laser-interference recording (holographic recording) in, for example, photorefractive crystals and polymer media. The resulting spatial modulation of refractive index (n) and absorption coefficient (α) is appropriate for the Kogelnik model.

As indicated in Fig. 8, an arbitrary grating slant is allowed. This is a key feature since in practical holographic recording, the angles of incidence of the recording waves are, in general, not symmetric with respect to the surface normal. The slant angle is denoted ϕ; $\phi=\pi/2$ defines an unslanted transmission grating, whereas $\phi=0$ for an unslanted reflection grating.

Table II summarizes the major features and limitations of Kogelnik's theory. The assumptions made yield simple analytical solutions that are easy to apply, provide valuable physical insight, and agree well with experiment in many cases. Consequently, these results are widely cited in the diffractive optics and holography literature.

The pertinent scalar wave equation (TE polarization) is

$$\nabla^2 E_y(x,z)+k^2(x,z)E_y(x,z)=0, \qquad (22)$$

TABLE Ib   Summary of Diffraction Efficiency Results from Transmittance Theory (Surface-Relief Phase Gratings)[a]

Grating profile		Efficiency (η_i)	$\eta_{\max}$
Sinusoidal		$J_i^2(g)$	33.8%
Rectangular ($0\leq f\leq 1$)		$\begin{cases}1-4f(1-f)\sin^2(g), & i=0 \\ [\sin(i\pi f)\sin(g)/(i\pi/2)]^2, & i\neq0\end{cases}$	40.5%
Triangular ($0\leq p\leq 1$)		$\begin{cases}p^2, & i=-\pi p/g \\ (1-p)^2, & i=\pi(1-p)/g \\ [g\sin(g+i\pi p)/(g+i\pi p)\{g-i\pi(1-p)\}]^2, & \text{otherwise}\end{cases}$	100%

[a] Remarks: $\tau(x)=\exp[-jk_0\Delta nd(x)]$, $g=\pi\Delta nd/\lambda_0$, $\Delta n=n_2-n_1$, and $\theta_{in}=0$.

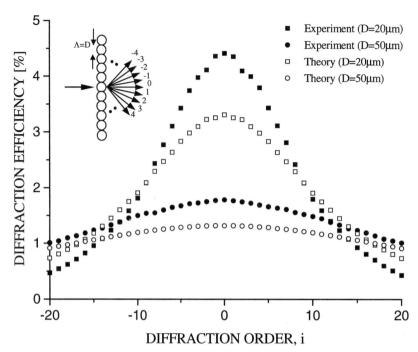

FIGURE 7 Comparison of experimental and theoretical efficiencies of diffracted orders from two ($\Lambda = 20$ and 50 μm) arrays of dielectric circular cylinders. The cylinder with a refractive index of 1.71 is surrounded by a slab with an index 1.47. A spatially filtered and collimated HeNe laser ($\lambda_0 = 632.8$ nm) is used as a source at normal incidence. [From Magnusson, R., and Shin, D. (1989). "Diffraction by periodic arrays of dielectric cylinders." *J. Opt. Soc. Am. A* **6**, 412–414.]

where $k(x, z)$ includes the spatial modulation of the optical constants for a mixed (phase and absorption modulated) grating and is given by

$$k^2(x, z) = \left(\frac{\omega}{c}\right)^2 \varepsilon(x, z) - j\omega\mu\sigma(x, z), \quad (23)$$

where $c = 1/\sqrt{\mu_0\varepsilon_0}$ is the vacuum (with permittivity ε_0 and permeability μ_0) speed of light, the dielectric constant is $\varepsilon(x, z) = n^2(x, z)$ and $\sigma(x, z)$ is the conductivity. For good dielectrics, $\sigma/\omega\varepsilon \ll 1$ and α can be approximated as (nonmagnetic media $\mu = \mu_0$) $\alpha \approx \sigma c\mu_0/2n$. Considering modulations as $n = n_0 + \Delta n(x, z)$ and $\alpha = \alpha_0 + \Delta\alpha(x, z)$ and assuming $\Delta n \ll 1$, $\Delta\alpha \ll 1$, and $\beta = 2\pi n_0/\lambda_0 \gg \alpha_0$, there results

$$k^2(x, z) = \beta^2 - 2j\alpha_0\beta$$
$$+ 2\beta\left(\frac{2\pi \Delta n(x, z)}{\lambda_0} - j\Delta\alpha(x, z)\right). \quad (24)$$

Taking sinusoidal modulations of the basic quantities ε and σ and converting to modulations of n and α leads to $\Delta n = n_1 \cos(\vec{\mathbf{K}} \bullet \vec{\mathbf{r}})$ and $\Delta\alpha = \alpha_1 \cos(\vec{\mathbf{K}} \bullet \vec{\mathbf{r}})$ where $\Delta n = \Delta\varepsilon/2n_0$, $\Delta\alpha = \mu_0 c\Delta\sigma/2n_0$, $\vec{\mathbf{K}}$ is the grating vector, and $\vec{\mathbf{r}} = x\hat{\mathbf{x}} + z\hat{\mathbf{z}}$ is the position vector in Fig. 8. Thus for sinusoidal $\Delta\alpha$ and Δn

$$k^2(x, z) = \beta^2 - 2j\alpha_0\beta + 2\beta\kappa[\exp(j\vec{\mathbf{K}} \bullet \vec{\mathbf{r}})$$
$$+ \exp(-j\vec{\mathbf{K}} \bullet \vec{\mathbf{r}})], \quad (25)$$

where $\kappa = \pi n_1/\lambda_0 - j\alpha_1/2$ is the coupling coefficient. The electric field inside the grating is expanded in coupled-wave form (similar to Eq. (8) in Section II.A retaining only two waves) as

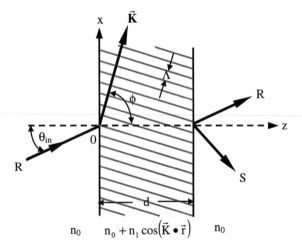

FIGURE 8 Geometry of diffraction by thick gratings (Kogelnik's model).

TABLE II Features and Limitations of Kogelnik's Theory

Features	Limitations
Sinusoidal modulation of n and α	Neglects second derivatives
Slanted gratings	No boundary reflections
Two-wave solution (plane waves)	Approximate boundary
Bragg or near-Bragg solutions	conditions (E only)
TE (H-mode, p) or TM (E-mode, p)	Neglects higher order waves
polarizations	
Transmission and reflection gratings	

$$E_y(x, z) = R(z) \exp(-j\vec{\rho} \bullet \vec{r}) + S(z) \exp(-j\vec{\sigma} \bullet \vec{r}) \quad (26)$$

with the wavevector of the diffracted wave expressed by the Floquet condition as $\vec{\sigma} = \vec{\rho} - \vec{K}$. The Bragg condition for the slanted grating is given by

$$\cos(\theta_{in} - \phi) = \frac{\lambda_0}{2n_0\Lambda} = \frac{K}{2\beta}. \quad (27)$$

By calculating $\nabla^2 E_y(x, z)$, including the modulations in $k^2(x, z)$, inserting both into Eq. (22), and collecting terms in $\exp(-j\vec{\rho} \bullet \vec{r})$ and $\exp(-j\vec{\sigma} \bullet \vec{r})$ leads to the coupled-wave equations

$$C_R \frac{\partial R}{\partial z} + \alpha_0 R = -j\kappa S$$

$$C_S \frac{\partial S}{\partial z} + (\alpha_0 + j\vartheta)S = -j\kappa R, \quad (28)$$

where κS, κR account for the coupling of the reference and signal waves, $\alpha_0 S$, $\alpha_0 R$ account for absorption, and ϑS defines dephasing of R and S on propagation through

the grating. To solve the coupled-wave Eqs. (28), the following boundary conditions are used:

Transmission type: $R(0) = 1$, $S(0) = 0$, with $S(d)$ giving the output amplitude of interest.
Reflection type: $R(0) = 1$, $S(d) = 0$, with $S(0)$ to be determined.

Diffraction efficiency is defined, as in Section II.A, by the ratio between the diffracted power density (along the z-direction) and the input power density and is given by $\eta = (|C_S|/C_R)|S|^2$, where $S = S(d)$ for a transmission grating and $S = S(0)$ for a reflection grating.

Table III summarizes key results for the practical case of lossless dielectric structures or pure-phase gratings. Example calculated results are presented in Fig. 9. The diffraction efficiency of a lossless transmission grating for three slightly different grating periods is shown in Fig. 9(a); these might represent three data pages in an angularly multiplexed holographic memory crystal. The reflectance spectra of several Bragg gratings are shown in Fig. 9(b); these might correspond to reflection gratings in an optical fiber used in a wavelength-division-multiplexed system.

D. Effective Medium Theory (EMT)

For gratings with sufficiently high spatial frequency ($\Lambda \ll \lambda_0$), only zero-order forward- and backward-diffracted waves propagate with all higher orders being evanescent, as seen from the grating equation. Such

TABLE III Summary of Main Results from Kogelnik's Theory (Pure-Phase Gratings)[a]

Type	Transmission gratings	Reflection gratings
Configuration		
Efficiency (η)	$\dfrac{\sin^2[(\nu^2 + \xi^2)^{1/2}]}{1 + \xi^2/\nu^2}$	$\left[1 + \dfrac{1 - \xi^2/\nu^2}{\sinh^2\{(\nu^2 - \xi^2)^{1/2}\}}\right]^{-1}$
ν	$\kappa d/(C_R C_S)^{1/2}$	$j\kappa d/(C_R C_S)^{1/2}$
ξ	$\dfrac{\vartheta d}{2C_S}, C_S > 0$	$-\dfrac{\vartheta d}{2C_S}, C_S < 0$
Unslanted case (On Bragg: $\vartheta = 0$)	$\eta = \sin^2\left(\dfrac{\pi n_1 d}{\lambda_0 \cos\theta_{in}}\right), \phi = \pi/2$	$\eta = \tanh^2\left(\dfrac{\pi n_1 d}{\lambda_0 \cos\theta_{in}}\right), \phi = 0$

[a] Parameters: $\kappa = \pi n_1/\lambda_0$, $C_R = \cos\theta_{in}$, $C_S = \cos\theta_{in} - K\cos\phi/\beta$, $\beta = 2\pi n_0/\lambda_0$, $\vartheta = K\cos(\theta_{in} - \phi) - K^2/2\beta$.

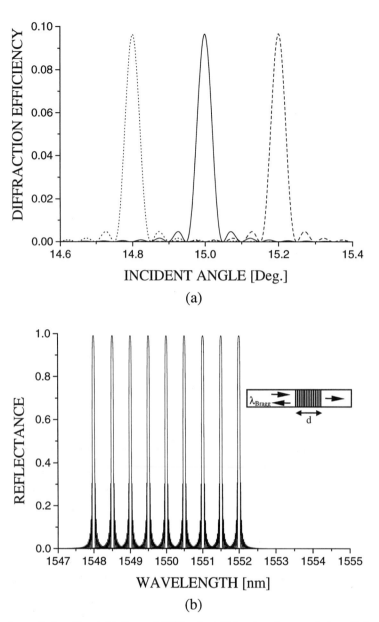

FIGURE 9 Example results from Kogelnik's theory. (a) Diffraction efficiencies of an angularly multiplexed transmission ($\phi = \pi/2$) hologram grating in LiNbO$_3$ as a function of incident angle. The parameters are $n_0 = 2.25$, $n_1 = 10^{-4}$, $d = 0.5$ mm, and $\lambda_0 = 514.5$ nm. Grating periods are 447.6 nm (dotted line), 441.8 nm (solid line), and 436.1 nm (dashed line). (b) Diffraction efficiencies from unslanted ($\phi = 0$) fiber Bragg gratings as a function of wavelength for several grating periods. The parameters are $n_0 = 1.45$, $n_1 = 5 \times 10^{-5}$, $d = 3$ cm, and $\Lambda = 534.5$ nm at $\lambda_0 = 550$ nm, for example.

gratings are referred to as subwavelength gratings or zero-order gratings. In this operating region, an incident electromagnetic wave cannot resolve the finely modulated layer and interacts with the structure, approximately, as if it were a homogeneous layer with its optical constants equal to the spatial average of those in the modulated layer. This is analogous to optical crystals which are periodic structures exhibiting strong diffraction effects in the x-ray region while acting as homogeneous media in the optical

spectral region. As crystals have a natural birefringence depending on the arrangement of atoms or molecules, subwavelength gratings also exhibit polarization dependence called form birefringence due to the spatial periodic modulation.

In the subwavelength region, a rectangular grating layer such as that in Fig. 1, may be approximately modeled as a homogeneous negative uniaxial layer with ordinary and extraordinary indices of refraction given by

$$n_O = \left[n_L^2 + f\left(n_H^2 - n_L^2\right) \right]^{1/2} \qquad (29)$$

$$n_E = \left[n_L^{-2} + f\left(n_H^{-2} - n_L^{-2}\right) \right]^{-1/2}. \qquad (30)$$

Similarly, a multilevel stair-step grating can be modeled as a stack of homogeneous layers with different effective indices. A grating with a continuously-varying dielectric constant profile $\varepsilon(x,z)$, such as a sinusoidal shape for example, can be approximated as a graded-index layer along the thickness direction as

$$n_O(z) = \left[\frac{1}{\Lambda} \int_{-\Lambda/2}^{\Lambda/2} \varepsilon(x, z)\, dx \right]^{1/2} \qquad (31)$$

$$n_E(z) = \left[\frac{1}{\Lambda} \int_{-\Lambda/2}^{\Lambda/2} \frac{1}{\varepsilon(x, z)}\, dx \right]^{-1/2}. \qquad (32)$$

For a wave incident from the cover (region C) as shown in Fig. 1, the ordinary and extraordinary waves correspond to TE- and TM-polarized waves, respectively. Classical (anisotropic) thin-film analysis can be applied to obtain the transmission and reflection coefficients (i.e., diffraction efficiencies) of the propagating zero-order waves.

The EMT expressions given above are independent of Λ/λ_0 and are called the zero-order effective indices. They provide accurate results for weakly modulated gratings ($\Delta\varepsilon = n_H^2 - n_L^2 \ll 1$) in the limit $\Lambda/\lambda_0 \ll 1$. The applicable range in Λ/λ_0 may be extended by using the Λ/λ_0-dependent second-order effective indices, given by

$$n_O^{(2)} = \left[n_O^2 + \frac{1}{3}\left\{ \pi \frac{\Lambda}{\lambda_0} f(1-f) \right\}^2 \left(n_H^2 - n_L^2\right)^2 \right]^{1/2} \qquad (33)$$

$$n_E^{(2)} = \left[n_E^2 + \frac{1}{3}\left\{ \pi \frac{\Lambda}{\lambda_0} f(1-f) \right\}^2 \right.$$
$$\left. \times \left(n_H^{-2} - n_L^{-2}\right)^2 n_E^6 n_O^2 \right]^{1/2}. \qquad (34)$$

More accurate results are found via higher-order EMT indices or by solving the Rytov eigenvalue equations numerically. Comparison of zero-order diffraction efficiencies found by EMT and RCWA is available in the literature.

Figure 10(a) shows the effective index of an etched GaAs grating, calculated using the effective medium theory, as a function of grating fill factor. This shows that an arbitrary effective index ranging from 1 (air) to 3.27 (bulk GaAs) is achievable for both polarizations by proper choice of the grating fill factor. Estimation using the second order EMT (dashed line) differs slightly from that using the zero-order EMT for this high-spatial-frequency ($\Lambda/\lambda_0 = 0.1$) grating. The maximum value of form bire-

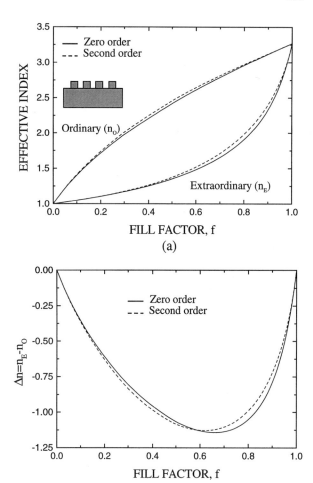

FIGURE 10 Effective medium properties of a one-dimensional, etched, GaAs subwavelength grating calculated by the zero-order (solid line) and second-order (dashed line) EMT. Parameters used are $\lambda_0 = 10.6\ \mu$m, $\Lambda = 0.1\lambda_0$, and $n_{GaAs} = 3.27$. (a) Effective refractive indices for ordinary and extra-ordinary waves as a function of fill factor. (b) Form birefringence of the GaAs subwavelength grating as a function of fill factor.

fringence, defined by $\Delta n = n_E - n_O$, is shown (Fig. 10(b)) to be -1.128 at fill factor of 0.62; this value is much larger than that of naturally birefringent materials such as calcite with $\Delta n = -0.172$ in the visible region.

III. APPLICATIONS

A. Optical Interconnects

Due to large bandwidths and inherent parallel-processing capability, optical interconnections are of interest in communication signal routing and optical computing. Figure 11 shows example optical interconnection devices made with diffractive components including interconnections in free space (Figs. 11(a)–(d)) and in integrated and

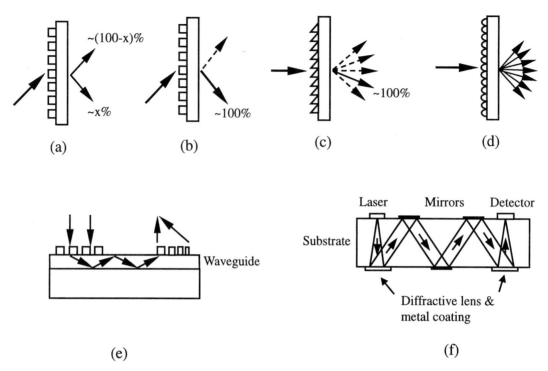

FIGURE 11 Optical interconnects: (a) beam divider (3 dB if $x = 50$), (b) beam deflector, (c) blazed grating, (d) array generator, (e) waveguide interconnect, and (f) substrate interconnect.

planar optical circuits (Figs. 11(e)–(f)). Gratings operating in a two-wave regime may be used as beam dividers (Fig. 11(a)). Under a Bragg incidence, these gratings generate two most efficient forward diffracted orders. The beam ratio (represented as x in the figure) may be controlled by the grating thickness; a beam deflector that transfers most of the energy into the first diffracted order, as shown in Fig. 11(b), is also achievable with this structure. Another type of beam deflector may be designed using a low-spatial-frequency blazed grating [Fig. 11(c)]. A low-frequency grating that generates multiple diffracted orders with similar efficiencies, as in Fig. 7, may be used as a channel divider (fanout element) [Fig. 11(d)]. In general, 1-to-N channel dividers with specified beam-division ratio can be achieved by Dammann gratings. Integrated optical input couplers are realizable using a diffractive phase matching element [Fig. 11(e)]. Similar gratings can also be used as output couplers. If a chirped grating (grating with variable local periods) is used, a focused output beam is obtainable, which is useful in optical memory readout devices. A substrate mode optical interconnect system is shown in Fig. 11(f).

B. Diffractive Lenses

Diffractive optical components are used to construct imaging and beam shaping devices. Compact, lightweight, and fast (high numerical aperture) lenses can be obtained via radially modulated diffractive optical components. Diffractive lenses may be constructed with reference to the refractive spherical lens as explained by Fig. 12(a). The phase profile of the spherical lens is transformed based on modulo 2π operation that eliminates bulk sections of 2π phase shifts, resulting in a Fresnel lens with a set of annular rings within which the surface has continuous profile. The maximum thickness of the lens is $d_{2\pi} = \lambda_0/[n(\lambda_0) - 1]$, which corresponds to a 2π phase shift at wavelength λ_0. These lenses may be fabricated by diamond turning, gray level masking, or direct laser writing lithography, depending on the feature size and aspect ratio. Alternatively, the continuous phase profile may be approximated by a discrete binary or multilevel profile as shown in Fig. 12(a). Fabrication of multilevel diffractive lenses is accomplished via photolithography employing accurately placed masks and sequential lithographic and etching steps.

Figure 12(b) illustrates schematically the operation of a diffractive lens using a ray approach. The lens may be viewed as a low-spatial frequency chirped grating with local grating periods decreasing in the radial direction. The Fresnel zone radii, r_h, are defined such that rays emanating from adjacent zones add constructively at the focal point. This means that the adjacent ray paths differ by λ_0, the design wavelength. Thus, the radius of h-th Fresnel zone is

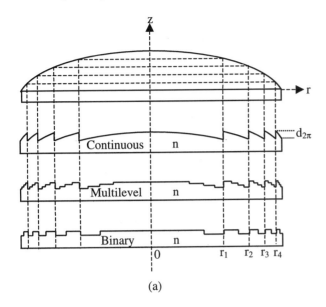

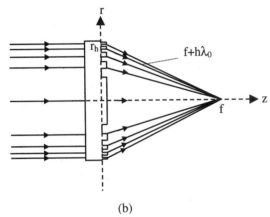

(a)

(b)

FIGURE 12 (a) A refractive spherical lens and analogous diffractive lenses. (b) Diffractive-lens focusing via ray tracing.

$$r_h = \left[(f + h\lambda_0)^2 - f^2\right]^{1/2}, \quad h = \text{integer}. \quad (35)$$

In the paraxial region ($f \gg h_{\max}\lambda_0$), it may be approximated as $r_h = [2h\lambda_0 f]^{1/2}$.

The wavelength-dependent focal length and diffraction efficiency of diffractive lenses with a continuous profile are given, through transmittance diffraction analysis, by

$$f_m(\lambda) = \frac{\lambda_0 f}{\lambda m}, \quad m = \text{integer} \quad (36)$$

and

$$\eta_m(\lambda) = \frac{\sin^2[\pi\{\alpha(\lambda) - m\}]}{[\pi\{\alpha(\lambda) - m\}]^2}, \quad (37)$$

where $\alpha(\lambda)$ is a detuning parameter defined by

$$\alpha(\lambda) = \frac{\lambda_0}{\lambda}\left[\frac{n(\lambda) - 1}{n(\lambda_0) - 1}\right]. \quad (38)$$

The expression for the focal length shows that the diffractive lens has an infinite number of focal points that correspond to the diffracted orders m, and that it is highly dispersive (i.e., f depends on λ). The diffraction efficiency can approach 100% for $\alpha(\lambda) = 1$. In contrast, the peak efficiency of a diffractive lens with a multilevel profile is given by

$$\eta = \frac{\sin^2[\pi/M]}{[\pi/M]^2} \quad (39)$$

at the operating wavelength λ_0 where M is the number of phase levels. Diffractive lenses with eight levels ($M = 8$), 4-levels ($M = 4$), and a binary phase profile ($M = 2$), for example, have maximum efficiencies of 95, 81, and 40.5%, respectively. Detailed discussion on diffractive lenses and their applications can be found in the literature.

C. Subwavelength Grating Devices

Due to advances in microfabrication technology, gratings with periods much smaller than the wavelength can be made. Various applications of subwavelength gratings are shown in Fig. 13. The operation of most subwavelength diffractive optical devices is based on the idea of an effective medium, i.e., the spatial averaging of the refractive index and the associated form birefringence described in Section II.D. Additional devices are based on combination of subwavelength gratings with thin-film and waveguide layers.

1. Antireflection (AR) Surfaces

Common AR coating techniques involve deposition of single or multiple homogeneous thin-film layers, whose refractive indices and thicknesses are chosen to reduce the Fresnel reflections at optical interfaces. Subwavelength gratings may be used as AR coatings (Fig. 13(a)) since the effective refractive index can be controlled by the grating fill factor as seen in Fig. 10(a). Single-layer AR structures can be etched into a variety of surfaces without additional thin-film deposition steps with improved adhesion and performance.

For example, an antireflection surface may be obtained by placing a rectangular subwavelength grating directly on the substrate. The commonly used quarter-wave matching technique (so called V-coating) requires that the grating layer have an effective refractive index equal to $n_{\text{eff}} = [n_c n_s]^{1/2}$ and layer thickness of $d = \lambda_0/4n_{\text{eff}}$ in order to be antireflective at the center wavelength of λ_0 for a normally incident wave. By connecting these with the zero-order effective index expressions (Eqs. (29) and (30)) with $n_L = n_C$ and $n_H = n_S$, an approximate grating fill factor required for antireflection is obtained as

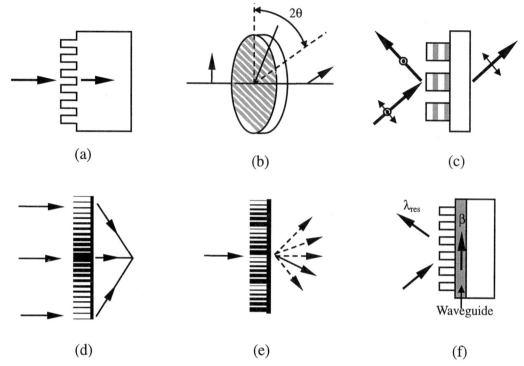

FIGURE 13 Applications of subwavelength gratings: (a) antireflection surface, (b) retardation plate, (c) polarizing beam splitter, (d) diffractive lens, (e) blazed grating, and (f) wavelength-selective reflector (filter).

$$f_{AR} = \begin{cases} n_C/(n_C + n_S) & \text{(TE)} \\ n_S/(n_C + n_S) & \text{(TM)} \end{cases}. \qquad (40)$$

Figure 14 shows an example spectral reflectance, calculated by RCWA, of a one-dimensional subwavelength antireflection grating formed by etching a GaAs substrate. The resulting reflectance (below ~0.1% at the center wavelength of 10.6 μm) with the grating layer is much

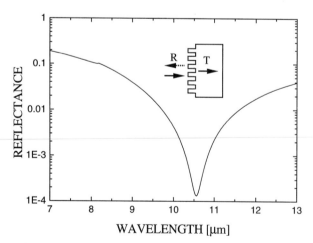

FIGURE 14 Spectral reflectance of a one-dimensional antireflection grating etched on a GaAs substrate under a normally incident wave at 10.6 μm. The device parameters are $n_{GaAs} = 3.27$, $\Lambda = 2.5$ μm, $f = 0.19$, and $d = 1.4655$ μm.

lower than the bulk GaAs substrate reflection (~28%). A multilevel or continuously varying grating profile obtains a wider antireflection bandwidth. For example, a triangular profile simulates a layer with a graded refractive index along the z-axis and provides an improved spectral and angular AR performance. Symmetric crossed gratings (grating modulation in two orthogonal directions) may be used for polarization-independent antireflection structures as shown in Fig. 15. Applications of subwavelength AR gratings include detectors, solar cells, and optical windows.

2. Polarization Devices

Subwavelength gratings may be used as polarization components (Figs. 13(b) and (c)). The metallic wire-grid polarizer is a well-known example. The form birefringence associated with dielectric subwavelength gratings enables construction of polarization components analogous to those made by naturally birefringent materials. For example, Fig. 13(b) shows a subwavelength retardation plate. The phase retardation experienced by a normally incident, linearly polarized wave passing through a subwavelength grating is given by

$$\Delta\phi = \phi_{TE} - \phi_{TM} = -k\Delta n d, \qquad (41)$$

where $\Delta n = n_{TM} - n_{TE}$ is the form birefringence and d is the thickness of the grating. Due to their large

$$f_x\Lambda_x \qquad \Lambda_x$$

FIGURE 15 Scanning electron micrograph of a two-dimensional Si AR surface designed for operation in the infrared (8- to 12-μm) band. The grating has a 2.45-μm period and fill factors $f_x = f_y = 0.7$. The grating is quarter wavelength thick at 10.6 μm. [From Raguin, D. H., Norton, S., and Morris, G. M. (1994). "Diffractive and Miniaturized Optics" (S. H. Lee, ed.), pp. 234–261, Critical review **CR 49**, SPIE Press, Bellingham.]

birefringence (Fig. 10(b)) compared with that provided by naturally birefringent materials, compact and light weight zero-order retardation plates such as quarter-wave plates, half-wave plates, polarization rotators, and polarization compensators may be fabricated. As this artificial birefringence can be obtained using isotropic materials, low-cost devices are feasible at any operating wavelength. Note that more compact devices are possible with higher index materials at the expense of higher insertion loss due to increased reflection. Design considerations may involve optimization to maximize the birefringence and minimize insertion loss.

Subwavelength gratings may be used to form multifunction polarization elements. The inset of Fig. 16 shows a diffractive polarizing mirror. A lateral subwavelength corrugation yields a polarization-dependent effective refractive index, while a stack of thin-film layers with alternating high/low refractive index simulates a distributed Bragg reflector. By taking each layer-thickness to be quarter wavelength for one polarization (TE polarization in this example), the structure reflects this polarization while it transmits the orthogonal polarization. The device in Fig. 16 can be used as a polarizing mirror at 1.55 μm. Similar structures can be designed to operate as polarizing beam splitter as shown in Fig. 13(c).

3. Distributed-Index Devices

The controllability of the effective refractive index (by the fill factor f) of a subwavelength grating enables applica-

tions such as phase plates. A refractive surface, such as shown in Fig. 17(a), may be modeled as a single, planar, laterally graded index layer with a fixed thickness if the x-dependent phase accumulation provided by the refractive surface profile (with fixed refractive indices n_1 and n_2) is translated into an equivalent refractive index

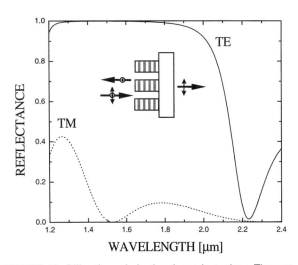

FIGURE 16 Diffractive polarization-dependent mirror. The grating consists of four pairs of quarter-wavelength thick Si–SiO$_2$ stack deposited on a fused silica substrate and etched with a period of 600 nm and a fill factor of 0.5. Refractive indices are $n_{Si} = 3.48$ and $n_{SiO2} = 1.44$. [From Tyan, R.-C., Salveker, A. A., Chou, H.-P., Cheng, C.-C., Scherer, A., Sun, P.-C., Xu, F., and Fainman, Y. (1997). "Design, fabrication, and characterization of form-birefringent multilayer polarizing beam splitter." *J. Opt. Soc. Am. A* **14**, 1627–1636.]

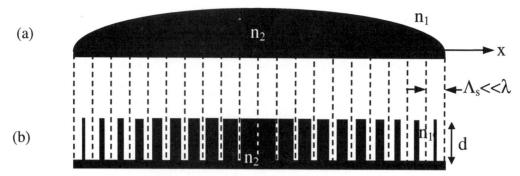

FIGURE 17 Construction of subwavelength distributed-index devices. (a) Refractive surface (center zone of diffractive lens). (b) Analogous structure realized by binary fill-factor modulated subwavelength gratings.

distribution. A subwavelength grating can be used to approximate this graded-index layer using the effective medium property, as shown in Fig. 17(b). That is, the structure is first replaced by a subwavelength binary grating layer with a sampling period of Λ_s. The grating thickness is set to $d = \lambda_0/(n_2 - n_1)$ to obtain a phase difference up to 2π by varying the fill factor in a local period. Then, in each local grating period, the fill factor is chosen, with the help of EMT, such that the overall effective index distribution approximates the graded-index profile required.

In general, an arbitrary phase function may be achieved by tailoring the local refractive index with two-dimensional fill-factor-modulated subwavelength gratings. Because the resulting structure is a single-layer binary-modulated surface, planar technology can be used for fabrication. Applications include diffractive lenses (Fig. 13(d)), blazed gratings (Fig. 13(e)), phase compensators, and other beam-shaping devices.

Figure 18 shows a scanning electron micrograph (SEM) of one period (along x) in a two-dimensional blazed

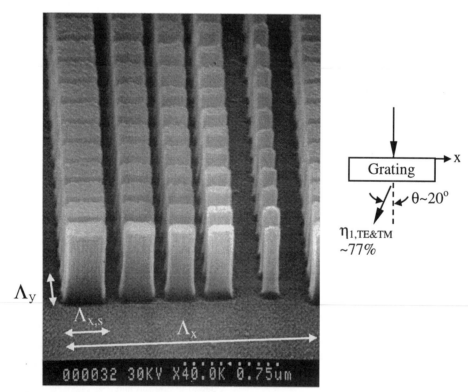

FIGURE 18 Scanning electron micrograph of a blazed grating operating at $\lambda_0 = 632.8$ nm realized by two-dimensional fill-factor-modulated subwavelength grating. The global grating periods are $\Lambda_x = 1900$ nm and $\Lambda_y = 380$ nm. The sampling period ($\Lambda_{x,s}$) is 380 nm. [From Lalanne, P., Astilean, S., Chavel, P., Cambril, E., and Launois, H. (1998). "Blazed binary subwavelength gratings with efficiencies larger than those of conventional echelette gratings." *Opt. Lett.* **23,** 1081–1083.]

grating made of TiO_2. The grating has global periods of $\Lambda_x = 1900$ nm and $\Lambda_y = 380$ nm such that an incident HeNe laser beam ($\lambda_0 = 632.8$ nm) is diffracted into five forward diffracted orders. Each x-period (Λ_x) is subdivided into five fill-factor modulated subwavelength gratings with a sampling period ($\Lambda_{x,s}$) equal to 380 nm to simulate the blazed grating profile. The two-dimensional area fill factors used in each local subwavelength grating are 0.31, 0.47, 0.53, 0.65, and 0.77. The grating was fabricated by e-beam lithography and reactive-ion etching. For a normally incident HeNe laser beam, this device redirect most of its energy into a first-order wave propagating at an angle $\sim 20°$ nearly independent of the polarization state. Measured diffraction efficiencies were 77 and 78% for TE and TM polarizations, respectively.

4. Guided-Mode Resonance (GMR) Devices

Thin-film structures containing waveguide layers and diffractive elements exhibit the guided-mode resonance effect. When an incident wave is phase-matched, by the periodic element, to a leaky waveguide-mode, it is reradiated in the specular-reflection direction as it propagates along the waveguide and constructively interferes with the directly reflected wave. This resonance coupling is manifested as rapid spectral or angular variations of the diffraction efficiencies of the propagating waves. When zero-order gratings are used, in particular, a high-efficiency resonance reflection can be obtained, which can be a basis for high-efficiency filtering. (Fig. 13(f))

In Fig. 19, the calculated (using RCWA) spectral reflectance from a double-layer waveguide-grating structure, SEM picture of which is shown in the inset, is given as a dotted line. The device includes HfO_2 waveguide layer and a holographically recorded photoresist grating. A high-efficiency resonant reflection with a linewidth of ~ 2.2 nm is shown to occur near 860 nm for a normally incident TE-polarized wave. This resonance is induced by phase matching to the TE_0 waveguide mode via the first evanescent diffracted order.

The resonance wavelength may be estimated by the phase matching condition

$$2\pi n_C \sin\theta_{in}/\lambda_0 - i2\pi/\Lambda = \beta(\lambda_0), \qquad (42)$$

where $\beta(\lambda_0)$ is the mode-propagation constant that can be obtained by solving the eigenvalue equation of the waveguide-grating structure. The resonance spectral linewidth is typically narrow (on the order of nanometers or less) and can be controlled by the modulation amplitude, fill factor, grating thickness, and the refractive-index contrast of the device layers. The resonance response (location and linewidth) is polarization dependent due to

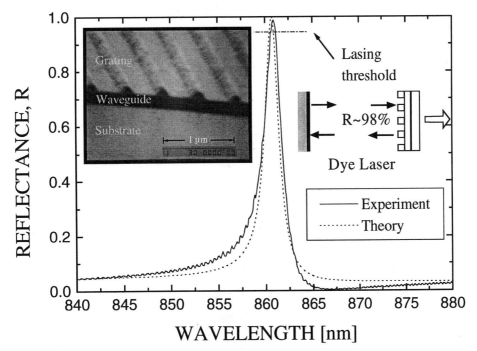

FIGURE 19 Theoretical (dotted line) and experimental (solid line) spectral response from a double-layer waveguide grating for a normally-incident TE-polarized wave. The device consists of holographically-recorded photoresist grating ($n = 1.63$, $\Lambda = 487$ nm, $f = 0.3$, $d = 160$ nm) on top of e-beam evaporated HfO_2 waveguide layer ($n = 1.98$, $d = 270$ nm) on a fused silica substrate ($n = 1.48$). [From Liu, Z. S., Tibuleac, S., Shin, D., Young, P. P., and Magnusson, R. (1998). "High-efficiency guided-mode resonance filter." *Opt. Lett.* **23**, 1556–1558.]

inherent difference in modal characteristics of excited TE- and TM-waveguide modes.

The experimentally obtained high efficiency of ~98% in Fig. 19 (solid line) confirms the feasibility of guided-mode resonance devices. Potential applications of these devices include laser resonator frequency-selective polarizing mirrors, laser cavity tuning elements, electro-optic modulators and switches, tunable filters, mirrors for vertical-cavity lasers, wavelength-division multiplexing, and chemical and biosensors. For example, the fabricated double-layer filter in Fig. 19 was used to realize a GMR laser mirror. The flat output mirror of the dye laser with broadband output (800–920 nm) was replaced with the GMR filter and the birefringent tuning element removed. Lasing occurred at a wavelength of ~860 nm. The laser power was ~100 mW when pumped with an Ar+ laser emitting a power of ~5 W at a 514-nm wavelength. The linewidth of the output laser beam was measured as ~0.3 nm. This linewidth was set by the GMR filter linewidth at the threshold reflectance for laser oscillation to occur; in this case at ~95% reflectance value in Fig. 19.

IV. FABRICATION

A. Photolithography

For diffractive components with features that are larger than the wavelength, ordinary photolithography may be applied for fabrication. Typically, several steps are needed including deposition of a photoresist layer on a thin film or substrate, UV light exposure of the photoresist through a prefabricated mask defining the diffractive element, development, and processing. The photoresist, a light-sensitive organic liquid, is spin-coated on the substrate resulting in a film with thickness (~1 μm) that depends on the spin rate. If the UV-exposed regions remain (vanish) after development, the resist is said to be negative (positive). Photoresist selection takes account of resolution, sensitivity, adhesion, and other factors. Typically, the resist is UV-exposed for a few seconds followed by development with appropriate chemical solutions. A postbake step at 100–200°C to increase etching resistance may follow.

B. Laser Interference Recording

Two-wave laser interference (Fig. 20(a)) may be used to record periodic elements with subwavelength features (Λ ~100 nm). The UV laser (for example helium cadmium, argon–ion, or excimer laser) beam is focused down with an objective lens and passed through a matching pinhole to spatially filter the beam. The central part of the emerging spherical wave, which has nearly planar phase fronts, illuminates the sample directly. A part of the wave is reflected towards the sample as shown. These two waves interfere to produce a standing, periodic intensity pattern with period

$$\Lambda = \lambda_0/(\sin\theta_1 + \sin\theta_2) \qquad (43)$$

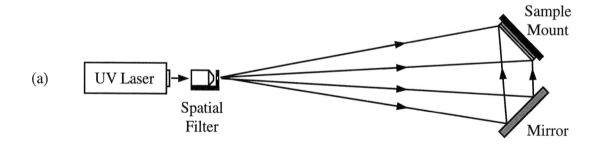

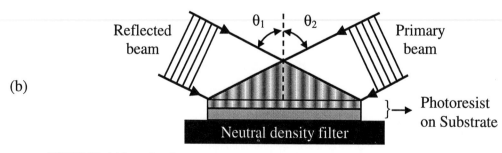

FIGURE 20 (a) Laser interference grating recording system. (b) Details of interference light pattern.

indicated by Fig. 20(b). For symmetrically incident ($\theta_1 = \theta_2 = \theta$) recording beams, the period is $\Lambda = \lambda_0/2\sin\theta$. The interference pattern has maximum contrast if the two recording beams are polarized in parallel (electric-field vector normal to the plane of incidence). This pattern must be kept stationary with respect to the photoresist during exposure. Heavy vibration-isolating, air-suspended optics tables are used for this; additional measures should be taken to screen off air currents and acoustic noise. The single beam system in Fig. 20 is often used in practice; two-beam systems with separate collimated beams are also in use.

From the expression for the period, it is seen that the minimum period obtainable with this approach is $\Lambda = \lambda_0/2$ for counterpropagating recording beams. It can be further reduced by increasing the refractive index of the input space by placing a glass block or a prism on the sample in which case the period is reduced by the refractive index of the prism. Laser interference recording, thus, allows fabrication of high-quality, large-area gratings with small periods. Noise patterns may arise from interface reflections during exposure. These can be minimized by use of index-matching liquids and neutral density filters (that absorb the light passing through preventing its return to the sample region) as indicated in Fig. 20(b).

C. Other Fabrication Techniques

Many additional fabrication methods are described in the references. Electron-beam writing is a high-resolution direct-write approach that is widely applied. Since the equipment is expensive and relatively slow, e-beam writing is particularly appropriate for making masks that can serve as masters for subsequent copying. Direct laser writing with a focused UV laser beam is another promising method. The generation of the spatial pattern defining the diffractive element is often followed by a variety of processing steps. These may include reactive-ion etching, wet chemical etching, ion-beam milling, etc. Finally, thin-film deposition may be needed with methods such as sputtering, thermal evaporation, e-beam evaporation, chemical vapor deposition, and molecular beam epitaxy being representative examples.

D. An Example Process

Figure 21 shows a process used for fabricating a buried waveguide grating. A thin film of silicon nitride (thickness $\sim$200 nm) is e-beam evaporated on a fused silica substrate followed by thermal evaporation of a thin aluminum layer (thickness $\sim$50 nm). A grating with period $\Lambda = 300$ nm is recorded using argon-ion ($\lambda_0 = 364$ nm) laser interference in photoresist. The developed photoresist grating is dry etched (RIE) to remove any residual resist in the grating troughs through which the Al layer is wet etched. The silicon nitride film is then dry etched through the metal mask and the Al grating removed. On sputtering deposition of a layer of silicon dioxide, a silicon nitride/silicon dioxide waveguide grating with a square-wave profile results; it is a waveguide grating as its average index of refraction exceeds that of the surrounding media.

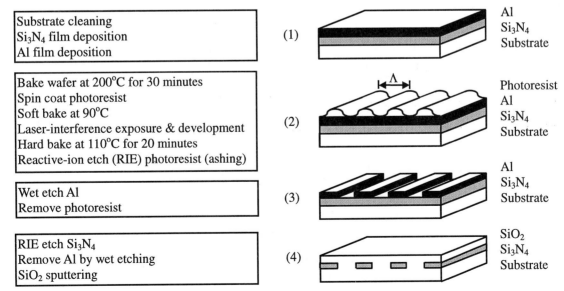

FIGURE 21 An example procedure to fabricate an embedded-grating device.

SEE ALSO THE FOLLOWING ARTICLES

HOLOGRAPHY • IMAGING OPTICS • LASERS • LASERS, OPTICAL FIBER • MICROOPTICS • OPTICAL DIFFRACTION • OPTICAL WAVEGUIDES (PLANAR) AND WAVEGUIDE DEVICES

BIBLIOGRAPHY

Born, M., and Wolf, E. (1980). "Principles of Optics" (5th edition), Pergamon, Oxford.

Gaylord, T. K., and Moharam, M. G. (1985). Analysis and applications of optical diffraction by gratings. *Proc. IEEE* **73,** 894–937.

Goodman, J. W. (1996). "Introduction to Fourier Optics" (2nd edition), McGraw Hill, New York.

Herzig, H. P. (ed.) (1997). "Micro Optics: Elements, Systems and Applications," Taylor and Francis, London.

Joannopoulos, J. D., Meade, R. D., and Winn, J. N. (1995). "Photonic Crystals: Molding the Flow of Light," Princeton University Press, Princeton, NJ.

Kogelnik, H. (1969). Coupled wave theory for thick hologram gratings. *Bell Sys. Tech. J.* **48,** 2909–2947.

Lee, S. H. (ed.). (1994). "Diffractive and Miniaturized Optics," Critical review **CR 49,** SPIE Press, Bellingham.

Mait, J. N., and Prather, D. W. (ed.) (2001). "Selected Papers on Subwavelength Diffractive Optics," Milestone series **MS 166,** SPIE Optical Engineering Press, Bellingham.

Martellucci, S., and Chester, A. N. (ed.) (1997). "Diffractive Optics and Optical Microsystems," Plenum, New York.

Moharam, M. G., Grann, E. B., Pommet, D. A., and Gaylord, T. K. (1995). Formulation of stable and efficient implementation of the rigorous coupled-wave analysis of binary gratings. *J. Opt. Soc. Am. A* **12,** 1068–1076.

Nishihara, H., Haruna, M., and Suhara, T. (1989). "Optical Integrated Circuits," McGraw-Hill, New York.

Petit, R. (ed.) (1980). "Electromagnetic Theory of Gratings," Springer Verlag, Berlin.

Sinzinger, S., and Jahns, J. (1999). "Micro-optics," Wiley-VCH, Weinheim.

Solymer, L., and Cooke, D. J. (1981). "Volume Holography and Volume Gratings," Academic Press, London.

Turunen, J., and Wyrowski, F. (ed.) (1997). "Diffractive Optics for Industrial and Commercial Applications," Akademie Verlag, Berlin.

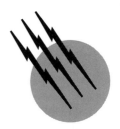

Digital Controllers

David M. Auslander
University of California, Berkeley

GLOSSARY

Actuator A means for modulating the application of power to a target system.

Aliasing When a signal is sampled at too low a frequency so that the high frequency content appears in the low frequency area of the sampled signal.

Analog Signals based on continuous values of a signal variable such as voltage; information content is limited by noise (*see also*, digital).

Cascade control A configuration of nested control loops for situations in which there are several measurements but only a single actuator.

Digital Signals based on discrete values of a signal variable such as voltage; normally binary, that is, two values per signal (*see also*, analog).

Drive Power amplifier for use with an electric motor.

Dynamic behavior Behavior for which the current output is dependent on the history of the inputs.

Eigenvalues Roots of the characteristic equation associated with an ordinary differential equation. These values give a compact characterization of system behavior (also called "poles").

Equilibrium A point at which a system is just "balanced," i.e., none of its state variables are changing.

Euler method A simple method for solving ordinary differential equations numerically.

Feedback control Explicit use of instruments, computation, and actuation to use observation of a system's output to control its actuation so as to influence the behavior toward a desirable outcome.

Feedforward Use of *a priori* knowledge of how a system will behave to apply an actuation that will produce a desired result.

Latency In real time software, the time between the occurrence of an event and the delivery of the response to that event.

Moore's law A prediction made by Gordon Moore (of Intel) that computing power per unit cost would double every 18 months.

ODE Ordinary differential equation.

PID control Control calculation based on the sum of terms proportional to: the error, the integral of the error, and the derivative of the error.

Poles Roots of the characteristic equation associated with an ordinary differential equation. These values give a compact characterization of system behavior (also called "eigenvalues").

Programmable logic controller (PLC) A form of computer designed to use Boolean logic in the solution of industrial control problems.

Real time software Software which must deliver a result at a specified time in order for the result to be usable.

Runga-Kutta methods A set of methods for numerical solution of ordinary differential equations.

Saturation When an actuator or instrument reaches the limit of its realizable range.

Setpoint The desired value for a system's output.

Simulation Use of a mathematical model of a physical system to predict its behavior.

Stability The behavior of a system when perturbed from its equilibrium; a stable system will return to the equilibrium, an unstable one will not.

Transfer function A means of expressing the Laplace transform of a dynamic system that emphasizes the dependence of the output on the input.

Windup Excess values that can accumulate in the integral portion of a PID controller when the actuator is saturated.

AN ENGINEERED SYSTEM is created according to a specification of what it should do. Many engineered systems involve the modulation of power in a manner that will meet those specifications. *Control* in the broadest sense refers to the means by which the power is modulated or manipulated. The art of engineering is to design systems that meet their specifications in the most cost-effective way possible. What is considered possible in creating a specification is very much a product of whatever technological and methodological base is generally known at the time it is created. Control has traditionally been exerted by humans who adjust the application of power according to their observation of the evolving result. In systems where the specification is relatively loose or the system very slow, the human can exert control on an intermittent basis, directing attention elsewhere in the interim. Examples of this approach abound: a woodworker with a traditional lathe, a cook, heating a home with a wood or coal stove, driving a vehicle (automobile or horse-drawn), innumerable sports activities, etc. Some of these activities involve continuous application of control (more-or-less), such as driving, others are distinctly intermittent such as heating a home with a wood stove.

I. FEEDBACK CONTROL

Prior to the 20th century, neither methodology nor technology existed for reproducing the human-mediated control process with machines, although both imagination (Jules Verne, for example) and a small number of notable examples (the Watt steam engine speed governor) whetted the appetite. It is not a coincidence that virtually every book on automatic control cites the Watt governor as a historic example; there was not much else available. The 20th century saw the development of feedback control methodology and technology to the point that it has become a ubiquitous component in engineered systems. Feedback control is the explicit use of the observe, decide, act cycle without human presence. It is also called automatic control to recognize the replacement of the human controller with a machine of some sort. It was primarily applied to large-scale industrial applications in the early part of the century, power generation using speed control derived from the Watt governor, and process control using newly invented pneumatic technology. Design methodology was weak in that period with most applications dependent on experience and experimentation to achieve satisfactory performance. Electronics and war fueled the mid-century developments of feedback control, largely for military applications. That electronics proved to be a far superior technology for feedback control implementation motivated major developments in design methodology as well.

As computer technology developed sufficiently for its application to feedback control in the last third of the century, however, both methodology and technology for feedback control exploded. Computers as the decision-making components were so far superior to previous technologies that the entire engineering domain of feedback control was liberated from restrictions on the nature of computations that could be carried out. While economics still limit what can be accomplished in any given system design, the uncanny accuracy of Moore's law—computing power will double every 18 months—motivates continuing widespread activity in design methodologies. It is the purpose of this article to highlight some of the critical factors needed for successful application of computers to feedback control. Beyond this article, there is an enormous instructional and research literature that can be used to provide insights into solution of a wide a variety of difficult problems.

A. Compensating for Ignorance

With perfect knowledge there would be no need for feedback control. It would be possible to know at all times exactly what input would cause the system to have the desired output. However, perfect knowledge is very expensive (infinitely expensive). The beauty of feedback control is that, at modest cost, it can coax maximal performance from imperfect hardware. In other words, compensating for ignorance (or imperfection) through the observe, decide, act feedback cycle is generally much less expensive than attempting a better approximation at perfection in order to reach the same level of performance.

Imagine, for example, controlling the speed of a rotating disk driven by an electric motor. This situation is typical of large numbers of applications requiring speed control. In the absence of feedback successful control of this system would require a combination of knowledge of exactly how the device operates, knowledge of the load being placed on the device at all times, and knowledge of disturbances from the environment that might cause the speed to vary. "Exact" in this context is defined as sufficient to allow operation within the specified tolerance for variation of the speed from its specifications. This would, for example, mandate knowledge of bearing operation and associated friction with changes in speed (the motor could be required to operate at various speeds or with a continuously varying speed) and changes in friction as a function of temperature and age of the machine. There would also have to be a means for knowing the precise characteristics of the load, which could change with device position (such as a robot) or as materials are picked up and dropped off. With the development of feedback control, control systems such as this that depend on prior knowledge and calibration have been called "open-loop" control systems, with the term "closed-loop" defined as synonymous with feedback control. Where the tolerance for speed control is very loose (as in a room-cooling fan, for example) open-loop control might work satisfactorily and economically. As the tolerances are tightened (in a precision grinding application, for example), achieving the desired control accuracy would get very expensive without feedback.

Why is feedback control for speed control a more economical solution? Primarily because achieving high accuracy with an open-loop system puts very stringent demands on the manufacture of the components. They have to be made so that their performance is impervious to most environmental changes such as temperature and humidity, as well as age of the machine. It is usually almost impossible to predict such changes as must be done if they are larger than the control tolerance so that design-manufacturing techniques to minimize them are necessary—and expensive. In a feedback control system these types of stringent requirements apply only to the instrument or *sensor* used to measure the variable to be controlled, in this example a speed sensor. Other components, the motor, the mechanics delivering power to the target system, etc., can be manufactured to more relaxed tolerances.

Building the sensor to tighter tolerances is much easier than building the power-delivery part of the system to the same tight tolerances. The primary reason is that measuring instruments do not have to deliver any significant amount of power—just enough to provide a readable signal to an amplifier. They can thus be much smaller and because they carry no load, are not subject to one of the main causes of speed change, load variation.

Thus, the use of feedback control is economically motivated: the main power delivery portion of the system can be built with short-term reproducibility that meets the performance specification; only the smaller, less expensive sensor need have long-term accuracy meeting those specifications. Moreover, in many situations it is impossible to know everything necessary to control a system without feedback.

The conclusion is that *if not for cost considerations* it would always be better to use open-loop control than feedback to meet a given performance specification. Open-loop control can never cause an otherwise stable system to become unstable. However, engineering cannot be separated from economics. Feedback is so much more cost effective that its use is essential in many practical situations.

B. Why Feedback is Hard: Dynamics

Feedback control has its limitations. There is a fundamental limitation imposed by the instrument: the overall performance of the control system cannot be any better than that of the sensor. There is another fundamental limit imposed by physical realizability: because no physical system can change state instantly, there is a limit to how fast a control system can respond to changes in either its command or to changes in disturbances. The "dynamic" behavior of a physical system describes how it changes in response to a change in its environment, purposeful change as when motor input power is changed, or unexpected change as when a "downdraft" causes sudden loss of altitude in a passenger airplane, much to the discomfort of the passengers.

The behavior of a dynamic system depends on the history of what has been done to it. It is not enough to know what its current inputs are—even if the airplane has moved past the downdraft, it will take some time to bring it back to its previous altitude. This is the crux of the feedback control design problem: react too energetically to a deviation from the desired output and the future consequences may be more than was bargained for; react too timidly and it will take altogether too long for the system to get to where it should be (maybe never). The essence of the mathematical side of control engineering is to devise methods for doing the feedback process just right. The practical side of control engineering is in understanding when and how to apply the mathematics and in designing the environment in which these mathematical methods are embedded, software and hardware, so as to build the most effective possible control systems.

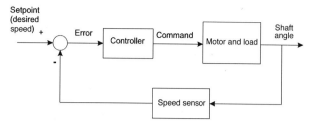

FIGURE 1 Block diagram of a speed control system.

C. Feedback Control System Structure

The classical representation of a control system is the "block" diagram. Figure 1 shows a block diagram for a speed control system. As is common in these diagrams, only the signals actually used by the control system are shown and the internal structure of the power delivery system is not shown. Environmental disturbances are sometimes shown on block diagrams, but showing them accurately is complicated because their power relationship to the system is often different than that of the control input. The block labeled "power delivery system" usually includes the power amplifier ("drive" for a motor), the actuator, and the object to which power is being delivered.

Each of the lines is a directed signal, with the direction shown by the arrow, and represents a single, scalar quantity. In the physical system, these lines represent wires between amplified points, with the arrow going from the output of one amplifier to the input of another. Double or extra wide lines are sometimes used to show vector quantities, with each of the vector quantities usually representing a single amplified signal. In cases where one or more of the signals are delivered via a computer network, there is no direct connection to any single wire in the physical system since the network carries different signals at different times (see the next section for more detail on amplification).

D. Amplification and Isolation

Amplification, and the isolation afforded by amplification, is one of the major enabling technologies for feedback control (the other big one is digital computation). Generally speaking, amplification is the control (or modulation) of a large power flow with a smaller one. The amplifier input (often also called the command) is the low power signal. It normally modulates the flow of power from a source to target device.

There are two unique domains of amplification: power amplification and signal amplification. Of these, power amplification has been known for much longer and was essential for any kind of control system, open-loop or simple feedback control systems. The Watt governor, for example, used a steam valve to modulate the flow of steam to an engine and thus control its speed. Prior to Watt, the same steam valve was used to manually control the engine speed. A steam valve is a form of power amplifier. The power source is the boiler, which contains high-pressure steam. Turning the valve stem operates some form of variable blockage that allows more or less steam to flow to the engine. The level of power needed to turn the valve stem is much lower than the power available in the flowing steam. Watt cleverly connected an instrument measuring the engine's speed to the speed valve so that as the speed increased the flow of steam was reduced, bringing the engine back to its desired speed. The instrument was a flyball, a vertical shaft that rotated with the main engine shaft. On it were a pair of weights hung from the top of the vertical shaft and connected to it so that they rotated with the shaft. As the shaft speeded up, the balls would fly outward. This outward motion was connected via a linkage to the steam valve so as to cause correct motion of the steam valve to regulate the shaft speed.

The Watt governor utilized the power amplifier, the steam valve, very well. But it suffered from lack of a signal amplifier. The power needed to operate the steam valve was substantial, even though much less than the power in the steam flow being modulated. That power was supplied from the main engine shaft, through the flyball measuring instrument. The flyball and its associated linkages thus played a triple role: measurement, computation, and power delivery to the steam valve. As a result, none of these functions could be optimized. All must be compromised by the need to meet all three requirements simultaneously.

II. COMPUTERS FOR CONTROL

The vacuum tube, and the applications of electronics to signal and power amplification, opened the door for feedback control in which each of the system elements could be isolated and optimized. On the decision-making side, electronics could be used to build a wide variety of linear, filter-like functions. These could be combined to yield a huge variety of functions. As compared to the Watt governor, for example, the computational flexibility was vastly improved. The problem of "droop," the inability of the control system to drive the error to within its tolerance, even if neither the setpoint nor the disturbances are changing, is difficult or impossible to solve with the Watt governor. Using an isolated, electronic computing element, however, makes that problem relatively easy to solve (see "integral mode" control, below).

Flexible as it is, electronics as a computing medium is severely limited. It is very effective at implementing linear operations (those representable by linear, ordinary differential equations) but it does not do well at nonlinear operations. Some nonlinear operations can be implemented by creative use of diodes—limits, function generators, peak holders, etc.—but these are difficult to work with and limited in scope.

Digital computers, on the other hand, have nearly unlimited computing complexity. If a way could be found to use them as the computing elements in feedback control systems that functionality could be optimized still further.

A. Computers and the Physical World

Using a computer as the computational element for control systems is a real challenge. First, the internal representation of numerical information is entirely abstract. That is, an internal coding is used that is arbitrary with respect to any physical realization of that number. Analog electronics used for control computation provided a relatively easy path in this regard because physical quantities are represented as voltages (or currents) in the computing circuit giving an easy analog to the external physical quantities they represent and easy conversion for input to and output from the computing circuit. There is no such easy path when a computer is used as the computing element in a feedback control system. Complex conversions must be done to connect the physical world to the computer and vice versa.

To make matters worse, computers are also sampling devices. Computer architecture is based on the idea of a powerful processing unit that is capable of executing instructions it gets from a memory. Each instruction is executed in turn, with the central processing unit (CPU) devoting its full attention to that instruction. The result is a one-thing-at-a-time device so that control of a system happens by a series of discrete operations: observation, computation, and actuation, with each of these operations requiring the execution of thousands of instructions. Thus control action is based on a momentary sample—the control object is ignored until the next time a measurement is made. This behavior is in contrast to standard electronic control systems in which all activities are carried out simultaneously. Modern computers contain some degree of parallelism, but not enough to change the sampling mode of behavior.

B. Signal Conversion

Measuring instruments designed for use in systems for which the human operator is providing the feedback control present their output information in human readable form, most commonly visual. Instruments designed for use in automatic feedback control systems (or in data logging systems) must present their outputs in forms that the control system can read. Because computers use electricity as their primary medium of operation, some form of electrical ouput is needed for measurements. The most common form of electrical output is an uncoded voltage, in which the instantaneous value of the voltage has a one-to-one functional relationship with the quantity being measured. Ideally, that relationship is linear, but linearity is not always achieved.

To be used in the computer, the voltage must be converted into a set of digital (binary) signals representing a binary integer (see Section II.V). This set of signals can then easily be copied into the computer's memory where it is usable as a "variable" by the control program. Generally, the value of the voltage scales linearly to the value of the integer. The device that does this is an "analog-to-digital converter" (A/D or A-to-D or ADC). While the details of A/D converter operation are not critical to this discussion, the general characteristics of the device are that it is fairly complex, consisting of digital logic and precision analog components, requires modest numbers of microseconds to complete a single conversion (which is slow on the time scale of a modern desktop computer), and typically has a conversion precision ranging from 8 to 16 bits (yielding resolutions from 1:256 to 1:65,536).

C. Discrete Control

Both numerical values and time are discrete in computer-based control. Numbers have binary digital signals as their basis and are thus normally finite precision. Time is discrete because the central processing unit can only handle one operation at a time. Both of these have severe consequences for feedback control. Finite precision arithmetic causes round-off error, which appears to the control system as noise. This is most severe in low-cost systems where economics forces the use of integer or fixed-point arithmetic.

Time discretization, however, is responsible for the most mischief. Because of the sampling nature of computer control, the control object is open loop most of the time. Disturbances that occur just after the computer has taken a sample are not detected until the next sampling instant. Control quality is improved as the sampling rate is increased, but faster sampling takes more computer power and is thus more expensive.

D. Computational Scale, Micro to Maxi

Personal computers have become a ubiquitous part of professional life—there is one on every desk to say nothing of the large numbers in homes. These computers are probably most people's major conscious image of a computer. Because applications tend to take advantage of the most computing power available, and because the cost of personal computers has dropped so much, the range of computing power is fairly narrow—there are not too many personal computers over five years old that are still in regular service.

The situation is very different with computers used for control of physical systems. These computers are normally embedded within a product and the economics are driven by the overall economics of the product. Thus, consumer products might impose limits of only $1 to $2 for the computer, while a machine used in a manufacturing process could allow for several thousand dollars or more for the control computer. Although all of these are computers and share the same general architecture, they impose vastly different constraints in terms of the types and amounts of computation that can be done within the time constraints of satisfactory control.

E. Number Representations, Precision and Scaling

How control computers do arithmetic has a significant effect on the control performance and on the productivity of the programmer writing the control software. In common applications such as spread sheets, simulation programs, stress analysis, etc., the users are isolated from the internal numerical representations because the programmers have made sure proper selections have been made. Programmers using common programming languages, C, C++, Java, Basic, etc., are given a wide choice of data types, but modern computers are capable of handling all of them with minimal penalty. Given the scale of computers used for control, however, some choices may be precluded because the performance penalties in these cases are significant.

The major considerations in selecting internal computer data representations to use in a control calculation are round-off error, dynamic range, and computing efficiency. The easiest number representation to use for control calculations is scientific notation, variously called "real" or "floating-point" in different computing environments. Computational scientific notation represents numbers by a mantissa and an exponent. The mantissa is a fixed-precision, fixed-point number. It is multiplied by a base to the "exponent" power to get the value of the floating-point number. The mantissa is always normalized, meaning that its highest order digit is always nonzero. In common notation, a scientific notation number is written as $1.2345*10^4$. The part before the asterisk is the mantissa; the exponent is the part after the ^. The major differences between the common paper notation and computer notations are (1) very few people pay much attention to the number of digits in the mantissa whereas that is critical in

computer representations and (2) in the computer the numbers are in base-2 (binary) format so the exponent raises 2 to a power. The major advantage to floating-point numbers is that because the mantissa is always normalized, the calculational precision is independent of the magnitude of the number (except for extreme numbers). The disadvantage is that processing of floating-point numbers is very complex and thus either expensive or slow (the usual trade-off with computers). Standard computer floating-point data types give from 5 to 15 decimal digit precision in the mantissa.

Low-cost applications cannot afford floating-point notation (given Moore's law, however, floating-point notation can be applied to more applications every year). They are restricted to some form of integer arithmetic. The simplest form is where the numbers represent whole integers, usually signed. In this case, the precision of the calculation depends strongly on the magnitudes of the numbers involved. "Precision" is defined as the ratio of the number to the difference between the number and its nearest neighbor. Take the number 1000, for example. Its nearest neighbors are 1001 and 999. In either case, the associated precision is 1000:1. On the other hand, take the number 2. Its precision is 2:1. The lower the precision, the larger the roundoff error in calculations. Roundoff errors appear as noise in computer control systems, so they can severely affect the quality of control. Maximizing the overall precision of calculations requires that numbers be scaled so they are large enough to have reasonable precision but small enough to avoid overflow problems (which have catastrophic results). This adds programming complexity to the job of creating control software.

Fixed-point numbers are somewhat more complicated than integers in that they allow for much more effective handling of quantities between zero and one. However, they suffer from the same precision problems as integers.

In summary, use floating-point if at all possible. Extra hardware cost may pay off in quicker time-to-market and a more reliable product.

F. Control System Software: Real Time

For successful control of a physical system, the events on the timeline described above (measure, compute, actuate) must happen at the correct times (within a specified tolerance). "Normal" software (word processors, numerical computation, spreadsheets, etc.) operate in ASAP mode. That is, the user would like the result as quickly as possible. Other than impatience if the operation takes too long, specific times are not a factor in standard software.

Software which must deliver results at specified times, or at specified time intervals after some external event, is designated "real-time" software. The added time dimension, plus the fact that a number of control activities must overlap, makes control system software substantially more of a challenge to deal with than standard software of approximately equal complexity.

Real-time software to implement feedback control has three major components: functions that service the instruments and actuators, a "main" section that implements the feedback algorithms, and the operator interface. Typically, the priorities of these components follow the same order—the instrument-actuator service is highest priority, the feedback next, and the operator interface is lowest. "Priority" for real time software refers primarily to the importance associated with a software module executing within a specified time of the event that triggers the module. That time interval is also called the "latency" specification for a software module. For control software, the general rule is that all software components must be activated within their latency specification, so that priority is based almost entirely on the latency specification. Although instruments and actuators fall into the highest priority class, there are very wide variations in latency requirements among different types. In most cases, the feedback control loops have similar latency requirements, so they can all be handled as a group. See Sections III.C and III.D for details on implementation of the feedback algorithm section. Finally, the operator interface must respond on human time scales, so it usually has the loosest latency specifications. It is also common to implement the operator interface on a separate computer to simplify the real-time structure. Operator interface concerns are very important to control system success. However, the subject is too vast to be treated here beyond this passing reference.

The mechanism used to realize a software system having components with substantially different latency requirements is the "interrupt" facility that is part of all computer hardware. Interrupts are triggered by events external to the computer and cause it to switch its attention from one software section (or "thread") to another. This switch takes place in a very short amount of time (normally less than a few microseconds). In this way, software modules with very short latency specifications can preempt activity by other software modules and respond very quickly to external events. The most common external event is a signal from a clock; others include switch closures, change in state of an electrical signal, etc. The computer's operating system also uses the interrupt system to service standard peripherals such as the disk drive, keyboard, and display.

The general principle in designing software for feedback control is to minimize the use of interrupts. Programs containing interrupts can be difficult to debug, harder to move from one computer to another, and require more maintenance. Assuming a computer does nothing but feedback control (perhaps with a number of active feedback loops), all of the feedback loops could operate in noninter-

rupt mode as could any of the sensors and actuators that do not need interrupt servicing. Examples of sensors that do not need interrupts are analog instruments where the signal comes into the computer via an A/D converter, instruments with their own digital interface that have substantial internal buffering, and on-off instruments. Likewise, on the actuator side, those that run from an analog command to a power amplifier, for example, do not require any interrupt servicing. On the other hand, a low-speed pulse width modulation (PWM) signal, which would be suitable for a large heater, would require interrupt servicing to maintain the accuracy of the PWM command.

Minimizing the use of interrupts requires only one major software design principle: keep all code nonblocking. This means that all decisions based on external events are computed immediately and completely when the events occur. Examples of this style of programming are given in Sections III.C and III.D.

G. Programmable Logic Controllers (PLC)

Programmable logic controllers are a class of computers designed specifically for industrial use and, traditionally, programmed using ladder logic, a schematic representation of relay logic. PLCs were originally designed as a replacement technology for relay logic, a means of implementing Boolean and sequential logic using sets of contacts and solendoids ("relays"). As such, the original device was suitable for solving problems suitable for digital logic representation, with sensors and actuators using logic (on/off) for input and output. As the use of PLCs expanded, facilities were added so that more general computation (using regular numerics rather than just logic values) could be mixed with the ladder programs. Eventually, a set of languages was established as a standard for these devices, ISO/IEC 1131 [International Organization for Standardization (www.iso.ch), International Electrotechnical Commission (www.iec.ch)]. The language set includes ladder programming and an algorithmic language, as well as other programming means.

Implementing feedback control using sequential logic is beyond the scope of this article, so the reader is referred to other texts on PLC programming. Using the algorithmic programming facility of IEC 1131 is the same as implementing feedback control with any other programming language, so all of the material here is relevant.

III. SIMULATION

If the physical system to be controlled or a reasonable physical model or prototype of it is not available (for example, not yet built) or if experimentation with it would be either too expensive, too time-consuming, or too dangerous, some form of control algorithm design and tuning based on mathematical formulations can be done. There are two main ways to start this process: simulation or linear systems. The issue here is not excluding the use of one technique or another, but of having a general notion of how to start. By the time the process is over, if the problem is a difficult one, every relevant technique in the book will be thrown at it.

In the past (i.e., when computing was very expensive) there was no question at all: the linear systems approach could produce results for many practical problems whereas the lack of generality of a simulation combined with the expense of getting results (which required a lot of computation) rendered it of little value except in extreme cases. Computational cost, however, is no longer a serious limitation and the cost per unit of computation continues to decline. With low-cost computation available, the advantages of simulation can be considered: much more generality in the nature of mathematical models that can be used and few limits on the types of performance measures that can be evaluated.

Most control classes are taught from the assumption that the linear systems approach is the better way to start a control system design problem. However, the computing power now available, along with effective mathematical software tools, makes the simulation approach extremely attractive. Thus, simulation is introduced first here as the preferred means for an initial approach to a control system design or analysis problem.

A. Models of the Control Object

What is a simulation? It is a numerical result obtained by solving a set of equations that purport to describe the desired aspects of the system-of-interest's behavior. The numerical result can be viewed in some appropriate form by a human analyst, who would then make some decision and, perhaps, do another simulation. It could also be used in further numerical procedures that analyze the result and make decisions that could also result in the execution of another simulation. As computing power has advanced to the point of being able to run multiple simulations in an automated process of analysis or optimization, simulation has risen in its importance as a primary design and analysis tool.

The "purports to" modifier in the above paragraph is a very important part of the mathematically based process—simulation, linear theory, or any other approach. While this article mainly addresses issues in producing and using the simulation results, a separate and equally important part of the job is the verification that the model (i.e., the set of equations) actually does describe the system-of-interest sufficiently well.

Most control objects to which computer control will be applied can be described with differential equations. Ordi-

nary differential equations (ODEs) suffice in many cases; the discussion here will be limited to those cases. There are, however, important control problems where the control object is more properly described with partial differential equations. The ODE representation takes the general form:

$$\frac{dx_1}{dt} = f_1(x_1, x_2, \ldots, x_n, t)$$
$$\frac{dx_2}{dt} = f_2(x_1, x_2, \ldots, x_n, t) \qquad (1)$$
$$\cdots$$

where the $x_1, x_2, \ldots$ are the state variables of the system.

B. Numerical Analysis

Producing a simulation result requires a numerical solution to the set of ODEs representing the system. Control problems are almost universally initial condition problems, meaning that the values of the state variables are known at time $t = 0$. Since the system equations, Eq. 1, are generally nonlinear, only aproximate solutions can be produced. Using the fact that initial conditions are always known, the method of solution is to define a small, but finite, time increment, Δt, and then to approximate the values of the derivatives of all of the state variables during that time interval. There are many ways to do that, each having different properties in terms of the accuracy, stability, and efficiency of the solution. The simplest approximation is that of Euler, in which the derivatives are computed at the beginning of the time interval, and assumed to be constant for the entire interval.

$$x_1(t + \Delta t) = f_1(x_1(t), x_2(t), \ldots, t)\Delta t + x_1(t)$$
$$x_2(t + \Delta t) = f_2(x_1(t), x_2(t), \ldots, t)\Delta t + x_2(t) \qquad (2)$$
$$\cdots$$

This format is intuitive and easy to understand as it basically expands the finite difference definition of the derivative. Since everything is known at zero-time, the right-hand side is completely determined and the solution is easy to start. However, except in cases where neither computing time nor stability of the numerical solution are very important, the Euler solution method is rarely used. Of the large number of possible methods, the most popular are the Runge-Kutta solutions. These are based on better approximations to the right-hand side obtained by expanding a Taylor series around the current point. Runge-Kutta solutions of various order are available; the fourth-order version is probably the most popular giving a good compromise between complexity of coding and efficiency of the solution. Most popular implementations also use an adaptive step size. In addition to the approximation to the one step ahead solution an estimate of the error is also produced. The step size is then adjusted until the error estimate falls below specified bounds. This is particularly important for nonlinear problems where the appropriate step size changes as the solution proceeds.

The Euler and Runge-Kutta methods and others like them are "explicit" ODE solvers. They are simple to implement but have the property that the maximum allowable step size is determined by the fastest behavioral mode of the system being simulated *regardless of whether the details of the fast modes are important or not*. For systems containing modes with widely separated characteristic times, this can lead to very inefficient solutions. "Implicit" methods handle these "stiff" equation sets much more efficiently if the fast details are not important. However, they are much more complex to code and take much more computing time per step.

C. Simulation of Digital Control Systems

Digital control systems are usually have a mixture of continuous-time and discrete-time behavior. The control object exists continuously in time (thus the differential equation representation) while the controller is active only for brief periods of time. The simulation software can recognize this situation by embedding a mini-event-manager into the main simulation loop.

The first part of the simulation program sets parameter values, initializes, states variables, etc. The working part of the program is the simulation loop. The main simulation loop has two parts: the event manager which controls execution of the discrete functions (those that the control computer would be doing) and the ODE solver for the continuous part of the system (normally the control object).

The event manager generally handles execution of one or more control (feedback) loops, data logging, external events that might affect operation, and when to terminate the simulation. On each pass through the event manager, code associated with all relevant events is executed and each of those events sets the time at which it will again need attention. The ODE simulation section then solves the system equations for the control object up to the time of the next event. Any ODE algorithm can be used; the sample software uses both fixed step size Euler solvers coded inline with the event loop and MATLAB solvers that use a separate file for the differential equation right-hand sides.

D. Control Software Implementation

As noted in Section II.F, the structure of the feedback control portion of the real-time program was deferred until after the simulation structure was discussed. The biggest difference between the simulation software and the actual control software is that no simulation section is needed for the actual implementation—"nature" solves the ODE. The other significant difference is that time being real must be determined by an external clock rather than through an internal calculation. The program determines the "current" time by a call to a system utility that reads the computer's clock (the clock is external to the central part of the computer, the CPU, but is normally part of the hardware set that makes up a functioning computer).

Any number of control loops and other events can be handled with this structure. The performance restriction is, as is the case with all real-time software, making sure that all control loops meet their latency restrictions. For feedback control loops, errors in when they run show up as noise in the control calculation. Large errors can affect stability and performance of the control loop. Since several control loops could conceivably be ready to run at the same time, the worst-case latency error is the sum of the execution times of all of the control loops. This can be calculated (based on computer performance figures) or measured experimentally and used to specify the computer speed needed to meet specifications for a given control configuration.

IV. BASIC FEEDBACK CONTROL

The dynamics of the control object is the largest source of difficulty in implementing control so most of the techniques devised to design controllers deal with dynamics and controller "tuning." Tuning refers to the process of setting parameter values so as to achieve satisfactory performance. In the digital control world, a "controller" is a section of software that implements an algorithm that takes the measurement of the control object's output as its input and produces as its output the actuation signal. In subsequent discussions, the term "control algorithm" will be used rather than "controller."

A. Small Signal Behavior: Tracking, Disturbances

The most fundamental behavior of a control system is exhibited by its response to small changes in either the setpoint (desired value) or in a disturbance. The behaviors that are important to observe are, for the short-term, characteristics (is the response monotonic or does it oscillate, does the oscillation persist or die away), and in the long-term change in the output for a given change in either the setpoint or disturbance (ideally, the long-term change should be equal to the setpoint change and be unaffected by disturbances).

B. Equilibrium, Stability, and State

These three terms, equilibrium, stability, and state, from system theory are important in understanding and characterizing the behavior of control systems. The "state" of a system is characterized by a set of variables (not surprisingly called state variables) such that if at any instant the values of those variables are known and the future inputs to the system are known then the complete future behavior of the system is determined. Classes of systems to which this characterization can be applied are called "state-determined systems." The state variables provide information about the independent energy storage modes in the system. The necessary number of state variables is unique, but the set of state variables is not unique, since any linearly independent set of state variables is an equally valid state variable.

A system is at "equilibrium" if for a constant input none of its state variables is changing. Thus, an equilibrium point in the state space is a point at which the rates-of-change for all of the state variables are zero (the state-space is the space for which each state variable is an axis). For example, a tank with liquid coming into it from above and liquid draining from the bottom can be characterized by a single state variable—liquid level or volume of liquid in the tank would be the most common choices. It is at an equilibrium point in its (one-dimensional) state space when the inflow is exactly balanced by the outflow.

"Stability" refers to what happens when an equilibrium is perturbed. A stable system will return to its equilibrium. In the tank example, if a bucket of water is suddenly thrown into the tank, the water level will rise above its equilibrium value. After a while, however, the level will go back to the equilibrium. Such a system is stable, in this case asymptotically stable, because it returns to the equilibrium exponentially. An inverted pendulum, on the other hand, which is carefully balanced in its upright position, will never return to the equilibrium position if it is given a little push.

An understanding of stability is of critical importance in control system design because the addition of a controller to a system which is otherwise stable can cause unstable behavior (an undesirable situation). A controller can also be used to stabilize an otherwise unstable system, such as an inverted pendulum.

C. On/Off Control

By far the easiest control algorithm is on/off: if the output value of the control object is above the setpoint turn the actuation to its minimum value (which is often off), otherwise turn it full on. This scheme also matches the common situation in which the actuation device is only capable of running in these modes. On/off actuators are usually considerably less expensive than those with a full range of operating values. Heating and cooling equipment, for example, often works this way. Figure 2 shows the response of a simulated single tank, liquid level control system using on/off control. The discrete nature of the digital control

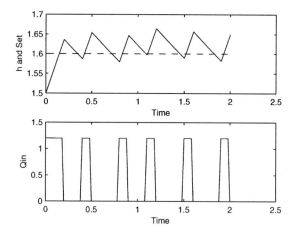

FIGURE 2 On/off control for a simple tank model.

shows up in the graph of inlet flow (actuation) which holds the same value from one sampling instant to the next.

The result is mixed. The output (liquid level) stays near the setpoint, but keeps oscillating as the actuator goes from full on to off. Thus, this kind of control is satisfactory where the tolerance on the output is fairly weak (as it is, for example, in a home heating and/or cooling system which uses this method).

D. Proportional Control

The simplest control algorithm beyond on/off, and one that is intuitively appealing, is to make the controller output proportional to the error,

$$m = k_p \varepsilon \qquad (3)$$

where, $\varepsilon = r - y$ is the error, r is the reference (setpoint) value, y is the output of control object, and m is the controller command (actuation) output. To be a bit more precise, the output of the system being controlled is not directly known; only its measurement, $\tilde{y}$, is actually available to the controller. This detail is often ignored in preliminary analysis because the sensor is commonly of much better quality than any other parts of the system so its imperfections can be ignored, at least initially.

Figure 3 shows the response for the same liquid level control system with proportional control. The behavior is much smoother than with on/off control, but, in this case, the output never quite reaches its desired value.

E. Proportional, Integral, Control (PI)

The most notable problem with using proportional control of liquid level is that the output never quite reaches the desired level (steady-state offset). This can be corrected by adding an additional mode to the control based on the integral of the error, the "I" mode. In this case the basic controller equation becomes

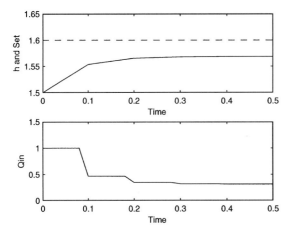

FIGURE 3 Proportional control, simple tank.

$$m = k_p \varepsilon + k_i \int \varepsilon \, dt \qquad (4)$$

This classic form of the PI control law cannot actually be realized with a computer controller because of its discrete operation. Instead, a summation as an approximation to the integration in usually used.

Figure 4 shows that the addition of integral action solves the problem of steady-state offset, with little change otherwise in the behavior.

F. Proportional, Integral, Derivative Control (PID)

When the problem in applying feedback control is that the response in overly oscillatory, or even divergent, the third control action, derivative, can be applied,

$$m = k_p \varepsilon + k_i \int \varepsilon \, dt + k_d \frac{d\varepsilon}{dt} \qquad (5)$$

The derivative term tends to be "anticipatory" so acts to modulate the controller output in advance of changes occurring in the output. As with integral action, the deriva-

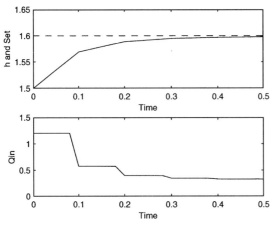

FIGURE 4 PI control, simple tank.

tive action cannot be realized exactly in control computers so difference based approximations are used. Examples of derivative action are given in the next section on controller tuning.

G. PID Tuning

Finding the best spot in the three-dimensional PID parameter space can be a daunting enterprise. A variety of techniques are available to accomplish this end. In many cases, the field tuning is the preferred environment. With the actual system built and available for experimentation and the control algorithm selected and implemented, the controller parameters are then selected based on experiments performed on the actual system. The problem with a purely experimental method is that even with only three parameters to tune, a full search of the space is rarely practical. A set of guidelines can reduce the dimensionality of the search to a manageable level. A simplified set of rules for hand-tuning a PID control that reduces the three-dimensional search to a series of one-dimensional searches is

1. With k_i and k_d set to zero, start with a very low P-gain (k_p) and increase it slowly until performance starts to deteriorate. Back off to a bit below this point. If the performance is satisfactory, the search is probably finished (additional control modes may improve the performance a bit, but at the expense of added complexity and possibly, robustness).
2. Determine whether the response obtained with the final P-gain suffers more from problems of oscillation and instability or from problems of steady-state error.
3. If stability problems dominated the search for a P-gain, proceed to tune the D-gain; if steady-state error problems predominate, proceed to tune the I-gain.
4. For D-gain increase k_d slowly and see if the performance improves. If it doesn't, keep the D-gain at zero. If it does, continue until the best performance is obtained. At this point, it may be possible to increase the P-gain and, possibly the I-gain to improve performance still further. Proceed to I-gain tuning if that has not yet been done and the steady-state error is still too large.
5. For I-gain increase k_i slowly to eliminate the steady-state error. When the best performance has been obtained, try reducing the P-gain a bit, and then retune the I-gain to try for better performance. Proceed to D-gain tuning if that has not been done yet.

This procedure assumes that the controller sampling interval has already been chosen. Although there can be some interaction with the tuning procedure, any choice of

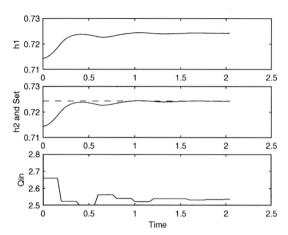

FIGURE 5 PID control of two-tanks, hand-tuned.

sampling interval based on performance will always select for faster sampling. Thus, by the time the controller tuning starts, the sample interval has probably been reduced to as low a value as is practical.

A system consisting of two tanks connected by a pipe allowing flow in either direction has more difficult dynamics than the single tank system used as an example above. Figure 5 shows a hand-tuned result, using the above rules, for small signal behavior of a two-tank system. As compared to the single-tank system, the use of derivative control is necessary to achieve reasonable stability, and the best (eyeballed) behavior still has some oscillation in it.

H. Controller Tuning by Optimization

The tuning procedure described in the previous section can be automated. The computing load to do that is quite high, but is not unreasonable for many problems. The most direct method is to define a scalar performance measure for the control system and then apply an optimization technique to find the controller parameter set that gives the best value of that performance measure.

The simplest performance criterion is the integral (or sum) of the squared error. This is a pure output-based criterion in that it does not weight the input effort at all. There are many variants of this that are in use, but the squared error will do for an example. Performance criteria that also weight the input effort can be used where energy is a consideration (as in a system that carries its own fuel) or where smoothness is important as well as how quickly the error is reduced. Although the definition of a quantitative performance index seems to add a degree of objectivity beyond the "eyeball" method so often used in hand-tuning, the procedure can actually be more arbitrary than one might imagine. Significantly different results can be obtained for different performance criteria,

all of which seem perfectly reasonable, different operating conditions (setpoint change, disturbance), even different length of simulation run.

The hand-tuned control system of Fig. 5 was optimized based on an error-squared performance criterion. The result is shown in Fig. 6. The results are very similar. In fact, the performance measure for the hand-tuned case is $J = 1.19$; the optimizer was only able to reduce it to 1.09 (the hand-tuning was done by eye without reference to any quantitative index). The optimizer used was the *fminsearch()* function of MATLAB. It uses the Nelder-Mead nonlinear, unconstrained simplex method. This method isn't necessarily the most efficient, but it tends to be quite robust. Several starting points, including the hand-tuned gains, were used to give some confidence that there were not any other peak points in the control parameter space that were missed (multimodal space). The optimized result had gains very similar to the hand-tuned gains, except for the integral gain which was nearly doubled.

Because much of the early part of the response necessarily has a large error, a common variant of the error-squared criterion is to multiply error-squared by time. This tends to weight the latter part of the response more heavily than the earlier part. A tuning optimization on this basis was also done, but did not result in a significantly different behavior.

I. Disturbance Rejection

The optimization in the section above was done for a change in setpoint. The purpose of feedback control is to insulate a system from changes in its environment. Setpoint change is only one possibility of many, although different from most in that it is changed purposely. The term "disturbance rejection" generally refers to how well the feedback control system can maintain the desired setpoint when external conditions change. A case in point

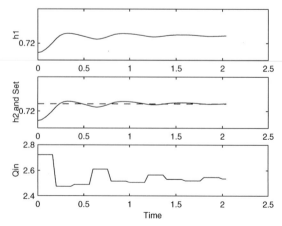

FIGURE 6 PID control of two-tanks, optimization-tuned.

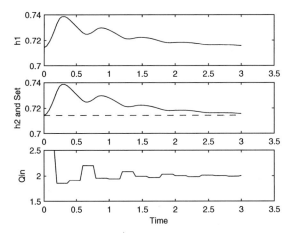

FIGURE 7 Disturbance rejection, Two-tank system, original tuning.

for the two-tank system is when the inflow is not what is expected. This could happen, for example, if a calibrated valve is used with no feedback measurement of the actual flow. In that case, a change in properties of the liquid (temperature, density, viscosity, etc.) could cause the flow to be different from the expected calibration. The difference between the actual and expected flow rates appears as an external disturbance.

Figure 7 shows the response to such a disturbance. The controller tuning used is the optimum tuning that was determined above for a change in setpoint.

Running the optimizer again for the disturbance rejection case resulted in a very different gain set. As the results in Fig. 8 show, a much higher integral gain was used to bring the tank height back toward the setpoint very quickly, but at the expense of more oscillatory behavior. This case is a good illustration of the dilemma of tuning: there is no one "best" tuning. What is best will depend entirely on what conditions of operation and performance measures are used.

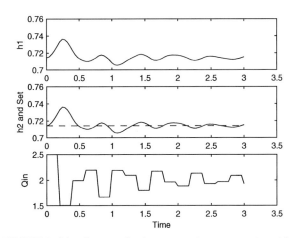

FIGURE 8 Disturbance rejection, two-tank system, retuned for disturbance.

V. CONTROL IMPLEMENTATION IN THE REAL WORLD

The material presented thus far gives the basics of control, but doesn't solve very many real problems. Probably the most restrictive aspect has been the assumption of small-signal behavior; real-world problems often require large changes in operating variables. This section examines some of these issues in an effort to bring the methodology closer to what is needed to build successful control applications.

A. Rule #1: Keep the Error Small

To some extent, the small-signal domain can be forced. As long as the controller error is small, even if the system is in the midst of a large change, many of the advantages of small-signal behavior are retained. The controller error comes about from two factors: changes in setpoint, which can be limited, and changes in disturbances, which are exogenous and cannot be controlled. Although the system operator may request a large change in setpoint, the whole change does not have to be passed to the controller at once. Imposing a gradual setpoint change, within the physical limits of the actuator-control object to respond, keeps the error very small during the change. If properly designed, the speed of the change can be almost as fast as other methods of dealing with large setpoint changes, and even as fast or faster in some cases.

B. Setpoint Profiling

There are two approaches to profiling the setpoint so as to keep the behavior within the physical limitations of the control object and thus keep the error small: (1) use program logic to specify certain functions that the setpoint will follow during changes (e.g., constant slope) or (2) pass the setpoint through a linear, low-pass filter to "soften" its behavior. The former approach (program logic) will be used here, primarily because it matches the actual physical limitations that actuators have better than the linear filter. Motors tend to have speed or current limits, pumps have flow rate limits, heaters have power limits, etc. These types of limitations generally impose limits on such geometric properties as slopes. A linear filter, on the other hand, because of its linear properties, scales its response to have the same shape regardless of the size of the change. If the filter parameters are set so that the maximum slope matches the physical limitation, the reponse will be too sluggish for smaller changes. Also, the program-logic-based methods have an identifiable "end" to the setpoint change process, which could be useful in coordination with other parts of the system. An advantage to using linear filters for softening is that the behavior of the entire

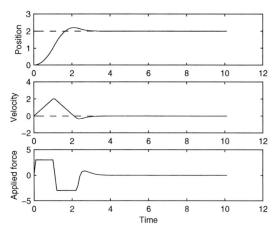

FIGURE 9 Positioning a mass: small move, no profile.

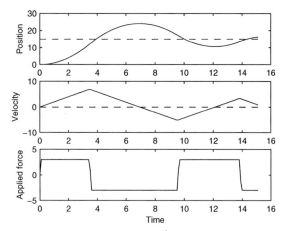

FIGURE 10 Positioning a mass: large move, no profile.

control system can be analyzed using linear system theory, which cannot be done when program logic is used for softening.

The use of setpoint profiling is illustrated for the case of motion control—moving an inertial load from one point to another. For the purposes of this problem, it is assumed that perfect measurements exist for the position and velocity of the mass (note this would apply equally to rotary positioning of a load with angular inertia). The control structure uses an interior control loop for the velocity control and an exterior loop for the position control. Each of these loops has just a P (proportional) control. This structure is called "cascade" control (see Section V.H).

Figure 9 shows the small-signal behavior of this system with nominal hand-tuning of the P gains for the velocity and position loops. These same controller gains are used for all examples in this section. The response shows a small overshoot and smooth behavior. The applied force stays within its limits.

When the distance to be moved is increased, the force reaches its limit immediately because of the large initial error. Figure 10, using exactly the same controller gains as in Fig. 9, shows an unsatisfactory response. The overshoot is large and the position has a slowly converging oscillation about the setpoint. Throughout this whole period, the force is either at its maximum or its minimum, behaving more like an on-off control than a proportioning control.

Figure 11 shows the behavior for a move of the same length but using a trapezoidal setpoint profile with the same controller gains. There is almost no overshoot and the response settles to its final position very quickly. Because of the setpoint profile, the error is always small and the applied force stays within its limits for the entire move. The trapezoidal profile has three sections: constant acceleration, constant velocity, and constant deceleration.

It is built around the two primary physical limitations of motor-driven systems: maximum acceleration due to current limitations and maximum velocity due to bearing and strength limitations.

An even smoother behavior can be obtained by using an S-shaped acceleration zone. The corners between the acceleration zones and the constant velocity zone in the trapezoidal profile cause sharp changes in the force. If the mechanism includes a structure with any flexibility, for example, these sharp changes could cause undesirable vibration. There are several ways to do this. Figure 12 shows the system behavior with a sinusoidally based profile. Note that it is substantially smoother than the response using the trapezoidal profile, although at the expense of a somewhat longer time to reach the final position. This profile is built using the same maximum acceleration, cruise velocity, and controller gains as were used for the trapezoidal profile.

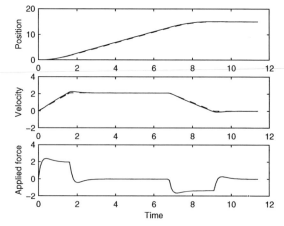

FIGURE 11 Positioning a mass: large move, trapezoidal profile.

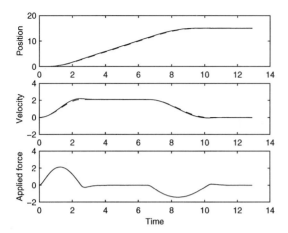

FIGURE 12 Positioning a mass: large move, sinusoidal profile.

C. Actuator Saturation

One consequence of not keeping error small is that the controller will ask for actuation output that cannot be attained, causing actuator "saturation," as was seen in the large move done without a setpoint profile in the previous section. There can be adverse effects on the actuator itself as well as issues associated with the performance of the control system if saturation occurs. The adverse effects range from damage to the actuator (for example, demagnetizing a motor) to changes in behavior during saturation such as actuator sticking where a certain amount of time or signal size is required to get it out of the saturation condition.

While an actuator is saturated, the feedback control is, in effect, deactivated. In many cases, the only problem caused by this is that the response time is lengthened and the feedback controller becomes active again when the saturation condition ends. If the system is open-loop unstable, however, the feedback control is required for stabilization. While the actuator is in saturation, this stabilizing influence cannot be exerted. In this case, the effect would be catastrophic and it would be very difficult or impossible to regain control.

D. Integrator Windup

A particular, well-known side effect of actuator saturation is integrator windup, which is a particular problem in digital control because of the large dynamic range of internal number representations. Briefly, what happens is that when the actuator goes into saturation, because of a large error, the integrator continues to accumulate. The integrator may have already become fairly large because the error is large. As long as the acutator stays saturated it cannot respond to larger controller output values. The integrator, however, seeing the large error, keeps adding

more and more to its accumulated value at each sample time. In computer control using some form of floating point number, there is no practical limit to how large that number can get.

Sooner or later, the error starts decreasing. However, the integrator term is by that time so large that the controller maintains the actuator in its saturated state. At some point, the error would reach and just cross zero. Until that time, the integrator cannot even begin to decrease. Thus, at the time the error is crossing zero, the controller is in a state at which it is still commanding maximum actuation output and will continue to do that for quite some time. The error thus changes sign, but continues in the same direction at a very high rate-of-change until the integrator actually comes down to a reasonable value. The net result of this is the controlled variable oscillates wildly. A completely unacceptable behavior. This is *not* a function of controller tuning. A controller with excellent small-signal behavior can exhibit integrator windup.

The two-tank system of Fig. 6 was tuned on the basis of small-signal behavior. Figure 13 shows the same system, but with a large change in setpoint and no setpoint profiling. It doesn't bear any resemblance to the small-signal response at all because saturation is a nonlinear effect so the shape of the response is not preserved as the size of the setpoint change. This response is a relatively mild case of integrator windup, but the windup can be seen clearly in the figure by noting that the level of the liquid in the second tank crosses the setpoint, but the controller remains saturated at its maximum value for a time after that as the integrator slowly winds down.

As demonstrated in Section V.B, setpoint profiling can be used to retain small-signal characteristics by keeping the error small. For the case of saturation, another (and simpler) method is to explicitly prevent the integrator

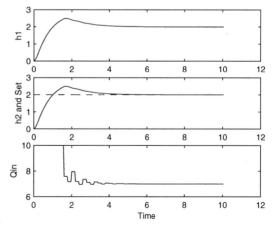

FIGURE 13 PID control of two-tanks, large setpoint change, no windup protection.

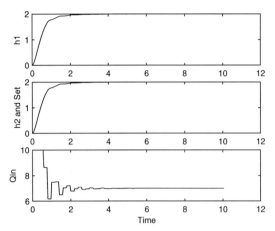

FIGURE 14 PID control of two-tanks, large setpoint change, with windup protection.

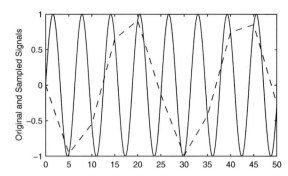

FIGURE 15 Original analog signal and aliased sampled signal.

windup but otherwise leave the control algorithm intact. The method used in this case to prevent windup is to freeze the value of the integrator as long as the controller remains in saturation. This takes only a couple of lines of code to implement. The result, even for this mild case of integrator windup, is dramatic. Figure 14, using this windup protection method, settles to the setpoint in approximately half the time taken by the system with no windup protection.

E. Noise and Aliasing

The world of digital computing is unique: there is no noise: That is why computer programs can be run over and over again, giving identical results each time. The design concept behind computing, digital, synchronous logic, assures that this state-of-affairs remains true within practical bounds (practical enough to base banking operations on it getting correct results). Control computers interact with the physical world. The interaction with the outside world is asynchronous so it breaks one of the basic rules by which noise-free operation is assured. Many of the signals used for control also originate as analog signals. Analog signals characterize the value of the quantity represented with a voltage (or current) value. Unlike a digital signal, whose information content is determined by the number of bits and is fixed, an analog signal's information content is determined by the magnitude of the noise mixed in with the signal. *All* analog signals contain noise.

In a computer, noise effects can be removed using digital filtering. This is only possible, however, if the noise is captured in the sampled signal. A problem unique to digital control arises under certain circumstances as an interaction between the noise present in an analog signal and the control computer's sampling process. For illustration, we can imagine that the "noise" is a purely sinusoidal

signal. If the sampling is at a frequency much lower than the frequency of the noise signal, the noise appears in the sampled signal at full amplitude, but at a frequency much lower than its original frequency. This is illustrated in Fig. 15, where the solid line is the original (analog) signal and the dashed line is the sampled version.

Once aliasing occurs, the standard method of removing noise, spectral filtering, can no longer be used because the noise signal is now in the same frequency range as the information. For those used to working with analog controllers and signal conditioning equipment, aliasing is counterintuitive since most analog equipment has natural low-pass filtering properties so that high-frequency noise tends to disappear in many cases and is easily filtered in others.

There is no easy solution to aliasing. The two relatively difficult solutions are to add an analog, low-pass filter between the noise source and the sampling device or to sample at a high enough rate so the noise can be removed after sampling. Both of these add considerable complexity to the control system. Another approach is to avoid the problem entirely by using digital sensors such as position encoders (incremental or absolute).

F. Computational Delay

Any delay in utilizing feedback information causes deterioration in the quality of control, up to and including instability. The classical example of this is using a television camera on the moon to show the operation of a vehicle or machine to an operator on earth. Because of the speed of light limitations on transmission speed, it takes over a second for the television signal from the moon to reach earth. The operator is thus seeing what was happening a second and some ago, not what is happening at the time the image is being viewed. A decision is made on an action (for example, with a joy stick) but it takes another second and a fraction for that signal to get back to the moon. If the system on the moon being controlled is very slow, the three seconds or so of total delay won't cause much of a

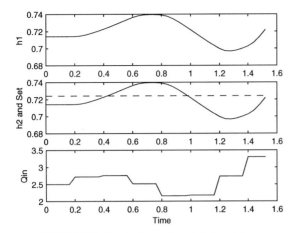

FIGURE 16 Two-tanks system: full step delay.

problem. If it is fast moving, however, the operator will be playing a losing game by always commanding actions that were appropriate for the time the signal was generated, but are no longer appropriate by the time the signal actually reaches the actuator. Most human operators adopt a wait-and-see strategy when faced with this type of situation. It works, but the overall performance is vastly slower than it would be in the absence of the delay.

The time a computer takes for computation between reading the instrument signal into the computer and generating the actuation output introduces such a delay. The worst case of such delay occurs when a small computer is dedicated to the control of a single loop and takes the full sample time to complete its computation. This introduces a full step delay into the system. For the two-tank system treated above, Fig. 16 shows that introduction of a full step delay in the small-signal (using the gains determined by the optimizer) results in an unstable and thus completely unacceptable behavior (note that in the original case the simulation was built with the implicit assumption that the computing time was a small fraction of the sample time and thus did not introduce any significant delay). In such circumstances, the controller must be retuned and will only be able to sustain substantially lower gains and thus more sluggish response.

G. Using Knowledge: Feedforward

"Knowledge is power" so don't waste it: While the emphasis thus far has been on feedback control as a means to compensate for ignorance, in many cases there is quite a lot of knowledge as to how the system operates. That knowledge can be used as a feedforward signal in addition to the command signal from the feedback controller. If it is reasonably good, the output of the controlled process or object will get fairly close to its desired value just

with the feedforward. The feedback will then have much less work to do since the degree of ignorance is much less.

The two-tank system offers two examples of potential use of feedforward. Knowing the setpoint, the input flow rate that will be needed for steady-state maintenance of that height in the second tank can be computed and added to the controller output. This does not change the observed behavior very much, but it does allow for controller gain tuning that nearly eliminates the integral gain. This makes the controller considerably more stable and robust as long as external unmeasured disturbances are not so large as to need the higher I-gain. On the other hand, if the flow disturbance described in Section IV.H could be measured, it could be compensated for with feedforward almost instantaneously, reducing the system excursion due to that disturbance to almost nothing. A measured flow disturbance in the second tank would fall in the middle since the feedforward flow correction could be applied well before a significant change in the level of the second tank would be detected.

H. Multiple Measurements: Cascade Control

Control systems with one measurement and one actuation point (SISO for single-input, single-output) are very common, but do not by any means encompass the full universe of feedback control systems. A full treatment of multiple-input, multiple-output (MIMO) systems is beyond the scope of this article, but there is a widely used control structure that deals with many systems having several measurements but only a single actuation (these might be called MISO but that acronym does not seem to be used). More measurements than actuations is common because the cost of measurement is generally much lower than the cost of actuation.

When the output of one component of a system becomes the input to the next a "cascade" is said to be present. Examples of this are the power amplifier providing power to a motor or the pump providing input flow to a tank. In these cases, it is tempting to add an instrument to the intermediate output, measuring the current output of the amplifier or the flowrate from the pump.

The motion control example used in Section V.B took advantage of two measurements, position and velocity, with a single actuation, force applied to the mass. A cascade control structure was utilized with velocity control as the inner loop and position control as the outer loop. Most motion systems also have a control loop inside the amplifier so that the amplifier command controls the current applied to the motor. The two-tank system could also use cascade control if the liquid height in the first tank were to be measured as well as the liquid height in the second tank. In that case, the liquid height control for the first tank would be the inner loop. This structure could be important if there were a need to make sure the first tank did not overflow since the single-loop structure does not put any limits on height of the liquid in the first tank.

I. Gain Scheduling

When tuning a feedback control system that has a large operating range, it is often impossible to find a set of controller parameters that will operate successfully across the entire range. Tunings that are always stable may be excessively sluggish at some parts of the operating range, and those that give lively performance may be unstable or highly oscillatory at other operating points. Probably the most dramatic example of this situation is in the control of high performance supersonic aircraft. The dynamic behavior changes so much as the plane goes from sub to supersonic that it is essentially impossible to use the same controller for both situations. Even in one domain or the other, the dynamic behavior remains a very strong function of altitude.

A solution to this problem is to schedule the gains so that they take on correct small-signal tunings at different parts of the operating range. Computer-based controllers offer the ability to store large tables of gains, and, most importantly, to carry out the transition computations that give a "bumpless" transfer from one gain set to another. Bumpless transfer assures that the control output does not make a sudden change when the gain set changes. If a pure P-control were being used, for example, as the gain changed the output would also change. The presence of an integral action, even if the I-gain is zero, allows the integrator value to be used to balance the output changes from other control modes.

VI. DESIGN BASED ON LINEAR MODELS

If the real world is entirely nonlinear (and it is) why study linear system analysis?

1. Small-signal behavior can often be adequately captured with linear models.
2. Linear analysis is incredibly powerful when it is applicable.

A. The Magic of Linearity

Systems that are linear obey additive superposition. In qualitatitive terms, that means that if the response to one particular input signal is known then so is the response

to another, the response to the sum of the two input signals is the sum of the corresponding outputs. In practice, this means that the solutions to all linear problems are expressed in series form, all using the *same* set of series: sines, cosines, and exponentials (which are all part of the same series family).

Because the sinusoidal-exponential family is so well understood, behavioral characteristics of linear systems can be determined parametrically, without having to perform full simulations. The origin of most of the linear system material predates computers so it is designed to extract the maximum amount of information about a system with minimum computation.

Linear system analysis for control is covered in a large number of texts as well as in thousands of research papers and articles. For that reason, only a few highlights will be pointed out here. The reader can refer to the extensive literature for details.

B. Linearization

Linear models can be constructed from a set of nonlinear differential equations, from simulations of those equations, or from experiments with the actual system. In all cases, a linear model is created that describes the system behavior near a specific operating point. When this process starts with data from experiments with the actual system, it is usually called system identification, but the end result is the same: a linear model.

When starting with a set of first-order (nonlinear) differential equations the object is to find the local slopes for the relationship between each of the system (state) variables in each equation and the rate of change of the corresponding state variable. This relationship can be expressed in matrix form; the matrix of local slopes is called the "Jacobian" matrix.

This procedure yields the classic "state space" form of the linear system equations for a control object,

$$\frac{d}{dt}\mathbf{x} = \mathbf{Ax} + \mathbf{Bu}; \qquad \mathbf{y} = \mathbf{Cx} + \mathbf{Du} \qquad (6)$$

where $\mathbf{x}$ is a vector of the system state variables, $\mathbf{u}$ is a vector of system inputs, $\mathbf{y}$ is a vector of system outputs, and $\mathbf{A}$, $\mathbf{B}$, $\mathbf{C}$, and $\mathbf{D}$ are matrices of (constant) coefficients. The sizes of the coefficient matrices depend on the number of states, inputs, and outputs.

While the equation is often written in this form in texts, for real systems this should be viewed as an incremental description of the behavior near an equilibrium point. The state, input, and output variables represent changes from the equilibrium values. This form is the basis for a large number of design and analysis methods based on linear algebra and calculus.

C. Transfer Functions

The most compact expression of a linear model is as an nth order differential equation. The matrix format often has many elements that are zero and a large number of relatively simple elements. The nth order form compacts all of the n^2 elements of the matrix format to n elements. The transfer function form originates as a Laplace transform, but is visually identical to the nth order differential equation format. The Laplace transform origin legitimizes manipulations using transfer functions as polynomials that make it very easy to combine linear elements and get a single transfer function for the combined system.

Transfer functions are freely convertable to state space matrix form and vice versa. The transfer function format is unique to a given physical system, but the state space format is not. A new set of state variables can be constructed from any linearly independent sum of other state variable sets. Like the state space format, a transfer function represents the incremental behavior near an equilibrium point.

"Classical" control design and analysis is based on the transfer function format. It focuses on the design of SISO control systems. Transfer functions represent SISO systems very well, even for high-order, complex dynamics.

D. Difference Equations

The state space and transfer function formats describe the continuous-time portion of the system, normally the control object. In computer-controlled systems, however, the controller is discrete-time and the control object is continuous-time. The discrete-time portion of the system is described by difference equations rather than the differential equations that describe the continuous-time portion. The discrete-time portion of the system description has two portions: a part that has been converted from the continuous-time model and a part for the controller. The first part is converted from a linear model, so will also be linear. The second part, the controller, is specified by the system designer. It is "linearized" by omitting the nonlinear parts (such as saturations).

The state space format for the discrete time equations is very similar in appearance to the continuous-time version,

$$\mathbf{x}(k + 1) = \mathbf{F}\mathbf{x}(k) + \mathbf{G}\mathbf{u}(k)$$

$$\mathbf{y}(k) = \mathbf{C}\mathbf{x} + \mathbf{D}\mathbf{u} \tag{7}$$

The coefficient matrices are sometimes given different names to avoid confusion, and sometimes given the same names to indicate similarity of function. The index k is the sample time counter.

Because the controller is normally the "open" part of the design, it is customary to convert the entire system to

the discrete-time domain. There are a number of methods for converting the continuous-time part to discrete-time. If the equations can be solved (or the transforms inverted) the conversion can be done exactly ("exactly" means that the responses will match at the sampling instants; nothing is said about the behavior between samples). There are also several approximations that are simpler to use, but only valid for specific ranges of sampling intervals.

E. Discrete-Time Transfer Functions

The Z-transform domain serves the same role for discrete-time systems that the Laplace domain does for continuous-time: it legitimizes the use of transfer functions to combine transfer functions for components and get a transfer function for overall system behavior. However, there are not as many design methods based on the Z-transform as there are for the Laplace transform-based transfer function. As with the Laplace transfer function, conversions can be made back and forth to the matrix format with the same uniqueness properties.

F. Equilibrium, Stability and Eigenvalues

Equilibrium, as defined above, becomes a very simple matter for linear systems: in most cases there is only a single equilibrium. In some cases there are an infinite number of equilibrium points, but, even with these systems, with the application of feedback control the final system normally has only one equilibrium point. It is very convenient to transform the state variables for the linear model of a system so that this equilibrium point is at the origin of the state space, so the equations are often written in this form. Because of the linear properties, this transformation has no effect on the results.

The other property of linear systems is that the stability properties at this equilibrium point are global, that is, they extend throughout the state space. The linear model, of course, only has a limited domain of applicability, so this result can be interpreted as extending the stability properties at the equilibrium point to the full range of the linear model's validity.

A major use of linear systems theory has been to parametrically define stability boundaries. Since unstable performance of a feedback control system is absolutely unacceptable, this boundary, expressed in terms of system parameters (such as controller gains), separates the region of unacceptable from the region of might-be-OK.

With modern computing tools, it is possible to learn much more than just whether a system is stable or not. Linear dynamic systems, discrete- or continuous-time, are described by a set of numbers characterizing their behavior. There are n of these "eigen" or "characteristic" values for a linear system, where n is the number of state variables. Examination of the set of values indicates whether a system will be stable or not, oscillatory or not. The numerical values give information about oscillatory frequency or speed of response. For systems of modest size (that is, modest number of state variables) computing these values is simple and straightforward in engineering computation systems such as MATLAB or in Fortran, C, etc.

G. Pole Placement Design

Since the eigenvalues of a linear system reveal so much about how a system will behave, it is natural to build a design method around placing them in desired locations. "Pole placement" (poles are another name for eigenvalues) is a means of doing just that. In some cases, the controller gains that will achieve the desired placement can be computed directly, in others optimization methods can be used to find the relevant gains.

H. Linear Quadratic Optimal Control

If the control system model with multiple measurements and a single actuator is carried to its extreme, there will be a measurement for every state variable. This represents the extreme because any additional measurements will not add any new dynamic information. Added measurements can be used for reliability or noise-reduction purposes, but from the perspective of linear controller design they do not add anything.

Feedback control algorithms can be designed to take advantage of all of these measurements. Linear quadratic optimal control (LQR for linear quadratic regulator) arises out of the much more general optimal control field. In general, an optimal control formulation will give the open loop input that is needed to optimize some specified performance of a dynamic system (it is closely related to dynamic programming). In the particular case of a quadratic performance index combining the square of the error and square of the actuation, the solution to the optimal control problem is a feedback control where the measurements used for the feedback are all of the state variables. In this formulation, each state variable is multiplied by a gain and the results are summed to get a single actuation value. The result of the LQR formulation is the set of gains, based on the relative weighting of the error and actuation in the performance index. The nice feature of LQR control as compared to pole placement is that instead of having to specify where n eigenvalues should be placed, a set of performance weightings are specified that could have more intuitive appeal. The result is a control that is guaranteed to be stable (to the extent, of course, that the model actually describes the real, physical system).

It should be noted that the particular form of the weighting function falls out of the optimal control theory. Although it does strike most people as a reasonable performance index, that is only coincidence. If any other index is desired an indirect design method such as a numerical optimizer must be used to achieve the design.

It is also possible to use LQR if the number of measurements is less than the number of state variables. This is accomplished through the use of an "estimator," a formulation that uses the linear system model to estimate all of the states based on measurement of only some of them. The derivative action in the PID control is a very primitive form of estimator that adds one estimated state variable to the measured variable that is used for SISO control.

SEE ALSO THE FOLLOWING ARTICLES

Baseband Digital Loops • Computer Architecture • Digital Speech Processing • Dynamic Programming • Human-Computer Interaction • Hydraulic Control Equipment • Numerical Analysis • Process Control Systems • Qualitative Simulation • Sensors for Control • System Theory

BIBLIOGRAPHY

Auslander, D. M., and Tham, C. H. (1990). "Real-Time Software for Control: Program Examples in C," Prentice-Hall, Englewood Cliffs, NJ.

Franklin, G. F., Powell, J. D., and Workman, M. L. (1990). "Digital Control of Dynamics Systems," Addison-Wesley, Reading, MA.

Kuo, B. C. (1995). "Automatic Control System, 7th Edition," Prentice-Hall, Englewood Cliffs, NJ.

Levine, W. S. (2000). "Control System Fundamentals," CRC Press, Washington, DC.

Ogata, K. (1995). "Discrete-Time Control Systems," Prentice-Hall, Englewood Cliffs, NJ.

Wittenmark, B., and Astrom, K. J. (1996). "Computer Controlled Systems: Theory and Design," Prentice-Hall, Englewood Cliffs, NJ.

Digital Electronic Circuits

David J. Comer
Donald T. Comer
Brigham Young University

GLOSSARY

Boolean algebra A type of algebra used to analyze logical relationships and logic circuits.

Digital code A collection on n digital signals that comprise an n-bit code.

Digital signal A voltage or current that can exist only at a finite number of levels or values. In many digital systems, the number of levels is two.

Integrated circuit or IC A small piece or chip of solid semiconductor material on or in which has been formed electrical components such as transistors, resistors, and capacitors. These components are interconnected within the chip to create a functioning electronic circuit or system.

Microprocessor An integrated circuit that includes the central processing unit of a computer (arithmetic unit plus the control unit) on a single IC chip.

Truth table A table that specifies the output values of a logic circuit as a function of all possible input combinations.

Voltage-controlled switch An electronic circuit that produces a high or a low level output, depending on the level of the input signal.

DIGITAL ELECTRONIC CIRCUITS are designed to respond to one or more input signals that have a finite number of voltage or current levels and produce one or more output signals that have a finite number of output current or voltage levels.

I. INTRODUCTION TO DIGITAL ELECTRONIC CIRCUITS

While the digital electronic circuit or digital circuit is used in other devices, the digital computer is the major consumer of these circuits. A single personal computer may contain several million transistors and associated electrical components to create the digital circuits that execute the computer's instructions. Examples of digital computer circuits are the flip-flop and logic gates such as the OR gate, NOR gate, AND gate, and NAND gate. These digital circuits are also used in noncomputer applications as are the monostable and astable multivibrators and other timing circuitry.

A. Digital Circuit Operation

The operation of digital circuits is predicated on the use of a finite number of voltage or current levels. The most common number of levels used is two, leading to the use of a binary or binary-related numbering system. A digital circuit may accept input signals with voltages that are near either the 0-V level or the 3-V level. The only two possible output levels are also approximately 0 V and 3 V. One of these voltage levels, often the high level, is associated with the binary number 1 and the other level is associated with the binary number 0. These digital circuits form the building blocks from which large digital systems such as computers can be constructed.

B. Digital Signals Compared to Analog Signals

Until the electronic digital computer was born, most electronic systems accepted analog input signals and generated analog output signals. These signals are continuous in time and can take on any value between the minimum level and the maximum level. The variation from one signal value to another is a smooth continuous change as shown in Fig. 1(a). The digital signal makes abrupt transitions

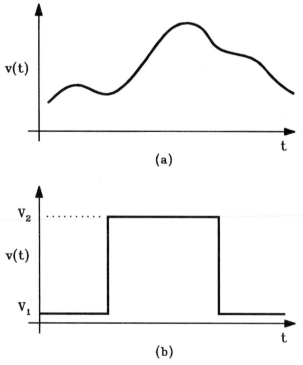

(a)

(b)

FIGURE 1   (a) Analog waveform. (b) Digital waveform.

from one level to another as shown in Fig. 1(b). Although there is a finite switching time of the digital circuit, the transition is almost instantaneous. The digital circuit is designed such that all acceptable input digital input signals will drive the output to either one of two levels. The output will never exist at any intermediate value between the two levels except during the very short transition from one level to the other.

In applications that require a signal to represent one of several numbers, the analog system can do so with much less circuitry. For example, an analog signal that is to represent numbers varying from 0 to 999 can be represented by 1000 different voltage levels. If the output signal can vary from a minimum of 0 to a maximum of 6 V, then each number will be represented by a voltage range of

$$\Delta V = \frac{6}{1000} = 0.006 \text{ V/number}. \tag{1}$$

The number 8 would correspond to a voltage that falls within the range 0.048 to 0.054 V.

While the output of a single analog circuit could represent 1000 different values, the very small voltage ranges for the different numbers would lead to accuracy problems. Most electronic circuits generate a small amount of electronic noise and some circuits pick up extraneous electronic noise from the electrical energy generated by communication circuits and other sources. Often this noise signal will exceed the voltage range of each number and cause the analog signal to be interpreted incorrectly. Furthermore, the transistors and other components used to construct circuits can cause the output voltage to depend on the temperature of the surroundings. The voltage change due to temperature may cause a shift in voltage greater than 0.006 V and the output number may again be misinterpreted.

The digital signal levels in a binary system may be 0 and 3 V. These levels are so widely separated that a small noise signal would cause no confusion in interpreting the signal level. Unfortunately, only two numbers can be represented by a single binary output. However, 10 binary outputs can be used to represent any binary number from 0 to 1023 by a digital code. Each output level can correspond to a different column of a 10-position binary number. The binary system is accurate and much more immune to noise problems, but requires more digital circuits than the analog system. Because integrated circuit technology can provide millions of circuits in a small volume, this disadvantage of digital circuits becomes less significant.

C. History of Digital Electronic Circuits

While mechanical or electromechanical computers were developed earlier, the first electronic digital computer was constructed at the University of Pennsylvania between 1943 and 1946. This system, called the ENIAC (Electronic Numerical Integrator and Computer), contained over 18,000 vacuum tubes, weighed 30 tons, and occupied a 30- by 50-ft room. The size and power consumption were very large compared to the computers of today, but the worst factor in ENIAC's inefficiency was the unreliable operation of the vacuum tubes making up the digital circuits. During 1952, for example, approximately 19,000 vacuum tubes had to be replaced to keep the computer functional. Although reliability was a problem, ENIAC demonstrated the feasibility of the electronic computer.

The point-contact transistor was invented toward the end of 1947 and the junction transistor in 1951 at Bell Labs. The present bipolar junction transistor (BJT) differs only in geometrical details from the device developed in 1951. Because vacuum tubes were firmly entrenched as the electronic device of choice, other companies developed computers using this component in the early 1950s. Toward the latter part of the decade, computers based on the more reliable BJT began to appear. These systems were still somewhat bulky and quite expensive and were generally purchased only by larger companies or institutions.

By 1964, the integrated circuit (IC) became commercially available. These small digital circuits might include 10 to 20 transistors along with resistors and capacitors in a package similar in size to that of a single transistor. The transistors, resistors, and capacitors are created within or on a single piece of silicon. These digital circuits are referred to as small-scale integrated (SSI) circuits. As IC technology improved, the medium-scale integrated (MSI) circuit, then the large-scale integrated (LSI) circuit became available. The LSI chip may contain over 10,000 electronic components on a single chip.

With the capacity to create this number of components on a single chip, the first microprocessor was developed in 1971. This chip was based on a 4-bit binary number, but implemented the circuitry for the arithmetic unit and control unit of a digital computer on a single silicon chip using BJTs.

The 8-bit microprocessor was introduced in 1972. Within a few years, very-large-scale integration (VLSI) was achieved and 16 and 32-bit microprocessors were then developed. The larger systems are now generally based on the metal–oxide–silicon field-effect transistor (MOSFET). This device was invented in the early 1960s and has several advantages over the BJT. More devices can be placed in a given area of the chip and less power is required to operate the MOSFET. Ultra-large-scale integration (ULSI) now creates tens of millions of MOSFETs on a single chip, allowing the construction of very powerful personal computers.

While computers use a large percentage of the total number of digital circuits fabricated, the digital circuit is also important in communications and other areas. Cellular phones, high-definition television, and computer modems apply digital techniques in signal transmission. Guidance control of space vehicles and missiles rely on digital circuits. Electronic instruments and automotive applications such as anti-skid braking systems employ digital circuits for implementation.

II. USING SWITCHES TO GENERATE BINARY VOLTAGE LEVELS

A major effort in the digital electronics field is spent in developing devices that emulate the behavior of a voltage-controlled switch. In these devices, an input voltage or current controls the opening and closing of an output switch. These devices are used to create switching or digital circuits.

Two important elements in the realization of switches are the BJT and the MOSFET. Some of the factors involved in selecting a device relate to how closely the device emulates the switch, how small the device can be made, how fast the switch can change states, and how many supporting components are required.

A. The Ideal Switch

In digital circuits, a voltage-controlled switch is used to generate two different output voltage levels; one when the switch is closed, the other when the switch is open. Figure 2 shows a configuration to generate these two levels using an ideal manual switch.

When the switch is closed, the output terminal is connected directly to ground forcing an output voltage of 0 V. When the switch is opened, the voltage is pulled to V_1 through the load resistor.

B. The Nonideal Switch

Figure 3 shows a nonideal switch that has some low resistance in the closed or "on" condition rather than a short circuit. In the open or "off" condition, the resistance of this switch is large, but finite rather than infinite. The closed or on-resistance of the switch is designated R_{sc} and the open or off-resistance R_{so}.

The low voltage level of the circuit occurs when the switch is closed and is

$$V_{low} = V_1 \frac{R_{sc}}{R_{sc} + R_L}. \tag{2}$$

If R_{sc} is much smaller than R_L, this voltage will be near the 0-V level. The high voltage occurs when the switch is open and is

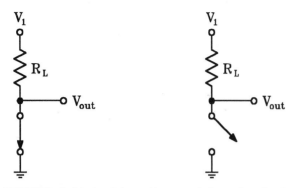

FIGURE 2 An ideal switch used to generate two voltage levels.

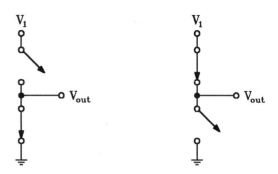

FIGURE 4 Complementary switch arrangement.

$$V_{high} = V_1 \frac{R_{so}}{R_{so} + R_L}. \qquad (3)$$

If R_{so} is much larger than R_L, this voltage will be near the voltage V_1.

The nonideal switch can be made to behave much like the ideal switch if the on-resistance is low and the off-resistance is high.

Another method of generating two different voltage levels replaces the resistor of Fig. 2 with a second switch, as shown in Fig. 4. This configuration is called a complementary arrangement of switches. One switch will always be set to the open position when the other is closed and vice-versa.

For ideal switches it is easy to see that when the top switch is open and the lower switch is closed, the output voltage will be 0 V. When the lower switch opens and the upper switch closes, the output voltage will be equal to V_1.

For nonideal switches, these same results will be approximated, depending on the values of on-resistance and off-resistance of the switches. In fact, the complementary switch will generate voltage levels nearer 0 V and V_1 than the single switching circuit of Fig. 3 when nonideal switch resistances are present.

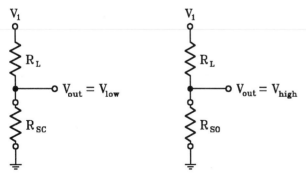

FIGURE 3 A nonideal switch.

The two voltage levels are

$$V_{low} = V_1 \frac{R_{sc}}{R_{sc} + R_{so}} \qquad (4)$$

and

$$V_{high} = V_1 \frac{R_{so}}{R_{so} + R_{sc}}. \qquad (5)$$

If the open switch resistance is a large factor, perhaps 10,000, times the closed switch resistance, the lower level will be almost zero and the higher level will be almost equal to V_1.

C. Voltage-Controlled Switches

In digital circuits, the switches are replaced by high-speed electronic devices that can switch between a high resistance state and a low resistance state under control of an input signal. The two types of semiconductor switches used today are the BJT and the MOSFET. In the 1960s and 1970s, the BJT was often used in a configuration that emulates the single switch with a resistive load. The MOSFET is often used in a configuration that emulates the complementary switching behavior of the two switches in Fig. 4. These are called complementary MOS or CMOS circuits. These switching circuits form the basis of the digital circuits that are used to construct computers and other digital systems.

At the present time, CMOS circuits are by far the most commonly used in microcomputers. This is primarily due to the huge number of MOS devices that can be fabricated on a single chip, lowering the cost of the chip. The continually improving performance achieved with CMOS circuits also contributes to the popularity of the MOS device.

1. The BJT Switch

The bipolar junction transistor or BJT is manufactured in two different arrangements. One is called the npn device,

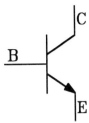

FIGURE 5 The symbol for an npn transistor.

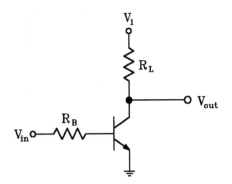

FIGURE 6 A switching circuit using an npn BJT.

the other is called the pnp device. The npn device is used most often in digital logic circuits.

The BJT consists of three different physical regions: the base, the emitter, and the collector as represented by the symbol for an npn device shown in Fig. 5. Each of the three distinct regions is electrically connected to a terminal, allowing connection of the regions to other circuit components. The emitter terminal is identified by an arrowhead that indicates the only possible direction of current flow for the npn device. The base terminal is the control for the device. If no voltage is applied from base to emitter, no current will flow from collector to emitter. If sufficient current is directed into the base terminal, leading to a voltage drop from base to emitter, a large current will flow from collector to emitter. These conditions allow the BJT to function as a controllable switch. The switch appears between the collector and emitter terminals. This switch is open if no current is applied to the base terminal and is closed when enough current is caused to flow into the base.

The BJT can operate in three different regions, the cutoff region, the active region, or the saturation region. For many switching applications, the device is either in the cutoff or saturation regions for most of the time with only short durations in the active region as a transition between the other two regions is made. Figure 6 shows an npn transistor in a common switching circuit configuration.

A high level voltage at the input saturates the transistor and the collector to emitter path approximates a closed switch. The output voltage is approximately zero. A low voltage level at the input leads to an approximate open switch between collector and emitter. The output is now at a voltage of V_1. This circuit is a voltage-controlled switch, emulating the configuration of Fig. 3, that is used in many different digital circuit applications.

Early BJT logic circuits such as the resistor–transistor–logic family (RTL) employed this basic configuration with little modification. Present day logic families such as transistor–transistor logic (TTL) and emitter-coupled logic (ECL) modify the basic configuration to speed the switching operation.

2. The MOSFET or CMOS Switch

The MOS device consists of four geometric regions called the gate, the source, the drain, and the substrate. A channel can exist between the source and drain regions. This device can be fabricated as an n-channel (nMOS) device or as a p-channel (pMOS) device. The symbols for both devices are shown in Fig. 7. The substrate is often tied to the source terminal in simple digital circuits, thus the substrate is not shown in the symbols of Fig. 7.

The resistance from drain to collector is controlled by the voltage applied between gate and source terminals. For the nMOS device, a zero or low value of V_{GS} causes a very large resistance to exist between source and drain. When V_{GS} exceeds some relatively small voltage called the threshold voltage, the resistance between source and drain becomes much smaller. For the pMOS device, a negative value of V_{GS} must be applied to lead to a low resistance between source and drain terminals. A very small negative value of V_{GS} leads to a high resistance between source and drain.

The MOS device approximates a switch better than the BJT does in the off state. It does not approximate a switch as closely as the BJT does in the on state. The resistance between source and drain in the off state may be in the range of hundreds of megohms for the MOS device. The resistance in the on state may be a few hundred to a few thousand Ohms for the MOSFET.

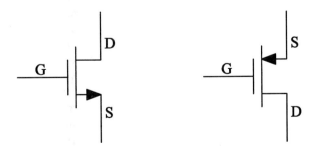

FIGURE 7 Symbols for the MOSFET: (a) nMOSFET; (b) pMOSFET.

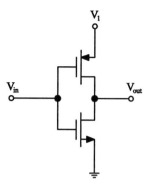

FIGURE 8 A complementary MOSFET (CMOS) switch.

The MOSFET is forced into the off state when the magnitude of the gate to source voltage is less than the threshold voltage of the device. This value is determined in the construction process and may be 0.8 V for the nMOS device and −0.8 V for the pMOS device. When the gate to source voltage for the nMOS device exceeds the threshold voltage, current flows from drain to source and the device is in the on state.

The basic switch configuration of the MOSFET switch corresponds to the complementary arrangement shown in Fig. 4. A pMOS device is used for the upper switch and an nMOS device is used for the lower switch as shown in Fig. 8. This is called a CMOS switch.

When the input voltage is at 0 V, the nMOS device is in the off state since the gate to source voltage is 0 V. The pMOS device has $-V_1$ V from gate to source and this device is in the on state. The output voltage is at V_1 V. This state is depicted in Fig. 9.

If the gate voltage changes to a high voltage, say V_1 V, the nMOS device will turn on and the pMOS device will turn off. This leads to the situation depicted in Fig. 10.

The key concept in digital BJT or CMOS circuits is that of a voltage controlled switch. The state of the switch or switches, as in the case of the CMOS circuit, can be controlled to generate either a high or low output voltage. The gate voltage controls the state of the CMOS circuit and draws no current at low and intermediate frequencies.

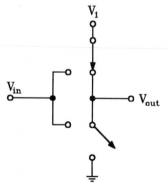

FIGURE 9 Model of the CMOS circuit for $V_{in} = 0$ V.

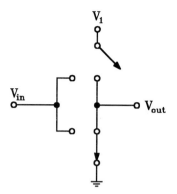

FIGURE 10 Model of the CMOS circuit for $V_{in} = V_1$ V.

III. TYPES OF DIGITAL CIRCUITS

The simple switching circuits discussed in the preceding paragraphs form the basis of many useful digital circuits. The digital computer applies digital circuits as building blocks to construct the subsystems that make up the computer.

An important group of circuits for the computer are called logic circuits. A simple logic circuit accepts several binary inputs and produces a binary output that depends on the combination of the binary input bits. Circuits such as inverters, AND gates, OR gates, NAND gates, and NOR gates are classified as combinational logic circuits. In addition, the bistable flip-flop circuit is also used in computer systems, often to act as a memory to store a binary number or bit. This circuit forms the basis of a sequential digital system. The outputs of n logic circuits can be grouped together to represent a digital code. In computers, these codes may represent numbers in some binary-related code or they may represent letters in some code such as the ASCII code.

Digital circuits are also used in noncomputer applications, for example, as controllers. These systems produce control signals that trigger certain events, based on the input signals and the time duration from some reference signal. One popular version of a controller is the digital state machine that is composed of logic gates and flip-flops.

Other digital circuits are used in a computer as well as in noncomputing digital systems. The monostable and astable multivibrators are examples of digital circuits used in both types of digital systems.

A. Digital Circuits for Computing Applications

1. Inverters

Essentially every personal computer now made is based on CMOS or closely related logic circuits. Computer chips are fabricated with tens of millions of MOSFET devices on each chip. These devices now operate at high switching speeds and with very little power consumption. For these reasons CMOS logic circuits have become the logic family of choice for personal computers and other digital systems. Consequently, BJT logic circuits such as transistor–transistor logic (TTL), emitter-coupled logic (ECL), and current-mode logic (CML) are not discussed here even though they are still used in several applications.

The CMOS switching circuit shown in Fig. 8 is often called an inverter. If the input voltage is low, the nMOS device is in cutoff and the pMOS device is in the conducting state with a low resistance between drain and source. The output voltage will be at the high voltage level. When the input voltage level moves positive to the high level, the pMOS device shuts off while the nMOS device now has a small resistance between drain and source. The output voltage is at 0 V for this input condition.

It has been mentioned earlier that the CMOS switch or inverter has a rail-to-rail swing at the output. For a 3-V supply, the high output level of the inverter is 3 V and the low level output is 0 V. This significant result guarantees that any circuit driven by the output of the inverter will have a high input level of 3 V and a low input level of 0 V. There is no degradation of voltage level through a CMOS inverter stage. This same result also applies to the CMOS NAND and NOR gates to be considered next.

2. Logic Gates

A key concept related to circuits is the use of Boolean variables to express inputs and outputs of logic gates. It is conventional to express logic operations in terms of Boolean variables and manipulate these expressions with Boolean algebra. For example, the two-input OR gate of Fig. 11 uses the variables A and B to represent inputs and X to represent the output. While the actual voltage levels of an input may take on the values of 0 and 3 V at different times, we associate the high voltage level with the binary number 1 and the low level with binary 0, assuming a positive logic convention. The inputs and output then take on values of 0 or 1.

The Boolean algebra expression used to define an OR gate is

$$X = A + B. \tag{6}$$

The "$+$" sign denotes the "OR" operation rather than the usual addition. This equation is read, "The output X is true

(or equal to 1) when either input A OR input B is true (or equal to 1).

The Boolean algebra expression used to define an AND gate is

$$X = A \cdot B \tag{7}$$

The "$\cdot$" sign denotes the "AND" operation rather than the usual multiplication. This equation is read, "The output X is true (or equal to 1) when both input A AND input B are true (or equal to 1). Often the "$\cdot$" symbol is deleted between the variables in the AND expression. There can be more than two inputs to a logic gate.

The output of a gate can be inverted (complemented) to provide a different logic function. If the output of the OR gate is inverted, it becomes a NOR gate. If the output of the AND gate is inverted, it becomes a NAND gate. The corresponding Boolean expressions for these gates are

$$X = \overline{A + B} \tag{8}$$

and

$$X = \overline{A \cdot B}. \tag{9}$$

a. The NOR gate. A truth table is often used to specify the output of a logic gate for all possible combinations of input signals. The truth table for a two-input NOR gate is

A	B	X
0	0	1
0	1	1
1	0	1
1	1	0

When either input is a logic 1, the output will be 0. The only combination of inputs that results in an output of 1 is $A = 0$ and $B = 0$.

The two-input NOR gate can be implemented in CMOS as shown in Fig. 12. The general logic symbol for the NOR gate is also indicated in this figure. The small circle at the output represents an inversion following the OR gate to create the NOR gate.

When both inputs are at 0 V or logic 0, both nMOS devices are off and both pMOS devices are on. This input condition leads to 0-V gate-to-source voltages for both M1 and M2 and -3-V gate-to-source voltages for M3 and M4. The output voltage will be very near 3 V or logic 1. Typical values of on resistance and off resistance are 10 kΩ and 200 MΩ, respectively.

When one or both inputs switch to logic 1 or 3 V, at least one nMOS device turns on while at least one pMOS device turns off. In either case, a very high resistance exists between the power supply voltage and output and a very low

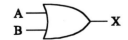

FIGURE 11 Symbol for an OR gate.

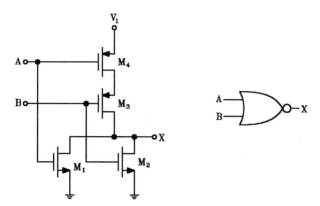

FIGURE 12 (a) Two-input CMOS NOR gate. (b) Symbol for the NOR gate.

resistance exists between output and ground. This forces the output to be at 0 V or logic 0.

The NOR gate can be extended to a higher number of inputs by adding a complementary pair of devices for each additional input. The additional pMOS device is placed in series with the other pMOS devices and the additional nMOS device is placed in parallel with the other nMOS devices.

b. The NAND gate. A two-input NAND gate satisfies the truth table

A	B	X
0	0	1
0	1	1
1	0	1
1	1	0

The logical expression for this gate is

$$X = \overline{A \cdot B} \tag{10}$$

For a positive logic system in which a high voltage level is defined as a logic 1, the NAND gate output is at the high level for all but one input combination. If both inputs are high, the output should be low. The CMOS gate of Fig. 13 satisfies this truth table. Also shown in the figure is the logic symbol for the NAND which is equivalent to an AND gate followed by an inversion.

For the combination of inputs $A = B = 1$, both M1 and M2 will be turned on and both M3 and M4 will be off. A very high impedance exists between the power supply and the output terminal while a low impedance exists between the output terminal and ground.

For any other input condition, the impedance from power supply to output will be low while the impedance from output to ground will be high.

The NOR gate and NAND gate make up a set of gates that are said to be functionally complete. This means that

any possible combinational logic function can be realized using nothing more than these two types of gates.

3. Flip-Flops

The circuits considered in the preceding paragraphs are combinational logic circuits. These systems have outputs that depend only on the applied input signals. For a given set of inputs, the outputs will always be the same. Combinational circuits are important, but so is another class of digital circuits referred to as sequential circuits.

A sequential circuit depends not only on the present inputs but also on past history of the inputs and time. The basis of sequential circuits is the flip-flop. The flip-flop is perhaps the most important single type of circuit in the digital field. The flip-flop belongs to a class of circuits called bistable multivibrators.

Flip-flops operate in one of two modes: direct or clocked. Direct-mode flip-flops respond directly to applied inputs. The outputs change as a direct result of the inputs. In clocked flip-flops, a change of input has no effect on the output until a clock signal is applied. When a clock transition from one voltage level to another occurs, the output changes to a value dictated by the input levels. The SR flip-flop and gated D flip-flop are examples of direct-mode operation devices. The toggle, clocked D, and JK flip-flops are clocked devices. Some clocked flip-flops also have direct inputs, allowing operation in either mode.

a. The direct input SR flip-flop. This form of flip-flop has two input lines and either one or two outputs. The output line that is always present is often labeled Q. Generally, a second output called $\bar{Q}$ will also be present. One input line is used to set the device to the $Q = 1$ state while the other input sets the device to the $Q = 0$ state. These inputs are called the SET or S input and the RESET or R input; hence the name SR flip-flop. The outputs Q

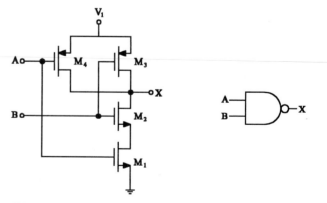

FIGURE 13 (a) Two-input CMOS NAND gate. (b) Symbol for the NAND gate.

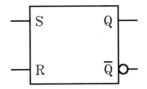

S	R	Q	$\bar{Q}$
0	0	Q_0	$\bar{Q}_0$
0	1	0	1
1	0	1	0
1	1	0	0

FIGURE 14 The SR flip-flop and characteristic table.

and $\bar{Q}$ make up what is called a double-rail output; that is, $\bar{Q}$ always equals the complement of Q except when both S and R are asserted simultaneously. This condition is normally avoided in actual circuit operation. The symbol for the SR flip-flop is shown in Fig. 14 along with the characteristic table. The notation Q_o and $\bar{Q}_o$ refers to the states of outputs Q and $\bar{Q}$ before the input conditions are applied.

From the characteristic table we see that Q may equal 0 or 1 for the input condition of $S = R = 0$, depending on the input conditions prior to setting both inputs low. This is a distinct difference from combinational circuits which always produce a unique output for a given set of inputs.

When both S and R are asserted, Q and $\bar{Q}$ are driven low. If both S and R are returned simultaneously to the low level, it is impossible to determine if Q will assume the 1 state or the 0 state. Generally the input condition of $S = R = 1$ is avoided unless the following input condition is either $S = 1$ and $R = 0$ or $S = 0$ and $R = 1$. In either of these cases, the output will be uniquely determined.

The SR flip-flop can be constructed from NOR gates as shown in Fig. 15. For this latch, applying a 1 on S (high level) and a 0 on R results in $\bar{Q}$ being asserted low. Since $\bar{Q}$ drives an input to the upper NOR gate, both inputs are low, resulting in a high value for Q. When S returns to a

low value, the high level of Q continues to drive the lower NOR gate to keep $\bar{Q}$ at a low level. If R is now asserted, the upper NOR gate drives Q low which now allows $\bar{Q}$ to return to a high level.

The SR flip-flop can function as a 1-bit memory. To store a 1 in this device, the S input is asserted. This input can then be deasserted, and the flip-flop remains in the $Q = 1$ state as long as the two inputs remain deasserted. The 0 bit can be stored by asserting R. Several SR flip-flops can be combined to form a storage register for several bits.

b. The clocked JK flip-flop. This bistable circuit has two gating inputs along with a clock input. The voltage level of the gates determine the output state to which the clock input will shift the flip-flop. Often direct set and reset inputs are provided that can override the clock input. For clocked operation the R and S inputs are set to the inactive level; but if a certain output condition is to be preset into the flip-flop before clocked operation takes place, these inputs can be used. The J and K inputs are called gates.

In the clocked mode of operation, the gating conditions are summarized in Fig. 16. The condition of both J and K low is called the "no change" condition. The toggle condition corresponds to both J and K high. The set condition occurs when $J = 1$ and $K = 0$ while the reset condition occurs when $J = 0$ and $K = 1$. It must be emphasized that the voltage levels applied to the J and K gates do not directly cause output changes. Levels on these gates determine to what level the clock transition drives the output. Changes on the J and K inputs cannot cause a change in Q if no clock pulse is applied.

There is a similarity between the function table of Fig. 16 and the SR table of Fig. 14. One obvious difference is the condition of $J = K = $ high leads to the toggling of the JK whereas the condition of $S = R = $ high should be avoided for the SR flip-flop. A more subtle difference is that the SR flip-flop operates in direct response to the S

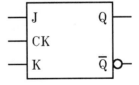

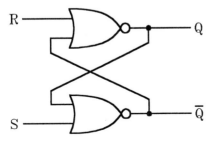

FIGURE 15 NOR gate configuration of an SR flip-flop.

S	R	Q	$\bar{Q}$	Condition
0	0	Q_0	$\bar{Q}_0$	No Change
0	1	0	1	Reset
1	0	1	0	Set
1	1	$\bar{Q}_0$	Q_0	Toggle

FIGURE 16 Gating conditions for a clocked JK flip-flop.

and *R* inputs while the levels on *J* and *K* determine the state that will exist after a clock transition occurs.

4. Frequency Limitations on Digital Circuits

A very significant factor in digital logic circuit performance is switching speed. Shorter switching times allow the execution of more operations per second by the computer. At the circuit level, rise and fall times and propagation delay times are measures of the speed of an individual circuit. The rise time is measured by applying a very abrupt input that will drive the output to the high voltage level. The time taken for the output to move from 10 to 90% of the total transition voltage is the rise time. The fall time measures the time to cross this same voltage interval when the output is driven from its high level to the low level. The propagation delay time is the time difference between the input transition and the output transition. It is measured between the 50% points of input and output transitions.

Although not the best measure of execution speed, the speed of the master clock of a computer gives an indication of the switching times required by the internal logic circuits. For a logic operation to be performed in one clock period, the clock cycle duration must exceed the propagation delay time and the rise time of the circuit performing the operation. By the year 2000, personal computer clocks reached frequencies of 1 GHz, improving from the low MHz range in two decades.

The switching speed of a circuit is limited by the parasitic capacitances associated with the transistors and connecting leads. The control of current flow through a BJT or a MOSFET is accomplished by means of a control voltage. As this voltage changes, a change in electronic charge takes place at various points within the device. Since capacitance can be defined as

$$C = \frac{dQ}{dV}, \tag{11}$$

these charge effects can be represented or modeled by capacitors. These capacitors are called parasitic because they are inadvertent results of the fabrication of the devices. In addition, the interconnections between devices on a chip and between chips introduce capacitance that also limits switching speeds.

In general, as MOSFET device size becomes smaller, the parasitic capacitance becomes smaller and higher switching speeds are possible. For years, as photolithographic technology improved to allow smaller devices, computer clock speeds continued to increase.

Another method of minimizing parasitic capacitances applies principles of physics rather than size reductions. The bipolar heterojunction transistor uses different but structurally similar materials, such as germanium and silicon, to decrease parasitic capacitance. Operation of these devices in the tens of GHz range was accomplished by the year 2000.

5. High Frequency Digital Circuits

Various configurations and devices have been developed to decrease the switching times of digital circuits. Most personal computers now use dynamic MOSFET circuits to achieve higher operating speeds. These circuits represent the two binary digits by the presence or absence of charge on a capacitor. To minimize the time to deposit or remove charge, only the small parasitic capacitance associated with the MOS devices is used.

In BJT high speed design using silicon BJTs, both emitter-coupled logic (ECL) and current mode logic (CML) are configurations that are popular. An improved device is the heterojunction BJT that allows very high speed operation in supercomputers.

B. Digital Circuits for Noncomputing Applications

Two important circuits used in timing applications are the monostable multivibrator and the astable multivibrator. The monostable multivibrator generates an output pulse of predetermined duration in response to a short trigger pulse at the input. This duration is set by the values of certain timing components in the circuit. The astable multivibrator produces a rectangular waveform at the output and requires no input signal. The frequency of the output signal is determined by the values of certain timing components in the circuit.

The astable is often referred to as a free-running multivibrator or as a clock circuit and the monostable is called a one-shot circuit. The flip-flop discussed earlier is classified as a bistable multivibrator. The basic difference among the types of multivibrator is the number of stable states associated with each type. The bistable has two stable states, the monostable has one stable state (and one quasi-stable state), and the astable has no stable states (two quasi-stable states). The waveforms of Fig. 17 show typical input and output signals that might be expected from the astable and the one-shot multivibrators.

The output of the one-shot remains at a specific level while in the stable state until an input signal initiates the quasi-stable state. The output voltage switches to another level in this state and remains at that level until the output reverts back to the original value. The time that the circuit remains in the quasi-stable state is called the period of the one-shot and is determined by circuit components. In the bistable and monostable multivibrators, the output signal levels are independent of the amplitudes of the input

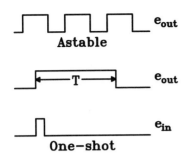

FIGURE 17 One-shot and astable waveforms.

signals, assuming these amplitudes are large enough to initiate switching.

The astable requires no input while the output continually switches between the two quasi-stable states. The length of time spent at each voltage level is determined by certain components of the circuit.

The one-shot circuit is used to create a pulse with a specific width in response to the input trigger signal. The pulse width may be as small as a few nanoseconds or as long as a thousand seconds. Such timed pulses are important in the control of machines or appliances.

The astable output waveform is used as a clock signal in low accuracy applications. For more precise applications, crystal controlled astables or oscillators are used.

SEE ALSO THE FOLLOWING ARTICLES

ANALOG ELECTRICAL FILTERS • ANALOG-SIGNAL ELECTRONIC CIRCUITS • BOOLEAN ALGEBRA • CIRCUIT THEORY • DIGITAL FILTERS • MATHEMATICAL LOGIC

BIBLIOGRAPHY

Comer, D. J. (1995). "Digital Logic and State Machine Design," (3rd edition, Saunders College Publishing, New York.

Martin, K. (2000). "Digital Integrated Circuit Design," Oxford University Press, New York.

Rabaey, J. M. (1996). "Digital Integrated Circuits," Prentice Hall, Upper Saddle River, NJ.

Uyemura, J. P. (2000). "A First Course in Digital Systems Design," Brooks/Cole, Pacific Grove, CA.

Wakerly, J. F. (2000). "Digital Design Principles and Practice," (3rd edition), Prentice Hall, Upper Saddle River, NJ.

Digital Filters

N. K. Bose
Pennsylvania State University

GLOSSARY

Filter banks Digital multirate filter banks, frequently used for compression of speech, image, and video signals, contain analysis and synthesis blocks for filtering, performing sampling rate conversion by decimation and interpolation, as well as coding, transmission, and decoding.

Multidimensional LSI filters Linear shift-invariant (LSI) digital filters for processing spatiotemporal signals in wavenumber–frequency space as well as for separating signals propagating with different velocities. Such types of filters are used, respectively, in image or video signal processing and in geophysical (including seismic) signal processing and span a variety of applications. Linear shift-varying as well as well as nonlinear filters (including the well-documented polynomial filters) can also be multidimensional.

Wave digital filter Low-sensitivity digital filter obtained from a doubly terminated analog passive prototype by replacing each type of analog network element by an appropriate type of digital realization.

Wavenumber Spatiotemporal angular frequency vector whose components are the variables in the Fourier transform of a spatiotemporal signal.

Velocity filter Filter capable of discriminating between signals collected by an array of sensors based upon their velocities of propagation.

I. INTRODUCTION

This article documents the present status of the topic of digital filters which, over the last three decades developed from a research discipline into an essential technology, spurred by the incredible advances in microminiaturization that facilitated the implementation of this digital signal processing workhorse in an expanding horizon of applications. Digital filters are now ubiquitous in the processing of spatial and temporal signals of diverse physical origin like in radio astronomy, seismology, bioengineering, speech and audio, image and video. The demands of the information age and the accompanying challenges for reliable, robust, high-speed communication of multimedia signals that include text, data, sound, image and video between remote locations assure the continuance of our dependence on digital filter technology well into the future.

The term, "digital filters," describing software or hardware designed to extract information from degraded data, appears in the title of several texts (Antoniou, 1993; Bose, 1993a; Hamming, 1989) and the subject it covers is treated extensively in works on digital signal processing. (Mitra and Kaiser, 1993; Bose and Rao, 1993). The heavy interest in this area relates to its importance in the attempt to attain robust, reliable, high-speed transmission and reception of information-bearing multimedia signals corrupted by noise or otherwise degraded during passage between multiusers and facilities. The increased sharing of spectrum and the need for multiplexing translate into a higher likelihood of user interference, and to accommodate a large number of users in the available spectrum, techniques for interference rejection become necessary. The design of an optimum filter for minimizing a chosen error criterion requires a priori information about the statistics of the data to be processed and an adaptive filter is needed when only partial knowledge of the signal characteristics is available. Interference-rejection techniques often need to be adapative because of the changing nature of interference and channel. The signals, weak or strong, could be spatial as well as temporal and may be of diverse physical origin.

The domain of application of digital filtering has expanded at a very rapid rate and the present and future needs of signal processing for communications in defense, surveillance, health care, and consumer electronics provide significant challenges to the objective of developing theory and practice, spurred by progress in microminiaturization and computing capabilities. For example, digital camera images often suffer from color noise and artifacts, which can be removed by filter plug-ins. The task of digital television bandwidth reduction involves two major data compression approaches: lossless (involving redundancy removal), which can compress most images in the range of 2:1 or 3:1, and lossy (discarding some information with reduced quality at higher compression), which can compress moving images greater than 100:1. Transform coding by discrete cosine transform (DCT), employed in the JPEG standard for still images, suffers from blocking effects due to the 8-by-8 block size at, say over 16:1 compression (because of the short basis functions for reconstruction). The blockiness is especially percieved in the blue and red channels in the flat regions of an image. On the other hand, artifacts like "mosquito noise" are conspicuous in areas of image with lots of detail. Most of the luminance information is carried in the green channel, while the noise and artifacts that are produced in digital cameras and JPEG compression occur primarily in the chrominance information that is dominant in the blue and red channels. After converting the RGB color images via RGB/YIQ transfor-

mation to chrominance (I, Q) and luminance components (Y), the chrominance is filtered so that the luminance details can be transferred to the blue and red channels, and then the image is converted back to RGB via the YIQ/RGB transformation. The so-called Quantum Mechanic filter uses an adaptive median-like filter (nonlinear) for removing large amounts of noise without destroying thin color details. Subband coding using wavelets (tree-structured filter banks) avoids blocking at medium bit rates because basis functions have variable lengths; at low bit rate, however, high wavenumbers (textures) are removed, causing ringing.

II. WAVE DIGITAL FILTERS

Considerable attention has been given to the problem of realizing low-sensitivity digital filters. As a rule of thumb, the simpler the realization procedure, the poorer the sensitivity of filter performance to elemental variations due, say, to finite arithmetic constraints like coefficient quantization with a specified number of bits. This is justified from the simultaneous increasing complexity of procedures for realization and improved passband sensitivity as one proceeds from the direct to the cascade and parallel structures culminating in the lattice-ladder realization of a digital filter transfer function. Fettweis (1971) considered the possibility of departure from the traditional digital domain approaches and showed how the richly documented synthesis techniques for analog circuits that can be used to generate a prototype analog filter structure could be transformed to a corresponding digital filter structure capable of capturing the low-sensitivity characteristics of the passive analog prototype. Expectation of low sensitivity to changes in multiplier coefficients was a major motivation for developing the principles of wave digital filtering.

A digital filter generated from an analog filter by the appropriate nesting of multiport filters (with associated port resistance at each port) called adaptors (realizable either in parallel or series form), along with elements (which could be one-port delays, reciprocal two-port unit elements, as well as quasi-reciprocal lines) and sources, are referred to as wave digital filters, motivated by the theory of wave propagation along a transmission line. The adaptors are built with adders and multipliers. A unconstrained (i.e., the port resistances are not constrained) series or a parallel n-port adaptor requires $n - 1$ multipliers and $3n - 3$ adders, while a constrained (made reflection-free at one port by appropriately relating the port resistances) adaptor rquires $n - 2$ multipliers and $3n - 5$ adders. The adaptors, which are the basic building blocks of wave digital filters, require some care in interconnection akin to that needed in the matching of resistors for maximum power transfer to a resistively terminated lossless two-port from a source with resistance.

The property of low sensitivity in the passband of a generic filter structure composed of a lossless two-port between a source with resistance and a resistive termination is expected to be carried over to a digital filter derived from the generic prototype. This expectation was confirmed by Fettweis (1972), who introduced the concepts of pseudopassivity and pseudolosslessness for digital filters as counterparts of the properties of passivity and losslessness associated with classical analog filter theory. The variety of distinct topologies in classical filter synthesis leads to several classes of digital filter structures that keep intact the excellent passband sensitivity characteristics of the analog doubly terminated lossless reference filters. Furthermore, the good stopband insensitivity to parameter variation of analog ladder filters motivated particular attention to this subclass of generic passive filter structures. It was shown that any *LC* ladder filter structure can be very simply translated into a corresponding digital structure where the number of multipliers exceeds the number of inductors and capacitors in the doubly terminated lossless two-port by one (Sedlmeyer and Fettweis, 1973). The adaptors used in the realization are three-ports, all unconstrained except one. Thus, the number of multipliers could be significantly larger than the order of the ladder filter, whose series or shunt arms are designed to realize the appropriate transmission zeros of the filter type. In wave digital filters designed from the symmetric lattice reference filters, the stopband insensitivity is poorer than ladder filters but the number of multipliers does not exceed the filter order and the passband insensitivity may even be better than for ladder structures. The wave digital counterpart of lattice-ladder filters is also possible. An excellent review of other aspects of wave digital filter theory and principles not mentioned here is available (Fettweis, 1996).

Before concluding this section, it is important to mention that a properly designed wave digital filter is free of zero-input limit cycles, does not exhibit zero-input parasitic oscillations even under looped conditions, offers satisfactory sensitivity properties in passbands as well as stopbands, and has good dynamic range. The principles of wave digital filtering generalize naturally to multidimensions, where, however, the absence of general synthesis techniques makes design of reference filters much more difficult in comparison to the one-dimensional case.

III. LSI FILTERING OF MULTIDIMENSIONAL DISCRETE FILTERS

Digital filtering is used to process discrete data obtained either from sampling continuous signals or in some other manner. Its range of application is extensive, including the processing of geophysical, biomedical, pictorial, sonar, and radar data. Most of the discussion in this section applies to multidimensional problems, though for the sake of brevity in exposition, the 2D case is emphasized.

Various strategies exist to sample band-limited multidimensional signals. Petersen and Middleton (1962) showed that rectangular sampling is a special case of a more general sampling strategy by which a band-limited waveform is sampled on a nonorthogonal sampling raster. Another special case of this general strategy is discussed in depth by Dudgeon and Mersereau (1984), who point out the advantages of hexagonal sampling. For more information on the subject of rectangular sampling along with reconstruction formulas for recovery of the analog signal from its discrete samples, see Zayed (1993).

The filtering of the discrete spatiotemporal multidimensional signals by linear shift-invariant (LSI) systems, defined next, will be considered in this section; the mathematical tools needed for the purpose will first be briefly summarized. For additional details, see Bose (1982).

A. Mathematical Tools

The notations and definitions, given for the 2D case for brevity, are directly generalizable to n dimensions. The reader should have no difficulty in extracting the corresponding 1D counterparts (Bose, 1993b).

Definition 1. A 2D discrete system whose input/output relationship $y[k_1, k_2] = T[x[k_1, k_2]]$ is characterized by the operator $T[\cdot]$ is said to be linear if and only if for arbitrary inputs $x_1[k_1, k_2]$ and $x_2[k_1, k_2]$ and arbitrary complex constants c_1 and c_2,

$$T[c_1 x_1[k_1, k_2] + c_2 x_2[k_1, k_2]]$$
$$= c_1 T[x_1[k_1, k_2]] + c_2 T[x_2[k_1, k_2]]. \quad (1)$$

The system characterized by $T[\cdot]$ is shift-invariant if and only if

$$y[k_1 - k_0, k_2 - l_0] = T[x[k_1 - k_0, k_2 - l_0]] \quad (2)$$

for all $x[k_1, k_2]$, with k_0, l_0 arbitrary integers.

A system satisfying both the above properties is linear shift-invariant (LSI).

The mathematical tool used in the study of multidimensional LSI systems is the multidimensional z-transform. In the case of 2D systems, the z-transform of a sequence $\{x[k_1, k_2]\}$ is defined to be

$$Z[x[k_1, k_2]] \stackrel{\triangle}{=} X(z_1, z_2)$$

$$\stackrel{\triangle}{=} \sum_{k_1=-\infty}^{\infty} \sum_{k_2=-\infty}^{\infty} x[k_1, k_2] z_1^{-k_1} z_2^{-k_2}. \quad (3)$$

An alternative definition replaces $z_1^{-k_1} z_2^{-k_2}$ with $w_1^{k_1} w_2^{k_2}$ in (3), but no essential conceptual difference arises due to these two possibilities. The expository advantage of working with the complex variables $w_1 \triangleq z_1^{-1}$ and $w_2 \triangleq z_2^{-1}$ is context dependent. In combinatorial studies, $X(z_1, z_2)$ is referred to as a generating function. The values of z_1, z_2 for which the double summation in (3) converges absolutely constitute the region of convergence, referred to as the *Reinhardt domain*. This domain is completely specified by the magnitudes $|z_1|$, $|z_2|$, respectively, of the complex variables z_1, z_2,

The inversion formula associated with (3) is

$$x[k_1, k_2] = \frac{1}{(2\pi j)^2} \oint_{C_1} \oint_{C_2} X(z_1, z_2) z_1^{-1-k_1} z_2^{-1-k_2} \, dz_1 \, dz_2. \quad (4)$$

In (4), the closed contours C_1, C_2 must be in the region of convergence of $X(z_1, z_2)$ in (1). The Reinhardt domain must be specified before the sequence $x[k_1, k_2]$ can be uniquely calculated from its z-transform $X(z_1, z_2)$ as in the one-dimensional case.

Any 2D LSI system is completely specified by its impulse response

$$h[k_1, k_2] = T[\delta[k_1, k_2]],$$

where the 2D unit impulse function is

$$\delta[k_1, k_2] = \begin{cases} 1 & \text{if} \quad k_1 = k_2 = 0, \\ 0 & \text{otherwise.} \end{cases}$$

The output $y[k_1, k_2]$ of any 2D LSI system with impulse response sequence $\{h[k_1, k_2]\}$ and specified input $\{x[i_1, i_2]\}$ is given by the 2D discrete convolution

$$y[k_1, k_2] = \sum_{i_1=-\infty}^{\infty} \sum_{i_2=-\infty}^{\infty} x[i_1, i_2] h[k_1 - i_1, k_2 - i_2]. \quad (5)$$

The input–output relationship of a 2D LSI system is defined by a difference equation of the form

$$\sum_{\substack{i_1 \\ (i_1,i_2) \in I_b}} \sum_{i_2} b[i_1, i_2] y[k_1 - i_1, k_2 - i_2]$$

$$= \sum_{\substack{i_1 \\ (i_1,i_2) \in I_a}} \sum_{i_2} a[i_1, i_2] x[k_1 - i_1, k_2 - i_2], \quad (6)$$

where I_b (output mask) and I_a (input mask) denote, respectively, the finite-arearegions of support for ar-

rays $\{b[i_1, i_2]\}$ and $\{a[i_1, i_2]\}$. With $b[0, 0] = 1$, we can write (6) as

$$y[k_1, k_2] = \sum_{\substack{i_1 \\ (i_1,i_2) \in I_a}} \sum_{i_2} a[i_1, i_2] x[k_1 - i_1, k_2 - i_2]$$

$$- \sum_{\substack{i_1 \\ (i_1,i_2) \in I_b \\ (i_1,i_2) \neq (0,0)}} \sum_{i_2} y[k_1 - i_1, k_2 - i_2]. \quad (7)$$

B. Two-Dimensional FIR Filter Principles

Finite impulse response (FIR) digital filters are extensively used in one-dimensional and multidimensional signal processing because of their inherent stability and linear phase (subject to appropriate symmetrization) properties. The importance of phase is especially underscored in multidimensions (Huang *et al.*, 1975). The respective generalizations of one-dimensional design techniques (Bose, 1993a; Hamming, 1989) are briefly summarized next.

1. Windowing Scheme

Perhaps the simplest and most straightforward design technique is the windowing technique, which is a judicious mixture of art and science. Various windows have been discovered and multidimensional design often exploits the documented one-dimensional case. Huang (1972) related the design of 2D and 1D windows in the following manner.

Fact: If $\omega(x)$ is a good symmetrical 1D window, then $\omega(\sqrt{x_1^2 + x_2^2})$ is a good circularly symmetrical 2D window.

The approximation is especially good if the Fourier transform of the window bisequence (this transform is called the window wavenumber response) has a support which is much smaller than the extent of bandlimitedness of the low-pass, 2D, circularly symmetric filter wavenumber response. Let the ideal 2D, circularly symmetric wavenumber response be defined by

$$H(\omega_1, \omega_2) = \begin{cases} 1 & \omega_1^2 + \omega_2^2 \leq B^2, \\ 0 & \text{otherwise.} \end{cases}$$

Let the Fourier transform of the 2D window $\omega[x_1, x_2]$ used to truncate the 2D impulse response, $h[x_1, x_2]$, be $W(\omega_1, \omega_2)$. Then, the wavenumber response of the truncated bisequence, denoted by $f[x_1, x_2] = h[x_1, x_2] w[x_1, x_2]$, is the 2D convolution

$$F(\omega_1, \omega_2) = H(\omega_1, \omega_2) * * W(\omega_1, \omega_2).$$

Clearly, if the "width" of $W(\omega_1, \omega_2)$ is much smaller than the "width" of $H(\omega_1, \omega_2)$, then $F(\omega_1, \omega_2)$ becomes essentially the convolution of $W(\omega_1, \omega_2)$ and a 2D step function. Therefore, if

$$\omega[x_1, x_2] = \omega\left(\sqrt{x_1^2 + x_2^2}\right),$$

then $F(\omega_1, \omega_2)$ is approximately of the functional form $F(\sqrt{\omega_1^2 + \omega_2^2})$.

2. Frequency-Sampling Scheme

In one dimension the Lagrange formula (Bose, 1993a) gives a polynomial $P(z)$ of degree n which interpolates over $n + 1$ distinct points $\{z_0, z_1, \ldots, z_n\}$. Specifically, if the frequency samples at $\{z_0, z_1, \ldots, z_n\}$ are specified by $P(z_k) = w_k, k = 0, 1, \ldots, n$, then

$$P(z) = \sum_{k=0}^{n} w_k \ell_k(z),$$

where

$$\ell_k(z) = \frac{\prod_{i \neq k}(z - z_i)}{\prod_{i \neq k}(z_k - z_i)}.$$

In 2D a similar interpolation is possible. Let $\{z_0, z_1, \ldots, z_n\}$ be $n + 1$ distinct points. Let $\{w_0, w_1, \ldots, w_m\}$ be a second such set of $m + 1$ points. Set (Davis, 1975)

$$P(z) = \prod_{i=0}^{n}(z - z_i),$$

$$Q(w) = \prod_{j=0}^{m}(w - w_j),$$

$$P_j(z) = \frac{P(z)}{(z - z_j)},$$

$$Q_k(w) = \frac{Q(w)}{(w - w_k)}.$$

The $(m + 1)(n + 1)$ polynomials

$$\ell_{jk}(z, w) = \frac{P_j(z)Q_k(w)}{P_j(z_j)Q_k(w_k)}$$

satisfy

$$\ell_{jk}(z_r, w_s) = \delta_{jr}\delta_{ks}.$$

Hence,

$$P(z, w) = \sum_{j=0}^{n}\sum_{k=0}^{m} \mu_{jk}\ell_{jk}(z, w)$$

is a bivariate polynomial of total degree $\leq m + n$ which satisfies the $(m + 1)(n + 1)$ interpolation conditions

$$P(z_j, w_k) = \mu_{jk}, \quad j = 0, 1, \ldots, n; \quad k = 0, 1, \ldots, m.$$

Thus, frequency sampling 2D FIR filters can be designed (Bose, 1982, pp. 253–255).

3. Minimax FIR Filters

The Parks–McClellan method for optimal (in the minimax sense) FIR filter design is based on the validity of a univalence condition referred to as the Haar condition (Bose, 1993a). This condition provides the uniqueness of the best approximation, a useful criterion (for convergence) in iterative implementation via the Remez exchange algorithm (Bose, 1993a). The Haar condition does not hold, in general, in the 2D case. Let the monomials in two variables z_1, z_2 be given as follows: $P_0(z_1, z_2) = 1, P_1(z_1, z_2) = z_1, P_2(z_1, z_2) = z_2, P_3(z_1, z_2) = z_1^2, P_4(z_1, z_2) = z_1 z_2, P_5(z_1, z_2) = z_2^2, P_6(z_1, z_2) = z_1^3, \ldots$. It is not always possible, having been given n arbitrary distinct points (z_{1i}, z_{2j}), to find a linear combination of $P_0, \ldots, P_{n-1}$ that takes on preassigned values at these points (Davis, 1975, p. 27).

4. Multidimensional FIR Filter via Transformation

To exploit the procedures richly documented for 1D digital filter design in the corresponding 2D case, a type of 1D-to-2D frequency/wavenumber transformation was proposed by McClellan (1973). This transformation maps a 1D zero-phase frequency response to a 2D frequency (wavenumber) respose and is suitable for the design and implementation of 2D digital filters whose wavenumber response magnitudes approximate either the low-pass circularly symmetric or fan characteristics (Dudgeon and Mersereau, 1984). The circularly symmetric low-pass design transformation was generalized to multidimensions through a simple but clever change of variables by Fettweis (1977).

C. Two-Dimensional IIR Filter Principles

The high implementational complexity of FIR filters is especially felt for those with sharp roll-off frequency (wavenumber) characteristics. The transition width of a 1D FIR filter is inversely proportional to the length of its impulse response sequence. For 2D FIR filters that approximate a diamond-shaped magnitude of wavenumber response, the filter support size is believed to be inversely proportional to the square of the transition width (Carrai *et al.*, 1994). The advantage of achieving reduction in computational complexity of implementation via IIR filtering is offset, however, by the difficulties with stabilization, nonlinearity of the phase response, and various problems encountered in implementation methods. Recursibility issues also have to be dealt with. In spite of these and other drawbacks a number of design techniques have been developed in both the traditional wavenumber domain as well as in the spatial domain (where local state-space models are used). Global state-space models for multidimensional filters are, in general, infinite dimensional. Only the important subclass of *recognizable* rational filter transfer functions (where the denominator polynomial is expressible as a product of univariate polynomials) have finite-dimensional global state-space models.

The wavenumber-domain IIR design methods can be grouped into two categories, those based on spectral transformations and those based on computer-aided optimization including iterative schemes. In the former approach a 2D IIR filter is designed from a 1D (or even a 2D) prototype (which may be either a stable analog or a stable digital filter) by transformations that besides keeping the stability property invariant also succeed in achieving the specified wavenumber response characteristics. In the computer-aided optimization approaches, a nonlinear optimization procedure iteratively adjusts the filter transfer function parameters untill a specified error criterion is minimized. Details on the various techniques are available in standard texts (e.g., Dudgeon and Mersereau, 1984; Bose, 1982).

The support of the impulse response of a recursive 2D digital filter need not be confined to a quarter-plane. In fact, wavenumber responses of quarter-plane filters are quite restricted (Bose, 1982). The condition for recursive computation of the output bisequence in Eq. (6) in response to a finite input bisequence can be met by cone (wedge) filters (to be described below) and the limiting case of cone (wedge) filters, namely half-plane filters. These filters are much more versatile than quarter-plane filters.

D. Recursibility

The conditions for recursibility in multidimensions have been discussed in sufficient generality (Dudgeon and Mersereau, 1984; Bose, 1982). Here the objective is to discuss important special cases which are not only easily understood and appreciated, but also bring out key features and methods in 2D recursive filter analysis such as unit impulse response computation of IIR filters and their stability. Consider the non-quarter-plane digital filter characterized by the transfer function

$$H(z_1, z_2) = \frac{Y(z_1, z_2)}{X(z_1, z_2)}$$

$$= \frac{1}{1 + 0.5 z_1 z_2^{-1} + z_1^{-1} + z_2^{-1} + z_1^{-2} z_2}, \quad (8)$$

where $Y(z_1, z_2)$, and $X(z_1, z_2)$ are, respectively, the transforms of the output and input bisequences. It is easy to

verify that the unit impulse response $\{h[k_1, k_2]\}$ is recursively computable with zero-boundary conditions from the 2D difference equation

$$h[k_1, k_2] = \delta[k_1, k_2] - \frac{1}{2} h[k_1 + 1, k_2 - 1] - h[k_1 - 1, k_2]$$

$$- h[k_1, k_2 - 1] - h[k_1 - 2, k_2 + 1]. \quad (9)$$

The support, $S_{p,q,r,t} \overset{\triangle}{=} \operatorname{supp}\{h[k_1, k_2]\}$, of the impulse response is in the causality cone $H_{1,1,1,2}$ defined by the intersection of the two half-planes,

$$H_{1,1}:\ x_1 + x_2 \geq 0 \qquad \text{and} \qquad H_{1,2}:\ x_1 + 2x_2 \geq 0.$$

It is important to note that if $\mathcal{Z}$ denotes the set of integers and $\mathcal{Z}^2 \overset{\triangle}{=} \mathcal{Z} \times \mathcal{Z}$, then there exists an index map

$$\phi:\ H_{p,q,r,t} \cap \mathcal{Z}^2 \to Q_1 \cap \mathcal{Z}^2$$

from the causality cone $H_{p,q,r,t} = H_{p,r} \cap H_{q,t}$ with support $S_{p,q,r,t}$ (p, q, r, t are nonnegative integers) to the first quadrant quarter-plane Q_1 with support $S_{1,0,0,1}$.

The map referred to is bijective, i.e., isomorphic (one-to-one and onto) and is described by the relation

$$[n_1, n_2] = \phi[k_1, k_2] = [pk_1 + rk_2, qk_1 + tk_2]. \quad (10)$$

The associated inverse map, with $pt - qr = 1$, is

$$[k_1, k_2] = \phi^{-1}[n_1, n_2] = [tn_1 - rn_2, -qn_1 + pn_2]. \quad (11)$$

Clearly,

$$\phi[t, -q] = [pt - qr, qt - qt] = [1, 0],$$

$$\phi[-r, p] = [-pr + pr, -qr + pt] = [0, 1],$$

which implies that points $[t, -q]$ and $[-r, p]$ in $H_{p,q,r,t}$ are mapped to points $[1, 0]$ and $[0, 1]$, respectively, in Q_1. The multidimensional counterpart of the 2D mapping and its inverse can be found in Bose (1985), Chapter 5.

The quarter-plane filter transfer function obtained from the image of the map is

$$\hat{H}(z_1, z_2) \overset{\triangle}{=} \sum_{n_1=0}^{\infty} \sum_{n_2=0}^{\infty} h[\phi^{-1}[n_1, n_2]] z_1^{-n_1} z_2^{-n_2},$$

where $\phi^{-1}[n_1, n_2]$ is defined in (11). Then

$$\hat{H}[z_1, z_2] = \sum_{n_1=0}^{\infty} \sum_{n_2=0}^{\infty} h[tn_1 - rn_2, -qn_1 + pn_2] z_1^{-n_1} z_2^{-n_2}$$

$$= \sum_{\text{supp}\{h[k_1,k_2]\}} \sum h[k_1, k_2] z_1^{-(pk_1+rk_2)} z_2^{-(qk_1+tk_2)}$$

$$= \sum_{\text{supp}\{h[k_1,k_2]\}} \sum h[k_1, k_2] \left(z_1^p z_2^q\right)^{-k_1} \left(z_1^r z_2^t\right)^{-k_2}$$

$$= H\left(z_1^p z_2^q, z_1^r z_2^t\right).$$

Notice that for notational convenience, below the z-transform involves positive powers of the indeterminates, $w_1 \overset{\triangle}{=} z_1^{-1}$, $w_2 \overset{\triangle}{=} z_2^{-1}$. Therefore,

$$\hat{H}[z_1, z_2] = \sum_{\text{supp}\{h[k_1, k_2]\}}\sum h[k_1, k_2]\big(z_1^p z_2^q\big)^{-k_1}\big(z_1^r z_2^t\big)^{-k_2}$$

$$= \sum_{\text{supp}\{h[k_1, k_2]\}}\sum h[k_1, k_2]\big(w_1^p w_2^q\big)^{k_1}\big(w_1^r w_2^t\big)^{k_2}.$$

The transfer function in the complex variables w_1, w_2 of the first quadrant quarter-plane filter obtained from $H(z_1, z_2)$ in (8) is

$$G[w_1, w_2]$$

$$= \frac{1}{1 + 0.5 z_1 z_2^{-1} + z_1^{-1} + z_2^{-1} + z_1^{-2} z_2}\Bigg|_{\substack{z_1 = w_1^{-1}w_2^{-1} \\ z_2 = w_1^{-1}w_2^{-2}}}$$

$$= \frac{1}{1 + 0.5 w_2 + w_1 w_2 + w_1 w_2^2 + w_1}.$$

Apply a first-quadrant quarter-plane stability test [see Chapter 3 in Bose (1982) for a comprehensive account] on $G(w_1, w_2)$, i.e., check whether or not $A(w_1, w_2) \neq 0$, $|w_1| \leq 1$, and $|w_2| \leq 1$, where $A(w_1, w_2) \overset{\triangle}{=} 1 + 0.5 w_2 + w_1 + w_1 w_2 + w_1 w_2^2$. Clearly $A(-1, 0) = 0$, which implies that $G(w_1, w_2)$ is BIBO unstable. Consequently, $H(z_1, z_2)$ in (8) is also a BIBO-unstable general support recursive digital filter. Indeed, from (9) it is straightforward to show that along the boundaries of the causality cone that provides the region of support of the filter,

$$h[-k, k] = \left(-\frac{1}{2}\right)^k, \qquad k \geq 0,$$

$$h[2k, -k] = (-1)^k, \qquad k \geq 0.$$

Subsequently, it is easy to show after some algebraic manipulation that

$$h[-k, k+1] = -\frac{3}{4}(k+1)\left(-\frac{1}{2}\right)^k, \qquad k \geq -1,$$

$$h[2k+1, -k] = \frac{1}{2}k(-1)^k, \qquad k \geq 0,$$

$$h[-k, k+2] = \left[-\frac{15}{2} + \frac{27}{4}(3+k) - \frac{9}{4}(3+k)^2\right]$$
$$\times \left(-\frac{1}{2}\right)^{k+3}, \qquad k \geq -2,$$

$$h[-k, k+3] = \left[-\frac{27}{2} + \frac{153}{8}(4+k) - \frac{27}{4}(4+k)^2\right.$$
$$\left. + \frac{9}{8}(4+k)^3\right]\left(-\frac{1}{2}\right)^{k+4}, \qquad k \geq -3,$$

etc. Better still, a plot of the impulse response $h[k_1, k_2]$ obtained by setting up a recursivecomputation on the pre-

ceding set of equations shows how it builds up, especially in the first quadrant.

IV. VELOCITY FILTERING

A two-dimensional signal often exhibits orientation in space called direction. When a signal has components along a direction, its Fourier transform also contains high-magnitude components along a well-defined direction. The so-called fan filter can be used as a directional filter for discriminating between signals along different directions. In array signal processing, the directional property of signals can be interpreted as velocity. Velocity filtering is essentially unique to seismology, at least to date, since the velocity of propagation is not constant for all waves. In radar, radio astronomy, and to a large extent also in sonar, the velocity is essentially constant.

Seismic waveforms consist of two distnct components. One is a signal reflected from subsurface formations and the other is a surface signal often referred to as ground roll. The information about subterranean deposits like oil and hydrocarbon in sedimentation layers is embedded in the reflected component, whereas the ground roll component transmitted along the surface of the earth is undesirable since it degrades the reflected signal of interest. The surface wave has a lower velocity of propagation than the reflected wave and can be targeted for attenuation by filtering (see Bose, 1993a, pp. 480–481).

Velocity filters provide a means for discriminating signals from noise or other undesired signals because of their different apparent (move-out) velocities. Thus, the use of velocity filters makes it possible selectively to enhance or attenuate signals occupying the same frequency ranges as noise or undesired signals. Velocity filtering can be performed by multidimensional filters included in the array processing scheme.

A. Signal and Noise Characteristics

Apparent velocity of arriving signals depends upon the wave type, source location, and structure beneath the recording site.

(a) Teleseismic signals: These are assumed to propagate coherently across the array. Observed apparent velocities usually lie above 8 km/sec (P-waves) and above 4 km/sec (S-waves). P-waves show predominant frequencies around 1 Hz and S-waves usually have lower peak frequency.

(b) Microseismic noise: This propagates with velocities from about 2.5 to 4 km/sec. Dominant frequencies occupy a broad low-frequency range. Depending upon the distance to the source and sensor spacing, some portions of microseismic noise may propagate coherently across the array.

(c) Incoherent noise: Various human activities (factories, traffic, construction), action of wind, smaller water basins, etc., generate high-frequency noise above 1 Hz. Corresponding apparent velocities vary for different sources, but in general have low values, around 1 km/sec. For sensor spacing larger than several kilometers, it is most likely that the high-frequency noise propagates incoherently across the array, irrespective of whether the location of the noise source is inside or outside the array pattern.

(d) Signal-generated noise: This is noise generated by the desired signal itself in the vicinity of the sensor as multiple reflections and mode conversions. It has a wide range of apparent velocities, depending upon the local structure, and it may propagate coherently across the array.

B. One-Dimensional Arrays for Velocity Filtering

In Fig. 1, P-waves with apparent velocities in the range from 8 km/sec to infinity occupy the triangular zone labeled "signal zone." All kinds of coherent noise with apparent velocities between 2.5 and 4 km/sec are distributed within the wedge marked "coherent noise."

Consider a 1D array and horizontal wave propagation parallel to the line of sensors. An adequate separation of the signal from the ambient noise may be achieved by velocity filtering, which is implemented via fan filters. For a fan filter the pass and rejection zones become defined as straight lines through the origin in the frequency–wavenumber, (f, k), plane. One of the major applications of fan filters is in directional filtering, where the filter can pass the signals along prescribed directions and reject

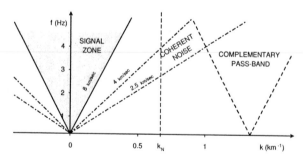

FIGURE 1 Idealized distribution of the signal and noise components in the frequency versus wavenumber plane. Space sampling period $\Delta x = 0.8$ km and time sampling period $\Delta t = 0.1$ sec are typical.

along others. Besides sampling there are no other limitations upon frequency and wavenumber intervals. Provided that the coherent noise occupies a region in the (f, k) plane that differs from that of the signal, the velocity filtering performs a perfect discrimination against the coherent noise.

Assume that it is desired to pass waveforms with wavenumbers within the range $-|f|/V \leq k \leq |f|/V$. Outside this wavenumber range all waveforms are rejected. The 2D transfer function is then defined as

$$H(f, k) = \begin{cases} 1 & -|f|/V \leq k \leq |f|/V \\ 0 & \text{elsewhere.} \end{cases} \quad (12)$$

The time–space impulse response of the filter may be expressed in terms of the inverse 2D Fourier transform of $H(f, k)$,

$$h(t, x) = \int_{-\infty}^{\infty} \int_{-\infty}^{\infty} H(f, k) \exp j2\pi(ft - kx)\, df\, dk. \quad (13)$$

Due to the periodicity of $H(f, k)$ in wavenumber, we must limit ourselves to the resolvable wavenumber band only:

$$-k_N \leq k \leq k_N. \quad (14)$$

It follows from the basic relation

$$f = Vk \quad (15)$$

that for a given apparent velocity V and frequencies $|f| > Vk_N$ in the (f, k) plane, we leave the primary and enter the complementary pass zone. Therefore, the frequency limits for a meaningful transfer function are

$$-f_N \leq f \leq f_N, \quad (16)$$

where, for a spatial period Δx,

$$f_N = Vk_N = \frac{V}{2\Delta x}. \quad (17)$$

Having established the value of the folding frequency f_N, we find that the temporal sampling period Δt which preserves all information within the frequency band $(-f_N, f_N)$ becomes

$$\Delta t = \frac{1}{2f_N}. \quad (18)$$

The last two equations also show that the apparent cutoff velocity satisfies

$$V = \frac{f_N}{k_N} = \frac{\Delta x}{\Delta t}. \quad (19)$$

For any signal to pass through this velocity filter, the apparent velocity must be equal to or higher than V, i.e., for a given sensor spacing the signal moveout τ must satisfy the condition

$$\tau \leq \Delta t. \tag{20}$$

Introducing the finite limits $\pm f_N$, $\pm k_N$, we have

$$h(t, x) = \int_{-f_N}^{f_N} \int_{-k_N}^{k_N} H(f, k) \exp j2\pi (ft - kx) \, df \, dk. \tag{21}$$

It may be expected intuitively that the sharpness between the pass and rejection zones increases and the amplitude of side lobes decreases with increasing number of array sensors.

C. Multidimensional Velocity Filters

Extensive documentation of design methods for multidimensional digital fan filters including the so-called cone and wedge filters is available in (Boll, 1992). For background information, see Dudgeon and Mersereau (1984).

V. FILTER BANKS

The journey from digital filters to banks of filters (filter banks) was motivated by applications that necessitated the decomposition of a discrete signal into frequency subbands (with the help of an analysis filter bank) prior to coding, transmission, and eventual reconstruction of the signal (with the help of a synthesis filter bank). The concept of subband decomposition originates from the expectation that if the input signal is decomposed into spectral bands, then it can be processed or manipulated more effectively (say, for topics of current interest like digital watermarking for copyright protection of images and even multimedia (Hartung and Kutter, 1999). The greater versatility of the filter bank can be further justified from the limitations inherent in a single filter. To wit, an IIR digital filter is unable to capture linearity of phase response, while a FIR filter cannot be made allpass. Thus, phase and amplitude cannot be manipulated simultaneously by either an FIR filter or an IIR filter. However, perfect reconstruction filter banks are capable of eliminating aliasing, phase, and amplitude distortions.

The idea of using multiband filter banks is crucial not only for encoding data from 1D signals like speech, but also including image and video. The generic analysis–synthesis system for perfect reconstruction consists of a bank of N filters each followed by a decimator or downsampler (with positive integer M as the downsampling factor in the uniform-filter-bank case) at the transmitting end and a corresponding bank of N synthesis filters each followed by an expander or upsampler (with upsampling

factor M in the uniform case) at the receiving end. The maximally decimated or critically sampled case occurs when $N = M$ and the discussions, unless mentioned otherwise, pertain to this case. For more efficient implementation, the polyphase representation is generated from the generic structure. In the polyphase representation, decimation of the signal precedes filtering at the transmitting end and upsampling follows filtering at the receiving end. The conditions for perfect reconstruction (up to a delay) of the original signal is straightforward to derive in terms of the matrices characterizing the multiband filter banks. These conditions can even be satisfied with FIR filters that are guaranteed to be stable. When the FIR analysis polyphase matrix is restricted to be paraunitary, the perfect reconstruction (PR) condition for the polyphase representation is automatically satisfied and the synthesis filters are the time-reversed conjugate version of the analysis filters. The perfect reconstruction property may, however, be lost in practical subband coding systems, where the signals in each subband are quantized. The possibility of PR even when the coefficients and the results of multiplications in the various stages of design of ladder structures are quantized was first discussed by Bruekers and van den Emden (1992). The existence of structures that keep invariant the paraunitary property after coefficient quantization is also known (Vaidyanathan, 1990).

Generalizations as well as limitations of recent results which incorporate the perfect reconstruction as well as linear phase (LP) constraints in multidimensional multiband digital filter banks are discussed with several examples and counterexamples in a recent paper (Charoenlarpnopparut and Bose, 1999). The PR and LP properties in filter banks are important for the subband coding of images because the human visual system is known to be more robust to symmetric distortions. When the class of filter banks is restricted to satisfy the conditions for LP as well as PR, the problem is completely solved for the two-band n-dimensional case using Gröbner bases (Buchberger, 1985). Although the problem is satisfactorily solved for the multiband 1D case in Basu and Choi (1999) and the two-band multidimensional case in Charoenlarpnopparut and Bose (1999), the more general problem associated with the multidimensional multiband counterpart is still open.

VI. CONCLUSIONS

Building on the solid foundation provided by the available literature on the filtering of continuous-time and discrete-time one-dimensional (1-D) temporal signals, this article gives state-of-the-art information on wave digital filters that exploit our knowledge of classical passive synthesis techniques in analog circuitry to offer digital simulations (accompanied with advantages of accuracy, robustness and dynamic range), multidimensional nonrecursive and recursive digital filters, and velocity discriminating filters (fan filters).

Digital filters are also used for time-frequency signal analysis and this has led to considerable research in the analysis and synthesis of digital filter banks. Multichannel filter banks have been used for signal compression, wavelet generation and related applications. This article provides a snapshot of the journey from filters to filter banks, the issues and implications of perfect reconstruction subject to constraints like linear phase and concludes with some frontiers of current research investigations in multidimensional filter banks and multidimensional wavelets.

The need for processing, coding, and transmitting multidimensional discrete signals in an expanding horizon of applications ensures the prominence of digital filters as an indispensable workhorse for electrical engineers. The emphasis placed on 2D digital filters is necessitated by the desired compromise between the extensive available documentation in the 1D case and the need for brevity in this presentation without sacrificing attention to some of the unique aspects of two- and higher dimensional digital filtering. Again, for the sake of brevity, attention here has been focused on linear shift-invariant filters. For an introduction to linear shift-variant multidimensional digital filters, see Bose (1985) [see in particular Valenzuela and Bose (1985), where other key references on the subject are available].

It is also important to point out that in spatiotemporal filtering, sometimes it is necessary to treat the temporal dimension in a manner different from the spatial dimensions. For example, in a baseball game the ball is thrown at a velocity higher than the motion of players. In that situation, a lowpass filter designed to preserve the details of the players and reduce noise would most likely have a cutoff frequency that would eliminate the baseball from the signal. Preservation of the baseball, however, would lead to unacceptable blurring of the players. To resolve the dilemma of temporal dependence on spatial variation, Dubois (1992) presented an approach to spatiotemporal filtering where the dimensions were treated unequally, unlike in conventional multidimensional filtering based, say, on least squares wavenumber domain-design.

Recent documentation of the theory for approximating magnitude only, group delay only, and both the magnitude and group delay of 1D as well as 2D recursive digital filters is available (Shenoi, 1999). There emphasis is placed on the designing of sharp-cutoff complexity digital filters for audio and image processing applications, where the need for approximating linear phase characteristics is crucial. Adaptive IIR filters remain as a potential tool for complex-

ity reduction in several different areas of signal processing (Regalia, 1995). The most demanding problem in adaptive IIR filtering currently is related to convergence characterization. Adaptive IIR filters have been used as notch filters for fast tracking of sinusoidal inputs, channel equalization for wired and wireless communications, and short training interfering echo cancellation.

Nonlinear digital filters have been widely used during the last couple of decades. For research conducted before 1990 see Pitas and Venetsanopoulos (1990). Detailed documentation of polynomial filters with emphasis on quadratic filters is available (Sicuranza, 1992). A type of nonlinear filter called a median filter is very suitable for removing impulsive noise. Vector median filters (Astola *et al.*, 1990) have been used for color images, and multidimensional nonlinear filters have been suggested for suppressing impulsive noise from video sequences with low signal-to-noise ratio (SNR) while keeping intact information pertaining to fine details and textures.

ACKNOWLEDGMENTS

The writing of this article was completed when the author was visiting Ruhr Universitaet, Lehrstuhl fuer Nachrichtentechnik, Fakultaet fuer Elektroechnik, Universitaetsstrassee 150, 44780 Bochum, Germany as an Humboldt Awardee. The author is very grateful to Professor Alfred Fettweis for several useful discussions and for supplying some references.

SEE ALSO THE FOLLOWING ARTICLES

ANALOG ELECTRICAL FILTERS • CIRCUIT THEORY • DIGITAL ELECTRONIC CIRCUITS • IMAGE PROCESSING • SYSTEM THEORY

BIBLIOGRAPHY

Antoniou, A. (1993). "Digital Filters: Analysis, Design, and Applications," McGraw-Hill, New York.

Astola, J., Haavisto, P., and Neuvo, Y. (1990). "Vector median filters," *Proc. IEEE* **78,** 678–689.

Basu, S., and Choi, H. M. (1999). "Hermite reduction methods for generation of complete class of linear phase perfect reconstruction filter banks: Part I—Theory," *IEEE Trans. Circuit Syst. II Analog Digital Signal Processing* **46**(2), 434–448.

Boll, R. (1992). "Entwurfsverfahren für mehrdimensionale discrete Faecherfilter," Dr. Ing. Dissertation, Fakultät für Elektrotechnik, Ruhr-Universität, Bochum, Germany.

Bose, N. K. (1982). "Applied Multidimensional Systems Theory," Van Nostrand Reinhold, New York.

Bose, N. K. (ed.). (1985). "Multidimensional Systems Theory: Progress, Directions, and Open Problems," Reidel, Dordrecht, Holland.

Bose, N. K. (1993a). "Digital Filters: Theory and Applications," Krieger, Malabar, Florida.

Bose, N. K. (1993b). *In* "Handbook for Digital Signal Processing," (S. K. Mitra and J. F. Kaiser, eds.), Chapter 3, Wiley, Ney York.

Bose, N. K., and Rao, C. R. (eds.). (1993). "Signal Processing and Its Applications," North-Holland, Amsterdam.

Bruekers, F. A. M. L., and van den Emden, A. W. M. (1992). "New networks for perfect inversion and perfect reconstruction," *IEEE J. Select. Areas Commun.* **10,** 130–137.

Buchberger, B. (1985). *In* "Multidimensional Systems Theory: Progress, Directions, and Open Problems" (N. K. Bose, ed.), Chapter 6, Reidel, Dordrecht, Holland.

Carrai, P., Cortelazzo, G. M., and Mian, G. A. (1994). "Characteristics of minimax FIR filters for video interpolationdecimation," *IEEE Trans. Circuits and Syst. Video Technol.* **4,** 453–467.

Charoenlarpnopparut, C., and Bose, N. K. (1999). "Multidimensional FIR filter bank design using Gröbner bases," *IEEE Trans. Circuits Syst. II Analog Digital Signal Processing* **46,** 1475–1486.

Davis, P. J. (1975). "Interpolation and Approximation," Dover, New York.

Dubois, E. (1992). "Motion-compensated filtering of time-varying images" *Multidimensional Syst. Signal Processing,* **3,** 211–239.

Dudgeon, D. E., and Mersereau, R. M. (1984). "Multidimensional Digital Signal Processing," Prentice-Hall, Englewood Cliffs, NJ.

Fettweis, A. (1971). "Digital filter structures related to classical filter networks," *Arch. Elek. Uebertragung.* **25,** 79–89.

Fettweis, A. (1972). "Pseudopassivity, sensitivity, and stability of wave digital filters," *IEEE Trans. Circuit Theory* **19,** 668–673.

Fettweis, A. (1977). "Symmetry requirements for multidimensional digital filters," *Int. J. Circuit Theory Appl.* **5,** 343–353.

Fettweis, A. (1996). "Wave digital filters: Theory and practice," *Proc. IEEE* **74,** 270–327.

Hamming, R. W. (1989). "Digital Filters," Prentice-Hall, Englewood Cliffs, NJ.

Hartung, F., and Kutter, M. (1999). "Multimedia watermarking techniques," *Proc IEEE* **87,** 1079–1107.

Huang, T. S. (1972). "Two-dimensional windows," *IEEE Trans. Audio Electroacoustics* **20,** 88–89.

Huang, T. S., Burnett, J. V., and Deczky, A. G. (1975). "The importance of phase in image processing filters," *IEEE Trans. Acoustics Speech Signal Processing* **23,** (December).

McClellan, J. H. (1973). "The design of two-dimensional filters by transformations." *In* "Proceedings Seventh Annual Princeton Conference on Information Systems," pp. 247–251.

Mitra, S. K., and Kaiser, J. F. (eds.). (1993). "Handbook for Digital Signal Processing," Wiley, New York.

Petersen, D. P., and Middleton, D. (1962). "Sampling and reconstruction of wavenumber-limited functions for N-dimensional Euclidean spaces," *Information Control* **5,** 279–323.

Pitas, I., and Venetsanopoulos, A. (1990). "Nonlinear Digital Filters," Kluwer, Dordrecht, The Netherlands.

Regalia, P. A. (1995). "Adaptive IIR Filtering in Signal Processing and Control," Marcel Dekker, New York.

Sedlmeyer, A., and A. Fettweis, A. (1973). "Digital filters with true ladder configuration," *Int. J. Circuit Theory* **1,** 5–10.

Shenoi, B. A. (1999). "Magnitude and Group Delay Approximation of 1-D and 2-D Digital Filters," Springer-Verlag, New York.

Sicuranza, G. L. (1992). "Quadratic filters for signal processing," *Proc. IEEE* **80,** 1263–1285.

Vaidyanathan, P. P. (1990). "Multirate digital filters, filter banks, polyphase networks, and applications: A tutorial," *Proc. IEEE* **78,** 56–93.

Valenzuela, H. M., and Bose, N. K. (1985). *In* "Multidimensional Systems Theory: Progress, Directions, and Open Problems" (N. K. Bose, ed.), Chapter 5, Reidel, Dordrecht, Holland.

Zayed, A. I. (1993). "Advances in Shannon's Sampling Theory," CRC Press, Boca Raton, FL.

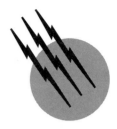

Digital Speech Processing

Biing Hwang Juang
M. Mohan Sondhi
Bell Laboratories, Lucent Technologies

Lawrence R. Rabiner
AT&T Laboratories

GLOSSARY

Adaptive coding A coding process that adapts its attributes and process parameters according to specific characteristics of the signal that is being encoded.

Continuous speech recognition Automatic recognition of speech utterances in which words are spoken continuously without enforced pauses between words.

Formant Resonance of the vocal tract manifested acoustically as a concentration of energy in the frequency spectrum.

Isolated word recognition Automatic recognition of words spoken as an isolated sequence of words with distinct pauses between individual words.

Keyword spotting Process of automatically detecting and recognizing a key word, or a key phrase, embedded in a naturally spoken sentence.

Predictive coding A coding process that involves using past (known) values of a signal to estimate the current value of the signal. The residual (i.e., the estimation error) has much lower variance than the original signal. This fact can be utilized to improve coding efficiency.

Speaker verification Process of authenticating a claimed identity based on spoken information.

Speech analysis The process of using computational algorithms to measure properties of the speech signal.

Speech coding Process of converting a sampled speech signal into a format suitable for storage or transmission over a network. The usual motivation for speech coding is storage or transmission efficiency.

Speech enhancement Processing of speech to make it less noisy, clearer, more intelligible, or easier to listen to.

Speech production model A model for the generation of speech by the human vocal apparatus.

Speech understanding High-level processing of a sequence of spoken words that extracts meaning from the spoken input.

Subjective quality Quality of a speech signal perceived by human listeners.

Text analysis Parsing and analyzing a sequence of words to make explicit the underlying syntactic and semantic structure so as to allow proper pronunciation and grouping of the words and to facilitate automatic understanding of the meaning in the given word sequence.

Text-to-speech synthesis Process of synthesizing speech from printed, usually unconstrained text.

Verbal information verification Process of verifying a claimed identity based on spoken user-specific information, such as a password, mother's maiden name, or a personal identification number (PIN).

SPEECH is the most fundamental form of communication among humans. Digital speech processing is the science and technology of transducing, analyzing, representing, transmitting, transforming, and reconstituting speech information by digital techniques. The technology is intended to improve communication between humans as well as to enable communications between humans and machines. Communications between humans can be enhanced by digital encoding and decoding of speech for efficient transmission, storage, and privacy and by suppressing noise, correcting distortion, and compensating for hearing loss. Communication between humans and machines comprises automatic recognition and understanding of speech, to give machines an "ear" with which to listen to spoken utterances, and digital synthesis of speech, to give machines a "mouth" with which to respond back to humans.

I. INTRODUCTION

Speech is the most sophisticated signal naturally produced by humans. A speech signal carries linguistic information for the sharing of information and ideas. It allows people to express emotions and verbally share feelings. It is the most fundamental form of communication among humans. The aim of digital speech processing is to take advantage of digital computing techniques to process the speech signal for increased understanding, improved communication, and increased efficiency and productivity associated with speech activities.

The field of speech processing includes speech analysis and representation, speech coding, speech synthesis, speech recognition and understanding, speaker verification, and speech enhancement. Speech is a complex signal that is characterized by varying distributions of energy in time as well as in frequency, depending on the specific sound that is being produced. The speech signal also possesses other characteristics that make it a very efficient means for carrying semantic (meaning) as well as pragmatic (task-dependent) information. The processes of *speech analysis and representation*, which lie at the technical basis of digital speech processing, attempt to use computational algorithms to "discover," measure, and represent the important properties of speech for many applications.

One of the most important applications of digital speech processing is *speech coding*, which is concerned with efficient and reliable communication between people who may be separated by geographical distance or by time. The former forms the basis of modern telephony, which enables people to converse regardless of their locations, and the latter forms the basis for applications such as "voice mail" that let people create verbal messages at arbitrary times, and retrieve them later. Speech coding enables both telephony and voice messaging by converting the speech signal into a digital format suitable for either transmission or storage. Relevant issues in speech coding are conservation of bandwidth (the rate of the voice coder), voice quality requirements, processing and transmission delay, and processing power, as well as techniques for privacy and secure communication.

Speech synthesis is concerned with providing a machine with the ability to talk to people in as intelligible and natural a voice as possible. A speech synthesis system can be as simple as a "prerecorded" announcement machine with a limited collection of utterances, or as complicated as a full text-to-speech conversion system, which automatically converts unconstrained printed text into speech. *Speech recognition*, often viewed as the counterpart to speech synthesis, is concerned with systems that have the ability to recognize speech (literally transcribe word-for-word what was spoken), and then understand the intended meaning of the spoken words (or more properly the recognized spoken words). Speech recognition systems range in sophistication from the simplest speaker-dependent, isolated-word or phrase recognizer to fully conversational systems that attempt to deal with virtually unlimited vocabularies and complex task syntax comparable to that of a natural language system.

Speaker recognition is concerned with machine verification or identification of individual talkers, based on their speech, for authorization of access to information, networks, computing systems, services, or physical premises. At times, as an important biometric feature, a talker's speech may also be used for forensic or criminal investigations.

Speech enhancement attempts to improve the intelligibility and quality of a speech signal that may have been corrupted with noise or distortion, causing loss of intelligibility or quality and compromising its effectiveness in communication. Speech enhancement methods are also useful in the design of hearing-aid devices, which transform the speech signal in such a way as to compensate for hearing loss in the auditory system of the wearer of the device.

Digital speech processing has continued to make substantial advances in the past decade, primarily due to the explosive growth in computational capabilities (both processing and storage), the introduction of key mathematical algorithms (see below), and the vast amount of real-world data that are being systematically made available through organized efforts among government agencies and research institutions. According to Moore's law, the computing capabilities of general-purpose processors grow at the rate of doubling the processor speed and memory every 18 months, without much increase in cost. Advances in very large scale integrated (VLSI) circuits have led to realization of extremely fast digital signal processors (DSPs) with very low power consumption. This has helped the proliferation of miniaturized devices such as cellular mobile phones. Today, a powerful DSP can run on as low as 0.9 volts, consumes a mere 0.05 mW per MIPS (million instructions per second), and has performance approaching 1000 MIPS (or GIPS). These advances in computing power are particularly beneficial in dealing with signals as complex as human speech. Finally, in the area of software, the convergence of programming languages (e.g., C and C++), the widespread use of command and interpretive languages for scientific computing such as MATLAB, and user-friendly environments for developing DSP assembly-code have made digital speech processing very accessible, practical, and easily realizable.

In the past decade, computing and digitization devices also have become ever more readily available for data recording and management. The availability of easily accessible databases has added a new direction to digital speech processing research, going beyond the traditional paradigm of hypothesis and observation made by experts in the field, to a data-driven mode in which computational algorithms are designed to learn from real-world data directly. This change of research paradigm is significant in making digital speech processing a useful practice because speech signals usually exhibit a wide range of variability, which can only be adequately observed (in a statistical sense) in a large collection of data. A human expert inherently lacks the ability to deal with such a large quantity of data. Realizing this, a number of government agencies and industrial research institutions have, in the past two decades, separately collected a number of task-specific speech databases. Such databases include speech sentences chosen for coder evaluations, the set of Naval Resource Management (RM) sentences for the study of continuous speech recognition, the set of Air Travel Information System (ATIS) sentences for speech recognition and dialog research, the set of North America News Broadcast (NAB) sentences for automatic speech transcription, and the Switchboard database for recognition of conversational telephone speech, just to name a few. These databases have guided many speech research programs and have been used for evaluating the progress in digital speech processing, thereby helping to make many system designs practically useful.

Many digital speech processing systems have been deployed for real-world services in the past decade. Major growth in the utility of voice systems exists in at least four major market sectors: telecommunications, business applications, consumer products, and government. In the *telecommunications* sector, speech coders are essential in digital and packet telephony, for traditional voice circuits as well as over the Internet. Specialized speech announcement systems use coded speech to provide timely information to customers, and speech recognition systems are used for automation of operator and attendant services as well as for retrieval of account information. In *business applications*, voice mail and store-and-forward messaging systems are in widespread use; automatic speech recognizers help people direct calls in a private branch exchange (PBX) arrangement; and voice interactive terminals let users dictate correspondence or issue voice commands for command and control of various devices or parts of a computer operating system. In the *consumer products* sector, toys incorporating speech synthesis and recognition have been available for years, and talking appliances and alarm announcement systems are also beginning to appear in household appliances. In the area of *government communications*, anticipated uses of speech processing include coding for secure communications, speech recognition for command and control of military systems, and automatic speech understanding and summarization for intelligence applications. These examples, by no means exhaustive, illustrate the burgeoning applications of digital speech processing and point to a growing market in the coming years.

II. SPEECH ANALYSIS AND REPRESENTATION

The traditional framework for analyzing speech is the source-tract model first proposed by Homer Dudley at

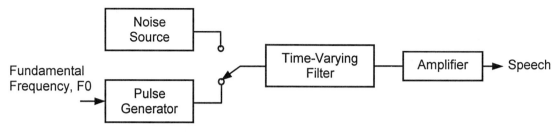

FIGURE 1 The speech production model, the basis for speech analysis.

Bell Laboratories in the 1930s. In this model, as depicted in Fig. 1, a speech excitation signal is produced by an excitation source and processed by a filter system that "modulates" the spectral characteristics of the excitation signal based on the shape of the vocal tract for the specific sound being generated. The excitation source has two components—a "buzz" source and a "hiss" source. The signal produced by the buzz source is a sequence of pulses with a controllable repetition rate (the fundamental frequency, or F0) and provides the carrier for voiced sounds such as /a/ in father, /e/ in met, and /o/ in hello. The hiss source produces a noise-like signal and provides the carrier for unvoiced sounds such as /s/ in sell, /sh/ in shout, and /k/ in kitten. The source signal is then filtered by a time-varying linear system with adjustable frequency response. The signal at the output of the time-varying filter is the final speech signal that has a time- and frequency-dependent distribution of energy. This time-varying filter system, which models the effects of the human vocal tract, is realized by a bank (ten channels in Dudley's original embodiment) of bandpass filters that span the range of speech frequencies. Any desired vocal tract frequency-response characteristic may be realized by adjusting the amplitudes of the outputs of the bandpass filters. Dudley was able to demonstrate (very successfully at the 1939 New York World's Fair) that a machine based on the production model of Fig. 1 can produce sounds that are very close in quality to that of natural speech.

Based on the work of Dudley, the source-tract model of speech production has been the dominant framework in speech analysis and representation for the past 60 years.

Key elements in the model—namely, specification of the frequency response of the slowly time-varying vocal tract filters and characteristics of the source signal—are thus the targets of speech analysis, which aims at finding optimal parameter values that best define the source-tract model to match a given speech signal.

To obtain the (time-varying) frequency-response characteristics of the vocal tract filters, the class of well-known methods of spectral estimation is normally employed. One such method uses a bank of bandpass filters, each followed by a nonlinearity, to measure the power level at the output of each channel. This method can be used to generate a local estimate of the spectral profile of the speech sound at a given time. Digital filter-bank design techniques are very well understood and are relatively straightforward to implement. Alternatively, one can use digital spectral analysis methods, such as the discrete Fourier transform (DFT) or fast Fourier transform (FFT) methods, to implement the filter banks according to a prescribed structure and spectral matching criterion. The resulting spectral profile of the time-varying speech signal, when plotted in the time–frequency plane using the density of the print to indicate the corresponding power level, is called a *sound spectrograph, sonogram,* or *spectrogram.* Figure 2 shows an example of a spectrogram corresponding to a speech segment of 2.34 seconds' duration.

Another important method for analyzing and representing the time-varying vocal tract frequency response is known as *linear prediction.* Linear prediction, or linear predictive coding (LPC), is motivated by the fact that the speech signal, at any time instant, can be approximated

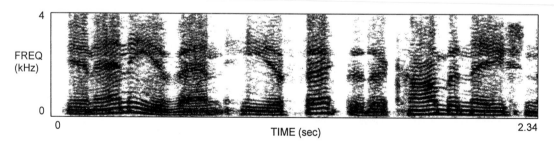

FIGURE 2 An example of a spectrogram for a speech segment.

by a linear combination of its past values. The difference between the predicted value and the true sample value is called the *residual, or prediction, error.* There are three important properties of LPC that make this method of analysis the most important one in most speech processing systems. First, the resulting set of optimal weights (the so-called linear prediction coefficients) that achieve the best prediction, in the sense of minimizing the residual for each short-time window of speech, defines an all-pole filter that best characterizes the behavior of the vocal tract. Second, the residual signal, when used as the excitation signal to drive the optimal tract filter, produces a signal essentially identical to the original signal. The residual signal has properties closely resembling those of the excitation source: It resembles a noise-like signal for unvoiced sounds and a pulse train for voiced sounds. Third, the optimal residual signal has a much smaller dynamic range than that of the original speech signal and is thus a preferred signal for coding (see Section III).

To model the time-varying characteristics of the speech signal, the LPC analysis procedure is repeated every few hundredths of a second. This process is generally referred to as *short-time spectral analysis.*

Another important property of the speech signal is the manner in which the vocal tract is excited. The excitation can be either voiced (in which case the vocal cords are vibrating and produce a periodic excitation) or unvoiced (in which case the vocal cords are not vibrating and a noise-like excitation is generated by air rushing through a narrow constriction somewhere in the tract). For voiced sounds, the key characteristic of the source is the fundamental frequency, which is the frequency of vocal-cord vibration. Another important speech parameter is the status of speech/non-speech activity (i.e., an indication of the presence or absence of speech).

There are several ways to determine whether a segment of speech is voiced and to determine the periodicity if it is voiced. In the time domain, one effective way is to compute the autocorrelation function of the speech signal. If the autocorrelation function exhibits no strong peak, then the speech segment is unvoiced. If a strong peak is observed, then the lag at which it occurs indicates the fundamental period of the signal (which is the reciprocal of the fundamental frequency). Another method for estimating the fundamental frequency of voiced sounds, which capitalizes on the regularity in the spectrum, is the so-called method of cepstral analysis. The cepstrum is the (inverse) Fourier transform of the log spectrum of the speech signal. Regularity in the log spectrum, such as repetitive spikes at essentially equal spacing (corresponding to the fundamental frequency), results in a clear spike in the cepstrum at a location (the "quefrency" index) corresponding to the pitch period. The cepstrum is also often used as a feature measurement in speech recognition algorithms.

III. SPEECH CODING

The goal of speech coding is to transform a sampled speech waveform into a digital representation that allows efficient transmission and storage of the signal. Such a transformation is performed by an algorithm or device called a *coder.* After storage or transmission, an approximation of the original signal may be resynthesized by means of a decoder. The process of coding and decoding, in general, introduces a processing delay and also results in a certain loss of fidelity. Generally, to achieve higher coding efficiency (i.e., lower bit rate) one must accept lower speech quality and longer processing delay. The three dimensions—coding efficiency, processing delay, and quality—are interrelated. The constraint of acceptable cost or algorithmic complexity also influences the achievable efficiency. The trade-off between these aspects must be adjusted according to the demands of a specific application. For instance, a delay of even a few seconds might be quite acceptable if the coded speech is to be stored and then retrieved at a later time. On the other hand, for telephony between humans, a total delay (due to processing plus that due to transmission) of over 200 msec would make fluent two-way communication quite difficult.

Quality of speech is a subjective measure, often expressed in terms of the so-called mean opinion score (MOS). The MOS is obtained by averaging the quality judgment from a pool of human listeners in response to a set of stimuli (speech samples) produced by the coder/decoder being tested. The subjective quality judgment is usually measured on a five-point scale ranging from 1 (unacceptable) to 5 (excellent). An MOS of 4 or above is generally regarded as high quality. A descriptive taxonomy of speech quality is as follows:

- Broadcast quality (FM bandwidth of ~7 kHz with MOS greater than 4)
- Toll quality (telephone bandwidth of ~3.2 kHz with MOS around 4)
- Communications quality (military and mobile radio with MOS around 3)

Speech intelligibility is another measure that may be used to evaluate coding/decoding algorithms, particularly those algorithms that attempt very low bit rates. Intelligibility is often tested by a diagnostic rhyme test (DRT) or diagnostic alliteration test (DALT). These tests measure errors made by human listeners between similar monosyllabic words. The DRT uses rhyming word pairs (e.g., *m*et vs. *n*et) and the DALT uses alliterative word pairs (e.g., fla*t* vs fla*k*). The words are chosen so as to uniformly cover six major linguistic attributes (e.g., voicing, nasality, etc.). Speech processing may introduce distortions that lead to increased confusions between such similar words as judged by human listeners. Except in some rare applications, speech processing algorithms need to be able to maintain a minimum DRT score of 90% to be acceptable.

Analysis methods in speech coding are generally based on two approaches. A *waveform coder* encodes the speech such that when the signal is reconstituted at the receiver for playback, the original waveform is reproduced as faithfully as possible. A *model-based speech coder*, or *vocoder*, transforms the speech into a representation that models human speech production. These two approaches represent the two ends of the technological spectrum that covers most speech coder designs.

The most widely used method of waveform coding is called pulse coded modulation (PCM), which was historically motivated by transmission applications (and hence the term modulation). In linear PCM, the amplitude of each sampled speech signal is approximated by a binary integer (i.e., an integer expressed in terms of 0s and 1s) that takes on one of a number of values uniformly spread across the amplitude range. To be able to faithfully reproduce a speech signal without detrimental perceptual effects, at least 4096 values are needed, which means a 12-bit linear PCM representation. To take advantage of the nonuniform amplitude distribution of the speech samples, a linear PCM system is generally preceded by amplitude compression during encoding and followed by amplitude expansion during decoding. Two such nonlinearities are commonly used in today's digital telephony—namely, the A-law and μ-law companders (compressor-expander). The resulting so-called A-law or μ-law PCM is more efficient than a straightforward linear PCM, and requires around 8 bits per sample to achieve a similar quality. For telephony applications, the speech signal is usually sampled at 8 kHz and encoded with 8-bit companded PCM, with a resulting transmission rate of 64 kb/sec.

Speech is a slowly varying signal and displays substantial correlation among the values of adjacent samples. A coding method that makes use of this correlation is thus generally more efficient than simple PCM encoding. The simplest way to utilize the adjacent-sample correlation in speech is the scheme of differential PCM (DPCM), which does not encode the speech signal directly but rather the difference between the current and the previous samples. When the coding scheme is made adaptive to certain time-varying properties of the signal (rather than remaining fixed for all time, independent of the signal characteristics), still higher efficiency is possible. The method of adaptive DPCM (or ADPCM) allows toll-quality transmission of speech at 32 kb/sec. Note that these simple differential schemes incur only a single sample (0.125 msec) of delay.

More sophisticated coding schemes, such as the adaptive predictive coding (APC) method, use more previous speech data samples to better predict the current speech sample value, resulting in a residual signal of much reduced variance that is easier to encode. The idea of using a linear weighted combination of the previous sample values to predict the current value gained tremendous attention in the late 1960s and early 1970s (see Section II). This is the method of linear predictive coding (LPC), in which a set of optimal weight coefficients (the so-called predictor coefficients) are calculated at a regular rate, say once every 20 msec, to minimize the variance (or energy) of the residual within a short window of speech samples. Studies in articulatory modeling were able to relate the predictor to the vocal tract shape and its resonance structure (or the formants), enabling LPC to become one of the most successful methods in digital speech processing.

The concept of model-based coding has its root in the original source-tract framework of Homer Dudley. In the source-tract model, key components are the parameters that define the tract filter, and the pitch and voicing information that defines the excitation signal. In the encoder, these components are obtained by applying speech analysis techniques to the signal and the resulting parameters are quantized with efficient coding schemes for transmission. The speech signal may be resynthesized from these parameters (the quantized spectrum and the quantized excitation signal), which vary relatively slowly with time. By transmitting these data at a slower rate (normally less than 30 data values once every 20 msec) than the original sampled waveform (normally 8 sample values every msec), the resulting vocoder thus provides a potential for speech transmission with a much higher efficiency. The advent of LPC makes analysis and representation of the time-varying spectral shape of the corresponding vocal tract straightforward and reliable. LPC-based vocoders operating at 2.4 kb/sec have been in military deployment for secure communications since the 1970s. At this bit rate, the coarse buzz–hiss model is usually used to represent the excitation, and the quality of the resynthesized speech is quite limited with an MOS normally below 3 and a DRT score in the low 90% range.

Many new digital speech coders fall into the category of a hybrid coder, in which the concept of vocal tract representation is integrated with a waveform-tracking scheme for the excitation signal to achieve a quality suitable for telephony applications. These include the multipulse excitation LPC coder, the code-excited LPC (CELP) coder, and the low-delay CELP (LD-CELP) coder. These coders normally operate at bit rates ranging from 4.8 to 16 kb/sec.

Another major advance in speech coding, namely vector quantization, took place in the late 1970s. The basic idea of vector quantization is to encode several samples (i.e., a vector), as opposed to a single sample (i.e., a scalar), at a time. Shannon's communication theorems provided strong motivations for the use of vector quantization. Furthermore, a number of algorithms were developed to automatically design the collection of reconstruction vectors (the codebook) according to the data distribution, so as to minimize the distortion incurred as a result of coding. Many recent speech coders employ vector quantization. When applied to model-based coding, vector quantization vocoders allow transmission of digital speech at 800 b/sec and below, with acceptable speech quality for special communications needs (e.g., in certain military applications).

For speech coding to be useful in telecommunication applications, coding systems have to be interoperable, thus the coding scheme has to be standardized before deployment. Speech coding standards are established by various standards organizations such as the International Telecommunications Union (ITU), the Telecommunications Industry Association (TIA), the Research and Development Center for Radio Systems (RCR) in Japan, the International Maritime Satellite Corporation (Inmarsat), the European Telecommunications Standards Institute (ETSI), and some government agencies.

Figure 3 illustrates the performance of various speech coders in terms of the perceptual quality of the decoded speech at the corresponding operating bit rate. The companded PCM coder at 64 kb/sec, the ADPCM coder at 32 kb/sec, the LD-CELP coder at 16 kb/sec and the algebraic-coded CELP (A-CELP) coder at 8 kb/sec are all capable of achieving an MOS of more than 4, thus are readily usable in telecommunication applications. The CELP at 4.8 kb/sec has an MOS of slightly less than 4 and is useful in several applications such as secure voice communication. The LPC vocoder at 2.4 kb/sec can only achieve an MOS of around 2.5 and is so far used only in military applications. A new multiband excitation LPC coder (ME-LPC), in which the characteristics of the excitation are determined for each frequency band separately, is able to operate at 2.4 kb/sec, and yet provides a perceptual quality approaching that of CELP at twice the bit rate. The chart also shows that the current challenge, in terms

of the rate-distortion performance of a coding scheme, lies at and below 4 kb/sec, where the perceptual quality drops radically.

The advent of the Internet and the possibility of packet telephony using the Internet Protocol (IP), the so-called voice-over IP (or VoIP), gives rise to another challenge in speech coding. The added dimensions from the new packet networks (as opposed to the traditional synchronous telephony network hierarchy) are

1. Network delay and delay jitter, which vary substantially depending on the network traffic conditions.
2. Packet loss due to congestion created by multiple services running on a common infrastructure.

These new applications call for a coding scheme that performs robustly even under these potentially adverse conditions.

IV. SPEECH SYNTHESIS

Modern speech synthesis is the product of a rich history of attempts to generate speech by mechanical means. The earliest known device to mimic human speech was constructed by Wolfgang von Kempelen over 200 years ago. His machine consisted of elements that mimicked various organs used by humans to produce speech—a bellows for the lungs, a tube for the vocal tract, a side branch for the nostrils, etc. Interest in such mechanical analogs of the human vocal apparatus continued well into the twentieth century. In the latter half of the nineteenth century, Helmholz and others began synthesizing vowels and other sonorants by superposition of harmonic waveforms with appropriate amplitudes. A significantly different direction was taken by Dudley in the 1930s with his discovery of the carrier nature of speech, and its corollary, the source-tract model shown in Fig. 1. By using a keyboard to control the time-varying filter and the choice of excitation for the system shown in that figure, he was able to synthesize quite good-quality, fluent speech.

The first *digital* speech synthesizer was demonstrated at Bell Laboratories around 1967. In some sense, that synthesizer was a throwback to von Kempelen's machine, with one major difference. Instead of *manipulating* a mechanical model, the synthesizer *computed* what a mechanical model would do if it were implemented. On the basis of rules derived from a study of human speech production, a computer program computed the sequence of shapes that a vocal tract would have to go through in order to generate speech corresponding to any text presented as an input. From these shapes, and the knowledge of the appropriate acoustic excitation (periodic pulses for voiced sounds,

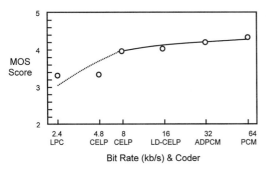

FIGURE 3 Speech quality in terms of MOS for various coders at typical bit rates.

noise-like excitation for unvoiced sounds), the program solved the wave equation with appropriate boundary conditions to compute the acoustic pressure at the lips. Finally, an electrical signal with the same waveform as the computed pressure was applied to a loudspeaker to produce the desired speech.

This method of speech synthesis has a strong appeal because it mimics the way a human produces speech. However, in spite of considerable effort, it has not yet proven possible to automatically generate good-quality speech from arbitrary text by this method. This is because it has not been possible to derive rules that work correctly in all circumstances. So far, good synthesis by rule seems to require too frequent *ad hoc* modification of the rules to be useful in practice. Also, it is difficult to generate certain speech sounds (e.g., bursts in sounds like /k/ of *k*itten) by rule.

Modern text-to-speech synthesis (TTS) is based on a much less fundamental but much more effective procedure called *concatenative synthesis.* Basically, a desired speech signal is assembled with "units" selected from an inventory compiled during a training phase of the synthesizer. The units are acoustical representations of small sub-word elements (e.g., phonemes, diphones, frequently occurring consonant clusters, etc.). The representation itself can take several alternative forms (e.g., the LPC along with the residual as described earlier). The complete process of concatenative synthesis, from text input to speech output, consists of several steps. These are outlined in Fig. 4 for one such system for the synthesis of English. The details would differ for other languages, but the general framework would be similar for a large class of languages.

Starting with the input text in some suitable format (e.g., ASCII) the text is first normalized. Normalization consists of detecting blank spaces, sentences, paragraphs, and capital letters and converting commonly occurring abbreviations to normal spelling. The abbreviations include symbols (such as $, %, &), as well as titles (e.g., Mr., Mrs., Dr.), abbreviations for months (Jan., Feb., etc.), and so on. Note that, as illustrated in the figure, the same abbreviation (Dr.) can give rise to different letter sequences (Doctor in the first example and Drive in the second), depending on the context.

The next step shown in Fig. 4 deals with the problem that a given string of letters must be pronounced differently depending on the part of speech or the meaning. So, for instance, "lives" as a verb is pronounced differently from "lives" as a noun, and "axes" is pronounced one way if it is the plural of "axe" and another way if it is the plural of "axis." There are many such confusions that the parser needs to disambiguate.

At the output of the syntactic/semantic parser, the text has been normalized and the intended pronunciation of all the words has been established. At this point, the pronunciation dictionary is consulted to determine the sequence of phonemes for each word. If a word does not exist in the dictionary (e.g., a foreign word or an unknown abbreviation), then the synthesizer falls back on letter-to-sound rules that make a best guess at the intended pronunciation.

Once the phoneme sequence has been established, the synthesizer specifies prosodic information—that is,

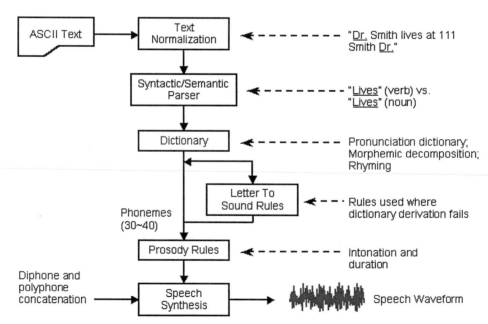

FIGURE 4 Block diagram of an English text-to-speech system.

information regarding the relative durations of various speech sounds and the intensity and pitch variations during the course of the utterance. Such information is critical to the generation of natural-sounding speech. If the wrong words are stressed in a sentence, or if the durations are inappropriate, the speech sounds unnatural. Not only that, inappropriate stress can alter the meaning of a sentence. Similarly, without a natural variation of pitch, the utterance might sound monotonous. Also, inappropriate pitch variation might, for instance, change a simple declaration into a question.

The final box in Fig. 4 does the actual synthesis of the speech. Its task is to select the appropriate sequence of units from the inventory; modify the pitch, amplitude, and/or duration of each unit; and concatenate these modified units to produce the desired speech waveform.

Currently, there are several directions in which concatenative text-to-speech synthesis is being extended and improved. One major effort is concerned with the collection of the inventory of units. Until recently, because of considerations of computational complexity and memory requirements, inventories had one (or at most a few) token for each needed unit. This token was then modified in pitch, amplitude, and duration as required by the context. Considerable effort was spent in the optimal selection of the tokens. More recently, the trend has been to have very large inventories in which most units might appear in many contexts, and with many different pitches and durations. With such an inventory, units can be concatenated after much less modification than was needed with the earlier inventories. Of course, much larger memory is required, and also the problem of searching the large inventory requires much more computation. Fortunately, both available memory and processor speed are increasing at a very rapid rate and becoming quite affordable.

Another direction is towards "multilingual" TTS. Clearly, the synthesis of a given language will have many features not shared with other languages. For example, word boundaries are not marked in Chinese; the grouping and order in which the digits of a given number are spoken in German are very different from those in English; the number and choice of the units inventory need to be different for different languages; and so on. However, rather than creating a collection of language-specific synthesizers, multilingual TTS aims at developing more versatile algorithms. The ultimate aim is that all language-specific information should be in tables of data, and all algorithms should be shared by all languages. Although this ideal is unlikely to be achieved in the foreseeable future, its pursuit focuses research more on language-independent aspects of synthesis.

Another direction of interest is that of adapting synthesis to mimic the speech of a specified person. Clearly,

speech synthesized in the manner described above will sound somewhat similar to that of the person who contributed the units in the inventory. Can one make it sound like the speech of any specified person? This appears to be highly unlikely. However, it might be possible to make the synthesizer sound like several different voices by manipulating the prosodic rules and making systematic modifications of the given inventory.

We conclude this section by noting that concatenative synthesis is by no means to be considered as the ultimate in speech synthesis. Although the quality of speech generated by this method is better than that of other methods, it is by no means an accurate mimic of human speech. Ultimately, we believe, accurate modeling of the human vocal apparatus might still be necessary to achieve a significantly higher level of speech quality.

V. SPEECH RECOGNITION

Speech recognition by machine in a limited and strict sense can be considered as a problem of converting a speech waveform into words. It requires analysis of the speech signal, conversion of the signal into elementary units of speech such as phonemes or words, and interpretation of the converted sequence in order to reconstruct the sentence or for other linguistic processing such as parsing and speech understanding. Applications of speech recognition include a voice typewriter, voice control of communication services and terminal devices, information services such as voice access to news and messages, and price inquiry and order entry in telecommerce, just to name a few. Sometimes the area of speech recognition is extended to include "speech understanding," because the utility of a speech recognizer often involves understanding of the spoken words in order to initiate a certain service action (for example, routing a call to an operator).

Until the 1970s, speech recognition was mostly considered to be a speech analysis problem. The fundamental belief was that if a proper analysis method were available that could reliably produce the identity of a speech sound, speech recognition would be readily attainable. Researchers in acoustic phonetics in the past advocated this deterministic view of the speech recognition problem, citing such examples as, "A stitch in dime saves nine" (in contrast to "A stitch in time saves nine"), which they believed can be recognized correctly only via the use of acoustic-phonetic features. This view may be appropriate in a microscopic sense (e.g., to distinguish the *isolated* distinction between /d/ and /t/ in the above example) but does not address the macroscopic question of how a recognizer should be designed such that on average (in dealing with all input sounds), it achieves the lowest error rate.

The introduction of the statistical pattern-matching approach to speech recognition, which matured in the 1980s and continues to flourish into the 21st century, helps set the problem on a solid analytical basis. Statistical pattern matching and recognition are motivated by Bayes' decision theory, which asserts that a pattern recognizer requires a knowledge of the statistical variation in the observations in order to be able to achieve the lowest recognition error probability. Error probability means, essentially, how likely it is that the recognizer will make a mistake, on average, over all unknown observations (i.e., samples within the recognition test set). The knowledge of statistical variation is expressed in the form of a model, whose parameters have to be learned from real data. These data, called the *training set*, are usually a large collection of utterances with known (or manually labeled) identities. A model is a probability measure and can be considered a typical "pattern" with associated statistical variances. The process of learning the statistical regularities as well as the associated variation from the data is often referred to as "training" the recognizer. Since the error probability, or error rate, is the most intuitive and reasonable measure of the performance of a recognition system, modern speech recognizers use Bayes' theory as the basic design principle.

The most successful model used for characterizing the statistical properties of speech today is the hidden Markov model (HMM). A hidden Markov model is a doubly stochastic process. Locally in time, a sequence of observations of a speech signal is described by one of a set of random processes. Globally, the statistics of the transitions between these random processes is described by a finite-state Markov chain. This model has been found to be quite suitable for the speech signal, which indeed displays two levels of variation, one due to the uncertainty in realization of a particular sound and the other due to the uncertainty in sequential changes from one sound to another. Today, most if not all of the automatic speech recognition systems in the field are based on the hidden Markov model technique.

Despite the advances of the last three decades, speech recognition research is still far from achieving its ultimate goal—namely, automatic recognition by machine of unconstrained speech from any speaker in any environment. There are several key reasons why this remains a challenging problem. One important factor is the difficulty of choosing the set of speech units used for recognition. The optimal choice seems to depend upon the task. Units as large as words and phrases are ideal for limited task environments (e.g., dialing telephone numbers, voice command words for text processing); however, these large speech units are totally unsuitable for speech recognition involving large vocabularies and/or continuous, naturally spoken sentences. There are two main reasons for this.

First, a prohibitively large amount of data needs to be collected and labeled for establishing the necessary statistical knowledge. (Consider the extreme case of using sentences as the units in continuous speech recognition. The number of possible sentences becomes astronomical even with a medium vocabulary size. Collecting all the realizations of these sentences is infeasible if not outright impossible.) Second, even if such a large amount of data were to be collected, the complexity of the search during the recognition phase would be prohibitive. For such applications, sub-word recognition units (e.g., dyad, diphone, syllable, fractional syllable, or even phoneme) become necessary, and techniques for composing words from such sub-word units (such as lexical access from stored pronouncing dictionaries) are needed. Since such units are often not well articulated in natural continuous speech, they can be decoded only with a high error rate. Grouping these error-prone sub-word sequences into sequences of words from a standard lexicon becomes a very difficult task.

Another important factor affecting performance is the size of the user population. So-called speaker-trained or speaker-dependent systems adapt to the voice patterns of an individual user via an enrollment procedure. For a limited number of frequent users of a particular speech recognizer, such an enrollment procedure works reasonably well and generally leads to good recognizer performance. However, for applications where the user population is large and the usage casual (e.g., users of automatic number dialers or order-entry services), it is infeasible to train the system on individual users. In such cases, the recognizer must be speaker independent and able to adapt to a broad range of accents and voice characteristics.

Other factors that affect recognizer performance include:

- Complexity of the vocabulary
- Transmission medium over which recognition is performed (e.g., over a telephone line, in an airplane cockpit, or using a hands-free speakerphone)
- Task limitations in the form of syntactic and semantic constraints on what can be spoken
- Cost and method of implementation

This partial list of factors illustrates why unconstrained automatic speech recognition by machine is unlikely to be achieved in the foreseeable future. Much rigorous research remains to be done.

Currently, speech recognition has achieved modest success by limiting the scope of its applications. Considerable success has been achieved with systems designed to recognize small- to moderate-size vocabularies (10 to 500 words) in a speaker-trained manner in a controlled environment with a well-defined task (e.g., order entry,

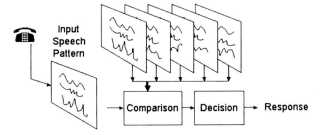

Stored Vocabulary Patterns/Models

Input Speech Pattern → Comparison → Decision → Response

FIGURE 5 Automatic speech recognition using a pattern recognition framework.

voice editing, telephone number dialing) and spoken cooperatively (as opposed to conversational utterances full of disfluencies such as partial words, incomplete sentences, and extraneous sounds like "uh" and "um"). Such systems have been, for the most part, based entirely on the techniques of statistical pattern recognition, as discussed above. In the simplest context, each word in the vocabulary is represented as a distinct pattern (or set of patterns) in the recognizer memory as shown in Fig. 5. (The pattern can be either a sample of the vocabulary word, stored as a temporal sequence of spectral vectors, or a statistical model of the spectrum of the word as a function of time.) Each time a word or sequence of words (either isolated or connected) is spoken, a match between the unknown pattern and each of the stored word vocabulary patterns is made, and the best match is used as the recognized string. In order to do the matching between the unknown speech pattern and the stored vocabulary patterns, a time-alignment procedure is required to register the unknown and reference patterns properly because speaking rate can vary between wide limits. Several algorithms based on the techniques of dynamic programming have been devised for optimally performing the match.

Tables I and II provide a summary of typical speech recognition performance for systems of the type shown in

Fig. 5. Table I shows average word error rate for context-free recognition of isolated words for both speaker-dependent (SD) and speaker-independent (SI) systems. (By context-free, we mean that no task syntax or semantics is available to help detect and correct errors.) It can be seen that, for the same vocabulary, SD and SI recognizers can achieve comparable performance. It is further seen that performance is more sensitive to vocabulary complexity than to vocabulary size. Thus, the 39-word alpha-digits vocabulary (letters A–Z, digits 0–9, three command words) with highly confusable word subsets—such as B, C, D, E, G, P, T, V, Z, and 3—has an average error rate of about 5 to 7%, whereas an 1109-word basic English vocabulary has an average error rate of ~4%.

Table II shows the typical performance of connected-word recognizers applied to tasks with various types of constraints. The highly accurate performance on connected digits (both speaker-trained and speaker-independent) is a result of significant advances in the training procedures for recognition. In the table, perplexity is defined as the average number of words that can follow an arbitrary word. It is usually substantially less than the size of the vocabulary because of the syntactic, grammatical, and semantic constraints of the task-specific language. It is a rough measure of the difficulty of the task and may be estimated from a statistical model of the language that defines the sequential dependence between words. For example, an N-gram model defines the probability of a sequence of N words (or units) as observed in a large collection of sentences in the language. Language models are obtained essentially by word counting procedures.

The Naval Resource Management task involves a particular kind of language used in naval duties and is highly stylized. Sentences in the Air Travel Information System are query utterances focusing on flight information such as flight time, fare, and the origin and destination of the intended travel. The vocabulary size reflects the number of city or airport names included in the database. The grammatical structure in the query sentence is quite limited compared to many other application domain languages. In the *Wall Street Journal* transcription task, the input is "read" speech recorded with a reasonably good-quality microphone. Read speech is known to be much easier to recognize than spontaneous, conversational speech. The Broadcast News transcription task involves signals often referred to as "found" speech such as radio or television news announcements. The signal may be degraded due to noise and distortion, and the spontaneity is generally somewhat higher than for "read" speech. The degradation in recognition accuracy is a clear indication that this is a more difficult task. These tasks, however, are all considered far easier than transcribing and understanding truly spontaneous conversational speech.

TABLE I Performance of Isolated Word Recognition Systems

Task/application	Vocabulary size	Mode	Word accuracy
Digits	10	SI	~100%
Voice dialer words	37	SD	100%
Alpha-digits plus command words	39	SD	96%
		SI	93%
Computer terms	54	SI	96%
Airline Words	129	SD	99%
		SI	97%
Japanese city names	200	SD	97%
Basic English	1109	SD	96%

TABLE II Performance Benchmark of HMM-Based Automatic Speech Recognition Systems for Various Connected and Continuous Speech Recognition Tasks or Applications

Task/application	Vocabulary size	Word Perplexity	Accuracy
Connected digit strings	10	10	~100%
Naval Resource Management	991	<60	97%
Air Travel Information System	1,800	<25	97%
Wall Street Journal transcription	64,000	<140	94%
Broadcast News transcription	64,000	<140	86%

The systems listed in Table II use phoneme-like statistical unit models to represent words, as do many speech recognition software packages offered on the market for personal computing (PC) applications such as dictating a letter. These phoneme-like unit models are also modified by the context in which they appear. For example, the /l/ in /e-l-i/ is different from the /l/ in /o-l-o/. Many systems employ thousands of such context-dependent unit models in continuous speech recognition. The benchmark results in Table II are based on very extensive speaker-independent training. On the other hand, PC-based software systems normally require speaker adaptation. The user of PC-based speech recognition software is asked to speak a designated set of sentences, ranging from 5 minutes to a few hours in duration, so that parameters of the baseline system can be modified for improved performance for the specific user.

As illustrated above, the performance of current systems is barely acceptable for large vocabulary systems, even with isolated word inputs, speaker training, and favorable talking environment. Almost every aspect of continuous speech recognition, from training to systems implementation, represents a challenge in performance, reliability, and robustness.

Another approach to machine recognition and understanding of speech, particularly for automated services in a limited domain, is the technique of keyword spotting. A word-spotting system aims at identifying a keyword (or a key phrase), which may be embedded in a naturally spoken sentence. This is very useful because it makes the interaction between the user and the machine more natural and more robust than with a rigid command-word recognition system. Experience shows that many (infrequent) users of a speech recognition system in telecommunication applications often speak words or phrases that are not part of the recognition vocabulary, thus creating so-called out-of-vocabulary (OOV) or out-of-grammar (OOG) errors. Rather than attempting to recognize every word in the utterance, a word-spotting system hypothesizes the presence of a keyword in appropriate portions of the speech utterance and verifies the hypothesis by computing two scores, one for the hypothesized portion of the speech signal and the keyword model, and the other for the former and a background speech model. These scores are subject to a ratio test against a threshold for the final decision. By avoiding forced recognition decisions on the unrecognizable and inconsequential regions of the speech signal, the system can accommodate natural command sentences with good results as long as the number of keywords is limited (say, less than 20). Today, a word-spotting system with five key phrases (*collect, credit card, third party, person-to-person*, and *operator*) has been extensively deployed. By automating billions of the telephone calls traditionally categorized as operator-assisted calls, the system provides tremendous savings in operating cost.

VI. SPEAKER VERIFICATION

The objective of speaker verification is authentication of a claimed identity from measurements on the voice signal. Applications of speaker verification include entry control to restricted premises, access to privileged information, funds transfer, credit card authorization, voice banking, and similar transactions.

There are two types of voice authentication: one verifies a talker's identity based on talker-specific voice characteristics reflected in spoken words or sentences, and the other based on the content of the spoken *password* or *passphrase*, such as a personal identification number (PIN), social security number, or mother's maiden name. In the former case, the test phrase may be in the open and even shared by talkers in the population, while in the latter case the password information is assumed to be known only to the authorized talker. We often refer to the former as speaker verification (SV) and the latter as verbal information verification (VIV).

Research in speaker verification has a much longer history than VIV and in the past has encompassed studies of acoustic and linguistic features in the speech signal that carry the characteristics of the talker. In recent years, the dominance of a statistical, data-driven pattern matching approach (see Section V) has substantially changed the research landscape. The assumption is that a powerful statistical model such as the hidden Markov model would be able to automatically "discover" talker-specific characteristics provided that sufficient spoken utterances from the particular talker are available to allow such learning. Indeed, the statistical modeling technique has made a similar impact on speaker verification as on speech recognition. It is based on a principle almost identical to the one illustrated in Fig. 5 in which the vocabulary models are replaced with the talker models or voice patterns. Each of the stored talker-specific models has to be trained according to an enrollment procedure before the system is put to use. In telephony applications these models may be stored in some remote location. For verification, the talker makes an identity claim (e.g., an account number) and speaks a test phrase (or simply the account number itself). The system compares the input speech with the stored model or pattern for the person with the claimed identity. On the basis of a similarity score and a carefully selected decision threshold, the system can accept or reject the speaker. The features useful for verification are those that distinguish talkers, independent of the spoken material. In contrast, the features useful for speech recognition are those that distinguish different words, independent of the talker. The decision threshold is often made dependent on the type of transaction that will occur as a result of the verification process. Clearly, a more stringent acceptance threshold is required for the transfer of money from one account to another than for reporting a current balance in a checking account. Key factors affecting the performance of speaker verification systems are the type of input string, the features that characterize the voice pattern, and the type of transmission system over which the verification system is used. Best performance is achieved when sentence-long utterances are used in a relatively noise-free speaking environment. A state-of-the-art system using a text-dependent test sentence is capable of achieving a 1 to 2% equal error rate (when the threshold is adjusted such that the probability of false acceptance and that of false rejection becomes equal). Conversely, poorer performance is achieved for short, unconstrained spoken utterances in a noisy environment (4 to 8% equal error rate using text-independent isolated words).

The challenge in speaker verification is to build adaptive talker models based on a small amount of training that perform well even for short input strings (e.g., one to four words). To achieve this goal, more research is needed in the area of talker modeling as well as in the area of robust analysis of noisy signals.

A VIV system stores the confidential information about a talker in a profile. When an identity claim is made, the talker is asked one or a series of questions based on the stored information; for example, "Please say your PIN" or "What is your birth date" or "What is your mother's maiden name?" The system verifies the information by scoring the spoken response with a composite speech model of the expected password or pass phrase constructed from the appropriate unit phone models. Such a system can achieve perfect (0 error) verification after three rounds of questions and answers.

VII. SPEECH ENHANCEMENT

During transmission from talker to listener, a speech signal may be degraded in a variety of ways. In this section, we will discuss some of these degradations and some of the methods that have been devised to deal with them.

Most speech enhancement algorithms are concerned with minimizing the perceptual effects of the degradation that results from additive noise. Noise can get mixed into a speech signal in several ways. For instance, speech may originate in a noisy environment, such as in a noisy airport, railway station, or shopping mall. A microphone placed in such an environment will pick up the sum of the desired speech and the ambient noise. Examples of practical devices that may be subjected to such degradation include cellular handsets, speakerphones, etc. Even low levels of ambient noise can become a problem in multipoint teleconferencing. This is because each participant in such a teleconference hears the ambient noise from all remote locations. Thus, assuming similar conditions at all locations, the noise would be about 10 dB higher in a teleconference among ten participants than in a two-way conference. Noise can also enter a conversation electrically in transmission lines, filters, amplifiers, etc., that are part of a telephone circuit.

In most such situations, all that is available to a speech enhancement algorithm is a single microphone signal representing the sum of the desired speech signal and the noise. Hence, to improve the quality of such degraded speech, the enhancement algorithm must operate "blindly." That is, it must operate in the absence of any prior knowledge of the properties of the interfering noise. It must estimate the noise component from the noisy signal and then attempt to minimize the perceptual effects of the noise.

There are, however, important applications in which additional information is available, besides the noisy speech signal. A prime example of such applications, and one that has been of interest for several decades, is echo cancellation on long-distance telephone circuits. Due to impedance mismatches on such circuits, an undesirable echo of a speech signal gets added to the desired signal being transmitted. This echo is the "noise" that needs to be eliminated from the "noisy" signal. If the speech signal that is responsible for the echo is observable, then the echo can be estimated and subtracted from the noisy signal. Another example of such applications might occur when a microphone is used to record the speech of a talker in a room with a noise source—say, an air-conditioning duct. If it is possible to place a secondary microphone very close to the noise source, then its output can be used to estimate and cancel the noise present at the recording microphone.

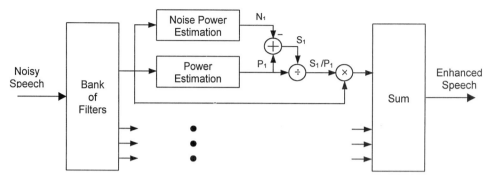

FIGURE 6 A speech enhancement system based on spectral subtraction.

A third type of speech enhancement is required to overcome the loss of intelligibility due to hearing loss that, in general, increases with age. Hearing loss is becoming more and more important in telephony, as life-expectancy increases and the percentage of older people in the population increases.

Finally, speech enhancement is of interest in situations where the speech may originate in a reasonably quiet environment but is to be received by a listener who is in a noisy environment. An example of such a situation is announcements over the public address (PA) system at a railway platform. In such cases, the noise itself is not under the control of a speech enhancer. However, if some estimate of the noise is available, then it is possible to preprocess the speech signal so as to improve its intelligibility as well as its spectral balance when listened to in the noisy environment.

Let us take a closer look at each of these four applications of speech enhancement.

In the case when only the noisy speech signal is available, speech enhancement is accomplished by systems that may be typified by the block diagram shown in Fig. 6. The noisy speech is passed through a bank of contiguous band-pass filters that span the frequency range of the speech signal (say, 200 to 3500 Hz for telephone speech). At the output of the ith filter, the short-term power, P_i, is an estimate of the power spectrum of the noisy speech at the center frequency of the filter. If the desired speech signal and the noise are uncorrelated, then $P_i = S_i + N_i$, where S_i is the power of the speech signal and N_i is the power of the noise at that frequency. Suppose for a moment that the noise power, N_i, can be estimated independently. (One way to estimate the noise power is to detect time intervals during which speech is present and measure the output power only during the other intervals. Several algorithms are known for detecting the presence of speech.) Then, $S_i = P_i - N_i$ is an estimate of the signal power. The ith channel signal is next adjusted by multiplying it with a gain factor S_i/P_i. Finally, these adjusted signals are all added together to yield the enhanced speech. It is seen that the output of a channel is attenuated more if it has more noise in it. This is what provides the enhancement.

Instead of introducing time-varying gains in the channels of a filter bank as shown in Fig. 6, an alternative approach derives the control signals of the type shown in Fig. 1 from the given noisy speech and then resynthesizes speech from these control signals. The first step as before is to estimate the signal power, S_i, in each channel and from it the signal amplitude (which is just the square root of the signal power). These estimated amplitudes characterize the time-varying filter of Fig. 1. In a parallel path, the given noisy signal is analyzed to determine intervals in which the signal is voiced and the intervals in which it is unvoiced. During voiced intervals, the fundamental frequency is determined. It turns out that such analysis can be performed reliably even in the presence of noise which is only 6 to 10 dB below the level of the speech signal. In this manner, the characteristics of the excitation source of Fig. 1 are determined. The enhanced signal is now synthesized as in Fig. 1.

All single-microphone enhancement systems in use today are variations of one of the two methods described above. They differ in the detailed characteristics of the band-pass filters, the method of estimating the noise power, the method of determining the fundamental frequency, etc.

These speech enhancement methods reduce the level of noise but unfortunately also introduce artifacts of their own. The main artifact is what is known as *musical noise*, which consists of bursts of periodic signals with randomly varying frequency. Many people prefer the original noisy signal to the processed signal because the musical noise is quite disconcerting to listen to. However, this type of artifact is almost inaudible when the noise in the original signal is at least 10 dB or more below the signal level. Also, enhancement system parameters can be adjusted so as to trade off this type of musical noise distortion against the residual noise. It should also be noted that except in

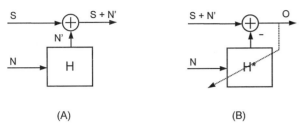

(A)   (B)

FIGURE 7 (A) Model for signal contamination; (B) Setup for canceling contamination from the desired signal.

one study with specialized test signals, these methods have *not* been shown to increase intelligibility. The main advantage of these speech enhancement methods is that they reduce fatigue that results from prolonged listening to noisy speech.

Algorithms for the second type of enhancement—echo cancellation or noise cancellation—are illustrated in the block diagram shown in Fig. 7. In Part A of this figure, the signal (S) is the desired signal to be recorded or transmitted. To that signal, an undesired signal (N') gets added. The signal N' is not known; however, it is known that N' is the output of a (hitherto unknown) linear filter (H), whose input is a known signal (N). In the echo cancellation example mentioned earlier, N' would be the echo generated by the signal N, and H would be the filter representing the transfer function of the echo path. In the other example, N' would be the noise signal at the recording microphone due to the noise N at the air-conditioning duct as picked up by the secondary microphone, and H would be the acoustic transfer function of the room from the air-conditioning duct to the recording microphone. An enhancement system for the echo or noise cancellation system is shown in part B of Fig. 7. The box marked H* denotes an estimate

of the filter (H). If we could make H* equal to H, then clearly the noise would be removed and the output signal (O) would be the same as the desired noise-free signal (S). To drive the transfer function of the filter H* towards that of H, an iterative gradient algorithm is used. During silent intervals of the speech signal (S), the algorithm adjusts H* in such a way as to drive output O towards zero. If O can be made exactly zero in those intervals, then H* provides a good estimate of H. If now the adaptive algorithm is turned off whenever the speech signal is present, then the noise portion of the signal, O, continues to be canceled, hence O approximates the desired noise-free signal S.

Such algorithms have been highly successful in the echo cancellation application. Indeed, several million such devices for echo cancellation have been deployed in various telephone systems. Their usage for the other application—reduction of room noise—is far less. First of all, there are far fewer occasions where this type of noise reduction is needed. Besides that, except at low frequencies, the noise at the recording microphone in a room cannot be accurately modeled as a single signal passed through a linear filter. In most cases, the noise at the microphone either lacks coherence with the noise reference (recorded at another location) or results from multiple, diffuse noise sources which cannot be represented by a single reference signal. Only in some special cases, e.g., in an airplane cockpit, or in a racing car, communication from the pilot or the driver can be somewhat improved by this technique.

Let us now turn to the third source of degradation mentioned previously, i.e., the degradation due to hearing loss. Fig. 8(a) shows the threshold of hearing as a function of frequency for a person with normal hearing. Also shown in that figure is the threshold of discomfort as a function of frequency. Any sound louder than this threshold would be

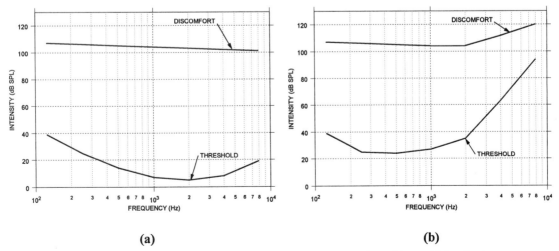

(a)   (b)

FIGURE 8 Typical thresholds of hearing and discomfort: (a) normal ear, (b) ear with hearing loss.

painful and can be harmful to the ear. Figure 8(b) shows the same two curves for a person with hearing loss. This figure is for illustration purposes only; there are many types of hearing loss with different characteristics.

The main noteworthy feature common to all types of hearing loss is that the threshold of hearing becomes higher than that for a normal ear, but the threshold of discomfort remains more or less unchanged. Also, in general at high enough intensities, the loudness at any frequency is the same for a person with normal hearing as for one with hearing loss. This last phenomenon is known as *loudness recruitment*.

A vast majority of hearing aids in use today provide only linear amplification, with frequency-dependent gain. From Figs. 8a and b, it is clear that this is not a satisfactory solution. If the gain is adjusted to be correct for a low-level sound, it will be too high for a high level sound, and vice versa. What is needed is to compress the dynamic range of sound at any frequency to fit the reduced dynamic range due to the hearing loss. Thus, an amplifier is needed such that its gain depends not only on the frequency but also on the signal level at that frequency so as to provide the appropriate compression. Hearing aids are now available that filter the speech signal into a number of bands (usually limited to two or three due to computational complexity) and provide a compression in each band that is appropriate for the hearing loss being corrected. This technique is called *multiband compression*.

Much improvement is still possible in this type of hearing aids. We do not yet know the optimal control strategy for the signal-dependent gain nor do we know the best procedures for fitting such a hearing aid. The fundamental reason for this is that the information summarized in Figs. 8a and b is obtained from tests with stationary signals and is inaccurate for a signal such as speech, which is continually changing. Also, note that the high gain required for low-level sounds also amplifies the noise. For this reason, hearing aids tend to be unsatisfactory in noisy environments. In the future, with better understanding of the hearing process, as well as with the availability of smaller and faster digital processors with low power requirements, hearing aids that combine noise reduction with signal-dependent gain will no doubt be available and will provide much better solutions to the problem of hearing loss.

Finally, let us consider the preprocessing of a speech signal to make it more intelligible in the presence of ambient noise. It turns out that the noise induces something akin to hearing loss in a normal hearing person. At any frequency, the threshold of hearing is elevated while a signal sufficiently above the noise level sounds almost as loud as in the absence of noise. Of course, the amount of loss depends on the level of noise at each frequency. Thus, if the clean speech signal is amplified with a gain that has the appropriate dependence on frequency and intensity, it can be made much more intelligible when listened to in the presence of noise. Of course, the speech will still be quite noisy, since the noise is not changed.

VIII. CONCLUDING REMARKS

This cursory overview of digital speech processing has been intended to highlight recent advances, current areas of research, and key issues for which new fundamental understanding is needed. Future progress in speech processing will surely be linked closely with advances in computation, microelectronics, and algorithm design.

SEE ALSO THE FOLLOWING ARTICLES

ARTIFICIAL INTELLIGENCE • COGNITIVE SCIENCE • DIGITAL CONTROLLERS • DIGITAL FILTERS • HUMAN-COMPUTER INTERACTION • INFORMATION THEORY • SOFTWARE ENGINEERING • SPEECH SYNTHESIS BASED ON LINEAR PREDICTION

BIBLIOGRAPHY

Flanagan, J. L. (1972). "Speech Analysis, Synthesis, and Perception," Springer-Verlag, New York.

Furui, S., and Sondhi, M. M., eds. (1991). "Advances in Speech Signal Processing," Dekker, New York.

Jayant, N. S., and Noll, P. (1984). "Digital Coding of Waveforms," Prentice-Hall, Englewood Cliffs, NJ.

Rabiner, L. R., and Gold, B. (1975). "Theory and Application of Digital Signal Processing," Prentice-Hall, Englewood Cliffs, NJ.

Rabiner, L. R., and Juang, B. H. (1993). "Fundamentals of Speech Recognition, Prentice-Hall, Englewood Cliffs, NJ.

Rabiner, L. R., and Schafer, R. W. (1978). "Digital Processing of Speech Signals," Prentice-Hall, Englewood Cliffs, NJ.

Direct Current Power Transmission, High Voltage

Narain G. Hingorani
Electric Power Research Institute

A. Figueroa
San Diego Gas & Electric Co.

GLOSSARY

Back-to-back tie Converter facility that ties two or more ac systems at one location, thus combines two or more converters.

Bipolar Having two poles; one positive and the other negative.

Converter Equipment for converting ac to dc or dc to ac.

Converter station High-voltage dc substation consisting of converters, transformers, filters, control equipment, and all other equipment needed by a facility for converting ac to dc or dc to ac or both.

Inverter Equipment for converting dc to ac.

Monopolar Having one pole, positive or negative.

Rectifier Equipment for converting ac to dc.

Thyristor Silicon switching device with anode, cathode, and gate, used as an element of a high-power valve.

Valve Assembly of thyristors in series and parallel

necessary to obtain the required voltage and current rating, along with necessary dividing circuits and thyristor level electronics.

HIGH-VOLTAGE DIRECT CURRENT

HIGH-VOLTAGE DIRECT CURRENT (HVDC) transmission in its simplest form is a means of transmission that involves conversion of alternating current (ac) to direct current (dc), transmission by dc, and conversion of dc back to ac, thus connecting two ac systems asynchronously. A more complex HVDC system, known as a multiterminal system, may interconnect more than two ac systems through converters. Another form of HVDC transmission, known as back-to-back tie, does not have a dc line, but merely connects two ac systems through converters. A variety of reasons make it necessary or economical to interconnect ac systems in this way. Early HVDC schemes involved the use of mercury arc valves, which have now been replaced by semiconductor devices known as thyristors. The HVDC transmission capacity ranges from a few tens of megawatts to several thousand megawatts.

I. THE BEGINNING OF HVDC

Use of direct current for power transmission dates back to the 1880s, the early days of electricity, when a controversy raged between Thomas Edison and his backers on the side of well-proven dc technology and George Westinghouse and his backers on the side of the new technology of alternating current. The latter was based on many innovative ideas, particularly those of Nicola Tesla.

Although there was an all-or-nothing struggle between ac and dc for a brief period during the 1880s and 1890s, it never really amounted to a complete switchover to ac; by the same token, dc will never assume a dominant role in energy transfer. Each has its unique advantages, depending on the application. The inventions in ac technology such as ac motors, ac generators, and transformers were truly revolutionary. The ac generators and motors proved to be economical and more reliable than dc generators and motors, but the invention responsible for the victory of ac was the transformer, a low-cost and reliable piece of equipment. Like magic, a hunk of steel and wire, to put it crudely, could transform ac power from one voltage level to another.

Eventually ac won, primarily because then, at low voltages, dc could not be efficiently transmitted over long distances. Just before the turn of the century, ac was selected for harnessing the energy at Niagara Falls because dc could not be transmitted economically to Buffalo, just 22 miles away. One of the principal reasons for the present use of dc, or HVDC as it is called, is that it can now be transmitted more economically than ac for long distances. It is necessary to understand the technological reasons behind this change to appreciate the scope of HVDC transmission.

Think of today's electrical system. The ac power generated at low voltage, say 20 to 30 kV, is transformed to higher voltages (hundreds of kilovolts) for interconnections and transmission to load centers; then it is transformed down to subtransmission level near cities, then down to tens of kilovolts in distribution systems of urban areas; and finally it is reduced to less than 10 kV at the street corners and to 230/110 V for home use. This is done so that power can be delivered to the user in a safe manner and at a low price. The transformer made possible the delivery of power cheaply via interconnections, as well as pooling of power generating plants and transmission of their output over significant distances.

The fact was, and still is, that sending power from one place to another by a transmission line, regardless of power level, is cheaper and more efficient by dc than by ac. If the power could be transformed from high-voltage ac to high-voltage dc and back to ac conveniently and economically, then HVDC could be used for transmission of power.

Many pioneering steps were taken during the early twentieth century in France, England, Germany, and the United States, using rotary converters and later thermionic valves. In 1926 a 17-mile HVDC transmission line was installed between Mechanicville and Schenectady, New York, transmitting 5.25 MW at 30 kV, connecting 40- and 60-Hz ac systems, using electronic valves. Important experiments were continued in Germany, Switzerland, and Sweden before and during World War II to improve the converter technology, particularly using mercury arc valves. In the German case, an experimental transmission system of 15 MW at 100 kV was built; this was intended to be a prototype of a 60-MW, 400-kV system with a transmission length of 110 km. This activity was interrupted at the end of the war. Then Sweden, under the leadership of Uno Lamm, regarded as the father of HVDC, initiated the modern era of HVDC by pioneering the world's first commercial HVDC transmission system, the Gotland Scheme, to transmit 20 MW at 100 kV over a distance of 100 km of submarine cable; this transmission could not have been achieved with ac. The Gotland Scheme was based on high-voltage mercury arc valves, using grading electrodes patented by Uno Lamm, and only one cable with current return through the earth. Based on this Swedish converter technology, several HVDC schemes were installed throughout the world; principal among them was the Pacific DC Intertie—850 miles, long rated 1440 MW, and operated at ±400 kV—which was energized in 1971 for transmission of hydropower from the Pacific Northwest to the Pacific Southwest.

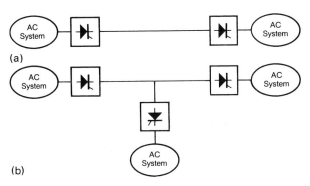

(a)

(b)

FIGURE 1 Concept of high-voltage direct-current (HVDC) power transmission. (a) Two-terminal scheme. (b) Three-terminal scheme.

Coming up in the wings was a new device called a silicon-controlled rectifier (SCR), now known as thyristor, a solid-state silicon chip device invented in the United States. Considerable advances in the thyristor device and HVDC valve by use of series-connected thyristors led to the last and most needed thrust in HVDC technology: cost reduction and increased reliability.

The HVDC technology, as it is known now, is a link connecting two or more ac substations through converters and dc lines (Fig. 1). Power is converted from ac to dc, or dc to ac as required, with power transmitted over the dc lines. A dc transmission line may be an overhead line, underground cable, or submarine cable, or any combination of these. In fact, in some cases, there may not be any transmission line. Such a link is referred to as back-to-back tie; the purpose of such a tie, which involves just the converters, is to interconnect two or more ac systems that cannot be connected otherwise by ac.

Realistically speaking, HVDC will not entirely replace ac transmission, but there is no doubt that it will play an important role for utilities, which are constantly striving to avail themselves of the lowest-cost energy.

To clarify the commonly used converter technology for HVDC systems: each converter has the capability to transfer power from ac to dc, in which case it operates as a rectifier, and it can also transfer power from dc to ac, in which case it operates as an inverter. The same converter, with full gate-firing angle control of valves, can operate either as a rectifier or as an inverter; the operating mode is simply a matter of the firing angle of the converter valves.

II. WHY USE HVDC?

Transmission, whether ac or dc, is built to serve several functions: to deliver power from power plants to load centers, to interconnect power plants and systems so that the least costly generation is used, to satisfy every system's

peak demand from minimum investment in generation, and to ensure a reliable power supply to the customers.

Once the need for transmission is established, the question is whether to use ac or dc. The following considerations would be taken into account; any one or more of these could be enough to decide in favor of dc.

1. Two ac systems may have different frequencies, different frequency controls, or enough stability problems that it is not possible to connect them with ac. Clearly, in such situations HVDC opens a way to exchange power, deferring the need for more power plants in either system. Power flow between Quebec and the United States, between Japan's 60- and 50-Hz systems, between England and France, and in similar regional and international connections provides opportunities, for HVDC. Similarly, the United States has three separate ac power systems: Western, Eastern, and Texas. These are interconnected by dc and would not otherwise be connected for a variety of reasons. For example, power transfer between the Western and Eastern ac systems by ac would require massive ac interconnections from the West Coast to the Midwest to keep the two systems running together without overloading the ac interconnections. With HVDC, practically any size link can be established where required.

2. A dc line is cheaper than an ac line (Fig. 2). Basically, dc transmission requires just one conductor if earth current is acceptable for the return path; otherwise it requires two conductors for bipolar dc (double circuit) transmission. With ac three conductors are needed for a single three-phase circuit and six conductors for a double three-phase circuit. Also, power flow on an ac line is not only the real (active) power, but includes a considerable amount of reactive power due to series inductance and shunt capacitance of the line. This means that full conductor capacity cannot be utilized for active power, and losses are increased by the flow of reactive power. This is particularly true of ac cables, which have several times more shunt capacitance

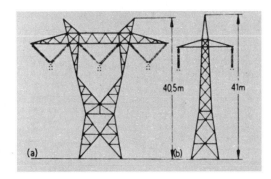

FIGURE 2 Comparison of towers for (a) an 800-kV ac line and (b) a ±500-kV dc line having the same transmission capacity.

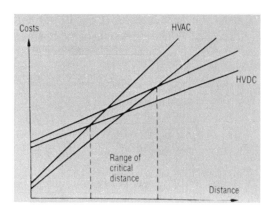

FIGURE 3 Cost comparison of ac and dc transmission.

than overhead lines, such that for a cable distance of about 30 miles, the reactive current could be enough to fully load the cable. It can be said with confidence that at any power level or transmission distance, using cable or overhead, line for line, dc is cheaper than ac. Thus, putting it hypothetically, if converter cost and losses were zero, all the transmission would be dc. Since the converter cost and system losses cannot be zero, there must be enough line length to compensate for the additional converter cost and system losses. Estimates for the so-called break-even distance (Fig. 3) between ac and dc range from 400 to 500 miles for overhead lines and 30 to 50 miles for underground cables.

3. One can transmit two to four times as much power through a given right-of-way with dc than with ac. Since rights-of-way are often difficult to obtain or expand, consideration should be given to building dc lines or converting ac lines to dc. For example, a double-circuit ac line (six conductors) can be converted to three bipolar dc circuits (six circuits) with more than triple the power capacity, higher reliability, and lower percentage losses.

4. Power through HVDC transmission is controlled by converter controls, which ensure that the power flows as required by the dispatcher. Power flow in an interconnected ac circuit is a complex matter depending on line impedances, phase angles, frequency, voltages at the buses, and so forth. Thus if one utility wants to interchange power with another utility, it may not be possible to do so just by building an ac line of the required capacity between the two utilities. Consideration and cooperation of all other interconnected utilities in the region are needed, since the power may actually flow through other lines. In addition, reactive power flow must be accounted for among various utilities. In general, any ac interconnected system requires long-term planning. If unexpected changes affect the required power and available sources or sociopolitical factors create uncertainties in long-term planning, then dc transmission can be a very useful tool to overcome unexpected power flow difficulties.

5. Use of dc can help to stabilize interconnected ac systems. Because the power flow over dc can be rapidly controlled, special controls can be designed to modulate or change the power flow on the dc tie in order to prevent overload and oscillations in the sending or receiving ac system. Thus a power system that incorporates some dc can be more stable than one without dc. Increasing the stability limit of a parallel ac line can, in turn, lead to an increase in the capacity of that line.

III. BASIC TYPES OF HVDC SYSTEMS

A. A Typical Converter

In a typical HVDC transmission system including a dc line, a converter station (Fig. 4) is made up of (1) a converter, which includes converter transformers, converter valves, and the cooling equipment; (2) an ac yard, which includes ac switchgear to connect converters to the ac lines, ac filters, and capacitor banks for reactive power compensation; (3) a dc yard, which includes smoothing reactors, dc filters, and dc switchgear connecting converters to the dc line; and (4) control and protection equipment. There are also various other components such as surge arresters and monitoring and measuring equipment, which are located throughout the converter station. A converter station may contain one or more 12 pulse converter units; as shown in Fig. 5, each converter unit consists of

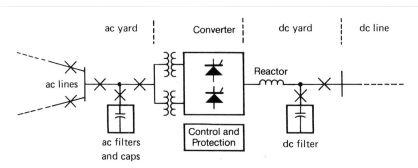

FIGURE 4 Simplified diagram of a typical converter station.

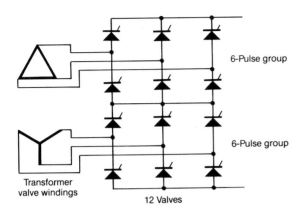

FIGURE 5 Twelve-pulse converter unit.

12 valves connected to two 3-phase secondaries (valve windings). Each set of six valves and a valve winding form a three-phase, full-wave bridge connection (its operation is discussed in Section IV). Two bridges, one with a delta-connected valve winding and the other with a wye-connected valve winding, form a 12-pulse converter group. Because of overall cost and performance considerations, this 12-pulse converter group has been universally adopted as a converter unit for thyristor-based HVDC schemes, in contrast to the adoption of the single-bridge (six-pulse) converter units used in mercury arc valve-based HVDC schemes.

For staged expansion, a converter station may include 12-pulse units in parallel or series, as shown in Fig. 6.

Valves and control and protection equipment are located in a building with a temperature- and humidity-controlled environment.

B. Back-to-Back Ties

In a back-to-back tie (Fig. 7) there is no dc line, and therefore both 12-pulse converters are located at the same site. There is no dc yard except the dc smoothing reactor. The

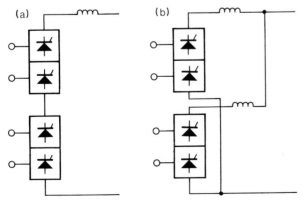

FIGURE 6 Two 12-pulse converter units in (a) series and (b) parallel.

valves and control and protection of both converters are located in one building. With no dc line, the line losses are not a factor; therefore the dc voltage is usually not very high (under 100 kV), and the dc current and voltage rating for the required power capacity is optimized, resulting in the lowest cost for the supplier. Of course, there may be more than one back-to-back tie in one complex, if reliability considerations require that loss of one tie will not lead to total loss of capacity; indeed, the remaining tie can accommodate most of the lost capacity through its short-time or winter overload capability. There may also be multiterminal back-to-back ties consisting of three or more converters connecting the corresponding number of ac systems, but these also require some dc switchgear.

C. Monopolar Earth Return System

In the monopolar earth return system (Fig. 8a), the dc line needs only one conductor because with dc the earth can be used as the current return conductor. Two earth electrodes are provided, one at each end, several miles away from the converter stations and connected to the converters with low-voltage lines called electrode lines. The electrodes are located several miles away from long pipelines to avoid the possibility of corrosion, and such pipelines are also provided with cathodic protection to eliminate corrosion, due not only to HVDC but also to natural earth currents and other electrolytic effects. The current does not follow the path of the line because the earth resembles a ball of complex resistances, and current is distributed in accordance with the resistances. Therefore, pipelines near the surface will not see any significant current level beyond the first few miles of the electrodes.

Usually, a monopolar system would be adopted only for submarine cable transmission because of the high cost of laying additional neutral submarine cable. In other cases, such as overhead transmission and land cable transmission, the monopolar earth return might be used only as the first stage in development of a bipolar scheme.

For power flow in one direction, say left to right in Fig. 8a, the conductor has positive voltage polarity to earth, and for power flow in the other direction the conductor current flows in the same direction, but the conductor has negative polarity to earth. Either side of the converter loop may have the conductor, and the other side would then be the earth path.

A dc reactor is needed at each end of the line, and if the HVDC transmission includes an overhead line, dc filters are also needed in order to decrease harmonic current flow in the dc line. In general, the dc reactors limit the rate of rise of current during faults, and both the dc reactors and the filters limit the harmonic currents in the dc line. In cable transmission, dc filters are not needed because there is no possibility of telephone interference from cable circuits.

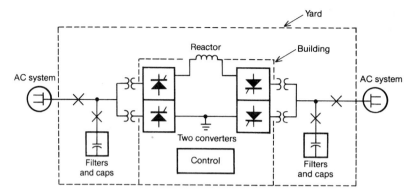

FIGURE 7 Simplified diagram of a back-to-back tie.

D. Bipolar HVDC System

A bipolar system combines two monopolar systems, with the dc conductor on opposite sides, as shown in Fig. 8b, for a two-terminal system. For power flow in one direction, one pole has positive polarity to earth and the other has negative polarity to earth. For power flow in the other direction, the two poles reverse their polarities. If the current in the two poles is equal, the current flow in the earth is zero; if not, then the difference flows through the earth.

This bipolar system is the most commonly used configuration. The advantage is that it has twice the capacity of a monopolar system, but the cost of building a bipolar overhead line should not be much more than the cost of a monopolar line of the same capacity. Equally important is the fact that no ground current flows when both poles are in operation with equal current; in effect, the current flows in the bipolar loop. When one pole is blocked (taken out of service) the other pole can continue to operate with return current through the earth. Since corrosion, if any, is

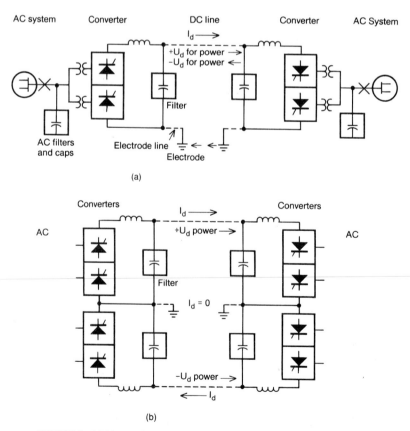

FIGURE 8 (a) Monopolar earth return system. (b) Bipolar HVDC system.

a long-term effect, occasional operation of one pole with earth return has not been objectionable to pipeline owners. The two poles may also be operated occasionally at different currents, if some unusual condition prevents operation of one pole with full current.

In some unusual cases, the use of ground for even a few hours a year may not be suitable. If the defective pole conductor still retains some low-voltage insulating capability, that conductor can be made to carry the return current. This monopolar operation is called the monopolar metallic-return operation.

E. Bipolar Metallic-Neutral System

If the earth current is not acceptable, or the distance between the system terminals is short, or an earth electrode location is not available, then the transmission line may be provided with a third conductor for a bipolar metallic-neutral system. Only one point of this conductor is grounded, and naturally this conductor requires a very low voltage insulation.

In all of the above, 12-pulse converters may be connected in series or parallel in each pole, for reliability considerations or staged development.

F. Multiterminal HVDC System

A variety of multiterminal HVDC systems can be used to meet different needs. An HVDC line may be tapped on the way to feed power along the way (see Fig. 1b). In another case, one can have the two-pole converter on one side of a bipolar system located at two different sites and/or connected to two different ac systems; this may be done when one power plant is required to feed power to two receiving locations. More complex multiterminal HVDC systems are shown in Fig. 9; these include the following:

1. Bulk power transmission—where low-cost energy from several power plants is transmitted over a long distance to different ac systems. A typical arrangement is shown in Fig. 9a, where power flow is usually in one direction.

2. AC network interconnection over a long or medium distance—where generation/load balancing and sharing of spinning reserves are of primary concern. As shown in Fig. 9b, dc systems may operate in conjunction with available ac ties.

3. Reinforcement of an ac network—where limited ac expansion possibilities exist. Energy from a new power plant is fed to different locations of an ac system, usually metropolitan. Power flow, as shown in Fig. 9c, is in one direction over short distances.

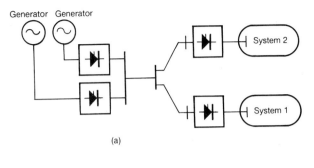

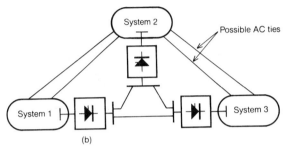

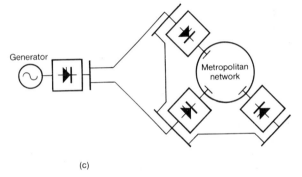

FIGURE 9 Similar potential applications for multiterminal and two-terminal dc systems. (a) Bulk power transmission, radial dc network. (b) AC network interconnection, mesh dc network. (c) Reinforcement of an urban network.

IV. EXPLANATION OF CONVERTER OPERATION

For conversion, the three-phase bridge connection (Fig. 10), also referred to as a six-pulse group, has been accepted as the best connection for HVDC converters because it provides full utilization of the transformer, valves, and other converter equipment.

To understand the operation of this six-pulse group as a rectifier, consider the circuit of Fig. 10a as consisting of two 3-phase half-wave circuits, shown in Fig. 10b. One circuit consists of three valves, 1, 3, and 5, with output between their common cathode bus and the neutral; the other consists of the other three valves, 2, 4, and 6, with output between their common anode bus and the neutral. The output voltage of the circuit with valves 1, 3, and 5 is

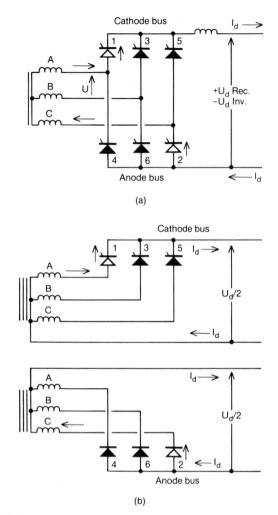

FIGURE 10 Six-pulse converter (a), which is the equivalent of two 3-phase half-wave converters (b).

Now when valve 3 fires at X_2, the current transfers from valve 1 to valve 3, and the time taken for this transfer (angle u) is called the commutation angle. The transfer of current is complete at X_3, and the output voltage thereafter follows the voltage of phase B. Valve 5 in phase C is fired in the same way at X_5, to take over current from valve 3 (phase B). Valve 1 then takes over from 5 and so on.

The other three-phase half-wave circuit consisting of valves 2, 4, and 6 (Fig. 10b) connected to phases C, A, and B, respectively, goes through the same cycle of commutation. But since the valves are reversed (cathodes connected to phases), the operation is on the opposite half-cycles. The full-wave circuit of Fig. 10a is obtained by combining the two half-wave circuits of Fig. 10b, eliminating one of the secondaries and the neutral connections.

For the whole circuit, therefore, the voltage between the anode and cathode busbars is represented by the voltage between the two thick lines in Fig. 11 and corresponds to six-pulse operation, as shown in Fig. 12a. The operating sequence of the group circuit as a whole is as follows.

Just before instant X_2, valves 1 and 2 are conducting through phases A and C. At instant X_2, valve 3 fires and takes over from valve 1, after which valves 3 and 2 conduct

shown by the thick line at the top in Fig. 11a, and similarly with current in valves 1, 3, and 5 in Fig. 11b. It is assumed that direct current is smooth.

Consider the operation from instant X_1 (Fig. 11), when valve 1 is conducting the direct current as shown by arrows in Fig. 10b. After instant X_1, the voltage of phase B becomes positive with respect to phase A, and with valve 1 conducting, this means that the anode of valve 3 is positive with respect to its cathode. Valve 3 can, therefore, fire any time after X_1, as soon as the firing pulse is given. In the wave form shown in Fig. 11, valve 3 is not fired at X_1, but at X_2, an angle α (called the delay angle) later. Up to that time valve 1 continues to conduct, with the instantaneous output voltage falling as shown by the thick line of phase A. It may be noted that firing of valve 3 could have been delayed up to X_6 (delay angle of 180°), after which the valve 3 voltage would become negative.

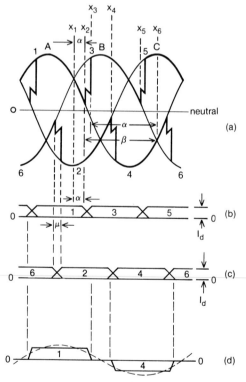

FIGURE 11 Operation of a bridge as a rectifier. (Numbers correspond to valves in Fig. 10.) (a) Output voltage; (b) current in valves 1, 3, and 5; (c) current in valves 2, 4, and 6; and (d) current in phase R.

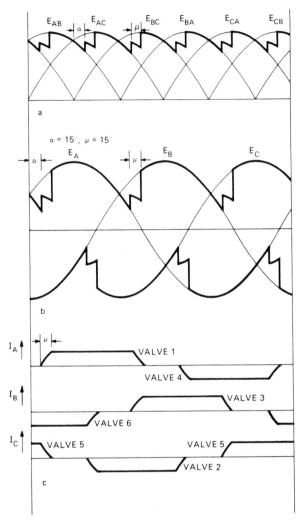

FIGURE 12 Voltages and currents of a 6-pulse group in rectifier operation. (a) DC line voltage constructed from line-to-line ac voltages. (b) DC line voltage constructed from line-to-neutral ac voltages. There is no scale change from (a) to (b). (c) Converter transformer valve-side ac currents.

through phases B and C. Then at instant X_4, 60° after X_2, valve 4 fires and takes over from valve 2, after which valves 3 and 4 conduct through phases B and A. After another 60°, valve 5 takes over from valve 3 and so on. The valve numbers correspond to the sequence of firing, and the valve currents are shown in Figs. 11b and 11c.

The average output voltage U_d, calculated by integrating and averaging the wave form, is given by

$$U_d = \frac{3\sqrt{2}}{2\pi} U (\cos \alpha + \cos(\alpha + u))$$

$$= \frac{3\sqrt{2}U}{\pi} \cos \alpha - \frac{3\omega L}{\pi},$$

where U is the phase–phase ac voltage, α the delay angle, u the commutation angle, and ωL is the transformer reac-

tance per phase. Output voltages with different values of delay angle α, neglecting commutation angle, are shown in Fig. 13.

The ac and dc current relationship is defined by

$$I_{ac} = \frac{\sqrt{6}}{\pi} I_d.$$

The solid line wave form in Fig. 11d shows the current wave form in phase A, which is made up of blocks of positive and negative current wave forms through valves 1 and 4, respectively. It is well known that any nonsinusoidal wave form is made of a series of sinusoidal wave forms of various frequencies. The nonsinusoidal wave form shown in Fig. 11d consists of the fundamental (60 Hz), shown by the dashed line, and the harmonics.

Currents in the other two phases, B and C, are the same except that they are phase-shifted by 120° with respect to each other (Fig. 12). Looking at the phase relationship of the fundamental current in phase A with respect to the voltage wave form of phase A (Fig. 11), it is noted that the current lags the voltage approximately by an angle of delay α, represented by the vector diagram in Fig. 13a.

If the dc current is maintained constant, the ac current wave form does not change in amplitude with change in the angle of delay, but simply changes in phase angle. As the angle of delay increases, the output voltage decreases as a cosine function of the angle of delay α. On the ac side, however, this is simply reflected by an increase in the power factor angle $\phi \simeq \alpha$ and a consequent reduction in the power component $I \cos \phi$ of the ac current, as shown in Fig. 13b.

It can be seen from Fig. 11 that commutation from, say, 1 to 3 can take place only as long as the voltage across valve 3 is positive, that is, phase B is positive with respect to A, a period of 180° between points X_1 and X_6. Angle $\beta = 180° - \alpha$ is called the angle of advance and angle $\gamma = 180° - \alpha - u = \beta - u$ is called the margin angle. As the angle of delay α is increased (angle of advance β decreased), the output voltage decreases, as seen in Fig. 13 (For simplicity, the commutation angle is neglected in Fig. 13.)

At $\alpha = 90°$, $\cos \alpha = 0$, and the average output voltage decreases to zero. Figure 13c shows the ac current vector when $\alpha = 90°$; its power component is zero and all the current is reactive. If α is increased even further, the negative areas of the output voltage exceed the positive areas and the average output voltage becomes negative. When α is sufficiently increased (Fig. 13d) there are no longer any positive areas in the output voltage. Figure 13d shows that when α is increased above 90°, the current lags voltage by an angle greater than 90°, and its power component $I \cos \phi$ is negative and the power flow is reversed.

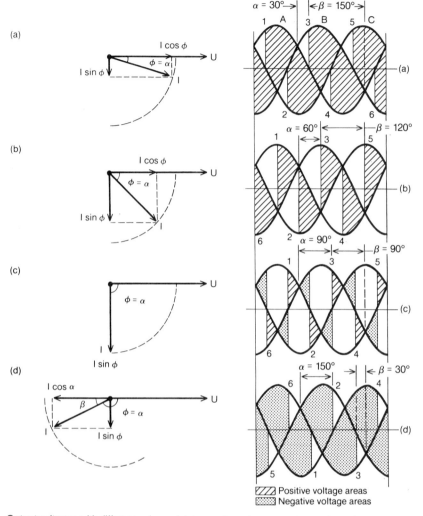

FIGURE 13 Output voltages with different values of delay angle α. Average output voltage is zero. (a and b) Rectifier; (c) zero-output voltage; and (d) inverter.

It is important to note that the current is still assumed to be flowing in the same direction, forced by another dc supply (the other converter working as a rectifier). Figure 14 shows the inverter output voltage along with commutation angles, and is seen to be the opposite of the rectifier wave form (Fig. 11). Valves 1, 3, and 5 conduct during the negative half of the voltage wave forms, whereas valves 2, 4, and 6 conduct during the positive half. Converter operation, when the converter output voltage is negative with current still flowing in the positive direction, is the inverter operation, and the power is being fed from the dc to the ac side.

Thus it is seen that the converter firing angle can be controlled over a wide range, and by this control the output voltage can also be changed over a wide range from positive to negative. The current direction is always the same, so when the output voltage is positive, the converter is a rectifier, and when the output voltage is negative, the converter is an inverter. Correspondingly, on the ac side (Fig. 13), when the ac current flowing from the ac system into the converter is in the fourth quadrant with respect to the ac voltage, the converter is a rectifier; and when the ac current is in the third quadrant, the converter is an inverter.

It is also seen from the vector diagrams of Fig. 13 that the reactive component of the ac current always lags the voltage. The reactive component is small for small angle of delay, increases to a maximum for 90° angle of delay (corresponding to zero output voltage), and decreases again for further increase in the angle.

V. VALVES

The basic element of a valve is a thyristor (Fig. 15), a solid-state device capable of switching a few thousand amperes and holding a few thousand volts 60 times a

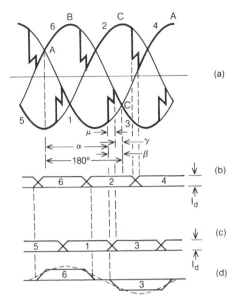

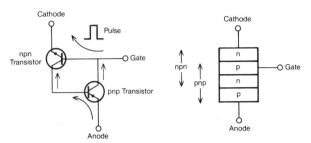

FIGURE 16 Basic structure of a thyristor.

FIGURE 14 Operation of a bridge as an inverter; voltage and current wave forms. (a) Output voltage; (b) current in valves 2, 4, 6; (c) current in valves 1, 3, 5; and (d) current in phase Y.

second for many tens of years, if operated within its rating. The thyristor is a hermetically sealed package enclosing the thyristor cell. Figure 15a shows both an electrical-fired thyristor (left) that has electrical gate-firing leads and a light-fired thyristor that has optical fibers connected to the internal gate. The cell is a silicon device with anode and cathode as the main electrodes

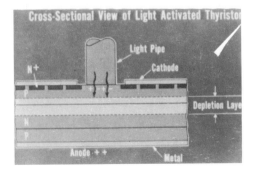

FIGURE 15 Light-fired thyristor.

(Fig. 15b) and a gate for triggering conduction. When the anode is positive with respect to the cathode, the thyristor is ready to turn on, but it turns on only when a small current pulse is passed from gate to cathode for a few microseconds.

The thyristor cell is composed of four layers—p, n, p, and n, as shown in Fig. 16. The end n layer is the cathode, the p layer next to the cathode is the gate, and the end p is the anode. This four-layer device is equivalent to two transistors, a p–n–p and an n–p–n transistor, connected together as shown in Fig. 16. When the anode is positive with respect to the cathode (with no gate signal), a very small leakage current flows from anode to cathode. But when a small positive current pulse is passed from the gate to the cathode, it turns on the first n–p–n transistor, whose anode current then becomes the gate current of the second transistor, which turns on, and the whole device acts like a closed switch from anode to cathode with only a small forward voltage drop ($1\frac{1}{2}$ to 2 V). Figure 15b shows structure of a light-fired thyristor, which is the same except that turn-on is accomplished by a light pulse in the gate region.

Once the thyristor turns on, it stays on until the current through it is brought to zero and a negative voltage is applied for a period of a few tens to a few hundred microseconds for recovery; after that, it can hold its rated positive voltage until the next gate pulse is applied. The thyristor can also withstand a high negative voltage up to its designed level with only a small leakage current flowing. If the positive or negative voltage or the rate of change of voltage exceeds the thyristor design level, the device could be damaged permanently.

In an HVDC valve rated for tens or hundreds of kilovolts, many thyristors are connected in series to provide the required voltage capability. In a 400-kV, 12-pulse converter, each of the 12 valves must withstand approximately 250 kV repetitively, along with frequent overvoltages. If a thyristor is rated for 5 kV, it will be used for a working voltage of about 2 kV of converter rating to allow for overvoltages and protection margins. Thus, a valve in a 400-kV, 12-pulse converter unit (200-kV six-pulse group) will have about 100 thyristors in series. With higher-voltage

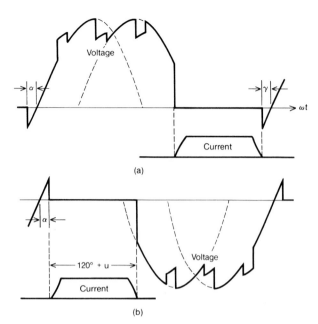

FIGURE 17 Voltage and current wave forms of a valve during (a) inverter and (b) rectifier operation.

thyristors, the number of series-connected thyristors decreases, leading to a less complex valve. When a thyristor fails, it becomes short-circuited and can stay in the valve as long as there are enough thyristors left to continue operation. Usually 2 to 5% extra thyristors are provided for such occurrences. Monitoring is provided to indicate which thyristor has failed, so that timely replacements can be made.

Figure 17 shows the current and voltage wave forms of a valve in an inverter and a rectifier operation. Current flows for a period of 120° plus the commutation angle u. When conducting, the valve voltage is very low, equal to the forward conducting voltage drop of all the series-connected thyristors (1.5. V per thyristor). During the nonconducting period, the valve must withstand a high voltage, which is the voltage between its own phase and one of the dc buses; it has a complex wave form made up of voltage jumps and different parts of the main frequency waves. While the converter is operating as a rectifier it must withstand mostly negative voltage, but when operating as an inverter it must withstand mostly positive voltage. At the start of conduction, the valve must be capable of a high rate of rise of current, and at the end of the conduction, it must withstand a high negative voltage jump with a high rate of rise.

Due largely to inductances and stray capacitances of the transformers, the voltage jumps result in oscillatory overvoltages, which must be damped. The valve design must also ensure that both positive and negative voltages are

shared equally among the thyristors. To ensure reasonably equal voltage division and damping of oscillations, voltage dividing/damping circuits consisting of capacitors and resistors are provided across the thyristor string. Voltage and current ratings must also take into account overcurrents and overvoltages that result from system disturbances.

Modern valves are very efficient; the total losses of all converter valves—consisting of losses due to forward voltage drop in thyristors, losses in damping–dividing circuits, and cooling losses—typically amount to about 0.25% of the converter station rating for all the converter station valves. Indeed, the whole converter station typically has only about 0.75% loss; one third of is in the valves and cooling, one third is in the transformers, and the remaining one third is in the filters, reactors, and so forth.

It is obviously important that when a valve is fired, all thyristors receive a firing pulse simultaneously. To protect against inadvertent uneven voltage distribution, special electronic circuits are provided to safely turn on the thyristor if its forward voltage or rate of change of voltage becomes too high, or thyristors are designed to self-fire safely.

In modern valves an optical firing pulse is sent from a control to each thyristor level through optical fibers (Fig. 18). If a thyristor is light-fired, designed to fire directly by light sent to its gate, then the optical fiber is placed directly over the thyristor gate and the light pulse from the

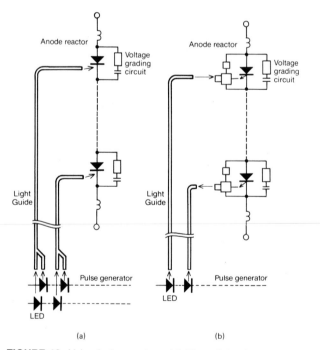

FIGURE 18 Valve fusing system. (a) Direct light triggering. (b) Indirect light triggering.

fiber turns on the thyristor directly. Alternatively, the optical pulse may fire a small pilot light-triggered thyristor connected between the anode and the gate of the main electrical-fired thyristor, so that the anode current of the pilot thyristor provides the electrical gate pulse for the main electrical-fired thyristor. Yet another approach is to convert the optical pulse to an electrical pulse by means of a photodiode and amplifier at the thyristor level and then apply the amplified electrical pulse to the main thyristor.

Structurally, there are a variety of ways in which thyristors are arranged in the valve structure and then in a valve hall to form the whole 12-pulse converter unit. Often a number of thyristors (5 to 10) are housed in a panel, and then several panels are housed in the valve structures. Finally, four valve structures are housed on top of each other, usually in the form of a quadruple valve (Figs. 19 and 20) making up one arm of the 12-pulse converter. Three such quadruple valves are housed in a valve hall to form a 12-pulse converter (Fig. 19).

The valves are cooled in a number of ways (Fig. 21). The simplest method is to cool them by recirculating air, which in turn is cooled in heat-exchangers by a second recirculating coolant, usually a water–glycol mixture; the latter is then cooled in evaporative water coolers. In another method, demineralized, deionized water (which has good electrical insulating qualities) is circulated through pipes passing through all the thyristors and resistors in series–parallel arrays; this water in turn is cooled directly or in heat exchangers by the water–glycol mixture, which

in turn is cooled by evaporative coolers. With water cooling, special measures must be provided to prevent corrosion. Also, measures are taken to minimize the possibility of any leaking water mixing with contamination on the valve structure, become conducting, and cause flashover.

VI. DC OVERHEAD LINES

Figure 22 shows a ±550-kV, bipolar dc line tower. It has two conductors with long insulators, one for each pole. Each conductor is a bundle of three subconductors. Above the main conductors are two ground wires (also known as shield wires), which capture most of the lightning strokes that would otherwise hit the main conductor. In areas where lightning and storms are less frequent, one ground wire is provided for shielding.

The size (cross section) of the pole conductor is determined by balancing the cost of the line against the cost of losses. Since there is no capacitive current flow as in the ac lines, the only losses are resistive ones due to the load current. A larger conductor means lower losses but a higher capital cost, and the final design is based on the least total cost. For this, it is essential to go to higher voltages for higher power unless the line length is only a few tens of miles. Generally, the optimum voltages corresponding to power levels are ±400 kV for 1000 MW, ±600 kV for 2000 MW, and ±800 kV for 4000 MW. Higher voltages, however, increase converter costs by about 1% for every 25-kV increase, so for short distances the optimum voltage will be lower for a given power.

Each pole conductor is divided into a number of subconductors (electrically connected but held a few inches apart by spacers) to reduce the voltage gradient at the conductor surface and hence reduce the radio interference, audible noise, corona losses, and ion generation from the conductors. Of course, the cost goes up with the number of subconductors. A single conductor should suffice for up to 300 kV, two subconductors for 300 to 500 kV, three subconductors for 500 to 600 kV, and four subconductors for 600 to 800 kV. The maximum gradient of 28 kV/cm at the conductor surface may be considered a satisfactory limit.

An optimally designed line conductor would usually have much more thermal capability than its nominal rating (up to twice as much), and this allows for a possible overload capability in an emergency.

Long insulators are of the polymeric type, reinforced with glass fibers, or they consist of a series of low-voltage porcelain or glass insulators locked together in a string. Because the conveters usually hold the short-time switching overvoltages to under 1.7 times the continuous line

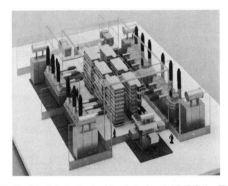

FIGURE 19 Model of a typical back-to-back HVDC tie. The building and walls are made transparent for clear view of the components. Inside the building are the two 12-pulse converters; each structure is a quadruple valve (four valve) converter; three quadruple valves together form a 12-pulse (12 valve) converter. Each 12-pulse converter is connected to converter transformers, located against the building walls as shown, for connections to the ac system on each side. The ac system connections and ac filters are not shown. On their dc side, the two converters are connected together through a smoothing reactor, shown by the near side of the valve hall. On the far side of the valve hall is the control room. (Courtesy of BBC.)

FIGURE 20 A quadruple valve (hanging structure) under test in a high-voltage laboratory. This valve is for the Intermountain Power Project ±550 kV, 1600 MW. (Courtesy of ASEA.)

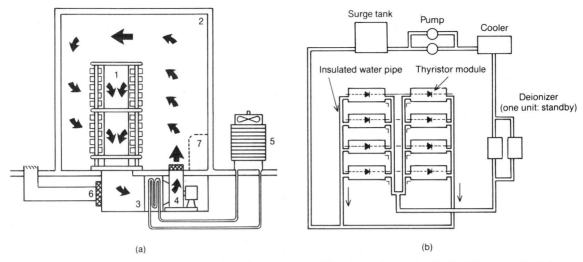

FIGURE 21 Schematic diagrams of (a) air-cooling system and (b) water-cooling system for thyristor valves. Parts in (a) are: 1, thyristor valve; 2, valve hall; 3, heat exchanger; 4, cooling fan; 5, cooling tower; 6, makeup air filter; and 7, aisle for watching the valves.

voltage, the length of the insulators is determined by the continuous line voltage and the type of insulator.

A dc potential tends to attract conducting dust particles, and therefore insulators tend to get dirty over a period of time. Therefore, dc insulators are designed with a long surface creepage path for a given length, hence the need for underskirts and/or larger-diameter insulators.

Adequate separation of each conductor from the tower is also necessary so that flashover does not occur across the air gap when lightning hits the tower or a maximum surge appears on the conductor. Tower height is determined by the minimum clearance required from the ground for safety considerations. Increased height also decreases the electric field and ion density near the ground. However, high towers are more visible and cost more.

VII. DC CABLES

Based on insulation and cooling, there are a variety of cable types: solid-dielectric paper tape, pressurized-oil paper tape, pressurized-gas paper tape, polypropylene–paper–polypropylene laminated tape, cross-linked polyethylene, extruded solid dielectric, and so forth. The cables may be self-contained with a grounded jacket and installed directly in the ground, or they may be of the pipe type with the cable pulled through a pipe through which oil is circulated for cooling. They may also be cooled with demineralized water flowing through the center of a hollow conductor. They are basically similar to ac cables but are specifically designed for dc application. The most economical type for dc is the simplest, the oil-impregnated solid-dielectric paper tape cable, self-contained or pipe

type. Figure 23 shows the cross section of a self-contained medium-pressure oil-cooled cable.

Factors that make dc cable design different from that of ac cables are the following:

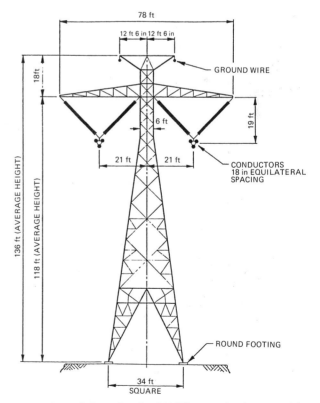

FIGURE 22 Schematic of ±500-kV tower for Intermountain Power Project in the United States. (Courtesy of Los Angeles Department of Water and Power.) Dimensions approximate.

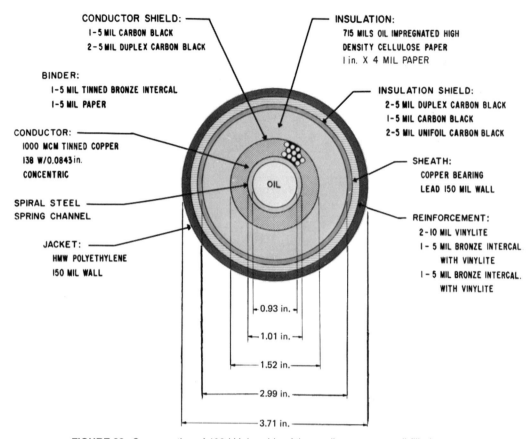

FIGURE 23 Cross section of 400-kV dc cable of the medium-pressure oil-filled type.

1. For ac the radial dielectric stress depends on the permittivity of the insulation, whereas for dc it depends on the resistivity. The permittivity does not change with temperature, and the stress distribution remains unchanged with temperature for ac cables and is always highest at the conductor surface. For dc cables, resistivity changes considerably with temperature. Initially, when cable is cold, the stress is highest next to the conductor. When the cable is hot the stress reverses and is highest near the sheath.

2. The insulation under dc stress retains its internal charge when the cable is de-energized. This charge dissipates with a long time constant of an hour or more. Thus, if the voltage polarity of a dc cable is reversed, the insulation is temporarily subjected to twice the stress, which gradually decreases as the original charge dissipates.

3. Insulation strength under dc is more than twice as high as under ac.

4. Because the voltage does not constantly change as for ac, the dielectric losses in a dc cable insulation are very low.

5. Unlike an ac cable, a dc cable does not have to carry a continuous charging current. Hence the dc cable may carry much more real power current for the same cooling and losses.

AC cable transmission for long distances (over 10 to 20 miles) is particularly difficult because it is necessary to provide shunt reactors every 10 miles or so; otherwise the capacitive charging current will be too large and will fully load the cable. This represents a high cost, and for submarine cable it is also virtually impossible to provide shunt reactors. With dc cable, distance is not a problem, which reduces the overall cost. As mentioned before, the savings for a dc cable over 30 to 50 miles can be enough to pay for the converter costs.

Since a small dc leakage current flows from the conductor to ground through the insulation, the cable sheath or the pipe can corrode over a long period of time. Therefore, appropriate cathodic protection must be provided for the cable with positive voltage polarity. The dc cable with negative voltage polarity will, of course, be self-protecting from corrosion because the leakage current goes in rather than out.

VIII. AC FILTERS AND REACTIVE POWER

AC current (see Fig. 11) resembles rectangular blocks of current made up by switching direct current from one phase to another in rotation. Apart from the main frequency component, the wave forms obviously contain harmonics. These harmonics are of the order of $n = 6k \pm 1$, where k is an integer, substitution for which gives 5, 7, 11, 13, ... The magnitude of each harmonic decreases with n and is approximately $1/n$th of the fundamental. Delta and wye windings have harmonics of equal magnitude; however, harmonics of order $n = 12k \pm 1$, that is, 11, 13, 23, 25, ..., are in phase, while the other harmonics, 5, 7, 17, 19, ..., are in phase opposition for the two windings. These two currents add up on the primary side; thus for a 12-pulse group, the harmonics entering the ac system are $12n + 1$ only. Since modern HVDC converters are large, these harmonics must be effectively filtered or else they may cause interference with open-wire telephone lines running parallel to the power lines and also cause unacceptable system voltage distortion. Therefore, ac filters are essential for any large ac–dc converter.

As discussed before, converters used for HVDC also consume reactive power, that is, current always lags ac voltage, whether they work as rectifiers or inverters. To reduce the reactive power flow to the ac line (in order to decrease ac line losses and voltage drop along the line), capacitor banks are provided at the terminal. In normal operation a converter consumes reactive power that is about 50% of the active power. Thus for a 1000-MW converter station, 450 to 500 megavolt ampere reactive (Mvar) of capacitor banks are needed to obtain a unity power factor on the ac side. Some of these capacitor banks can be converted to filters so that they not only supply 60-Hz reactive power but also filter out harmonics.

Figure 24 shows one possible arrangement of filters and capacitors for a 1000-MW bipolar converter station.

A total of 480 Mvar of the active power supply is divided into two 240-Mvar banks. Each 240-Mvar bank is divided into 30 Mvar of 11th filter, 30 Mvar of 13th filter, 60 Mvar of high-pass filter (for all higher harmonics), and four capacitor banks of 30 Mvar each.

When starting up the dc system from zero power, first only one filter bank of 11, 13, and high pass is connected; then as the load increases, other filters and capacitor banks are switched in one at a time to compensate for the increased reactive power consumption as well as increase the filtering of harmonics.

If the ac system is weak and liable to large voltage fluctuations, a small synchronous condenser or a thyristor-controlled reactive power compensator may be provided. Since the reactive power consumed by the converter is a function of its delay angle, this angle can be varied over a short range to regulate the ac bus voltage in order to smooth out the step changes caused by switching of capacitor banks.

There are many possible variations on these filter–capacitor bank arrangements. AC systems are complex and are frequently changing with line outages, switching, and so forth; they must be properly studied for response to all harmonics, and filters must be designed so that the voltage distortion is kept below 1 or 2% and the telephone interference factor is kept below the required level.

IX. DC FILTERS

The direct output voltage of a converter (see Fig. 13) is not smooth, but consists of 60°-wide sections of ac voltage

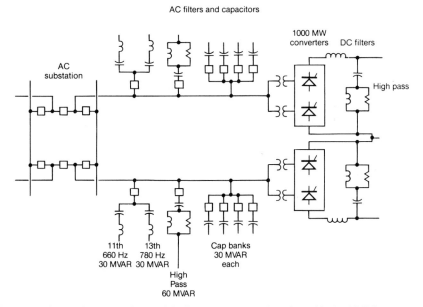

FIGURE 24 A typical ac filter/capacitor and dc filter arrangement for a large bipolar HVDC converter station.

in sequence. In the six-pulse converter there are six such sections per cycle, 360 per second for a 60-Hz system. Analysis shows that the wave form contains dc voltage plus harmonics of order $n = 6k$, where k is an integer, i.e. 6, 12, 18, 24, ... The harmonic amplitude is less for higher harmonics and increases with increase in the delay angle, reaching a maximum at about 90° and decreasing again with delay angles greater than 90° (i.e., within inverter operating range).

In a 12-pulse converter, the 60°-wide sections in the two six-pulse converters, one with delta-connected secondary and the other with wye-connected secondary, are phase-shifted by 30°, because the two secondary voltages are phase-shifted by 30°. When the two voltages are added, the net voltage sections are 30° wide, and the harmonics are of the order $n = 12k$ only, that is, 12, 24, ... The remaining harmonics, 6, 18, ... that are present in the two 6-pulse converters are out of phase and therefore cancel each other out, while the 12, 24, ... harmonics add up for the two 6-pulse groups. Thus the total harmonics for a 12-pulse group are greatly reduced compared to 6-pulse groups for the same dc voltage. Harmonics are also influenced by commutation angle, and they decrease with increasing commutation angle (i.e., increasing load). The maximum amplitude of each harmonic occurs when the delay angle is 90° and the commutation angle is zero. Under normal operation, with a delay angle of about 15°, the 12th harmonic will be about 3% of the dc voltage, the 24th harmonic about 1.5%, with higher harmonics decreasing to very low levels.

These harmonic voltages cause corresponding harmonic currents to flow in the line. The harmonics from the two converter stations of the dc system add up, and in the worst case add up in phase. The current flow is limited by the smoothing inductors and is also largely bypassed by the filters. DC filters, like ac filters, are necessary for overhead lines in order to limit possible inductive interference in any open-wire telephone lines running parallel to the overhead lines (modern communications, i.e., microwave and optical-fiber links, do not have this problem, but the old open-wire communications still exist in some places). DC cable transmission and back-to-back ties do not need filters because the question of interference does not arise.

In bipolar operation, the harmonic currents return through the other pole, largely canceling out the induced interference. In monopolar ground-return operation, the harmonics return through the ground.

In most schemes, large enough filters are provided so that the induced voltage in a parallel open-wire telephone line 1 km long and 1 km from the dc line will be less than 10 mV, an arbitrary practice that seems to have served well so far.

A typical dc filter (Fig. 24) is a single high-pass filter tuned to the 12th harmonic. For improved filtering, a tuned filter for the 12th and a high-pass filter for the 24th and higher harmonics are provided.

X. HVDC CONTROL

Basically, an ac–dc converter has the ability to control dc voltage from positive maximum to negative maximum in about 10 to 12 msec while the current flows in one direction. Since power is the product of voltage and current, it is the direction of voltage polarity that determines whether the converter is a rectifier or an inverter. With two converters connected in a loop, as done in two-terminal dc transmission (Fig. 25a), current flow is the difference between the two voltages divided by the resistance of the loop

$$I_d = (U_{dR} - U_{dI})/R.$$

Clearly, by varying the output voltages, the direct current and therefore the power can be changed, and very rapidly if required. The rectifier power is $I_d \times U_{dR}$ and the inverter power $I_d \times U_{dI}$; the difference is due to the losses in the dc transmission. The dc output voltage U_d of each converter is

$$U_d = (3\sqrt{2}U/\pi)\cos\alpha - (3\omega L/\pi)I_d,$$

where α is the delay angle, U the ac phase-to-phase voltage, and ωL the transformer reactance per phase.

The dc voltage can be changed rapidly by changing the delay angle α; it can also be changed by changing the ac voltage U with the transformer tap changer, but rather slowly. Both angle control and tap changer control are used in HVDC converters, the former for fast control and the latter to arrive slowly at the most desirable steady-state operating condition.

Primarily, the system is controlled to maintain a constant current set on the controller, which compares the actual current with the set current (Fig. 25a). Converter characteristics for achieving this are shown in Fig. 25b for the two converters A and B. The X axis is the dc current and the Y axis the dc voltage. It is seen that characteristic A has three parts. The upper part is the maximum dc voltage that converter A can produce with minimum allowed delay angle. The vertical part is the constant-current control, which controls the output voltage from maximum positive to the third part of the characteristic, which is the maximum inversion voltage. The voltage is changed as necessary to maintain the current at its set value, I_{ds}. The maximum rectification and inversion voltage characteristics will move up and down with the change in ac voltage, which is controlled by the tap changer.

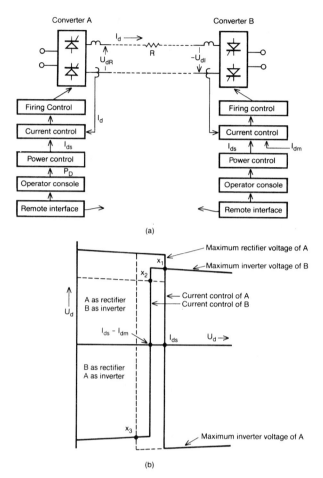

FIGURE 25 (a) Simplified block diagram of control. (b) Converter characteristics for achieving control.

Converter B's characteristic in Fig. 25b is shown upside down since in a system loop one converter is inverted compared to the other. Converter A has a constant-current setting of I_{ds}, while converter B has a setting slightly smaller than I_{ds} by a margin setting of I_{dm} (I_{dm} is usually about 5% of rated current). The system current is I_d, corresponding to the only crossing point X_1 of the two characteristics, and results from the fact that converter B has a somewhat lower current setting than converter A. Converter A works on constant-current control with the necessary delay angle, while converter B works as an inverter with maximum inversion voltage because the actual current is larger than its set current. Now if, due to some drop in ac voltage, the characteristic of converter A drops (shown by the horizontal dashed line), then the current corresponds to X_2, converter A continues to work as a rectifier with its maximum rectification voltage, and converter B's constant-current control takes over and does not allow the current to drop below $I_{ds} - I_{dm}$ (a drop of about 5%). Thus even if the rectifier as voltage drops to a low

level, the current is maintained by the inverter backing off. If the inverter voltage drops, the rectifier delay angle α increases and prevents the current from rising above I_{ds}. This basic characteristic, with several improvisations, is used by all manufacturers because it represents the best control strategy.

If the negative current margin is applied to converter A instead of converter B, thus lowering the current setting of converter A below that of converter B, shown by the dashed vertical line in Fig. 25b, the operating point now becomes X_3; that is, dc voltage polarity reverses, current is maintained in the same direction, and the power flow reverses. Thus complete reversal of power can be achieved in 10 to 20 msec simply by reversing the margin setting signal.

In steady-state operation it is normal to operate the inverter with maximum inversion voltage and the rectifier with constant-current control (operating point X_1). This is achieved by assigning the inverter tap changer control the function of maintaining constant dc voltage to cover for its ac voltage fluctuations, and assigning the rectifier tap changer the function of maintaining constant-current control within a narrow upper range of delay angle of about 10 to 15°.

The power flow is changed by changing the current setting (Fig. 25a). Constant power control is provided, which determines the current setting for the constant-current control. The set power is divided by the actual dc voltage, and the calculated current order is fed into the current controller as the set current order.

Power can also be modulated as desired by modulating the set power or the set current. The purpose of modulation may be to stabilize one or both ac systems. The power flow may also be changed to control the frequency of either system. The ability to change dc power, dc current, and dc voltage offers the possibility of many other special controls. Firing angle can be modulated on either side in order to damp subsynchronous oscillations in large ac machines. Or firing angle can be changed to change the consumed reactive power in order to compensate for sudden changes in ac voltage or to maintain a constant ac voltage.

To take advantage of these possibilities, extensive system control studies are made to ensure the compatibility of various requirements. It is clear that power can be controlled precisely, almost at will, a capability that is well appreciated by the system operators and planners.

XI. PROTECTION

Like any power system, HVDC equipment is protected from overvoltages by surge arresters across each valve, ac bus to ground, dc bus to ground, across each six-pulse converter, and across various reactors. It is normal to provide a 10 to 15% insulation margin above the protection level of arresters.

In an ac system, protective relays are provided which, on detection on a fault, trip the appropriate circuit breaker to isolate the faulted section. In HVDC the converter control can be used to bring current rapidly to zero, and it is used to do so unless the fault is in the converter itself, in which case the firing pulses are stopped and the ac side breaker is opened. Appropriate detection circuits are provided to detect converter faults, including transformer faults, valve short circuits, failure of a valve to fire, repetitive commutation failures, and converter-to-ground faults.

When a dc line fault occurs, it is detected by rate of fall of voltage and reduced level of voltage. Detection takes about 5 to 25 msec, and the rectifier firing angle is immediately increased to more than 90°, the inverter at the other end is prevented from going into rectification, and the current comes to zero in 10 to 20 msec. Current is then held to zero for 100 to 200 msec to allow the fault area to deionize and recover insulation strength. Then the system is restarted, and since most faults are temporary, the event is over in 120 to 250 msec in more than 90% of fault cases. If the fault restrikes, it is suppressed again, but this time the restart is made at a reduced dc voltage (about 75%) by holding the rectifier firing angle to a larger than normal level. This allows the line to recover even if the line insulation is significantly damaged. If two or three restart attempts fail, the system is shut down for operator action.

The same protection strategy can be adapted for multiterminal systems (i.e., systems without any dc breakers). If the fault is permanent in one section, fast disconnect switches can be provided to remove the faulty section while the current is held to zero, and then the remaining system is restarted. However, this strategy means that the unfaulted sections would remain shut down for 100 to 200 msec. If this time is too long, dc breakers are provided, in which case the unfaulted system can be back in service in less than 50 msec.

Figure 26a shows electrical circuits of a dc breaker module. The interrupter is a typical ac breaker (air blast, SF_6 puffer, etc.), internally modified to enhance arc instability. In parallel are a capacitor and a powerful, highly nonlinear resistor (such as a gapless zinc oxide arrester) with a knee voltage about $1\frac{1}{2}$ times the dc voltage for energy dissipation. When the interrupter opens, a current oscillation builds up between the interrupter and the capacitor. When the oscillation current equals the direct current, the interrupter current comes to zero long enough for it to interrupt. The capacitor is left with the dc current, which then charges up to the resistor's knee voltage, and the current

FIGURE 26 (a) An HVDC breaker concept. (b) A 500-kV HVDC breaker with four interrupter modules.

finally transfers to the resistor. Since the resistor presents a back emf of 1.5 times the dc line voltage, the current quickly goes to zero. Modules are connected in series to obtain required breaker rating. Figure 26b shows a 500-kV breaker with four modules in series.

XII. TYPICAL LAYOUT OF CONVERTER STATIONS

Figures 27 and 28 show a typical layout and site view of a bipolar HVDC converter station. The layout, as seen from Fig. 27, is quite straightforward. Two ac lines come into the two ac yards, one for each pole. The ac yards on the two sides of the building consist of the necessary shunt capacitor banks, 11th, 13th, and high-pass filters, and switching equipment. The two ac busbars are brought to the transformers located on the sides of the building, which contains mainly the converter valves and the control room. Six single-phase transformers are shown on each side, although they could well be two 3-phase transformers or three single-phase 3-winding transformers required for a 12-pulse converter unit. Three quadruple valves (not shown) on each side of the building make up the 12 valves needed for each pole. The transformers have nearly horizontal bushings, which are slid into the valve hall through openings. These openings are sealed with flexible removable covers around the bushings. The building would also contain the control room, heat exchangers, and much of the auxiliary equipment. To the left of the building are the two smoothing reactors, one for each pole, and outgoing dc busbars for the two poles. The dc yard containing the dc filters and the necessary switchgear is shown on the left side of the building. The site view in Fig. 28 also shows an ac substation on the right side of the converter station.

The valve hall building is screened by use of metal sheet structure and screened windows and doors in order to prevent leakage of radio noise from the valve hall. The

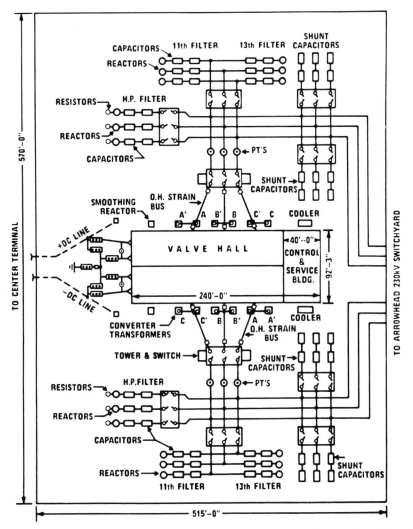

FIGURE 27 Typical layout of a bipolar HVDC converter station (actual layout of the Square Butte Project in Minnesota. (Courtesy of GE.)

building is also temperature- and humidity-controlled to avoid over-heating of the valves and moisture condensation in the valve structures. The layout is such that either pole can be shut down for maintenance while the other pole continues to operate. The valve hall building is divided down the middle for the same reason. The building would also include offices, meeting room, storage area, workshop, capability for lifting and moving any failed equipment, and other facilities. The yard also includes space for various spares, such as a transformer and a reactor, as shown in the left lower corner of the site view. Necessary facilities for communication with the remote terminal and dispatcher would also be provided.

Figure 19 shows the inside of a 250-kV valve hall housing three quadruple valve structures making up the 12 valves of a 12-pulse converter unit. There are other ways to arrange these valves, of course. For example, a

small project may have a single-valve structure for all 12 valves.

The layout shown in Figs. 27 and 28 would do just as well for a back-to-back tie, in which case, the two poles would become the two terminals and there would be no dc busbars and no dc yard. In a back-to-back scheme,

FIGURE 28 Photograph of the Square Butte converter station (Layout in Fig. 27). (Courtesy of GE.)

however, the dc voltage would be rather low (less than 100 kV), since no dc transmission is involved.

SEE ALSO THE FOLLOWING ARTICLES

POWER TRANSMISSION, HIGH VOLTAGE • TRANSFORMERS, ELECTRICAL

BIBLIOGRAPHY

CIGRE Symposium S09-87 (1987). "AC/DC Transmission Interaction and Comparisons," Boston, M.

CIGRE International Colloquim on "HVDC Power Transmission," (1989). Recife, Brazil.

Electrobras *et al.* of Brazil. (1983). Proc. Int. Symp. HVDC Technol—Sharing Brazilian Experience, 1983, Rio de Janeiro. Sponsored by Electrobras *et al.* of Brazil.

IEE (1981). *Proc. Int. Conf. Thyristor Variable Static Equipment ACDC Transmission, 1981, London.* Organized by Institute of Electrical Engineers, United Kingdom.

IEE (1985). *Int. Conf. AC DC Power Transmission, 4th, 1985, London,* Institute of Electrical Engineers, United Kingdom.

IEEE (1984). *Proc. Panel Session Conf. Exposition, 9th, 1984, Kansas City, Missouri.* "Basis of Selection of HVDC for Recent Transmission Projects in North America," IEEE Publication No. 85TH0122-2 PWR.

IEEE (1986). *Proc. Panel Session Conf. Exposition, 10th, 1986, Anaheim, California.* "Physical Layout of Recent HVDC Transmission Projects in North America," IEEE Publication No. 87TH01776-PWR.

IREQ and IEEE (1984). *Proc. Int. Conf. DC Power Transmission, 1984, Montreal.* Sponsored by IREQ and IEEE, Montreal.

Litzenberger W. H., and Rajiv V. (eds.) (1998). "An Annotated Bibliography of HVDC Transmission and FACTS Devices, 1996–1997," Bonneville Power Administration and EPRI. This material may be accessed on the Internet at http://www.transmission.bpa.gov and may be downloaded from http://www.bpa.gov

U.S. Department of Energy (1980). *Proc. Symp. Incorporating HVDC Power Transmission System Planning, 1980, Phoenix, Arizona.* Sponsored by U.S. Department of Energy.

U.S. Department of Energy (1983). *Proc. Symp. Urban Applications HVDC Power Transmission, 1983, Philadelphia.* Sponsored by U.S. Department of Energy.

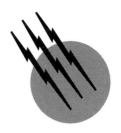

Discrete Mathematics and Combinatorics

Douglas R. Shier

Clemson University

I. Nature of Combinatorics
II. Basic Counting Techniques
III. Recurrence Relations and Generating Functions
IV. Inclusion–Exclusion Principle
V. Existence Problems

GLOSSARY

Algorithm Systematic procedure or prescribed series of steps followed in order to solve a problem.

Binary Pertaining to the digits 0 and 1.

Event Set of occurrences defined with respect to some probabilistic process.

Identity Mathematical equation that always holds.

Integers The numbers 0, 1, 2, . . . and their negatives.

List Ordered sequence of elements.

Mutually exclusive Events that cannot occur simultaneously.

Prime Integer greater than 1 that cannot be evenly divided by any integer other than itself and 1.

Set Unordered collection of elements.

String Ordered sequence of letters taken from some alphabet.

Universal set Set that contains all elements relevant to the current discussion.

COMBINATORICS is a branch of discrete mathematics that involves the study of arrangements of various objects.

Typically, the focus of combinatorics is on determining whether arrangements can be found that satisfy certain properties or on counting all possible arrangements of such objects. While the roots of combinatorics extend back several thousands of years, its relevance to modern science and engineering is increasingly evident.

I. NATURE OF COMBINATORICS

Combinatorics constitutes a rapidly growing area of contemporary mathematics and is one with an enviable repertoire of applications to areas as diverse as biology, chemistry, physics, engineering, communications, cryptography, and computing. Of particular significance is its symbiotic relationship to the concerns and constructs of computer science. On the one hand, the advent of high-speed computers has facilitated the detailed study of existing combinatorial patterns as well as the discovery of new arrangements. On the other hand, the design and analysis of computer algorithms frequently require the insights and tools of combinatorics. It is not at all surprising, then, that computer science, which is ultimately concerned with

Encyclopedia of Physical Science and Technology, Third Edition, Volume 4
Copyright © 2002 by Academic Press. All rights of reproduction in any form reserved.

the manipulation of finite sets of symbols (e.g., strings of binary digits), and combinatorial mathematics, which provides tools for analyzing such patterns of symbols, have rapidly achieved prominence together. Moreover, since the symbols themselves can be abstract objects (rather than simply numerical quantities), combinatorics supports the more abstract manipulations of symbolic mathematics and symbolic computer languages.

Combinatorics is at heart a problem-solving discipline that blends mathematical techniques and concepts with a necessary touch of ingenuity. In order to emphasize this dual nature of combinatorics, the sections that follow will first present certain fundamental combinatorial principles and then illustrate their application through a number of diverse examples. Specifically, Sections II–IV provide an introduction to some powerful techniques for counting various combinatorial arrangements, and Section V examines when certain patterns can be guaranteed to exist.

II. BASIC COUNTING TECHNIQUES

A. Fundamental Rules of Sum and Product

Two deceptively simple, but fundamentally important, rules allow the counting of complex patterns by decomposition into simpler patterns. The first such principle states, in essence, that if we slice a pie into two nonoverlapping portions, then indeed the whole (pie) is equal to the sum of its two parts.

Rule of Sum. Suppose that event E can occur in m different ways, that event F can occur in n different ways, and that the two events are mutually exclusive. Then, the compound event where at least one of the two events happens can occur in $m + n$ ways.

The second principle indicates the number of ways that a menu of choices (one item chosen from E, another item chosen from F) can be selected.

Rule of Product. Suppose that event E can occur in m different ways and that subsequently event F can occur in n different ways. Then, a choice from E followed by a choice from F can be made in $m \times n$ ways.

EXAMPLE 1. A certain state anticipates a total of 2,500,000 registered vehicles within the next ten years. Can the current system of license plates (consisting of six digits) accommodate the expected number of vehicles? Should there instead be a change to a proposed new system consisting of two letters followed by four digits?

Solution. To analyze the current situation, there are ten possibilities (0–9) for each of the six digits, so application of the product rule yield $10 \times 10 \times 10 \times 10 \times$

$10 \times 10 = 1,000,000$ possibilities, not enough to accommodate the expected number of vehicles. By contrast, the proposed new system allows (again by the product rule) $26 \times 26 \times 10 \times 10 \times 10 \times 10 = 6,760,000$ possibilities, more than enough to satisfy the anticipated demand.

EXAMPLE 2. DNA (deoxyribonucleic acid) consists of a chain of nucleotide bases (adenine, cytosine, guanine, thymine). How many different three-base sequences are possible?

Solution. For each of the three positions in the sequence, there are four possibilities for the base, so (by the product rule) there are $4 \times 4 \times 4 = 64$ such sequences.

EXAMPLE 3. In a certain computer programming language, each identifier (variable name) consists of either one or two alphanumeric characters (A–Z, 0–9), but the first character must be alphabetic (A–Z). How many different identifier names are possible in this language?

Solution. In this case, analysis of the compound event can be broken into counting the possibilities for event E, a single-character identifier, and for event F, a two-character identifier. The number of possibilities for E is 26, whereas (by the product rule) the number of possibilities for F is $26 \times (26 + 10) = 936$. Since the two events E and F are mutually exclusive, the total number of distinct identifiers is $26 + 936 = 962$.

B. Permutations and Combinations

In the analysis of combinatorial problems, it is essential to recognize when order is important in the arrangement and when it is not. To emphasize this distinction, the set $X = [x_1, x_2, \ldots, x_n]$ consists of n elements x_i, assembled without regard to order, whereas the list $X = [x_1, x_2, \ldots, x_n]$ contains elements arranged in a prescribed order.

In the previous examples, the order of arrangement was clearly important so lists were implicitly being counted. More generally, arrangements of objects into a list are referred to as *permutations*. For example, the objects a, b, c can be arranged into the following permutations: $[a, b, c]$, $[a, c, b], [b, a, c], [b, c, a], [c, a, b], [c, b, a]$. By the product rule, n distinct objects can be arranged into:

$$n! = n \times (n - 1) \times (n - 2) \times \cdots \times 2 \times 1$$

different permutations. (The symbol $n!$, or n factorial, denotes the product of the first n positive integers.) A permutation of size k is a list with k elements chosen from the n given objects, and there are exactly

$$P(n, k) = n \times (n - 1) \times (n - 2) \times \cdots \times (n - k + 1)$$

such permutations.

EXAMPLE 4. In a manufacturing plant, a particular product is fabricated by processing in turn on four different machines. If any processing sequence using all four machines is permitted, how many different processing orders are possible? How many processing orders are there if only two machines from the four need to be used?

Solution. Each processing order corresponds to a permutation of the four machines, so there are $P(4, 4) = 4! = 24$ different orders. If processing on any two machines is allowable then there are $P(4, 2) = 4 \times 3 = 12$ different orders.

When the order of elements occurring in the arrangement is not pertinent, then a way of arranging k objects, chosen from n distinct objects, is called a *combination* of size k. For example, the objects a, b, c can be arranged into the following combinations, or sets, of size 2: $\{a, b\}, \{a, c\}, \{b, c\}$. The number of combinations of size k from n objects is given by the formula:

$$C(n, k) = P(n, k)/k!$$

EXAMPLE 5. A group of ten different blood samples is to be split into two batches, each consisting of five "pooled" samples. Further chemical analysis will then be performed on the two batches. In how many ways can the samples be split in this fashion?

Solution. Any division of the samples $S_1, S_2, \ldots, S_{10}$ into the two batches can be uniquely identified by those samples belonging to the first batch. For example, $\{S_1, S_2, S_5, S_6, S_8\}$ defines one such division. Since the order of samples within each batch is not important, there are $C(10, 5) = 252$ ways to divide the original samples.

EXAMPLE 6. Suppose that 12 straight lines are drawn on a piece of paper, with no two lines being parallel and no three meeting at a single point. How many different triangles are formed by these lines?

Solution. Any three lines form a triangle since no lines are parallel. As a result, there are as many triangles as choices of three lines selected from the 12, giving $C(12, 3) = 220$ such triangles.

EXAMPLE 7. How many different solutions are there in nonnegative integers x_i to the equation $x_1 + x_2 + x_3 + x_4 = 8$?

Solution. We can view this problem as an equivalent one in which eight balls are placed into four numbered boxes. For example, the solution $x_1 = 2, x_2 = 3, x_3 = 2, x_4 = 1$ corresponds to placing 2, 3, 2, 1 balls into boxes 1, 2, 3, 4. This solution can also be represented by the string $* * | * * * | * * | *$ which shows the number of balls residing in the four boxes. The number of solutions is then the number of ways of constructing a string of 11 symbols (eight stars and three bars); namely, we can select the three bars in $C(11, 3) = 165$ ways.

C. Binomial Coefficients

Ways of arranging objects can also be viewed from an algebraic perspective. To understand this correspondence, consider the product of n identical factors $(1 + x)$, namely:

$$(1 + x)^n = (1 + x)(1 + x) \cdots (1 + x)$$

The coefficient of x^k in the expansion of this product is just the number of ways to select the symbol x from exactly k of the factors. However, the number of ways to select these k factors from the n available is $C(n, k)$, meaning that:

$$(1 + x)^n = C(n, 0) + C(n, 1)x$$
$$+ C(n, 2)x^2 + \cdots + C(n, n)x^n \quad (1)$$

Because the coefficients $C(n, k)$ arise in this way from the expansion of a two-term expression, they are also referred to as *binomial coefficients*. These coefficients can be conveniently placed in a triangular array, called *Pascal's triangle*, as shown in Fig. 1. Row n of Pascal's triangle contains the values $C(n, 0), C(n, 1), \ldots, C(n, n)$. Several patterns are apparent from this figure. First, the binomial coefficients are symmetrically placed within each row: namely, $C(n, k) = C(n, n - k)$. Second, the coefficient appearing in any row equals the sum of the two coefficients appearing in the previous row just to the left and to the right. For example, in row 5 the third entry, 10, is the sum of the second and third entries, 4 and 6, from the previous row. In general, the binomial coefficients satisfy the identity:

$$C(n, k) = C(n - 1, k - 1) + C(n - 1, k)$$

The binomial coefficients satisfy a number of other interesting and useful identities. To illustrate one way of

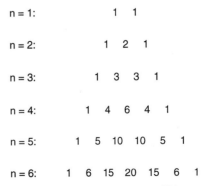

n = 1:						1		1					
n = 2:						1	2	1					
n = 3:					1		3		3		1			
n = 4:				1		4		6		4		1		
n = 5:			1		5		10		10		5		1	
n = 6:		1		6		15		20		15		6		1

FIGURE 1 Arrangement of binomial coefficients $C(n, k)$ in Pascal's triangle.

discovering such identities, formally substitute the value $x = 1$ into both sides of Eq. (1), yielding the identity:

$$2^n = C(n, 0) + C(n, 1) + C(n, 2) + \cdots + C(n, n)$$

In other words, the binomial coefficients for n must sum to the value 2^n. If instead, the value $x = -1$ is formally substituted into Eq. (1), the following identity results:

$$0 = C(n, 0) - C(n, 1) + C(n, 2) - \cdots + (-1)^n C(n, n)$$

This simply states that the alternating sum of the binomial coefficients in any row of Fig. 1 must be zero.

A final identity involving the numbers in Fig. 1 concerns a string of coefficients progressing from the left-hand border along a downward sloping diagonal to any other entry in the figure. Then, the sum of these coefficients will be found as the value just below and to the left of the last such entry. For instance, the sum $C(2, 0) + C(3, 1) + C(4, 2) + C(5, 3) = 1 + 3 + 6 + 10 = 20$ is indeed the same as the binomial coefficient $C(6, 3)$. In general, this observation can be expressed as the identity:

$$C(n + k + 1, k) = C(n, 0) + C(n + 1, 1)$$
$$+ C(n + 2, 2) + \cdots + C(n + k, k)$$

This identity can be given a pleasant combinatorial interpretation, namely, consider selecting k items (without regard for order) from a total of $n + k + 1$ items, which can be done in $C(n + k + 1, k)$ ways. In any such selection, there will be some item number r so that items $1, 2, \ldots, r$ are selected but $r + 1$ is not. This then leaves $k - r$ items to be selected from the remaining $n + k + 1 - (r + 1) = n + k - r$ items, which can be done in $C(n + k - r, k - r)$ ways. Since the cases $r = 0, 1, \ldots, k$ are mutually exclusive, the sum rule shows the total number of selections is also equal to $C(n + k, k) + C(n + k - 1, k - 1) + \cdots + C(n + 1, 1) + C(n, 0)$. Thus, by counting the same group of objects in two different ways, one can verify the above identity. This technique of "double counting" provides a powerful tool applicable to a number of other combinatorial problems.

D. Discrete Probability

Probability theory is an important area of mathematics in which combinatorics plays an essential role. For example, if there are only a finite number of outcomes $S_1, S_2, \ldots, S_m$ to some process, the ability to count the number of occurrences of S_i provides valuable information on the likelihood that outcome S_i will in fact be observed. Indeed, many phenomena in the physical sciences are governed by probabilistic rather than deterministic laws; therefore, one must generally be content with assessing the probability that certain desirable (or undesirable) outcomes will occur.

EXAMPLE 8. What is the probability that a hand of five cards, dealt from a shuffled deck of cards, contains at least three aces?

Solution. The population of 52 cards can be conveniently partitioned into set A of the 4 aces and set N of the 48 non-aces. In order to obtain exactly three aces, the hand must contain three cards from set A and two cards from set N, which can be achieved in $C(4, 3) C(48, 2) = 4512$ ways. To obtain exactly 4 aces, the hand must contain four cards from A and one card from N, which can be achieved in $C(4, 4) C(48, 1) = 48$ ways. Since the total number of possible hands of five cards chosen from the 52 is $C(52, 5) = 2,598,960$, the probability of the required hand is $(4512 + 48)/2,598,960 = 0.00175$, indicating a rate of occurrence of less than twice in a thousand.

EXAMPLE 9. In a certain state lottery, six winning numbers are selected from the numbers $1, 2, \ldots, 40$. What are the odds of matching all six winning numbers? What are the odds of matching exactly five? Exactly four?

Solution. The number of possible choices is the number of ways of selecting six numbers from the 40, or $C(40, 6) = 3,838,380$. Since only one of these is the winning selection, the odds of matching all six numbers is $1/3,838,380$. To match exactly five of the winning numbers, there are $C(6, 5) = 6$ ways of selecting the five matching numbers and $C(34, 1) = 34$ ways of selecting a nonmatching number, giving (by the product rule) $6 \times 34 = 204$ ways, so the odds are $204/3,838,380 = 17/319,865$ for matching five numbers. To match exactly four winning numbers, there are $C(6, 4) = 15$ ways of selecting the four matching numbers and $C(34, 2) = 561$ ways of selecting the nonmatching numbers, giving $15 \times 561 = 8415$ ways, so the odds are $8415/3,838,380 = 561/255,892$ (or approximately 0.0022) of matching four numbers.

EXAMPLE 10. An alarm system is constructed from five identical components, each of which can fail (independently of the others) with probability q. The system is designed with a certain amount of redundancy so that it functions whenever at least three of the components are working. How likely is it that the entire system functions?

Solution. There are two states for each individual component, either good or failed. The state of the system can be represented by a binary string $x_1 x_2 x_3 x_4 x_5$, where x_i is 1 if component i is good and is 0 if it fails. A functioning state for the system thus corresponds to a binary string having at most two zeros. The number of states with exactly two zeros is $C(5, 2) = 10$, so the probability of two failed and three good components is $10q^2(1-q)^3$. Similarly, the probability of exactly one failed component is $C(5, 1)q^1(1-q)^4 = 5q(1-q)^4$, and the probability of no failed components is $C(5, 0)(1-q)^5 = (1-q)^5$. Altogether, the probability that the system functions is given by $10q^2(1-q)^3 + 5q(1-q)^4 + (1-q)^5$. For example, when $q = 0.01$, the system will operate with probability 0.99999015 and thus fail with probability 0.00000985; this shows how adding redundancy to a system composed of unreliable components (1% failure rate) produces a highly reliable system (0.001% failure rate).

III. RECURRENCE RELATIONS AND GENERATING FUNCTIONS

A. Recurrence Relations and Counting Problems

Not all counting problems can be solved as readily and as directly as in Section II. In fact, the best way to solve specific counting problems is often to solve instead a more general, and presumably more difficult, problem. One technique for doing this involves the use of recurrence relations.

Recall that the binomial coefficients satisfy the relation:

$$C(n, k) = C(n-1, k-1) + C(n-1, k)$$

Such an expression shows how the value $C(n, k)$ can be calculated from certain "prior" values $C(n-1, k-1)$ and $C(n-1, k)$. This type of relation is termed a recurrence relation, since it enables any specific value in the sequence to be obtained from certain previously calculated values.

EXAMPLE 11. How many strings of eight binary digits contain no consecutive pair of zeros?

Solution. It is easy to find the number f_1 of such strings of length 1, since the strings "0" and "1" are both acceptable, yielding $f_1 = 2$. Also, the only forbidden string of length 2 is "00" so $f_2 = 3$. There are three forbidden strings of length 3 ("001," "100," and "000"), whereupon $f_3 = 5$. At this point, it becomes tedious to calculate subsequent values directly, but they can be easily found by noticing that a certain recurrence relation governs the sequence f_n. In an acceptable string of length n, either the first digit is a 1 or it is a 0. In the former case, the remaining digits can be any acceptable string of length $n-1$ (and there are f_{n-1} of these). In the latter case, the second digit must be a 1 and then the remaining digits must form an acceptable string of length $n-2$ (there are f_{n-2} of these). These observations provide the recurrence relation:

$$f_n = f_{n-1} + f_{n-2} \tag{2}$$

Using the initial conditions $f_1 = 2$ and $f_2 = 3$, the values $f_3, f_4, \ldots, f_8$ can be calculated in turn by substitution into Eq. (2):

$$f_3 = 5, \qquad f_4 = 8, \qquad f_5 = 13,$$
$$f_6 = 21, \qquad f_7 = 34, \qquad f_8 = 55$$

Therefore, 55 binary strings of length 8 have the desired property.

In this problem it was clearly expedient to solve the general problem by use of a recurrence relation that stressed the interdependence of solutions to related problems. The particular sequence obtained for this problem, [1, 2, 3, 5, 8, 13, 21, 34, ...], with $f_0 = 1$ added for convenience, is called the *Fibonacci sequence*, and it arises in numerous problems of mathematics as well as biology, physics, and computer science.

EXAMPLE 12. Suppose that ten straight lines are drawn in a plane so that no two lines are parallel and no three intersect at a single point. Into how many different regions will the plane be divided by these lines?

Solution. As seen in Fig. 2, the number of regions created by one straight line is $f_1 = 2$, the number created by two lines is $f_2 = 4$, and the number created by three lines is $f_3 = 7$. The picture becomes excessively complicated with more added lines, so it is prudent to seek a general solution for f_n, the number of regions created by n lines in the plane. Suppose that $n-1$ lines have already been drawn and that line n is now added. Because the lines are all mutually nonparallel, line n must intersect each existing line exactly once. These $n-1$ intersection points divide the new line into n segments and each segment serves to subdivide an existing region into two regions. Thus, the n segments increase the number of regions by exactly n, producing the recurrence relation:

$$f_n = f_{n-1} + n$$

Given the initial condition $f_1 = 2$, application of this recurrence relation yields the values $f_2 = f_1 + 2 = 4$ and $f_3 = f_2 + 3 = 7$, as previously verified. In fact, such a recurrence relation can be explicitly solved, giving:

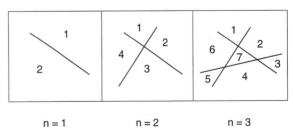

n = 1   n = 2   n = 3

FIGURE 2 Number of regions created by the placement of *n* lines.

$$f_n = (n^2 + n + 2)/2$$

indicating that there are $f_{10} = 112/2 = 56$ regions bounded by ten lines.

In representing mathematical expressions, the placement of parentheses can be crucial. For example, $((x - y) - z)$ does not in general give the same result as $(x - (y - z))$. Moreover, the placement of parentheses must be syntactically valid as in $((x - y) - z)$ or $(x - (y - z))$, whereas $(x) - y) - (z$ is not syntactically valid.

EXAMPLE 13. How many valid ways are there of parenthesizing an expression involving n variables?

Solution. Notice that there is only one valid form for a single variable x, namely (x), giving $f_1 = 1$; similarly, it can be verified that $f_2 = 1$ and $f_3 = 2$. More generally, suppose that there are n variables. Then any valid form must be expressible as (AB), where A and B are themselves valid forms. Notice that if A contains k variables, then B must contain $n - k$ variables, so in this case the f_k valid forms for A can be combined with the f_{n-k} valid forms for B to yield $f_k f_{n-k}$ valid forms for the whole expression. Since the different possibilities for k ($k = 1, 2, \ldots, n - 1$) are mutually exclusive, we obtain the recurrence relation:

$$f_n = f_1 f_{n-1} + f_2 f_{n-2} + \cdots + f_{n-2} f_2 + f_{n-1} f_1$$

This equation can be explicitly solved for f_n in terms of binomial coefficients:

$$f_n = C(2n - 2, n - 1)/n$$

The numbers in this sequence, $[1, 1, 2, 5, 14, 42, 132, \ldots]$, are called *Catalan numbers* (Eugène Catalan, 1814–1894), and they occur with some regularity as solutions to a variety of combinatorial problems (such as triangulating convex polygons, counting voting sequences, and constructing rooted binary trees).

B. Generating Functions and Counting Problems

The examples in Part A of this section serve to illustrate the theme that solving a specific problem can frequently be aided by relating the given problem to other, often simpler, problems of the same type. For example, a problem of size n might be related to a problem of size $n - 1$ and to another problem of size $n - 3$. Another way of pursuing such interrelationships among problems of different sizes is through use of a generating function. As a matter of fact, this concept has already been previewed in studying the binomial coefficients. Specifically, Eq. (1) shows that $f(x) = (1 + x)^n$ can be viewed as a generating function for the binomial coefficients $C(n, k)$:

$$f(x) = C(n, 0) + C(n, 1)x + C(n, 2)x^2 + \cdots + C(n, n)x^n$$

The variable x simply serves as a formal symbol and its exponents represent placeholders for carrying the coefficient information. More generally, a generating function is a polynomial in the variable x:

$$f(x) = a_0 + a_1 x + a_2 x^2 + \cdots + a_n x^n + \cdots$$

and it serves as a template for studying the sequence of coefficients $[a_0, a_1, a_2, \ldots, a_n, \ldots]$.

Recall that the binomial coefficients $C(n, k)$ count the number of combinations of size k derived from a set $\{1, 2, \ldots, n\}$ of n elements. In this context, the generating function $f(x) = (1 + x)^n$ for the binomial coefficients can be developed by the following reasoning. At each step $k = 1, 2, \ldots, n$ a decision is made as to whether or not to include element k in the current combination. If x^0 is used to express exclusion and x^1 inclusion, then the factor $(x^0 + x^1) = (1 + x)$ at step k compactly encodes these two choices. Since each element k presents the same choices (exclude/include), the product of the n factors $(1 + x)$ produces the desired enumerator $(1 + x)^n$. This reasoning applies more generally to cases where the individual choices at each step are not identical, so the factors need not all be the same (as in the case of the binomial coefficients). The following examples give some idea of the types of problems that can be addressed through the use of generating functions.

EXAMPLE 14. In how many different ways can change for a dollar be given, using only nickels, dimes, and quarters?

Solution. The choices for the number of nickels to use can be represented by the polynomial:

$$(1 + x^5 + x^{10} + \cdots) = (1 - x^5)^{-1}$$

where x^i signifies that exactly i cents worth of nickels are used. Similarly, the choices for dimes are embodied in:

$$(1 + x^{10} + x^{20} + \cdots) = (1 - x^{10})^{-1}$$

and the choices for quarters in:

$$(1 + x^{25} + x^{50} + \cdots) = (1 - x^{25})^{-1}$$

Multiplying together these three polynomials produces the required generating function:

$$f(x) = (1 - x^5)^{-1}(1 - x^{10})^{-1}(1 - x^{25})^{-1}$$

The coefficient of x^n in the expanded form of this generating function indicates the number of ways of making change for n cents. In particular, there are 29 different ways of making change for a dollar ($n = 100$). This can be verified by using a symbolic algebra package such as *Mathematica* or *Maple*.

A *partition* of the positive integer n is a set of positive integers (or "parts") that together sum to n. For example, the number 4 has the five partitions: $\{1, 1, 1, 1\}$, $\{1, 1, 2\}$, $\{1, 3\}$, $\{2, 2\}$, and $\{4\}$.

EXAMPLE 15. How many partitions are there for the integer n?

Solution. The choices for the number of ones to include as parts is represented by the polynomial:

$$(1 + x + x^2 + \cdots) = (1 - x)^{-1}$$

where the x^i term means that 1 is to appear i times in the partition. Similarly, the choices for the number of twos to include is given by:

$$(1 + x^2 + x^4 + \cdots) = (1 - x^2)^{-1}$$

the choices for the number of threes is given by:

$$(1 + x^3 + x^6 + \cdots) = (1 - x^3)^{-1}$$

and so forth. Therefore, the number of partitions of n can be found as the coefficient of x^n in the generating function:

$$f(x) = (1 - x)^{-1}(1 - x^2)^{-1}(1 - x^3)^{-1} \cdots$$

EXAMPLE 16. Find the number of partitions of the integer n into *distinct* parts.

Solution. Since the parts must be distinct, the choices for any integer i are whether to include it (x^i) or not (x^0) in the given partition. As a result, the generating function for this problem is

$$f(x) = (1 + x)(1 + x^2)(1 + x^3) \cdots$$

For example, the coefficient of x^8 in the expansion of $f(x)$ is found to be 6, meaning that there are six partitions of 8 into distinct parts: namely, $\{8\}$, $\{1, 7\}$, $\{2, 6\}$, $\{3, 5\}$, $\{1, 2, 5\}$, $\{1, 3, 4\}$.

IV. INCLUSION–EXCLUSION PRINCIPLE

Another important counting technique is based on the idea of successively adjusting an initial count through systematic additions and subtractions that are guaranteed to produce a correct final answer. This technique, called the *inclusion–exclusion principle*, is applicable to many instances where direct counting would be impractical.

As a simple example, suppose we wish to count the number of elements that are *not* in some subset A of a given universal set U. Then, the required number of elements equals the total number of elements in U, denoted by $N = N(U)$, minus the number of elements in A, denoted by $N(A)$. Expressed in this notation,

$$N(A') = N - N(A)$$

where $A' = U - A$ designates the set of elements in U that do not appear in A. Figure 3a depicts this relation using a Venn diagram (John Venn, 1834–1923), in which the enclosing rectangle represents the set U, the inner circle represents A, and the shaded portion represents A'. The quantity $N(A')$ is thus obtained by excluding $N(A)$ elements from N.

EXAMPLE 17. The letters a, b, c, d, e are used to form five-letter words, using each letter exactly once. How many words do *not* contain the sequence *bad*?

Solution. The universe here consists of all words, or permutations, formed from the five letters, so there are $N = 5! = 120$ words in total. Set A consists of all such words containing *bad*. By treating these three letters as a new "megaletter" x, the set A equivalently contains all words formed from x, c, e so $N(A) = 3! = 6$. The number of words not containing *bad* is then $N - N(A) = 120 - 6 = 114$.

Figure 3b shows the situation for two sets, A and B, contained in the universal set U. As this figure suggests,

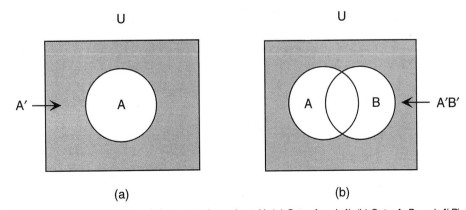

FIGURE 3 Venn diagrams relative to a universal set U. (a) Sets A and A'; (b) Sets A, B, and $A'B'$.

the number of elements in either A or B (or both) is then the number of elements in A plus the number of elements in B, minus the number of elements in both:

$$N(A \cup B) = N(A) + N(B) - N(AB) \qquad (3)$$

Here $A \cup B$ denotes the elements either in A or in B, or in both, whereas AB denotes the elements in both A and B. Since the sum $N(A) + N(B)$ counts the elements of AB twice rather than just once, $N(AB)$ is subtracted to remedy the situation. The number of elements $N(A'B')$ in neither A nor B is $N - N(A \cup B)$, thus an alternative form of Eq. (3) is

$$N(A'B') = N - [N(A) + N(B)] + N(AB) \qquad (4)$$

This form shows how terms are alternately included and excluded to produce the desired result.

EXAMPLE 18. In blood samples obtained from 50 patients, laboratory tests show that 20 patients have antibodies to type A bacteria, 29 patients have antibodies to type B bacteria, and 8 patients have antibodies to both types. How many patients have antibodies to neither of the two types of bacteria?

Solution. The given data state that $N = 50$, $N(A) = 20$, $N(B) = 29$, and $N(AB) = 8$. Therefore, by Eq. (4):

$$N(A'B') = 50 - [20 + 29] + 8 = 9$$

meaning that nine patients are immune to neither type of bacteria.

The foregoing equations generalize in a natural way to three sets A, B, and C:

$$N(A \cup B \cup C) = N(A) + N(B) + N(C) - [N(AB)$$
$$+ N(AC) + N(BC)] + N(ABC) \qquad (5)$$
$$N(A'B'C') = N - [N(A) + N(B) + N(C)] + [N(AB)$$
$$+ N(AC) + N(BC)] - N(ABC) \qquad (6)$$

In each of these forms, the final result is obtained by successive inclusions and exclusions, thus justifying these as manifestations of the inclusion–exclusion principle.

EXAMPLE 19. An electronic assembly is comprised of components 1, 2, and 3 and functions only when at least two components are working. If all components fail independently of one another with probability q, what is the probability that the entire assembly functions?

Solution. Let A denote the event in which components 1 and 2 work, let B denote the event in which components 1 and 3 work, and let C denote the event in which components 2 and 3 work. Of interest is the probability that at least one of the events A, B, or C occurs. The analogous form of Eq. (5) in the probabilistic case is

$$Pr(A \cup B \cup C) = Pr(A) + Pr(B) + Pr(C) - [Pr(AB)$$
$$+ Pr(AC) + Pr(BC)] + Pr(ABC)$$

Here, $Pr(A)$ is the probability of event A occurring, $Pr(AB)$ is the probability of event AB occurring, and so forth. Notice that $Pr(A) = Pr(B) = Pr(C) = (1-q)^2$ and $Pr(AB) = Pr(AC) = Pr(BC) = Pr(ABC) = (1-q)^3$, so the assembly functions with probability:

$$Pr(A \cup B \cup C) = 3(1-q)^2 - 3(1-q)^3 + (1-q)^3$$
$$= 3(1-q)^2 - 2(1-q)^3$$

Two positive integers are called *relatively prime* if the only positive integer evenly dividing both is the number 1. For example, 7 and 15 are relatively prime, whereas 6 and 15 are not (they share the common divisor 3).

EXAMPLE 20. How many positive integers not exceeding 60 are relatively prime to 60?

Solution. The appropriate universe here is $U = \{1, 2, \ldots, 60\}$ and the (prime) divisors of $N = 60$ are 2, 3, 5. Relative to U, let A be the set of integers divisible by 2, let B be the set of integers divisible by 3, and let C be the set of integers divisible by 5. The problem here is to calculate $N(A'B'C')$, the number of integers that share no divisors with 60. Because every other integer is divisible by 2, we have $N(A) = 60/2 = 30$. Similarly, $N(B) = 60/3 = 20$ and $N(C) = 60/5 = 12$. Because any integer divisible by both 2 and 3 must also be divisible by 6, we have $N(AB) = 60/6 = 10$. Likewise, $N(AC) = 60/10 = 6$, $N(BC) = 60/15 = 4$, and $N(ABC) = 60/30 = 2$. Substituting these values into Eq. (6) gives:

$$N(A'B'C') = 60 - [30 + 20 + 12] + [10 + 6 + 4] - 2$$
$$= 16$$

so there are 16 positive numbers not exceeding 60 that are relatively prime to 60.

EXAMPLE 21. Each package of a certain product contains one of three possible prizes. How likely is it that a purchaser of five packages of the product will get at least one of each prize?

Solution. The number of possible ways in which this event can occur will first be calculated, after which a probabilistic statement can be deduced. Let the prizes be denoted a, b, c, and let the contents of the five packages be represented by a string of five letters (where repetition is allowed). For example, the string *bacca* would represent one possible occurrence. Define A to be the set of all such strings that do not include a. In similar fashion, define B (respectively, C) to be the set of strings that do not include b (respectively, c). It is required then to calculate

$N(A'B'C')$, the number of instances in which a, b, and c all occur. The approach will be to use the inclusion–exclusion relation, Eq. (6), to calculate this quantity. Here, N is the number of strings of five letters over the alphabet $\{a, b, c\}$, so (by the product rule) $N = 3^5 = 243$. Also, $N(A)$ is the number of strings from the alphabet $\{b, c\}$, whereupon $N(A) = 2^5 = 32$. By similar reasoning, $N(B) = N(C) = 32$, $N(AB) = N(AC) = N(BC) = 1^5 = 1$, and $N(ABC) = 0$. As a result,

$$N(A'B'C') = 243 - 3(32) + 3(1) - 0 = 150$$

In summary, the total number of possible strings is 243 and all three prizes are obtained in 150 of these cases. If all strings are equally likely (i.e., any package is just as likely to contain each prize), then the required probability is $150/243 = 0.617$, indicating a better than 60% chance of obtaining three different prizes in just five packages. A similar analysis shows that for six packages, the probability of obtaining all three prizes increases to 0.741; for seven packages, there is a 0.826 probability.

V. EXISTENCE PROBLEMS

A. Pigeonhole Principle

In certain combinatorial problems, it may be exceedingly difficult to count the number of arrangements of a prescribed type. In fact, it might not even be clear that any such arrangement actually exists. What is needed then is some guiding mathematical assurance that configurations of the desired type do indeed exist. One such principle is the so-called *pigeonhole principle*. While its statement is overwhelmingly self-evident, its applications range from the simplest to the most challenging problems in combinatorics.

The pigeonhole principle states that if there are more than k objects (or pigeons) to be placed in k locations (or pigeonholes), then some location must house two (or possibly more) such objects. A simple illustration of this principle assures that at least two residents of a town with 400 inhabitants have the same birthday. Here, the objects are the residents and the locations are the 356 possible birthdays. Since there are more objects than locations, some location must contain at least two objects, meaning that some two residents (or more) must share the same birthday. Notice that this principle only guarantees the *existence* of two such residents; it does not give any information about finding them.

EXAMPLE 22. In any selection of five different elements from $\{1, 2, \ldots, 8\}$, there must be some pair of selected elements that sum to 9.

Solution. Here the objects are the numbers $1, 2, \ldots, 8$ and the locations are the four sets $A_1 = \{1, 8\}$, $A_2 = \{2, 7\}$, $A_3 = \{3, 6\}$, and $A_4 = \{4, 5\}$. Notice that for each of these sets its two elements sum to 9. According to the pigeonhole principle, placing five numbers into these four sets results in some set A_i having both its elements selected, so the sum of these two elements (by the construction of set A_i) must equal 9.

EXAMPLE 23. In a room with $n \geq 2$ persons, there must be two persons having exactly the same number of friends in the room.

Solution. The number of possible friendships for any given person ranges from 0 to $n - 1$. However, if $n - 1$ occurs then that person is a friend of everyone else, and (assuming that friendship is a mutual relation) no other person can be without friends. Thus, both 0 and $n - 1$ cannot simultaneously occur in a group of n persons. If $1, 2, \ldots, n - 1$ are the possible numbers of friendships, then using these $n - 1$ numbers as locations for the n persons (objects), the pigeonhole principle assures that some number in $\{1, 2, \ldots, n - 1\}$ must appear twice. A similar result can be established for the case $\{0, 1, \ldots, n - 2\}$, demonstrating there must always be at least two persons having the same number of friends in the room.

Two strings $x = x_1 x_2 \cdots x_n$ and $y = y_1 y_2 \cdots y_m$ over the alphabet $\{a, b, \ldots, z\}$ are said to be disjoint if they share no common letter and are said to overlap otherwise.

EXAMPLE 24. In any collection of at least six strings, there must either be three strings that are mutually disjoint or three strings that mutually overlap.

Solution. For a given string x, let the $k \geq 5$ other strings in the collection be divided into two groups, D and O; D consists of those strings that are disjoint from x, and O consists of those that overlap with x. By a generalization of the pigeonhole principle, one of these two sets must contain at least three elements. Suppose that it is set D. Then, either D contains three mutually overlapping strings or it contains two disjoint strings y and z. In the first case, these three strings satisfy the stated requirements. In the second case, the elements x, y, z are all mutually disjoint, so again the requirements are met. A similar argument can be made if O is the set containing at least three elements. In any event, there will either be three mutually disjoint strings or three mutually overlapping strings.

This last example is a special case of *Ramsey's theorem* (Frank Ramsey, 1903–1930), which guarantees that if there are enough objects then configurations of certain types will always be guaranteed to exist. Not only does this theorem (which generalizes the pigeonhole principle) produce some very deep combinatorial results, but it has also been applied to problems arising in geometry, the design of communication networks, and information retrieval.

B. Combinatorial Designs

Combinatorial designs involve ways of arranging objects into various groups in order to meet specified requirements. Such designs find application in the planning of statistical experiments as well as in other areas of mathematics (number theory, coding theory, geometry, and algebra).

As one illustration, suppose that an experiment is to be designed to test the effects of five different drugs using five different subjects. One clear requirement is that each subject should receive all five drugs, since otherwise the results could be biased by variation among the subjects. Each drug is to be administered for one week, so that at the end of five weeks the experiment will be completed. However, the order in which drugs are administered could also have an effect on their observed potency, so it is also desirable for all drugs to be represented on any given week of the experiment. One way of designing such an experiment is depicted in Fig. 4, which shows one source of variation—the subjects ($S_1, S_2, \ldots, S_5$)—appearing along the rows and the other source of variation—the weeks ($W_1, W_2, \ldots, W_5$)—appearing along the columns. The entries within each row show the order in which the drugs ($A, B, \ldots, E$) are administered to each subject on a weekly basis. Such an arrangement is termed a *Latin square*, since the five treatments (drugs) appear exactly once in each row and exactly once in each column.

Figure 4 clearly demonstrates the existence of a 5×5 Latin square; more generally, Latin squares of size $n \times n$ exist for each value of $n \geq 1$. There are also occasions when it is desirable to superimpose certain pairs of $n \times n$ Latin squares. An example of this arises in testing the effects of n types of fertilizer and n types of insecticide on the yield of a particular crop. Suppose that a field on which the crop is grown is divided into an $n \times n$ grid of plots. In order to minimize vertical and horizontal variations in the composition and drainage properties of the soil, each fertilizer should appear on exactly one plot in each "row" and exactly one plot in each "column" of the grid. Likewise, each insecticide should appear once in each row and once in each column of the grid. In other words, a Latin square design should be used for each of the two treatments. Figure 5a shows a Latin square design for four fertilizer types (A, B, C, D), and Fig. 5b shows another Latin square design for four insecticide types (a, b, c, d).

In addition, the fertilizer and insecticide treatments can themselves interact, thus an ideal design would ensure that each of the n^2 possible combinations of the n fertilizers and n insecticides appear together once. Figure 5c shows that the two Latin squares in Figs. 5a and b (when superimposed) have this property; namely, each fertilizer–insecticide pair occurs exactly once on a plot. Such a pair of Latin squares is called *orthogonal*.

A pair of orthogonal $n \times n$ Latin squares need not exist for all values of $n \geq 2$. However, it has been proved that the only exceptions occur when $n = 2$ and $n = 6$. In all other cases, an orthogonal pair can be constructed.

Latin squares are special instances of *complete* designs, since every treatment appears in each row and in each

	W_1	W_2	W_3	W_4	W_5
S_1	A	B	C	D	E
S_2	B	C	D	E	A
S_3	C	D	E	A	B
S_4	D	E	A	B	C
S_5	E	A	B	C	D

FIGURE 4 Latin square design for drug treatments ($A, \ldots, E$) applied to subjects ($S_1, \ldots, S_5$) by week ($W_1, \ldots, W_5$).

A	B	C	D
C	D	A	B
D	C	B	A
B	A	D	C

(a)

a	b	c	d
d	c	b	a
b	a	d	c
c	d	a	b

(b)

Aa	Bb	Cc	Dd
Cd	Dc	Ab	Ba
Db	Ca	Bd	Ac
Bc	Ad	Da	Cb

(c)

FIGURE 5 Orthogonal Latin squares. (a) Latin square design for fertilizers ($A, \ldots, D$); (b) Latin square design for insecticides ($a, \ldots, d$); (c) superimposed Latin squares.

column. Another useful class of combinatorial designs is one in which not all treatments appear within each test group. Such *incomplete* designs are especially relevant when the number of treatments is large relative to the number of tests that can be performed on an experimental unit.

As an example of an incomplete design, consider an experiment in which subjects are to compare $v = 7$ brands of soft drink $(A, B, \ldots, G)$. For practical reasons, every subject is limited to receiving $k = 3$ types of soft drink. Moreover, to ensure fairness in the representation of the various beverages, each soft drink should be tasted by the same number $r = 3$ of subjects, and each pair of soft drinks should appear together the same number $\lambda = 1$ of times. It turns out that such a design can be constructed using $b = 7$ subjects, with the soft drinks compared by each subject i given by the set B_i below:

$$B_1 = \{A, B, D\}; \qquad B_2 = \{A, C, F\};$$
$$B_3 = \{A, E, G\}; \qquad B_4 = \{B, C, G\};$$
$$B_5 = \{B, E, F\}; \qquad B_6 = \{C, D, E\};$$
$$B_7 = \{D, F, G\}$$

The sets B_i are referred to as *blocks*, and such a design is termed a (b, v, r, k, λ) *balanced incomplete block design*. In any such design, the parameters b, v, r, k, λ must satisfy the following conditions:

$$bk = vr, \qquad \lambda(v - 1) = r(k - 1)$$

In the previous example, these relations hold since $7 \times 3 = 7 \times 3$ and $1 \times 6 = 3 \times 2$. While the above conditions must hold for any balanced incomplete block design, there need not exist a design corresponding to every set of parameters satisfying these conditions.

SEE ALSO THE FOLLOWING ARTICLES

COMPUTER ALGORITHMS • PROBABILITY

BIBLIOGRAPHY

Bogart, K. P. (2000). "Introductory Combinatorics," 3rd ed. Academic Press, San Diego, CA.

Cohen, D. I. A. (1978). "Basic Techniques of Combinatorial Theory," Wiley, New York.

Grimaldi, R. P. (1999). "Discrete and Combinatorial Mathematics," 4th ed. Addison-Wesley, Reading, MA.

Liu, C. L. (1985). "Elements of Discrete Mathematics," 2nd ed. McGraw–Hill, New York.

McEliece, R. J., Ash, R. B., and Ash, C. (1989). "Introduction to Discrete Mathematics," Random House, New York.

Roberts, F. S. (1984). "Applied Combinatorics," Prentice Hall, Englewood Cliffs, NJ.

Rosen, K. H. (1999). "Discrete Mathematics and Its Applications," 4th ed. McGraw–Hill, New York.

Rosen, K. H., ed. (2000). "Handbook of Discrete and Combinatorial Mathematics," CRC Press, Boca Raton, FL.

Tucker, A. (1995). "Applied Combinatorics," 3rd ed. John Wiley & Sons, New York.

Discrete Systems Modeling

Egon Börger

Università di Pisa

GLOSSARY

Abstract state machine (ASM)   A generalization of finite state machines to arbitrary structures, instead of finitely many control states, and to distributed runs, instead of sequential computations that transform input to output. The two major subclasses of abstract state machines are the sequential ASMs and the distributed (multi-agent) ASMs. Such machines encompass any known form of virtual machines and real computers in a way that can be made rigorous (see the explanation of the ASM thesis below, where a precise definition of ASMs is also given).

Algorithm   A procedure (also the finite text describing it) that can be mechanized, at least in principle. Examples are electronic booking procedures, the operating system of a computer, or protocols that govern the interaction of multiple computer programs on the Internet. An important special group of algorithms includes those procedures that compute functions (i.e., procedures which, started with any given argument, eventually terminate and yield the value of the function for that argument as output). Examples are the well-known methods for computing the four elementary arithmetical functions or, more generally, computer programs that compute numerical functions.

Computer program   A finite text, satisfying the syntactical conditions of the (so-called programming) language in which it is formulated, which can be executed on a computer. A computer program refers to the specific data structures of the programming language to which it belongs. A particularly important class of computer programs is reactive programs whose role is to continuously execute the actions associated with them, as distinguished from sequential transformational programs that compute an input–output relation.

Finite state machine (FSM)   A machine that executes a set of instructions that determine, for each of a finite number of internal (or control) states and for each of a finite number of possible finite inputs, the next internal state and the (equally finite) output. A more precise definition of these transformational sequential FSMs is given in the text below.

Logic   The branch of mathematics that develops rigorous systems of reasoning, which are also called logics, and studies their mathematical properties. Usually a system of logic is defined syntactically by a formal language that establishes the set of legal expressions (also called formulae) of this logic, together with a system of rules allowing us to deduce from given formulae (axioms) their logical consequences. Semantically, a logic is defined by characterizing the intended interpretations of its formulae, yielding classes of models where the axioms and the consequences deduced from them by the rules are "true." For first-order logic, such models are structures consisting of finitely many domains with functions and relations defined over them.

DISCRETE SYSTEMS are dynamic systems that evolve in discrete steps due to the abrupt occurrence of internal or external events. The system evolution is typically modeled as resulting from firing state transforming rules, which are triggered when certain (internal and/or external) conditions become true. Such systems encompass sequential (also called transformational) algorithms and their implementations as computer programs, but also systems of distributed (asynchronous concurrent) processes such as telecommunication systems, operating on physically or logically separate architectures and typically triggered by external (discrete or continuous physical) events. Methods for modeling sequential or distributed systems are intimately related to methods for validating and verifying these systems against their models.

I. SEQUENTIAL SYSTEMS

Sequential discrete systems are dynamic systems with a law that determines how they evolve in discrete steps, producing for every initial state a sequence of states resulting from firing state transforming rules, which are triggered when certain conditions become true. Such sequences S_0, S_1, ... are determined by the evolution law (usually rules that model the dynamics of the system) and by the initial state S_0. This sequence constitutes what is also called a computation of the algorithmic or transformational process defined by the rules. There is a great variety of such algorithmic systems, including manufacturing, transportation, business, administrative, and information handling processes, as well as computers, for which the rules are described by computer programs. Numerous general purpose or specialized languages are used to model and implement such systems.

A widespread criterion to classify different modeling languages is whether the system definitions they allow to formulate are state-transformation based (also called operational) or purely logico-functional (also called declarative). This dichotomy came into use around 1970 with the denotational and algebraic approaches to system descriptions which were biased to modeling systems by sets of algebraic equations or axioms. A representative example is the Vienna development method (VDM) for constructing models of software systems in early design stages. Also, declarative set-theoretic modeling (e.g., in Z) belongs here, as well as numerous forms of system modeling by

axioms of specific logics, such as temporal logics, logics of action, logics of belief, etc. The stateless form of modeling also underlies functional programming (e.g., pure LISP or ML) and logic programming (e.g., PROLOG), as opposed to imperative or object-oriented programming (e.g., FORTRAN, C, JAVA).

The fundamental problem, to whose solution the distinction between declarative and operational system models has been set out to contribute, is the necessity to provide the system designer with abstraction methods that enable one to cope with the ever increasing complexity of software and hardware systems. The idea has been to tackle this problem by developing system definitions at levels of abstraction higher than the level of the machines where these systems are implemented, so that the resulting models can be analyzed before the detailed design, coding, and testing are started. Unfortunately, formalizations by logical axioms inherently lead to global descriptions; in fact, in addition to formalizing dynamic changes of mostly local system elements, the overwhelming majority of the sets of axioms usually is concerned with stating that, under the specified conditions, all the other system parameters do not change. This phenomenon is known as the frame problem and yields high-level system models that are considerably larger than the final system. For example, in the SCR method, which uses the temporal logic x/x' notation within table-based modeling of systems, the frame problem yields numerous "no change" (NC) clauses. Furthermore, it is usually difficult to reliably relate the (stateless) logical descriptions by a sequence of stepwise refinements to imperative code, which is executed and tested on state-transforming machines. Consider, for instance, the difficulties encountered in attempts to extend logical characterizations of passive databases by descriptions of the inherently reactive behavior of active databases, in particular when it comes to modeling multiple database transactions that may happen during the execution of an active rule. Thus, logico-functional modeling methods contribute to the gap encountered in the practice of large-scale system design, between high-level system models and their executable counterpart on virtual or physical machines.

This same idea (namely, to abstract from the machine details that are only relevant for the implementation to obtain better understandable and verifiable high-level system models) underlies the development since the end of the 1960s of numerous notions and specimens of abstract machines, also called virtual machines. Famous examples, developed for the implementation of programming languages, are the class concept of Simula67; Wirth's P-machine (for executing PASCAL programs); Warren's abstract machine (for efficient implementations of PROLOG programs); Peyton Jones' spineless, tagless G-machine (for efficient implementations of functional programs);

and the Java virtual machine (for platform independent execution of JAVA programs). Such machines split the implementation of programs into two steps, compilation into intermediate byte code and interpretation of byte code by an abstract machine or its further compilation into runnable code. Virtual machines also appear in other areas; consider, for example, the numerous forms of database machines, instruction set architectures, etc. This development paralleled for over 20 years the rich deployment of logico-algebraic specification methods, without being connected to it, so that no general notion of abstract machine came out which could be used as a uniform and practical basis for high-level system modeling together with methods for refining the abstract models to executable and thereby testable ones.

Gurevich's notion of an abstract state machine (ASM) resolves the impractical dichotomy between declarative and operational modeling methods and yields a simple universal concept of virtual machines. The intended abstractions are realized by:

1. Appropriate choice of the relevant, *a priori* arbitrary, data structures that can be tailored to the system under investigation to make up the needed notion of states
2. Providing corresponding abstract machine instructions which describe the intended evolution (transformation) of states, at the desired level of detailing

A. Sequential Abstract State Machines

Sequential abstract state machines capture the notion of a sequential algorithm, in the sense that for every sequential algorithm, there exists an equivalent sequential ASM (i.e., with the same set of states, the same set of initial states, and the same state transformation law). This sequential ASM thesis relies upon the following three postulates for sequential algorithms from which it can be proved. The *sequential time postulate* expresses that every sequential algorithm is associated with a set of states, a subset of initial states, and a state transformation law (which is a function from states to states). The *abstract-state postulate* requires that the states of a sequential algorithm are first-order structures, with fixed domain and signature, and are closed under isomorphisms (respecting the initial states and the state transformation law). The *bounded exploration postulate* states that for every sequential algorithm, the transformation law depends only upon a finite set of terms over the signature of the algorithm, in the sense that there exists a finite set of terms such that, for arbitrary states X and Y which assign the same values to each of these terms, the transformation law triggers the same state changes for X and Y.

A sequential ASM is defined as a set of transition rules of the form:

If condition then updates

which transforms first-order structures (the states of the machine). The guard *condition*, which has to be satisfied for a rule to be applicable, is a variable free first-order formula, and *updates* is a finite set of function updates (containing only variable free terms) of the form:

$$f(t_1, \ldots, t_n) := t$$

The execution of these rules is understood as updating, in the given state and in the indicated way, the value of the function f at the indicated parameters, leaving everything else unchanged. This proviso avoids the frame problem of declarative approaches. In every state, all the applicable rules are simultaneously applied (if the updates are consistent) to produce the next state. If desired or useful, declarative features can be built into an ASM by integrity constraints and by assumptions on the state, environment, and applicability of rules.

Computations of sequential ASMs formalize the so-called transformational character of sequential systems, namely that each step of the system consists of an ordered sequence of two substeps of the environment and the transformational program of the system. First the environment prepares the input on which the program is expected to be run, then the program fires its rules on the given input without further intervention from the environment. An example can be found in current implementations of active databases, where the occurrence of external events and the processing of active rules alternate. Part of the second substep may be that the program produces some output, which may be used by the environment for preparing the next input. This is often considered a third substep in this sequence of interactions between the environment and the transformational program, which together constitute the sequential system.

In a sequential ASM, M, this separation of the action of the environment from that of the transformational program is reflected in the following classification of functions that make up the state of the system. A function f is called static or rigid if its values do not change in any of the states of M; f is called dynamic or flexible if its values may change from one to the next state of M. Dynamic functions are further classified into input functions, controlled functions, output functions, and shared functions. Input functions for M are those functions that M can only read, which means that these functions are determined entirely by the environment of M. These functions are often also called monitored functions, because they are used to reflect the occurrence of events that trigger rules. Controlled functions of M are updated by some of the rules of M and are never changed by the environment. This means that M and only M can update these functions. Output functions for M are functions which M can only update but not read, whereas the environment can read them (without updating them). Shared functions are functions that can be read and updated by both M and the environment, which means that their consistency has to be guaranteed by special protocols.

This function classification includes a fundamental distinction for selection functions, which are an abstract form of scheduling algorithms. In fact, static selection functions, such as Hilbert's ε-operator, whose values depend only upon the value of the set they choose from, are distinguished from dynamic selection functions, the choice of which may depend upon the entire computation state. There is a standard notation for not-furthermore-specified selection functions which is often used for modeling the phenomenon of nondeterminism in discrete systems, namely for applying a rule for an element which satisfies a given condition:

$$\text{Choose } x \text{ satisfying cond } (x) \text{ in rule}(x).$$

The above-defined abstract machines provide a practical (and, by the sequential ASM thesis, most general) framework for high-level modeling of complex discrete systems and for refining high-level models to executable system models. Important examples are given below. There are various ways to refine high-level ASM models to executable models, using various tools for executing ASMs (e.g., ASMGofer, ASM Workbench, XASM), coming with implementations of natural ASM definitions for standard programming constructs, such as sequentialization, iteration, and submachine calls.

B. Specialized Modeling Languages for Sequential Systems

Most of the numerous specification and programming languages are tailored to the particular modeling needs of the given domain of application or of the given framework for modeling (read: defining and implementing) the class of systems of interest.

The best-known examples include programming languages belonging to the major (non-concurrent) programming paradigms—namely, functional, logical, imperative, and object-oriented programming. All these modeling approaches can be viewed naturally, and in a rigorous way, as producing instantiations of ASMs at the corresponding level of abstraction. This has been proved explicitly for numerous programming languages (e.g., PROLOG, C, JAVA) by defining their semantics and their implementations on virtual machines as ASMs. In this way, ASMs have also been used for modeling and analyzing proto-

cols, architectures, ASICS, embedded software systems, etc.

Active databases are special forms of ASMs, essentially systems of rules with event, condition, and action components. The event and condition parts together form the ASM rule guard; the event describes the trigger which may result in firing the rule; the condition extracts, from the context in which the event has taken place, that part which must be satisfied to actually execute the rule by performing the associated action. This action part describes the task to be carried out by the database rule, if the event did occur and the condition was true; it corresponds to the updates of ASM rules. Different active databases result from variations of:

1. Underlying notions of state, as constituted by the signature of events, conditions, and actions, and of their relation to the database states
2. Scheduling of the evaluation of condition and action components relative to the occurrence of events (e.g., using coupling modes and priority declarations)
3. Rule ordering (if any)

Abstract state machines provide a rigorous and flexible semantical basis for reflecting and analyzing these different models for the interaction between active rules and the database proper and to classify their implementations. The sequential ASM thesis guarantees, for instance, that the intuitive notion of active database "actions," which may range from performing simple operations to the execution of arbitrary programs, is captured completely by the rigorous concept of the ASM rule.

For some other widely used forms of system models, we now provide their explicit definition in terms of ASMs. One of the historically first and most important types of system models is finite automata, also called Moore automata or finite state machines (FSMs), which are sequential ASMs where all the rules have the form:

$$\text{If } ctl = i \text{ and in} = a \text{ then ctl} := j$$

with functions in for input, and ctl for internal (control) states, which assume only a finite number of values. FSMs that also yield output (also called Mealy automata) are equipped with an additional function out for output, which assumes only a finite number of values and is controlled by an additional update out $:= b$ in the ASM rule above. By analogy, Mealy-ASMs are ASMs whose rules are like those of Mealy automata but with the output update replaced by a fully blown ASM rule. In a similar fashion, one can define all the classical models of computation (e.g., the varieties of Turing machines, Thue systems, Markov algorithms, Minsky machines, recursive functions, Scott machines, Eilenberg's X machines, push down automata) and their modern extensions, such as relational database machines, Wegner's interaction machines, or timed automata or a discrete event system (DES).

Discrete event systems are tailored to describe the control structure for the desired occurrence of events, in the system to be modeled, by specifying allowed event sequences as belonging to languages generated by finite automata. Time can be incorporated into discrete event systems by timer variables, whose update rules relate event occurrences and the passage of time, thus constituting one among many other timed extensions of finite automata for modeling real-time systems.

Wegner's interacting machines add to Turing machines that each computation step may depend upon and also influence the environment, namely by reading input and by yielding an output that may affect the choice of the next input by the environment. This results in the following instantiation of ASMs, where *ctl* represents the current internal state, *headpos* the current position of the reading head, and *mem* the function that yields the current content (read: the memory value) of a tape position. In the traditional notation of Turing machines by sets of instructions, the functions *nextstate*, *move*, and *print* correspond to the machine program. They determine how the machine changes its control state, the content of its working position, and this position itself, depending on the current values of these three parameters; *output* represents an abstract output action, which depends on the same parameters:

$$ctl := nextstate(ctl, mem(headpos), input)$$

$$headpos := move(ctl, mem(headpos), input)$$

$$mem(headpos) := print(ctl, mem(headpos), input)$$

$$output \ (ctl, mem(headpos), input)$$

Considering the output as written on the in–out tape results in defining the output action as follows, where *out* is a function determined by the program instructions:

$$output := input * out(ctl, mem(headpos), input)$$

Similarly, viewing the input as a combination of preceding outputs and the new user input comes down to defining *input* as follows, where *user-input* is a monitored function, and *combine* formalizes the way the environment mixes the past output with the new input:

$$input = combine \ (output, user\text{-}input)$$

The differentiation between single and multiple stream interacting Turing machines is only a question of instantiating *input* to a tuple $(inp_1, \ldots, inp_n)$ of independent input functions, each representing a different environment agent.

The classical models of computation come with simple data structures, typically integers or strings, into which other structures have to be encoded to guarantee the universality of the computational model. Modern modeling languages offer more and richer data structures, but nevertheless the level of abstraction is usually fixed und thereby leads to the necessity of encoding when structures have to be modeled whose data types are not directly supported by the modeling language. For example, VDM models are instances of sequential ASMs with a fixed abstraction level, which is described by the VDM-SL ISO standard. It is obtained by restricting sets to VDM-SL types (built from basic types by constructors), functions to those with explicit or implicit definitions, operations to procedures (with possible side effects), and states to records of read/write variables (0-ary instead of arbitrary functions). Similarly, CSP models, as well as their refinements to OCCAM programs that can be executed on the TRANSPUTER, are known to be instances of ASMs equipped with agents and mechanisms for communication and nondeterministic choice. Parnas tables, which underlie the SCR approach to system modeling, classify functions into monitored or controlled, without providing support for system modularization by auxiliary external functions. The table notation used in SCR for updating dynamic functions of time reflects particular instances of ASMs. For example, normal Parnas tables, which specify how a function value $f(x, y)$ will change to, say, $t_{i,j}$ when a row condition $r_i(x)$ and a column condition $c_j(y)$ are true, represent the ASM with the following rules (for all row indices i and all column indices j):

$$\text{If } r_i(x) \quad \text{and} \quad c_j(y), \quad \text{then } f(x, y) := t_{i,j}$$

Relational machines add to Turing machines arbitrary, but fixed relational structures.

The unified modeling language (UML) shares with the ASM approach a general first-order view of data structures, although UML documents do not formulate anything which goes as far as the abstract-state postulate. Instead, they describe the general first-order view of data structures only vaguely by declaring that the intended models consist of (read: are universes of) "things" ("abstractions that are first-class citizens in a model") together with their relationships. UML comes with a set of graphical notations, which are proposed as "visual projections" into the textual, fully detailed specifications of system models. This set includes activity diagrams, state diagrams, and use-case diagrams (supported by collaboration and sequence diagrams), which are offered to model the dynamic aspects of discrete systems, although no precise meaning is defined for these diagram types by the UML documents. The dynamic semantics of these diagrams has been defined rigorously using particular classes

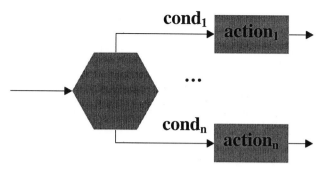

FIGURE 1 ASM normal form of sequential UML activity diagrams.

of ASMs— sequential ASMs for all diagrams that describe purely sequential system behavior. For instance, each sequential UML activity diagram (that is, activity diagrams with only synchronous concurrent nodes) can be built from alternating branching and action nodes, as shown in Fig. 1.

The meaning of alternating branching and action nodes is a generalization of the above ASM rules for FSMs. In moving through such a diagram along the arrows to perform in succession the actions inscribed in the rectangles, being placed on an arc i visualizes being in the control state i (i.e., $ctl = i$); changing the position from one to another arc, as indicated by the arrows, visualizes changing the control state correspondingly. The input reading condition $in = a$ in the FSM rules above is generalized to an arbitrary guard $cond_j$, which labels an arc exiting a branching node and leading to an action node $action_j$; it shows that when the control is positioned on the branching node, one is allowed to perform $action_j$ and to move the control correspondingly—namely, from the arc entering the branching node, along the arc guarded by $cond_j$, to the arc, say j, which exits the action node $action_j$—if the condition $cond_j$ is true in the current state. The output action $out := b$ is generalized to an arbitrary ASM *rule*, which provides a rigorous semantical definition for what in UML is called and described as an (atomic) action. Due to the simultaneous firing of all rules of an ASM, in any state, this rational reconstruction of an atomic UML action as one step of an ASM also provides, within sequential systems, an interpretation for synchronous concurrent UML nodes.

Therefore Fig. 1 depicts the ASM with the following set of rules (for all $0 < j < n + 1$):

$$\text{If } ctl = i \text{ and } cond_j \text{ then } action_j$$
$$ctl := j$$

where i, j stand for the control states defined by the corresponding arcs.

Using this ASM normal form representation of sequential UML activity diagrams, the experimental evidence provided by the experience of 50 years of computing

(namely, that these diagrams represent a notation for a most general modeling language) is theoretically confirmed by the above-mentioned proof for the sequential ASM thesis.

The remaining UML diagrams, in particular class (object, component, deployment) diagrams, are used for modeling the static aspects of discrete systems—namely, the signature of the model, together with the constraints imposed on its structure.

Sequential ASMs also naturally reflect the computational model of synchronous programming languages, such as ESTEREL, that are widely used for modeling parallel synchronous systems. This is due to the ASM concept of simultaneous execution of all applicable rules, which is enhanced by the following operator that applies a rule for all elements satisfying a given condition:

forall x satisfying *cond (x)* do *rule (x)*

A typical use of such a rule is illustrated in the ASM model for Conway's Game of Life, which consists of the following rule:

forall c in cell: if aliveNeighb(c) = 3 then resume(c)

if aliveNeighb(c) < 2

or aliveNeighb(c) > 3

then suspend(c)

where resume(c) stands for alive(c) := true and suspend(c) for alive(c) := false. The rule expresses how at each step, for all cells in the system, their life status may change from non-alive to alive or vice versa.

II. DISTRIBUTED SYSTEMS

Distributed systems consist of physically or logically separate but concurrently executed sequential systems, which interact with each other via communication and synchronization and typically lack global control while manifesting causal dependencies. From the point of view of each sequential component of a distributed system, the other components can be regarded in their entirety as environment, which continuously interacts with this component, concurrently and without any *a priori* known order of component and environment actions.

Distributed systems can be modeled as sets of cooperating sequential systems, independently of whether the transfer of information between the interacting agents is based upon shared memory, or upon a form of (synchronous or asynchronous) message-passing mechanism or remote procedure calls. For distributed ASMs, the notion of run, which is defined for sequential systems as a sequence of computation steps of a single agent, is replaced by the notion of a partial order of moves of finitely many agents, such that the following three conditions are satisfied:

1. Co-finiteness: Each move has only finitely many predecessors.

2. Sequentiality of single agents: The moves of every agent are linearly ordered.

3. Coherence: Each finite initial segment X corresponds to a state $\sigma(X)$, interpreted as the result of executing all moves in X, which for every maximal element $x \in X$ is obtainable by applying move x in state $\sigma(X - \{x\})$. The moves of the single agents can be atomic or durative, but for simplicity the preceding definition of distributed runs assumes actions to be atomic.

This definition of distributed ASMs, which are also called multi-agent ASMs, provides a theoretical basis for a coherent "global" system view for concurrent sequential computations of single agents, each executing its own sequential ASM, at its own pace and with atomic actions applied in its own "local" states, including input from the environment as monitored functions. The definition guarantees that, given any finite initial segment of a distributed run, all linearizations of this segment, also called interleavings, yield the same global view of the state resulting from the computation. In other words, if moves of different agents are independent of each other in a given run, they can be scheduled relative to each other in an arbitrary manner, without influencing the overall view of the state resulting from the run.

The notion of distributed ASM runs comes with no recipe for constructing such runs which, in applications, may turn out to be a challenging problem. However, it is this generality of the concept that provides the freedom to design and analyze models for distributed systems without any *a priori* commitment to special synchronization or communication concepts, an important consequence for the practice of modeling. This includes the abstraction from any particular conditions on the timing of moves of the single agents. The ordering of moves reflects only their causal dependency, which is a before-and-after relation, without further details on the precise timing of the moves.

Another consequence is that numerous concepts of distributed systems, which are tailored to specific modeling needs, can be naturally viewed as instances of classes of distributed ASMs. See, for example, the modeling approaches which work with the method of interleaving of atomic actions, where, in runs of the given system of parallel processes, at each moment exactly one atomic action is chosen for execution. This is a special way to abstract from possible simultaneity of atomic actions, or from overlapping of durative actions, which may happen in concurrent computations on different processors. Interleaving is often viewed as reducing concurrency to nondeterminism, in the sense that a given concurrent run represents a random choice among the corresponding interleaving orders yielding a sequential run. This is particularly convenient for executable models of concurrency. In other approaches, the notion of run is refined by imposing additional fairness constraints on the scheduling of enabled transitions; for example, in the language Unity it is assumed that in infinite runs each assignment is executed infinitely many times. In fact, interleaving alone does not guarantee that the choices made for scheduling are fair. Many specification or programming languages for distributed systems provide special coordination constructs, such as semaphores, monitors, critical regions, rendezvous, or handshaking, which support programming the desired scheduling of allowed single actions in distributed runs.

Although at present there is no proof for a distributed ASM thesis, experimental evidence indicates that, for distributed systems, ASMs constitute a most general computational framework, naturally encompassing the major paradigms used for modeling distributed systems. An industrially relevant example is SDL, the Specification and Description Language for distributed systems, which is widely used for modeling telecommunication systems and whose recent standard for version SDL-2000 has been defined in terms of distributed ASMs by the international standardization body for telecommunications (International Telecommunication Union).

Another important modeling approach for distributed systems is known as Petri nets, and incorporates a broad spectrum of different languages. All the known types of Petri nets turn out to be instances of distributed ASMs, namely agents which are equipped with rules of the form:

If condition (pre-places) then updates (post-places)

where pre- and post-places are finite, not necessarily disjoint, sets of so-called places which appear as rule parameters, and where updates $(p_1, \ldots, p_n)$ is a set of function updates $f(p_i) := t_i$. The ASM rules correspond to what in the traditional phrasing of Petri nets is called the information flow relation, which leads along arcs from pre-places via transitions to post-places. The rule guards represent the "passive" net components (places, graphically represented by circles), and the updates are the "active" components (transitions, graphically represented by bars or boxes). The conditions and the values to which the functions are updated are usually about some marking of places by weighted sets of tokens, about their coloring, etc. In the above characterization of Petri nets as particular classes of distributed ASMs, the ASM notion of state as an arbitrary structure naturally encompasses the generalized notions of Petri nets, where tokens can be arbitrary data.

In the famous example of the Dining Philosophers, the value function is about ownership of a left and right fork; each agent, below referred to as *me*, has as his program the instance of each of the following two sequential ASM rules:

If Owner $(LF (me)) =$ Owner $(RF (me)) =$ none

then Owner $(LF (me)) :=$ me

Owner $(RF (me)) :=$ me

If Owner $(LF (me)) =$ Owner $(RF (me)) =$ me

then Owner $(LF (me)) :=$ none

Owner $(RF (me)) :=$ none

These rules allow an agent to possibly fetch the pair of his left fork $LF(me)$ and his right fork $RF(me)$, if they are free (i.e., not owned by any other agent), or to release this pair for use by one of the neighbor agents. In every distributed run (of dining philosophers), the partial order reflects the way competing agents have solved their conflicts about accessing the common resources (in the example, forks); in general, the partial order avoids introducing forms of pairwise synchronization.

The states of Petri nets are sometimes also defined as first-order logical predicates, associated to places. This is a peculiar logical view of states as structures (where those predicates can be true or false) and of actions (Petri net transitions) as changing the truth value of these predicates. The concept of local states with associated atomic actions, which is characteristic for Petri net transitions and underlies their graphical representation, is reflected in the above ASM definition by the fact that the single agents are sequential ASMs, firing their transition rules in their own "substate" of the "global" state resulting from the distributed computation. This locality of causation and effect avoids the difficulties of global-event-based modeling, where a single action may depend upon tracing an event in the entire system, as is familiar from CCS and from the event-driven, run-to-completion scheme of UML state machines (also called statecharts). The rich variety of different synchronization mechanisms, which has been developed for Petri nets, defines special patterns for synchronizing the corresponding distributed ASMs runs.

Co-design finite state machines (CFSMs), which are used for modeling in hardware–software co-design, are distributed ASMs whose agents are equipped with Mealy ASMs as defined above. The data structure capabilities of Mealy ASMs are needed for including arbitrary combinational (external and instantaneous) functions in the definition of CFSMs. These machines often come with an additional global scheduler agent that controls the interaction of the other system components and/or with timing conditions in case the agents are able to perform durative instead of atomic actions. Due to the parallelism of simultaneous execution of multiple rules that is incorporated into sequential ASMs, the sequential Mealy ASM components of CFSMs have a locally synchronous behavior, a fact that facilitates their validation and verification, whereas the partial order of runs of the CFSMs reflects the flexibility of the globally asynchronous system features.

III. VALIDATION AND VERIFICATION OF SYSTEMS

An intrinsic goal of modeling systems is that the model, which as an algorithmic description has to obey the laws of logic and mathematics, corresponds in the intended way to the desired physical system it was built to design. To show that this goal has been reached implies various engineering and mathematical methods, depending on the phase of the modeling process.

A. Validation and Verification of Ground Models

At the beginning of the modeling work, the so-called specification or requirements engineering phase, a model has to be defined which reflects the informally described requirements for the system to be built. Such models, also called ground models, have the role of mediating between, on the one hand, the formulation of the application domain view of the desired system, which usually is expressed in natural language interspersed with symbolic elements (tables, formulae, and the like), and, on the other hand, its formalization through the design. The ground model itself is of mathematical character, but usually comes with a large number of interfaces to the natural language description of the given requirements. This hybrid character of ground models reflects the fact that they represent the link between the reality (e.g., physical) to be modeled and the formal (mathematical) models.

For the designer the ground models capture the requirements in such a way that, due to their mathematical character, they can serve as a starting point for producing the executable system by rigorously verifiable and validatable refinements of the requirements model. For the designer the ground models document the relevant application domain knowledge in a traceable way, which allows one to link the requirements to the detailed design. For the customer the ground models provide intersubjective means for requirements inspection, which make the correctness and completeness of the ground model checkable with respect to the intentions of the requirements. The particular character of such inspections is that they cannot have the status

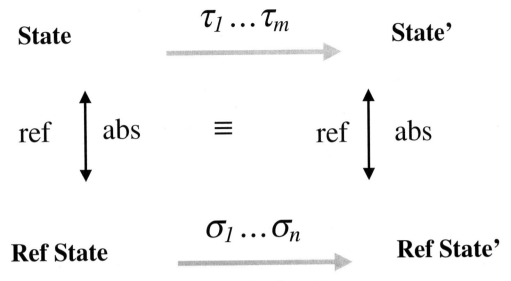

FIGURE 2 Commutative refinement diagrams.

of mathematical verifications, since the informal requirements are not given with mathematical rigor (and by an Aristotelian argument there is no infinite regress to push the limits of formalizability and verifiability). But this does not exclude that certain system properties required by the informal descriptions can be proved or disproved in the ground model, once their formalization there is recognized as faithful to the intentions. It also does not exclude that the ground model itself is validated (or falsified, in the Popperian sense) by executing the ground model for relevant experiments (e.g., use cases or scenarios that appear as part of the requirements), either mentally or by machines. The ground model also provides the possibility of formulating rigorous test plans before the beginning of the design which can serve as bases for testing the final code.

B. Validation and Verification of Model Refinements

During the phase of detailed design, starting with the ground model, a hierarchy of stepwise refined models is developed which links the ground model to the implementation of the system. The justification problem during this phase is of a mathematical nature, given that all models involved are mathematical objects which can be subject to verification. Here the problem is that of crossing system levels by rigorous but practical notions of abstraction and refinement which have to reflect good principles of system composition or decomposition. Typically, at each refinement step, for a given level of abstraction (notion of state and computation), a more detailed level of abstraction has to be found, together with a notion of refinement and ab-

straction, which links the abstract and the refined states of interest, as well as the relevant segments of abstract and refined computations. Then the locations of interest that are related to each other have to be determined, and it has to be defined in which sense they are supposed (and then proved) to be equivalent. This situation is illustrated by Fig. 2. Such refinement hierarchies yield good system documentation which reports the results of the modeling and of its analysis. This helps design engineers to communicate design patterns. It prepares the ground for effective reuse of models by exploiting the orthogonality of different hierarchical levels. It also helps the maintenance engineer by providing faithful models for different system components and their interplay.

The main problem during this refinement phase of the modeling process is to find and formulate the correct refinements and abstractions that:

1. Faithfully reflect the intended design decision (or reengineering idea, in case the modeling starts from the detailed model and has to be abstracted into a less detailed one)
2. Can be justified as faithful by verification and validation (through experiments with executing the models)

C. Separating Different Design and Analysis Concerns

There is an intimate relation between modeling and verification problems and methods. Clearly, every modeling language affects the comprehension of the models expressed in the language, as well as the potential for analysing their properties. Most modeling approaches come with fixed schemes for hierarchical definition and corresponding proof principles. Nevertheless, from the practical engineering point of view, it is usually advantageous to separate design (modeling) concerns from analysis (verification and validation) concerns as much as possible, to make both tasks achievable even for large designs.

The separation of design from analysis supports letting orthogonal design decisions stand out as independent of each other by reflecting them in different models. This helps to keep the design space open, thus realizing the two important modeling principles:

1. Specify for change
2. Avoid premature design decisions

It also helps to structure the design space, providing rigorous interfaces for system composition and decomposition and a basis for experimentation with different design decisions. A design effect of ASMs being unstructured (i.e., without any *a priori* fixed structuring or decomposition principle) is to provide the freedom for the system designer to develop compositional and hierarchical design techniques for ASMs as the necessity arises, tailored to and guided by the kind of system being investigation, thus taking advantage of the possibly simplifying particularities of the system under study. This may indeed also simplify the verification and validation task by supporting the *a posteriori* development of modular proof techniques that are tuned to the application.

The systematic separation of modeling from verification and validation means in particular that the validation (debugging) of a system, through simulation based upon the executability of the model, is separated from the proof-oriented verification of system properties of interest. A further separation of concerns comes with the distinction between different verification levels, characterized by corresponding degrees of detailing the arguments. It is one thing to justify the design by reasoning made for human inspection and understanding; it is another to provide a proof in a rule-based reasoning system. Furthermore, such proofs may be produced by a human or by an interactive computerized system or by an automatic procedure (for instance, model checkers or fully automated theorem provers).

A widespread prejudice identifies rigor, stratified in over 2000 years of mathematics, with the very particular degree of formality characteristic of modern systems of logic which come with a machine-oriented (usually algorithmic, or mechanizable) syntax, semantics, and proof rules to carry out verifications by deductions. Such a logical degree of rigor may, in the ideal case and under the assumption that the formalization is in-

tuitively correct with respect to the system under study, yield machine-checkable verifications that—under the additional assumptions that the prover design is logically correct, that its implementation is semantically correct, and that the environment running the code executes the prover's implementation correctly—exclude any sources of error in the verified model (although not necessarily in the real system). However, the price to be paid for such full-fledged logical formalizations is that usually they enforce decisions about single details which may be required by the particular proof system and not by the problem under study or which may unnecessarily restrict the implementor's freedom to choose among the many design options, against good engineering practice. Furthermore the cost of such formalized verifications is typically very high, in terms of labor and time, and is usually prohibitive. The often criticized "formal explosion" of logical formalizations belongs here, as it yields model descriptions that are longer and more difficult to understand than the programs they are intended to specify. In practice, the most intriguing problem of system design and verification is to provide for the human reader, (1) an insight into the structure of the system to be modeled, and (2) a sufficient understanding of the reasons why the model and its refinements to an implementation are correct, intrinsically and with respect to the real system. Such an understanding is supported by rigorous modeling and proof techniques, where the degree of rigor varies with the given reliability requirements.

The ASM approach to the verification of models makes no *a priori* commitment to a special logical notation, to specific proof rules, or to specific implementations of machine-supported verification assistants. As a consequence, it allows one to use any mathematical notation and argument for justifying a design; however, it fits any specific implementation of the state-of-the-art algorithmic or deductive approaches to verification, namely through model checkers or by proof systems such as PVS, HOL, ISABELLE, KIV, Z-EVE, ACL2, OTTER, or Coq, to mention only a few among dozens of such systems. Model checkers are automatic procedures, but they are applicable only to finite-state systems and in real-life applications quickly face the intrinsic "state explosion" problem. In contrast, proof systems have the advantage of being applicable to infinite as well as to finite systems, but their fully automatic versions are usually rather restricted in applicability. Interactive proof systems open a wider range of applicability but require the expertise of a verification specialist to exploit their verification space.

In combined system design and verification approaches, some of which work well for the development of small-sized designs, the proof rules correspond to program constructs, mappable to restricted classes of programs, for developing programs and proofs of their desired properties hand-in-hand. This holds in particular for declarative approaches that are based upon specialized logics, often proven to be complete for a particular class of properties. Numerous examples can be found in the axiomatic semantics approaches that are based on so-called Floyd or Hoare "assertion calculi" for program verification.

Another set of examples is constituted by temporal logics, coming with operators dealing with the future or with the past of computation states, which are widely used as languages for specifying properties of reactive programs (i.e., programs that are specified in terms of their ongoing behavior). Logical constructs are classified and are then used to obtain classifications of program properties into hierarchies. An example is the hierarchy of safety, guarantee, obligation, response, persistence, and reactivity properties for reactive systems. Another example is provided by the classification into so-called safety, liveness, and progress properties. Safety properties are defined as properties that always hold throughout the entire computation (i.e., in each of the states of a computation), so they are also called state properties (e.g., global or local state invariants, deadlock freedom, fault freedom, mutual exclusion, event ordering principles, etc.). Liveness properties are defined as properties that will hold eventually. Progress properties are defined as conjunctions of safety and liveness properties (e.g., termination and total correctness, freedom from individual starvation and livelocks, accessibility, fairness requirements, etc.).

Petri nets are *a priori* unstructured like most transition systems. Nevertheless, they usually are tightly linked to obtain a combination of modeling and analysis techniques. In those approaches, local states are viewed as logical formulae, and actions are introduced as predicate transformers. In other words, a net transition (a rule) is characterized by logical formulae, describing the relation between pairs of a state and its next state with respect to the given transition. This logical rule description is usually split into the enabling condition, which formalizes the rule guard, and the modification statement, which formalizes the result of the updates, similar to what has been explained above for the event and action part of active database rules.

Another method that advocates linking the construction of models to their verification in corresponding proof systems is the so-called rigorous approach to industrial software engineering, known under the acronym RAISE. Another combined design and verification method that has been tailored to the development of sequential executable code is Abrial's B method. It improves upon its predecessor, the axiomatic set-theoretic Z method, by achieving the executability of models which are described using basically syntactical logical means. The semantics of B machines is defined using Dijkstra's weakest precondition notion. Atomic machine actions are formalized syntac-

tically, by a generalized substitution, so that the fundamental (local) notion of assignment appears as a specialization of the global notion of substitution. The intended meaning of the logical definition of the semantics of B machines is provided by set-theoretic models, in contrast to the ASM approach where the assignment operator constitutes the not-furthermore-analyzed semantical basis of arbitrary machine operations (local function updates). The construction of B machines is linked to proving their properties by corresponding proof rules, tailored in particular to provide termination proofs. As a consequence, the resulting notion of refinement is restricted to refining single operations. This reflects that the underlying computation model of B is that of a pocket calculator, with only finite states (that is, finite sets and finitely many variables ranging over them) and a purely sequential operation mode.

SEE ALSO THE FOLLOWING ARTICLES

COMPUTER ALGORITHMS • DATA STRUCTURES • DISTRIBUTED PARAMETER SYSTEMS • REAL-TIME SYSTEMS • SELF-ORGANIZING SYSTEMS • SYSTEM THEORY

BIBLIOGRAPHY

Abrial, J.-R. (1996). "The B-Book: Assigning Programs to Meanings," Cambridge University Press, London.

Börger, E. (1999). "High level system design and analysis using abstract state machines," *In* "Current Trends in Applied Formal Methods (FM-Trends 98)" (D. Hutter, W. Stephan, P. Traverso, and M. Ullmann, eds.), pp.1–43, LNCS Vol. 1641, Springer–Verlag, Heidelberg.

Börger, E., Cavarra, A., and Riccobene, E. (2000). "Modeling the dynamics of UML state machines," *In* "Abstract State Machines: Theory and Applications" (Y. Gurevich, P. W. Kutter, M. Odersky, and L. Thiele, eds.), pp. 231–241, LNCS Vol. 1912, Springer–Verlag, Berlin.

Fitzgerald, J., and Gorm Larsen, P. (1998). "Modeling Systems: Practical Tools and Techniques in Software Development," Cambridge University Press, London.

Gurevich, Y. (2000). "Sequential abstract-state machines capture sequential algorithms," *ACM Trans on Computational Logic* **1,** 77–111.

ITU-T. (2000). "Specification and Description Language (SDL)," Recommendation Z.100 Annex F. ITU-T (Telecommunication Standardization Sector of International Telecommunication Union).

Manna, Z., and Pnueli, A. (1992). "The Temporal Logic of Reactive and Concurrent Systems: Specification," Springer–Verlag, Berlin.

Manna, Z., and Pnueli, A. (1995). "Temporal Verification of Reactive Systems: Safety," Springer–Verlag, Berlin.

Mosses, P. D. (1992). "Action Semantics," Cambridge University Press, London.

Ramadge, P. J., and Wonham, W. M. (1988). "The control of discrete-event systems," *IEEE Proc*, **77,** 81–98.

Reisig, W. (1998). "Elements of Distributed Algorithms. Modeling and Analysis with Petri Nets," Springer–Verlag, Berlin.

Rumbaugh, J., Jacobson, I., and Booch, G. (1999). "The Unified Modeling Language Reference Manual," Addison–Wesley, Reading, MA.

Spivey, J. M. (1992). "The Z Notation: A Reference Manual," 2nd ed., Prentice Hall, New York.

Stärk, R., Schmid, J., and Börger, E. (2001). "Java and the Java Virtual Machine, Definition, Verification, Validation," Springer–Verlag, Berlin.

Wirsing, M. (1990). "Handbook of algebraic specifications," *In* "Handbook of Theoretical Computer Science," (J. van Leeuwen, ed.), pp. 675–788, Elsevier, Amsterdam.

ADDITIONAL READING

Apt, K. R., and Olderog, E.-R. (1991). "Verification of Sequential and Concurrent Programs," Springer–Verlag, Berlin.

Börger, E., and Schmid, J. (2000). "Composition and submachine concepts for sequential ASMs," *In* "Proc. 14th International Workshop CSLL" (P. Clote and H. Schwichtenberg, eds.), pp. 41–60, LNCS Vol. 1862, Springer–Verlag, Heidelberg.

Brandin, B. A., and Wonham, W. M. (1994). "The supervisory control of timed DES," *IEEE Trans. Automatic Control* **39**, 329–342.

Clarke, E. M., Grumberg, O., and Peled, D. (1998). "Model Checking," MIT Press, Cambridge, MA.

Gurevich, Y. (1995). "Evolving algebras 1993: Lipari Guide," *In* "Specification and Validation Methods" (E. Börger, ed.), pp. 9–36, Oxford University Press, London.

Harel, D., and Politi, M. (1998). "Modeling Reactive Systems with Statecharts," McGraw–Hill, New York.

Distillation

M. R. Resetarits
Koch-Glitsch, Inc.

M. J. Lockett
Praxair, Inc.

I. Distillation Equipment
II. Distillation Theory
III. Distillation Column Design
IV. Applications of Distillation
   Including Energy Considerations

GLOSSARY

Azeotrope Mixture that does not change in composition on distillation and usually has a boiling point higher or lower than any of its pure constituents.

Column (tower) Vertical cylindrical vessel in which distillation is carried out.

Distillate Product of distillation formed by condensing vapor.

Efficiency (overall column efficiency) Ratio of the number of theoretical stages required to effect a distillation separation to the number of actual trays.

Height of a theoretical plate (HETP) Height of packing in a distillation column that gives a separation equivalent to one theoretical stage.

K value Ratio of the concentration of a given component in the vapor phase to its concentration in the liquid phase when the phases are in equilibrium.

Packing Specially shaped metal, plastic, or ceramic material over which the liquid trickles to give a large surface area for contact with the vapor.

Reflux ratio Ratio of the flow rate of the liquid that is returned to the top of the column (the reflux) to the flow rate of the overhead product.

Relative volatility Ratio of the K values of two components; a measure of the ease with which the two components can be separated by distillation.

Theoretical stage Contact process between vapor and liquid such that the exiting vapor and liquid streams are in equilibrium.

Trays (plates) Perforated metal sheets, spaced at regular intervals within a column, on which intimate contact of vapor and liquid occurs.

Vapor pressure Pressure at which a liquid and its vapor are in equilibrium at a given temperature.

DISTILLATION is a physical process for the separation of liquid mixtures that is based on differences in the boiling points of the constituent components. The art of distillation is believed to have originated in China around 800 BC. Early applications of the process were concerned

FIGURE 1 Large distillation column for the production of styrene. [Courtesy of Shell Chemical Canada, Ltd.]

with alcoholic beverage production and the concentration of essential oils from natural products. Over the centuries the technique spread widely, and the first book on the subject, *Das kleine Distillierbuch*, by Brunswig, appeared in 1500. Originally, distillation was carried out in its simplest form by heating a liquid mixture in a still pot and condensing the vapor that boiled off. Condensation was simply carried out by air cooling and later in water-cooled condensers. The origin of the word *distillation* is the Latin *destillare*, which means "dripping down," and it is related to the dripping of condensed vapor product from the condenser.

I. DISTILLATION EQUIPMENT

A. General Description

Distillation is the dominant separation process in the petroleum and chemical industries. It is carried out continuously more often than batchwise, in large, vertical, hollow cylindrical columns (or towers). Figure 1 shows a large distillation column with its associated piping, heat exchangers, vessels, ladders, platforms, and support structures. Figure 2 shows a simple schematic representation.

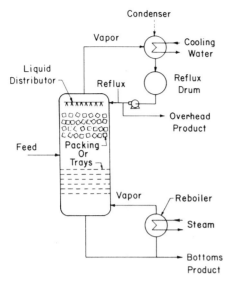

FIGURE 2 Schematic representation of a distillation column.

The process of distillation begins with a feed stream that requires treatment. It is usually necessary to separate the feed into two fractions: a low-boiling fraction (the light product) and a high-boiling fraction (the heavy product). The feed can be in a vapor or a liquid state or a mixture of both. Assuming the feed in Fig. 2 is a liquid, after entering the column it flows down through a series of trays or a stack of packing (see Section 1.B). Liquid leaving the bottom of the column is split into a bottoms product and a fraction that is made available for boiling. The bottoms product, which is rich in low-volatility components, is sometimes called the *tails* or the *bottoms*. A heat exchanger (the reboiler) is employed to boil the portion of the bottoms liquid that is not drawn off as product. The vapor produced flows up through the column (through the trays or packing) and comes into intimate contact with the downflowing liquid. After the vapor reaches and leaves the top of the column, another heat exchanger (the condenser) is encountered where heat is removed from the vapor to condense it. The condensed liquid leaving the condenser passes to a reflux drum and from there is split into two streams. One is the overhead product, which is rich in high-volatility components and is usually called the *distillate* or sometimes the *make* or the *overheads*. The other liquid stream is called the *reflux* and is returned to the top of the column. As the reflux liquid flows down the column, it comes into intimate contact with upflowing vapor. Approximately halfway down the column the reflux stream meets the liquid feed stream and both proceed down the column.

The reflux liquid returned to the top of the column has a composition identical to that of the overhead product. As the reflux, which is rich in high-volatility compo-

nents, encounters upflowing vapor, which is not as rich in these components, the difference in composition, or lack of equilibrium between the two phases, causes high-volatility components to transfer from the liquid to the vapor and low-volatility components to transfer from the vapor to the liquid. The upflowing vapor is made richer in high-volatility components and vice versa for the liquid.

Refluxing improves the separation that is achieved in most distillation columns. Any reflux rate increase, however, requires an increase in the rate of vapor production at the bottom of the column and hence an increase in energy consumption.

Contrary to the implication of Fig. 2, the condenser is usually not located at the top of the column and instead is often located some 3 to 6 m above the ground on a permanent scaffold or platform. The reflux drum is located beneath the condenser. A pump sends the reflux liquid to the top of the column and the distillate to storage or further processing.

The average distillation column at a typical refinery or petrochemical plant is probably 1 to 4 m in diameter and 15 to 50 m tall. Some columns, however, are 15 m in diameter and can extend to a height of 100 m. Columns taller than this are unfeasible to construct and erect. In addition, column height-to-diameter ratios greater than 30 are uncommon because of the support problems encountered with tall, thin columns. Most distillation columns in industrial service are bolted onto thick concrete slabs. Tall, thin columns can employ guy wires for extra support when shell thicknesses are insufficient to prevent excessive sway in the face of high winds.

Elliptical or spherical heads are employed at the top and bottom of the column. Whenever possible, industrial columns are fabricated from carbon steel, but when corrosive chemicals are encountered, columns can be made from, or lined with, more expensive materials such as stainless steel, nickel, titanium, or even ceramic materials. Operation at low temperatures also requires the use of more expensive materials. Shell thickness is generally between 6 and 75 mm. Large-diameter, high-pressure columns require thick shells to prevent shell rupture. Hoop-stress considerations alone dictate a shell thickness of 70 mm for a carbon steel column that is 3 m in diameter and operating at a pressure of 35 bars. At a height of 30 m such a vessel would weigh approximately 180,000 kg. Fortunately, most distillations are run at pressures much less than 35 bars, and thinner and less expensive columns can be employed. Column height also affects shell thickness. Height increases require shell thickness increases to combat wind forces. In addition, columns that are operated below atmospheric pressure require extra shell thickness and/or reinforcement rings to prevent column deformation or collapse. Most columns are wrapped with about

TABLE I Typical Steam Pressures Available for Distillation

Designation	Pressure (bars)	Condensation temperature (°C)
Low pressure	2.5	127
Medium pressure	15	198
High pressure	40	250

75 to 150 mm of insulation to prevent heat gain or loss, since distillation fluids are often at temperatures other than ambient.

Some distillation columns must handle two or more feed streams simultaneously. Furthermore, alternative feed nozzles are often provided to allow the actual feed-point locations to be altered. By optimizing the feed-point locations, energy consumption in the reboiler can often be minimized.

The most common energy source used in reboilers is steam. Most refineries and petrochemical plants have several steam pressure levels available. Some examples are listed in Table I. The condensation temperature of the steam used in the reboiler must be approximately 15°C greater than the boiling temperature of the bottom product. Other common heat sources used in reboilers are hot oil, hot water, and direct firing by burning oil or gas. In contrast, low-temperature columns, in ethylene plants, for example, often use propylene in a refrigeration circuit as the heating and cooling medium.

B. Column Internals

Sieve trays (Fig. 3) and valve trays (Fig. 4) are the two types of distillation trays most commonly used. In recent years these have supplanted previously widely used bubble-cap trays except when very large flow-rate range-abilities are needed. Figure 5 shows that liquid flows across the tray deck over the outlet weir and passes down the downcomer to the next tray.

Vapor passes through holes in the tray deck where it comes into contact with the liquid to form a froth, foam, or spray. Columns operating at high pressures typically must handle large volumetric liquid flow rates per unit cross-sectional column area. Under such conditions, multiple liquid flow passes are used. Figure 6 shows two- and four-pass arrangements. Compared with a single-pass tray (Fig. 5), multipass trays have more downcomer area and a longer total outlet weir length and are capable of handling higher liquid rates. However, the number of liquid flow passes is usually minimized since multipass trays are prone to liquid and vapor maldistribution and, because they are structurally more complex, they are more expensive.

Recently there has been an increasing trend to replace the conventional trays depicted in Fig. 5 by trays having receiving pans that terminate some 15 cm above the tray deck. This provides more column cross-sectional area for vapor flow and allows increased vapor capacity. Even greater vapor capacity can be obtained from trays that utilize localized, upward co-current flow of vapor and liquid. But, as each tray then requires a vapor–liquid separation device, they are more expensive and are used only in specialized applications.

As an alternative to trays, especially at low volumetric liquid-to-vapor ratios, packing can be used to promote vapor–liquid contact. One approach is to dump specially shaped pieces of metal, glass, or ceramic material into the column, wherein they are supported on a grid. An example of dumped or random packing is shown in Fig. 7.

FIGURE 3 Sieve tray. [Courtesy of Koch–Glitsch, Inc.]

FIGURE 4 Valve tray. [Courtesy of Koch–Glitsch, Inc.]

Another approach is to fabricate and install a precisely defined packing structure, which is carefully placed to fill the column. An example of a structured packing is shown in Fig. 8. Both types of packing are most commonly made from stainless steel. The surface area per unit volume is a key variable. Large surface area packings have lower efficiencies, higher capacities, lower pressure drops, and lower costs than small surface area packings.

Liquid is introduced into a packed column via a distributor (Fig. 9), which causes a large number of liquid streams to trickle over the surface of the packing. The design of the distributor is often critical for successful packed-column operation. Structured packing generally has a higher capacity for vapor–liquid flow than dumped packing when compared under conditions of identical mass-transfer performance, but is usually more expensive. In general, packing has a lower pressure drop than trays, although it is often more expensive and less reliable in operation. Structured packing has proven to be particularly advantageous in vacuum and air separation columns.

II. DISTILLATION THEORY

The process of distillation depends on the fact that the composition of the vapor that leaves a boiling liquid mixture is different from that of the liquid. Conversely, drops of liquid that condense from a vapor mixture differ in composition from the vapor.

A key physical property in distillation theory is the vapor pressure. Each pure component has a characteristic vapor pressure at a particular temperature, and vapor pressure increases with temperature and generally with a reduction in molecular weight. Vapor pressure is defined as the pressure at which a liquid and its vapor can coexist in equilibrium at a particular temperature.

The vapor pressure of a liquid mixture is given by the sum of the partial pressures of the constituents. Raoult's law is

$$p_i = p_i^{\circ} x_i \qquad (1)$$

where p_i is the partial pressure of component i, p_i° the vapor pressure of pure component i, and x_i the mole fraction of component i in the liquid. For a vapor mixture, Dalton's law is

$$p_i = y_i \pi \qquad (2)$$

where y_i is the mole fraction of component i in the vapor, and π is the total pressure. Combining Raoult's and Dalton's laws,

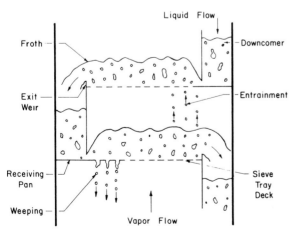

FIGURE 5 Single-pass distillation trays.

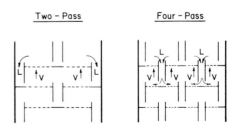

FIGURE 6 Multipass distillation trays.

FIGURE 7 Dumped packing. [Courtesy of Sulzer Chemtech, Ltd.]

$$y_i = (p_i°/\pi)x_i \qquad (3)$$

Equation (3) relates the composition of a liquid to the composition of its equilibrium vapor at any pressure and temperature (since $p_i°$ depends on temperature). Equation (3) is often written:

$$y_i = K_i x_i \qquad (4)$$

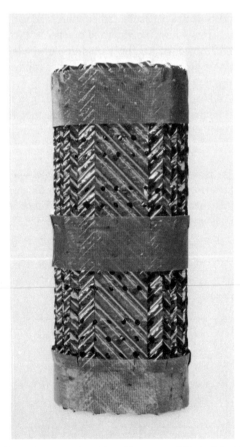

FIGURE 8 Structured packing. [Courtesy of Koch–Glitsch, Inc.]

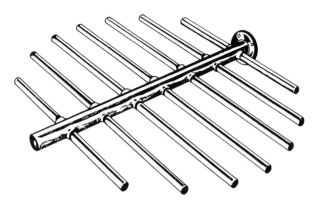

FIGURE 9 Packed column distributor.

where the equilibrium K value,

$$K_i = p_i°/\pi \qquad (5)$$

Mixtures that obey Eq. (5) exactly are termed *ideal mixtures.*

Deviations from ideality often occur, and the K_i value depends not only on temperature and pressure but also on the composition of the other components of the mixture. A more detailed discussion of vapor–liquid equilibrium relationships for nonideal mixtures is outside the scope of this article.

The relative volatility α of components 1 and 2 is obtained from Eq. (4) as:

$$\alpha_{12} = K_1/K_2 = p_1°/p_2° = (y_1/x_1)(x_2/y_2) \qquad (6)$$

For a binary mixture,

$$x_1 + x_2 = 1 \quad \text{and} \quad y_1 + y_2 = 1 \qquad (7)$$

Substituting into Eq. (6) gives:

$$y_1 = \alpha_{12}x_1/[1 + (\alpha_{12} - 1)x_1] \qquad (8)$$

Figure 10 shows the relationship between y_1 and x_1 for different values of α_{12} calculated from Eq. (8). When two components have close boiling points, by implication they have similar vapor pressures, so that α_{12} is close to unity. Separation of mixtures by distillation becomes more difficult as α_{12} approaches unity. Figure 11 indicates some of the x, y diagrams that can be obtained for distillation systems. Also shown are corresponding temperature–composition diagrams. The saturated vapor or dewpoint curve is determined by finding the temperature at which liquid starts to condense from a vapor mixture. Similarly, the saturated liquid or bubble-point curve corresponds to the temperature at which a liquid mixture starts to boil. For ideal mixtures, the dewpoint and bubble-point curves can be calculated as follows. From Eq. (3), at the dew point, since

$$\sum_{i=1}^{n} x_i = 1$$

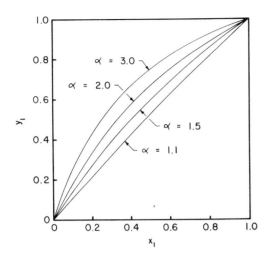

FIGURE 10 Vapor (y_1) versus liquid (x_1) concentration as a function of relative volatility.

where there are n components in the mixture,

$$\sum_{i=1}^{n}(y_i\pi/p_i^{\circ}) = 1 \qquad (9)$$

Similarly, at the bubble point,

$$\sum_{i=1}^{n} y_i = 1$$

Therefore,

$$\sum_{i=1}^{n}(p_i^{\circ} x_i/\pi) = 1 \qquad (10)$$

Since p_i° is a function of temperature, the dewpoint and bubble-point temperatures for an ideal vapor or liquid mixture can be determined as a function of the total pressure π from Eq. (9) or (10), respectively. An analogous procedure can be used for real mixtures, but the nonidealities of the liquid and vapor phases must be accounted for.

Azeotropes occur when $x_1 = y_1$, as indicated in Figs. 11c and d. Distillation of a mixture having the composition of an azeotrope is not possible since there is no difference in composition between vapor and liquid. Figure 11c shows how the azeotrope composition is affected as the pressure is changed.

When complex multicomponent mixtures are distilled, particularly those associated with oil refining, it is difficult to characterize them in terms of their components. Instead, they are characterized in terms of their boiling range, which gives some indication of the quantities of the components present. The true boiling point distillation (TBP) is probably the most useful, in which the percent distilled is recorded as a function of the boiling temperature of the mixture. For the TPB distillation, a 5 : 1 reflux ratio is often used with 15 theoretical stages in a laboratory characterization column (see Section III).

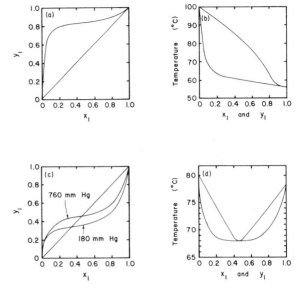

FIGURE 11 x, y and corresponding temperature–composition diagrams. (a, b) Acetone (1)–water at 1.0 bar; (c) ethanol (1)–benzene at 1.0 and 0.24 bar; (d) ethanol (1)–benzene at 1.0 bar.

III. DISTILLATION COLUMN DESIGN

It is convenient to perform calculations for both packed and trayed distillation columns in terms of theoretical equilibrium stages. A theoretical equilibrium stage is a contact process between liquid and vapor in which equilibrium is achieved between the streams leaving the theoretical stage. Figure 12 shows a representation of a theoretical stage. The compositions of y_{out} and x_{out} are in equilibrium, and the temperature and pressure of V_{out} and L_{out} are identical. The composition of y_{out} is related to x_{out} by an equilibrium relationship such as Eq. (4) or, for a binary mixture, Eq. (8). For calculation purposes, a distillation column can be modeled as a series of theoretical stages stacked one above the other. The design of a new distillation column to achieve a target separation can be broken down into a sequence of steps:

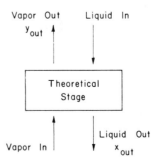

FIGURE 12 Theoretical stage concept.

1. Fix the pressure of operation of the column.
2. Determine the number of theoretical stages necessary to achieve the required separation as a function of the reflux ratio R.
3. Estimate the optimum value for R.
4. Relate the required number of theoretical stages to the actual height of the column needed.
5. Determine the necessary column diameter.
6. Refine steps 1 to 5 to achieve an optimum design.

The following sections deal with steps 1 to 5 in more detail.

A. Column Operating Pressure

The condensing temperature of the overhead vapor is reduced by lowering the column pressure. Very often, cooling water is used for condensation, and typically it has a temperature of ~35°C. Consequently, the condensing vapor must have a temperature of not less than ~50°C, and this sets the lower limit of the column operating pressure.

The boiling temperature of the bottoms product increases as the column pressure increases. Typically, medium-pressure steam, which has a temperature of ~200°C, is used in the reboiler.

When this steam is used for heating, the bottoms product cannot have a boiling temperature greater than ~185°C which sets an upper limit on the column operating pressure.

Other heating and cooling arrangements can be employed, such as the use of a refrigerant in the condenser or higher pressure steam in the reboiler, but they increase costs and are avoided whenever possible. An additional consideration that often limits the maximum temperature of the bottoms product is polymerization and product degradation at high temperatures (and therefore at high pressures). Furthermore, at lower pressures the relative volatility tends to increase so fewer theoretical stages are required, but at the same time the column diameter tends to increase.

As a result of these factors the distillation pressure varies widely. Typically, the distillation pressure falls as the molecular weight of the feed increases. Some typical operating pressures and temperatures are shown in Table II.

B. Calculation of the Required Number of Theoretical Stages

Figure 13 shows a McCabe–Thiele diagram, which can be used when the mixture to be distilled consists of only two components or can be represented by two components. Starting at the required overhead product composition x_D, an upper-section operating line is drawn hav-

TABLE II   Typical Operating Conditions in Distillation

	Pressure (bars), top	Temperature (°C)		Theoretical stages
		Top	Base	
Demethanizer	33	−94	−8	32
Deethanizer	28	−18	72	40
Ethane–ethylene splitter	21	−29	−45	80
Propane–propylene splitter	18	45	60	150
Isobutane-*n*-butane splitter	7	45	65	60
Deisohexanizer	1.6	55	120	60
Oxygen–nitrogen separation	1.1	−194	−178	70
Ethylbenzene–styrene separator	0.06	55	115	85
Crude oil distillation	0.03	93	410	—

ing a slope $R/(R + 1)$. The operating line for the lower section below the feed is drawn by joining the required bottom composition to a point located by the intersection of the upper section operating line and the q line. The q line of Fig. 13 represents a liquid feed at its bubble point, but the slope of the q line differs for other thermal conditions of the feed. The number of theoretical stages required is determined by stepping off between the operating lines and the equilibrium line, as shown in Fig. 13. Each step on the diagram represents a theoretical stage. For the example shown, only nine theoretical stages are required, but usually many more are needed in industrial columns.

In practice, feeds rarely consist of only two components, and the McCabe–Thiele diagram cannot be used.

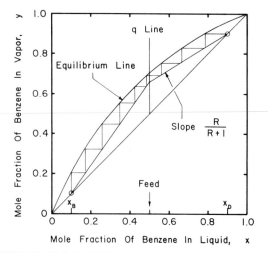

FIGURE 13 McCabe–Thiele diagram for benzene and toluene (top column pressure, 1.0 bar).

For multicomponent mixtures, the approach is to solve a complex system of matrix equations involving vapor and liquid compositions, flow rates from each theoretical stage, and temperature and pressure distributions through the column. This procedure, known as *tray counting* or *column simulation*, usually gives the required reflux ratio for specified product compositions and number of theoretical stages. Several commercial computer programs are available for tray counting.

C. Optimum Reflux Ratio

By using the procedures outlined in Section III.B, it is possible to determine the number of theoretical stages required to achieve the desired separation as a function of the reflux ratio (Fig. 14). Two limits are apparent: the minimum reflux ratio at which an infinite number of theoretical stages is necessary and the minimum number of theoretical stages that would be needed as the reflux ratio tends toward infinity. (A column operating with no feed and no product withdrawals operates at total reflux.) The optimum reflux ratio depends mainly on a balance between the investment cost of extra stages, hence extra column height, which results as R is reduced, and the operating cost of the heating medium used in the reboiler, which increases as R is increased. Generally, the optimum reflux ratio is about 1.2 to 1.5 times the minimum value.

D. Column Height

The number of actual trays required in a column can be determined from the calculated number of theoretical stages by invoking an efficiency. Various definitions of efficiency are used, but the simplest is an overall column efficiency E_o for which

$$\text{Actual trays} = \text{Theoretical stages}/E_o \quad (11)$$

For distillation, E_o is typically in the range 0.5 to 0.9. The vertical spacing between trays ranges from 200 to 900 mm. In some trayed columns, an undesirable bubbly foam can form above the liquid–vapor mixture. Antifoam chemicals must be added to such columns or diameters or tray spacings must be increased. Packed columns foam less often than trayed columns.

The required height of a packed column is determined from:

$$\text{Packed height} = \text{Theoretical stages} \times \text{HETP}$$

where HETP is the height equivalent of a theoretical plate. Note that the terms *plate, stage*, and *tray* tend to be used interchangeably. HETP varies with the packing size and is typically in the range of 250 to 800 mm.

E. Column Diameter

The column diameter is sized to suit the maximum anticipated rates of vapor and liquid flow through the column. Usually, the diameter is determined primarily by the vapor flow rate, and a rough estimate can be obtained from:

$$D = 4.5 Q_V^{0.5} [\rho_V/(\rho_L - \rho_V)]^{0.25} \quad (12)$$

where D is the column diameter in meters, Q_V is the vapor flow rate in cubic meters per second, and ρ_V and ρ_L are the vapor and liquid densities, respectively, in kilograms per cubic meter.

Columns operated at vapor and liquid flow rates greater than those for which they were designed become "flooded." Unexpected foaming can also cause flooding. In a flooded column, liquid cannot properly descend against the upflowing vapor. Poor separation performance results, the overhead condensation circuit fills with process liquid, the reboiler is starved of process liquid, and the column quickly becomes inoperable.

IV. APPLICATIONS OF DISTILLATION INCLUDING ENERGY CONSIDERATIONS

A. Flash Distillation

In contrast to the description of distillation given earlier, which dealt with multistage distillation, flash distillation (Fig. 15) is carried out in a single stage. Liquid flows continuously through a heater, across a valve, and into a flash vessel. By heating the liquid and reducing its pressure across the valve, partial vaporization occurs in the flash

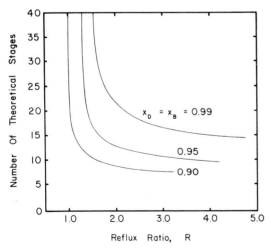

FIGURE 14 Theoretical stages versus reflux ratio (benzene–toluene at 1.0 bar). x_D, mole fraction benzene in overhead; x_B, mole fraction toluene in bottoms.

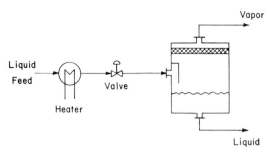

FIGURE 15 Flash distillation.

vessel. The temperature and pressure of the liquid entering the flash vessel are adjusted to achieve the required degree of vaporization. The compositions of the product streams leaving the flash vessel are different and are a function of the extent to which vaporization occurs.

Although the flash vessel itself is simple, care must be taken to ensure that the resultant vapor and liquid phases are separated completely from one another. To this end, the entering feed is often introduced tangentially rather than at a 90-degree angle to the vessel wall. An annular baffle directs the liquid droplets that are created by the flash toward the bottom of the vessel. By installing a wire mesh (approximately 75 mm thick) near the top of the vessel, fine liquid drops are prevented from leaving the top of the vessel as entrainment in the high-velocity vapor stream.

At best, only one theoretical stage is achieved by a flash distillation; however, it is used frequently in cryogenic and petroleum processing applications, where its simplicity is often attractive for nondemanding separations. Flashing often occurs in conventional distillation columns as feed and reflux streams enter. This flashing must be considered when column entrance devices and distributors are being designed.

B. Batch Distillation

Batch distillation (Fig. 16) is often preferable to continuous distillation when small quantities of feed material are processed. A liquid feed is charged to a still pot and heated until vaporization occurs. Vapor leaves the top of the column, and after condensation, part is removed as product and the rest returned to the column as reflux. As distillation proceeds, the contents of the still pot and the overhead product become richer in less volatile components. When operated at a fixed reflux ratio, an overhead product *cut* is collected until the product composition becomes unaccceptable. As an alternative, the reflux ratio can be gradually increased to hold the product composition constant as the cut is taken. For a fixed rate of heat addition to the still pot, the latter option results in a steadily declining product flow rate. After the first cut, subsequent cuts can be taken to obtain lower volatility products. Intermediate cuts of mixed composition are sometimes taken between each product cut, and these are saved and later returned to the still pot for inclusion in the next batch.

C. Extractive and Azeotropic Distillation

Conventional distillation tends to be difficult and uneconomical because of the large number of stages required when the relative volatility between the components to be separated is very low. In the extreme case, in which an unwanted azeotrope is formed, distillation past the azeotrope becomes impossible. Extractive or azeotropic distillation can sometimes be used to overcome these difficulties.

Both processes involve the addition of a new material, the *solvent*, to the mixture. The solvent is chosen so as to increase the relative volatility of the components to be separated. During extractive distillation, the solvent is generally added near the top of the column, and because it has a low volatility it is withdrawn with the product at the bottom. In azeotropic distillation, the solvent is withdrawn as an azeotrope with one or more of the components to be separated—usually in the overhead product. If the ratio of the components to be separated is different in the withdrawn azeotrope from their ratio in the feed to the column, then at least a partial separation has been achieved. In both processes it is necessary to separate the solvent from the product. This can be accomplished, for example, by distillation, solvent extraction, or even gravity settling, depending on the characteristics of the components involved.

D. Reactive Distillation

Many distillation columns reside upstream or downstream of catalytic reactors. Over the last decade, catalysts have

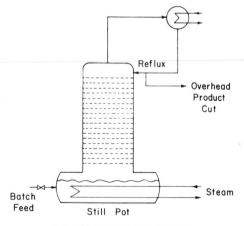

FIGURE 16 Batch distillation.

been increasingly employed inside distillation columns to simultaneously effect distillation and reaction. Oxygenates such as methyl-tert-butyl-ether (MTBE) and tertiary-methyl-ether (TAME) are produced in this manner for utilization within reformulated gasolines (RFGs). In reactive distillation, catalysts can be employed between the sheets of structured packings, on the decks or inside the downcomers of trays, or in dedicated beds between packed or trayed column sections. It is expected that reactive distillation will be used even more extensively in the future.

E. Energy Consumption

Approximately 30% of the energy used in U.S. chemical plants and petroleum refineries is for distillation, and it accounts for nearly 3% of the total U.S. annual energy consumption. The energy usage associated with some specific distillation products is shown in Table III. The cost of energy for distillation can be reduced by using waste heat such as is available from quench water in ethylene plants, for example, or exhaust steam from mechanical drivers such as compressors.

Energy costs can also be reduced by thermally linking neighboring distillation columns, as shown in Fig. 17. The overhead vapor from column 1 is condensed in an integrated condenser–reboiler, and the latent heat of condensation is used to boil the bottoms of column 2. In some cases, it may be necessary to operate columns 1 and 2 at different pressures so as to achieve the necessary temperature difference in the condenser–reboiler. The same strategy can be adopted for two columns performing identical separations in parallel. By raising the pressure of column 1, overhead vapors from column 1 can be used to drive column 2. The total energy consumption can be reduced by as much as half in this way.

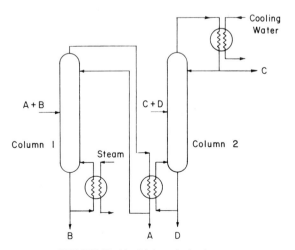

FIGURE 17 Heat-integrated columns.

TABLE III Distillation Energy Consumption

Component classification	Total U.S. distillation energy consumption (quads/yr)[a]	Specific distillation energy consumption (Btu/lb product)
Petroleum fuel fractions		
Crude distillation	0.36115	193
Vacuum distillation	0.08990	132
Catalytic hydrotreating/ hydrorefining	0.07726	101
Catalytic cracking fractionator	0.06803	112
Naphtha fractionator	0.06105	132
Catalytic hydrocracking	0.05964	632
Catalytic reforming	0.04988	132
Thermal operations	0.00936	60
Ethylene primary fractionator (naphtha/gas oil cracking)	0.00205	352
Total	0.77832	331
Light hydrocarbons		
Natural gas processing	0.07495	827
Ethylene and propylene	0.04821	1517
Alkylation HF	0.04701	1046
Alkylation H_2SO_4	0.03065	570
Light ends processing	0.01729	699
Isomerization	0.01312	803
Butadiene	0.01024	3151
Cyclohexane	0.00021	98
Total	0.24168	928
Water-oxygenated hydrocarbons		
Ethylene glycols	0.01065	2795
Ethanol	0.01063	9008
Phenol	0.00947	4344
Adipic acid	0.00739	4862
Methanol	0.00733	1175
Vinyl acetate (monomer)	0.00710	4797
Acetic acid	0.00701	2885
Isopropanol	0.00651	3785
Ethylene oxide	0.00554	1325
Methyl ethyl ketone	0.00481	9431
Terephthalic acid	0.00425	1756
Acetone	0.00417	2172
Dimethyl terephthalate	0.00412	1567
Formaldehyde	0.00412	733
Acetic anhydride	0.00267	1669
Propylene oxide	0.00219	1217
Glycerine	0.00202	14,870
Acetaldehyde	0.00174	1081
Total	0.10172	2366
Aromatics		
BTX[b]	0.02437	933
Styrene	0.01554	2467

continues

TABLE III (*continued*)

Component classification	Total U.S. distillation energy consumption (quads/yr)[a]	Specific distillation energy consumption (Btu/lb product)
Ethylbenzene	0.01388	2264
o-Xylene	0.00638	6019
Cumene	0.00390	1450
Total	0.06407	1515
Water-inorganics		
Sour water strippers	0.02742	240
Sodium carbonate	0.01398	1875
Urea	0.01030	133
Total	0.05170	411
Others		
Vinyl chloride (monomer)	0.01256	2188
Oxygen and nitrogen	0.00846	158
Acrylonitrile	0.00826	5434
Hexamethylenediamine	0.00612	8164
Total	0.03540	567
Remaining 30% of chemicals		
Production	0.10869	1973
Total for all component classifications	1.38158	623

From Mix, T. J., Dweck, J. S., and Weinberg, M. (1978). *Chem. Engr. Prog.* **74** (4), 49–55. Reproduced by permission of the American Institute of Chemical Engineers.

[a] 1 quad = 10^{15} Btu.

[b] Benzene-toluene-xylene.

A technique for energy reduction that has received considerable attention since 1970 is vapor recompression, or heat pumping. Vapor recompression takes advantage of the fact that when a vapor is compressed its temperature is simultaneously increased. Figure 18 shows typical temperatures and pressures associated with the use of heat pumping for splitting C_4 hydrocarbons. Through the use of a compressor, vapor leaving the top of the column is compressed from 3.8 bars and 27°C to 10.7 bars and 69°C. The compressed vapor is then hot enough to be used to boil the liquid at the bottom of the column, where the temperature is 46°C.

Vapor recompression eliminates the need for a conventional heat source, such as steam, to drive the reboiler. There is, however, an electrical energy requirement to drive the compressor which is not present in conventional distillation. The key advantage of vapor recompression is that the cost of running the compressor is often lower than the cost of driving a conventional reboiler. Under ideal conditions, the operating cost of a vapor recompression

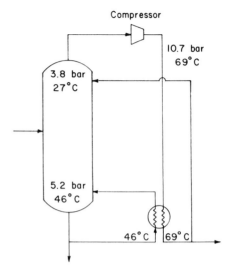

FIGURE 18 Vapor recompression.

system can be one-sixth of that associated with conventional distillation. As the temperature difference between the top and bottom of the column increases, compression costs become prohibitive. Vapor recompression is rarely used if the temperature difference exceeds 30°C.

F. Distillation Column Control

A typical control scheme for a distillation column is shown in Fig. 19. Flow controllers (FCs) regulate the flow rates of the feed and overhead products. Each flow rate is measured by a device such as an orifice plate placed upstream

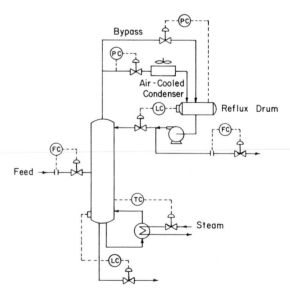

FIGURE 19 Typical distillation column control scheme.

of the control valve. The flow controller is used to open or close the control valve in response to differences between the measured flow rate and the target flow rate (the flow controller's set point). The rate of steam flow to the reboiler is regulated by measuring the temperature (usually with a thermocouple) at a point in the column and comparing this temperature to the set point of the temperature controller (TC). The rate of flow of the bottoms product is regulated by measuring the level of liquid in the column sump and opening or closing a control valve using a level controller (LC) to keep the level steady and at its set point. Similarly, the liquid level in the reflux drum is controlled by regulating the flow of reflux back to the column. Column pressure is controlled via a pressure controller (PC) acting on the condenser inlet valve, and the reflux drum pressure is controlled by a valve in the bypass line around the condenser.

The control scheme described is just one of a wide variety. In the past few years, the art and science of column control have developed rapidly, and now control system design tends to be the prerogative of the specialist control engineer.

SEE ALSO THE FOLLOWING ARTICLES

CHEMICAL THERMODYNAMICS • FLUID DYNAMICS, CHEMICAL ENGINEERING • FLUID MIXING • MEMBRANES, SYNTHETIC • PETROLEUM REFINING

BIBLIOGRAPHY

Billet, R. (1995). "Packed Towers," VCH, Weinheim.

Kister, H. Z. (1990). "Distillation Operation," McGraw–Hill, New York.

Kister, H. Z. (1992). "Distillation Design," McGraw–Hill, New York.

Lockett, M. J. (1986). "Distillation Tray Fundamentals," Cambridge University Press, Cambridge, U.K.

Luyben, W. L. (1992). "Practical Distillation Control," Van Nostrand–Reinhold, New York.

Seader, J. D., and Henley, E. J. (1998). "Separation Process Principles," Wiley, New York.

Shinskey, F. G. (1984). "Distillation Control for Productivity and Energy Conservation," McGraw–Hill, New York.

Stichlmair, J. G., and Fair, J. R. (1998). "Distillation," Wiley, New York.

Strigle, R. F. (1994). "Packed Tower Design and Applications," Gulf Pub., Houston, TX.

Taylor, R., and Krishna, R. (1993). "Multicomponent Mass Transfer," Wiley, New York.

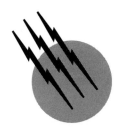

Distributed Parameter Systems

N. U. Ahmed
University of Ottawa

GLOSSARY

Banach space Normed vector space in which every Cauchy sequence has a limit; normed space complete with respect to the norm topology.

Cauchy sequence Sequence $\{x_n\} \in X$ is called a *Cauchy sequence* if $\lim_{n,m \to \infty} \|x_n - x_m\| = 0$; an element $x \in X$ is its limit if $\lim_{n \to \infty} \|x_n - x\| = 0$.

Normed vector space $X \equiv (X, \|\cdot\|)$ Linear vector space X furnished with a measure of distance between its elements $d(x, y) = \|x - y\|$, $\|x\| = d(0, x)$ satisfying: (a) $\|x\| \geq 0$, $\|x\| = 0$ iff $x = 0$, (b) $\|x + y\| \leq \|x\| + \|y\|$, (c) $\|\alpha x\| = |\alpha| \|x\|$ for all $x, y \in X$, and α scalar.

Reflexive Banach space Banach space X is reflexive if it is equivalent to (isomorphic to, indistinguishable from) its bidual $(X^*)^* \equiv X^{**}$.

Strictly convex normed space Normed space X is said to be strictly convex if, for every pair $x, y \in X$ with $x \neq y$, $\|\frac{1}{2}(x + y)\| < 1$ whenever $\|x\| \leq 1$, $\|y\| \leq 1$.

Uniformly convex normed space Normed space X is said to be uniformly convex if, for every $\varepsilon > 0$, there exists a $\delta > 0$, such that $\|x - y\| > \varepsilon$ implies that $\|\frac{1}{2}(x + y)\| < 1 - \delta$; for $\|x\| \leq 1$, $\|y\| \leq 1$; a uniformly convex Banach space is reflexive.

Weak convergence and weak topology Sequence $\{x_n\} \in X$ is said to be weakly convergent to x if, for every $x^* \in X^*$, $x^*(x_n) \to x^*(x)$ as $n \to \infty$; the notion of weak convergence induces a topology different from the norm topology and is called the weak topology; the spaces L_p, $1 \leq p < \infty$, are weakly (sequentially) complete.

Weak* convergence and w^*-topology Sequence $\{x_n^*\} \in X^*$ is said to be w^*-convergent to x^* if, for every $x \in X$, $x_n^*(x) \to x^*(x)$ as $n \to \infty$; the corresponding topology is called w^*-*topology*.

Weak and weak* compactness Set K in a Banach space X is said to be weakly (sequentially) compact if every sequence $\{x_n\} \in K$ has a subsequence $\{x_{n_k}\}$ and an $x_0 \in K$ such that, for each $x^* \in X^*$, $x^*(x_{n_k}) \to x^*(x_0)$; weak* compactness is defined similarly by interchanging the rolls of X^* and X.

DISTRIBUTED PARAMETER SYSTEMS are those whose state evolves with time distributed in space. More precisely, they are systems whose temporal evolution of state can be completely described only by elements of ∞-dimensional spaces. Such systems can be described by partial differential equations, integro-partial differential equations, functional equations, or abstract differential or integro-differential equations on topological spaces.

I. SYSTEM MODELS

Mysteries of the universe have baffled scientists, poets, philosphers, and even the prophets. Scientists have tried to learn the secrets of nature by building and studying mathematical models for various systems. It is a difficult task to give a precise definition of what is meant by a *system*. But, roughly speaking, you may think of any physical or abstract entity that obeys certain systematic laws, deterministic or probabilistic, as being a system.

Systems can be classified into two major categories: (a) lumped parameter systems, and (b) distributed parameter systems. Lumped parameter systems are those whose temporal evolution can be described by a finite number of variables. That is, they are systems having finite degrees of freedom and they are governed by ordinary differential or difference equations in a finite-dimensional space. On the other hand, distributed parameter systems are those whose temporal evolution can be described only by elements of an ∞-dimensional space called the *state space*. These systems are governed by partial differential equations or functional differential equations or a combination of partial and ordinary differential equations.

A. Examples of Physical Systems

In this section we present a few typical examples of distributed parameter systems. The Laplace equation in $\mathbb{R}^n$ is given by:

$$\Delta\phi \equiv \sum_{i=1}^{n} \frac{\partial^2 \phi}{\partial x_i^2} = 0 \qquad (1)$$

For $n = 3$, this equation is satisfied by the velocity potential of an irrotational incompressible flow, by the gravitational field outside the attracting masses, by electrostatic and magnetostatic fields, and also by the temperature of a body in thermal equilibrium. In addition to its importance on its own merit, the Laplacian is also used as a basic operator in diffusion and wave propagation problems. For example, the temperature of a body is governed by the so-called heat equation:

$$\frac{\partial T}{\partial t} = k\Delta T + f(t, x),$$

$$(t, x) \in I \times \Omega, \qquad \Omega \subset \mathbb{R}^3 \quad (2)$$

where k is the thermal conductivity of the material and f is an internal source of heat. The classical wave equation in R^n is given by:

$$\frac{\partial^2 \psi}{\partial t^2} = c^2 \Delta\psi \qquad (3)$$

where c denotes the speed of propagation. It may represent the displacement of a vibrating string, or propagation of acoustic or light waves and surface waves on shallow water.

Maxwell's equation, for a nonconducting medium with permeability μ and permittivity ε describing the temporal evolution of electric and magnetic field vectors H and E, respectively, in $\mathbb{R}_3$, is given by:

$$\mu\frac{\partial H}{\partial t} + \nabla \times E = 0,$$

$$\varepsilon\frac{\partial E}{\partial t} - \nabla \times H = 0$$

$$\text{div } E \equiv \nabla \cdot E = 0, \qquad (4)$$

$$\text{div } H \equiv \nabla \cdot H = 0$$

Under the assumptions of small displacement, the classical elastic waves in $\mathbb{R}^3$ are governed by the system of equations,

$$\rho\frac{\partial^2 y_i}{\partial t^2} = \mu\,\Delta y_i + (\lambda + \mu)\frac{\partial}{\partial x_i}(\text{div } y)$$

$$i = 1, 2, 3,$$

$$y = (y_1, y_2, y_3)$$

$$y_i = y_i(t, x_1, x_2, x_3) \quad (5)$$

where y represents the displacement of the elastic body from its unstrained configuration, ρ is the mass density, and λ, μ, are Lamé constants. In $\mathbb{R}^3$, the state of a single particle of mass m subject to a field of potential $v = v(t, x_1, x_2, x_3)$ is given by the Schrödinger equation:

$$i\hbar\frac{\partial \psi}{\partial t} = -\frac{\hbar^2}{2m}\Delta\psi + v\psi \qquad (6)$$

where $\hbar = 2\pi h$ is Planck's constant.

In recent years the nonlinear Schrödinger's equation has been used to take into account nonlinear interaction of particles in a beam by replacing $v\psi$ by a suitable nonlinear function $g(t, x, \psi)$.

The equation for an elastic beam allowing moderately large vibration is governed by a nonlinear equation of the form:

$$\rho A\frac{\partial^2 y}{\partial t^2} + \beta\frac{\partial y}{\partial t} + \frac{\partial^2}{\partial x^2}\left(EI\frac{\partial^2 y}{\partial x^2}\right) + N\frac{\partial^2 y}{\partial x^2} = f(t, x)$$

$$N = \frac{EA}{2l}\int_0^l \left(\frac{\partial y}{\partial x}\right)^2 dx \qquad (7)$$

$$x \in (0, l), \qquad t \geq 0$$

where E denotes Young's modulus, I the moment of area of the cross section A of the beam, l the length, ρ the mass density, N the membrane force, and f the applied force. For small displacements, neglecting N and β, one obtains the Euler equation for thin beams. The dynamics of Newtonian fluid are governed by the Navier–Stokes equations:

$$\rho\left(\frac{\partial v}{\partial t} + v \cdot \nabla v\right) - \nu \,\Delta v - (3\lambda + \nu)\, \text{grad div } v$$

$$+ \text{grad } p = f \tag{8}$$

$$\frac{\partial \rho}{\partial t} + \text{div}(\rho v) = 0, \qquad t \geq 0, \qquad x \in \Omega \subset \mathbb{R}^n$$

obtained from momentum and mass conservation laws, where ρ is the mass density, v the velocity vector, p the pressure, f the force density, and ν, λ are constant parameters.

Magnetohydrodynamic equations for a homogeneous adiabatic fluid in $\mathbb{R}^3$ is given by:

$$\frac{\partial v}{\partial t} + v \cdot \nabla v + \rho^{-1} \nabla p + (\mu\rho)^{-1}(B \times \text{rot } B) = f$$

$$\frac{\partial \rho}{\partial t} + \text{div}(\rho v) = 0, \qquad \frac{\partial s}{\partial t} + v \cdot \nabla s = 0 \tag{9}$$

$$\frac{\partial B}{\partial t} - \text{rot}(v \times B) = 0, \qquad \text{div } B = 0$$

$$p = g(\rho, s), \qquad \text{rot } B \equiv \nabla \times B$$

where v, ρ, p, and f are as in Eq. (8); s is the entropy; B, the magnetic induction vector; and μ, the permeability.

In recent years semilinear parabolic equations of the form,

$$\frac{\partial \phi}{\partial t} = D\,\Delta\phi + f(t, x; \phi), \qquad t \geq 0, \qquad x \in \Omega \subset \mathbb{R}^n \tag{10}$$

have been extensively used for modeling biological, chemical, and ecological systems, where D represents the migration or diffusion coefficient.

The dynamics of a spacecraft with flexible appendages is governed by a coupled system of ordinary and partial differential equations.

In the following two sections we present abstract models that cover a wide variety of physical systems including those already mentioned.

B. Linear Systems

A general spatial differential operator used to construct system models is given by:

$$A(x, D)\phi \equiv \sum_{|\alpha| \leq 2m} a_\alpha(x)\, D^\alpha \phi$$

$$x \in \Omega \equiv \text{open subset of } \mathbb{R}^n \tag{11}$$

Under suitable smoothness conditions on the coefficient a_α and the boundary $\partial\Omega$ of Ω one can express Eq. (11) in the so-called divergence form:

$$A(x, D)\phi \equiv \sum_{|\alpha|, |\beta| \leq m} (-1)^{|\alpha|} D^\alpha \left(a_{\alpha\beta} D^\beta \phi\right) \tag{12}$$

The operator A is said to be strongly uniformly elliptic on Ω if there exists a $\gamma > 0$ such that

$$(-1)^m \, \text{Re}\left\{\sum_{|\alpha|=2m} a_\alpha(x)\xi^\alpha\right\} \geq \gamma|\xi|^{2m}$$

$$\text{for all} \quad x \in \Omega \tag{13}$$

or

$$\text{Re}\left\{\sum_{|\alpha|=|\beta|=m} a_{\alpha\beta}(x)\xi^\alpha\xi^\beta\right\} \geq \gamma|\xi|^{2m} \tag{14}$$

Many physical processes in steady state (for example, thermal equilibrium or elastic equilibrium) can be described by elliptic boundary value problems given by:

$$A(x, D)\phi = f \qquad \text{on} \quad \Omega$$

$$B(x, D)\phi = g \qquad \text{on} \quad \partial\Omega \tag{15}$$

where $B = \{B_j,\ 0 \leq j \leq m-1\}$ is a system of suitable boundary operators. For example, the boundary operator may be given by the Dirichlet operator,

$$B \equiv \left\{\frac{\partial^j}{\partial \nu^j}, 0 \leq j \leq m - 1\right\}$$

where $\partial/\partial\nu$ denotes spatial derivatives along the outward normal to the boundary $\partial\Omega$. The boundary operators $\{B_j\}$ cannot be chosen arbitrarily; they must satisfy certain compatibility conditions with respect to the operator A. Only then is the boundary value problem (15) well posed; that is, one can prove the existence of a solution and its continuous dependence on the data f and $g = \{g_j,\ 0 \leq j \leq m-1\}$. The order of each of the operators B_j is denoted by m_j.

Evolution equations of parabolic type arise in the problems of heat transfer, chemical diffusions, and also in the study of Markov processes. The most general model describing such phenomenon is given by:

$$\frac{\partial \phi}{\partial t} + A\phi = f, \qquad t, x \in I \times \Omega = Q$$

$$B\phi = g, \qquad t, x \in I \times \partial\Omega \tag{16}$$

$$\phi(0) = \phi_0, \qquad x \in \Omega$$

where f, g, and ϕ_0 are the given data.

Second-order evolution equations of hyperbolic type describing many vibration and wave propagation problems have the general form:

$$\frac{\partial^2 \phi}{\partial t^2} + A\phi = f, \qquad t, x \in I \times \Omega$$

$$B\phi = g, \qquad t, x \in I \times \Omega$$

$$\phi(0) = \phi_0, \qquad\qquad\qquad (17)$$
$$x \in \Omega$$
$$\dot{\phi}(0) = \phi_1,$$

Schrödinger-type evolution equations are obtained if A is replaced by iA in Eq. (16).

In the study of existence of solutions of these problems, Garding's inequality is used for *a priori* estimates. If A is strongly uniformly elliptic, the principal coefficients satisfy Hölder conditions on Ω uniformly with respect to $t \in I$, and the other coefficients are bounded measurable, then one can prove the existence of a $\lambda \in \mathbb{R}$ and $\alpha > 0$ such that

$$a(t, \phi, \phi) + \lambda |\phi|_H^2 > \alpha \|\phi\|_V^2, \qquad t \in I \qquad (18)$$

where

$$a(t, \phi, \psi) \equiv \sum_{|\alpha|, |\beta| \leq m} \langle a_{\alpha, \beta}(t, \cdot) D^\beta \phi, D^\alpha \psi \rangle_\Omega \qquad (19)$$

and V is any reflexive Banach space continuously and densely embedded in $H = L_2(\Omega)$ where $L_p(\Omega)$ is the equivalence classes of pth-power Lebesgue integrable functions on $\Omega \subset R^n$, with the norm given by:

$$\|f\|_{L_{p(\Omega)}} \equiv \begin{cases} \left(\int \Omega |f(x)|^p \, dx \right)^{1/p} & \text{for} \quad 1 \leq p < \infty \\ \text{ess sup}\{f(x), x \in \Omega\} & \text{for} \quad p = \infty \end{cases}$$

For example, V could be $W_0^{m,p}$, $W^{m,p}$, or $W_0^{m,p} \subset V \subset W^{m,p}$ for $p \geq 2$ where $W_0^{m,p}(\Omega)$ is the closure of C^∞ functions with compact support on Ω in the topology of $W^{m,p}$, and

$$W^{m,p}(\Omega) \equiv \left\{ f \in L_p(\Omega) : D^\alpha f \in L_p(\Omega), \ |\alpha| \leq m \right\}$$

furnished with the norm topology:

$$\|f\|_{W^{m,p}} \equiv \sum_{|\alpha| \leq m} \|D^\alpha f\|_{L_{p(\Omega)}}, \qquad p \geq 1$$

$$D^\alpha \equiv D_1^{\alpha_1} D_1^{\alpha_2} \cdots D_n^{\alpha_n}, \qquad D_i^{\alpha_i} \equiv \frac{\partial^{\alpha_i}}{\partial x_i^{\alpha_i}}$$

$$\|\alpha\| = \sum_{i=1}^n \alpha_i, \qquad \alpha_i \equiv \text{nonnegative integers}$$

C. Nonlinear Evolution Equations and Differential Inclusions

Nonlinear systems are more frequently encountered in practical problems than the linear ones. There is no clear-cut classification for these systems. We present here a few basic structures that seem to cover a broad area in the field.

1. Elliptic Systems

The class of elliptic problems which have received considerable attention in the literature is given by:

$$A\psi = 0 \qquad \text{on} \quad \Omega$$
$$D^\alpha \psi = 0 \qquad \text{on} \quad \partial\Omega, \qquad |\alpha| \leq m - 1 \qquad (20)$$

where

$$A\psi \equiv \sum_{|\alpha| \leq m} (-1)^{|\alpha|} D^\alpha(|D^\alpha \psi|^{p-2} D^\alpha \psi) + \sum_{|\beta| \leq m-1} (-1)^{|\beta|}$$
$$\times D^\beta \big(b_\beta(x, \psi, D^1 \psi, \ldots, D^m \psi) \big) \qquad (21)$$

2. Semilinear Systems

The class of nonlinear evolution equations that can be described in terms of a linear operator and a nonlinear operator containing lower order derivatives has been classified as semilinear. These systems have the form:

$$\frac{d\psi}{dt} + A(t)\psi + f(t, \psi) = 0$$
$$\psi(0) = \psi_0 \qquad (22)$$

and

$$\frac{d^2\psi}{dt^2} + A(t)\psi + f(t, \psi) = 0$$
$$\psi(0) = \psi_0, \qquad (23)$$
$$\dot{\psi}(0) = \psi_1$$

where $A(t)$ may be a differential operator of the form (11) and the nonlinear operator f may be given by:

$$f(t, \psi) \equiv \sum_{|\alpha| \leq 2m-1} (-1)^{|\alpha|} D^\alpha$$
$$\times \big(b_\alpha(t, \cdot; \psi, D^1 \psi, \ldots, D^{2m-1} \psi) \big) \qquad (24)$$

For example, a second-order semilinear parabolic equation is given by:

$$\frac{\partial \phi}{\partial t} + \sum_{i, J} a_{ij}(t, x)\phi_{x_i x_j} + a(t, x, \phi, \phi_x) = 0 \qquad (25)$$

which certainly covers the ecological model, Eq. (10). In short, Eq. (22) is an abstract model for a wide variety of nonlinear diffusions. It covers the Navier–Stokes equation (8) and many others including the first-order semilinear hyperbolic system:

$$\frac{\partial y}{\partial t} + \left(\sum_{i=1}^{n} A_i(t, x) \frac{\partial y}{\partial x_i} + B(t, x)y \right) + f(t, x, y) = 0 \tag{26}$$

The second-order abstract semilinear equation (23) covers a wide variety of nonlinear vibration and wave propagation problems.

3. Quasilinear Systems

The most general form of systems governed by quasilinear evolution equations that have been treated in the literature is given by:

$$\frac{d\psi}{dt} + A(t, \psi)\psi + f(t, \psi) = 0 \tag{27}$$
$$\psi(0) = \psi_0$$

It covers the quasilinear second-order parabolic equation or systems of the form:

$$\frac{\partial \psi}{\partial t} + \sum_{i,J}^{n} a_{ij}(t, x; \psi, \psi_x)\psi_{x_i x_j} + a(t, x, \psi, \psi_x) = 0 \tag{28}$$

where, for parabolicity, one requires that

$$a_{ij}(t, x, p, q)\xi_i\xi_j \geq \gamma|\xi|^2$$
$$\gamma > 0, \qquad t, x \in Q, \qquad p \in \mathbb{R}, \qquad q \in \mathbb{R}^n$$

It covers the quasilinear hyperbolic systems of the form:

$$A_0(t, x, y)\frac{\partial y}{\partial t} + \sum_{i}^{n} A_i(t, x, y)\frac{\partial y}{\partial x_i} + B(t, x, y) = 0 \tag{29}$$

including the magnetohydrodynamic equation (9).

4. Differential Inclusions

In recent years abstract differential inclusions have been used as models for controlled systems with discontinuities. For example, consider the abstract semilinear equation (22). In case the operator f, as a function of the state ψ, is discontinuous but otherwise well defined as a set-valued function in a suitable topological space, one may consider Eq. (22) as a differential inclusion:

$$-\dot{\psi}(t) \in A(t)\psi(t) + f(t, \psi(t)) \qquad \text{a.e.} \tag{30}$$
$$\psi(0) = \psi_0$$

Such equations also arise from variational inequalities.

D. Stochastic Evolution Equations

In certain situations because of lack of precise knowledge of the system parameters arising from inexact observation or due to gaps in our fundamental understanding of the physical world, one may consider stochastic models to obtain most probable answers to many scientific questions. Such models may be described by stochastic evolution equations of the form:

$$d\psi = (A(t)\psi + f(t, \psi)) \, dt + \sigma(t, \psi) \, dW \tag{31}$$
$$\psi(0) = \psi_0$$

where A is a differential operator possibly of the form (11), f is a nonlinear operator of the form (24), and σ is a suitable operator-valued function. The variable $W = \{W(t), t \geq 0\}$ represents a Wiener process with values in a suitable topological space and defined on a probability space. The uncertainty may arise from randomness of ψ_0, the process W, and even from the operators A, f, and σ. This is further discussed in Sections II and III.

II. LINEAR EVOLUTION EQUATIONS

In this section, we present some basic results from control theory for systems governed by linear evolution equations.

A. Existence of Solutions for Linear Evolution Equations and Semigroups

A substantial part of control theory for distributed parameter systems has been developed for first- and second-order evolution equations of parabolic and hyperbolic type of the form:

$$\frac{d\phi}{dt} + A(t)\phi = f, \qquad \phi(0) = \phi_0 \tag{32}$$

$$\frac{d^2\phi}{dt^2} + A(t)\phi = f, \qquad \phi(0) = \phi_0 \tag{33}$$
$$\dot{\phi}(0) = \phi_1$$

A fundamental question that must be settled before a control problem can be considered is the question of existence of solutions of such equations.

We present existence theorems for problems (32) and (33) and conclude the section with a general result when the operator A is only a generator of a c_0-semigroup (in case of constant A) or merely the generator of an evolution operator in a Banach space (in case A is variable). The concepts of semigroups and evolution operators are introduced at a convenient place while dealing with the questions of existence.

For simplicity we consider time-invariant systems, although the result given below holds for the general case.

Theorem 1. Consider system (33) with the operator A time invariant and self-adjoint and suppose it satisfies the conditions:

(a1) $|\langle A\phi, \psi\rangle| \le c\|\phi\|_V \|\psi\|_V$

$\qquad c \ge 0, \qquad \phi, \psi \in V$

(a2) $\langle A\phi, \phi\rangle + \lambda|\phi|_H^2 \ge \alpha\|\phi\|_V^2$

$\qquad \lambda \in R, \qquad \alpha > 0, \qquad \phi \in V$

Then, for every $\phi_0 \in V$, $\phi_1 \in H$, and $f \in L_2(I, H)$, system (33) has a unique solution ϕ satisfying:

(c1) $\phi \in L_\infty(I, V) \cap C(\bar{I}, V)$

(c2) $\dot{\phi} \in L_\infty(I, V) \cap C(\bar{I}, H)$

and

(c3) $(\phi_0, \phi_1, f) \to (\phi, \dot{\phi})$

is a continuous mapping from $V \times H \times L_2(I, H)$ to $C(\bar{I}, V) \times C(\bar{I}, H)$.

Proof (Outline). The proof is based on Galerkin's approach which converts the infinite-dimensional system (33) into its finite-dimensional approximation, then by use of *a priori* estimates and compactness arguments one shows that the approximating sequence has a subsequence that converges to the solution of problem (33). $\qquad\square$

Note that the second-order evolution equation (33) can be written as a first-order equation $d\psi/dt + \tilde{A}\psi = \tilde{f}$ where

$$\psi = \begin{bmatrix} \phi \\ \dot{\phi} \end{bmatrix}, \qquad \tilde{A} = \begin{bmatrix} 0 & -I \\ A & 0 \end{bmatrix}$$

$$\tilde{f} = \begin{bmatrix} 0 \\ f \end{bmatrix} \qquad (34)$$

Defining $X = V \times H$, with the product topology, as the state space, it follows from Theorem 1(c3) that there exists an operator-valued function $S(t)$, $t \ge 0$, with values $S(t) \in \mathcal{L}(X)$ so that

$$\psi(t) = S(t)\psi_0 + \int_0^t S(t-\theta)\tilde{f}(\theta)\, d\theta \qquad (35)$$

The family of operators $\{S(t), t \ge 0\}$ forms a c_0-semigroup in X and $\psi \in C(\bar{I}, X)$ where $C(I, X)$ is the space of continuous functions on I with values in the Banach space X with the norm (topology):

$$\|f\| = \sup\{|f(t)|_X, t \in I\}$$

For system (32) one can prove the following result using the same procedure.

Theorem 2. Consider system (32) and suppose that the operator A satisfies assumptions (a1) and (a2) (see Theorem 1). Then, for each $\phi_0 \in H$ and $f \in L_2(I, V^*)$, system (32) has a unique solution:

$$\phi \in L_2(I, V) \cap L_\infty(I, H) \cap C(\bar{I}, H)$$

and further

$$(\phi_0, f) \to \phi$$

is a continuous map from $H \times L_2(I, V^*)$ to $C(\bar{I}, H)$.

As a consequence of this result there exists an evolution operator $U(t, s)$, $0 \le s \le t \le \infty$, with values $U(t, s) \in \mathcal{L}(H)$ such that

$$\phi(t) = U(t, 0)\phi_0 + \int_0^t U(t, \theta)f(\theta)\, d\theta \qquad (36)$$

is a (mild) weak solution of problem (32).

We conclude this section with some results on the question of existence of solutions for a class of general time-invariant linear systems on Banach space.

Let X be a Banach space, and $S(t)$, $t \ge 0$, a family of bounded linear operators from X to X satisfying the properties:

(a) $S(0) = I$ (identity operator)

(b) $S(t + \tau) = S(t)S(\tau), \qquad t, \tau \ge 0$ $\qquad (37)$

(c)   Strong $\lim_{t\downarrow 0^+} S(t)\xi = \xi \in X$

The operator $S(t)$, $t \ge 0$, satisfying the above properties is called a *strongly continuous semigroup* or, in short, a c_0-semigroup. Let A be a closed, densely defined linear operator with domain $D(A) \subset X$ and range $R(A) \subset X$. Suppose there exist numbers $M > 0$ and $\omega \in R$ such that

$$\|(\lambda I - A)^{-1}\|_X \le M/(\lambda - \omega)$$

for all real $\lambda > \omega$. Then, by a fundamental theorem from the semigroup theory known as the Hille–Yosida theorem, there exists a unique c_0-semigroup $S(t)$, $t \ge 0$, with A as its infinitesimal generator. The semigroup $S(t)$, $t \ge 0$, satisfies the properties:

(a) $\|S(t)\|_{\mathcal{L}(X)} \le Me^{\omega t}, \qquad t \ge 0$

(b)   for $\xi \in D(A)$

$\qquad S(t)\xi \in D(A) \qquad$ for all $t \ge 0$ $\qquad (38)$

(c)   for $\xi \in D(A)$

$$\frac{d}{dt}S(t)\xi = AS(t)\xi = S(t)A\xi, \qquad t \ge 0$$

If $\omega = 0$, we have a *bounded semigroup*; for $\omega = 0$ and $M = 1$ we have a *contraction semigroup*, and for $\omega < 0$, we have a *dissipative semigroup*. In general, the abstract Cauchy problem:

$$\frac{dy}{dt} = Ay, \qquad y(0) = \xi \qquad (39)$$

has a unique solution $y(t) = S(t)\xi$, $t > 0$, with $y(t) \in D(A)$, provided $\xi \in D(A)$. If $\xi \in D(A)$ and f is any strongly continuously differentiable function with values $f(t) \in X$, then the inhomogeneous problem,

$$\frac{dy}{dt} = Ay + f, \qquad y(0) = \xi \qquad (40)$$

has a unique continuously differentiable solution y given by:

$$y(t) = S(t)\xi + \int_0^t S(t-\theta) f(\theta) \, d\theta$$

$$\text{with} \quad y(t) \in D(A) \qquad \text{for all} \quad t \geq 0 \quad (41)$$

A solution satisfying these conditions is called a *classical solution*. For control problems these conditions are rather too strong since, in general, we do not expect the controls, for example $f(t)$, to be even continuous. Thus, there is a need for a broader definition and this is provided by the so-called mild solution.

Any function $y: I \to X$ having the integral representation (41) is called a mild solution of problem (40). In this regard we have the following general result.

Theorem 3. Suppose A is the generator of a c_0-semigroup $S(t)$, $t \geq 0$, in X and let $y_0 \in X$ and $f \in L_p(I, X)$, $1 \leq p \leq \infty$, where $L_p(I, X)$ is the space of strongly measurable functions on I taking values in a Banach space X with norm:

$$\|f\|_{L_p(I,X)} = \begin{cases} \left(\int_I |f(t)|_X^p \, dt \right)^{1/p}, & 1 \leq p < \infty \\ \text{ess sup } \{|f(t)|_X, t \in I\}, & p = \infty \end{cases}$$

Then, evolution Eq. (40) has a unique mild solution y given by:

$$y(t) = S(t)y_0 + \int_0^t S(t-\theta) f(\theta) \, d\theta \qquad (42)$$

In this case $y(t)$ does not necessarily belong to $D(A)$. Another special but important class of strongly continuous semigroups $S(t)$, $t \geq 0$, that satisfies the property:

$$S(t)X \subset D(A) \qquad \text{for all} \quad t > 0$$

is called the *analytic* (holomorphic, parabolic) *semigroup*. An analytic semigroup has the following properties:

(a) $S(t)X \subset D(A^n)$ for all integers $n \geq 0$ and $t > 0$,

(b) $S(t)$, $d^n S(t)/dt^n = A^n S(t)$ are all bounded operators for $t > 0$.

One reason for calling the analytic semigroups parabolic semigroups is that $S(t)$, $t \geq 0$, turns out to be the fundamental solution of certain parabolic evolution equations. Consider the differential operator,

$$L(x, D) = \sum_{|\alpha| \leq 2m} a_\alpha(x) D^\alpha$$

and suppose it is strongly elliptic of order $2m$ and

$$B_j(x, D) = \sum_{|\alpha| \leq m_j} b_{j,\alpha} D^\alpha, \qquad 0 \leq j \leq m - 1$$

is a set of normal boundary operators as defined earlier. Define

$$D(A) = W^{2m,p}(\Omega) \cap W_0^{m,p}(\Omega) \qquad (43)$$

$$(A\phi)(x) = -L(x, D)\phi(x), \qquad \phi \in D(A)$$

Then, under certain technical assumptions, A generates an analytic semigroup $S(t)$, $t \geq 0$, in the Banach space $X = L_p(\Omega)$. The initial boundary value problem,

$$\frac{\partial \phi}{\partial t}(t, x) + L(x, D)\phi(t, x) = f(t, x)$$

$$t \in (0, T), \qquad x \in \Omega$$

$$(D^\alpha \phi)(t, x) = 0 \qquad (44)$$

$$|\alpha| \leq m - 1, \qquad t \in (0, T), \qquad x \in \partial\Omega$$

$$\phi(0, x) = \phi_0(x), \qquad x \in \Omega$$

can be considered to be an abstract evolution equation in $X = L_p(\Omega)$ and be written as:

$$\frac{d\phi}{dt} = A\phi + f$$

$$\phi(0) = \phi_0$$

with mild solution given by:

$$\phi(t) = S(t)\phi_0 + \int_0^t S(t-\theta) f(\theta) \, d\theta \qquad (45)$$

where $f \in L_p(I, X)$.

B. Stability

The question of stability is of signnificant interest to system scientists, since every physical system must be stable to function properly.

In this section, we present a Lyapunov-like result for the abstract evolution equation,

$$\frac{dy}{dt} = Ay \qquad (46)$$

where A is the generator of a strongly continuous semigroup $S(t)$, $t \geq 0$ on H. Let $\mathscr{L}^+(H)$ denote the class of symmetric, positive, self-adjoint operators in H; that is,

$T \in \mathscr{L}(H)$, $T = T^*$, and $(T\xi, \xi) > 0$ for $\xi \ (\neq 0) \in H$. We can prove the following result.

Theorem 4. A necessary and sufficient condition for the system $dy/dt = Ay$ to be exponentially stable in the Lyapunov sense (i.e., there exist $M \geq 1$, $\beta > 0$ such that $\|y(t)\| \leq Me^{-\beta t}$) is that the operator equation,

$$A^*Y + YA = -\Gamma \tag{47}$$

has a solution $Y \in \mathscr{L}^+(H)$ for each $\Gamma \in \mathscr{L}^+(H)$ with $(\Gamma\xi, \xi) \geq \gamma |\xi|_H^2$ for some $\gamma > 0$.

REMARK. Equation (47) is understood in the sense that the equality,

$$0 = (\Gamma\xi, \eta) + (A\xi, Y\eta) + (Y\xi, A\eta)$$

holds for all $\xi, \eta \in D(A)$.

Corollary 5. If the autonomous system (46) is asymptotically stable, then the system,

$$\frac{dy}{dt} = Ay + f, \qquad t \geq 0 \tag{48}$$

is stable in the L_p sense; that is, for every $y_0 \in H$ and input $f \in L_p(0, \infty; H)$, the output $y \in L_p(0, \infty; H)$ for all $1 \leq p \leq \infty$ and in particular, for $1 \leq p < \infty$, $y(t) \to 0$ as $t \to \infty$.

We conclude this section with a remark on the solvability of Eq. (47). Let $\{\lambda_i\}$ be the eigenvalues of the operator A, each repeated as many times as its multiplicity requires, and let $\{\xi_i\}$ denote the corresponding eigenvectors complete in H. Consider Eq. (47) and form

$$\left(A^*Y\xi_i, \xi_j\right) + (YA\xi_i, \xi_j) = -(\Gamma\xi_i, \xi_j) \tag{49}$$

for all integers $i, j \geq 1$. Clearly, from this equation, it follows that

$$(Y\xi_i, \xi_j) = -(\Gamma\xi_i, \xi_j)/(\lambda_i + \bar{\lambda}_j) \tag{50}$$

Hence, if $\lambda_i + \bar{\lambda}_j \neq 0$ for all i, j, Γ determines Y uniquely, and if $\lambda_i + \bar{\lambda}_j = \mathrm{Re}\ \lambda_i < 0$ for all i, then $Y \in \mathscr{L}^+(H)$. In other words, if system (46) is asymptotically stable then the operator equation (47) always has a positive solution.

C. System Identification

A system analyst may know the structure of the system—for example, the order of the differential operator and its type (parabolic, hyperbolic, etc.)—but the parameters are not all known. In that case, the analyst must identify the unknown parameters from available information. Consider the natural system to be given by $\dot{y}^* = A(q^*)y^*$. Assume that q^* is unknown to the observer but the observer can observe certain data z^* from a Hilbert space K, the output space, which corresponds to the natural history y^*. The

observer chooses a feasible set Q, where q^* may possibly lie, and constructs the model system,

$$\dot{y}(q) = A(q)y(q), \qquad y(q)(0) = y_0$$
$$z(q) = Cy(q) \tag{51}$$

where C is the observation or output operator, an element of $\mathscr{L}(H, K)$. The analyst may choose to identify the parameter approximately by minimizing the functionl,

$$J(q) = \frac{1}{2} \int_0^T |z(q) - z^*|_K^2 \, dt \qquad \text{over } Q \tag{52}$$

Similarly, one may consider identification of an operator appearing in system equations. For example, one may consider the system,

$$\frac{d^2y}{dt^2} + Ay + B^*y = f$$
$$y(0) = y_0, \qquad \dot{y}(0) = y_1 \tag{53}$$
$$z = Cy$$

where the operator A is known but the operator B^* is unknown. One seeks an element B from a feasible set $P^0 \subset \mathscr{L}(V, V^*)$ so that

$$J(B) = \int_0^T g(t, y(B), \dot{y}(B)) \, dt \tag{54}$$

is minimum, where g is a suitable measure of discrepancy between the model output, $z(B) = Cy(B)$, and the observed data z^* corresponds to the natural history y^*.

In general, one may consider the problem of identification of all the operators A, B including the data y_0, y_1, and f. For simplicity, we shall consider only the first two problems and present a couple sample results. First, consider problem (51) and (52).

Theorem 6. Let the feasible set of parameters Q be a compact subset of a metric space and suppose for each $q \in Q$ that $A(q)$ is the generator of a strongly continuous contraction semigroup in H. Let

$$(\gamma I - A(q_n))^{-1} \to (\gamma I - A(q_0))^{-1} \tag{55}$$

in the strong operator topology for each $\gamma > 0$ whenever $q_n \to q_0$ in Q. Then there exists $q^0 \in Q$ at which $J(q)$ attains its minimum.

Proof. The proof follows from the fact that under assumption (55) the semigroup $S_n(t)$, $t \geq 0$, corresponding to q_n strongly converges on compact intervals to the semigroup $S_0(t)$, $t \geq 0$, corresponding to q_0. Therefore, J is continuous on Q and, Q being compact, it attains its minimum on Q. $\qquad \square$

The significance of the above result is that the identification problem is well posed.

For the second-order evolution equation (53) we have the following result.

Theorem 7. Consider system (53). Let P^0 be a compact (in the sense of strong operator topology) subset of the ball,

$$P_b \equiv \left\{ B \in \mathcal{L}(V, V^*) \colon \|B\|_{\mathcal{L}(V,V^*)} \leq b \right\}$$

Then, for each g defined on $I \times V \times H$ which is measurable in the first variable and lower semi continuous in the rest, the functional $J(B)$ of Eq. (54) attains its minimum on P^0.

The best operator B^0 minimizing the functional $J(B)$ can be determined by use of the following necessary conditions of optimality.

Theorem 8. Consider system (53) along with the functional,

$$J(B) \equiv \frac{1}{2} \int_0^T |Cy(B) - z^*(t)|_K^2 \, dt$$

with the observed data $z^* \in L_2(I, K)$, the observer $C \in \mathcal{L}(H, K)$, $f \in L_2(I, H)$, $y_0 \in V$, $y_1 \in H$ and P^0 as in Theorem 7. Then, for B^0 to be optimal, it is necessary that there exists a pair $\{y, x\}$ satisfying the equations,

$$\ddot{y} + Ay + B^0 y = f$$

$$\ddot{x} + A^* x + (B^0)^* x = C^* \Lambda_K (Cy(B^0) - z^*)$$

$$(56)$$

$$y(0) = y_0, \qquad x(T) = 0$$

$$\dot{y}(0) = y_1, \qquad \dot{x}(T) = 0$$

and the inequality

$$\int_0^T \langle B^0 y(B^0), x \rangle_{V^*, V} \, dt \geq \int_0^T \langle By(B^0), x \rangle_{V^*, V} \, dt$$

$$(57)$$

for all $B \in P^0$, where Λ_K is the canonical isomorphism of K onto K^* such that $\|\Lambda_K e\|_{K^*} = \|e\|_K$.

By solving Eqs. (56) and (57) simultaneously one can determine B^0. In fact, a gradient-type algorithm can be developed on the basis of Eqs. (56) and (57).

D. Controllability

Consider the controlled system,

$$\dot{\phi} = A\phi + Bu, \qquad t \geq 0 \qquad (58)$$

with ϕ denoting the state and u the control. The operator A is the generator of a strongly continuous semigroup $S(t)$, $t \geq 0$, in a Banach space X and B is the control operator with values in $\mathcal{L}(E, X)$ where E is another Banach space. Let $\mathcal{U}$ denote the class of admissible controls, possibly a proper subset of $L_p^{loc}(E) \equiv$ locally pth power summable E-valued functions on $\mathbb{R}_0 = [0, \infty)$. For a given initial state $\phi_0 \in \mathcal{U}$,

$$\phi(t) = S(t)\phi_0 + \int_0^t S(t - \theta)Bu(\theta) \, d\theta, \qquad t \geq 0$$

denotes the mild (weak) solution of problem (58).

Given $\phi_0 \in X$ and a desired target $\phi_1 \in X$, is it possible to find a control from $\mathcal{U}$ that transfers the system from state ϕ_0 to the desired state ϕ_1 in finite time? This is the basic question of controllability. In other words, for a given $\phi_0 \in X$, one defines the attainable set:

$$\mathcal{A}(t) \equiv \left\{ x \in X \colon x = S(t)\phi_0 \right.$$

$$\left. + \int_0^t S(t - \theta)Bu(\theta) \, d\theta, \ u \in \mathcal{U} \right\}$$

and inquires if there exists a finite time $\tau \geq 0$, such that $\phi_1 \in \mathcal{A}(\tau)$ or equivalently,

$$\phi_1 - S(\tau)\phi_0 \in R(\tau) \equiv \mathcal{A}(\tau) - S(\tau)\phi_0$$

The set $\mathcal{R}(\tau)$, given by:

$$\mathcal{R}(\tau) \equiv \left\{ \xi \in X \colon \xi = L_\tau u \right.$$

$$\left. \equiv \int_0^\tau S(\tau - \theta)Bu(\theta) \, d\theta, \ u \in \mathcal{U} \right\}$$

is called the *reachable set*. If $S(t)B$ is a compact operator for each $t \geq 0$, then $\mathcal{R}(\tau)$ is compact, hence the given target may not be attainable. A similar situation arises if $BE \subset D(A)$ and $\phi_0 \in D(A)$ and $\phi_1 \notin D(A)$. As a result, an appropriate definition of controllability for ∞-dimensional systems may be formulated as follows.

Definition. System (58) is said to be controllable (exactly controllable) in the time interval $[0, \tau]$ if $R(\tau)$ is dense in $X[\mathcal{R}(\tau) = X]$ and it is said to be controllable (exactly controllable) in finite time if $\cup_{\tau>0} R(\tau)$ is dense in $X[\cup_{\tau>0} R(\tau) = X]$. Note that for finite-dimensional systems, $X = R^n$, $E = R^m$, and controllability and exact controllability are all and the same. It is only for ∞-dimensional systems that these concepts are different.

We present here a classical result assuming that both X and E are self-adjoint Hilbert spaces with $\mathcal{U} = L_2^{loc}(E)$.

Theorem 9. For system (58) the following statements are equivalent:

(a) System (58) is controllable in time τ

(b) $(L_\tau L_\tau^*) \in \mathcal{L}^+(X)$

(c) $\text{Ker } L_\tau^* \equiv \{\xi \in X \colon L_\tau^* \xi = 0\} = \{0\}$

where L_τ^* is the adjoint of the operator L_τ and $L_\tau u \equiv \int_0^\tau S(\tau - \theta) Bu(\theta) d\theta$.

Note that by our definition, here controllability means approximate controllability; that is, one can reach an arbitrary neighborhood of the target but never exactly at the target itself. Another interesting difference between finite- and infinite-dimensional systems is that in case $X = R^n$, $E = R^m$, condition (b) implies that $(L_\tau L_\tau^*)^{-1}$ exists and the control achieving the desired transfer from ϕ_0 to ϕ_1 is given by:

$$u = (L_\tau L_\tau^*)^{-1}(\phi_1 - S(\tau)\phi_0)$$

For ∞-dimensional systems, the operator $(L_\tau L_\tau^*)$ does not in general have a bounded inverse even though the operator is positive.

Another distinguishing feature is that in the finite-dimensional case the system is controllable if and only if the rank condition,

$$\text{rank}(B, AB, \ldots, A^{n-1}B) = n$$

holds. In the ∞-dimensional case there is no such condition; however, if $BE \subset \bigcap_{n=1}^\infty D(A^n)$ then the system is controllable if

$$\text{closure} \left\{ \bigcup_{n=0}^\infty \text{range}(A^n B) \right\} = X \qquad (59)$$

This condition is also necessary and sufficient if $S(t)$, $t \geq 0$, is an analytic semigroup and $BE \subset \bigcup_{t>0} S(t)X$.

In recent years, much more general results that admit very general time-varying operators $\{A(t), B(t), t \geq 0\}$, including hard constraints on controls, have been proved. We conclude this section with one such result. The system,

$$\dot{y} = A(t)y + B(t)u, \qquad t \geq 0$$

$$y(0) = y_0 \qquad (60)$$

is said to be globally null controllable if it can be steered to the origin from any initial state $y_0 \in X$.

Theorem 10. Let X be a reflexive Banach space and Y a Banach space densely embedded in X with the injection $Y \subset X$ continuous. For each $t \geq 0$, $A(t)$ is the generator of a c_0-semigroup satisfying the stability condition and $A \in L_1^{\text{loc}}(0, \infty; \mathscr{L}(Y, X))$ where $\mathscr{L}(X, Z)$ is the space of bounded linear operators from a Banach space X to a Banach space Z; $\mathscr{L}(X) \equiv \mathscr{L}(X, X)$, $B \in L_q^{\text{loc}}(0, \infty; \mathscr{L}(E, X))$, and $\mathscr{U} = \{u \in L_p^{\text{loc}}(E); u(t) \in \Gamma \text{ a.e.}\}$ where Γ is a closed bounded convex subset of E with $o \in \Gamma$ and $p^{-1} + q^{-1} = 1$. Then a necessary and sufficient condition for global null controllability of system (60) is that

$$\int_0^\infty H_\Gamma(B^*(t)\psi(t)) dt = +\infty \qquad (61)$$

for all nontrivial weak solutions of the adjoint system

$$\dot{\psi} + A^*(t)\psi = 0, \qquad t \geq 0$$

where $H_\Gamma(\xi) = \sup\{(\xi, e)_{E^*, E}, e \in \Gamma\}$.

E. Existence of Optimal Controls

The question of existence of optimal controls is considered to be a fundamental problem in control theory. In this section, we present a simple existence result for the hyperbolic system,

$$\ddot{\phi} + A\phi = f + Bu, \qquad t \in I \equiv (0, T)$$

$$\phi(0) = \phi_0, \qquad \dot{\phi}(0) = \phi_1 \qquad (62)$$

Similar results hold for parabolic systems. Suppose the operator A and the data ϕ_0, ϕ_1, f satisfy the assumptions of Theorem 1. Let E be a real Hilbert space and $\mathscr{U}_0 \subset L_2(I, E)$ the class of admissible controls and $B \in \mathscr{L}(E, H)$. Let $S_0 \subset V$ and $S_1 \subset H$ denote the set of admissible initial states. By Theorem 1, for each choice of $\phi_0 \in S_0$, $\phi_1 \in S_1$, and $u \in \mathscr{U}_0$ there corresponds a unique solution ϕ called the *response*. The quality of the response is measured through a functional called the *cost functional* and may be given by an expression of the form,

$$J(\phi_0, \phi_1, u) \equiv \alpha \int_0^T g_1(t, \phi(t), \dot{\phi}(t)) dt$$

$$+ \beta g_2(\phi(T), \dot{\phi}(T)) + \lambda g_3(u) \qquad (63)$$

$\alpha + \beta > 0$; $\alpha, \beta, \gamma \geq 0$, where g_1, g_2, and g_3 are suitable functions to be defined shortly. One may interpret g_1 to be a measure of discrepancy between a desired response and the one arising from the given policy $\{\phi_0, \phi_1, u\}$. The function g_2 is a measure of distance between a desired target and the one actually realized. The function g_3 is a measure of the cost of control applied to system (62). A more concrete expression for J will be given in the following section. Let $P \equiv S_0 \times S_1 \times \mathscr{U}_0$ denote the set of admissible policies or controls. The question is, does there exist a policy $p^0 \in P$ such that $J(p^0) \leq J(p)$ for all $p \in P$? An element $p^0 \in P$ satisfying this property is called an *optimal policy*. A set of sufficient conditions for the existence of an optimal policy is given in the following result.

Theorem 11. Consider system (62) with the cost functional (63) and let S_0, S_1, and $\mathscr{U}_0$ be closed bounded convex subsets of V, H, and $L_2(I, E)$, respectively. Suppose for each $(\xi, \eta) \in V \times H$, $t \to g_1(t, \xi, \eta)$ is measurable on I and, for each $t \in I$, the functions $(\xi, \eta) \to g_1(t, \xi, \eta)$ and $(\xi, \eta) \to g_2(\xi, \eta)$ are weakly lower semicontinuous on $V \times H$ and the function g_3 is weakly lower semicontinuous on $L_2(I, E)$. Then, there exists an optimal policy,

$$p^0 = (\phi_0^0, \phi_1^0, u^0) \in P$$

Another problem of considerable interest is the question of existence of time-optimal controls. Consider system (33) in the form (34) with

$$f = Bu, \qquad \tilde{f} = \begin{pmatrix} 0 \\ Bu \end{pmatrix}$$

and solution given by:

$$\psi(t) = S(t)\psi_0 + \int_0^t S(t - \theta)\tilde{f}(\theta)\, d\theta \qquad (35')$$

where $S(t)$, $t \geq 0$, is the c_0-semigroup in $X \equiv V \times H$ with the generator $-\tilde{A}$ as given Eq. (34). Here, one is given the initial and the desired final states ψ_0, $\psi_1 \in X$ and the set of admissible controls $\mathcal{U}_0$. Given that the system is controllable from state ψ_0 to ψ_1 in finite time, the question is, does there exist a control that does the transfer in minimum time? A control satisfying this property is called a *time-optimal control*. We now present a result of this kind.

Theorem 12. If $\mathcal{U}_0$ is a closed bounded convex subset of $L_2(I, E)$ and if systems (34) and (35') are exactly controllable from the state ψ_0 to $\psi_1 \in X$, then there exists a time-optimal control.

Proof. Let $\psi(u)$ denote the response of the system corresponding to control $u \in \mathcal{U}_0$; that is,

$$\psi(u)(t) = S(t)\psi_0 + \int_0^t S(t - \theta)\begin{pmatrix} 0 \\ Bu(\theta) \end{pmatrix}\, d\theta$$

Let $\mathcal{U}_{00} \subset \mathcal{U}_0$ denote the set of all controls that transfer the system from state ψ_0 to state ψ_1 in finite time. Define

$$\mathcal{I} = \{t \geq 0 : \psi(u)(0) = \psi_0,\ \psi(u)(t) = \psi_1,\ u \in \mathcal{U}_{00}\}$$

and $\tau^* = \inf\{t \geq 0 : t \in \mathcal{I}\}$. We show that there exists a control $u^* \in \mathcal{U}_{00}$ having the transition time τ^*. Let $\{\tau_n\} \in \mathcal{I}$ such that τ_n is nonincreasing and $\tau_n \to \tau^*$. Since $\tau_n \in \mathcal{I}$ there exists a sequence $u_n \in \mathcal{U}_{00} \subset \mathcal{U}_0$ such that $\psi(u_n)(0) = \psi_0$ and $\psi(u_n)(\tau_n) = \psi_1$. Denote by f_n the element $(_B^0 u_n)$. By virtue of our assumption, $\mathcal{U}_0$ is weakly compact, B is bounded, and there exists a subsequence of the sequence $\{f_n\}$ relabeled as $\{f_n\}$ and

$$f^* = \begin{pmatrix} 0 \\ Bu^* \end{pmatrix} \in L_2(I, X)$$

with $u^* \in \mathcal{U}_0$ such that $f_n \to f^*$ weakly in $L_2(I, X)$. We must show that $u^* \in \mathcal{U}_{00}$. Clearly, by definition of $\{\tau_n\}$ we have

$$\psi_1 = \psi(u_n)(\tau_n) \equiv S(\tau_n)\psi_0 + \int_0^{\tau_n} S(\tau_n - \theta)f_n(\theta)\, d\theta$$

$$\text{for all} \quad n$$

Let $x^* \in X^* \equiv V^* \times H$, where X^* is the dual of the Banach space X which is the space of continuous linear functionals on X; for example,

$$X = L_p, \qquad 1 \leq p < \infty$$

$$X^* = L_q \qquad \text{with} \quad p^{-1} + q^{-1} = 1$$

Then,

$$x^*(\psi_1) = x^*(S(\tau_n)\phi_0) + x^*\left(\int_0^{\tau_n} S(\tau_n - \theta)f_n(\theta)\, d\theta\right) \tag{64}$$

By virtue of the c_0-property of the semigroup $S(t)$, $t \geq 0$,

$$\lim_{n \to \infty} x^*(S(\tau_n)\psi_0) = x^*(S(\tau^*)\psi_0) \tag{65}$$

Splitting the integral in Eq. (64) into two parts, we have

$$x^*\left(\int_0^{\tau_n} S(\tau_n - \theta)f_n(\theta)\, d\theta\right)$$

$$= x^*\left(\int_0^{\tau^*} S(\tau_n - \tau^*)S(\tau^* - \theta)f_n(\theta)\, d\theta\right)$$

$$+ x^*\left(\int_{\tau^*}^{\tau_n} S(\tau_n - \theta)f_n(\theta)\, d\theta\right)$$

$$= \left\langle \int_0^{\tau^*} S(\tau^* - \theta)f_n(\theta)\, d\theta,\ S^*(\tau_n - \tau^*)x^* \right\rangle_{X,X^*}$$

$$+ x^*\left(\int_{\tau^*}^{\tau_n} S(\tau_n - \theta)f_n(\theta)\, d\theta\right)$$

where S^* is the dual of the operator S. $\mathcal{L}_u(X, Y)$ is the space of unbounded linear operators from X into Y. Let $\{x^*, y^*\} \in X^* \times Y^*$ be the set of all pairs for which

$$\langle y^*, Sx \rangle_{Y^*,Y} = \langle x^*, x \rangle_{X^*,X}$$

for all $x \in D(S) \subset X$ where

$$x^*(x) = \langle x^*, x \rangle_{X^*,X} = \langle x, x^* \rangle_{X,X^*}$$

is the duality pairing between the elements $x \in X$ and $x^* \in X^*$ or the value of x^* at x. If $D(S)$ is dense in X (i.e., closure of $D(S) = X$), then the above relation determines uniquely the dual S^* of S and its domain $D(S^*) \subset Y^*$. If $D(S) = X$, then $S \in \mathcal{L}(X, Y)$ and $S^* \in \mathcal{L}(Y^*, X^*)$. Clearly,

$$\int_0^{\tau^*} S(\tau^* - \theta)f_n(\theta)\, d\theta$$

$$\xrightarrow{w} \int_0^{\tau^*} S(\tau^* - \theta)f^*(\theta)\, d\theta \qquad \text{in} \quad X \tag{66}$$

and since V and hence V^* are all reflexive Banach spaces S^* is a c_0-semigroup in X^* and consequently

$$S^*(\tau_n - \tau^*)x^* \xrightarrow{s} x^* \qquad \text{in} \quad X^* \tag{67}$$

Further, by the c_0-property of $S(t)$, $t \geq 0$ there exists a finite $M > 0$ such that

$$\left| x^* \left(\int_{\tau^*}^{\tau_n} S(\tau_n - \theta) f_n(\theta) \, d\theta \right) \right|$$

$$\leq M \|x^*\|_{X^*} \left(\int_{\tau^*}^{\tau_n} \|f_n(\theta)\|_X^2 \, d\theta \right)^{1/2} (\tau_n - r^*)^{1/2} \quad (68)$$

Since $\mathcal{U}_0$ is bounded, it follows from this that

$$\lim_{n \to \infty} x^* \left(\int_{\tau^*}^{\tau_n} S(\tau_n - \theta) f_n(\theta) \, d\theta \right) = 0$$

Using Eqs. (65) to (67) in (64) we obtain:

$$x^*(\psi_1) = x^* \left(S(\tau^*) \psi_0 \right) + x^* \left(\int_0^{\tau^*} S(\tau^* - \theta) f^*(\theta) \, d\theta \right)$$

for all $x^* \in X^*$. Hence,

$$\psi_1 = S(\tau^*) \psi_0 + \int_0^{\tau^*} S(\tau^* - \theta) f^*(\theta) \, d\theta$$

and $u^* \in \mathcal{U}_{00}$. This completes the proof. $\qquad \square$

The method of proof of the existence of time-optimal controls presented above applies to much more general systems.

F. Necessary Conditions of Optimality

After the questions of controllability and existence of optimal controls are settled affirmatively, one is faced with the problem of determining the optimal controls. For this purpose, one develops certain necessary conditions of optimality and constructs a suitable algorithm for computing the optimal (extremal) controls. We present here necessary conditions of optimality for system (62) with a quadratic cost functional of the form,

$$J(u) \equiv \alpha \int_0^T (C\phi(t) - z_1(t), C\phi(t) - z_1(t))_{H_1} \, dt$$

$$+ \beta \int_0^T (D\dot{\phi}(t) - z_2(t), D\dot{\phi}(t) - z_2(t))_{H_2} \, dt$$

$$+ \gamma \int_0^T (N(t)u, u)_E \, dt \quad (69)$$

where $\alpha, \beta, \gamma > 0$. The output spaces, where observations are made, are given by two suitable Hilbert spaces H_1 and H_2 with output operators $C \in \mathcal{L}(H, H_1)$ and $D \in \mathcal{L}(H, H_2)$. The desired trajectories are given by $z_1 \in L_2(I, H_1)$ and $z_2 \in L_2(I, H_2)$. The last integral in Eq. (69) gives a measure of the cost of control with $N(t) \geq \delta I$ for all $t \in I$, with $\delta > 0$. We assume that $N(t) = N^*(t), t \geq 0$. Our problem is to find the necessary and sufficient conditions an optimal control must satisfy. By Theorem 11, we know that, for the cost functional (69) subject to the dynamic

constraint (62), an optimal control exists. Since in this case J is strictly convex, there is, in fact, a unique optimal control. For characterization of optimal controls, the concept of Gateaux differentials plays a central role. A real-valued functional f defined on a Banach space X is said to be Gateaux differentiable at the point $x \in X$ in the direction $h \in X$ if

$$\lim_{\varepsilon \to 0} \left\{ \frac{f(x + \varepsilon h) - f(x)}{\varepsilon} \right\} = f'(x, h) \quad (70)$$

exists. In general, $h \to f'(x, h)$ is a homogeneous functional, and, in case it is linear in h, we write:

$$f'(x, h) = (f'(x), h)_{X^*, X} \quad (71)$$

with the Gateaux derivative $f'(x) \in X^*$. Since the functional J, defined on the Hilbert space $L_2(I, E)$, is Gateaux differentiable and strictly convex, and the set of admissible controls $\mathcal{U}_0$ is a closed convex subset of $L_2(I, E)$, a control $u^0 \in \mathcal{U}_0$ is optimal if and only if

$$(J'(u^0), u - u^0) \geq 0 \qquad \text{for all} \quad u \in \mathcal{U}_0 \quad (72)$$

Using this inequality, we can develop the necessary conditions of optimality.

Theorem 13. Consider system (62) with the cost functional (69) and $\mathcal{U}_0$, a closed bounded convex subset of $L_2(I, E)$. For $u^0 \in \mathcal{U}_0$ to be optimal, it is necessary that there exists a pair

$$\{\phi^0, \psi^0\} \in C(\bar{I}, V) \times C(\bar{I}, V)$$

with

$$\{\dot{\phi}^0, \dot{\psi}^0\} \in C(\bar{I}, H) \times C(\bar{I}, H)$$

satisfying the equations:

$$\ddot{\phi}^0 + A\phi^0 = f + Bu^0$$

$$\phi^0(0) = \phi_0, \quad (73a)$$

$$\dot{\phi}^0(0) = \phi_1$$

$$\ddot{\psi}^0 + A^*\psi^0 + \int_t^T g_1 \, d\theta + g_2 = 0$$

$$\psi^0(T) = 0, \qquad \dot{\psi}^0(T) = 0$$

$$g_1 = 2\alpha C^* \Lambda_1 (C\phi^0 - z_1) \quad (73b)$$

$$g_2 = 2\beta D^* \Lambda_2 (D\dot{\phi}^0 - z_2)$$

and the inequality

$$\int_0^T \left(u - u^0, 2\gamma N u^0 + \Lambda_E^{-1} B^* \dot{\psi}^0\right)_E dt \geq 0 \qquad (74)$$

for all $u \in \mathcal{U}_0$.

Proof. By taking the Gateaux differential of J at u^0 in the direction w we have

$$(J'(u^0), w) \equiv \int_0^T dt \left\{ 2\alpha \left(C\hat{\phi}(u^0, w), C\phi(u^0) - z_1\right)_{H_1} \right.$$
$$+ 2\beta \left(D\dot{\hat{\phi}}(u^0, w), D\dot{\phi}(u^0) - z_2\right)_{H_2}$$
$$\left. + 2\gamma (w, Nu^0)_E \right\} \qquad (75)$$

where $\hat{\phi}(u^0, w)$ denotes the Gateaux differential of ϕ at u^0 in the direction w, which is given by the solution of

$$\ddot{\hat{\phi}}(u^0, w) + A\hat{\phi}(u^0, w) = Bw$$
$$\qquad (76)$$
$$\hat{\phi}(u^0, w)(0) = 0, \qquad \dot{\hat{\phi}}(u^0, w)(0) = 0$$

Introducing the duality maps,

$$\Lambda_1: H_1 \rightarrow H_1^*, \qquad \Lambda_2: H_2 \rightarrow H_2^*$$

in expression (75) and defining

$$g_1(t, \phi(u^0)) \equiv 2\alpha C^* \Lambda_1 \left(C\phi(u^0) - z_1(t)\right)$$
$$g_2(t, \phi(u^0)) \equiv 2\beta D^* \Lambda_2 \left(D\dot{\phi}(u^0) - z_2(t)\right)$$

we obtain:

$$(J'(u^0), w) = \int_0^T dt \left\{ \left(\hat{\phi}(u^0, w), g_1\right)_H + \left(\dot{\hat{\phi}}(u^0, w), g_2\right)_H \right.$$
$$\left. + (w, 2\gamma Nu^0)_E \right\} \qquad (77)$$

for all $w \in \mathcal{U}_0$.

Defining $\hat{\phi}_1 = \hat{\phi}$, $\hat{\phi}_2 = \dot{\hat{\phi}}$ one can rewrite Eq. (76) as a first-order evolution equation:

$$\frac{d}{dt} \begin{pmatrix} \hat{\phi}_1 \\ \hat{\phi}_2 \end{pmatrix} = \begin{pmatrix} 0 & I \\ -A & 0 \end{pmatrix} \begin{pmatrix} \hat{\phi}_1 \\ \hat{\phi}_2 \end{pmatrix} + \begin{pmatrix} 0 \\ Bw \end{pmatrix}$$
$$\qquad (78)$$
$$\begin{pmatrix} \hat{\phi}_1(0) \\ \hat{\phi}_2(0) \end{pmatrix} = \begin{pmatrix} 0 \\ 0 \end{pmatrix}$$

Then, by introducing the adjoint evolution equation,

$$\frac{d}{dt} \begin{pmatrix} p_1 \\ p_2 \end{pmatrix} = -\begin{pmatrix} 0 & -A^* \\ I & 0 \end{pmatrix} \begin{pmatrix} p_1 \\ p_2 \end{pmatrix} + \begin{pmatrix} g_1 \\ g_2 \end{pmatrix}$$
$$\qquad (79)$$
$$\begin{pmatrix} p_1(T) \\ p_2(T) \end{pmatrix} = \begin{pmatrix} 0 \\ 0 \end{pmatrix}$$

one can easily verify from Eqs. (78) and (79) that

$$\int_0^T \left\{(\hat{\phi}_1, g_1)_H + (\hat{\phi}_2, g_2)_H\right\} dt = -\int_0^T (Bw, p_2)_H dt$$
$$\qquad (80)$$

Using Eq. (80) in Eq. (77) and the duality map Λ_E: $E \rightarrow E^*$, we obtain, for $w = u - u^0$,

$$(J'(u^0), u - u^0) = \int_0^T \left(2\gamma Nu^0 - \Lambda_E^{-1} B^* p_2, u - u^0\right)_E dt$$
$$u \in \mathcal{U}_0 \qquad (81)$$

Defining $\psi^0(t) = \int_t^T p_2(\theta) d\theta$, one obtains the adjoint equation (73b), and the necessary inequality (74) follows from Eqs. (72) and (81).

REMARK 1. In case $\beta = 0$, the adjoint equation is given by a differential equation rather than the integro-differential equations (73b). That is, p_2 satisfies the equation:

$$\ddot{p}_2 + A^* p_2 + g_1 = 0$$
$$p_2(T) = 0, \qquad \dot{p}_2(T) = 0$$

and in Eq. (74) one may replace $\dot{\psi}^0$ by $-p_2$.

REMARK 2. In case of terminal observation, the cost functional (69) may be given by:

$$J(u) \equiv \alpha \|C\phi(T) - z_1\|_{H_1}^2 + \beta \|D\dot{\phi}(T) - z_2\|_{H_2}^2$$
$$+ \gamma \int_0^T (N(t)u, u)_E \, dt \qquad (82)$$

where $z_1 \in H_1$ and $z_2 \in H_2$.

In this case, the necessary conditions of optimality are given by:

$$\int_0^T \left(u - u^0, 2\gamma Nu^0 - \Lambda_E^{-1} B^* p_2\right) dt \geq 0$$
$$u \in \mathcal{U}_0 \qquad (83)$$

where p_2 satisfies the differential equation,

$$\frac{d^2 p_2}{dt^2} + A^* p_2 = 0$$
$$p_2(T) = g_2 \equiv 2\beta D^* \Lambda_2 \left(D\dot{\phi}^0(T) - z_2\right) \qquad (84)$$
$$\dot{p}_2(T) = -g_1 \equiv -2\alpha C^* \Lambda_1 \left(C\phi^0(T) - z_1\right)$$

In recent years several interesting necessary conditions of optimality for time-optimal control problems have been reported. We present here one such result. Suppose the system is governed by the evolution equation,

$$\frac{dy}{dt} = Ay + u$$

in a Banach space X, and let

$$\mathcal{U} \equiv \left\{u \in L_p^{\text{loc}}(R_0, X) : u(t) \in B_1\right\}$$

with $p > 1$, $B_1 = $ unit ball in X, denote the class of admissible controls.

Theorem 14 (Maximum Principle.) Suppose A is the generator of a c_0-semigroup $S(t)$, $t \geq 0$, in X and there exists a $t > 0$ such that $S(t)X = X$. Let y_0, $y_1 \in X$, and suppose u^0 is the time-optimal control, with transition time τ, that steers the system from the initial state y_0 to the final state y_1. Then there exists an $x^* \in X^*$ ($=$dual of X) such that

$$\langle S^*(\tau - t)x^*, u^0(t) \rangle_{X^*, X}$$
$$= \sup \left\{ \langle S^*(\tau - t)x^*, e \rangle, e \in B_1 \right\}$$
$$= |S^*(\tau - t)x^*|_{X^*} \tag{85}$$

Suppose X is a reflexive Banach space and there exists a continuous map $v: X^* \backslash \{0\} \to X$ such that, for $\xi^* \in X^*$,

$$|v(\xi^*)|_X = 1$$

and

$$\langle \xi^*, v(\xi^*) \rangle_{X^*, X} = |\xi^*|_{X^*}$$

then the optimal control is bang-bang and is given by $u^0(t) = v(S^*(\tau - t)x^*)$, and it is unique if X is strictly convex.

Maximum principle for more general control problems are also available.

G. Computational Methods

In order to compute the optimal controls one is required to solve simultaneously the state and adjoint equations (73a) and (73b), along with inequality (74). In case the admissible control set $\mathcal{U}_0 \equiv L_2(I, E)$, the optimal control has the form:

$$u^0 = -(1/2\gamma)N^{-1}\Lambda_E^{-1}B^*\psi^0$$

Substituting this expression in Eqs. (73a) and (73b), one obtains a system of coupled evolution equations, one with initial conditions and the other with final conditions specified. This is a two-point boundary-value problem in an ∞-dimensional space, which is a difficult numerical problem. However, in general one can develop a gradient-type algorithm to compute an approximating sequence of controls converging to the optimal. The required gradient is obtained from the necessary condition (74). The solutions for the state and adjoint equations, (73a) and (73b), can be obtained by use of any of the standard techniques for solving partial differential equations, for example, the finite difference, finite element, or Galerkin method.

In order to use the result of Theorem 13 to compute the optimal controls, one chooses an arbitrary control $u^1 \in \mathcal{U}_0$ and solves Eq. (73a) to obtain ϕ^1, which is then used in Eq. (73b) to obtain ψ^1. Then, on the basis of Eq. (74) one takes

$$J'(u^1) = 2\gamma Nu^1 + \Delta_E^{-1}B^*\psi^1$$

as the gradient at u^1 and constructs a new control $u^2 = u^1 - \varepsilon_1 J'(u^1)$, with $\varepsilon_1 > 0$ sufficiently small so that $J(u^2) \leq J(u^1)$. This way one obtains a sequence of approximating controls,

$$u^{n+1} = u^n - \varepsilon_n J'(u^n)$$

with $J(u^{n+1}) \leq J(u^n)$, $n = 1, 2, \ldots$. In practical applications, a finite number of iterations produces a fairly good approximation to the optimal control.

H. Stochastic Evolution Equations

In this section we present a very brief account of stochastic linear systems. Let $(\Sigma, \mathcal{A}, P)$ denote a complete probability space and $\mathcal{F}_t$, $t \geq 0$, a nondecreasing family of right-continuous, completed subsigma algebras of the σ-algebra $\mathcal{A}$; that is, $\mathcal{F}_s \subset \mathcal{F}_t$ for $0 \leq s \leq t$. Let H be a real separable Hilbert space and $\{W(t), t \geq 0\}$ an H-valued Wiener process characterized by the properties:

(a) $P\{W(0) = 0\} = 1$,

(b) $W(t)$, $t \geq 0$, has independent increments over disjoint intervals, and

(c) $E\{e^{i(W(t),h)}\} = \exp[-t/2(Qh, h)]$,

where $E\{\cdot\} \equiv \int_\Sigma \{\cdot\} dP$ and $Q \in \mathcal{L}^+(H)$ is the space of positive self-adjoint bounded operators in H. Symbolically, a stochastic linear differential equation is written in the form

$$dy = Ay\,dt + \sigma(t)\,dW(t), \qquad t \geq 0, \qquad y(0) = y_0$$

where A is the generator of a c_0-semigroup $S(t)$, $t \geq 0$, in H and $\sigma \in \mathcal{L}(H)$ and y_0 is an H-valued random variable independent of the Wiener process. The solution y is given by:

$$y(t) = S(t)y_0 + \int_0^t S(t - \theta)\sigma(\theta)\,dW(\theta), \qquad t \geq \theta$$

Under the given assumptions one can easily show that $E|y(t)|_H^2 < \infty$ for finite t, whenever $E|y_0|_H^2 < \infty$, and further $y \in C(I, H)$ a.s. (a.s. $\equiv$ with probability one). In fact, $\{y(t), t \geq 0\}$ is an $\mathcal{F}_t$–Markov random process and one can easily verify that

$$E\{y(t) | \mathcal{F}_\tau\} = S(t - \tau)y(\tau)$$

for $0 \leq \tau \leq t$. The covariance operator $C(t)$, $t \geq 0$, for the process $y(t)$, $t \geq 0$, defined by:

$$(C(t)h, g) = E\{(y(t) - Ey(t), h)(y(t) - Ey(t), g)\}$$

is then given by:

$$(C(t)h, g) = \int_0^t (S(t-\theta)\sigma(\theta)Q\sigma^*(\theta)$$
$$\times S^*(t-\theta)h, g)\, d\theta + (S(t)C_0 S^*(t)h, g)$$

Denoting the positive square root of the operator Q by $\sqrt{Q}$ we have

$$(C(t)h, h) = \int_0^t |\sqrt{Q}\sigma^*(\theta)S^*(t-\theta)h|_H^2\, d\theta$$
$$+ (C_0 S^*(t)h, S^*(t)h)$$

This shows that $C(t) \in \mathcal{L}^+(H)$ if and only if the condition,

$$S^*(t)h = 0, \qquad \text{or}$$

$$\sigma^*(\theta)S^*(t-\theta)h \equiv 0, \qquad 0 \le \theta \le t$$

implies $h = 0$. This is precisely the condition for (approximate) controllability as seen in Theorem 9. Hence, the process $y(t)$, $t \ge 0$, is a nonsingular H-valued Gaussian process if and only if the system is controllable. Similar results hold for more general linear evolution equations on Banach spaces with operators A and σ both time varying.

Linear stochastic evolution equations of the given form arise naturally in the study of nonlinear filtering of ordinary Ito stochastic differential equations in R^n. Such equations are usually written in the weak form,

$$d\pi_t(f) = \pi_t(Af)\, dt + \pi_t(\sigma f)\, dW_t$$

$$t \ge 0, \qquad f \in D(A)$$

where A is a second-order partial differential operator and $\{W(t), t \ge 0\}$ is an $\mathbb{R}^d$-valued Wiener process $(d \le n)$. One looks for solutions π_t, $t \ge 0$, that belong to the Banach space of bounded Borel measures satisfying $\pi_t(f) \ge 0$ for $f \ge 0$. Questions of existence of solutions for stochastic systems of the form,

$$dy = (A(t)y + f(t))\, dt + \sigma(t)\, dW$$

$$y(0) = y_0, \qquad t \in (0, T)$$

$$dz = -((A^*(t) - B(t))z + g)\, dt + \sigma(t)\, dW$$

$$z(T) = 0, \qquad t \in (0, T)$$

have been studied in the context of control theory. A fundamental problem arises in the study of the second equation and it has been resolved by an approach similar to the Lax–Milgram theorem in Hilbert space. Questions of existence and stability of solutions of nonhomogeneous boundary value problems of the form,

$$\frac{\partial y}{\partial t} + A(t)y = f(t), \qquad \text{on} \quad I \times \Omega$$

$$By = g(t), \qquad \text{on} \quad I \times \partial\Omega$$

$$u(0) = y_0, \qquad \text{on} \quad \Omega$$

have been studied where f and g have been considered as generalized random processes and y_0 as a generalized random variable. Stability of similar systems with g replaced by a generalized white noise process has been considered recently. Among other things, it has been shown that $y(t)$, $t \ge 0$, is a Feller process on H, a Hilbert space, and there exists a Feller semigroup T_t, $t \ge 0$, on $C_b(H)$, a space of bounded continuous functions on H, whose dual U_t determines the flow $\mu_t = U_t\mu_0$, $t \ge 0$, of the measure induced by $y(t)$ on H. μ_t, $t \ge 0$, satisfies the differential equation:

$$\frac{d}{dt}\mu_t(f) = \mu_t(Gf), \qquad t \ge 0$$

for all $f \in D(G) \subset C_b(H)$ where G is the infinitesimal generator of the semigroup T_t, $t \ge 0$. Optimal control problems for a class of very general linear stochastic systems of the form,

$$dy = (A(t)y + B(t)u)\, dt + \sigma(t)\, dW(t)$$

$$J(u) = E\left\{ \int_0^T \left[|Cy - z_d|^2 + (Nu, u) \right] dt \right\}$$

$$\equiv \min$$

have been considered in the literature, giving results on the existence of optimal controls and necessary conditions of optimality, including feedback controls. In this work, A, B, σ, C, and N were considered as operator-valued random processes.

III. NONLINEAR EVOLUTION EQUATIONS AND DIFFERENTIAL INCLUSIONS

The two major classes of nonlinear systems that occupy most of the literature are the semilinear and quasilinear systems. However, control theory for such systems has not been fully developed; in fact, the field is wide open. In this section, we shall sample a few results.

A. Existence of Solutions, Nonlinear Semigroup

We consider the questions of existence of solutions for the two major classes of systems, semilinear and quasilinear. In their abstract form we can write them as

$$\frac{d\phi}{dt} + A(t)\phi = f(t, \phi) \qquad \text{(semilinear)} \quad (86)$$

$$\frac{d\phi}{dt} + A(t, \phi)\phi = f(t, \phi) \qquad \text{(quasilinear)} \quad (87)$$

and consider them as evolution equations on suitable state space which is generally a topological space having the structure of a Banach space, or a suitable manifold therein.

For example, let us consider the semilinear parabolic equation with mixed initial and boundary conditions:

$$\frac{\partial \phi}{\partial t} + L\phi = g(t, x; \phi, D\phi, \ldots, D^{2m-1}\phi)$$

$$(t, x) \in I \times \Omega \equiv Q$$

$$\phi(0, x) = \phi_0(x), \qquad x \in \Omega \qquad (88)$$

$$D_\nu^k \phi = 0, \qquad 0 \leq k \leq m - 1$$

$$(t, x) \in I \times \partial\Omega$$

where

$$(L\phi)(t, x) \equiv \sum_{|\alpha| \leq 2m} a_\alpha(t, x) D^\alpha \phi \qquad (89)$$

and $D_\nu^k \phi = \partial^k \phi / \partial \nu^k$ denotes the kth derivative in the direction of the normal ν to $\partial\Omega$. We assume that L is strongly elliptic with principal coefficients a_α, $|\alpha| = 2m$, in $C(\bar{Q})$ and the lower order coefficients a_α, $|\alpha| \leq 2m - 1$, $L_\infty(Q)$, and further they are all Hölder continuous in t uniformly on $\bar{\Omega}$. Let $1 < p < \infty$ and define the operator-valued function $A(t)$, $t \in I$, by:

$$D(A(t)) \equiv \{\psi \in X = L_p(\Omega) : (L\psi)(t, \cdot) \in X$$

$$\text{and } D_\nu^k \psi \equiv 0 \text{ on } \partial\Omega, 0 \leq k \leq m - 1\} \qquad (90)$$

The domain of $A(t)$ is constant and is given by:

$$D \equiv W^{2m,p} \cap W_0^{m,p} \qquad (91)$$

Then, one can show that for each $t \in I$, $-A(t)$ is the generator of an analytic semigroup and there exists an evolution operator $U(t, \tau) \in \mathcal{L}(X)$, $0 \leq \tau \leq t \leq T$, that solves the abstract Cauchy problem:

$$\frac{dy}{dt} + A(t)y = 0$$

$$y(0) = y_0 \qquad (92)$$

for each $y_0 \in X$; that is, $y(t) = U(t, 0)y_0$, $t \in I$, with $y \in C(\bar{I}, X)$ and $\dot{y} \in C((0, T], X)$. In general, if $f \in L_p(I, X)$, then y, given by:

$$y(t) = U(t, 0)y_0 + \int_0^t U(t, \tau) f(\tau) \, d\tau \qquad (93)$$

is a mild solution of the Cauchy problem,

$$\frac{dy}{dt} + A(t)y = f$$

$$y(0) = y_0 \qquad (94)$$

This would be the generalized (weak) solution of the parabolic initial boundary value problem (88) if g were replaced by $f \in L_p(I, X)$. In order to solve problem (88), one must introduce an operator f such that

$$f(t, u) \equiv g(t, \cdot; u, D^1 u, \ldots, D^{2m-1} u) \in X \qquad \text{a.e.}$$

for u in a suitable subspace Y with $D(A) \subset Y \subset X$. Problem (88) can then be considered as an abstract Cauchy problem,

$$\frac{d\phi}{dt} + A(t)\phi = f(t, \phi)$$

$$\phi(0) = \phi_0 \qquad (95)$$

In view of Eq. (93), a mild solution of Eq. (95) is given by a solution of the integral equation,

$$\phi(t) = U(t, 0)\phi_0 + \int_0^t U(t, \theta) f(\theta, \phi(\theta)) \, d\theta \qquad (96)$$

if one exists. Defining the operator G by:

$$(G\phi)(t) \equiv U(t, 0)\phi_0 + \int_0^t U(t, \theta) f(\theta, \phi(\theta)) \, d\theta$$

one then looks for a fixed point for G, that is, an element ϕ such that $\phi = G\phi$. Using *a priori* estimates, the most difficult part of the program, one can establish the existence of a solution by use of a suitable fixed-point theorem—for example, Banach, Schauder, or Leray–Schauder fixed-point theorems. We state the following result without proof.

Theorem 15. Consider the semilinear parabolic problem (88) in the abstract form (95) and suppose A generates the evolution operator $U(t, \tau)$, $0 \leq \tau \leq t \leq T$, and f satisfies the properties:

(F1) $\qquad \|f(t, u)\|_X \leq c\{1 + \|A^\beta(t)u\|_X\}$

$\qquad\qquad t \in I$

$\qquad\qquad$ for constants $\quad c > 0, \quad 0 \leq \beta < 1$

$\qquad\qquad$ and $u \in D(A^\beta)$, $\qquad\qquad\qquad\qquad (97)$

(F2) $\qquad \|f(t, u) - f(t, w)\|_X$

$\qquad\qquad \leq C\{\|A^\beta(t)v - A^\beta(t)w\|_X^\rho\}$

$\qquad\qquad t \in I$

$\qquad\qquad$ for some $\quad 0 < \rho \leq 1$,

$\qquad\qquad u, w \in D(A^\beta) \qquad\qquad\qquad\qquad\qquad (98)$

Then, Eq. (95) has a mild solution $\phi \in C(I, X)$, hence the semilinear parabolic equation (88) has a generalized solution. The solution is unique if $\rho = 1$.

REMARK 1. Condition (F1) is satisfied if the function g satisfies the growth condition:

$$|g(t, x, u, Du, \ldots, D^{2m-1}u)| \leq k\left\{1 + \sum_{j=0}^{2m-1} |D^j u|_j^r\right\}$$

for $0 \leq r_j \leq (2m + n/q)/(j + n/q)$, $1 < q < \infty$, and a number β satisfying $(2m - 1)/2m < \beta < 1$. In the case

$\rho = 1$, condition (F2) is satisfied if the function g is Lipschitz in the last $2m$ variables uniformly with respect to $(t, x) \in \bar{I} \times \bar{\Omega}$.

If the coefficients $\{a_\alpha, |\alpha| \leq 2m\}$ in Eq. (89) are also dependent on ϕ so that $\{a_\alpha = a_\alpha(t, x; \phi, D^1\phi, \ldots, D^{2m-1}\phi)$ then system (88) becomes a quasilinear system and parabolic if

$$(-1)^m \operatorname{Re}\left\{ \sum_{|\alpha|=2m} a_\alpha(t, x; \eta)\xi^\alpha \right\} \geq c|\xi|^{2m}$$

$$c > 0 \quad (99)$$

for $(t, x) \in \bar{Q}$ and $\eta \in R^N$, where $N = \sum_{j=0}^{2m-1} N_j$ with N_j denoting the number of terms representing derivatives of order exactly j appearing in the arguments of $\{a_\alpha\}$. System (88) then takes the form:

$$\frac{d\phi}{dt} + A(t, \phi)\phi = f(t, \phi)$$

$$\phi(0) = \phi_0 \quad (100)$$

This problem is again solved by use of *a priori* estimates and a fixed-point theorem under the following assumptions:

(A1) The operator $A_0 = A(0, \phi_0)$ is a closed operator with domain D dense in X and

$$\left\|(\lambda I - A_0)^{-1}\right\| \leq k/(1 + |\lambda|)$$

$$\text{for all } \lambda, \qquad \operatorname{Re} \lambda \leq 0.$$

(A2) A_0^{-1} is a completely continuous operator that is; it is continuous in X and maps bounded sets into compact subsets of X.

(A3) There exist numbers ε, ρ satisfying $0 < \varepsilon \leq 1$, $0 < \rho \leq 1$, such that for all $t, \tau \in I$,

$$\|(A(t, u) - A(\tau, w))A^{-1}(\tau, w)\|$$

$$\leq k_R\left\{|t - \tau|^\varepsilon + \left\|A_0^\beta u - A_0^\beta w\right\|^\rho\right\}$$

for all u, w such that $\|A_0^\beta u\|, \|A_0^\beta w\| < R$ with k_R possibly depending on R.

(F1)   For all $t, \tau \in I$,

$$\|f(t, v) - f(t, w)\| \leq k_R\left\{\left\|A_0^\beta v - A_0^\beta w\right\|^\rho\right\}$$

for all $v, w \in X$ such that $\|A_0^\beta v\|, \|A_0^\beta w\| \leq R$.

Theorem 16. Under assumptions (A1) to (A3) and (F1) there exists a $t^* \in (0, T)$ such that Eq. (100) has at least one mild solution $\phi \in C([0, t^*], X)$ for each $\phi_0 \in D(A_0^\beta)$ with $\|A_0^\beta \psi_0\| \leq R$. Further, if f also satisfies the Hölder condition in t, the solution is C^1 in $t \in (0, t^*)$. If $\rho = 1$, the solution is unique.

Proof. We discuss the outline of a proof. The differential equation (100) is converted into an integral equation and then one shows that the integral equation has a solution. Let $v \in C([0, t^*], X)$ and define:

$$A^v(t) \equiv A\left(t, A_0^{-\beta} v(t)\right),$$

$$f^v(t) \equiv f\left(t, A_0^{-\beta} v(t)\right)$$

and consider the linear system,

$$\frac{dy}{dt} + A^v(t)y = f^v(t), \quad t \in [0, t^*]$$

$$y(0) = \phi_0 \quad (101)$$

By virtue of assumptions (A1) to (A3), $-A^v(t)$, $t \in [0, t^*]$, is the generator of an evolution operator $U^v(t, \tau)$, $0 \leq t \leq t^*$. Hence, the system has a mild solution given by:

$$y^v(t) = U^v(t, 0)\phi_0 + \int_0^t U^v(t, \theta)f^v(\theta)\, d\theta$$

$$0 \leq t \leq t^* \quad (102)$$

Defining an operator G by setting

$$(Gv)(t) \equiv A_0^\beta U^v(t, 0)\phi_0 + \int_0^t A_0^\beta U^v(t, \theta)f^v(\theta)\, d\theta$$

$$(103)$$

one then looks for a fixed point of the operator G, that is, an element $v^* \in C([0, t^*], X)$ such that $v^* = Gv^*$. In fact, one shows, under the given assumptions, that for sufficiently small $t^* \in I$, there exists a closed convex set $K \subset C([0, t^*], X)$ such that $GK \subset K$ and GK is relatively compact in $C([0, t^*], X)$ and hence, by the Schauder fixed-point theorem (which is precisely as stated) has a solution $v^* \in K$. The solution (mild) of the original problem (100) is then given by $\phi^* = A_0^{-\beta} v^*$. This is a genuine (strong) solution if f is also Holder continuous in t. If $\rho = 1$, G has the contraction property and the solution is unique. $\square$

According to our assumptions, for each $t \in [0, t^*]$ and $y \in D(A_0^\beta)$, the operator $A(t, y)$ is the generator of an analytic semigroup. This means that Theorem 16 can handle only parabolic problems and excludes many physical problems arising from hydrodynamics and wave propagation phenomenon including the semilinear and quasilinear symmetric hyperbolic systems discussed in Section I. This limitation is overcome by allowing $A(t, y)$, for each t, y in a suitable domain, to be the generator of a c_0-semigroup rather than an analytic semigroup. The fundamental assumptions required are:

(H1) X is a reflexive Banach space with Y being another Banach space which is continuously and densely embedded in X and there is an isometric isomorphism S of Y onto X.

(H2) For $t \in [0, T]$, $y \in W \equiv$ an open ball in Y, $A(t, y)$ is the generator of a c_0-semigroup in X.

(H3) For $t, y \in [0, T] \times W$,

$$(SA(t, y) - A(t, y)S)S^{-1} = B(t, y) \in \mathcal{L}(Y, X)$$

and $\|B(t, y)\|_{\mathcal{L}(Y, X)}$ is uniformly bounded on $I \times Y$.

Theorem 17. Under assumptions (H1) to (H3) and certain Lipschitz and boundedness conditions for A and f on $[0, T] \times W$, the quasilinear system (100) has, for each $\phi_0 \in W$, a unique solution,

$$\phi \in C([0, t^*), W) \cap C^1([0, t^*), X)$$

$$\text{for some} \quad 0 < t^* \leq T$$

The proof of this result is also given by use of a fixed-point theorem but without invoking the operator A_0. Here, for any $v \in C([0, t^*), W)$, one defines $A^v(t) = A(t, v(t))$, $f^v(t) = f(t, v(t))$ and $U^v(t, \tau)$, $0 \leq \tau \leq t \leq T$, the evolution operator corresponding to $-A^v$ and constructs the operator G by setting

$$(Gv)(t) = U^v(t, 0)\phi_0 + \int_0^t U^v(t, \tau)f^v(\tau)\,d\tau$$

where the expression on the right-hand side is the mild solution of the linear equation (101). Any v^* satisfying $v^* = Gv^*$ is a mild solution of the original problem (100).

From the preceding results it is clear that the solutions are defined only over a subinterval $(0, t^*) \subset (0, T)$ and it may actually blow up at time t^*. Mathematically this is explained through the existence of singularities, which physically correspond to the occurrence of, for example, turbulence or shocks in hydrodynamic problems. However, global solutions are defined for systems governed by differential equations with monotone operators. We present a few general results of this nature.

Let X be a Banach space with dual X^* and suppose A is an operator from $D(A) \subset X$ to X^*. The operator A is said to be monotone if

$$(Ax - Ay, x - y)_{X^*, X} \geq 0 \qquad \text{for all} \quad x, y \in D(A) \tag{104}$$

It is said to be demicontinuous on X if

$$Ax_n \overset{w}{\to} Ax_0 \qquad \text{in } X^*$$

$$\text{whenever} \quad x_n \overset{s}{\to} x_0 \quad \text{in } X \tag{105}$$

And, it is said to be hemicontinuous on X if

$$A(x + \theta y) \overset{w}{\to} Ax \qquad \text{in } X^*$$

$$\text{whenever} \quad \theta \to 0 \tag{106}$$

Let H be a real Hilbert space and V a subset of H having the structure of a reflexive Banach space with V dense in H. Let V^* denote the (topological) dual of V and suppose H is identified with its dual H^*. Then we have $V \subset H \subset V^*$. Using the theory of monotone operators we can prove the existence of solutions for nonlinear evolution equations of the form,

$$\frac{d\phi}{dt} + B(t)\phi = f, \qquad t \in I = (0, T)$$

$$\phi(0) = \phi_0 \tag{107}$$

where $B(t)$, $t \in I$, is a family of nonlinear monotone operators from V to V^*.

Theorem 18. Consider system (107) and suppose the operator B satisfy the conditions:

(B1) $B: L_p(I, V) \to L_q(I, V^*)$ is hemicontinuous:

$$\|B\phi\|_{L_q(I, V^*)} \leq K_1\left(1 + \|\phi\|_{L_p(I, V)}^{p-1}\right) \tag{108}$$

where $K_1 > 0$, and $1 < p, q < \infty$, $1/p + 1/q = 1$.

(B2) For all $\phi, \psi \in L_p(I, V)$,

$$(B\phi - B\psi, \phi - \psi)_{L_q(I, V^*), L_p(I, V)} \geq 0 \tag{109}$$

That is, B is a monotone operator form $L_p(I, V)$ to $L_q(I, V^*)$.

(B3) There exists a nonnegative function $C: \mathbb{R} \to \bar{R}$ with $C(\xi) \to +\infty$ as $\xi \to \infty$ such that for $\psi \in L_p(I, V)$,

$$(B\psi, \psi)_{L_q(I, V^*), L_p(I, V)} \geq C(\|\psi\|)\|\psi\| \tag{110}$$

Then for each $\phi_0 \in H$ and $f \in L_q(I, V^*)$ system (107) has a unique solution $\phi \in L_p(I, V) \cap C(\bar{I}, H)$ and $\dot{\phi} \in L_q(I, V^*)$. Further, ϕ is an absolutely continuous V^*-valued function on $\bar{I}$.

It follows from the above result that, for $\phi_0 \in H$, $\phi \in C(\bar{I}, H)$ and hence $\phi(t) \in H$ for $t \geq 0$. For $f \equiv 0$, the mapping $\phi_0 \to \phi(t)$ defines a nonlinear evolution operator $U(t, \tau)$, $0 \leq \tau \leq t \leq T$, in H. In case B is time invariant, we have a nonlinear semigroup $S(t)$, $t \geq 0$, satisfying the properties:

(a) $\quad S(0) = I$ and, as $t \to 0$,

(b) $\quad S(t)\xi \to {}^s\xi$ in H and, due to uniqueness,

(c) $\quad S(t + \tau)\xi = S(t)S(\tau)\xi, \xi \in H$.

Further, it follows from the equation $\dot{\phi} + B\phi = 0$ that $(\dot{\phi}(t), \phi(t)) + (B\phi(t), \phi(t)) = 0$; hence,

$$|\phi(t)|_H^2 = |\phi_0|_H^2 - 2\int_0^t (B\phi(\theta), \phi(\theta))\,d\theta, \qquad t \geq 0$$

Thus, by virtue of (B3), $|\phi(t)|_H \leq |\phi_0|_H$; that is, $|S(t)\phi_0|_H \leq |\phi_0|_H$. Hence the semigroup $\{S(t), t \geq 0\}$ is a family of nonlinear contractions in H, and its generator is $-B$.

A classical example is given by the nonlinear initial boundary-value problem,

$$\frac{\partial \phi}{\partial t} - \sum_i \frac{\partial}{\partial x_i}\left(\left|\frac{\partial \phi}{\partial x_i}\right|^{p-2}\frac{\partial \phi}{\partial x_i}\right)$$

$$= f \quad \text{in} \quad I \times \Omega = Q$$

$$\phi(0,x) = \phi_0(x) \qquad\qquad (111)$$

$$\phi(t,x) = 0 \quad \text{on} \quad I \times \partial\Omega$$

For this example, $V = W_0^{1,p}(\Omega)$ with $V^* = W^{-1,q}$ and $H = L_2(\Omega)$ and

$$Bu = -\sum \frac{\partial}{\partial x_i}\left(\left|\frac{\partial v}{\partial x_i}\right|^{p-2}\frac{\partial v}{\partial x_i}\right), \qquad p \geq 2$$

$$f \in L_q(I, W^{-1,q}), \qquad \phi_0 \in L_2(\Omega)$$

We can rewrite problem (111) in its abstract form,

$$\frac{d\phi}{dt} + B\phi = f, \qquad t \in I$$

$$\phi(0) = \phi_0$$

noting that V absorbs the boundary condition.

We conclude this section with one of the most general results involving monotone operators. Basically we include a linear term in model (107) which is more singular than the operator B. For convenience of presentation in the sequel we shall write this model as:

$$\frac{d\phi}{dt} = A(t)\phi + f(t,\phi), \qquad \phi(0) = \phi_0 \qquad (112)$$

where f represents $-B + f$ of the previous model (107) and $A(t)$, $t \in I$, is a family of linear operators more singular than f in the sense that it may contain partial differentials of higher order than that in f and hence may be an unbounded operator.

For existence of solutions we use the following assumptions for A and f:

(A1) $\{A(t), t \in I\}$ is a family of densely defined linear operators in H with domains $D(A(t)) \subset V$ and range $R(A(t)) \subset V^*$ for $t \in I$.

(A2) $\langle A(t)e, e\rangle_{V^*,V} \leq 0$ for all $e \in D(A(t))$.

(F1) The function $t \to \langle f(t,e), g\rangle$ is measurable on I for $e, g \in V$, and $f: I \times V \to V^*$ is demicontinuous in the sense that for each $e \in V$

$$\langle f(t_n, \xi_n), e\rangle_{V^*,V} \to \langle f(t,\xi), e\rangle$$

whenever $t_n \to t$ and $\xi_n \to \xi$ in V.

(F2) $\langle f(t,\xi) - f(t,\eta), \xi - \eta\rangle_{V^*,V} \leq 0$ for all $\xi, \eta \in V$.

(F3) There exists an $h \in L_q(I, R_+)$, $R_+ = [0, \infty)$, and $\alpha \geq 0$ such that

$$|f(t,\xi)|_{V^*} \leq h(t) + \alpha(\|\xi\|_V)^{p/q} \qquad \text{a.e.}$$

for each $\xi \in V$.

(F4) There exists an $h_1 \in L_1(I, R)$ and $\beta > 0$ such that

$$\langle f(t,\xi), \xi\rangle_{V^*,V} \leq h_1(t) - \beta(\|\xi\|_V)^p \qquad \text{a.e.}$$

for each $\xi \in V$.

Theorem 19. Consider system (112) and suppose assumptions (A1) to (A2) and (F1) to (F4) hold. Then, for each $\phi_0 \in H$, Eq. (112) has a unique (weak) solution $\phi \in L_p(I, V) \cap C(\bar{I}, H)$ and further the solution ϕ is an absolutely continuous V^*-valued function.

The general result given above also applies to partial differential equations of the form:

$$\frac{\partial \phi}{\partial t} + \sum_{|\alpha| \leq m+1} (-1)^{|\alpha|} D^\alpha\left(a_{\alpha\beta}(t,x)D^\beta \phi\right)$$

$$+ \sum_{|\alpha| \leq m} (-1)^{|\alpha|} D^\alpha F_\alpha(t,x;\phi,D^1\phi,\ldots,D^m\phi)$$

$$= 0 \qquad \text{on} \quad I \times \Omega$$

$$\phi(0,x) = \phi_0(x) \qquad \text{on} \quad \Omega$$

$$(D^\alpha \phi)(t,x) = 0, \qquad 0 \leq |\alpha| \leq m \qquad (113)$$

$$\text{on} \quad I \times \partial\Omega$$

where the operator $A(t)$ in Eq. (112) comes from the linear part and the boundary conditions, and the nonlinear operator f comes from the nonlinear part in Eq. (113). The space V can be chosen as $W_0^{m,p}$, $p \geq 2$, with $V^* \equiv W^{-m,q}$ where $1/p + 1/q = 1$.

REMARK 2. Again, if both A and f are time invariant, it follows from Theorem 19 that there exists a nonlinear semigroup $S(t)$, $t \geq 0$, such that $\phi(t) = S(t)\phi_0$, $\phi_0 \in H$.

In certain situations the function $\{F_\alpha, |\alpha| \leq m\}$ may be discontinuous in the variables $\{\phi, D^1\phi, \ldots, D^m\phi\}$ and as a result the operator f, arising from the corresponding Dirichlet form, may be considered to be a multivalued function. In other words $f(t,\xi)$, for $t, \xi \in I \times V$, is a nonempty subset of V^*. In that case, the differential equation becomes a differential inclusion $\dot{\phi} \in A(t)\phi + f(t,\phi)$. Differential inclusions may also arise from variational evolution inequalities. Define the operator S by:

$$S_t g = U(t,0)\phi_0 + \int_0^t U(t,\theta)g(\theta)\,d\theta$$

$$t \in I, \qquad g \in L_1(I, V^*) \qquad (114)$$

where U is the transition operator corresponding to the generator A and ϕ_0 is the initial state. Then, one questions the existence of a $g \in L_1(I, V^*)$ such that $g(t) \in f(t, S_t g)$ a.e. If such a g exists, then one has proved the existence of a solution of the initial value problem:

$$\dot{\phi}(t) \in A(t)\phi(t) + f(t, \phi(t))$$
$$\phi(0) = \phi_0 \tag{115}$$

These questions have been considered in control problems.

B. Stability, Identification, and Controllability

We present here some simple results on stability and some comments on the remaining topics.

We consider the semilinear system,

$$\frac{d\phi}{dt} = A\phi + f(\phi) \tag{116}$$

in a Hilbert space $(H, \|\cdot\|)$ and assume that f is weakly nonlinear in the sense that (a) $f(0) = 0$, and (b) $\|f(\xi)\| = o(\|\xi\|)$, where

$$\lim_{\|\xi\| \to 0} \left\{ \frac{o(\|\xi\|)}{\|\xi\|} \right\} = 0 \tag{117}$$

Theorem 20. If the linear system $\dot{\phi} = A\phi$ is asymptotically stable in the Lyapunov sense and f is a weakly nonlinear continuous map from H to H, then the nonlinear system (116) is locally asymptotically stable near the zero state.

The proof is based on Theorem 4.

For finite-dimensional systems, Lyapunov stability theory is most popular in that stability or instability of a system is characterized by a scalar-valued function known as the Lyapunov function. For ∞-dimensional systems a straight forward extension is possible only if strong solutions exist.

Consider the evolution equation (116) in a Hilbert space H, and suppose that A is the generator of a strongly continuous semigroup in H and f is a continuous map in H bounded on bounded sets. We assume that Eq. (116) has strong solutions in the sense that $\dot{\phi}(t) = A\phi(t) + f(\phi(t))$ holds a.e., and $\phi(t) \in D(A)$ whenever $\phi_0 \in D(A)$. Let T_t, $t \geq 0$, denote the corresponding nonlinear semigroup in H so that $\phi(t) = T_t(\phi_0)$, $t \geq 0$. Without loss of generality we may consider $f(0) = 0$ (if necessary after proper translation in H) and study the question of stability of the zero state. Let Ω be a nonempty open connected set in H containing the origin and define $\Omega_D \equiv \Omega \cap D(A)$, and $B_a(D) \equiv \{\xi \in H : |\xi|_H < a\} \cap D(A)$ for each $a > 0$. The system is said to be stable in the region Ω_D if, for each ball $B_R(D) \subset \Omega_D$, there exists

a ball $B_r(D) \subset B_R(D)$ such that $T_t(\phi_0) \in B_R(D)$ for all $t \geq 0$ whenever $\phi_0 \in B_r(D)$. The zero state is said to be asymptotically stable if $\lim_{t \to \infty} T_t(\phi_0) = 0$ whenever $\phi_0 \in \Omega_D$.

A function $V : \Omega_D \to [0, \infty]$ is said to be positive definite if it satisfies the properties:

(a) $V(x) > 0$ for $x \in \Omega_D \setminus \{0\}$, $V(0) = 0$.
(b) V is continuous on Ω_D and bounded on bounded sets.
(c) V is Gateaux differentiable on Ω_D in the direction of H, in the sense that, for each $x \in \Omega_D$ and $h \in H$,

$$\lim_{\varepsilon \to 0} \left\{ \frac{V(x + \varepsilon h) - V(x)}{\varepsilon} \right\} \equiv V'(x, h)$$

exists, and for each $h \in H$, $x \to V'(x, h)$ is continuous.

The following result is the ∞-dimensional analog of the classical Lyapunov stability theory.

Theorem 21. Suppose the system $\dot{\phi} = A\phi + f(\phi)$ has strong solutions for each $\phi_0 \in D(A)$ and there exists a positive definite function V on Ω_D such that along any trajectory $\phi(t)$, $t \geq 0$, starting from $\phi_0 \in \Omega_D$,

$$V'(\phi(t), A\phi(t) + f(\phi(t))) \leq 0 \quad (<0) \tag{118}$$

for all $t \geq 0$. Then the system is stable (asymptotically stable) in the region Ω_D.

If the system admits only mild solutions, Theorem 21 must be modified by using positive definite functions which have Gateaux derivatives in the directions $\{h\}$ in spaces larger than H.

We conclude this section with a result for systems governed by monotone nonlinear operators as in Eqs. (107) and (112).

Theorem 22. Consider system (112) with the operators A and f satisfying the assumptions of Theorem 19 for all $t \geq 0$, and suppose $h_1 \in L_1(0, \infty; R)$ and the injection $V \subset H$ is continuous. Then, the system is globally asymptotically stable with respect to the origin in H.

The questions of identification of parameters appearing in any of the system equations treated above can be dealt with in a similar way as in the linear case. In fact, an identification problem may be considered as a special case of a control problem with controls appearing in the system coefficients. Such classes of problems have been covered well in the literature. However, controllability questions for the general systems are more difficult and almost nothing is known.

C. Existence of Optimal Controls

Existence of optimal controls for strongly nonlinear parabolic systems and more general nonlinear evolution equations of the form (86) have been treated in the literature. The technical details are rather involved and long. We shall limit ourselves to a brief summary of some results.

Consider system (107) with controls denoted by u:

$$\frac{d\phi}{dt} + B(t)\phi = f(t, u) \qquad (119)$$

where the operator B is nonlinear and may be given by the expression:

$$B(t)\phi = \sum_{|\alpha| \le m} (-1)^{|\alpha|} D^\alpha F_\alpha(t, x; \phi, D^1\phi, \dots, D^m\phi),$$

$$t, x \in I \times \Omega \qquad (120)$$

Under quite general assumptions on the function F_α, the operator B has properties (B1) to (B3) of Theorem 18. For the space V one may choose $W_0^{m,p}$, $p \ge 2$, or any closed subspace of $W^{m,p}$ so that $W_0^{m,p} \subset V \subset W^{m,p}$. Here V is a reflexive Banach space. For admissible controls we choose any reflexive Banach space E of functions defined on Ω and Γ, a closed bounded convex subset of E, and consider $\mathcal{U}$ to be the class of admissible controls which are strongly measurable functions defined on $I = (0, T)$, with values in Γ. Let $f: I \times \Gamma \to V^*$, so that for each t, $f(t, \cdot)$ is weakly continuous (or more generally demi-continuous) on Γ; for each $v \in \Gamma$, $f(\cdot, v)$ is measurable on I (or continuous), and for each $u \in \mathcal{U}$, $f(u) \in L_q(I, V^*)$ where $f(u)(t) \equiv f(t, u(t))$.

The system,

$$\frac{d\phi}{dt} + B(t)\phi = f(t, u(t)), \qquad t \in I$$

$$\phi(0) = \phi_0, \qquad u \in \mathcal{U} \qquad (121)$$

is written in its weak form:

$$(L\phi, \psi) + b(\phi, \psi) = (f(u), \psi)$$

$$\text{for all} \quad \psi \in L_p(I, V) \cap C(\bar{I}, H) \qquad (122)$$

$$\phi(0) = \phi_0, \qquad u \in \mathcal{U}$$

where L denotes the extension of d/dt as an operator from the space $L_p(I, V)$ to the space $L_q(I, V^*)$ and b is the Dirichlet form given by:

$$b(\phi, \psi) \equiv \int_I \sum_{|\alpha| \le m} \int_\Omega F_\alpha(t, x, \phi, D^1\phi, \dots, D^m\phi)$$

$$\cdot D^\alpha \psi \, dx \, dt \qquad (123)$$

We consider the following control problems:

(P1) *Terminal control.* Let $J(u) = Z(\phi(T))$, where Z is a real-valued function on H and the pair $\{u, \phi\}$ is subject to the dynamic constraint (122). The problem is to find a control $u \in \mathcal{U}$ that minimizes the functional J.

(P2) *Time-optimal control.* Let M, a subset of H, be the target set. The requirement is to find a control $u \in \mathcal{U}$ that transfers the system from the state $\phi_0 \in H$ to the target set M in minimum time.

The existence of optimal controls depends on the properties of admissible trajectories, attainable sets, and the cost functionals. Let $\mathscr{X}$ denote the set of admissible trajectories, that is, the set of all $\{\phi\} \in L_p(I, V) \cap C(\bar{I}, H)$ such that ϕ is a solution of Eq. (122) corresponding to some control $u \in \mathcal{U}$. Similarly, the attainable set may be defined as:

$$\mathcal{A}(t) \equiv \{\xi \in H : \xi = \phi(t) \quad \text{for some} \quad \phi \in \mathscr{X}\}$$

Under a number of technical assumptions on the operators B and f and the control set Γ one can prove the following result.

Theorem 23. (a) The set of admissible trajectories $\mathscr{X}$ is a weakly closed and weakly sequentially compact subset of $L_p(I, V)$. (b) For each $t \in [0, T]$, the attainable set $\mathcal{A}(t)$, is a weakly compact subset of H.

Using the preceding result, one can prove the existence of optimal controls for problems (P1) and (P2).

Theorem 24. Let Z be a weakly lower semicontinuous functional defined on H and bounded from below. Then there exists an optimal control solving problem (P1).

Theorem 25 (P2). Suppose the given target set M is a weakly closed subset of H and the system is controllable in the sense that there exists an admissible u, and $\tau \in \bar{I}$, such that $\phi(u)(\tau) \in M$. Then there exists an optimal control that steers the systems from state ϕ_0 to the target set M in minimum time.

Optimal control problems for the more general system (112) recently have been studied in several papers giving existence results for measurable controls and measure-valued controls. Systems governed by differential inclusions of the form (115) and their associated control problems also have been studied recently. The technical details are too long for presentation here. Interested readers may consult the Bibliography.

D. Necessary Conditions of Optimality

For completeness we shall present a result on the necessary conditions of optimality. Consider system (122) along with the cost functional given by:

$$J(u) = Z(\phi(T)) + \int_0^T f^0(t, \phi(t), u(t)) \, dt \qquad (124)$$

The problem is to find a control $u^0 \in \mathcal{U}$ that minimizes the functional J subject to constraint (122). Let $\{F_\alpha(t, x, \xi)\}$,

t, $x \in I \times \Omega, \xi = \{\xi_\alpha, |\alpha| \leq m\} \in R^N$, denote the functions defining the operator B (see Eqs. (119) and (120)) and $\{F_\alpha^\beta, |\beta| \leq m\}$ their directional derivatives with respect to $\xi \in R^N$. We assume that for fixed t, $x \in I \times \Omega$, the functions $\xi \to F_\alpha^\beta(t, x, \xi)$ are continuous on R^N for all α, β satisfying $|\alpha|, |\beta| \leq m$, and, for fixed $\xi \in R^N$, $(t, x) \to F_\alpha^\beta$ (t, x, ξ) are measurable on $I \times \Omega$. For any fixed $\phi \in L_p(I, V)$, the bilinear form,

$$b_\phi(\psi, v) \equiv \sum_{|\alpha|, |\beta| \leq m} \int_I \langle D^\alpha \psi, F_\beta^\alpha(t, x; \phi,$$

$$D^1\phi, \ldots, D^m\phi)D^\beta v \rangle_\Omega \, dt \qquad (125)$$

is well defined on $L_p(I, V) \times L_p(I, V)$. Let

$$f_1^0: I \times V \times E \to V^*$$
$$f_2^0: I \times V \times E \to E^* \qquad (126)$$

denote the linear Gateaux differentials of f^0 with respect to the state and control variables, respectively, and let

$$F: I \times E \to \mathcal{L}(E, V^*) \qquad (127)$$

denote the linear Gateaux differential of f with respect to the control variable.

Under a number of technical assumptions on the functions $\{F_\alpha, |\alpha| \leq m\}$, f, f^0, and Z and the control constraint set $\Gamma \subset E$, one can prove the following necessary conditions of optimality.

Theorem 26. Consider system (122) with the cost functional (124). For the pair $\{u^0, \phi^0\} \in \mathcal{U} \times \mathcal{X}$ to be optimal it is necessary that there exists a $\psi^0 \in L_p(I, V) \cap C(\bar{I}, H)$ so that the triple $\{u^0, \psi^0, \phi^0\}$ satisfy the conditions:

(a) $(L\phi^0, v) + b(\phi^0, v) = (f(u^0), v)$, $\phi^0(0) = \phi_0$

for all

$$v \in \mathscr{F}_1 \equiv \{v \in L_p(I, V) \cap C(\bar{I}, H): v(T) = 0\}$$

(b) $-(L\psi^0, v) + b_{\phi^0}(\psi^0, v) + (f_1^0, v) = 0$,

$$\psi^0(T) = -Z'(\phi^0(T))$$

for all

$$v \in \mathscr{F}_0 \equiv \{v \in L_p(I, V) \cap C(\bar{I}, H): v(0) = 0\}$$

(c) $\int_I \langle f_2^0(t, \phi^0(t), u^0(t))$

$$-F^*(t, u^0(t))\psi^0(t), \, w(t) - u^0(t) \rangle_{E^*, E} \, dt \geq 0 \qquad (128)$$

for all $w \in \mathcal{U}$, where F^* denotes the dual of the operator $F(t, u^0(t)) \in \mathcal{L}(E, V^*)$ and Z' the Gateaux derivative of Z on H.

For time-optimal controls, similar necessary conditions exist. In this case, the optimal control is also characterized by inequality (c), with the exceptions that $f_2^0 \equiv 0$ and the upper limit of the integral is the optimal time t^0 instead of T.

A number of interesting observations can be made from the above result. For example, if $f(t, u) = T(t)u$ and $f^0(t, \phi, u) = \tilde{f}^0(t, \phi) + \langle Nu, u \rangle_{E^*, E}$ and the control set $\Gamma = E$, then it follows from inequality (c) that

$$(N + N^*)u^0(t) = T^*(t)\psi^0(t) \qquad (129)$$

Hence, if E is reflexive and $N = N^*$ and N is invertible, then $u^0(t) = \frac{1}{2}N^{-1}T^*(t)\psi^0(t)$. This is precisely the form of optimal controls for linear systems with quadratic cost functionals.

E. Nonlinear Stochastic Evolution Equations

We present here a brief account of nonlinear stochastic systems. The simplest nonlinear stochastic evolution equation may be given by:

$$dy = Ay \, dt + f(y) \, dW(t), \qquad t \geq 0, \qquad y(0) = y_0 \qquad (130)$$

where W is the Wiener process with covariance operator $Q \in \mathcal{L}^+(H)$ as in the linear case. A is the generator of a c_0-semigroup $S(t)$, $t \geq 0$, on H, and $f: H \to \mathcal{L}(H)$ satisfying:

(a) $\|f(x)\|_{\mathcal{L}(H)}^2 \leq K^2\left(1 + |x|_H^2\right)$

(b) $\|f(x) - f(y)\|_{\mathcal{L}(H)}^2 \leq K^2|x - y|_H^2$ (131)

Define the nonlinear operator G by:

$$(Gx)(t) \equiv S(t)y_0 + \int_0^t S(t - \theta)f(x(\theta)) \, dW(\theta) \quad (132)$$

for $x \in X \equiv C(I, L_2(\Omega, H))$, where X is a Banach space with respect to the topology given by:

$$\|x\|_X \equiv \sqrt{\sup\left\{E|x(t)|_H^2, t \in I\right\}}$$

Under the above assumptions one can prove the existence of an integer n, such that the nth iteration of G, denoted by G^n, is a contraction in X. Hence, by the Banach fixed-point theorem, there exists $y \in X$ such that $y = Gy$. In other words, the integral equation in X,

$$y(t) = S(t)y_0 + \int_0^t S(t - \theta)f(y(\theta)) \, dW(\theta)$$

$$t \in [0, T]$$

has a unique solution $y \in X$ whenever $y_0 \in L_2(\Sigma, H)$, hence y is a mild solution of system (130).

Existence theory for semilinear stochastic evolution equations of the form,

$$dy = (A(t)y + f(t, y)) \, dt + \sigma(t) \, dW \qquad (133)$$

has been developed under much weaker hypotheses on the operators A and f using the Leray–Schauder degree theory. There are also results in which A has been considered to be a strongly measurable function from $(\Sigma, \mathcal{A}, P)$ to $\mathcal{L}_c(H)$, the space of closed densely defined linear (not necessarily bounded) operators in H. In this case, A is assumed to generate a strongly measurable (random) c_0-semigroup in H. Existence theory for nonlinear stochastic boundary value problems of the form,

$$\frac{\partial \phi}{\partial t} + A(t)\phi = f + F(\phi) \qquad \text{on} \quad I \times \Omega$$

$$B\phi = g + G(\phi) \qquad \text{on} \quad I \times \partial\Omega \qquad (134)$$

$$\phi(0) = \phi_0 \qquad \text{on} \quad \Omega$$

has been considered with f, g being generalized random processes, ϕ_0 a generalized random variable, and F, G nonlinear accretive operators. Stability problems for systems of the form (134) with $f = 0$; $g = N$, a generalized white noise; $G = 0$; and F being a monotone operator have been studied. It has been shown that the system is asymptotically stable with respect to a ball around the origin with radius determined by the trace of the covariance operator of the associated Wiener process.

It appears from the literature that for nonlinear systems the theory of optimal control, filtering, identification, and controllability is far from satisfactory. This is a difficult but fascinating field and certainly a challenging subject of the future.

IV. RECENT ADVANCES IN INFINITE-DIMENSIONAL SYSTEMS AND CONTROL

In this section we discuss some recent advances in the theory and applications of distributed parameter systems since the time of first publication of this encyclopedia. Details of these new developments can be found in the references. There has been substantial theoretical development of distributed parameter systems, as indicated in References 11 to 36. These include general boundary control problems, control of deterministic and stochastic evolution inequalities and differential inclusions, uncertain systems, and the so-called B-evolutions. Due to space limitations, we cannot include the details. Here we shall give only a brief outline of the major new concepts introduced in recent years. On the theoretical front, there are three new major developments:

1. The first one is in the area of control theory for systems governed by m-times integrated semigroups or distribution semi groups.
2. The second one is in the area of fundamental concepts, in particular, the notion of solution, a new notion of solutions, called *measure solutions*, has been introduced very recently and has been used in the theory of control of distributed parameter systems.
3. The third front extends the concept of impulsive systems to infinite dimensional Banach spaces; we shall discuss briefly these new developments and their implication in mathematical sciences.
4. The fourth front represents recent applications of the theory of distributed parameter systems to the physical sciences.

A. *m*-Times Integrated Semigroups

The classical semigroup theory, as seen in Section II, is based on the assumptions that A is closed and $D(A)$ is dense in X and that the Hille–Yosida inequality,

$$\|R(\lambda, A)\| \equiv \|(\lambda I - A)^{-1}\| \leq M/(\lambda - \omega),$$

$$\lambda \in \rho(A) \supset (\omega, \infty) \qquad (135)$$

holds for some $M \geq 0$ and $\omega \in R$. These are the necessary and sufficient conditions for the existence of a C_0-semigroup $S(t)$, $t \geq 0$, and hence the existence of a solution of the Cauchy problem,

$$\dot{x} = Ax, \qquad x(0) = \xi \qquad (136)$$

in X. The solution is given by $x(t) = S(t)\xi$, $t \geq 0$. No doubt this covers a very large class of partial differential operators with given boundary conditions and hence a large class of distributed parameter systems. However, there are classes of operators A which do not satisfy the Hille–Yosida theorem, yet such systems have solutions in some generalized sense. According to the Hille–Yosida theorem, $R(\lambda, A)$ is the Laplace transform of some operator-valued function $S(t)$, $t \geq 0$. It is now known that the Cauchy problem stated above has a solution in some generalized sense even if only $R_m(\lambda, A) \equiv R(\lambda, A)/\lambda^m$, $\lambda \in \rho(A)$, is the Laplace transform of an operator-valued function $T(t)$, $t \geq 0$. In this case, $T(t)$, $t \geq 0$, is said to be the m-times integrated semigroup and A is said to be its infinitesimal generator. The classical solution for the Cauchy problem as stated above is now given by:

$$x(t) = (d^m/dt^m)T(t)\xi, t \geq 0, \qquad \text{for} \quad \xi \in D(A^{m+1})$$

In general, if $\xi \in X$ there is no classical solution but we may admit generalized derivatives and hence generalized solutions. For example, if $D(A^{m+1})$ is dense in X, one can choose a sequence $\{\xi_k\}$ converging strongly to ξ and consider the entity x as the generalized solution if it satisfies the identity:

$$\int_0^\tau \langle x(t), \phi(t) \rangle_{X, X^*} \, dt$$

$$= (-1)^m \lim_{k \to \infty} \int_0^\tau \langle T(t)\xi_k, D^m\phi(t) \rangle_{X, X^*} \, dt \quad (137)$$

for all ϕ in a class of test functions. A suitable class of test functions for this problem is the Sobolev class $W_0^{m,1}(I, X^*)$ which consists of X^*-valued functions whose derivatives up to order $m - 1$ vanish on the boundary of the set $I \equiv (0, \tau)$ and belong to $L_1(I, X^*)$. By duality, the solution $x \in W^{-m, \infty}(I, X)$. For further details on m-times integrated semigroups, generalized solutions, stochastic systems, and optimal controls of systems involving operators that generate such semigroups, the reader may consult References 13, 21, and 24.

B. Measure Solution

Consider the system,

$$\dot{x} = f(x), \qquad x(0) = \xi \in E \quad (138)$$

It is well known that if $E = R^n$ is a finite-dimensional space and f is merely continuous, the system has at least one local solution in the sense that there exists a maximal interval of time $(0, \tau_m)$ and an absolutely continuous function $x^*(t)$, $t \in (0, \tau_m)$, that satisfies the differential equation along with the initial condition $x^*(0) = \xi$, with the possibility of blow-up at time τ_m. That is,

$$\lim_{t \to \tau_m} \|x^*(t)\| = \infty \quad (139)$$

An elementary example is $\dot{y} = y^2$, $y(0) = \gamma$. If $\gamma > 0$, the blow-up time is $\tau_m = (1/\gamma)$.

In contrast, if E is an infinite-dimensional Banach space, mere continuity of $f : E \to E$ is no more sufficient to guarantee even a local solution. Even continuity and boundedness of this map do not guarantee existence. Compactness is necessary. However, under mere continuity and local boundedness assmptions, one can prove the existence of generalized (to be defined shortly) solutions. This is what prompted the concept and development of measure-valued solutions. Consider the semilinear system,

$$\dot{x} = Ax + f(x), t \geq 0,$$

$$x(0) = x_0 \quad (140)$$

and suppose A is the infinitesimal generator of a C_0-semigroup in E. Let $BC(E)$ denote the Banach space of bounded continuous functions on E with the standard topology induced by the sup norm $\|\phi\| \equiv \sup\{|\phi(\xi)|, \xi \in E\}$. The dual of this space is the space of regular,

bounded, finitely additive measures denoted by $\sum_{rba}(E)$ and defined on the algebra of sets generated by closed subsets of E. This is a Banach space with respect to the topology induced by the total variation norm. Let $\Pi_{rba}(E) \subset \sum_{rba}(E)$ denote the space of regular, finitely additive probability measures. For a $\nu \in \sum_{rba}(E)$ and $\phi \in BC(E)$, the pairing

$$(\nu, \phi) \equiv \nu(\phi) \equiv \int_E \phi(\xi)\nu(d\xi)$$

is well defined. Letting $D\phi$ denote the Frechet derivative of $\phi \in BC(E)$, we introduce the class,

$$\mathcal{F} \equiv \{\phi \in BC(E) : D\phi \text{ continuous with } \phi, D\phi$$

$$\text{having bounded supports}\}$$

for test functions. Define the operator $\mathcal{A}$ with domain:

$$D(\mathcal{A}) \equiv \{\phi \in F : \mathcal{A}\phi \in BC(E^+)\}$$

where for $\phi \in D(\mathcal{A})$,

$$\mathcal{A}\phi \equiv \langle A^*D\phi(\xi), \xi \rangle_{E^*, E} + \langle D\phi(\xi), f(\xi) \rangle_{E^*, E} \quad (141)$$

and E^+ is a suitable compactification of E that makes E^+ a compact Hausdorff space containing E as a dense subspace. The new notion of a solution for the Cauchy problem (140) can be stated as follows.

Definition: A measure function μ_t, $t \geq 0$, with values in $\Pi_{rba}(E)$, is said to be a generalized solution of the semilinear evolution equation if, for each $\phi \in D(\mathcal{A})$, the following identity holds:

$$\mu_t(\phi) = \phi(x_0) + \int_0^t \mu_s(\mathcal{A}\phi) \, ds, \qquad t \geq 0 \quad (142)$$

The concepts of measure solutions and stochastic evolution equations have been extended. For example, consider the infinite-dimensional stochastic systems on a Hilbert space H, governed by Eq. (130) or, more generally, the equation:

$$dx = Ax \, dt + f(x) \, dt + \sigma(x) \, dW$$

$$x(0) = x_0 \quad (143)$$

Again, if f and ω are merely continuous and bounded on bounded sets, all familiar notions of solutions (strong, mild, weak, martingale) fail. However, measure solutions are well defined. In this case, expression (142) must be modified as follows:

$$\mu_t(\phi) = \phi(x_0) + \int_0^t \mu_s(\mathcal{A}\phi) \, ds$$

$$+ \int_0^t \langle \mu_s(\mathcal{B}\phi), dW(s) \rangle, \qquad t \geq 0 \quad (144)$$

where now the operator $\mathcal{A}$ is given by the second-order partial differential operator,

$$\mathcal{A}\phi \equiv (1/2)Tr(D^2\phi\sigma\sigma^*) + \langle A^* D\phi(\xi), \xi\rangle_H$$
$$+ \langle D\phi(\xi), f(\xi)\rangle_H \qquad (145)$$

and the operator $\mathcal{B}$ is given by:

$$\mathcal{B}\phi \equiv \sigma^* D\phi$$

The last term is a stochastic integral with respect to a cylindrical Brownian motion (for details, see Reference 31). The operators $\mathcal{A}$ and $\mathcal{B}$ are well defined on a class of test functions given by:

$$\mathcal{F} \equiv \{\phi \in BC(E) : \phi, D\phi, D^2\phi \text{ continuous having}$$

bounded supports and

$$Tr(D^2\phi\sigma\sigma^*(\xi)) < \infty, \xi \in H\} \qquad (146)$$

A detailed account of measure solutions and their optimal control for semilinear and quasilinear evolution equations can be found in References 20, 22, 28, and 30–32.

C. Impulsive Systems

In many physical problems, a system may be subjected to a combination of continuous as well as impulsive forces at discrete points of time. For example, in building construction, piling is done by dropping massive weights on the top of vertically placed steel bars. This is an impulsive input. In the management of stores, inventory is controlled by an agreement from the supplier to supply depleted goods. An example from physics is a system of particles in motion which experience collisions from time to time, thereby causing instantaneous changes in directions of motion. In the treatment of patients, medications are administered at discrete points of time, which could be considered as impulsive inputs to a distributed physiological system.

The theory of finite dimensional impulsive systems is well known.[12] Only in recent years has the study of infinite dimensional impulsive systems been initiated.[27,33–36] Here, we present a brief outline of these recent advances. Let $I \equiv [0, T]$ be a closed bounded interval of the real line and define the set $D \equiv \{t_1, t_2, \cdot, \cdot, t_n\} \in (0, T)$. A semilinear impulsive system can be described by the following system of equations:

$$\dot{x}(t) = Ax(t) + f(x(t)),$$
$$t \in I\backslash D, \qquad x(0) = x_0,$$
$$\Delta x(t_i) = F_i(x(t_i)), \qquad (147)$$
$$0 = t_0 < t_1 < t_2, \cdots < t_n < t_{n+1} = T$$

where, generally, A is the infinitesimal generator of a C_0-semigroup in a Banach space E, the function f is a continuous nonlinear map from E to E, and $F_i: E \to E$, $i = 1, 2, \cdot, \cdot, n$, are continuous maps. The difference operator $\Delta_x(t_i) \equiv x(t_i + 0) - x(t_i - 0) \equiv x(t_i + 0) - x(t_i)$ denotes the jump operator. This represents the jump in the state x at time t_i with F_i determining the jump size at time t_i. Similarly, a controlled impulsive system is governed by the following system of equation:

$$dx(t) = [Ax(t) + f(x(t))]\, dt + g(x(t))\, dv(t),$$
$$t \in I\backslash D, x(0) = x_0,$$
$$\Delta x(t_i) = F_i(x(t_i)), \qquad (148)$$
$$0 = t_0 < t_1 < t_2, \cdots < t_n < t_{n+1} = T$$

where $g : E \to \mathcal{L}(F, E)$ and the control $v \in BV(I, F)$. For maximum generality, one can choose the Banach space $BV(I, F)$ of functions of bounded variation on I with values in another Banach space F as the space of admissible controls. This class allows continuous as well as jump controls. For each $r > 0$, let $B_r \equiv \{\xi \in E : \|\xi\| \leq r\}$ denote the ball of radius r around the origin. Let $PWC_\ell(I, E)$ denote the Banach space of piecewise continuous functions on I taking values from E, with each member being left continuous, having right-hand limits. For solutions and their regularity properties we have the following.

Theorem 27. Suppose the following assumptions hold:

(A1) A is the infinitesimal generator of a C_0-semigroup in E.

(A2) The maps g, F_i, $i = 1, 2, \cdot, \cdot, n$ are continuous and bounded on bounded subsets of E with values in $\mathcal{L}(F, E)$ and E, respectively,

(A3) The map f is locally Lipschitz having at most linear growth; that is, there exist constants $K > 0$, $K_r > 0$, such that:

$$\|f(x) - f(y)\|_E \leq K_r\|x - y\|_E, \qquad x, y \in B_r$$
$$\text{and} \quad \|f(x)\| \leq K(1 + \|x\|)$$

Then, for every $x_0 \in E$ and $v \in BV(I, F)$, system (148) has a unique mild solution, $x \in PWC_l(I, E)$.

Using this result one can construct necessary conditions of optimality. We present here one such result of the author.[35] Consider $\mathcal{U} \in BV(I, F)$ to be the class of controls comprised of pure jumps at a set of arbitrary but prespecified points of time $J \equiv \{0 = t_0 = s_0 < s_1 < s_2, \cdot, \cdot, < s_{m-1} < s_m, m \geq n\} \subset [0, T)$. Clearly $\mathcal{U}$ is isometrically isomorphic to the product space $\mathcal{F} \equiv \prod_{k=1}^{m+1} F$, furnished with the product topology. We choose a closed convex subset $\mathcal{U}_{ad} \subset \mathcal{U}$ to be the class of admissible controls. The basic control problem is to find a control policy that imparts a minimum to the following cost functional,

$$J(v) = J(q) \equiv \int_0^T \ell(x(t), v(t)) \, dt + \varphi(v) + \Phi(x(T))$$

$$\equiv \int_0^T \ell(x(t), q) \, dt + \varphi(q) + \Phi(x(T))$$

Theorem 28. Suppose assumptions (A1) to (A3) hold, with $\{f, F_i, g\}$ all having Frechet derivatives continuous and bounded on bounded sets, and the functions ℓ, φ, Φ are once continuously Gateaux differentiable on $E \times F, \mathcal{F}, E$, respectively. Then, if the pair $\{v^o (or q^o), x^o\}$ is optimal, there exists a $\psi \in PWC_r(I, E^*)$ so that the triple $\{v^0, x^0, \psi\}$ satisfies the following inequality and evolution equation:

(a) $$\sum_{i=0}^m \left\langle \left\{ g^*\big(x^o(s_i)\big) \psi(s_i) + \varphi_{q_i}(q^o) \right. \right.$$

$$\left. \left. + \int_{s_i}^T \ell_u(x^o(t), v^o(t)) \, dt \right\}, q_i - q_i^o \right\rangle_{F^*, F} \geq 0$$

for all $q = \{q_0, q_1, \cdot, \cdot, q_m\} \in \mathcal{U}_{ad}$.

(b) $$d\psi = -\big(A^*\psi + f_x^*(x^o(t))\psi - \ell_x(x^o(t), v^o(t)) \, dt$$

$$- g_x^*(x^o(t), \psi(t)) \, dv^o, \qquad t \in I \backslash D$$

$$\psi(T) = \Phi_x(x^o(T))$$

$$\Delta_r \psi(t_i) = -F_{i,x}^*\big(x^o(t_i)\big)\psi(t_i), \qquad i = 1, 2, \cdot, \cdot, n$$

where the operator Δ_r is defined by $\Delta_r f(t_i) \equiv f(t_i - 0) - f(t_i)$.

(c) The process x^o satisfies the system equation (148) (in the mild sense) corresponding to the control v^o.

In case a hard constraint is imposed on the final state requiring $x^o(T) \in \mathcal{K}$ where $\mathcal{K}$ is a closed convex subset of E with nonempty interior, the adjoint equation given by (b) requires modification. The terminal equality condition is replaced by the inclusion $\psi(T) \in \partial I_\mathcal{K}(x^o(T))$, where $I_\mathcal{K}$ is the indicator function of the set $\mathcal{K}$.

Recently a very general model for evolution inclusions has been introduced:[36]

$$\dot{x}(t) \in Ax(t) + F(x(t)),$$

$$t \in I \backslash D, \qquad x(0) = x_0, \qquad (2.3)$$

$$\Delta x(t_i) \in G_i(x(t_i)),$$

$$0 = t_0 < t_1 < t_2, \cdots < t_n < t_{n+1} \equiv T$$

Here, both F and $\{G_i, i = 1, 2, \cdot, \cdot, n\}$ are multivalued maps. This model may arise under many different situations. For example, in case of a control problem where

one wishes to control the jump sizes in order to achieve certain objectives, one has the model,

$$\dot{x}(t) \in Ax(t) + F(x(t)),$$

$$t \in I \backslash D, \qquad x(0) = x_0,$$

$$\Delta x(t_i) = g_i(x(t_i), u_i)),$$

$$0 = t_0 < t_1 < t_2, \cdots < t_n < t_{n+1} \equiv T$$

where the controls u_i may take values from a compact metric space U. In this case, the multis are given by $G_i(\zeta) = g_i(\zeta, U)$. For more details on impulsive evolution equations and, in general, inclusions, the reader may consult References 35 and 36.

D. Applications

The slow growth of application of distributed systems theory is partly due to its mathematical and computational complexities. In spite of this, in recent years there have been substantial applications of distributed control theory in aerospace engineering, including vibration suppression of aircraft wings, space shuttle orbiters, space stations, flexible artificial satellites, and suspension bridges. Several papers on control of fluid dynamical systems governed by Navier–Stokes equations have appeared over the past decade with particular reference to artificial heart design. During the same period, several papers on the control of quantum mechanical and molecular systems have appeared. Distributed systems theory has also found applications in stochastic control and filtering. With the advancement of computing power, we expect far more applications in the very near future.

SEE ALSO THE FOLLOWING ARTICLES

CONTROLS, LARGE-SCALE SYSTEMS • DIFFERENTIAL EQUATIONS, ORDINARY • DIFFERENTIAL EQUATIONS, PARTIAL • TOPOLOGY, GENERAL • WAVE PHENOMENA

BIBLIOGRAPHY

Ahmed, N. U., and Teo, K. L. (1981). "Optimal Control of Distributed Parameter Systems," North-Holland, Amsterdam.

Ahmed, N. U. (1983). "Properties of relaxed trajectories for a class of nonlinear evolution equations on a Banach space," *SIAM J. Control Optimization* **2**(6), 953–967.

Ahmed, N. U. (1981). "Stochastic control on Hilbert space for linear evolution equations with random operator-valued coefficients," *SIAM J. Control Optimization* **19**(3), 401–403.

Ahmed, N. U. (1985). "Abstract stochastic evolution equations on Banach spaces," *J. Stochastic Anal. Appl.* **3**(4), 397–432.

Ahmed, N. U. (1986). "Existence of optimal controls for a class of systems governed by differential inclusions on a Banach space," *J. Optimization Theory Appl.* **50**(2), 213–237.

Ahmed, N. U. (1988). "Optimization and Identification of Systems Governed by Evolution Equations on Banach Space," Pitman Research Notes in Mathematics Series, Vol. 184, Longman Scientific/Wiley, New York.

Balakrishnan, A. V. (1976). "Applied Functional Analysis," Springer–Verlag, Berlin.

Butkovskiy, A. G. (1969). "Distributed Control Systems," Elsevier, New York.

Curtain, A. F., and Pritchard, A. J. (1978). "Infinite Dimensional Linear Systems Theory," Lecture Notes, Vol. 8, Springer–Verlag, Berlin.

Lions, J. L. (1971). "Optimal Control of Systems Governed by Partial Differential Equations," Springer–Verlag, Berlin.

Barbu, V. (1984). "Optimal Control of Variational Inequalities," Pitman Research Notes in Mathematics Series, Vol. 246, Longman Scientific/Wiley, New York.

Lakshmikantham, V., Bainov, D. D., and Simeonov, P. S. (1989). "Theory of Impulsive Differential Equations," World Scientific, Singapore.

Ahmed, N. U. (1991). "Semigroup Theory with Applications to Systems and Control," Pitman Research Notes in Mathematics Series, Vol. 246, Longman Scientific/Wiley, New York.

Ahmed, N. U. (1992). "Optimal Relaxed Controls for Nonlinear Stochastic Differential Inclusions on Banach Space," Proc. First World Congress of Nonlinear Analysis, Tampa, FL, de Gruyter, Berlin, pp. 1699–1712.

Ahmed, N. U. (1994). "Optimal relaxed controls for nonlinear infinite dimensional stochastic differential inclusions," *Lect. Notes Pure Appl. Math., Optimal Control of Differential Equations* **160,** 1–19.

Ahmed, N. U. (1995). "Optimal control of infinite dimensional systems governed by functional differential inclusions," *Discussiones Mathematicae (Differential Inclusions)* **15,** 75–94.

Ahmed, N. U., and Xiang, X. (1996). "Nonlinear boundary control of semilinear parabolic systems," *SIAM J. Control Optimization* **34**(2), 473–490.

Ahmed, N. U. (1996). "Optimal relaxed controls for infinite-dimensional stochastic systems of Zakai type," *SIAM J. Control Optimization* **34**(5), 1592– 1615.

Ahmed, N. U., and Xiang, X. (1996). "Nonlinear uncertain systems and necessary conditions of optimality," *SIAM J. Control Optimization* **35**(5), 1755–1772.

Ahmed, N. U. (1996). "Existence and uniqueness of measure-valued solutions for Zakai equations," *Publicationes Mathematicae* **49**(3–4), 251–264.

Ahmed, N. U. (1996). "Generalized solutions for linear systems governed by operators beyond Hille–Yosida type," *Publicationes Mathematicae* **48**(1–2), 45–64.

Ahmed, N. U. (1997). "Measure solutions for semilinear evolution equations with polynomial growth and their optimal control," *Discussiones Mathematicae (Differential Inclusions)* **17,** 5–27.

Ahmed, N. U. (1997). "Stochastic *B*-evolutions on Hilbert spaces," *Nonlinear Anal.* **30**(1), 199–209.

Ahmed, N. U. (1997). "Optimal control for linear systems described by *m*- times integrated semigroups," *Publicationes Mathematicae* **50**(1–2), 1–13.

Xiang, X., and Ahmed, N. U. (1997). "Necessary conditions of optimality for differential inclusions on Banach space," *Nonlinear Anal.* **30**(8), 5437–5445.

Ahmed, N. U., and Kerbal, S. (1997). "Stochastic systems governed by *B*-evolutions on Hilbert spaces," *Proc. Roy. Soc. Edinburgh* **127A,** 903–920.

Rogovchenko, Y. V. (1997). "Impulsive evolution systems: main results and new trends," *Dynamics of Continuous, Discrete, and Impulsive Systems* **3**(1), 77–78.

Ahmed, N. U. (1998). "Optimal control of turbulent flow as measure solutions," *IJCFD* **11,** 169–180.

Fattorini, H. O. (1998). "Infinite dimensional optimization and control theory," *In* "Encyclopedia of Mathematics and Its Applications," Cambridge Univ. Press, Cambridge, U.K.

Ahmed, N. U. (1999). "Measure solutions for semilinear systems with unbounded nonlinearities," *Nonlinear Anal.* **35,** 478–503.

Ahmed, N. U. (1999). "Relaxed solutions for stochastic evolution equations on Hilbert space with polynomial nonlinearities," *Publicationes Mathematicae* **54**(1–2), 75–101.

Ahmed, N. U. (1999). "A general result on measure solutions for semilinear evolution equations," *Nonlinear Anal.* **35**.

Liu, J. H. (1999). "Nonlinear impulsive evolution equations: dynamics of continuous, discrete, and impulsive systems," **6,** 77–85.

Ahmed, N. U. (1999). "Measure solutions for impulsive systems in Banach space and their control," *J. Dynamics of Continuous, Discrete, and Impulsive Systems* **6,** 519–535.

Ahmed, N. U. (2000). "Optimal impulse control for impulsive systems in Banach spaces," *Int. J. of Differential Equations and Applications* **1**(1).

Ahmed, N. U. (2000). "Systems governed by impulsive differential inclusions on Hilbert space," *J. Nonlinear Analysis.*

DNA Testing in Forensic Science

Moses S. Schanfield

Monroe County Public Safety Laboratory

GLOSSARY

Alleles Alternate forms of an inherited trait. Such as the "A," "B," or "O" alleles of the ABO blood group. Simple polymorphisms will have as few as two alleles, while hypervariable polymorphisms will have five or more alleles, such that the majority of the population is heterozygous.

Amplified fragment length polymorphisms (AFLPs or AmpFLPs) Polymorphism in length of DNA segments detected using the PCR process to amplify a specific segment of DNA, followed by separation using electrophoresis. These are normally Variable Number Tandem Repeat (VNTR) regions that can have different size repeats. AFLPs are classified by the size of the repeat, repeats consisting of 10 or more bases are referred to as large tandem repeat (LTR) AFLP, while those containing seven or fewer base

pairs per repeat are referred to as short tandem repeats (STRs).

Gene A sequence of DNA which codes for the production of a specific protein or part of a specific protein, such as a glactosyl transferase (the B allele of the ABO blood group).

Genotype The genetic type of a person determined either by the fact that they have two different alleles at a locus, or that the individuals parents have been tested to determine what alleles the person has genetically inherited.

Heterozygotes An individual with two different alleles at a given locus.

Homozygotes An individual with two alleles that are the same.

Locus The place (location) where a specific gene resides in the human genome. The names of genetic regions have been standardardized by the International System of Gene Nomenclature (Shows *et al.*, 1987). Inherited traits that code for genes usually have a name that reflects its biological function, such as THO1 for the first tyrosine hydroxylase locus. Some identified inherited traits do not code for proteins, but are regions that show variation (polymorphism) in length or sequence of DNA. These areas are referred to as "DNA loci," such as D1S7, which represents and DNA inherited trait (**D**), located on chromosome **1**, "**S**" indicates it is a single copy region and "**7**" indicates it was the seventh polymorphism found on that chromosome.

Phenotype The observed results of a genetic test. If a person has two different alleles (e.g., is heterozygous), the phenotype and the genotype are the same. However, if a person only has one allele detected, without testing the parents we cannot be certain that the person has two alleles the same (i.e., homozygous) since the person could have two alleles the same, or one detected allele and one undetected allele, for what ever reason. In the case of RFLP loci this is usually referred to as a single band pattern. For AFLP- and sequence-based PCR-based systems this is referred to as a homozygous phenotype. In general since we do not know the genetic type of homozygous individuals it is better to refer to homozygous and heterozygous phenotypes, unless the genetic type of the individual is determined by pedigree analysis.

Polymorphism Genetically inherited variation with two or more forms, the least common of which occurs at a frequency of greater than 1%.

Restriction fragment length polymorphism (RFLP) Polymorphism in length of DNA segments detected using a restriction enzyme, followed by separation using electrophoresis.

Variable number tandem repeat (VNTR) regions of inherited variation that consist of alleles which contain different numbers of repeating segments. It is easiest to think of them as freight trains with different numbers of box cars. Different loci will have different numbers of repeats. Though RFLP loci are also VNTR loci the number of repeats is generally not known because of the detection technology.

FORENSIC SCIENCE is applied science. That is to say that the scientific methodologies used in forensic science were developed by biologists, chemists, and geneticists and then taken over by forensic scientists to help them solve problems. The first inherited trait to be used in forensic testing was the ABO blood group on red blood cells and secreted blood group substance found in saliva and other body fluids of individuals called "Secretors." Research in the 1950s, 1960s, and 1970s identified proteins that had genetic variation or were "polymorphic." Some of these were enzymes such as acid phosphatase [ACP, referred to as erythrocyte acid phosphatase (EAP) by forensic scientists], esterase D (ESD), phosphoglucomutase (PGM1) and some transport and functional proteins such as group specific component [GC, now known as Vitamin D binding globulin [VDBG]], haptoglobin (HP), the immunoglobulin allotypes (GM and KM), and transferrin (TF) were used forensically. These markers were used in the late 1970s and early 1980s by forensic science in the United States and abroad to individualize blood stains. Although, some of these markers did not last long in bloodstains they made it possible to often individualize bloodstains with greater than a 99% certainty. Semen stains, saliva, and urine had relatively few markers that could be detected, making it difficult to provide information in cases with these types of evidence.

In the mid-1970s two independent areas of research would change the future of forensic science. Research on bacterial enzymes that cut DNA at specific places led to the development of restriction fragment length polymorphisms (RFLPs). Initially these were only useful for the diagnosis of genetic diseases such as Sickle Cell Anemia caused by a mutation in the hemoglobin gene. In 1980 with the identification of the first hypervariable DNA polymorphism (D14S1), detected by restriction length polymorphism technology (RFLP), the door was opened to the possibility that DNA technology could be applied to forensic evidence. In the next several years, the search for new markers lead to the identification of many forensically useful markers, detected by RFLP, some of which are still used routinely in forensic DNA testing. In the meantime, hypervariable minisatellite regions which identified many genetic regions at one time (multilocus probes) were found. The term "DNA fingerprinting" was used to describe these bar code-like patterns. Although these regions proved to be highly informative for parentage testing, they did not have the sensitivity needed for forensic testing. Though multilocus probes were used for paternity testing and some forensic applications, they are rarely used at the present time. Using a battery of five to seven RFLP loci made it possible to individualize samples into the 100s of millions and billions. This means that the chance of any two unrelated individuals matching was very unlikely.

At the same time the revolution in RFLP was beginning in the mid 1970s, an early version of copying DNA using repair enzymes called polymerases was being explored. It would not be until the early 1980s when the modern Polymerase Chain Reaction (PCR) tests was developed. The role of the polymerase enzymes in copying DNA had been known since the 1970s. The use of high temperature Taq polymerase allowed for the automation of thermal cycling and the introduction of modern PCR. Tests were developed to identify human leukocyte antigens (HLA) for the transplantation community. The first marker that they developed was to the HLA region called DQα (now called

DQA1). This marker looked at a genetic marker that had originally been tested at the protein level at the DNA level. This type of marker looked at DNA-sequence-based differences. This became the first PCR based tested to be used forensically. Other sequence-based tests were developed but they did not provide the same level of identification produced by RFLP based testing. Using the available sequence based tests only allowed individualization in the 100s to 100,000s.

The FBI established a standardized system for publicly funded crime laboratories in the United States. Similar work was going on in Canada, the United Kingdom, and Europe. The introduction of DNA restriction fragment length polymorphism (RFLP) technology revolutionized the field of forensic identity testing. This is especially true for the area of sexual assault evidence that historically has been an area of limited information. With DNA-based testing sperm DNA could be separated from the victims type allowing for the first-time regular direct testing of the sperm donor. Though RFLP technology has been a tremendous aid, it has several problems. It is expensive to implement, labor intensive, expensive to test, and is limited by both quantity and quality of DNA obtained. Further, because the process involved measuring the movement of bands and not directly the DNA product there were many statistical problems with representing the data. The technical feasibility of amplifying specific segments of DNA using the polymerase chain reaction (PCR) had the potential to overcome the shortcomings of RFLP technology.

PCR-based technology is much less expensive to implement, since it does not require a laboratory capable of handling radioactive isotopes. It has higher through put since each worker can do more cases in the same amount of time. PCR by its nature works with smaller amounts of DNA and with DNA that has been environmentally abused. Finally, since the DNA product can be identified as different alternative forms the statistical manipulation of data was similar to that for other genetic polymorphisms.

In 1989 at the same time that the RFLP-based testing was being converted to PCR-based testing creating the first amplified fragment length polymorphism or AFLP, new polymorphisms were being found directly using PCR. The converted RFLP loci which consisted of different numbers of repeated segments, much like freight trains with different numbers of box cars were referred to as "variable number of tandem repeat" or VNTR regions or loci. These repeats consisted of 15 to 70 base pairs and were referred to at "large tandem repeat loci or LTRs. The new regions being found had much smaller repeats consisting of two, three, four, or five bases in a repeat unit. These new markers were called "short tandem repeats" or STRs for short. The four base pair or tetra nucleotide repeats became the

genetic markers used to map the human genome. With this large number of markers available it became possible to pick sets of these markers to make highly discriminatory multiplexes (multiple regions amplified at the same time). The use of fluorescent detection of DNA fragments on automated DNA sequencers from the human genome project made it possible to create fluorescent multiplexes with automated detection. These methodologies are now the methods of choice for use in the forensic testing of DNA samples.

I. WHAT IS DNA?

DNA stands for deoxyribonucleic acid. It is the biological blueprint of life. DNA is made up of a double-stranded structure consisting of sugar (deoxyribose) and phosphate back bone, cross linked with two types of nucleic acids referred to as purines (adenine and guanine) and pyrimidines (thymine and cytosine) (Fig. 1). The cross linking nucleic acids always pair a purine with a pyrimidine, such that adenine always pairs with thymine and guanine always pairs with cytosine.

DNA can be found in several areas of a cell. The majority of DNA is located in the nucleus of cells (Fig. 2) organized in the form of chromosomes (22 pairs of autosomes and a set of sex chromosomes (X and Y)). Each nucleated cell normally has 46 chromosomes that represent the contribution from both parents. In the formation of gametes (eggs and sperm) one chromosome of each pair is randomly separated and placed in the gamete. The separation of chromosomes is referred to as segregation. The transmission of half of our chromosomes to our children in the form of gametes is the basis of Mendelian inheritance. This DNA is referred to as nuclear or genomic DNA. With the exception of identical twins, no two people share the same genomic DNA sequence.

Another source of DNA is found in the mitochondria in the cytoplasm of cells (Fig. 2). Unlike nuclear DNA, which only has two copies of each genetic region, mitochondrial DNA is involved in energy production within the cell and can have between 100 and 10,000 copies per cell. Structurally, instead of a linear arrangement of DNA within chromosomes, mitochondrial DNA has a circular structure. Mitochondrial DNA is inherited from the mother because it is found in the cytoplasm which comes from the egg (ova).

A. Where Is DNA Found?

Nuclear or genomic DNA is found in all nucleated cells as well as in the reproductive cells (eggs and sperm). The amount of DNA we can expect to find in different cells and types of evidence are found in Table I. DNA has been

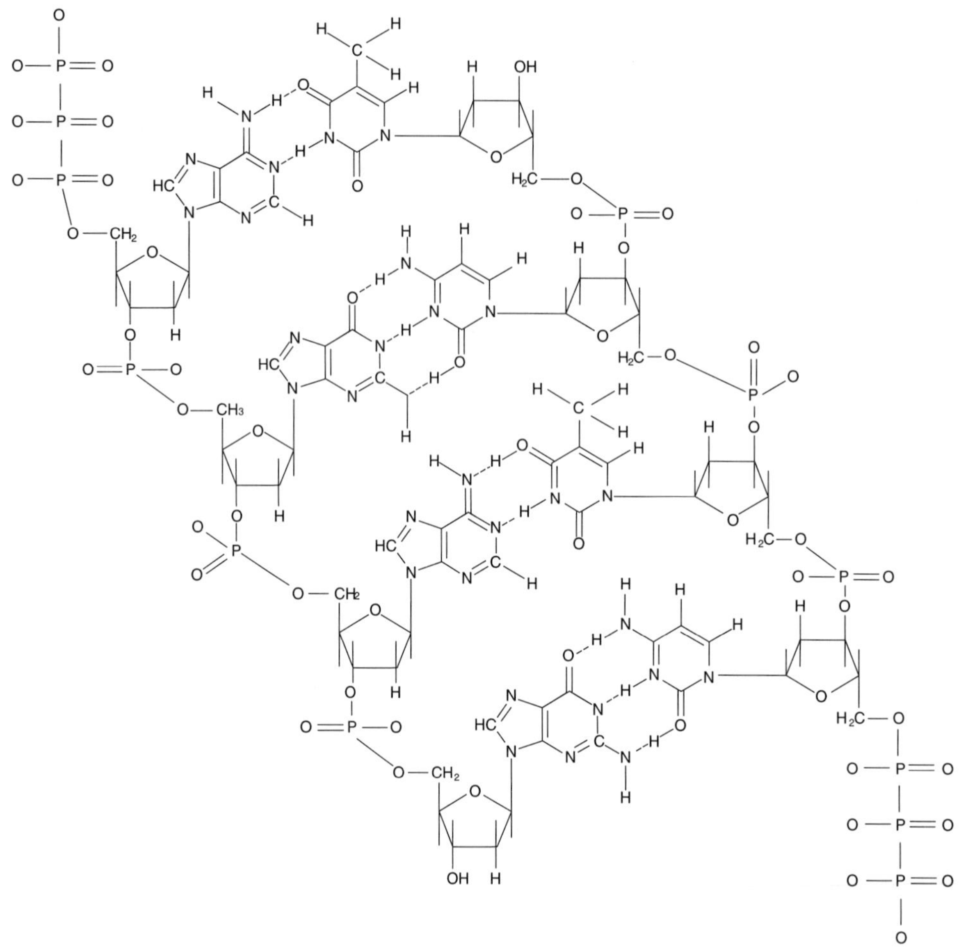

FIGURE 1 Molecular structure of DNA. From top to bottom: Adenine–Thymine, Guanine–Cytosine, Adenine–Thymine and Guanine–Cytosine. (From Schanfield, M. S. (2000). Deoxyribonucleic Acid/Basic Principles. *In* "Encyclopedia of Forensic Sciences" (Siegel, J. A., Saukko, P. J., and Knupfer, G. C., eds.), Academic Press, London, p. 481.)

successfully obtained from blood and bloodstains, vaginal and anal swabs, oral swabs, well-worn clothing, bone, teeth, most organs, and to some extent urine. It is less likely to obtain DNA from some types of evidence than others. Blood or semen stains on soil and leather are historically not good sources of evidenciary DNA. Saliva, per se, has few nucleated cells, but, beer and wine bottles, drinking glasses, beer cans, soda cans, cigarettes, stamps and envelope flaps have all been found to provide varying amounts of DNA.

B. How Much DNA Is Needed for Forensic Testing?

The amount of DNA needed to perform testing depends on the technology used. RFLP technology usually needs at least 50 ng of intact high-molecular-weight DNA. In contrast PCR-based testing can use as little as 250 pg. Most

PCR based tests are set up to use between 1 and 10 ng of genomic DNA.

C. Destruction of DNA

Biological materials are going to be affected by their environment. Enzymes lose activity over time and type of storage conditions. DNA has been found to be relatively robust when it is in the form of dry stains. Initial environmental studies indicated some of the limitations of DNA based on the material it is deposited upon and the environmental conditions. Environmental insult to DNA does not change the results of testing, you will either obtain results, or if the DNA has been too badly affected by the environment (i.e., the DNA is degraded) you do not get RFLP results. One report on the success rate of obtaining RFLP results and noted that depending on the substrate or condition of the stain, results were obtained

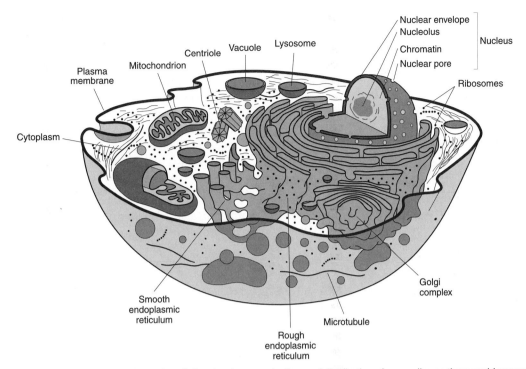

FIGURE 2 A generalized eukaryotic cell showing the organization and distribution of organelles as they would appear in transmission electron microscope. The type, number, and distribution of organelles are related to cell function. (From Schanfield, M. S. (2000). Deoxyribonucleic Acid/Basic Principles. *In* "Encyclopedia of Forensic Sciences" (Siegel, J. A., Saukko, P. J., and Knupfer, G. C., eds.), Academic Press, London, p. 481.)

between 0 (carpet stains or putrefied samples) and 61.5% (scrapped dried stains) of the time with and average of 52% for the 100 items of evidence tested.. Thus, the material DNA is deposited on and the degree of further insult can markedly affect the ability to obtain RFLP DNA results.

All of the published studies on environmental insult were done on prepared dried stains. Since biological fluids are liquid the effects of ultraviolet radiation on liquid DNA have been evaluated. The results of exposing 100-μl samples of a standard DNA solution to fluorescent light in the laboratory, a UV germicidal light (254 nm), midday sunlight in January, and early sunset light in January in

15-min increments, up to 1 hr are presented in Fig. 3. There is a linear decrease in high-molecular-weight DNA with the UV germicidal light, such that after an hour about 96% of the high-molecular-weight DNA has been lost. Even in the weak midday light in January, over 60% of the high-molecular-weight DNA was lost. In contrast, the fluorescent lighting in the laboratory and the after sunset light had

TABLE I DNA Content of Various Tissues

1 sperm	3 pg
1 cell	6 pg
1 shed hair	1 ng[a]
1 plucked hair	300 ng[b]
1 drop of blood	1,500 ng

[a] The success rate for PCR on shed hairs is 30 to 50%, so this average is optimistic.

[b] There is a great deal of variation among hair roots. Fine blond hair will tend to have much less DNA, while course dark hair with large roots more.

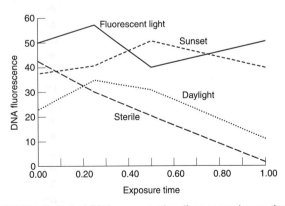

FIGURE 3 Plot of DNA concentration (frorescence) over time after exposure to different light sources. (From Schanfield, M. S. (2000). Deoxyribonucleic Acid/Basic Principles. *In* "Encyclopedia of Forensic Sciences" (Siegel, J. A., Saukko, P. J., and Knupfer, G. C., eds.), Academic Press, London, p. 482.)

no effect on the amount of high-molecular-weight DNA. This was not a rigorous experiment, but the effects are dramatic enough to demonstrate the effect of ultra violet light exposure to DNA before stains dry.

II. EXTRACTION OF DNA

As stated previously DNA exists inside of cells. Because most evidence is in the form of dry stains, the DNA must be removed from the stain before it can be tested. The process of removing DNA from the cells on the evidence and dissolving it is referred to as extraction. There are several procedures available for removing DNA from evidence so that it can be used. They are referred to as either "organic" extraction or "nonorganic" extraction based on the nature of the chemicals used. Further, there are two special types of extraction. The first, called differential extraction, was developed for sexual assault evidence to separate the cells that come from the victim (epithelial cells from the vagina, rectum, or mouth) from those of the perpetrator (male sperm cells). The second method is a specialized "nonorganic" extraction using Chelex beads. Chelex beads can only be used when PCR-based DNA testing is going to be used. The basic DNA extraction procedures, whether organic or nonorganic, can be adapted for special circumstances such as hair or tissue.

A. Chloroform–Phenol Extraction

This is the oldest procedure available for extracting DNA from blood and it has been extended to include hair, tissue, and semen stains. The basic procedure consists of opening up cells with a buffer and an enzyme, usually Protease K, and then denaturing and separating the proteins from the DNA. The latter part is done using a mixture of chloroform (24:1 chloroform:isoamyl alcohol) and phenol (buffered). The phenol–chloroform mixture denatures proteins liberated by the first stage. The major disadvantage of this procedure is the fact that phenol–chloroform is a hazardous waste and could theoretically pose a risk to pregnant employees. A modern protocol for phenol–chloroform extraction of various types of evidence can be found in the literature.

B. Nonorganic Extraction

In nonorganic extraction the hazardous phenol–chloroform protein denaturation step was replaced by a salting out of proteins. This allowed for the same chemistry to be used for the initial phase of DNA extraction, and replacement of the hazardous elements of the procedure with a nonhazardous alternative. The salting-out procedure has several advantages over the phenol–chloroform extraction. The first is that instead of having two liquid phases (organic and nonorganic) that can occasionally trap the DNA in the wrong phase (organic) phase, by precipitating the proteins (e.g., the proteins become insoluble and become a solid), there are liquid and solid phases with the DNA only in the liquid phase (nonorganic). The second advantage is that the hazardous phenol–chloroform is replaced with a harmless salt solution. Comparison of the organic and nonorganic procedures for blood, and semen indicate that the nonorganic extraction is on the average as good or better than organic extraction, whether quantitated by yield gel or slot blot (Table II).

Either method of DNA extraction described earlier can be used for both RFLP- or PCR-based DNA testing. Organic DNA extraction is widely used in laboratories doing criminal casework while nonorganic DNA extraction is

TABLE II Comparison of Organic and Nonorganic Extraction of DNA from Blood and Semen Stains[a]

Quantitation method	Blood organic	Blood nonorganic	Semen organic	Semen nonorganic
Yield gel				
Mean	185 ng	258 ng	175 ng	207 ng
N	21	8	22	8
p	.054		.122	
Slot blot				
Mean	515 ng	908 ng	627 ng	1175 ng
N	22	8	27	8
p	.022		.008	

[a] Data taken from Tables 1 and 2 of Laber *et al.* (1992). Differences in means tested by Kruskall–Wallace nonparametric analysis of variance, H statistic with 1 df, uncorrected p values presented.

widely used in laboratories performing paternity testing, research, and diagnostics. On a worldwide basis nonorganic DNA extraction is the more prevalent. With the shift to PCR-based testing this choice in extraction is increasingly common.

C. Chelex Extraction

In 1991, a method of DNA extraction was described that was specifically aimed at the extraction of small amounts of dilute DNA for PCR-based testing using Chelex beads. The method is simple, relatively fast, and biohazard free. It is widely used by forensic laboratories doing PCR-based typing which has increased the number of laboratories using nonorganic, biohazard-free DNA extraction. The only limitations of Chelex extraction is that it produces a dilute solution of DNA that may need to be concentrated before it can be used with some of the newer high-resolution PCR-based typing systems and it cannot be used for RFLP testing.

III. QUANTITATION OF DNA

A. Yield-Gel Quantitation

Whether RFLP- or PCR-based testing is performed it is necessary to know how much DNA is present. One of the earliest methods of quantitating small amounts of DNA is the use of a yield gel. A small gel is made using a salt solution to carry electrical current and a supporting medium made of agarose (a complex carbohydrate made from seaweed). Much like gelatin, the agarose dissolves in water that is heated to near boiling and the liquid is cooled slightly and poured into a casting tray. A plastic comb or well former with rectangular teeth is placed in the liquid agarose. Once the agarose gels, the comb is removed leaving behind rectangular wells in the agarose gel. The DNA to be tested is mixed with loading buffer, and placed in the wells. Loading buffer is a mixture of a large amount of sugar and dye. The high concentration of sugars makes the mixture heavier than the salt solution so that the DNA sinks to the bottom of the well. The dye allows the migration of the DNA to be monitored. The agarose was melted in water containing salt. When electrical current is applied to the gel, electricity flows through the gel because of the salt and moves (migrates) from the negative electrode (cathode) toward the positive electrode (anode). Since all DNA has a negative charge, and was placed in the wells at the cathodal end of the gel, the negatively charged DNA will migrate out of the wells toward the positive end of the gel. If the DNA is broken into pieces that are different sizes, the smaller pieces will move through the gel faster than the larger pieces and will

be separated based on size. This process of separating DNA using an electric current is called electrophoresis, which simply means separation (phoresis) by means of electricity (electro).

Since DNA is colorless it is not possible to see the DNA after it has been separated without the use of special dyes that bind to it. One of the earliest dyes used was ethidium bromide, which fluoresces pink when bound to double-stranded DNA and exposed to ultraviolet light. Figure 4 is an ethidium bromide stained yield gel. To quantitate the amount of DNA in the DNA extracts, a set of DNA quantitation standards are placed on the gel. By visual comparison of the unknown DNA to the known DNA the amount of DNA can be approximated. This test provides information about the relative amount of DNA and whether it is degraded (i.e., the DNA is broken down so that different-size pieces of DNA are present). It does not indicate if the DNA is human, however, since all DNA will fluoresce. Thus the DNA present may be bacterial as well as human DNA. For RFLP testing the total amount of DNA in the sample is the important determinant of how the samples migrate in the gel. Therefore yield-gel electrophoretic quantitation of DNA is an appropriate method. Yield-gel quantitation of DNA for RFLP testing was considered to be such an integral part of quality assurance that it was included in the National Institute of Standards, Standard Reference Material 2390, "DNA Profiling Standard."

As with the extraction of DNA using the organic method, ethidium bromide is potentially hazardous because the dye is associated with an increased cancer risk. Though ethidium bromide is still widely used for the identification of DNA it is currently being replaced by a new dye called Sybr® green which is much less carcinogenic and can detect smaller amounts of DNA than ethidium bromide.

B. Slot-Blot Quantitation

In contrast to RFLP, for PCR-based testing, the amount of human DNA and not the total amount of DNA is an important determinant in how likely it will be to obtain results. A slot blot does not rely on electrophoresis to separate the DNA but rather on the ability of denatured (separated DNA strands) DNA to bind to homologous complementary sequences. The ability to quantitate human DNA requires sequences of DNA that are common in the human genome so that a single DNA sequence can recognize them and bind to them. The repeated DNA sequence called D17Z1 is the basis for all human DNA slot-blot quantitation systems. There are several of these procedures commercially available. In one of the most widely used tests, the quantitation requires that denatured DNA is applied to a membrane using a slotted plastic apparatus. The denatured DNA binds to the membrane. The

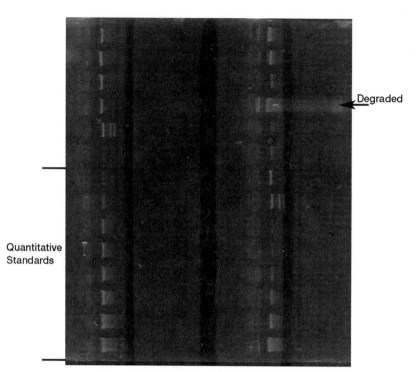

FIGURE 4 Ethidium bromide stained yield gel. Bottom left samples are quantitative standards. Other samples represent various samples of DNA. Upper right sample is degraded DNA. (From Schanfield, M. S. (2000). Deoxyribonucleic Acid/Basic Principles. *In* "Encyclopedia of Forensic Sciences" (Siegel, J. A., Saukko, P. J., and Knupfer, G. C., eds.), Academic Press, London, p. 484.)

membrane is exposed to a solution of denatured DNA fragments that recognizes a repeating sequence of human or primate DNA. Pieces of DNA that recognize a specific region of DNA are referred to as a "probe." The probe will bind to complementary DNA fragments stuck on the membrane. The probe has an indicator attached to it so that the binding of the DNA to the probe can be detected. The unbound probe is washed off and the probe is detected using either chemicals that change color (colorimetric detection) or chemicals that give off light (chemiluminscent detection). To be able to quantitate the amount of human DNA present, standards with different amounts of human DNA are also placed on the membrane so that it is possible to determine the approximate amount of DNA bound to the membrane by visual comparison to the known standards. More precise quantitation can be obtained by scanning the membrane with a scanning densitometer and determining the amount of color associated with each band. Most forensic laboratories use visual comparison.

IV. CURRENT FORENSIC DNA TESTING

At this point in time, at the beginning of the 21st century, forensic DNA testing has moved away from RFLP test-

ing and is replacing sequence-based PCR strip technology with fluorescent STR-based testing. Thus it is important to understand the nature of PCR-based testing and the power it provides to forensic scientists.

A. Definition and Description of PCR

The Polymerase Chain Reaction (PCR) is based on biochemical processes within cells to repair damaged DNA and to make copies of the DNA as the cells replicate. In the repair mode, if a single strand of DNA is damaged, the damaged area is removed so that there is a single-stranded section of DNA with double-stranded sections at either end. The polymerase enzyme fills in the missing complementary DNA. In the copy mode an entire strand is copied during DNA replication. Figure 5 illustrates a polymerase enzyme copying a portion of a strand of DNA.

In a cell a specific gene is copied or translated from DNA to RNA because the polymerase has specific start and stop signals coded into the DNA. To copy a sequence of DNA *in vitro*, artificial start and stop signals are needed. These signals can only be made once the sequence of the region to be amplified is known. Once a sequence is known, the area to be copied or amplified can be defined by a

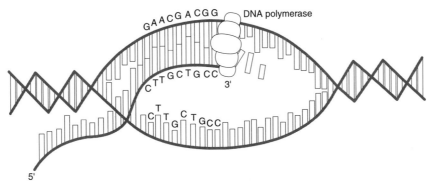

FIGURE 5 DNA polymerase copying one strand of a portion of double-stranded DNA. (From Schanfield, M. S. (2000). Deoxyribonucleic Acid/Polymerase Chain Reaction. *In* "Encyclopedia of Forensic Sciences" (Siegel, J. A., Saukko, P. J., and Knupfer, G. C., eds.), Academic Press, London, p. 516.)

unique sequence of DNA. For a primer to recognize a unique sequence in the genome it must be long enough for no other sequence to match it by chance. This can usually be achieved with a sequence of 20 to 25 nucleotides. These manufactured pieces of DNA are called "primers" and they are complementary to the start and stop areas defined earlier. The "forward primer" is complementary to the beginning sequence on one strand of DNA, usually called the positive strand. The "reverse primer" is complementary to the stop sequence on the opposite or negative strand of DNA.

B. Multiplexing PCR Reactions

One of the advantages of PCR is that more than one region can be amplified at a time. Although it is necessary to select carefully primers that cannot bind to each other, the only limitation is how many pairs of primers can be placed together is the ability to detect the amplified product.

C. PCR Process

To perform a PCR reaction several ingredients are needed. They include PCR reaction buffer, which is basically a salt solution at the right pH for the enzyme being used, the four nucleotides (DNA building blocks), primers, a thermostable DNA polymerase (Taq, Pfu, Vent, Replinase, etc), and template DNA. These reactants are placed in small plastic reaction tubes. The process consists of heating a solution of DNA to greater than 90°C. Double-stranded DNA comes apart or melts to form single-stranded DNA at this temperature. This is called the *denaturation* step. The solution is then cooled down to between 50 and 65°C the primers will bind to their complementary locations. This is called the *annealing* or probe hybridization step. Finally, the solution temperature is raised to 72°C at which point the polymerase makes a copy of

the target DNA defined by the primers. This is called the *extension* step. This completes one cycle of the PCR process. To make enough copies of the target DNA to detect the process is repeated from 25 to 40 times. This is done using a device called a thermalcycler. The process is illustrated in Fig. 6. If the process were perfect 30 cycles would create over a billion copies of the original target DNA.

The heating and cooling of the tubes are done in an electromechanical device call a "thermalcycler," which in general, consists of an aluminum blocks with wells designed to fit the plastic PCR reaction tubes. The aluminum block has heating and cooling elements controlled by a microprocessor that can raise and lower the temperature of the block and the plastic PCR reaction tubes in the block. In the thermal cyclers that were first made, the plastic reaction tubes extended above the thermal block. This allowed cooling to take place above the reaction. The water in the reaction mixture would evaporate and condense at the top of the tube, changing the concentration of reactants and affecting the success of the amplification. To limit the evaporation, mineral oil was placed on top of the reaction mixture. New thermal cyclers have heated lids on top of the block to prevent or minimize evaporation. The microprocessor can store many sets of instructions such that different programs can be kept in the microprocessor to amplify different sequences of DNA.

D. Detection of PCR Products

There are many methods for detecting PCR products. Since large amounts of product are produced there is no need to use techniques such as radioactive detection, although it has been used in some clinical setting. In forensic testing, one of the advantages of PCR-based testing is that it does not require the use of hazardous materials to detect it. There is normally enough product so that if the PCR products are run on a yield gel and stained with ethidium bromide or Cyber green, there is normally enough DNA

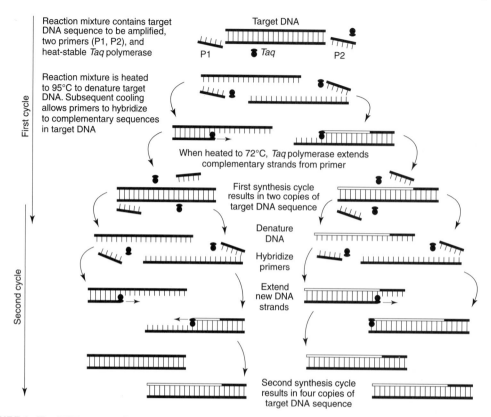

Reaction mixture contains target DNA sequence to be amplified, two primers (P1, P2), and heat-stable *Taq* polymerase

Reaction mixture is heated to 95°C to denature target DNA. Subsequent cooling allows primers to hybridize to complementary sequences in target DNA

First cycle

Second cycle

Target DNA

P1   *Taq*   P2

When heated to 72°C, *Taq* polymerase extends complementary strands from primer

First synthesis cycle results in two copies of target DNA sequence

Denature DNA

Hybridize primers

Extend new DNA strands

Second synthesis cycle results in four copies of target DNA sequence

FIGURE 6 The PCR process. Courtesy of PE Cetus Instruments. (From Schanfield, M. S. (2000). Deoxyribonucleic Acid/Polymerase Chain Reaction. *In* "Encyclopedia of Forensic Sciences" (Siegel, J. A., Saukko, P. J., and Knupfer, G. C., eds.), Academic Press, London, p. 517.)

to be detected. This is a suggested method to verify if the PCR amplification was successful and that there is a PCR product to detect.

V. STRs USED FORENSICALLY

At this point in time with the large number of STR loci, a demand for a standardized panel in the United States, and a need for there to be at least some sharing of loci with forensic counterparts in Canada, England, and Europe, the Technical Working Group on DNA Analysis Methods (TWGDAM) implemented a multi-laboratory evaluation of those STR loci available in kits in the United States. The loci chosen would be the PCR-based core of a national sex offender file required under the 1994 DNA Identification Act. The national program is called Combined DNA Indexing System or CODIS for short.

The TWGDAM/CODIS loci were announced at the Promega DNA Identification Symposium in the fall of 1997 and at the American Academy of Forensic Science meeting in February 1998. The following loci were chosen to be part of what was originally called the CODIS 13 loci: CSF1PO, D3S1358, D5S818, D7S820,

D8S1179, D13S317, D16S539, D18S51, D21S11, FGA, THO1, TPOX, and VWA03. These loci overlapped with the Forensic Science Services multiplexes and the Interpol multiplexes. The 13 loci can be obtained in two amplifications using Profiler Plus and Cofiler or in a single amplification using a kit in development called Identifiler from PE Biosystems, or in two amplifications using Powerplex 1 and Powerplex 2 from Promega (Table II), or in a single reaction with Powerplex 16 by Promega.

A. Fluorescent Dyes

Before discussing the equipment used to detect fluorescent STRs some understanding of fluorescent dyes are necessary. Fluorescent dyes or minerals when subjected to light at one wave length, such as untraviolet light or black light, will give off colored light at a slightly different wavelength or color. A characteristic of fluorescent dyes or materials is that the compound is excited at one frequency of light, referred to as its absorption or excitation peak, and emits or gives off light a different frequency, referred to as its emission peak.

To label a PCR primer with a fluorescent dye, the dye is attached to the 5′ end of the molecule. Since DNA is translated from 5 to 3′ it is at the very end of the primer and should not affect the amplification or binding of the primer if made correctly. One of the oldest fluorescent dyes is fluoroscein. Many devices have been made that will detect fluoroscein, and it has been used extensively to label antibodies and other materials. Many of these dyes have been used for a long time and are in the public domain. Others have been developed for specific projects. The dyes used by PE Biosystems were originally proprietary and part of a patented four color DNA sequencing system (Blue, Green, Yellow, Red). These dyes are now becoming more readily available.

B. Fluorescent Detection Equipment

The equipment used to detect the product of fluorescently labeled STR tests fall into two categories. The first are devices that scan a gel after the DNA products have been separated by a process called electrophoreses. Examples of post electrophoresis scanners are the Hitachi FMBIO® Fluorescent Scanner, the Molecular Dynamics Fluorimager™ and the Beckman Genomics SC scanner. The Hitachi and Molecular Dynamics use a laser as a light with filters to identify the proper frequency and a CCD camera to capture the image. The Beckman Genomics SC uses a monochromatic Xenon light source and uses filters to detect the appropriate light for the CCD camera. The CCD camera scans back and forth over the gel as it is exposed to the light source and detects the various fluorescent colors using filters that change. This type of equipment has flexibility because different formats of electrophoresis gels can be used and scanned. The output is in the form of an electronic image with bands, that look much like a set of RFLP bands. Figure 7A is an example of actual images

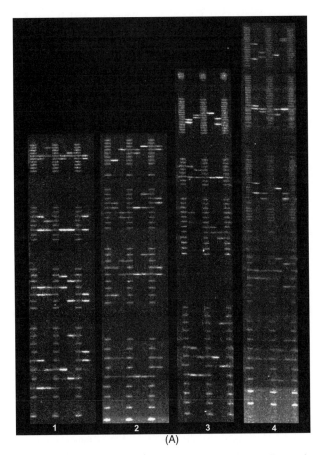

(A)

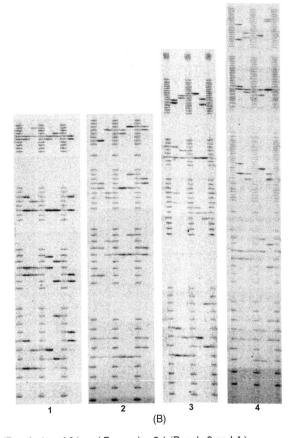

(B)

FIGURE 7 (A) is the original CCD images of Powerplex 1.1 (Panels 1 and 2) and Powerplex 2.1 (Panels 3 and 4). (B) is the reverse image of the same images. (B) Panel 1 is the TMR loci from top to bottom CSF1PO, TPOX, THO1, and VWA03. (B) Panel 2 is the fluoroscein loci, from top to bottom D16S539, D7S820, D13S317, and D5S818. (B) Panel 3 is the TMR loci from top to bottom FGA, TPOX, D8S1179, and VWA03. (B) Panel 4 is the fluoroscein loci from top to bottom Penta E, D18S51, D21S11, THO1, and D3S1358. The same samples are loaded between the ladders. The sample format is the TWGDAM format when an internal size standard is not used. The overlapping loci for sample confirmation are THO1, TPOX and VWA03. (From Schanfield, M. S. (2000). Deoxyribonucleic Acid/Polymerase Chain Reaction-Short Tandem Repeats. *In* "Encyclopedia of Forensic Sciences" (Siegel, J. A., Saukko, P. J., and Knupfer, G. C., eds.), Academic Press, London, p. 530.)

(light on dark) recorded by the CCD camera, and Fig. 7B is the reverse image (dark on light) that is reminiscent of an RFLP autoradiograph or lumigraph. It should be noted that the scanning time is added onto the electrophoresis time, with increased time for each color read.

The second type of imaging system is a real time system, in which the DNA fragments, after the bands have been resolved, pass beneath a light source scanner that recovers the spectrum of light from the different fluorophors. This is the ABI Prism® system from PE Biosystems. It includes the older Model 373 DNA Sequencer and 377 DNA Sequencer, which use slab acrylamide electrophoresis to separate the DNA fragments, and the 310 and 3100 Genetic Analyzers which use capillary electrophoresis to separate the DNA fragments. Capillary electrophoresis is a technology in which a fine glass capillary is filled with a proprietary separation polymer. The sample is pulled into the capillary by applying an electric current to it, then using high-voltage electrophoresis (12,000 V), and the DNA fragments are separated over the length of the column and move past a laser detector. The 377 can put approximately 60 samples on a gel at one time, and with modifications, 96. In contrast, the 310 CE system does one sample at a time, with a separation time of approximately 20 min. However, as this is automated, a cassette can be filled with samples for testing and left to run unattended. The 3100 uses 10 capillary tubes with higher throughput. The output of these devices is not a CCD image, but a series of electropherograms with a profile for each color scanned (nominally Blue, Green, Yellow, and Red). Since these are difficult to interpret the computer software provides decomposed single color graphs. Figure 8 contains an electropherogram of the single amplification Promega PowerPlex 16.2 System run on an ABI PRISM 310 Genetic Analyzer.

One of the extremely useful characteristics of fluorescent imaging devices is that the amount of light read by the detection device is quantified. The electropherograms produced by the real time scanners is quantitative. However, the CCD image can also be used as a scanning densitometer to determine the amount of light in each peak. Since there is one fluorescent molecule per band the amount of fluorescence is linear with the number of molecules in a band. This allows for many different types of analysis to be performed on the data generated. One of the more important of these is the ability to detect mixtures (see following).

VI. PCR SETUP AND DETECTION

The manufacturers of the kits have done forensic validations of the kits; however, each laboratory is responsible for the individualized validation required before test-

ing. The guidelines for those validations for laboratories in the United States are governed by the DNA Advisory Board Guidelines (as of October 1, 1998) as implemented by ASCLD-LAB. Other areas of the world are regulated by other guidelines, unless they are also ASCLD-LAB accredited.

A. STR Detection

The major difference in the typing of the STR loci is the ability to include an internal size standard if the detection device used has multicolor capability. Under the TWG-DAM Guidelines forensic samples are to be placed adjacent to an allele ladder, as seen in Fig. 7 (PCR–STR). Since the Beckman Genomyx SC only has two filters (fluoroscein and TMR) an internal ladder could not be used, so the adjacent ladder format is used. In this situation there is no special preparation for detection. When the four-color Hitachi FMBIO II Fluorescent Scanner, ABI Prism 377 or 310 is used, an internal standard is used to size the DNA fragments. As part of the electrophoresis setup a ROX ladder is added to PE Biosystems amplified products while a CRX ladder is added to Promega kits. (See Figure 8 for example.) Amplified products including the

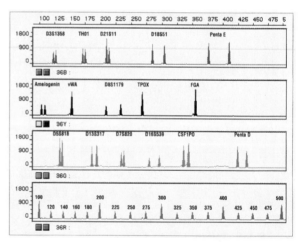

FIGURE 8 Electropherogram of a single DNA sample amplified using the 16-locus prototype Powerplex™ 16.2 System detected with the ABI PRISM® 310 Genetic Analyzer. All 16 loci were amplified in a single reaction and detected in a single capillary. The fluorescein-labeled loci (D3S1358, THO1, D21S11, D18S51, and Penta E) are displayed in blue, the TMR labeled loci (Amelogenin, VWA03, D8S1179, TPOX, and FGA) are displayed in black, and the loci labeled with a new dye (D5S818, D13S317, D7S820, D16S539, CSF1PO, and Penta D) are displayed in green. The fragments of the prototype ILS-500 size marker are labeled with CXR and are shown in red. (Taken from Promega promotional literature.) (From Schanfield, M. S. (2000). Deoxyribonucleic Acid/Polymerase Chain Reaction-Short Tandem Repeats. *In* "Encyclopedia of Forensic Sciences" (Siegel, J. A., Saukko, P. J., and Knupfer, G. C., eds.), Academic Press, London, p. 533.)

alllelic size ladders. The internal size standard is used to size all fragments within a lane by detector supplied software and assigns repeat numbers from the allele ladder sizings.

B. The Use of Internal Size Standards

It was previously demonstrated that within gel variation in DNA migration could be compensated for by placing a size ladder within the lane and measuring each fragment with the internal size standard. This allows for highly precise measurements of fragments. This is necessary since the electrophoresis systems used to detect the STR loci must have the capability of resolving differences between alleles as little as one base pair, to make sure that the fragment sizes can be accurately converted to repeat numbers. This would not be critical if all STR were regular. That is to say always four repeats for the tetra nucleotide STRs. However, this is not the case. Some STR loci have one or more common alleles that differ by only a single base pair (THO1, FGA, and D21S11). An example of this is seen in Fig. 7B, Panel 1, third sample from the left, in the third locus from the top.

VII. USEFULNESS IN DETECTING MIXTURES

One of the major problems in the analysis of forensic evidence is posed by samples containing biological material from more than one source. The large number of discrete alleles at multiple loci make STR multiplexes and excellent tool for identifying components of mixtures.

VIII. INDIVIDUALIZATION

In the United States, the FBI has started releasing reports indicating that biological material originated from a specific source, much as fingerprint examiners have done for many years. The FBI has decided that if the population frequency exceeds 1 in 230 billion the sample is individualized. Other laboratories have chosen high thresold levels such as 1 in 500 billion. Whatever the level chosen it is estimated that the average power of exclusion for these 13 CODIS loci exceeds 1 in a million billion, and though it is possible to obtain a frequency more common than that required for individualization it will occur infrequently.

SEE ALSO THE FOLLOWING ARTICLES

HYDROGEN BOND • MASS SPECTROMETRY IN FORENSIC SCIENCE • PROTEIN STRUCTURE • SPECTROSCOPY IN FORENSIC SCIENCE • TOXICOLOGY IN FORENSIC SCIENCE • TRANSLATION OF RNA TO PROTEIN

BIBLIOGRAPHY

Adams, D. E., Presley, L. A., and Baumstark, A. L., *et al.* (1991). "Deoxyribonucleic Acid (DNA) analysis by restriction fragment length polymorphism of blood and other body fluid stains subjected to contamination and environmental insults," *J. Forensic Sci.* **36,** 1284–1298.

Von Beroldingen, C. H., Blake, E. T., Higuchi, R., Sensabaugh, G., and Ehrlich, H. (1989). "Application of PCR to the analysis of biological evidence. *In* "PCR Technology: Principles and Applications for DNA Amplification" (H. A. Ehrlich, ed.), Stockton Press, New York, pp. 209–223.

Blake, E., Mihalovich, J., Higuchi, R., Walsh, P. S., and Ehrlich, H. (1992). "Polymerase chain reaction (PCR) amplification of human leukocyte antigen (HLA)-DQ oligonucleotide typing on biological evidence samples: casework experience," *J. Forensic Sci.* **37,** 700–726.

Budowle, B., Moretti, T. R., Baumstark, A. L., Defebaugh, D. A., and Keys, K. (1999). "Population data on the thirteen CODIS core short tandem repeat loci in African Americans, US Caucasians, Hispancics, Bahamians, Jamaicans and Trinidadians," *J. Forensic Sci.* **44,** 1277–1286.

Dieffenbach, C. W., and Dveksler, G. S. (1993). "Setting up a PCR laboratory," *PCR Methods Appl.* **3,** 2–7.

Frégeau, C. J., and Fourney, R. M. (1993). "DNA typing with fluorescently tagged short tandem Repeats: A sensitive and accurate approach to human identification," *BioTechniques* **15,** 100–119.

Haugland, R. P. (1996). "Handbook of Fluorescent Probes and Research Chemicals, 6th Edition," Molecular Probes, Eugene, OR, pp. 144–156.

Jeffreys, A., Wilson, V., and Thein, S. L. (1985). "Hypervariable 'minisatellite' regions in human DNA," *Nature* **314,** 67–73.

Kimpton, K. P., Gill, P., Walton, A., Urquhart, A., Millinan, E. A., and Adams, M. (1993). "Automated DNA profiling employing multiplex amplification of short tandem repeat loci," *PCR Methods Appl.* **3,** 13–22.

Laber, T. L., O'Connor, J. M., Iverson, J. T., Liberty, J. A., and Bergman, D. L. (1992). "Evaluation of four Deoxyribonucleic Acid (DNA) extraction protocols for DNA yield and variation in restriction fragment length polymorphism (RFLP) sizes under varying gel conditions," *J. Forensic Sci.* **37,** 404–424.

Maniatis, T., Fritsch, E. F., and Sambrook, J. (1982). "Molecular Cloning: A Laboratory Manual," Cold Spring Harbor Laboratory, NY, pp. 458–459, 469.

McNally, L., Shaler, R. C., Baird, M., Balazs, I., Kobilinsky, L., and De Forest, P. (1989). "The effects of environment and substrata on deoxyribonucleic acid (DNA): The use of casework samples from New York city," *J. Forensic Sci.* **34,** 1070–1077.

Miller, S. A., Dykes, D. D., and Polesky, H. F. (1988). "A simple salting out procedure for extracting DNA from human nucleated cells," *Nucl. Acids Res.* **16,** 1215.

Mullis, K. B., Faloona, F. A., Scharf, S. J., Saiki, R. K., Horn, G. T., and Erlich, H. A. (1986). "Specific enzymatic amplification of DNA *in vitro*: The polymerase chain reaction," *Cold Spring Harbor Symp. Quant. Biol.* **51,** 263–273.

Shows, T. B., McAlpine, P. J., and Bouchix, C., *et al.* (1987). "Guidelines for human gene nomenclature. An international system for human gene nomenclature (ISGN,1987)," *Cytogenet. Cell Genet.* **46,** 11–28.

Walsh, P. S., Metzger, D. A., and Higuchi, R. (1991). "Chelex® 100 as a medium for simple extraction of DNA for PCR based typing from forensic material," *BioTechniques* **10,** 506–513.

Walsh, P. S., Varlaro, J., and Reynolds, R. (1992). "A rapid chemiluminescent method for quantitation of human DNA," *Nucl. Acids Res.* **20,** 5061–5065.

Waye, J. S., and Willard, H. F. (1986). "Structure, organization and sequence of alpha satellite DNA from human chromosome 17: evidence for evolution by unequal crossing-over and an ancestral pentamer repeat shared with the human X chromosome," *Mol. Cell. Biol.* **6,** 3156–3165.

Wyman, A., and White, R. (1980). "A highly polymorphic locus in human DNA," *Proc. Nat. Acad. Sci. (USA)* **77,** 6754–6758.

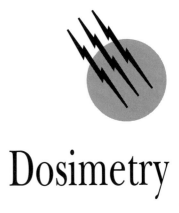

Dosimetry

J. W. Poston, Sr.

Texas A. & M. University

I. Sources of Radiation
II. Quantities and Units
III. Interactions of Radiation with Matter
IV. Dosimetric Techniques, External
V. Dosimetric Techniques, Internal

GLOSSARY

This glossary is intended to define a few words in common use in the United States at this time. It is not intended to be exhaustive. Many other quantities and terms are defined in appropriate locations in the text. For additional definitions and concepts the reader is referred to publications of the National Council on Radiation Protection and Measurements (NCRP) and federal agencies such as the U.S. Nuclear Regulatory Commission (USNRC).

Absorbed dose Amount of energy deposited by ionizing radiation in a material per unit mass of the material. Usually expressed in the special radiological unit rad or in the SI unit gray.

Dose equivalent Product of the absorbed dose, quality factor, and other modifying factors. This quantity is used to express the effects of radiation absorbed dose from the many types of ionizing radiation on a common scale. The special radiological unit is the rem and the equivalent SI unit is the sievert.

Dosimeter Any device worn or carried by an individual to establish the total exposure, absorbed dose, or dose equivalent (or the rates) in the area or to the individual worker while he or she is occupying the area.

Exposure Quantity defined as the charge produced in air by photons interacting in a volume of air with a known mass. A general term used to indicate any situation in which an individual is in a radiation field.

Ionization Process of removing (or adding) one or more electrons from (or to) an atom or a molecule.

Isotope One of two or more atoms with the same number of protons but a different number of neutrons in their nuclei. A radio-isotope is an isotope of an element that is unstable and transforms by the emission of nuclear particles or electromagnetic radiation to reach a more stable state. A radio-isotope may also be called a radionuclide.

Radiation Used here to indicate ionizing radiation; that is, nuclear particles or electromagnetic radiation with sufficient energy to cause ionization of the atoms and molecules composing the material in which the radiation is interacting. Directly ionizing radiations are charged particles that interact directly with the electrons through coulombic interactions. These radiations include, for example, alpha particles, beta particles, electrons, and protons. Indirectly ionizing radiations are uncharged radiations (e.g., X-rays, gamma rays, neutrons) that must interact with the material, producing a charged particle, which then causes further ionization in the material.

DOSIMETRY is the theory and application of principles and techniques associated with the measurement of ionizing radiation. In practice, dosimetry is divided into two primary categories: external dosimetry and internal dosimetry. External dosimetry usually encompasses the use of radiation-detection instrumentation that can be used to establish, or measure, the characteristics of the radiation field. These characteristics may include the type of radiation; the energy (or energy distribution) of the radiation; its intensity, angular dependence, time dependence, and location within an area; and many other parameters. From this knowledge, and the use of other devices (dosimeters), the absorbed dose or dose equivalent to individuals, or samples, in the radiation field can also be established. In external dosimetry the source of radiation is outside the body and the absorbed dose (or dose rate) depends on the distance from the source, time spent in the vicinity of the source, and influence of materials interposed between the source and the individual. These concepts are called simply time, distance, and shielding, but they play an important role in reducing personnel exposure to external sources of ionizing radiation.

Internal dosimetry is a process of measurement and calculation that results in an estimate of the absorbed dose (or dose equivalent) to tissues of the body from an intake of radioactive material. In this case, the radioactive material is assumed to be taken into the body whereupon it becomes subject to the control of normal body processes in terms of where it is deposited and the length of time the material remains in the body, for example. Absorbed dose estimates are based on measurements of radioactivity in material excreted from the body or on measurements by sensitive radiation detectors placed near the body that indicate the amount of radioactivity in the body. In either case, the dose estimate must rely on a mathematical model that has been derived to describe retention of the radioactive material in the body. The ability of this model to reflect the actual situation in any particular individual is always of concern in performing internal dose assessments.

I. SOURCES OF RADIATION

In this section, the term *radiation* is used to mean ionizing radiation. Only those radiations that can produce ionization in the atoms or molecules with which an interaction

occurs are discussed in this section. Ionization is simply the process of removing (or adding) one or more electrons from (or to) an atom or a molecule. Therefore, ionizing radiation is usually considered to be any radiation that can displace an electron(s) from an atom or a molecule, thereby producing ions. The remaining atom (or molecule) and the liberated electron are called an ion pair; these play an important role in the detection of ionizing radiation.

Ionizing radiation is generally divided into two categories: directly ionizing radiation and indirectly ionizing radiation. The first category encompasses those radiations that possess an electrical charge and have sufficient kinetic energy to produce ionization by "collision." Actually, these radiations interact with the matter through which they are passing primarily by coulombic interactions; thus, the word *collision* can be misleading. Primary radiations that constitute this category are alpha particles, beta particles (both negatively and positively charged species), electrons and protons.

Indirectly ionizing radiations are those that have no charge and interact in a way that produces directly ionizing radiations. Some or all of the kinetic energy of the radiation is transferred by the interaction, and the directly ionizing radiations produced usually can cause additional ionization as this kinetic energy is dissipated in the medium. In this discussion, only two radiations are considered: photons (X-rays and gamma rays) and neutrons.

A. Types of Radiation

The discovery of X-rays, by Wilhelm Konrad Roentgen, and the discovery of natural radioactivity, by A. Henri Becquerel, occurred within a few months of each other. On November 8, 1895, Roentgen discovered X-rays, and translations of his work appeared in *Nature* on January 23, 1896. Becquerel noticed a fogging of photographic plates (similar to that reported by Roentgen) in his studies with uranium-based phosphorescent materials. Becquerel concluded that the fogging must be due to emissions from the uranium. Even though Becquerel continued his experiments for a number of years, his original observation (reported on February 24, 1896) formed the basis for the studies of many investigators. Thus, it is generally agreed that these two discoveries opened the door to the study of radiation and radioactivity (the term introduced by Marie and Pierre Curie).

A summary of the entire history of research into radiation and radioactivity is not presented here; instead, this very brief history sets the stage for the discussion of the characteristics of the types of radiations considered in this section. The information that follows was obtained over a number of years and encompasses the research of an untold number of scientists throughout the world. The importance of this effort and its magnitude are often lost when simple, concise statements are made regarding the characteristics of ionizing radiations.

1. Alpha Particles

Radioactive decay (or transformation) of nuclei by the emission of alpha particles was determined by Ernest Rutherford in 1899. However, it was not until 1911 that Boltwood and Rutherford identified the properties of these radiations. These researchers concluded that an alpha particle is basically a helium nucleus consisting of two protons and two neutrons bound together in a stable configuration. This is a massive particle on a nuclear scale relative to the other radiations. Alpha particles, each with a mass of 4 units and a positive charge of 2, are emitted primarily in the radioactive transformation of heavy nuclei (e.g., uranium, thorium). These radiations are monoenergetic and usually possess a large amount of kinetic energy, typically in the range of 4–8 million electron volts (MeV). Because of their massive size and positive charge, alpha particles do not travel far in most media. The distance that alpha radiation travels in the air (i.e., the range) depends on the energy, but typical values range from 2.5 to 7.5 cm. In more dense materials (e.g., human tissue) the range of alpha particles is measured in micrometers.

2. Beta Particles

Radioactive decay by beta-particle emission confounded the scientific world for a number of years. In 1896, Becquerel noted the emission of energetic radiation from the salts of uranium that had penetrating powers similar to those of X-rays. In Rutherford's research on alpha particles, he also noted the emission of other radiations that had a penetration equal to the "average X-ray bulb." However, a full explanation of beta decay was not provided until the mid-1930s. In the early 1930s, Wolfgang Pauli postulated that the emission of beta radiation involved the release of not one but two radiations, one of which must be electrically neutral. This postulation allowed the conservation of energy and satisfied the accepted concept of discrete energy nuclear levels. Enrico Fermi proposed his theory of beta decay in 1934, in which he incorporated the postulates of Pauli. Fermi proposed that the radiation emitted along with the beta particle must have essentially no mass and no charge. He named this uncharged radiation the neutrino.

Radionuclides that have an excess of neutrons in the nucleus (neutron-rich) usually emit beta radiation when they transform. The beta particles emitted have all the characteristics of electrons. That is, each particle has a mass equivalent to an electron and a single negative charge.

Some scientists call these radiations negatrons to distinguish the negative species from the positively charged species emitted in the decay of certain radionuclides. In contrast to alpha particles, beta particles do not have discrete energies. Instead, the energies are distributed in a continuum up to a maximum energy, which is equivalent to the total energy available in the transformation. The available energy is shared between the beta particle and the neutrino. However, in a discussion of dosimetry, it is common to characterize a certain beta-emitting radionuclide by the "average energy" of the beta particles. A rule of thumb is to assume that the average energy of a beta-emitter is one-third of the maximum energy.

Some radionuclides may be proton-rich (i.e., have an excess of protons in the nucleus) and will transform by the emission of a positively charged beta particle. These radiations are usually called positrons. Again, these radiations have the same mass as electrons and differ only in the fact that each possesses a single positive charge. When a positron comes to rest, it will combine with a free electron and "annihilate": the electron and positron combine, and their rest mass is converted into energy by the production of two photons (called annihilation radiation). These latter radiations are important in dosimetry and must not be ignored when an individual is dealing with positron-emitting radionuclides.

The penetrating power of beta particles, as indicated in early experiments, is much greater than that of alpha particles. Although it is not completely correct to speak of the "range" of beta particles, it is instructive to consider the range of this type of radiation in air. A rule of thumb in common use is that a 1-MeV beta particle has a range in air of ~3.6 m.

3. Photons

In this section, the term *photons* is used to describe either X-rays or gamma rays. A photon has been described as a "bundle" or "particle" of radiation. This is because photons possess both particle- and wavelike properties; a photon possesses energy but it is assumed to have no mass. Both X-rays and gamma rays are electromagnetic radiation and differ only in their origin. X-rays originate from rearrangements in the electron structure of the atom. Gamma rays, however, originate from within the nucleus and are associated with the radioactive transformation of many radionuclides. In addition, gamma rays are usually assumed to have shorter wavelengths than X-rays have.

Photons can carry large amounts of energy and can have great penetrating powers. The degree of penetration is a function of the material; hence, dense materials such as lead are excellent shields against photon radiation.

4. Neutrons

Neutrons are relatively massive particles that are one of the primary constituents of the nucleus. However, neutrons can be produced in a number of ways and can represent a significant source of indirectly ionizing radiation. Generally, neutrons are segregated into several categories on the basis of their energy. Thermal neutrons are those that are in thermal equilibrium with matter and, in special cases, have a Maxwellian distribution of velocities. In this distribution, the most probable velocity at 295 K is 2200 m/sec, corresponding to an energy of 0.025 eV.

Neutrons in the energy range 0.5–10 keV are called intermediate neutrons. These neutrons may also be called resonance or epithermal neutrons. Fast neutrons are those in the energy range 10 keV to 10 MeV. In this energy range, neutrons interact with matter through elastic collisions (i.e., billiard-ball–type collisions). Neutrons with energies >10 MeV are called relativistic neutrons.

B. Natural Radioactivity

Every creature on earth is exposed continuously to ionizing radiation from natural sources. These sources can be divided into two basic categories: extraterrestrial and terrestrial. That is, some of the radiation originates from sources in space, whereas other radiation results from exposure to the naturally radioactive substances that constitute a portion of the earth's crust.

Extraterrestrial radiation sources can be further divided into two classes: cosmic radiation and cosmogenic radionuclides. The term *cosmic radiation* is used to mean both the primary energetic particles that interact in the earth's atmosphere and the secondary particles that result from these interactions. Primary cosmic radiation arises from two sources: galatic radiation, which is incident on our solar system, and solar radiation, which is emitted by our sun.

The components of galactic radiation are protons (87%), alpha particles (11%), and a few heavier nuclei and electrons. The energy of these radiations can exceed 10^{20} eV, but most of the radiation has energies in the range of 10^8 to 10^{11} eV. It is believed that these radiations originate from high-energy cosmic processes such as supernova explosions as well as other explosive phenomena. It is also believed that the higher energy radiations may actually originate outside our own galaxy.

The sun represents a continuous source of charged particles. However, these particles are of such a low energy level (∼1 keV) that it is not possible for them to penetrate the earth's magnetic field and reach the atmosphere. Solar flares (large magnetic disturbances) can, however, generate large quantities of particles with energies approaching several GeV. Normally, these radiations are in the energy range of 1–100 MeV.

Primary galactic and solar radiations are attenuated by the earth's atmosphere and secondary radiations are produced. The first generation of secondary particles consists mainly of neutrons, protons, and pions. Decay of the pions results in the production of electrons, photons, and muons. It has been estimated that cosmic radiation contributes between 30 and 50% of the total dose from all external environmental radiation exposure.

Cosmogenic radionuclides are numerous although in most cases the atmospheric concentrations are quite small. Many of these radionuclides are produced by cosmic radiation interaction with extraterrestrial dust. Typical radionuclides include ^{7}Be, ^{22}Na, ^{26}Al, ^{60}Co, and many more. The major source of cosmogenic radionuclides is interactions with atmospheric gases. Primary radionuclides produced in these processes are ^{3}H and ^{14}C. Estimated equilibrium activities of these radionuclides are 28 million curies (MCi) for ^{3}H and 230 MCi for ^{14}C (with only ∼2% of this in the atmosphere). Because of previous nuclear weapons tests in the atmosphere, this equilibrium has been disturbed. As of 1963, an estimated 1900 MCi of ^{3}H had been injected into the atmosphere of the Northern Hemisphere. Weapons tests have increased the ^{14}C concentration to approximately twice the pre-1950 concentration.

Cosmogenic radionuclides may also be produced by interactions in the upper 1–2 m of the earth's crust. It is estimated that ∼70% of the earth's inventory of ^{36}Cl results from activation of ^{35}Cl, an abundant nuclide in rocks and seawater.

C. Nuclear Reactors

Nuclear reactors provide copious quantities of neutrons and gamma rays for many research applications. In addition, reactors provide many challenges from the dosimetry point of view. The leakage spectrum from a reactor may vary widely depending on the reactor type and its intended use, as well as the interposition of moderator, coolant, or shielding between the core of the reactor and the point of dosimetric interest. For example, the leakage spectrum from an unshielded fast reactor closely approximates that of the fission spectrum. Before definitive dosimetry measurements can be made in the mixed neutron- and gamma-radiation field present around a nuclear reactor, measurements must be made to difine as carefully as possible the radiation environment.

In addition, a reactor facility may pose other dosimetry problems caused by the activation of materials passing through the core (carried by the coolant) or by releases of fission products normally contained within the fuel itself.

Thus, in a reactor facility, there may exist a need to assess not only the neutron- and gamma-radiation environment, but also the radiation field produced by beta-emitting radionuclides at various locations throughout the facility.

D. Typical Radiation Environments

Recently, a significant effort has been devoted to characterizing radiation environments in and around nuclear power facilities and evaluating the response of personnel-monitoring devices and other radiation detectors in these radiation fields. The results of these studies have been published in a number of documents issued by the USNRC. This section summarizes the results of these studies, as the findings have a significant impact on dosimeter selection and the evaluation of the measured dose received by workers in these radiation fields.

1. Photon-Radiation Fields

Spectral and dosimetric measurements have been made in seven commercial nuclear power facilities in the United States. Both pressurized water reactors (PWRs) and boiling water reactors (BWRs) were included and surveys were made both while the reactors were operating and while they were shut down. Results of these studies showed that the photon-radiation fields in these facilities could be classified in four categories: (1) radiation fields dominated by photons emitted in the decay of radioactive materials associated with neutron-activated or fission-product deposits; (2) radiation fields dominated by scattered photons, represented as a continuum of energies; (3) radiation fields containing short-lived noble gases; and (4) radiation fields dominated by high-energy photons.

Each of these radiation fields is discussed briefly below. First, it is necessary to define the meanings of low-, medium-, and high-energy photons. For the purposes of this discussion, photons with energies <200 keV are referred to as low energy, medium-energy photons are those with energies in the range 200 keV to 3 MeV, and high-energy photons comprise all those with energies >3 MeV.

Dose rates in most areas of the nuclear plants were dominated by lightly shielded radioactive sources in neutron-activated or fission-product deposits. Measurements of photon spectra in various locations showed the expected radionuclides. These included the typical activation products ^{58}Co, ^{60}Co, ^{54}Mn, ^{51}Cr, ^{59}Fe, and ^{65}Zn. Primary fission products identified in the photon spectra included only ^{134}Cs, ^{137}Cs, and Zr-Nb-95. The average energy of the photon field strongly depends on the radionuclide mix. The photon field in some areas was found to be composed of only one or two radionuclides and the average energy was easily obtained from a knowledge of the decay

schemes for the radionuclides. For example, one of the most prevalent radionuclides found through these measurements was ^{60}Co. In many cases this radionuclide was the only one present and, obviously, the average energy of the photon field was 1.25 MeV. In other cases, the average energy of the photon field may be low (200–300 keV) because of the complex mixture of radionuclides constituting the field.

The scattered photon field (a continuum of energies) varied with location in the facilities. The continuum usually had a maximum at ~120 keV, with a half-maximum range from 70 to 250 keV. The distribution was skewed on the high-energy side with a high-energy "tail" exceeding 500 keV. However, the conclusion that this continuum is caused entirely by scattered photons can be misleading. As discussed above, for complex mixtures of activation and fission products, the average energy of the photon field also may fall into this energy range.

Radioactive noble gases were measured inside the containments at operating PWRs. The presence of photons with energies near or greater than 1 MeV obscured the effects of the low-energy photons from the xenon isotopes (i.e., 81 and 249 keV).

High-energy photons (≤ 8 MeV) were measured inside the containments at operating PWRs. Even though these photons contributed significantly to the radiation field, the measurements also showed contributions from medium- and low-energy photons. In turbine rooms at BWRs, ~80% of the dose was due to the high-energy photons from ^{16}N (6.1 MeV). Annihilation radiation (0.511 MeV) also contributed significantly to the total dose in the turbine areas.

The study concluded that the potential for inaccurate dosimetry results is greater for high-energy photon-radiation fields than for low-energy fields. It was estimated that some dosimeters may overrespond by as much as 60%. This overresponse occurs in film dosimeters and those that use filters manufactured from high–atomic-numbered materials and is due to differences in the pair production cross sections. However, as this study pointed out, the dose estimates obtained with these dosimeters are always conservative.

2. Neutron-Radiation Fields

A series of measurements of neutron energy spectra, neutron dose-equivalent rates, and personnel neutron-dosimeter responses also have been made at six commercial nuclear power plants. In this study, five of the plants were PWRs designed by the three reactor manufacturers. The sixth plant was an operating BWR. These measurements showed that most dose-producing neutrons had energies from 25 to 500 keV with an average between 50 and 100 keV in the containments of the PWRs. In general, spectral measurements revealed no significant numbers of neutrons with energies >1 MeV. At the BWR, measurements were resticted to the areas outside shield penetrations. In these areas, the neutron energy spectrum was somewhat more energetic with average energies from 150 to 250 keV.

Another set of measurements inside the containments of two PWRs operating at full power showed the neutron spectrum to be "soft" with few neutrons with energies >700 keV. In fact, in some locations, the average neutron energies were found to range from 0.9 to 90 keV. Thus, a dosimeter selected for use in nuclear power plant environments must be sensitive to neutrons with energies in the intermediate energy range or on the lower end of the fast-neutron energy range (typically fast neutrons are assumed to have energies from 10 keV to 10 MeV). This requirement limits the choice of dosimeters because some of the more conventional dosimeters (e.g., film badges) do not respond to neutrons in this energy range. The response of dosimeters to neutron radiation is discussed in Section IV.A.6.

3. Beta-Radiation Fields

Beta-radiation fields in and around a commercial nuclear power facility are extremely difficult to characterize. In actual plant situations, the mixture of beta-emitting radionuclides and the ratio of beta- to gamma-radiation intensity may change with time. In general, many of the radionuclides just discussed contribute significantly to the beta-radiation field in the nuclear power environment. However, the components of the radiation field at any time depend on the operating history of the plant, integrity of the fuel cladding, quality of the reactor coolant chemistry, and status of the plant (operating or shut down), Measurements in selected areas in a PWR showed a wide variation in the average beta energy. For example, at the steam generator manway and diaphragm, the average energy was 76 keV, whereas measurements in the reactor coolant system (gas) area showed an average energy of 561 KeV. In general, the average beta energy in the 12 reactor areas surveyed ranged from 100 to 300 KeV. Data obtained at a BWR, from area smears and analysis of resin samples, gave an average beta energy of ~240 keV.

E. Accelerators

Accelerator-produced radiations are also of dosimetric concern although most accelerators are heavily shielded and personnel are not normally exposed to these radiations. However, accelerators are used as sources of neutrons and other radiations in dosimetry research. The use of a Van de Graaff accelerator allows the production of monoengergetic neutrons over a wide range of energies. Such monoenergetic sources are used for the calibration and intercomparison of neutron dosimetry systems. Linear accelerators are also used with the most common reactions being the $D(d, n)^3$He reaction that produces 3-MeV neutrons and the $T(d, n)^4$He reaction that produces 14.3-MeV neutrons.

Accelerators are also used extensively in radiation therapy, and careful dosimetry is a requirement for effective treatment of disease. In this application, Van de Graaff generators, linear accelerators, and betatrons have been used. Dosimetry for purposes of radiation therapy (including treatment planning) is not discussed here.

F. Isotopic Sources

A large number of isotopic radiation sources are available for use in the calibration of dosimeters. This section describes typical sources used in dosimetry.

1. Beta-Radiation Sources

The number of beta-radiation sources available for dosimeter calibration is limited. Few of the radionuclides that decay by beta emission are actually "pure" beta-emitters; that is, the radionuclide emits only beta radiation in the process of transforming. In addition, only a very few of these pure beta-emitters have radioactive half-lives that are sufficiently long to make them suitable for use in dosimeter calibration. These radionuclides, and some general characteristics of each, are listed in Table I.

2. Gamma-Radiation Sources

Many more radionuclides that also emit gamma radiation in their decay are available and many have very long half-lives. Two of the most widely used radionuclides are ^{60}Co (with a half-life of 5.27 yr) and ^{137}Cs (with a half-life of 30.0 yr). Both these radionuclides decay by beta emissions. In the case of ^{60}Co two high-energy gamma rays

TABLE I Comparison of Some Common Beta Sources

Radionuclide	Half-life	Average energy (MeV)	Maximum energy (MeV)
^{147}Pm	2.6 yr	0.062	0.225
^{204}Tl	3.8 yr	0.267	0.765
^{90}Sr	29.12 yr	0.200	0.544
^{32}P	14.3 d	0.694	1.709
^{90}Y	64.0 hr	0.931	2.245

TABLE II Comparison of Some Common Gamma Sources

Radionuclide	Half-life	Energy (MeV)	Specific gamma-ray ray constant (R-cm²/hr-mCi)
^{60}Co	5.27 yr	1.17, 1.33	13.2
^{137}Cs	30.0 yr	0.661 (^{137m}Ba)	3.3
^{24}Na	15.0 hr	1.37, 2.75	18.4
^{54}Mn	303.0 d	0.835	4.7
^{22}Na	2.6 yr	1.275	12.0

are emitted in cascade in >99% of the transitions. The radionuclide ^{137}Cs is actually a pure beta-emitter and the gamma ray associated with the decay of this radionuclide is actually due to the decay of ^{137m}Ba (2.55 min). Characteristics of selected gamma-radiation sources are presented in Table II.

3. Neutron Sources

Isotopic neutron sources have been available for a long time. Much of the early research that led to the discovery of fission and the possibility of producing a slef-sustaining nuclear chain reaction used such isotopic sources. In general, most isotopic sources rely on a radionuclide that emits either alpha radiation or gamma radiation in combination with beryllium. Irradiation of beryllium with alpha particles results in complex nuclei formed by absorption of the alpha particles by ^{9}Be nuclei. These complex nuclei are highly excited and a neutron is emitted almost instantaneously (within approximately one billionth of a second). Characteristics of these sources, called (α, n) sources are presented in Table III. Neutrons emitted by these sources span a wide range of energies, yet it is common to assign some average energy to the neutrons emitted from a particular source. However, this point should not be ignored when an individual is calibrating neutron dosimeters.

Neutrons can also be produced when gamma rays interact in beryllium. These sources, often called photoneutron sources, require reasonably high-energy gamma radiation and are characterized by low neutron yields. In addition, with the exception of ^{226}Ra, the gamma-radiation sources

used for photoneutron sources have short half-lives. Characteristics of selected sources are given in Table IV.

The availability of spontaneously fissioning radionuclides has had a strong influence on neutron dosimety since the early 1970s. Approximately 30 radionuclides decay by spontaneous fission, usually in competition with alpha decay. It is now possible to produce sufficient quantities of these radionuclides so that the sources are useful in the calibration of dosimeters. These sources have neutron- and gamma-ray energy spectra that are basically equivalent to the fission spectra and are extremely useful for calibration of dosimeters to be used in mixed-field dosimetry around certain types of nuclear reactors. In addition, moderators have been designed to enclose these sources, making them suitable for standardization of many types of dosimeters. A summary of some of the pertinent characteristics is given in Table V.

II. QUANTITIES AND UNITS

A. Basic Definitions

Measurements with radiation dosimeters may yield results in a number of different units. These include count rate, exposure rate, absorbed-dose rate, and dose-equivalent rate to name only a few. Each of these units may or may not be appropriate for the particular instrument and the measurement being made. However, it is extremely important that the user understand the meaning of these results and the fundamental quantities that are represented by these data. For this reason, the fundamental definitions of the appropriate quantities and their associated units are presented before a detailed discussion of radiation dosimetry systems.

Even though the International Commission on Radiation Units and Measurements (ICRU) has issued new definitions for many of the dosimetric concepts, the United States has been slow to adopt these new concepts. Thus, the definitions that follow reflect those still in common use. When possible, the SI unit has been included along with the traditional unit.

1. Activity

The activity A of an amount of radioactive nuclide in a particular energy state at a given time is the quotient of dN by dt, where dN is the expectation value of the number of spontaneous nuclear transitions from that energy state in the time interval dt. That is,

$$A = dN/dt.$$

TABLE III Comparison of Some (α, n) Neutron Sources

Source	Half-life	Average E_n (MeV)	Maximum E_n (MeV)	Output for 1 Ci (neutrons/sec)
^{210}Po-B	138.4 d	2.8	5.0	2.0×10^5
^{210}Po-Be	138.4 d	4.0	10.8	2.5×10^6
^{241}Am-Be	458.0 yr	4.3	11.0	2.0×10^6
^{226}Ra-Be	1622.0 yr	4.5	13.2	1.5×10^7
^{239}Pu-Be	2.44×10^4 yr	4.1	10.6	1.5×10^6

TABLE IV Comparison of Some (γ, n) Neutron Sources

Source	Half-life	E_γ (MeV)	Energy calculated (MeV)	Energy measured (MeV)	Yield
^{24}Na + Be	15.0 hr	2.757	0.966	0.83	1.3×10^5
^{24}Na + D_2O	15.0 hr	2.757	0.261	0.22	2.7×10^5
^{88}Y + Be	104.0 d	1.853	0.166	0.158	1.0×10^5
^{124}Sb + Be	60.0 d	1.70	0.031	0.0248	1.9×10^5
^{226}Ra + Be	1622.0 yr	Many	—	0.7 max	1.2×10^4
^{226}Ra + Be	1622.0 yr	Many	—	0.12	0.1×10^4

The traditional unit of activity is the curie with 1 Ci = 3.7 E10/sec. The SI unit for activity is the becquerel and 1 Bq = 1/sec; therefore, a Ci = 3.7 E10 Bq. In the above definition, the "particular energy state" is the ground state of the nuclide unless otherwise specified. The activity of an amount of a radionuclide is equal to the product of the decay constant and the number of nuclei in that particular state.

2. Exposure

The exposure X is the quotient of dQ by dm. where dQ is the absolute value of the total charge of the ions of one sign produced in air when all electrons liberated by photons in a volume element of air having a mass dm are completely stopped in air; that is,

$$X = dQ/dm.$$

This definition indicates that exposure may be measured by collecting the charge produced in a known volume (mass) of air produced by the interaction of photons in the air.

The special unit of exposure is the roentgen, named after the discoverer of X-rays. One roentgen is equivalent to 2.58×10^{-4} C/kg. Under the SI units, the roentgen will no longer be used and exposure will simply be measured in units of C/kg (charge produced per unit mass).

TABLE V Comparison of Some Spontaneous Fission Sources

Nuclide	Half-life, spontaneous fission (Yr)	Half-life (α-decay)	Neutrons/g-sec
^{236}Pu	3.5×10^9	2.7 yr	3.1×10^4
^{238}Pu	4.9×10^{10}	89.6 yr	2.3×10^3
^{240}Pu	1.3×10^{11}	6600.0 yr	7.0×10^2
^{242}Cm	7.2×10^6	162.5 d	1.8×10^9
^{244}Cm	1.4×10^7	18.4 yr	1.0×10^7
^{252}Cf	85.5	2.7 yr	2.3×10^{12}

This definition is very restrictive in that it applies only to photons (i.e., X-rays and gamma rays) interacting in air; exposure to other radiations should not be expressed in units of exposure. In addition, radiation energy interacting with other types of matter (e.g., tissue) cannot be expressed in units of exposure. An additional factor often forgotten is that the quantity exposure is not defined for photons with energies >3 MeV. Nevertheless, this quantity is still widely used in radiation protection, and most pocket dosimeters (direct and indirect reading) "read out" in units of exposure, as do many portable, air-filled ionization chamber survey instruments (more correctly, exposure rate). With the introduction of SI units, some pocket ionization chamber manufacturers supply their dosimeters with an internal scale that is read directly in units of coulombs per kilogram. At the present time, these direct-reading dosimeters are not sold in the United States.

A More useful quantity is specified in the definition of the absorbed dose. Absorbed dose D is the quotient of $d\varepsilon$ by dm, where $d\varepsilon$ is the mean energy imparted by ionizing radiation to matter in a volume element and dm is the mass of the matter in that volume element. Mathematically,

$$D = d\varepsilon/dm.$$

This fundamental definition simply states that absorbed dose is the energy absorbed per unit mass of the material being irradiated. the special unit of absorbed dose is the rad and 1 rad = 0.01 J/kg. The newer SI unit for absorbed dose is the gray and 1 Gy = 1 J/kg. In other words, 1 Gy = 100 rad.

In occupational radiation protection the quantity of interest is the *dose equivalaent*. The dose equivalent H is the product of the absorbed dose, the quality factor, and any other modifying factors that may be appropriate for the exposure situation. that is,

$$H = D \times Q \times N.$$

In this equation, D represents the absorbed dose from the equation above, Q the quality factor, and N the product of all modifying factors. For external exposure situations,

N is always assumed to be equal to 1. (More recently, the ICRU has decided to drop the factor N because there have been no cases in which it has been assigned a value other than unity.) Values of the quality factor depend on the type of radiation and the linear energy transfer (LET) of the radiation. However, for radiation protection purposes, it is usually assumed that the values of Q are constants for particular types or radiation. The currently accepted values are

$Q = 1$ for electrons, beta radiation, X-rays, gamma radiation, and bremsstrahlung.

$Q = 20$ for alpha particles, fission fragments, and recoil nuclei, although none of these pose an external radiation hazard.

$Q = 3$ for thermal and intermediate neutrons (energies <10 keV).

$Q = 10$ for fast neutrons (energies >10 keV) and protons, although the International Commission on Radiological Protection (ICRP) has recently recommended that Q be increased to 20 for fast neutrons. This change has not been widely incorporated into national recommendations on radiation protection, nor has it been implemented in federal regulations.

It should be noted that neutron flux dose-equivalent data presented in tabular form in Title 10 Code of Federal Regulations Part 20.4, "Units of Radiation Dose," can be used to derive a quality factor as a function of neutron energy.

The traditional unit for dose equivalent is the rem. Under the traditional definition of the dose equivalent, the unit rem has no cgs (centimeter-gram-second) or mks (meter-kilogram-second) equivalent. Reserving a special unit for use with the quantity dose equivalent and associating no other conventional units with it was a way of indicating that the dose equivalent was the product of a physical quantity (absorbed dose), an empirical factor applied to account for the differences in biological response for equal absorbed doses from different radiations (quality factor), and a factor that took into account any other effects that might be noted for particular exposure situations (other modifying factors).

More recently, the ICRP has chosen to replace the term *quality factor, Q* with the term *radiation weighting factor, w_R*. However, for the purposes of this discussion, the quality factor and the radiation weighting factor are conceptually equivalent.

The SI unit for the dose equivalent is called the sievert, where 1 Sv = 1 J/kg and, therefore, 1 Sv = 100 rem. In this case, the ICRU decided that the quality factor and the distribution factor were dimensionless and, therefore, it was clear that both the absorbed dose and the dose equivalent must have the same units in the SI system. Assigning the sievert the identical units as the gray has caused a great deal of confusion and discussion in the radiation protection community. However, this need not cause trouble at this point because the SI units have not yet been adopted for use in radiation protection purposes in the United States.

3. Kerma

The kerma K is the quotient of $dE(tr)$ by dm, where $dE(tr)$ is the sum of the initial kinetic energies of all charged ionizing particles liberated by uncharged ionizing particles in a material of mass dm. That is,

$$K = dE(tr)/dm.$$

The traditional unit for the absorbed dose (i.e., rad) may also be used as the unit for kerma. However, the SI unit for kerma is the gray. The extent to which absorbed dose and kerma are equal depends on the degree of charged particle equilibrium and bremsstrahlung is negligible.

4. Linear Energy Transfer

The linear energy transfer L of a material for charged particles is the quotient of dE by dl, where dE is the energy lost by a charged particle traversing a distance dl; that is,

$$L = dE/dl.$$

Many scientists use the notation LET to represent the linear energy transfer. In conventional units, L is usually expressed in keV/μm (i.e., kiloelectron volts per micrometer); however, the SI unit is joules per meter.

B. The Bragg-Gray Principle

Measurements of absorbed dose due to the interactions of ionizing radiation in matter rest on the Bragg-Gray principle, a fundamental concept in dosimetry. This principle states that the energy absorbed from secondary electrons per unit volume of a solid medium is equal to the product of the ionization per unit volume in a small gas-filled cavity in the medium, the mean energy expended in the gas, and the ratio of the mass stopping powers of the secondary electrons in the medium and the gas. Stated more simply, the principle indicates that the amount of ionization produced in a gas-filled cavity serves as a measure of the energy deposited in the surrounding medium.

For this statement to be true, several conditions must be met. These include

1. The cavity must be of such dimensions that only a small fraction of the particle energy is dissipated in it. This

condition means that only a small fraction of the particles contributing to the ionization will enter the cavity with a range that is less than the cavity dimensions.

2. Radiation interactions in the gas in the cavity should contribute only a negligible proportion of the total ionization in the cavity. This condition is usually satisfied if the first condition is met.

3. The cavity must be surrounded with an equilibrium thickness of the solid medium. This is the thickness that will result in electronic equilibrium. Electronic equilibrium is a situation that exists when electrons (produced by radiation interacting in a volume) escape from the volume but are replaced by electrons produced outside the volume that enter the volume and dissipate a portion of their energy. In other words, this equilibrium thickness is the thickness of material equal to the range of the most energetic secondary electrons produced by the primary radiation.

4. Energy deposition by the ionizing radiation must be essentially uniform throughout the solid medium immediately surrounding the cavity.

If these conditions are met, the energy absorbed per gram of the solid material is related to the ionization per gram of the gas in the cavity by

$$E_m = J_g \times W \times s_m,$$

where

J_g = the number of ion pairs formed per unit mass of the gas (usually expressed in units of grams).

W = the average energy required to produce an ion pair.

s_m = the ratio of the mass stopping power of the medium to that of the gas in the cavity for the secondary electrons.

The value of W in air for X or gamma radiation is $\sim$34 eV per ion pair. The factor s_m can be expressed as

$$s_m = \frac{[N_m \times S_m]}{[N_g \times S_g]},$$

where

N_m = the number of electrons per gram of the medium.

N_g = the number of electrons per gram of the gas.

S_m = the stopping power (for electrons) for the medium.

S_g = the stopping power (for electrons) for the gas.

Basically, the factor s_m indicates how much more frequently ionization will occur in the medium than in the

gas in the cavity. Thus, measurement of J_g, the ionization per unit mass of the gas in the cavity, coupled with a knowledge of the values of s_m and W, makes it possible to determine the absorbed energy (absorbed dose) in the medium. If the medium is tissue, then the Bragg-Gray principle allows the measurement of the absorbed dose to the irradiated tissue.

III. INTERACTIONS OF RADIATION WITH MATTER

A. Alpha Radiation

As charged particles, such as alpha particles, move through material, energy is transferred from the radiation to the atoms or molecules that make up the material. The major energy-loss mechanisms are electronic excitation and ionization. The alpha particle has a high electrical charge but a low velocity due to its large mass, and interactions are frequent. These interactions are with the loosely bound, outer electrons of the atoms in the material and should not be considered collisions. Since the particle is positively charged, it exerts an attractive force on the oppositely charged electron. In some cases, this force is not sufficient to separate the electron from the atom, but the electron is raised to a higher energy state and the atom is said to be "excited." In other cases, the attractive force is sufficient to remove the electron from the atom (ionization). The closer an alpha particle passes near an electron the stronger the force and the higher the probability an ionizing event will occur. In these situations, the electron may be imagined as being "ripped" from its orbit as the alpha particle passes nearby.

The number of ion pairs created per unit length of travel is called the specific ionization. The specific ionization of alpha particles is, of course, dependent on the energy of the radiation. Only $\sim$34 eV of energy is required to produce an ionizing event in a gas such as air. It should be clear then that a typical alpha particle, with perhaps 5 MeV of energy, will cause a large amount of ionization and it is safe to say that alpha particles have a high specific ionization. In air, the specific ionization may be $\sim$10,000 ion pairs per centimeter or more. As the alpha particle gives up its energy, it slows and therefore spends more time in the vicinity of atoms. For this reason, the specific ionization increases near the end of the alpha particle's travel. Near the very end of the travel, the specific ionization decreases to zero as the particle acquires two electrons and becomes a neutral atom.

Alpha particles can be characterized as having straight paths and discrete ranges. In describing the movement of alpha particles through matter, the term *mean range* is used. The mean range is the absorber thickness traversed by an "average" alpha particle. Empirical equations have been derived that can be used to calculate the range of alpha particles in materials. Usually, the range is specified in air and, if necessary, this range is used to convert to a range in any other material. For alpha particles in the energy range 4–8 MeV one such equation is

$$R_{cm} = 1.24 E_{MeV} - 2.62,$$

where

$R =$ the range in air in centimeters.
$E =$ the energy of the alpha particle in million electron volts.

The range in tissue is obtained by using the equation

$$R_{air} \times \rho_{air} = R_{tissue} \times \rho_{tissue},$$

where

$\rho =$ the density of the materials.
$R =$ the range in centimeters.

Since the density of tissue is assumed to be 1 g/cm^3 the equation reduces to

$$R_{tissue} = R_{air} \times \rho_{air}.$$

B. Beta Radiation

As mentioned above, beta particles have the same mass as electrons and may differ only in the charge on the particle (i.e., positron). These radiations also interact with the matter through which they are passing by means of excitation and ionization. However, in the case of electrons and beta particles (i.e., the negatively charged species), the interaction processes involve scattering (i.e., inelastic collisions) rather than attraction. Since the particles have like charges to those of the orbital electrons, there is a repulsive force exerted between the two. The net effect is the same, however, because electrons may be moved to higher energy states or sufficient force may be exerted to ionize the atom. Positron interactions are also considered to be scattering reactions even though the radiation is positively charged.

In contrast to alpha particles, as beta particles move through material, many scattering events occur and the path is far from straight. Some researchers have described this path as "tortuous," resulting from multiple scattering events with atoms along the particle's path. Two terms are used to describe electron absorption in material. The range of a beta particle is the linear thickness of a material required to absorb the particle. The path length is the actual distance traveled before all the particle's kinetic energy is lost. It should be clear that the path length is much greater than the range.

Beta particles may also lose energy by radiative collisions or bremsstrahlung production. The word bremsstrahlung means "braking radiation," and it is the term used to describe energy that is radiated when the beta particle (electron) is accelerated due to the presence of the nucleus. Bremsstrahlung is usually important only at high energy and in high–atomic-numbered absorbers.

Range energy relations have also been derived for beta particles. One such equation is

$$R = 0.542E - 0.133 \quad \text{for} \quad E > 0.8 \text{ MeV}.$$

In this equation, E has units of MeV and the range (R) is given in units of g/cm^2 (grams per square centimeter).

Beta particles that have lost their kinetic energy can exist in nature as electrons. However, this is not true for positrons. When a positron comes to rest, it combines with a free electron and annihilates. That is, the electron and the positron neutralize each other and convert their combined rest mass into energy. This rest mass is released in the form of two photons, each with 0.511 MeV of energy. The production of these energetic photons must be considered when an individual is performing dosimetry or designing shielding to protect against positron radiation.

C. X and Gamma Radiation

Photons interact with matter through three primary mechanisms: the photoelectric effect, Compton scattering, and pair production. The probability of each of these interactions occurring depends on the energy of the radiation and the material through which it is passing.

The photoelectric effect occurs primarily at low photon energies and in high–atomic-number (Z) materials. This interaction should be considered to occur with the entire atom even though the energy transfer is between the photon and an orbital electron. In this interaction a photon strikes a tightly bound electron and transfers its entire energy to the electron. If this energy is greater than the binding-energy of the electron to the atom, then the electron will be knocked out of the atom. The electron (a photoelectron) may possess kinetic energy as a result of this interaction. This energy is the difference between the initial energy of the photon and the binding energy of the electron.

Photoelectric interactions are most probable with the most tightly bound electrons (K shell), and the loss of an electron from the inner shell(s) leaves a vacancy that must be filled. An electron from a higher orbit will drop into the vacancy, but it in turn leaves another vacancy. There is in effect a cascading of electrons as they drop into lower energy states to fill the existing vacancies. As each electron fills a vacancy, a photon is emitted whose energy is equal to the difference between the initial and final energy levels. These photons are called characteristic X-rays because the energy differences between the electron orbits are unique for an atom and the photons are characteristic of the element from which they originate.

As stated previously, photoelectric interactions are most probable at low photon energies. The interaction is relatively unimportant for photons with energies >1 MeV, except in very heavy elements.

Compton scattering is an interaction that occurs between a photon and an essentially "free" electron. That is, the electron is in one of the outer orbits and its binding energy is significantly less than the energy of the photon. In Compton scattering, the requirements for the conservation of momentum and energy make it impossible for complete transfer of the photon energy to the electron. Basically, the photon has a collision with the electron and transfers only a portion of its energy to the electron. The photon is deflected from its original path (scattered) and has less energy (longer wavelength) than the incident photon. The Compton electron has kinetic energy equivalent to the difference between the initial photon and the Compton-scattered photon.

The probability of Compton scattering decreases with increasing photon energy and with increasing Z of the absorber. This interaction is, therefore, more probable in the middle photon energy range (i.e., 0.1–1 MeV) and with light materials.

The third interaction, pair production, may be considered the opposite of the production of annihilation radiation. In this case, a high-energy photon comes into the near vicinity of the nucleus of an atom and has a coulombic interaction in which the photon disappears and two charged particles are produced in its place. These charged particles, a positron and an electron, share (as kinetic energy) any available energy of the photon over and above the threshold energy for the reaction. The rest-mass energy of each of these charged particles is equivalent to 0.511 MeV and, therefore, pair production is not possible below a "threshold" of 1.022 MeV. Even though the threshold for this reaction is just >1 MeV, pair production does not become important until a photon energy of ~4 MeV is reached.

When the positron has expended its kinetic energy in the medium, it will annihilate with a free electron, as described previously.

D. Neutrons

The type of neutron interaction depends strongly on the kinetic energy of the neutron. For thermal neutrons, the most important interaction with matter is capture. That is, the neutron is captured by the nucleus and the nuclear structure is transformed. In most situations, this transformation results in an unstable nucleus and energy is emitted by the nucleus as radiation. For example, in tissue the important reaction at low energy is the neutron–gamma reaction with hydrogen. This reaction produces a gamma ray with 2.2 MeV of energy. Another reaction in tissue is the neutron–proton reaction with nitrogen producing a 0.6-MeV proton. Both these reactions are of concern in dosimetry.

For intermediate-energy neutrons, the neutron slowing-down process is the important interaction with matter. Capture and nuclear reactions may also occur in this region.

In dosimetry, fast neutron interactions are the most important, especially those occurring in tissue. The most important dose-depositing interaction of fast neutrons with tissue is elastic scattering with hydrogen. Collision of a neutron with a nucleus results in deflection of the incident particle, along with the transfer of a portion of the neutron energy to the struck nucleus. Energy losses by elastic scattering depend on the size of the colliding nucleus and the collision angle (glancing or head-on).

Inelastic scattering becomes important as the neutron energy increases, first occurring for most nuclei at an energy of ~1 MeV. At energies >10 MeV, inelastic scattering may be as probable as elastic scattering. The most important inelastic reactions in soft tissue are those with nuclei of carbon, nitrogen, and oxygen.

Cross sections for inelastic processes of interest in tissue become significant at >5 MeV and increase generally, but not always, monotonically with neutron energy to ~15 MeV. Most of these reactions are accompanied by deexcitation gamma rays, but proton- and alpha-producing reactions are of special importance because of the higher LET of the particles and the total absorption of the particle energy very near the reaction site.

In the relativistic energy range, especially >20 MeV, inelastic scattering is more important than elastic scattering. For high–atomic-number materials, the elastic cross section may be neglected entirely. However, for hydrogenous materials, such as tissue, elastic processes are still important.

IV. DOSIMETRIC TECHNIQUES, EXTERNAL

The basic requirement of any dosimetric device is that it measure (register) the dose received with sufficient reproducibility and reasonable accuracy over the entire range of energies, doses, and dose rates expected during its use. The dosimeter may be a standard device used to establish or characterize a particular radiation field or it may be a monitoring device worn by radiation workers to establish their occupationally related dose. The accuracy of a dosimetry system may vary depending on the intent of the dosimetry and the dose levels to which the dosimeter is exposed. National and international guidance on personnel monitoring indicate that an accuracy of ±50% is acceptable for those exposures classed as "routine occupational exposure." This term refers to the exposure that radiation workers receive during normal work activities; the term is also used to describe exposures that are well below any legal limit. As exposure levels increase, the desired accuracy of a personnel monitoring system becomes more restrictive. Both the National Council on Radiation Protection and Measurements (NCRP) and the International Commission on Radiological Protection (ICRP) recommend that for exposures approaching the permissible levels the accuracy be approximately ±30%. The International Atomic Energy Agency (IAEA), in its latest Code of Good Practice on personnel monitoring, recommends even more accuracy. For doses approaching those that could have clinical significance, the IAEA document recommends a desired accuracy of ±25%. The term *clinical significance* includes those exposures that may be life-threatening; however, it is important to note that the level of interest extends down to acute exposures in excess of ∼1 rad.

In other situations, the requirements on the accuracy of the dosimetry may be much more severe. For example, in dosimetry for radiation-therapy treatment planning a much higher degree of accuracy (perhaps a few percent) is required because of the potential for severe harm if overexposure occurs. A clear understanding of the requirements and the ability of a particular dosimetry system to meet these requirements is an important facet of "good dosimetry."

Individuals responsible for a dosimetry program regardless of its intent must understand the performance characteristics and limitations of the particular dosimetry system in use at their facility. It should be understood that quantitative measurements made with the system depend on many factors. These include

1. Variation of the dosimeter response from the ideal. Factors such as radiation quality (LET), radiation intensity, energy dependence, and angular dependence may influence the indicated dose.

2. Reliability with which the dosimeter maintains its calibration or retains the recorded dose. Terms such as *fading* or *leakage* are normally used to describe the loss of information originally recorded by the dosimeter.

3. Influence of environmental factors. Temperature, humidity, dust, vapors, light, and many other factors may affect the dosimeter response or the ability of the dosimeter to retain the recorded dose information. In addition, other factors, such as rough treatment or contamination, may lead to invalid monitoring results.

The adequacy of a personnel monitoring system cannot be fully evaluated unless some consideration is given to these factors.

To evaluate fully any dosimetry system, an individual must also understand the desired characteristics of an "ideal" system. A list of these characteristics would include

1. Adequate sensitivity over the anticipated exposure range.

2. Adequate reproducibility.

3. Stability before, during, and after exposure. There should be a minimum loss of information before use, during the measurement period, and during any waiting period between collection and evaluation of the dosimeters.

4. An energy-independent response.

5. A linear relation between response and dose. This characteristic is not completely necessary as long as the relationship is well-known and unambiguous.

6. LET-independent response.

7. A response independent of dose rate.

8. Minimum sensitivity to environmental factors.

9. A response independent of the angle of incidence of the radiation.

10. No response to unwanted radiations. The dosimeter should record the dose due to all types of radiation equally well, or it should be sensitive to only one type of radiation.

11. If the dosimeter is a personnel monitoring device, no interference with the worker's ability to perform their routine tasks. Primary considerations here are the size and weight of the dosimeter.

12. Easy identification to facilitate issue, collection, and proper assignment of the recorded doses.

No dosimetry system (especially a personnel monitoring system) is capable of meeting all the ideal characteristics just listed. However, a knowledge of the strengths and weaknesses of available systems can do much to ensure the proper use and interpretation of these dosimeters. The desired characteristics should be kept in mind as the available personnel monitoring systems are discussed.

A. Gas-Filled Detectors

Gas-filled detectors represent probably the most widely used class of radiation detectors. All detectors in this class are based on the collection of ions produced in the sensitive volume of the detector due to the passage of ionizing radiation. Ionization chambers, proportional counters, and Geiger-Müller counters are the primary detectors in this class. However, ionization is also important in a number of other detector systems.

The popularity of gas-filled detectors stems from a number of important factors. The detectors are relatively simple to construct, are easy to operate, and require only a minimum of equipment. This is especially true for some types of ionization chambers. A large number of gases, even air, may be used to fill the detectors. Finally, detectors in this class may be constructed in a variety of shapes and sizes to fit virtually any application and to provide a wide range of sensitivities. Most of these detectors are available commercially.

1. General Considerations

Radiation interacting in a gas-filled volume produces ion pairs in the volume through the process called ionization. Excitation of the molecules also occurs. The ion pairs (electrons and positive ions) are the result of interactions of the incident radiation with orbital electrons of the gas molecules.

The incident radiation may have a majority of its interactions in the gas or in the material making up the wall of the detector. Directly ionizing particles in the gas-filled region result from some combination of these interactions. This depends on the type and energy of the radiations being detected. Alpha particles may not have sufficient energy to enter the sensitive volume of the detector unless the walls of the detector are very thin and, therefore, ionization is produced directly in the gas. Beta radiation is much more penetrating, and there is a higher probability that the particles can enter the detector volume. Both of these radiations are classed as directly ionizing radiations and have the ability to cause ionization in the gas volume.

Gamma radiation and neutrons are common examples of indirectly ionizing radiations. These radiations interact with molecules of the gas or those composing the chamber wall, producing charged particles that result in ionization and excitation of the gas. In gas-filled detectors, gamma-ray interactions occur with a much higher probability in the dense wall of the detector than in the less dense, gas-filled volume. Thus, the primary source of charged particles in the gas-filled volume due to gamma-ray interactions are electrons released from (or knocked out of) the wall by these interactions.

Neutrons may be detected by supplying the radiation detector with a material that has a high–neutron-absorption cross section. In this case, the neutron interaction produces an excited nucleus, which emits a charged particle that ionizes the gas. This technique, as well as others, is discussed in more detail later in this section.

Charged-particle interactions in the gas produce a region around the particle track in which the ion pairs exist.

The ion pair density depends on the specific ionization of the particle. For a heavily charged particle (e.g., an alpha particle), the track is very straight and the density of ion pairs along the track is quite high. The specific ionization of an alpha particle may exceed 10,000 ion pairs per centimeter. Electron tracks are not straight because of the large number of scattering interactions these particles may experience. In addition, the specific ionization is much less (perhaps 100 ion pairs per centimeter) and thus the ion-pair density is less than that for alpha particles or protons.

Electrons produced in ionizing events may make many collisions with the gas molecules as they move through the detector volume. For most common gases the mean free path (i.e., the average distance between collisions) is in the range of 10^{-4} to 10^{-5} cm. Often, electron collisions with gas molecules result in electron attachment to the molecule, which forms a negative ion. The probability of attachment to a neutral molecule per electron collision is called the electron attachment coefficient. The value of the coefficient depends on the electron energy and the type of gas. Values range from 10^{-6} for gases such as argon to 10^{-3} for the halogen gases. The best gases for use in radiation detectors are those that have a low electron attachment coefficient.

Since the positive and negative ions and electrons exist in proximity to each other, recombination may be a common occurrence; that is, the ions simply recombine to form neutral molecules. Of course, the number of recombinations is proportional to the density of positive and negative charges present. Therefore, the potential for recombination along the track of an alpha particle is much higher than that along a beta-particle or electron track. If no electric field is present in the detector, then the above effects predominate and tend to erase the effects of the ionizing radiation.

2. Ionization Chamber Regime

Consider a radiation detector shown schematically in Fig. 1. The diagram shows a gas-filled volume with a central electrode insulated from the outer walls of the chamber. A voltage is applied, through a resistor shunted by a capacitor, between the outer wall and the central electrode.

Ionizing radiation passing through the sensitive volume of the detector will produce ion pairs within the volume due to interactions either in the walls or in the gas-filling. The positive and negative charges, under the influence of the applied electric field, will move toward their respective electrodes. The number of ion pairs reaching the collecting electrodes depends on the chamber design, gas-filling, and applied voltage. A typical plot of the relationship between the number of ion pairs collected and the applied voltage is shown in Fig. 2.

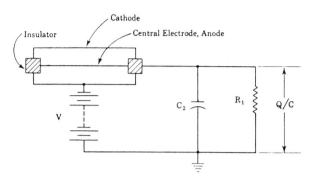

FIGURE 1 Diagram of a gas-filled detector.

In this figure there are four very different regions. In region I, at low applied voltages, several mechanisms are active and compete for the ion pairs produced. The primary factors are a loss of ion pairs due to recombination and the removal of the charges from the volume by the collecting electrodes. At low voltages there is a net drift of the charges in a direction parallel to field lines of force. The drift velocity is a complex relation dependent on the type of gas. The drift velocity, however, is directly proportional to the electric field strength and inversely proportional to the gas pressure.

As the applied voltage is increased, the drift velocity of the ions increases significantly and there is a corresponding decrease in the time available for recombination. Thus, the number of ion pairs collected increases with increasing voltage. In addition, the force acting on the ions accelerates the electrons faster than the heavier positive ions, and hence the drift velocity of the electrons is much higher than that of the positive ions.

The number of ion pairs collected increases with increasing voltage. Because the probability of reaching the collecting electrodes significantly exceeds that of meeting an oppositely charged ion, it is possible to reach a voltage such that all ion pairs produced in the volume by the primary ionizing event are swept from the volume and collected. At this point, further increases in applied voltage result in no increase in the number of ion pairs collected. The voltage at which this occurs is called the saturation voltage and the region (i.e., region II in Fig. 2) is called the ionization region. Gas-filled detectors that operate in this region are called ionization chambers. The voltage required to produce saturation in the ionization region depends on the type of gas used, gas pressure, and chamber dimensions.

Since the charge per ion is only 1.6×10^{-19} C, the ionization current caused by primary ion pairs produced in the chamber is small. This results in a current or a pulse from the chamber that is very small. Usually an electrometer or an electrometer tube with amplification stages is required to detect this small current. The events that occur in an ionization chamber are summarized schematically in Fig. 3.

3. Proportional Counter Region

In region II of Fig. 2 a voltage was reached at which all primary ions were collected; that is, the collecting voltage was sufficient to sweep all ions from the detector volume before significant recombination occurred. In this situation the drift velocity of the electrons may reach 10^5 to 10^7 cm/sec as they acquire kinetic energy from the accelerating force of the electric field. As these ions move toward the electrodes, they may collide with gas molecules. In each collision a fraction of the kinetic energy is transferred to the molecules, exciting but not ionizing them. Energy is also lost in each scattering collision, but more energy is acquired as each ion moves on under the influence of the electric field.

The broad area of Fig. 2 marked region III presents a regime in which a phenomenon called gas multiplication begins to predominate. In the regime the voltage has been increased so much that the ions acquire sufficient energy to create additional ionization in the gas themselves.

Under normal conditions, electrons produced by a primary ionizing event will drift under the influence of the electric field toward the anode, whereas positive ions move outward (toward the cathode). In the arrangement shown in Fig. 1, the electric field intensity is strongest surrounding

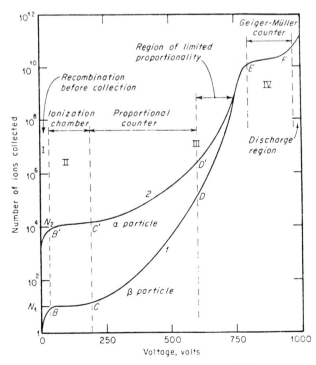

FIGURE 2 Counting characteristics of gas-filled detectors.

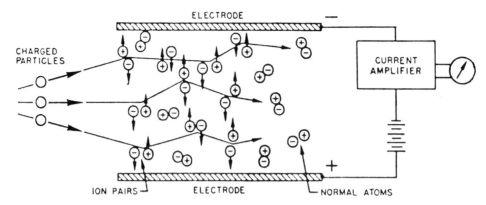

FIGURE 3 Operation of an ionization chamber.

the central electrode. When electrons enter this steeply increasing electric field, which is perhaps only a few mean free paths away from the central electrode, the result of their interactions with the gas molecules changes significantly. In this region the electrons receive sufficient kinetic energy to produce secondary ionizations in the gas; that is, the electrons, which resulted from a primary ionizing event, become directly ionizing particles. The electrons released in these ionizing events are called secondary electrons. These electrons are also under the influence of the electric field and ultimately may possess enough kinetic energy to produce additional ionizations. (The electrons produced in these tertiary ionizing events may also produce ionization.) Thus, each primary ionizing event produces an avalanche of electrons all moving toward the anode. This is the phenomenon of gas multiplication that is common to both proportional counters and Geiger-Müller counters.

As the applied voltage is increased, the gas volume in which gas multiplication can occur will expand. In a cylindrical chamber such an increase in applied voltage implies an increase in the radius of a cylindrical "multiplication" region around the anode, which results in a growth in the number of electrons produced per primary ionizing event. Under suitable conditions the effects of gas multiplication will become significant and the charge collected may be increased by several orders of magnitude (e.g., a gas multiplication factor, the number of ions collected per primary event, of $\geq 10^4$ is typical).

In the detector itself there may be several primary ionization events produced by the passage of a single ionizing particle. The electron avalanche, often called a Townsend avalanche, produced by each primary interaction moving paralel to the collector is confined to a small length of the wire. If there are only a few ionizations produced in the detector volume, there will be no interaction between the avalanches caused by the primary electrons. Within certain limitations, all the avalanches produced by individual electrons (from primary ionizations) are approximately the same size; that is, the total charge collected per primary event is uniform and independent of location within the chamber of the passing particle. Thus, the resultant output pulse, which is the sum of all these avalanches, is proportional to the number of primary electrons produced by the passage of the ionizing particle. The region designated region III in Fig. 2 is called the proportional region, and radiation detectors that operate in this region are called proportional counters, This process is illustrated schematically in Fig. 4.

Electrons are more mobile and hence are collected much more quickly than the positive ions. In many cases the avalanche region may extend only a fraction of a millimeter into the gas. This rapid collection of electrons from the gas volume leaves a positive-ion sheath around the central electrode that, in the time required for electron collection, appears to be essentially stationary. In the proportional counter, these positive ion sheaths remain localized. The counter may receive another pulse at another location on the electrode even while the original ion sheath remains in position on the electrode. Thus, the ability to resolve the passage of many ionizing particles is quite good in a proportional counter. This means that the counter is responsive to new ionizing events as soon as existing ions have been swept out of the main counter volume by the applied voltage.

If the electrode potential is increased further, there is a corresponding increase in the gas multiplication factor and the positive-ion sheath spreads along the electrode. This spreading causes a corresponding increase in the resolving time of the detector. (The resolving time is the minimum time that can elapse between the interactions of two successive particles within the detector if they are to produce two counts.) Since the avalanche in a proportional counter is terminated by the collection at the anode of the

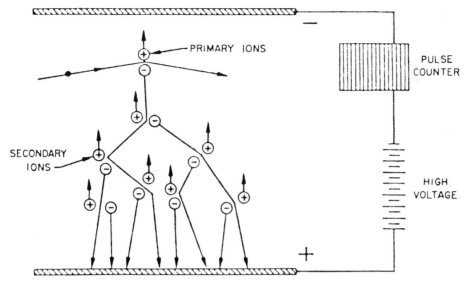

FIGURE 4 Operation of a proportional counter.

liberated electrons, the rise time of the pulse is governed primarily by the time required for electrons produced at the extreme end of the primary ion track to travel to the electrode. Usually this time is very short. In some applications a resolving time of 0.2–0.5 μsec may be obtained. However, such applications usually involve only the detection of radiation. In many other applications the actual resolving time may be limited, not by the detector, but by the external electronic circuits associated with the detector system.

In the region in which gas multiplication occurs, the multiplication for a given applied voltage is independent of the initial ionizing event since it depends only on the energy of the positive and negative ions drifting through the gas volume. In this way, the proportionality of pulse sizes is maintained; that is, the total number of ions is still proportional to the total energy lost by the incident particle in the gas volume. As the voltage is increased further, the strict proportionality no longer holds because not all primary ions generate an identical number of secondaries. In the upper portion of region III in Fig. 2, the pulse size is independent of the initial ionization and this is called the region of limited proportionality.

Before considering the last region, note that Fig. 2 also illustrates the difference in the number of ion pairs produced by densely ionizing particles (curve 2 for alpha particles) and more sparsely ionizing radiation (curve 1 for beta particles). In the region of limited proportionality it is still possible to distinguish between the two types of particles, but the ratio of the pulse heights provides no useful information. As the voltage is increased further even this ability disappears and in region IV the ability to distinguish

between particles with different ionizing power does not exist.

4. Geiger-Müller Regime

The last region, region IV, presents one in which the charge collected is independent of the initial ionization because each primary event causes an "avalanche" of secondary ions extending throughout the whole counter volume. Thus, only the number of events is detected, not the energy transferred. Actually, the degree of gas multiplication is what differentiates this region from the proportional region. In region IV gas multiplication is limited only by the characteristics of the detector and the external circuit. This regime is commonly called the Geiger-Müller, or simply the GM, mode of operation.

In a GM counter the gas-multiplication process produces a pulse of a uniform size regardless of the number of ion pairs formed by the primary ionization. The avalanche continues until a certain number of ion pairs are produced in the detector volume, typically 10^9 electrons.

In the GM counter the initial Townsend avalanche builds up rapidly. It is terminated when all the electrons produced in the avalanche reach the central electrode because the positive-ion sheath causes a reduction in the field strength near the wire. The initial ionization avalanche is followed by successive avalanches, each one triggered by the preceding one. In a GM counter the effects of the Townsend avalanche are thought to be propagated by the excitation of neutral atoms within the avalanche region. These atoms may deexcite by the emission of photons in the UV region that may initiate further avalanches. In counters filled with a mixture of gases, avalanches are propagated by ionization of other atoms that have a lower ionization potential than that of the original excited atom.

The UV radiation may also initiate the emission of photoelectrons by interactions in the cathode. The time sequence to these secondary avalanches may be delayed significantly when compared with that of the original discharge. This is due to metastable excitation states of the noble gases, which serve to delay the emission of the UV radiation and the subsequent production of photoelectrons and another avalanche.

In pure gases, positive ions may initiate photoelectrons as they are neutralized at the cathode. Figure 5a illustrates schematically this process. Positive ions migrate to the cathode and are neutralized by combining with electrons from the wall. This process (i.e., neutralization) may produce an additional electron by two separate mechanisms. First, a photon may be radiated from the wall and may produce more photoelectrons. The energy of this photon is initially the energy difference between the ionization potential of the ion and the work function of the cathode material.

The second mechanism that produces an additional electron occurs if the positive ion has an ionization potential that is at least twice the work function of the cathode. In this case, a second electron, rather than a photon, may be ejected from the cathode. In either case, the emission of this excess energy results ultimately in another avalanche.

The latter two processes are obviously governed by the transit time of the positive ions to the cathode. This implies a significant delay after the initial discharge before additional discharges are produced. In a typical situation the time interval may approach 200 μsec; this, in turn, introduces a long "dead time" during which the detector will not respond to new ionizing particles.

In all cases discussed here, the production of another electron, by whatever process, results in another discharge in the detector. This propagation of avalanches will continue unless steps are taken to prevent these occurrences. This is called quenching the discharge. Quenching consists of the introduction of an electronegative impurity that can absorb some of the excitations without further ionization. Thus, it limits further spreading of the avalanche region and permits early recovery of the ionized gas.

Quenching is not important in proportional counters because the number of excited atoms and ions formed is small. Thus, successive avalanches become smaller and result in no further charge generation. The need for quenching is another characteristic of the difference between proportional and GM counters.

A number of methods have been used to quench the self-perpetuating avalanches in GM counters. External circuits were used with older GM counting systems. These circuits were designed to reduce the voltage across the detector to less than the value required to maintain the discharge. The voltage is reduced only momentarily but this is sufficient to terminate the discharge of the counter. External quenching results in long resolving times and for this reason self-quenching GM counters are commonly used.

Self-quenching GM counters contain a quenching gas that is added as an impurity to the major gas-filling. Two types of self-quenching GM tubes are available: those quenched by the addition of a halogen gas and those organically quenched. In self-quenching GM counters, as positive ions move toward the cathode, there is a transfer of charge to the molecules of the quenching gas. When the charged molecules reach the cathode they dislodge electrons from the chamber wall, but this process results in the dissociation of the quenching gas molecules rather than the production of additional electrons (see Fig. 5b).

Typical organic quenching gases are polyatomic gases, including ethyl alcohol, ethyl formate, and amyl acetate. For organic quenching gases, the dissociation is irreversible and, therefore, the life of an organically quenched GM counter is limited. The useful lifetime is typically 10^8 to 10^{10} pulses. Bromine and chlorine gases are commonly used in halogen quenching GM counters. In this type of counter the quenching gas molecules recombine after a finite period and, thus, the counters have essentially an unlimited useful lifetime. However, all gas-filled counters have a practical limitation on their lifetime set by leakage of counter seals.

GM counters are typically cylindrical in shape and filled with a mixture of argon and quenching gas (<1%) at a pressure of 40 Torr (5.3 kPa). The collecting electrode is usually a tungsten wire; the outer electrode is a thin stiffened steel or aluminum shell coated with graphite to ensure uniform field distribution. For alpha or beta detection, counters have been designed with thin mica windows or very thin walls. Such counters require very careful handling and the window should never be touched by hand.

Since the Geiger regime is such that all ions produced by a single initiating event are collected as a single pulse

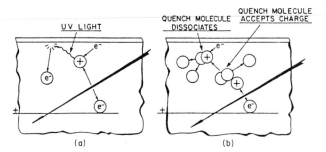

FIGURE 5 Operation of a GM counter: (a) formation of initial ion pair; (b) effect of quench molecules.

event, the output pulses have a constant amplitude regardless of the energy of the initiating event. Hence, a GM counter is a very sensitive radiation detector whose large output pulses may be of the order of volts in amplitude, which makes it convenient to count them directly. However, since it is not possible to correlate pulse amplitude with energy deposition in the detector, GM counters have found only limited use in dosimetry.

5. Ionization Chambers

Two broad classifications of ionization chambers are usually assigned in the discussion of radiation detectors of this type: passive detectors and active detectors. The term *passive* implies a situation in which several steps must be taken to use the detector. That is, the detector must be prepared for use, exposed to radiation, and evaluated. The information provided by the detector may be obtained at the end of the exposure or after the detector has been removed from the radiation field. Therefore, passive detectors have to be integrating devices since the detector reading gives the total exposure accumulated over the entire exposure period, unless the situation arises in which the detector is overexposed. In this case, the integrating (passive) dosimeter gives no useful information.

Active detectors are those that provide an immediate indication of the exposure received. This information is obtained through the use of an external electronic circuit that is connected (semipermanently) directly to the detector. The electronic system may be close to the detector, as in certain portable radiation detectors, or it may be located at some distance from the detector, as in the case of radiation detectors used in the operation of a nuclear reactor.

a. Passive ionization chambers.
One of the earliest radiation detectors was the gold-leaf electroscope. This detector, although no longer used, remains as the basis for some radiation detectors used in personnel monitoring.

The quartz-fiber electroscope, developed originally by Lauritsen, replaced the gold-leaf design because it is more compact and portable.

In this detector there are two metallized quartz fibers, one movable and the other stationary, mounted on a central support. These fibers are insulated from the outer case and represent the sensitive element of the detector.

To use the detector, the operator must charge it through an external circuit; usually 200 V is required. Since the two fibers have the same polarity, the repulsion of like charges displaces the movable fiber from the stationary fiber. The electroscope is equipped with an eyepiece, a transparent scale, and a window to illuminate the fiber. When fully charged, the movable quartz fiber casts a shadow at the zero position on the scale. When the electroscope is exposed to radiation, ionization of the gas in the volume reduces the charge on the electroscope and the repulsive force between the fibers is reduced. The movement of the fiber toward the fixed fiber is reflected on the scale as the total amount of radiation exposure.

The above principle is applied in the self-reading pocket ionization chamber. In this device the fibers are in the form of loops and all components are enclosed in a rugged metal case. The chamber is small and can fit easily in a shirt pocket. Typically the range of the chambers is 200 mR, but others are available with full-scale readings up to 50 R.

The second major type of passive ionization chambers are those called condenser-type chambers, also called indirect-reading dosimeters. These devices are simpler than the electroscope type but require an external circuit not only for charging but also for evaluation. The detector consists of a cylindrical outer case constructed of a material, such as Bakelite, and a central metallic electrode coaxial with, but insulated from, the case. The detector is charged from a battery pack (or in some cases a stationary unit called a minometer), removed from the charger, and exposed to radiation. The exposure is evaluated by replacing the detector in the charger, which also serves as a reader, and evaluating the reduction in total charge that is proportional to the total ionization produced in the chamber.

Usually the charger–reader is calibrated to read directly in units of exposure, that is, roentgens. However, the exposure can be evaluated by the relation.

$$C \times V = Q.$$

In this relation, C represents the electrical capacitance of the chamber, V the change in voltage before and after exposure, and Q the charge collected during the exposure. If the chamber volume is known, then the exposure can be calculated on the basis of the definition of exposure.

Certain types of condenser chambers may serve as secondary standard devices for use in calibrating the output of X-ray machines, radio isotope sources used to calibrate survey instruments, and so forth. These detectors are often referred to as R chambers. These chambers are manufactured with walls of different thicknesses and compositions for use in a wide range of photon fields. Usually, the wall material is air-equivalent and the thickness of the wall is adjusted for a particular photon energy. For example, R chambers are available for use in the energy range 6–35 keV and also for photons in the energy range 0.25–1.4 MeV. In addition, the total exposure range may be selected by choosing a detector of a certain volume. Typical detectors in use have sensitivities from 0.025 R to 250 R. Other sensitivities are available (e.g., 0.001 R) for use in special exposure situations, such as environmental measurements or for the measurement of "stray" radiation aroun heavy shields. However, it must be remembered for all indirect-reading chambers that the calibration is valid only for energies near that of the calibration source.

R chambers require a charge–reader for use. This charger–reader gives an output directly in units of exposure (i.e., roentgens). Normally the complete system, usually called an R meter, is purchased with calibration certificates from the National Institute for Standards and Technology (NIST) or from a regional calibration laboratory.

The simplicity of passive ionization chambers is such that there are a variety of detector types and arrangements that have been designed for special applications. Usually, the walls of the chambers are constructed of material that is air-equivalent; that is, the material has photon scattering and absorption properties similar to those of air. Materials such as Bakelite, graphite, polystyrene, nylon, aluminum, or any other material that for a selected photon energy has similar characteristics are commonly used.

For other applications, chambers may need a wall material that is tissue-equivalent of muscle-equivalent. Chambers have also been manufactured with bone-equivalent materials. A large number of equivalent materials are available for the construction of the chambers. The choice depends on the application, materials available to the designer–builder, and expected accuracy of the intended use. For example, small condenser chambers can be made sensitive to thermal neutrons by the addition of a material such as boron, either as a linear or as part of the wall material. Condenser pocket chambers used as neutron dosimeters are commercially available and be used for personnel monitoring. However, since the chamber responds to photon irradiation as well as thermal neutrons, the user obtains no information as to the relative contribution of each radiation to the measured exposure when both are present.

This disadvantage can be overcome by the use of two chambers, one sensitive only to photons and the other sensitive to neutrons as well as photons. Both detectors must be properly calibrated since even the detector that responds only to photons has a finite (but small) response to thermal neutrons. Although this presenets no real problems, the simple example illustrates the complexities encountered with measurements in mixed-radiation fields, even with the simplest radiation detectors.

b. Active ionization detectors. In an active (or current-type) ionization chamber, electrons collected at the anode compose a direct current that can be amplified in an electrometer tube and measured with a microammeter. For situations requiring high accuracy, the small current may be measured with a vibrating reed electrometer that transforms a small direct current into pulses that can be amplified. Essentially all ionization chambers are evaluted in one of these ways.

Probably the simplest of the active ionization chambers, yet the type least familiar in terms of widespread use, is the free-air ionization chambers. This detector is used as a primary standard in national standardization laboratories throughout the world. The chamber is a parallel-plate design that satisfies the operational definition of exposure in units of roentgens. The photon beam is collimated as it enters the chamber and interacts in a volume of air defined by the collimator aperture and the electric field between the collecting electrodes. The chamber features a guard ring and guard wires to maintain straight lines of force between the two electrodes. The entire device is enclosed, usually with a lead-lined material. Ions produced in the chamber volume due to photon interactions are collected at the plates. The current flow is measured by an external circuit, and from it the number of ions produced in the volume and, ultimately, the exposure can be calculated.

For this measurement to be valid, electronic equilibrium must exist in the detector. In other words, all the energy of primary electrons produced in the sensitive volume of the chamber must be dissipated in the chamber. Obviously, many electrons produced in the detector volume by photon interactions will leave the sensitive volume. Electronic equilibrium is maintained by making the entire chamber larger than the maximum range of the primary electrons in air. In this situation, primary electrons produced in the sensitive volume that leave the volume are replaced by primary electrons that were produced outside the sensitive volume but enter it. Thus, electronic equilibrium is obtained as an electron of equal energy enters into the sensitive volume for every electron that leaves.

The thickness of air between the entrance port and the collecting volume needed to provide electronic equilibrium increases with increasing photon energy. For example, 9 cm of air is required for highly filtered, 250-kV X-rays, whereas for 500-kV X-rays, the air thickness is 40 cm. This fundamental requirement limits the use of free-air chambers since the size of the chamber for higher photon energies is extremely large. For example, the NIST has three free-air ionization chambers. The chambers are intended to cover the X-ray generating potentials of 10–60 kV, 20–100 KV and 60–250 kV. These chambers were manufactured at the NIST, but similar chambers are commercially available with a useful range up to ~300 keV. At ranges greater than this, photon energy, operational difficulties, chamber size, and so forth limit this detector's usefulness even in a standards laboratory.

The surface dose due to beta-emitters may be determined by use of the extrapolation chamber. This special ionization chamber is a parallel-plate detector similar in design to the free-air chamber described above. However, in this design the distance between the plates can be varied. Usually, one plate that acts as a thin window is placed as close as possible to the source to be measured. A series of measurements is obtained while the spacing between the plates is decreased. The results of these measurements are plotted and extrapolated to zero spacing. This gives the dose at the surface of the beta source and eliminates secondary gas or wall effects.

The use of an extrapolation chamber is a good example of the application of the Bragg-Gray principle to the measurement of absorbed dose (discussed earlier in this article). The chamber, introduced by Failla, is quite useful because it recognizes the fundamental requirement that the detector cavity be small compared with the electron ranges. The chamber has also been used for measurements in areas where no electronic equilibrium exists, for example, at interfaces between two dissimilar materials. Currently, there is increased interest in the use of extrapolation chambers for measurements in beta-radiation fields found in nuclear utilities. However, this is a special application of this detector system since the chamber is not sufficiently rugged to survive in routine use in these environments. Tissue-equivalent extrapolation chambers have also been designed and used in a number of dosimetry research activities. However, there has been no widespread application of this system to routine dosimetry.

Ionization chambers have found wide use in surveys for radiation protection purposes. The ionization chamber is the only gas-filled detector that allows the direct determination of the absorbed dose. This is because the measured current is directly proportional to the ionization produced in the sensitive volume and that in turn is directly proportional to the energy deposited in the detector.

A number of active detectors have been designed with characteristics similar to the condenser R chambers discussed in the previous section. One such system is a precision instrument designed specifically for the measurement of ionizing radiation used in medical diagnostic and therapeutic procedures. The individual chambers have walls constructed of air-equivalent materials, and the sensitive volume is filled with air. A preamplifier located close to the detector allows a reasonably long run of cable between the detector and the readout. The readout system functions either as a rate-meter or as an integrating device. In addition, a high-voltage supply for the chamber is an integral part of readout. The entire system is very stable and accurate, and state-of-the-art solid-state electronics makes it easy to operate. As with the condenser R chambers, the detectors may be purchased with a calibration traceable directly to the NIST. These systems have found wide use in instrument calibration facilities in many utilities. The systems provide immediate and accurate indications of the exposure rates, which lends confidence to portable calibration procedures. In addition, the systems can be used in the integrate mode to monitor standard exposures of pocket chambers or TLD badges.

A multitude of special detectors have been designed and used for dosimetry in mixed-radiation fields. These systems use paired chambers, one of which is sensitive essentially to only one components of the field and the other detector which is sensitive to both components of the radiation field. If the detectors have been properly calibrated, the exposure rate (or dose rate) of the radiation field for each component can be obtained by subtraction. One such system uses a chamber constructed of tissue-equivalent material through which a tissue-equivalent gas is flowing. This detector is sensitive to both gamma and neutron radiation. The other detector is constructed of graphite, and the filling-gas is carbon dioxide. The graphite is sensitive only to gamma radiation, whereas the tissue-equivalent chamber is sensitive to both neutron and gamma radiation. The neutron component of the radiation field can be obtained by substracting the dose rate indicated by the graphite detector from that indicated by the tissue-equivalent detector.

6. Proportional Counters

Proportional counters in common use are usually of two types: gas-flow and sealed tubes. In the gas-flow device, the counting gas is continuously circulated at a slow rate through the detector volume. These detectors have essentially an infinite life because the counting gas is constantly replenished and molecules degraded by the ion-formation process are continually removed. Gas-flow detectors are standard equipment in most laboratories. However, these detectors are not usually applied to dosimetry. Sealed proportional counters have a finite life since there is usually no mechanism for replenishing the filling-gas that will accumulate impurities due to corrosion of sealing materials in bombardment effects.

a. Sealed proportional counters. The widest use of sealed-tube proportional counters is in the detection of thermal neutrons. A material with a high-neutron cross section (e.g., boron) is introduced into the chamber and the counter functions by detecting the charged particles liberated by the (n, α) reaction in ^{10}B. The reaction has a high–thermal-neutron cross section ($\sim$3800 b) and exhibits a simple-energy dependence over a wide neutron-energy range, from the thermal energy to $\sim$30 keV. The reaction is exothermic and an energy release of up to 2.78 MeV is shared between the alpha particle and the recoil lithium nucleus. There is also an excited state of ^7Li in which a 0.48-MeV gamma ray is emitted and, in this case, the shared energy is 2.30 MeV. This latter reaction is the most probable.

The boron may be incorporated in the detector in two ways. First, a thin layer of boron may be used to coat the inside of the cathode. These boron layers are usually ≤ 0.5 mg/cm^2 because a coating thicker than the maximum range of the alpha particles (~ 0.75 mg/cm^2) would actually reduce the sensitivity of the detector. This type of detector has the advantage that a conventional counting gas can be used. However, alpha particles produced in the boron lining lose some of their energy in escaping the wall and produce a wide distribution of pulse heights. The count rate versus operating voltage curve obtained has a plateau with a slope greater than that obtained in other proportional counters.

An alternative method of introducing the neutron absorber is to combine it with the counting gas. In this detector the gas-filling is boron trifluoride (BF$_3$) enriched in the isotope ^{10}B, typically at a pressure of 120–600 Torr (16.3–81.6 kPa). This detector has the advantage that the (n, α) reaction is produced in the gaseous volume, and, therefore, the entire energy of the alpha particle will be used in producing ionization in the sensitive volume. Usually the plateau in the operating curve is very flat.

Proportional counters designed to detect neutrons have found wide use in mixed neutron–gamma-radiation fields that may exist around reactors, accelerators, or certain isotopic neutron sources. The detectors not only have good discrimination characteristics against gamma radiation but also may be used to detect both thermal and fast neutrons. This is accomplished by the use of an additional moderator. For example, in normal use a BF$_3$ proportional counter is sensitive only to thermal neutrons. However, if a thick moderator, such as paraffin or polyethylene, is placed around the detector, it becomes sensitive to fast neutrons. Fast neutrons incident on the moderator are thermalized and are detected by the proportional counter as thermal neutrons. The moderator may be any convenient shape: spheres, right-cylinders, and rectangular moderators are in common use. Thus, with the BF$_3$ counter outside the moderator, the thermal-neutron component of the radiation field is detected. Inside the moderator the fast-neutron component of the field is detected. In practice, these detectors must be carefully calibrated before use because often no precise information is available on the sensitivity of a specific detector, the energy spectrum of the fast-neutron field or the percentage of the fast-neutron component that is thermalized. In addition, the response of the detector system depends on the thickness of the moderator materials and the incident neutron energy.

Many types of proportional counters have been designed for use as dosimeters, especially for the detection of neutrons. There has been wide use of ^{3}He-filled detectors and detectors filled with methane. Proportional counters have also been lined with hydrogenous material, paraffin or polyethylene, and used as fast-neutron detectors. One such detector is the Hurst absolute fast-neutron dosimeter. This cylindrical proportional counter is lined with polyethylene and the gas-filling is cyclopropane (ethylene has also been used). The gas pressure is typically 500 Torr (68 kPa) for cyclopropane and 750 Torr (102 kPa) for ethylene. In either case, the gas and the walls have the same atomic composition, which satisfies one of the requirements of the Bragg-Gray relationship. In this case, ionization produced by proton recoils (due to fast-neutron interactions in the wall) produces relatively large pulses. Using this detector allows an individual to measure the energy absorbed per unit mass of the gas. From this measurement, the energy absorbed per gram of tissue can be determined since, over the energy range 0.01–20 MeV, the ratio of the absorbed dose in ethylene to that in tissue is ~ 1.45.

The Hurst counter has a low response to gamma radiation caused by the differences in pulse heights between those produced by proton recoils and secondary electrons. In addition, the associated electronics usually feature a discriminator that allows the measurement of a fast-neutron dose of 0.001 rad/hr in a ^{60}Co–gamma-radiation field of 50–100 R/hr.

Many other specialized proportional counters have been designed and used in radiation measurements. One of these is the Rossi LET chamber. This proportional counter consists of a sphere of tissue-equivalent material featuring a helical field-defining wire around the central electrode. The intent of the detector is to simulate small volumes of tissue, by varying the filling pressure of the tissue-equivalent gas. Absorbed dose may be determined through a measurement of the pulse-height distributions from the detector and a calculation of the LET distribution from these data. These detectors are available commercially, in sizes up to ~ 2 in. (5 cm) in diameter, with quick-disconnect fittings for gas-filling and built-in preamplifiers. However, after an initial flurry of interest, this type of proportional counter has not found wide use in neutron dosimetry research.

Tissue-equivalent proportional counters (TEPCs) have been constructed in various sizes and shapes for a variety of applications. One application of the TEPC has been its experimental use to monitor the radiation exposure of flight crews on commercial airliners and crews on space shuttle missions. These specially designed and constructed detectors use a tissue-equivalent plastic (Shonka A-150) as the wall of the chamber and propane as the filling-gas. The detectors are portable but can be placed in mountings in the spacecraft to provide an indication of the dose rate and total dose associated with each mission.

It is anticipated that these TEPC detectors will be used on the International Space Station and also will be present in the spacecraft destined for the Mars mission. On this mission, two detectors will be used, one of tissue-equivalent material and another constructed of graphite. These two detectors will allow separation of the radiation field into two components.

Another method for detecting thermal neutrons with proportional counters is to use a fission counter. This detector uses a thin coating of a fissionable material, for example, ^{235}U, on the electrodes to detect thermal neutrons. Thermal-neutron capture in the ^{235}U results in a fission event and the fission fragments produce a high density of ion pairs in the detector gas. The advantages of a fission counter are its low sensitivity to very high gamma-radiation fields and the large amplitude of pulses due to the fission fragments, which allows easy discrimination against other radiation, including alpha particles emitted in the normal radioactive decay of ^{235}U. Fission chambers are used widely as part of the start-up instrumentation on nuclear reactors but not as dosimeters.

B. Solid-State Detectors

Although gas-filled detectors satisfactorily fulfill many detection tasks, they have a number of inherent shortcomings for many applications. The low density of the gas medium makes the interaction with incident radiations inherently inefficient, especially for photons and high-energy radiations, where large detector volumes or high gas pressures would be required for good sensitivity. Also, the finite drift mobility of the ions produced in the gas results in a slow response time, of the order of milliseconds in most cases. This leads to an appreciable dead time, during which the detector may not respond to fresh-incident radiation, and to a slow rise time in the charge collection pulses, which again limits the rate of detection. One way of overcoming these limitations is to use condensed-state or solid-state detectors, in which signal generation processes occur more rapidly, and where the high atomic density results in a high probability of interaction over a relatively short range. There are several phenomena that meet these conditions, though no one detection mechanism is ideal for all applications.

1. Scintillation Detectors

The first solid-state detection process discussed here is known as scintillation detection. It was one of the earliest to be used, if only in a rather primitive form when Rutherford and his collaborators observed the emission of alpha particles by means of a spinthariscope in the 1920s. The spinthariscope consisted of a screen of zinc sulfide that could fluoresce or "scintillate" when struck by an alpha particle. An observer, sitting in a fully darkened room (after several hours in which the observer's eyes adjusted to the dark) could count these scintillations, which could then be used to estimate alpha activities. Modern scintillation counting dates from 1947 when Coltman and Marshall developed the photomultiplier tube and Kallmann and Broser combined the scintillating material with a photomultiplier, enabling more efficient light detection while eliminating problems of eye fatigue.

The general principle of scintillation counting depends on interaction of the incident radiation with a suitable fluorescent material, called the scintillator or phosphor. On absorbing energy from the incident radiation the phosphor under-goes excitation to a higher electron state. This is followed by a prompt (or delayed) return to the ground state, accompanied by emission of electromagnetic radiation (light) of a wavelength appropriate to the difference in energy levels. Provided the material is transparent to light of that wavelength, this light may be observed outside the phosphor material. Otherwise, only light emission from the surface layer can be seen. In many materials the emitting transition also corresponds to an absorption transition, so that the energy cannot escape, that is, the material is opaque at that frequency. For this reason it is usually necessary to introduce an impurity into the scintillator to which the exciton can be transferred. The impurity gives rise to trapping levels from which transfer to the ground state can occur at a wavelength significantly different from the absorption wavelength; therefore, the light is emitted in a region for which the material is transparent.

Another reason for introducing impurities into phosphor materials is to serve as wavelength shifters. Many materials, especially organic materials, will fluoresce readily when excited. However, the transitions associated with light emission tend to give rise to wavelengths in the near or far UV regions, where detection of this light is relatively inefficient and many materials require replacement of glass envelopes by quartz. By introducing small concentrations of impurities into the scintillator enough characteristic trapping centers can be provided to ensure a high probability for exciton transfer leading to a high proportion of fluorescence emission from these now lower energy levels. The net effect of this is that emission may then occur in the blue or green regions of the visible spectrum where sensitive photomultiplier detectors are readily available and at a much lower cost.

The fluorescent light emitted by the excited scintillator is guided through a suitable optical medium to a photomultiplier tube. In the photomultiplier each incident light photon is photoelectrically converted to one or more electrons. These photoelectrons are then accelerated by an electric field and hit a low–work-function electrode, called a dynode, where each energetic electron causes the emission of two or more secondary electrons. Repetition of this secondary electron emission process at successive, more positively charged dynodes enables each initial electron to give rise to a large swarm of electrons as the electrons move down the "multiplying" chain. At the last dynode, the resulting charge pulse can be collected and further amplified, if desired. In general the magnitude of this pulse will be proportional to the number of photons reaching the photocathode and, hence, to the energy of the incident radiation. The number of pulses, representing the number of separate (in time) exciting events in the phosphor will be proportional to the intensity of the incident radiation. After amplification, the pulse rate can be counted to obtain intensity information. With suitable treatment of the electronic pulse output from the photomultiplier, the distribution of pulse amplitudes can be used to obtain energy spectra, particularly for gamma radiation.

The key component in scintillation counting is the scintillator, where conversion of incident radiation into light occurs. For efficient detection. the scintillator, or phosphor, should have a high density and a high fluorescent-light yield, be transparent to the emitted light with little internal scattering, have a refractive index compatible with the light guide and photomultiplier window, and have adequate physical and mechanical stability. The dimensions of the scintillator should be sufficient to stop most of the incident radiation and be readily contained in a light-tight shield. Not all of these criteria can be met by any one scintillator material and a variety of phosphors have been developed for different applications. Scintillator materials may be categorized as inorganic or organic and may be solid or liquid. Various gases, particularly the noble gases, also exhibit fluorescence phenomena, but they offer no practical advantages over gas-filled ion collection systems and, thus, are not considered further here.

Although scintillation detectors are used extensively in radiation detection, their use in dosimetry has been limited. For dosimetry, LiI(Eu) is the major inorganic scintillator of interest. This scintillator is used primarily for the detection of fast and thermal neutrons. Inorganic scintillators have found wide use in dosimetry; however, the ability to produce reasonably pure crystals of some of the materials has limited the application of organic scintillators to dosimetry to only a few specific phosphors. Of these, the most popular are plastic scintillators, in particular NE-102 and NE-213.

2. Thermoluminescence Dosimetry

The most popular method of personnel dosimetry at commercial nuclear power plants (as well as many other facilities) in the United States involves the use of thermolumi-

nescence dosimetry (TLD). TLDs have many of the same characteristics as those required for an "ideal" dosimeter. However, TLDs also have certain characteristics that influence their response and the resulting dose estimates obtained with the dosimeter. These characteristics must be known and appreciated if this system is to be used and evaluated properly as a personnel monitor. In this section, a general discussion of the mechanism of thermoluminescence and the characteristics of the most popular TLD materials will be presented.

3. Theory

Thermoluminescence (TL) has been observed for centuries; it occurs whenever certain fluorites and limestones are heated. It is reported that Sir Robert Boyle and his colleagues studied TL in the early 1660s and that Boyle presented a paper on TL to the Royal Society in London in 1663. Some investigators have gone even further in examining the history of TL; one investigator has postulated that the early cavemen and early alchemists often observed the phenomenon of TL, even though neither possessed an explanation of the mechanism. TL has been studied extensively; such well-known scientists as Henri Becquerel (and his father before him) have mentioned the phenomenon in their scientific papers. However, it was not until about 1950 that Daniels proposed the use of this phenomenon as a radiation detector; more specifically, it was his suggestion that TL could be used as a radiation dosimeter. It is interesting that this suggestion was so late in coming because the relation between TL and exposure to X-rays was observed as early as 1904.

In TLD the absorbed dose is determined simply by observing the emitted light from the crystal as the crystal is heated under a controlled manner. The amount of light emitted is directly proportional to the radiation energy deposited in the TL material. However, TLD is not an absolute dosimetry system, and therefore the system must be properly calibrated to establish the relationship between the amount of light emitted and the deposited energy (i.e., the absorbed dose). Detailed discussions of the chemical and physical theories of TL have been offered by many scientists, but some would contend the actual phenomenon is not completely understood. However, the basic phenomenon is qualitatively understood and this is sufficient for the purposes of discussing TL applied to radiation dosimetry.

Normally, TL is explained by referring to a hypothetical energy-level diagram of an insulating crystal. Although the model is greatly simplified, it serves to illustrate the fundamental process. If a crystal exhibiting TL (sometimes called a phosphor) is exposed to ionizing radiation, interactions in the crystal free electrons from their respective atoms (ionization). In the energy-level diagram, electrons are released from the valence band and move to the conduction band. The loss of electrons in the valence band creates positively charged atoms (or sites) called holes. The electrons and holes may migrate through the crystal until they recombine or are trapped in metastable states. The "traps" prevent the electrons from returning to the valence band, and therefore the radiation energy is in effect stored in the crystal. The metastable states are thought to be associated with defects in the crystal or with impurity sites (actually, impurities are introduced intentionally into crystals used as TLDs). These impurity sites are called trapping sites, and it should be remembered that these sites may exist at many energy levels in the crystal; it should not be assumed that all the electrons are trapped at exactly the same energy level. The importance of this point will be made clear in the discussion of the characteristics of typical TLD materials.

Basically, the crystal has stored the energy that caused the electrons to be released. If the crystal can be stimulated in some way so that the stored energy is released and that released energy can be measured in some way, then the material can be used as a radiation dosimeter. Usually, the energy to release the electrons is supplied by thermally heating the crystal (thermo); the stored energy is then released in the form of visible light (luminescence). Thermally heating the crystal was one of the first approaches taken to release the energy stored in the TLD. There are a number of possible ways to release this stored energy and there are even variations in the design of systems to thermally heat the crystal. For example, optical stimulation has been studied extensively. Nevertheless, in the discussions that follow, it will be assumed that the TLD material has been heated conventionally using thermal heating.

At this point, there are two possible ways the stored energy can be released. First, as the crystal is heated, sufficient energy may be given to the trapped electrons to release them for the trapping sites and raise them into the conduction band. The electron may recombine with a hole, returning to the valence band, and giving up the excess energy in the form of light (a luminescence photon). Light photons are released with energies proportional to the difference between the excited and stable electron energy levels. On the other hand, the hole trap may be less stable than the electron trap, and when the crystal is heated the hole receives sufficient energy to wander until it combines with a trapped electron and a luminescence photon is released. Usually, since the two processes are similar only the first possibility is presented in discussions of TL.

The energy gap between the valence and conduction bands is related to the temperature required to release the electrons and produce luminescence photons. In practical situations, many trapped electrons and holes are produced during irradiation of the TLD. As the temperature of the crystal is increased, the probability of releasing any electron is increased. At some sufficiently high temperature, there is virtual certainty that all electrons will be released. Thus, the emitted light from the crystal may be weak at low temperatures, pass through one or more maxima at higher temperatures, and then decrease again to zero.

4. TLD Glow Curves

A plot of the light emitted by the TLD as a function of temperature (or time) is called a glow curve. The most usual case is to plot the temperature of the crystal on the abscissa versus the light emitted by the phosphor on the ordinate. However, literature on commercially available systems typically presents the latter (i.e., time versus light emitted). Glow curves are obtained by electronically plotting the signals from a thermocouple in close contact with the container holding the TL material and the current from a photomultiplier tube (PMT) viewing the light emitted by the material. More modern systems now use an electronic system called a glow-curve analyzer, instead of plotting these curves on a plotter. These newer systems allow extensive analysis of the structure of the individual peaks constituting the glow curve.

Typical glow curves will show one or more peaks (maxima) as traps at various energy levels are emptied. The relative amplitudes of the peaks indicate approximately the relative populations of electrons in the various traps.

Either the total light emitted during part or all of the heating cycle, or the height of one or more of the peaks, may be used as a measure of the absorbed dose in the phosphor (or the exposure in air, depending on the calibration technique). When the peak height technique is used to determine the absorbed dose, the heating cycle must be extremely reproducible to avoid causing peak height fluctuations. These fluctuations influence strongly the accuracy and reproducibility of the dosimetric measurements. For this reason, most commercial TLD systems use a technique that simply integrates the light output (i.e., the PMT current) over part or all of the heating cycle. The results obtained with this technique are much more reproducible because only the heating rate and the maximum temperature of the sample must be controlled.

One of the disadvantages of most TLD systems is the fact that there is no permanent record of the exposure as is possible with other methods of dosimetry (e.g., film badges). However, it is possible to record the glow curve for each individual TLD as it is read and to use this as the permanent record of the exposure. This is actually done in some dosimetry programs. In such cases, the reproducibility of the heating cycle is extremely important.

After the traps have been emptied by heating to a high temperature for a sufficient length of time, and the phosphor has been cooled, the TLD is ready to be reused. In some cases, the heat applied during the reading cycle is sufficient to prepare the TLD for reuse. In other cases, special heat treatments (called annealing) are required before the TLD can be put back into service. The exact prescription for the heat treatment of a TLD material depends on the material itself, form of the material, and intended use (e.g., the anticipated exposure level).

5. TLD Readers

The basic components of a device to evaluate (or "read") TLDs are extremely simple. These consist of a system to heat the material, a detector sensitive to the light emitted by the TLD, and a measuring or recording instrument. In a research situation, it is common to use both a digital system to record pulses from the PMT and an X–Y plotter to record glow curves. Commercially available systems usually give only the integrated current but some can be calibrated and adjusted to read-out directly in units of exposure (mR or R). Some of the more sophisticated systems may be coupled directly to a computer-based, record-keeping system in which the measured exposures are entered directly into the employee's exposure record. The characteristics of commercially available systems vary widely and the choice of vendor and the complexity of the reader are highly dependent on the application of the TLD system.

There are a number of factors that may affect the shape of the glow curve. In addition to heating rate, these include the size, shape, and thermal conductivity of the sample; the irradiation and annealing history of the sample; the recording instrument selected for use; and other spurious effects that may appear. The term *tribothermoluminescence* is normally used to describe many of these spurious effects, some of which are unexplained.

6. TL Materials

There are a large number of TL materials. In fact. most materials thermoluminesce to some extent and with careful use and proper calibration many common materials can be used as a dosimeter. For example, approximately 17 years after the bombing of Hiroshima, the TL in roof tiles taken from houses in the city was used to provide a check on dose estimates made for the survivors. TL has been used for dating meteorites, minerals, and ancient pottery as well as for personnel radiation monitoring and other dosimetry uses.

To be useful for most dosimetric applications, the TL material should have a relatively strong light output and be able to retain trapped electrons for reasonable periods of time at the temperatures expected to be encountered in

the particular application. This requirement limits useful TL phosphors to those with traps $\geq 80°C$. More detailed information is presented in the sections that follow.

a. Calcium sulfate. Manganese-activated calcium sulfate ($CaSO_4 : Mn$) has a long history as a TL dosimeter. In the late 1960s it held the distinction of being the most widely studied phosphor of calcium sulfate. Reports of studies of the TL properties of this phosphor can actually be found in the literature as early as 1895.

This phosphor has a glow curve with a single peak occurring in the range 80–120°C. Commercially available $CaSO_4 : Mn$ is listed as having the temperature of the main TL glow peak at 110°C. There have been a wide variety of fading rates reported for this phosphor (typical fading rate of 50% in the first 24 hr), and this represents one of the major disadvantages in the use of the material. This disadvantage is offset partially by the high sensitivity and the wide usable exposure range exhibited by this phosphor. In the evaluation of TLD materials, LiF (TLD-100) is considered the standard; all other materials are compared to it. At ^{60}Co energies, the light output of $CaSO_4$: Mn per unit exposure is 70 higher than that for LiF. This means that the TLD material should be useful for measurements at very low dose rates perhaps approaching environmental levels. The TLD can be used routinely to measure in the 1- to 10-mR range and, with care, can be extended down into the μR range. The dosimeter has been used for measurements in the exposure range of 20 μR with a standard deviation of $\pm 50\%$.

The response of the phosphor as a function of exposure is linear, but there is some disagreement as to the upper limit of usability. At least one investigator found the upper limit of linear response to be $\sim 10,000$ R. Another investigator reported a value slightly less than 5000 R, while a third reported that the behavior of the phosphor was nonlinear at exposures >400 R. The commercially available TLD material is specified to be usable up to 10,000 R. However, results such as these serve as good examples of the possible influence of fading during exposure or storage, variations between batches of the TLDs due to phosphor preparation techniques, environmental factors such as UV light, and so forth.

Other characteristics of this TLD material influence its usefulness as a dosimeter. The material has a density of 2.6 g/cm^3 and an effective atomic number (effective Z or Z_{eff}) of 15.5. The dosimeter is not tissue-equivalent and shows a marked overresponse at low photon energies (due to the high Z_{eff}). Another measure of usefulness of a TLD material is obtained by taking the ratio of the response per unit exposure of the TLD at 30 keV to the response per unit exposure of the dosimeter at ^{60}Co energies. For $CaSO_4$: Mn, this ratio is ~ 10.

Commercially available calcium sulfate doped with dysprosium is sensitive to thermal neutrons. During Comparison of the neutron sensitivity of TLD materials it is common practice to compare the response to a thermal-neutron fluence of 10^{10} n/cm^2 and relate this exposure to a gamma-equivalent exposure. For commercially available calcium sulfate doped with dysprosium, such an exposure would produce a light emission equivalent to a gamma-ray exposure of 0.5 R. No data were available on the commercially available manganese-doped material. However, for the same neutron fluence, a manganese–lithium-fluoride doped material would yield a light output equivalent to an exposure of 100 R. If ^{6}Li fluoride is included in the manganese-doped material, neutron exposure at the same fluence would be equivalent to a gamma-ray exposure of >1000 R. The thulium-doped material is the least thermal neutron–sensitive material with the standard thermal-neutron exposure corresponding to an equivalent gamma-ray exposure of only ~0.2 R.

It should be pointed out that calcium sulfate is extremely light-sensitive (UV), which enhances fading considerably. The vendor recommends that these materials should be handled, used, and stored in opaque containers to reduce fading due to light exposure. Some procedures even call for handling this material in reduced and nondirect light. If these materials are used in environmental-monitoring programs, the light sensitivity must be taken into account in the handling, packaging, use, and evaluation of the dosimeters.

b. Calcium fluoride. Calcium fluoride (CaF_2) exists in nature as the mineral fluorite. Fluorite exhibits a strong radiation-induced TL and, after special treatment, can be used satisfactorily for radiation dosimetry purposes. The first reported use of the "radiothermoluminescence" of natural calcium fluoride occurred in 1903.

This phosphor exhibits three principal peaks in the glow curve that occur in the temperature ranges 70–100°C, 150–190°C, and 250–300°C. The material has shown serious fading characteristics as a function of storage time, most likely due to the low-temperature glow peak. The response of calcium fluoride as a function of gamma-ray exposure is linear from a few mR to ~500 R with a SD of ±2%. Enclosing the material in a metal filter (e.g., lead) can make the dosimeter response constant, within ±20–30%, over the gamma-ray energy range of 80 keV to 1.2 MeV.

The response of the natural phosphor to fast neutrons is negligible. However, the response to thermal neutrons is about the same as for gamma rays, per rem in tissue. Some investigators have studied the use of calcium fluoride as a dosimeter for mixed-radiation fields of thermal neutrons and gamma rays. However, for routine personnel monitoring, this technique is not used.

Synthetic calcium fluoride materials are available commercially. One of these phosphors is activated with manganese and shows only a single glow peak located at ~260°C. In some cases, this phosphor has exhibited a spurious luminescence, but proper techniques and some innovative dosimeter devices have reduced the significance of this spurious effect.

CaF_2:Mn has a density of 3.18 g/cm^3 and an effective atomic number (Z_{eff}) of 16.3. The efficiency of the material relative to LiF (at ^{60}Co energies) is 10 and the energy response at 30 keV relative ^{60}Co is ~13. The dosimeter can be used to measure over the exposure range 100 μR to 300,000 R. Fading is quoted as being ~10% in the first 24 hr and ~15% (total) in the first 2 weeks. Thermal-neutron sensitivity of the phosphor has been reported by many investigators. For the standard thermal-neutron fluence of 10^{10} n/cm^2, the equivalent gamma-ray exposure ranges from 0.1 to 0.6 R.

Calcium fluoride doped with dysprosium is one of the most popular TL materials in use today. This material has a complex glow curve that is composed of six peaks. The commercial source of this material in the United States lists 180°C as the temperature of the main glow. As with the manganese-doped material, this phosphor has a density of 3.18 g/cm^3 and an effective Z of 16.3. Efficiency of the phosphor relative to LiF at ^{60}Co energies is 30 and the response ratio (30 keV to ^{60}Co energies) is ~12.5. The material is usable over the exposure range 10 μR to 1,000,000 R and fading is reported as 10% during the first 24 hr and 16% (total) in the first 2 weeks. Usually, a postirradiation, preevaluation anneal at 100°C for 20 min will stabilize the material and eliminate further fading. However, fading must be considered when the dosimeter is used in long-term environmental-monitoring applications.

Thermal-neutron sensitivity of the commercially available phosphor has been reported to be in the range 0.5–0.7 R, equivalent gamma-ray exposure for a thermal-neutron fluence of 10^{10} n/cm^2.

As with calcium sulfate, the calcium fluoride materials are extremely light-sensitive, and users are cautioned about handling, use, and storage of the materials.

c. Lithium fluoride. Lithium fluoride was first studied as a TLD material in about 1950. It was studied in the form of pellets of pressed LiF powder because this form solved many of the problems associated with handling the powder. However, these studies were abandoned soon after they began, primarily due to problems encountered with the material. These difficulties were caused principally by a low-temperature glow peak at ~120°C and the inability of standard techniques to eliminate the problem. Later, Cameron studied the crystals of lithium fluoride from which the pellets had been made. Working closely with the Harshaw Chemical Co. (in Solon, Ohio), he was able to develop the TL grade lithium fluoride material activated with magnesium and titanium, called TLD-100.

LiF has received a great deal of attention due to its many advantageous characteristics. A survey of the literature on TLD investigations made several years ago indicated that approximately half of all the publications on TLD were concerned with LiF. The characteristics of LiF (primarily TLD-100) that sparked this great interest are

1. A nearly constant ("flat") energy response per unit exposure over a wide range of energies. In the previous discussion it was indicated that significant overresponses were observed for most TLD materials in the 30- to 40-keV energy range. For LiF, the response at 30 keV is only ~25% higher than that at ^{60}Co energies. In some cases, this overresponse can be reduced through the use of a suitable energy compensation shield.

2. An effective atomic number much closer to that of tissue. LiF has a density of 2.64 g/cm^3 and an effective Z of 8.2. Tissue is normally assumed to have an effective Z in the range of 7.4–7.6.

3. The light emitted by the main glow peak (190°C) shows little fading with storage time at room temperature. Cameron has reported the value of 5% per year but other investigators have reported fading of up to 15% within the first 3 weeks. Harshaw Chemical Co. recommends a postirradiation. preevaluation anneal of 100°C for ~10 min to stabilize the phosphor. If this treatment is used, the fading is reported to be negligible.

In addition to the preceding benefits, the phosphor is usable over a wide range of exposures, typically tens of mR up to ~300,000 R. However, these data are those quoted by the vendor, and others have reported a usable range extending only to ~700 R with saturation occurring before 100,000 R is reached. The lower limit of usefulness depends on controlling spurious effects (tribothermoluminescence, perhaps). Typically, TLD readers flow dry nitrogen gas into the heating/reading volume to reduce these effects. Under these conditions LiF can be used to measure exposures in the 10-mR range.

In addition to TLD-100, two other LiF phosphors are available for measuring exposures to neutrons. Of the three phosphors, TLD-100 has the natural, isotopic abundance of the lithium isotopes, ^{6}Li and ^{7}Li. The phosphor TLD-600 is highly enriched in ^{6}Li, and TLD-700 is made essentially from pure ^{7}Li.

The thermal-neutron cross section of ^{7}Li for the most likely reaction is only 0.033 b. This cross section is considered negligible when compared with the 945-b cross section for the (n, α) reaction in ^{6}Li. Thus, TLD-700 has

essentially no response to thermal neutrons when it is compared with either TLD-600 or TLD-100. Either of these materials can be used in conjunction with TLD-700 for measurements in mixed thermal-neutron–gamma-ray fields. For example, TLD-600 will respond to both the thermal-neutron and the gamma-ray exposure whereas TLD-700 will respond only to the gamma-ray component of the radiation field. Subtraction of the dose indicated by the TLD-700 detector from the dose indicated by the TLD-600 detector gives the dose due to thermal neutrons.

Care must be exercised in the calibration and use of these detectors for measurements in mixed-radiation fields. For example, nine different values have been reported for the thermal-neutron sensitivity of the Harshaw TLD-100. Assuming an exposure to the standard thermal-neutron fluence of 10^{10} n/cm^2, the equivalent gamma-ray response ranged from 65 to 535 R. For TLD-600 the equivalent response for four measurements ranged from 870 to 2190 R. Even though TLD-700 contains only a trace amount of ^{6}Li, it is incorrect to assume the phosphor has no thermal-neutron sensitivity. Six values of an equivalent response have been reported in the literature, ranging from 0.7 to 2.5 R. However, it should be obvious that the wide difference in thermal-neutron sensitivity between TLD-600 and TLD-700 (when combined with the appropriate calibration) makes the technique useful for most monitoring situations.

d. Lithium borate. Another TL phosphor that has received much attention lately is lithium borate (or more properly lithium tetraborate). Usually manganese is added as the impurity to this material, but one commercial dosimetry system uses lithium borate doped with copper. Characteristics of the manganese system are described first because this element appears to be the most promising of the lithium-based phosphors. In addition, this phosphor exhibits many of the "ideal" dosimeter characteristics listed at the beginning of this section.

Lithium borate (TLD-800) has many characteristics that make it attractive as a radiation dosimeter. It is essentially tissue-equivalent (effective Z of 7.4) even though its density is ~2.4 g/cm^3. The phosphor compares favorably with the "standard" LiF. The low-temperature peak in the lithium borate curve decays very rapidly (these data were obtained 10 min after exposure), but the main glow peak at 200°C is relatively stable. Harshaw Chemical Co. quotes the fading at <5% in 3 months. Some researchers have reported that the main peak actually consists of more than one component and the prominent peak depends on the magnitude of the exposure. For example, for exposures of <1000 R, the peak appears at 180°C, whereas for exposures >1000 R (extending to ~1,000,000 R), the main peak appears at 210°C. This point is not particularly important when an individual is considering exposures in the personnel-monitoring range. However, it serves to illustrate that information provided in sales literature may be a condensation of typical characteristics, and actual performance of a dosimetry system depends on the application. In addition, these results point out the necessity that those responsible for personnel monitoring using TLD badges be familiar with the characteristics of the materials used in the system.

The literature indicates that lithium borate ($Li_2B_4O_7$) is not as sensitive to gamma-ray exposure as are other TLD materials. For example, the efficiency relative to LiF at ^{60}Co energies is only ~0.15. This limits the lower useful exposure range to ~50 mR. However, this reduced sensitivity allows use of the dosimeter for exposures >100,000 R. This lack of sensitivity is actually due to the design of the TLD readers and not entirely to the gamma-ray sensitivity of the lithium borate. The photomultiplier tubes used in standard readers are sensitive to light over a range best suited for LiF (i.e., 3500–6000 Å). The emission spectrum of $Li_2B_4O_7$ is in the range 5300–6300 Å, where the standard photomultiplier tubes have reduced sensitivity.

Characteristics of the material are highly dependent on the concentration of the manganese impurity. Data on lithium borate indicate that the optimum manganese concentration is ~0.4%. For this concentration, the response per unit exposure is essentially energy-independent. Actually the response ratio 30 keV to ^{60}Co energies is 0.9 for a concentration of 0.4% manganese impurity. But this dosimeter has response characteristics that closely approximate the "ideal."

Lithium borate has some sensitivity to thermal neutrons. Again, if the material is exposed to the standard thermal-neutron fluence of 10^{10} n/cm^2, the equivalent gamma-ray response is in the range 230–390 R. This response is approximately the same as that of TLD-100. These data are for the material available commercially from Harshaw Chemical Co. Other investigators have reported equivalent gamma-ray responses ranging from 300 to 670 R.

e. Aluminum oxide.

Aluminum oxide (Al_2O_3) has long been investigated for use as a TLD material for use in personnel dosimetry and for environmental measurements. Daniels' early investigations led him to believe that this material displayed more favorable TL characteristics than those of LiF. However, extensive research on the applicability of Al_2O_3 as a TLD was curtailed in 1960 when Cameron and the Harshaw Chemical Co. developed LiF TLD-100. Recently, an Al_2O_3 dosimeter has been introduced uses a carbon dopant (i.e., Al_2O_3 : C). These dosimeters exhibit an extremely high sensitivity to gamma radiation (about 60 times that of LiF TLD-100), which

makes them useful for monitoring low-radiation exposures in the environment and to workers in controlled areas.

Aluminum oxide exhibits a sensitivity comparable to that of calcium sulfate doped with manganese ($CaSO_4$: Mn) but has many more desirable characteristics than the other high-sensitivity materials. Al_2O_3 : C has a Z_{eff} of 10.2 and exhibits a relatively flat energy response from 150 keV to 1.5 MeV. Data on the energy response at 30 keV relative to that at ^{60}Co energies has been determined to be in the range of 3. The relation of indicated exposure to actual exposure is linear from about 0.05 mrad to 100 rad and fading is quoted at 3% per year (under suitable conditions).

This phosphor has three glow peaks that occur at approximately 100°C, 185°C, and 250°C. The main glow peak at 185°C contains more than 99.5% of the TL signal emitted. The low temperature peak is extremely unstable and is not discernible 15 min after exposure to ionizing radiation. Nevertheless, a pre-read anneal at 100°C is recommended to ensure the removal of this low-temperature peak. The recommended readout cycle consists of raising the phosphor temperature to a maximum of 270°C at a rate of 10°C per second. As with other high-sensitivity phosphors, this material is extremely light-sensitive. Exposure to sunlight or normal laboratory light will cause the phosphor to lose almost 100% of the stored TL signal in less than 24 hr.

7. Beta Sensitivity of TLD Materials

All TLD materials in common use are sensitive to beta radiation. Theoretically, it would be possible to use TLDs for routine beta dosimetry. However, accurate assessment of beta-radiation absorbed dose is difficult to achieve with the personnel-monitoring devices currently available. Most of the conventional dosimeters were designed to detect the penetrating component of the radiation field and estimates of the nonpenetrating component radiation field and estimates of the nonpenetrating component are often obtained through the use of algorithms derived from calibrations in standard radiation fields.

The inability to perform accurate beta dosimetry can be attributed to a number of factors. These include the spectral energy distribution of the radiation, the low penetrating nature of the radiation, the wide energy range of beta-emitters encountered in the work environment, the influence of backscatter and attenuation in the badge components, and the lack of suitable calibration sources and techniques. An additional and very important factor is the steeply sloped energy-response curve exhibited by most TLD materials. For example, if the relative response per unit exposure of LiF TLD-100 to the 2.2-MeV beta particles from ^{90}Y is assumed to be 1.0, the response is ~0.2 for beta radiation from ^{204}Tl (0.76 MeV) and ^{90}Sr (0.55 MeV). The decrease in response continues, reaching ~0.08 for ^{99}Tc (0.29 MeV) and ~0.04 for ^{35}S (0.17 MeV).

Within the last few years a number of personnel-monitoring badges have been designed that are intended to measure the beta component of the radiation field more accurately. These dosimeters are usually multielement (i.e., four or more TLDs) and require fairly sophisticated algorithms to obtain an estimate for the beta dose. Many attempts to produce an ultrathin TLD have been reported. Often, a TLD powder is mixed with a polyethylene base so that a thin but flexible dosimeter can be produced. At least one commercial badge has a thin (0.015-in.-thick) TLD incorporated in it strictly as a beta dosimeter. However, enclosing the dosimeter in some sort of badge or holder usually defeats the purpose of the beta dosimeter.

At this point, the development of TLD dosimeters specifically for use in beta-radiation fields has not progressed past these techniques described. Other, more sophisticated techniques are under study. These include the implantation of materials such as carbon into the crystal to alter its response and new reading techniques using lasers. However, it can be concluded that the errors associated with personnel beta-radiation monitoring may be quite large and improvements urgently needed.

8. Fast-Neutron Dosimetry with TLDs

As indicated above, neutron-radiation fields in and around nuclear power reactors are typically composed of neutrons with energies <500 keV. Therefore, detailed consideration of dosimetry for fast neutrons is not necessary since this is not a problem in light-water–moderated reactors. However, this discussion on a typical system is presented to illustrate the use of TLDs for fast-neutron dosimetry. The system chosen is the Hoy thermoluminescence neutron dosimeter (TLND). The TLND is an albedo-type detector using TLD-600 and TLD-700 chips embedded in a hemisphere of polyethylene.

Basically, an albedo dosimeter is designed to measure the fluence of thermal neutrons that escape from the body when an individual is exposed to fast neutrons. Fast neutrons incident on the body are moderated and scattered by the body and many escape as thermal neutrons. Detection of these thermal neutrons with a carefully calibrated dosimeter will provide a reasonable estimate of the incident fast-neutron dose (or dose equivalent). Typical albedo dosimeters use various combinations of TLDs, primarily TLD-600 and TLD-700 (see earlier discussion of lithium fluoride TLDs).

If the system is to function as an effective albedo-type dosimeter, the incident thermal-neutron component of the radiation field must be removed. This is normally accomplished by placing a material that captures thermal neutrons over the TLDs. Many different designs have been reported that use materials such as cadmium, polyethylene, and boron-loaded plastic.

The TLND badge was designed specifically to respond to albedo neutrons. Under most conditions, this component of the radiation field can be related to total fast-neutron exposure (badges are typically calibrated to give a neutron–dose-equivalent response). Albedo neutrons and associated gamma rays are detected by a pair of TLDs (TLD-600 and TLD-700) placed at the center of a moderating hemisphere of polyethylene (~2 in. in diameter). The hemisphere is covered by a 0.03-in.-thick cadmium dome to shield against incident thermal neutrons. Another pair of TLD chips, in a small compartment near the top of the dome, detects a portion of the incident radiation and provides a correction factor for overresponse of the dosimeter when it is exposed to low-energy neutron spectra. The badge components are enclosed in a protective stainless steel case in the shape of a 2-in.-diameter hemisphere. A belt is provided to ensure that the dosimeter is in close contact with the body (dosimeter response depends on this fact).

The Holy TLND is probably one of the largest of the albedo dosimeters and, because of its size and weight, is worn on a belt at the waist. However, because of the large amount of polyethylene used in the badge, it is one of the most sensitive albedo-neutron dosimeters in use. The dosimeter has exhibited characteristics far superior to those of nuclear emulsions for routine personnel-monitoring purposes. It can be used to measure in the dose-equivalent range 10 mrem to 50,000 rem with no observed rate dependence. In addition, the dosimeter is not affected by exposure to gamma radiation except when used to measure very low neutron doses.

The Hoy dosimeter has been studied extensively in a number of neutron-exposure situations. These neutron sources include plutonium sources thought to be representative of the production material at the Savannah River Plant (Aiken, South Carolina), several configurations using a ^{252}Cf source and moderators, a number of reactors, and accelerator-produced monoenergetic neutrons. For sources other than the accelerator-produced neutrons, and in situations in which the neutron spectrum is known, the neutron dose-equivalent indicated by the TLND is within a factor of 2 of the actual neutron dose equivalent. More specifically, for the majority of exposure situations, the indicated neutron dose equivalent was within 50% of the actual neutron dose equivalent.

Data obtained with monoenergetic neutron sources indicate that the dosimeter exhibits a marked under response for neutrons in the energy range 1–10 MeV. However, the dosimeter response is relatively constant over this energy range at ~30% of the actual dose equivalent. At levels <1 MeV, there is a marked overresponse with decreasing neutron energy. This overresponse reaches a maximum of ~70% greater than the actual dose equivalent for neutrons of ~300 keV. These data indicate that scattered neutrons will produce an overresponse in the dosimeter that could lead to a significant overestimate of the neutron dose equivalent.

V. DOSIMETRIC TECHNIQUES, INTERNAL

A. Basis for Internal Dosimetry

The term *internal dosimetry* has always held a certain mystery about it that has confused and confounded health physicists for a long time. It is an unfortunate term that, historically, was intended to serve in contrast to the term *external dosimetry*. Generally, the term *external dosimetry* simply means the measurement of radiation exposure due to sources located outside the body. These sources of radiation, which are usually located in well-defined positions in an area, have the ability to penetrate into the body, depositing energy and potentially causing harm to the person being irradiated.

In general, the health physicist and radiation workers are more comfortable dealing with external sources of radiation exposure. If work is to be done in an area, the health physicist can make a radiation survey with portable instruments and obtain an estimate of the anticipated exposure from this survey. "Hot spots" or other areas with high-exposure rates can be identified; areas in which the dose rates are low can also be identified where workers can wait when not needed in a particular operation, and "stay times" for the work can be calculated, if necessary.

Most workers have become comfortable working in controlled areas where the radiation presents only an external hazard. In addition, workers are aware of the usual methods of controlling exposure to external radiation sources by using *time, distance*, and *shielding*, and the work can be performed with these methods in mind. These techniques are taught in all general employee-training courses for radiation workers. Use of stay times, low–dose-rate waiting areas, and temporary shielding are examples of how these techniques are practiced in keeping all exposures as low as is reasonably achievable (ALARA).

For some high-risk work, it may be necessary for the health physicist responsible for monitoring the work to control, as closely as possible, the exposure of the workers. This is usually accomplished through continuous monitoring of the work as it progresses. Nevertheless, the point remains the same; radiation workers enter radiation and high-radiation areas hundreds of times per day and usually give little thought to this type of exposure.

All workers entering the radiation area are required to wear personnel-monitoring devices, which give accurate estimates of the doses received. Many dosimeters can be used to monitor the radiation where the field varies significantly over the total body or to monitor certain parts of the body (e.g., the lens or the gonads). Some of these dosimeters can be evaluated almost immediately after the workers exit the area to provide estimates of the whole-body dose and the dose to other important parts of the body, for example, the extremities or the lens of the eye. These data can be used to evaluate the effectiveness of the radiation survey, control measures taken to keep exposures ALARA, and future job assignments of the individual workers.

If the potential for internal exposure is present, then the situation facing the health physicist, and the attitudes of the workers, is entirely different. It is still possible to perform prework surveys of the area to determine the potential hazard. However, these surveys are a different type. In this case, the health physicist must make measurements of airborne radioactivity concentrations. This requires drawing a known amount of air through a filter (or other collection device) and analyzing the sample to determine total ("gross") activity or to determine the radionuclides in the sample and their individual activities. In addition, surface-contamination surveys might be made to determine the potential for resuspension of other radioactive materials into the air. Although the work environment can be defined with some degree of confidence, exposure of the workers to radioactive materials that may be deposited internally is not so easy to predict. These measures of the potential internal exposure hazard cannot be correlated directly with the exposure the workers may receive.

Once the work has begun, use of the control methods—time, distance, and shielding—does not play an important role in preventing an internal exposure (although limiting the time in the area certainly reduces the probability of an internal exposure). In addition, if an internal exposure should occur, then the radiation protection staff and the individual worker have little control over the time the material remains in the body; distance is no longer a protective technique and neither is the use of temporary shielding. The radioactive material is inside the body, where almost all the radiation energy emitted in the decay of the radionuclide will be absorbed in tissues of the body.

Monitoring during the work is difficult and not very effective in predicting the accumulated exposure as the work proceeds. Local or area air-monitoring systems do not give a true indication of the concentrations of radioactive material in the breathing zones of individual workers. Personal air samplers may be used to obtain breathing zone samples, but low flow rates, high-failure rates, and worker acceptance can be a problem. Also, care must be taken to ensure that these samplers do not interfere with the worker so that the actual exposure period (to external or internal radiation sources) is extended.

At the completion of the work, no direct measurement techniques are available to evaluate the exposure. Samples of mucus from the nasal passages may be taken (either through nasal swabs or blowing of the nose) to determine if radioactive material has deposited in the upper nasal passages. However, these samples must be obtained soon after exposure as the effectiveness of such samples in indicating an internal exposure is limited to a very short period after exposure (~15–30 min). Bioassay is the only other method available to evaluate an internal exposure. This term is used to include both the measurement of radioactive material in the body and the measurement of radioactive material excreted from the body. Bioassay is clearly an "after the fact" evaluation technique.

Therefore, the most effective method of controlling exposure to internally deposited radioactive material is to prevent the exposure. There are many methods to accomplish this goal and all of them are in use at most nuclear facilities. Simple restrictions such as controlling smoking, eating, and drinking in many areas can prevent the inadvertent intake of radioactive materials. On a larger scale, the first line of defense is containment of the radioactive material so that it cannot become airborne. Good housekeeping plays an important, but often overlooked, role in preventing material from becoming airborne and keeping exposures to such material ALARA. Engineering controls, that is, the design of equipment to move, exchange, filter, and clean air, are effective in keeping airborne concentrations low in most areas. When these methods are not effective, respiratory protective devices and protective clothing are used to prevent (or limit) exposure to airborne radioactive material.

If an internal exposure should occur, then internal dosimetry (the historical name) is not really the process that is followed. The term *dosimetry* literally means "dose measurement" and this is *not* possible when the radioactive material is inside the body. In keeping with traditional usage, the term *internal dosimetry* is used here, but it is defined as a process of measurement and calculation that results in an estimate of the dose-equivalent to tissues of the body due to the intake of radioactive material. The term *measurement* applies to bioassay techniques in which the quantity (activity) of radioactive material in the body is measured by using very sensitive radiation detectors located outside the body (called direct bioassay). It also applies to the measurement of the concentration of radioactive materials excereted from the body, usually in the urine and feces (called indirect bioassay). These

data are combined with a mathematical model that has been derived to explain the uptake, deposition, movement, metabolism, retention, and excretion of the particular element in the human body. This combination results in a series of "calculations" that produce "estimates" of the dose equivalent (over a specific period of time) to certain organs of the body from this intake of radioactive material.

From this short discussion it should be clear that internal dosimetry really means internal dose assessment. The sections that follow show just how these assessments are performed. The discussions that follow begin with the "old techniques" for internal exposure assessment (in use in the United States until January 1, 1994). These are presented in some detail in order to preserve the recent past. In addition, an understanding of the previous approach provides some insight into the more recent formulations for internal exposure assessment. This discussion is followed by a more detailed discussion of the current techniques used in the United States. Finally, a short discussion of the more recent recommendations of the ICRP is presented.

1. Internal Dose Control

The current concept of controlling internal exposure to radioactive materials in most nuclear facilities is based on limiting the concentrations of these materials in air and attempting to limit oral intake. In earlier recommendations, radionuclides were controlled by establishing maximum allowable concentrations in both air and water.

Inhalation and ingestion of radioactive material are considered to be the most likely pathways of entry into the body in the work situation. Usually, no consideration is given in the internal dosimetry regulations to accidents such as intakes through wounds (injection) or absorption of radioactive material through the intact skin. The development of this systematic set of concentration limits is based on a four-step process that forms the basis for the establishment of all radiation protection standards in current use. These steps are

1. Establish limits, which should not be exceeded, for radiation exposure, based on a careful review of available biological data.

2. Calculate the maximum allowable amount of each radionuclide (and its daughters) that can be in the body without exceeding the dose limits established in step 1.

3. Establish possible routes of entry into the body for each element (or radionuclide) and derive an allowable intake rate that will satisfy both steps 1 and 2.

4. On the basis of the physiological parameters established for the routes of entry (e.g., inhalation and ingestion), calculate the allowable concentrations of the

radionuclide in air or water that will satisfy steps 1, 2, and 3.

These steps have resulted in a terminology well-known in health physics, but the logical basis for the establishment of these terms has been lost. Recommendations of the ICRP and the NCRP published in 1959 gave each of these logical steps a name. The process taking place in step 1 is the establishment of the maximum permissible dose equivalent. In step 2, the maximum permissible body burden is calculated based on the maximum permissible dose equivalent, and in step 3 the maximum allowable intake has been established. The final step, and the one used daily to control exposure to radioactive materials for many years, is the establishment of the maximum permissible concentration values for a particular radionuclide. These four simple steps form the basis for all internal exposure recommendations and regulations in use throughout the world. More recently, the ICRP introduced a new approach to the assessment of internal exposure. However, although the terms have been changed (see Section V.B.2), the four basic steps presented above remain unchanged.

2. Regulatory Requirements

Until January 1, 1994, U.S. guidance and regulations related to external radiation exposure and the control of the internally deposited radionuclides were based on recommendations of the ICRP and the NCRP published between 1953 and 1969. Other publications on specific groups of radionuclides, for example, the alkaline earths and the actinides, were issued, but these had little impact on the federal regulations. The ICRP published more recent recommendations on a risk-based approach to radiation protection (in 1977), and completely revised the internal dosimetry scheme (1979), but these approaches took a long time to find their way into the federal regulations and radiation protection practice in the United States. Subsequently, the ICRP and the NCRP have published additional recommendations on radiation-exposure limits and the approaches to radiation protection. These newer recommendations have not yet been incorporated into federal guidance. Nevertheless, any consideration of internal dose assessment must begin with a discussion of the old ICRP and NCRP formulations, published originally in 1959. The following material establishes the foundation for these considerations.

3. Basic Definitions

Several definitions and a discussion of the concepts and assumptions embodied in the internal dosimetry recommendations are necessary before proceeding. These are presented below (definitions related to the newer publications of the ICRP are given in Section V.B.2. The following definitions are from ICRP Publication 2 (ICRP-2).

The *maximum permissible dose equivalent (MPDE)* for an individual is that dose, accumulated over a long period of time or resulting from a single exposure, which (in light of present knowledge) carries a negligible probability of severe somatic or genetic injuries. The ICRP expands this definition by stating, "Furthermore, it is such a dose that any effects that ensue more frequently are limited to those of a minor nature that would not be considered unacceptable by the exposed individual and by competent medical authorities." Even though the emphasis has been, and is today, placed on keeping all exposures as low as achievable, this definition of the MPDE has formed the foundation of the internal dosimetry guidance for many years.

The *critical body organ* is that organ of the body whose damage by radiation results in the greatest damage to the body. It should be noted that in most discussions, the ICRP simply calls this the critical organ. Actually, a number of factors must be considered in designating the critical organ(s) for use in internal dose calculations. The following are included in the selection criteria: (a) the organ that accumulates the greatest concentration of radioactive material; (b) the importance of the organ to the well-being of the entire body; (c) the radiosensitivity of the organ (i.e., the organ damaged by the lowest dose); and (d) the organ damaged by the route of entry of the radionuclide into the body. This consideration is particularly important for highly insoluble materials that have low transfer coefficients across barriers such as the lung and the lining of the gastrointestinal tract. In these cases, the *only* organ that may be damaged is the organ in which the material is contained or through which the material passes as it enters and exits the body.

Even though all these factors should play an important role in the selection of the critical organ, the primary factor usually is the organ that accumulates the greatest concentration of material, that is, criterion (a). But, the ICRP has not left the task of selecting the critical organ to the individual. In most cases, the ICRP has selected one or more critical organs for each element and, in some cases, up to 14 organs have been selected. In all cases, one of the critical organs selected is the total body. This critical organ is very useful when the health physicist is dealing with mixtures of radionuclides.

The *body burden* of a particular radionuclide in an individual is the activity of the radionuclide present in the individual's body at a particular time.

The *maximum permissible body burden (MPBB)* is the activity of a particular radionuclide that delivers an MPDE to the whole body or to one or more organs of the body. The MPBB is computed on the assumption that the particular radionuclide is the only radionuclide in the body. According to the definitions, an exposure less than the MPBB is not thought to produce an observable biological change.

The *maximum permissible concentration (MPC)* for any radionuclide is that concentration of material, in air or water, for which continuous exposure may occur without exceeding the MPDE. Actually, the ICRP recommendations include two exposure situations. MPC values are given for a normal 40-hr work week (called occupational exposure) and for a 168-hr week (called continuous occupational exposure).

Publication 10 of the ICRP, issued in 1968, gave several additional definitions that were intended to clarify some of the problems with the understanding of ICRP-2. *Intake* is defined as the amount entering the body by nose or mouth; *uptake* as the amount absorbed into extracellular fluid; *deposition* as the amount present in the organ of reference; and *transportable* as the property of a radionuclide-containing compound that results in its ready transfer across body membranes.

For the purposes of this discussion, it is convenient to substitute the word *blood* for the term *extracellular fluid*. Although the term *blood* is not completely correct, it is simpler to think of the blood as the transport mechanism for radioactive material in the body; that is, the blood carries (transports) the material to the organs of the body in which it will deposit. The definition for *transportable* was suggested by the ICRP as a substitute for the earlier term *soluble*, which was used widely in ICRP-2. In this latter publication, materials were classed as soluble and insoluble. In this discussion these terms are used in keeping with the MPC values specified in the early ICRP and NCRP recommendations.

4. Internal Dosimetry Scheme

The internal dosimetry scheme given in Fig. 6 is essentially universal in its utility in explaining the factors that must be considered in establishing limits and protecting workers from exposure to internally deposited radionuclides. The primary concern is the absorbed dose (or dose equivalent) to organs or tissues of the body due to the intake of a radionuclide. This statement is true whether concentration limits are being derived or an actual exposure is being evaluated. The dose is normally obtained through a series of calculations, which rely on a large number of factors. These factors, shown at the bottom of the figure, include the dosimetric concepts discussed above, the radiological parameters associated with the particular radionuclide, and anatomical and physiological data necessary to describe the intake and deposition of the radionuclide in the body. These data must be coupled with dosimetric models and calculational techniques to derive an estimate of the dose from an intake of a particular radionuclide. The dosimetric models are specific for a particular element and describe mathematically the metabolism, retention, and excretion of the element.

It is important that this scheme be kept in mind in the sections that follow. In addition, there is one other simple fact that must be remembered for an individual to gain full appreciation for internal dosimetry. To determine the absorbed dose from the intake of a radionuclide, the health physicist must determine only two things: (1) the number

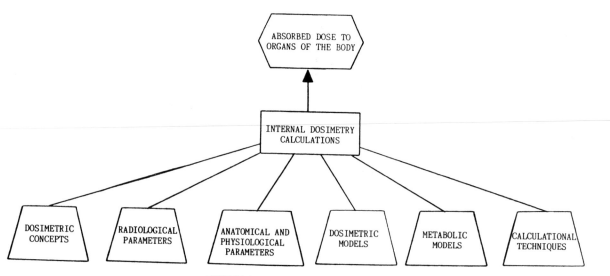

FIGURE 6 Internal dosimetry scheme.

of disintegrations of the radionuclide that take place in the organ over the time period of interest and (2) the energy per unit mass (i.e., per gram) deposited in the organ per disintegration of the radionuclide. Other factors, such as the names of quantities and their units and the quality factors chosen, may change, but this fundamental fact remains unaltered by the recommendations and regulations of national or international bodies.

a. Dosimetric concepts. The dosimetric concepts necessary for an internal dose calculation under the old system are those defined in Section V.A.3. These include the MPDE, MPBB, MPC, critical organ, and many others. These constitute the language used in performing these calculations. As will be seen, changes in this language do not cause any significant change in the techniques used in the calculation.

b. Radiological parameters. The radiological parameters are those that are associated with the radionuclide itself. The ICRP formulation assumes that only one radionuclide enters the body and that, at the time of intake, the radionuclide is "pure." That is, the assumption is always made that the radionuclide contains no radioactive daughter products at the time of intake. However, as we will see, the calculation includes the daughter(s) in an interesting way. The term *radiological parameters* includes the half-life of the radionuclide, types of radiation emitted in the decay, energies of these radiations, and frequency at which each of the radiations is emitted in the decay of the radionuclide. Other parameters include the quality factors associated with the radiations and the properties of any daughter radionuclides. If the radionuclide has a radioactive daughter(s), then the same parameters must be determined for each daughter in the decay chain. All of this information is found in a "decay scheme" for the radionuclide.

c. Anatomical and physiological data. Early in the history of internal dose calculations it was recognized that a need existed for a standard set of biological parameters that would be accepted for calculating permissible levels for radioactive materials. The first set of data was presented at a conference at Chalk River. Canada, in 1949. The data adopted at this conference included values for the mass of certain organs and data on the chemical composition of selected organs and of the total body. In addition, data on patterns of intake and excretion, water balance, respiration rates, and retention of particulate matter in the lung were accepted. The data were called Standard Man and were intended to represent the average occupationally exposed radiation worker. The Standard Man values were modified several times and additional data were included;

TABLE VI Organs of Standard Man Including Mass and the Effective Radius[a]

Organ or tissue	Mass, m (g)	% of total body	Effective radius, X (cm)
Total body[b]	70,000	100.0	30.0
Muscle	30,000	43.0	30.0
Skin and subcutaneous tissue[c]	6,100	8.7	0.1
Fat	10,000	14.0	20.0
Skeleton			
Without bone marrow	7,000	10.0	5.0
Red bone marrow	1,500	2.1	—
Yellow bone marrow	1,500	2.1	—
Blood	5,400	7.7	—
Gastrointestinal tract[b]	2,000	2.9	30.0
Contents of the gastrointestinal tract			
Lower large intestine	150	—	5.0
Stomach	250	—	10.0
Small intestine	1,100	—	30.0
Upper large intestine	135	—	5.0
Liver	1,700	2.4	10.0
Brain	1,500	2.1	15.0
Lungs (2)	1,000	1.4	10.0
Lymphoid tissue	700	1.0	—
Kidneys (2)	300	0.43	7.0
Heart	300	0.43	7.0
Spleen	150	0.21	7.0
Urinary bladder	150	0.21	—
Pancreas	70	0.10	5.0
Testes (2)	40	0.057	3.0
Thyroid gland	20	0.029	3.0
Thymus	10	0.014	—
Ovaries (2)	8	0.011	3.0

[a] Adapted from International Commission on Radiological Protection (1959). "Report of Committee II on Permissible Dose for Internal Radiation," ICRP Publication 2, Pergamon, Oxford.
[b] Does not include the contents of the gastrointestinal tract.
[c] The mass of the skin alone is taken to be 2000 g.

for example, the rate of passage and the mass of the contents of the gastrointestinal tract were added in 1953. Data on Standard Man form the basis for the recommendations contained in ICRP-2. The mass and effective radius of organs in Standard Man are given in Table VI. Intake and excretion parameters for water and an "air balance" are shown in Table VII. These data are taken directly from ICRP Publication 2.

d. Dosimetric models. The primary models used in dose calculations are those associated with the lung and the gastrointestinal tract. Other models include situations such

TABLE VII Intake and Excretion of Standard Man[a]

Water balance			
Intake (cm³/d)		Excretion (cm³/d)	
Food	1000	Urine	1400
Fluids	1200	Sweat	600
Oxidation	300	From lungs	300
		Feces	200
Total	2500	Total	2500

Air balance			
	Oxygen (vol %)	Carbon dioxide (vol %)	Nitrogen + others (vol %)
Inspired air	20.94	0.03	79.03
Expired air	16.0	4.0	80.0
Alveolar air (inspired)	15.0	5.6	—
Alveolar air (expired)	14.0	6.0	—

Vital capacity of lungs	3–4 L (men)
	2–3 L (women)
Air inhaled during 8-hr workday	$1.0\,E+7\ cm^3/d$
Air inhaled during 16-hr workday	$1.0\,E+7\ cm^3/d$
Total	$2.0\,E+7\ cm^3/d$
Interchange area of lungs	50 m²
Area of upper respiratory tract, trachea, bronchi	20 m²
Total surface area of respiratory tract	70 m²
Total water in body	$4.3\,E+4\ g$
Average life span of man	70 yr
Occupational exposure time of humans	8 hr/d; 40 hr/week; 50 weeks/yr; 50 yr total time

[a] Adapted from International Commission on Radiological Protection (1959). "Report of Committee II on Permissible Dose for Internal Radiation," ICRP Publication 2, Pergamon, Oxford.

as exposure in a cloud of a radioactive material or special exposure situations such as the deposition of a radionuclide in the skeleton. The distribution of inhaled material in Standard Man is shown in Table VIII and the model of the gastrointestinal tract is summarized in Table IX. These models will form the basis for the old ICRP internal dose calculations.

e. Metabolic models. Each element that enters the body is transported, deposited, retained, and excreted in a manner that can be described by a set of mathematical equations. In the current ICRP scheme these equations generally have the form of an exponential equation. That is, the retention and excretion of an element in the body can be "modeled" by using equations that have a similar form to those very familiar equations used in radioactive decay.

f. Calculation techniques. The last entry in this internal dosimetry scheme is included to serve as a reminder that a series of calculations must be made in any internal dose assessment. In the old ICRP recommendations these calculations were relatively simple and did not require a computer. Howerver, any discussion of the newer internal dosimetry techniques shows that the methods are more sophisticated, and for some of the calculations very complex computer codes are required.

B. Generally Accepted Techniques

1. The ICRP Publication 2 Formulation (A Review)

The techniques of the ICRP-2 formulation and the limits derived from their use are based on a very old system proposed originally in 1959. The system establishes limits for the dose-equivalent rates to organs of the body (or

TABLE VIII Particulates in the Respiratory System of Standard Man[a]

Distribution[b]	Readily soluble compounds (%)	Other compounds[c] (%)
Exhaled	25	25
Deposited in upper respiratory, passages and subsequently swallowed	50	50
Deposited in lungs lower (respiratory passages)	25	25[d]

[a] Adapted from International Commission on Radiological Protection (1959). "Report of Committee II on Permissible Dose for Internal Radiation," ICRP Publication 2, Pergamon, Oxford.

[b] Retention of particulate matter in the lungs depends on many factors, such as the size, shape, and density of particles; the chemical form; and whether the person is a mouth-breather; however, when specific data are lacking it is assumed that the distribution is as shown here.

[c] In the tables of MPC values, these compounds are called insoluble.

[d] Of this, half is eliminated from the lungs and swallowed in the first 24 hr, which makes a total of 62.5% swallowed. The remaining 12.5% is retained in the lungs with a half-life of 120 d, it being assumed that this portion is taken into body fluids.

the total body), and allowable concentrations of radionuclides in the organ(s) are calculated based on these limits. To control exposure to a radionuclide that may be deposited in the body, the health physicist must calculate concentrations in air and water such that, assuming standard inhalation and ingestion rates, the allowable organ concentration (and, therefore, the dose-equivalent rate) is not exceeded.

This discussion focuses on the ICRP recommendations rather than the NCRP recommendations contained in Handbook 69. These recommendations are nearly identical and the differences insignificant. The adoption of NCRP recommendations by the Atomic Energy Commission, and later its successor, the NRC, in 10CFR20 was essentially an adoption of the ICRP recommendations on an internal exposure control.

TABLE IX Gastrointestinal Tract of Standard Man[a]

Portion of the gastrointestinal tract that is critical tissue	Mass of contents (g)	Time food remains (d)	Fraction from lung to tract	
			Soluble	Insoluble
Stomach (S)	250	1/24	0.50	0.625
Small intestine (SI)	1100	4/24	0.50	0.625
Upper large intestine (ULI)	135	8/24	0.50	0.625
Lower large intestine (LLI)	155	18/24	0.50	0.625

[a] Adapted from International Commission on Radiological Protection (1959). "Report of Committee II on Permissible Dose for Internal Radiation," ICRP Publication 2, Pergamon, Oxford.

a. Radiation exposure standards. Limits on radiation exposure recommended by the ICRP in Publication 2 are different from those found in the old 10CFR20. However, since these form the basis for the calculation of the MPBB, the ICRP recommendations must be discussed here. This section concentrates on the recommended limits for occupational exposure; those recommendations appropriate to the general population are mentioned briefly in this section.

i. Occupational exposure. The ICRP recommended a limit for the total body and gonands of 3 rem in any 13 consecutive weeks with the total exposure not to exceed $5(N - 18)$ rem if the exposure begins at age 18 yr. If the exposure begins before age 18, occupational exposure must not exceed 5 rem/yr and should not exceed a total of 60 rem at age 30.

The dose equivalent to the skeleton is based on a knowledge of the dose to the skelton from ^{226}Ra. The ICRP recommends that in any 13 consecutive weeks this dose should not exceed the dose due to the deposition in the skeleton of 1 μCi of ^{226}Ra. The dose-equivalent rate from this amount of ^{226}Ra is 0.56 rem/week.

The thyroid gland and the skin had a dose-equivalent limit of 8 rem in any 13 consecutive weeks, not to exceed 30 rem/yr.

For any single organ, except the gonads, skeleton, skin, and thyroid, the dose equivalent should not exceed 4 rem in any 13-week period and should not exceed 15 rem/yr.

ii. External versus internal exposure. These limits apply to both external and internal exposure. In other words, the ICRP recommended that radiation exposure to both external and internal sources conform to these limits: external and internal exposure should be added and the total dose equivalent should not exceed these limits. In addition, the following guidelines are given by the ICRP to be used if both external and internal exposure occur.

If no external exposure has occurred, then the appropriate MPC value may be used to control internal radiation exposure. If a combined exposure has occured (i.e., both external and internal), then the appropriate MPC value must be modified by the factor $(D - E)/D$, where D is the quarterly dose-equivalent limit for the organ and E is the dose equivalent due to external exposure.

If exposure to an airborne radionuclide lasted for only 1 hr, then the appropriate MPC value for occupational exposure may be increased by a factor of 40. The same applies to a radionuclide in water, but this situation is not discussed further here because this intake pathway is very unlikely in an occupational exposure situation. It should be remembered that the rules on concurrent exposure still apply, and the allowable MPC value may require modification so that permissible exposure limits are not exceeded.

In special situations, especially those in which sufficient monitoring was available, the recommendations allowed a person to work for 1 hr in an area in which the concentration in air of a radionuclide (with the total body as the critical organ) exceeded the 40-hr MPC value by a factor of 1200 (i.e., $40 \times 13 \times \frac{12}{5} = 1200$). This factor was determined on the assumption that the $5(N - 18)$ formula had not been exceeded. In addition, if such an exposure had been allowed, no additional exposure would have been permitted during the remainder of the 13-week period.

b. Assumptions, restrictions, and explanations.

Values of MPBB and MPC were based on a number of assumptions. In addition, these data had certain restrictions placed on them that influenced their interpretation and use in controlling exposure to internally deposited radionuclides. These assumptions and restrictions are presented and discussed in this section.

ICRP-2 presented MPBB and MPC data for approximately 240 radionuclides. These data were given for soluble and relatively insoluble compounds, and no consideration was given to the chemical structure of the compound. In other words, these inexact terms were used in place of the more conventional descriptions, such as oxides and hydroxides, that might have been expected when exposure to airborne radionuclides is discussed. In addition, ingestion and inhalation were considered as the only pathways into the body. Injection through wounds and absorption through the intact skin were not considered in the recommendations. In several situations, exposure by submersion in a radioactive cloud was considered, where important.

All the calculations assumed a standard worker (i.e., Standard Man); individual variations were not considered. These individual variations, in reality, could have had a significant impact on the evaluation of an exposure, but the ICRP had no other choice but to ignore them.

Below is a list of the other principal features, assumptions, and conditions placed on the MPBB and MPC calculations.

1. In all cases, values are given for both soluble and insoluble compounds of the radionuclide. Usually, a number of critical organs are listed as well as the total body.

2. The values are computed for occupational exposure at the rate of 40 hr/week, 50 weeks/yr for a continuous working period of 50 yr. In addition, a 50-yr continuous exposure period (i.e., 168 hr/week) is also considered.

3. The calculated dose rate takes into account the amount of radionuclide actually present in the critical organ (or the body) rather than an assumed equilibrium value. MPC values based on the critical organ must meet the requirement that the dose rate (rem/week) after 50 yr of occupational exposure must not exceed the limits specified by the ICRP. Because of the short effective half-lives of most radionuclides in the body, a majority will reach a state of equilibrium in the body during a 50-yr exposure period. Most radionuclides that did not satisfy this statement were assigned a biological half-life of 200 yr.

4. If the radionuclide has a radioactive daughter, the calculation assumes that only the parent radionuclide enters the body. That is, it is always assumed that the exposure is to a "pure" radionuclide. Certainly this is not a realistic assumption for most exposure situations. However, the estimated dose rate from the radionuclide does include a consideration of the energy released by daughter radionuclides formed inside the body. Two exceptions to this general statement are the radionuclides ^{220}Rn and ^{222}Rn. For these radionuclides, a state of equilibrium usually attained in air is assumed.

5. A compartment model is assumed for use in the calculations. In this model each organ is assigned a biological half-life, and elimination of the radionuclide from the organ is assumed to occur at a constant rate. In general, this model is assumed to be a single compartment described by a single exponential function. Values used in the equations are selected to produce, *in 50 yr of constant level exposure*, the radionuclide retention indicated by a more detailed model. The ICRP cautions that these models should be used with great care when exposures of shorter duration are considered because they may not accurately represent the actual situation.

For exposures >50 yr, the dose rate in the critical organ must not exceed the permissible levels set by the ICRP. However, if the radionuclide has not reached equilibrium in the organ during the 50-yr period, the dose rate will continue to increase.

6. The average breathing rate of Standard Man was 1×10^7 cm^3 per 8-hr workday. This value was assumed to be one-half the amount of air breathed in 24 h (i.e., 2×10^7 cm^3).

7. The average rate of water consumption was 1100 cm^3 per 8-hr workday. This was assumed to be one-half the water consumed in a 24-h period (i.e., 2200 cm^3)

8. The dose from submersion in a cloud of inert gas, with radiation of sufficient energy to penetrate the epidermal layer (7 mg/cm^2), results from an external exposure in the cloud and not from exposure due to the inhalation of the radioactive gas into the body. Only two critical organs are considered for submersion in a cloud of noble or inert gas. These are the skin (nonpenetrating radiation) and the total body (penetrating radiation). Again, the radons are exceptions to this general assumption. In this latter case, the lung is considered the critical organ.

9. In general, the ICRP calculations do not consider the chemical toxicity of the element in establishing the MPBB or MPC values. The exception to this rule is the element uranium. In the case of uranium, chemical toxicity was considered (the kidneys are the critical organ), and it was the limiting criterion for certain of the long-lived isotopes of uranium.

These features are the stated assumptions used in the calculations of MPBB and MPC values. However, other assumptions embodied in this technique are often unclear or are lost in the discussion of the equations. Before proceeding, these assumptions must be presented.

Radioactive material inhaled or ingested by a worker is assumed to be either soluble or insoluble. If soluble material is inhaled, then the amount deposited in the lung (lower respiratory passages) is assumed to go directly into the blood and, subsequently, to be deposited in one or more organs of the body. The important point is that a dose to the lungs is not considered. Soluble material deposited in the gastrointestinal tract may also enter the blood across the lining of this tract. The model assumes that this transfer occurs only in the small intestine. Again, any material entering the blood through the gastrointestinal tract may be deposited in one or more organs of the body. Unless all the material deposited in the tract enters the body, the sections of the tract may be critical organs.

Radioactive material deposited in an organ is assumed to be uniformly distributed throughout the entire organ. The dose calculation proceeds by calculating the dose to the organ from radioactive material contained in the organ. If the radioactive material decays by emitting alpha or beta radiation, then it is assumed that all this energy is deposited in the organ containing the radioactive material. If the decay involves the emission of gamma radiation, a correction is applied to account for that portion of the energy escaping the organ. Only the energy deposited in the critical organ is considered. No consideration is given to the ultimate deposition of the radiation escaping the single critical organ.

If the material is insoluble, it is assumed that there is no transfer of material to the blood in either the lung or the gastrointestinal tract. In this case, the number of critical organs are limited to the lung and the four sections of the tract.

i. Calculation of maximum permissible exposure values. A calculation of maximum permissible exposure values (specifically MPC values) begins with establishing the MPBB for a particular radionuclide. This calculation can proceed along two paths depending on the nature of the particular radionuclide. For bone-seeking radionuclides that primarily emit alpha and beta radiation, the MPBB is based on a comparison with ^{226}Ra. For other radionuclides, the MPBB is determined by establishing the activity of the radionuclide that can be present without exceeding the permissible weekly dose-equivalent rate for the critical organ.

ii. Body burden based on comparison with ^{226}Ra. This method is the result of a calculation that attempts to determine (1) the amount of a radionuclide deposited in bone that will deliver the same effective dose equivalent as delivered by the deposition of 1 μCi of ^{226}Ra and its daughter products, and (2) the amount of a radionuclide deposited in bone that will result in damage comparable to that observed from known deposits of ^{226}Ra in bone. At the time these recommendations were formulated, there was an extensive body of knowledge regarding ^{226}Ra; this information was used to establish limits for other radionuclides depositing in bone.

Thus, the first method used to establish the MPBB is based on a comparison with ^{226}Ra and is applicable only to alpha- and beta-emitting radionuclides that deposit in bone (i.e., in the skeleton). To simplify the equations, the symbol q is used for the MPBB in the equations that follow. The MPBB for a radionuclide deposited in bone is given by

$$q = \frac{q' \times f(2)'}{f(2)} \times \frac{\varepsilon'}{\varepsilon}, \tag{1}$$

where

q = the MPBB for the radionuclide of interest.
q' = the MPBB for ^{226}Ra, 0.1 μCi.
$f(2)$ = the fraction of the radionuclide in the critical organ (i.e., skelton) of that in the total body.
$f(2)'$ = for ^{226}Ra, f(2) = 0.99.
ε = effective absorbed energy per disintegration of the radionuclide.
ε' = for ^{226}Ra, the effective absorbed energy is 110 MeV per disintegration.

Substituting these values into the equation gives

$$q = \frac{0.1(0.99)}{f(2)} \times \frac{110}{\varepsilon}$$

$$q = \frac{11}{f(2)\varepsilon}. \tag{2}$$

iii. Body burden based on dose-equivalent rate to a critical organ. In general, for radionuclides that do not distribute in the bone, there is a lack of specific information that can be used to set acceptable body burdens. For this reason, MPBB values are calculated under the assumption that the MPBB is the amount of a radionuclide, distributed throughout the body, that will result in the maximum permissible dose-equivalent rate to the critical organ. The equation to be used is derived as follows. Let R be the permissible dose-equivalent rate, usually in rem/week, to a critical organ or the total body. Then,

$$R \propto \frac{q \times f(2) \times \varepsilon}{m}, \tag{3}$$

or

$$R = k \times \frac{q \times f(2) \times \varepsilon}{m}, \tag{4}$$

where

> $k =$ a constant to be evaluated later.
> $q =$ MPBB.
> $f(2) =$ the fraction of the radionuclide in the critical organ of that in the total body.
> $\varepsilon =$ the effective absorbed energy per disintegration.
> $m =$ the mass of the target organ.

Before the constant k is evaluated, the equation should be examined in more detail. First, note that it specifies the dose-equivalent rate in an organ; that is, R has units of rem/week. Second, notice that the right side of the equation has the basic components mentioned earlier for calculating dose. The product of q and $f(2)$ gives the activity present in the critical organ; that is, $qf(2)$ is the disintegration rate of the radionuclide in the critical organ. If the disintegration rate is known, then dose rate is simply the product of the disintegration rate and the energy deposited in the critical organ per disintegration per gram of the critical organ. The quotient of ε by m gives the absorbed energy per disintegration of the radionuclide per unit mass of the critical organ. Thus, the equation can be put in any acceptable units of absorbed dose or dose equivalent if the proper conversion factors are collected in the proportionality constant.

The units of each component of the equation are as follows:

> q is in microcuries
> $f(2)$ is a fraction and has no units
> ε is in units of MeV/disintegration
> m is in units of grams

Thus, the constant required must bring all the units together such that the resultant is in units of dose-equivalent rate (i.e., in units of rem/week). In this case the constant has a value of 358. The equation becomes

$$R = \frac{358 \times q \times f(2) \times \varepsilon}{m}. \tag{5}$$

But the object of this exercise was to derive an equation to be used to calculate the MPBB. Thus, solving for q gives

$$q = \frac{m \times R}{358 \times f(2) \times \varepsilon}, \tag{6}$$

or

$$q = 2.8 \times 10^{-3} \frac{m \times R}{f(2) \times \varepsilon}. \tag{7}$$

The MPBB for any organ can be calculated by substituting the appropriate values for the parameters and establishing the permissible dose-equivalent rate for the critical organ. For example, suppose the critical organ is the liver. This organ is not listed specifically in the ICRP limits so the limit for "other organs" (15 rem/yr) is applied for the calculation. The permissible dose-equivalent rate is 0.3 rem/week (it is always assumed that there are 50 working weeks in a year). Substituting this value into the equation gives

$$q = 2.8 \times 10^{-3} \frac{m \times 0.3}{f(2) \times \varepsilon}, \tag{8}$$

or

$$q = 8.4 \times 10^{-4} \frac{m}{f(2) \times \varepsilon}. \tag{9}$$

iv. MPCs in air and water. The next step in the process is to calculate MPC values for air and water for use in controlling exposure to internally deposited radionuclides. This section gives the equations to be used to calculate MPCs for all organs of the body except the gastrointestinal tract. It is assumed that radioactive material is taken into the body at a constant rate of $P \, \mu Ci/d$ and that biological elimination from the critical organ follows a simple exponential relationship. Under these assumptions, the rate of change of radioactive material in the critical organ can be expressed easily in a word equation:

$$\begin{bmatrix} \text{rate of change} \\ \text{of material in} \\ \text{critical organ} \end{bmatrix} = \begin{bmatrix} \text{rate of intake} \\ \text{of material into} \\ \text{critical organ} \end{bmatrix}$$
$$- \begin{bmatrix} \text{rate of loss} \\ \text{of material from} \\ \text{critical organ} \end{bmatrix}.$$

Replacing the words with the appropriate symbols gives

$$\frac{d(qf(2))}{dt} = P - \lambda_e(qf(2)). \tag{10}$$

In this equation, λ_e is the effective decay constant. The solution with $qf(2) = 0$ when $t = 0$ is

$$qf(2) = \frac{P}{\lambda_e}(1 - \exp(-\lambda_e t)). \tag{11}$$

Note that q is the "allowed activity" (or the MPBB) of radioactive material in the body and that $qf(2)$ is the allowed activity in the critical organ (i.e., the maximum permissible organ burden, MPOB). Thus, P must be the allowed intake rate. Solving for P gives

$$P = \frac{\lambda_e q f(2)}{(1 - \exp(\lambda_e t))}. \tag{12}$$

Remembering that $\lambda_e = 0.693/T$ (where T is the effective half-life) and rewriting to eliminate the decay constant yields

$$P = \frac{0.693(q f(2))}{T(1 - \exp(-0.693 t/T))}. \tag{13}$$

The allowable intake rate is the product of three factors:

$$P = S \times C \times f,$$

where

S = the rate of intake of either air or water in cm^3/d.

C = the concentration of radioactive material in either air or water in $\mu Ci/cm^3$.

f = the fraction of the material that reaches the critical organ by either inhalation or ingestion.

Substituting for P in Eq. (13) gives

$$S \times C \times f = \frac{0.693(q f(2))}{T(1 - \exp(-0.693 t/T))}. \tag{14}$$

The maximum allowed rate of intake occurs when the concentration of the radioactive material is at a maximum. This is true because the inhalation and ingestion parameters of Standard Man (i.e., S) are constant and therefore do not change. In addition, f, the fraction of material inhaled or ingested that reaches the critical organ, is assumed to be constant for each particular element. The only variable in the equation is the concentration of the radioactive material in air or water (i.e., C). In other words, if the allowable rate of intake is to be controlled, this equation indicates that the concentration of the radioactive material must be controlled in either air or water. Therefore, we can substitute into this equation a term called the maximum allowable concentration, in effect, the MPC. This done, we arrive at a general equation for the MPC of radioactive material such that the maximum allowable intake is not exceeded and, more important, the maximum permissible dose-equivalent rate in the critical organ is not exceeded. The general equation has the form

$$MPC = \frac{0.693(q f(2))}{S \times f \times T(1 - \exp(1 - 0.693 t/T))}. \tag{15}$$

In this equation t and T are expressed in days, and t is the time period of exposure (that is always 50 yr expressed in days).

These equations can be solved for the MPC in air or water by substituting the proper values for S and f. For air, these parameters are designated as S_a and f_a and, for water, they are S_w and f_w. Actually, there are two values for S_a and two values for S_w (see Section V.B. 1). For air, the parameters are

$$S_a = 6.9 \times 10^6 cm^3/d \text{ (occupational exposure)}$$

and

$$S_a = 2.0 \times 10^7 cm^3/d$$

(continuous occupational exposure).

The parameter f_a is specific to the particular element and is called the fraction inhaled that reaches the critical organ. The first value of f_a given here does not agree with the values given earlier for the assumed inhalation rate. The ICRP modified this value to provide for the fact that an occupational worker spends only 5 of 7 days per week and 50 of 52 weeks per year on the job. Thus, $(1 \times 10^7 cm^3/d) \times (5/7) \times (50/52) = 6.9 \times 10^6 cm^3/d$. The MPC equations for inhalation become

$$MPC_a = \frac{1.0 \times 10^{-7}(q f(2))}{f_a \times T(1 - \exp(-0.693 t/T))} \tag{16}$$

for occupational exposure, and

$$MPC_a = \frac{3.5 \times 10^{-8}(q f(2))}{f_a \times T(1 - \exp(-0.693 t/T))} \tag{17}$$

for continuous occupational exposure.

For ingestion, the equations are

$$MPC_w = \frac{9.2 \times 10^{-4}(q f(2))}{f_w \times T(1 - \exp(-0.693 t/T))} \tag{18}$$

for occupational exposure, and

$$MPC_w = \frac{3.2 \times 10^{-4}(q f(2))}{f_w \times T(1 - \exp(-0.693 t/T))} \tag{19}$$

for continuous occupational exposure. In these last two equations, S_w has the following values:

$$S_w = 750 \text{ cm}^3/d \quad \text{(occupational exposure)}$$

and

$$S_w = 2200 \text{ cm}^3/d$$

(continuous occupational exposure).

Note that f_w is the fraction ingested that reaches the critical organ and that t and T are expressed in days. The first value of S_w has been modified as described above.

This discussion presents the general equations used for calculation of MPC values under the ICRP-2 system.

Several special cases exist, for example, for the gastrointestinal tract and for exposure to semi-infinite clouds of noble gases. However, these are not discussed here.

2. ICRP Publication 26

The recommendations contained in ICRP-26 were the first real pronouncements of the Commission on radiation protection since the early 1960s. Recommendations were made in a number of areas including protection from external and internal radiation sources, exposures of population groups, exposure of pregnant women, and planned special exposures. This discussion focuses primarily on those recommendations and techniques that have an impact on methods to be used for internal dosimetry.

First, the ICRP restated the objectives of radiation protection that formed the basis for the new dose-limitation system they proposed. These objectives were

1. No practice should be adopted unless it provides a positive net benefit.
2. All exposures should be kept as low as reasonably achievable, economic and social factors being taken into account.
3. The dose equivalent to individuals should not exceed the limits recommended for the appropriate circumstances by the Commission.

The first statement reflects the Commission's commitment to the benefit–risk philosophy and the second states the ALARA philosophy adopted much later. In all situations, the Commission warns that their recommended limits should not be exceeded.

The Commission also provided several new concepts that must be defined and explained if these recommendations are to be understood and used effectively. The most significant change in the new recommendations is the introduction of the terms *stochastic* and *nonstochastic* effects of radiation. *Stochastic effects* are those for which the probability, rather than the severity, of an effect occuring is regarded as a function of dose without threshold. *Nonstochastic effects* are those for which the severity of the effect varies with the dose, and for which a threshold may therefore occur.

Some somatic effects of radiation are considered to be stochastic. The most important of these effects is carcinogenesis, and it is considered to be the chief somatic risk of radiation exposure at low doses (i.e., at the doses encountered in radiation protection). For this reason, cancer is the main concern when stochastic effects of radiation are being considered. The Commission states that, for the dose range involved in radiation protection, hereditary effects of radiation are also considered to be stochastic.

It is tempting to substitute the words *linear* and *threshold* effects of radiation for the terms just defined. However, the definition of stochastic effects is slightly different from the concept of a linear dose response curve because of the word *probability*. In addition, the ICRP includes certain effects of radiation in the two categories that may not have been included previously. For example, nonstochastic effects include cataracts of the lens of the eyes, nonmalignant damage to the skin, cell depletion in the bone marrow causing certain blood deficiencies, and gonadal cell damage leading to impairment of fertility. In a recent review, Upton identified a large number of nonstochastic effects of radiation. This review included estimates of the threshold doses for nonstochastic effects in >30 tissues of the adult.

Thus, the goals of radiation protection are to prevent the detrimental nonstochastic effects of radiation exposure and to limit the probability of stochastic effects to levels deemed "acceptable." Prevention of nonstochastic effects can be achieved if the dose-equivalent limits are selected such that a threshold is never reached. The ICRP goal was to select a level such that a threshold could not be reached even if the exposure lasted for an entire lifetime. The limitation on stochastic effects was selected based on a consideration of the benefit–risk relation with the ALARA philosophy in mind.

In addition, the limit on stochastic effects was selected by comparing the risks of occupationally exposed workers to those of workers in other "safe" industries. That is, the limit was selected such that the risk of producing a fatal cancer per unit exposure was essentially equivalent to the risk of an occupationally related death in other industries.

The Commission also formally defined the term *committed dose equivalent*, even though this term is in common use in many segments of the nuclear industry. *Committed dose equivalent* is the dose equivalent to a given organ or tissue that will be accumulated over a period of 50 yr, representing a working lifetime, after a single intake of radioactive material into the body. Mathematically, the committed dose equivalent is defined by

$$H_{50,\mathrm{T}} = \int_t^{t+50y} H(t)\,dt, \qquad (20)$$

where $H(t)$ is the relevant dose-equivalent rate and t is the time of intake. The ICRP states that this quantity may be considered a special case of the dose-equivalent commitment, but this distinction is not particularly important to this discussion of internal dosimetry.

The ICRP recommended a dose-equivalent limit for stochastic effects based on the total risk of all tissues irradiated. A single dose-equivalent limit is set for uniform irradiation of the whole body and a dose-limitation system is established to ensure that the total risk from irradiation of parts of the body does not exceed that from uniform irradiation of the whole body. In addition, no single tissue should be irradiated in excess of the dose-equivalent limit set to prevent nonstochastic damage.

Thus, for stochastic effects of radiation, the dose-limitation system is based on the principle that the risk should be equal whether the whole body is being irradiated uniformly or whether there is nonuniform irradiation. The ICRP concludes that this condition will be met if

$$\sum_{T} w_T H_T \leq H_{WB,L}, \tag{21}$$

where

w_T = a weighting factor representing the proportion of the stochastic risk resulting from irradiation of tissue (T) to the total risk, when the whole body is irradiated uniformly.

H_T = the annual dose equivalent in tissue T.

$H_{WB,L}$ = the recommended annual dose-equivalent limit for uniform irradiation of the whole body.

Later, at their 1978 meeting, the ICRP decided to call the quantity on the left side of Eq. (21) the *effective dose equivalent*. Thus,

$$H_E = \sum_{T} w_T H_T. \tag{22}$$

These conditions stated will be met, in the opinion of the ICRP, if the limits on radiation exposure are the following:

Stochastic effects are 5 rem/yr for uniform irradiation of the whole body, that is, 0.05 Sv/yr under the new SI system of units. The ICRP recommendations are written using only the new units.

Nonstochastic effects are 50 rem/yr (0.5 Sv/yr) to all tissues except the lens of the eye. The lens was finally assigned a limit of 15 rem/yr (0.15 Sv/yr), but some of the early publications contained the value of 30 rem/yr (0.30 Sv/yr). To accommodate the use of the concept of effective dose equivalent into the new scheme for internal dose assessment and for establishing secondary limits for use in controlling the work area, the ICRP found it necessary to define another term. This term is called the *committed effective dose equivalent* (later given the symbol $H_{E,50}$); that is,

$$H_{E,50} = \sum_{T} w_T H_{T,50}.$$

This concept requires that the committed dose equivalent (see Eq. (20)) be calculated for each organ or tissue in the body that is significantly irradiated. Then, the committed effective dose equivalent is obtained by multiplying each of these values by the respective stochastic tissue weight-ing factor (called the weighted committed dose equivalent) and summing over all the organs and tissues.

a. Tissues at risk. The internal dosimetry system described in ICRP-2 introduced the concept of a critical organ, and all the calculations of MPBB and MPC were made under the assumption that there was a need to control the dose-equivalent rate to this critical organ to the limit set by the ICRP (i.e., the MPDE). The new ICRP concept takes into account the total risk that can be attributed to the exposure of all tissues irradiated. If this concept is to be implemented, then it is necessary to specify the organs and tissues of the body that should be considered at risk and to establish some measure of this risk.

The most visible change in the new ICRP recommendations is that the critical organ concept has been discarded. This change was necessary because the concept of a single critical organ did not fit into the scheme of specifying an effective dose equivalent (H_E) relative to a uniform whole body irradiation. This scheme requires that the committed effective dose equivalent be the sum of the weighted committed dose equivalent ($w_T H_{T,50}$) to each organ in the body, each with a specific sensitivity to radiation effects. This sensitivity is given by the weighting factor shown in Eqs. (21) and (22).

The description of tissues at risk that follows is used to derive the weighting factors needed to calculate the effective dose equivalent. The derivation of risk factors is based on an average risk to a particular tissue from irradiation. No consideration is given to the effects of age-dependent or sex-dependent differences. The tissues considered and the risk factors derived are based on (1) a review of the suceptibility of the tissue to radiation damage, (2) a review of the seriousness of this radiation-induced damage, and (3) a consideration of the extent to which this damage is treatable. In addition, only the likelihood of inducing fatal malignant disease, nonstochastic changes, or substantial genetic defects is considered.

i. Gonads. Irradiation of the gonads can cause effects in three different ways. First, there is the probability of tumor induction. However, the gonads appear to have a low sensitivity to radiation and no carcinogenic effects have been documented. Impairment of fertility is also a possible effect but such an effect is clearly age-dependent. Again, the ICRP did not consider this an important radiation effect. The major effect considered for irradiation of the gonads is the production of hereditary effects over the first two generations. On the basis of an evaluation of hereditary effects over the first two subsequent generations from the irradiation of either parent, the risk appears to be $\sim 1.0 \times 10^{-2}$ Sv^{-1} (1.0×10^{-4} rem^{-1}). This value was obtained by considering the proportion of exposures that were likely to be genetically significant. The ICRP concluded that the genetic risk must be less than the mortality risk from fatal cancers. Thus, the risk estimate was reduced by the ratio of the mean reproductive life to the total life expectancy (i.e., ~0.40).

ii. Red bone marrow. Irradiation of the red bone marrow is clearly linked with the induction of leukemia. Other blood-forming tissues are thought to play only a very minor role in leukemia induction. For radiation protection purposes, the risk coefficient for leukemia was taken to be 2.0×10^{-3} Sv^{-1} (2.0×10^{-5} rem^{-1}).

iii. Bone. ICRP-11 (published in 1968) identified the radiosensitive cells in bone as the endosteal cells and the epithelial cells on bone surfaces. The sensitive cells are assumed to lie within 10 μm of the bone surfaces. The primary radiation-induced effect in these cells is cancer; however, the bone seems to be much less sensitive to radiation than other organs and tissues are. For this reason, the risk coefficient assigned by the ICRP was 5.0×10^{-4} Sv^{-1} (5.0×10^{-6} rem^{-1}).

iv. Lung. For the lung, the major radiation-induced effect is lung cancer. The evidence examined by the ICRP indicated that the risk of cancer was of the same order of magnitude as for the development of leukemia. Therefore, the risk coefficient assigned to the lung was 2.0×10^{-3} Sv^{-1} (2.0×10^{-5} rem^{-1}). In addition to considering the threat of lung cancer from radiation exposure, the ICRP again dismissed the thought that a "hot particle" in the lung would present a higher risk situation than that for material distributed uniformly in the lung.

v. Thyroid gland. The thyroid gland has a high sensitivity to cancer induction due to radiation exposure. In fact, it seems to be higher than that for the induction of leukemia. However, mortality from these thyroid cancers is quite low primarily due to the success in the treatment (e.g., in the United States, thyroid cancer is almost 100% survivable). The risk coefficient assigned to the thyroid gland was 5.0×10^{-4} Sv^{-1} (5.0×10^{-6} rem^{-1}).

vi. Breast. During reproductive life, the female breast may be one of the most radiosensitive tissues in the human body. For radiation protection purposes, the ICRP assigned a risk coefficient of 2.5×10^{-3} Sv^{-1} (2.5×10^{-5} rem^{-1}) to the breast.

vii. All other tissues. There is evidence that radiation is carcinogenic in many other organs and tissues of the body. However, there was not sufficient data available to the ICRP to allow the assignment of individual risk factors. Nevertheless, there was sufficient data to conclude that the risk factor for all other tissues was lower than those specified here. On the basis of that review the ICRP assigned a combined risk coefficient for all remaining unspecified tissues of 5.0×10^{-3} Sv^{-1} (5.0×10^{-5} rem^{-1}). The ICRP assumed that no single tissue was responsible for more than one-fifth of this value. In the discussion

of the ICRP calculations, we will see that these unspecified tissues are called the "remainder." However, this simple designation of the remainder is confusing to apply in most dose calculations. A more detailed explanation of the interpretation of the weighting factors for the remainder will be given later in Section V.B.3.a.

3. ICRP Publication 30

Publication of ICRP-30 in 1979 brought with it an entirely new dosimetry scheme for calculating the dose equivalent due to the uptake of radionuclides in the body. The scheme was based on the material discussed above. In the first part of ICRP-30, these basic concepts are reviewed hurriedly, but the dosimetry scheme is discussed in great detail. Although some of the following discussion was initially introduced in ICRP-26, it seems much more appropriate to discuss it in the context of the ICRP-30 formulation for internal dosimetry.

a. Determination of the tissue weighting factors. Up to this point the progress of the ICRP has been traced to lead up to deriving the tissue weighting factors needed in Eq. (21). The susceptible tissues have been identified and a risk coefficient has been assigned to each tissue based on the available biological evidence. The next step is to calculate the individual weighting factors. This is accomplished by taking the ratio of the individual risk for a tissue to the sum of all the risk coefficients. In other words, the weighting factor is given by

$$w_{\mathrm{T}} = \frac{\text{risk coefficient for tissue (T)}}{\text{sum of all risk coefficients}}. \quad (23)$$

This calculation is summarized in Table X where the tissues at risk, the radiation effects, the risk coefficients, and the weighting factors are given. The "remainder" category is assigned to the five tissues, other than those in Table X, which receive the highest dose equivalents. A weighting factor of 0.06 is assigned to each of five tissues. If the gastrointestinal tract is irradiated, each section of the tract is considered to be a separate tissue.

The remainder, according to the ICRP, consists of those organs or tissues not mentioned in (a) the metabolic model for the element, (b) the gastrointestinal tract model, and (c) the table of weighting factors (i.e., Table X). As stated above, the weighting factor assigned to any single organ cannot exceed 0.06, and no more than five organs may be considered when the health physicist is applying these factors. A complication arrives when the gastrointestinal tract or organs mentioned in the metabolic model are irradiated to a significant extent. What weighting factors are to be applied to these organs?

TABLE X Calculation of Tissue Weighting Factors of Stochastic Risks

Tissue (T)	Radiation effect	Risk coefficient (Sv^{-1})	W_T
Gonads	Hereditary	0.4×10^{-2}	0.25
Breast	Cancer	2.5×10^{-3}	0.15
Red bone marrow	Leukemia	2.0×10^{-3}	0.12
Lungs	Cancer	2.0×10^{-3}	0.12
Thyroid gland	Cancer	5.0×10^{-4}	0.03
Bone surfaces	Cancer	5.0×10^{-4}	0.03
Remainder[a]	Cancer	5.0×10^{-3}	0.3
		1.65×10^{-2} (Total)	

[a] Assigned to any five organs and tissues not designated here. See Section V.B.3.a for a more detailed explanation.

The ICRP has introduced a scheme to account for this situation. The weighting factor used for each of the individual organs is chosen to be 0.06. The weighting factor to be applied to the remainder is reduced by subtracting the sum of the weighting factors applied to these organs from the total weighting factor available for use (i.e., 0.30). The following example clarifies this point.

Suppose that the organs considered to be irradiated significantly are the liver, small intestine, lower large intestine, adrenals, pancreas, and uterus. The liver has the highest committed dose equivalent and the uterus the lowest. The other organs are arranged in rank order according to the value of the committed dose equivalent for each. In the calculation of the weighted committed dose equivalent, the committed dose equivalent values for the liver, small intestine, and lower large intestine are each multiplied by the factor 0.06. Based on the recommended ICRP scheme, the weighting factor for the other three organs cannot exceed 0.12. In this case, the ICRP recommends that the organ with the highest committed dose equivalent be selected and that the entire weighting factor be applied to this single organ. In recommending this procedure, they are assuming that the weighted committed dose equivalent for all three organs will be overestimated.

This procedure may seem confusing and it is, but the ICRP procedure is relatively straightforward and exceptions are usually noted in supplemental material published by the Commission to support the basic limits.

b. Secondary and derived limits. The Commission has defined new terms and recommended new limits for use in radiation protection. Use of the words *maximum permissible* has been discontinued due to the misinterpretation of the intent of the concept and the misuse of the limits recommended by the ICRP. To meet the basic ICRP limits on radiation exposure of workers, intakes of radioactive material in any one year must be limited to satisfy the following conditions:

$$I \sum_T w_T H_{50,T} \leq 0.05 \text{ Sv/yr} \qquad (24)$$

and

$$I H_{50,T} \leq 0.5 \text{ Sv/yr}. \qquad (25)$$

Note that Eq. (24) applies a limit to stochastic effects whereas Eq. (25) limits nonstochastic effects from the intake of radioactive materials. The ICRP also emphasizes that it is sufficient to limit the intake of radioactive materials in any one year to the recommended limits and there is no need to specify a limit on the rate of intake.

A secondary limit has been defined to meet the basic conditions for occupational exposure stated in Eqs. (24) and (25). This limit is called the *annual limit on intake (ALI)* and is defined as the activity of a radionuclide that if taken in alone, would irradiate a person, represented by Reference Man, to the limit set by the ICRP for each year of occupational exposure.

More specifically, the ALI is the greatest value of the annual intake *I* that satisfies both the following inequalities:

$$I \sum_T w_T (H_{50,T} \text{ per unit intake}) \leq 0.05 \text{ Sv/yr} \qquad (26)$$

and

$$I(H_{50,T} \text{ per unit intake}) \leq 0.5 \text{ Sv/yr}, \qquad (27)$$

where *I* (in Bq) is the annual intake of the specified radionuclide either by ingestion or inhalation and the other parameters are as identified earlier.

In ICRP-2 exposure to radionuclides was controlled by applying recommended MPCs in air or water to the specific exposure situation. Even though the basic recommendation of the Commission under the new formulation is based on the ALI, the Commission chose to include another quantity for convenience in controlling exposure to airborne radionuclides. This quantity was called the derived air concentration (DAC).

i. Derived air concentration. That concentration of a radionuclide in air, which if breathed by Reference Man for a working year of 2000 hr under conditions of "light activity," would result in the ALI by inhalation. That is,

$$\text{DAC (Bq/m}^3) = \frac{\text{ALI (Bq/yr)}}{(2000 \text{ hr/yr})(1.2 \text{ m}^3/\text{hr})}, \qquad (28)$$

or

$$\text{DAC (Bq/m}^3) = \frac{\text{ALI (Bq/yr)}}{2.4 \times 10^{-3} \text{ m}^3/\text{yr}}. \qquad (29)$$

In Eq. (28) the time of 2000 hr is obtained from the assumption of a 40-hr work week for 50 weeks per year. The quantity 1.2 is the volume of air, in units of cubic meters, assumed to be breathed by Reference Man per hour under conditions of light activity.

c. Other definitions.
Several other definitions are necessary before proceeding with a detailed discussion of the ICRP internal dosimetry scheme. As mentioned earlier, the ICRP discarded the critical organ concept because it did not fit the new scheme. However, there is still a need to call the organs and tissues of the body by some name. The ICRP has introduced the terms *source tissue* and *target tissue* to describe these tissues. *Source tissue (S)* is a tissue (which may be a body organ) that contains a significant amount of a radionuclide after intake of that radionuclide into the body. *Target tissue (T)* is a tissue (which may be a body organ) in which radiation energy is absorbed.

It should be clear that each source tissue is also a target tissue because some, if not all, of the radiation emitted by the radionuclide in the source tissue will be absorbed in the source tissue. It should also be clear that every target tissue is not necessarily a source tissue. This is true because some tissues may lie a significant distance from the source tissue but be irradiated because the radionuclide may emit X-rays or gamma rays, which can travel large distances in tissue.

Before the details of the calculation of committed dose equivalent are discussed, there are two more definitions to introduce in this section. These definitions are important in discussing the bone model used in the new calculations.

Volume seekers are radionuclides that tend to be distributed throughout the bone volume.

Surface seekers are radionuclides that tend to remain preferentially on bone surfaces.

d. Calculation of the committed dose equivalent.
The committed dose equivalent is defined as the total dose equivalent to an organ or tissue over the period of 50 yr after intake of a radioactive material. That is, the dose equivalent (or the committed dose equivalent) is proportional to the product of the total number of transformations occurring in the tissue over the time period of interest and the energy absorbed per gram of tissue per transformation of the radionuclide. In other words,

$$\begin{bmatrix} \text{committed dose} \\ \text{equivalent} \end{bmatrix} \propto \begin{bmatrix} \text{total number of transformations} \\ \text{in tissue over period of interest} \end{bmatrix}$$
$$\times \begin{bmatrix} \text{energy absorbed per gram of target} \\ \text{tissue per transformation} \end{bmatrix}$$

And, in the new symbolism used by the ICRP, the preceding word equation becomes

$$H_{T,50} = k \times U_S \times \text{SEE}, \tag{30}$$

where U_S represents the total number of spontaneous nuclear transformations of a radionuclide occurring in the source tissue (S) over a period of 50 yr after intake and SEE stands for the specific effective energy imparted per gram of the target tissue (T) from a transformation occurring in the source tissue.

The total number of transformations in the organ is obtained by integrating (or summing) over time an equation that describes the way material is retained in the organ. This equation includes loss by radioactive decay as well as loss through biological elimination. In early discussions of internal dose, the integrated activity was often called the cumulated activity and had units of microcurie-days. In the new ICRP formulation, the units on the quantity total number of nuclear transformations over a period of 50 yr after intake of the radionuclide are transformations per becquerel.

The specific effective energy is obtained from a consideration of the radiological characteristics of the nuclide deposited in the organ. These parameters, except for one, may be obtained from a review of the decay scheme of the particular radionuclide. The equation for SEE is

$$\text{SEE (T} \leftarrow \text{S)} = \sum_i \frac{Y_i E_i AF(\text{T} \leftarrow \text{S}) Q_i}{M_T}, \tag{31}$$

where

$$\begin{aligned} Y_i &= \text{the yield of the radiations of type} \\ &\quad i \text{ per transformation of the} \\ &\quad \text{radionuclide } j. \\ E_i &= \text{the average or unique energy} \\ &\quad \text{of radiation } i \text{ in units of million} \\ &\quad \text{electron volts.} \\ AF(T \leftarrow S) &= \text{the fraction of energy absorbed in} \\ &\quad \text{target organ T per emission of} \\ &\quad \text{radiation } i \text{ in source organ S.} \\ Q_i &= \text{the appropriate quality factor for} \\ &\quad \text{radiation of type } i. \\ M_T &= \text{the mass of the target organ in grams.} \end{aligned}$$

The factor $AF(\text{T} \leftarrow \text{S})$ is called the absorbed fraction of energy and is simply the ratio of the energy absorbed in a target organ to the total energy emitted by the radionuclide in the source organ. The symbols S and T and the arrow appear in the equation as reminders of this relationship. Before proceeding further with the development of the ICRP equations, the concept of the absorbed fraction of energy should be reviewed. For alpha and beta radiation (i.e., nonpenetrating radiation), all radiation energy is assumed to be absorbed in the organ containing the radionuclide. In this case, the absorbed fraction in the source organ (i.e., S ← S) is equal to 1.0. The absorbed fraction in all other target organs (i.e., T ← S) is assumed to be 0. There are two exceptions to this general rule for alpha and beta radiation. These are special situations in which the source is the bone or the contents of the gastrointestinal tract and the targets are cells in the bone or in the walls of the gastrointestinal tract.

It should be clear then that the absorbed fraction concept really applies in situations where the radionuclide has a significant penetrating radiation component (i.e., X-ray or gamma-ray emission). In the discussion of the ICRP-2 techniques, the only organ considered in the dose calculation was the organ containing the radionuclide (i.e., the critical organ). In the new ICRP internal dose formulation, the absorbed fraction concept allows the source organ to irradiate another target organ, which may be located some distance away (e.g., liver, thyroid). The British have nicknamed this concept "cross-fire."

There is one more complication of committed dose equivalent. The ICRP has not published data that give the absorbed fractions of energy for radionuclides that can be used in dose calculations. Instead, the ICRP published data on the specific absorbed fractions (SAF). The SAF is defined as the absorbed fraction divided by the mass of the target organ. In other words,

$$SAF = \frac{AF(T \leftarrow S)}{M_T}.$$ (32)

In this equation, both parameters on the right side are included in the specification of the SEE.

To return to the equation for committed dose equivalent, we evaluate the constant in Eq. (30). In the new ICRP formulation, the committed dose equivalent has units of sieverts per unit intake (1 Sv = 1 J/kg). Therefore, the quantities on the right side of the equation must be multiplied by a constant to bring both sides into agreement. The total number of transformations has units of transformations per becquerel and the specific effective energy has units of million electron volts per gram per transformation. In addition, the quality factor is "hidden" in the calculation of the SEE. Therefore, to bring both sides of the equation into agreement, we must multiply by 1000 g/kg and 1.6×10^{-13} J/MeV. Eq. (30) then becomes

$$H_{T,50}(T \leftarrow S) = 1.6 \times 10^{-10} U_S \times SEE(T \leftarrow S).$$ (33)

The subscript T,50 on the committed dose equivalent indicates that the committed dose equivalent was calculated for a single organ or tissue of the body (target tissue) and that the time period of concern is 50 yr after the intake. This

equation also applies to the situation in which a source organ is irradiating a single target organ. There may be cases in which the target organ is irradiated by several source organs. In this case, the committed dose equivalent contributed by each source organ must be summed to obtain the total committed dose equivalent for a particular target organ.

As stated previously, these ICRP concepts formed the basis for the current federal regulations on radiation exposure (even though the discussion focused primarily on internal dosimetry). More recent recommendations of the ICRP and the NCRP were issued in 1990 and 1993, respectively. These new recommendations are based on a re-evaluation of the risk coefficients used in ICRP-26 and include new concepts and exposure limits. However, these revised recommendations have not been adopted into the federal regulations in the United States and thus are not be discussed here. The interested reader is referred to publications of the ICRP and NCRP listed in the Bibliography.

Summary

In this section, the concepts of the ICRP for internal dose assessment (i.e., internal dosimetry) have been introduced. However, a great deal of the detailed information regarding the new techniques has not been explained fully. The intent was simply to introduce these concepts, and further study of the concepts, models, and detailed calculations are left to the reader.

These concepts appear to be extremely complicated, and the calculations seem to have many complexities. However, this is not completely the case. Remember that the concepts may be called by different "names" but the basic idea of calculating the energy deposited in an organ containing radioactive material *has not changed*. This statement is true with regard to the "old ICRP Publication 2" techniques as well as the techniques introduced in the early 1940s. It is still necessary only to determine the total number of nuclear transformations occurring in the tissue over the time period of interest and multiply this quantity by the total energy deposited in the tissue per transformation per gram of tissue.

SEE ALSO THE FOLLOWING ARTICLES

COSMIC RADIATION • HEALTH PHYSICS • NUCLEAR POWER REACTORS • NUCLEAR RADIATION DETECTION DEVICES • NUCLEAR REACTOR THEORY • NUCLEAR SAFEGUARDS • RADIATION PHYSICS • RADIATION SHIELDING AND PROTECTION • RADIOACTIVITY • SOLAR PHYSICS

BIBLIOGRAPHY

Attix, F. H., and Roesch, W. C. (1968). "Radiation Dosimetry," Vol. 1, 2nd ed., Academic Press, Orlando, Fla.

Cember, H. (1983). "Introduction to Health Physics," 2nd ed., Pergamon, Oxford.

Eichholz, G. G., and Poston, J. W. (1985). "Principles of Nuclear Radiation Detection," Lewis, Chelsea, Mississippi.

Greening, J. R. (1985). "Fundamentals of Radiation Dosimetry," 2nd ed., Medical Physics Handbooks 15, Adam Hilger Ltd., Accord, Mass.

International Commission on Radiological Protection (1959). "Report of Committee II on Permissible Dose for Internal Radiation," ICRP Publication 2, Pergamon, Oxford.

International Commission on Radiological Protection (1977). "Recommendations of the ICRP," ICRP Publication 26, Pergamon, Oxford.

International Commission on Radiological Protection (1979). "Limits for Intakes of Radionuclides by Workers," ICRP Publication 30, Annals of the ICRP, 2, Pergamon, Oxford.

International Commission on Radiological Protection (1990). "Recommendations of the ICRP," ICRP Publication 60, Pergamon, Oxford.

Kase, K. R., Bjarngard, B. E., and Attix, F. H. (1985, 1987, 1990). "The Dosimetry of Ionizing Radiation," Vols. 1, 2, and 3, Academic Press, Orlando, Fla.

Loevinger, R., Budinger, T. F., and Watson, E. E. (1988). "MIRD Primer for Absorbed Dose Calculations," Society of Nuclear Medicine, New York.

National Council on Radiation Protection and Measurements (1993a). "Risk Estimates for Radiation Protection," NCRP Report No. 115, Bethesda, Md.

National Council on Radiation Protection and Measurements (1993b). "Limitation of Exposure to Ionizing Radiation," NCRP Report No. 116, Bethesda, Md.

Drinking Water Quality and Treatment

Dan Askenaizer

Montgomery Watson Engineers

GLOSSARY

Disinfection Water systems add disinfectants to destroy microorganisms that can cause disease in humans. Primary methods of disinfection include chlorination, chloramines, chlorine dioxide, ozone, and ultraviolet light.

Disinfection byproducts Side reactions can occur in water when chemical oxidants such as chlorine and ozone are used to control potentially pathogenic microorganisms. These reactions can form low levels of disinfection byproducts, several of which have been regulated for potential adverse human health effects.

Filtration Filtration is the process of removing suspended solids from water by passing the water through a permeable fabric or porous bed of material. The most common filtration process employs a granular media (e.g., sand, anthracite coal). Filtration is usually a combination of physical and chemical processes.

Membrane filtration Membrane separation processes use semipermeable membranes to separate impurities from water. The membranes are selectively permeable to water and certain solutes. A driving force is used to force the water to pass through the membrane, leaving the impurities behind as a concentrate. The amount and type of material removed depends upon the type of membrane, the type and amount of the driving force and the characteristics of the water.

Ozone Ozone is a colorless gas that is extremely unstable and is a strong oxidizing agent that is capable or reacting with a wide variety of organic and inorganic solutes in water. Effectiveness of ozone disinfection is a function of the pH, temperature of water and method for ozone application.

Ultraviolet light Ultraviolet light is electromagnetic energy that is located in the electromagnetic spectrum at wavelengths between those of X-rays and visible light. UV light that is effective is destroying microbial entities in located in the 200- to 310-nm range of the energy spectrum. Most typical applications of UV at water treatment plants apply UV light in the wavelength range of 250 to 270 nm.

WATER is the most abundant compound on the surface of the earth. The physical and chemical properties of water are important issues with regard to water supply, water quality, and water treatment processes. With the increasing growth of urban areas and activities that can possibly introduce contaminants into drinking water sources, source water protection has become an increasingly important facet of providing safe drinking water. Water treatment professionals deal with a wide range of water qualities and they have a growing array of treatment methods at their disposal. During the 1970s and 1980s there was a growing interest and concern with groundwater contamination due to organic chemicals such as solvents and pesticides. In the early 1990s and in to the new millennium, there is a growing awareness of the need to balance the risks of the need to disinfect the water to reduce the threat of disease from microorganisms against the potential health risks from disinfection byproducts that are formed as a results of adding a disinfectant. Microorganisms such as *Giardia* and *Cryptosporidium* present challenges to regulators and water treatment engineers. The purpose of this chapter is to provide an overview of drinking water quality and treatment methods.

I. OVERVIEW OF DRINKING WATER SOURCES

The vast majority of fresh water in the world is provided by precipitation resulting from the evaporation of seawater. This transfer of moisture from the sea to land and back to the sea is referred to as the hydrologic cycle. Figure 1 presents a depiction of the hydrologic cycle.

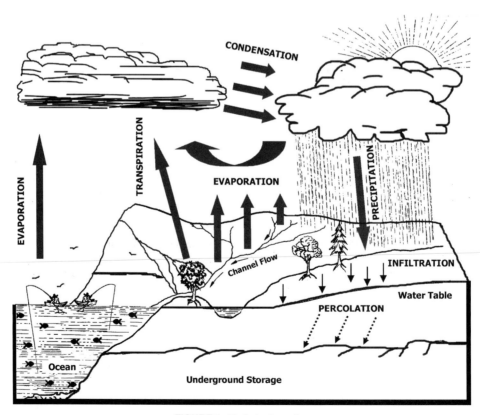

FIGURE 1 Hydrologic cycle.

About two thirds of the precipitation which reaches land surface is returned to the atmosphere by evaporation from water surfaces, soil, and vegetation, and through transpiration by plants. The remainder of the precipitation returns ultimately to the ocean through surface or underground channels.

The following section presents a brief overview of three sources of water: surface water sources, groundwater sources, and the use of reclaimed water. In the United States around 35% of the population served by community water systems drink groundwater while nearly 65% of the population served by community water systems receive water taken primarily from surface water sources. Increasingly, communities are looking to reclaimed water to meet a portion of their needs.

A. Surface Water

Surface waters sources for drinking water include lakes, rivers, canals, runoff, and impounding reservoirs. Water quality conditions in streams and rivers can change dramatically over a short period of time. Surface water sources such as streams and rivers are susceptible to chemical spills and accidental releases.

There are three stages (trophic levels) in the life cycle of a body of water. These are oligotrophic (low nutrients, minimal microbiological activity), mesotrophic (moderate nutrients, moderate microbiological activity), and eutrophic (high nutrients, high microbiological activity).

Water quality in lakes and reservoirs can change throughout the year as the water can stratify during warmer months. Thermal stratification can be a significant process in many lakes. During the warmer months of the year the warmer water (and therefore less dense water) will stay near the surface of the water body while the cooler and more dense water is trapped below. In the absence of strong winds there will be little mixing of the colder, denser water below with the warmer, less dense water near the surface. Under certain conditions where adequate nutrients are present this can lead to a depletion of oxygen in the lower parts of the water body and can thus cause water quality issues such as taste and odor problems and problems with iron and manganese (which will have increased solubility under the reducing conditions). As temperatures cool and the temperature of the surface of the lake cools, this together with wind action can cause mixing throughout the reservoir.

Depending upon nutrient, temperature, and carbonate conditions the upper regions of a lake or reservoir can be susceptible to algal blooms (which can cause changes in source water turbidity, alkalinity, taste, odor, and pH) and can make it difficult to treat the water near the surface of the lake.

B. Groundwater

Groundwater sources are beneath the land surface and include springs and wells. As can be seen from the hydrologic cycle, when rain falls to the ground, some water flows along the land to streams or lakes, some water evaporates into the atmosphere, some is taken up by plants, and some seeps into the ground. As water seeps into the ground, it enters a zone referred to as the unsaturated zone or vadose zone. Water moves through the unsaturated zone into the saturated zone, where the interconnected openings between rock particles are filled with water.

Groundwater quality is typically constant over time; however, changes in hydrogeological conditions can lead to differences in water quality over a relatively short distance. The chemistry of the groundwater is influenced by the composition of the aquifer and by the chemical and biological processes that occur as water infiltrates through the aquifer.

C. Reclaimed Water

An additional source of water that must be included in any discussion of sources is the use of reclaimed water for nonpotable and potable uses. Reclaimed water sources include desalination of brackish water or seawater, and reuse or recycling of wastewater through the application of appropriate treatment technology. In some cities, reclaimed wastewater has been used to irrigate golf courses and parks. In addition, several communities have seriously evaluated and studied the use of reclaimed wastewater to augment the drinking water supply (e.g., adding highly treated wastewater meeting drinking water standards under the Safe Drinking Water Act to a reservoir that is the source for raw water for the community's drinking water treatment plant). This drive toward utilizing a greater amount of reclaimed water comes from growing urban populations and constraints on the development of new water sources. Some public health authorities are reluctant to support or endorse the planned use of reclaimed water to augment a drinking water supply. However, it is already the case that there are many surface water sources (river and streams) that are subject to sewage contamination prior to their use as a potable drinking water supply. In these instances, in effect, the cities are practicing unplanned indirect potable reuse.

II. PROTECTION OF SOURCE WATER

Natural waters acquire their chemical characteristics by dissolution and by chemical reactions with solids, liquids, and gases with which they have come into contact during the various parts of the hydrological cycle. An example

TABLE I Water Quality Results for Three Different Sources[a] (from Snoeyink, Jenkins, 1980)

Constituent (mg/L)	Source 1[b] (reservoir)	Source 2[c] (river water)	Source 3[d] (groundwater)
SiO_2	9.5	1.2	10
Fe(III)	0.07	0.02	0.09
Ca^{2+}	4.0	36	92
Mg^{2+}	1.1	8.1	34
Na^+	2.6	6.5	8.2
K^+	0.6	1.2	1.4
HCO_3^-	18.3	119	339
SO_4^{2-}	1.6	22	84
Cl^-	2.0	13	9.6
NO_3^-	0.41	0.1	13
Total dissolved solids	34	165	434
Total hardness as $CaCO_3$	14.6	123	369

[a] (From Snoeyink, V. L., Jenkins, D. (1980). "Water Chemistry," John Wiley and Sons, New York.)

[b] (Source 1. Pardee Reservoir. East Bay Municipal Utility District, Oakland, CA, 1976.)

[c] (Source 2. Niagara River, Niagara Falls, New York.)

[d] (Source 3. Groundwater, Dayton, OH.)

of this would be weathering reactions, which are caused by the interaction of water and atmosphere with the crust of the earth. Table I presents examples of water quality for three different sources

Beginning in the early 1970s there was a growing interest in the presence of and increased understanding of health effects associated with low levels of organic compounds in water. Organic compounds in water can occur due to (1) degradation of naturally occurring organic material (i.e., leaves), (2) human activities such as handling and disposal of chemicals, and (3) chemicals reactions during the treatment of water (i.e., the production of disinfection byproducts).

Protecting sources of drinking water has become an increasingly important aspect of providing safe drinking water. Source water quality management is the first step toward ensuring an adequate supply of safe drinking water.

Potential sources of contamination or water quality problems in source water include the following:

- Climate (primarily precipitation causing increased levels of sediment, turbidity, and other contaminants)
- Temperature (can affect biological activity, oxygen saturation, and mass transfer coefficients)
- Watershed characteristics (steep slopes, vegetative cover, wildlife)
- Geology (e.g., mineral content)
- Presence of nutrients (can stimulate microbiological growth)
- Saltwater intrusion (increased salinity)
- Wastewater discharges (bacterial contamination, depletion of dissolved oxygen increased levels of inorganic and organic contaminants)
- Industrial discharges (accidental or planned discharges of chemical contaminants)
- Hazardous waste facilities (release of toxic, reactive, corrosive, or flammable contaminants)
- Mine drainage (acid discharges, increases in sediments, turbidity, color)
- Agricultural runoff (pesticides, herbicides, fertilizers)
- Livestock (microbial contamination, increased erosion increased nitrates)
- Urban Runoff (petroleum products, metals, salts, silts, and sediments)
- Land development (increased erosion and sediment loading, increased human activities that can release contaminants to the environment)
- Atmospheric deposition (acid rain)
- Recreational activities (swimming, boating, camping)

Activities to protect and enhance the quality of surface water sources include conducting sanitary surveys, programs to monitor source water quality and activities to provide watershed control. Other activities that can be undertaken to protect source water include storm-water management, development and implementation of emergency response procedures (to contain and clean up spills to prevent contamination of source water).

A sanitary survey is an on-site review of the water source, facilities, equipment, operation, and maintenance of a public water system to evaluate the adequacy of the source, facilities, equipment, operation, and maintenance for producing and distributing drinking water.

Monitoring programs can be conducted of both chemical and microbiological parameters at locations throughout a source of supply. A thorough monitoring program can provide valuable information toward an understanding and identification of possible changes in source water quality.

For groundwater sources, the potential effectiveness of a groundwater management program depends upon the degree to which potential contamination is accurately identified and the practicality of the response, remediation, and protection measures that are developed. All residential and commercial development and industrial and agricultural activities within the well field zone of influence and upstream of the general direction of groundwater flow should be investigated, and monitoring programs can be implemented to detect and control contaminants that could be introduced into the groundwater. For surface water sources, source protection can involve such activities as storm water management and controls on activities in a watershed that could impact the quality of the source water, including activities such as fishing, boating, swimming, hunting, and camping.

III. OVERVIEW OF BASIC DRINKING WATER TREATMENT PROCESSES

The amount and type of treatment applied by a given public water system will vary depending upon the source type and quality. Many, if not most, groundwater systems can provide adequate treatment that involves little or no treatment of the source. Surface waters, however, are exposed to the atmosphere and surface runoff and are more likely to contain contaminants. Surface water systems, therefore, must implement a greater level of treatment to provide safe and potable drinking water.

Water utilities can use a variety of treatment processes together at a single treatment plant to remove contaminants, remove turbidity, and provide disinfection. The most common physical processes used at public water systems with surface water supplies include coagulation, flocculation, sedimentation, and filtration. The following sections provide descriptions of physical and chemical processes that can be used to treat drinking water. In addition to a description of some basic water treatment methods (coagulation, flocculation, sedimentation and filtration, slow sand filtration, lime-soda softening, granular activated carbon), the following section also provides description of some typical water quality issues that can be addressed through treatment (iron and manganese removal, taste and odor problem, corrosion control) and additional treatment methods (membranes and ion exchange).

A. Coagulation

Colloidal suspended particles in water have like electrical charges that tend to keep them in suspension. Coagulation is defined as the destabilization of the charge on colloids and suspended solids, including bacteria and viruses, by use of a coagulant. The most commonly used coagulants are metal salt coagulants such as aluminum sulfate, ferric chloride, and ferric sulfate. In water, metal salts undergo hydrolysis. The products of this hydrolysis readily adsorb to colloid particles and cause the destabilization of their electrical charge. An important parameter in determining the effectiveness of a given coagulant is the pH of the water.

Synthetic polymers such as polydiallyl dimethyl ammonium (PDADMA) are also used as coagulants. Organic polymers can be used as the primary coagulant or as a co-agulant aid. Polymers are classified as anionic, cationic, or nonionic. Anionic polymers ionize in water to form negative sites along the polymer molecule. Cationic polymers ionize to form positively charged sites, while nonionic polymers exhibit only slight ionization.

Flash or rapid mixing is an important part of coagulation. The purpose of flash mixing is to quickly and uniformly disperse water treatment chemicals throughout the water. Effective flash mixing is especially important when using metal salt coagulants, since their hydrolysis occurs within a second and subsequent adsorption to colloidal particles is almost immediate. Rapid mix processes can typically be accomplished in just a few minutes.

B. Flocculation

Flocculation is a gentle mixing phase that follows the initial rapid mix step. During the flocculation step the chemically treated water is sent into a basin where the suspended particles can collide and form heavier particles called floc. Gentle agitation and appropriate detention times are used to allow this process to occur. Typical time for the flocculation step could be on the order of 15–30 min. After the flocculation step the water can then move into the sedimentation step.

C. Sedimentation

Also known as clarification, the purpose of sedimentation is to remove a majority of the settleable solids by gravitational settling. By removing the majority of the settleable solids in the sedimentation step this will maximize downstream unit processes such as filtration. During the sedimentation step, the velocity of water is decreased so that suspended material can settle out of the water stream by gravity. The key to effective sedimentation is proper coagulation and flocculation of suspended material in the raw water. Removal and disposal of the sludge from the sedimentation basin are important parts of the treatment process.

D. Filtration

Filtration is the process of removing suspended solids from water by passing the water through a permeable fabric or porous bed of material. The most common filtration process employs a granular media (e.g., sand, anthracite coal). Filtration is usually a combination of physical and chemical processes. Mechanical straining removes some particles by trapping them between the grains of the filter medium (such as sand). Adhesion is an equally important process by which suspended particles stick to the surface of filter grains or previously deposited material. The average filtration rate in the United States is 5–6 gal/min/ft^2 of filter area. At a conventional treatment plant the filters are

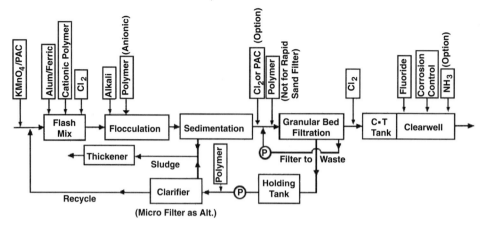

FIGURE 2 Conventional treatment process.

preceded by coagulation, flocculation, and sedimentation. At a direct filtration plant, the filters are preceded by coagulation and flocculation only; with the floc being removed directly by the filters. Figure 2 presents a schematic for a conventional treatment plant together with examples of potential application points for chemical addition. Figure 3 presents the schematic for a direct filtration plant.

1. Slow Sand Filters

Slow sand filters are operated at very low filtration rates without the use of coagulation. Slow sand filters are a simple, reliable and easy to operate system. The filtration rate for slow sand filters is typically 50–100 times slower than that of granular media filters. Therefore, a much larger area is needed for the filter bed to produce an equivalent amount of water. Contaminants are removed

from water through a combination and physical straining and microbiological processes in a slow sand filter. When in operation, the surface of the filter bed is covered by a thin layer of medium, known as the "schmutzdecke." This layer contains a large variety of microorganisms and enables these filters to remove large numbers of bacteria. Slow sand filters do not require highly trained operators, have minimal power requirements, and can tolerate reasonable hydraulic and solids shock loadings. Some of the disadvantages of slow sand filters include the large amount of land they require; the filters can be easily clogged by excessive amounts of algae; they are not very effective at removing color; and intermittent operation of the filters may degrade the quality of the filter effluent by promoting anaerobic conditions within the filter bed. The filters must be periodically cleaned by scraping off a thin layer of sand from the surface of the filter bed.

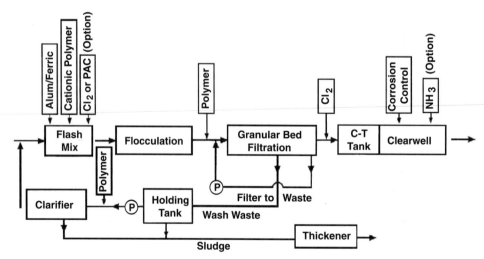

FIGURE 3 Direct filtration process.

TABLE II   General Classification of Water Hardness[a]

Level of hardness (mg/L)	Classification
0–75	Soft water
75–150	Moderately hard water
150–300	Hard water
>300	Very hard water

[a] Hardness is sometimes expressed as grains per gallon, where 1 grain of $CaCO_3$/gallon is equivalent to 17.1 mg/L as $CaCO_3$.

E. Lime-Soda Softening

The hardness of water is defined as the concentrations of divalent metallic cations in water and is expressed as mg/L of $CaCO_3$. Table II presents a general classification of water hardness. The principal divalent metallic cations that contribute to the hardness of a water are calcium and magnesium, with contributions from iron, manganese, and strontium. The main purpose of water softening is to reduce the levels of calcium and magnesium in the water to reduce the hardness of the water. In the United States there are over 1000 treatment plants were softening is practiced. Historically, softening was important due to the high consumption of soap by hard water. With today's synthetic detergents this is no longer a major issue. However, there are other benefits to softening including removal of heavy metals, metallic elements and organic compounds; effective destruction of bacteria, viruses and algae; and improvement in boiler feed water and cooling waters.

While there are many variations, the primary method of softening a water is the addition of lime (calcium hydroxide) and soda ash (sodium carbonate) to the water. The purpose of adding these compounds to a water is to change the hardness compounds such that they become insoluble and precipitate (e.g., calcium and magnesium are converted, respectively, to calcium carbonate ($CaCO_3$, partially soluble) and magnesium hydroxide ($Mg(OH)_2$). In addition to chemical treatment, ion exchange resins and membranes can be used to soften a water.

F. Granular Activated Carbon

Granular activated carbon (GAC) has been used as a substitute for granular filter media and as an additional process in conventional treatment plants for the removal of organic compounds including compounds producing taste and odors, pesticides and other synthetic organic compounds. GAC can be manufactured from a large variety of materials including wood, nuts, shells, coal, peat, or petroleum residues. GAC used in water treatment plants is typically manufactured from bituminous or lignite coal by heating the coal under anaerobic conditions. The product of this process is then activated by exposure to a mixture of steam and air at a temperature of 1500°F, which oxidizes the surface of the carbon pores and allows the surface to attract and hold organic compounds. After being used at a water treatment plant, the spent GAC can be regenerated through steam, thermal regeneration and chemical means.

G. Iron and Manganese Removal

The presence of iron and manganese is drinking water have been associated with unpleasant taste and odors, staining of laundry and fixtures as well as causing the formation of mineral deposits. Iron and manganese are commonly found in soil in insoluble forms. When water contains carbon dioxide (or is an acidic water) then the ferric iron can be reduced to the ferrous form (which is soluble in water) and manganese is reduced to a form that is also soluble in water. Treatment to remove iron and manganese includes oxidation (aeration, chlorination, chlorine dioxide, potassium permanganate, ozone) followed by clarification and filtration; ion exchange; stabilization through use of a sequestering agent, and lime softening.

H. Taste and Odor Control

The most frequent causes of taste and odor in a drinking water are metabolites of algae (most commonly blue-green algae), actinomycetes (filamentous bacteria), and decaying vegetation. Other potential causes of taste and odor issues are hydrogen sulfide, agricultural runoff, industrial chemical spills, and sewage pollution.

The most common odor-producing compounds are geosmin and 2-methlyisoborneal (MIB) which can impart objectionable odor at very low concentrations. These compounds are responsible for the earthy-musty odors in water and have been isolated from actinomycetes (*Actinomyces, Nocardia, Streptomyces*) and from blue-green algae (e.g., *Anabaena* and *Oscillatoria*). Control methods for taste and odor include: prevention at the source (reservoir mixing, aquatic plant control, reservoir management), removal of a particular constituent at the treatment plant (aeration, oxidation, adsorption), and control within the distribution system (minimization of dead-ends, use of blow-off and cleanout assemblies, distribution system flushing).

I. Corrosion Control

All waters are corrosive to some degree. Corrosion can reduce the life of a pipe by reducing wall thickness until there are leaks, it can result in encrustations that reduce the effective carrying capacity and can result in corrosion by-products at the consumer's tap that have public health implications (i.e., lead and copper). The tendency of a water to be corrosive will depend on its physical and chemical characteristics as well as on the nature of the material it comes into contact with. The most common types of materials used in distribution systems include cast iron, ductile iron (cast iron containing a small amount of alloying elements such as magnesium), asbestos-cement, steel, copper, galvanized iron, and plastics.

The mechanisms of corrosion in a water distribution system are typically a complex and interrelated combination of physical, chemical, and even biological processes. The basic principles that affect corrosion of materials include solubility, described by chemical equilibria among materials and constituents in the water; and the rate of dissolution, which is described by chemical and electrochemical kinetics. Electrochemical corrosion occurs where two different metals have an electropotential between them are immersed in a common body of water. All waters can act as an electrolyte, but the degree to which they do so depends on the dissolved chemicals present. High-velocity water flow can cause pitting and erosion of surfaces due to cavitation. Certain types of bacteria, sulfate-reducing bacteria, and iron bacteria can also can cause internal corrosion in piping material. Other factors that can influence the corrosivity of a given water include the concentration of dissolved salts in the water, level of dissolved gases in the water, water temperature, and stress and fatigue.

In terms of lead solubility, the most important water quality parameters are pH, alkalinity, dissolved inorganic carbonate and orthophosphate levels. In general, low pH levels have been associated with higher lead levels at the tap. Soft waters that are low in pH and alkalinity are often corrosive toward lead and other metals. Water treat-ment for controlling and/or reducing lead leaching include pH adjustment, carbonate adjustment, use of corrosion inhibitors, calcium carbonate deposition, as well as painting, coating, and usage of cathodic protection systems.

J. Membrane Filtration Technology

Membrane separation processes use semipermeable membranes to separate impurities from water. The membranes are selectively permeable to water and certain solutes. A driving force is used to force the water to pass through the membrane, leaving the impurities behind as a concentrate. The amount and type of material removed depends upon the type of membrane, the type and amount of the driving force, and the characteristics of the water. Important issues involved with the operation of membrane systems include membrane fouling and disposal of the concentrate.

There are two classes of membrane treatment systems. These include low-pressure membrane systems (such as microfiltration (MF) and ultrafiltration (UF)), and high-pressure membrane systems (such as nanofiltration (NF) and reverse osmosis (RO)). Low-pressure membranes are operated at pressures ranging from 10 to 30 lb/in.^{2} (psi), whereas high-pressure membranes, including nanofiltration are operated at pressures ranging from 75 to 250 psi. Figure 4 presents a general description of various membranes types and their ability to remove impurities from water.

MF can remove particles that are greater than 0.5 μm in diameter. UF is capable of removing colloids, bacteria, viruses, and high-molecular-weight organic compounds.

Some advantages of using low-pressure membranes include small waste stream, limited chemical usage, a

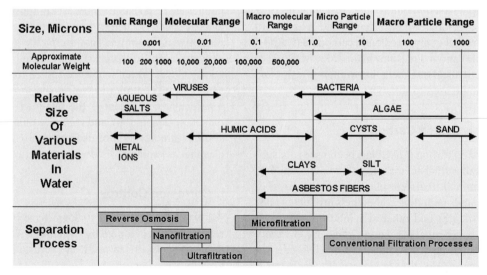

FIGURE 4 Membrane size ranges.

relatively small area needed for equipment, good pathogen reduction, and no disinfection by-product formation. Low-pressure membranes, however, are ineffective for the removal of dissolved organic matter. Therefore, color-causing organic matter, taste-and-odor causing compounds such as Geosmin and methylisoborneol (MIB), and man-made chemicals can pass through the membranes into the treated water.

NF membranes have been used successfully for groundwater softening since they achieve greater than 90% rejection of divalent ions such as calcium and magnesium. NF membranes are also capable of removing greater than 90% of natural organic matter present in the water. Therefore, they are also excellent candidates for the removal of color, and also DBP precursor material.

RO membranes have long been used for the desalination of seawater around the world. These membranes can consistently remove about 99% of the total dissolved solids (TDS) present in the water, including monovalent ions such as chloride, bromide, and sodium.

K. Ion-Exchange Technology

Ion-Exchange (IX) technology removes undesirable ions from raw water and exchanges them for desirable ions. The two most common applications of ion exchange are for water softening (Ca^{2+} and Mg^{2+} removal), either at the water treatment plant or as a point-of-entry (POE) treatment process, and for industrial applications, such as the production of fully demineralized water.

Examples of ions that can be removed using IX include nitrate, arsenic, selenium, barium, radium, lead, fluoride, and chromate.

In the IX process water passes through a resin bed where contaminant ions present in the water are exchanged with ions on the resin surface, thus removing the contaminant ions from the water and concentrating them on the resin. The resin is frequently regenerated to remove the contaminant from the resin surface and replenish the resin with the original exchange ion. There are four primary types of IX resins:

1. Strong Acid Cationic (SAC) Resin
2. Weak Acid Cationic (WAC) Resin
3. Strong Base Anionic (SBA) Resin
4. Weak Base Anionic (WBA) Resin

SAC and WAC resins are used to remove cations from water (e.g., Ca^{2+}, Mg^{2+}, Ra^{2+}, Ba^{2+}, Pb^{2+}), while SBA and WBA resins are used to remove anions from water (e.g., NO_3^-, SO_4^{2-}, ClO_4^-, $HAsO_4^{2-}$, SeO_3^{2-}). During water softening, SAC resins can remove both carbonate and noncarbonate hardness, whereas WAC resins can only re-move carbonate hardness. On the other hand, WAC resins are easier to regenerate than SAC resins and do not result in sodium concentration increase as SAC resins do.

An important issue for the application of IX technology is the waste stream produced by the process. The volume of the waste stream is not large on the order of 2 to 5% of the water volume treated. However, the waste stream contains a high concentration of acid (HCl), base (NaOH), or salt (NaCl). The waste stream also can contain a high concentration of the contaminant removed from the water (e.g., NO_3^-, Pb^{2+}, etc. . .).

IV. DISINFECTION

Disinfection is an important step in ensuring that water is safe to drink. Water systems add disinfectants to destroy microorganisms that can cause disease in humans. Primary methods of disinfection include chlorination, chloramines, chlorine dioxide, ozone, and ultraviolet light.

From a historical perspective, the Chick–Watson model has been the predominant model used to describe the kinetics of using disinfectants to inactivate microorganisms. Chick's law (1908) expresses the rate of destruction of microorganisms using the relationship of a first-order chemical reaction. Watson (1908) refined the equation to produce an empirical relationship that reflected changes in the disinfectant concentration. The Chick–Watson model can be expressed as follows:

$$\log N/N_o = -kC^n T,$$

where N_o = the initial concentration of bacteria, N = the concentration of surviving bacteria at time T, k = rate constant (coefficient of specific lethality), C = concentration of the disinfectant, and n = concentration of the dilution (empirically derived). The Chick–Watson model leads to a correlation between the level of inactivation and the product of the concentration of the disinfectant (C) and the contact time (T). The United States Environmental Protection Agency's (USEPA) Surface Water Treatment Rule includes tables that associate specific CT values (mg*min/L) with a given level of inactivation of *Giardia* and viruses.

A. Chlorine

Chlorination of potable water has been practiced in the United States since 1903. Chlorine can be applied by drinking water treatment plants as chlorine gas, sodium hypochlorite solutions, or as solid calcium hypochlorite. Free chlorine reacts rapidly with many substances in the water, including microorganisms. The effectiveness of chlorine to provide disinfection is affected by many variables including amount of oxidizable substances in the water (that exert a demand on the chlorine), concentration of particulate matter, pH, temperature, contact time, and the level of residual chlorine. The formation of disinfection byproducts including trihalomethanes (THMs) is one of the major concerns with regard to the use of chlorine disinfection.

1. Chlorine Gas

Elemental chlorine is a toxic, yellow-green gas at normal pressures. At high pressures, it is a liquid. Chlorine gas is typically supplied as a liquid in high pressure cylinders. Chlorine gas is released from the liquid chlorine cylinder by a pressure reducing and flow control valve operating at pressures less than atmospheric. The gas is led to an injector in the water supply pipe where highly pressurized water is passed through a venturi orifice creasing a vacuum that draws the chlorine into the water stream. Adequate mixing and contact time must be provided after injection to ensure complete disinfection of pathogens. Gaseous chlorine, when added to water, rapidly hydrolyzes to hypochlorous acid (HOCl) and hydrochloric acid (HCl) as follows:

$$Cl_2 + H_2O \rightleftarrows HOCl + H^+ + Cl^-$$

Hypochlorous acid is subject to further reaction including disinfection, reactions with various organic and inorganic compounds or dissociation to hydrogen and hypochlorite ion (OCl^-) as follows:

$$HOCl \rightleftarrows H^+ + OCl^-$$

The relative concentrations of hypochlorous acid and hypochlorite ion are dependent on the pH and the temperature. Generally, hypochlorous acid is a better disinfecting agent than is hypochlorite ion.

2. Sodium Hypochlorite Solution

Sodium hypochlorite is available as a solution in concentrations of 5 to 15% chlorine. Sodium hypochlorite is easier to handle than chlorine gas or calcium hypochlorite. It is, however, extremely corrosive and should be kept away from equipment that could be damaged by corrosion. Hypochlorite solutions decompose and should not be stored for more than 1 month and must be stored in a cool, dark, dry area.

The sodium hypochlorite solution is diluted with water in a mixing/holding tank. The diluted solution is then injected by a chemical pump in to the water supply pipe at a controlled rate. Adequate mixing and contact time must be provided. Sodium hypochlorite can be generated on site by using electrolysis of sodium chloride solution. Hydrogen gas is given off as a by-product and must be safely dispersed.

3. Calcium Hypochlorite

Calcium hypochlorite is a white solid that contains 65% available chlorine and dissolves easily in water. Calcium hypochlorite is very stable and can be stored for an extended period of time. Calcium hypochlorite is a corrosive material with a strong odor. Reactions between calcium hypochlorite and organic material can generate enough heat to cause a fire or explosion. It must be kept away from organic materials such as wood, cloth, and petroleum products. Calcium hypochlorite readily absorbs moisture, forming chlorine gas.

Calcium hypochlorite can be dissolved in a mixing/ holding tank and injected in the same manner as sodium hypochlorite. Alternatively, where the pressure can be lowered to atmospheric, such as at a storage tank, tablets of calcium hypochlorite can be directly dissolved in the free flowing water.

V. DISTRIBUTION SYSTEM

While beyond the scope of this chapter, a vital component toward ensuring the delivery of safe drinking water is the series of transmission mains, fire hydrants, valves, pump stations, booster chlorination stations, storage reservoirs, standpipes, and service lines that constitute the distribution system. The proper design, construction material, and maintaining the integrity of the distribution system and the individual components are important to maintain the microbiological safety of the drinking water.

VI. DISINFECTION BY-PRODUCTS

In 1974 trihalomethanes (THMs), were first identified in finished drinking water. In 1975 the USEPA conducted the National Organics Reconnaissance Survey of 80 cities in the United States and observed that the occurrence of THMs was widespread in chlorinated drinking water and was associated with the practice of using chlorine to disinfect the water. Later studies demonstrated that THMs continued to form in the distribution system.

THMs are a class of organic compounds where three hydrogen atoms in the methane molecule have been replaced with three halogen atoms (chlorine or bromine). The four THMs identified were chloroform, bromodichloromethane, dibromochloromethane, and bromoform. THMs were important to regulators initially as suspected human carcinogens. Recent information suggests that some disinfection by-products (DBPs) may have adverse developmental and reproductive impacts.

The production of THMs can be shown simply as follows:

Free Chlorine + Natural Organics (precursors)
+ bromide → THMs + Other byproducts

The natural organic precursors in the raw water are generally humic and fulvic acids. The humic and fulvic acids enter water from the degradation of algae, leaves, bark, wood, and soil. Humic and fulvic acids comprise the major fraction of dissolved organic carbon in most natural waters. Factors that can influence the concentration of THMs in a given water include the concentration and type of precursor material, the concentration of free chlorine residual, contact time with chlorine, water temperature, water pH, and the concentration of bromide ion. The brominated species of THMs has been attributed to the presence of bromide in the raw water and the fact that hypochlorous acid can oxidize the bromide ion to hypobromous acid. Hypobromous acid can undergo addition and substitution reactions with various types of organic compounds in water to produce halogenated organics such as the THMs.

Since the initial observation on the occurrence and formation of THMs, additional research has shown that all chemical disinfectants (chlorine, monochloramines, ozone, and chlorine dioxide) can form various types of disinfection byproducts (DBPs).

Table III presents a list of disinfection byproducts that can be formed through the use of chemical disinfectants.

Approaches for controlling or limiting the formation of THMs include the following: use of an alternative source (such as groundwater versus a surface water with higher concentration of organic precursor material), improved flocculation and sedimentation (enhanced coagulation) to remove the DBP precursor material, moving the point of chlorine application within the treatment plant to minimize contact time, and the use of alternative disinfectants.

VII. ALTERNATIVES TO CHLORINATION

Over the years, public water systems have used various alternatives to chlorine disinfection. The more common alternatives include (1) chloramines, (2) Ozone, (3) Chlorine Dioxide, and (4) Ultraviolet irradiation.

A. Chloramines

Chloramines (referred to as combined chlorine) are formed when water containing ammonia is chlorinated. There are three inorganic chloramine species: monochloramine (NH_2Cl), dichloramine ($NHCl_2$), and trichloramine (NCl_3). The species of chloramines that are formed depends on factors such as the ratio of chlorine to ammonia–nitrogen, chlorine dose, temperature, pH, and alkalinity.

The principal reactions for chloramine formation are presented below:

$$NH_3(aq) + HOCl \rightleftharpoons NH_2Cl + H_2O \text{ (monochloramine)}$$

$$NH_2Cl + HOCl \rightleftharpoons NHCl_2 + H_2O \text{ (dichloramine)}$$

$$NHCl_2 + HOCl \rightleftharpoons NCl_3 + H_2O \text{ (trichloramine)}$$

At a typical water treatment plant, the dominant chloramine species will be monochloramines. Chloramine generating reactions are 99% complete within a few minutes. Chloramines are a weak disinfectant that are less effective against viruses or protozoa than free chlorine but produce fewer disinfection by-products. The use of chloramines as a DBP control strategy is well established in the United States. Chloramines are generated onsite at the treatment plant. Anhydrous ammonia and ammonia sulfate are examples of ammonia containing chemicals used by water systems to form chloramines. In most situations in the United States, chloramines are used as a secondary disinfectant to maintain a residual in the distribution system.

B. Ozone

Ozone is a colorless gas that is extremely unstable and is a strong oxidizing agent that is capable or reacting with a wide variety of organic and inorganic solutes in water. Ozone can also undergo a rapid, autocatalytic

TABLE III Disinfection Byproducts

Trihalomethanes	Haloacetonitriles
Chloroform	Dichloroacetonitriles
Bromodichloromethane	Trichloroacetonitrile
Dibromochloromethane	Dibromoacetonitrile
Bromoform	Tribromoacetonitrile
	Bromochloroacetonitrile
Haloacetic acids	
Monochloroacetic acid	Halopicrins
Dichloroacetic acid	Chloropicrin
Trichloroacetic acid	Bromopicrin
Monobromoacetic acid	
Dibromoacetic acid	Chloral hydrate
Tribromoacetic acid	
Bromochloroacetic acid	Oxyhalides
Bromodichloroacetic acid	Chlorite
	Chlorate
Aldehydes	Bromate
Formaldehyde	
Acetaldehyde	Haloketones
Glyoxal	1,1-Dichloroacetone
Methyl glyoxal	1,1,1-Trichloroacetone

decomposition to form a variety of oxidant species, with the hydroxyl radical (^-OH) being the most notable. Both molecular ozone and the free hydroxyl radical are powerful oxidants. Ozone does not provide a lasting residual, therefore a secondary disinfectant, chlorine or chloramines, are typically used to provide a stable residual within the public water system distribution system.

The actual ozone dose utilized at a water treatment plant is determined by the quality of the raw water (i.e., how much ozone demand will be exerted) and what is the ultimate objective for using ozone. Plants using ozone for iron and manganese oxidation will need to apply a different dose than plants using ozone as a disinfectant for inactivation of protozoans such as *Cryptosporidium*.

Ozone must be manufactured and used onsite. Ozone is manufactured by passing air or oxygen through two electrodes with a high, alternating potential difference. The use of ozone to provide disinfection at a water treatment plant involves taking into account the concentration of ozone applied, competing ozone demands, and providing a minimum contact time. As indicated, ozone is a strong oxidant and requires a shorter reaction time as compared to free chlorine to accomplish the same results. The basic components of an ozone system at a water treatment plant include ozone generation equipment, feed gas preparation, ozone contacting in the water and a means to destroy ozone off-gas.

While ozone does not form many of the byproducts formed through the use of chlorine (e.g., trihalomethanes, haloacetic acids, etc.) it is important to note that ozone can produce a byproduct, bromate (BrO_3^-), which is a potential human carcinogen, when the water being treated contains bromide. In general, bromide concentrations in raw water greater than 50 μg/L may result in bromate formation at levels greater than the current USEPA maximum contaminant level (MCL) of 10 μg/L in the United States. Additional byproducts of ozonation include aldehydes, ketones, and carboxylic acids. However, these later compounds have as yet not been considered a concern for public health protection at the levels produced in drinking water treatment using ozone.

Ozone off-gas destruction is the final major component in the ozonation process. This system is required to remove ozone from spent off-gas streams, which are collected and treated prior to discharge into the atmosphere. Both catalytic and thermal destruction devices are used for this purpose.

C. Chlorine Dioxide

Chlorine dioxide (ClO_2) has been used as a primary disinfectant throughout Europe but only in a limited number of situations in the United States. Chlorine dioxide is a strong oxidant and has been used primarily to control taste and odors but it is also an effective bactericide, equal to or slightly better than hypochlorous acid. Chlorine dioxide is also used for iron and manganese oxidation. The use of chlorine dioxide can lead to the formation of chlorite (ClO_2^-) and chlorate (ClO_3^-) ions that have been a concern to regulators due to possible adverse health effects. When applied to surface water or groundwater, ClO_2 reacts rapidly with naturally occurring, oxidizable organic material. Up to 70% of the applied dose is reduced to ClO_2^- with the remainder being ClO_3^- or Cl^-. Once produced, chlorate ion, is not removed by conventional treatment. Clorite ion (ClO_2^-) can be minimized by optimizing the operation of the chlorine dioxide generator or by using a reducing agent such as ferrous iron.

Chlorine dioxide is unstable at high concentrations and must be generated on site at the water treatment plant. Chlorine dioxide is highly soluble and hydrolyzes very slowly in water.

D. Ultraviolet (UV) Irradiation Technology

UV light is electromagnetic energy that is located in the electromagnetic spectrum at wavelengths between those of X-rays and visible light. UV light that is effective is destroying microbial entities in located in the 200- to 310-nm range of the energy spectrum. Most typical applications of UV at water treatment plants apply UV light in the wavelength range of 250 to 270 nm. Most lamps emit UV irradiation by passing an electrical arc between filaments in a pressurized gas or vapor (typically mercury vapor). Ultraviolet dosage is commonly measured as milliWatt-second per square centimeter ($mW\text{-}s/cm^2$) or milliJoule per square centimeter (mJ/cm^2).

Typically a UV process is designed such that water flows in a narrow region around a series of UV lamps. Microorganisms in water are inactivated through exposure to the UV light. In general, a molecule in the ground state absorbs electromagnetic energy from the UV source and the bonds in the molecule are transformed to an excited state, and chemical and physical processes become thermodynamically possible. The process works on the principle that UV energy disrupts the DNA of the microorganisms and prevents it from reproducing.

There are four types of UV technologies of interest to the water industry. They include (1) low-pressure, low-intensity UV technology; (2) low-pressure, medium-intensity UV technology; (3) medium-pressure, high-intensity UV technology; and (4) pulsed-UV technology. Unlike using disinfectants such as ozone, chlorine, or chlorine dioxide, UV irradiation does not provide oxidation for color, taste, and odor control because UV light is not a strong oxidant.

A UV treatment process is comprised of a series of UV lamps enclosed inside a quartz sleeve. The UV light passes through the quartz sleeve and into the water. Due to the high energy emitted by the UV lamps, the temperature of the quartz sleeve can rise substantially causing the precipitation of various scales on the surface of the sleeve, thus blocking the passage of the UV light into the water and dramatically reducing the efficiency of the process. The scales are commonly caused by the precipitation of calcium, iron, or magnesium salts. Preventing the buildup of this scale is a major operational challenge for the use of UV. One of the current problems facing the use of UV irradiation is determining the actual UV dose the water receives, because measuring a residual is not possible.

VIII. ADVANCED TREATMENT METHODS

The term Advanced Oxidation Processes (AOPs) was first used to describe a process that produces hydroxyl radicals (^-OH) for the oxidation of organic and inorganic water impurities. AOPs can have multiple uses in water treatment. Examples include oxidation of synthetic organic chemicals (SOCs), color, taste-and-odor causing compounds, sulfide, iron, and manganese, and destruction of DBP precursor material prior to the addition of chlorine.

A. Peroxone

As an example of an AOP process, the following discussion on ozone with hydrogen peroxide is presented (referred to as "Peroxone"). The reactions of ozone can be divided into two types, direct ozonation reactions and free radical decomposition reactions. When hydrogen peroxide (H_2O_2) is added to ozonated water, it reacts with molecular ozone to accelerate the formation of hydroxyl radicals. Therefore, in an ozone–H_2O_2 process, the goal is to increase the concentration of hydroxyl radicals, which are a stronger oxidizer than molecular ozone, and reduce the concentration of molecular ozone. Therefore, hydrogen peroxide is added to an ozone process if it is used as an oxidation process, but not if ozone is intended to be used in a disinfection process that relies on the prevalence of a high concentration of molecular ozone. The Peroxone process is used for the destruction of taste and odor causing compounds, color removal, and the destruction of micropollutants, such as VOCs, pesticides, and herbicides.

Currently, the conventional design of an ozone–H_2O_2 treatment process is one where hydrogen peroxide is fed as a liquid to the influent water and an ozone-rich gas is fed through fine-bubble diffusers at the bottom of a contactor.

IX. NEW EMERGING ISSUES FOR DISINFECTION

For many years public water systems and regulatory agencies, such as the USEPA, have approached the idea of changing water treatment practices to limit the formation of DBPs with caution due to the possible implications for control of microbial risks in drinking water. Balancing acute risks from microbial pathogens such as viruses and protozoans against adverse health risks from DBPs has been a challenging situation to address. Selecting the appropriate disinfectant involves not only consideration of its ability to form potentially harmful DBPs but also its ability to effectively inactivate waterborne pathogens among other water quality considerations. For example, the use of chloramines can be an effective strategy for limiting the production of THMs, however, they are a weaker disinfectant than chlorine.

Giardia lamblia and *Cryptosporidium parvum* are two protozoan pathogens that have emerged to present significant challenges to public water systems and regulators for the delivery of *safe*, potable drinking water.

A. Giardia

Giardia lamblia is a flagellated protozoan that has a trophozoite form and a cyst form. The length of the trophozoite form ranges from 9 to 21 μm, the width from 5 to 15 μm and a thickness from 2 to 4 μm. In an unfavorable environment the parasite encysts, and the cysts are characteristically oval or ellipsoid in shape (slightly asymmetric) with a length of 8–14 μm and a width of 7–10 μm.

Giardiasis is an acute, self-limiting diarrheal disease caused by ingestion of Giardia cysts. Giardiasis is one of the most commonly identified waterborne intestinal diseases in the U.S. *Giardia lamblia* has been identified as the causative agent of numerous outbreaks involving public and private drinking water systems.

Outbreaks of Giardiasis have occurred due to the use of untreated surface water, contaminated water distribution systems and treatment deficiencies. Wild and domestic mammals can be significant sources of *Giardia* cysts. Human wastewater is another source of *Giardia* contamination. A person infected with *Giardia lamblia* may shed on the order of 10^8 cysts/day. Giardia cysts can be effectively removed from drinking water through the use of physical treatment such as filtration. The cysts can also be inactivated using disinfectants such as chlorine.

B. Cryptosporidium

Cryptosporidium is a protozoan parasite that reproduces within the gut of an animal host. The life cycle of

Cryptosporidium is extremely complex and includes the formation of resistant oocysts. *Cryptosporidium* oocysts can originate from a number of animal hosts, including cattle, swine, horses, deer, chicken, ducks, fish, turtles, guinea pigs, cats, and dogs. *Cryptosporidium* is transmitted by ingestion of oocysts that have been excreted in the feces of infected humans or animals. *Cryptosporidium parvum* has been recognized as a human pathogen since 1976 when the first case of cryptosporidiosis (the disease caused by *Cryptosporidium*) was diagnosed.

Like Giardiasis, Cryptosporidiosis is an acute self-limiting diarrheal disease. The oocysts can remain infective in water and moist environments for several months and are resistant to high concentrations of chlorine. This, together with the small size of the oocysts (4–6 μm in diameter) can lead to oocysts occasionally passing through conventional water treatment plants. Not all infected individuals develop symptoms but the illness can be serious or life-threatening for individuals with a compromised immune system (e.g., AIDS patients). Cryptosporidiosis outbreaks have been documented in the United States and around the world.

Table IV presents an overview of cryptosporidiosis outbreaks and the suspected source of *Cryptosporidium*

TABLE IV Waterborne Outbreaks of Cryptosporidiosis[a]

Location	Date	Water source	Number of cases (estimated)	Treatment	Suspected source of contamination
Braun Station, TX	July 1984	Well	47–117 (2006)	Chlorine	Raw Sewage
Sheffield, England	May–June 1986	Surface water	49 (537)	Unfiltered	Agricultural, nonpoint source pollution during heavy rainfall
Bernalillo County, NM	July–Oct. 1986	Surface water	78	Untreated	Runoff from livestock grazing areas
Carrolton, GA	Jan.–Feb. 1987	River	(13,000)	Conv., chlorine	Raw sewage and runoff from cattle grazing areas
Ayrshire, Scotland	Apr. 1988	Reservoir	(13,000)	N/A	Cross-connection to sewage contaminated source in a distribution system tank
Swindon and Oxfordshire, England	Mar. 1989	River	516	N/A	Agricultural nonpoint source pollution during heavy rainfall event
Berks County, PA	Aug. 1991	Well	(551)	Chlorine	Septic tank influence
Jackson County, OR (Talent and Medford)	Jan.–June 1992	Spring/river	(15,000)	Conv., chlorine	Surface water, treated wastewater or runoff from agricultural areas
Milwaukee, WI	March 1993	Lake	(403,000)	Conv., chlorine	Cattle wastes, slaughterhouse wastes and sewage carried by tributary rivers
Yakima County, WA	April 1993	Well	3 (7)	Untreated	Infiltration of runoff from cattle, sheep or elk grazing areas
Cook County, MN	Aug. 1993	Lake	5 (27)	Pressure filter, chlorine	Backflow of sewage or septic tank effluent into distribution, raw water inlet lines or both
Clark County, NV	Dec. 1993	Lake	78	Direct, chlorine	Treated wastewater, sewage from boats
Walla Walla, WA	Aug. 1994	Well	86	Chlorine	Treated wastewater
Alachua County, FL	July 1995	N/A	72	N/A	Backflow from a wastewater line in a camp's drinking water system
Collingwood, CANADA	1996	39(150)	N/A	N/A	Source of contamination unclear. 100 year storm prior to outbreak

[a] (Craun, G. F., Hubbs, S. A., Frost, F., Calderon, R. L., Via, S. H. (1998). "Waterborne Outbreaks of Cryptosporidiosis," *J. Am. Water Works Assoc.* **90**, 81–91.)

Solo-Gabriele, H., Neumeister, S. (1996). "US Outbreaks of Cryptosporidiosis," *J. Am Water Works Assoc.* 76–86.

Frey, M. M., Hancock, C., Logsdon, G. S. (1997). "*Cryptosporidium*: Answers to Questions Commonly Asked by Drinking Water Professionals," Amer. Water Works Assoc. Research Foundation.

that occurred between 1984 and 1996. The following presents brief description of two of those outbreaks of cryptosporidiosis that were significant in the United States (Milwaukee, WI; Las Vegas, NV).

1. Milwaukee Outbreak

The Milwaukee outbreak occurred during March and April 1993. Estimates are that 403,000 people had diarrhea (estimates are that around 100 people died) due to ingestion of *Cryptosporidium parvum*.

At the time of the outbreak, the city of Milwaukee operated two treatment plants. Both of the plants treated Lake Michigan water using conventional treatment. The intake for the treatment plant associated with the outbreak was located 42 ft below the surface of the lake. Chlorine and/or potassium permanganate were occasionally added at the intake to control taste and odors and to control zebra mussels.

Historically, alum had been used as the coagulant. In September 1992, the facility switched to polyaluminum chloride (PACl) as the coagulant. The switch was made after consulting with other water agencies treating Lake Michigan water, regulatory authorities and the chemical manufacturer. By converting to PACl, the intent was to achieve a higher finished water pH (for corrosion control), reduce sludge volume and improve coagulation effectiveness in cold water conditions.

After the rapid mix stage, the water passed into one of four 1 million-gallon parallel baffled coagulation basins. The water then passed through eight sand and gravel media filters. Flow rates and head loss (i.e., the pressure drop between two point along the path of a flowing liquid) were monitored across each filter and were the basis for initiating each filter cleaning procedure. Coagulant dosages were adjusted as needed on the basis of laboratory jar tests.

Between March 18 and April 8, the turbidity of the raw water into the treatment plant associated with the outbreak ranged from 1.5 nephelometric turbidity units (NTU) to 44 NTU. Levels of total coliforms ranged from less than 1 colony forming unit (cfu) per 100 mL to about 3200 cfu per 100 mL. Plant staff responded to the fluctuations in turbidity throughout the treatment processes by adjusting coagulant dosages. Coagulation dosage adjustments were also made to compensate for coagulation demands resulting from treatment to control taste and odor. The coagulant dosages were based on jar-test data and consultation with the chemical supplier.

Several times during the period of the incident, filter effluent turbidity levels exceeded turbidity values achieved in previous months. On April 2, 1993, the plant switched back to alum as the primary coagulant. From March 18 through April 8 (the day the plant was shut down) the efflu-

ent turbidity was highly variable and ranged between 0.1 and 2.7 NTU. The ultimate source of the *Cryptosporidium* oocysts was not determined. As indicated previously, nearly 400,000 people had diarrhea during the outbreak and estimates are that nearly 100 people died.

2. Las Vegas Outbreak

Between January and May 1994, Las Vegas experienced an outbreak of cryptosporidiosis in the severely immunocompromised population of the city. The treatment plant in question had a capacity of 400 million gallons per day and employed direct filtration. At this treatment plant, filter backwash water (e.g., the water used to clean the filter beds) was recycled to the beginning of the treatment plant after sedimentation in clarifiers. The source water was Lake Mead and the intake was located 130 ft below the surface. The average raw water turbidity from January 1993 through June 1995 was 0.14 NTU, with a high of 0.3 NTU and a low of 0.1 NTU in December 1993 and April 1994.

In March 1994, the local health department observed an increase in cryptosporidiosis in the HIV-infected community in the city. From June 28 through December 31, 1993, there were 9 cases of cryptosporidiosis reported. From January 1 through March 19, 1994, there were 49 cases of cryptosporidiosis reported. In June 1994, the outbreak ceased. There had been no changes in the treatment process of the distribution system.

In March 1995 the U.S. Centers for Disease Control and Prevention (CDC) released the results of their investigation. The data indicated that HIV-infected patients who drank tap water were at greater risk of contracting cryptosporidiosis than if they drank bottled or filtered water. CDC found no deficiencies in the treatment plant or in the distribution system. CDC was unable to locate a specific source of *Cryptosporidium* or evidence of *Cryptosporidium* in the water supply. During January 1 through April 30, 1994, seventy-eight people became infected, 65 adults and 13 children. Of the 65 adults, 61 had HIV infections. Of the 13 children, 2 were HIV infected. There was no geographic clustering of the cases of cryptosporidiosis observed.

X. DRINKING WATER REGULATIONS

A. United States Regulations

The Safe Drinking Water Act (SDWA) was enacted by the United States Congress in 1974. Through the SDWA, the federal government gave the USEPA the authority to set standards for contaminants in drinking water supplies. The EPA was required to establish primary regulations for the control of contaminants which affect public health and secondary regulations for compounds which affect the taste or aesthetics of drinking water.

The first step taken by the EPA to establish MCLs is to determine a maximum contaminant level goal (MCLG) for the target compound. The MCLG represents the estimated concentration in which no known or anticipated adverse health effects occur, including an adequate margin of safety. MCLGs are strictly a health-based number and do not take into account issues associated with analytical methods, treatment technology or economics. The MCLs, however, are then set as close to the MCLG as is technically and economically feasible. Under the SDWA, EPA can regulate compounds by requiring utilities to institute specified treatment techniques to limit its concentration in drinking water (the Surface Water Treatment Rule, the Lead and Copper Rule are examples of treatment techniques). The SDWA was reauthorized and amended by Congress in 1986 and again in 1996. Under the SDWA, individual states are allowed to adopt and implement drinking water regulations. To do so, states must adopt drinking water regulations that are no less stringent than the regulations published by USEPA (state regulations can be more stringent than federal regulations).

1. THM Regulation

The THM regulation was promulgated in 1979 and applied to all public water systems serving populations greater than 10,000. The THM regulation established an MCL of 100 μg/L for total trihalomethanes (TTHMs) in the distribution system. [Total trihalomethanes include the sum of chloroform, bromodichloromethane, dibromochloromethane, and bromoform.] Systems must collect a minimum of four distribution system samples per treatment plant on a quarterly basis. Compliance with the MCL is based on a running annual average concentration of four quarterly monitoring periods.

In 1998 USEPA published a DBP regulation that revised the 100-μg/L standard for THMs to 80 μg/L as well as establishing MCLs for HAAs (60 μg/L), bromate (10 μg/L), and chlorite (1000 μg/L).

EPA has also established maximum residual disinfectant levels (MRDLs) to limit the applied dose of chlorine, chloramines and chlorine dioxide during drinking water treatment:

Chlorine	4.0 mg/L
Chloramines	4.0 mg/L as total chlorine
Chlorine Dioxide	0.8 mg/L

a. Enhanced coagulation. Since only approximately 50% of the total organic halides produced during

TABLE V Enhanced Coagulation TOC Removal Requirements

Source Water TOC	Source Water Alkalinity, mg/l		
	0 to 60	>60 to 120	>120
>2.0 to 4.0 mg/l	35%	25%	15%
>4.0 to 8.0 mg/l	45%	35%	25%
>8.0 mg/l	50%	40%	30%

disinfection of drinking water have been identified, the EPA established a requirement for conventional treatment plants to remove total organic carbon (TOC) as a surrogate for DBP precursor material. The reasoning was that by removing TOC this would reduce the formation of unidentified DBPs with potential adverse health effects.

Performance criteria for TOC removal based on raw water TOC levels and alkalinity have been established to serve as a definition of enhanced coagulation. Table V presents a matrix identifying the percent TOC removal required based on raw water TOC and alkalinity levels. If a conventional treatment plant can not achieve the percent reductions listed in Table V, the system can use an alternative criteria for evaluating enhanced coagulation. This alternative criteria requires a system to conduct bench- or pilot-scale studies to identify optimum coagulation conditions.

2. Surface Water Treatment Rule and Interim Enhanced Surface Water Treatment Rule

The Surface Water Treatment Rule (SWTR) was published in June 1989. The SWTR was promulgated to control the levels of turbidity, *Giardia lamblia*, viruses, *Legionella*, and heterotrophic plate count bacteria in U.S. drinking waters. In 1998, the USEPA published an Interim Enhanced Surface Water Treatment Rule (ESWTR) that added additional restrictions to the SWTR and addressed the issue of *Cryptosporidium* removal.

The SWTR requires all utilities utilizing a surface water supply or a ground water supply under the influence of a surface water supply, to provide adequate disinfection and under most conditions, to provide filtration. (Surface water supplies can avoid the requirement to filter their supply if they meet specific requirements for water quality and the utility can demonstrate control of the watershed.)

a. General requirements. The SWTR and the Interim ESWTR includes the following general requirements in order to minimize human exposure to microbial contaminants in drinking water.

- Utilities are required to achieve at least 99.9% removal and/or inactivation of *Giardia lamblia* cysts (3-log removal), a minimum 99.99% removal and/or inactivation of viruses (4-log removal), and a 2-log removal of *Cryptosporidium*. The required level of removal/inactivation must occur between the point where the raw water is not longer subject to surface water runoff and the point at which the first customer is served.
- The disinfectant residual entering the distribution system must not fall below 0.2 mg/L for more than 4 hr during any 24-hr period.
- A disinfectant residual must be detectable in 95% of distribution system samples. A heterotrophic plate count (HPC) concentration of less than 500 colonies per milliliter can serve as a detectable residual if no residual is measured.
- Each utility must perform a watershed sanitary survey at least every 5 years.
- The combined filtered water turbidity must be less than or equal to 0.3 NTU in 95% of samples taken during each month (measured every four hours), never to exceed 1 NTU.
- Systems are required to achieve 2-log reduction in *Cryptosporidium*.
- If a system meets the turbidity performance standard of 0.3 NTU in 95% of the samples taken each month (never to exceed 1 NTU) than the system gets credit for achieving the 2-log reduction in *Cryptosporidium*.
- Systems must monitor the turbidity for individual filters, and any filter that is not performing per specified criteria must be evaluated.

b. Removal credit.

The level of removal credit given a utility for both *Giardia lamblia* and viruses is determined by the type of treatment process used. For a conventional water treatment plant the SWTR provides a 2.5-log removal credit for *Giardia lamblia* and a 2.0-log removal credit for viruses. As described above, water treatment plants meeting the turbidity performance standard of 0.3 NTU in 95% of the monthly measurements also get credit for a 2-log reduction in *Cryptosporidium*.

c. Disinfection credit.

Disinfection during conventional treatment (assuming all operational criteria and performance standards are met), must achieve 0.5-log inactivation of *Giardia lamblia* and 2.0-log inactivation of viruses. As a substitute to actually measuring the inactivation of *Giardia lamblia* and viruses achieved at a treatment plant, the SWTR established the concept of CT to evaluate inactivation. CT is the product of the concentration of disinfectant remaining at the end of a treatment process ("C" in milligrams per liter) and the contact time in which

10% of the water passes through the treatment process ("T" or "T$_{10}$" in minutes). The SWTR provides tables which identify the log removal of both *Giardia lamblia* and viruses achieved for a calculated CT value based on the type of disinfectant, the water temperature, and pH.

3. Total Coliform Rule

The Total Coliform Rule (TCR) was promulgated by the EPA in June 1989. A specific MCL value was not established for total or fecal coliforms under the TCR. Instead, there are three potential scenarios in which an MCL is violated that depend on the presence and/or absence of total coliforms (and fecal coliforms or E. *coli*). Public water systems monitor for total coliforms throughout the distribution system to determine compliance with the TCR.

TABLE VI WHO Guidelines for Bacteriological Quality of Drinking Water[a]

Organisms	Guideline value
All water intended for drinking	
E. *coli* or thermotolerant coliform bacteria[b,c]	Must not be detectable in any 100-mL sample
Treated water entering the distribution system	
E. *coli* or thermotolerant coliform bacteria[b]	Must not be detectable in any 100-mL sample
Total coliform bacteria	Must not be detectable in any 100-mL sample
Treated water in the distribution system	
E. *coli* or thermotolerant coliform bacteria[b]	Must not be detectable in any 100-mL sample
Total coliform bacteria	Must not be detectable in any 100-mL sample. In the case of large supplies, where sufficient samples are examined, must not be present in 95% of samples taken throughout any 12-month period.

[a] Immediate investigative action must be taken if either E. *coli* or total coliform bacteria are detected. The minimum action in the case of total coliform bacteria is repeat sampling; if these bacteria are detected in the repeat sample, the cause must be determined by immediate further investigation.

[b] Although E. *coli* is the more precise indicator of fecal pollution, the count of thermotolerant coliform bacteria is an acceptable alternative. If necessary, proper confirmatory tests must be carried out. Total coliform bacteria are not acceptable indicators of the sanitary quality of rural water supplies, particularly in tropical areas where many bacteria of no sanitary significance occur in almost all untreated supplies.

[c] It is recognized that, in the great majority of rural water supplies in developing countries, fecal contamination is widespread. Under these conditions, the national surveillance agency should set medium-term targets for the progressive improvement of water supplies.

TABLE VII WHO and USEPA Limits for Inorganic Chemical Constituents[a]

Constituent	WHO guideline value (mg/L)	USEPA limit (mg/L)
Antimony	0.005 (P)[b]	0.006
Arsenic	0.01 (P)	0.05
Asbestos	—	7 MFL
Barium	0.7	2
Beryllium	—	0.004
Boron	0.3	—
Cadmium	0.003	0.005
Chromium	0.05 (P)	0.1
Copper	2 (P)	TT
Cyanide	0.07	0.2
Fluoride	1.5[c]	4.0
Lead	0.01	TT
Manganese	0.5 (P)	—
Mercury	0.001	0.002
Molybdenum	0.07	—
Nickel	0.02	—
Nitrate (as NO_3^-)	50	45
Nitrite (as NO_2^-)	3 (P)	—
Selenium	0.01	0.05
Radium 226 and 228	—	5 pCi/L[d]
Gross alpha particle activity	0.1 Bq/L[c]	15 pCi/L
Beta particle and photon activity	1 Bq/L	4 mrem (annual dose equivalent)[e]

[a] MFL = million fibers per liter greater than 10 μ in length.
[b] (P) = Provisional.
[c] Bq/L = Becquerel per liter.
[d] pCi/L = picoCuries per liter.
[e] mrem = millirem.

4. Lead and Copper Rule

The Lead and Copper Rule (LCR) was published by the EPA on June 7, 1991. The objective of the LCR is to minimize the corrosion of lead and copper-containing plumbing materials in public water systems (PWS) by requiring utilities to optimize treatment for corrosion control. The LCR establishes "action levels" in lieu of MCLs for regulating the levels of both lead and copper in drinking water. The action level for lead was established at 0.015 mg/L while the action level for copper was set at 1.3 mg/L. An action level is exceeded when greater than 10% of samples collected from the sampling pool contain lead levels above 0.015 mg/L or copper levels above 1.3 mg/L.

5. Arsenic

Arsenic is the 20th most abundant element in nature and the 12th most abundant in the human body. Arsenic occurs in both the organic and inorganic forms. The inorganic forms are arsenite (As^{+3}) and arsenate (As^{+5}). These are the most toxic forms of arsenic, and arsenite appears to be the most toxic. Human exposure to arsenic comes primarily from food sources (shellfish and grain raised in arsenic-laden soils) and is mostly in the organic forms. EPA is in the process of lowering the MCL for arsenic, perhaps significantly lower than the current standard. Historically, arsenic has been regulated in drinking water based on its potential to cause skin cancer (usually a nonfatal disease). Information indicates that arsenic may be associated with internal cancers such as lung, kidney and bladder.

Potential treatment processes for reducing arsenic levels include activated alumina, activated carbon, ion exchange, adsorption/coprecipitation with Fe/Al oxides, and membrane processes.

B. WHO Regulations

The World Health Organization (WHO) has published guidelines for drinking water quality. According to WHO,

TABLE VIII USEPA and WHO Limits for Volatile Organic Chemicals in Drinking Water

Contaminant	Federal MCL (mg/L)	WHO guideline (mg/L)
1,1-Dichloroethylene	0.007	0.030
1,2-Dichloroethylene	—	0.050
1,1,1-Trichloroethane	0.2	2 (P)
1,1,2-Trichloroethane	0.005	—
1,2-Dichloroethane	0.005	0.030
1,2-Dichloropropane	0.005	0.020 (P)
1,3-Dichloropropylene	—	0.020
1,2,4-Trichlorobenzene	0.07	—
Benzene	0.005	0.010
Carbon tetrachloride	0.005	0.002
cis-1,2-Dichloroethylene	0.07	—
Dichloromethane	0.005	0.020
Ethylbenzene	0.7	0.30
Monochlorobenzene	0.1	0.30
1,2-Dichlorobenzene	—	1
1,3-Dichlorobenzene	0.6	—
1,4-Dichlorobenzene	0.075	0.30
Trichlorobenzene (total)	—	0.020
Styrene	0.1	0.020
Tetrachloroethylene	0.005	0.040
Toluene	1	0.70
trans-1,2-Dichloroethylene	0.1	—
Trichloroethylene	0.005	0.070 (P)
Vinyl chloride	0.002	0.005
Xylenes (total)	10	0.50

TABLE IX USEPA and WHO Limits for Synthetic Organic Chemical in Drinking Water[a]

Contaminant	Federal MCL (mg/L)	WHO guideline value (mg/L)
2,3,7,8-TCDD (Dioxin)[b]	3×10^{-8}	—
2,4-D	0.07	0.030
2,4,5-T	—	0.009
2,4,5-TP (Silvex)	0.05	—
Acrylamide	TT	0.0005
Alachlor	0.002	0.020
Aldicarb	(a)	0.010
Aldrin/dieldrin	—	0.00003
Atrazine	0.003	0.002
Bentazone	—	0.030
Benzo(a)pyrene	0.0002	0.0007
Carbofuran	0.04	0.005
Chlordane	0.002	0.0002
Chlorotoluron	—	0.030
Dalapon	0.2	—
Di(2-ethylhexyl)adipate	0.4	0.080
Di(2-ethylhexyl)phthalate	0.006	0.008
Dibromochloropropane (DBCP)	0.0002	0.001
Dichlorprop	—	0.10
Dinoseb	0.007	—
Diquat	0.02	—
DDT	—	0.002
Edetic acid (EDTA)	—	0.20
Endothal	0.1	—
Endrin	0.002	—
Epichlorohydrin	TT	0.0004
Ethylene dibromide	0.00005	—
Fenoprop	—	0.009
Glyphosate	0.7	—
Heptachlor	0.0004	0.00003
Heptachlor epoxide	0.0002	0.00003
Hexachlorobenzene	0.001	0.001
Hexachlorobutadiene	—	0.0006
Hexachlorocyclopentadiene	0.05	—
Isoproturon	—	0.009
Lindane	0.0002	0.002
MCPA	—	0.002
Mecoprop	—	0.010
Methoxychlor	0.04	0.020
Metolachlor	—	0.010
Molinate	—	0.006
Nitrilotriacetic acid	—	0.20
Oxamyl (vydate)	0.2	—
Pendimethalin	—	0.020
Pentachlorophenol	0.001	0.009 (P)
Permethrin	—	0.020
Picloram	0.5	—

continues

TABLE IX (*continued*)

Contaminant	Federal MCL (mg/L)	WHO guideline value (mg/L)
Polychlorinated biphenyl (PCB)	0.0005	—
Propanil	—	0.020
Pyridate	—	0.10
Simazine	0.004	0.002
Tributyltin oxide	—	0.002
Trifluralin	—	0.020
Toxaphene	0.003	—

[a] P = Provisional guideline value. This term is used for constituents for which there is some evidence of a potential hazard but where the available information on health effects in limited or where an uncertainty factor greater than 1000 has been used in the derivation of the tolerable daily intake. Provisional guideline values are also recommended (1) for substances for which the calculated guideline value would be below the practical quantification level, or below the level that can be achieved through practical treatment methods; or (2) where disinfection is likely to result in the guideline value being exceeded.

TT = Treatment Technique.

[b] The USEPA MCLs for aldicarb (MCL = 0.01 mg/L), aldicarb sulfoxide (MCL = 0.01 mg/L), aldicarb sulfone (MCL = 0.04 mg/L) are not enforced.

these guidelines are intended to be used as the basis for the development of national standards that will ensure the safety of drinking water supplies.

According to WHO, the judgement of safety—or what is an acceptable level of risk in particular circumstances—is a matter in which society as a whole has a role to play. The final judgement as to whether the benefit resulting from the adoption of any of the guideline values given here as standards justifies the cost is for each country to decide.

The guideline values presented by WHO represent the concentration of a given constituent that does not result in any significant risk to the health of the consumer over a lifetime of consumption. Short-term deviations above the guideline values do not necessarily mean that the water is unsuitable for consumption. The amount by which, and the period for which, any guideline values can be exceeded

TABLE X USEPA and WHO Limits for Disinfectants

Disinfectant	USEPA MRDL[a] (mg/L)	WHO guideline value (mg/L)
Chorine[b]	4.0	
Chloramine[b]	4.0	
Monochloramine	—	3
Di- and trichloramine	—	5
Chlorine dioxide	0.8	

[a] MRDL = maximum residual disinfectant level.

[a] Measured as free chlorine.

without affecting public health depends upon the specific substance involved. In developing national drinking water standards based on these guideline values, it will be necessary to take account of a variety of geographical, socio-economic, dietary, and other conditions affecting potential exposure. This may lead to national standards that differ appreciably from the guideline values.

C. Current USEPA Regulations and WHO Guideline Values

Table VI presents the WHO guidelines for bacteriological quality of drinking water. Tables VII, VIII, and IX contain the USEPA standards and WHO Guideline Values for inorganic, volatile organic and synthetic organic chemicals, respectively. Table X and XI presents the USEPA and WHO Guidelines Values for disinfectants and disinfection byproducts. Table XII presents the USEPA limits (secondary standards that are not enforceable at the federal level) and the WHO Guideline Values for constituents affecting the taste and odor of the water.

TABLE XI USEPA and WHO Limits for Disinfection By-Products[a]

	USEPA MCL (mg/L)	WHO guideline value (mg/L)
Constituent		
Bromate	0.010	0.025 (P)
Chlorite	1.0	0.20
2,4,6-chlorophenol	—	0.20
Formaldehyde	—	0.90
Total THMs	0.080	—
Bromoform	—	0.10
Dibromochloromethane	—	0.10
Bromodichloromethane	—	0.06
Chloroform	—	0.20
HAA5	0.060	—
Dichloroacetic acid	—	0.050 (P)
Trichloroacetic acid	—	0.10 (P)
Chloral hydrate	—	0.010 (P)
Dichloroacetonitrile	—	0.090 (P)
Dibromoacetonitrile	—	0.10 (P)
Trichloroacetonitrile	—	0.001 (P)
Cyanogen chloride	—	0.070

[a] P = Provisional guideline value. This term is used for constituents for which there is some evidence of a potential hazard but where the available information on health effects in limited or where an uncertainty factor greater than 1000 has been used in the derivation of the tolerable daily intake. Provisional guideline values are also recommended (1) for substances for which the calculated guideline value would be below the practical quantification level, or below the level that can be achieved through practical treatment methods; or (2) where disinfection is likely to result in the guideline value being exceeded.

TABLE XII USEPA and WHO Limits for Constituents Affecting Taste and Odor

	USEPA limit (mg/L)	WHO guideline value (mg/L)[a]
Color	15 CU	15 TCU
Turbidity	—	5 NTU
Aluminum	0.05–0.2	0.2
Ammonia	—	1.5
Chloride	—	250
Copper	1.0	1
Fluoride	2.0	—
Hydrogen sulfide	—	0.05
Iron	0.3	0.3
Manganese	0.05	0.1
Sodium	—	200
Sulfate	250	250
Total dissolved solids	500	1000
Zinc	5	3
Touene	—	0.024–0.170
Xylene	—	0.020–1.8
Ethylbenzene	—	0.002–0.2
Styrene	—	0.004–2.6
Monochlorobenzene	—	0.010-0.12
1,2-dichlorobenzene	—	0.001–0.010
1,4-Dichlorobenzene	—	0.0003–0.030
Trichlorobenzene (total)	—	0.005–0.050
Chlorine	—	0.6–1.0
2-Chlorophenol	—	0.0001–0.010
2,4-Dichlorophenol	—	0.0003–0.040
2.4.6-Trichlorophenol	—	0.002–0.3

[a] The levels indicated are not precise numbers. Problems may occur at lower or higher values according to local circumstances. A range of taste and odor threshold concentrations is given for organic constituents. TCU = time color unit. CU = color unit. NTU = nephelometric turbidity unit.

SEE ALSO THE FOLLOWING ARTICLES

DRINKING WATER QUALITY AND TREATMENT ● ENVIRONMENTAL MEASUREMENTS ● ENVIRONMENTAL TOXICOLOGY ● POLLUTION, AIR ● POLLUTION, ENVIRONMENTAL ● SOIL AND GROUNDWATER POLLUTION ● TRANSPORT AND FATE OF CHEMICALS IN THE ENVIRONMENT ● WASTEWATER TREATMENT AND WATER RECLAMATION ● WATER POLLUTION ● WATER RESOURCES

BIBLIOGRAPHY

American Water Works Association (1990). Pontius, F. (technical ed.). "Water Quality and Treatment," McGraw-Hill, New York.
Faust, S. D., and Aly, O. M. (1998). "Chemistry of Water Treatment," Ann Arbor Press, MI.

Fox, K. R., and Lytle, D. A. (1996). "Milwaukee's crypto outbreak: Investigation and recommendations," *J. Am. Water Works Assoc.* **88,** 87–94.

Glaze, Wh. H., Kang, H. W., and Chapin, H. D. (1987). "The chemistry of water treatment processes involving ozone, hydrogen peroxide and ultraviolet radiation," *Ozone Sci. Engrg.* **9,** 335.

Goldstein, S. T., *et al.* (1996). Cryptosporidiosis: An Outbreak Associated with Drinking Water Despite State-of-the-Art Treatment. *Ann. Int. Med.* **124,** 459–468.

Karamura, S. (1991). "Integrated Design of Water Treatment Facilities," Wiley, New York.

MacKenzie, W. R., *et al.* (1994). "A massive outbreak in Milwaukee of *Cryptosporidium* infection transmitted through the public water supply," *New Engl. J. Med.* **331** (3), 161–167.

Montgomery, J. M. (Consulting Engineer) (1985). "Water Treatment Principles and Design," Wiley, New York.

National Research Council (1998). "Issues in Potable Reuse. The Viability of Augmenting Drinking Water Supplies with Reclaimed Water," National Academy Press. Washington, D.C.

Singer, P. (1999). "Formation and Control of Disinfection By-Products in Drinking Water," American Water Works Association. Denver, CO.

Twort, A., Ratnayaka, D., and Brandt, M. (2000). "Water Supply," (5th edition), Arnold and IWA Publishing, London.

USEPA (1999). National Primary Drinking Water Regulations. "Code of Federal Regulations," **40,** Sections 141.1–141.175.

USEPA (1998). National Secondary Drinking Water Regulations. "Code of Federal Regulations," **40,** Sections 143.1–143.5.

WHO (1993). "Guidelines for Drinking-Water Quality," (2nd edition), Volume 1, Recommendations, World Health Organization, Geneva.

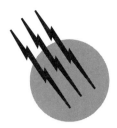

Dynamic Programming

Martin L. Puterman
The University of British Columbia

I. Sequential Decision Problems
II. Finite-Horizon Dynamic Programming
III. Infinite-Horizon Dynamic Programming
IV. Further Topics

GLOSSARY

Action One of several alternatives available to the decision maker when the system is observed in a particular state.

Decision rule Function that determines for the decision maker which action to select in each possible state of the system.

Discount factor Present value of a unit of currency received one period in the future.

Functional equation Basic entity in the dynamic programming approach to solving sequential decision problems. It relates the optimal value for a $(t + 1)$-stage decision problem to the optimal value for a t-stage problem. Its solution determines an optimal decision rule at a particular stage.

Horizon Number of stages.

Markov chain Sequence of random variables in which the conditional probability of the future is independent of the past when the present state is known.

Markov decision problem Stochastic sequential decision problem in which the set of actions, the rewards, and the transition probabilities depend only on the current state of the system and the current action selected; the history of the problem has no effect on current decisions.

Myopic policy Policy in which each decision rule ignores the future consequence of the decision and uses the action at each stage that maximizes the immediate reward.

Policy Sequence of decision rules.

Stage Point in time at which a decision is made.

State Description of the system that provides the decision maker with all the information necessary to make future decisions.

Stationary Referring to a problem in which the set of actions, the set of states, the reward function, and the transition function are the same at each decision point or to a policy in which the same decision rule is used at every decision point.

IN ALL AREAS of endeavor, decisions are made either explicitly or implicitly. Rarely are decisions made in isolation. Today's decision has consequences for the future because it could affect the availability of resources or limit the options for subsequent decisions. A sequential decision problem is a mathematical model for the problem faced by a decision maker who is confronted with a sequence of interrelated decisions and wishes to make them in an optimal fashion. Dynamic programming is a collection of mathematical and computational tools for analyzing sequential decision problems. Its main areas of application are operations research, engineering, statistics, and resource management. Improved computing capabilities will lead to the wide application of this technique in the future.

I. SEQUENTIAL DECISION PROBLEMS

A. Introduction

A system under the control of a decision maker is evolving through time. At each point of time at which a decision can be made, the decision maker, who will be referred to as "he" with no sexist connotations intended, observes the state of the system. On the basis of this information, he chooses an action from a set of alternatives. The consequences of this action are two-fold; the decision maker receives an immediate reward or incurs an immediate cost, and the state that the system will occupy at subsequent decision epochs is influenced either deterministically or probabilistically. The problem faced by the decision maker is to choose a sequence of actions that will optimize the performance of the system over the decision-making horizon. Since the action selected at present affects the future evolution of the system, the decision maker cannot choose his action without taking into account future consequences.

Dynamic programming is a procedure for finding optimal policies for sequential decision problems. It differs from linear, nonlinear, and integer programming in that there is no standard dynamic programming problem formulation. Instead, it is a collection of techniques based on developing mathematical recursions to decompose a multistage problem into a series of single-stage problems that are analytically or computationally more tractable. Its implementation often requires ingenuity on the part of the analyst, and the formulation of dynamic programming problems is considered by some practitioners to be an art. This subject is best understood through examples. This section proceeds with a formal introduction of the basic sequential decision problem and follows with several examples. The reader is encouraged to skip back and forth between these sections to understand the basic ingredients of such a problem. Dynamic programming methodology is discussed in Sections II and III.

B. Problem Formulation

Some formal notation follows. Let T denote the set of time points at which decisions can be made. The set T can be classified in two ways; it is either finite or infinite and either a discrete set or a continuum. The primary focus of this article is when T is discrete. Discrete-time problems

are classified as either finite horizon or infinite horizon according to whether the set T is finite or infinite. The problem formulation in these two cases is almost identical; however, the dynamic programming methods of solution differ considerably. For discrete-time problems, T is the set $\{1, 2, \ldots, N\}$ in the finite case and $\{1, 2, \ldots\}$ in the infinite case. The present decision point is denoted by t and the subsequent point by $t + 1$. The points of time at which decisions can be made are often called stages. Almost all the results in this article concern discrete-time models; the continuous-time model is briefly mentioned in Section IV.

The set of possible states of the system at time t is denoted by S_t. In finite-horizon problems, this is defined for $t = 1, 2, \ldots, N + 1$, although decisions are made only at times $t = 1, 2, \ldots, N$. This is because the decision at time N often has future conseqences that can be summarized by evaluating the state of the system at time $N + 1$. This is analogous to providing boundary values for differential equations. If at time t the decision maker observes the system in state $s \in S_t$, he chooses an action a from the set of allowable actions at time t, $A_{s,t}$. As above, S_t and $A_{s,t}$ can be either finite or infinite and discrete or continuous. This distinction has little consequence for the problem formulation.

As a result of choosing action a when the system is in state s at time t, the decision maker receives an immediate reward $r_t(s, a)$. This reward can be positive or negative. In the latter case it can be thought of as a cost. Furthermore, the choice of action affects the system evolution either deterministically or probabilistically. In the deterministic case, the choice of action determines the state of the system at time $t + 1$ with certainty. Denote by $w_t(s, a) \in S_{t+1}$ the state the system will occupy if action a is chosen in state s at time t; $w_t(s, a)$ is called the transfer function. When the system evolves probabilistically, the subsequent state is random and choice of action specifies its probability distribution. Let $p_t(j|s, a)$ denote the probability that the system is in state $j \in S_{t+1}$ if action a is chosen in state s at time t; $p_t(j|s, a)$ is called the transition probability function. When S_t is a continuum, $p_t(j|s, a)$ is a probability density. Such models are discussed briefly in Section IV. A sequential decision problem in which the transitions from state to state are governed by a transition probability function and the set of actions and rewards depends only on the current state and stage is called a Markov decision problem.

The deterministic model is a special case of the probabilistic model obtained by choosing $p_t(j|s, a) = 1$ if $j = w_t(s, a)$ and $p_t(j|s, a) = 0$ if $j \neq w_t(s, a)$. Even though there is this equivalence, the transfer function representation is more convenient for deterministic problems.

A decision rule is a function $d_t : S_t \to A_{s,t}$ that specifies the action the decision maker chooses when the system is

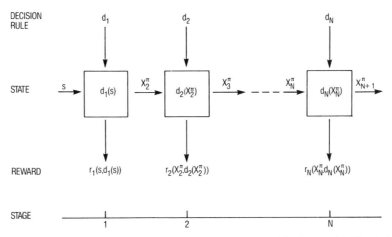

FIGURE 1 Evolution of the sequential decision model under the policy $\pi = (d_1, d_2, \ldots, d_N)$. The state at stage 1 is s.

in state s at time t; that is, $d_t(s)$ specifies an action in $A_{s,t}$ for each s in S_t. A decision rule of this type is called Markovian because it depends only on the current state of the system. The set of allowable decision rules at time t is denoted by D_t and is called the decision set. Usually it is the set of all functions mapping S_t to $A_{s,t}$, but in some applications, it might be a proper subset.

Many generalizations of deterministic Markovian decision rules are possible. Decision rules can depend on the entire history of the system, which is summarized in the sequence of observed states and actions observed up to the present, or they can depend only on the initial and current state of the system. Furthermore, the decision rule might be randomized; that is, in each state it specifies a probability distribution on the set of allowable actions so that by using such a rule the decision maker chooses his action at each decision epoch by a probabilistic mechanism. For the problems considered in this article, using deterministic Markovian decision rules at each stage is optimal so that the generalizations referred to above will not be discussed further.

A policy specifies the sequence of decision rules to be used by the decision maker over the course of the planning horizon. A policy π is a finite or an infinite sequence of decision rules; that is, $\pi = \{d_1, d_2, \ldots, d_N\}$, where $d_t \in D_t$ for $t = 1, 2, \ldots, N$ if the horizon is finite, or $\pi = \{d_1, d_2, \ldots\}$, where $d_t \in D_t$ for $t = 1, 2, \ldots$ if the horizon is infinite. Let Π denote the set of all possible policies; $\Pi = D_1 \times D_2 \times \cdots \times D_N$ in the finite case and $\Pi = D_1 \times D_2 \times \cdots$ in the infinite case.

In deterministic problems, by specifying a policy at the start of the problem, the decision maker completely determines the future evolution of the system. For each policy the sequence of states the system will occupy is known with certainty, and hence the sequence of rewards the decision maker will receive over the planning horizon is

known. Let X_t^π denote the state the system occupies at time t if the decision maker uses policy π over the planning horizon. In the first period the decision maker receives a reward of $r_1(s, d_1(s))$, in the second period a reward of $r_2(X_2^\pi, d_2(X_2^\pi))$, and in the tth period $r_t(X_t^\pi, d_t(X_t^\pi))$. Figure 1 depicts the evolution of the process under a policy $\pi = \{d_1, d_2, \ldots, d_N\}$ in both the deterministic and stochastic cases. The quantity in each box indicates the interaction of the incoming state with the prespecified decision rule to produce the indicated action $d_t(X_t^\pi)$. The arrow to the right of a box indicates the resulting state, and the arrow downward the resulting reward to the decision maker. The system is assumed to be in state s before the first decision.

The decision maker evaluates policies by comparing the value of a function of the policy's income stream. Many such evaluation functions are available, but it is most convenient to assume a linear, additive, and risk-neutral utility function over time, which leads to using the total reward over the planning horizon for evaluation. Let $v_N^\pi(s)$ be the total reward over the planning horizon. It is given by the expression

$$v_N^\pi(s) = \sum_{t=1}^{N+1} r_t\left(X_t^\pi, d_t\left(X_t^\pi\right)\right), \tag{1}$$

in which it is implicit that $X_1^\pi = s$. For deterministic problems, evaluation formulas such as Eq. (1) always depend on the initial state of the process, although this is not explicitly stated below.

In probabilistic problems, by specifying a policy at the start of the problem, the decision maker determines the transition probability functions of a nonstationary Markov chain. The sequence of states the system will occupy is not known with certainty, and consequently the sequence of rewards the decision maker will receive over the planning horizon is not known. Instead, what is known is the joint probability distribution of system states and rewards. In this case, expectations with respect to the joint probability distributions of the Markov chain conditional on the state at the first decision epoch are often used to evaluate policy performance. As in the deterministic case, let X_t^π denote the state the system occupies at time t if the decision maker uses policy π over the planning horizon. For finite-horizon problems, let $v_N^\pi(s)$ equal the total expected reward over the planning horizon. It is given by the expression

$$v_N^\pi(s) = E_{\pi,s}\left\{\sum_{t=1}^{N+1} r_t\left(X_t^\pi, d_t\left(X_t^\pi\right)\right)\right\}, \quad (2)$$

where $E_{\pi,s}$ denotes the expectation with respect to probability distribution determined by π conditional on the initial state being s.

In both the deterministic and stochastic problems the decision maker's problem is to choose at time 1 a policy π in Π to make $v_N^\pi(s)$ as large as possible and to find the maximal reward,

$$v_N^*(s) = \sup_{\pi \in \Pi} v_N^\pi(s) \qquad \text{for all} \quad s \in S_1. \quad (3)$$

Frequently the problem is such that the supremum in Eq. (3) is attained—for example, when both $A_{s,t}$ and S_t are finite for all $t \in T$. In such cases the decision maker's objective is to maximize $v_N^\pi(s)$ and find its maximal value.

For infinite-horizon problems, the total reward or the expected total reward the decision maker receives will not necessarily be finite; that is, the summations in Eqs. (1) and (2) usually will not converge. To evaluate policies in infinite-horizon problems, decision makers often use discounting or averaging. Let λ represent the discount factor, usually $0 \le \lambda < 1$. It measures the value at present of one unit of currency received one period from now. Let $v_\lambda^\pi(s)$ equal the total discounted reward in deterministic problems or the expected total discounted reward for probabilistic problems if the system is in state s, before choosing the first action. For deterministic problems it is given by

$$v_\lambda^\pi(s) = \sum_{t=1}^{\infty} \lambda^{t-1} r_t\left(X_t^\pi, d_t\left(X_t^\pi\right)\right), \quad (4)$$

and for stochastic problems it is given by

$$v_\lambda^\pi(s) = E_{\pi,s}\left\{\sum_{t=1}^{\infty} \lambda^{t-1} r_t\left(X_t^\pi, d_t\left(X_t^\pi\right)\right)\right\}. \quad (5)$$

In this setting the decision maker's problem is to choose at time 1 a policy π in Π to make $v_\pi^\lambda(s)$ as large as possible and to find the supremal reward,

$$v_\lambda^*(s) = \sup_{\pi \in \Pi} v_\lambda^\pi(s) \qquad \text{for all} \quad s \in S_1. \quad (6)$$

Alternatively in the infinite-horizon setting, the decision maker might not be willing to assume that a reward received in the future is any less valuable than a reward received at present. For example, if decision epochs are very close together in real time, then all rewards the decision maker receives would have equal value. In this case the decision maker's objective might be to choose a policy that maximizes the average or expected average reward per period. This quantity is frequently called the gain of a policy. For a specified policy it is denoted by $g^\pi(s)$ and in both problems is given by

$$g^\pi = \lim_{N \to \infty} \frac{1}{N} v_N^\pi(s), \quad (7)$$

where $v_N^\pi(s)$ is defined in Eqs. (1) and (2).

In this setting the decision maker's problem is to choose at time 1 a policy π in Π to make $g^\pi(s)$ as large as possible and to find the supremal average reward,

$$g^*(s) = \sup_{\pi \in \Pi} g^\pi(s) \qquad \text{for all} \quad s \in S_1. \quad (8)$$

Dynamic programming methods with the average reward criteria are quite complex and are discussed only briefly in Section III. The reader is referred to the works cited in the Bibliography for more details.

Frequently in infinite-horizon problems, the data are stationary. This means that the set of states, the set of allowable actions in each state, the one-period rewards, the transition or transfer functions, and the decision sets are the same at every stage. When this is the case, the time subscript t is deleted and the notation S, A_s, $r(s, a)$, $p(j|s, a)$ or $w(s, a)$, and D is used. Often stationary policies are optimal in this setting. By a stationary policy is meant a policy that uses the identical decision rule in each period; that is, $\pi = (d, d, \ldots)$. Often it is denoted by d when there is no possible source of confusion.

C. Examples

The following examples clarify the notation and formulation described in the preceding sections. The first example illustrates a deterministic, finite-state, finite-action, finite-horizon problem; the second a deterministic, infinite-state, infinite-action, finite-horizon problem; and the third a stochastic, finite-state, finite-action problem with both finite- and infinite-horizon versions. In Sections II and III, these examples will be solved by using dynamic programming methodology.

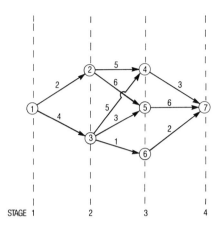

STAGE 1   2   3   4

FIGURE 2 Network for the longest-route problem.

1. A Longest-Route Problem

A finite directed graph is depicted in Fig. 2. The circles are called nodes, and the lines connecting them are called arcs. On each arc, an arrow indicates the direction in which movement is possible. The numerical value on the arc is the reward the decision maker receives if he chooses to traverse that arc on his journey from node 1 to node 7. His objective is to find the path from node 1 to node 7 that maximizes the total reward he receives on his journey. Such a problem is called a longest-route problem. A practical application is determining the length of time needed to complete a project. In such problems, the arc length represents the time to complete a task. The entire project is not finished until all tasks are performed. Finding the longest path through the network gives the minimum amount of time for the entire project to be completed since it corresponds to that sequence of tasks that requires the most time. The longest path is called a critical path because, if any task in this sequence is delayed, the entire project will be delayed.

In other applications the values on the arcs represent lengths of road segments, and the decision maker's objective is to find the shortest route from the first node to the terminal node. Such a problem is called a shortest-route problem. All deterministic, finite-state, finite-action, finite-horizon dynamic programming problems are equivalent to shortest- or longest-route problems. Another example of this will appear in Section II.

An important assumption is that the network contains no directed cycle; that is, there is no route starting at a node that returns to that node. If this were the case, the longest route would be infinite and the problem would not be of interest.

The longest-route problem is now formulated as a sequential decision problem. This requires defining the set of states, actions, decision sets, transfer functions, and rewards. They are as follows:

Decision points:

$$T = \{1, 2, 3\}$$

States (numbers correspond to nodes):

$$S_1 = \{1\}; \qquad S_2 = \{2, 3\}; \qquad S_3 = \{4, 5, 6\};$$
$$S_4 = \{7\}$$

Actions (action j selected in node i corresponds to choosing to traverse the arc between nodes i and j; the first subscript on A is the state and the second the stage):

$$A_{1,1} = \{2, 3\}$$
$$A_{2,2} = \{4, 5\}; \qquad A_{3,2} = \{4, 5, 6\}$$
$$A_{4,3} = \{7\}; \qquad A_{5,3} = \{7\}; \qquad A_{6,3} = \{7\}$$

Rewards:

$$r_1(1, 2) = 2; \qquad r_1(1, 3) = 4$$
$$r_2(2, 4) = 5; \qquad r_2(2, 5) = 6;$$
$$r_2(3, 4) = 5, \qquad r_2(3, 5) = 3; \qquad r_2(3, 6) = 1$$
$$r_3(4, 7) = 3; \qquad r_3(5, 7) = 6; \qquad r_3(6, 7) = 2$$

Transfer function:

$$w_t(s, a) = a$$

The remaining ingredients in the sequential decision problem formulation are the decision set, the set of policies, and an evaluation formula. The decision set at stage t is the set of all arcs emanating from nodes at stage t. A policy is a list of arcs in which there is one arc starting at each node (except 7) in the network. The policy set contains all such lists. Each policy contains a route from node 1 to node 7 and some superfluous action selections. The value of the policy is the total of the rewards along this route, and the decision maker's problem is to choose a policy that maximizes this total reward.

The structure of a policy is described in more detail through an example. Consider the policy $\pi = \{(1, 2), (2, 4), (3, 4), (4, 7), (5, 7), (6, 7)\}$. Implicit in this definition is a sequence of decision rules $d_t(s)$ for each state and stage. They are $d_1(1) = 2$, $d_2(2) = 4$, $d_2(3) = 4$, $d_3(4) = 7$, $d_3(5) = 7$, and $d_3(6) = 7$. This policy can be formally denoted by $\pi = \{d_1, d_2, d_3\}$. The policy contains one unique routing through the graph, namely, $1 \rightarrow 2 \rightarrow 4 \rightarrow 7$ and several unnecessary decisions. We use the formal notation $X_1^\pi = 1$, $X_2^\pi = 2$, $X_3^\pi = 4$, and $X_4^\pi = 7$ so that

$$v_3^\pi(1) = r_1(1, d_1(1)) + r_2(2, d_2(2)) + r_3(4, d_3(4))$$
$$= r_1(1, 2) + r_2(2, 4) + r_3(4, 7)$$
$$= 2 + 5 + 3 = 10.$$

In such a small problem, one can easily evaluate all policies by enumeration and determine that the longest route through the network is $1 \rightarrow 2 \rightarrow 5 \rightarrow 7$ with a return of 14. For larger problems this is not efficient; dynamic programming methods will be seen to offer an efficient means of determining the longest route.

The reader might note that the formal sequential decision process notation is quite redundant here. The subscript for stage does not convey any useful information and the specification of a policy requires making decisions in nodes that will never be reached. Solution by dynamic programming methods will require this superfluous information. In other settings this information will be useful.

2. A Resource Allocation Problem

A decision maker has a finite amount K of a resource to allocate between N possible activities. Using activity i at level x_i consumes $c_i(x_i)$ units of the resource and yields a reward or utility of $f_i(x_i)$ to the decision maker. The maximum level of intensity for activity i is M_i. His objective is to determine the intensity for each of the activities that maximizes his total reward. When any level of the activity is possible, this is a nonlinear programming problem. When the activity can operate only at a finite set of levels, this is an integer programming problem. In the special case that the activity can be either utilized or not ($M_i = 1$ and x_i is an integer) this is often called a knapsack problem. This is because it can be used to model the problem of a camper who has to decide which of N potential items to carry in his knapsack. The value of item i is $f_i(1)$ and it weighs $c_i(1)$. The camper wishes to select the most valuable set of items that do not weigh more than the capacity of the knapsack.

The mathematical formulation of the resource allocation problem is as follows:

Maximize $f_1(x_1) + f_2(x_2) + \cdots + f_N(x_N)$
subject to

$$c_1(x_1) + c_2(x_2) + \cdots + c_N(x_N) \leq K \qquad (9)$$

and

$$0 \leq x_t \leq M_t, \qquad t = 1, 2, \ldots, N. \qquad (10)$$

The following change of variables facilitates the sequential decision problem formulation. Define the new variable $s_i = c_i(x_i)$ and assume that c_i is a monotone increasing function on $[0, M_i]$. this assumption says that the more intense the activity level, the more resource utilized. Define $g_i(s_i) = f_i(c_i^{-1}(s_i))$ and $m_i = c_i^{-1}(M_i)$. This change of variables corresponds to formulating the problem in terms of the quantity of resource being used. In this notation the formulation above becomes

Maximize $g_1(s_1) + g_2(s_2) + \cdots + g_N(s_N)$
subject to

$$s_1 + s_2 + \cdots + s_N \leq K \qquad (11)$$

and

$$0 \leq s_t \leq m_t, \qquad t = 1, 2, \ldots, N. \qquad (12)$$

It is not immediately obvious that this problem is a sequential decision problem. The formulation is based on treating the problem as if the decision to allocate resources to the activities were done sequentially through time with allocation to activity 1 first and activity N last. Decisions are coupled, since successive allocations must take the quantity of the resource allocated previously into account. That is, if $K - s$ units of resource have been allocated to the first t activities, then s units are available for activities $t + 1, t + 2, \ldots, N$.

The following sequential decision problem formulation is based on the second formulation above:

Decision points (correspond to activity number):

$$T = \{1, 2, \ldots, N\}$$

States (amount of resource available for allocation in remaining stages): For $0 \leq t \leq N$,

$$S_t = \begin{cases} \{s : 0 \leq s \leq m_t\} \\ \quad \text{if resource levels are continuous} \\ \{0, 1, 2, \ldots, m_t\} \\ \quad \text{if resource levels are discrete} \end{cases}$$

For $t = N + 1$,

$$S_t = \begin{cases} \{s : 0 \leq s \leq K\} \\ \quad \text{if resource levels are continuous} \\ \{0, 1, 2, \ldots, K\} \\ \quad \text{if resource levels are discrete} \end{cases}$$

Actions (s is amount of resource available for stages t, $t + 1, \ldots, N$):

$$A_{s,t} = \begin{cases} \{u : 0 \leq u \leq \min(s, m_t)\} \\ \quad \text{if resource levels are continuous} \\ \{0, 1, 2, \ldots, \min(s, m_t)\} \\ \quad \text{if resource levels are discrete} \end{cases}$$

Rewards:

$$r_t(s, a) = g_t(a)$$

Transfer function:

$$w_t(s, a) = s - a$$

The decision set at stage t is the set of all functions from S_t to $A_{s,t}$, and a policy is a sequence of such functions, one for each $t \in T$. A decision rule specifies the amount of resource to allocate to activity t if s units are available for allocation to activities $t, t + 1, \ldots, N$, and a policy specifies which decision rule to use at each stage of the sequential allocation. As in the longest-route problem, a policy specifies decisions in many eventualities that will not occur using that policy. This may seem wasteful at first but is fundamental to the dynamic programming methodology. The quantity X_t^π is the amount of remaining resource available for allocation to activities $t, t+1, \ldots, N$ using policy π. Clearly, $X_1^\pi = K$, $X_2^\pi = K - d_1(K)$, and so forth. The decision maker compares policies through the quantity $v_N^\pi(K)$, which is given by

$$v_N^\pi(K) = g_1(d_1(K)) + g_2(d_2(X_2^\pi)) + \cdots + g_N(d_N(X_N^\pi)).$$

When the set of activities is discrete, the resource allocation problem can be formulated as a longest-route problem, as can any discrete state and action sequential decision problem. This is depicted in Fig. 3 for the following specific example:

Maximize $\quad 3s_1^2 + s_2^3 + 4s_3$

subject to

$$s_1 + s_2 + s_3 \le 4,$$

$s_1, s_2,$ and s_3 are integers, and

$$0 \le s_1 \le 2; \quad 0 \le s_2 \le 2; \quad 0 \le s_3 \le 2.$$

In the longest-route formulation, the node labels are the amount of resource available for allocation at subsequent stages. A fifth stage is added so that there is a unique destination and all decisions at stage 4 correspond to moving from a node at stage 4 to node 0 with no reward. This is because an unallocated resource has no value to the decision maker in this formulation. The number on each arc is the reward, and the amount of resource allocated is the difference between the node label at stage t and that at stage $t+1$. For instance, if at stage 2, there are 3 units of resource available for allocation over successive stages

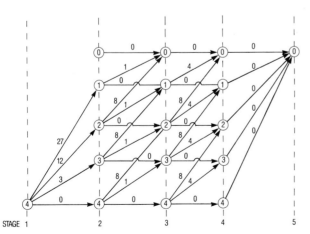

FIGURE 3 Network for the resource allocation problem.

and the decision maker decides to allocate 2 units, he will receive a reward of $2^3 = 8$ units and move to node 1.

When the resource levels form a continuum, the network representation is no longer valid. The problem is reduced to a sequence of constrained one-dimensional nonlinear optimization problems through dynamic programming. Such an example will be solved in Section II by using dynamic programming methods.

3. A Stochastic Inventory Control Problem

Each month, the manager of a warehouse must determine how much of a product to keep in stock to satisfy customer demand for the product. The objective is to maximize expected total profit (sales revenue less inventory holding and ordering costs), which may or may not be discounted. The demand throughout the month is random, with a known probability distribution. Several simplifying assumptions make a concise formulation possible:

a. The decision to order additional stock is made at the beginning of each month and delivery occurs instantaneously.

b. Demand for the product arrives throughout the month but is filled on the last day of the month.

c. If demand exceeds the stock on hand, the customer goes elsewhere to purchase the product.

d. The revenues and costs and the demand distribution are identical each month.

e. The product can be sold only in whole units.

f. The warehouse capacity is M units.

Let s_t denote the inventory on hand at the beginning of month t, a_t the additional product ordered in month t, and D_t the random demand in month t. The demand has a known probability distribution given by $p_d = P\{D_t = d\}$, $d = 0, 1, 2, \ldots$. The cost of ordering u units in any month is $O(u)$ and the cost of storing u units for 1 month is $h(u)$. The ordering cost is given by

$$O(u) = \begin{cases} K + c(u), & \text{if} \quad u > 0 \\ 0, & \text{if} \quad u = 0, \end{cases} \quad (13)$$

where $c(u)$ and $h(u)$ are increasing functions of u. For finite-horizon problems, if u units of inventory are on hand at the end of the planning horizon, its value is $g(u)$. Finally, if u units of product are demanded in a month and the inventory is sufficient to satisfy demand, the manager receives $f(u)$. Define $F(u)$ to be the expected revenue in a month if the inventory before receipt of customer orders is u units. It is given in period t by

$$F(u) = \sum_{j=0}^{u-1} f(j)p_j + f(u)P\{D_t \ge u\}. \quad (14)$$

Equation (14) can be interpreted as follows. If the inventory on hand exceeds the quantity demanded, j, the revenue is $f(j)$; p_j is the probability that the demand in a period is j units. If the inventory on hand is u units and the quantity demanded is at least u units, then the revenue is $f(u)$ and $P\{D_t \geq u\}$ is the probability of such a demand. The combined quantity is the probability-weighted, or expected, revenue.

This is a stochastic sequential decision problem (Markov decision problem), and its formulation will include a transition probability function instead of a transfer function. The formulation follows:

Decision points:
$$T = \{1, 2, \ldots, N\};$$
N may be finite or infinite

States (units of inventory on hand at the start of a month):
$$S_t = \{0, 1, 2, \ldots M\}, \qquad t = 1, 2, \ldots, N+1$$

Actions (the amount of additional stock to order in month t):
$$A_{s,t} = \{0, 1, 2, \ldots, M - s\}$$

Expected rewards (expected revenue less ordering and holding costs):
$$r_t(s, a) = F(s + a) - O(a) - h(s + a),$$
$$t = 1, 2, \ldots, N$$

Value of terminal inventory (no actions are possible):
$$r_{N+1}(s, a) = g(s)$$

Transition probabilities (see explanation below):
$$p_t(j|s, a) = \begin{cases} 0, & \text{if } j > s + a \\ p_j, & \text{if } j = s + a - D_t, \\ & \quad s + a \leq M, \text{ and} \\ & \quad s + a > D_t \\ q_{s+a}, & \text{if } j = 0, \quad s + a \leq M, \\ & \quad \text{and } s + a \leq D_t \end{cases}$$

where
$$q_{s+a} = P\{D_t \geq s + a\} = \sum_{d=s+a}^{\infty} p_d.$$

A brief explanation of the transition probabilities might be helpful. If the inventory on hand at the beginning of period t is s units and an order is placed for a units, the inventory before external demand is $s + a$ units. If the demand of j units is less than $s + a$ units, then the inventory at the beginning of period $t + 1$ is $s + a - j$ units. This occurs with probability p_j. If the demand exceeds $s + a$ units, then the inventory at the start of period $t + 1$ is 0

units. This occurs with probability q_{s+a}. Finally, the probability that the inventory level ever exceeds $s + a$ units is zero, since this demand is nonnegative.

The decision sets consist of all rules that assign the quantity of inventory to be ordered each month to each possible starting inventory position in a month. A policy is a sequence of such ordering rules. Unlike deterministic problems, in which a decision rule is specified for many states that will never be reached, in stochastic problems such as this, it is necessary for the decision maker to determine the decision rule for *all* states. This is because the evolution of the inventory level over time is random, which makes any inventory level possible at any decision point. Consequently the decision maker must plan for each of these eventualities.

An example of a decision rule is as follows: Order only if the inventory level is below 3 units at the start of the month and order the quantity that raises the stock level to 10 units on receipt of the order. In month t this is given by
$$d_t(s) = \begin{cases} 10 - s, & s < 3 \\ 0, & s \geq 3. \end{cases}$$

The evaluation method for a policy depends on the time horizon under consideration. For finite-horizon problems, the total expected cost conditional on the initial stock level is a convenient summary. Assuming the stock level at time 1 is s, the expected total reward for policy π is
$$v_N^\pi(s) = E_{\pi,s} \left\{ \left[\sum_{t=1}^N F\left(X_t^\pi + d_t^\pi(X_t^\pi)\right) - O\left(d_t^\pi(X_t)\right) \right. \right.$$
$$\left. \left. - h\left(X_t^\pi + d_t(X_t^\pi)\right) \right] - g\left(X_{N+1}^\pi\right) \right\}.$$

If, instead, the decision maker wishes to discount future profit at a monthly discount rate of λ, $0 \leq \lambda < 1$, the term λ^{t-1} is inserted before each term in the above summation and λ^N before the terminal reward g. For an infinite-horizon problem, the expected total discounted profit is given by
$$v_\lambda^\pi = E_{\pi,s} \left\{ \sum_{t=1}^{\infty} \lambda^{t-1} \left[F\left(X_t^\pi + d_t(X_t^\pi)\right) \right. \right.$$
$$\left. \left. - O\left(d_t^\pi(X_t^\pi)\right) + h\left(X_t^\pi + d_t(X^\pi)\right) \right] \right\}.$$

The decision maker's problem is to choose a sequence of decision rules to maximize expected total or total discounted profits.

Many modifications of this inventory problem are possible; for example, excess demand in any period could be backlogged and a penalty for carrying unsatisfied demand could be charged, or there could be a time lag between placing the order and its receipt. The formulation herein can easily be modified to include such changes; the interested reader is encouraged to consult the Bibliography for more details.

A numerical example is now provided in complete detail. It will be solved in subsequent sections by using dynamic programming methods. The data for the problem are as follows: $K = 4$, $c(u) = 2u$, $g(u) = 0$, $h(u) = u$, $M = 3$, $N = 3$, $f(u) = 8(u)$, and

$$
p_d = \begin{cases} \frac{1}{4}, & \text{if} \quad d = 0 \\ \frac{1}{2}, & \text{if} \quad d = 1 \\ \frac{1}{4}, & \text{if} \quad d = 2. \end{cases}
$$

The inventory is constrained to be 3 or fewer units, and the decision maker wishes to consider the effects over three periods. All the costs and revenues are linear. This means that for each unit ordered the per unit cost is 2, for each unit held in inventory for 1 month the per unit cost is 1, and for each unit sold the per unit revenue is 8. The expected revenue when u units of stock are on hand before receipt of an order is given by

u	$F(u)$
0	0
1	$0 \times \frac{1}{4} + 8 \times \frac{3}{4} = 6$
2	$0 \times \frac{1}{4} + 8 \times \frac{1}{2} + 16 \times \frac{1}{4} = 8$
3	$0 \times \frac{1}{4} + 8 \times \frac{1}{2} + 16 \times \frac{1}{4} = 8$

Combining the expected revenue with the expected shortage and holding costs gives the expected profit in period t if the inventory level is s at the start of the period and an order for a units is placed. If $a = 0$, the ordering and holding cost equals s, and if a is positive, it equals $4 + s + 3a$. It is summarized in the tabulations below, where an X corresponds to an action that is infeasible. Transition probabilities depend only on the total inventory on hand before the receipt of orders. They are the same for any s and a that have the same total $s + a$. So that redundant information is reduced, transition probabilities are presented as functions of $s + a$ only. The information in the following tabulations defines this problem completely:

$r_t(s, a)$

s	a = 0	a = 1	a = 2	a = 3
0	0	−1	−2	−5
1	5	0	−3	X
2	6	−1	X	X
3	5	X	X	X

$p_t(j \mid s, a)$

s + a	j = 0	j = 1	j = 2	j = 3
0	0	0	0	0
1	$\frac{3}{4}$	$\frac{1}{4}$	0	0
2	$\frac{1}{4}$	$\frac{1}{2}$	$\frac{1}{4}$	0
3	0	$\frac{1}{4}$	$\frac{1}{2}$	$\frac{1}{4}$

II. FINITE-HORIZON DYNAMIC PROGRAMMING

A. Introduction

Dynamic programming is a collection of methods for solving sequential decision problems. The methods are based on decomposing a multistage problem into a sequence of interrelated one-stage problems. Fundamental to this decomposition is the *principle of optimality*, which was developed by Richard Bellman in the 1950s. Its importance is that an optimal solution for a multistage problem can be found by solving a functional equation relating the optimal value for a $(t + 1)$-stage problem to the optimal value for a t-stage problem.

Solution methods for problems depend on the time horizon and whether the problem is deterministic or stochastic. Deterministic finite-horizon problems are usually solved by backward induction, although several other methods, including forward induction and reaching, are available. For finite-horizon stochastic problems, backward induction is the only method of solution. In the infinite-horizon case, different approaches are used. These will be discussed in Section III. The backward induction procedure is described in the next two sections. This material might seem difficult at first; the reader is encouraged to refer to the examples at the end of this section for clarification.

B. Functional Equation of Dynamic Programming

Let $v^t(s)$ be the maximal total reward received by the decision maker during stages $t, t + 1, \ldots, N + 1$, if the system is in state s immediately before the decision at stage t. When system transitions are stochastic, $v^t(s)$ is the maximal expected return. Recall that decisions are made at stages $1, 2, \ldots, N$ and not at time $N + 1$; however, a reward might be received at stage $N + 1$ as a consequence of the decision in stage N. In most deterministic problems, S_1 consists of one element, whereas in stochastic problems, solutions are usually required for all possible initial states.

The basic entity of dynamic programming is the functional equation, or Bellman equation, which relates $v^t(s)$ to $v^{t+1}(s)$. For deterministic problems it is given by

$$
v^t(s) = \max_{a \in A_{s,t}} \{ r_t(s, a) + v^{t+1}(w_t(s, a)) \},
$$

$$
t = 1, \ldots, N \tag{15}
$$

and

$$
v^{N+1}(s) = 0, \tag{16}
$$

where Eq. (15) is valid for all $s \in S_t$ and Eq. (16) is valid for all $s \in S_{N+1}$. Equation (15) is the basis for the backward induction algorithm for solving sequential decision problems. This equation corresponds to a one-stage problem in which the decision maker observes the system in state s and must select an action from the set $A_{s,t}$. The consequence of this action is that the decision maker receives an immediate reward of $r_t(s, a)$ and the system moves to state $w_t(s, a)$, at which he receives a reward of $v^{t+1}(w_t(s, a))$. Equation (15) says that he chooses the action that maximizes the total of these two rewards. This is exactly the problem faced by the decision maker in a one-stage sequential decision problem when the terminal reward function is v^{t+1}. Equation (16) provides a boundary condition. When the application dictates, this value 0 can be replaced by an arbitrary function that assigns a value to the terminal state of the system. Such might be the case in the inventory control example.

Equation (15) emphasizes the dynamic aspects of the sequential decision problem. The decision maker chooses that action which maximizes his immediate reward *plus* his reward over the remaining decision epochs. This is in contrast to the situation in which the decision maker behaves myopically and chooses the decision rule that maximizes the reward only in the current period and ignores future consequences. Some researchers have given conditions in which such a myopic policy is optimal; however, in almost all problems dynamic aspects must be taken into account.

The expression "max" requires explanation because it is fundamental to the dynamic programming methodology. If $f(x, y)$ is any function of two variables with $x \in X$ and $y \in Y$, then

$$g(x) = \max_{y \in Y}\{f(x, y)\}$$

if for each $x \in X$, $g(x) \geq f(x, y)$ for all $y \in Y$ and there exists a $y^* \in Y$ with the properties that $f(x, y^*) \geq f(x, y)$ for all $y \in Y$ and $g(x) = f(x, y^*)$. Thus, Eq. (15) states that the decision maker chooses $a \in A_{s,t}$ to make the expression in braces as large as possible. The quantity $v^t(s)$ is set equal to this maximal value.

In stochastic problems, the functional equation (15) is modified to account for the probabilistic transition structure. It is given by

$$v^t(s) = \max_{a \in A_{s,t}}\left\{r_t(s, a) + \sum_{j \in S_{t+1}} p_t(j|s, a)v^{t+1}(j)\right\},$$

$$t = 1, \ldots, N. \tag{17}$$

The stochastic nature of the problem is accounted for in Eq. (17) by replacing the fixed transition function $w_t(s, a)$ by the random state j, which is determined by the probability transition function corresponding to selecting action a. The second expression in this equation equals the expected reward received over the remaining periods as a consequence of choosing action a in period t.

C. Backward Induction and the Principle of Optimality

Backward induction is a procedure that uses the functional equation in an iterative fashion to find the optimal total value function and an optimal policy for a finite-horizon sequential decision problem. That this method achieves these objectives is demonstrated by the principle of optimality. The principle of optimality is not a universal truth that applies to all sequential decision problems but a mathematical result that requires formal proof in each application. For problems in which the (expected) total reward criterion is used, as considered in this article, it is valid. A brief argument of why it holds in such problems is given below.

To motivate backward induction, the following iterative procedure for finding the total reward of some specified policy $\pi = (d_1, d_2, \ldots, d_N)$ is given. It is called the policy evaluation algorithm. To simplify notation, assume that $p_t(j|s, a) = p(j|s, a)$ and $r_t(s, a) = r(s, a)$ for all s, a, and j.

a. Set $t = N + 1$ and $v^{N+1}(s) = 0$ for all $s \in S_{N+1}$.
b. Substitute $t - 1$ for $t(t - 1 \rightarrow t)$ and compute $v^t(s)$ for each $s \in S_t$ in the deterministic case by

$$v^t(s) = r(s, d_t(s)) + v^{t+1}(w_t(s, d_t(s))), \tag{18}$$

or in the stochastic case by

$$v^t(s) = r(s, d_t(s)) + \sum_{j \in S_{t+1}} p(j|s, d_t(s))v^{t+1}(j). \tag{19}$$

c. If $t = 1$, stop; otherwise, return to step b.

This procedure inductively evaluates the policy by first fixing its value at the last stage and then computing its value at the previous stage by adding its immediate reward to the previously computed total value. This process is repeated until the first stage is reached. This computation process yields the quantities $v^1(s)$, $v^2(s)$, $\ldots$, $v^{N+1}(s)$. The quantity $v^1(s)$ equals the expected total value of policy π, which in earlier notation is given by $v_\pi^N(s)$. The quantities $v^1(s)$ correspond to the value of this policy from stage t onward. This procedure is extended to optimization by iteratively choosing and evaluating a policy consisting of the actions that give the maximal return from each stage to the end of the planning horizon instead of just evaluating a fixed prespecified policy.

The backward induction algorithm proceeds as follows:

a. Set $t = N + 1$ and $v^{N+1}(s) = 0$ for all $s \in S_{N+1}$.

b. Substitute $t - 1$ for $t (t - 1 \rightarrow t)$ and compute $v^t(s)$ for each $s \in S_t$ using Eq. (15) or (17) depending on which is appropriate. Denote by $A^*_{s,t}$ the set of actions a^* for which in the deterministic case,

$$v^t(s) = r(s, a^*) + v^{t+1}\big(w_t(s, a^*)\big), \qquad (20)$$

or in the stochastic case,

$$v^t(s) = r(s, a^*) + \sum_{j \in S_{t+1}} p(j|s, a^*)v^{t+1}(j). \qquad (21)$$

c. If $t = 1$, stop; otherwise, return to step b.

By comparing this procedure with the policy evaluation procedure above, we can easily see that the backward induction algorithm accomplishes three objectives:

a. It finds sets of actions $A^*_{s,t}$ that contain all actions in $A_{s,t}$ that obtain the maximum in Eq. (15) or (17).

b. It evaluates any policy made up of actions selected from the sets $A^*_{s,t}$.

c. It gives the total return or expected total return $v_t(s)$ that would be obtained if a policy corresponding to selecting actions in $A^*_{s,t}$ were used from stage t onward.

Thus, if the decision maker had specified a policy that selected actions in the sets $A^*_{s,t}$ before applying the policy evaluation algorithm, these two procedures would be identical. It will be argued below that any policy obtained by selecting an action from $A^*_{s,t}$ in each state at every stage is optimal and consequently $v^1(s)$ is the optimal value function for the problem; that is, $v^1(s) = v^*_N(s)$.

In deterministic problems, specifying a policy often provides much superfluous information, since if the state of the system is known before the first decision, a policy determines the system evolution with certainty and only one state is reached at each stage. Since all deterministic problems are equivalent to longest-route problems, the objective in such problems is *only* to find a longest route. The following route selection algorithm does this. The system state is known before decision 1, so S_1 contains a single state.

a. Set $t = 1$ and for $s \in S_t$ define $d_t(s) = a^*$ for some $a^* \in A^*_{s,t}$. Set $u = w_t(s, a^*)$.

b. For $u \in S_{t+1}$, define $d_{t+1}(u) = a^*$ for some $a^* \in A^*_{u,t+1}$. Replace u by $u = w_{t+1}(u, a^*)$.

c. If $t + 1 = N$, stop; otherwise, $t + 1 \rightarrow t$ and return to step b.

In this algorithm, the choice of an action at each node determines which arc will be traversed, and decisions are necessary only at nodes that can be reached from the initial state. the algorithm traces forward through the network along the path determined by decisions that obtain the maximum in Eq. (15). It produces *one* route through the network with longest length. If at any stage the set $A^*_{s,t}$ contains more than one action, then several optimal routings exist, and if all are desired, the procedure must be carried out to trace each path.

A problem closely related to the longest-route problem is that of finding the longest route from *each* node to the final node. When there is only one action in $A^*_{s,t}$ for each s and t, then specifying the decision rule that in each state is equal to the unique maximizing action produces these routings. This is closer in spirit to the concept of a policy than a longest route.

That the above procedure results in an optimal policy and optimal value function is due to the additivity of rewards in successive periods. A formal proof of these results is based on induction, but the following argument gives the main idea. The backward induction algorithm chooses maximizing actions in reverse order. It does not matter what happened before the current stage. The only important information for future decisions is the current state of the system. First, for stage N the best action in each state is selected. Clearly, $v^N(s)$ is the optimal value function for a one-stage problem beginning in stage s at stage N. Next, in each state at stage $N - 1$, an action is found to maximize the immediate reward plus the reward that will be obtained if, after reaching a state at stage N, the decision maker chooses the optimal action at that stage. Clearly, $v^{N-1}(s)$ is the optimal value for a one-stage problem with terminal reward $v^N(s)$. Since $v^N(s)$ is the optimal value for the one-stage problem starting at stage N, no greater total reward can be obtained over these two stages. Hence, $v^{N-1}(s)$ is the optimal reward from stage $N - 1$ onward starting in state s. Now, since the sets $A^*_{s,N}$ and $A^*_{s,N-1}$ have been determined by the backward induction algorithm, choosing any policy that selects actions from these sets at each stage and evaluating it with the policy evaluation algorithm above will also yield $v^{N-1}(s)$. Thus, this policy is optimal over these two stages since its value equals the optimal value. This argument is repeated at stages $N - 2, N - 3, \ldots, 1$ to conclude that a policy that selects an action from $A^*_{s,t}$ at each stage is optimal.

The above argument contains the essence of the principle of optimality, which appeared in its original form on p. 83 of Bellman's classic book, "Dynamic Programming," as follows:

An optimal policy has the property that whatever the initial state and initial decision are, the remaining decisions must constitute an optimal policy with regard to the state resulting from the first decision.

The functional equations (15) and (17) are mathematical statements of this principle.

It might not be obvious to the reader why the backward induction algorithm is more attractive than an enumeration procedure for solving sequential decision problems. To see this, suppose that there are N stages, M states at each stage, and K actions that can be chosen in each state. Then there are $(K^M)^N$ policies. Solving a deterministic problem by enumeration would require $N(K^M)^N$ additions and $(K^M)^N$ comparisons. By backward induction, solution would require NMK additions and NK comparisons, a potentially astronomical savings in work. Solving stochastic problems requires additional M multiplications at each state at each stage to evaluate expectations. Enumeration requires $MN(K^M)^N$ multiplications, whereas backward induction would require NKM multiplications. Clearly, backward induction is a superior method for solving any problem of practical significance.

D. Examples

In this section, the use of the backward induction algorithm is illustrated in terms of the examples that were presented in Section I. First, the longest-route problem in Fig. 2 is considered.

1. A Longest-Route Problem

Note that $N = 3$ in this example.

a. Set $t = 4$ and $v^4(7) = 0$.
b. Since $t \neq 1$, continue. Set $t = 3$ and
$$v^3(4) = r_3(4, 7) + v^4(7)$$
$$= 3 + 0 = 3,$$
$$A^*_{6,3} = \{7\},$$
$$v^3(5) = r_3(5, 7) + v^4(7)$$
$$= 6 + 0 = 6,$$
$$A^*_{5,3} = \{7\},$$
$$v^3(6) = r_3(6, 7) + v^4(7)$$
$$= 2 + 0 = 2,$$
$$A^*_{4,3} = \{7\}.$$

c. Since $t \neq 1$, continue. Set $t = 2$ and
$$v^2(2) = \max\{r_2(2, 4) + v^3(4), r_2(2, 5) + v^3(5)\}$$
$$= \max\{5 + 3, 6 + 6\} = 12,$$
$$v^2(3) = \max\{r_2(3, 4) + v^3(4), r_2(3, 5) + v^3(5),$$
$$r_2(3, 6) + v^3(6)\}$$
$$= \max\{5 + 3, 3 + 6, 1 + 2\} = 9,$$
$$A^*_{2,2} = \{5\}; \qquad A^*_{3,2} = 5.$$

d. Since $t \neq 1$, continue. Set $t = 1$ and
$$v^1(1) = \max\{r_1(1, 2) + v^2(2), r_1(1, 3) + v^2(3)\}$$
$$= \max\{2 + 12, 4 + 9\} = 14,$$
$$A^*_{1,1} = \{2\}.$$

e. Since $t = 1$, stop.

This algorithm yields the information that the longest route from node 1 to node 7 has length 14, the longest route from node 2 to node 7 has length 12, and so on. To find the choice of arcs that corresponds to the longest route, we must apply the route selection algorithm:

a. Set $t = 1$, $d_1(1) = 2$, and $u = 2$.
b. Set $d_2(2) = 5$ and $u = 5$.
c. Since $t + 1 \neq 3$, continue. Set $t = 2$, $d_3(5) = 7$, and $u = 7$.
d. Since $t + 1 = 3$, stop.

This procedure gives the longest path through the network, namely, $1 \to 2 \to 5 \to 7$, which can easily be seen to have length 14. Note that choosing the myopic policy at each node would not have been optimal. At node 1, the myopic policy would have selected action 3; at node 3, action 4; and at node 4, action 7. The path $1 \to 3 \to 4 \to 7$ has length 12. By not taking future consequences into account at the first stage, the decision maker would have found himself in a poor position for subsequent decisions.

The backward induction algorithm has also obtained an optimal policy. It is given by

$$\pi^* = \left(d_1^*, d_2^*, d_3^*\right),$$

where

$$d_1^*(1) = 2, \qquad d_2^*(2) = 5, \qquad d_2^*(3) = 5,$$
$$d_3^*(4) = 7, \qquad d_3^*(5) = 7, \qquad \text{and} \qquad d_3^*(6) = 7.$$

This policy provides the longest route from each node to node 7, as promised by the principle of optimality. In the language of graph theory, this corresponds to a maximal spanning tree. The longest route from each node to node 7 is depicted in Fig. 4.

2. A Resource Allocation Problem

The backward induction algorithm is now applied to the continuous version of the resource allocation problem of Section 1. Computation in the discrete problem is almost identical to that in the longest-route problem and will be left to the reader.

The bounds on the s_i's are changed to simplify exposition. The problem that will be solved is given by

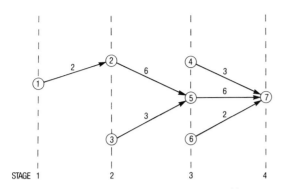

FIGURE 4 Solution to the longest-route problem.

Maximize $\quad 3s_1^2 + s_2^3 + 4s_3$

subject to

$$s_1 + s_2 + s_3 \leq 4$$

and

$$0 \leq s_1 \leq 3; \qquad 0 \leq s_2; \qquad 0 \leq s_3.$$

The backward induction is applied as follows:

a. Set $t = 4$ and $v^4(s) = 0, 0 \leq s \leq 4$.
b. Since $t \neq 1$, continue. Set $t = 3$ and

$$v^3(s) = \max_{0 \leq a \leq s} \left\{ r_3(s, a) + v^4(s - a) \right\}$$

$$= \max_{0 \leq a \leq s} \{4a\} = 4s$$

and $A_{s,3}^* = \{s\}$.
c. Since $t \neq 1$, continue. Set $t = 2$ and

$$v^2(s) = \max_{0 \leq a \leq s} \left\{ r_2(s, a) + v^3(s - a) \right\}$$

$$= \max_{0 \leq a \leq s} \{a^3 + 4(s - a)\},$$

$$v^2(s) = \begin{cases} 4s, & 0 \leq s \leq 2 \\ s^3, & 2 \leq s \leq 4, \end{cases}$$

and

$$A_{s,2}^* = \begin{cases} \{0\}, & 0 \leq s \leq 2 \\ \{s\}, & 2 \leq s \leq 4. \end{cases}$$

d. Since $t \neq 1$, continue. Set $t = 1$. A solution is obtained for $v^1(4)$ only. Obviously it is optimal to allocate all resources in this problem. In most problems one would obtain a $v^1(s)$ for all s. That is quite tedious in this example and unnecessary for solution of the original problem.

$$v^1(4) = \max_{0 \leq a \leq 3} \left\{ r_1(s, a) + v^2(4 - a) \right\}$$

$$= \max \left\{ \max_{\substack{0 \leq a \leq 3 \\ 0 \leq 4 - a \leq 2}} \{3a^2 + 4(4 - a)\}, \right.$$

$$\left. \max_{\substack{0 \leq a \leq 3 \\ 2 \leq 4 - a}} \{3a^2 + 4(4 - a)^3\} \right\}$$

$$= \max \left\{ \max_{2 \leq a \leq 3} \{3a^2 + 4(4 - a)\}, \right.$$

$$\left. \max_{0 \leq a \leq 2} \{3a^2 + (4 - a)^3\} \right\}$$

$$= \max\{31, 30\} = 31$$

and $A_{4,1}^* = \{3\}$.
e. Since $t = 1$, stop.

The objective function value for this constrained resource allocation problem is 31. To find the optimal resource allocation, we must use the second algorithm above:

a. Set $t = 1$, $d_1(4) = 3$, and $u = 1$.
b. Set $d_2(1) = 0$ and $u = 1$.
c. Since $t + 1 \neq 3$, continue. Set $t = 2$, $d_3(1) = 1$, and $u = 0$.
d. Since $t + 1 = 3$, stop.

This procedure gives the optimal allocation, namely, $s_1 = 3$, $s_2 = 0$, and $s_3 = 1$, which corresponds to the optimal value of 31.

3. The Inventory Example

The backward induction algorithm is now applied to the numerical stochastic inventory example of Section I. Since the data are stationary, the time index will be deleted. Also, the following additional notation will be useful. Define $v^t(s, a)$ by

$$v^t(s, a) = r(s, a) + \sum_{j \in S} p(j|s, a)v^{t+1}(j). \qquad (22)$$

a. Set $t = 4$ and $v^4(s) = 0$, $s = 0, 1, 2, 3$.
b. Since $t \neq 1$, continue. Set $t = 3$, and for $s = 0, 1, 2, 3$,

$$v^3(s) = \max_{a \in A_s} \left\{ r(s, a) + \sum_{j \in S} p(j|s, a)v^4(j) \right\}$$

$$= \max_{a \in A_s} \{r(s, a)\}. \qquad (23)$$

It is obvious from inspecting the values of $r(s, a)$ that in each state the maximizing action is 0; that is, do not order. Thus,

s	$v^3(s)$	$A^*_{s,3}$
0	0	0
1	5	0
2	6	0
3	5	0

s	$d^*_1(s)$	$d^*_2(s)$	$d^*_3(s)$	$v^*_3(s)$
0	3	2	0	$\frac{67}{16}$
1	0	0	0	$\frac{129}{16}$
2	0	0	0	$\frac{194}{16}$
3	0	0	0	$\frac{227}{16}$

c. Since $t \neq 1$, continue. Set $t = 2$ and

$$v^2(s) = \max_{a \in A_s}\{v^2(s, a)\},$$

where, for instance,

$$v^2(0, 2) = r(0, 2) + p(0|0, 2)v^3(0) + p(1|0, 2)v^3(1)$$
$$+ p(2|0, 2)v^3(2) + p(3|0, 2)v^3(3)$$
$$= -2 + \left(\tfrac{1}{4}\right) \times 0 + \left(\tfrac{1}{2}\right) \times 5 + \left(\tfrac{1}{4}\right) \times 6 +$$
$$0 \times 5 = 2.$$

The quantities $v^2(s, a)$, $v^2(s)$, and $A^*_{s,2}$ are summarized in the following tabulation, where Xs denote infeasible actions:

		$v^2(s,a)$				
s	$a=0$	$a=1$	$a=2$	$a=3$	$v^2(s)$	$A^*_{s,2}$
0	0	$\tfrac{1}{4}$	2	$\tfrac{1}{2}$	2	2
1	$6\tfrac{1}{4}$	4	$2\tfrac{1}{2}$	X	$6\tfrac{1}{4}$	0
2	10	$4\tfrac{1}{2}$	X	X	10	0
3	$10\tfrac{1}{2}$	X	X	X	$10\tfrac{1}{2}$	0

d. Since $t \neq 1$, continue. Set $t = 1$ and

$$v^1(s) = \max_{a \in A_s}\{v^1(s, a)\}.$$

The quantities $v^1(s, a)$, $v^1(s)$, and $A^*_{s,1}$ are summarized in the following tabulation, where Xs denote infeasible actions:

		$v^1(s,a)$				
s	$a=0$	$a=1$	$a=2$	$a=3$	$v^1(s)$	$A^*_{s,1}$
0	2	$\frac{33}{16}$	$\frac{66}{16}$	$\frac{67}{16}$	$\frac{67}{16}$	3
1	$\frac{129}{16}$	$\frac{98}{16}$	$\frac{99}{16}$	X	$\frac{129}{16}$	0
2	$\frac{194}{16}$	$\frac{131}{16}$	X	X	$\frac{194}{16}$	0
3	$\frac{227}{16}$	X	X	X	$\frac{227}{16}$	0

e. Since $t = 1$, stop.

This procedure has produced the optimal reward function $v^*_3(s)$ and optimal policy $\pi^* = (d^*_1(s), d^*_2(s), d^*_3(s))$, which are as follows:

This policy has a particularly simple form: If at decision point 1 the inventory in stock is 0 units, order 3 units; otherwise, do not order. If at decision point 2 the inventory in stock is 0 units, order 2 units; otherwise, do not order. And at decision point 3 do not order. The quantity $v^*_3(s)$ gives the expected total reward obtained by using this policy when the inventory before the first decision epoch is s units.

A policy of this type is called an (s, S) policy. An (s, S) policy is implemented as follows. If in period t the inventory level is s^t units or below, order the number of units required to bring the inventory level up to S^t units. Under certain convexity and linearity assumptions on the cost functions, Scarf showed in an elegant and important 1960 paper that (s, S) policies are optimal for the stochastic inventory problem. His proof of this result is based on using backward induction to show analytically that, for each t, $v^t(s)$ is K-convex, which ensures that there exists a maximizing policy of (s, S) type. This important result plays a fundamental role in stochastic operations research and has been extended in several ways.

III. INFINITE-HORIZON DYNAMIC PROGRAMMING

A. Introduction

Solution methods for infinite-horizon sequential decision problems are based on solving a stationary version of the functional equation. In this section, the state and action sets, the rewards, and the transition probabilities are assumed to be stationary and only the stochastic version of the problem is considered. Reward streams are summarized using the expected total discounted reward criterion, and the objective is to find a policy with maximal expected total discounted reward as well as this value. The two main solution techniques are value iteration and policy iteration. The former is the extension of backward induction to infinite-horizon problems and is best analyzed from the perspective of obtaining a fixed point for the Bellman equation. Policy iteration corresponds to using a generalization of Newton's method for finding a zero of the functional equation. It is not appropriate for finite-horizon problems. Other solution methods include modified policy iteration, linear programming, Gauss-Seidel iteration, successive overrelaxation, and extrapolation. Of these, only modified policy iteration and linear programming will be discussed.

B. Basic Results

The basic data of the stationary, infinite-horizon, stochastic sequential decision model are the state space S; the set of allowable actions in state s, A_s; the one-period expected reward if action a is selected in state s, $r(s, a)$; the probability the system is in state j at the next stage if it is in state s and action a is selected, $p(j|s, a)$; and the discount factor λ, $0 \leq \lambda < 1$. Both S and A_s are assumed to be finite. Let M denote the number of elements in S.

For a policy $\pi = (d_1, d_2, \ldots)$, the infinite-horizon expected total discounted reward is denoted by $v_\lambda^\pi(s)$. Let $p_\pi^m(j|s) = P\{X_{m+1}^\pi = j | X_1^\pi = s\}$. For $m = 2$ it can be computed by

$$p_\pi^2(j|s) = \sum_{k \in S} p(j|k, d_2(k)) p(k|s, d_1(s)).$$

This is the matrix product of the transition probability matrices corresponding to the decision rules $d_1(s)$ and $d_2(s)$. In general $p_\pi^m(j|s)$ is given by the matrix product of the matrices corresponding to $d_1, d_2, \ldots, d_m$. Using this notation, we can compute $v_\lambda^\pi(s)$ by

$$v_\lambda^\pi(s) = r(s, d_1(s)) + \sum_{j \in S} \lambda p(j|s, d_1(s)) r(j, d_2(j))$$

$$+ \sum_{j \in S} \lambda^2 p_\pi^2(j|s) r(j, d_3(j)) + \cdots. \quad (24)$$

Equation (24) cannot be implemented since an arbitrary nonstationary infinite-horizon policy cannot be completely specified. However, if the rewards are bounded, the above infinite series is convergent, because $\lambda < 1$. When the policy is stationary, Eq. (24) simplifies so that a policy can be evaluated either inductively or by solution of a system of linear equations. Let d denote the stationary policy that uses decision rule $d(s)$ at each stage and $p_d^m(j|s)$ its corresponding m-step transition probabilities. Then,

$$v_\lambda^d(s) = r(s, d(s)) + \sum_{j \in S} \lambda p(j|s, d(s)) r(j, d(j))$$

$$+ \sum_{j \in S} \lambda^2 p_d^2(j|s) r(j, d(j))$$

$$+ \sum_{j \in S} \lambda^3 p_d^3(j|s) r(j, d(j)) + \cdots \quad (25)$$

$$= r(s, d(s)) + \sum_{j \in S} \lambda p(j|s, d(s)) v_\lambda^d(j). \quad (26)$$

Equation (26) is derived from Eq. (25) by explicitly writing out the matrix powers, factoring out the term $p(j|s, d(s))$ from all summations, and recognizing that

the remaining terms are exactly $v_\lambda^d(s)$ Equation (26) can be rewritten as

$$\sum_{j \in S} [\delta(j, s) - \lambda p(j|s, d(s))] v_\lambda^d(j) = r(s, d(s)),$$

where $\delta(j, s) = 1$ if $j = s$ and 0 if $j \neq s$. This equation can be re-expressed in matrix terms as

$$(I - \lambda P_d) \mathbf{v_\lambda^d} = \mathbf{r_d}, \quad (27)$$

where P_d is the matrix with entries $p(j|s, d(s))$, I is the identity matrix, and $\mathbf{r_d}$ is the vector with elements $r(s, d(s))$. Equation (27) has a unique solution, which can be obtained by Gaussian elimination, successive approximations, or any other numerical method. A consequence of this equation is the following convenient representation for $v_\lambda^d(s)$:

$$\mathbf{v_\lambda^d} = (I - \lambda P_d)^{-1} \mathbf{r_d}. \quad (28)$$

The inverse in Eq. (28) exists because P_d is a probability matrix, so that its spectral radius is less than or equal to 1 and $0 \leq \lambda < 1$.

The functional equation of infinite-horizon discounted dynamic programming is the stationary equivalent of Eq. (17). It is given by

$$v(s) = \max_{a \in A_s} \left\{ r(s, a) + \sum_{j \in S} \lambda p(j|s, a) v(j) \right\}$$

$$\doteq T v(s). \quad (29)$$

Equation (29) defines the nonlinear operator T on the space of bounded M vectors or real-valued functions on S. Since T is a contraction mapping (see Section III.C.1), this equation has a unique solution. This solution is the expected discounted reward of an optimal policy. To see this, let $v^*(s)$ denote the solution of this equation. Then for any decision rule $d_t(s)$,

$$v^*(s) \geq r(s, d_t(s)) + \sum_{j \in S} \lambda p(j|s, d_t(s)) v^*(j). \quad (30)$$

By repeatedly substituting the above inequality into the right-hand side of Eq. (30) and noting that $\lambda^n \to 0$ as $n \to \infty$, we can see that what follows from Eq. (24) is that $v^*(s) \geq v_\lambda^\pi(s)$ for any policy π. Also, if $d^*(s)$ is the decision rule that satisfies

$$r(s, d^*(s)) + \sum_{j \in S} \lambda p(j|s, d^*(s)) v^*(j)$$

$$= \max \left\{ r(s, a) + \sum_{j \in S} \lambda p(j|s, a) v^*(j) \right\}, \quad (31)$$

then the stationary policy that uses $d^*(s)$ each period is optimal. This follows because if d^* satisfies Eq. (31), then

$$v^*(s) = r(s, d^*(s)) + \sum_{j \in S} \lambda p(j|s, d^*(s)) v^*(j),$$

but since this equation has a unique solution,

$$v^*(s) = v_\lambda^{d^*}(s) = v_\lambda^*(s).$$

These results are summarized as follows. There exists an optimal policy to the infinite-horizon discounted stochastic sequential decision problem that is stationary and can be found by using Eq. (31). Its value function is the unique solution of the functional equation of discounted dynamic programming.

This result plays the same role as the principle of optimality in finite-horizon dynamic programming. It says that for us to solve the infinite-horizon dynamic programming problem, it is sufficient to obtain a solution to the functional equation. In the next section, methods for solving the functional equation will be demonstrated.

C. Computational Methods

The four methods to be discussed here are value iteration, policy iteration, modified policy iteration, and linear programming. Only the first three are applied to the example in Section III.D. Value iteration and modified policy iteration are iterative approximation methods for solving the dynamic programming functional equation, whereas policy iteration is exact. To study the convergence of an approximation method, we must have a notion of distance. If $\mathbf{v}$ is a real-valued function on S (an M vector), the norm of $\mathbf{v}$, denoted by $\|\mathbf{v}\|$ is defined as

$$\|\mathbf{v}\| = \max_{s \in S} |v(s)|.$$

The distance between two vectors $\mathbf{v}$ and $\mathbf{u}$ is given by $\|\mathbf{v} - \mathbf{u}\|$. This means that two vectors are ε units apart if the maximum difference between any two components is ε units. This is often called the L^∞ norm.

A policy π is said to be ε-optimal if $\|\mathbf{v}_\lambda^\pi - \mathbf{v}_\lambda^*\| < \varepsilon$. If ε is specified sufficiently small, the two iterative algorithms can be used to find policies whose expected total discounted reward is arbitrarily close to optimum. Of course, the more accurate the approximation, the more iterations of the algorithm that are required.

1. Value Iteration

Value iteration, or successive approximation, is the direct extension of backward induction to infinite-horizon problems. It obtains an ε-optimal policy d^ε as follows:

a. Select $\mathbf{v}^0$, specify $\varepsilon > 0$, and set $n = 0$.
b. Compute $\mathbf{v}^{n+1}$ by

$$v^{n+1}(s) = \max_{a \in A_s} \left\{ r(s, a) + \sum_{j \in S} \lambda p(j|s, a) v^n(j) \right\}. \quad (32)$$

c. If $\|\mathbf{v}^{n+1} - \mathbf{v}^n\| < \varepsilon(1 - \lambda)/2\lambda$, go to step d. Otherwise, increment n by 1 and return to step b.

d. For each $s \in S$, set $d^\varepsilon(s)$ equal to an $a \in A_s$ that obtains the maximum on the right-hand side of Eq. (32) at the last iteration and stop.

This algorithm can best be understood in vector space notation. In Eq. (29), the operator T is defined on the set of bounded real-valued M vectors. Solving the functional equation corresponds to finding a fixed point of T, that is, a $\mathbf{v}$ such that $\mathbf{Tv} = \mathbf{v}$. The value iteration algorithm starts with an arbitrary $\mathbf{v}^0$ (0 is usually a good choice) and iterates according to $\mathbf{v}^{n+1} = T\mathbf{v}^n$. Since T is a contraction mapping, that is,

$$\|T\mathbf{v} - T\mathbf{u}\| \leq \lambda \|\mathbf{v} - \mathbf{u}\|$$

for any $\mathbf{v}$ and $\mathbf{u}$, the iterative method is convergent for any v^0. This is because

$$\|\mathbf{v}^{n+1} - \mathbf{v}^n\| \leq \lambda^n \|\mathbf{v}^1 - \mathbf{v}^0\|,$$

and the space of bounded real-valued M vectors is a Banach space (a complete normed linear space) with respect to the norm used here. Since a contraction mapping has a unique fixed point, $\mathbf{v}_\lambda^*$, $\mathbf{v_n}$ converges to it. The rate of convergence is geometric with parameter λ, that is,

$$\|\mathbf{v}^n - \mathbf{v}^*\| \leq \lambda^n \|\mathbf{v}^0 - \mathbf{v}^*\|.$$

The algorithm terminates with a value function $\mathbf{v_{n+1}}$ and a decision rule $\mathbf{d}^\varepsilon$ with the following property:

$$v^{n+1}(s) = r(s, d^\varepsilon(s)) + \sum_{j \in S} \lambda p(j|s, d^\varepsilon(s)) v^n(j). \quad (33)$$

The stopping rule in step c ensures that the stationary policy that uses $\mathbf{d}^\varepsilon$ every period is ε-optimal.

The sequence of iterates $\mathbf{v^n}$ have interesting interpretations. Each iterate corresponds to the optimal expected total discounted return in an n-period problem in which the terminal reward equals $\mathbf{v}^0$. Alternatively, they correspond to the expected total discounted returns for the policy in an n-period problem that is obtained by choosing a maximizing action in each state at each iteration.

2. Policy Iteration

Policy iteration, or approximation in the policy space, is an algorithm that uses the special structure of infinite-horizon stationary dynamic programming problems to find all optimal policies. The algorithm is as follows:

a. Select a decision rule $d^0(s)$ for all $s \in S$ and set $n = 0$.

b. Solve the system of equations

$$\sum_{j \in S} [\delta(j, s) - \lambda p(j|s, d^n(s))] v^n(j) = r(s, d^n(s)) \quad (34)$$

for $v^n(s)$.

c. For each s, and each $a \in A_s$, compute

$$r(s, a) + \sum_{j \in S} \lambda p(j|s, a) v^n(j). \quad (35)$$

For each s, put a in $A_{n,s}^*$ if a obtains the maximum value in Eq. (35).

d. If, for all s, $d^n(s)$ is contained in $A_{n,s}^*$, stop. Otherwise, proceed.

e. Set $d^{n+1}(s)$ equal to any a in $A_{n,s}^*$ for each s in S, increment n by 1, and return to step b.

The algorithm consists of two main parts: step b, which is called policy evaluation, and step c, which is called policy improvement. The algorithm terminates when the set of maximizing actions found in the improvement stage repeats, that is, if the same decision obtains the maximum in step b on two successive passes through the iteration loop.

This algorithm terminates in a *finite* number of iterations with an optimal stationary policy and its expected total discounted reward. This is because the improvement procedure guarantees that $v^{n+1}(s)$ is strictly greater than $v^n(s)$, for some $s \in S$, until the termination criterion is satisfied, at which point $v^n(s)$ is the solution of the dynamic programming functional equation. Since each $\mathbf{v^n}$ is the expected total discounted reward of the stationary policy $\mathbf{d^n}$, and there are only finitely many stationary policies, the procedure must terminate in a finite number of iterations.

If only an ε-optimal policy is desired, a stopping rule similar to that in step c of the value iteration procedure can be used.

3. Modified Policy Iteration

The evaluation step of the policy iteration algorithm is usually implemented by solving the linear system

$$(I - \lambda P_{d^n}) \mathbf{v^{d^n}} = \mathbf{r_{d^n}} \quad (36)$$

by using Gaussian elimination, which requires $\frac{1}{3} M^3$ multiplications and divisions. When the number of states is large, exact solution of Eq. (34) can be computationally prohibitive. An alternative is to use successive approximations to obtain an approximate solution. This is the basis of the modified policy iteration, or value-oriented successive approximation, method. The modified policy iteration algorithm of order m is as follows:

a. Select $\mathbf{v^0}$, specify $\varepsilon > 0$, and set $n = 0$.

b. For each s and each $a \in A_s$, compute

$$r(s, a) + \sum_{j \in S} \lambda p(j|s, a) v^n(j). \quad (37)$$

For each s, put a in $A_{n,s}^*$ if a obtains the maximum value in Eq. (37).

c. For each s in S, set $d^n(s)$ equal to any a in $A_{n,s}^*$.

(i) Set $k = 0$ and define $u^0(s)$ by

$$u^0(s) = \max_{a \in A_s} \left\{ r(s, a) + \sum_{j \in S} \lambda p(j|s, a) v^n(j) \right\}. \quad (38)$$

(ii) If $\|\mathbf{u^0} - \mathbf{v^n}\| < \varepsilon(1 - \lambda)/2\lambda$, go to step d. Otherwise, go to step (iii).

(iii) Compute u^{k+1} by

$$u^{k+1}(s) = r(s, d^n(s)) + \sum_{j \in S} \lambda p(j|s, d^n(s)) u^k(j). \quad (39)$$

(iv) If $k = m$, go to step (v). Otherwise, increment k by 1 and return to step (i).

(v) Set $\mathbf{v^{n+1}} = \mathbf{u^{m+1}}$, increment k by 1, and go to step b.

d. For each $s \in S$, set $d^\varepsilon(s) = d^n(s)$ and stop.

This algorithm combines features of both policy iteration and value iteration. Like value iteration, it is an iterative algorithm that terminates with an ε-optimal policy; however, value iteration avoids step c above. The stopping criterion used in step (ii) is identical to that of value iteration, and the computation of $\mathbf{u^0}$ in step (i) requires no extra work because it has already been determined in step b. Like policy iteration, the algorithm contains an improvement step, step b, and an evaluation step, step c. However, the evaluation is not done exactly. Instead, it is carried out iteratively in step c, which is repeated m times. Note that m can be selected in advance or adaptively during the algorithm. For instance, m can be chosen so that $\|\mathbf{u^{m+1}} - \mathbf{u^m}\|$ is less than some prespecified tolerance that can vary with n. Recent studies have shown that low orders of m work well, while adaptive choice is better.

The modified policy iteration algorithm serves as a bridge between value iteration and policy iteration. When $m = 0$, it is equivalent to value iteration, and when m is infinite, it is equivalent to policy iteration. It will converge in fewer iterations than value iteration and more iterations than policy iteration; however, the computational effort per iteration exceeds that for value iteration and is less than that for policy iteration. When the number of states, M, is large, it has been shown to be the most computationally efficient method for solving Markov decision problems.

4. Linear Programming

The stationary infinite-horizon discounted stochastic sequential decision problem can be formulated and solved by linear programming. The primal problem is given by

Minimize

$$\sum_{j \in S} \alpha_j v(j)$$

subject to, for $a \in A_s$ and $s \in S$,

$$v(s) \geq r(s, a) + \sum_{j \in S} \lambda p(j|s, a) v(j),$$

and $v(s)$ is unconstrained.

The constants α_j are positive and arbitrary. The dual problem is given by

Maximize

$$\sum_{s \in S} \sum_{a \in A_s} x(s, a) r(s, a)$$

subject to, for $J \in S$,

$$\sum_{a \in A_s} x(j, a) - \sum_{s \in S} \sum_{a \in A_s} \lambda p(j|s, a) x(s, a) = \alpha_j,$$

and $x(j, a) \geq 0$ for $a \in A_j$ and $j \in S$.

Using a general-purpose linear programming code for solving dynamic programming problems is not computationally attractive. The dynamic programming methods are more efficient. The interest in the linear programming formulation is primarily theoretical but allows inclusion of side constraints. Some interesting observations are as follows:

  a. The dual problem is always feasible and bounded. Any optimal basis has the property that for each $s \in S$, $x(s, a) > 0$ for only one $a \in A_s$. An optimal stationary policy is given by $d^*(s) = a$ if $x(s, a) > 0$.
  b. The same basis is optimal for all α_j's.
  c. When the dual problem is solved by the simplex algorithm with block pivoting, it is equivalent to policy iteration.
  d. When policy iteration is implemented by only changing the action that gives the maximum improvement over all states, it is equivalent to solving the dual problem by the simplex method.

D. Numerical Examples

In this section, an infinite-horizon version of the stochastic inventory example presented earlier is solved by using value iteration, policy iteration, and modified policy iteration. The data are as analyzed in the finite-horizon case;

however, the discount rate λ is chosen to be .9. The objective is to determine the stationary policy that maximizes the expected total infinite-horizon discounted reward.

1. Value Iteration

To initiate the algorithm, we will take $\mathbf{v}^0$ to be the zero vector; ε is chosen to be .1. The algorithm will terminate with a stationary policy that is guaranteed to have an expected total discounted reward within .1 of optimal. Calculations proceed as in the finite-horizon backward induction algorithm until the stopping criterion of

$$\|\mathbf{v}^{n+1} - \mathbf{v}^n\| \leq \frac{\varepsilon(1 - \lambda)}{2\lambda} = \frac{.1 \times .1}{2 \times .9} = .0056$$

is satisfied. The value functions $\mathbf{v}^n$ and the maximizing actions obtained in step b at each iteration are provided in the tabulation on the following page. The above algorithm terminates after 58 iterations, at which point $\|\mathbf{v}^{58} - \mathbf{v}^{57}\| = .0054$. The .1-optimal stationary policy is $\mathbf{d}^\varepsilon = (3, 0, 0, 0)$, which means that if the stock level is 0, order 3 units; otherwise, do not order. Observe that the optimal policy was first identified at iteration 3, but the algorithm did not terminate until iteration 58. In larger problems such additional computational effort is extremely wasteful. Improved stopping rules and more efficient algorithms are described in Section III.E.

2. Policy Iteration

To initiate policy iteration, choose the myopic policy, namely, that which maximizes the immediate one-period reward $r(s, a)$. The algorithm proceeds as follows:

  a. Set $d^0 = (0, 0, 0, 0)$ and $n = 0$.
  b. Solve the evaluation equations:

$$\begin{aligned}
(1 - .9 \times 1)v^0(0) &&&= 0, \\
(-.9 \times .75)v^0(0) &+ (1 - .9 \times 25)v^0(1) &&= 5, \\
(-.9 \times .25)v^0(0) &+ (-.9 \times .5)v^0(1) && \\
&&+ (1 - .9 \times 25)v^0(2) &= 6,
\end{aligned}$$

and

$$\begin{aligned}
(-.9 \times .25)v^0(1) &+ (-.9 \times .50)v^0(2) & \\
&+ (1 - .9 \times .25)v^0(3) &= 5.
\end{aligned}$$

These equations are obtained by substituting the transition probabilities and rewards corresponding to policy $\mathbf{d}^0$ into Eq. (34). The solution of these equations is $\mathbf{v}^0 = (0, 6.4516, 11.4880, 14.9951)$.

  c. For each s, the quantities

$$r(s, a) + \sum_{j=0}^{3} \lambda p(j|s, a) v^0(j)$$

are computed for $a = 0, \ldots, 3 - s$, and the actions that achieve the maximum are placed into $A_{0,s}^*$. In this example

	$v^n(s)$				$d^n(s)$			
n	$s=0$	$s=1$	$s=2$	$s=3$	$s=0$	$s=1$	$s=2$	$s=3$
0	0	0	0	0	0	0	0	0
1	0	5.0	6.0	5.0	2	0	0	0
2	1.6	6.125	9.6	9.95	2	0	0	0
3	3.2762	7.4581	11.2762	12.9368	3	0	0	0
4	4.6632	8.8895	12.6305	14.6632	3	0	0	0
5	5.9831	10.1478	13.8914	15.9831	3	0	0	0
10	10.7071	14.8966	18.6194	20.7071	3	0	0	0
15	13.5019	17.6913	21.4142	23.0542	3	0	0	0
30	16.6099	20.7994	24.5222	26.6099	3	0	0	0
50	17.4197	21.6092	25.3321	27.4197	3	0	0	0
56	17.4722	21.6617	25.3845	27.4722	3	0	0	0
57	17.4782	21.6676	25.3905	27.4782	3	0	0	0
58	17.4736	21.6730	25.3959	27.4836	3	0	0	0

there is a unique maximizing action in each state, and it is given by

$$A_{0,0}^* = \{3\}; \qquad A_{0,1}^* = \{2\};$$
$$A_{0,2}^* = \{0\}; \qquad A_{0,3}^* = \{0\}.$$

d. Since $d^0(0) = 0$, it is not contained in $A_{0,0}^*$, so continue.

e. Set $\mathbf{d}^1 = (3, 2, 0, 0)$ and $n = 1$, and return to the evaluation step.

The detailed step-by-step calculations for the remainder of the algorithm are omitted. The value functions and corresponding maximizing actions are presented below. Since there is a unique maximizing action in the improvement step at each iteration, $A_{n,s}^*$ is equivalent to $d^n(s)$ and only the latter is displayed.

The algorithm terminates in three iterations with the optimal policy $\mathbf{d}^* = (3, 0, 0, 0)$. Observe that an evaluation was unnecessary at iteration 3 since $\mathbf{d}^2 = \mathbf{d}^3$ terminated the algorithm before the evaluation step. Unlike value iteration, the algorithm has produced an optimal policy as well as its expected total discounted reward $\mathbf{v}^3$, which is the optimal expected total discounted reward. This computation shows that the .1-optimal policy found by using value iteration is in fact optimal. This information could not be obtained by using value iteration unless the action elimination method described in Section III.E were used.

3. Modified Policy Iteration

The following illustrates the application of modified policy iteration of order 5. The first pass through the algorithm is described in detail; calculations for the remainder are presented in tabular form below.

a. Set $\mathbf{v}^0 = (0, 0, 0, 0)$, $n = 0$, and $\varepsilon = .1$.

b. Observe that

$$r(s, a) + \sum_{j=0}^{3} \lambda p(j|s, a)v^0(j) = r(s, a),$$

so that for each s the maximum value occurs for $a = 0$. Thus, $A_{n,s}^* = \{0\}$ for $s = 0, 1, 2, 3$ and $\mathbf{d}^n = (0, 0, 0, 0)$.

(i) Set $k = 0$ and $\mathbf{u}^0 = (0, 5, 6, 5)$.

(ii) Since $\|\mathbf{u}^0 - \mathbf{v}^0\| = 6 > .0056$, continue.

(iii) Compute $\mathbf{u}^1$ by

$$u^1(s) = r(s, d^0(s)) + \sum_{j=0}^{3} \lambda p(j|s, d^0(s))u^0(j)$$

$$= r(s, 0) + \sum_{j=0}^{3} \lambda p(j|s, 0)u^0(j) \tag{40}$$

$$= 0 + .9 \times 1 \times 0 = 0, \qquad \text{for} \quad s = 0,$$
$$= 5 + .9 \times \tfrac{3}{4} \times 0 + .9 \times \tfrac{1}{4} \times 5 = 6.125,$$
$$\qquad \qquad \qquad \qquad \text{for} \quad s = 1,$$
$$= 6 + .9 \times \tfrac{1}{4} \times 0 + .9 \times \tfrac{1}{2} \times 5$$
$$\quad + .9 \times \tfrac{1}{4} \times 6 = 9.60, \qquad \text{for} \quad s = 2,$$
$$= 6 + 9 \times \tfrac{1}{4} \times 5 + .9 \times \tfrac{1}{2} \times 6$$
$$\quad + .9 \times \tfrac{1}{4} \times 5 = 10.95, \qquad \text{for} \quad s = 3,$$

so that $\mathbf{u}^1 = (0, 6.125, 9.60, 10.95)$.

(iv) Since $k = 1 < 5$, continue.

	$v^n(s)$				$d^n(s)$			
n	$s=0$	$s=1$	$s=2$	$s=3$	$s=0$	$s=1$	$s=2$	$s=3$
0	0	6.4516	11.4880	14.9951	0	0	0	0
1	10.7955	12.7955	18.3056	20.7955	3	2	0	0
2	17.5312	21.7215	25.4442	27.5318	3	0	0	0
3	X	X	X	X	3	0	0	0

n	$v^n(s)$				$d^n(s)$			
	$s=0$	$s=1$	$s=2$	$s=3$	$s=0$	$s=1$	$s=2$	$s=3$
0	0	0	0	0	0	0	0	0
1	0	6.4507	11.4765	14.9200	3	2	0	0
2	7.1215	9.1215	14.6323	17.1215	3	0	0	0
3	11.5709	15.7593	19.4844	21.5709	3	0	0	0
4	14.3639	18.5534	22.2763	24.3639	3	0	0	0
5	15.8483	20.0377	23.7606	25.8483	3	0	0	0
10	17.4604	21.6499	25.3727	27.4604	3	0	0	0
11	17.4938	21.6833	25.4062	27.4938	3	0	0	0

The loop is repeated four more times to evaluate $\mathbf{u}^2$, $\mathbf{u}^3$, $\mathbf{u}^4$, and $\mathbf{u}^5$. Then $\mathbf{v}^1$ is set equal to $\mathbf{u}^5$ and the maximization in step b is carried out. The resulting iterates are shown at the bottom of the page. In step (ii), following iteration 11, the computed value of $\mathbf{u}^0$ is (17.4976, 21.6871, 25.4100, 27.4976), so that $\|\mathbf{u}^0 - \mathbf{v}^{11}\| = .0038$, which guarantees that the policy (3, 0, 0, 0) is ε-optimal with $\varepsilon = .1$.

4. Comparison of the Algorithms

Value iteration required 58 iterations to obtain a .1-optimal solution. At each iteration a maximization was required so that each action had to be evaluated to determine Eq. (32) at each iteration. Modified policy iteration of order 5 required 11 iterations to determine a .1-optimal policy so that the maximization step was carried out only 11 times. However, at each iteration the inner loop of the algorithm in step c was invoked five times. Thus, modified policy iteration required far fewer maximizations than did value iteration. This would lead to considerable computational savings when the action sets or the problem had many states.

Policy iteration found an optimal policy in three iterations; however, each iteration required both the evaluation of all actions in each state and the solution of a linear system of equations. In this small example, it is not time consuming to solve the linear system by using Gaussian elimination, but when the number of states is large, this can be prohibitive. In such cases, modified iteration is preferred over both policy iteration and value iteration; however, which order is the best to use is an open question.

E. Bounds on the Optimal Total Expected Discounted Reward

At each stage of the value iteration, policy iteration, and modified policy iteration, the computed value of $\mathbf{v}^n$ can be used to obtain upper and lower bounds on the optimal expected discounted reward. These bounds can be used to terminate any of the iterative algorithms, eliminate suboptimal actions at each iteration of an algorithm, and develop improved algorithms.

Bounds are given for value iteration; however, they have also been obtained for policy iteration and modified policy iteration. First, define the following two quantities:

$$L^n = \min_{s \in S}\{v^{n+1}(s) - v^n(s)\}$$

and

$$U^n = \max_{s \in S}\{v^{n+1}(s) - v^n(s)\}.$$

Then, for each n and $s \in S$,

$$v^n(s) + \frac{1}{1-\lambda}L^n \le v^{d^n}(s) \le v^*(s) \le v^n_\lambda(s) + \frac{1}{1-\lambda}U^n. \tag{41}$$

We can easily see, using the two extreme bounds, that if $U^n - L^n < \varepsilon(1 - \lambda)$, then

$$0 \le v^*_\lambda(s) - \left(v^n(s) + \frac{1}{1-\lambda}L^n\right) \le \varepsilon. \tag{42}$$

This provides an alternative stopping criterion for value iteration and can be modified for any of the above algorithms. In particular in value iteration, if the algorithm is terminated when $U^n - L^n < \varepsilon(1 - \lambda)$, then the quantity $v_n(s) + (1 - \lambda)^{-1}L^n$ is within ε of the optimal value function. If this had been implemented in the value iteration algorithm, in Section III.C., it would have terminated after 9, as opposed to 58, iterations, when $\varepsilon(1 - \lambda) = .1 \times (1 - .9) = .01$.

These bounds can be used at each iteration to eliminate actions that cannot be part of an optimal policy. This is important computationally because the maximization in the improvement step can be made more efficient if all of A_s need not be evaluated at each iteration. Elimination is based on the result that action a is suboptimal in state s, if

$$r(s, a) + \sum_{j \in S} \lambda p(j|s, a)v^*_\lambda(j) < v^*_\lambda(s). \tag{43}$$

Of course, $v^*_\lambda(s)$ is not known in Eq. (43), but by substituting an upper bound for it on the left-hand side and a lower bound on the right-hand side, one can use the result to eliminate suboptimal actions in state s. The bounds in Eq. (41) can be used.

Action elimination procedures are especially important in approximation algorithms such as value iteration and modified policy iteration that produce only ε-optimal policies. If a unique optimal policy exists, by eliminating suboptimal actions at each iteration, one obtains an optimal policy when only one action remains in each state. This will occur in finitely many iterations.

F. Turnpike and Planning-Horizon Results

Infinite-horizon sequential decision models are usually approximations to long finite-horizon problems with many decision points. A question of practical concern is, *When is the optimal policy for the infinite-horizon model optimal for the finite-horizon problem?* An answer to this question is provided through planning-horizon, or turnpike, theory. The basic result is the following. There exists an N^*, called the planning horizon, such that, for all $n \geq N^*$, the optimal decision when there are n periods remaining is one of the decisions that is optimal when the horizon is infinite. This result means that if there is a unique optimal stationary policy d^* for the infinite-horizon problem, then in an n-period problem with $n \geq N^*$, it is optimal to use the stationary strategy for the initial $n - N^*$ decisions and to find the optimal policy for the remaining N^* periods by using the backward induction methods of Section II. The optimal infinite-horizon strategy is called the turnpike, and it is reached after traveling N^* periods on the nonstationary "side roads." The term *turnpike* originates in mathematical economics, where it refers to the policy path that produces optimal economic growth.

Another interpretation of the above result is that it is optimal to use d^* for the first decision in any finite-horizon problem in which it is known that the horizon exceeds N^*. Thus, it is not necessary to specify the horizon, only to know that there are at least N^* decisions to be made. Bounds on N^* are available, and this concept has been extended to nonstationary infinite-horizon problems and the expected average reward criteria. This is referred to as a rolling horizon approach.

The computational results for the value iteration algorithm in Section III.D give further insight into the planning-horizon result. There, $N^* = 3$, so that in any problem with horizon greater than 3, it is optimal to use the decision rule $(3, 0, 0, 0)$ until there are three decisions left to be made, at which point it is optimal to use the decision rule $(2, 0, 0, 0)$ for two periods and $(0, 0, 0, 0)$ in the last period.

G. The Average Expected Reward Criteria

In many applications in which infinite-horizon formulations are natural, the total discounted reward criterion is not relevant. For instance, in a large telecommunications network, millions of packet and call routing decisions are made every second, so that discounting the consequences of latter decisions makes little sense. An alternative is the expected average reward criterion defined in Eq. (7). Using this criterion means that rewards received at each time period receive equal weight.

Computational methods for the average reward criterion are more complex than those for the discounted reward problem. This is because the form of the average reward function, $g^\pi(s)$, depends on the structure of the Markov chain corresponding to the stationary policy π. If the policy is unichain, that is, the Markov chain induced by the policy has exactly one recurrent class and possibly several transient classes, then $\mathbf{g}^\pi$ is a constant vector. In this case, the functional equation is

$$v(s) = \max_{a \in A_s} \left\{ r(s, a) - g + \sum_{s \in S} p(j|s, a)v(j) \right\}. \quad (44)$$

Solving this equation uniquely determines g and determines $v(s)$ up to an additive constant. The quantity g is the optimal expected average reward. For us to specify $\mathbf{v}$ uniquely, it is sufficient to set $v(s_0) = 0$ for some s_0 in the recurrent class of a policy corresponding to choosing a maximizing action in each state. If this is done, $v(s)$ is called a relative value function and $v(j) - v(k)$ is the difference in expected total reward obtained by using an optimal policy and starting the system in state j as opposed to state k.

As in the discounted case, an optimal policy is found by solving the functional equation. This is best done by policy iteration. The theory of value iteration is quite complex in this setting and is not discussed here. The policy iteration algorithm is given below. It is assumed that all policies are unichain.

a. Select a decision rule $\mathbf{d^0}$ and set $n = 0$.

b. Solve the system of equations

$$\sum_{j \in S} [\delta(j, s) - p(j|s, d^n(s))]v^n(j) - g^n = r(s, d^n(s)),$$
$$(45)$$

where $v^n(s_0) = 0$ for some s_0 in the recurrent class of d^n.

c. For each s, and each $a \in A_s$, compute

$$r(s, a) + \sum_{j \in S} p(j|s, a)v^n(j). \quad (46)$$

For each s, put a in $A_{n,s}^*$ if a obtains the maximum value in Eq. (46).

d. If for each s, $d^n(s)$ is contained in $A_{n,s}^*$, stop. Otherwise, proceed.

e. Set $d^{n+1}(s)$ equal to any a in $A_{n,s}^*$ for each s in S, increment n by 1, and return to step b.

Note that this algorithm is almost identical to that for the discounted case. The only difference is the linear system of equations solved in step b. If the assumption that all policies are unichain is dropped, solution of the functional equation [Eq. (44)] is no longer sufficient to determine an optimal policy. Instead, a nested pair of optimality equations is required.

IV. FURTHER TOPICS

A. Historical Perspective

The development of dynamic programming is usually credited to Richard Bellman. His numerous papers in the 1950s presented a formal development of this subject and numerous interesting examples. Most of this pioneering work is summarized in his book, "Dynamic Programming." However, many of the themes of dynamic programming and sequential decision processes are scattered throughout earlier works. These include studies that appeared between 1946 and 1953 on water resource management by Masse; sequential analysis in statistics by Wald; games of pursuit by Wald; inventory theory by Arrow, Blackwell, and Girshick; Arrow, Harris, and Marshak; and Dvoretsky, Kiefer, and Wolfowitz; and stochastic games by Shapley.

Although Bellman coined the phrase "Markov decision processes," this aspect of dynamic programming got off the ground with Howard's monograph, "Dynamic Programming and Markov Processes" in 1960. The first formal theoretical treatment of this subject was by Blackwell in 1962. In 1960, deGhellinck demonstrated the equivalence between Markov decision processes and linear programming. Other major contributions are those of Denardo in 1968, in which he showed that value iteration can be analyzed by the theory of contraction mappings; Veinott in 1969, in which he introduced a new family of optimality criteria for dynamic programming problems; and Federgruen and Schweitzer between 1978 and 1980, in which they investigated the properties of the sequences of policies obtained from the value iteration algorithm. Modified policy iteration is usually attributed to Puterman and Shin in 1978; however, similar ideas appeared earlier in works of Kushner and Kleinman and of van Nunen. In 1978, Puterman and Brumelle demonstrated the equivalence of policy iteration to Newton's method. Puterman's book "Markov Decision Processes" provides a comprehensive overview of theory, application, and calculations.

B. Applications

Dynamic programming methods have been applied in many areas. These methods have been used numerically to compute optimal policies, as well as analytically to determine the form of an optimal policy under various assumptions on the rewards and transition probabilities. A brief and by no means complete summary of applications appears in Table I. Only stochastic dynamic programming is considered; however, in many cases, the problems have also been analyzed in the deterministic setting. In these applications, probability distributions of the random quantities are assumed to be known before solution of the problem; adaptive estimation of parameters is not necessary.

A major limitation in the practical application of dynamic programming has been computational. When the set of states at each stage is large—for example, if the state description is vector-valued—then solving a sequential decision problem by dynamic programming requires considerable storage as well as computationl time. Bellman recognized the difficulty early on and referred to it as the "curse of dimensionality." Research in the 1990s addressed this issue by developing approximation methods for large-scale applications. These methods combined concepts from stochastic approximation, simulation, and artificial intelligence and are sometimes referred to as reinforcement learning. A comprehensive treatment of this line of research appears in "Neuro-Dynamic Programming" by Bertsekas and Tsitsiklis.

C. Extensions

In the models considered in this article, it has been assumed that the decision maker knows the state of the system before making a decision, that decisions are made at discrete time points, that the set of states is finite and discrete (with the exception of the example in Section II.D.2), and the model rewards and transition probabilities are known. These models can be modified in several ways: the state of the system may be only partially observed by the decision maker, the sets of decision points and states may be continuous, or the transition probabilities or rewards may not be known. These modifications are discussed briefly below.

1. Partially Observable Models

This model differs from the fully observable model in that the state of the system is not known to the decision maker at the time of decision. Instead, the decision maker receives a signal from the system and on the basis of this signal updates his estimate of the probability distribution of the system state. Updating is done using Bayes' theorem. Decisions are based on this probability distribution, which is a sufficient statistic for the history of the process.

When the set of states is discrete, these models are referred to as partially observable Markov decision processes. Computational methods in this case are quite

TABLE I Stochastic Dynamic Programming Applications

Area	States	Actions	Reward	Stochastic aspect
Capacity expansion	Size of plant	Maintain or add capacity	Costs of expansion and production at current capacity	Demand for product
Cash management	Cash available	Borrow or invest	Transaction costs less interest	External demand for cash
Catalog mailing	Customer purchase record	Type of catalog to send to customer, if any	Purchases in current period less mailing costs	Customer purchase amount
Clinical trials	Number of successes with each treatment	Stop or continue the trial, and if stopped, choose best treatment if any	Costs of treatment and incorrect decisions	Response of a subject to a treatment
Economic growth	State of the economy	Investment or consumption	Utility of consumption	Effect of investment
Fisheries management	Fish stock in each age class	Number of fish to harvest	Value of the catch	Population size
Football	Position of ball	Play to choose	Expected points scored	Outcome of play
Forest management	Size and condition of stand	Harvesting and reforestation activities	Revenues less harvesting costs	Stand growth and price fluctuations
Gambling	Current wealth	Stop or continue playing the game	Cost of playing	Outcome of the game
Hotel and airline reservations	Number of confirmed reservations	Accept, wait-list, or reject new reservations	Profit from satisfisfied reservations less overbooking penalties	Demand for reservations and number of arrivals
Inventory control	Stock on hand	Order additional stock	Revenue per item sold less ordering, holding, and penalty costs	Demand for items
Project selection	Status of each project	Project to invest in at present	Return from investing in project	Change of project status
Queuing control	Number in the queue	Accept or reject arriving customers or control service rate	Revenue from serving customer less delay costs	Interarrival times and service times
Reliability	Age or status of equipment	Inspect and repair or replace if necessary	Inspection and repair costs plus failure cost	Failure and deterioration
Scheduling	Activites completed	Next activity to schedule	Cost of activity	Length of time to complete activity
Selling an asset	Current offer	Accept or reject the offer	The offer less the cost of holding the asset for one period	Size of the offer
Water resource management	Level of water in each reservoir in river system	Quantity of water to release	Value of power generated	Rainfall and runoff

complex, and only very small problems have been solved numerically. When the states form a continuum, this problem falls into the venue of control theory. An extremely important result in this area is the Kalman filter, which provides an updating formula for the expected value and covariance matrix of the system state. Another important result is the separation theorem, which gives conditions that allow decomposition of this problem into separate problems of estimation and control.

2. Continuous-State, Discrete-Time Models

The resource allocation problem in Section I is an example of a continuous-state, discrete-time, deterministic model.

Its solution using dynamic programming methodology is given in Section II. When transitions are stochastic, only minor modifications to the general sequential decision problem are necessary. Instead of a transition probability function, a transition probability density is used, and summations are replaced by integrations throughout. This modification causes considerable theoretical complexity; the main issues concern measurability and integrability of value functions.

Problems of this type fall into the realm of stochastic control theory. Although dynamic programming is used to solve such problems, the formulation is quite different. Instead of explicitly giving a transition probability function for the state, the theory requires use of a dynamic equation to relate the state at time $t+1$ to the state at time t. A major result in this area is that when the state dynamics are linear in the state, action, and random component and the cost is quadratic in the state and the action, then a closed-form solution is available for the optimal decision rule and it is linear in the system state. These problems have been studied extensively in the engineering literature.

3. Continous-Time Models

Stochastic continuous-time models are categorized according to whether the state space is continuous or discrete. The discrete-time model has been widely studied in the operations research literature. The stochastic nature of the problem is modeled as either a Markov process, a semi-Markov process, or a general jump process. The decision maker can control the transition rates, transition probabilities, or both. The infinite-horizon versions of the Markov and semi-Markov decision models are analyzed in a similar fashion to the discrete-time Markov decision process; however, general jump processes are considerably more complex. These models have been widely applied to problems in queuing and inventory control.

When the state space is continuous and Markovian assumptions are made, diffusion processes are used to model the transitions. The decision maker can control the drift of the system or can cause instantaneous state transitions or jumps. The discrete-time optimality equation is replaced by a nonlinear second-order partial differential equation and is usually solved numerically. These models are studied in the stochastic control theory literature and have been applied to inventory control, finance, and statistical modeling.

4. Adaptive Control

When transition probabilities and/or rewards are unknown, the decision maker must adaptively estimate them to control the system optimally. The usual approach to analysis of such systems is to assume that the rewards and transition probabilities depend on an unknown parameter, such as the arrival rate to a queuing system, and then use the observed sequence of system states to adaptively estimate this parameter. In a Bayesian analysis of such models, uncertainty about the parameter value is described through a probability distribution which is periodically updated as information becomes available. The classical approach to such models uses maximum likelihood theory to estimate the parameter and derive its statistical properties.

ACKNOWLEDGMENT

Preparation of this article was supported by Natural Sciences and Engineering Research Council Grant A-5527.

SEE ALSO THE FOLLOWING ARTICLES

LINEAR OPTIMIZATION • NONLINEAR PROGRAMMING • OPERATIONS RESEARCH

BIBLIOGRAPHY

Bellman, R. E. (1957). "Dynamic Programming," Princeton University Press, Princeton, N.J.

Bertsekas, D. P. (1995). "Dynamic Programming and Optimal Control," Vols. 1 and 2, Athena Scientific, Belmont, Mass.

Bertsekas, D. P., and Tsitsiklis, J. M. (1995). "Neuro-Dynamic Programming," Athena Scientific, Belmont, Mass.

Blackwell, D. (1962). *Ann. Math. Stat.* **35**, 719–726.

Denardo, E. V. (1967). *SIAM Rev.* **9**, 169–177.

Denardo, E. V. (1982). "Dynamic Programming, Models and Applications," Prentice-Hall, Englewood Cliffs, N.J.

Fleming, W. H., and Rishel, R. W. (1975). "Deterministic and Stochastic Optimal Control," Springer-Verlag, New York.

Howard, R. A. (1960). "Dynamic Programming and Markov Processes," MIT Press, Cambridge, Mass.

Puterman, M. L. (1994). "Markov Decision Processes," Wiley, New York.

Ross, S. M. (1983). "Introduction to Stochastic Dynamic Programming," Academic Press, New York.

Scarf, H. E. (1960). *In* "Studies in the Mathematical Theory of Inventory and Production," (K. J. Arrow, S. Karlin, and P. Suppes, eds.), Stanford University Press, Stanford, Calif.

Veinott, A. F., Jr. (1969). *Ann. Math. Stat.* **40**, 1635–1660.

Wald, A. (1947). "Sequential Analysis," Wiley, New York.

White, D. J. (1985). *Interfaces* **15**, 73–83.

White, D. J. (1988). *Interfaces* **18**, 55–61.

Dynamics of Elementary Chemical Reactions

H. Floyd Davis
Hans U. Stauffer
Cornell University

GLOSSARY

Complex-mediated reaction An elementary bimolecular reaction proceeding via a long-lived collision complex having lifetimes ranging from several vibrational periods (100 fs) to many rotational periods (>10 ps). If complex lifetimes exceed several rotational periods, product angular distributions from crossed beam reactions exhibit forward-backward symmetry in the center-of-mass frame of reference.

Direct reaction An elementary bimolecular reaction proceeding via direct passage through the transition state region. The absence of long-lived intermediates in such reactions leads to anisotropic center-of-mass product angular distributions that often provide insight into the most favorable geometries for reaction.

Free radical An atom or molecule possessing one or more unpaired electrons. Free radicals may either be stable molecules (e.g., NO, O_2, or NO_2), or highly reactive transitory chemical intermediates (e.g., H, Cl, CH_3) that react on essentially every collision with stable molecules.

Ionization The process by which one or more valence electrons are removed from an atom or molecule. Most often achieved by electron impact or absorption of one or more ultraviolet photons.

Laser-induced fluorescence (LIF) A spectroscopic technique usually employing visible or UV laser light, in which the fluorescence emission from a gaseous, liquid, or solid sample is monitored. A fluorescence excitation spectrum is a plot of the total emitted fluorescence vs. excitation wavelength and provides information similar to an absorption spectrum.

Molecular beam Collimated stream of gaseous molecules produced by expansion of a gas through an orifice into an evacuated chamber. A supersonic molecular beam is characterized by a velocity distribution much narrower than a Boltzmann distribution.

Potential energy surface Schematic two- or three-dimensional representation of the total potential energy of a chemical system as a function of internuclear coordinates.

Transition state Region of the PES corresponding to the critical geometry through which a reacting system must pass for reactants to become products.

AN ELEMENTARY CHEMICAL reaction is any process involving bond fission and/or bond formation following a single collision between two reactants. Chemical reactions occur in all three phases of matter (gas, liquid, and solid), and at their interfaces. Under the experimental conditions most commonly used to carry out reactions, the overall reaction usually consists of a sequence of two or more *elementary reactions*. For example, the reaction of gaseous hydrogen with chlorine forming hydrogen chloride is represented by the following balanced chemical equation:

$$H_2 + Cl_2 \rightarrow 2HCl. \quad (1)$$

This reaction proceeds by a chain mechanism involving a repetitive sequence of elementary reactions involving the three stable molecules listed in Eq. (1), as well as two short-lived free radical intermediates, i.e., chlorine atoms (Cl) and hydrogen atoms (H). The most important elementary steps in the overall reaction mechanism are:

$$\text{Initiation:} \quad Cl_2 \rightarrow 2Cl. \quad (2)$$

$$\text{Propagation:} \quad Cl + H_2 \rightarrow HCl + H \quad (3a)$$
$$H + Cl_2 \rightarrow HCl + Cl. \quad (3b)$$

$$\text{Termination:} \quad H + Cl + M \rightarrow HCl + M \quad (4a)$$
$$Cl + Cl + M \rightarrow Cl_2 + M \quad (4b)$$
$$H + H + M \rightarrow H_2 + M. \quad (4c)$$

The chain reaction is initiated by dissociation of Cl_2, a stable molecule, to form two highly reactive chlorine atoms (Cl). Since chemical bond fission requires the input of energy, initiation may be achieved by heating the sample (see Lindemann mechanism) or by ultraviolet irradiation (photodissociation). Following initiation, the two elementary bimolecular propagating reactions will continue until either or both of the reactants (H_2 and Cl_2) are consumed, at which time the termination steps end the chain reaction. Termination typically involves termolecular recombination of two radicals to form a stable molecule in the presence of a third body (M), which may be a molecule or the wall of the container. Note that the two elementary propagating reactions may be added like mathematical equations, yielding the overall chemical reaction (1).

For an overall reaction such as that in Eq. (1) involving a sequence of elementary steps, the overall rate of formation of products may be a complex function of reactant concentrations, because products are formed by several different elementary processes. In the previous example, the HCl products are formed by reactions (3) and (4), each of which has its own rate law and rate constant. Thus, for a complex multistep process such as reaction (1), the rate law can only be determined through experiment.

For an *elementary* bimolecular reaction $A + B \rightarrow C + D$, the reaction rate is proportional to the concentrations (denoted by [A], [B], etc.) of the reactants:

$$\frac{d[C]}{dt} = \frac{d[D]}{dt} = -\frac{d[A]}{dt} = -\frac{d[B]}{dt} = k[A][B]. \quad (5)$$

The proportionality constant, k, is called the reaction rate constant. Since an elementary reaction involves a single bimolecular collision between A and B, the maximum possible rate constant is usually the frequency of collisions between reactants. A few simple atom-transfer reactions (e.g., $F + H_2 \rightarrow HF + H$) actually do occur on nearly every collision, and are said to proceed at or near the "gas kinetic limit."

I. KINETICS AND COLLISION THEORY

In order to estimate the frequency of collisions between gaseous A and B molecules, consider a beam of molecules of incident flux I_A (molecules/cm$^2 \cdot$ s) impinging on a static cell containing molecules at a concentration [B] (molecules/cm^3). The particles interact in a volume element V. The collision rate per unit time, Z, is given by

$$Z = \sigma I_A [B] V. \quad (6)$$

Here, σ is the collision cross section, which may be estimated using a simple hard sphere model for colliding particles (Fig. 1). Two particles collide with a relative velocity vector, $\mathbf{g}$, the magnitude of which is denoted by g, and impact parameter b, also known as the "aiming error" of the collision. A hard sphere collision will occur provided $0 \le b \le (r_A + r_B)$. The collision cross section is therefore the area of a circle of radius $d_{AB} = r_A + r_B$, i.e., $\sigma_{h.s.} = \pi d_{AB}^2$. The incident flux, $I_A = [A]g_A$, may then be substituted into Eq. (6). If the rate of reaction between A and B is simply the collision rate, then

$$-\frac{d[A]}{dt} = Z = \sigma_{h.s.} g_A [A][B]. \quad (7)$$

$b < d_{AB} = r_A + r_B$

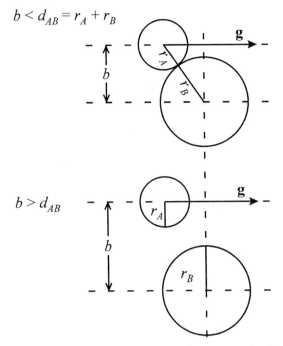

$b > d_{AB}$

FIGURE 1 Depiction of hard sphere collision cross section. Note that collisions only occur for impact parameters $b < d_{AB}$.

Comparison of Eqs. (5) and (7) indicates that the collision rate constant, k, is related to the collision cross section and relative velocity of colliding particles by

$$k(g) = \sigma g. \qquad (8)$$

In deriving Eq. (8), it is assumed that molecules A and B collide with a single relative velocity g. In a real gaseous sample containing both A and B molecules at thermal equilibrium, the distribution of relative velocities is described by the Maxwell–Boltzmann Distribution Law:

$$f(g) = 4\pi g^2 \left(\frac{\mu}{2\pi k_B T}\right)^{3/2} e^{-\frac{\mu g^2}{2k_B T}}, \qquad (9)$$

where $\mu = m_A m_B/(m_A + m_B)$ is the reduced mass of the colliding particles, k_B is Boltzmann's constant, and T is the temperature in Kelvin. The hard sphere collision rate constant, $k_{h.s.}$, is thus temperature dependent, and may be evaluated explicitly by integrating over all possible relative velocities, g:

$$k_{h.s.}(T) = \int_{g=0}^{\infty} \sigma_{h.s.} g f(g) \, dg = \pi d_{AB}^2 \left(\frac{8k_B T}{\pi \mu}\right)^{1/2}. \qquad (10)$$

Using typical molecular values of $d_{AB} \approx 0.35$ nm and $\mu = 14$ amu for room temperature collisions between N_2 molecules, one observes the magnitude of a hard sphere collision rate constant to be on the order of 2.6×10^{-10} cm^3/molecule · s.

Whereas the hard sphere cross section depends only on the sum of the radii of colliding particles, a reaction cross section may depend strongly on the energy of the collision and therefore on the relative velocity of the colliding particles, the magnitude of which is given by g. As illustrated in Fig. 2, the relative velocity vector, $\mathbf{g}$, may be decomposed into two perpendicular components. The first is a radial component, $g_r = \dot{r}$, i.e., the time derivative of r, the distance between the particle centers. Perpendicular to the radial axis is a tangential component, $g_\perp = r\dot{\theta}$, where $\dot{\theta}$ is the time derivative of θ, the angle between $\mathbf{g}$ and the radial axis. As a result, the total kinetic energy of collision, $E_{kin} = {}^1/_2\mu g^2$, can be thought of as a sum of a radial kinetic energy,

$$E_r = {}^1/_2\mu g_r^2 \qquad (11)$$

and an energy associated with the perpendicular velocity component,

$$E_\perp = {}^1/_2\mu g_\perp^2 = {}^1/_2\mu r^2 \dot{\theta}^2 = \frac{L^2}{2\mu r^2}, \qquad (12)$$

where $L = \mu r^2 \dot{\theta} = \mu g b$ is the magnitude of the angular momentum associated with the colliding pair. Thus, for interaction potentials that depend solely on the distance between the colliding pair, $V(r)$, only g_r is effective in surmounting potential energy barriers such as those associated with the energy required to break and form bonds during reaction; $g_\perp$ is associated purely with rotational motion of the two particles.

One way to model the energy dependence of σ is to assume that reaction can only occur if the component

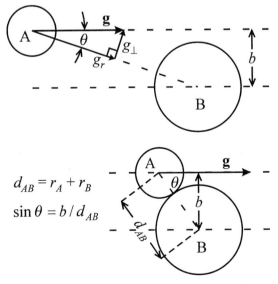

$d_{AB} = r_A + r_B$

$\sin \theta = b / d_{AB}$

FIGURE 2 Decomposition of relative velocity vector, $\mathbf{g}$, into radial (g_r) and perpendicular ($g_\perp$) components, with θ defined as the angle between $\mathbf{g}$ and the internuclear axis. At the moment of a hard sphere collision, $\sin \theta = b/d_{AB}$.

of kinetic energy along the axis of the collision (i.e., E_r), exceeds a critical energy, V_c. Note that during the course of a collision, θ varies from 0 to 180° as r changes from $-\infty$ to $+\infty$. However, if the particles are treated as hard spheres, the internuclear potential, $V(r)$, is zero for $r > d_{AB}$, and there is no interaction between them until collision, at which time $(\sin\theta) = b/d_{AB}$. Using the fact that $g_r = g\cos\theta$, the radial kinetic energy at the moment of contact is

$$E_r = \tfrac{1}{2}\mu g^2 \cos^2\theta = E_{kin}(1 - \sin^2\theta) = E_{kin}\left(1 - \frac{b^2}{d_{AB}^2}\right). \tag{13}$$

In order for the reaction to be successful, $E_r \geq V_c$. For a given collision energy, E_{kin}, this implies that the impact parameter must be smaller than a critical impact parameter, b_c, defined such that

$$E_{kin}\left(1 - \frac{b_c^2}{d_{AB}^2}\right) = V_c. \tag{14}$$

Thus, the cross section for a reaction involving a critical energy, V_c, given by

$$\sigma = \pi b_c^2 = \pi d_{AB}^2\left[1 - \frac{V_c}{E_{kin}}\right] \tag{15}$$

is expected to increase with energy, as illustrated in Fig. 3. Note that this model predicts that the threshold for reaction occurs at $E_{kin} = V_c$, and that the cross section reaches half the hard sphere value at $E_{kin} = 2V_c$, asymptotically approaching the hard sphere value as $E_{kin} \to \infty$.

In many cases, reaction cross sections for real systems differ considerably from that shown in Fig. 3. For example, some processes, such as charge exchange (e.g., $A^+ + B^- \to A + B$), proceed with reaction cross sections far exceeding the hard sphere limit. Here, the long-range Coulomb potential causes reactants to be attracted to one another at large distances, considerably increasing the reaction cross section. Even for neutral–neutral interactions, the interaction potential, $V(r)$, often differs substantially from that of hard spheres. Long-range induced dipole-

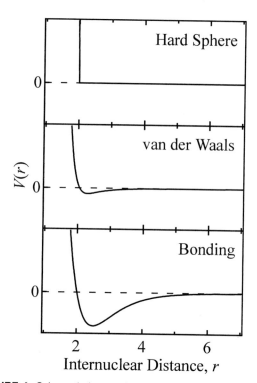

FIGURE 4 Schematic internuclear potentials for different models of atomic and molecular interactions. The hard sphere model exhibits only a repulsive component at small r; more realistic potentials exhibit attractive and repulsive components.

induced dipole interactions (van der Waals' interactions) result in an attractive region of the potential surface at longer bond distances even in cases when formal bonds between the interacting pair cannot be formed. When strong bonds can be formed between colliding particles, as in the case of two halogen atoms like Cl, a strongly attractive component of the potential is present over a wide range of internuclear separations. In such cases, the hard sphere potential would only be useful in modeling the interaction potential at small distances where electron–electron repulsion becomes dominant, as demonstrated in Fig. 4. However, the magnitudes of most reaction cross sections are controlled predominantly by the form of the attractive component of the potential at longer internuclear separations.

We now discuss a relatively simple model for reactions involving gaseous particles interacting through a potential $V(r)$ operating at long range. Recall from Eq. (11) and (12) that E_{kin} can be written as a sum of radial energy, E_r, and energy associated with rotational motion of the interacting particles. Thus, the total energy, $E = E_{kin} + V(r)$, of the colliding partners is

$$E = E_r + \frac{(\mu g b)^2}{2\mu r^2} + V(r), \tag{16}$$

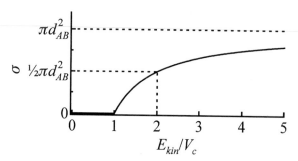

FIGURE 3 Cross section (σ) dependence on kinetic energy, E_{kin}, for the hard sphere model requiring an energetic threshold, V_c.

making it convenient to conceptualize an effective potential,

$$V_{eff}(r) = \frac{(\mu g b)^2}{2\mu r^2} + V(r), \qquad (17)$$

that governs the radial motion of the particles. At large internuclear distances, the total energy of the system, $E = \frac{1}{2}\mu g^2$, is simply the radial kinetic energy, since $\lim_{r\to\infty} V_{eff}(r) = 0$. However, at smaller values of r, where $V_{eff}(r) > 0$, some of the initial kinetic energy is converted into energy associated with the effective potential. The distance of closest approach, or turning point, $r_{t.p.}$, is reached when the magnitude of the effective potential is equal to the initial radial kinetic energy, and the radial velocity becomes zero:

$$E = \frac{1}{2}\mu g^2 = \frac{(\mu g b)^2}{2\mu r_{t.p.}^2} + V(r_{t.p.}). \qquad (18)$$

This may be rearranged to solve for $b = b_{max}$, the maximum impact parameter for a given relative velocity, g, that allows the collision partners to reach a fixed critical distance for reaction, r_c:

$$b_{max}(g) = r_c \left[1 - \frac{2V(r_c)}{\mu g^2} \right]^{1/2}, \qquad (19)$$

which holds for $g \geq g_{min}$, where, in order to be physically meaningful,

$$g_{min} = [2V(r_c)/\mu]^{1/2} \quad \text{if } V(r_c) > 0$$
$$= 0 \quad \text{if } V(r_c) \leq 0. \qquad (20)$$

The reaction rate constant may in both of these cases be determined analytically by integration over all relative velocities exceeding g_{min}:

$$k(T) = \int_{g_{min}}^{\infty} \pi \{b_{max}(g)\}^2 g f(g) \, dg$$

$$= \pi r_c^2 \left(\frac{8k_B T}{\pi \mu} \right)^{1/2} e^{-\frac{V(r_c)}{k_B T}} \quad \text{if } V(r_c) > 0 \qquad (21)$$

$$= \pi r_c^2 \left(\frac{8k_B T}{\pi \mu} \right)^{1/2} \left(1 - \frac{V(r_c)}{k_B T} \right) \quad \text{if } V(r_c) \leq 0. \qquad (22)$$

In cases where $V(r_c)$ is positive, the critical distance rate constant expression [Eq. (21)], is similar to the hard sphere collision rate constant [Eq. (10)]; however, an additional exponential term is present. This term represents the fraction of molecules at temperature T having sufficient energy to react. This temperature dependence is thus similar

to the empirical Arrhenius expression found to satisfactorily model a large number of chemical reactions:

$$k = A e^{-E_a/k_B T}, \qquad (23)$$

where A is the Arrhenius preexponential factor, and E_a is the Arrhenius activation energy. These quantities are most readily determined by plotting $\ln k$ vs. $1/T$, which should be linear with a slope $-E_a/k_B$ and intercept $\ln A$. Note that this purely empirical relationship often holds for elementary as well as multistep reactions. The obvious similarity between Eqs. (21) and (23) suggests that E_a is at least loosely related to the height of the potential energy barrier for the rate-limiting step in the reaction. However, the Arrhenius parameters are only phenomenological quantities derived from the temperature dependence of reaction rate constants. In fact, Arrhenius plots are in many cases found to be markedly nonlinear, suggesting the occurrence of a multistep reaction mechanism or a mechanism that changes at different temperatures.

The critical distance model can be used to derive an explicit formula for the temperature dependence of the reaction rate constant for charge transfer reactions of the form $A^+ + B^- \to A + B$. Such interactions are subject to long-range Coulomb attractions of the form $V(r) = -q^2/4\pi\varepsilon_0 r$, where q is the charge of an electron. Taking the critical distance, r_c, to be the ionic–covalent curve crossing radius (R), which corresponds to the distance at which the Coulomb attraction between ions balances the energy required for electron transfer, one obtains by substitution into Eq. (22) the following expression for the charge exchange rate constant:

$$k(T) = \pi R^2 \left(\frac{8k_B T}{\pi \mu} \right)^{1/2} \left(1 + \frac{q^2}{4\pi\varepsilon_0 R k_B T} \right). \qquad (24)$$

This reaction rate constant expression bears some similarity to a hard-sphere rate constant; however, an additional term ($q^2/4\pi\varepsilon_0 R k_B T$) results from the long range attractive interaction, and is in general the dominant contribution to the reaction rate constant for reactions of this type.

For attractive potentials of the form $V(r) = -a/r^s$, V_{eff} is given by

$$V_{eff}(r) = \frac{L^2}{2\mu r^2} - \frac{a}{r^s} = \frac{(\mu g b)^2}{2\mu r^2} - \frac{a}{r^s}. \qquad (25)$$

Provided $s \geq 3$, V_{eff} has a local maximum for a given impact parameter, b, at a radial distance, r_{max}, determined by

$$r_{max} = \left(\frac{sa}{\mu g^2 b^2} \right)^{2-s}. \qquad (26)$$

For close approach required for reaction, the two particles must overcome this maximum, $V_{max} = V_{eff}(r_{max})$, as

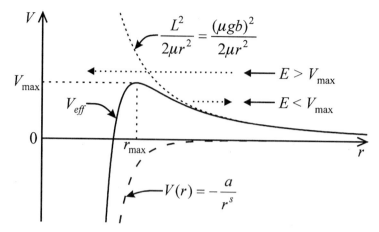

FIGURE 5 Relationship between $V_{eff}(r)$ and $V(r)$ for a given impact parameter, b. Close approach necessary for reaction requires $E > V_{max}$, where V_{max} is the maximum in $V_{eff}(r)$.

illustrated in Fig. 5. Classically, if the energy of the colliding reactants is exactly equal to V_{max}, all of the initial kinetic energy is converted into V_{eff} when $r = r_{max}$, and "orbiting" will occur; i.e., the interacting pair will rotate together for an infinite amount of time since both the radial kinetic energy and the centrifugal force acting on the particles $(-dV_{eff}/dr)$ are zero. The orbiting impact parameter, b_{orb}, defined as the impact parameter for a given initial collision energy, $E = \frac{1}{2}\mu g^2$, at which orbiting occurs, can be determined by setting $V_{max} = E$ and rearranging to arrive at

$$b_{orb}^2 = \left(\frac{a(s-2)}{\mu g^2}\right)^{2/s}\left(\frac{s}{s-2}\right). \quad (27)$$

Since smaller impact parameter collisions will result in a smaller value of V_{max}, only collisions occurring with impact parameters less than b_{orb} will lead to a close collision and reaction. Thus, the reaction cross section, $\sigma = \pi b_{orb}^2$, which depends on the relative velocity g, may be integrated over all relative velocities to derive the rate constant temperature dependence:

$$k(T) = \int_0^\infty \left(\pi b_{orb}^2\right) g f(g)\, dg$$

$$= 2^{\frac{3s-4}{2s}}\left(\frac{\pi}{\mu}\right)^{1/2}(s-2)^{2/s}a^{2/s}(k_BT)^{\frac{s-4}{2s}}\Gamma\left(\frac{s-2}{s}\right), \quad (28)$$

where the gamma function, Γ, is available in mathematical tables. This equation predicts that reactions involving quenching of an electronically excited state, which can be modeled using $s = 3$, will show weak inverse temperature dependence, $k(T) \propto T^{1/6}$. Ion–molecule reactions, having $s = 4$, are predicted to have rate constants independent of temperature. Reactions dominated by

van der Waals interactions (i.e., $s = 6$) are expected to show a small positive temperature dependence ($k \propto T^{1/6}$).

The profound effect of the exact form of the internuclear potential on the interaction between particles can be observed in elastic scattering experiments. These studies allow determination of the angle of deflection of a particle from its original direction upon interaction with the second particle. Conceptually, the scattering process can be understood by considering the effect of a particle of mass μ colliding with an infinitely massive particle fixed in space. The deflection angle, χ, defined as the angle between the initial and final relative velocity vectors of the colliding pair, will depend on the form of potential, $V(r)$, and, based on the impact parameter at which a given collision occurs, the region of $V(r)$ that is sampled by the colliding pair.

Recall from Eq. (17) that the effective potential, V_{eff}, governs the radial motion of the colliding particles, and therefore determines the radial turning point, $r_{t.p.}$, for a given magnitude of initial collision energy, E. Figure 6 shows three effective potentials (for impact parameters $b = 0$, b_1, b_2, and b_3, where $0 < b_1 < b_2 < b_3$) for an internuclear potential, $V(r)$, with both long-range attractive and short-range repulsive components. If the two particles collide with a fixed collision energy (e.g., that denoted by E_1), the radial turning point, and therefore the regions of $V(r)$ accessed during the collision, will depend strongly on the magnitude of b. For very large values of b, ($b \sim b_3$), the turning point lies at very large r, and the particle experiences little deflection upon approach ($\chi \sim 0$). For smaller values of b ($b \sim b_2$), the turning point moves to smaller values of r, and the interacting particles therefore sample more of the long-range attractive part of $V(r)$, resulting in a deflection toward more negative angles. As b becomes even smaller, however, the effects of the repulsive part of the potential begin to play a role, and χ

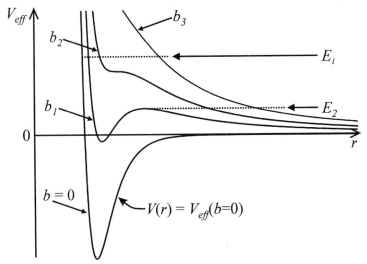

FIGURE 6 Dependence of V_{eff} on impact parameter, b, where $0 < b_1 < b_2 < b_3$. Larger magnitudes of b result in larger values of the radial turning point for a given collision energy.

reaches a minimum value, called the "rainbow" angle, χ_r, by analogy to the optical rainbow resulting from scattering in small water droplets. The impact parameter that results in this most negative degree of deflection is referred to as the rainbow impact parameter, b_r. For impact parameters less than b_r ($b \sim b_1$), short-range repulsion begins to dominate, and the deflection angle becomes less negative, reaching zero at the impact parameter at which the attractive and repulsive forces during the collision exactly offset, the so-called "glory" impact parameter, b_g. For impact parameters smaller than b_g, repulsion dominates, and the particle is scattered to positive angles, reaching a max-

imum of $\chi = \pi$ for direct head-on collisions ($b = 0$). A pictorial depiction of the dependence of χ on the impact parameter, known as the deflection function, is shown in Fig. 7, where particle trajectories during the course of a collision are shown for a wide range of impact parameters. Note that this figure depicts the dependence of the deflection angle, χ, on $b^* = b/r_e$, where r_e is the internuclear separation where $V(r)$ reaches a minimum.

Although collision theory has provided considerable insight beyond simple hard spheres, it cannot properly address questions such as what fraction of collision geometries are likely to lead to reaction. Such issues cannot be

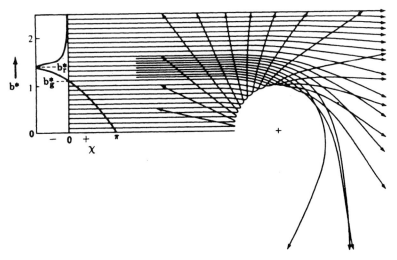

FIGURE 7 Deflection angle, χ, dependence on impact parameter. Trajectories depict the degree of deflection of impinging particle resulting from attractive and repulsive components of the interaction potential. (From Levine, R. D., and Bernstein, R. B. (1987). "Molecular Reaction Dynamics and Chemical Reactivity," Oxford University Press, New York.)

properly accounted for unless the details of the molecular structure of reactants are considered. Often, the rate constant calculated from collision theory must be reduced by inclusion of an *ad hoc* "steric factor" which represents the fraction of collisions that have proper geometry for reaction. This additional factor is an empirical factor used to bring experimental observation into line with collision theory, and accounts for those important factors not addressed by collision theory.

II. ACTIVATED COMPLEX THEORY

Many models of chemical reactions are based on the concept of an "activated complex," or "transition state," which corresponds to the nuclear configuration with the highest potential energy of the system traversed during the course of the reaction. The transition state corresponds to a critical geometry of the reacting system marking the boundary between reactants and products. Consider the elementary reaction:

$$A + HB \rightarrow AH + B. \tag{29}$$

The overall rate for this elementary reaction is given by

$$\frac{d[AH]}{dt} = k[A][HB]. \tag{30}$$

The reaction coordinate for this H-atom transfer reaction may be considered to be translational motion of the H atom from B to A. A potential energy diagram for this process may be represented in 2D or in 3D, as shown in Fig. 8.

According to transition state theory, a fast equilibrium exists between reactants and molecules at the transition state, denoted by $AHB^{\ddagger}$:

$$A + HB \underset{k_{-1}}{\overset{k_1}{\rightleftarrows}} AHB^{\ddagger} \overset{\omega}{\rightarrow} AH + B. \tag{31}$$

The equilibrium concentration of molecules at the transition state is given by

$$[AHB^{\ddagger}] = K^{\ddagger}[A][HB], \tag{32}$$

where $K^{\ddagger} = k_1/k_{-1}$. The overall rate of product formation depends on the rate constant, ω, with which the activated complexes cross over to products:

$$\frac{d[AH]}{dt} = \omega[AHB^{\ddagger}] = \omega K^{\ddagger}[A][B]. \tag{33}$$

A comparison of Eqs. (30) and (33) indicates that the rate constant $k = \omega K^{\ddagger}$. The equilibrium constant for production of activated complexes $K^{\ddagger}$ is related to molecular partition functions (Q), calculated using statistical mechanics:

$$K^{\ddagger} = \frac{[AHB^{\ddagger}]}{[A][HB]} = \frac{Q_{AHB\ddagger}}{Q_A Q_{HB}} e^{-\Delta E/k_B T}. \tag{34}$$

In the above equation, $\Delta E = E_{AHB\ddagger} - E_A - E_{HB}$ is the energy difference between the reactants and activated complex. Assuming that the reaction involves translational motion of the hydrogen atom from B to A, the rate of passage through the transition state is given by

$$\omega = \frac{v}{\delta} = \left(\frac{k_B T}{2\pi\mu}\right)^{1/2} \frac{1}{\delta}, \tag{35}$$

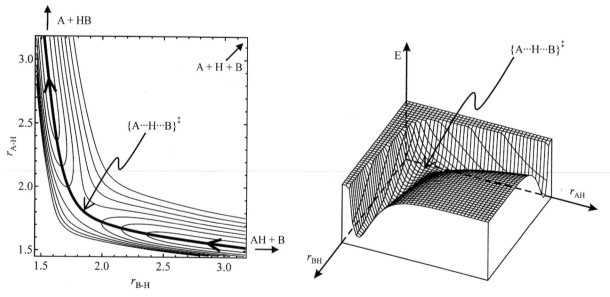

FIGURE 8 Schematic potential energy surface of $A + HB \rightarrow AH + B$ reaction, shown as a two-dimensional contour plot (left) and a three-dimensional surface plot (right). Trajectory on contour plot corresponds to lowest energy pathway from reactants to products and traverses the region corresponding to the reaction transition state, $AHB.^{\ddagger}$

where v is the velocity through the transition state region and δ is the "width" of transition state along the reaction coordinate.

The contribution to the partition function of the activated complex corresponding to motion along the reaction coordinate, in this case translational motion, is factored out of the transition state partition function, $Q_{AHB\ddagger}$, to yield:

$$Q_{AHB\ddagger} = q_{trans,AHB\ddagger} Q'_{AHB\ddagger} = \frac{(2\pi\mu k_B T)^{1/2}\delta}{h} Q'_{AHB\ddagger}. \tag{36}$$

Note that the width parameter δ appears again in Eq. (36). This parameter ultimately factors out in the transition state theory rate constant, k_{TST}, which is given by:

$$k_{TST} = \omega K^{\ddagger} = \frac{k_B T}{h} \frac{Q'_{AHB\ddagger}}{Q_A Q_{HB}} e^{-\Delta E/k_B T}. \tag{37}$$

Although we have derived this equation assuming that the reaction coordinate for atom transfer corresponds to translational motion, the same expression is obtained if the reaction coordinate is assumed to be vibrational motion. According to Eq. (37), the reaction rate constant may be calculated using the relevant molecular partition functions, known from statistical mechanics, remembering that $Q'_{AHB\ddagger}$ does not include the translational motion contribution to the transition state partition function.

III. UNIMOLECULAR VERSUS BIMOLECULAR REACTIONS

Many early experiments showed that thermal decomposition of a molecule A, forming products P_1 and P_2, often exhibits first order kinetics at high pressure, and second-order kinetics at low pressure. As first proposed by Lindemann, the mechanism involves collisional excitation of the reactant A:

$$A + M \underset{k_{-1}}{\overset{k_1}{\rightleftharpoons}} A^* + M, \qquad A^* \overset{k_2}{\rightarrow} P_1 + P_2. \tag{38}$$

Under steady-state conditions, the concentration of the collisionally activated molecule, A^* is constant, i.e., the rate of its formation is exactly balanced by the rate of its destruction:

$$\frac{d[A^*]}{dt} = k_1[A][M] - k_{-1}[A^*][M] - k_2[A^*] = 0. \tag{39}$$

Rearranging:

$$[A^*] = \frac{k_1[A][M]}{k_{-1}[M] + k_2}. \tag{40}$$

The overall rate of product formation is

$$\frac{d[P_1]}{dt} = \frac{d[P_2]}{dt} = -\frac{d[A]}{dt} = k_2[A^*] = \frac{k_1 k_2[A][M]}{k_{-1}[M] + k_2}. \tag{41}$$

At high pressure, $k_{-1}[M] \gg k_2$, and $d[P_1]/dt = (k_1 k_2/k_{-1})[A]$, yielding first-order kinetics. Under low-pressure conditions, $k_{-1}[M] \ll k_2$ and $d[P_1]/dt = k_1[A][M]$ resulting in second-order behavior. The important result is that reactions that appear to be *unimolecular*, exhibiting first-order kinetics at high pressures, actually involve *bimolecular* processes.

An important class of elementary bimolecular reactions are those that involve formation of persistent collision complexes, denoted AB^*, that may ultimately form products C + D, or decay back to reactants:

$$A + B \underset{k_{-1}}{\overset{k_1}{\rightleftharpoons}} AB^* \overset{k_2}{\rightarrow} P_1 + P_2. \tag{42}$$

Following formation of the AB^* intermediate by bimolecular collision, the reaction dynamics are somewhat analogous to those in the Lindemann mechanism: an internally excited molecule may decay to products or reform reactants. In the absence of collisions, the dynamics are solely determined by the unimolecular dynamics of the complex. However, at high pressures, particularly if the lifetime of the AB complex is long, it may undergo collision with another body, possibly carrying away excess energy resulting in formation of stable AB:

$$AB^* + M \overset{k_3}{\rightarrow} AB + M. \tag{43}$$

IV. STATISTICAL THEORIES OF UNIMOLECULAR DECOMPOSITION

The Lindemann mechanism as well as reactions occurring via formation of long-lived complexes involve participation of highly internally excited intermediate species that may ultimately dissociate by one or more chemical channels. For example, the intermediate complex AB^* in reaction (42) may form new products $P_1 + P_2$, or decay back to reactants, A + B. The total rate constant for decay of AB^* is the sum of the two rate constants, $k_{-1} + k_2$, and the relative importance of these competing processes is defined as the product branching ratio k_2/k_{-1}. Of key importance in understanding these reactions are the reaction rate constants k_{-1} and k_2. A number of theories have been developed to quantitatively predict rate constants for unimolecular reactions.

Rice, Ramsperger, and Kassel developed a simple theory, now known as RRK theory, which contains many fundamental elements underlying most modern theories of unimolecular reaction. According to RRK theory, the energized molecule produced by bimolecular collision may be considered to consist of a group of s identical harmonic oscillators, each of frequency v. If any oscillator accumulates sufficient energy $E_o = mhv$, where m is an integer and h is Planck's constant, the energized molecule will dissociate. An underlying assumption of RRK (as well as other related theories) is that energy may flow freely between the oscillators in the molecule. The total energy of all of the oscillators is denoted by $E = nhv$. The number of ways that n quanta can be placed in a molecule consisting of s oscillators is:

$$w_n = \frac{(n+s-1)!}{n!(s-1)!}.$$ (44)

The number of ways that n quanta can be placed in a molecule such that at least m quanta are in one oscillator is given by

$$w_m = \frac{(n-m+s-1)!}{(n-m)!(s-1)!}.$$ (45)

The probability, P, of dissociation is the ratio of these quantities:

$$P = \frac{w_m}{w_n} = \frac{(n-m+s-1)!n!}{(n-m)!(n+s-1)!}.$$ (46)

If n and m are large, then Sterling's approximation $(x! \approx x^x/e^x)$ may be applied, and, if s is small relative to $(n-m)$, then $(n-m+s-1) \approx n-m$. The probability of decomposition then reduces to

$$P = \left(\frac{n-m}{n}\right)^{s-1}.$$ (47)

The rate constant for unimolecular decomposition is the probability P multiplied by a frequency factor v:

$$k_{RRK} = v\left(\frac{n-m}{n}\right)^{s-1} = v\left(\frac{E-E_o}{E}\right)^{s-1}.$$ (48)

For a given value of s, the reaction rate constant increases with increasing energy E above threshold, E_o. If $E \gg E_o$, the reaction rate constant approaches the frequency factor v. On the other hand, for reactions involving two similar but different-sized molecules having the same E and E_o, since the larger molecule has a greater number of oscillators, s, the reaction rate constant k is smaller (since $(E-E_o)/E < 1$). In practice, to obtain agreement with experiment, it is necessary to use values of s in Eq. (48) which are approximately one half of the actual number of vibrational modes in the molecule. Of course, because RRK theory treats all oscillators as having the same vibrational frequency, the theory employs very simple equations that represent qualitatively but not quantitatively the behavior of real molecules.

With the development of computers, accurate calculations using theoretical models better able to represent the behavior of real molecules has become widespread. A very important extension of the original theory, due to Marcus, is known as RRKM theory. Here, the real vibrational frequencies are used to calculate the density of vibrational states of the activated molecule, $N(E)$. The number of ways that the total energy can be distributed in the activated complex at the transition state is denoted $W(E')$. Note that the geometry of the transition state need not be known, but the vibrational frequencies must be estimated in order to calculate $W(E')$. In calculating the total number of available levels of the transition state, explicit consideration of the role of angular momentum is included. The RRKM reaction rate constant is given by:

$$k_{RRKM} = \frac{W(E')}{hN(E)},$$ (49)

where h is Planck's constant.

V. REACTIONS IN SOLUTION

The density of molecules is substantially higher in liquids than in the gas phase. However, for reactions carried out in solution under relatively dilute conditions, the concentrations of reactants are not appreciably different from in the gas phase. Since reactant molecules A and B must undergo collision in solution in order to react, many of the same principles developed for gas-phase reactions also apply in solution. However, the presence of solvent molecules leads to important differences between reactions in solution and in the gas phase. In solution, the rate of diffusion often limits the rate of approach of molecule A to within a sufficient distance to B for reaction to occur. Once an encounter pair AB is formed, however, the solvent may act as a "cage," effectively holding them in close proximity, thereby increasing the probability of reaction.

In solution, the overall reaction may again be broken down into a sequence of elementary steps:

$$A + M \underset{k_{-d}}{\overset{k_d}{\rightleftarrows}} AB, \qquad AB \overset{k_r}{\rightarrow} P,$$ (50)

where, AB is an encounter pair, k_d and k_{-d} are rate constants for approach and separation of the reactant molecules by diffusion, and k_r is the rate of conversion of encounter pairs to products. Applying the steady-state approximation to the concentration of encounter pairs, we obtain

$$\frac{d[P]}{dt} = \frac{k_d k_r}{k_{-d} + k_r}[A][B] = k[A][B].$$ (51)

If the activation energy for the reaction is large, $k_r \ll k_d$ and the reaction rate constant $k \approx k_d k_r / k_{-d}$. Alternatively, if $k_r \gg k_d$, then $k \approx k_d$, and the reaction rate constant is the rate of diffusion of reactants to sufficiently close proximity to facilitate reaction. The rate of a diffusion controlled reaction is determined by the magnitude of the diffusion coefficients, D, for the reactants A and B:

$$k = 4\pi(D_A + D_B)R. \qquad (52)$$

The diffusion coefficients are related to the viscosity of the solvent, η, and hydrodynamic radius, r, of the diffusing species by the Stokes–Einstein Law:

$$D = \frac{k_B T}{\pi \beta \eta r}, \qquad (53)$$

where β is a constant typically ranging from 4 to 6. Note that this derivation assumes that diffusing particles are equally likely to move in any direction. However, if the reactants are oppositely charged ions, Coulomb attraction substantially increases reaction rates relative to the rate of diffusion.

VI. EXPERIMENTAL TECHNIQUES

It is well known from the empirical Arrhenius expression that reaction rate constants often increase with reactant temperature. One goal of molecular reaction dynamics is to understand the roles that different forms of reactant energy (translational, electronic, vibrational, and rotational) play in chemical reaction. Also, if a reaction *is* successful, how is the total available energy channeled into the various available degrees of freedom of the product molecules?

Many of the microscopic details of a chemical reaction are related to the nature of the transition state. Most insight into the transition state region has been obtained through experiments focusing on the asymptotic limits of the reaction; i.e., reactants and products. Four general categories of experiments will be discussed here, all focusing on elementary gas-phase reactions. In the first type of experiment, the total cross sections for various chemical reaction channels are measured, usually as a function of collision energy. In the second type, the angular and velocity distributions of products from single reactive encounters are measured. In the third type of experiment, the quantum state distributions (vibrational and rotational) of products are measured using spectroscopic techniques. In this latter approach, velocity distributions may also be obtained using Doppler or velocity imaging methods. Finally, a relatively recent development is "transition state spectroscopy," which focuses directly on the transition state itself, usually through spectroscopic measurements.

A. Cross Section Measurements

The reaction cross section may be determined experimentally as a function of collision energy using a variety of methods. For the reaction $H + D_2 \rightarrow HD + D$, the reaction has been carried out in a flow cell containing mixtures of D_2 and a stable H atom precursor such as HBr. The H atoms are produced by photolysis of HBr at various UV wavelengths. Reaction cross sections are determined by measuring the concentration of products relative to reactants directly. In the present case, this involves monitoring the relative concentrations of H and D atoms (i.e., reactants and products) as a function of time following photodissociation of reactant precursor. Both H and D atoms may be detected by laser induced fluorescence (LIF) excitation spectroscopy near 121 nm (Lyman-α). By choosing different photolysis wavelengths and H atom precursor molecules, reaction rate constants may be determined for different collision energies. In this case, reaction cross sections are determined directly from Eq. (8), since the relative velocity is well defined using photolytic reactants. In Fig. 9, experimental data from several different laboratories are shown as solid points surrounded by rectangles, and theoretical values are connected by a solid line. The reaction cross section is zero below 0.4 eV due to the presence of a potential energy barrier for reaction, as discussed in detail in Section VI.B. The reaction cross section increases with increasing energy above threshold, with behavior qualitatively similar to that predicted by Eq. (15).

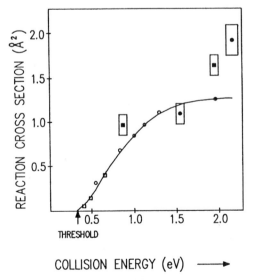

FIGURE 9 Cross section for $H + D_2 \rightarrow HD + D$ reaction vs. collision energy. Solid points surrounded by rectangles are experimental data and open points connected by solid line is theoretical calculation. (From Gerlach-Meyer, U., Kleinermanns, K., Linnebach, E., and Wolfrum, J. (1987). *J. Chem. Phys.* **86**, 3047–3048.)

In some cases, reaction may lead to more than one type of chemical product. For example, the reaction of chromium cations, Cr^+, with methane, CH_4, may lead to production of $CrCH_2^+ + H_2$ or $CrH^+ + CH_3$. In order to study ion-neutral reactions such as these, a beam of mass-selected reactant ions, e.g., Cr^+, is accelerated to a well-defined laboratory kinetic energy. The ion beam encounters target molecules held in a gas cell, where bimolecular reaction occurs. The product ions are then extracted from the reaction volume, mass selected, and counted. A wide variety of reactions have been studied using such techniques. In Fig. 10, the cross sections for reactions of ground state Cr^+ are shown using solid symbols. Due to the endoergicity of reaction, the formation of $CrCH_2^+$ only occurs at energies above ~2.3 eV, and CrH^+ formation has a threshold just under 3.0 eV. Reactions of an electronically excited state of Cr^+ with CH_4, on the other hand, have a large cross section for production of $CrCH_2^+$ even down to zero collision energy, as indicated by open symbols. Experiments such as this provide insight into the role of electronic state on chemical reactivity. Furthermore, by measuring energetic thresholds for reaction, thermodynamic quantities such as bond dissociation energies may be determined directly.

B. Angular and Velocity Resolved Studies

Crossed molecular beam reactive scattering facilitates experimental determination of the angular and velocity distributions of chemical products from elementary bimolecular reactions. The technique involves production of two molecular beams containing the reactants, initially moving at right angles relative to one another, in an evacuated

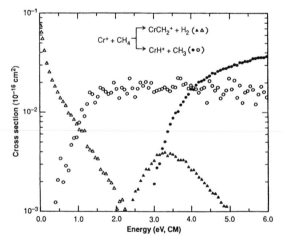

FIGURE 10 Experimental cross sections for $Cr^+ + CH_4$ reaction. Solid points denote reaction of ground electronic state Cr^+ and open points denote reactions of electronically excited Cr^+. (From Armentrout, P. B. (1991). *Science* **251,** 175–179.)

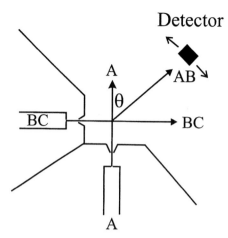

FIGURE 11 Schematic diagram of crossed molecular beam apparatus. Beams cross at right angles; products are detected by a detector that may be rotated with respect to the reactant beams.

($<10^{-6}$ Torr) chamber (Fig. 11). The beams, denoted A and BC here, are usually generated by a supersonic expansion, in which a gas at a pressure near an atmosphere or greater passes through a pinhole into an evacuated chamber. Due to adiabatic cooling of the molecules during expansion, rotational temperatures are reduced to less than 100 K. Vibrational excitation is also cooled, although less efficiently. Velocity distributions of molecules in a supersonic beam are typically less than 10% full width at half maximum (FWHM); much narrower than in a Boltzmann distribution. One or more collimating devices (skimmers) are used to define the beams to angular divergences of several degrees. Since molecules within the beam move together at nearly the same velocity in the evacuated chamber, the probability of collisions within a beam or with a background gas molecule is negligible. A small fraction of reactants in one beam undergo a single collision with molecules in the other, forming chemical products, e.g., $A + BC \rightarrow AB + C$. In a typical experiment, the beams are several millimeters in diameter and the detector is a mass spectrometer with electron impact ionizer that may be rotated in the plane of the beams. By measuring the arrival times of products at the detector at different angles, the laboratory angular and velocity distributions of the products are determined.

Reactive and nonreactive scattering data are analyzed with the aid of a Newton diagram, illustrated in Fig. 12 for the hypothetical reaction $A + BC \rightarrow AB + C$. This vector diagram in velocity space facilitates the transformation between the laboratory and center-of-mass (CM) reference frames. Laboratory angles are denoted by Θ, with $\Theta = 0$ and $90°$ defined by the directions of the A and BC beams, respectively. All laboratory velocities are denoted by the variable v, and have their origins in the lower left corner of

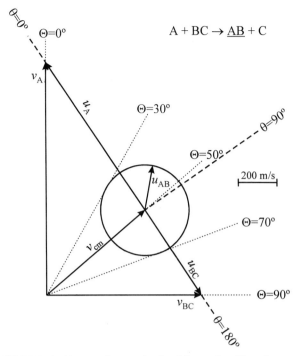

$\Theta=0°$

$A + BC \rightarrow \underline{AB} + C$

$\Theta=30°$

$\theta=90°$

$\Theta=50°$

200 m/s

$\Theta=70°$

$\Theta=90°$

$\theta=180°$

FIGURE 12 Newton diagram for $A + BC$ reaction. Note that velocity scale is indicated on right side of diagram. See text for details.

the Newton diagram. The relative velocity vector is given by $\mathbf{v}_{rel} = \mathbf{v}_A - \mathbf{v}_{BC} = \mathbf{u}_A - \mathbf{u}_{BC}$, with velocities in the CM reference frame denoted by u and angles in the CM frame denoted by θ.

The laboratory velocity of the center-of-mass of the colliding partners, $\mathbf{v}_{cm}$, which in this case is oriented along $\Theta = 50°$, may be calculated from the masses and initial velocities of the $A + BC$ reactants. Note that AB products scattered from the collision volume will be traveling with a velocity in the laboratory reference frame that is simply $\mathbf{v}_{cm} + \mathbf{u}_{AB}$. Thus, a circle is shown in Fig. 12, centered at the tip of $\mathbf{v}_{cm}$. The radius of this circle corresponds to the maximum possible velocity of AB products in the CM reference frame, u_{AB}, for all possible orientations of the vector $\mathbf{u}_{AB}$. From energy conservation, this maximum velocity corresponds to production of internally cold AB + C products. Thus, if the thermodynamics of the reaction are known or can be estimated, u_{AB} may be calculated using energy and momentum conservation laws. According to Fig. 12, product AB molecules may be observed at laboratory angles Θ ranging from 30 to 70°; no products can be observed outside of this range. Products formed with progressively greater internal energies, and hence smaller translational energies, are constrained to smaller Newton circles, which are always centered on the tip of $\mathbf{v}_{cm}$. A single CM recoil velocity vector $\mathbf{u}_{AB}$ is shown in this figure, at $\theta = 45°$, with $\theta = 0$ and 180° corresponding to the ve-

locities of A and BC reactants, respectively, in the CM reference frame. However, depending upon the dynamics of the reaction, θ can range from 0 to 360°. For randomly oriented reactant molecules, CM angular distributions must be symmetric about the relative velocity vector, i.e., about $\theta = 0$ and 180°.

Among the earliest reactions to be studied using crossed molecular beams were those producing alkali atoms and alkali–halide molecules, due to their ease of detection using surface ionization methods. The reaction of potassium atoms (K) with molecular bromine (Br_2) leads to formation of KBr + Br via the so-called "harpoon mechanism." Cuts of the two relevant potential energy surfaces are illustrated in Fig. 13. In this reaction, the neutral $K + Br_2$ reactants approach on the covalent potential energy surface. At a distance r_c, termed the ionic–covalent curve crossing distance, the energy required to transfer an electron from K to Br_2, given by $IP_K - EA_{Br_2}$, is exactly compensated by the Coulomb energy gained by formation of the $K^+Br_2^-$ ion pair. Reaction is initiated by electron transfer, which in the present case occurs at long range (~ 7.5 Å), and is analogous to the throwing of a harpoon during a whale hunt. Because the electron is transferred into an antibonding orbital, the equilibrium bond length of Br_2^- is much greater than for Br_2 and rapid Br–Br$^-$ stretching in the presence of K^+ leads to immediate formation of the $K^+Br^- + Br$ products.

This reaction has been termed a "spectator stripping" reaction, by analogy with some nuclear reactions. Here, the spectator is the bromine atom *not* involved in formation of the KBr product. Because of the rapid dissociation of Br–Br$^-$ while the K^+ is still at relatively long range, the motion of the spectator Br atom remains essentially unperturbed throughout the course of the collision.

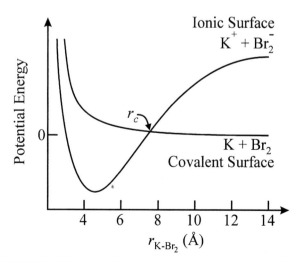

FIGURE 13 Ionic and covalent potential energy curves for $K + Br_2$ reaction.

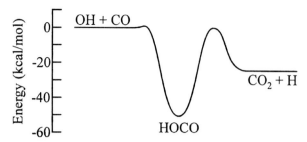

FIGURE 14 Energy level diagram for OH + CO reaction.

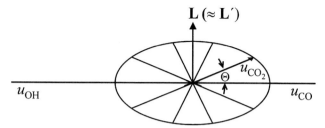

FIGURE 15 Schematic of one OH + CO collision event showing reactant velocities in CM frame. CO_2 product velocities are constrained by angular momentum conservation to lie in plane near that of reactants.

Consequently, this spectator atom is produced with a CM velocity vector that is essentially the same as that of the initial Br_2 molecule reactant. This leads to a highly anisotropic spatial distribution of products, with the Br fragment scattered at CM angles near $\theta = 180°$ relative to the direction of the K atom reactant, whereas the KBr counterfragment is scattered preferentially near $\theta = 0°$. The KBr product is thus "forward scattered" with respect to the incoming K atom in the CM reference frame. Note that $r_c \approx 7.5$ Å for the K + Br_2 system (Fig. 13). Since a large fraction of those collisions involving electron transfer lead to reaction, the reaction cross section is essentially the electron transfer cross section, $\sigma = \pi r_c^2 \approx 177$ Å^2, which is much larger than typical hard sphere cross sections.

The elementary reaction OH + CO $\rightarrow$ CO_2 + H is the primary mechanism for formation of CO_2 in combustion. This reaction, as well as its reverse, have been studied extensively using a variety of methods, including the crossed molecular beams method. The reaction is thought to involve the participation of long-lived HOCO intermediates, as illustrated in Fig. 14. Note that although OH + CO may form the intermediate complex with at most only a small potential energy barrier, a substantial barrier exists for H-atom loss forming CO_2.

In a crossed molecular beams experiment using supersonic beams in which the OH and CO are in their lowest few rotational levels, the total angular momentum of the HOCO intermediate is primarily orbital angular momentum, $L = \mu v_{rel} b$, with the vector $\mathbf{L}$ lying perpendicular to the plane containing the colliding species (Fig. 15). In many reactions, including this one, a relatively small fraction of total angular momentum appears as product rotational angular momentum. Consequently, the final orbital angular momentum, L', for the H + CO_2 products is approximately the total angular momentum, L, and the product CM velocities $\mathbf{u}_{AB}$ and $\mathbf{u}_C$, must lie in a plane nearly identical to that of the incident reactants. Because the intermediate HOCO lifetime is long compared to its rotational period, $\mathbf{u}_{AB}$ may lie in any direction in the plane of the collision.

The experimentally measured CO_2 angular distribution, shown in Fig. 16, is broad with products scattered to both sides of the center-of-mass of the system. Such angular distributions are said to exhibit "forward–backward symmetry" in the center-of-mass frame of reference. From analysis of product translational energy distributions (not shown), the degree of CO_2 vibrational excitation has been inferred for this reaction.

Perhaps the most extensively studied elementary bimolecular reaction is the simplest: the hydrogen exchange reaction and its isotopic variants. For example, studies of reactions such as H + D_2 $\rightarrow$ HD + D provide direct insight into reactive collisions. This reaction is slightly endothermic and exhibits a large ($\sim$0.4 eV) potential energy barrier to product formation (Fig. 17). The reaction is direct, and proceeds via a near-collinear H–D–D transition state on a timescale short relative to rotation, leading to anisotropic D atom angular distributions.

The experimental apparatus used in a recent H + D_2 study is shown in Fig. 18. Two molecular beams containing HI and D_2 are generated in separately pumped

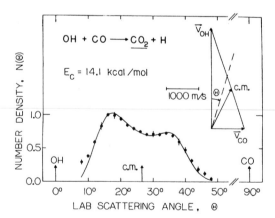

FIGURE 16 CO_2 laboratory angular distribution from crossed beams OH + CO reaction. (From Alagia, M., Balucani, N., Casavecchia, P., Stranges, D., and Volpi, G. G. (1993). *J. Chem. Phys.* **98**, 8341–8344.)

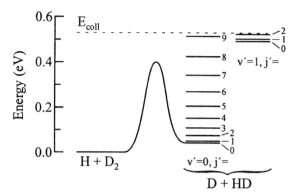

FIGURE 17 Energy level diagram for $H + D_2$ reaction.

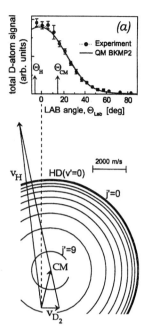

FIGURE 19 Lower figure: Newton diagram for $H + D_2$ reaction. Circles show maximum CM velocities for differerent rotational levels of HD ($v' = 0$) products. Upper figure: D atom laboratory angular distribution; solid line shows theoretical simulation. (From Casavecchia, P., Balucani, N., and Volpi, G. G. (1999). *Annu. Rev. Phys. Chem.* **50**, 347–376.)

regions and allowed to propagate parallel to one another, 30 mm apart. The HI molecules are photodissociated by a pulsed 266-nm laser, producing ground state H atoms with a very narrow translational energy distribution. Some of these H atoms collide with the D_2 molecules in the second beam, leading to chemical reaction forming HD + D. The D atom products from the reaction are "tagged" before they leave the interaction region by excitation to high-n Rydberg states using two tunable "probe" lasers. The first laser, operating at 121.6 nm, is tuned to the $n = 1 \rightarrow 2$ transition, and the second laser, at 365 nm, excites from $n = 2 \rightarrow \sim 40$. The lifetimes of these highly excited D atoms are in excess of 100 μs. Since the velocities of the atoms are essentially unchanged during the tagging step, they evolve spatially with their nascent velocities to a detector located approximately 30 cm away from the reaction zone, where they are field ionized and collected. The detector may be rotated with respect to the fixed beams in order to map out the angular distribution of the products. By measuring the time of arrival at the detector following time zero for reaction, the velocity and hence translational energy distribution is obtained.

At a collision energy, E_{coll}, of 0.53 eV, the HD product may be produced in either the ground or first excited vibrational states (Fig. 17). A Newton diagram for H-atom products recoiling from HD in various rotational levels of the ground vibrational state ($v' = 0$) is shown in Fig. 19. The largest circle corresponds to the velocity of D atoms recoiling from internally cold HD (i.e., $v' = 0$, $j' = 0$). Smaller circles correspond to recoil from HD molecules born with progressively greater internal energies. A measurement of the D atom velocity distribution at different

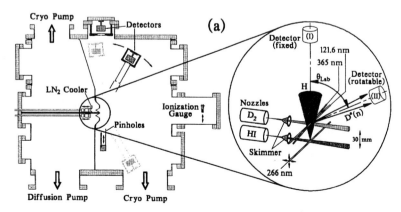

FIGURE 18 Experimental apparatus for studies of $H + D_2$ reaction in crossed molecular beams. (From Schnieder, L., Seekamp-Rahn, K., Wrede, E., and Welge, K. H. (1997). *J. Chem. Phys.* **107**, 6175–6195.)

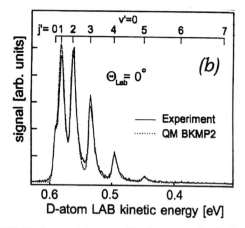

FIGURE 20 D-atom laboratory kinetic energy distribution from $H + D_2$ reaction at 0.53 eV. (From Casavecchia, P., Balucani, N., and Volpi, G. G. (1999). *Annu. Rev. Phys. Chem.* **50**, 347–376.)

angles thus provides a direct measure of the internal state distribution of the HD product.

The D atom angular distribution is highly anisotropic, peaking at laboratory angles near zero degrees (Fig. 19). In Fig. 20, the translational energy distribution for the D product, obtained by time of flight measurements for the D atoms, is shown at a laboratory angle, Θ, of $0°$. Structure corresponding to varying degrees of HD $(v' = 0)$ rotational excitation (denoted by j') is observed in the D atom kinetic energy distribution. Note that in addition to the experimental distributions, a calculated kinetic energy distribution, also shown, is found to almost exactly reproduce the degree of HD rotational excitation. This calculated distribu-

tion is based on exact trajectory calculations on high level *ab initio* potential energy surfaces. This close agreement between state-of-the-art theory and experiment is a particularly impressive achievement, demonstrating that the hydrogen exchange reaction is presently very well understood from a theoretical standpoint. Other related three atom systems that have been extensively studied include halogen atom reactions $X + H_2 \rightarrow HX + H; X = F, Cl$, and their reverse. Like the hydrogen exchange reaction, these reactions proceed via direct mechanisms over substantial potential energy barriers.

C. Product Detection Using Spectroscopic Methods

Many different spectroscopic techniques have been applied to studies of elementary bimolecular reactions. Measurements of the infrared chemiluminescence emitted from reaction provides a direct measure of the product state (rotational and vibrational) distributions. Studies of many different types of reactions have led to the development of fundamental principles that facilitate an understanding of how various forms of energy are partitioned between reactants and products. Recall that the transition state is a region of the potential energy surface that must be traversed in order for reaction to occur. The location of the transition state relative to reactants and products has a significant effect on the dynamics of the reaction. As shown in Fig. 21, an analogy may be drawn between the motion of atoms during a chemical reaction and the motion of a ball rolling on a three-dimensional potential energy surface. If

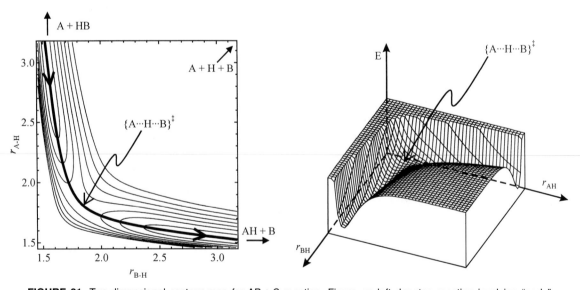

FIGURE 21 Two-dimensional contour map for $AB + C$ reaction. Figure on left denotes reaction involving "early" transition state, in which translation is more effective in promoting reaction than is vibration; figure on right denotes reaction having "late" transition state, where vibrational energy promotes reaction.

the transition state geometry more closely resembles the reactants (AB + C), than the products (A + BC), the transition state is termed "early." A "late" transition state, on the other hand, corresponds to the case where the transition state more closely resembles the products. For reactions involving an early barrier, reactant translational energy is more effective than vibrational energy in surmounting the barrier, and much of the available energy is deposited into product vibration. Note that the opposite is true for reactions involving late barriers. In this case, vibrational energy of the reactants is more effective than translational energy, and the products from the reaction tend to be vibrationally cold, with a large fraction of available energy appearing in translational energy.

To illustrate these ideas, consider the reaction $H + Cl_2 \rightarrow HCl + Cl$. Studies of the infrared chemiluminescence emitted from the diatomic product HCl indicate that although the reaction is highly exothermic, only a small fraction of the total available energy is channeled into HCl vibrational and rotational energy. Since electronically excited Cl and HCl products are not energetically accessible, a large fraction of the available energy must be partitioned into relative translational energy. This reaction is therefore an example of one involving a late transition state.

Measurement of product state distributions using various spectroscopic techniques has been applied to a wide variety of reactions, including $H + CO_2$. The $H + CO_2 \rightarrow OH + CO$ reaction, the reverse of the reaction shown in Fig. 14, is substantially endoergic. However, UV photodissociation of a diatomic molecule such as HI or HBr can produce H atoms with more than enough kinetic energy to react. The OH products may be readily detected using fluorescence excitation spectroscopy in which the total OH fluorescence intensity is monitored as a function of OH excitation wavelength. This provides a direct measure of the OH quantum state distribution from the bimolecular reaction. In the upper panel of Fig. 22, the rotational distribution of OH ($v' = 0$, 1) from photodissociation of HBr in the presence of CO_2 molecules is shown. For both OH vibrational levels, the most probable rotational level is near $j' = 8$. In the lower panel, a similar experiment is carried out, but in this case the reaction is initiated within weakly bound $HBr \cdots CO_2$ van der Waals complexes, produced by coexpanding HBr and CO_2 in a dilute He mixture in a supersonic molecular beam. In this case, the HBr moiety of the complex is photodissociated, producing H atoms that react with CO_2 to form $OH + CO$. As indicated in the lower panel, the OH rotational distributions are substantially "colder" for products from reactions initiated within complexes, with the most probable rotational level near $j' = 4$. The different behavior is believed to result from the presence of the halogen counterfragment

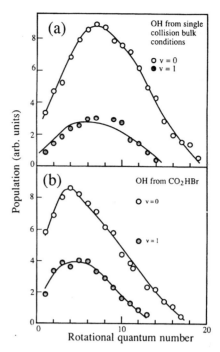

FIGURE 22 OH ($v = 0$ and 1) rotational distributions from $H + CO_2$ reaction. Upper panel indicates distribution obtained from gas phase reaction. Lower panel indicates distribution obtained from reaction photoinitiated in $CO_2 \cdots HBr$ van der Waals complexes. (From Wittig, C., Engel, Y. M., and Levine, R. D. (1988). *Chem. Phys. Lett.* **153**, 411–416.)

near the reacting $H + CO_2$ pair in the complex-initiated reaction.

Spectroscopic methods can also be used to determine rates of elementary reactions. As noted earlier, the existence of HOCO intermediates persisting longer than their picosecond rotational timescales during the $OH + CO$ reaction has been inferred from CO_2 product angular distributions in crossed molecular beam experiments. More direct determinations of bimolecular reaction timescales are often very difficult because molecules must first collide before they can react. Due to the "random" nature of bimolecular collisions, both in time and in space, it is not usually possible to define the "time zero" for a bimolecular reaction to sufficient precision to permit a direct study of reaction timescales. These difficulties, however, have been surmounted by initiating the bimolecular reaction via photodissociation of van der Waals complexes such as $HI \cdots CO_2$. Since the $H + CO_2$ reactants are already in close proximity, and because the time zero for reaction can be accurately defined using an ultrashort laser pulse (<1 ps), it is possible to measure the reaction rate in real time by measuring the buildup of OH products following photodissociation of HI within a binary complex. Rate constants for $H + CO_2 \rightarrow OH + CO$ obtained at different HI photodissociation wavelengths, and hence

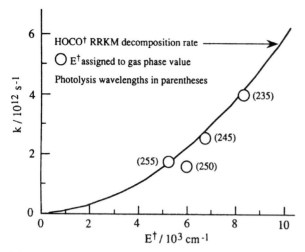

FIGURE 23 Experimentally determined bimolecular reaction rate constants for $H + CO_2$ reaction photoinitiated in $HI \cdots CO_2$ van der Waals complexes. Solid line denotes calculated RRKM decomposition rate for HOCO reaction intermediates. (From Ionov, S. I., Brucker, G. A., Jaques, C., Valachovic, L., and Wittig, C. (1992). *J. Chem. Phys.* **97,** 9486–9489.)

collision energies, are shown in Fig. 23. Note that the rate constant, k, for this elementary reaction, which is the inverse of the HOCO lifetime (τ), increases monotonically with increasing energy, $E^\dagger$, in excess of the potential energy barrier for $H + CO_2$ production. This behavior is in close agreement with that predicted by statistical theories such as RRKM theory, illustrated by a solid curve.

Through the use of direct optical excitation of reactants, it has been possible in several cases to direct the course of reaction to favor certain product channels or product state distributions. Using an intense pulsed near-infrared laser, the very weakly absorbing OH vibrational overtones in small molecules like water may be excited relatively efficiently, depositing a substantial amount of vibrational energy into well-defined vibrational modes of reactants. Using this approach, the role of reactant vibrational excitation in bimolecular reactivity has been studied. Because of the local mode character of OH bonds in water, the vibrational excitation initially deposited into the molecule remains localized in a well-defined oscillator. For example, the reaction of $Cl + HOD$ ($4\nu_{OH}$) (having four quanta in the OH stretch with zero in the OD stretch), leads to relatively efficient production of $HCl + OD$, but essentially no $DCl + OH$, as determined by LIF detection of OH and OD products (Fig. 24). In this example of *bond specific* chemistry, the vibrationally excited OH bond is highly reactive, whereas the unexcited OD bond is relatively inert. In other related studies of reactions of H_2O, *mode specific* chemistry has been observed, in which different excited H_2O vibrational states at nearly the same energy displayed somewhat different reaction dynam-

ics. For example, the reactivity of H_2O $|04\rangle$, having four quanta in one OH bond, was compared to that for H_2O $|13\rangle$, which has three quanta in one OH bond and one in the other. It was found that $Cl + H_2O$ $|04\rangle$ produced primarily $HCl + OH$ ($v = 0$), whereas reaction of $Cl + H_2O$ $|13\rangle$ yielded primarily $HCl + OH$ ($v = 1$). This illustrates that the OH bond *not* involved in the reaction acts much like a spectator, retaining its initial level of vibrational excitation in the chemical products.

D. Spectroscopy of the Transition State

In the experiments described previously, much of the information about the reaction pathway is obtained through measurements of the velocity, angular, or internal state distributions of products, or through measurements of their cross sections or timescales for formation. In some cases, the effect of different forms of reactant energy in promoting chemical reactivity also provided insight into the thermodynamics of the reaction or on the nature of the transition state. A conceptually different and more recent development involves direct measurement of the transition state itself. Since, in most reactions, the transition state corresponds to an energy maximum along the reaction coordinate, this region is only accessed for a very short time, typically on timescales comparable to a vibrational period ($<10^{-13}$ s). Consequently, the fraction of molecules in a reacting sample near the transition state at any given time

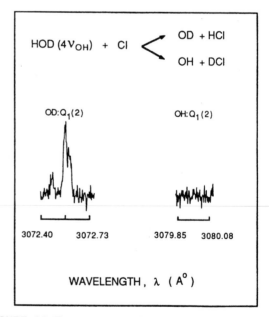

FIGURE 24 Fluorescence excitation spectrum for OD and OH products from $HOD(4\nu_{OH}) + Cl$ reaction. (From Sinha, A., Thoemke, J. D., and Crim, F. F. (1991). *J. Chem. Phys.* **96,** 372–376.)

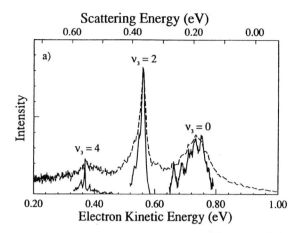

Scattering Energy (eV)

FIGURE 25 Photoelectron spectra at two different resolutions from IHI⁻ photodetachment. Peaks correspond to asymmetric stretching levels of neutral IHI product. (From Metz, R. B., and Neumark, D. M. (1992). *J. Chem. Phys.* **97**, 962–977.)

is vanishingly small, making it difficult to study such species directly. Much of the recent progress in the area of transition state spectroscopy has therefore utilized photoinitiation qualitatively similar to that described above for HI $\cdots$ CO$_2$ to produce molecules near the transition state.

For example, the transition state regions for a number of hydrogen exchange reactions have been studied by first producing molecular beams containing stable negative ions, such as IHI⁻. The geometries of these stable negative ions often resemble that of the transition state for the neutral reaction, e.g., I + HI′ → HI + I′. Photodetachment of the electron produces the neutral IHI species. As in conventional photoelectron spectroscopy, by measuring the kinetic energy distribution of the ejected electron, the vibrational energy levels of the molecular counterfragment, in this case IHI at nuclear geometries near the transition state, are obtained. In Fig. 25, two spec-

tra, recorded at different experimental resolutions, show structure corresponding to antisymmetric stretching vibrational levels of the neutral IHI molecule. This approach has also been applied to studies of a wide variety of other reactive systems involving transfer of a light atom, such as F + CH$_3$OH → HF + CH$_3$O, and F + H$_2$ → HF + H.

SEE ALSO THE FOLLOWING ARTICLES

ATOMIC AND MOLECULAR COLLISIONS • COHERENT CONTROL OF CHEMICAL REACTIONS • ENERGY TRANSFER, INTRAMOLECULAR • ION KINETICS AND ENERGETICS • KINETICS, CHEMICAL • MOLECULAR BEAM EPITAXY, SEMICONDUCTORS

BIBLIOGRAPHY

Baer, T., and Hase, W. L. (1996). "Unimolecular Reaction Dynamics, Theory and Experiments," Oxford Univ. Press, New York.

Crim, F. F. (1999). "Vibrational state control of bimolecular reactions: Discovering and directing the chemistry," *Acc. of Chem. Res.* **32**, 877–884.

Johnston, H. S. (1966). "Gas Phase Reaction Rate Theory," Ronald Press.

Lee, Y. T. (1987). "Molecular beam studies of elementary chemical processes," *Science* **236**, 793–798.

Levine, R. D., and Bernstein, R. B. (1987). "Molecular Reaction Dynamics and Chemical Reactivity," Oxford Univ. Press, New York.

Neumark, D. M. (1992). "Transition State Spectroscopy of Bimolecular Chemical Reactions" In Annual Review of Physical Chemistry (H. L. Strauss, G. T. Babcock, and S. R. Leone, eds.), Vol. 43, pp. 153–176, Annual Reviews Inc., Palo Alto, CA.

Pilling, M. J., and Seakins, P. W. (1995). "Reaction Kinetics," Oxford Science Publications, New York.

Polanyi, J. C. (1987). "Some concepts in reaction dynamics," *Science* **236**, 680–690.

Polanyi, J. C., and Zewail, A. H. (1995). "Direct observation of the transition state," *Acc. Chem. Res.* **28**, 119–132.

Scoles, G. (1988). "Atomic and Molecular Beam Methods, Vol. 1," Oxford Univ. Press, New York.

Earthquake Engineering

Robert V. Whitman

Massachusetts Institute of Technology

I. Scope of Earthquake Engineering
II. Basic Concepts Concerning
   Structural Response
III. General Principles of Design
   Against Earthquakes
IV. Analysis of Buildings
V. Lifeline Components and Systems
VI. Identification and Evaluation
   of Earthquake Hazards

GLOSSARY

Ductility A property of a structure that allows it to continue to support its load after the structure has yielded or cracked.

Geotechnical engineering Engineering that deals with construction on, within, and with soil and rock (e.g., foundations and earth dams).

Intensity The strength of ground shaking at a site. One measure of intensity is the peak ground acceleration.

Liquefaction A loss of soil strength or stiffness. It occurs when cyclic straining causes large pressures to build up in the water phase of a soil.

Magnitude A measure of the strain energy released by an earthquake.

Nonstructural The parts of a building system that do not support loads (e.g., partitions, lighting fixtures).

Seismology A branch of earth science dealing with earthquakes.

EARTHQUAKE ENGINEERING is a multidisciplinary field that assesses threats to people and their natural and constructed environment that occur as a result of earthquakes, and that develops and puts into action plans and designs that will reduce the consequences of earthquakes. Earthquake engineering overlaps with earth sciences on one hand and public policy issues on the other. Membership in the professional society for earthquake engineering (Earthquake Engineering Research Institute, or EERI) is primarily drawn from the structural and geotechnical engineering fields, but it also includes architects, building code officials, public safety officials, transportation officials, regulators of utilities, city planners, land use planners, medical personnel, economists, political scientists, and other professionals and scientists.

I. SCOPE OF EARTHQUAKE ENGINEERING

A. Damage Caused by Earthquakes

Large, damaging earthquakes are the result of a sudden slip along a fault. If the fault break reaches the surface of the earth and passes directly beneath a building, bridge, highway, or pipeline, for example, or through a tunnel, the damage to these structures can be enormous. However, usually the greatest damage from earthquakes is caused by the resulting release of strain energy stored in the adjacent rock. The release causes stress waves to spread out through the earth from the epicenter, which causes the ground at any location to shake back and forth as the waves pass. Special instruments—strong-motion seismographs—have been developed to record the very strong motions that are of primary interest for earthquake engineering. Figure 1 shows horizontal acceleration measured at the base of a building during a strong ground motion. The motion changes erratically with time, as a consequence of many different frequencies present in the motion. Although the vertical component of ground motions can be important in some engineering problems, the horizontal components are usually of the greatest consequence. The number of such records has been increasing dramatically since about 1970, and they have become essential for understanding the ground motions that are caused by earthquakes.

Ground motions such as those shown in Fig. 1 will likely cause severe damage and even collapse of poorly constructed buildings, resulting in serious injuries and fatalities. After major earthquakes, especially those occurring in developing countries, newspapers feature pictures of collapsed houses made of adobe bricks, rounded field stones, or even brick masonry. However, such damage can also happen to modern-looking structures that have been poorly constructed. Even when buildings have been designed with some seismic resistance and constructed properly, there may be expensive-to-repair damage—especially to nonstructural features such as exterior walls, partitions, ceilings, mechanical equipment, contents, etc.—accompanied by serious injuries. Business or occupancy interruption can add significantly to the cost of all such damage.

The same potential problems and challenges apply to the many other parts of our built infrastructure. It is especially important that *essential facilities*—hospitals, fire stations, and police stations—be able to function in the hours and days following a damaging earthquake. Damage that impairs the functioning of *lifelines*—highways and other transportation systems, as well as electrical, water supply, and waste disposal systems—can seriously disrupt the functioning of a community.

Strong ground shaking resulting from earthquakes can also induce other geological hazards, which in turn may cause damage to buildings and other structures.

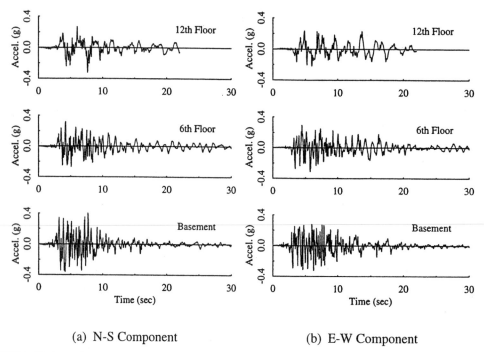

(a) N-S Component   (b) E-W Component

FIGURE 1 Horizontal accelerations measured in two directions—(a) vertical and (b) horizontal—at various levels in a 12-story building. Accelerations are normalized to the acceleration of gravity, *g*. [From Uang, C. M. *et al.* (1997). "Earthquake Spectra," Earthquake Engineering Research Institute, Oakland, CA.]

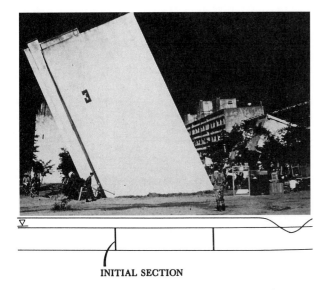

INITIAL SECTION

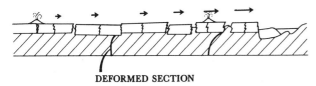

DEFORMED SECTION

FIGURE 2 Consequences of liquefaction. (a) Settlement and tipping of an apartment building in Niigata, Japan, as a consequence of the 1964 earthquake. [Courtesy of George Housner.] (b) Schematic diagram of lateral spreading.

Landslides, rock falls, and avalanches may be triggered by earthquakes and in extreme cases have been known to bury entire towns or to dam valleys and thus create new lakes. When loose, saturated soils are shaken strongly, they may momentarily lose stiffness and strength and thus lose the capacity to support the weight of structures (Fig. 2a). This phenomenon, known as *liquefaction,* may also cause lateral earth movements of many feet, even in nearly horizontal ground (Fig. 2b), and landslides in steeper ground. Liquefaction effects can be severe along the edges of oceans, lakes, and rivers and may be especially damaging to port facilities and to highways, railroads, and pipelines.

Extreme damage to buildings and other facilities such as gas pipelines, fuel storage sites, and factories may be the cause of secondary hazards such as fires and releases of hazardous and toxic substances. Fires following earthquakes have often been a major cause of damage and loss of life. If dams storing water fail because of faulting, ground shaking, liquefaction, or landslides, downstream flooding may potentially submerge large areas, causing great damage and many casualties. Dams formed by landslides are often later overtopped, which causes downstream flooding months after the causative earthquake. All these secondary hazards have the potential to cause greater losses than those occurring during the original ground shaking.

Finally, ocean-front properties may experience *tsunamis,* which are long and deep waves generated when a portion of the ocean's bottom drops or rises suddenly as a result of motion along a submerged fault. Tsunamis can travel large distances across oceans at great speeds. When the waves meet shallow water at a shore, they can rise up to immense heights and surge inland, causing enormous devastation. A similar type of phenomenon can also occur in large lakes.

B. Reduction and Prevention of Damage

Prevention of damage caused by geological hazards—faulting, liquefaction, landslides, and tsunamis—requires, first, identification of these hazards and, second, land use planning and regulation to avoid siting of structures in such locations. However, when private property is involved, taking such actions in the political arena is never simple and easy. Efforts to avoid damage from geological hazards become much more difficult when structures of any type already exist in areas identified as susceptible to these hazards.

The design of new buildings and other structures to resist the effects of ground shaking is a matter of good engineering plus the willingness of an owner to pay for whatever additional costs are required. Building codes in many cities and states require design against earthquakes. Engineers and architects implement these code provisions. However, there still are communities without adequate code requirements in regions subject to earthquakes. Convincing such communities to adopt suitable code requirements is a continuing task.

Building codes are aimed primarily at protecting occupants from injury. A building can be designed to meet code and fulfill the mission of protecting life safety and yet experience significant damage during an earthquake. Owners must bear the cost of expensive repairs, and buildings may be unusable until repairs are made. An increasing number of owners want to ensure continued usability of their buildings following an earthquake, and they are enlisting engineers and seismologists for advice.

However, existing buildings and structures designed before there were serious concerns about earthquakes constitute by far the greatest risk to life and property. To bring all such buildings just up to code requirements would involve prohibitive costs. Hence there are challenging questions: Which buildings are so dangerous that some actions should be required? What types of strengthening actions are most cost-effective? Often the most dangerous buildings are those occupied by the poorest portion of the population. What happens to these people while buildings are strengthened? Higher rents charged for the improved buildings may force the occupants to seek shelter and business space elsewhere. The "existing building problem" thus involves social, economic, and political considerations in addition to engineering considerations.

This discussion of earthquake-related hazards and the challenges involved in dealing with these problems has emphasized the broad, multidisciplinary nature of earthquake engineering. It suggests the breadth of knowledge and skills required to assess the potential hazards and risks and their consequences and to develop and put into action measures to mitigate the earthquake threat. Buildings, housing residences, and offices represent the large bulk of the earthquake hazard mitigation challenge; therefore, the following discussion focuses primarily on the analysis and design of buildings. There are limited comments concerning other types of structures.

II. BASIC CONCEPTS CONCERNING STRUCTURAL RESPONSE

The theory of structural dynamics is fundamental to the understanding and analysis of the response of structures subject to earthquake ground motions. This theory identifies a number of basic concepts, the most important of which are as follows:

- A structural system has a number of *degrees-of-freedom*. The idealized building in Fig. 3a has three degrees-of-freedom in the plane of the paper that correspond to horizontal motions of the three floors where mass is concentrated. The number of important degrees-of-freedom typically increases with the height (or size) of a building and with the flexibility of the structural members.
- The structure possesses a number of *natural frequencies (or natural periods)* equal to the number of degrees-of-freedom. The smallest of these natural frequencies is the *fundamental frequency*. The fundamental frequency (and of course other natural frequencies as well) decreases with increasing building height and flexibility. If a structure is disturbed and allowed to vibrate freely without further application of force or displacement (*free vibrations*), the vibrations will occur at these natural frequencies.

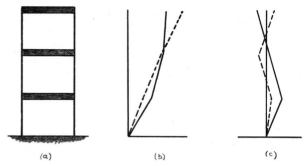

FIGURE 3 (a) Idealized three-degrees-of-freedom structure. (b) Typical shapes for fundamental mode. (c) Typical shapes for second (solid lines) and third (dashed line) modes.

- When a structure is vibrating at a natural frequency, there will be a characteristic pattern (*mode shape*) for the displacements. Figure 3b shows typical shapes of fundamental modes, which depend upon the distribution of mass and stiffness. Typical shapes for second and third modes are sketched in Fig. 3c.
- When a building is *forced* by a base motion such as that in Fig. 1, the response of the structure is a superposition of the mode shapes, with the relative importance of each mode shape depending upon the ratio of the predominant frequency in the base motion to the natural frequency of that mode.

When the earth beneath a building moves rapidly, the building attempts to move with the ground. Force must develop between the building and the ground to make it move and within the building to make the upper portions move with the lower part. It is these forces, and the associated deformations, that can cause the structural system and nonstructural features to be damaged or even to fail. If a building were completely rigid, the horizontal force at each level would equal the mass of the overlying building times the ground acceleration. Actual buildings, however, are flexible to some degree; as a result, accelerations within a building may be either greater or smaller than those at its base, and, as seen in Fig. 1, will have different frequencies.

The important role of flexibility is captured by a *response spectrum* (Fig. 4a), which for a particular ground motion plots the *spectral acceleration, S_a* (approximately the peak acceleration), experienced by the mass of a single-degree-of-freedom mass–spring–dashpot system versus the undamped natural period T of that system. T is given by

$$T = 2\pi (M/k)^{1/2}, \qquad (1)$$

where M = mass and k = spring constant for the simple system. The product MS_a gives the force in the spring.

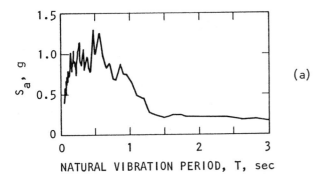

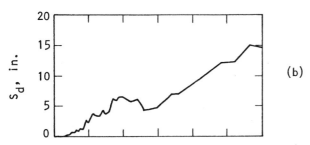

FIGURE 4 Response spectra. (a) Spectral acceleration, S_a. (b) Spectral displacement, S_d.

The case $T = 0$ corresponds to a truly rigid building, and in such a case the peak acceleration of the mass does equal the peak acceleration of the ground. As T increases, the peak acceleration of the mass generally becomes larger than that of the ground. There typically is some range of T for which the mass experiences accelerations several times greater than those of the ground. The peaked portion of the spectrum corresponds to the periods of ground motion that are strongest and most sustained. With further increase of T (which implies further increase of flexibility), the ratio of peak accelerations at mass to those at ground decreases and eventually becomes significantly less than unity.

Figure 4b is a related plot of maximum distortion in the spring (*spectral displacement, S_d*), which is related to spectral acceleration by

$$S_d = MS_a/k. \qquad (2)$$

Comparing Figs. 4a and 4b reveals that stiff structures (small T) experience large forces but small distortions, whereas flexible structures (large T) may feel small forces but at the expense of large distortions.

These concepts remain useful for understanding and analyzing the response of actual buildings having several degrees-of-freedom. The acceleration and deformations experienced by a structure are also influenced by the

damping present in the structure, and hence a series of curves (*response spectra*) for different values of damping are often plotted. Spectra are calculated from actual ground motions. Spectra from different motions have often been combined to produce "smoothed spectra" for use in design studies. Response spectra are engineering tools and are related to, but basically different from, the Fourier spectra that seismologists often use to represent the characteristics of ground motions.

A final, but very important, point concerns the duration of ground shaking. Such duration is primarily a function of the magnitude of an earthquake, being short (a few seconds) for earthquakes of magnitude less than 6 but long (several minutes) for great earthquakes of magnitude 8 or higher. Increasing duration tends to cause somewhat larger ordinates in response spectra. However, the greatest consequence occurs when the strength of a soil or structure decreases dramatically as shaking continues.

III. GENERAL PRINCIPLES OF DESIGN AGAINST EARTHQUAKES

A. Ductility

A very important, basic concept is that reaching the limiting strength of a structure during a rapid shaking does not necessarily mean that the structure as a whole will "fail" in the sense of collapsing. A structure is said to be "ductile" if it continues to hold together and support its weight after its strength has been momentarily reached. Conversely, a "brittle" structure that lacks ductility may come apart and collapse as a result of a shaking that causes its strength to be reached.

The ductility of a building depends not only upon the material used for beams, columns, floors, and walls, but also upon the way in which these elements are connected. Wood-framed walls with adequate bracing or sheathing are ductile. However, wood-framed houses inadequately connected to a foundation may collapse during an earthquake. Structural steel is a ductile material; a beam may yield and experience plastic deformations without cracking. However, connections between beams and columns must be specially designed to achieve ductile behavior. Concrete is inherently brittle; once its strength is reached, it begins to crack and split. Nevertheless, providing sufficient reinforcing steel within concrete allows a degree of ductility to be achieved. With precast concrete construction, having ductility depends upon the manner in which beams are connected to columns and the way in which walls are tied to floors and roofs. Unreinforced masonry construction is very brittle. Some ductility can be achieved by inserting steel bars into masonry walls.

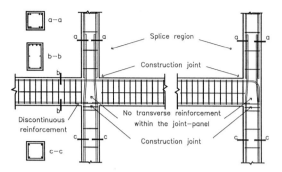

FIGURE 5 Elevation view of interior and exterior reinforced concrete beam–column joint regions. The reinforcement is typical of buildings designed without regard for earthquakes. [From Beres, A. *et al.* (1996). "Earthquake Spectra," Earthquake Engineering Research Institute, Oakland, CA.]

Past earthquakes have provided invaluable lessons concerning adequate and inadequate ductility. Following any major earthquake, teams of experts examine damaged and undamaged structures so as to understand the reasons for good or bad performance. "Learning from earthquakes" has been the key to achieving ever better design for buildings. In addition, many large-scale tests have been performed in laboratories. Various structural systems have been examined in this way, including reinforced concrete

members and assemblages of steel and concrete members. These experiences form the basis for rules concerning adequate reinforcement for concrete members and adequate connections for steel members.

As an example of the challenge in providing ductility, Fig. 5 depicts test specimens representing portions of a building. The double lines within the concrete represent steel reinforcing bars, which give tensile strength to the concrete. The solid narrow lines are hoop steel that restrain the reinforcing bars from buckling outward and participate in resisting shear. This configuration is typical of a reinforced concrete structure that has been designed without regard to earthquakes. When these assemblies are tested, sustained vertical loads representing the weight of an overlying structure are applied, then cyclic horizontal loads are superimposed. Figure 6 shows typical results. Figure 6c illustrates inadequate behavior: After the second large cycle, the resistance falls off rapidly with subsequent cycles. Thus, failure of this beam–column assembly would occur during an earthquake with duration long enough to impose more than a few cycles of shaking. The behavior charted in Figs. 6a and 6b is better, but still inadequate for resisting intense ground shaking. Such results plus experiences during actual earthquakes show the need for additional reinforcement at and near beam–column junctions.

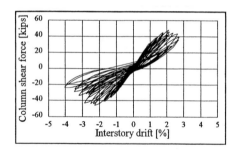

a. Interior joint, continuous positive bars

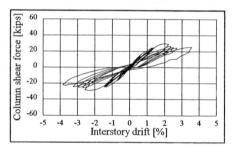

b. Interior joint, discont. positive bars

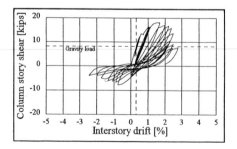

c. Exterior joint

FIGURE 6 Results from cyclic loading tests on the assemblages shown in Fig. 5: (a) interior joint, continuous positive bars; (b) interior joint, discontinuous positive bars; (c) exterior joint. Interstory drift is the relative displacement between adjacent floors. [From Beres, A. *et al.* (1996). "Earthquake Spectra," Earthquake Engineering Research Institute, Oakland, CA.]

B. Layout of Building

In addition to choosing materials and details that provide ductility, paying close attention to the layout of a building and structural system is essential. Irregularities in the layout of a building often cause difficulties. A change in the structural framing system at some elevation within a building can introduce a weak spot. A very common problem, and the cause of many collapses, is a "soft first story," where walls have been omitted so as to accommodate offices, stores, or garaging of vehicles. Irregularity in plan can also be the cause of severe damage and even collapse, in part by causing a building to twist. If one building is placed too close to another structure, they may pound against each other, which will likely cause damage and even partial collapse.

C. Controlling Damage

Although providing ductility can make it possible for buildings to undergo large deformations and resulting damage without collapsing, there are several reasons why large motions and deformations with associated damage may be undesirable. Large interstory deformations may cause the exterior cladding of a building, including windows, to fall from the building and threaten lives on the ground below. Ceilings may fall upon occupants of rooms. Elevators may become inoperable, and stairways blocked by fallen partitions, which would hinder escape from the damaged building. For these reasons, building codes generally limit the permissible interstory displacement. In addition, care in the design of nonstructural components is needed to reduce losses.

If essential facilities are to perform their critical roles following an earthquake, electrical and mechanical components must remain functional. Some elementary examples are the large doors of fire stations and the electrical supply to hospitals. There are enormous implications to the loss of large computerized databases in banks, and increasingly manufacturers are concerned about protecting production lines from disruption. Such key nonstructural features must themselves be robust enough to endure strong shaking without failing, but they must also be mounted properly.

At still another level, the release of toxic and poisonous substances as a result of earthquake shaking can threaten large populations in circumstances where escape from them is hampered by clogged roads and failed bridges. Nuclear power plants are an extreme example. In such cases, drastic and costly measures must be taken to ensure that containment vessels and structures do not experience significant damage and that control and safety equipment survive an earthquake.

Protecting against damage that is socially or economically unacceptable first suggests designing buildings to be stronger and stiffer. However, there are limits to the economic viability of such an approach. One possible alternative approach is *base isolation*. Conceptually, this approach means placing below a structure a weak surface that limits the shear force that can be transmitted upward into the structure from moving ground. The building as a whole may "slide" over the ground, with permanent displacement at the end of a major earthquake, but the structure above that level and its contents are undamaged. An immediately obvious requirement is to provide flexible connections where utilities enter the structure. Indeed many details need attention if the base isolation scheme is to function reliably and effectively. A number of proprietary base isolation systems have been developed and have been applied to key governmental buildings, manufacturing plants, banks, and so forth. They have also been used to achieve life safety requirements in existing, hazardous, historic buildings. In addition, special energy-absorbing devices, such as large dampers, have been developed for placement within structural systems to reduce deformations during earthquakes.

IV. ANALYSIS OF BUILDINGS

A variety of approaches may be used to analyze an existing structure or a trial design for a new structure, depending upon its nature and importance.

A. Equivalent Static Analysis

The great majority of buildings and other structures are simply designed in accordance with codes, which permit the use of horizontal static loads to simulate the forces associated with earthquakes. The procedure, which takes into account key aspects of dynamic response, is as follows:

- The horizontal shear, V, at the base of a building is computed from

$$V = ZC(T)W/R, \qquad (3)$$

where W is the weight of the building, $C(T)$ is a seismic coefficient depending upon the fundamental period T of the building, Z is a zone factor, and R is the force reduction factor dependent upon the structural system used for the building.
- The function $C(T)$ is a simplified, generic version of a response spectrum. Codes give simple equations for computing T depending upon the height of the building and the type of structural framing.
- Z is a zone factor that accounts for the appropriate local seismicity. It is read from a map of the nation. In the United States, Z is largest for states bordering the

Pacific Ocean and markedly less for many states in the eastern part of the country.

- Forces that sum to the base shear are distributed over the height of the building following procedures set forth in the code as a way of taking into account the distribution of mass and the expected shapes of the first several modes of the building.
- These forces are then distributed to the members of the beams, columns, and walls of the building in accordance with common structural engineering procedures.
- These pseudodynamic forces are combined with dead loads and other live loads, and, along with suitable safety factors, are then used to design the structural system.

If R were equal to 1 and no safety factors were to be used, then the ground shaking during a very strong (for the location) earthquake would bring a building essentially to the point of yielding. However, since such earthquakes are rare and the cost of designing against such a strong shaking can be significant, codes permit the use of values of $R > 1$; just how much greater depends upon using structural systems that will continue to support dead loads after being damaged by the strong shaking. For very ductile structural systems (e.g., steel frames with moment-resisting beam-to-column connections), $R = 8$ is allowed, whereas for building systems with poor ductility (e.g., unreinforced masonry or reinforced concrete lacking in adequate reinforcing steel bars), R is permitted to exceed 1 only slightly.

This simple approach, which is adequate for many buildings, has evolved since the 1930s. Values of $C(T)$ and R have been adjusted and readjusted to reflect experiences with buildings during actual earthquakes. Other factors—S, to account for the effects of soil conditions at the building site, and I, to reflect the importance of a building—have also been added, which leads to Eq. (4):

$$V = ZSIC(T)W/R. \qquad (4)$$

The most important improvements in this equivalent static approach have been in the requirements that a designer must fulfill in order to qualify the building for values of R such as 4–8. Particular attention must be given to beam–column connections in steel structures and to the details of reinforcing steel in concrete or masonry structures. Codes also require attention to some nonstructural features, especially to ensure that the exterior cladding and windows will not fall off a building as a result of ground shaking.

B. Linear Dynamic Analysis

Today, computer codes for performing dynamic analysis of buildings subjected to ground motions are readily available and are found in the offices of nearly all structural engineering firms—at least in states with significant earthquake problems. It is necessary only to input the geometry of the framing system and the properties of the members, and often this can be done automatically by linking to a computer program used as part of the development and design of the building. The dynamic analysis code will compute natural periods and mode shapes and, once the motion at the base is specified, will determine forces in all members of the building for comparison with allowable forces. Using graphical interfaces, the user can watch the building move and deform as the shaking takes place.

Since the 1930s, the methods for performing dynamic analysis have evolved through many stages. A first major step was the theory for modal analysis. In this procedure, the response of each mode is determined separately and then these modal responses are superimposed. Although a full time-history of response can be obtained in this way, often it suffices to focus only upon the peak forces in members, and these can be estimated by using rules to combine the peak responses for each mode. The peak modal responses can be obtained by reading the response spectrum for ground motion at the natural periods of the several modes. Except for tall buildings, often it suffices to consider only the first three to five modes.

With today's powerful computer programs, response will more likely be calculated by direct integration of the simultaneous equations of motion. Fourier analysis is also sometimes used.

The use of linear dynamic analysis becomes important for buildings with unusual framing systems, very tall buildings, and especially buildings that are irregularly shaped in either plan or elevation. These are situations for which the approximate analysis following a building code may give misleading forces in some members. Often analyses will be made by using two different levels of shaking. The first is a shaking that can reasonably be expected to occur during the life of the structure. For this shaking, all stresses should remain within allowable stresses as specified by codes. The second shaking is the largest credible shaking, and it may lead to predicted forces that exceed the yield strength of the materials. Even though a linear analysis then is, strictly speaking, incorrect, experience has shown that the displacements and deformations predicted by the analysis will be approximately correct. These displacements and deformations can be compared with those that members can sustain and still remain intact.

C. Pushover Analysis

Pushover analysis is a static analysis used to investigate how far into the inelastic range a building can go before it is on the verge of a total or a partial collapse. A model for the building is assembled on a computer, with all

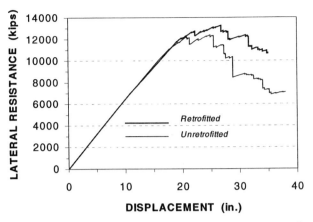

FIGURE 7 Pushover curves for a building before and after retrofit to increase ductility. [Courtesy of EQE, Oakland, CA.]

load-resisting elements together with their force–deformation relationships both before and after yielding and with dead loads plus average live loads. Then a small set of horizontal forces is applied so as to simulate the effects of ground motions, and deformations are calculated. The forces are then increased in steps so as to develop a plot of base motion versus deformation. Examination of this plot reveals the largest base motion that the building can resist.

This approach was developed to permit analysis of existing buildings and to study the effectiveness of schemes for strengthening these buildings and giving them greater ductility. Figure 7 shows typical plots for an existing building both before and after it is remediated. The existing building begins to lose resistance rapidly once its peak resistance is exceeded, and it would likely collapse during a very strong shaking. The strengthened building exhibits some ductility that will make collapse much less likely. Pushover analysis is now also used frequently to evaluate the expected performance of designs for new buildings.

D. Nonlinear Dynamic Analysis

There now exist computer programs that are capable of calculating nonlinear response in detail, including the effects of repeated yielding during successive cycles of ground motion. Analyses using these programs can consume large amounts of computer time. Such programs are used to evaluate designs for large and important buildings and those for which base isolation devices are to be used. The biggest challenge in the use of such programs is establishing realistic force–deformation relationships for all of the load-resisting elements, especially when these members are of reinforced concrete. Results from nonlinear analyses can be quite sensitive to the details of applied ground shakings. For this reason, analysts select a set of different "scenario earthquakes" that could cause severe shaking of the building so as to adequately check the performance of the building.

E. Response of Nonstructural Systems

A building may house and provide support for sensitive equipment and piping, the failure of which can have a severe impact on the functioning and safety of the building. Examples are hospitals, banks, and, at the extreme, the containment structure for a nuclear power plant. A dynamic analysis for the structure itself can be carried out in such a way as to provide dynamic motions at the points of support for such sensitive equipment. Again it is necessary to use multiple "scenario earthquakes" to account for potential uncertainties in earthquake ground motion and in the response of the basic structure. In effect, the task is to establish suitable support-point response spectra.

F. Soil–Structure Interaction

The response of a building to ground shaking can be affected by flexibility in the ground supporting or surrounding the base of the building. Rocking is an obvious example. The soil–structure interaction may have important consequences for especially tall or massive structures. Special methodologies for analyzing such interaction effects have been developed.

V. LIFELINE COMPONENTS AND SYSTEMS

A. Transportation Lifelines

Much of what has been said concerning the analysis and design of buildings also applies to bridges. Failures of highway bridges and viaducts in California during earthquakes in the late twentieth century have led to considerable research, testing, and development of new procedures for designing new bridges and strengthening existing ones. One of the first challenges has been to design supports that allow for temperature-caused changes in the length of girders and yet keep the girders on these supports during a major earthquake. A particular problem with analyzing very long bridges is that ground motions can be significantly out of phase at the several points of support. California and some other states have undertaken programs to replace or strengthen the most hazardous existing highway bridges.

Except where they cross faults, tunnels generally perform well provided that the liners are flexible enough to adjust to distortions experienced by the ground as it shakes.

Major attention has been given to underwater tunnels, such as the crossing of the Bay Area Rapid Transit system between San Francisco and Oakland, which performed well during a serious (but not severe) shaking in 1989.

Lateral spreading or landslides that take out sections of pavement can seriously interrupt highways and other transportation systems.

B. Dams and Earth Retaining Structures

Failure of a dam, with rapid release of stored water, can be a catastrophic event. Fortunately, there were no earthquake-caused failures of large dams during the last half of the twentieth century, but there were two near misses and several failures of smaller dams. One near miss, near Los Angeles in 1971, spotlighted the susceptibility of earth dams constructed by hydraulic filling techniques. This event led to the development of techniques for analyzing such dams and to continuing efforts to upgrade the performance of existing dams constructed in that manner. Similarly, a near failure of a concrete arched dam in India in 1967 stimulated study of such dams. It is worthy of note that Pacoima Dam, a large arch dam near Los Angeles, has twice experienced very strong earthquake shaking without failure-threatening damage.

Walls to retain highway cuts and fills have generally performed well during earthquakes, unless there has been liquefaction-susceptible soil in the backfill or foundation. As might be expected, there have been widespread problems with retaining walls along rivers and especially in seaports because of liquefaction. Failures during many recent earthquakes have stimulated the use and development not only of techniques for *in situ* improvement of liquefaction-susceptible soils, but also of improved methods for analyzing and constructing retaining structures.

C. Electrical Systems

While electrical generating plants have proven to be robust during earthquake shaking, transformers, circuit breakers, and other equipment in switchyards have often failed, which has caused widespread disruption of electrical service. At least in more earthquake-prone regions, electrical utilities have been taking steps to install more robust equipment.

D. Water and Waste Systems

Although pipes of water distribution systems can be broken just by stress waves passing through soil, most such failures are the result of lateral spreading, landslides, or fault slips intersecting pipes. The resulting system outages are of special concern to a community striving to recover from an earthquake-caused disaster. These problems are addressed by use of pipes that are more tolerant of lateral movements, automatic shutoff valves, increased redundancy in distribution systems, and special steps at fault crossings.

E. Other Lifelines

The seismic resistance of gas pipelines, and other gas-handling facilities, has recently been given serious attention, partly because of economic consequences but also because of the potential for explosions. Communication utilities have generally acted to protect their facilities from disruption by earthquake shaking, so that user overload rather than equipment failure is the primary reason for postearthquake difficulties.

VI. IDENTIFICATION AND EVALUATION OF EARTHQUAKE HAZARDS

For the great majority of buildings and other structures that are simply designed in accordance with codes, a national zonation map gives the ground-shaking hazard based upon a major effort by seismologists of the U.S. Geological Survey.

For some structures, the owner may want assurance that a building will experience little damage during an earthquake that will likely occur during the life of the structure, plus very strong assurance that life safety is protected even during the largest credible earthquake. In such cases, a seismologist will typically become involved in recommending ground motions that meet these specifications. The owner will then expect careful engineering of the structure and, for a building, careful attention to the design of nonstructural features.

When the failure of a structure would threaten many lives (e.g., a large dam or a nuclear power plant), a regulatory agency may become involved in the approval of ground motions assumed for analysis and design. If this happens, geologists will then be asked to assess the possibility that very large earthquakes might occur within a large area structure, and seismologists will be asked to study how large the resulting motions might be at the site. Very thorough and detailed engineering analyses will then be performed.

A. Effects of Local Soil Conditions

Ground motions atop a deposit of soil generally are stronger than those at underlying rock. Figure 8 shows spectra computed from motions measured at nearby soil (upper left recording) and rock (lower left recording) sites

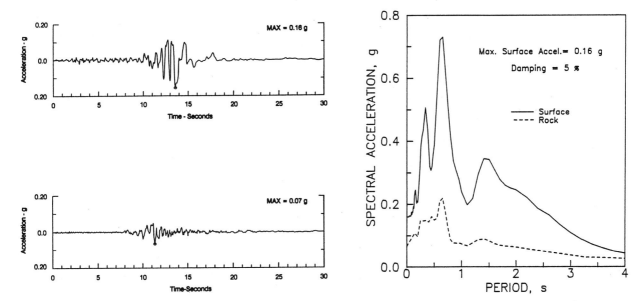

FIGURE 8 Accelerations measured on nearby islands in San Francisco Bay during the 1989 Loma Prieta earthquake. (Bottom left) Yerba Buena Island consists of rock, whereas (top left) Treasure Island is manmade fill over soft clay. The response spectra (far right) were calculated from the recorded accelerations. [From Seed *et al.* (1990). "EERC-90/05," University of California at Berkeley.]

in San Francisco Bay. The significant effect of soil is shown by the corresponding response spectra (Fig. 8, far right). The earthquake that caused great damage and loss of life in Mexico City in 1985 provides a striking example of the potential importance of local soil effects. Figure 9 shows the unusual response spectra recorded in the central part of the city. The large peak at the SCT site unfortunately coincided with the fundamental period of several large buildings, which led to very large deformations and then collapse.

Geotechnical engineers have developed theories for predicting the modifying effects of soil. In the simplest analysis, a column of soil is analyzed in much the same way as the idealized structure in Fig. 3. The stiffness of soils is represented by shear modulus, related to measured or estimated *in situ* shear wave velocity. Since soil is nonlinear, curves giving shear modulus reduction and damping ratio as a function of strain have been developed. Examples appear in Fig. 10. Usually linear analysis is used, with modulus and damping adjusted iteratively so as to be consistent with computed strains. Such calculations from theory have been reasonably well validated by the still scarce information from recordings during earthquakes. The calculations show that low-frequency components of ground motion are increased on soil, in accordance with Figs. 8 and 9. However, because of the large damping that occurs during very strong shaking, calculations suggest that high-frequency components of ground motion may be reduced between the bottom and the top of a soft soil.

Figure 11 summarizes peak ground accelerations measured on soft soils, plus theoretical predictions indicating that peak ground accelerations may be less atop soil than at rock. This aspect of the effect of site conditions is still somewhat controversial.

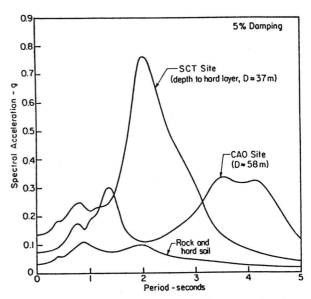

FIGURE 9 Response spectra computed from ground motions recorded in Mexico City during the earthquake of 1985. The SCT and CAO sites are two locations within the city, underlain by different depths of the very soft Mexico City clay. [From Romo, M., and Seed, H. B. (1987). "Int. Conf. Mexican Earthquake," American Society of Civil Engineers, Reston, VA.]

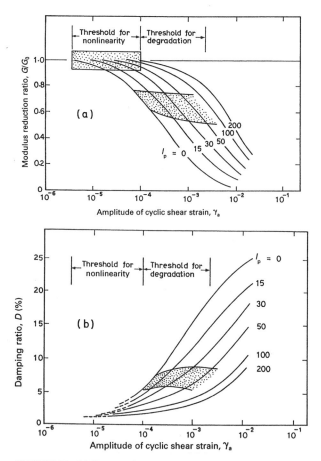

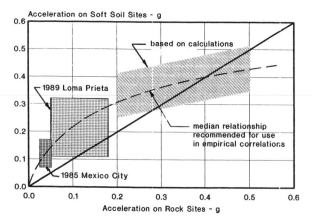

FIGURE 10 (a) Secant shear modulus during cyclic straining in relation to the shear modulus G_0 at very small strains. (b) Damping ratio, which is essentially the same as percentage of critical damping. I_p denotes the plasticity index, with $I_p = 0$ corresponding to cohesionless soil and $I_p = 200$ to a very plastic clay. At the threshold of degradation, the reduction factor and the damping ratio begin to be affected by the number of cycles of straining. [From Ishihara, K. (1996). "Soil Behavior in Earthquake Geotechnics," Clarendon Press, Oxford, after data by M. Vulcetic and R. Dobry.]

FIGURE 11 Relation of peak acceleration atop soft clay to peak acceleration at underlying rock. [From Idriss, I. M. (1990). "Seed Memorial," Vol. 2, BiTech Publishers, Vancouver, BC.]

Theoretical analyses and field experiences have been used to establish the soil factors S prescribed by building codes, and for important buildings, site-specific calculations are often made by geotechnical engineers. The fact that motions may be different at various elevations within a soil deposit has important implications concerning the behavior of tunnels or other structures embedded within soil, as well as for the dynamic response of buildings having deep foundations. In addition, it is the presence of dynamic stresses within soil and rock during ground shaking that can cause "failure," either liquefaction or landslides.

B. Liquefaction

Liquefaction is the consequence of the tendency for soils to become denser as a result of cyclic straining. When a soil is saturated, it cannot densify if water does not have time to flow out of the soil. This can cause the pressure in the fluid phase of soil to increase until there is little

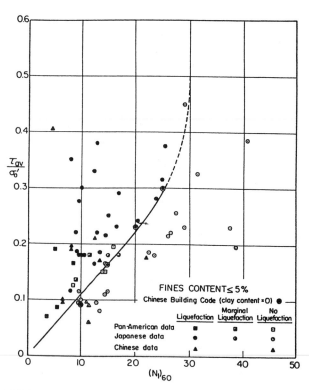

FIGURE 12 Instances of liquefaction and nonliquefaction observed in earthquakes, in relation to intensity of ground shaking (vertical axis) and resistance of soil (horizontal axis). τ_{av}/σ_0' is the ratio of earthquake-caused shear stress in the soil (generally calculated from estimated or measured accelerations) to effective overburden stress. $(N_1)_{60}$ is blows per foot of penetration during a standard penetration test, modified to account for the effective overburden pressure and the energy used to cause penetration. [From Seed, H. B. et al. (1985). J. Geotech. Eng., American Society of Civil Engineers, Reston, VA.]

or no stress through the soil skeleton, and the soil loses stiffness and strength. Liquefaction is most likely to occur in fine sands, but it can also occur in gravels confined by less permeable layers and in silts.

Although liquefaction has always occurred during earthquakes, the phenomenon was not identified as such until the Niigata, Japan, earthquake of 1964. It has since been thoroughly studied in the laboratory in an attempt to understand the process and the factors that affect it. It is very difficult to measure directly the liquefaction resistance of a particular soil, unless very specialized sampling techniques (e.g., *in situ* freezing) are used. Hence, in practice, resistance is typically evaluated indirectly by measuring resistance to penetration as rods are pushed or driven into ground following standardized procedures. Figure 12 shows the form for assembling experience from past earthquakes. Recorded instances of liquefaction or nonliquefaction are plotted versus penetration resistance and peak acceleration. A curve separates the zones in which liquefaction is or is not expected at a new site.

Potential consequences of liquefaction—lateral spreading, landsliding, settlement, and so forth—are difficult to analyze and predict by theoretical analysis, but constitutive models and computer codes have been developed and used in connection with special projects. Centrifuge model tests have proved very useful in achieving better knowledge of such phenomena and in validating theoretical methods. These are specialized tests in which a small model of a soil mass is spun at high speeds so as to properly scale the effects of gravity and then is subjected to base shaking while it is spinning. Some useful empirical predictions have also been developed.

SEE ALSO THE FOLLOWING ARTICLES

DAMS, DIKES, AND LEVEES • EARTHQUAKE MECHANISMS AND PLATE TECTONICS • EARTHQUAKE PREDICTION • MECHANICS OF STRUCTURES • SEISMOLOGY, ENGINEERING

BIBLIOGRAPHY

Bolt, B. A. (1999). "Earthquakes," Freeman, San Francisco.
Chopra, A. K. (1995). "Dynamics of Structures," Prentice-Hall, Englewood Cliffs, N.J.
Federal Emergency Management Agency (1997a). "NEHRP Guidelines for the Seismic Rehabilitation of Buildings," FEMA 273, Washington, D.C.
Federal Emergency Management Agency (1997b). "NEHRP Recommended Provisions for the Seismic Regulation of New Buildings," FEMA 222A, Washington, D.C.
Ishihara, K. (1996). "Soil Behavior in Earthquake Geotechnics," Clarendon Press, Oxford.
Kramer, S. L. (1996). "Geotechnical Earthquake Engineering," Prentice Hall, Upper Saddle River, N.J.
Naeim, F., and Kelly, J. M. (1999). "Design of Seismic Isolated Structures," Wiley, New York.

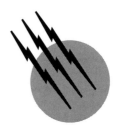

Earthquake Mechanisms and Plate Tectonics

Seth Stein
Eryn Klosko
Northwestern University

GLOSSARY

Euler vector Rotation vector describing the relative motion between two plates. The magnitude of the Euler vector is the rotation rate, and its intersection is the Euler pole. The linear velocity at any point on the plate boundary is the vector product of the Euler vector and the radius vector to that point.

Global plate motion model A set of Euler vectors specifying plate motions. Such models can be derived for spans of millions of years using rates from seafloor magnetic anomalies and directions of motion from the orientations of transform faults and the slip vectors of earthquakes on transforms and at subduction zones. They can also be derived for spans of a few years using space-based geodesy.

Space-based geodesy Techniques using space-based technologies to measure the positions of geodetic monuments to accuracies of better than a centimeter, even for sites thousands of kilometers apart. Measurements of positions over time yield relative velocities to very high precision.

Plate boundary zone The diffuse zone of deformation, often marked by a distribution of seismicity, active faulting, and topography, within which relative plate motion is accommodated. Plate boundary zones are typically narrow in oceanic lithosphere and broad in continental interiors.

I. INTRODUCTION

Earthquake seismology has played a major role in the development of our current understanding of global plate tectonics. Because earthquakes occur primarily at the boundaries between lithospheric plates, their distribution

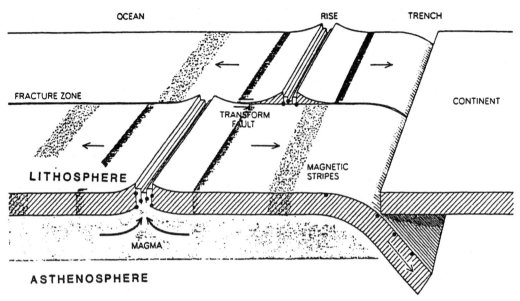

FIGURE 1 Plate tectonics at its simplest. Plates are formed at ridges and subducted at trenches. At transform faults, plate motion is parallel to the boundries. Each boundary type has typical earthquakes.

is used to map plate boundaries, and their focal mechanisms provide information about the motion at individual boundaries.

Plate boundaries are divided into three types (Fig. 1). Oceanic lithosphere is formed at *spreading centers*, or mid-ocean ridges, and is destroyed at *subduction zones* or trenches. Thus, at spreading centers plates move away from the boundary, whereas at subduction zones the subducting plate moves toward the boundary. At the third boundary type, *transform faults*, plate motion is parallel to the boundary. The *slip vectors* of the earthquakes on plate boundaries, which show the motion on the fault plane, reflect the direction of relative motion between the two plates.

The basic principle of plate kinematics is that the relative motion between any two plates can be described as a rotation on a sphere about an *Euler pole* (Fig. 2). Specifically, at any point along the boundary between plates i and j, with latitude λ and longitude μ, the linear velocity of plate j with respect to plate i is

$$\mathbf{v}_{ji} = \omega_{ji} \times \mathbf{r},$$

the usual formulation for rigid body rotations in mechanics. The vector $\mathbf{r}$ is the position vector to the point on the boundary, and ω_{ji} is the rotation vector or *Euler vector*. Both are defined from an origin at the center of the earth.

The direction of relative motion at any point on a plate boundary is a small circle, a parallel of latitude *about the Euler pole* (not a geographic parallel about the North Pole!). For example, in Fig. 3a, the pole shown is for the motion of plate 2 with respect to plate 1. The first named plate ($j=2$) moves counterclockwise about the pole with

respect to the second plate ($i = 1$). The segments of the boundary, where relative motion is parallel to the boundary, are transform faults. Thus, transforms are small circles about the pole, and earthquakes occurring on them should have pure strike-slip mechanisms. Other segments have relative motion away from the boundary and are thus spreading centers. Figure 3b shows an alternative case. The pole here is for plate 1 ($j = 1$) with respect to plate 2 ($i = 2$), so plate 1 moves toward some segments of the boundary, which are subduction zones. Note that the ridge and subduction zone boundary segments are not small circles.

The magnitude, or rate, of relative motion increases with distance from the pole, since

$$|\mathbf{v}_{ji}| = |\omega_{ji}||\mathbf{r}|\sin\gamma,$$

where γ is the angle between the Euler pole and the site (corresponding to a colatitude about the pole.) Thus, although all points on a plate boundary have the same angular velocity, the linear velocity varies.

If we know the Euler vector for any plate pair, we can write the linear velocity at any point on the boundary between the plates in terms of the local E-W and N-S components by a coordinate transformation. With this, the rate and azimuth of plate motion become

$$\text{rate} = |\mathbf{v}_{ji}| = \sqrt{\left(\mathbf{v}_{ji}^{NS}\right)^2 + \left(\mathbf{v}_{ji}^{EW}\right)^2}$$

$$\text{azimuth} = 90 - \tan^{-1}\left(\left(\mathbf{v}_{ji}^{NS}\right/\left(\mathbf{v}_{ji}^{EW}\right)\right)$$

such that azimuth is measured in degrees clockwise from north.

Given a set of Euler vectors with respect to one plate, those with respect to others are found by vector arithmetic.

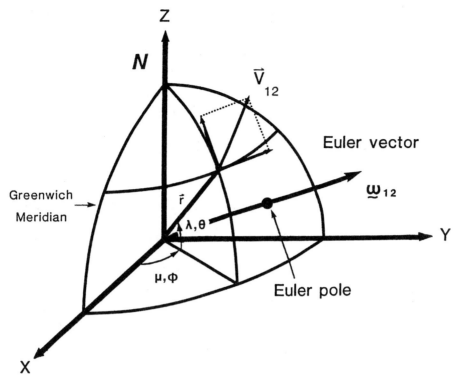

FIGURE 2 Geometry of plate motions. The linear velocity at point **r** is given by $V_{ji} = \omega_{ji} \times r$. The Euler pole is the intersection of the Euler vector ω_{ji} with the earth's surface.

For example, the Euler vector for the reverse plate pair is the negative of the Euler vector,

$$\omega_{ij} = -\omega_{ji}.$$

Euler vectors for other plate pairs are found by addition,

$$\omega_{jk} = \omega_{ji} + \omega_{ik},$$

so, given a set of vectors all with respect to plate i, any Euler vector needed is found from

$$\omega_{jk} = \omega_{ji} - \omega_{ki}.$$

For further information on plate kinematics, see an introductory text like Cox and Hart (1986). As discussed there, motions between plates can be determined by combining three different types of data from different boundaries. The rate of spreading at ridges is given by seafloor magnetic anomalies, and the directions of motion are found from the orientations of transform faults and the slip vectors of earthquakes on transforms and at subduction zones. As is evident, earthquake slip vectors are only one of three types of plate motion data available. Euler vectors are determined from the relative motion data, using geometrical conditions. Since slip vectors and transform faults lie on small circles about the pole, the pole must lie on a line at right angles to them (Fig. 3). Similarly, the rates of plate motion increase with the sine of

the distance from the pole. These constraints make it possible to locate the poles. Determination of Euler vectors for all the plates can thus be treated as an overdetermined least-squares problem, and the best solution can be found using the generalized inverse to derive a *global plate motion model* (DeMets *et al.*, 1990). Because these models use magnetic anomaly data, they describe plate motion averaged over the past few million years.

New data has become available in recent years due to the rapidly evolving techniques of space-based geodesy. These techniques [Very Long Baseline radio Interferometry (VLBI), Satellite Laser Ranging (SLR), the Global Positioning System (GPS), and DORIS (similar to GPS, but using ground transmitters)] use space-based technologies to measure the positions of geodetic monuments to accuracies of better than a centimeter, even for sites thousands of kilometers apart. Hence, measurements of positions over time yield relative velocities to precisions almost unimaginable during the early days of plate tectonic studies. A series of striking results, first with VLBI and SLR (e.g., Robbins, Smith, and Ma, 1993), and now with GPS (Argus and Heflin, 1995; Larson, Freymueller, and Philipsen, 1997), show that plate motion over the past few years is generally quite similar to that predicted by the global plate motion model NUVEL-1A. This agreement is consistent with the prediction that episodic

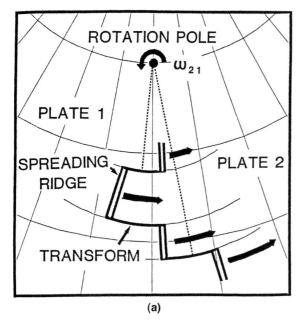

(a)

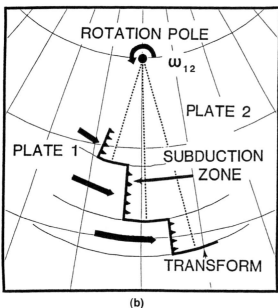

(b)

FIGURE 3 Relationship of motion on plate boundaries to the Euler pole. Relative motion occurs along small circles about the pole; the rate increases with distance from the pole. Note the difference the sense of rotaation makes: ω_{ji} is the Euler vector corresponding to the rotation of plate j counterclockwise with respect to i.

motion at plate boundaries, as reflected in occasional large earthquakes, will give rise to steady motion in plate interiors due to damping by the underlying viscous asthenosphere (Elsasser, 1969). As a result, the earthquake mechanisms can be compared to the plate motions predicted by both global plate motion models and space-based geodesy.

II. OCEANIC SPREADING CENTER FOCAL MECHANISMS

Earthquake mechanisms from the mid-ocean ridge system reflect the spreading process. Figure 4 schematically shows a portion of a spreading ridge offset by transform faults. Because new lithosphere forms at the ridges and then moves away, the relative motion of lithosphere on either side of a transform is in opposing directions. The direction of transform offset, not the spreading direction, determines whether there is right or left lateral motion on the fault. This relative motion, defined as transform faulting, is not what produces the offset of the ridge crest. In fact, if the spreading at the ridge is symmetric (equal rates on either side), the length of the transform will not change with time. This is a very different geometry from a transcurrent fault, where the offset is produced by motion on the fault and the length of the offset between ridge segments would increase with time.

The model is illustrated by focal mechanisms. Figure 5 (top) shows a portion of the Mid-Atlantic Ridge composed of north-south trending ridge segments and offset by transform faults, such as the Vema Transform, which trend approximately east-west. Both the ridge crest and the transforms are seismically active. The mechanisms show that the relative motion along the transform is right lateral. Seafloor spreading on the ridge segments produces the

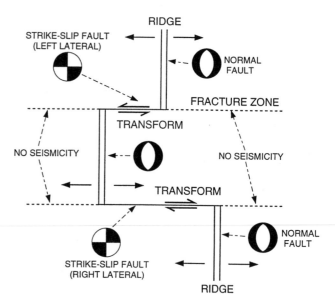

FIGURE 4 Possible tectonic settings of earthquakes at an oceanic spreading center. Most events occur on the active segments of the transform and have strike-slip mechanisms consistent with transform faulting. On a slow spreading Ridge, like the Mid-Atlantic, ridge normal fault earthquakes occur. Very few events occur on the inactive fracture zone.

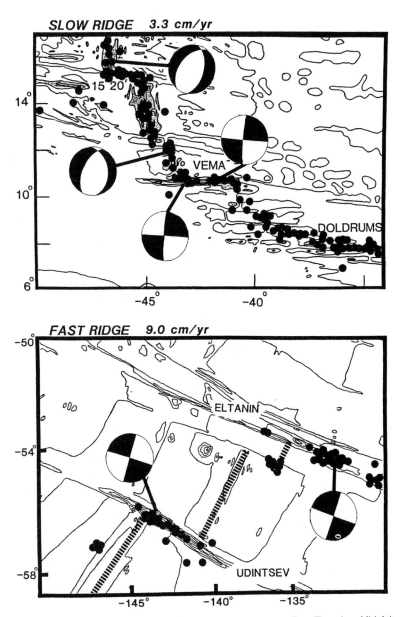

FIGURE 5 Maps contrasting faulting on slow and fast spreading centers. Top: The slow Mid-Atlantic Ridge has earthquakes both on the active transform and on the ridge segment. Strike-slip faulting on a plane parallel to the transform azimuth is characteristic. On the ridge segments, normal faulting with nodal planes parallel to the ridge trend is seen. Bottom: The fast East Pacific Rise has only strike-slip earthquakes on the transform segments. Mechanisms form Engeln, Wiens, and Stein (1986), Huang *et al.* (1986), and Stewart and Okal (1983).

observed relative motion. For this reason, earthquakes occur almost exclusively on the active segment of the transform fault between the two ridge segments, rather than on the inactive extension, known as a *fracture zone*. Although no relative plate motion occurs on the fracture zone, it is often marked by a distinct topographic feature, due to the contrast in lithospheric ages across it. Unfortunately, some transform faults named before this distinction became clear, such as the Vema, are known as "fracture zones" along their entire length. Earthquakes also occur on the

spreading segments. Their focal mechanisms show normal faulting, with nodal planes trending along the ridge axis.

The seismicity is different on fast spreading ridges. Figure 5 (bottom) shows a portion of the Pacific-Antarctic boundary along the East Pacific Rise. Here, strike-slip earthquakes occur on the transforms, but we do not observe the ridge crest normal faulting events. These observations can be explained by the thermal structure of the lithosphere, because fast spreading produces younger and thinner lithosphere than slow spreading. The axis of a fast

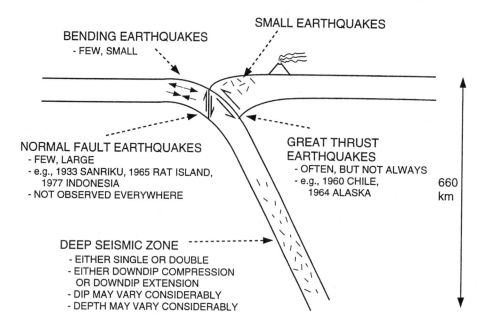

BENDING EARTHQUAKES
- FEW, SMALL

SMALL EARTHQUAKES

NORMAL FAULT EARTHQUAKES
- FEW, LARGE
- e.g., 1933 SANRIKU, 1965 RAT ISLAND, 1977 INDONESIA
- NOT OBSERVED EVERYWHERE

GREAT THRUST EARTHQUAKES
- OFTEN, BUT NOT ALWAYS
- e.g., 1960 CHILE, 1964 ALASKA

660 km

DEEP SEISMIC ZONE
- EITHER SINGLE OR DOUBLE
- EITHER DOWNDIP COMPRESSION OR DOWNDIP EXTENSION
- DIP MAY VARY CONSIDERABLY
- DEPTH MAY VARY CONSIDERABLY

"COMPOSITE" SUBDUCTION ZONE

FIGURE 6 Schematic of some of the features observed at subduction zones. Not all features are seen at all subduction zones.

ridge has a larger magma chamber than the slow ridge, and the lithosphere moving away from a fast spreading ridge is more easily replaced than for a slow ridge. Thus, in contrast to the axial valley and normal faulting earthquakes on a slow ridge, a fast ridge has an axial high and an absence of earthquakes.

The mechanisms are consistent with the predictions of plate kinematics. The area in Fig. 5 is a portion of the boundary between the South American and Nubian (West African) plates. An Euler vector for Nubia with respect to South America with a pole at 62°N, 37.8°W and a magnitude of 0.328 degrees/Myr predicts that at 0°N, 20°W Africa is moving N81°E, or almost due east, at 33 mm/year with respect to South America. The Vema is a boundary segment parallel to this direction, and so is a transform fault characterized by strike-slip earthquakes with directions of motion along the trace of the transform. The short segments at essentially right angles to the direction of relative motion are then spreading ridge segments. The spreading rate determined from magnetic anomalies, and thus the slip rate across the transform, is described by the Euler vector.

III. SUBDUCTION ZONE FOCAL MECHANISMS

Both the largest earthquakes and the majority of large earthquakes occur at subduction zones. Their focal mech-

anisms reflect various aspects of the subduction process. Figure 6 is a composite cartoon showing some of the features observed in different subduction zones.

Most of the large, shallow, subduction zone earthquakes indicate thrusting of the overriding plate over the subducting lithosphere. The best such examples are the two largest ever recorded: the 1960 Chilean (M_0 2.7×10^{30}, M_s 8.3) and 1964 Alaskan (M_0 7.5×10^{29}, M_s 8.4) earthquakes. These were impressive events; in the Chilean earthquake 24 m of slip occurred on a fault 800-km long along strike and 200-km long downdip. Smaller, but large, thrust events are characteristic. For example, Fig. 7 (top) shows the focal mechanisms of large, shallow earthquakes along a portion of the Peru-Chile trench, where the Nazca plate is subducting beneath the South American plate. The mechanisms along the trench show thrust faulting on fault planes with a consistent geometry: parallel to the coast, which corresponds to the trench axis, with shallow dips to the northeast.

These thrust events directly reflect the plate motion. At a point on the trench (17°S, 75°W), the global plate motion model NUVEL-1A (DeMets *et al.*, 1994) predicts motion of the Nazca plate with respect to South America at a rate of 68 cm/year and an azimuth of N76°E. The direction of motion is toward the trench, as expected at a subduction zone. The major thrust earthquakes at the interface between subducting and overriding plates thus directly reflect the subduction, and slip vectors from their

NAZCA - SOUTH AMERICA PLATE BOUNDARY ZONE

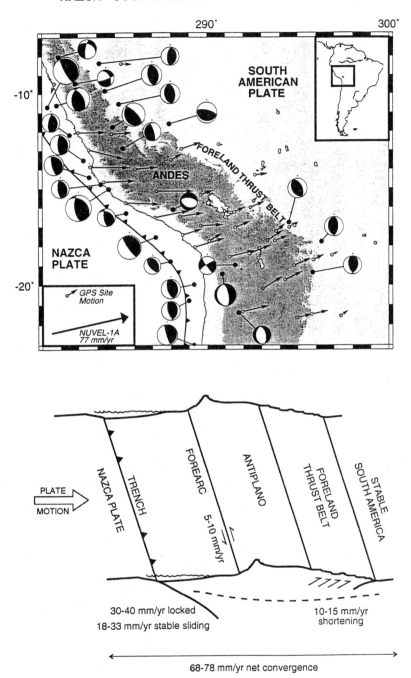

FIGURE 7 Top: GPS site velocities relative to stable South America (Norabuena *et al.*, 1998), and selected earthquake mechanisms in the boundary zone. Rate scale is given by the NUVEL-1A vector. Bottom: Cross-section across Andean orgenic system showing velocity distribution inferred from GPS data.

focal mechanisms can be used to determine the direction of plate motion. The rate of subduction is harder to assess. Although the rate can be computed from global plate motion models or space geodesy, all of the plate motion is not always released seismically in earthquakes (Kanamori,

1977). In this case, the seismic slip rate estimated from seismic moments can be only a fraction of the real plate motion. Nonetheless, it is useful to determine the seismic slip rate to assess the fraction of seismic slip, as it reflects the mechanics of the subduction process. It is also interesting to know how this seismic slip varies as a function of time and position along a subduction zone.

Figure 6 also shows other types of shallow, subduction zone earthquakes. An interesting class of subduction zone earthquakes results from the flexural bending of the downgoing plate as it enters the trench. Precise focal depth studies show a pattern of normal faulting in the upper part of the plate to a depth of 25 km and thrusting in its lower part, between 40 and 50 km. These observations constrain the position of the neutral surface separating the upper extensional zone from the lower flexural zone and, thus, provide information on the mechanical state of the lithosphere. Occasionally, trenches are the sites of large, normal fault earthquakes (e.g., Sanriku, 1933; Indonesia, 1977). There has been some controversy as to whether to interpret these earthquakes as bending events in the upper flexural sheet or as "decoupling" events showing rupture of the entire downgoing plate due to "slab pull."

The deeper earthquakes, which form the Wadati-Benioff zone, go down to depths of 700 km within the downgoing slab. Their mechanisms provide important information about the physics of the subduction process. The essence of the process is the penetration and slow heating of a cold slab of lithosphere in the warmer mantle. This temperature contrast has important consequences. The subducting plate is identified by the locations of earthquakes in the Wadati-Benioff zone below the zone of thrust faulting at the interface between the two plates. Earthquakes occur to greater depths than elsewhere because the slab is colder than the surrounding mantle. The mechanisms of earthquakes within the slab similarly reflect this phenomenon. The thermal evolution of the downgoing plate and its surroundings is controlled by the relation between the rate at which cold slab material is subducted to that at which it heats up, primarily by conduction as it equilibrates with the surrounding mantle. In addition, adiabatic heating due to the increasing pressure with depth and phase changes contribute.

Numerical temperature calculations show that the downgoing plate remains much colder than the surrounding mantle until considerable depths, where the downgoing slab heats up to the ambient temperature. Comparison of calculated temperatures, the observed locations of seismicity, and images from seismic tomography show that earthquakes occur in the cold regions of the slab. The thermal structure also helps explain their focal mechanisms. The force driving the subduction is the integral over the slab of the force due to the density contrast between the denser subducting material and the density of "normal" mantle material outside. This force, known as slab pull, is the plate driving force due to subduction. Its significance for stresses in the downgoing plate and for driving plate motions depends on its size relative to the resisting forces

at the subduction zone. There are several such forces. As the slab sinks into the viscous mantle, material must be displaced. The resulting force depends on the viscosity of the mantle and the subduction rate. The slab is also subject to drag forces on its sides and resistance at the interface between the overriding and downgoing plates. The latter, of course, is often manifested as the shallow thrust earthquakes.

One way to study the relative size of the negative buoyancy and resistive forces is to use focal mechanisms to examine the state of stress in the downgoing slab. Earthquakes above 300 km show generally downdip tension, whereas those below 300 km show generally downdip compression. A proposed explanation is that there are two basic processes operating: near the surface the slab is being extended by its own weight; at depth the slab begins to "run into" stronger material and downdip compression occurs. Numerical models of stress in downgoing slabs are consistent with this interpretation as they can reproduce the shallow downdip tension and the deep downdip compression.

Finally, it is worth noting that not all features shown in Fig. 6 have been observed at all places. For example, the dips and shapes of subduction zones vary substantially. Some show double planes of deep seismicity, and some do not. Even the very large thrust earthquakes, considered characteristic of subduction zone events, are not observed in all subduction zones. In recent years, considerable effort has been made to understand such variations.

IV. DIFFUSE PLATE BOUNDARY EARTHQUAKE FOCAL MECHANISMS

Although the basic relationships between plate boundaries and earthquakes apply to continental as well as oceanic lithosphere, the continents are more complicated. The continental crust is much thicker, less dense, and has very different mechanical properties from the oceanic crust. Because continental crust and lithosphere are not subducted, the continental lithosphere records a long, involved tectonic history. In contrast, the oceans record only the past 200 million years. One major result of these factors is that plate boundaries in continents are often diffuse, rather than the idealized narrow boundaries assumed in the rigid plate model, which are a good approximation to what we see in the oceans. The initial evidence for this notion comes from the distribution of seismicity and the topography, which often imply a broad zone of deformation between the plate interiors. This effect is especially evident in continental interiors, such as the India-Eurasia collision zone in the Himalayas or the Pacific-North America boundary zone in the western United States. Plate boundary zones

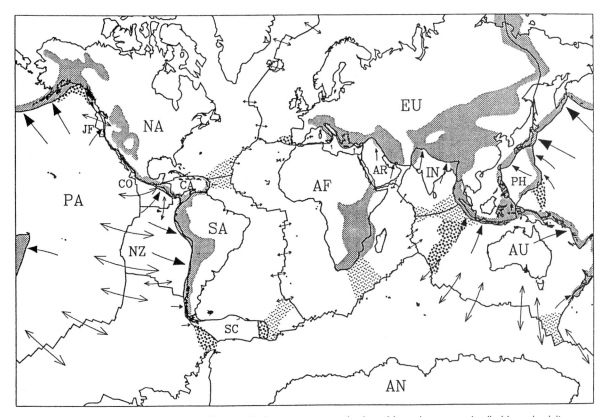

FIGURE 8 Comparison of the idealized rigid plate geometry to the broad boundary zones implied by seismicity, topography, or other evidence of faulting. Fine stipple shows mainly subaerial regions where the deformation has been inferred from seismicity, topography, other evidence of faulting, or some combination of these. Medium stipple shows mainly submarine regions where the nonclosure of plate circuits indicates measurable deformation; in most cases these zones are also marked by earthquakes. Coarse stipple shows mainly submarine regions where the deformation is inferred mainly from the presence of earthquakes. These deforming regions from wide plate boundary zones, which cover about 15% of the earth's surface. The precise geometry of these zones, and in some cases their existence, in under investigation. Plate motions shown are for the NUVEL-1 global relative plate motion model. Arrow lengths are proportional to the displacement if plates maintain their present relative velocity for 25 Myr. Divergence across mid-ocean ridges is shown by diverging arrows. Convergence is shown by single arrows on the underthrust plate. [After Gordon, R. G., and Stein, S. (1992). "Global tectonics and space geodesy." *Science* **256**, 333.]

(Fig. 8), indicated by earthquakes, volcanism, and other deformation, appear to cover about 15% of the earth's surface (Gordon and Stein, 1992; Stein, 1993).

Insight into plate boundary zones is being obtained by combining focal mechanisms with geodetic, topographic, and geological data. Although plate motion models predict only the integrated motion across the boundary, GPS, geological, and earthquake data can show how this deformation varies in space and time. Both variations are of interest. Possible spatial variations include a single fault system taking up most of the motion (e.g., Prescott, Lisowski, and Savage, 1981), a smooth distribution of motion (e.g., England and Jackson, 1989), or motion taken up by a few relatively large microplates or blocks (e.g., Acton, Stein, and Engeln, 1991; Thatcher, 1995). Each of these possibilities appears to occur, sometimes within the same boundary zone. The distribution of the motion

in time is of special interest because steady motion between plate interiors gives rise to episodic motion at plate boundaries, as reflected in occasional large earthquakes and, in some cases, steady creep (Fig. 9). The detailed relation between plate motions and earthquakes is complicated and poorly understood and hence forms a prime target of present studies.

For example, Fig. 7 (top) shows focal mechanisms and vectors derived from GPS illustrating the distribution of motion within the boundary zone extending from the stable interior of the oceanic Nazca plate, across the Peru-Chile trench to the coastal forearc, across the high Altiplano and foreland thrust belt, and into the stable interior of the South American continent. The GPS site velocities are relative to stable South America, so if the South American plate were rigid and all motion occurred at the boundary, they would be zero. Instead, they are highest

Plate Boundary Zone Slip Distribution

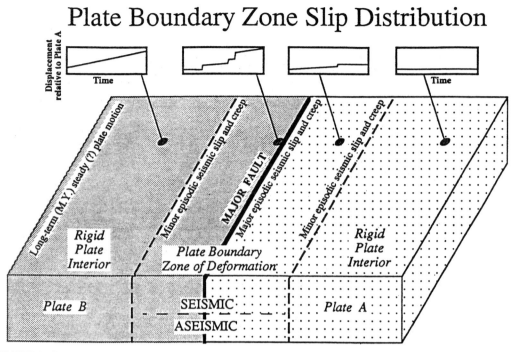

FIGURE 9 Schematic illustration of the distribution of motion in space and time for a strike-slip boundary zone between two major plates (Stein, 1993).

near the coast and decrease relatively smoothly from the interior of the Nazca plate to the interior of South America. Figure 7 (bottom) shows an interpretation of these data. In this, about half of the plate convergence (30–40 mm/year) is locked at the plate boundary thrust interface, causing elastic strain that is released in large interplate trench thrust earthquakes. Another 18–30 mm/year of the plate motion occurs aseismically by smooth stable sliding at the trench. The rest occurs across the sub-Andean fold-and-thrust belt, causing permanent shortening and mountain building, as shown by the inland thrust fault mechanisms. Comparison of strain tensors derived from GPS and earthquake data shows that the shortening rate inferred from earthquakes is significantly less than indicated by the GPS, implying that much of the shortening occurs aseismically. The focal mechanisms also indicate some deformation within the high Andes themselves.

Another broad plate boundary zone is the Pacific–North America boundary in western North America. Figure 10 shows the boundary zone in a projection about the Euler pole. The relative motion is parallel to the small circle shown. Thus, the boundary is extensional in the Gulf of California, essentially a transform along the San Andreas fault system, and convergent in the eastern Aleutians and the western Aleutians. The focal mechanisms reflect these changes. For example, in the Gulf of California we see strike-slip motion along oceanic transforms and normal faulting on a ridge segment. The San Andreas has both pure strike-slip earthquakes (Parkfield) and earthquakes with some dip-slip motion (Northridge, San Fernando, and Loma Prieta) when it deviates from pure transform behavior. The plate boundary zone is also broad, as shown by the distribution of seismicity. Although the San Andreas fault system is the locus of most of the plate motion and hence large earthquakes, seismicity extends as far eastward as the Rocky Mountains. For example, the Landers earthquake shows strike-slip motion east of the San Andreas, and the Borah Peak earthquake illustrates Basin and Range faulting. The diffuse nature of the boundary is also illustrated by vectors showing the motion of GPS and VLBI sites with respect to stable North America. Net motion across the zone is essentially that predicted by the global plate motion model NUVEL-1A. The site motions show that most of the strike-slip motion occurs along the San Andreas fault system, but significant motions occur for some distance eastward.

V. INTRAPLATE DEFORMATION AND INTRAPLATE EARTHQUAKES

A final important use of earthquake mechanisms is to study the internal deformation of major plates. Although idealized plates would be purely rigid, the existence of

PACIFIC - NORTH AMERICA PLATE BOUNDARY ZONE

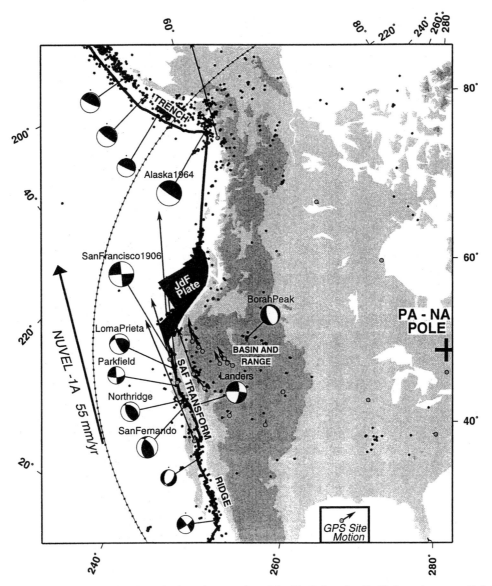

FIGURE 10 Geometry and focal mechanisms for a portion of the North America-Pacific boundary zone. Dot-dash line shows small circle, and thus direction of plate motion, about the Pacific-North America Euler pole. The variation in the boundary type along its length from extension, to transform, to convergence is shown by the focal mechanism. The diffuse nature of the boundary zone is shown by seismicity (small dots), focal mechanisms, topogaphy (1000-m contour shown), and vectors showing the motion of GPS and VLBI sites with respect to stable North America (Bennett, Davis, and Wernicke, 1999; Newman *et al.*, 1999).

intraplate earthquakes reflects the important and poorly understood tectonic processes of intraplate deformation. One such example is the New Madrid area in the central United States which had very large earthquakes in 1811–1812. The seismicity of such regions is generally thought to be due to the reactivation of pre-existing faults or weak zones in response to intraplate stresses. Because motion in

these zones is at most a few millimeters per year, compared to the generally much more rapid plate boundary motions, seismicity is much lower (Fig. 10). Similarly, major intracontinental earthquakes occur substantially less frequently than plate boundary events; recurrence estimates for 1811–1812 type earthquakes average 500–1000 years. Efforts are being made to combine geodetic data, which indicate deviations from rigidity, to the earthquake data. For example, comparison of the velocities for permanent GPS sites in North America east of the Rocky Mountains to velocities predicted by modeling these sites as being on a single rigid plate shows that the interior of the North American plate is rigid at least to the level of the average velocity residual, less than 2 mm/year (Dixon, Mao, and Stein, 1996; Newman *et al.*, 1999). Similar results emerge from geodetic studies of other major plates, showing that plates thought to have been rigid on geological timescales are quite rigid on decadal scales. Moreover, geological data suggest that such intraplate seismic zones may be active for only a few thousands of years, even though plate motions have been steady for millions of years. As a result, understanding how these intraplate seismic zones operate is a major challenge. A special case of this phenomenon occurs at passive margins, where continental and oceanic lithosphere join. Although these areas are, in general, tectonically inactive, magnitude 7 earthquakes can occur, as on the eastern coast of North America. Such earthquakes are thought to be associated with stresses at the continental margin, including those due to the removal of glacial loads, which reactivate the faults remaining along the continental margin from the original rifting.

SEE ALSO THE FOLLOWING ARTICLES

CONTINENTAL CRUST • EARTHQUAKE ENGINEERING • EARTHQUAKE PREDICTION • EARTH'S CORE • EARTH'S MANTLE • OCEANIC CRUST • PLATE TECTONICS • SEISMOLOGY, ENGINEERING • SEISMOLOGY, OBSERVATIONAL • SEISMOLOGY, THEORETICAL • TSUNAMIS

BIBLIOGRAPHY

Acton, G. D., Stein, S., and Engeln, J. F. (1991). "Block rotation and continental extension in Afar: a comparison to oceanic microplate systems," *Tectonics* **10,** 501–526.

Argus, D. F., and Heflin, M. B. (1995). "Plate motion and crustal deformation estimated with geodetic data from the Global Positioning System," *Geophys. Res. Lett.* **22,** 1973–1976.

Bennett, R. A., Davis, J. L., and Wernicke, B. P. (1999). "Present-day pattern of Cordilleran deformation in the western United States," *Geology* **27,** 371–374.

Cox, A., and Hart, R. B. (1986). "Plate Tectonics: How it Works," Black-well Sci, Palo Alto.

DeMets, C., Gordon, R. G., Argus, D. F., and Stein, S. (1990). "Current plate motions," *Geophys. J. Int.* **101,** 425–478.

DeMets, C., Gordon, R. G., Argus, D. F., and Stein, S. (1994). "Effect of recent revisions to the geomagnetic reversal time scale on estimates of current plate motion," *Geophys. Res. Lett.* **21,** 2191–2194.

Dixon, T. H., Mao, A., and Stein, S. (1996). "How rigid is the stable interior of the North American plate?" *Geophys. Res. Lett.* **23,** 3035–3038.

Elsasser, W. M. (1969). "Convection and stress propagation in the upper mantle," *In* "The Application of Modern Physics to the Earth and Planetary Interiors," (S. K. Runcorn, ed.), pp. 223–246, Wiley, New York.

Engeln, J. F., Wiens, D. A., and Stein, S. (1986). "Mechanisms and depths of Atlantic transform earthquakes," *J. Geophys. Res.* **91,** 548–577.

England, P., and Jackson, J. (1989). "Active deformation of the continents," *Annu. Rev. Earth Planet. Sci.* **17,** 197–226.

Gordon, R. G., and Stein, S. (1992). "Global tectonics and space geodesy," *Science* **256,** 333–342.

Huang, P. Y., Solomon, S. C., Bergman, E. A., and Nabelek, J. L. (1986). "Focal depths and mechanisms of Mid-Atlantic Ridge earthquakes from body waveform inversion," *J. Geophys. Res.* **91,** 579–598.

Kanamori, H. (1977). "Seismic and aseismic slip along subduction zones and their tectonic implications," *In* "Island Arcs, Deep-Sea Trenches and Back-Arc Basins, Maurice Ewing Ser., 1," (M. Talwani, and W. C. Pitman, III, eds.), pp. 163–174, Am. Geophys. Union, Washington, DC.

Larson, K. M., Freymueller, J. T., and Philipsen, S. (1997). "Global plate velocities from the Global Positioning System," *J. Geophys. Res.* **102,** 9961–9981.

Newman, A., Stein, S., Weber, J., Engeln, J., Mao, A., and Dixon, T. (1999). "Slow deformation and lower seismic hazard at the New Madrid Seismic Zone," *Science,* **284,** 619–621.

Norabuena, E., Leffler-Griffin, L., Mao, A., Dixon, T., Stein, S., Sacks, I. S., Ocala, L., and Ellis, M. (1998). "Space geodetic observations of Nazca-South America convergence along the Central Andes," *Science* **279,** 358–362.

Prescott, W. H., Lisowski, M., and Savage, J. C. (1981). "Geodetic measurements of crustal deformation on the San Andreas, Hayward, and Calaveras faults, near San Francisco, California," *J. Geophys. Res.* **86,** 10,853–10,869.

Robbins, J. W., Smith, D. E., and Ma, C. (1993). "Horizontal crustal deformation and large scale plate motions inferred from space geodetic techniques," *In* "Contributions of Space Geodesy to Geodynamics: Crustal Dynamics, Geodynamics Series 23," (D. E. Smith, and D. L. Turcotte, eds.), pp. 21–36, Am. Geophys. Union, Washington, DC.

Stein, S. (1993). "Space geodesy and plate motions," *In* "Space Geodesy and Geodynamics, Geodynamics Ser. Vol. 23," (D. E. Smith, and D. L. Turcotte, eds.), pp. 5–20, Am. Geophys. Union, Washington, DC.

Stewart, L. M., and Okal, E. A. (1983). "Seismicity and aseismic slip along the Eltanin Fracture Zone," *J. Geophys. Res.* **88,** 10,495–10,507.

Thatcher, W. (1995). "Microplate versus continuum descriptions of active tectonic deformation," *J. Geophys. Res.* **100,** 3885–3894.

Earthquake Prediction

Tsuneji Rikitake
Kazuo Hamada

Association for the Development of Earthquake Prediction

GLOSSARY

Active fault Fault that has moved in historic or recent geological time.
Earthquake precursor Anomalous phenomenon preceding an earthquake.
Macroscopic precursor Earthquake precursor detected by human sense organs, for example, anomalous animal behavior, gush of well water, earthquake light, and rumbling.
Precursor time Time span between onset of anomalous phenomenon and occurrence of main shock.
Seismic gap Area in which seismic activity is extremely low. Such an area often becomes the seat of a large earthquake later.

EARTHQUAKE PREDICTION is the human effort to predict the time, location, and magnitude of a future earthquake. Earthquake prediction programs have been promoted in Japan, China, the United States, the former Soviet Union, and other countries. Scientifically, earthquake prediction relies on anomalous phenomena that precede an earthquake. Although there have been a few successful cases of earthquake prediction, most of them in China, it will be some time before we have the capability of issuing an earthquake warning based on accurate earthquake prediction information.

I. LONG-TERM EARTHQUAKE PREDICTION

A. Historical, Archeological, and Geological Seismicity

Historical data on earthquake occurrence are sometimes useful for assessing the probability of future seismic activity, especially in countries having a long history of earthquake activity, such as China, Japan, and Turkey.

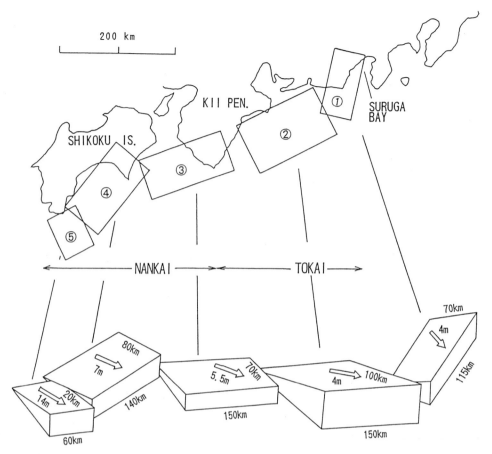

FIGURE 1 Seismic fault model for the 1707 Hoei earthquake of $M = 8.4$. The dimensions of the faults, along with the fault slips, are shown for the five segments considered.

It is well known that great earthquakes having a magnitude (M) of 8 or thereabout on the Richter scale have been repeatedly occurring, with a return period of 100–150 yr, off the pacific coast of the Tokai–Nankai (east sea–south sea) zone of Japan. These earthquakes occur along the Nankai trough, a deep sea canyon. It is believed that the 1707 Hoei earthquake ($M = 8.4$), the largest one in history, was caused by the fault breaks at the five segments shown in Fig. 1. The fault models are obtained from the tsunami data.

Although all the segments seem to have moved almost simultaneously at the time of the 1707 Hoei earthquake, there have been other cases in which only one or two segments have moved. For instance, it is believed that segments 3 and 4, off Kii Peninsula and Shikoku Island, moved during the 1946 Nankai earthquake ($M = 8.0$). Since segment 1 has not moved since the 1854 Ansei–Tokai earthquake, it is feared that a great earthquake will occur there soon.

Examination of historical records in Japan has revealed the occurrence periods of great earthquakes in the Tokai–

Nankai zone, as shown in Table I. These data have been supplemented by paleoseismological studies on ground liquefaction that have become popular in recent years, the results of which are also indicated in Table I. It is thus possible to obtain a frequency distribution of the return periods of great earthquakes in the Tokai–Nankai area.

According to current plate tectonics, it is believed that an ocean plate called the philippine sea plate, which is moving in a northwest direction, is subducting at the Nankai trough. When the land plate, which is compressed and pulled down by the subduction of the ocean plate, is deformed so strongly that the deformation exceeds a certain limit, rupture takes place at the land–sea interface. When this happens, the land plate rebounds because it becomes free, and the accumulated strain energy is radiated in the form of seismic waves.

Given the above mechanism, we may be able to understand why great earthquakes have recurred many times at almost the same area in association with the trough. We can apply a Weibull distribution analysis to obtain further statistics, as follows. It is assumed that the probability that

TABLE I Great earthquakes in the Tokai–Nankai Zone[a]

Earthquake		Tokai zone				Nankai zone			
		Suruga Bay (segment 1)		Off Tokai (segment 2)		Off Kii Peninsula (segment 3)		Off Shikoku Island (segment 4)	
Year	Magnitude	Evidence	Interval (yr)	Evidence	Interval (yr)	Evidence	Interval (yr)	Evidence	Interval (yr)
684	8.3			(●)		(●)		○	
887	8–8.5							○	
1096	8–8.5			○					
1099	8–8.3							○	
1361	8.3–8.5							○	
									137
1498	8.2–8.4			○		○		●	
					107		107		107
1605	7.9			○		○		○	
					102		102		102
1707	8.4	○		○		○		○	
			147		147		147		147
1854	8.4	○		○		○		○	
					90				
1944	7.9			○			92		92
1946	8.0					○		○	

[a] Open circles show firmly established epochs in history. Solid circles represent ground liquefaction evidence. Those with parentheses are a little less reliable.

the return period will lie between t and $t + \Delta t$ is given by $\lambda(t)$, on the condition that no great earthquake occurs before t and that

$$\lambda(t) = Kt^{m+1}, \qquad (1)$$

where $K > 0$ and $m > -1$.

Let us denote a cumulative probability for the recurrence of a great earthquake during the period between 0 and t by $F(t)$, the last earthquake being assumed to have occurred at $t = 0$. Putting

$$R(t) = 1 - F(t), \qquad (2)$$

We assume a Weibull distribution function to be

$$R(t) = \exp[-Kt^{m+1}/(m+1)]. \qquad (3)$$

Parameters K and m, which govern the distribution, can be determined by analysis of the actual data. When the parameters are known, the mean return period and its standard deviation are readily calculated. The mean return period and its standard deviation for the Nankai–Tokai zone shown in Table I are thus estimated to be 109 yr and 33 yr, respectively.

With $F(t)$, which is called the cumulative probability, evaluated with the aid of the parameters thus obtained, we can evaluate a conditional probability $FS(t)$, which, according to the terminology of quality control engineering, is called the hazard rate. On the condition that no

earthquake occurs in the time range between 0 and t, the probability of having an earthquake between t and $t + s$ is defined as $FS(t)$, which can be written as

$$FS(t) = [F(t + s) - F(t)]/[1 - F(t)]. \qquad (4)$$

Changes in $F(t)$ and $FS(t)$ as time goes on are calculated, as shown in Fig. 2, in which $s = 10$ yr is assumed. Time origin $t = 0$ is taken at the year 1854, when the last Tokai earthquake occurred. We can observe from Fig. 2 that $F(t)$ and $FS(t)$ steadily increase, reaching fairly high values such as

$$F(t) = 0.868, \qquad FS(t) = 0.435 \qquad (5)$$

at $t = 146$ yr, or in the year 2000.

Although we do not know exactly how to appraise the probabilities thus obtained, one of the writers (T.R.) believes that these values are so high that we should anticipate having a great Tokai earthquake within 10–20.

In addition to the analysis of plate-boundary earthquakes mentioned above, similar work can be performed for inland active faults that have caused large earthquakes in the past. In Japan, a great many active faults have been identified (Fig. 3) on the basis of geomorphological studies. These faults are classified according to their slip rate, and so it is clear that many large earthquakes are associated with first-class faults. The accumulation rate of crustal strain is also high over an active fault.

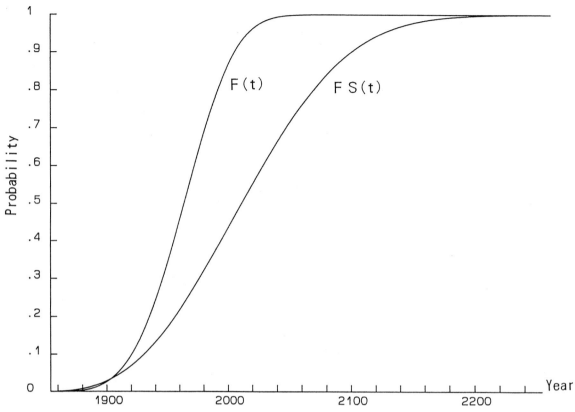

FIGURE 2 Changes in the cumulative probability $F(t)$ and the hazard rate $FS(t)$ for a great earthquake in the Tokai area as evaluated from the recurrence tendency.

It is now standard practice to dig into faults and look for evidence of earthquakes in the past, a procedure called trenching. It is believed that the approximate recurrence interval of an earthquake can be inferred from such studies as long as the dating is accurate. For instance, the recurrence interval is estimated to be ~1000 yr for the Tanna fault, ~100 km southwest of Tokyo, the site of an earthquake of magnitude 7.3 in 1930. Likewise, the recurrence interval is estimated to be ~150 yr for portions of the San Andreas fault in California.

In a manner similar to that used to analyze the subduction zone mentioned in the previous paragraphs, researchers can evaluate to some extent the probability that an active fault, of which the return period is somehow estimated, will generate the next earthquake. For instance, the probability of the Tanna fault's moving in the 30 yr since the mid-1990s has been estimated to be nearly 0%. Conversely, a U.S. Geological Survey working group estimates that the probability that a damaging earthquake of $M \geq 7$ will hit San Francisco between 1990 and 2020 is 67%. Their conclusion was obtained as the synthetic probability of four faults in the San Francisco Bay area.

The point that long- and medium-term probabilities of earthquake occurrence have been evaluated in specific cases should be appreciated as one of the achievements of the earthquake prediction program in recent years. Such evaluation is also possible from monitoring crustal strain, as is discussed next.

B. Crustal Strain

Japanese and other programs on earthquake prediction place much emphasis on monitoring crustal deformation in order to achieve long-term earthquake prediction. Figure 4 shows the changes in distance between the first-order triangulation stations in the South Kanto area to the southwest of Tokyo during 1925–1991. It is amazing that the distance between Izu Oshima Island and the Izu Peninsula decreased by more than 100 cm during the 66-yr period, whereas the distances between the island and the Boso and Miura Peninsulas increased by scores of centimeters.

In general, the crustal movement shown in Fig. 4 indicates that one-third of the deformation that took place at the time of the 1923 Kanto earthquake of magnitude 7.9 has recovered. Such deformations of the earth's crust must have been caused by the northwestward motion of the ocean plate. Maximum shearing strain for a triangle formed by any combination of three triangular stations

FIGURE 3 Conspicuous active faults in Japan.

can be calculated from the survey results shown in Fig. 4. The mean strain rate is then estimated for the triangles covering Sagami Bay as 4.7×10^{-7} minus per year.

If we assume that accumulated crustal strain was released at the time of the great 1923 Kanto earthquake, which killed more than 140,000 citizens, and that crustal strain has been stored again since then, the amount of crustal strain accumulated by a certain epoch can readily be calculated by multiplying the above strain rate by the time span in years since 1923. On the other hand, the ultimate strain leading to crustal rupture is known through data collected during intermittent geodetic surveys. It is known that the ultimate strain is distributed around a mean value of $\sim 5 \times 10^{-5}$.

By comparing the accumulated strain with the distribution of ultimate strain, researchers can estimate the probability that the crust will break in a specified time period. The calculation is similar to that performed in Section I.A.

The hazard rate, or the probability that crustal break will occur within 10 yr from 1999, is estimated at 10% or so. This value is not so high as to cause major concern about a recurrence of the great 1923 earthquake at the moment. However, the probability that another earthquake will hit the Tokyo–Yokohama area by the end of the twenty-first century is appreciably high.

A similar probability estimate can be made on the basis of the crustal strain accumulated in the Tokai area. The increase in the probabilities resembles that shown in Fig. 2, which indicates that the occurrence of the suspected Tokai earthquake is fairly imminent.

Global positioning System (GPS) monitoring of crustal strain has become very popular in recent years. In Japan, more than 1000 fixed stations for GPS observation were distributed across the country at the end of the 1990s. Real-time observation data from these stations are telemetered to a center in the Geographical Survey Institute (GSI) at

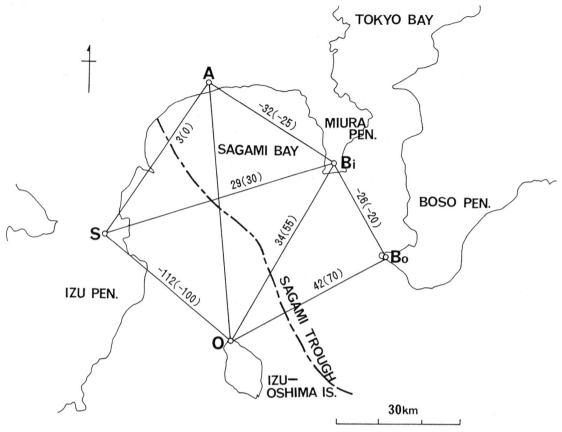

FIGURE 4 Changes in the lengths, in centimeters, between the neighboring triangulation stations in the Sagami Bay area for the 1925–1991 period. Similar changes for the 1925–1971 period are shown in parentheses.

Tsukuba. It is hoped that the relation between crustal strain and earthquake occurrence will be defined more clearly by this system in the near future.

A similar GPS network has been operating in California, where the seismicity is fairly high. Such a dense network of GPS observation does not seem to exist in other countries.

C. Seismic Gap

A large earthquake occurs when the earth's crust ruptures after storing strain energy to a certain limit. Thus, it is understandable that an area where a large earthquake is going to occur is free of conspicuous seismic activity for some time while strain energy steadily accumulates there without being released by small-scale earthquakes. Such a quiet area is called a seismic gap and is useful for identifying an area of seismic potential.

At the moment, remarkable gaps are observed at the Tokai area of Japan, a portion of the San Andreas fault to the northeast of Los Angeles, the Peru–Chile border, and other areas.

It is interesting to note that a number of large earthquakes at plate boundaries have been predicted on the basis of the concept of seismic gap. Most of them were foretold several years before the earthquake occurred, although no exact date and magnitude could be indicated.

For instance, since the 1939 Erzincan earthquake of magnitude 8.0 that occurred on the eastern portion of the North Anatolian fault in Turkey, a number of magnitude 7–8 earthquakes have occurred along the fault. The epicenters have moved westward in order. Considering the circumstances, it was strongly surmised that the next large earthquake would occur at the western-most portion of the fault. This hypothesis was confirmed by the occurrence of the 1999 Izmit, Turkey, earthquake of magnitude 7.4, which killed more than 15,000 people. In fact, in the early 1980s, the United Nations Educational, Scientific, and Cultural Organization (UNESCO) proposed to nominate the epicenter area of the 1999 earthquake as one of the candidate sites of an international earthquake prediction experiment. However, no actual work materialized as a result of financial difficulty and other reasons. Bilateral cooperative work between Turkey, and Japan, Germany, the United States, England, and others has been conducted over the area, though.

II. EARTHQUAKE PRECURSORS AND SHORT-TERM PREDICTION

A. Types of Earthquake Precursors

1. Land Deformation

One of the most powerful means of making short-term predictions is to detect anomalous land deformations. A fair number of examples of anomalous land uplift preceding a large earthquake have been reported.

One of the most typical examples of precursory land deformation was reported in association with the 1964 Niigata, Japan, earthquake of magnitude 7.5. As a result of repeated leveling surveys, it was discovered that anomalous uplift at a number of benchmarks had taken place 7–8 yr before the earthquake occurred. The distances between the epicenter and the benchmarks amounted to a few tens of kilometers.

In another instance, it was fortunate that a leveling survey was being conducted along a leveling route around Kakegawa City near Point Omaezaki, which is a promontory in central Japan (shown in the inset of Fig. 5) that projects into the Pacific Ocean, on the very day of the 1944 Tonankai earthquake of magnitude 7.9. The distance from the epicenter was ~150 km. Because of the unusually large closing error of the survey, it was concluded that a northward tilting had occurred during the survey period. A more

detailed analysis of the survey data suggested that the land around Kakegawa began to upheave a few days before the earthquake and that the speed of uplift was enormously accelerated a few hours before the earthquake.

Point Omaezaki had been subsiding for a long time before the earthquake. Such pre-earthquake subsidence and co-seismic upheaval were also reported at the time of the 1854 Ansei–Tokai earthquake of magnitude 8.4 that occurred off the point. Because it is thought that the up-and-down movement of Point Omaezaki may be an important factor in the short-term prediction of the coming Tokai earthquake, the leveling route between the two benchmarks A and B (shown in the inset of Fig. 5) is now being surveyed every 3 months. As can be seen in Fig. 5, the results clearly indicate that benchmark A near Point Omaezaki has been subsiding relative to benchmark B at a rate of ~5 mm/yr during the past 30 yr. Seasonal, zigzag changes are superimposed on the general trend, although the reason we observe such changes is not clear.

Close examination of Fig. 5 suggests that the rate of secular subsidence has slowed in the past 5 yr or so. If the tendency of subsidence would reverse, Japanese seismologists would assume that the next catastrophe was approaching. To catch such a change in crustal movement, the GSI set up 25 fixed stations of GPS observation along the leveling route, and real-time observation was initiated in 1999.

Tide-gauge observation also sometimes discloses local uplift or subsidence relative to sea level, although much care must be taken to eliminate noise of oceanographic and meteorological origin.

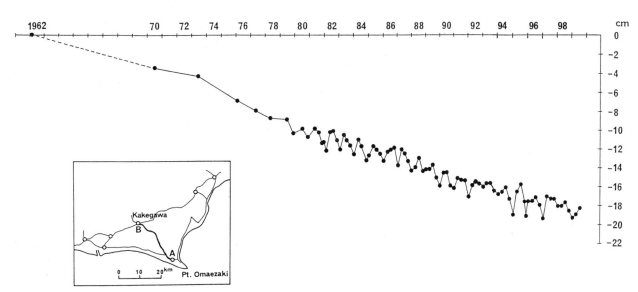

FIGURE 5 Changes in height at benchmark A relative to benchmark B.

Continuous monitoring of the length of a baseline ranging from several to tens of kilometers is another method believed to be useful for investigating land deformation. Electro-optical devices are used for this purpose, especially on the San Andreas fault in California and around and across Suruga Bay in the Tokai district of Japan.

Such observations will gradually be replaced by GPS observations.

2. Tilt and Strain

Until GPS observation was introduced, geodetic surveys were carried out intermittently. To observe the crustal movement between two survey epochs, researchers set up crustal-movement observatories equipped with tiltmeters and strainmeters over earthquake areas. There are various types of tiltmeters and strainmeters. Among the former are the horizontal pendulum type, the water-tube type, and the bubble type. Many precursory changes in tilt and strain have been reported, notably in China, Japan, the United States, and the former Soviet Union. Typical examples can be seen in textbooks of earthquake prediction.

It is currently standard practice to make array-type observations with many instruments, which are installed in deep boreholes. For instance, the Japan Meteorological Agency (JMA), which is responsible for the short-term or imminent earthquake prediction of the feared Tokai earthquake, set up 31 borehole volume strainmeters along several hundred kilometers of the Pacific coast in the hope of detecting changes in the crustal strain state. The signals taken by these strainmeters are telemetered to JMA headquarters in Tokyo on a real-time, on-line basis.

3. Seismic Activity

It is certain that seismic activity provides basic data for earthquake prediction, although so far there is no solid confirmation of the relation between preseismic activity and the occurrence of the main shock.

It has been reported that intense foreshock activity sometimes precedes a main shock. The 1975 Haicheng, China, earthquake of magnitude 7.3 was preceded by conspicuous foreshocks, and so local governments were able to officially issue an earthquake warning, which saved many inhabitants' lives. There are many instances, however, in which few or almost no foreshocks are observed before the main shock.

It is interesting to examine the precursor time of foreshocks, which is defined as the time span between the commencement of foreshock activity and the occurrence of the main shock, in relation to the magnitude of the main shock. With regard to 122 pieces of Japanese foreshock data collected by one of the writers (T.R.), no systematic

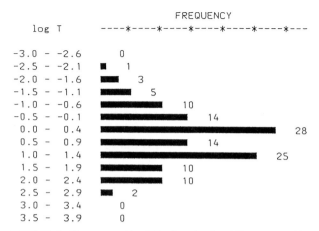

FIGURE 6 Histogram of log T for foreshocks. T is measured in units of days.

correlation between precursor time and main shock magnitude was seen.

The distribution of logarithmic precursor time T in units of days is shown as the histogram in Fig. 6. A Weibull distribution analysis of the foreshock data leads to a mean value of precursor time amounting to 2.0 days. Judging from the spread of the foreshock precursor times shown in Fig. 6, it is difficult to deterministically fix the date of main shock occurrence, although the probability of main shock occurrence time within a specified range may be evaluated, as shown later.

It is well known that a relation

$$\log N = a - bM$$

approximately holds for N, which is the number of earthquakes of magnitude M that occur in a certain area during a certain period, a and b being constants (b usually takes on a value around 1). It is often pointed out that b takes on a smaller value for foreshocks and that it decreases well before a large earthquake occurs. It has become clear that the larger the magnitude of the main shock, the longer the period of the b-value decrease.

A similar relation has occasionally been found between seismic quiescence and occurrence of a major earthquake. Seismic quiescence is different from seismic gap, which occurs over a very long period, in that seismic activity becomes weak in an area at a certain period and a large earthquake occurs in that area sometime later. The dependence of precursor time on main shock magnitude is similar to that for the b value. The successful prediction of the 1978 Oaxaca, Mexico, earthquake of magnitude 7.8 was based on the seismic quiescence, which is sometimes called the seismic gap of the second kind.

Other aspects of seismic activity such as microseismicity, change in source mechanism, and hypocentral migration are sometimes useful for predicting large earthquakes. However, these factors have not been exploited to the same extent as those discussed previously.

It was once thought that changes in the velocity of seismic waves passing through a region that becomes a future focal region of a large shock provide a powerful means of making earthquake predictions. Such phenomena were first detected in middle Asia and tested in many seismically active areas. Despite the fact that marked changes in seismic-wave velocity seem likely to precede an earthquake, as originally reported, subsequent high-precision research has proved that no conspicuous changes in seismic-wave velocity are usually observed. Therefore, this criterion does not seem to hold much promise.

4. Geoelectricity, Geomagnetism, and Electromagnetic Emission

It has been reported from time to time that an anomalous change in earth currents is observed preceding an earthquake. Observations are usually performed by measuring the electric potential between two electrodes buried in the ground and separated by a distance ranging from 100 to 1000 m. Unfortunately, however, no clear-cut relation between earth-current anomalies and earthquake occurrence has been established.

Changes in ground resistivity have also been extensively observed in relation to earthquake occurrence, notably in the Soviet Union and China. Marked decreases in resistivity have often been reported before large earthquakes, although the reason for such decreases is by no means clear.

In recent years, however, a powerful method for earthquake prediction is said to have been put foward by a Greek group. This method is called the VAN method (V, A, and N are the first letters of the last names of the three scientists involved). These scientists reported a number of successful cases of earthquake prediction. Although the VAN method basically differs little from traditional earth-current observation, it does involve selecting a measuring station, which records precursory signals of earth potential from earthquakes occurring in a particular area. The VAN researchers claimed that they succeeded in foretelling fairly many earthquakes of moderately large magnitude in Greece. International evaluation of the VAN method is now under way, as is mentioned later in Section IV.

Seismomagnetic study has a long history, and an anomalous change in the geomagnetic field preceding an earthquake has often been reported. It is believed, however, that real change of this kind has become measurable only recently because of the noise involved in observation. Since proton precession magnetometers were put to practical use around 1960, a number of highly reliable geomagnetic changes preceding earthquakes have been stored in the data file on earthquake precursors. A premonitory change usually amounts to several nanoteslas, that is, only about 1/10,000 of the geomagnetic field itself. It appears that such a change is caused by the piezomagnetic effect of magnetized rock.

Electromagnetic radiation of some sort is also said to be observed occasionally before an earthquake. Reports on such radiation have come mostly from Asia and Japan in recent years. Although the detailed nature of the radiation is yet unknown, it is hoped that electromagnetic radiation and precursory change in the ionosphere will become a powerful means of short-term earthquake prediction. Fairly clear-cut examples of the anomaly were reported in association with the 1995 Kobe earthquake of magnitude 7.2.

5. Underground Water and Hot Springs

There is little doubt that some underground water anomalies sometimes precede an earthquake. There have been many reports on changes in water level and chemical composition. This is also the case with hot springs. Even gushes of well water are often reported. The monitoring of underground water is believed to be a powerful means of making earthquake predictions, notably in China.

Geochemical observations are now widely made for the purpose of earthquake prediction as a result of the marked successes with this approach in middle Asia, the former Soviet Union, and China. Much effort has been made to monitor the radon concentration in well water in the United States, Japan, and Turkey, as well as the nations mentioned above. Monitoring of other geochemical elements such as helium and chlorine has also been attempted.

B. Nature of Earthquake Precursors

1. When do We Observe a Precursor?

As the amount of precursor data increases, it becomes possible to classify precursors on the basis of precursor time. It has gradually been recognized that, for certain types of precursors, the longer the precursor time, the larger the magnitude of the main shock. Figure 7 shows the $\log T - M$ plots for the 202 precursors studied by one of the writers (T.R.), including land deformation, tilt observed by a water-tube tiltmeter, strain observed by an extensometer, b value, change in seismic wave velocity, anomalous seismicity, seismic quiescence, geomagnetic field, and so forth. As before, T is measured in days. Each alphabetical letter in the figure indicates a discipline of precursor.

Despite considerable scattering of the data, we can see that the aforementioned tendency approximately holds good for the group for which $\log T$ is larger than 0, and the averaged $\log T - M$ relation is given by

$$\log T = -1.01 + 0.60M. \qquad (6)$$

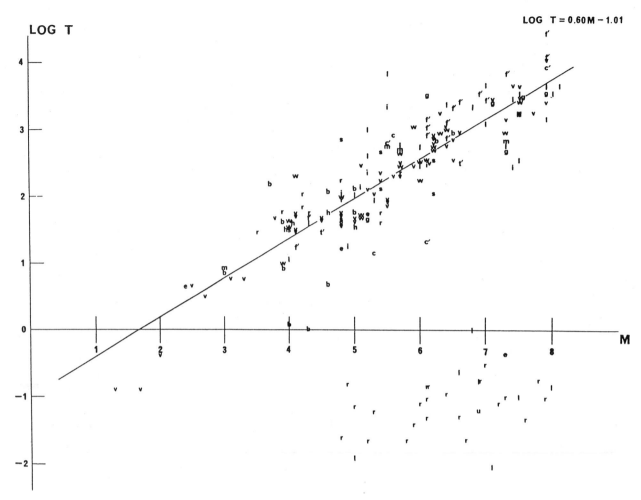

FIGURE 7 Logarithm of precursor time (*T*) in days versus main shock magnitude (*M*) for the selected precursor data.

In Fig. 7, we can also see that some of the precursors, such as electromagnetic radiation, resistivity change measured by a variometer, and the like, are scattered around log *T* = −1 irrespective of magnitude. Foreshocks and tilt and strain, which are not include in Fig. 7, also seem to belong to this class. As was shown in Fig. 6 for foreshocks, log *T* is distributed around its mean value. No dependence of *T* on *M* is seen in this case. Let us hereafter call the precursor that is governed by Eq. (6) the precursor of the first kind, and the precursor for which precursor time does not depend on main shock magnitude the precursor of the second kind.

2. Where do We Observe a Precursor?

To demonstrate the extent of the area over which we observe a precursor, we present plots of magnitude *M* versus logarithmic epicentral distance *D* (measured in kilometers) in Fig. 8. The plots represent data for 180 precursors studied in Japan. The types of precursors involved were land deformation, tilt and strain, microseismicity,

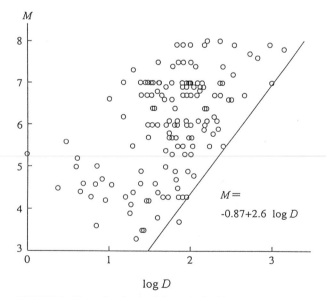

FIGURE 8 Plots of main shock magnitude *M* versus logarithmic epicentral distance *D* in kilometers for precursor data in Japan. The types of precursor phenomena involved are given in the text.

geomagnetic field, earth currents, resistivity, electromagnetic radiation, radon concetration, and underground water.

As is clear from Fig. 8, for earthquakes of large magnitude, there is a tendency for a precursor to be observed at a location more distant from the epicenter. The threshold line beyond which we observe no precursor is also shown in the figure. It is known empirically that a contour line on which a co-seismic step associated with an earthquake takes on a certain value can be drawn on the $M - \log D$ plane. It is interesting that the straight line in Fig. 8 roughly corresponds to the contour line for 10^{-9} in strain. It may be said that no precursor smaller than 10^{-9} can practically be observed when its magnitude is interpreted in terms of crustal strain.

Of the plots in Fig. 8, those for land deformation as detected by geodetic surveys and tide-gauge observation are scattered around a line corresponding to 10^{-6} in strain. This is also the case for changes in the geomagnetic field. The plots for microseismicity, earth current, long-distance resistivity, radon, and underground water are scattered around the 10^{-7} line. The sensitivity of tilt and strain varies, but it is close to 10^{-7}. The plots for resistivity by a high-sensitivity variometer are located farthest to the right on the graph, corresponding to 10^{-8} to 10^{-9}.

It is thus apparent how sensitive an observation of various types of precursors is for detecting a specific earthquake precursor. For instance, geodetic surveys detect a precursory land deformation of the order of 10^{-7} in strain or larger, so that the sensitivity of detection is not high. Once an anomalous land deformation is observed by geodetic means, however, its reliability must be high. This is why so much emphasis has been placed on geodetic work since the beginning of earthquake prediction research.

It has been known that the larger the earthquake magnitude, the wider the crustal deformation area associated with that earthquake. Precursors seem to be observed in an area wider than that over which land deformation directly connected with an earthquake takes place, or the epicentral area. It is expected that precursors for an earthquake of magnitude 8 would appear over an area having a diameter of 100 km or more.

3. What Causes Precursors?

No physical mechanism for earthquake precursors has been determined so far. One hypothesis is based on the concept of dilatancy. When crustal stress increases, many small cracks may be produced in rocks forming the earth's crust, the result of which is in an increase in volume. Such a state is called the dilatant state. When this happens, land deformation, such as anomalous uplift, takes place. The velocity of a seismic P wave being propagated in a di-

latancy region may be reduced. Because of high stress, some piezomagnetic effect gives rise to a change in the geomagnetic field. At the same time, precursory seismic activity may occur in the stressed portion of the crust. It is thus possible to interpret the occurrence of precursors, of which the precursor time can be correlated to the magnitude of the main shock by Eq. (4).

It appears to the writers, however, that such a hypothesis does not account for all the aspects of earthquake precursors. The study of the physical mechanisms of earthquake precursors is still incomplete. Much effort should be made to establish these mechanisms.

C. Macroscopic Precursors

Anomalous phenomena preceding an earthquake that can be sensed by humans without the aid of sophisticated instruments have often been reported. They include rumbling and detonation, unusual animal behavior, earthquake light and fireball, and so on. It is difficult to give credence to all these reports because they are usually vague and sometimes rather fantastic. Yet we cannot utterly rule out the possibility that there is some truth to these reports. These phenomena, called macroscopic precursors by Chinese seismologists, are now undergoing scientific analysis, although no detailed aspect of the study is presented here.

As a result of an intensive search for macroscopic precursors, more than 200 incidents of anomalous animal behavior were recorded for the 1923 Kanto earthquake of magnitude 7.9, the worst shock in Japan's history. For example, there were numerous reports that starting a month before the quake, many sardines swam upstream in river shoals flowing into Tokyo Bay. Because the reports were independent of one another, it is certain that this anomalous behavior actually took place before the catastrophe, although there is no guarantee that such an anomaly is firmly connected to earthquake occurrence. Fish may swim upstream for some reason even if no earthquake is going to occur.

Nonetheless, the number of animal reports tends to increase about 10 days before an earthquake, reaching a maximum about 1 day before the event. There appears to be another peak of report frequency a few hours before the main shock. Since this distribution of precursor time data is common for a number of large earthquakes, attention should be drawn to macroscopic data in any overall approach to earthquake prediction, although it would be difficult to predict an earthquake on the basis of these precursors alone.

Recent intensive investigations on macroscopic precursors indicate that the general characteristics of these precursors are more or less the same as those of geoscientific precursors. The macroscopic precursor seems to be essentially a short-term one resembling foreshock and other geoscientific precursors of the second kind. In addition, as with geoscientific precursors, the area over which anomalies are observed becomes wider as the main shock magnitude becomes larger.

III. PRACTICAL PREDICTION BASED ON PRECURSOR APPEARANCE

One of the most common approaches to practical prediction is to specify an area where a large earthquake is likely to occur on the basis of long-term prediction and to set up a highly dense network of observations of various precursory phenomena over the area in the hope of achieving a short-term or an imminent prediction. Such networks have been developed over the San Andreas fault in California, the North Anatolian fault in Turkey, and the Tokai–Kanto area in Japan, as well as faults in Garm and associated areas in the former Soviet Union and the Beijing–Tianjin–Tangshan–Zhangjiakou area in China.

To achieve a short-term and possibly imminent prediction, which can be converted to an earthquake warning, individuals responsible for monitoring earthquake data should send any anomalous signals recorded by observation instruments set up over the area to prediction headquarters on a real-time basis. Such a well-organized telemetering system is in operation in the Tokai area, Japan. Because of fear of the impending great earthquake, observational data of various types at more than 180 sites are telemetered by telephone lines to the JMA in Tokyo.

Someone monitors the data day and night at the JMA, and if an anomaly exceeding a certain prescribed limit is observed, the monitor is expected to notify the Prediction Council, consisting of six university professors.

It should be emphasized that an ocean-bottom seismograph system, which is located ~110 km SSW off Point Omaezaki (see Fig. 5), provides a powerful means of monitoring offshore seismicity. Seismograms taken by the system are sent to the JMA by means of ocean-bottom cables and telephone lines.

When a number of precursors are successively observed, the following approach is possible. Earthquake precursor data so far accumulated in Japan show that the larger the main shock magnitude, the larger the distance between the epicenter and an observation point where a precursor is observed. It is possible to establish empirically an approximate relationship between maximum detectable distance. D_{max}, and main shock magnitude, M, the relationship being different from discipline to discipline of precursor.

We may then draw a circle with a radius equal to D_{max} for each precursor discipline, centered at the respective observation point, on the condition that M takes on a certain value. The epicenter of a future earthquake should be located in the area that is common to all the circles. If M is too small, some of the circles will not overlap. If it is too large, the epicentral area will be too wide to be realistic. In this way, an approximate epicenter location and a rough value of the main shock magnitude can be assessed. Applying this procedure to the precursors of the Izu Oshima Kinkai, Japan, earthquake ($M = 7.0$, 1978) yields remarkable success, as can be seen in Fig. 9.

Probabilities of an earthquake's occurring in a specified time interval can be evaluated as a function of time when a precursor is observed. For such an evaluation, we rely either on the $\log T - M$ relationship, with a prescribed value of M, or on the frequency distribution of $\log T$, which is empirically obtained depending on precursor disciplines, T being the precursor time.

When a number of precursors are observed one by one, changes in the synthetic probability of earthquake occurrence can be estimated. In Fig. 10, such changes in the probability of having the main shock within 100 days from a specified epoch are shown for the Izu Oshima Kinkai earthquake, along with the precursor appearances, each indicated by a bar with a number. If one predicts that an earthquake will occur within 100 days of when the probability reaches 95%, the prediction turns out to be correct.

The on-line, real-time observation system had not been fully developed over the earthquake area of the Izu Oshima Kinkai earthquake by the time of earthquake occurrence. The evaluation of magnitude, epicenter, and occurrence period presented in the previous paragraphs therefore had to rely on the data collected after the earthquake. This is not a real earthquake prediction, although a much improved approach is certainly possible in the future because of completion of real-time observation, if another earthquake of fairly large magnitude occurs there.

How to cope with spurious signals in analyses of precursors is a serious problem. Although no established way of distinguishing a false signal (one that results from a source other than the coming earthquake) from a genuine one is known, it is possible to introduce a precursory signal reliability parameter in evaluating the occurrence probability. Such a parameter somehow be estimated from the experience.

The D_{max} method and probability evaluation of occurrence time can also be applied to the data set of macroscopic anomalies. Although no real-time analysis has so far been achieved, it has been proved that epicenter location, main shock magnitude, and occurrence probability of an earthquake may be surmised for a number of large earthquakes in Japan, such as the 1891 Nobi earthquake of magnitude 8.0, the 1923 Kanto earthquake of magnitude 7.9, the 1944 Tonankai earthquake of magnitude 7.9, and

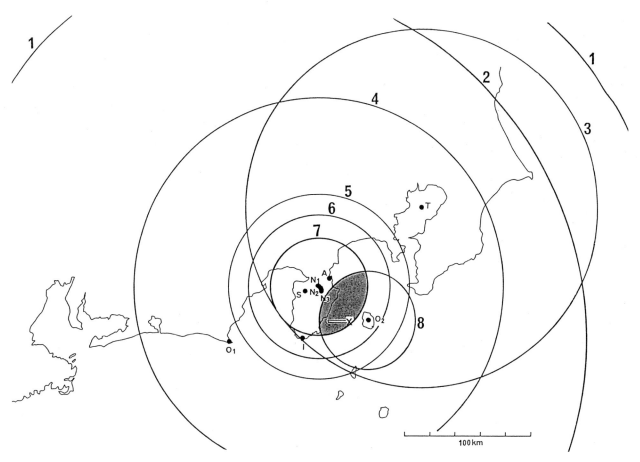

FIGURE 9 The shaded area represents the epicentral area of the Izu Oshima Kinkai earthquake as determined by the D_{max} method. $M = 7.0$ is assumed. The cross and the rectangle represent the actual epicenter and the horizontal projection of the source fault, respectively. A, I, N_1, N_2, N_3, O_1, O_2, S, and T are the observation points. Circles 1 to 8 are drawn with D_{max}'s for the following precursory signals at the respective points: 1, underground water level at S; 2, the same at O_1; 3, microearthquake at T; 4, earth currents at N_1; 5, anomalous land uplift centered at N_2; 6, geomagnetic field at N_1; 7, resistivity at N_1; 8, resistivity at O_2.

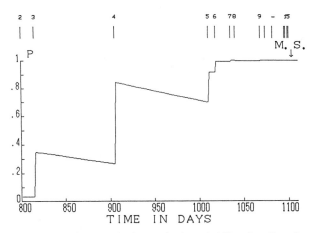

FIGURE 10 Changes in the synthetic probability of earthquake occurrence within 100 days' time from a specified epoch. The occurrence times of respective precursors are shown by vertical bars at the top, while the occurrence time of the main shock of the Izu Oshima Kinkai earthquake is indicated by an arrow. The abscissa denotes the time (in days) after the appearance of precursor 1.

the 1978 Izu Oshima Kinkai earthquake of magnitude 7.0. It has become clear through a test of some kind that the reliability parameter takes on a small value around 0.2 for macroscopic anomalies. It is hoped that real-time monitoring of macroscopic anomalies through the Internet or other advanced means of communication will become a highly useful method for pre-earthquake monitoring of such anomalies and evaluation of the time-dependent occurrence probability.

IV. PRESENT STATUS OF EARTHQUAKE PREDICTION RESEARCH AND EVALUATION OF PREDICTION TECHNIQUE

The first national program on earthquake prediction was launched in Japan in 1965. This program has affected many other countries since then. As a result, earthquake prediction research is now carried out in numerous countries in the world. Observation, surveying, experimentation, and research have been conducted in many places with the aim of realizing people's dream of earthquake prediction. These works are multidisciplinary, including not only seismology, but geology, geodesy, electromagnetism, geochemistry, and so forth, as can be seen in the earlier part of this article. Despite such work, however, earthquake prediction technology is not established but still only but still only at a developmental stage. The optimism about prediction that was seen at an early stage of earthquake prediction research, some tens of years ago, seems to have disappeared, although there has been steady progress in earthquake-related sciences and technological developments useful for observation, surveying, experimentation, and computation. Earthquake prediction is a long-term challenge to nature.

A. Recent Research Trends

1. Characteristic Features of Earthquake Precursors

Almost all methods of predicting earthquakes are experiential and based on precursors to earthquakes. These precursors include not only simple precursors observed by one instrument, but also special patterns of seismic activities such as seismic gap. There are many reports on possible precursors in the world.

Precursors appear before some earthquakes. However, many earthquakes occur without a precursor. In addition, precursor that appeared before earthquakes in the past does not always accompany later earthquakes. This means that there is a phenomenon that comes from other sources that is apparently similar to the precursor. Therefore, of course, there is no precursor that always appears before an earthquake. Detectable precursors are essential in predicting earthquakes as real problems, even if there are reliable theoretical, experimental, or empirical bases for earthquakes prediction. The practical approach to prediction is not deterministic but probabilistic owing to the nature of precursors.

Figure 11 illustrates a large scattering of $M - \log T$ plots for short-term precursors that were studied by one of the writers (K.H.). As before, M is the magnitude of main shock and T is the precursor time in days. The short-term precursors include foreshock sequences, continuous measurements of crustal deformation, earth current, resistivity, radon anomalies, groundwater level, and so forth. Table II clearly shows the characteristic feature that large earthquakes tend to be accompanied by more precursors than the number accompanying small earthquakes. Since the prediction of large earthquakes is more important, this

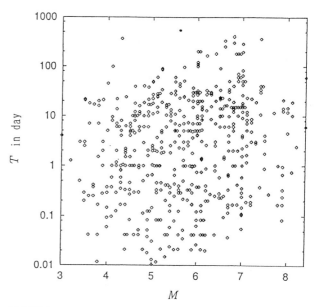

FIGURE 11 $M - \log T$ plots for all the short-term precursors that were observed in and around Japan from 1971–1993. M is the magnitude of main shock. T is the precursor time. [From Hamada, K. (1995). "Precursory phenomena to earthquakes occurring in and around Japan," *Report of the Coordinating Committee for Earthquake Prediction* **53**, 682–693. (in Japanese)]

is a favorable nature of the precursor. Table III shows disciplines and the number of precursors that were observed in and around Japan. The precursors related to seismic activities, more than half of which were foreshock sequences, account for approximately 70% of all precursors. To understand these statistics in the two tables, we should pay attention to the fact that these statistics are strongly dependent on the detection capability of observational systems. Therefore, these statistics cannot be easily generalized. Also, most precursors were reported after the earthquake. Therefore, there is the remaining problem of

TABLE II   Statistics of Earthquakes and Precursors in and around Japan (1971–1993)[a,b]

	Magnitude range		
	$5 \leqq M < 6$	$6 \leqq M < 7$	$7 \leqq M$
A. Total number of earthquakes	1100	115	15
B. Total number of precursors	145	118	81
C. Number of earthquakes with precursors	88	40	13
B/C	1.6	3.0	6.2
C/A	8%	35%	87%

[a] From Hamada, K. (1995). "Precursory phenomena to earthquakes occurring in and around Japan," *Report of the Coordinating Committee for Earthquake Prediction* **53**, 682–693. (in Japanese)

[b] Area: 29.3–45.7° N, 128.0–147.0° E. Depth: <90 km.

TABLE III Disciplines and Numbers of Precursors to Earthquakes Occurring in and around Japan up to 1993[a]

Discipline	Number	Percentage of Total
Related to geodesy	31	3.4
Observed by the instruments for continuous measurement of crustal movement	106	11.5%
Related to seismic activities	632	68.3
Related to geoelectromagnetism	75	8.1
Related to geochemistry and groundwater	72	7.8
Others	925	1.0

[a] From Hamada, K. (1995). "Precursory phenomena to earthquakes occurring in and around Japan," *Report of the Coordinating Committee for Earthquake Prediction* **53,** 682–693. (in Japanese)

how to recognize precursors before the occurrence of the earthquake.

2. Experimental Law of Precursor Appearance with Respect to Time

On the basis of the data given in Fig. 11, it is possible to evaluate the probability that a precursor will appear at a certain time. Considering the probability that T is between T and $T + dT$ for a precursor randomly selected from the population given in Fig. 11, the probability is denoted by $Pd(T)\,dT$, where $Pd(T)$ is called the probability density function. The probability that a precursor will appear between 0 and T is then obtained as

$$Pc(T) = \int_0^T Pd(T)\,dT, \qquad (7)$$

which is called the probability distribution function.

Figure 12 illustrates the logarithmic probability density function (Pd) for all the short-term precursors illustrated in Fig. 11. The plot is for the probability density averaged for all the magnitudes. The figure shows an obvious linear relationship between log Pd and log T in the approximate range $0.1 \leq T \leq 30$. The figure also shows the best-fit straight line for the five plots from the left. The straight line is approximated by

$$\log Pd = -0.901 - 0.850 \log T. \qquad (8)$$

The empirical formula represents a drastic increase in the precursor appearance probability with respect to the time approaching the main shock. Approximately, the probability density at 1 day before the main shock is 10 times larger than that at 10 days before the main shock. When the probability density function Pd is given, according to the definition, the probability distribution function Pc can be derived. Figure 13 shows an empirical curve of Pc thus derived, along with observed data. There is excel-

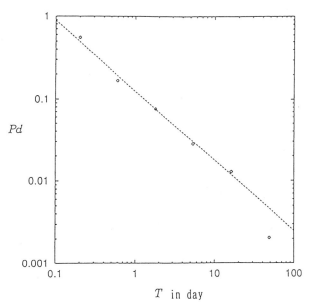

FIGURE 12 Probability density function for all the short-term precursors illustrated in Fig. 11. The dotted line is the regression line determined by the five data points from the left.

lent agreement between the calculated Pc and the observed ones for $0.1 \leq T \leq 30$ days.

3. Recent Research Trends

Recently, there has been increased interested and a number of achievements in electromagnetics associated with earthquakes, as briefly mentioned in Section II.A.4. Earthquake phenomena occur predominantly in the lithosphere of the earth. However, during the past few decades, there have been many convincing reports of the presence of electromagnetic disturbances and atmospheric and ionosphric perturbations associated with earthquakes. The lithosphere–atmosphere–ionosphere coupling seems to be a new, challenging scientific field. Some of the more recent observations suggest that electromagnetic effects also occur to an earthquake.

If there is a reasonable physical process that explains the observed results, electromagnetic methods might be used as a tool for prediction. However, scientific investigation must first address the following needs: (1) clear definition of anomalies supposedly related to earthquakes, (2) development of experiments to test causality between electromagnetic fields and earthquakes, (3) identification of physical mechanisms for the generation of electromagnetic fields that might explain observed data, (4) identification of the mechanisms for impending earthquakes, (5) identification of coupling mechanisms that could transform electromagnetic fields generated at the earthquake source into large–scale electromagnetic disturbances, and

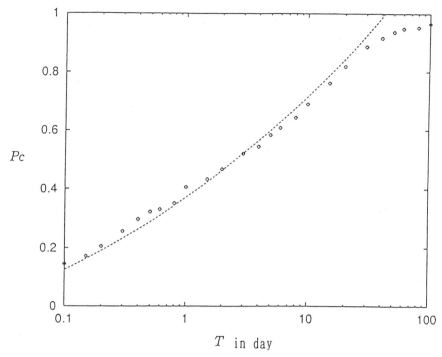

FIGURE 13 Probability distribution function for all the short-term precursors, which is calculated from the regression line in Fig. 12, and observed data plots. [From Hamada, K. (1995). "Precursory phenomena to earthquakes occurring in and around Japan," *Report of the Coordinating Committee for Earthquake Prediction* **53,** 682–693. (in Japanese)]

(6) rigorous statistical analysis of the significance of possible relations between earthquakes and suggested anomalies.

Besides the work in electromagnetics, there are two trends of recent earthquake prediction research. One is quantitative testing of hypotheses of the precursor. In many cases in the past, it was insufficient to select precursors quantitatively from the observed data, and quantitative definition of successful prediction was insufficient. For scientists to develop the prediction method, an appropriate evaluation for the method is essential, which can be accomplished only when data processing is quantitative and objective. The other trend is to put stress on deepening our understanding of the source process, such as from the earthquake nucleation to dynamic rupture, in other words, modeling of the comprehensive physical process of the earthquake source. There is no clear boundary between the fundamental research on the earthquake source and the prediction research of this trend.

4. Recent Social Environment Regarding Earthquake Prediction in Japan

Considering the huge amount of damage caused by the 1995 Southern Hyogo Prefecture earthquake (Kobe earthquake) in Japan, the Special Measures Law for Earthquake Disaster Prevention was enacted in July 1995 to enhance earthquake disaster prevention measures. This law reviewed the past system and attempted to unify the system of responsibility by establishing the Headquarters for Earthquake Research Promotion (the Headquarters) in the Prime Minister's Office under the leadership of the Ministry of Science and Technology Agency, abolishing the existing Headquarters for Earthquake Prediction Promotion. Stress has been shifted from the realization of earthquake prediction technology to earthquake hazard reduction, although earthquake prediction research is still included.

The following was recognized as the background of such political and organizational changes.

- Regarding the earthquake prediction that specifies place, time, and magnitude of an earthquake with appropriate accuracy, the realization of the prediction technology is difficult in general at the present research level.
- Short-term prediction of earthquakes is more difficult than long-term prediction and/or forecasting.
- It is important to promote earthquake prediction research because prediction technology has a great potential to reduce earthquake hazards.
- The prediction of the "Tokai earthquake" is probably, possible, considering that the 1944 Tonankai earthquake, which occured in an area adjacent to the

Tokai area, was associated with highly possible precursors.

B. Methods for Evaluating Prediction

There are not many documented examples in which earthquakes were actually predicted that can withstand scientific criticism, largely because methods for evaluating earthquake prediction are not established. One famous debate occurred at the meeting. "A Critical Review of Van," held jointly by the International Council of Scientific Union (ICSU) and the Royal Society at the Society's London premises in 1995. The proceedings of the meeting (Lighthill, 1996) not only reflect a wide diversity of views on the VAN method, which involves prediction based on seismic electric signals in Greece, but also demonstrate the present state of earthquake prediction evaluation.

1. Statistical Evaluation Methods

Even random predictions of earthquakes sometimes succeed by chance. For this reason, an appropriate method for evaluating a prediction is necessary. Evaluation methods are used not only to evaluate the result of the prediction that has already been made, but also to develop incomplete prediction methods. One of the writers (K.H.) would like to propose a statistical evaluation method for predictions, which can be commonly used for different types of predictions based on various kinds of precursors. This stastical method evaluates only the result of the prediction and seems to be the one and only impartial method at present.

Following is the procedure for the statistical evaluation method. First a space–time domain must be selected for evaluation and an earthquake catalog for that selected space–time domain (SSTD) prepared, in which all earthquake with magnitude equal to or greater than a specified value are included. Next, the definition of successfully predicted earthquakes must be described quantitatively and objectively. This is a very important description. For example, the predicted event must be located within the predicted space–time domain (PSTD) that is exactly determined by the adopted criterion. Also, the magnitude of the predicted event has to be located within the predicted range.

In general, two or more events might be selected as the predicted event. In such cases, additional rules to select the predicted event are necessary. These conditions for successful prediction regarding space, time, and magnitude are not easily determined in fact. However, they have to be set for statistical evaluation. When prediction information is compared with the earthquake catalog, on the basis of these conditions for successful prediction, it will become clear whether the issued predictions are successful or false. Judgment should be fair during the comparison. Then, a

success rate of the prediction, a success rate by chance, a probability gain, and a rate of predicted events are derived. These rates, which are fundamental parameters for statistical evaluation, are described next.

a. Success rate of the prediction. The success rate of the prediction is defined here as the following: (the number of successful predictions)/(the number of predictions issued). In general, if the PSTD and the magnitude range are small, the success rate of the prediction might be low; however, it might be easy to consider countermeasures. If the PSTD and the magnitude range are large, the success rate of the prediction might be high; however, it might be difficult to consider countermeasures.

b. Success rate by chance. The success rate by chance can be derived by assuming a random model of seismic activity. The random model has the same SSTD that is selected for evaluation at the beginning and has the same number of earthquakes with the same magnitudes that are given in the earthquake catalog for the SSTD. The point of the random model is random occurrence of these events with respect to space and time. The success rate by chance is the probability that the target earthquake whose magnitude is within the predicted range is included within the PSTD based on the random model.

c. Probability gain. The probability gain is defined here as the following: (the success rate of the prediction)/(the success rate by chance). The probability gain shows the statistical significance of the prediction. The success rate of the prediction without any scientific bases does not differ from that of the prediction by chance. We can increase the success rate of the prediction by using a larger PSTD and/or a larger predicted magnitude range. However, the probability gain cannot be increased by the same way because the success rate by chance would also increase by the same way. The benefit of the prediction is expressed by the probability gain, not by the apparent success rate.

d. Rate of predicted events. The rate of predicted events is defined here as the following: (the number of predicted events)/(the number of all target events). This rate helps determine the usefulness of the prediction. Total usefulness of the prediction is shown by the combination of the rate of predicted events and the success rate of the prediction. These two rates are probably dependent on each other in many cases. For example, if the prediction is issued only when many possible precursors are observed, the success rate might be high; however the rate of predicted events might be low. If the prediction is issued whenever possible precursors are observed, the success rate might be low; however the rate of predicted events might be high.

2. Statistical Method to Check the Statistical Significance of a Prediction

Assumptions:

- Target earthquakes occur i times in an area A in a j-day period.
- The prediction is always issued for all of area A in that period.
- The predicted time interval is the k-day length from the issuance of the prediction. That is, if the target event occurs in that area within k days after the prediction, the prediction is successful; otherwise, it is false.

On the basis of a random model of earthquake occurrence, the success rate by chance for a single trial of the prediction is estimated as

$$1 - \exp(-ik/j). \tag{9}$$

Assuming that an event E occurs with a probability p in a single trial, then the probability p_{sum} that the event E will occur r times or more in n trials is derived as follows:

$$p_{sum} = \sum_{r=r}^{n} Pr; \tag{10}$$

$$p_r = C_{nr} p^r q^{n-r}, \qquad q = 1 - p, \qquad C_{nr}: \text{combination}.$$

The probability that the successful prediction occurs r times or more by chance in n trials can be calculated by using Eq. (10), where p is replaced with Eq. (9). When the calculated result p_{sum} is less than 0.05 or 0.01, with a confidence level larger than 95% or 99%, it is rejected that r successful predictions out of n predictions can be explained by chance.

SEE ALSO THE FOLLOWING ARTICLES

Dams, Dikes, and Levees • Earthquake Mechanisms and Plate Tectonics • Electromagnetics • Global Seismic Hazards • Seismology, Engineering • Seismology, Observational • Seismology, Theoretical

BIBLIOGRAPHY

Hamada, K. (1995). "Precursory phenomena to earthquakes occurring in and around Japan," *Report of the Coordinating Committee for Earthquake Prediction* **53,** 682–693. (in Japanese)

Lighthill, J., ed. (1996). "A Critical Review of VAN Earthquake Prediction from Seismic Electrical Signals," World Scientific, Singapore.

Rikitake, T. (1976). "Earthquake Prediction," Elsevier, Amsterdam.

Rikitake, T. (1982). "Earthquake Forecasting and Warning," Center for Academic Publications Japan, Tokyo/Reidel, Dordrecht.

Rikitake, T. (1999). "Probability of a great earthquake to recur in the Tokai district, Japan: Reevaluation based on newly-developed paleoseismology, plate tectonics, tsunami study, micro-seismicity and geodetic measurements," *Earth Planets Space* **51,** 147–157.

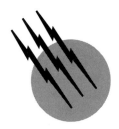

Earth Sciences, History of

Naomi Oreskes

University of California, San Diego

I. Earth History and Geological Time
II. Extinction, Evolution, and the Age of the Earth
III. Continental Drift and Plate Tectonics
IV. Oceanography, Meteorology, Seismology, and Planetary Sciences
V. Historiographic Reflections

GLOSSARY

Asthenosphere A region of the earth's upper mantle at a depth of approximately 75–150 km, characterized by low mechanical strength, attenuation of seismic shear waves, and partial melting. The term was coined in the 1910s by geologist Joseph Barrell to describe the zone in which isostatic adjustment occurs and basaltic magmas are generated. With the development of plate tectonics in the 1960s, the asthenosphere is now understood as the plastic zone over which the rigid plates move.

Bathymetry Measurement of ocean depths.

Benioff zones Zones of intermediate and deep-focus earthquakes, dipping 30°–45° from the ocean toward the continents, described by seismologist Hugo Benioff in the 1950s. While Benioff interpreted them as fault planes along continental margins, these zones are now understood to mark the locations where oceanic crust is subducted beneath continental margins at convergent plate margins.

Fennoscandian rebound In the early nineteenth century, farmers and fishermen observed that the shorelines and coastal islands of Scandinavia were rising. In the early twentieth century, this uplift was recognized as a regional phenomena encompassing Norway, Sweden, and Finland, resulting from landscape adjustment to post glacial conditions. During the Pleistocene period, the land had sunk under the weight of glacial ice. When the ice melted, the land began to rebound in a gradual process, still continuing today.

Geognosy A term introduced in the eighteenth century by German mineralogist Abraham Gotlob Werner to denote factual knowledge of rocks, minerals, and their spatial relations, without reference to theoretical interpretation.

Geosynclines Regions bordering the continental margins where thick sequences of shallow-water sedimentary strata accumulated, and subsequently were compressed and folded into mountain belts. The concept was developed in the nineteenth century by geologists James Dana and James Hall, Jr., to account for the features of the American Appalachians, and in the United States it became the generally accepted explanation for the origins of mountains prior to plate tectonics. However, it was never widely accepted elsewhere.

Isostasy The theory that the earth's crust floats in hydrostatic equilibrium within a denser, fluid or plastic substrate.

Orogenesis The theory or study of mountains and their origins.

Uniformitarianism The principle, articulated by nineteenth century British geologist Sir Charles Lyell, that geological history can be explained by reference to presently observable processes. In his now-classic work, *Principles of Geology* (1830), Lyell argued that past geological processes were the same in rate, intensity, and kind as those currently in operation, and earth history was not progressive (*substantive uniformitarianism*). Many geologists were unconvinced by Lyell's arguments for constant rates and intensity, but enthusiastically adopted uniformitarianism as a methodological program of interpreting the geological record by analogy with modern environments and events (*methodological uniformitarianism*). In the twentieth century, uniformitarianism was often said to be the fundamental guiding principle of modern geology.

ALTHOUGH THE TERM "geology" did not come into widespread use until the nineteenth century, interest in geological phenomena dates back at least to the ancient Greeks. Historian Mott T. Greene has suggested that the battle between Zeus and Kronos in Hesiod's *Theogony* is a description of the eruption of the volcano at Thera in the fifteenth century BC. No doubt the inhabitants of the ancient world were fearful of the volcanoes and earthquakes that intermittently threatened their existence. No doubt also they paid heed to the earth materials used for the construction of their cities, art, pottery, and weaponry.

Explicit speculations on the causes of earthquakes, volcanoes, lightning, and floods are found in the writings of Ionian philosophers such as Thales, Anaximander, and Xenophanes (c. sixth century BC). Earthquakes and floods feature in Plato's *Timaeus* in the account of the disappearance of Atlantis; in *Meteorologica* Aristotle (384–322 BC) discusses the processes responsible for changing the earth's surface. In the Roman world, the natural historian Pliny the Elder was killed in the 79 AD eruption of Vesuvius that destroyed Pompeii. In the medieval period, interest in rocks and minerals is recorded in the lapidaries of European and Arabic scholars.

Interest in geological matters flourished in the late fifteenth and early sixteenth centuries, when the growing commercial and intellectual activity of Renaissance Europe stimulated demand for minerals and building materials and curiosity about natural objects encountered through travel and trade. Among Leonardo da Vinci's (1452–1519) multitude of interests was a fascination with earth materials. As an architect and engineer supervising excavations, he recognized the similarity of fossils with living forms and interpreted rock strata as evidence of a progressive earth history in which terrestrial regions were previously inundated. Da Vinci also advanced one of the earliest recorded estimates of the duration of a geological process: 200,000 years for the Po River to lay down its alluvial plain.

Leonardo's interest in strata exposed in building sites was paralleled by Georgius Agricola's (1494–1555) interest in the rocks of the mining districts of Saxony. Best known for *De Re Metallica*, his influential treatise on ore deposits, mining techniques, and mineral processing, Agricola also wrote extensively on physical geology, subsurface fluid flow, mineralogy, fossils, and the causes of geological features. He proposed that mineral veins were precipitated from circulating fluids, and suggested that sequences of geological events could be interpreted from the structural relations of rocks and minerals.

The realization that materials could be introduced into rocks after their formation advanced the study of ore deposits but confounded the study of fossils. Renaissance naturalists used the term *fossil* to refer to any object dug up from rocks, and there was considerable dispute over whether they were the remains of past life or had grown *in situ*. Much of the debate hinged on the degree of similarity between fossil and living forms. Conrad Gesner's (1516–1565) *On Fossil Objects* was the first text to include systematic woodcuts explicitly comparing fossils with living organisms. In the late sixteenth century, the advent of copper engraving enabled naturalists to produce illustrations of great clarity and detail. Together with the development of museum collections and private "cabinets" of specimens, this innovation facilitated communication among specialists, who increasingly acknowledged the similarity between some fossil and living forms. The organic origins of fossils remained disputed, however, in part for lack of an account of how the fossils got inside the rock.

I. EARTH HISTORY AND GEOLOGICAL TIME

A. From Stratigraphy to Earth History

The first scientific account of sedimentation and stratigraphy is credited to Niels Stensen, better known as Nicholas Steno (1638–1686). As physician to the Grand Duke of Tuscany, Ferdinand II, Steno dissected a shark that fisherman brought ashore in 1666 and was struck by the similarity between the shark's teeth and the "tongue-stones" of naturalists' discussions. He concluded that tongue-stones were petrified teeth, preserved in soft mud, later hardened into rock. From this insight he developed the law of superposition: that stratified rocks are precipitated in horizontal layers, oldest at the bottom; tilted rocks have been disrupted by later events. Therefore, a sequence of historical events could be inferred from the structural relations of rocks. In England, Steno's work was furthered by Robert Hooke (1635–1703). In one of the first uses of microscopy in paleontology, *Micrographia* (1665), Hooke had compared the fine structure of fossil and living wood as demonstration of the organic origins of fossils. Drawing on Steno, he argued persuasively for the preservation of organic remains by sedimentary processes. But how long did such processes take?

Traditionally, scholars had assumed that earth history and human history were coextensive, and texts were reliable sources of evidence about natural historical events such as the Biblical flood. Most saw little reason to doubt the work of Archbishop James Ussher, who placed the origins of the world at 4004 BC on the basis of textual evidence, or Thomas Burnet (1635–1715) whose *Sacred Theory of the Earth* (1680) integrated natural and supernatural explanation to account for the physical features of the earth's surface and the literary record provided by scripture. This view was challenged by George Buffon (1707–1788), who argued for human history as only the last of seven epochs preserved in the stratigraphic record. Buffon argued that the study of human history had to be decoupled from earth history, and rocks rather than texts were the relevant source of information about the latter. Inspired by Leibniz, he experimented with cooling globes to obtain an estimate of the age of the earth of several tens of thousands of years.

The eighteenth century scholar who most advanced naturalistic study of earth history was the German mineralogist Abraham Gottlob Werner (1749–1817). Hailing from a family associated with ironworks, Werner was educated at the Bergakademie Freiberg, where he then built his career. Students came from across Europe to hear his famed lectures (including later luminaries Leopold von Buch, Alexander von Humboldt, and Friedrich Mohs), and Freiberg became the most celebrated school of geology in the world. In his systematic mineralogy, Werner laid the foundations of the system of identifying minerals by their external characteristics still used by field geologists today. He also introduced the concepts of geognosy—knowledge of rock masses, their mineral contents, and their spatial relations—and formations. Challenging taxonomic tradition, he argued that rocks should be grouped by their mode and time of formation, even if they looked different, thereby providing a basis for regional correlation.

Werner divided the rock record into five basic formations: primitive, transition, flötz, alluvial, and volcanic. At the start of geological time, perhaps one million years ago, the earth was entirely fluid. The first three units formed by progressive precipitation, with the least soluble materials forming the primitive rocks, the most soluble forming the flötz (which included such obviously soluble materials as salt and gypsum). The fossil content became more abundant and complex with time, suggesting that plant and animal life began to flourish as the waters gradually receded. Structural dislocations were attributed to periods of stormy conditions, and lithological variations were attributed to spatial and temporal heterogeneity in the ocean from which the materials precipitated. Above the three aqueous formations were the alluvial materials, formed by erosion and redeposition of the earlier sequences, and the volcanic rocks, pumice and lava, formed from burning coal beds in the flötz. Not all rocks were given a chemical interpretation, but all were given superficial origins.

Neptunism was subsequently discredited in the debate over the origins of basalt, leading later writers to caricature Werner's theory and dismiss his contributions. This is a mistake. By enabling correlation between distant locales, Werner's formation concept made it possible to talk about earth history in a global manner. Formations became the organizing principle of stratigraphy, inspiring scores of geologists in the early nineteenth century to take up systematic field mapping. One example is William Maclure, who produced the first geological map of the United States in 1809.

The proof of the igneous origins of basalt is properly credited to the French geologist, Nicolas Desmarest (1725–1815), who, in explaining the basalts of Auvergne, articulated the principle of uniformity generally associated with his Scottish contemporary, James Hutton (1726–1797). In fact, many French and German geologists in the late eighteenth century interpreted geological patterns by reference to observable processes of sedimentation, erosion, and volcanism; by late century the habit of examining modern environments for interpretive analogues was established. Hutton, however, elevated this practice into a unifying methodological precept, linked to a steady-state theory of earth history.

A polymathic founding member of the Royal Society of Edinburgh and avid Newtonian, Hutton believed that rocks recorded the "natural history" of the earth: as monuments recorded human history and their functions could be understood by comparison with contemporary edifices, so ancient rocks could be understood by comparison with processes presently operating on the earth's surface. Sedimentary rocks formed by erosion and redeposition of terrestrial materials, but this implied that an earlier generation of rocks had been exposed at the surface. How? Hutton's answer, presented in his *Theory of the Earth* (1795), was a geological cycle of deposition, burial, heating, melting, expansion, uplift, erosion, and deposition again. But erosion and deposition were scarcely detectable; their rates of operation were nearly vanishingly slow. So Hutton concluded that geological time must be "indefinite," prompting his famous description of earth history as bearing "no vestige of a beginning, no prospect of end."

Hutton and Werner have commonly been cast as antagonists, in part because their theories diverged on the question of whether earth history was progressive or steady state. But there was an important commonality: a shared emphasis on the primacy of fieldwork as the methodological tool to unravel earth history through the structural relations of rocks. In the nineteenth century geological field work came to prominence, culminating in the establishment of the geological time scale.

B. The Geological Time Scale

In 1835, the British Geological Survey was founded under Henry Thomas De la Beche (1796–1855), reflecting an increased governmental interest in the materials fueling the industrial revolution: coal, limestone, iron, tin, copper, and lead. The Survey was preceded and complemented by the work of private land surveyors, such as William Smith (1769–1839), who pioneered the use of fossil assemblages to distinguish lithologically similar units, and produced the first geological map of England in 1815. Complementing economic motivation, academic natural theologians such as William Buckland (1784–1856) and Adam Sedgwick (1785–1873) pursued field geology as a means to know God through his works, while gentlemen-scientists such as Roderick Murchison (1792–1871) adopted natural history an appropriate vocation for men of independent means. The task of geological reconstruction was facilitated by the British Isles' remarkably complete and largely undeformed sequences. Field mapping flourished; by midcentury geologists had reconstructed the sequence of rock units across Great Britain, Europe, and portions of North America, and defined the geological time scale in use today.

Geological mapping established that earth history was marked by distinctive rocks with characteristic fossil assemblages. However, in any one place only fragments of the whole record were found, and gaps were commonly associated with structural dislocations where rocks were tilted or folded like putty. It seemed logical, even obvious, to interpret these gaps and dislocations as the result of cataclysmic events such as the Lisbon earthquake of 1756, the Etna eruption of 1669, or the Biblical flood. Challenging this view, Charles Lyell (1797–1875) revived Hutton's interpretive reliance on observable processes and promoted it as the basis for geological explanation. In *Principles of Geology* (1830), he argued the view later dubbed uniformitarianism: that the accumulation of incremental change, operating in the past at the same rate and intensity as in the present, was sufficient to explain the transformations of the geological record. It was not merely that the laws of nature were temporally and spatially uniform—any Newtonian would have accepted that—but also that the processes themselves were uniform. Study of modern analogues was therefore both necessary and sufficient for understanding past geological process. While the details of Lyell's position were hotly argued by his contemporaries, particularly the assertion of constant rates of change, the uniformitarian framework was widely accepted. Many would later call it the central methodological principle of geology.

The uniformitarian reading of geological time was far greater than any literal or even semiliteral interpretation of the Old Testament might allow, thus contributing to a more flexible interpretation of religious doctrine. Geological knowledge also facilitated the exploitation of earth materials, warranting the expansion and institutionalization of the science. By the late nineteenth century, academic departments and government surveys had been established across Europe and the United States.

II. EXTINCTION, EVOLUTION, AND THE AGE OF THE EARTH

A. The Fact of Extinction

With the establishment of geological time, historian Paolo Rossi notes, fossils "become the clocks that can measure the long time periods of natural history." The newly established geological time periods were partly defined by their distinctive fossil assemblages. Most notably, the ends of geological eras—Paleozoic, Mesozoic, Cenozoic—as well as the ends of the periods—Cambrian, Permian, Cretaceous, etc.—were marked by the disappearance of one or more major species.

Extinction had often been addressed in the context of divine providence: Why would God destroy his own work? Why would he create imperfect forms? Initially, many paleontologists thought that seemingly missing creatures might still be lurking in the depths of oceans and the interiors of uncharted terrains. However, as increasing numbers of fossils were discovered and their relations to living forms scrutinized, this explanation became increasingly untenable. In his 1796 paper "*On the Species of Living and Fossil Elephants,*" the French paleontologist and comparative anatomist George Cuvier (1769–1832) demonstrated that the fossil "elephants" (mammoths) of Siberia were distinct from living forms. The fact of the extinction of mammoths and many other species, from belemnites to ammonites, was soon accepted.

This led to a new scientific question: Why did species go extinct? And where did new species come from? It was increasingly clear that new species appeared at intervals in the fossil record as surely as some old ones had disappeared, and these changes provided a metric by which to measure geological time. Cuvier recognized three explanatory options: destruction, modification, or migration. Seeing no direct evidence of modification in the fossil record, he opted for destruction followed by migration: rapid marine transgressions had destroyed the faunal assemblages of earlier periods, after which species migrated in from elsewhere. Drawing on the political events of his own time, he called these periods revolutions. These natural revolutions were not necessarily sudden or violent, but they were major transformations in which old species vanished, new ones appeared.

B. From Revolution to Evolution

The question of modification of species was taken up by Jean-Baptiste Lamarck (1744–1829) and later more successfully by Charles Darwin (1809–1882). A follower and later a friend of Lyell, Darwin was deeply influenced by his argument for the efficacy of small changes accumulated over geological time. This became the foundation for his theory of the origin of species by natural selection. By analogy with a presently observable process—artificial selection by breeders—Darwin conceptualized a mechanism of natural evolutionary change. Uniformitarianism provided theoretical justification for a gradualist interpretation of faunal change in the absence of direct evidence. Darwin famously attributed the "missing links" to gaps in the fossil record, and most geologists concurred. By the end of the nineteenth century, the fact of evolution (if not the mechanism of it) was widely accepted. This fact hinged in turn on a very old earth.

But how old was old? Most geologists spoke in qualitative terms. Werner had described the great duration of geological time "in contrast to which written history in only a point." Hutton had called geological time "indefinite"; others spoke of its "limitless stores." Uniformitarianism inspired attempts at quantification: Darwin used modern sedimentation rates to estimate the time for the deposition of the Weald (a unit of the lower Cretaceous) at 300 million years—implying an earth age of hundreds of billions. Irish geologist John Joly (1857–1933) used the salt concentration of modern rivers to calculate how long it would take to salinize the world's oceans. His result: 90 million. While the rigor of Joly's calculations appealed to some, many geologists intuitively felt it was too low. Most preferred billions to millions.

C. Millions or Billions?

Geologists' conviction in a prodigiously old earth was radically challenged in the late nineteenth century by British physicist William Thomson, later Lord Kelvin (1824–1907). Kelvin revived an idea earlier proposed by Buffon that the earth's age could be deduced on the principle of a cooling sphere. Drawing on the Kant–LaPlace nebular hypothesis, that the solar system had formed by condensation from a gaseous cloud, and applying Fourier's theory of heat transfer, Kelvin calculated the time required for the earth to cool to its present temperature at no more than 20–40 million years. If uniformitarianism implied otherwise, then uniformitarianism must be wrong, and geological practice based on a fallacy. Moreover, Darwin's theory of the origin of species by natural selection must also be wrong, resting as it did on the premise of eons of time.

This was a profound challenge. Some geologists tried to adjust themselves to Kelvin's numbers; some sought a compromise or defended the geological view. Darwin suggested that the rate of evolution might be faster than he had supposed. The debate was resolved in the early twentieth century on the geological side by the discovery of radiogenic heat, which mooted Kelvin's starting premises. (Today it is further mooted by the theory of cold accretion of the terrestrial planets.) By 1913, British geologist Arthur Holmes (1890–1965) had used U–Pb ratios to demonstrate that Precambrian rocks were at least 2 billion years old. In the 1930s, the development of the mass spectrometer by Alfred Nier (1911–1994) made it possible to differentiate lead isotopes, pushing calculations upward toward 3 billion. In 1953, American geochemist Clair Patterson (b. 1922) used iron–nickel meteorites to determine primeval lead ratios, producing the presently accepted earth age of 4.55 billion years.

III. CONTINENTAL DRIFT AND PLATE TECTONICS

A. The Origin of Mountains: Thermal Contraction

Geologists in the eighteenth and nineteenth century made great advances in documenting earth history and explaining the origins of sedimentary rocks; they were less successful in accounting for the origins of mountains, earthquakes, and igneous processes. In Europe, mountains had often been viewed as frightening and dangerous places, but they also contained many valuable mineral deposits. In the early nineteenth century, they attracted increasing artistic attention as sites of great natural beauty, cultural attention as potential sites of human conquest, and political attention as a barriers to expansion. They attracted scientific attention as well. How were they formed? What caused their spectacular structural features? What made the earth move?

Most nineteenth-century theories of orogenesis invoked terrestrial contraction as a causal force. In Europe, Austrian geologist Edward Suess (1831–1914) popularized the image of the earth as a drying apple. Like Kelvin, Suess built on the premise of secular cooling. Drawing on the earlier theories of Léonce Elie de Beaumont (1798–1874) and Henry De la Beche, he proposed that the earth contracted as it cooled, and mountains resulted from crustal wrinkling in response. Initially, the crust was continuous, but it broke apart as the earth's interior shrunk, and the collapsed portions formed the ocean basins. With further cooling, the remaining elevated portions (the continents) became unstable and collapsed to form the next generation of ocean floor; what had formerly been ocean now became dry land. This explained the presence of marine deposits on land (which had puzzled da Vinci), and the interleaving of marine and terrestrial materials in the stratigraphic record. Suess's theory also explained the widely known similarities of fossil assemblages in parts of India, Africa, and South America by attributing them to an early period when these continents were still contiguous. He called this ancient supercontinent Gonawanaland.

In North America, a different version of contraction theory was developed by James Dwight Dana (1813–1895), famous at the age of 24 for his *System of Mineralogy*, which comprehensively organized minerals according to their chemical affiliations. (First published in 1837, a version is still in print today.) Drawing on his understanding of the chemical properties of minerals, Dana suggested that the earth's continents had formed first, when minerals with relatively low fusion temperatures such as quartz and feldspar had solidified. Then the globe continued to cool and contract, until the high temperature minerals such as olivine and pyroxene finally solidified: on the moon, to form the lunar craters, on earth, to form the ocean basins. Continued contraction after solidification caused surface deformation. The greatest pressure was experienced at the boundaries between the oceanic and continental blocks, explaining the localization of mountains (particularly the Appalachians). Because continents and oceans were understood as permanent features of the globe, Dana's version of contraction came to be known as permanence theory.

In North America, permanence was linked to the theory of geosynclines, developed by Dana and James Hall (1811–1889), State Paleontologist of New York and the first President of the Geological Society of America (1889). Hall suggested that materials eroded off the continents accumulated in the adjacent marginal basins, causing the basin to subside. Subsidence allowed more sediments to accumulate, causing more subsidence, until finally the weight of the pile caused the sediments to be heated, lithified, and by some mechanism not elucidated, uplifted into mountains. Dana modified Hall's view by arguing that thick sedimentary piles were not the cause of subsidence but the result of it. Either way the theory provided a concise explanation of how thick sequences of shallow-water rocks could form.

B. Continental Drift as Alternative to Contraction Theory

In the early twentieth century, contraction theory was refuted by three independent lines of evidence. First, field mapping in the Swiss Alps and the North American Appalachians demonstrated hundreds of miles of shortening of strata, which would require impossibly huge amounts of terrestrial contraction to explain. Second, geodesists studying the problem of surface gravitational effects showed that the surface mass associated with mountains was counterbalanced by a mass deficit within or beneath them. Mountains were held aloft not by their internal strength, but by floating—a concept called isostasy. Continents and oceans were not interchangeable, because continents could not sink to form ocean basins. Third, physicists discovered radiogenic heat, which refuted the premise of secular cooling. With contraction no longer axiomatic, earth scientists were motivated to search for other driving forces of deformation. Many did; Alfred Wegener (1880–1930) is the most famous.

Primarily known as a meteorologist and author of a pioneering textbook on the thermodynamics of the atmosphere (1911), Wegener realized that paleoclimate change could be explained if continents had migrated across climate zones, and the changing configurations of continents and oceans periodically altered climate patterns. However, continental drift was more than a theory of paleoclimate change. It was an attempt at unification of disparate elements of earth science: on one hand, the paleontological evidence that the continents had once been connected; on the other, the geodetic evidence that they could not be connected in the way Suess had imagined. Wegener's answer was to reconnect the continents by moving them laterally.

Wegener's theory was widely discussed in the 1920s and early 1930s. It was also hotly rejected, particularly by Americans who labeled it bad science. The standard explanation for the rejection of Wegener's theory is its lack of a causal mechanism. But this explanation is false. There was a spirited and rigorous international debate over the possible mechanisms of continental migration. Much of it centered on the implications of isostasy: if continents floated in a denser substrate, then this substrate had to be plastic or fluid, and continents should be able to move through it. The Fennoscandian rebound—the progressive uplift of central Scandinavia since the melting of Pleistocene glacial ice—provided empirical evidence that they did, at least in the vertical direction and at least in the Pleistocene. But here the cause of motion was known: the weight of glacial ice and the pressure release upon its removal. What force would cause horizontal movement? And would the substrate respond comparably to horizontal as to vertical movement? Debate over the mechanisms of drift therefore concentrated on the long-term behavior of the substrate, and the forces that could cause continents to move laterally. Various proposals emphasized the earth's layered structure, which allowed for decoupling of the continental layer from the one beneath.

John Joly linked the problem of continental drift to the discoveries in radioactivity. He had demonstrated that pleochroic haloes in mica were caused by radiation damage from tiny inclusions of U- and Th-bearing minerals, proving that radioactive elements were ubiquitous in rocks. Then radiogenic heat was also ubiquitous, and as it built up it would melt the substrate. During these molten periods the continents could move under the influence of small forces that would otherwise be ineffectual.

Joly's theory responded to a geophysical complaint against a plastic substrate: that the propagation of seismic waves indicated a fully solid and rigid earth. More widely credited was the suggestion of Arthur Holmes that the substrate was partially molten or glassy. Underscoring arguments made by Wegener, Holmes emphasized that the substrate need not be liquid, only plastic. Furthermore, it might be rigid under high strain rates (during seismic events) yet be ductile under the low strain rates prevailing under most geological conditions. If it were plastic in response to long-term stress, then continents could move within it. Holmes's driving force was convection currents in the mantle. He argued that the midocean ridges were the sites of upwelling convection currents, where continents had split, and the ocean deeps (geosynclines) were the sites of downwelling currents, where continents were deformed as the substrate descended. Between the ridges and the trenches, continents were dragged along in conveyor-like fashion.

C. The Rejection of Continental Drift

Arthur Holmes's papers were widely read and cited; many thought he had solved the mechanism problem. However, opposition to drift was nonetheless for that, particularly in the United States, where reaction to Wegener's theory was harshly negative, even vitriolic. Evidently more was at stake than a matter of scientific fact.

Three factors contributed to the American animosity to continental drift. One, Americans were widely committed to the method of multiple working hypotheses, and Wegener's work was interpreted as violating it. For Americans, right scientific method was empirical, inductive, and involved weighing observational evidence in light of alternative explanatory possibilities. Good theory was also modest, holding close to the objects of study. Most closely associated with the University of Chicago geologist T. C. Chamberlin (1843–1928), who named it, the method of multiple working hypotheses reflected American ideals expressed since the eighteenth century linking good science to good government: Good science was antiauthoritarian, like democracy. Good science was pluralistic, like a free society. If good science provided an exemplar for good government, then bad science threatened it. To American eyes Wegener's work was bad science. It put the theory first, and then sought evidence for it. It settled too quickly on a single interpretive framework. It was too large, too unifying, too ambitious. Features that were seen as virtues of plate tectonics were attacked as flaws of continental drift.

Continental drift was also incompatible with the version of isostasy to which Americans subscribed. In the late nineteenth century, two accounts of isostatic compensation had been proposed: John Henry Pratt (1809–1871) attributed it to density variations, George Biddell Airy (1801–1892) attributed it to differences in crustal thickness. Until the early twentieth century, there had been no empirical confirmation of the concept beyond the original evidence that had inspired it, nor any means to differentiate the two explanations. Then American geodesists John Hayford (1868–1925) and William Bowie (1872–1940) used the Pratt model to demonstrate that isostatic compensation was a general feature of the crust. By making the assumption of a uniform depth of compensation, they were able to predict the surface effects of isostasy to a high degree of precision throughout the United States. At first, their work was hailed as proof of isostasy in general, but in time, it was viewed as confirmation of the Pratt model in particular. However, if continental drift were true, then the large compressive forces involved would squeeze the crust, generating thickness differentials. Continental drift seemed to refute Pratt isostasy, which had worked for Americans so well, and because it had worked, they had come to believe was true. Rather then reject Pratt isostasy, they rejected continental drift.

Finally, Americans rejected continental drift because of the legacy of uniformitarianism. By the early twentieth century, the methodological principle of using the present to interpret the past was deeply entrenched in the practice of historical geology. Many believed this to be the *only* way to interpret the past, and that uniformitarianism made geology a science, for without it what proof was there that God had not made the Earth in seven days, fossils and all? Historical geologists routinely used faunal assemblages to make inferences about climate zones, but on drift-based reconstructions, continents placed in the tropics did not necessarily have tropical faunas, because the reconfiguration of continents and oceans might change things altogether. Wegener's theory raised the specter that the present was not the key to the past—it was just a moment in earth history, no more or less important than any other. This was not an idea Americans were willing to accept.

In North America, the debate over continental drift was quelled by an ad hoc explanation of the faunal evidence. In 1933, geologists Charles Schuchert (1858–1942) and Bailey Willis (1857–1949) proposed that the continents had been intermittently connected by isthmian links, just as the isthmus of Panama connects North and South America and the Bering Land Bridge once connected North America to Asia. The isthmuses had been raised up by orogenic forces, subsided under the influence of isostasy. This explanation was patently ad hoc—there was no evidence of isthmian links other than the paleontological data they were designed to explain. Nevertheless, the idea was widely accepted, and a major line of evidence of continental drift undercut. In 1937, South African geologist Alexander du Toit (1878–1948) published *Our Wandering Continents*, a comprehensive synthesis of the geological evidence of continental drift, but it had little impact in North America.

D. Plate Tectonics

In the late 1930s, Dutch geodesist Felix Vening Meinesz (1887–1966) and American Harry Hess (1906–1969) applied the idea of convection currents to explain downwarpings of the oceanic crust associated with gravity anomalies in the Caribbean and the Dutch East Indies. However, this work was cut short by World War II. In the 1950s, continental drift was revived by British geophysicists working on rock magnetism as a means to investigate the earth's magnetic field, one group at Imperial College led by P. M. S. Blackett (1897–1974), and one at Cambridge (later at Newcastle) led by S. Keith Runcorn (1922–1995).

Both groups found evidence that rocks had had moved relative to the earth's magnetic poles, so either the continents or the poles had moved. Initially geophysicists were more receptive to the idea of polar wandering, but by the late 1950s comparative evidence from India and Australia pointed in the direction of differentially moving continents. Inspired by these results, Harry Hess revisited convection currents as driving force for continental motion, and proposed the hypothesis Robert Dietz (1914–1995) dubbed sea floor spreading. Hess suggested that mantle

convection drives the crust apart at midocean ridges and downward at ocean trenches, forcing the continental migrations in their wake. Hess interpreted the oceanic crust as a hydration rind on serpentinized mantle; Dietz modified this to generate oceanic crust by submarine basalt eruptions. Dietz's interpretation was later confirmed by direct examination of the sea floor.

Meanwhile American researchers, Richard Doell (b. 1923), Brent Dalrymple (b. 1937) and Allan Cox (1923–1987) were studying a different aspect of rock magnetism: the record of reversals in the earth's magnetic field. Detailed field studies of basaltic lava flows convinced them that reversals were not an artifact of cooling or laboratory procedures. In fact, one could construct a chronology of paleomagnetic reversals—a geomagnetic time scale. Magnetic reversals plus seafloor spreading added up to a testable hypothesis, proposed independently by Canadian Lawrence Morley and British geophysicists Frederick Vine (b. 1939) and Drummond Matthews (1931–1997): If the seafloor spreads while the earth's magnetic field reverses, then the basalts forming the ocean floor will record these events in the form of a series of parallel "stripes" of normal and reversely magnetized rocks. Since World War II, the United States Office of Naval Research had supported seafloor studies for military purposes, and large volumes of magnetic data had been collected. American and British scientists quickly set to work examining these data, and by 1966 the Vine and Matthews hypothesis had been confirmed.

At this point, many workers turned to the problem. Among the most important were J. Tuzo Wilson (1908–1993), who thought of another test of the theory. The midocean ridges were repeatedly offset by faults; the slip direction on these faults would be one direction if seafloor spreading were taking place, the opposite if it were not. Wilson called the latter transform faults, as they transformed one segment of a spreading ridge into another. Seismic data analyzed by Lynn Sykes (b. 1937) at the Lamont–Doherty Geological Observatory confirmed the existence of transform faults.

Sykes's Lamont coworkers, Walter Pitman (b. 1931) and James Heirtzler (b. 1925), used paleomagnetic data to refine the reversal time scale, and confirm that the patterns on either side of the mid-Atlantic ridge were symmetrical. In 1967–1968, these various lines of evidence were independently synthesized by Daniel P. McKenzie (b. 1942) and Robert L. Parker (b. 1942) working at the Scripps Institution of Oceanography, and by Jason Morgan at Princeton University. Both showed that existing data could be used to analyze crustal motions as rigid body rotations on a sphere. The result became known as plate tectonics, which by the early 1970s had become the unifying theory of the earth sciences.

IV. OCEANOGRAPHY, METEOROLOGY, SEISMOLOGY, AND PLANETARY SCIENCES

A. Historical Background

Oceanography, meteorology, seismology, and planetary science have only recently received sustained historical attention. One reason is itself historical: these are young sciences. The earth's surface permitted scientific study sooner than the other parts. Anyone could pick up a rock or a fossil and begin to develop a collection, but it required instruments, financing, and often national or international cooperation to study the oceans, atmosphere, and earth's interior. The distant planets were even less accessible.

Prior to the late nineteenth century, oceanographic and meteorological questions were commonly understood as issues of geography, and investigations were piggybacked on expeditions that were not primarily scientific. Seismology began to flourish around the same time in conjunction with earthquake studies and the development of precise seismographs. Early in the twentieth century, it grew substantially, first though its application to petroleum exploration, later more dramatically through its application to nuclear test ban verification. Planetary science similarly developed in a limited way in the early century with the construction of powerful reflecting telescopes, and then much more fully in conjunction with the U.S. and Soviet space programs.

B. Oceanography

Oceanography is the oldest of the ancillary earth sciences, with its links to navigation and exploration, colonialism and trade. In the seventeenth century, members of the British Royal Society compiled sea temperatures; Isaac Newton (1642–1727) and Edmund Halley (1656?–1743) theorized the astronomical cause of tides. In the eighteenth century, British sea captain James Cook (1728–1779) recorded tides and temperatures during his voyages.

Currents were obviously important to navigation, and most early workers thought they were driven by density differences—hence attention to surface temperature. In the early nineteenth century, scientists began to consider other factors. English surveyor James Rennell (1742–1830) suggested the role of winds in driving surface currents, while the German polymath, Alexander von Humboldt (1769–1859), promoted the idea that sinking of cold water in high latitudes could drive deep circulation. Many ship voyages therefore included some measurements of salinity, temperature, surface currents, and soundings. The founding of systematic bathymetry is generally credited to Matthew Fontaine Maury (1806–1873), who pioneered

this work at the U.S. Navy Hydrographic Office, but resigned from federal service during the American Civil War.

The voyage of H. M. S. *Challenger* (1872–1876) was the first explicitly oceanographic expedition. Lasting three and a half years and covering 110,000 km, the expedition took thousands of soundings, measurements of temperature and salinity, and biological and bottom sediment samples. The results (published in 50 volumes) provided the first systematic description of the oceans. The data demonstrated the existence of the abyssal plains at approximately 5 km depth, the constancy of the proportion of salts in seawater, and the variation of deep-water temperature. The detection of a systematic temperature difference between the eastern and western half of the Atlantic Ocean was first suggestion of the existence of a mid-Atlantic Ridge. Dredging the seafloor revealed the predominance of calcareous oozes and red clays, quite unlike terrestrial sediments, providing an early argument against the interchangeability of continents and oceans. Among the biological results, 3225 new species were discovered, including many at great depths, disproving the idea that ocean depths were devoid of life.

Toward the end of the century, interest in oceanography received impetus from Arctic explorers, particularly Scandinavians Fritjof Nansen (1861–1930) and Roald Amundsen (1872–1928?). In the *Fram* expedition (1893–1896), Nansen completed the first drift across the Arctic, discovering that the ice drift diverged 45° from prevailing wind directions. Suspecting the Coriolis force [articulated by French natural philosopher Gaspard Coriolis (1792–1843), but as yet unapplied to ocean dynamics], Nansen proposed the problem to a young Swedish oceanography student, Wagn Ekman (1874–1954), who demonstrated mathematically that the drift was indeed consistent with wind-driven circulation modified by the Coriolis effect—dubbed the "Ekman spiral." Amundsen repeated the Arctic drift on the *Maud* (1918–1925) with improved instruments permitting better measurement of wind and current directions, and confirmed Ekman's results.

Ekman's work became a cornerstone of the "dynamic oceanography" developed by Bjørn Helland-Hansen (1877–1957), Johan Sändström (1874–1947), and Harald Sverdrup (1888–1957). Using temperature and salinity data to determine a density field, one deduced a pressure field, and from this, ocean currents. However, mathematical treatment relied on accurate data; thus the development of dynamic oceanography also depended on improved instrumentation. Chief among these was the "Nansen bottle." In response to problems encountered on the *Fram*, Nansen invented a self-sealing insulated bottle to collect deep-water samples and measure their temperatures *in situ* with a pressure-protected thermometer. The

Nansen bottle was widely used until the development of the bathythermograph for temperature measurement in the 1940s, and electronic techniques for salinity measurement in the 1960s.

Oceanography also benefited from international cooperation. Alarmed by declining fish stocks, Swedish oceanographer Otto Pettersson (1848–1941) spearheaded the creation of *The International Council for the Exploration of the Seas* (ICES) in 1902, in the hope of stemming further losses through improved scientific understanding. From Germany, the *Meteor* expedition (1925–1927) confirmed the existence of deep-water circulation from north to south in the northern hemisphere and middepth flow south to north in southern hemisphere, while Albert Defant (1884–1974) published the first German textbook of the dynamic method, *Dynamische Ozeanographie* (1929) and George Wust (1890–1977) demonstrated that calculations of the Gulf Stream based on dynamic method matched empirical measurements.

Dynamic oceanography was brought to the United States by Harald Sverdrup, who had sailed with Amundsen on the *Maud*. Together with Martin Johnson (1893–1984) and Richard Fleming (1909–1990), Sverdrup authored the first comprehensive textbook of oceanography in English, *The Oceans* (1942). Perhaps the single most influential book in the history of oceanography, it framed the agenda for oceanographic research for several decades. Together with Walter Munk (b. 1917), Sverdrup developed the methods used to predict surf conditions for amphibious landings during World War II.

After World War II, oceanography flourished in the United States as the U.S. Office of Naval Research (ONR) provided abundant funding for research relevant to subsurface warfare and communication. Munk and Henry Stommel (1920–1992) improved prediction of ocean circulation by adding friction and detailed wind data. This work explained how wind-driven circulation leads to intensification of currents on the western sides of oceans in the northern hemisphere (eastern sides in the southern hemisphere), helping to explain phenomena such as upwelling currents and El Niño events. Stommel also predicted the existence of abyssal circulation involving the sinking of cold water at high latitudes, as suggested by Humboldt, now thought to be of major importance in controlling the earth's climate. The ONR also supported investigations of the structure of the seafloor and oceanic crust, including the paleomagnetic studies that demonstrated seafloor spreading.

Postwar advances in isotope geochemistry, headed by American chemist Harold Urey (1893–1981) and geochemist Harmon Craig (b. 1926), led to the recognition that O^{18}/O^{16} ratios in fossil foraminifera could be used to measure paleotemperatures. The results revealed many more temperature fluctuations than previously recognized from lithological evidence. Cesare Emiliani (1922–1995) linked these fluctuations to Milankovitch cycles—small variations in earth's orbit on a time scale of 20–100,000 years—providing the first widely accepted theoretical account of the cause of the ice ages.

C. Meteorology

Natural philosophers in the seventeenth and eighteenth century sporadically investigated atmospheric phenomena, but systematic meteorology developed in the nineteenth century when forecasters began to organize weather data to predict storms. This advance hinged on the telegraph, which permitted forecasters to integrate geographically dispersed information. It also hinged on the military value of weather forecasts, which in the United States were sent alongside military communications.

In the late nineteenth century this empirical tradition was criticized by researchers who hoped to reduce meteorological systems to physics and hydrodynamics. Leading the effort was Norwegian physicist Vilhelm Bjerknes (1862–1951), whose polar front concept gave physical interpretation to the behavior of storm systems and the interaction of air masses. Bjerknes was an inspired teacher; among his students he counted the pioneers of dynamic oceanography, who applied Bjerknes's principles and methods to the movement of water masses. As oceanography received support in part for its application to fisheries and navigation, so meteorology grew with the development of the aeronautics industry.

Bjerknes believed that the behavior of weather systems could be deterministically calculated, much as one could compute advance positions of planets by knowing their orbits and initial conditions. This problem was taken up in midcentury by British mathematician Lewis Fry Richardson (1881–1953), who pioneered the use of digital computers in weather forecasting. Richardson achieved great improvements in short-term forecasting, but the hope of a deterministic science of meteorology was dashed by the work of Edward Lorenz (b. 1917), who discovered that small changes in initial conditions could create very large perturbations in meteorological models. This realization—dubbed the "butterfly effect"—was a key element in the development of chaos theory.

In the late twentieth century attention shifted from weather to climate, as researchers developed General Circulation Models to understand the effect of altered atmospheric chemistry caused by burning of fossil fuels. Just as telegraph communication was critical to nineteenth-century empirical forecasting, the development of high-speed digital computers has been critical to twentieth-century climate modeling. However, given the complexity of the climate system and its interrelation with ocean circulation, the goal of accurate prediction of climate change remains a task for the future.

D. Seismology

As in meteorology and oceanography, scientific seismology required accurate instrumentation and synthesis of geographically dispersed data, which began to develop in the midnineteenth century. Early seismic investigations focussed primarily on earthquake occurrences; one of the earliest comprehensive catalogs was compiled by Irish engineer, Robert Mallet (1810–1881), who also coined the term *seismology* and developed an early design for an electromagnetic seismograph, later developed in Italy by Luigi Palmieri (1867–1896). In 1862, Mallet undertook a detailed investigation of the 1857 Basilicata (Naples) earthquake ($M \cong 7$), mapping the distribution of ground motion based on damage, and in the process inventing the term and concept of isoseismal lines.

Fear of earthquake damage provided the major incentive for government support of seismology in the nineteenth century; in Japan, the Meiji government recruited several British engineers to study seismology and seismic hazard. In 1878, John Milne (1850–1913) was hired at the Imperial College of Engineering, where he began detailed investigations of earthquake ground motions, improved seismograph design, and helped to establish the Seismological Society of Japan, the first national society of its kind. He also developed a rivalry with fellow expatriates Thomas Gray (1850–1908) and James Ewing (1850–1935), engineering professors at Tokyo University. Using a modified version of Milne's horizontal-pendulum seismograph, Gray and Ewing obtained high-quality records of earthquake ground motion. In 1895 Milne returned to Britain to establish the first world-wide seismic recording network, with financial support from the British Association for the Advancement of Science. With 30 stations world-wide, Milne's network remained the most extensive until the creation of the World Wide Standard Seismic Network (WWSSN) in 1960 (see below).

Seismograph design was also pursued in Germany, where Emil Wiechert (1861–1928) developed the inverted-pendulum sensor in 1904—widely used throughout the first half of the twentieth century—and in Russia, where B. B. Golicyn (Galitzin) (1862–1916) developed a sensitive design based on electrodynamic sensors and photographically recording galvanometers, in 1906. The development of high accuracy instruments enabled seismologists to differentiate between various "phases" of earthquake arrivals, which were soon related to different kinds of earthquake waves with different velocities and travel paths. Seismology now began to develop along two distinct lines: determining the internal structure of the earth based on arrival times, and understanding the mechanism of earthquakes and wave propagation.

The initial use of arrival times to deduce internal structure is often credited to R. C. Oldham (1858–1936), the Director of the Geological Survey of India, who discovered in 1900 that shock waves from earthquakes came in two pulses: an early (P or Primary) wave and a later (S or secondary) wave. Because they were traveling at different rates, the difference in their arrival times must be proportional to the distance to the source, and one could triangulate to determine the earthquake epicenter. In addition, Oldham found that the velocities increased with distance traveled, suggesting that the waves were traveling faster in the earth's deeper portions. Therefore, one could use seismic velocities to determine the earth's interior density. On this basis, he suggested that the earth might contain an iron core.

This line of investigation was taken up vigorously by Wiechert and his students, particularly Beno Gutenberg (1889–1960). Weichert is credited with the first accurate solution to the problem of deducing seismic wave velocities from travel times; Gutenberg used this to determine the radius of the core: 2900 km. Oldham had suggested the core might be liquid; proof of this came from Sir Harold Jeffreys (1891–1989), who provided the rigorous mathematical treatment to rule out alternative interpretations, and Danish seismologist Inge Lehmann (1888–1993), who demonstrated that seismic data were most consistent with a twofold structure: a liquid outer core and a solid inner core.

With the confirmation of the earth's core, it became clear that the earth had a three-layer structure—crust, mantle and core—but the depth to the mantle was not known. In the 1950s, Anglophone scientists realized it had already been identified by Croatian seismologist Andrija Mohorovicic (1857–1936), who in 1909 discovered the discontinuity in seismic waves under the continents at a depth of about 45 km that now bears his name. However, the core–mantle distinction was confounded by Gutenberg's discovery of the attenuation of S-waves in a region below the crust, at a depth of about 80 km. Gutenberg suggested this was a of partial melting, which he called the asthenosphere—borrowing the term coined by geologist Joseph Barrell (1869–1919) to describe the weak zone in which isostatic compensation occurred. It was not until the 1960s that this idea was generally accepted, when the distinction between the lithosphere and asthenosphere became a foundation of plate tectonics.

Seismologists in the twentieth century also studied the mechanisms of earthquakes themselves. As in Japan, seismology in the United States was motivated primarily by seismic hazard. Drawing on evidence from the devastating

1906 San Francisco quake, Johns Hopkins professor Harry Fielding Reid (1859–1944) developed elastic rebound theory: that strain builds up slowly in rocks, which then break suddenly along faults. While the association of earthquakes with surface faults seems obvious in retrospect, this connection was not always accepted prior to Reid's work. Meanwhile Gutenberg had moved to the United States, where he collaborated with Caltech colleague Charles Richter (1900–1985) to develop the magnitude scale that bears the latter's name.

Attention to earthquakes themselves led to renewed interest in their spatial distribution, and in the 1930s, Japanese seismologist Kiyoo Wadati (1902–1995) discovered that deep focus earthquakes are almost entirely restricted to narrow zones dipping steeply under the continental margins. In the 1950s, this was rediscovered by Hugo Benioff (1899–1968), and "Benioff zones" were soon recognized as slabs of oceanic lithosphere subducting under continental lithosphere at plate margins.

Two other concerns contributed to the development of seismology in the twentieth century: petroleum exploration and nuclear test ban verification. In the first half of the century, seismic refraction and reflection techniques were developed for the analysis of shallow crustal structure to identify petroleum traps. In the second half of the century, seismology was greatly boosted by the demand for identification of nuclear weapons tests. In 1960, the United States established the World Wide Standard Seismic Network, with over 100 stations world wide, under project Vela-Uniform, to distinguish nuclear blasts from natural seismic events. This program was further boosted by the 1963 Limited Test Ban treaty, which made seismic verification essential to treaty enforcement. While developed for political reasons, the WWSSN produced greatly improved location of earthquake foci and analysis fault-plane mechanisms, which proved critical to the modern understanding of earthquakes as indicators of plate motions.

E. Planetary Science

The principal development of planetary science came with the U.S and Soviet space programs. While information had been gleaned through telescopic study going back to Galileo's study of the moons of Jupiter, space probes and manned space flight yielded a cornucopia of information about the other planets. Among the many important results of the space program were the confirmation of composition of the Moon and its age (the same as the earth), the multiple moons of Jupiter, the rings of Neptune, and the discovery that Pluto may not be a planet at all. From the perspective of the earth sciences, the two most important conceptual developments are the recognition of the geo-

logical importance of impact cratering, and the absence of plate tectonics on other planets.

The Apollo lunar missions greatly increased scientific understanding of the lunar craters, proving after decades of debate that they were impact rather than volcanic craters. In addition, the various space craft returned showing the effects of thousands of tiny impacts from cosmic dust and micrometeorites. The resulting resurgence of interest in meteorite impacts on earth led to the suggestion by Eugene Shoemaker (1928–1997) and Robert Dietz that meteorite impacts had been important forces in geological and evolutionary history. In the early 1980s, Walter Alvarez (b. 1911) and coworkers discovered the existence of a thin layer of Ir-rich sediment at the Cretaceous–Tertiary (K–T) boundary. Interpreting this as the remains of an impact event, they proposed that the mass extinction of dinosaurs, ammonites, and other species at the K–T boundary could be thus explained. While impacts as a general cause of mass extinction has not been proven, most geologists now accept that a massive meteorite impact—and its associated effects on ocean and atmospheric chemistry—was the primary cause of the demise of the dinosaurs. Similar large impacts may well have played a role in the other mass extinctions of earth history.

While the Apollo missions showed that lunar features were relevant to understanding earth processes, studies of Venus and Mars have shown that tectonics there are very different than on earth. Venus has enormous volcanoes, far greater than any on earth, and appears to be dominated by plume tectonics rather than by plate tectonics. Mars likewise has volcanoes and rifts, but no evidence of lateral crustal motion. In our solar system, it appears that only earth has plate tectonics.

V. HISTORIOGRAPHIC REFLECTIONS

Given the importance of the earth sciences for industry, transportation, communication, and warfare, and for understanding the planet we live on, it is remarkable that historians have paid so little attention. Perhaps the utility of the earth sciences has made them seem less worthy to historians enamored of a pure science ideal. Perhaps the highly international and cooperative nature of advances in oceanography and seismology have made these sciences harder to idealize by those searching glory through national scientific accomplishments. In addition, the military connections of these sciences, particularly in the twentieth century, have made the documentary record difficult to uncover.

Nor do the earth sciences fit models of scientific advance that celebrate the role of individual genius and accomplishment, crucial experiments, conceptual revolutions, or advance through reductionism. Meteorology is a case in point: Bjerknes went far in his efforts to reduce forecasting to principles of physics, but he also reached a roadblock that involved more than lack of computational power. Advances in the earth sciences have been as much a function of improved communication and data collection as of improved theory. Often theory has followed advances in instrumentation and data-processing rather than the other way around.

The enormous need for data collection in these sciences has in the past confused historians who have mistaken their methods for "naïve empiricism," and dismissed them as primitive and uninteresting. But the earth sciences are less primitive than they are complex; like the systems they study, they are not easily reduced to simple formulations. Rather than understand them as naïve, we might understand them as big science—requiring substantial institutional support—before the term "big science" existed. Today, large-scale data collection remains central to the earth sciences. As remote sensing permits the creation of synoptic images of earth systems, and as computer models give us new means of investigating these systems, the traditional methods of field mapping and data analysis find renewed importance as "ground-truth" for these images and models.

The rise of computer modeling in the earth sciences has also raised the question of prediction in the earth sciences. Many people believe that the goal of science is prediction and control, and from this perspective the earth sciences look like failures. But prediction and control have only rarely been viewed as primary purposes in the earth sciences. When prediction and control were on the scientific agenda—as in earthquake prediction or weather control—often they were placed there by government patrons, such as military officials wishing to manipulate weather to strategic advantage. Scientists may or may not have shared these aspirations; whether they did is a historical question. Rather than assume all sciences share the goals of prediction and control, it would be better to ask, What were the goals of the earth sciences? Often, the primary goal was to develop an explanatory framework of processes that were clearly too large and too powerful to control. From this perspective, the earth sciences in the twentieth century have proved a conspicuous success.

SEE ALSO THE FOLLOWING ARTICLES

CLIMATOLOGY • EARTH'S CORE • EARTH'S MANTLE • GEOLOGIC TIME • METEOROLOGY, DYNAMIC • PLATE TECTONICS • PHYSICAL OCEANOGRAPHY • PLANETARY GEOLOGY • SEISMOLOGY, OBSERVATIONAL • SEISMOLOGY, THEORETICAL • SOLAR SYSTEM, GENERAL • VOLCANOLOGY

BIBLIOGRAPHY

Agnew, D. C. (in press). *In* "IASPEI International Handbook of Earthquake and Engineering Seismology," Chap. 1. Academic Press, San Diego.

Bowler, P. J. (1992). "The Norton History of the Environmental Sciences," New York, W. W. Norton.

Brush, S. G. (1996). "A History of Modern Planetary Physics: Vol. 1, Nebulous Earth; Vol. 2, Transmuted Past; Vol. 3 , Fruitful Encounters," Cambridge University Press, New York.

Burchfield, J. D. (1990). "Lord Kelvin and the Age of the Earth," 2nd ed., The University of Chicago Press, Chicago.

Doel, R. E. (1996). "Solar System Astronomy in America: Communities, Patronage and Inter-Disciplinary Research," Cambridge University Press, New York.

Ellenberger, F. (1996). "Histoire de la Geologie," A. A. Balkema, Rotterdam.

Fleming, J. R. (1998). "Historical Perspectives on Climate Change," Oxford University Press, New York.

Friedman, R. M. (1989). "Appropriating the Weather: Vilhelm Bjerknes and the Construction of a Modern Meteorology," Cornell University Press, Ithaca, NY.

Greene, M. T. (1992). "Natural Knowledge in Preclassical Antiquity," The Johns Hopkins University Press, Battimore.

Laudan, R. (1987). "From Mineralogy to Geology: The Foundations of a Science, 1650–1830," The University of Chicago Press, Chicago.

Le Grand, H. E. (1988). "Drifting Continents and Shifting Theories," Cambridge University Press, New York.

Oldroyd, D. (1996). "Thinking about the Earth: A History of Ideas in Geology," Athlone Press, London.

Oreskes, N. (1999). "The Rejection of Continental Drift: Theory and Method in American Earth Science," Oxford University Press, New York.

Rossi, P. (1984). "The Dark Abyss of Time: The History of the Earth and the History of Nationals from Hooke to Vico" (L. G. Cochrane, trans.), The University of Chicago Press, Chicago.

Rudwick, M. J. S. (1985). "The Meaning of Fossils: Episodes in the History of Paleontology," 2nd ed., The University of Chicago Press, Chicago.

Rudwick, M. J. S. (1985). "The Great Devonian Controversy: The Shaping of Scientific Knowledge among Gentlemanly Specialists," The University of Chicago Press, Chicago.

Earth's Core

David Loper

Florida State University

GLOSSARY

Convection Motion driven within a fluid body due to density differences. The density differences may be of thermal or compositional origin.

Core–mantle boundary The boundary between the outer core and the mantle: a nearly spherical surface 3480 km in radius.

Inner core The central, solid spherical region of the core having a radius of 1200 km.

Inner-core boundary The boundary between the inner core and the outer core: a nearly spherical surface 1200 km in radius.

Mush A mixed region of solid and liquid phases.

Outer core The molten part of the core lying between 1200 and 3480 km from the center.

THE EARTH'S METALLIC CORE occupies the central portion of the planet and is surrounded by the rocky mantle. To a first approximation the core is a sphere having a radius of 3480 km and a mean density of 10,480 kg/m^3. It comprises 16% by volume of earth and 32% by mass. The core is composed principally of iron, but contains a small percentage of nickel and a small but significant amount ($\sim$10%) of one or more nonmetallic elements. The central portion of the core, called the inner core, is solid, while the remainder, called the outer core, is liquid. Earth has had a core since it formed some 4.5 billion years ago. It is likely that the entire core was initially molten, that the solid inner core crystallized from the outer core as earth cooled over earth history and that this process is continuing. The outer core is very likely to be convecting vigorously, driven principally by the release of latent heat and compositionally buoyant material at the inner-core boundary. These convective motions provide energy to the geomagnetic field by means of dynamo action, sustaining it against ohmic decay. Direct evidence of core motions is found in the secular variations of the geomagnetic field. It is a remarkable fact that these variations occur on times of human scale (decades), whereas continental drift takes roughly a million times longer. This suggests that the core is far different, both dynamically and structurally, from the mantle and may be likened more to the atmosphere or oceans.

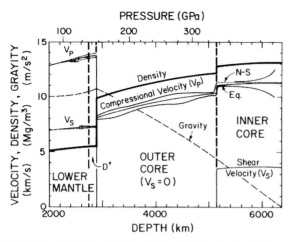

FIGURE 1 Seismologically determined elastic-wave speed, density and gravity as functions of depth in the core. Uncertainties of wave speeds are indicated by the multiple lines. [From Jeanloz, J. A. (1990). *Annu. Rev. Earth Planet. Sci.* **18**, 357–386. Copyright 1990 by Annual Reviews Inc.]

I. STRUCTURE

The internal properties of earth must be determined by means of external observations, which is a difficult inverse problem. Seismology is the principal source of information about most of the earth's interior, particularly the mantle. However, seismology provides relatively little information about the liquid outer core, and the secular variations of the geomagnetic field are used to probe its topmost layer. Additional evidence regarding the core comes from measurements of the length of day, gravity, and moment of inertia and from high-pressure experimentation (see Fig. 1).

To a good approximation, the core is a spherically symmetric, self-gravitating body in hydrostatic equilibrium. The pressure, p, density, ρ, and local acceleration of gravity, g, of any such body are related by

$$\frac{dp}{dr} = -\rho g, \qquad g = G\frac{m}{r^2},$$

$$\frac{dm}{dr} = 4\pi\rho r^2 \qquad \text{and} \qquad \rho = \rho(p),$$

where $G = 6.67 \times 10^{-11}$ N m²/kg² is the gravitational constant and m is the amount of mass within a sphere of radius r. The last of these equations is an equation of state for density; the functional form of this relation determines the internal structure of a given planetary body. A good approximation to the equation of state for earth's core is that the incompressibility is a linear function of pressure:

$$\rho\frac{dp}{d\rho} = k^1(p + p_1),$$

with $p_1 = 68$ GPa and $k^1 \approx 3.23$. [Remarkably, this equation with the same constants fits the lower mantle as well.]

The pressure at the center of earth is about 364 GPa (i.e., several million atmospheres) and 135 GPa at the top of the core. Iron compressibility is important at these pressures; core material exceeds the density of iron at zero pressure by 25% at the top of the core and by 65% at earth's center. However, these densities are smaller than those of pure iron, strongly suggesting alloying with lighter elements.

A. Core–Mantle Boundary

The outer boundary of the core is often denoted by its acronym: CMB (core–mantle boundary). Above the CMB is the low-density and electrically insulating elastic silicate mantle, while below is the high-density and electrically conducting fluid outer core. The CMB is identified by a strong and sharp change in seismic and electromagnetic properties. Primary or compressive (P) wave speeds drop from 13.6 km/s in the lower mantle to 10.0 km/s at the top of the core, while secondary or shear (S) wave speeds drop from 7.3 km/s in the mantle to zero in the core. (Fluids cannot sustain shear waves.) Reflections of seismic phases such as *PcP*, *ScS*, and *PnKP* at the CMB indicate that the CMB is less than 2 km thick. The seismically determined mean CMB radius, as codified in the Preliminary Reference Earth Model, is 3480 km. This radius may be estimated independently from geomagnetism, assuming that the mantle is an electrical insulator and the core is perfectly conducting. The geomagnetic CMB is the depth at which the unsigned magnetic flux does not vary with time. The radius of the CMB determined by this method depends on the magnetic-field model used, but is in good agreement with the seismically determined value.

The shape of the CMB deviates from spherical due to the centrifugal force of rotation, making the polar radius about 9 km less than the equatorial. Smaller scale deviations of the CMB from its mean radius can be estimated in principle by seismic tomography, but observations are confused by strong heterogeneities in the lower mantle and there is no clear consensus on the shape or magnitude of the deviations. Estimates of the root mean square (rms) deviations typically lie in the range from 0.5 to 5.0 km.

B. Inner-Core Boundary

The inner-core boundary (ICB) is believed to be a phase-change boundary between the metallic liquid outer core above and the metallic solid inner core below. Since the outer core is electromagnetically opaque, we must rely on seismic studies for direct information about this feature. The mean radius of the ICB is 1220 km.

Geographical deviations of the ICB from its mean radius are difficult to quantify, due to the small number and poor geographical distribution of relevant seismic measurements. Indirect information regarding the structure of the ICB comes from metallurgy; see Section III.B.

C. Outer-Core Structure

The primary evidence for existence of the outer core comes from the shadow zone for direct P waves at angular distances between 100 and 143 degrees from the seismic source. There is no reliable evidence that the outer core is other than a well-mixed homogeneous liquid. The relatively rapid secular variations of earth's magnetic field observed at the surface are strong, but indirect, evidence of vigorous motions within the core which maintain this state. Furthermore, the dynamo process believed to be operating in the outer core (see Section V) provides further strong, but indirect, argument for vigorous motions throughout most of the outer core.

D. Inner-Core Structure

The first evidence of structure within the inner core came from studies of seismic attenuation, which found relatively high levels of attenuation in the uppermost 200–300 km of the inner core. More recently, it has been determined that the inner core is seismically anisotropic, with P-wave speeds being larger in the direction parallel to earth's rotation axis than in the perpendicular direction. There is also evidence of seismic anisotropy in the western hemisphere of the inner core. The most likely cause of these structures is crystal alignment, induced by asymmetrical core growth and deformation by relaxation toward equilibrium ellipticity.

II. COMPOSITION AND PROPERTIES

A. Overall Composition

Strong, but indirect, evidence that the core is composed principally of iron comes from cosmochemistry, from the existence of iron meteorites, and from high-pressure experimentation. There is also strong evidence from the last of these that the core must contain a significant percentage of light, nonmetallic material. The nature of this material is uncertain, but the most likely elements are sulfur, silicon, and oxygen.

An important, but undetermined, issue is whether the core contains significant amounts of heat-producing radioactive elements. This has bearing on the energetics of the core, the age of the inner core, and the energy source for the geodynamo.

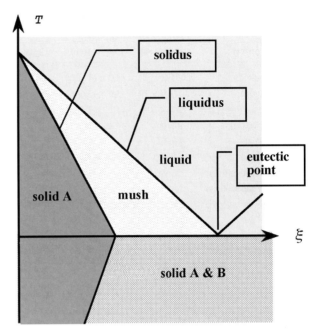

FIGURE 2 A typical phase diagram for a binary eutectic alloy. T is temperature and ξ is mass fraction of constituent B. Within the region labeled mush, solid having the solidus composition and liquid having the liquidus composition coexist at a common temperature.

B. Radial Variations of Composition

It is very likely that the outer core is homogeneous to a high degree of precision; the relative density differences associated with convective motions are on the order of 10^{-9}. The largest and most significant change of composition is that which occurs across the ICB. In general, the solid which forms by solidification of an alloy has a composition that differs from the parent liquid (see Fig. 2). The outer core is dominantly iron and the inner core is likely to be composed of iron crystals containing less of the nonmetallic elements. The density contrast attributed to the compositional difference is roughly 0.5 g/cc $= 500$ kg/m^3.

C. Physical Properties and State

The physical properties and state of the core are summarized in Table I. The first entry in columns 2 and 3 is that at the top and the second is that at the bottom. For more detail, see Appendices F and G of Stacey (1992).

III. EVOLUTION OF THE CORE

A. Initial Formation

The existence of stony and iron meteorites provides strong evidence that planetary cores, in general, and earth's core, in particular, formed by separation of less dense silicate phases and more dense metallic phases as they accreted

TABLE I Physical Properties and State of the Core

Property	Outer core (top–bottom)	Inner core (top–bottom)	Units
P-wave speed	8,065–10,356	11,028–11,266	m/s
S-wave speed	0–0	3,504–3,668	m/s
Incompressibility	644–1,304	1,343–1,425	GPa
Poisson's ratio	0.5–0.5	0.444–0.441	—
Specific heat	707–659	641–637	J/kg K
Thermal expansivity	15.6–7.8	6.7–6.4	10^{-6}/K
Thermal conductivity	28–36.8	49.5–50.9	W/m K
Electrical conductivity	3	4	10^5 S/m
Pressure	136–329	329–364	GPa
Density	9,903–12,166	12,764–13,088	kg/m^3
Gravity	10.7–4.4	4.4–0.0	m/s^2
Temperature	3,750–4,960	4,960–5,100	K

during the formation of the solar system some 4.5 billion years ago. This was a strongly exothermic process; the gravitational potential energy released by this process is sufficient to heat earth by some 2000°. It follows that earth likely was very hot soon after its formation and has been cooling since then.

B. Formation and Growth of Inner Core

It is very likely that the inner core has grown by solidification from the outer core as earth has cooled during the past 4.5 billion years and that solidification and growth is continuing. The inner core may well be a relatively recent feature; in some models of the evolution of the core it begins to grow roughly 2 billion years ago.

The core is cooled by transfer of heat to the mantle, and the rate of cooling is largely controlled by the thermal structure of the lowermost mantle (the D″ layer). The outer core is coolest at the top, near the CMB, but freezing proceeds from the center outward because the increase of the freezing (liquidus) temperature with pressure is greater than the adiabatic gradient:

$$\frac{dT_L}{dp} > \frac{dT_A}{dp}.$$

As the inner core grows, both latent heat and buoyant material are released at the base of the outer core. These work in parallel to drive convective motions in the outer core.

Solidification of outer-core material at the ICB is similar to the metallurgical process of unidirectional solidification of molten metallic alloys; the mathematical model is called a Stefan problem. The simplest solution to the Stefan problem involves the steady advance of a planar solidification front into a quiescent liquid. This simple solution has two known forms of instability. If the freezing process involves a change of composition (see Fig. 2) and the material rejected by the solid phase is buoyant compared with the parent liquid, the static state is prone to a compositional convective instability. It is very likely that this instability occurs in the outer core and that the resulting convective motions participate in the dynamo process which sustains earth's magnetic field.

Solidification of an alloy at a planar interface is prone to a second, morphological instability. The material rejected by the solid phase accumulates on the liquid side of the freezing interface, depressing the liquidus and making that liquid compositionally (or constitutionally) supercooled. This causes the flat freezing interface to be unstable and become convoluted. These convolutions can become extreme, forming a so-called mushy zone. Again, it is very likely that this instability occurs in the core and that the inner core is, in fact, an intimate mixture of solid and liquid. Dynamic processes cause the fraction of liquid phase to be small, so that the inner core acts structurally as a solid even though, thermodynamically, it behaves as a solid-liquid mixture.

IV. CORE–MANTLE INTERACTIONS

The core and the mantle may exchange heat, material, and angular momentum. The geodynamo operating in the outer core requires transfer of heat from core to the mantle. In addition, the heat conducted down the adiabatic gradient must be transferred to the mantle. It is an open question whether the rate of heat conduction down the adiabat is greater or less than the rate of transfer from core to mantle. If greater, then the top of the outer core may be thermally stratified. In this case, compositional buoyancy has the capacity to maintain the adiabat all the way to the top.

Four types of material exchanges across the CMB are possible: silicate from core to mantle, silicate from mantle to core, metal from core to mantle, and metal from mantle to core; but which occurs, if any at all, remains uncertain. During the accretion of earth, silicates and metals were chemically equilibrated at low pressure. It is an open question whether silicates and metals are equilibrated across the CMB. If metals are leaching into the mantle and/or silicates into the core, the top of the core may be compositionally stratified.

Angular momentum transfers between the core and the mantle are responsible for the long-term (decade and longer) changes in the length of day. The principal mechanism of transfer is unclear; possible coupling mechanisms include electromagnetic, topographic, and gravitational torques. Electromagnetic torques require significant electrical conductivity in the lowermost mantle, topographic torques require variations in the shape of the CMB, and gravitational torques require density anomalies in the mantle plus a nonspherical ICB. The existence of the requisite electrical conductivity within the lower mantle is uncertain, but it appears that irregularities of the shape of the CMB and of the density of the lower mantle are sufficient to produce topographic and gravitational torques of the required magnitude.

V. CORE ENERGETICS

The core is cooling by transfer of heat to the mantle. The rate of transfer is controlled by the thermal structure of the lowermost mantle (i.e., the D″ layer). The possible sources of energy within the core include sensible heat (i.e., the heat capacity of the core plus gravitational energy released by thermal contraction) released by the slow cooling of the core, latent heat of fusion released by the progressive solidification of the inner core (plus gravitational energy released by the volume change), gravitational potential energy released by the selective solidification of the denser metallic constituents in the core, and radioactive heating. The first three of these are linked to the cooling of the core and have released approximately 3.0×10^{29} J of energy since the inner core formed. Gravitational energy has supplied about 13% of this total.

The magnitude of radioactive heating is difficult to estimate, as the partition coefficients of the relevant elements (U, Th, K) between the core and the mantle are very poorly known. Due to the finite half-lives of the isotopes ^{238}U, ^{235}U, ^{232}Th, and ^{40}K, radioactive heating was more significant early in earth history than it is now. If these elements contribute significantly to the present heat budget, then the core would have been heating for much of its history, and it would not have been possible to form a solid inner core by cooling. It is quite likely that the present amount of radioactive heating in the core is relatively insignificant.

If the outer core is well mixed by convective motions, as appears very likely, the temperature decreases significantly with increasing radius due to adiabatic decompression. The adiabatic gradient is given by

$$\frac{dT}{dr} = \frac{\alpha T g}{C_p},$$

where α is the coefficient of thermal expansion and C_p is the specific heat; see Table I. The rate, $\dot{Q}$, that heat is conducted radially outward along this adiabat is quantified by

$$\dot{Q} = 4\pi r^2 k \frac{dT}{dr} = 4\pi r^2 k \frac{\alpha T g}{C_p},$$

where k is the thermal conductivity. If $\dot{Q}$ is less than the rate of transfer of heat to the mantle across the CMB, then thermal buoyancy contributes to convective motions at that level. Conversely, if $\dot{Q}$ exceeds the rate of transfer, then the

top of the outer core is thermally stably stratified. Current estimates of the properties of the outer core, particularly the thermal conductivity, are not known with sufficient accuracy to determine which possibility in fact occurs. Using values from Table I, $\dot{Q} \approx 3.5 \times 10^{12}$ W. [If thermal conductivity is as high as 47 W/m K, then $\dot{Q} \approx 6 \times 10^{12}$ W.]

It is important to distinguish between thermal and compositional (i.e., gravitational) energy sources, because, as noted above, thermal energy is "short circuited" by conduction down the adiabat. On the other hand, molecular diffusion is ineffective in redistributing matter, and compositional convection is much more likely than thermal convection in the outer core.

VI. CORE DYNAMICS

A. Oscillations

The outer core may sustain oscillations involving inertial (Coriolis), magnetic (Lorentz), and/or buoyancy forces. Oscilations involving all three are referred to as MAC waves (M = magnetic, A = Archimedian, C = Coriolis). The role of buoyancy forces in sustaining oscillations may, in fact, be negligible, in which case MC waves result. Ideal MC waves, involving the fluid velocity, $\mathbf{u}$, pressure, p, and perturbation magnetic field, $\mathbf{b}$, are governed by the momentum, mass, and magnetic-diffusion equations:

$$\frac{\partial \mathbf{u}}{\partial t} + 2\Omega \times \mathbf{u} = -\nabla p + \frac{1}{\rho\mu}(\mathbf{B} \cdot \nabla)\mathbf{b},$$

$$\nabla \cdot \mathbf{u} = 0 \quad \text{and} \quad \frac{\partial \mathbf{b}}{\partial t} = (\mathbf{B} \cdot \nabla)\mathbf{u},$$

where Ω is the rotation rate of earth, ρ is the density of core fluid, μ is the magnetic permeability, and $\mathbf{B}$ is the magnetic field (assume locally constant). Plane-wave solutions obey the dispersion relation

$$\left[\omega^2 - \frac{(\mathbf{B} \cdot \mathbf{k})^2}{\rho\mu}\right]^2 = 4\frac{(\Omega \cdot \mathbf{k})^2}{k^2}\omega^2.$$

The solutions are of two distinct types; one is the same as classic, nonmagnetic rotational oscillations to dominant order, and the second is a strongly modified Alfven wave which has a slow phase and group speeds. The phase speed of this latter type of wave is consistent with the speed of motions at the top of the outer core inferred from secular variations of the magnetic field.

B. Convection

Convective motions in the outer core are driven by sources of buoyancy at the ICB or sinks at the CMB; in the absence of forcing, thermal conduction drives the outer core toward an isothermal state, which is strongly stable. The sources and sinks may be compositional or thermal.

A compositional sink of buoyancy at the CMB results from the transfer of silicate to the mantle or metal to the core. A thermal sink of buoyancy at the CMB results from the transfer of heat from core to mantle at a rate greater than the rate heat is conducted radially outward within the outer core to the CMB. It is uncertain whether any significant transfer of material occurs at the CMB and whether the rate of transfer of heat from core to mantle exceeds that conducted down the adiabat.

Compositional and thermal sources of buoyancy at the ICB result from the growth of the inner core; latent heat is released at too great a rate to be conducted down the adiabat, and molecular diffusion is quite ineffective in redistributing the buoyant material released by solidification. It is very likely that both thermal and compositionally buoyant material is released at the ICB and that this material drives the convective motions in the bulk of the outer core.

C. Outer-Core Stratification

Given that the only plausible explanation for the existence of the geomagnetic field is a convective dynamo in the outer core, the bulk of the outer core must be convecting and hence unstratified. However, the outer core might be stratified at the top. If the rate of heat conduction down the adiabat were greater than the rate of transfer from core to mantle, then thermal buoyancy forces would tend to stratify the top of the outer core. Similarly, if silicate material were leaking into the core and/or metallic material were leaking into the mantle, then the top of the outer core would be compositionally stratified. The rates of transfer of heat and material at the CMB are not known with sufficient accuracy to determine whether the top of the outer core is stratified. Any possible stratification is too weak to be detected seismically. The best observational evidence of the dynamic state of the top of the outer core comes from geomagnetic secular variation, which can be inverted to give velocity fields. Current models of core motion do not show any tendency for stratification. If the top layer of the outer core were stably stratified and if the rate of transfer of heat from core to mantle were geographically variable (as seems likely), then strong thermal winds would be generated at the top of the outer core. Such winds are not seen in the models of core surface motion.

D. Inner-Core Rotation

There are strong dynamical reasons to believe that to a first approximation the inner core is corotating with the mantle and with the bulk of the outer core. If the inner core were rotating about a different axis or at a different rate, enormous electromagnetic torques would be generated which would restore the state of corotation. In the late 1990s, several seismic studies produced evidence that the inner core is rotating slightly (from 0.2 to 3%) faster than the mantle. This conclusion is controversial, as other studies find no significant difference in rotation rates of the inner core and the mantle.

VII. THE GEODYNAMO

Given the rapid secular variation of earth's magnetic field and its episodic reversals of polarity, the only plausible explanation of its origin is the dynamo action of convective motions in the outer core. The so-called geodynamo problem has proved to be one of the most difficult of mathematical geophysics. Early results in the 1930s were negative, in the form of anti-dynamo theorems. Further progress on this problem was slow until the 1970s when it was shown that certain velocity fields were capable of sustaining a magnetic field. This kinematic dynamo problem required solution of the magnetic diffusion equation

$$\eta \nabla^2 \mathbf{B} + \nabla \times (\mathbf{u} \times \mathbf{B}) = \frac{\partial \mathbf{B}}{\partial t}$$

in some spatial domain (e.g., a sphere) with suitable boundary conditions (e.g., insulating surroundings having a potential field). Here, η is the magnetic diffusivity and $\mathbf{u}$ is a specified velocity. This is in effect a vector eigenvalue problem.

Generalization of this kinematic problem to the dynamic case has proved to be difficult. In the full problem the velocity and pressure are determined by the momentum and continuity equations, e.g.,

$$\frac{\partial \mathbf{u}}{\partial t} + 2\Omega \times \mathbf{u} = -\nabla p + C\mathbf{g} + \frac{1}{\rho\mu}(\mathbf{B} \cdot \nabla)\mathbf{B} + \nu\nabla^2\mathbf{u}$$

and

$$\nabla \cdot \rho\mathbf{u} = 0,$$

while the fractional density perturbation, C, which is the driving force for the convective motions, obeys an advective-diffusion equation

$$\frac{\partial C}{\partial t} + \mathbf{u} \cdot \nabla C = D\nabla^2 C.$$

This problem is too complex for analytic solution, and successful numerical simulation of dynamo action in a spherical body was first achieved by Glatzmaier and Roberts in 1995. As seen in Fig. 3, the output of this and similar models can appear quite realistic. The limitations in size and speed of current computers require the diffusivities of magnetic field (η), momentum (ν), and buoyancy (D) to

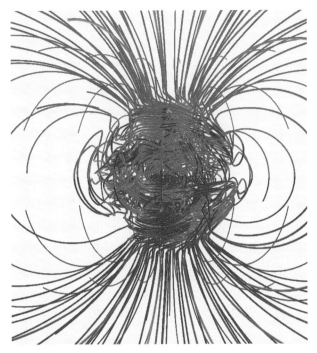

FIGURE 3 A representation of the magnetic field produced by the Glatzmaier-Roberts dynamo model. The structure of the field changes abruptly at the CMB. (Figure courtesy of Gary Glatzmaier.)

be parameterizations of small-scale turbulence rather than assuming their molecular values.

The full dynamo problem is driven through the boundary conditions on the density perturbation, C, which pro-

vide gravitational potential energy to the system. This gravitational energy is converted to kinetic energy by means of convective instabilities and then to magnetic energy through magnetic induction. Next, ohmic dissipation converts the magnetic energy to heat, principally within the core. This heat is conducted and convected to the CMB and transferred to the mantle.

SEE ALSO THE FOLLOWING ARTICLES

CONTINENTAL CRUST • GEOMAGNETISM • HEAT FLOW • HIGH-PRESSURE SYNTHESIS (CHEMISTRY) • MANTLE CONVECTION AND PLUMES • OCEANIC CRUST • SEISMOLOGY, THEORETICAL

BIBLIOGRAPHY

Jacobs, J. A. (1975). "The Earth's Core," Academic Press, San Diego.

Jacobs, J. A. (1992). "Deep Interior of the Earth," Chapman & Hall, London.

Jeanloz, J. A. (1990). "The nature of the Earth's core," *Annu. Rev. Earth Planet. Sci.* **18,** 357–386.

Merrill, R. T., McElhinny, M. W., and McFadden, P. L. (1996). "The Magnetic Field of the Earth: Paleomagnetism, the Core, and the Deep Mantle," Academic Press, San Diego.

Poirier, J. P. (1994). "Light elements in the Earth's outer core: a critical review," *Phys. Earth Planet. Inter.* **85,** 319–337.

Stacey, F. D. (1992). "Physics of the Earth," 3rd ed., Brookfield Press, Kenmore, Brisbane.

Stixrude, L., and Brown, J. M. (1998). "The Earth's core," *Rev. Mineral.* **37,** 261–282.

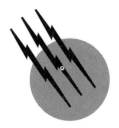

Earth's Mantle (Geophysics)

Raymond Jeanloz

University of California, Berkeley

I. Structure and Physical Properties
II. Composition and Mineralogy
III. Dynamics and Temperature

GLOSSARY

Asthenosphere Region in the upper mantle immediately beneath the lithosphere. The asthenosphere is thought to exhibit relatively low viscosity, and it corresponds approximately in depth to the seismologically observed, low-velocity, attenuating layer between 80 and 250 km depth.

D″ layer Layer approximately 200 (±200) km thick at the base of the mantle, characterized by anomalous seismological, properties: decreased average velocity gradients with depth, complex velocity structure, and enhanced scattering of elastic waves.

Geotherm Profile of temperature with depth in the earth.

Komatiite Rock solidified from a high-temperature lava and characterized by magnesium-rich olivine exhibiting a special texture formed by rapid quenching from the melt. Ultramafic komatiites, indicative of especially high internal temperatures of more than 1900 K, are observed only in Archean (older than 2.5 billion years) rock sequences.

Lithosphere Elastic layer making up the tectonic plates observed at the surface. The lithosphere includes the crust and the uppermost part of the mantle that are observed as the "lid," which transmits seismic waves with relatively high velocity and little attenuation. Because of its relatively low temperature, the lithosphere is rigid compared with the underlying mantle, which is hot and ductile. Cooling of the lithosphere occurs mainly by upward conduction of heat; hence, it corresponds closely to the thermal boundary layer at the top of the mantle.

Peridotite Rock consisting mainly of olivine [$(Mg, Fe)_2SiO_4$] and pyroxene [$(Mg, Fe, Ca)SiO_3$]. Other minerals, such as garnet, can be present as well. A synthetic analog, pyrolite, has been extensively used for experiments simulating the formation of basalt in the mantle.

Perovskite Crystal structure of the dense, high-pressure phase of pyroxene [$(Mg, Fe, Ca)SiO_3$] that is stable only above ~20 GPa (200 kbar).

Transition zone Region between the upper and lower mantle that is bounded by the seismologically observed discontinuities at depths of 400 and 670 km beneath the surface. Large increases in density and elastic properties with depth are ascribed mainly to high-pressure mineral transformations in this region.

Xenolith Rock fragment brought up volcanically from depth. Commonly no more than a few centimeters across, xenoliths originating from depths as great as 150–200 km, and possibly deeper, have been documented. Occasionally, xenoliths from the mantle are associated with occurrences of diamonds.

THE EARTH'S MANTLE and crust make up the solid, rocky part of the planet surrounding the largely molten, metallic core at the center. The mantle is a 2900-km thick shell that includes the bulk of the earth's interior (Table I). In comparison, the crust, which surrounds the mantle and is the outermost layer of the planet, is only 25 km thick on average. Thus, the core (diameter of 6960 km), mantle, and crust represent 16.3, 82.6, and 1.1% of the earth's volume, respectively. Magnesium-rich silicate minerals are

the primary constituents of the mantle, in contrast to the predominantly aluminum silicate minerals making up the crust. The minerals of the upper mantle are transformed to denser crystal structures with depth, due to the increase in pressure toward the earth's center. Pressures as high as 136 GPa (1.36 million atmospheres) are achieved at the base of the mantle, but the mineral transformations occur almost entirely at pressures of 12–25 GPa, corresponding to depths of 400–700 km beneath the surface. In this depth range, high-pressure mineral phases, such as the spinel form of Mg_2SiO_4 and $MgSiO_3$ in the perovskite structure, first appear. Because the major minerals of the upper mantle break down to this latter structure under pressure, the silicate perovskite phase is considered to be the dominant mineral of the lower mantle. As a result of both self-compression and high-pressure transformations, the physical properties of the mantle (e.g., density and elastic moduli) vary mainly with depth rather than laterally. Temperatures, however, vary by comparable amounts in the horizontal and vertical directions. For example, the temperature at a given depth beneath oceanic trenches and ridges can differ by 1000 K or more. Similarly, the average temperature increases from ~1500 K at the top of the mantle to ~3500 K at its base. Although the mantle is almost entirely solid, the temperatures are sufficiently high that it deforms by solid-state creep of the constituent minerals. As a result, the mantle is ultimately weak, and it behaves like a fluid over geological time scales of 10^5 to 10^9 years. The fluid-like convection or flow of the mantle is responsible for many of the geological processes observed at the earth's surface, such as the large-scale horizontal movements of crustal plates and the associated volcanism, the occurrence of earthquakes, the formation of sedimentary basins, and the uplift of mountains. Variations in density at a given depth in the mantle, caused by horizontal differences in temperature (and possibly composition), produce the buoyancy forces that cause mantle convection and near-surface tectonic activity. In addition, flow of the mantle is the predominant means by which heat is transported outward from the deep mantle and core. Thus, mantle convection has determined the cooling and chemical evolution of the interior over the earth's 4.5-billion-year history.

TABLE I Bulk Properties of Earth's Mantle[a]

Property	Value
Volume	9.06×10^{20} m^3
Mass	4.06×10^{24} kg
Fraction of earth	
Mass	67%
Atomic[b]	84%
Average density	4.48 Mg m^{-3}
Moment of inertia[c]	7.02×10^{37} kg m^{-2}
Gravitational acceleration[d]	10.3 ($\pm$0.4) m sec^{-2}
Outward heat loss (present values)	
Total	37 ($\pm$5) TW
Average flux	72 ($\pm$10) mW m^{-2}
Estimated radioactive heat production (present value)[e]	24 ($\pm$10) TW
Viscosity (upper mantle)	5×10^{19} to 10^{21} Pa sec
Electrical conductivity (upper mantle)	10^{-3} to 10^{-1} S m^{-1}
Age	1.43×10^{17} sec (4.54×10^9 year)

[a] Sources:

Basaltic Volcanism Studies Project (1981). "Basaltic Volcanism on the Terrestrial Planets," Pergamon, New York.

Cathles, L. M., III (1975). "The Viscosity of the Earth's Mantle," Princeton Univ. Press, Princeton, NJ.

Davies, G. F. (1980), *J. Geophys, Res.* **85**, 2517–2530.

Dziewonski, A. M., and Anderson, D. L. (1981). *Phys. Earth Planet. Int.* **25**, 297–356.

Garland, G. D. (1981). *Annu. Rev. Earth Planet. Sci.* **9**, 147–174.

Jeanloz, R. (1987). *In* "Mantle Convection" (W. R. Peltier, ed.), Gordon and Breach, New York.

O'Connell, R. J., and Hager, B. H. (1980). *In* "Physics of the Earth's Interior" (A. M. Dziewonski and E. Boschi, eds.), pp. 270–317, Elsevier, New York.

Peltier, W. R. (1980). *In* "Physics of the Earth's Interior" (A. M. Dziewonski and E. Boschi, eds.), pp. 362–431, Elsevier, New York.

Sclater, J. G., Jaupart, C., and Galson, D. (1980). *Rev. Geophys. Space Phys.* **18**, 269–311.

Verhoogen, J. (1980). "Energetics of the Earth," Natl. Acad. Sci., Washington, DC.

[b] Value depends on the assumed composition of the core.

[c] Value for the entire mantle.

[d] Value at any depth in the mantle.

[e] Values for the mantle alone, excluding contributions from the crust or core.

I. STRUCTURE AND PHYSICAL PROPERTIES

A. Seismology and Elastic Properties

The basic structure of the mantle consists of concentric layers or shells that are defined by variations of physical properties with depth (Fig. 1). The most detailed

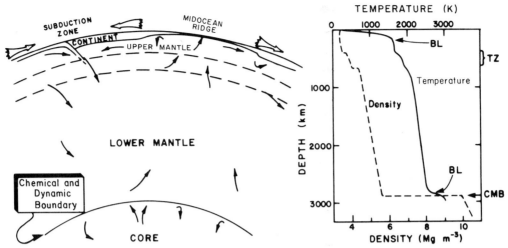

FIGURE 1 Summary of the basic structure of the mantle, schematically illustrating the observed motion of lithospheric plates at the surface and a possible flow pattern in the convecting mantle. New lithosphere is formed at mid-ocean ridges, moves horizontally along the surface (large arrows), and sinks into the mantle along subduction zones that originate from oceanic trenches at the surface. The seismologically determined density profile (Table II) is shown on the right, along with the average temperature with depth corresponding to the flow pattern shown. The presence of thermal boundary layers (BL) at the top and bottom of the mantle are indicated along the temperature profile. The depths of the core–mantle boundary (CMB) and of the transition zone (TZ) located between the 400- and 670-km discontinuities (dashed lines between upper mantle and lower mantle on left) are given at the far right. In this figure, the mantle is assumed to be of uniform average composition and, hence, to be thoroughly stirred by convection. [From Jeanloz, R. (1981). *In* "Proceedings of the American Chemical Society 17th State-of-the-Art Symposium," p. 74, Am. Chem. Soc., Washington, DC.]

information on the present structure of the mantle is obtained from seismology. Specifically, the density and elastic-wave velocities are determined as functions of depth by examining the waves generated by earthquakes or large explosions. Away from the source (e.g., the earthquake hypocenter), these are almost perfectly elastic waves that travel through and around the globe: body waves and surface waves, respectively. Several types of measurement are involved. (1) The travel times from the earthquake or explosion source to receivers (seismometers at the surface) are recorded for body waves that are refracted and reflected to the surface by the internal structure; (2) similarly, the travel times for surface waves encircling the globe are obtained as a function of frequency; (3) waveform analysis of the three-dimensional displacement observed as a function of time at the seismometer complements the travel-time measurements, yielding considerable detail in determinations of mantle structure; and (4) the frequencies of the earth's free oscillations are determined by measuring over an extended time (e.g., several hours or days) the standing waves that are generated across the entire globe by the source displacements. The frequencies involved span the approximate range from 10 Hz for body waves to 10^{-3} Hz or less for free oscillations, with surface waves being at intermediate frequencies. The higher frequency waves yield more detail

about the internal structure but tend to be scattered or attenuated more severely than those of lower frequency; for example, waves exceeding 100 Hz are only observed over relatively short distances across the crust.

From the combination of free-oscillation, body-wave, and surface-wave measurements, the longitudinal or compressional-wave velocity V_p, the transverse or shear-wave velocity V_s, and the density are ρ obtained as functions of depth. The longitudinal and transverse waves involve distortions that are, respectively, parallel and perpendicular to the direction of propagation of the wave. These velocities are related to the average elastic moduli, the adiabatic incompressibility or bulk modulus K_s and the rigidity or shear modulus μ, in the following way:

$$V_p = \sqrt{\left(K_s + \tfrac{4}{3}\mu\right)/\rho} \tag{1}$$

$$V_s = \sqrt{\mu/\rho}. \tag{2}$$

Normally, the density, elastic moduli, and wave velocities of a crystal increase continuously with pressure. Thus, anomalously rapid or discontinuous changes in the wave velocities with depth indicate unexpected changes in the average elastic moduli or density of the mantle rock. These are ascribed either to changes in rock type (i.e., changes in bulk composition) or to high-pressure transformations (changes in mineral structures) at a given depth, and they

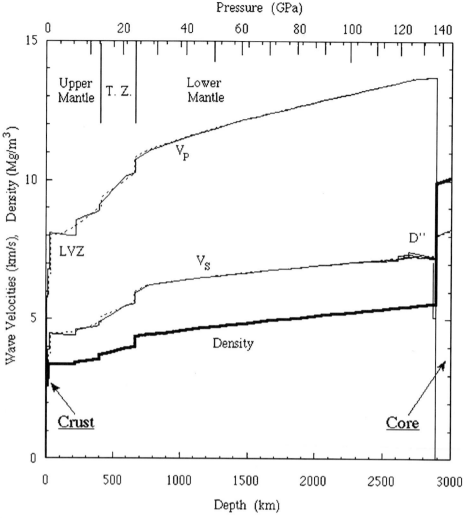

FIGURE 2 Seismologically determined properties as a function of depth in the outer 3000 km of the earth. Horizontally averaged compressional- and shear-wave velocities (V_P and V_S; thin curves) and density (bold curve) as a function of depth (bottom scale) and pressure (top) through the upper mantle, the transition zone (TZ), and the lower mantle according to the Preliminary Reference Earth Model (PREM: solid curves) and iasp91 (dotted curves). The low-velocity zone (LVZ) is indicated in the upper mantle, and several illustrative V_S profiles are shown for the D'' region at the base of the mantle. [Data from Dziewonski, A. M., and Anderson, D. L. (1981). *Phys. Earth Planet. Int.* **25,** 297–356. Kennett, B. L. N., and Engdahl, E. R. (1991). *Geophys. J. Int.* **105,** 429–465. Wysession, M. E., *et al.* (1998). *In* "The Core–Mantle Boundary Region," Am. Geophys. Union, Washington, DC.]

define the layered structure of the mantle (Figs. 1 and 2; Table II).

One way to determine whether the change of density with depth is anomalous is to compare the observed variation of density with that expected from the elastic moduli obtained from the wave velocities. Assuming adiabatic conditions (see Section III), the expected change in density with pressure P is given by the bulk modulus

$$K_s = \rho(\partial P/\partial \rho)_s = \rho\left(V_p^2 - (4/3)V_p^2\right) \equiv \rho\varphi, \qquad (3)$$

in which the second equality is derived from Eqs. (1) and (2) and φ is defined as the seismic parameter. The actual

change in pressure with depth z is described by the hydrostatic formula

$$dP = \rho g\, dz, \qquad (4)$$

with g being the acceleration of gravity at depth z. Equation (4) is valid for the mantle because the rock is weak and it therefore sustains only small nonhydrostatic stresses (estimated to be less than ~ 0.1 GPa, a value that is negligible compared with the pressure). From Eq. (4), the variation of pressure and density in the mantle is therefore

$$K_E \equiv \rho\, dP/d\rho = \rho^2 g/(d\rho/dz). \qquad (5)$$

TABLE II Vertical Structure and Properties of Earth's Mantle[a]

Region	Depth z (km)	Pressure P (GPa)	Density ρ (Mg m^{-3})	Incompressibility K_s (GPa)	Rigidity[b] μ (GPa)	Quality factor Q (in shear)	Inhomogeneity parameter η
Upper mantle							
Lid	24	0.6	3.38	131.5	68.2	600	−0.13
Low-velocity zone	80	2.5	3.37	130.3	67.4	80	−0.13
	220	7.1	3.36	127.0	65.6	80	−0.12
Mass fraction of mantle, 15.4%	220	7.1	3.44	152.9	74.1	140	0.78
Temperatures, 1200–1700 K	400	13.4	3.54	173.5	80.6	140	0.83
400-km discontinuity							
Transition zone	400	13.4	3.72	189.9	90.6	140	1.73
	500	17.1	3.85	218.1	105.1	140	1.86
Mass fraction of mantle, 11.2%	600	21.0	3.98	248.9	121.0	140	0.37
Temperatures, 1400–2300 K	670	23.8	3.99	255.6	123.9	140	0.37
670-km discontinuity							
	670	23.8	4.38	299.9	154.8	310	0.98
Lower mantle	770	28.3	4.44	313.3	173.0	310	0.97
	1000	38.8	4.58	351.9	185.6	310	0.98
	1250	50.3	4.72	393.2	200.5	310	0.99
Mass fraction of mantle, 73.4%	1500	62.3	4.86	434.1	215.3	310	0.99
Temperatures, 2000–3600 K	1750	74.4	4.99	473.4	229.4	310	0.99
	2000	87.1	5.12	513.8	243.7	310	1.00
	2250	99.9	5.25	554.0	257.5	310	1.00
	2500	113.6	5.37	597.6	272.5	310	1.00
D″ layer	2750	127.6	5.50	642.1	287.1	310	1.01
	2890	135.8	5.57	655.6	293.8	310	0.99

[a] Sources:

Dziewonski, A. M., and Anderson, D. L. (1981). *Phys. Earth Planet. Int.* **25**, 297–356.

Jeanloz, R., and Morris, S. (1986). *Annu. Rev. Earth Planet. Sci.* **14**, 377–415.

[b] Values at a reference frequency of 1 Hz.

Here, dp/dz is the observed density gradient with depth and K_E is an effective bulk modulus describing the actual pressure–density relation of the mantle. The ratio of the expected and observed pressure–density relations is defined as the inhomogeneity parameter $\eta = K_s/K_E$. Combining Eqs. (3) and (5) yields

$$\eta = (\varphi/\rho g)(d\rho/dz). \tag{6}$$

Thus, if η deviates from 1, the observed density profile in the mantle deviates from the expected density variation based on the velocities.

The inhomogeneity parameter can be determined at each depth because all of the terms on the right-hand side of Eq. (6) are obtained from seismological measurements. To find the gravitational acceleration, which is required to determine either the pressure or η, the density profile must be integrated,

$$g(r) = \frac{4\pi G}{r^2} \int_0^r r'^2 \rho(r') \, dr', \tag{7}$$

with $G = 6.67 \times 10^{-11}$ m^3 kg^{-1} sec^{-2} being the gravitational constant and $r = 6371$ km $-z$ being the radial distance from the earth's center (6371 km is the mean radius of the planet). Because of the high density of the core (average value of 10.6 Mg m^{-3}), g is almost independent of depth or radial distance through the mantle: $g \cong 10$ m^2 sec^{-1} (Table I). Given the observed densities, pressure increases with depth by about 35–45 MPa km^{-1} (0.35–0.45 kbar km^{-1}) throughout the mantle [Eq. (4); Table II].

A seismological earth model, such as that summarized in Table II, is based on the observed velocity and density profiles through the globe. From these profiles, the gravitational acceleration, pressure, elastic moduli, and inhomogeneity parameter are derived for each depth from Eqs. (1)–(7). The result is that the density and elastic properties are found to increase smoothly with depth just as a consequence of self-compression through most of the mantle. Specifically, the inhomogeneity parameter is close to 1.0 across nearly the entire lower mantle.

B. Mantle Structure

The most significant deviations from homogeneity occur at the four seismological discontinuities that define the basic layering of the mantle (Fig. 2): the Mohorovičić discontinuity (or Moho) at the top of the mantle, the discontinuity at 400 km depth, the discontinuity at 670 km depth, and the discontinuity at the base of the mantle. The largest changes in properties occur at the first and last of these, the crust–mantle and core–mantle boundaries, but the mid-mantle discontinuities separating the upper and lower mantle represent the predominant internal structure of the mantle. Understanding the causes of the 400- and 670-km discontinuities and the nature of the transition zone between them is a central goal in geophysical and geochemical studies of the mantle.

The Moho is characteristically observed as a sudden increase in longitudinal velocity, from V_p less than 7 km sec^{-1} in the crust to a value exceeding 8 km sec^{-1} at the top of the mantle. Also, the average density increases from ~2.8 Mg m^{-3}, typical of the lower crust, to 3.4 Mg m^{-3} in the upper mantle. The depth of the Moho beneath the surface is quite variable. That is, the thickness of the crust ranges from less than 10 km beneath oceans or in regions of crustal extension to as much as 60 km in areas of continental collision. In contrast, there is no evidence for such large ($\gtrsim$10–30 km) variations in the depths of the other discontinuities.

The upper mantle appears to be the most complex region between the crust and the core: properties vary considerably with depth, resulting in significant deviations from homogeneity; there are major horizontal variations in wave velocity; and this is the largest part of the mantle that is known to be anisotropic in its elastic properties. The complexity of the upper mantle is not altogether surprising in light of other geophysical information. For example, it is known from laboratory studies that the minerals of the upper-most mantle react and are transformed to denser phases at the pressures corresponding to depths of 100–700 km. More will be said about this in Section II, but the transformations expected to occur within the upper mantle can explain most of the observed deviations from $\eta = 1$ at these depths (Table II). Similarly, the evidence for plate tectonics at the surface, and hence convection within the mantle, demonstrates that cold crust sinks into the mantle at subduction zones and hot mantle rises beneath mid-ocean ridges (Fig. 1; see Section III). The resulting temperature variations at a given depth in the mantle would be expected to produce several percent horizontal variations in velocity, as is observed.

In many regions of the upper mantle, seismic waves appear to travel systematically faster in the horizontal than in the vertical direction at a given depth. This transverse anisotropy in elastic properties is not well understood but may be due to a fine-scale layering that is below the limit of seismic resolution. One suggestion has been that the large-scale horizontal flow associated with plate tectonics at the surface preferentially aligns minerals in the uppermost mantle; horizontal alignment of olivine, which is strongly anisotropic in its elastic properties, could explain the transverse anisotropy. Also characteristic of the topmost mantle is the presence of a low-velocity zone: a decrease in wave velocity over a limited depth range. On average, velocities do not increase significantly with depth throughout the upper 200 km (see Table II), but detailed regional studies indicate considerable variation in the exact amount of decrease and in the depths involved (Fig. 2). In general, areas characterized by tectonic activity and high heat flow at the surface exhibit the most pronounced low-velocity zones. This observation supports the conventional interpretation that the decrease in velocity is associated with an especially rapid increase in temperature with depth. As increases in pressure and temperature, respectively, cause increases and decreases in the wave velocities, temperature gradients locally dominate the velocity profiles in the low-velocity zone. Throughout the rest of the mantle down to the D'' layer, the increase in pressure dominates that in temperature with depth, and the velocities systematically increase with depth.

Much of the upper mantle, the entire transition zone, and the top 100 km of the lower mantle exhibit anomalous velocity gradients with depth and, thus, strong deviations from homogeneity (Table II; Fig. 2). To a large extent, these anomalous velocity profiles, including the 400- and 670-km discontinuities, can be explained in terms of phase transformations that are expected at these depths. Less certain is the degree to which the anomalous variations in properties are due to changes in bulk composition with depth. Although not considered likely for dynamic reasons (convection acts to stir the mantle), the seismologically observed variations could be entirely ascribed to variations in composition with depth. Alternatively, because most (and perhaps all) of the anomalous changes in elastic properties with depth can be explained in terms of mineral transformations alone, it is often assumed that bulk composition remains constant with depth throughout the mantle. Whether compositional variations are required to explain the seismological structure of the deep mantle is a major topic of geophysical research. As described in Section III, changes in composition with depth could control the thermal evolution of the mantle.

The apparent homogeneity of the lower mantle makes this the largest region of the interior, comprising 61% of the earth on an atomic basis. Except for the top 100 km and the bottom ~200 km (D'' layer), the lower mantle is remarkably uniform in its seismologically observed properties. Large-scale horizontal variations in velocity are within about 1–2% of the mean values, as compared with variations two to three times larger observed in the upper mantle. This homogeneity is taken as an indication that the lower mantle is thoroughly mixed by convection. Also, the homogeneity is compatible with the fact that no significant phase transformations have been experimentally found in mantle minerals at pressures exceeding 30 GPa (300 kbar).

In contrast with the transition zone, high-pressure phase transformations are usually not invoked to explain the anomalous velocity gradients in the D'' layer at the base of the mantle. The most common explanation is that heat flow out of the core produces an increased temperature gradient, or thermal boundary layer, and hence decreased velocity gradients near the core–mantle boundary (see Section III). This is completely analogous to the explanation for the low-velocity zone as being due to a rapid increase in temperature beneath the surface. Detailed studies, however, indicate that the D'' layer is characterized by a complicated velocity profile, with lateral velocity variations causing seismic waves to be strongly scattered. Thus, in addition to possible temperature effects, there may be a chemical or physical intermixing between the mantle and the core that produces the observed properties of the D'' region.

The core–mantle boundary at a depth of 2890 km involves the largest changes of properties observed within the earth. The transverse-wave velocity vanishes across this boundary, and the longitudinal velocity drops by 41%, demonstrating the fluidity (i.e., complete loss of rigidity) of the outer core [see Eqs. (1) and (2)]. The 72% increase in density from mantle to core reflects a fundamental change in composition, from oxide and silicate minerals in the outer part of the earth to a metallic iron alloy toward the center.

C. Anelasticity

In addition to the structure and elastic properties of the mantle, seismology provides information on the anelastic properties as a function of depth. That is, the mantle is not quite perfectly elastic. What is observed is that the amplitudes of seismic waves decrease with distance and the amplitudes of the free oscillations decrease with time because of damping caused by internal friction processes. Such processes include the movement of dislocations or other defects within minerals and the sliding of neighboring crystals along grain boundaries: the damping is associated mainly with shear deformations. The result is that deformation energy is dissipated as heat, thus diminishing the amplitude of the seismic wave. Therefore, a quantitative measure of the degree of anelasticity is given by the relative decrease in amplitude $\delta A/A$ per cycle of the seismic wave:

$$\delta A/A = \tfrac{1}{2}(\delta E/E) \equiv \pi/Q, \tag{8}$$

In Eq. (8), $\delta E/E$ is the relative decrease in strain energy per cycle and Q is termed the quality factor (a smaller

value of Q implies greater anelasticity than a larger value of Q).

With increasing temperature, internal friction processes are generally activated or enhanced, so Q is found to decrease as the melting point of a rock or crystal is approached. Hence, the maximum shear dissipation, or minimum Q, found in the low-velocity zone (Table II) supports the interpretation that temperature increases rapidly beneath the surface, approaching the melting point of the mantle at 100–200 km depth. The same interpretation was given for the low values of seismic wave velocity in this region. Below 220 km depth, the effect of pressure is apparently to increase the melting point of the mantle more rapidly than the average temperature increases with depth. As a result, the quality factor is relatively high through most of the mantle. Similarly, the relatively cold temperatures near the surface lead to high Q, or more perfectly elastic behavior in the lid above the low-velocity zone (Table II). The seismologically defined lid corresponds roughly to the mantle portion of the lithosphere, the region that behaves elastically in response to tectonic stresses exerted over geological time periods. The lithosphere, containing crust and uppermost mantle, comprises the tectonic plates at the surface.

II. COMPOSITION AND MINERALOGY

A. Upper-Mantle Composition

The bulk composition and mineralogical content of rocks in the uppermost mantle are essentially determined by direct observations. For most of the mantle, however, the main constraints on composition and mineralogy come from the interpretation of geophysical observations, most notably the seismological data. In this approach, the compositions of broad regions, such as the entire upper mantle, are treated as uniform; lateral variations in composition, in particular, are for the most part considered to be of secondary importance. Given the large variations in rock types observed at the surface, this coarse description of mantle composition may seem simplistic, and it certainly reflects the indirect nature of the observations involved. Nevertheless, it is more justifiable for the mantle than for the crust because of the high temperatures of the interior, near the melting point of the minerals in the mantle. For example, the effects of partial melting on a small scale and of convection on a large scale are expected to smooth out in the mantle many of the compositional variations that characterize the cold crust at the surface. In fact, variations observed among rocks of the upper mantle are generally subtle compared with those of the crust and mainly involve differences in the contents of minor or trace elements. Very locally, mantle compositions can differ substantially from the average, as exemplified by the eruption of diamonds in kimberlite, but the occurrences are rare.

Mantle rocks are brought to the surface by two types of processes: tectonic and volcanic. The main example of the former is the occurrence of ophiolites, slivers of oceanic crust that over geological time periods are thrust upward and onto preexisting continental or oceanic crust. The process involved is not well understood, but the product has been extensively studied. Ophiolites characteristically exhibit a rock sequence of marine sediments overlying a thin basaltic layer typical of oceanic crust. Beneath the basalt are found the ultramafic rocks (rocks dominated by magnesium silicate minerals) that are thought to be common in the upper mantle. These rocks are mainly peridotite and dunite, which consist of the minerals olivine $[(Mg, Fe)_2SiO_4]$ plus pyroxene $[(Mg, Fe, Ca)SiO_3]$ and of olivine alone, respectively. It is worth pointing out that the wave velocities of ophiolite rocks measured in the laboratory match the seismologically observed profile of wave velocities through sediments, crust, and mantle underneath oceans.

Rock fragments brought up volcanically from the interior, termed xenoliths, often include samples from the lower crust and upper mantle. Among these, ultramafic xenoliths found in certain basalt flows are considered to originate in the mantle. In addition, some ultramafic fragments are brought up along with diamonds in kimberlite, a gas-rich fluid that is explosively erupted through the crust. Diamond is a highpressure form of carbon, which rapidly turns to graphite if heated at pressures less than ~ 5 GPa (50 kbar). Therefore, kimberlite eruptions containing diamonds must originate at depths of at least 150 km; the accompanying peridotite xenoliths are considered to be among the deepest samples available from the mantle.

On the basis of these observations, garnet peridotite, consisting of approximately 50–60% olivine, 20–40% pyroxene, and 10% garnet $[(Mg, Fe, Ca)_3Al_2Si_3O_{12}]$, is thought to be the primary rock making up the upper mantle. Indirect evidence for this conclusion comes from the study of basalt, which originates by partial melting of the mantle. Because basalt is by far the most voluminous lava erupted from the interior, solidifying to form the crust beneath the oceans, its source is considered representative of the upper mantle. Experimental melting studies demonstrate that partial melting of a peridotitic composition at plausible temperatures and pressures for the upper mantle does yield basaltic liquid. Because iron and aluminum enter preferentially into the liquid, continued melting depletes the source rock in these constituents. Mineralogically, the result is that garnet (the main aluminum-bearing phase) and ultimately pyroxene are lost with increased

melting: refractory dunite can be the end product. Garnet-bearing periodotite xenoliths, being relatively undepleted, are sometimes termed fertile. The implication is that fertile rock can yield basalt on melting, whereas infertile (garnet-poor, depleted) xenoliths cannot. Thus, considerable effort is made to identify the least depleted compositions in the examination of mantle rocks brought to the surface. These correspond most closely to primary, unmelted mantle.

The estimated composition of the uppermost mantle is summarized in Table III. Typically, rock analyses are presented in weight percentages of oxide components (left column), but a translation to atomic percentages is presented for clarity (right side). First, this emphasizes that the primary atomic constituent of the rock is oxygen, with magnesium and silicon being of secondary importance. Second, the uncertainties are minor for the abundances of the first three elements, comprising 94% of the rock. The minor-element abundances that follow are less certain because the percentages are small and, with the effects of partial melting on depletion of these elements being more severe, they exhibit more scatter in natural samples. From the atomic fractions, the ratio of magnesium to iron plus magnesium components is near $X_{Mg} = 0.9$; for comparison, typical olivine and pyroxene compositions in mantle peridotites are roughly $(Mg_{0.9}Fe_{0.1})_2SiO_4$ and $(Mg_{0.9}Fe_{0.1})SiO_3$. Finally, the ratio of olivine to olivine and pyroxene constituents is given as $Ol/(Ol + Px) \cong 0.54$, in accord with the description of garnet peridotite given above.

B. Mineral Transformations

Up to this point, mineral names have been used to designate compositions. In addition, the names often indicate a particular phase or crystal structure (Table IV). Thus, *olivine* and *garnet* refer to specific crystallographic structures that are orthorhombic and cubic, respectively. There are actually two types of pyroxenes that are commonly observed. These are orthopyroxene and clinopyroxene, with orthorhombic and monoclinic crystal structures, respectively. Nevertheless, because the compositions overlap, the high-pressure transitions are similar, and the bulk physical properties are nearly identical; the two types of pyroxene are not distinguished here. In general, whether a mineral name designates a composition or a crystal structure should be clear from the context in which it is used.

Xenoliths provide samples of the mantle to depths of 150–200 km, with fragments originating from greater depths having been less conclusively documented. Hence, the nature of the deeper mantle is inferred on the basis of comparing the seismologically observed densities and elastic moduli with the results of high-pressure

TABLE III Models of Upper-Mantle Composition[a]

Oxide component	Weight fraction (%)[b]	Element	Atomic fraction (%)
SiO_2	45.0 ($\pm$1.4)	O	58.3
TiO_2	0.18 ($\pm$0.05)	Mg	20.1 ($\pm$1.0)
Al_2O_3	4.5 ($\pm$1.5)	Si	15.8 ($\pm$0.5)
Cr_2O_3	0.4 ($\pm$0.1)		
FeO[c]	7.6 ($\pm$1.7)	Fe	2.2 ($\pm$0.5)
MnO	0.11 ($\pm$0.05)	Al	1.9 ($\pm$0.6)
NiO	0.23 ($\pm$0.05)	Ca	1.2 ($\pm$0.4)
MgO	38.4 ($\pm$2.0)	Na	0.27 ($\pm$0.14)
CaO	3.3 ($\pm$1.0)	Cr	0.11 ($\pm$0.03)
Na_2O	0.4 ($\pm$0.2)		
K_2O	0.01 ($\pm$0.1)	Ni	0.06 ($\pm$0.01)
		Ti	0.05 ($\pm$0.01)
Total	100.1	Mn	0.03 ($\pm$0.01)
Atomic or molar proportions		K	0.004 ($\pm$ 0.04)

$$X_{Mg} = \frac{Mg}{Fe + Mg} = 0.90 \, (\pm 0.05)$$

$$\frac{Ol}{Ol + Px} = \frac{Mg + Fe + Ca - Si}{Si} - Ca = 0.54 \, (\pm 0.08)$$

[a] Sources:

Aoki, K. (1984). *In* "Materials Science of the Earth's Interior" (I. Sunagawa, ed.), pp. 415–444, Terra Scientific, Tokyo.
Basaltic Volcanism Studies Project (1981). "Basaltic Volcanism on the Terrestrial Planets," Pergamon, New York.
Green, D. H., Hibberson, W. O., and Jacques, A. L. (1979). *In* "The Earth: Its Origin, Structure and Evolution" (M. W. McElhinny, ed.), pp. 265–299, Academic Press, New York.
Ringwood, A. E. (1975). "Composition and Petrology of the Earth's Mantle," McGraw-Hill, New York.
Yoder, H. S. (1976). "Generation of Basaltic Magma," Natl. Acad, Sci., Washington, DC.

[b] Average values and uncertainties are representative of the range of most published values.

[c] All iron listed as Fe^{2+}.

experiments on garnet peridotite and its constituent minerals. The most important effect of increasing pressure is that the minerals of the uppermost mantle are transformed to denser phases, with major changes in physical properties being observed. The mineral phases involved are summarized in Table IV, but to simplify the discussion they are considered in two separate groupings: (1) olivine and its high-pressure phases; (2) pyroxenes and garnet and their high-pressure phases.

Aside from being the most abundant mineral of the upper mantle, olivine exhibits structural characteristics that are typical of minerals occurring near the earth's surface. Specifically, four oxygen atoms are coordinated around each silicon atom, and six oxygen atoms are coordinated around each of the other cations (e.g., Mg and Fe). The packing of these oxygen polyhedra, the tetrahedra around silicon and the octahedra around magnesium or iron, is commonly not very efficient or space saving in crustal minerals, thus resulting in relatively large volumes or low densities. Such is the case for olivine, but a more efficient packing is achieved in its highpressure forms; as is required thermodynamically, the high-pressure polymorphs are systematically denser.

At pressures corresponding to the bottom of the upper mantle, olivine is trasformed to two related crystal structures: that of spinel (γ phase), a cubic structure, and that of a related spinelloid structure, which is orthorhombic (β phase). There is no change in oxygen coordinations around the cations (Table IV), but the enhanced structural packings leas to significant increase in density and elastic moduli compared with olivine. Consequently, the 400-km seismic discontinuity is ascribed to an experimentally observed high-pressure reaction between the olivine, β-phase, and γ-spinel polymorphs, with the changes in elastic moduli being similar to what is observed seismologically in the mantle. Because the pressure at 400 km depth is known (Table II) and the effect of temperature on the pressure of the reaction among the olivine polymorphs has been measured in the laboratory, this reaction has been used to estimate the average temperature to be $\sim$1700 ($\pm$300) K at the top of the transition zone (400-km depth).

The reason that olivine, on the one hand, and pyroxenes and garnet, on the other, can be considered separately is that these mineral groups tend not to react, but are transformed separately as the garnet peridotite is taken to higher

TABLE IV Properties of Mantle Minerals[a,b]

Mineral phase	Crystal chemical formula[c]	Pressure range of existence (GPa)	Density[d] ρ (Mg m^{-3})	Adiabatic bulk modulus K_s (GPa)	Shear modulus μ (GPa)	Thermal diffusivity[e,f] κ (10^{-6} m^2 sec^{-1})	Thermal expansion coefficient[e] α (10^{-5} K^{-1})
Olivine	^{VI}(Mg, Fe)$_2^{IV}$ SiO$_4$	0–13	3.22	129	81	0.7	4.0
Pyroxene[g]	^{VI}(Mg, Fe, Ca)^{IV} SiO$_3$	0–13	3.21	108	76	1–2	4.2
Garnet–	^{VIII}(Mg, Fe, Ca)$_3^{VI}$ Al$_2^{IV}$ Si$_3$O$_{12}$	0–30	3.56	174	89	1–2	2.9
majorite	^{VIII}(Mg, Fe, Ca)$_3^{VI}$ [(Mg, Fe)Si]^{IV}Si$_3$O$_{12}$		3.51	220	—[h]	—[h]	—[h]
β Phase	^{VI}(Mg, Fe)$_2^{IV}$ SiO$_4$	13–17	3.47	174	114	—[h]	3.4
γ Spinel	^{VI}(Mg, Fe)$_2^{VI}$ SiO$_4$	17–26	3.55	184	119	1	2.7
Magnesio-wüstite	^{VI}(Mg, Fe)O	0 to >100	3.58	163	131	2.6	4.8
Perovskite	$^{VIII–XII}$(Mg, Fe, Ca)^{VI}(Si, Al)O$_3$	20 to >70	4.10	265	185	—[h]	4.0

[a] Unless otherwise stated, these values were experimentally determined at zero pressure and room temperature.

[b] Sources:

Jeanloz, R., and Thompson, A. B. (1983). *Rev. Geophys. Space Phys.* **21,** 51–74.

Knittle, E., Jeanloz, R., and Smith, G. L. (1985). *Nature* **319,** 214–216.

Knittle, E., and Jeanloz, R. (1987). *Science* **235,** 668–670.

Weidner, D. J., Sawamoto, H., Sasaki, S., and Kumazawa, M. (1984). *J. Geophys. Res.* **89,** 7852–7860.

Yeganeh-Haeri, A., Weidner, D. J., and Ito, E. (1989). *Science* **243,** 787–789.

[c] Coordination number of oxygen around cation sites is indicated by superscript roman numerals.

[d] Values for magnesium-end-member composition.

[e] High-temperature values, measured at ∼1000 K.

[f] Thermal diffusivity is related to thermal conductivity k by $k = \kappa\rho C_p$, with C_p($\sim$1kJ K^{-1} kg^{-1} for mantle minerals at high temperature) being the specific heat at constant pressure.

[g] Both clinopyroxene and orthopyroxene are included under this heading.

[h] Unmeasured value.

presures. In general, the olivine phases maintain a constant composition, with aluminum, calcium and many of the trace elements remaining concentrated in the pyroxene–garnet constituents. Like olivine, the pyroxenes consist of relatively inefficiently packed oxygen tetrahedra (around silicon) and octahedra (around magnesium, iron, and calcium). In contrast, garnet is a more densely packed cubic structure, with magnesium, iron, or calcium cations being eightfold coordinated by oxygen. Although it exists in the crust, garnet exhibits a remarkable stability under pressure and can in many respects be considered a high-pressure phase. Thus, with increasing depth the pyroxene constituents, of composition (Mg, Fe, Ca)SiO$_3$, dissolve into the garnet structure. The pure silicate garnet of this composition is a high-pressure phase of pyroxene called majorite; in the mantle some aluminum is undoubtedly present, but in a lesser amount than in garnets common to the crust (Table IV). What is interesting about the majorite garnet is that some of the silicon is octahedrally coordinated by oxygen, a coordination that is characteristic of high-pressure silicate phases.

At pressures of the transition zone and top of the lower mantle, above ∼20 GPa (200 kbar), both constituents of garnet periodtite, γ-spinel and garnet at these depths, are transformed to the dense perovskite structure of approx-

imate composition (Mg, Fe, Ca)(Si, Al)O$_3$. In addition, lesser amounts of oxide [(Mg, Fe)O, magnesiowüstite] are produced from the breakdown of the olivine phases. The silicate perovskite is orthorhombic in symmetry, being a distorted form of the ideal perovskite structure. The coordinations involved are octahedral (6-fold) for oxygen around the silicon or aluminum and 8- to 12-fold for the distorted polyhedron of oxygen around the other cations (Table IV).

Experimental and theoretical work demonstrates that the changes in seismological properties from about 600 to 800 km depth can be ascribed largely to the appearance of perovskite over this depth range.

Befitting its high density, silicate perovskite is exceptionally stable under pressure; it is known to exist to at least 120 GPa (1.2 Mbar), and it is probably stable to the 136 GPa (1.38 Mbar) pressure at the base of the mantle. Along with the fact that the predominant minerals of the upper mantle are ultimately transformed to this phase, the high-pressure stability of the high pressure stability of the perovskite ensures that this is the most abundant mineral of the mantle.

Broadly speaking, the density and elastic moduli of the lower mantle can be explained in terms of the measured properties of high-pressure phases (Table IV) corresponding to the composition of the upper mantle (Table III). In detail, however, there may be discrepancies between observed and expected properties of the lower mantle, suggesting the possibility of compositions that are slightly ($\lesssim 10\%$) enriched in iron or silicon relative to the upper mantle. No consensus exists on the magnitude of these enrichments or even on whether the deep mantle differs in bulk composition from the upper mantle. This question of the uniformity (or alternatively the variation with depth) of the mantle composition is particularly significant for models of the dynamics and thermal state of the earth's interior.

III. DYNAMICS AND TEMPERATURE

A. Mantle Viscosity

The high temperature of the planetary interior cause the mantle to convect over geological time periods. The reason this is possible is that the ductility of crystalline rocks increases rapidly with temperature: the movement of grain boundaries, dislocations, and other defects allow the minerals to deform by creep in the solid state. In contrast, the cool outer layer of the earth is strong and behaves elastically; this is the lithosphere that makes up the tectonic plates at the surface. The thickness of the lithosphere is ~100 km, encompassing the entire crust and the top part of the uppermost mantle. The ductile region immediately underlying the lithosphere, termed the asthenosphere, occurs at about the depth of the sesismological low-velocity zone (see Table II).

The ductility of the mantle is directly inferred from observations of the rate at which the earth's surface is responding to the disappearance of the Pleistocene ice sheets. The load of the ice that covered much of the Northern Hemisphere during the Pleistocene ice age depressed the surface by hundreds to thousands of meters. In response to the melting of the ice over the past 100,000 years, the lithosphere has been rebounding upward. By dating and measuring the elevation of ancient shorelines, the amount of uplift relative to sea level is documented as a function of time. Because the lithosphere can move upward only to the degree that the mantle flows back under the glacially induced depression of the surface, the flow rate or viscosity of the mantle can be determined from the observed rate of uplift.

The measurement of uplift rates and regional gravity anomalies caused by the glacial unloading has been complemented by the determination of $\dot{J}_2$, the time dependence of the second harmonic of the earth's gravity field. This quantity is obtained from precise tracking of satellite orbits over several years. The combined mea-

surements yield values ranging between 10^{19} and 10^{22} Pa sec for the viscosity through the upper mantle and into the lower mantle. Compared with laboratory experience, such values are large for the viscosity of a fluid: five to eight orders of magnitude higher than the conventional value at which glass is considered to be solid, for example. The difference is that over geological time scales of 10^{13} to 10^{17} sec, the viscosity of the crystalline mantle is relatively small. Since shear-deformation rates are $\sim 10^{-16}$ sec^{-1} for the mantle (see Section III.B), a viscosity of 10^{22} Pa sec or less implies that shear stresses of only 10^6 Pa (10 bars) are sufficient to cause large-scale deformation of the mantle. As noted previously, shear stresses in the mantle are negligible compared with the pressures achieved.

The glacial-rebound data suggest that the viscosity of the mantle varies with depth, with a region of low viscosity occuring at about 100–250 km beneath the surface. The depth dependence of the viscosity is obtained by Fourier analyzing the uplift measurements as a function of location: the longer spatial wavelengths provide a deeper sampling of mantle viscosity than the shorter wavelengths. Modeling the data in terms of an elastic plate, the lithosphere at the surface, overlying a fluid of variable viscosity yields a value $v \cong 5 \times 10^{19}$ Pa sec for the viscosity in the asthenosphere, the low-viscosity channel beneath the lithosphere. Below ~200 km the viscosity is thought to increase to $\sim 10^{21}$ Pa sec, remaining constant thereafter to depths exceeding 700 or 800 km (the uncertainty increases rapidly beyond ~700 km depth). The lithosphere includes the entire crust and the part of the upper mantle corresponding to the seismologically observed, high-Q lid (Table II). The astheno-sphere is defined in terms of its low viscosity over geological time periods, but it occurs at about the same depth as the seismological low-velocity, low-Q zone. The low viscosity is qualitatively explained by the high, near-melting temperatures at this depth, just as was the case for explaining the observed minimum in Q and velocity. That temperatures are close to the melting point in the asthenosphere is confirmed by the occurrence of mantle xenoliths, which are brought up by volcanic processes and therfore indicate that partial melting occurs locally at depths of 100–150 km below the surface.

B. Mantle Convection

Several lines of evidence demonstrate that the mantle convects vigorously. Since the mind-1960s and 1970s, the tectonic processes or geological deformations observed at the surface have been widely accepted as being a consequence of convective flow of the mantle. Plate tectonics, in particular, is directly attributable to the dynamics of the underlying mantle. In addition, virtually all of the heat lost from the deep mantle and core is transported by convection. That is, flow of the mantle cools the planetary interior and results in a highly nonuniform pattern of outward heat flux observed at the surface.

The plate tectonic cycle of oceanic lithosphere involves the creation of the lithospheric plate at the mid-ocean ridges, horizontal movement of the plate at 1–10 cm year^{-1}, and sinking of the lithosphere into the mantle at subduction zones (Fig. 1). This process is well documented by the pattern of earthquake sources around the globe, by the symmetry of magnetic anomalies recorded in the oceanic crust, and by geological evidence for large-scale displacements along plate boundaries, such as the San Andreas fault. Since the mid-1980s, very long baseline interferometric (VLBI) and Global Positioning System (GPS) measurements at the surface have directly recorded the relative movements of plates at the centimeter per year velocities expected from the geological and magnetic-anomaly observations. Along with the horizontal movement of plates, the upwelling of mantle rock beneath the mid-ocean ridges where the plates separate and the downwelling associated with the converging plates at oceanic trenches (i.e., along subduction zones) define the convective flow pattern of the upper mantle. The rate of shear deformation associated with mantle flow is roughly given by the ratio of the observed flow velocity ($u \sim 1$ cm year$^{-1} \sim 3 \times 10^{-10}$ m sec^{-1}) to the length scale over which flow occurs. With the depth of the mantle being $D \cong 3 \times 10^6$ m, the deformation rate is $\sim 10^{-16}$ sec^{-1}.

The fact that most of the earth's internal heat emerges through mid-ocean ridges and the adjoining young oceanic crust provides additional documentation of convection in the mantle. Quantitatively, more than 50% of the total heat lost from the interior emerges through oceanic crust less than 50 million years old. In comparison, the age of the earth is ~ 100 times greater ($4.5 \sim 10^9$ years), and the average age of old oceanic crust is ~ 120 million years upon subduction. This means that the most intense transfer of heat out of the planetary interior is associated with the upwelling mantle beneath the mid-ocean ridges.

There are three possible mechanisms of heat transfer within the mantle: conduction and radiation, in which heat is transmitted *through* the rock, and convection, by which hot rock is bodily transported to a cooler environment (e.g., from the interior to the surface). Laboratory measurements demonstrate that radiative transfer, which is limited by the opacity of the minerals, is only a fraction of the heat conducted through the rock. Thus, both radiative and conductive transport of heat are combined into an effective conductivity; the corresponding thermal diffusivity of mantle rock is experimentally observed to be near $\kappa \sim 10^{-6}$ m^2 sec^{-1} (see Table IV). With this value, a quantitative comparison between the relative effectiveness of heat transport by convection, as opposed to conduction plus radiation, is offered by the Peclet number:

$$\text{Pe} = lu/\kappa. \tag{9}$$

The Peclet number is simply the ratio of time needed for heat to be transmitted a distance l by convection vs conduction. Because vertical (upwelling and downwelling) velocities in the mantle must be comparable to plate tectonic velocities at the surface, $u \sim 3 \times 10^{-10}$ m sec^{-1}, the Peclet number is in the range of 10^2–10^3 for distances greater than 10^2–10^3 km. That is, for upward heat transfer on a global scale, convection predominates over conduction and radiation by a factor of 100–1000.

One of the main consequences of mantle flow is the formation of lithosphere at the surface by the upwelling of hot rock from depth at mid-ocean ridges. From the mid-ocean ridge, heat is transported horizontally as the plate moves toward the subduction zone. At the same time, upward conduction of heat allows the plate to cool from the top. As it cools, the plate becomes denser because of the effect of thermal expansion with the thermal expansion coefficient

$$\alpha = -(1/\rho)(\partial\rho/\partial T)_\text{P} \tag{10}$$

being of the order of 10^{-5} K^{-1} (Table IV), the density increases by a small percentage in response to a temperature decrease of 1000 K.

The amount of cooling of the lithospheric plate is determined by the thermal diffusivity of the rock. Over a time period Δt, cooling reaches a characteristic depth

$$\delta = \sqrt{\kappa \Delta t}. \tag{11}$$

Therefore, the density of the plate increases in proportion to the square root of its age Δt. An analysis of the balance of forces involved demonstrates that the increasing density of the plate causes it to sink into the mantle; that is, the bathymetric depth beneath the surface of the ocean increases with the square root of age as the plate moves away from the mid-ocean ridge and cools. Moreover, the cooling is predicted to cause the heat flux out of the top of the plate to decrease inversely with the square root of age. The quantitative agreement between the predicted $\sqrt{\Delta t}$ behavior and the observed bathymetry and heat flow through the oceanic crust provides a powerful justification for associating plate tectonics at the surface with convective heat transfer in the mantle (Fig. 3).

In addition to these surface observations, the upper and lower mantle are expected to convect vigorously, based on the viscosities inferred inferred from the postglacial-rebound measurements. To demonstrate this conclusion, the balance of forces driving convection against the resistive and dissipative forces opposing convection must be considered. Imagine that heating is applied at depth D

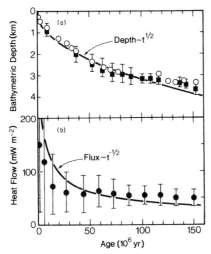

FIGURE 3 (a) Average values of bathymetric depth below the mid-ocean ridge crest as a function of age for crust beneath the North Pacific (○) and North Atlantic (■). (b) Average value of heat flow as a function of age for crust beneath all oceans; the large error bars are due to effects of hydrothermal circulation, especially in young crust. The theoretically expected dependencies, which are directly and inversely proportional to the square root of age, are shown by solid lines. [Data from Sclater, J. G., Anderson, R. N., and Bell, M. L. (1971). *J. Geophys. Res.* **76**, 7888–7915. Parsons, B., and Sclater, J. G. (1977). *J. Geophys. Res.* **82**, 803–827. Sclater, J. G., Jaupart, C., and Galson, D. (1980). *Rev. Geophys. Space Phys.* **18**, 269–311.]

within a fluid body, such that the temperature at this depth exceeds that at the surface by an amount ΔT. As a result, a buoyancy force $\alpha g \rho \, \Delta T$ arises from the temperature-induced decrease in density at depth; the hot, less dense fluid attempts to rise toward the surface [see Eqs. (4) and (10)]. This buoyancy would lead immediately to convection, except that it is resisted by the viscosity v and thermal diffusivity κ of the fluid. Thus, a nondimensional ratio of forces, the Rayleigh number, describes the likelihood and vigor of convection:

$$\mathrm{Ra} = \alpha g \rho \, \Delta T \, D^3 / \kappa v. \qquad (12)$$

Convection occurs if the Rayleigh number exceeds ~1000, and it is extremely vigorous for Ra above 10^6 (the exact values depend on the specific pattern or flow and are established by experiment or detailed numerical analysis). As will be seen in Section III. D, the base of the mantle ($D \cong 3 \times 10^6$ m) is at least $\Delta T = 2500$ K hotter than the surface, so the Rayleigh number of the mantle is in the range of 10^5–10^9 based on the observed viscosity. That is, despite the fact that the mantle is crystalline and highly viscous on a laboratory time scale, it is expected to undergo vigorous convection over geological time periods. The surface observations of plate movements and heat flow confirm this expectation.

The overall flow pattern throughout the mantle is not well known, but is most directly imaged by seismological methods. Seismic tomography, akin to ultrasonic imaging used in medicine, makes it possible to determine horizontal variations in wave velocity at depth within the mantle. Figure 4, for example, showing evidence for the cold (fast seismic velocity) lithosphere sinking westward into the mantle underneath Japan, suggests that the subducting material extends well into the lower mantle to at least 800 km depth, and perhaps 1800 km if not 2800 km depth. If the latter, the schematic illustration in Fig. 1 is correct in showing a flow pattern extending across the entire mantle, which would therefore be completely mixed and of uniform composition.

The resolution of seismic tomography is currently inadequate to prove that such a simple, through-going pattern of flow is completely appropriate for the mantle. To what degree are high-velocity regions fortuitously lined up in such cross-sections as Fig. 4? Could the pattern of flow have changed over geological history? These and related questions have kept open the debate about how uniformly mixed the mantle really might be.

This is significant because if the deep mantle differs in composition from the upper mantle (see Section II), these regions may convect separately (Fig. 5). The reason that compositional layering of the mantle can modify the flow pattern is that the density differences associated with thermal convection are small. That is, the thermally induced buoyancy force driving convection in the mantle is caused by a density variation $\alpha \rho \Delta T$ [Eq. (12)] that amounts to only a few percent. Since comparable changes in density are produced by small variations in composition, the thermal buoyancy force can be readily counteracted by a change in composition. Specifically, if the deeper mantle is intrinsically denser than the upper mantle, due, for example, to an enrichment in iron content by a small percentage, then a layered system of convection would be expected (Fig. 5; in this case, the Rayleigh number would be calculated using $D = 7 \times 10^5$ m and $D = 2.2 \times 10^6$ m for the upper mantle and the lower mantle, respectively). The fact that such a small change is within the uncertainty of mantle composition estimates (see Table III) emphasizes the importance of both compositional and thermal variations in controlling the dynamics of mantle flow. It is currently thought that if there exists a change in bulk composition within the mantle, it is likely to occur at variable depths and possibly deep in the lower mantle (rather than the single "boundary" shown in Fig. 5).

C. Heat Sources

Because the amount of heat lost through the surface is relatively well known, an estimate of the heat being produced

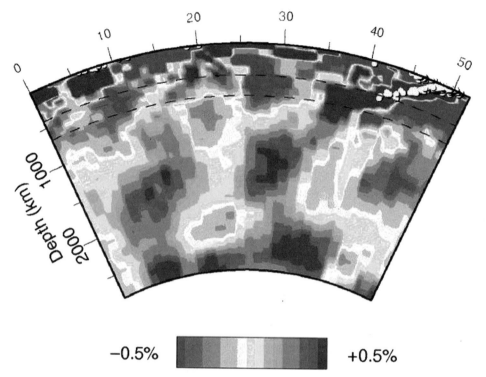

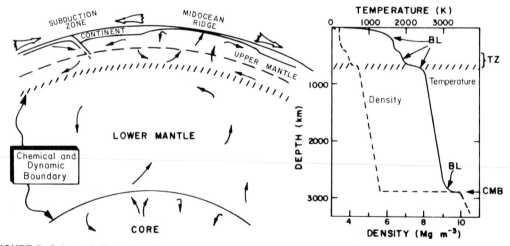

FIGURE 4 East-west cross-section through the mantle beneath Asia, at about 40°N latitude and extending 52° in longitude, shows horizontal variations in V_p (see Fig. 2): blue and red indicate regions having velocities that are, respectively, faster and slower than average at that depth (see scale at bottom). Earthquakes, indicated by open circles, are especially evident along the subduction zone beneath Japan (upper right). Because increasing temperature typically reduces the seismic-wave velocity in rock blue (seismically fast) regions within the mantle are inferred to be colder than average, hence probably sinking. [After Van de Voo, R., Spakman, W., and Bijward, H. (1999). *Nature* **397,** 246–249.]

FIGURE 5 Schematic illustration of an alternative model to Fig. 1. Here, the lower-mantle and upper-mantle compositions differ, and these regions convect separately. The resulting temperature profile involves thermal boundary layers (BL) at the 670-km discontinuity and therefore temperatures in the lower mantle that are higher than those shown in Fig. 1. CMB, core–mantle boundary; TZ, transition zone. [From Jeanloz, R. (1981). *In* "Proceedings of the American Chemical Society 17th State-of-the-Art Symposium," p. 74, Am. Chem. Soc., Washington, DC.]

internally is required to determine the energy balance associated with convection. The planet is believed to be cooling, for example, and this implies that more heat is being lost than is being produced. The thermal evolution of the interior reflects this balance of heat sources and heat sinks through time. In addition, because the heat sources ultimately provide the energy that drives convection in the mantle, the tectonic and thermal evolution of the earth are directly linked.

Radioactive decay is thought to provide most of the heat currently produced in the mantle. The isotopes $^{238,235}U$, ^{40}K, and ^{235}Th are the most significant radiogenic sources; other isotopes either exist in too low an abundance or produce too little energy over too long a time to be important. The present concentrations of uranium, potassium, and thorium are inferred from the abundances measured in mantle rocks: typical estimates are about 0.02, 400, and 0.1 ppm by weight. Partial melting can affect the concentrations of these trace elements, but the values are close to those observed in chondritic meteorites. This suggests that the relative elemental abundances in the solar system largely determine the amount of radioactive heat production in the earth and other planets. The resulting heat production in the mantle is estimated to be near 2×10^{13} W, with a probable uncertainty of nearly 50% (Table 1). Since the half-lives of the isotopes are known, this value can be extrapolated back to an estimated radioactive heat production of about 1 to 3×10^{14} W at the origin of the earth, 4.5 billion years ago. The decrease in radioactive heat production with time is an inherent consequence of the decay process, by which the abundance of radiogenic isotopes decreases with time.

Not only was radioactive heating more intense early in earth history, but a number of other heat-producing processes are thought to have been especially important at or near the time the planet formed. These primordial heat sources include, most notably, the energies of accretion and differentiation of the earth. The former basically involves the shock-induced impact heating created as particles and planetesimals aggregate to form the growing planet. Differentiation refers to the chemical unmixing of the planetary interior, the most important example being the separation of the metallic core from the silicate and oxide mantle. The gravitational energy released in such an unmixing process is enormous: estimates of how much energy could have been released during the formation of the earth's core are $\sim 10^{31}$ J. This would correspond to an average heat production of 7×10^{13} W throughout geological history (twice the present heat loss), if the heat were released continuously. Differentiation is thought to have been more sudden, however, contributing mainly in the first 10^8–10^9 years after formation. Still,

the possibility that some amount of chemcial differentiation is presently underway cannot be ruled out; the complex nature of the D″ layer, for example, suggests ongoing chemical interaction between the mantle and the core (see Section I).

In addition to the contributions of radioactive and primordial heat sources within the mantle, similar energy sources in the core are expected to heat the base of the mantle. The amount of heat lost from the core to the mantle is estimated, in part, on the basis of thermodynamic models, in part, on the basis of thermodynamic process that generates the earth's magnetic field inside the core. Either radioactive or chemical differentiation sources are plausibly sufficient to provide the energy required to produce the magnetic field. The resulting estimates of heat loss into the mantle range between about 2 and 12 TW but again must be considered uncertain.

To summarize, the present heat production inside the earth is estimated to be roughly one-half of the heat lost at the surface. The main heat sources are thought to be radioactive at present, but both radioactive and primordial sources were significant during the early part of the earth's history. Heat production has probably decreased by about one order of magnitude in 4.5 billion years, and it is thought to have been less than contemporaneous heat losses. That is, the earth has been cooling throughout geological history. This conclusion is supported by geological observations summarized in the next section.

D. Temperature Distribution

Estimates of the current, average temperature as a function of depth (the geotherm) through the mantle are summarized in Fig. 6. The main constraints on estimates of the temperature at depth arise from experimental determinations of melting temperatures. The mantle is known to be crystalline, based on the seismological evidence for a finite rigidity and the fact that xenoliths are crystalline. Therefore, the average temperature in both the upper mantle and the lower mantle must be below the melting point of peridotite and of the silicate perovskite phase, as shown in the figure. In contrast, the iron alloy making up the outer core is molten, thus providing a lower bound for the temperature at the base of the mantle (Fig. 6). In addition, the temperature at the 400-km discontinuity is obtained from the olivine-β-phase-γ-spinel reaction, indicated by $\alpha \rightarrow \beta$ in the figure (see Section II). All of these solid–liquid and solid–solid phase equilibria have been determined from high-pressure experiments. Finally, the temperatures and pressures at which xenoliths were formed in the upper mantle can be obtained from detailed measurements of the constituent olivine, pyroxene, and garnet compositions.

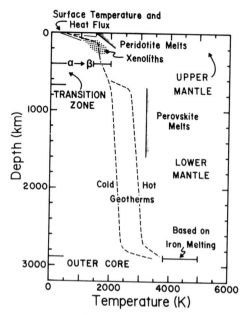

FIGURE 6 Summary of experimental constraints on the present, horizontally averaged temperature as a function of depth (geotherm) through the crust and the mantle. Melting temperature for peridotite and perovskite define upper limits for the geotherm through much of the mantle. At the base of the mantle, high-pressure determinations of melting temperatures for iron and iron alloys provide a lower limit for the geotherm (the estimated uncertainty is shown). Temperatures are relatively well determined in the crust and the upper mantle from the surface heat flux, the xenolith compositions, and the depth of the olivine–spinel reaction ($\alpha \rightarrow \beta$, with an uncertainty as indicated). A characteristic geotherm is illustrated by the dashed line in this region. Two possible geotherms are shown for the lower mantle ("hot" and "cold" dashed lines), reflecting greater uncertainties in the average temperature at these depths (see Figs. 1 and 4). [From Jeanloz, R., and Morris, S. (1986). *Annu. Rev. Earth Planet. Sci.* **14**, 380.]

The compositions of the coexisting minerals depend on pressure and temperature in a way that has, again, been reproduced in the laboratory.

The sigmoidal shape of the geotherm through the mantle is characteristic of what would be expected for a convecting system. In the interior of the mantle, heat is transported adiabatically in the vertical direction by the convective flow. The adiabatic change of temperature with pressure (or depth) is experimentally found to be small for minerals, amounting, for example, to less than 1000 K across the entire lower mantle. (The adiabatic gradient is somewhat larger through the transition zone because of the contribution of heats of reaction associated with the mineral transformations; Fig. 6.) At the top and bottom, however, the flow of the mantle is entirely horizontal, so heat can be transferred outward (vertically) only by conduction. Thermal conduction is a relatively inefficient process of heat transfer for minerals, which characteristically have low values of thermal conductivity k (see Table IV).

The vertically conducted heat flux is given by Fourier's law,

$$\text{Flux} = k(dT/dz), \qquad (13)$$

in one dimension. The observed flux at the surface ($\sim 7 \times 10^{-2}$ W m^{-2}) and the known thermal conductivity of rock (~ 3 W K^{-1} m^{-1}) require that the average temperature gradient with depth z be ~ 20 K km^{-1} at shallow depths, ~ 50 times larger than the adiabatic gradient deeper in the mantle.

The regions in which flow is predominantly horizontal, and hence vertical heat transfer is mainly conductive, are known as thermal boundary layers. Most of the temperature change along the mantle geotherm occurs in the thermal boundary layers, in particular the one at the surface and the one thought to exist at the base of the mantle (the latter, on the assumption that some heat enters the mantle from the core). These are associated with the lithosphere at the top and the D″ layer at the bottom, both regions differing significantly in seismic wave velocity and attenuation Q from the mantle nearby (see Section I). The characteristic thickness of the top thermal boundary layer is given by δ, the depth to which conductive cooling penetrates from the surface [Eq. (11)]. With a typical age of ~ 120 million years ($\Delta t \sim 4 \times 10^{15}$ sec) at subduction, δ is ~ 60 km; values of the order of 10^5 m are thought to be typical for boundary layer thicknesses in the mantle. The corresponding temperature increase across the thermal boundary layer, $\Delta T \sim \delta$ flux/k [Eq. (13)], is $\sim 10^3$ K.

The significance of boundary layers in determining the thermal state of the interior is well illustrated by the consequences of compositional layering, if it is present in the mantle. Because a change in composition with sufficient density increase produces a barrier to convection (Fig. 5), thermal boundary layers would result at the bottom of the transition zone. That is, the geotherm through the lower mantle would be $\sim 10^3$ K higher with compositional layering (hot geotherm in Fig. 6.) than without (cold geotherm in Fig. 6). Moreover, through-going connection is hindered, and the shallower mantle effectively acts as a thermal blanket, tending to retain heat within the deeper mantle and core over the age of the earth.

Although the average temperature with depth through the mantle is relatively well constrained, the lateral variations are not known in great detail. These variations are large and are directly linked with the flow field, which changes with time. At the surface, for example, it is known that hot lava emerges at the mid-ocean ridges at up to 1500 K and cold (~ 300 K) lithosphere sinks into the mantle at subduction zones. The corresponding temperature difference between upwelling and downwelling regions at the same depth in the mantle can reach 1200 K, comparable to the vertical change in temperature across the bulk of the interior of the mantle. The horizontal variations of temperature are significant not only in magnitude, but also becasue these produce the buoyancy forces that drive convection, as noted above.

The fact that heat production within the earth was higher in the past than at present suggests that interior temperatures may have decreased with time. One geological observation does indicate that the mantle has cooled, namely, the occurrence of high-temperature komatiites exclusively in Archean time, more than 2.5 billion years ago. Komatiites are extremely refractory ultramafic lavas, consisting mainly of olivine. It is known from melting experiments that these lavas must have been erupted onto the surface at temperatures of at least 1900 K. Therefore, maximum temperatures associated with upwelling in the mantle have likely decreased by ~400–500 K over 3–4 billion years. This amount of cooling, less than 25% of the interior temperature over earth history, is small compared with the order-of-magnitude decrease in heat production over the same time period. The reason for this apparent contradiction is that the diminished heat production over geological history is accommodated by less vigorous convection and, hence, decreased heat loss at the surface. That is, the thermal evolution of the interior has caused the intensity of convection in the mantle and of tectonic activity at the surface to evolve with time.

SEE ALSO THE FOLLOWING ARTICLES

CONTINENTAL CRUST • EARTH SCIENCES, HISTORY OF • EARTH'S CORE • MINERALOGY AND INSTRUMENTA-TION • OCEANIC CRUST • ORE PETROLOGY • PLATE TECTONICS • ROCK MECHANICS • SEISMOLOGY, ENGINEERING • SEISMOLOGY, OBSERVATIONAL • SEISMOLOGY, THEORETICAL

BIBLIOGRAPHY

Ahrens, T. J., ed. (1995), "Global Earth Physics, A Handbook of Physical Constants," Am. Geophys. Union, Washington, DC.

Basaltic Volcanism Studies Project (1981). "Basaltic Volcanism on the Terrestrial Planets," Pergamon, New York.

Bott, M. H. P. (1982). "The Interior of the Earth: Its Structure, Constitution and Evolution," Elsevier, New York.

DeBremaecker, J. C. (1985). "Geophysics: The Earth's Interior," Wiley, New York.

Decker, R., and Decker, B. (1982). "Volcanoes and the Earth's Interior," Freeman, San Francisco.

Jeanloz. R. (1990). *Annu. Rev. Earth Planet. Sci.* **18,** 357–386.

Jeanloz, R., and Morris. S. (1986). *Annu. Rev. Earth Planet. Sci.* **14,** 377–415.

Jeanloz, R., and Thompson, A. B. (1983). *Rev. Geophys. Space Phys.* **21,** 51–74.

Kellogg, L. H., Hager, B. H., and van der Hilst, R. D. (1999). "Compositional stratification in the deep mantle." *Science* **283**(5409), 1881–1884.

Ozima, M. (1981). "The Earth: Its Birth and Growth," Cambridge Univ. Press, New York.

Ringwood, A. E. (1975). "Composition and Petrology of the Earth's Mantle," McGraw-Hill, New York.

Silver, P., Carlson, R. W., and Olson, P. (1988). *Annu. Rev. Earth Planet. Sci.* **16,** 477–541.

Turcotte, D. L., and Schubert, G. (1982). "Geodynamics," Wiley, New York.

Yoder, H. S., Jr. (1976). "Generation of Basaltic Magma," Natl. Acad. Sci., Washington, DC.

Elasticity

Herbert Reismann

State University of New York at Buffalo

GLOSSARY

Anisotropy A medium is said to be anisotropic if the value of a measured, physical field quantity depends on the orientation (or direction) of measurement.

Eigenvalue and eigenvector Consider the matrix equation $\mathbf{AX} = \lambda\mathbf{X}$, where $\mathbf{A}$ is an $n \times n$ square matrix, and $\mathbf{X}$ is an n-dimensional column vector. In this case, the scalar λ is an eigenvalue, and $\mathbf{X}$ is the associated eigenvector.

Isotropy A medium is said to be isotropic if the value of a measured, physical field quantity is independent of orientation.

ELASTICITY THEORY is the (mathematical) study of the behavior of those solids that have the property of recovering their size and shape when the forces that cause the deformation are removed. To some extent, almost all solids display this property. In this article, most of the discussion will be limited to the special case of linearly elastic solids, where deformation is proportional to applied forces. This topic is usually referred to as classical elasticity theory. This branch of mathematical physics was formulated during the nineteenth century and, since its inception, has been developed and refined to form the background and foundation for disciplines such as structural mechanics; stress analysis; strength of materials; plates and shells; solid mechanics; and wave propagation and vibrations in solids. These topics are fundamental to solving present-day problems in many branches of modern engineering and applied science. They are used by structural (civil) engineers, aerospace engineers, mechanical engineers, geophysicists, geologists, and bioengineers, to name a few. The deformation, vibrations, and structural integrity of modern high-rise buildings, airplanes, and high-speed rotating machinery are predicted by applying the modern theory of elasticity.

I. ONE-DIMENSIONAL CONSIDERATIONS

If we consider a suitably prepared rod of mild steel, with (original) length L and cross-sectional area A, subjected

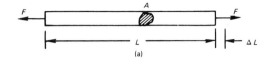

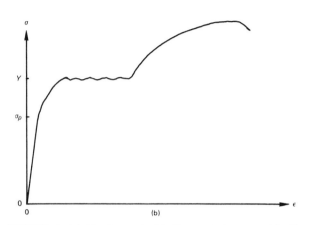

FIGURE 1 (a) Tension rod. (b) Stress–strain curve (ductile material).

to a longitudinal, tensile force of magnitude F, then the rod will experience an elongation of magnitude ΔL, as shown in Fig. 1a. So that we can compare the behavior of rods of differing cross section in a meaningful manner, it is convenient to define the (uniform) axial stress in the rod by $\sigma = F/A$ and the (uniform) axial strain by $\varepsilon = \Delta L/L$. We note that the unit of stress is force per unit of (original) area and the unit of strain is change in length divided by original length. If, in a typical tensile test, we plot stress σ versus strain ε, we obtain the curve shown in Fig. 1b. In the case of mild steel, and many other ductile materials, this curve has a straight line portion that extends from $0 < \sigma < \sigma_p$, where σ_p is the proportional limit. The slope of this line is $\sigma/\varepsilon = E$, where E is known as Young's modulus (Thomas Young, 1773–1829). When $\sigma_p < \sigma$, the stress–strain curve is no longer linear, as shown in Fig. 1b. When the rod is extended beyond $\sigma = \sigma_p$ (the proportional limit), it suffers a permanent set (deformation) upon removal of the load F. At $\sigma = Y$ (the yield point), the strain will increase considerably for relatively small increases in stress (Fig. 1b). For the majority of structural applications, it is desirable to remain in the linearly elastic, shape-recoverable range of stress and strain ($0 \leq \sigma \leq \sigma_p$). The mathematical material model that is based on this assumption is said to display linear material characteristics. For example, an airplane wing will deflect in flight because of air loads and maneuvers, but when the loads are removed, the wing reverts to its original shape. If this were not the case, the wing's lifting capability would not be reliably predictable, and, of course, this would not be desirable. In addition, if the load is doubled, the deflection will also double.

Within the context of the international system of units (*Système International*, or SI), the unit of stress is the pascal (Pa). One pascal is equal to one newton per square meter (N m^{-2}). The unit of strain is meter per meter, and thus strain is a dimensionless quantity. We note that 1 N m$^{-2} = 1$ Pa $= 1.4504 \times 10^{-4}$ psi and 1 psi $= 6894.76$ Pa.

Typical values of the Young's (elastic) modulus E and yield stress in tension Y for some ductile materials are shown in Table II in Section IV. The tension test of a rod and naive definitions of stress and strain are associated with one-dimensional considerations. Elasticity theory is concerned with the generalization of these concepts to the general, three-dimensional case.

II. STRESS

Elastic solids are capable of transmitting forces, and the concept of stress in a solid is a sophistication and generalization of the concept of force. We consider a material point P in the interior of an elastic solid and pass an oriented plane II through P with unit normal vector $\mathbf{n}$ (see Fig. 2). Consider the portion of the solid which is shaded. Then on a (small) area ΔA surrounding the point P, there will act a net force of magnitude $\Delta \mathbf{F}$, and the stress vector at P is defined by the limiting process

$$\mathbf{T}(\mathbf{n}) = \lim_{\Delta A \to 0} \frac{\Delta \mathbf{F}}{\Delta A}. \tag{1}$$

It is to be noted that the magnitude as well as the direction of the stress vector $\mathbf{T}$ depends upon the orientation of $\mathbf{n}$. If we resolve the stress vector along the (arbitrarily chosen) $(x, y, z) = (x_1, x_2, x_3)$ axes, then we can write

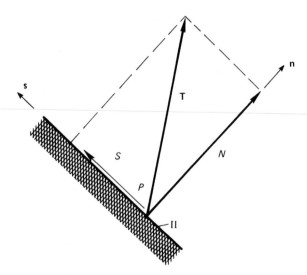

FIGURE 2 Stress vector and components.

$$\mathbf{T}_1 = \tau_{11}\mathbf{e}_1 + \tau_{12}\mathbf{e}_2 + \tau_{13}\mathbf{e}_3$$

$$\mathbf{T}_2 = \tau_{21}\mathbf{e}_1 + \tau_{22}\mathbf{e}_2 + \tau_{23}\mathbf{e}_3 \qquad (2)$$

$$\mathbf{T}_3 = \tau_{31}\mathbf{e}_1 + \tau_{32}\mathbf{e}_2 + \tau_{33}\mathbf{e}_3,$$

where $\mathbf{T}_i = \mathbf{T}(\mathbf{e}_i)$ for $i = 1, 2, 3$; that is, the $\mathbf{T}_i$ are stress vectors acting upon the three coordinate planes and $\mathbf{e}_i$ are unit vectors associated with the coordinate axes $(x, y, z) = (x_1, x_2, x_3)$. We note that here and in subsequent developments, we use the convenient and common notation $\tau_{12} \equiv \tau_{xy}, \mathbf{T}_1 \equiv \mathbf{T}_x, \mathbf{T}_2 \equiv \mathbf{T}_y$, etc. In other words, the subscripts 1, 2, 3 take the place of x, y, z. We can also write

$$T_1 = \tau_{11}n_1 + \tau_{12}n_2 + \tau_{13}n_3$$

$$T_2 = \tau_{21}n_1 + \tau_{22}n_2 + \tau_{23}n_3 \qquad (3)$$

$$T_3 = \tau_{31}n_1 + \tau_{32}n_2 + \tau_{33}n_3,$$

where

$$\mathbf{n} = \mathbf{e}_1 n_1 + \mathbf{e}_2 n_2 + \mathbf{e}_3 n_3$$

and

$$\mathbf{T}(\mathbf{n}) = \mathbf{e}_1 T_1 + \mathbf{e}_2 T_2 + \mathbf{e}_3 T_3$$

$$= \mathbf{T}_1 n_1 + \mathbf{T}_2 n_2 + \mathbf{T}_3 n_3. \qquad (4)$$

This last expression is known as the lemma of Cauchy (A. L. Cauchy, 1789–1857). The stress tensor components

$$[\tau_{ij}] = \begin{bmatrix} \tau_{11} & \tau_{12} & \tau_{13} \\ \tau_{21} & \tau_{22} & \tau_{23} \\ \tau_{31} & \tau_{32} & \tau_{33} \end{bmatrix} \qquad (5)$$

can be visualized with reference to Fig. 3, with all stresses shown acting in the positive sense. We note that τ_{ij} is the stress component acting on the face with normal $\mathbf{e}_i$, in the direction of the vector $\mathbf{e}_j$.

With reference to Fig. 2, it can also be shown that relative to the plane II, the normal component N and the shear component S are given by

$$T_n \equiv N = \mathbf{T} \cdot \mathbf{n} = \sum_{i=1}^{3} T_i n_i$$

$$= \sum_{i=1}^{3}\sum_{j=1}^{3} n_i n_j \tau_{ij} \qquad (6a)$$

and

$$T_s \equiv S = \mathbf{T} \cdot \mathbf{s} = \sum_{i=1}^{3} T_i s_i$$

$$= \sum_{i=1}^{3}\sum_{j=1}^{3} n_i s_j \tau_{ij}, \qquad (6b)$$

where $\mathbf{n} \cdot \mathbf{s} = 0$ and $\mathbf{n}, \mathbf{s}$ are unit vectors normal and parallel to the plane II, respectively.

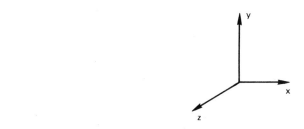

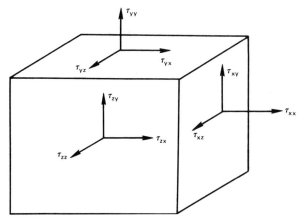

FIGURE 3 Stress tensor components.

At every interior point of a stressed solid, there exist at least three mutually perpendicular directions for which all shearing stresses $\tau_{ij}, i \neq j$, vanish. This preferred axis system is called the principal axis system. It can be found by solving the algebraic eigenvalue–eigenvector problem characterized by

$$\begin{bmatrix} \tau_{11} - \sigma & \tau_{12} & \tau_{13} \\ \tau_{21} & \tau_{22} - \sigma & \tau_{23} \\ \tau_{31} & \tau_{32} & \tau_{33} - \sigma \end{bmatrix} \begin{Bmatrix} n_1 \\ n_2 \\ n_3 \end{Bmatrix} = \begin{Bmatrix} 0 \\ 0 \\ 0 \end{Bmatrix}, \qquad (7)$$

where n, n_2, and n_3 are the direction cosines of the principal axis system such that $n_1^2 + n_2^2 + n_3^2 = 1$; and σ_1, σ_2, and σ_3 are the (scalar) principal stress components. The necessary and sufficient condition for the existence of a solution for Eq. (7) is obtained by setting the coefficient determinant equal to zero. The result is

$$\sigma^3 + I_1\sigma^2 + I_2\sigma - I_3 = 0, \qquad (8)$$

where the quantities

$$I_1 = \tau_{11} + \tau_{22} + \tau_{33}, \qquad (9a)$$

$$I_2 = \begin{vmatrix} \tau_{11} & \tau_{12} \\ \tau_{21} & \tau_{22} \end{vmatrix} + \begin{vmatrix} \tau_{22} & \tau_{23} \\ \tau_{32} & \tau_{33} \end{vmatrix} + \begin{vmatrix} \tau_{33} & \tau_{31} \\ \tau_{13} & \tau_{11} \end{vmatrix}, \qquad (9b)$$

and

$$I_3 = \begin{vmatrix} \tau_{11} & \tau_{12} & \tau_{13} \\ \tau_{21} & \tau_{22} & \tau_{23} \\ \tau_{31} & \tau_{32} & \tau_{33} \end{vmatrix} \qquad (9c)$$

are known as the first, second, and third stress invariants, respectively. For example, we consider these stress tensor components at a point P of a solid, relative to the x, y, z axes:

$$[\tau_{ij}] = \begin{bmatrix} 3 & 1 & 1 \\ 1 & 0 & 2 \\ 1 & 2 & 0 \end{bmatrix}. \tag{10}$$

Thus,

$$I_1 = 3, \qquad I_2 = -6, \qquad I_3 = -8$$

and

$$\sigma^3 - 3\sigma^2 - 6\sigma + 8 = (\sigma - 4)(\sigma - 1)(\sigma + 2) = 0.$$

Consequently, the principal stresses at P are $\sigma_1 = 4$, $\sigma_2 = 1$, and $\sigma_3 = -2$. With the aid of Eq. (7), it can be shown that the principal directions at P are given by the mutually perpendicular unit vectors

$$\mathbf{n}^{(1)} = \mathbf{e}_1 \frac{2}{\sqrt{6}} + \mathbf{e}_2 \frac{1}{\sqrt{6}} + \mathbf{e}_3 \frac{1}{\sqrt{6}}$$

$$\mathbf{n}^{(2)} = \mathbf{e}_1 \left(-\frac{1}{\sqrt{3}} \right) + \mathbf{e}_2 \frac{1}{\sqrt{3}} + \mathbf{e}_3 \frac{1}{\sqrt{3}} \tag{11}$$

$$\mathbf{n}^{(3)} = \mathbf{e}_1 (0) + \mathbf{e}_2 \left(-\frac{1}{\sqrt{2}} \right) + \mathbf{e}_3 \left(\frac{1}{\sqrt{2}} \right).$$

When the Cartesian axes are rotated in a rigid manner from x_1, x_2, x_3, to $x_{1'}, x_{2'}, x_{3'}$, as shown in Fig. 4, the components of the stress tensor transform according to the rule

$$\tau_{p'q'} = \sum_{i=1}^{3} \sum_{j=1}^{3} a_{p'i} a_{q'j} \tau_{ij}, \tag{12}$$

where $a_{p'i} = \cos(x_{p'}, x_1) = \cos(\mathbf{e}_{p'}, \mathbf{e}_i)$ are the nine direction cosines that orient the primed coordinate system relative to the unprimed system.

For example, consider the rotation of axes characterized by the table of direction cosines

$$[a_{i'j}] = \begin{bmatrix} \frac{2}{3} & -\frac{2}{3} & -\frac{1}{3} \\ \frac{1}{3} & \frac{2}{3} & -\frac{2}{3} \\ \frac{2}{3} & \frac{1}{3} & \frac{2}{3} \end{bmatrix}. \tag{13}$$

The stress components τ_{ij} in Eq. (10) relative to the x, y, z axes will become

$$[\tau_{p'q'}] = \begin{bmatrix} 0.889 & 0.778 & 0.222 \\ 0.778 & -1.444 & 1.444 \\ 0.222 & 1.444 & 3.556 \end{bmatrix} \tag{14}$$

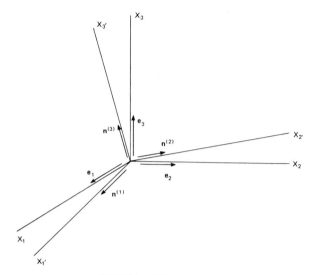

FIGURE 4 Principal axes.

when referred to x', y', z' axes, according to the law of transformation [Eq. (12)]. The extreme shear stress at a point is given by $\tau_{\max} = \frac{1}{2}(\sigma_1 - \sigma_3)$ and this value is $\tau_{\max} = \frac{1}{2}(4 + 2) = 3$ for the stress tensor [Eq. (10)]. It should be noted that the principal stresses are ordered, that is, $\sigma_1 \geq \sigma_2 \geq \sigma_3$, and that $\sigma_1(\sigma_3)$ is the largest (smallest) normal stress for all possible planes through the point P.

If we now establish a coordinate system coincident with principal axes then in "principal stress space," the normal stress N and the shear stress S on a plane characterized by the outer unit normal vector $\mathbf{n}$ are, respectively,

$$N = n_1^2 \sigma_1 + n_2^2 \sigma_2 + n_3^2 \sigma_3 \tag{15a}$$

and

$$S^2 = n_1^2 n_2^2 (\sigma_1 - \sigma_2)^2 + n_2^2 n_3^2 (\sigma_2 - \sigma_3)^2 + n_3^2 n_1^2 (\sigma - \sigma_1)^2, \tag{15b}$$

where σ_1, σ_2, and σ_3 are principal stresses. We now visualize eight planes, the normal to each of which makes equal angles with respect to principal axes. The shear stress acting upon these planes is known as the octahedral shear stress τ_0, and its magnitude is

$$\tau_0 = \frac{1}{3} \left[(\sigma_1 - \sigma_2)^2 + (\sigma_2 - \sigma_3)^2 + (\sigma_3 - \sigma_1)^2 \right]^{1/2} \geq 0. \tag{16}$$

It can be shown that the octahedral shear stress is related to the average of the square of all possible shear stresses at the point, and the relation is

$$\tfrac{3}{5}(\tau_0)^2 = \langle S^2 \rangle. \tag{17}$$

It can also be shown that

$$9\tau_0^2 = 2I_1^2 - 6I_2, \qquad (18)$$

where I_1 and I_2 are the first and second stress invariants, respectively [see Eqs. (9a) and (9b)]. We also note the bound

$$1 \le \sqrt{\frac{3}{2}}\frac{\tau_0}{\tau_{\max}} \le \frac{2}{\sqrt{3}} \qquad (19)$$

and the associated implication that $\frac{3}{2}\tau_0 \cong 1.08\,\tau_{\max}$ with a maximum error of about 7%. Returning to the stress tensor [Eq. (10)], we have

$$\tau_{\max} = 3$$

and

$$9\tau_0^2 = 2I_1^2 - 6I_2 = (2)(9) + (6)(6) = 54,$$

or

$$\tau_0 = \sqrt{6} = 2.4495,$$

and

$$1 \le \frac{\sqrt{3/2}\tau_0}{\tau_{\max}} = \frac{\sqrt{3/2}(2.4495)}{3} \le \frac{2}{\sqrt{3}},$$

or

$$1 = 1 \le 1.1547.$$

III. STRAIN

In our discussion of the concept of stress, we noted that stress characterizes the action of a "force at a point" in a solid. In a similar manner, we shall show that the concept of strain can be used to quantify the notion of "deformation at a point" in a solid.

We consider a (small) quadrilateral element in the unstrained solid with dimensions dx, dy, and dz. The sides of the element are taken to be parallel to the coordinate axes. After deformation, the volume element has the shape of a rectangular parallelepiped with edges of length $(dx+du)$, $(dy+dv)$, $(dz+dw)$. With reference to Fig. 5, the material point P in the undeformed configuration is carried into the point P' in the deformed configuration. A projection of the element sides onto the x–y plane, before and after deformation, is shown in Fig. 5. We note that all changes in length and angles are small, and they

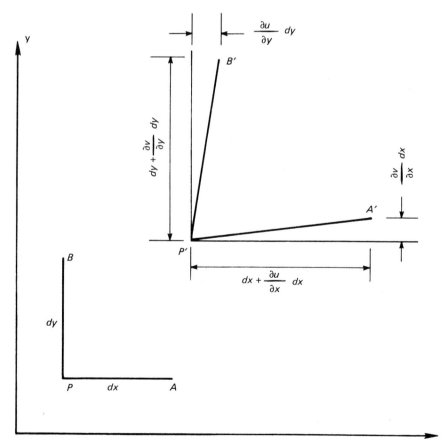

FIGURE 5 Strain.

have been exaggerated for purposes of clarity. We now define extensional strain $\varepsilon_{xx} = \varepsilon_{11}$ as change in length per unit length, and therefore for the edge PA (in Fig. 5), we have

$$\varepsilon_{xx} = \frac{[dx + (\partial u/\partial x)\,dx] - dx}{dx} = \frac{\partial u}{\partial x}$$

and

$$\varepsilon_{yy} = \frac{[dx + (\partial v/\partial y)\,dy] - dy}{dy} = \frac{\partial v}{\partial y},$$

and a projection onto the y–z plane will result in

$$\varepsilon_{zz} = \frac{[dz + (\partial w/\partial z)\,dz] - dz}{dz} = \frac{\partial w}{\partial z}.$$

The shear strain is defined as one-half of the decrease of the originally right angle APB. Thus, with reference to Fig. 5, we have

$$2\varepsilon_{xy} = 2\varepsilon_{yx} = \frac{(\partial v/\partial x)\,dx}{dx + (\partial u/\partial x)\,dx} + \frac{(\partial u/\partial y)\,dy}{dy + (\partial v/\partial y)\,dy}$$

$$= \frac{\partial v/\partial x}{1 + (\partial u/\partial x)} + \frac{\partial u/\partial y}{1 + (\partial v/\partial y)} = \frac{\partial v}{\partial x} + \frac{\partial u}{\partial y}$$

because it is assumed that

$$1 \gg \partial u/\partial x; \qquad 1 \gg \partial v/\partial y \qquad \text{(small rotations)}.$$

In a similar manner, using projections onto the planes y–z and z–x, we can show that

$$2\varepsilon_{yz} = \frac{\partial v}{\partial z} + \frac{\partial w}{\partial y}$$

and

$$2\varepsilon_{zx} = \frac{\partial w}{\partial x} + \frac{\partial u}{\partial z}.$$

Consequently, the complete (linearized) strain-displacement relations are given by

$$
\begin{bmatrix}
\varepsilon_{xx} & \varepsilon_{xy} & \varepsilon_{xz} \\
\varepsilon_{yx} & \varepsilon_{yy} & \varepsilon_{yz} \\
\varepsilon_{zx} & \varepsilon_{zy} & \varepsilon_{zz}
\end{bmatrix}
$$

$$
= \begin{bmatrix}
\dfrac{\partial u}{\partial x} & \dfrac{1}{2}\left(\dfrac{\partial u}{\partial y} + \dfrac{\partial v}{\partial x}\right) & \dfrac{1}{2}\left(\dfrac{\partial u}{\partial z} + \dfrac{\partial w}{\partial x}\right) \\[2mm]
\dfrac{1}{2}\left(\dfrac{\partial v}{\partial x} + \dfrac{\partial u}{\partial y}\right) & \dfrac{\partial v}{\partial y} & \dfrac{1}{2}\left(\dfrac{\partial v}{\partial z} + \dfrac{\partial w}{\partial y}\right) \\[2mm]
\dfrac{1}{2}\left(\dfrac{\partial w}{\partial x} + \dfrac{\partial u}{\partial z}\right) & \dfrac{1}{2}\left(\dfrac{\partial w}{\partial y} + \dfrac{\partial v}{\partial z}\right) & \dfrac{\partial w}{\partial z}
\end{bmatrix}.
$$

$$(20)$$

Equation (20) characterizes the deformation of the solid at a point. If we define the mutually perpendicular unit vectors $\mathbf{n}$ and $\mathbf{s}$ with reference to a plane II through a

point P in a solid (see Fig. 2), then it can be shown that the extensional strain N in the direction $\mathbf{n}$ is given by the formula

$$N = \sum_{i=1}^{3}\sum_{j=1}^{3} \varepsilon_{ij} n_i n_j \qquad (21a)$$

and the shear strain relative to the vectors $\mathbf{n}$ and $\mathbf{s}$ is

$$S = \frac{1}{2}\sum_{i=1}^{3}\sum_{j=1}^{3} \varepsilon_{ij} n_i s_j. \qquad (21b)$$

Equation (21a) expresses the extensional strain and Eq. (21b) expresses the shearing strain for an arbitrarily chosen element; therefore, we can infer that the nine (six independent) quantities $\varepsilon_{ij}(i = 1, 2, 3; j = 1, 2, 3)$ provide a complete characterization of strain associated with a material point in the solid. It can be shown that the nine quantities ε_{ij} constitute the components of a tensor of order two in a three-dimensional space, and the appropriate law of transformation under a rotation of coordinate axes is

$$\varepsilon_{p'q'} = \sum_{i=1}^{3}\sum_{j=1}^{3} a_{p'i}a_{q'j}\varepsilon_{ij}; \qquad (22)$$

$$p = 1, 2, 3; \qquad q = 1, 2, 3,$$

where the $a_{p'i}$ are direction cosines as in Eq. (12). As in the case of stress, there will be at least one set of mutually perpendicular axes for which the shearing strains vanish. These axes are principal axes of strain. They are found in a manner that is entirely analogous to the determination of principal stresses and axes. (See Section II.)

It should be noted that a single-valued, continuous displacement field for a simply connected region is guaranteed provided that the six equations of compatibility of A. J. C. Barré de Saint-Venant (1779–1886) are satisfied:

$$\frac{\partial^2 \varepsilon_{xx}}{\partial y^2} + \frac{\partial^2 \varepsilon_{yy}}{\partial x^2} = 2\frac{\partial^2 \varepsilon_{xy}}{\partial x \partial y}, \qquad (23a)$$

$$\frac{\partial^2 \varepsilon_{xx}}{\partial y \partial z} = \frac{\partial}{\partial x}\left(-\frac{\partial e_{yx}}{\partial x} + \frac{\partial \varepsilon_{xz}}{\partial y} + \frac{\partial \varepsilon_{xy}}{\partial z}\right),$$

$$(23b)$$

and there are two additional equations for each of Eqs. (23a) and (23b), which are readily obtained by cyclic permutation of x, y, z.

IV. HOOKE'S LAW AND ITS LIMITS

The most general linear relationship between stress tensor and strain tensor components at a point in a solid is given by

$$\tau_{ij} = \sum_{k=1}^{3} \sum_{l=1}^{3} C_{ijkl}\varepsilon_{kl}; \qquad i = 1, 2, 3;$$

$$j = 1, 2, 3, \qquad (24)$$

where the $3^4 = 81$ constants C_{ijkl} are the elastic constants of the solid. If a strain energy density function exists (see Section V), and in view of the fact that the stress and strain tensor components are symmetric, the elastic constants must satisfy the relations

$$C_{ijkl} = C_{ijlk}, \qquad C_{ijkl} = C_{jikl}, \qquad C_{ijkl} = C_{klij}, \qquad (25)$$

and therefore the number of independent elastic constants is reduced to $\frac{1}{2}(6^2 - 6) + 6 = 21$ for the general anisotropic elastic solid. If, in addition, the elastic properties of the solid are independent of orientation, the number of independent elastic constants can be reduced to two. In this case of an isotropic elastic solid, the relation between stress and strain is given by

$$E\varepsilon_{xx} = \tau_{xx} - \nu(\tau_{yy} + \tau_{zz})$$

$$E\varepsilon_{yy} = \tau_{yy} - \nu(\tau_{zz} + \tau_{xx})$$

$$E\varepsilon_{zz} = \tau_{zz} - \nu(\tau_{xx} + \tau_{yy})$$

$$2G\varepsilon_{xy} = \tau_{xy} \qquad (26)$$

$$2G\varepsilon_{yz} = \tau_{yz}$$

$$2G\varepsilon_{zx} = \tau_{zx},$$

where $G = E/2(1 + \nu)$ is the shear modulus, E is Young's modulus (see Section I), and ν is Poisson's ratio (S. D. Poisson, 1781–1840). Equation (26) is known as Hooke's law (Robert Hooke, 1635–1693) for a linearly elastic, isotropic solid. A listing of typical values of the elastic constants is provided in Table I.

Many failure theories for solids have been proposed, and they are usually associated with specific classes of

TABLE I Typical Values of Elastic Constants[a]

Material	ν	E (Pa)[b]	G (Pa)[b]
Aluminum	0.34	6.89×10^{10}	2.57×10^{10}
Concrete	0.20	0.76×10^{10}	1.15×10^{10}
Copper	0.34	8.96×10^{10}	3.34×10^{10}
Glass	0.25	6.89×10^{10}	2.76×10^{10}
Nylon	0.40	2.83×10^{10}	1.01×10^{10}
Rubber	0.499	1.96×10^{6}	0.654×10^{6}
Steel	0.29	20.7×10^{10}	8.02×10^{10}

[a] Adapted from Reismann, H., and Pawlik, P. S. (1980). "Elasticity: Theory and Applications," Wiley (Interscience), New York.
[b] Note that $1\ Pa = 1\ N\ m^{-2} = 1.4504 \times 10^{-4}\ lb\ in.^{-2}$.

TABLE II Some Material Properties for Ductile Materials[a]

Material	Yield point stress, σ_Y (tension, Pa)	Young's modulus, E (Pa)	Strain at yield point, ε_Y (tension)
Aluminum alloy (2024 T 4)	290×10^6	7.30×10^{10}	0.00397
Brass	103×10^6	10.3×10^{10}	0.00100
Bronze	138×10^6	10.3×10^{10}	0.00134
Magnesium alloy	138×10^6	4.50×10^{10}	0.00307
Steel (low carbon, structural)	248×10^6	20.7×10^{10}	0.00120
Steel (high carbon)	414×10^6	20.7×10^{10}	0.00200

[a] Adapted from Reismann, H., and Pawlik, P. S. (1980). "Elasticity: Theory and Applications," Wiley (Interscience), New York.

materials. In the case of a ductile material with a well-defined yield point (see Fig. 1b), there are at least two failure theories that yield useful results.

A. The Hencky-Mises Yield Criterion

This theory predicts failure (yielding) at a point of the solid when $9\tau_0^2 \geq 2Y^2$, where τ_0 is the octahedral shear stress [see Eq. (16)] and Y is the yield stress in tension (see Fig. 1b). In this case, the ratio of yield stress in tension Y to the yield stress in pure shear τ has the value $Y/\tau = \sqrt{3}$.

B. The Tresca Yield Criterion

This theory postulates that yielding occurs when the extreme shear stress τ_{max} at a point attains the value $\tau_{max} \geq Y/2$. We note that for this theory the ratio of yield stress in tension to the yield stress in pure shear is equal to $Y/\tau = 2$. A listing of the values of Y for some commonly used materials is given in Table II.

V. STRAIN ENERGY

We now consider an interior material point P in a stressed, elastic solid. We can construct a Cartesian coordinate system x, y, z with origin at P, which is coincident with principal axes at P. The point P is enclosed by a small, rectangular parallelepiped with sides of length dx, dy, and dz. The areas of the sides of the parallelepiped are $dA_z = dx\,dy, dA_x = dy\,dz, dA_y = dz\,dx$, and the volume is $dV = dx\,dy\,dz$. The potential (or strain) energy stored in the linearly elastic solid is equal to the work of the external forces. Consequently, neglecting heat generation, if W is the strain energy per unit volume (strain energy density), we have

$$WdV = \tfrac{1}{2}(\tau_{xx}A_x)(dx\varepsilon_{xx}) + \tfrac{1}{2}(\tau_{yy}A_y)(dy\varepsilon_{yy})$$
$$+ \tfrac{1}{2}(\tau_{zz}A_z)(dz\varepsilon_{zz})$$
$$= \tfrac{1}{2}(\tau_{xx}\varepsilon_{xx} + \tau_{yy}\varepsilon_{yy} + \tau_{zz}\varepsilon_{zz})\,dV,$$

and therefore the strain energy density referred to principal axes is

$$W = \tfrac{1}{2}(\tau_{xx}\varepsilon_{xx} + \tau_{yy}\varepsilon_{yy} + \tau_{zz}\varepsilon_{zz}).$$

In the general case of arbitrary (in general, nonprincipal) axes, this expression assumes the form

$$W = \tfrac{1}{2}(\tau_{xx}\varepsilon_{xx} + \tau_{xy}\varepsilon_{xy} + \tau_{xz}\varepsilon_{xz})$$
$$+ \tfrac{1}{2}(\tau_{yx}\varepsilon_{yx} + \tau_{yy}\varepsilon_{yy} + \tau_{yz}\varepsilon_{yz})$$
$$+ \tfrac{1}{2}(\tau_{zx}\varepsilon_{zx} + \tau_{zy}\varepsilon_{zy} + \tau_{zz}\varepsilon_{zz}),$$

or, in abbreviated notation,

$$W = \frac{1}{2}\sum_{i=1}^{3}\sum_{j=1}^{3}\tau_{ij}\varepsilon_{ij}. \qquad (27)$$

In view of the relations in Eqs. (24) and (27), the expression for strain energy density can be written in the form

$$W = \frac{1}{2}\sum_{i=1}^{3}\sum_{j=1}^{3}\sum_{k=1}^{3}\sum_{l=1}^{3}C_{ijkl}\varepsilon_{ij}\varepsilon_{kl}. \qquad (28a)$$

In the case of an isotropic elastic material [see Eq. (26)], this equation reduces to

$$W = \frac{1}{2}\left[\lambda(\varepsilon_{11} + \varepsilon_{22} + \varepsilon_{33})^2 + 2G\sum_{i=1}^{3}\sum_{j=1}^{3}\varepsilon_{ij}\varepsilon_{ij}\right],$$
$$(28b)$$

where

$$\lambda = E/(1 + v)(1 - 2v).$$

Thus, with reference to Eq. (28), we note that the strain energy density is a quadratic function of the strain tensor components, and W vanishes when the strain field vanishes. Equation (28) serves as a potential (generating) function for the generation of the stress field, that is,

$$\tau_{ij} = \frac{\partial W(\varepsilon_{ij})}{\partial\varepsilon_{ij}} = \sum_{k=1}^{3}\sum_{l=1}^{3}C_{ijkl}\varepsilon_{kl}; \quad i = 1, 2, 3;$$
$$j = 1, 2, 3 \quad (29)$$

[see Eq. (24)]. The concept of strain energy serves as the starting point for many useful and important investigations in elasticity theory and its applications. For details, the reader is referred to the extensive literature, a small selection of which can be found in the Bibliography.

VI. EQUILIBRIUM AND THE FORMULATION OF BOUNDARY VALUE PROBLEMS

External agencies usually deform a solid by two distinct types of loadings: (a) surface tractions and (b) body forces. Surface tractions act by virtue of the application of normal and shearing stresses to the surface of the solid, while body forces act upon the interior, distributed mass of the solid. For example, a box resting on a table is subjected to (normal) surface traction forces at the interface between tabletop and box bottom, whereas gravity causes forces to be exerted upon the contents of the box.

Consider a solid body B bounded by the surface S in a state of static equilibrium. Then at every internal point of B, these partial differential equations must be satisfied:

$$\frac{\partial\tau_{xx}}{\partial x} + \frac{\partial\tau_{xy}}{\partial y} + \frac{\partial\tau_{xz}}{\partial z} + F_x = 0$$
$$\frac{\partial\tau_{yx}}{\partial x} + \frac{\partial\tau_{yy}}{\partial y} + \frac{\partial\tau_{yz}}{\partial z} + F_y = 0 \qquad (30)$$
$$\frac{\partial\tau_{zx}}{\partial x} + \frac{\partial\tau_{zy}}{\partial y} + \frac{\partial\tau_{zz}}{\partial z} + F_z = 0,$$

where $\tau_{xy} = \tau_{yx}$, $\tau_{yz} = \tau_{zy}$, $\tau_{zx} = \tau_{xz}$, and $\mathbf{F} = F_x\mathbf{e}_x + F_y\mathbf{e}_y + F_z\mathbf{e}_z$ is the body force vector per unit volume.

The admissible boundary conditions associated with Eq. (30) may be stated in the form:

$$\mathbf{T} \equiv (T_1, T_2, T_3) \qquad \text{on} \qquad S_1$$

and

$$\mathbf{u} \equiv (u, v, w) \qquad \text{on} \qquad S_2, \qquad (31)$$

where $\mathbf{T}$ is the surface traction vector [see Eq. (4)], $\mathbf{u}$ is the displacement vector, and $S = S_1 + S_2$ denotes the bounding surface of the solid.

The solution of a problem in (three-dimensional) elasticity theory requires the determination of

$$\left.\begin{array}{l}\text{the displacement vector field } \mathbf{u} \\ \text{the stress tensor field } \tau_{ij} \\ \text{and the strain tensor field } \varepsilon_{ij}\end{array}\right\} \quad \text{in } B. \quad (32)$$

This solution is required to satisfy the equations of equilibrium [Eq. (30)], the equations of compatibility [Eq. (23)], the strain-displacement relations [Eq. (20)], and the stress–strain relations [Eq. (26) or (24)], as well as the boundary conditions [Eq. (31)]. This is a formidable task, even for relatively simple geometries and boundary conditions, and the exact or approximate solution requires extensive use of advanced analytical as well as numerical mathematical methods in most cases.

VII. EXAMPLES

A. Example A

We consider an elastic cylinder of length L with an arbitrary cross section. The cylinder is composed of a linearly elastic, isotropic material with Young's modulus E and Poisson's ratio v. The cylinder is inserted into a perfectly fitting cavity in a rigid medium, as shown in Fig. 6, and subjected to a uniformly distributed normal stress $\tau_{zz} = T$ on the free surface at $z = L$. We assume that the bottom of the cylinder remains in smooth contact with the rigid medium, and that the lateral surfaces between the cylinder and the rigid medium are smooth, thus capable of transmitting normal surface tractions only. Moreover, normal displacements over the lateral surfaces are prevented. Thus, we have the displacement field

$$u = v = 0, \qquad w = (\delta/L)z,$$

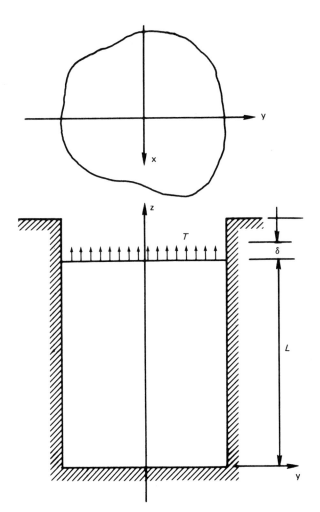

FIGURE 6 Transversely constrained cylinder.

where δ is the z displacement of the top of the cylinder. With the aid of Eq. (20), we obtain the strain field

$$\varepsilon_{xx} = \varepsilon_{yy} = 0; \qquad \varepsilon_{zz} = \delta/L;$$
$$\varepsilon_{ij} \equiv 0, \qquad i \neq j. \tag{33}$$

In view of Eqs. (26) and (33), we have

$$\tau_{xx} - v(\tau_{yy} + \tau_{zz}) = 0,$$
$$\tau_{yy} - v(\tau_{xx} + \tau_{zz}) = 0,$$

and

$$\tau_{zz} - v(\tau_{yy} + \tau_{xx}) = E(\delta/L),$$

and therefore,

$$\tau_{xx} = \tau_{yy} = \frac{v}{1-v}\tau_{zz}$$
$$\tau_{zz} = E\frac{\delta}{L}\frac{(1-v)}{(1-2v)(1+v)} = T, \tag{34}$$
$$\tau_{ij} = 0 \qquad \text{for} \quad i \neq j.$$

In the case of a copper cylinder, we have (see Table I) $v = 0.34$, $E = 8.96 \times 10^{10}$ Pa; and for an axial strain $\varepsilon_{zz} = \delta/L = 0.0005$, we readily obtain

$$\tau_{xx} = \tau_{yy} = 35.53 \times 10^6 \text{ Pa}$$

and

$$\tau_{zz} = 68.9 \times 10^6 \text{ Pa}.$$

Thus, when we compress the copper cylinder with a stress $\tau_{zz} = T = -68.9 \times 10^6$ Pa, there will be induced a lateral compressive stress $\tau_{xx} = \tau_{yy} = -35.53 \times 10^6$ Pa. We note that the strain field [Eq. (33)] satisfies the equations of compatibility [Eq. (23)] and the stress field [Eq. (34)] satisfies the equations of equilibrium [Eq. (30)] provided the body force vector field $\mathbf{F}$ vanishes (or is negligible).

B. Example B

We consider the case of plane, elastic pure bending (or flexure) of a beam by end couples as shown in Fig. 7. In the reference state, the z axis and the beam longitudinal axis coincide. The cross section of the beam (normal to the z axis) is constant and symmetrical with respect to the y axis. Its area is denoted by the symbol A, and the centroid of A is at $(0, 0, z)$. The beam is acted upon by end moments $M_x = M$ such that

$$M_x = \int_A \tau_{zz} y \, dA = M$$

and

$$M_y = \int_A \tau_{zz} x \, dA = 0.$$

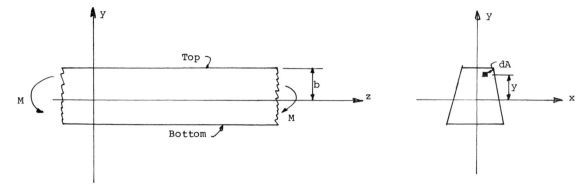

FIGURE 7 Pure bending of a beam.

The present situation suggests the stress field

$$\begin{bmatrix} \tau_{xx} & \tau_{xy} & \tau_{xz} \\ \tau_{yx} & \tau_{yy} & \tau_{yz} \\ \tau_{zx} & \tau_{zy} & \tau_{zz} \end{bmatrix} = \begin{bmatrix} 0 & 0 & 0 \\ 0 & 0 & 0 \\ 0 & 0 & \dfrac{My}{I} \end{bmatrix}, \qquad (35)$$

where $I = \int_A y^2\, dA$, on account of physical reasoning and (elementary) Euler-Bernoulli beam theory. Upon substitution of Eq. (35) into Eq. (26), and in view of Eq. (20), we obtain

$$\varepsilon_{xx} = -\frac{\nu}{E}\tau_{zz} = -\frac{\nu}{E}\frac{M}{I}y = \frac{\partial u}{\partial x}$$

$$\varepsilon_{yy} = -\frac{\nu}{E}\tau_{zz} = -\frac{\nu}{E}\frac{M}{I}y = \frac{\partial v}{\partial y} \qquad (36)$$

$$\varepsilon_{zz} = \frac{\tau_{zz}}{E} = \frac{M}{EI}y = \frac{\partial w}{\partial z},$$

and all shearing strains vanish.

We now integrate the partial differential equations in (36), subject to the following boundary conditions: At $(x, y, z) = (0, 0, 0)$ we require $u = v = w = 0$ and

$$\frac{\partial u}{\partial z} = \frac{\partial y}{\partial z} = \frac{\partial u}{\partial y} = 0.$$

Thus, the beam displacement field is given by

$$u = -\frac{M\nu}{EI}xy$$

$$v = -\frac{M}{2EI}[z^2 + \nu(y^2 - x^2)] \qquad (37)$$

$$w = \frac{M}{EI}yz.$$

We note that the strain field (36) satisfies the equation of compatibility (23) and the stress field (35) satisfies the equations of equilibrium (30) provided the body force vector field **F** vanishes (or is negligible).

With reference to Fig. 7, in the reference configuration, the top surface of the beam is characterized by the plane $y = b$. Subsequent to deformation, the top surface of the beam is characterized by

$$v = -\frac{M}{2EI}(z^2 - \nu x^2) - \frac{\nu Mb^2}{2EI}, \qquad (38)$$

and for $(x, y, z) = (0, b, 0)$ we have

$$v(0, b, 0) = -\frac{\nu Mb^2}{2EI}.$$

We now write Eq. (38) in the form

$$V = v + \frac{\nu Mb^2}{2EI} = -\frac{M}{2EI}(z^2 - \nu x^2), \qquad (39)$$

and we note that V denotes the deflection of the (originally) plane top surface of the beam. The contour lines $V = $ constant of this saddle surface are shown in Fig. 8a. We note that the contour lines consist of two families of hyperbolas, each having two branches. The asymptotes are straight lines characterized by $V = 0$, so that $\tan \alpha = z/x = \sqrt{\nu}$.

An experimental technique called holographic interferometry is uniquely suited to measure sufficiently small deformations of a beam loaded as shown in Fig. 7. In Fig. 8b we show a double-exposure hologram of the deformed top surface of a beam loaded as shown in Fig. 7. This hologram was obtained by the application of a two (light) beam technique, utilizing Kodak Holographic 120-02 plates. The laser was a 10-mW He-Ne laser, 632.8 nm, with beam ratio 4:1. The fringe lines in Fig. 8b correspond to the contour lines of Fig. 8a. The close correspondence between theory and experiment is readily observed. We also note that this technique results in the nondestructive, experimental determination of Poisson's ratio ν of the beam.

C. Example C

We wish to find the displacement, stress field, and strain field in a spherical shell of thickness $(b - a) > 0$ subjected to uniform, internal fluid (or gas) pressure p. The shell is

(a)

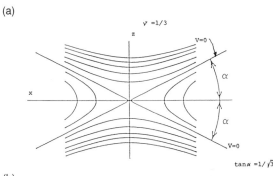

(b)

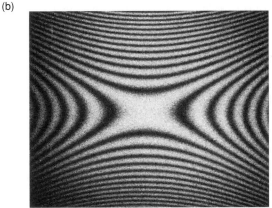

FIGURE 8 (a) Contour lines; V, constant. (b) Double-exposure hologram of deformed plate surface. (Holographic work was performed by P. Malyak in the laboratory of D. P. Malone, Department of Electrical Engineering, State University of New York at Buffalo.) [This hologram is taken from Reismann, H., and Pawlik, P. S. (1980). "Elasticity: Theory and Applications," Wiley (Interscience), New York.]

bounded by concentric spherical surfaces with outer radius $r = b$ and inner radius $r = a$, and we designate the center of the shell by O. In view of the resulting point-symmetric displacement field, there will be no shear stresses acting upon planes passing through O and upon spherical surfaces $a \le r \le b$. Consequently, at each point of the shell interior, the principal stresses are radial tension (or compression) τ_{rr} and circumferential tension (or compression) $\tau_{\theta\theta}$, the latter having equal magnitude in all circumferential directions.

To obtain the pertinent equation of equilibrium, we consider a volume element (free body) bounded by two pairs of radial planes passing through O, each pair subtending a (small) angle $\Delta\theta$, and two spherical surfaces with radii r and $r + \Delta r$. Invoking the condition of (radial) static equilibrium, we obtain

$$(\tau_{rr} + \Delta\tau_{rr})[(r + \Delta r)\Delta\theta]^2 - \tau_{rr}(r\Delta\theta)^2$$

$$= 2\tau_{\theta\theta}\left(r + \frac{\Delta r}{2}\right)\Delta r(\Delta\theta)^2.$$

We now divide this equation by $(\Delta\theta)^2$ and Δr then take the limit as $\Delta r \to 0$ and $\Delta\tau_{rr}/\Delta_r \to d\tau_{rr}/dr$. The result of these manipulations is the stress equation of equilibrium

$$\frac{d\tau_{rr}}{dr} + \frac{2}{r}(\tau_{rr} - \tau_{\theta\theta}) = 0. \tag{40}$$

In view of the definition of strain in Section III, the strain-displacement relations for the present problem are

$$\varepsilon_{rr} = \frac{(dr + du) - dr}{dr} = \frac{du}{dr}$$

$$\varepsilon_{\theta\theta} = \frac{2\pi(r + u) - 2\pi r}{2\pi r} = \frac{u}{r}, \tag{41}$$

where the letter u denotes radial displacement. For our present purpose, we now write Hooke's law (26) in the following form:

$$\tau_{rr} = (\lambda + 2G)\varepsilon_{rr} + 2\lambda\varepsilon_{\theta\theta}$$

$$\tau_{\theta\theta} = 2(\lambda + G)\varepsilon_{\theta\theta} + \lambda\varepsilon_{rr}, \tag{42}$$

where

$$\lambda = \frac{Ev}{(1 + v)(1 - 2v)} = \frac{2Gv}{(1 - 2v)}.$$

If we substitute Eq. (41) into Eq. (42) and then substitute the resulting equations into Eq. (40), we obtain the displacement equation of equilibrium

$$\frac{d^2u}{dr^2} + \frac{2}{r}\frac{du}{dr} - \frac{2}{r^2}u = 0. \tag{43}$$

The spherical shell has a free boundary at $r = b$ and is stressed by internal gas (or liquid) pressure acting upon the spherical surface $r = a$. Consequently, the boundary conditions are

$$\tau_{rr}(a) = -p, \tag{44}$$

where $p \ge 0$ and $\tau_{rr}(b) = 0$. The solution of the differential equation (43) subject to the boundary conditions (44) is

$$u = \frac{pa^3r}{3K(b^3 - a^3)} + \frac{pa^3b^3}{4G(b^3 - a^3)r^2}, \qquad a \le r \le b, \tag{45}$$

where $K = E/[3(1 - 2v)] = (3\lambda + 2G)/3$ is the modulus of volume expansion, or bulk modulus. Upon substitution of Eq. (45) into Eq. (41), we obtain the strain field

$$\varepsilon_{rr} = \frac{pa^3}{3K(b^3 - a^3)} - \frac{pa^3b^3}{2G(b^3 - a^3)r^3}$$

$$\varepsilon_{\theta\theta} = \frac{pa^3}{3K(b^3 - a^3)} + \frac{pa^3b^3}{4G(b^3 - a^3)r^3}, \tag{46}$$

and upon substitution of Eq. (46) into Eq. (42), we obtain the stress field

$$\tau_{rr} = \frac{pa^3}{(b^3 - a^3)}\left[1 - \left(\frac{b}{r}\right)^3\right] = \sigma_2 = \sigma_3 \le 0$$

$$\tau_{\theta\theta} = \frac{pa^3}{(b^3 - a^3)}\left[1 + \frac{1}{2}\left(\frac{b}{r}\right)^3\right] = \sigma_1 \ge 0. \tag{47}$$

We also note the following relations:

$$\tau_{rr} + 2\tau_{\theta\theta} = \frac{3pa^3}{(b^3 - a^3)}, \quad \varepsilon_{rr} + 2\varepsilon_{\theta\theta} = \frac{pa^3}{K(b^3 - a^3)},$$

$$\frac{\tau_{rr} + 2\tau_{\theta\theta}}{\varepsilon_{rr} + 2\varepsilon_{\theta\theta}} = 3K. \tag{48}$$

With reference to Eq. (16), the octahedral shear stress is

$$\tau_0 = \frac{1}{3}\left[(\sigma_1 - \sigma)^2 + (\sigma_2 - \sigma_3)^2 + (\sigma_3 - \sigma_1)^2\right]^{1/2}$$

$$= \frac{\sqrt{2}}{2}\frac{pa^3}{(b^3 - a^3)}\left(\frac{b}{r}\right)^3, \tag{49}$$

and the maximum shear stress (as a function of r) is

$$\tau_{\max} = \frac{1}{2}(\sigma_1 - \sigma_3) = \frac{3}{4}\frac{pa^3}{(b^3 - a^3)}\left(\frac{b}{r}\right)^3, \tag{50}$$

and we note that for the present case we have $\tau_0/\tau_{\max} = (2\sqrt{2})/3 \cong 0.9428$ and [see Eq. (19)]

$$1 < \sqrt{\frac{3}{2}}\frac{\tau_0}{\tau_{\max}} = \frac{2}{\sqrt{3}}. \tag{51}$$

We now apply the failure criterion due to Hencky-Mises (see Section IV): Yielding will occur when $3\tau_0 = \sqrt{2}Y$, where Y denotes the yield stress in simple tension of the shell material. Upon application of this criterion and with the aid of Eq. (49), we obtain

$$p = \frac{2}{3}\frac{(b^3 - a^3)}{a^3}\left(\frac{r}{b}\right)^3 Y, \tag{52}$$

and the smallest value of p results when $r = a$. Thus we conclude that the Hencky-Mises failure criterion predicts yielding on the surface $r = a$ when

$$p = \frac{2}{3}\left[1 - \left(\frac{a}{b}\right)^3\right]Y. \tag{53}$$

The criterion due to Tresca (see Section IV) predicts failure when $\tau_{\max} = Y/2$. With the aid of Eq. (50), this results again in Eq. (53), and we conclude that for the present example, the failure criteria of Hencky-Mises and Tresca predict the same pressure at incipient failure of the shell given by the formula (53).

SEE ALSO THE FOLLOWING ARTICLES

ELASTICITY, RUBBERLIKE • FRACTURE AND FATIGUE • MECHANICS, CLASSICAL • MECHANICS OF STRUCTURES • NUMERICAL ANALYSIS • STRUCTURAL ANALYSIS, AEROSPACE

BIBLIOGRAPHY

Boresi, A. P., and Chong, K. P. (1987). "Elasticity in Engineering Mechanics," Elsevier, Amsterdam.

Brekhovskikh, L., and Goncharov, V. (1985). "Mechanics of Continua and Wave Dynamics," Springer-Verlag, Berlin and New York.

Filonenko-Borodich, M. (1963). "Theory of Elasticity," Peace Publishers, Moscow.

Fung, Y. C. "Foundations of Solid Mechanics," Prentice-Hall, Englewood Cliffs, NJ.

Green, A. E., and Zerna, W. (1968). "Theoretical Elasticity," 2nd ed., Oxford Univ. Press, London and New York.

Landau, L. D., and Lifshitz, F. M. (1970). "Theory of Elasticity" (Vol. 7 of Course of Theoretical Physics), 2nd ed., Pergamon, Oxford.

Leipholz, H. (1974). "Theory of Elasticity," Noordhoff-International Publications, Leyden, The Netherlands.

Lur'e, A. I. (1964). "Three-Dimensional Problems of the Theory of Elasticity," Wiley (Interscience), New York.

Novozhilov, V. V. (1961). "Theory of Elasticity," Office of Technical Services, U.S. Department of Commerce, Washington, D.C.

Parkus, H. (1968). "Thermoelasticity," Ginn (Blaisdell), Boston.

Parton, V. Z., and Perlin, P. I. (1984). "Mathematical Methods of the Theory of Elasticity," Vols. I and II, Mir Moscow.

Reismann, H., and Pawlik, P. S. (1974). "Elastokinetics," West, St. Paul, Minn.

Reismann, H., and Pawlik, P. S. (1980). "Elasticity: Theory and Applications," Wiley (Interscience), New York.

Solomon, L. (1968). "Elasticité Linéaire," Masson, Paris.

Southwell, R. V. (1969). "An Introduction to the Theory of Elasticity," Dover, New York.

Timoshenko, S. P., and Goodier, J. M. (1970). "Theory of Elasticity," 3rd ed., McGraw-Hill, New York.

Elasticity, Rubberlike

Jean-Pierre Queslel
James E. Mark
University of Cincinnati

GLOSSARY

Cross-linking Process by which macromolecular chains are connected by chemical bonds to form a network.

Elasticity Property by which a body resists and recovers from deformation produced by a force.

Entanglement Physical interaction between chains due to their uncrossability.

Freely jointed chain Ideal chain without bond angle restrictions and hindered rotations.

Gaussian chain Chain for which the density distribution of end-to-end chain vectors is Gaussian.

Hookean solid Solid that when deformed in uniaxial tension (or compression) has a stress proportional to strain.

Mechanical testing Determination of mechanical properties.

Modulus of elasticity Ratio of stress to the corresponding strain.

Poisson's ratio Absolute value of the ratio of transverse strain to the corresponding axial strain resulting from uniformly distributed axial stress.

Rubber Material that is capable of recovering from large deformations and that can swell but is insoluble in all solvents.

Strain Change, due to force, in the size or shape of a body relative to its original size or shape.

Strength, tensile Maximum tensile stress that a material is capable of sustaining.

Stress, nominal Ratio of a force to the initial area on which this force acts.

RUBBERLIKE ELASTICITY is a unique property exhibited by polymeric materials called elastomers. These materials can undergo large deformations (up to 10 times their original length) under tension without rupturing. Moreover, when the deforming force is removed, they spontaneously recover their original dimensions. The presence of long chains is a necessary condition, but not a sufficient one. Permanent network structure, high chain flexibility, and weak interchain interactions are also required.

I. INTRODUCTION

In 1496, Columbus and his companions reported that natives in Haiti were playing with small elastic balls. Much later, in 1839, Goodyear discovered that natural rubber cured with sulfur (vulcanization) exhibits elastic properties. In this curing process, chains are joined together by sulfur cross-linkages to form a network, and the ability to recover after deformation originates in the existence of this permanent structure. Gough had already reported, in 1805, the important thermoelastic properties of (unvulcanized) natural rubber: (1) a stressed piece of rubber shrinks upon heating, and (2) the temperature of rubber increases during adiabatic extension. These first observations initiated the thermodynamic analysis of thermoelasticity by Kelvin, supported by a set of experiments on vulcanized natural rubber carried out by Joule. In 1932, Meyer, von Susich, and Valkó were the first to attribute high elasticity to the tendency of the deformed, coiling chain molecule to come back to its original random shape. The field of rubberlike elasticity became widely investigated and has now been covered by extensive reviews.

High elastic deformability is a virtue found exclusively in materials consisting of long macromolecular chains with a large number of accessible configurations (high chain flexibility). Polymeric substances that are neither glassy nor crystalline are capable of exhibiting this property if their chains are cross-linked together into a permanent network and if the interchain interactions are weak. A glassy polymer can possess the same number of conformations as does a strip of rubber. However, it is incapable of spontaneous reversibility because it lacks sufficient mobility to utilize the geometric possibilities.

A large variety of elastomers exist. They differ by their chemical nature, by the microstructure and macrostructure of their chains, and by the cross-linking process used to prepare them. So that specific application requirements can be fulfilled, rubbers are mixed with several ingredients before curing. For example, rubber compounds are frequently reinforced by addition of fillers such as carbon black or silica. The physical properties obtained are very sensitive to small compounding or processing changes. Mechanical testing is used to measure and understand the elastomer's intrinsic properties such as its shear modulus or Young's modulus.

The thermodynamic behavior of rubber is very similar to that of a gas. Specifically, compression of a gas is also governed by entropy changes. On the other hand, deformation of metals is primarily enthalpic. Kelvin pioneered the analysis of rubberlike elasticity by classical thermodynamics, which became relevant through the establishment of the second law of thermodynamics. Thermoelastic properties are directly connected to polymer chain dimensions.

A condition required for high elasticity is weak interchain interactions; therefore the study of the behavior of a single chain is of importance. As a way to elucidate the molecular mechanism of elasticity, statistical mechanics is applied to an ensemble of macromolecules. Molecular theories are still actively developed and are useful to characterize the molecular structure of elastomeric networks.

Besides the usual ways of testing mechanical properties, techniques that directly measure chain orientation contribute to understanding elastomer behavior.

Rupture, filler reinforcement, and strain-induced crystallization are other topics of considerable importance in this area.

II. PHYSICAL TESTING OF RUBBERS

A. Stress

It is easy to realize that the force that causes rupture of a uniaxially stretched strip of rubber depends on its cross-sectional area. So to be independent of specimen geometry, the stress (or engineering stress or nominal stress) σ is defined as the ratio of a force f to the initial area A on which this force acts. This area may vary with deformation, and a true stress, defined as the ratio of force to the real deformed surface, is sometimes used instead.

An elementary volume of sample is shown in Fig. 1. The nominal stress is the ratio of force to its original cross-sectional area:

$$\sigma_i = \sigma_{ii} = f_i/A_i, \tag{1}$$

where f_i and A_i are as represented in Fig. 1a. Shear stress is the ratio of force to tangential surface area. For example, shear stress σ_{ij}, as represented in Fig. 1b, is defined as

$$\sigma_{ij} = \sigma_j/A_i. \tag{2}$$

The square shown in Fig. 1b does not rotate if $\sigma_{ij} = \sigma_{ij}$. More generally it is possible to define a symmetric stress tensor

$$\sigma = \begin{bmatrix} \sigma_1 & \sigma_{12} & \sigma_{13} \\ \sigma_{12} & \sigma_2 & \sigma_{23} \\ \sigma_{13} & \sigma_{23} & \sigma_3 \end{bmatrix}. \tag{3}$$

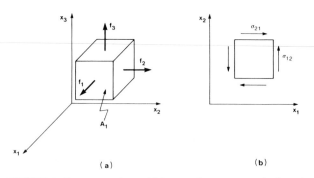

FIGURE 1 Representations of (a) normal stress $\sigma_{ii} = f_i/A_i$ and (b) shear stress $\sigma_{ij} = f_j/A_i$.

The preferred units of stress are newtons per square meter (N/m^2). Other units are often used and are easily converted by the following relationships among megapascals (MPa), newtons per square millimeter (N/mm^2), pounds per square inch ($lb/in.^2$), and kilogram force per square centimeter (kgf/cm^2):

$$MPa \equiv N/mm^2 \equiv lb/in.^2 \times 0.00689$$

and

$$MPa \equiv kgf/cm^2 \times 0.09806.$$

B. Deformation

Forces acting on a body change its dimensions. The deformation tensor ϵ is a matrix with three normal and three tangential components. Upon deformation, portions of the body are moved from an initial position defined by the Cartesian coordinates x_i ($i = 1, 2, 3$) to a final position defined likewise by x_i' (translation or rotation is not deformation since dimensions do not change). Deformation is characterized by the initial coordinate gradients λ_{ij}, displacements d_i, and displacement gradients e_{ij} (i, j assuming in turn the values 1, 2, 3) defined by the following relationships:

$$\lambda_{ij} = \partial x_i' / \partial x_j, \tag{4}$$

$$d_i = x_i' - x_i, \tag{5}$$

and

$$e_{ij} = \partial d_i / \partial x_j. \tag{6}$$

The infinitesimal strain tensor (Cauchy) is symmetric and is defined in terms of displacement gradient or coordinate gradient components. Specifically,

$$\varepsilon_{ij} = (e_{ij} + e_{ji})/2 \tag{7}$$

and

$$\varepsilon_{ij} = (\lambda_{ij} + \lambda_{ji})/2 - \delta_{ij}, \tag{8}$$

where δ_{ij} is the Kronecker symbol. Since deformation or strain is a displacement per unit original dimension, it is dimensionless.

C. Strain Energy and Stress–Strain Relations

From the theory of linear elasticity under homogeneous material conditions, the stress can be expressed as the derivative of the strain energy W with respect to strain as

$$\sigma_{ij} = \partial W / \partial \varepsilon_{ij}. \tag{9}$$

The most general anisotropic form of linear elastic stress–strain relations is given by

$$\sigma_{ij} = C_{ijkl}\varepsilon_{kl}, \tag{10}$$

where C_{ijkl} is the fourth-order tensor of elastic moduli, called the stiffness tensor. Rectangular Cartesian coordinates are employed, with the usual Cartesian tensor notation involving summation on repeated indices. W is defined as

$$W = (\tfrac{1}{2})C_{ijkl}\varepsilon_{ij}\varepsilon_{kl}. \tag{11}$$

It follows from Eqs. (9) through (11) that

$$C_{ijkl} = C_{klij}. \tag{12}$$

The number of independent components C_{ijkl} is reduced to 21. In the case of isotropy, the stress–strain relations can be written as

$$\sigma_{ij} = \lambda_L \varepsilon_{kk} + 2\mu_L \varepsilon_{ij}, \tag{13}$$

where λ_L and μ_L are the Lamé elastic constants. A way to define shear and bulk moduli involves deviatoric components of stress and strain, respectively:

$$a_{ij} = \sigma_{ij} - (\tfrac{1}{3})\delta_{ij}\sigma_{kk} \tag{14}$$

and

$$b_{ij} = \varepsilon_{ij} - (\tfrac{1}{3})\delta_{ij}\varepsilon_{kk}. \tag{15}$$

Equation (13), combined with Eqs. (14) and (15), leads to

$$a_{ij} = 2\mu_L b_{ij} \tag{16}$$

and

$$\sigma_{kk} = 3K\varepsilon_{kk}, \tag{17}$$

where μ_L is the shear modulus and K the bulk modulus. Expressions for these will be developed subsequently.

D. Mechanical Properties of Rubbers

As mentioned earlier, the physical properties of rubbers are highly sensitive to specimen preparation. They are dependent also on testing method and conditions (strain rate, temperature), but standardization has been achieved through the American Society for Testing and Materials (ASTM). Other important standard methods are those of the International Standards Organization (ISO) and the British Standards Institution (BSI). The preparation of materials (ASTM D15-59T) involves mixing of rubber and several ingredients such as cross-linking agent, accelerator, activator antioxidant, and processing aid, and then molding and curing. A typical compound of natural rubber containing the essential ingredients is described in Table I. Values of the stress cited have been measured at 500% deformation. As defined by ASTM designation D412-62T, tensile strength is the force at the time of the rupture of a

TABLE I Typical Example of a Natural Rubber Compound and Its Main Physical Properties[a]

Formula		
The elastomer	SMR5 (natural rubber)[b]	100[c]
Processing aid	VANFRE AP-2	2
Activators		
A fatty acid	Stearic acid	2
A metallic oxide	Zinc oxide	5
Antioxidant	AGERITE STALITE S	1
Vulcanizing agent	Sulfur	2.75
Accelerators of vulcanization		
Primary	ALTAX	1
Secondary	METHYL TUADS	0.1
		113.85

Physical properties			
at 153°C Time of cure (min)	Stress at 500% ($\times 10^{-6}$ N/m^2)	Tensile ($\times 10^{-6}$ N/m^2)	Elongation (%)
5	1.31	25.51	780
10	1.45	23.79	750
15	1.38	22.62	750

[a] Reprinted with permission from Babbit, R. O. (Ed.) (1978). "The Vanderbilt Rubber Handbook," R. T. Vanderbilt, Norwalk, Conn.

[b] SMR, standard Malaysian rubber.

[c] Part per hundred parts of rubber (in weight).

specimen divided by original cross-sectional area. It is calculated by dividing the breaking force by the cross section of the unstressed specimen. As mentioned earlier, elongation at rupture is generally high, for example, ~750% in Table I.

A great variety of tests exist. A few that involve simple states of deformation of the specimen enable one to determine the three interrelated moduli, namely, Young's, shear, and bulk moduli, which are characteristic ratios of stress and strain for isotropic elastic materials.

1. Young's Modulus

The simplest deformation that can be applied to a sample is uniaxial tension (or compression), as shown in Fig. 2a. The application of a stress σ to a real body will result in a strain ε:

$$\sigma = f/AB \qquad (18)$$

and

$$\varepsilon = \Delta C/C, \qquad (19)$$

where f, A, B, C, and ΔC are as defined in Fig. 2a.

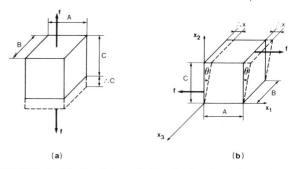

FIGURE 2 (a) Tension and (b) simple shear in a three-dimensional body. [Reprinted with permission from Aklonis, J. J., and MacKnight, W. J. (1983). "Introduction to Polymer Viscoelasticity," 2nd ed., Wiley, New York.]

When a material that obeys Hooke's law is deformed in uniaxial tension (or compression), the stress σ, at any given strain ε, is given by

$$\sigma = E\varepsilon, \qquad (20)$$

where E is called Young's modulus. The situation is less clear in the case of vulcanized rubber since it does not obey Hooke's law. As can be seen in Fig. 3, its stress–strain relation is markedly nonlinear. However, it is generally assumed to be linear over small strains, and hence Young's modulus E is usually defined as the slope of the stress–strain curve over small tensile or compressive strains. It must be emphasized that this discussion is limited to

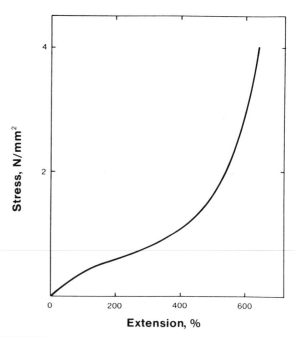

FIGURE 3 Load–elongation curve for a typical vulcanized rubber. [Reprinted with permission from Hearle, J. W. S. (1982). "Polymers and Their Properties," Vol. 1, Wiley, New York.]

time-independent phenomena. Elastomers are not purely elastic materials; they also present viscous effects, which are time dependent. Intrinsic properties such as Young's modulus must be determined from a stress–strain curve pertaining to equilibrium; that is, measurement of stress is made after relaxation at each state of deformation.

It is not easy to measure Young's modulus since strains must be kept small in order for them to be in the linear region of the stress–strain curve. However, they must be large enough to enable reasonably precise measurements. One may, therefore, prefer to determine Young's modulus from hardness measurements. The hardness test consists of measuring the difference between the depths of penetration of a rigid ball into the rubber under a small initial load and a large final load (ASTM designation D1415-62T). It is noteworthy that the slopes of the stress–strain curve in the tensile and compressive regions are slightly different. It has also been reported that moduli in tension and compression increase significantly as zero deformation is approached, but this is probably due to experimental errors.

Tensile tests are standardized (ASTM 1414-56T, 412-62T), and commercial apparatus are available. Two shapes of test pieces are usually used: rings and dumbbells. A typical apparatus to record the stress–strain curve at uniaxial extension stress equilibrium is shown in Fig. 4. Force is measured by an electrical transducer, and deformation is calculated after determination of initial and deformed distances between two ink marks drawn on the sample.

2. Shear Modulus

Simple shear deformation is illustrated in Fig. 2b. The application of the force f will result in the deformation

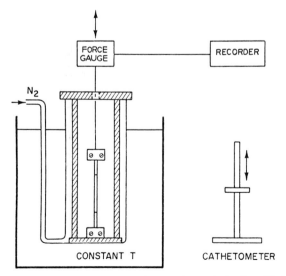

FIGURE 4 Extensometer. [Reprinted with permission from Mark, J. E. (1976). *J. Polymer Sci., Macromol. Rev.* **11**, 135.]

shown by the dashed lines. It is important that the height C is kept constant. Here the stress is given by

$$\sigma_s = f/AB, \qquad (21)$$

where the subscript s denotes shear. The shear strain is given by

$$\gamma = \Delta x/C = \tan\theta, \qquad (22)$$

where angle θ, Δx, and C are as defined in Fig. 2b. The shear modulus G is the ratio of shear stress to shear strain:

$$G = \sigma_s/\gamma. \qquad (23)$$

In Cartesian coordinates (x_1, x_2, x_3), shown in Fig. 2b,

$$\sigma_s = \sigma_{12} = \sigma_{21} \qquad (24)$$

and

$$\varepsilon_{12} = \varepsilon_{21} = \gamma/2. \qquad (25)$$

Other stress and strain tensor components are zero. Again the compressive strength is the maximum compressive stress that a material is capable of sustaining (ASTM E6-62).

Testing in shear is not more difficult than testing in tension, the only practical difficulty being the necessity to bond the rubber test piece to rigid members to provide attachment for applying the shearing force (ASTM E143-61). The stress–strain curve in simple shear is approximately linear up to relatively large strains.

According to the statistical theory of rubberlike elasticity, the stress–strain relation up to 50% extension or compression for an incompressible body is given to a sufficient approximation by

$$\sigma = G(\lambda - \lambda^{-2}), \qquad (26)$$

where λ is the extension ratio [Eq. (8)]:

$$\lambda = \varepsilon + 1. \qquad (27)$$

By substituting $(\varepsilon + 1)$ for λ and expanding, one can rewrite Eq. (26) as

$$\sigma = G(3\varepsilon - 3\varepsilon^2 + 4\varepsilon^3 - 5\varepsilon^4 \cdots). \qquad (28)$$

For infinitesimal strains, comparison of Eqs. (20) and (28) leads to

$$E = 3G. \qquad (29)$$

3. Bulk Modulus

The bulk modulus is the reciprocal of the compressibility. It is defined as the ratio of the hydrostatic pressure p to the volume strain $\Delta V/V_o$ (volume change per unit original volume):

$$K = p/(\Delta V/V_o). \qquad (30)$$

The bulk modulus can be determined not only by static compression experiments, but also (and more easily) by measurement of the velocity of longitudinal elastic waves, that is, sound.

4. Poisson's Ratio

According to ASTM E6-62, Poisson's ratio is the absolute value of the ratio of transverse strain to the corresponding axial strain. It can be measured by setting extensometers in the axial and transverse directions of the test piece (ASTM E132-61). Poisson's ratio μ_P is related to the change of volume dV accompanying infinitesimal deformation $d\varepsilon$ by the expression

$$\mu_P \equiv (\tfrac{1}{2})[1 - (1/V)(dV/d\varepsilon)]; \qquad (31)$$

for an incompressible material, $(dV/d\varepsilon) = 0$ and $\mu_P = 0.5$. This value is the maximum possible value for any substance. Violation of this rule, however, may occur as a result of a phase change (i.e., crystallization) during stretching. Poisson's ratio of rubber is ~ 0.49; therefore unfilled elastomers are nearly incompressible.

It can be shown that the three elastic moduli are interrelated by the equations

$$E = 2G(1 + \mu_P) \qquad (32)$$

and

$$K = (E/3)/(1 - 2\mu_P). \qquad (33)$$

For an incompressible body, K is infinite and μ_P equals 0.5. Thus Eq. (29) is recovered ($E = 3G$).

5. Other Tests

a. Torsion test. Torsional behavior has also been investigated. In principle, the shear modulus could be measured by using a test piece strained in torsion. However, it is more generally used as a low-temperature test. For a strip, force and deflection are related by

$$\tau = kbt^2 G\theta/l, \qquad (34)$$

where τ is the applied torque, k the shape factor, b the width of the test piece, t the thickness of the test piece, G the shear modulus, θ the angle of twist, and l the effective length of the test piece.

b. Tear test. Tensile stress is a rupture stress measurement carried out by using a flawless test piece. In a tear test, however, the force is not applied evenly, but is concentrated on a deliberate flaw or sharp discontinuity, and the force required to continuously produce a new surface is measured (ASTM D654-54). Tear strength is not, however, a fundamental property of a material. Rivlin

and Thomas have developed the concept of the characteristic energy of tearing, which is the energy required to form a unit area of new surface by tearing. This field of strong interest in engineering practice is being extensively investigated.

III. THERMODYNAMICS AND THERMOELASTICITY

Gough, Kelvin, and Joule investigated the behavior of rubber under various thermodynamic circumstances (e.g., the effect of temperature in adiabatic extensions). Thermodynamic studies are important for engineering purposes since it is often necesary to predict the effects of environmental changes (temperature, pressure). Moreover, it can be shown that molecular information such as change of end-to-end distance of chains with temperature (which can then be related to conformational characteristics) and the contribution of internal energy to the elastic force can be obtained from thermoelasticity experiments.

A. Thermodynamic Equation of State

In simple elongation, the change in the Helmholtz free energy A of the system caused by change in temperature T, volume V, and length L of the system is given by

$$dA = -p\,dV + f\,dL - S\,dT, \qquad (35)$$

where p is the external pressure and f the external force of extension at equilibrium. From Eq. (35), the retractive force f due to change in length L at constant temperature and volume is the derivative of A versus L:

$$f = (\partial A/\partial L)_{T,V}. \qquad (36)$$

The relationships among entropy, pressure, and Helmholtz energy can also be deduced from Eq. (35) as

$$p = -(\partial A/\partial V)_{T,L} \qquad (37)$$

and

$$S = -(\partial A/\partial T)_{V,L}. \qquad (38)$$

However, A may be expressed as

$$A \equiv U - TS, \qquad (39)$$

where U is the internal energy. The change in U accompanying the stretching of an elastic body may be written as

$$dU = dQ - dW, \qquad (40)$$

where dQ is the element of heat absorbed by the system and dW the element of work done by the system on the surroundings:

$$dW = p\,dV - f\,dL \qquad (41)$$

and

$$dQ = T\,dS. \qquad (42)$$

Equation (36) may be transformed to

$$f = (\partial U/\partial L)_{T,V} - T(\partial S/\partial L)_{T,V}. \qquad (43)$$

This shows that f is the sum of a contribution f_e from the internal energy and a contribution f_s from the entropy:

$$f = f_e + f_s, \qquad (44)$$

with

$$f_e = (\partial U/\partial L)_{T,V} \qquad (45)$$

and

$$f_s = -T(\partial S/\partial L)_{T,V}. \qquad (46)$$

The second derivative obtained by differentiating $(\partial A/\partial L)_{T,V}$ with respect to T at constant V and L is identical to that obtained by differentiating $(\partial A/\partial T)_{V,L}$ with respect to L at constant V and T [application of Maxwell's relation to Eqs. (36) and (38)]:

$$(\partial f/\partial T)_{V,L} = -(\partial S/\partial L)_{T,V}. \qquad (47)$$

This leads to

$$f_s = T(\partial f/\partial T)_{V,L}. \qquad (48)$$

Equation (48) is the fundamental relation of thermoelasticity that relates the entropy component of the retractive force to its temperature coefficient at constant volume and length. The fractional contribution to f due to internal energy can be written in several alternative forms:

$$
\begin{aligned}
f_e/f &= 1 - f_s/f \\
&= 1 - (T/f)(\partial f/\partial T)_{V,L} \\
&= -T[\partial \ln(f/T)/\partial T]_{V,L} \\
&= (1/T)[\partial \ln(f/T)/\partial(1/T)]_{V,L}. \qquad (49)
\end{aligned}
$$

The contributions f_e and f_s in Eqs. (45), (46), (48), and (49) are related to experiments carried out at constant length and at constant volume. So that this latter condition can be fulfilled, a hydrostatic pressure has to be applied to nullify the increase in volume due to thermal expansion. Other thermodynamic relationships can be obtained in case of measurements at constant temperature and pressure.

The Gibbs free energy F is defined by

$$F \equiv H - TS, \qquad (50)$$

where H is the heat content or enthalpy:

$$H \equiv U + pV. \qquad (51)$$

The differential of the Gibbs free energy follows from Eqs. (41), (42), (50), and (51):

$$dF = V\,dP + f\,dL - S\,dT. \qquad (52)$$

It follows from Eq. (52) that

$$f = (\partial F/\partial L)_{T,p}, \qquad (53)$$

$$f = (\partial H/\partial L)_{T,p} - T(\partial S/\partial L)_{T,p}, \qquad (54)$$

and

$$S = -(\partial F/\partial T)_{p,L}. \qquad (55)$$

Again, application of Maxwell's relation to Eqs. (53) and (55) leads to

$$(\partial f/\partial T)_{p,L} = -(\partial S/\partial L)_{T,p}. \qquad (56)$$

By making use of Eq. (56), one can transform Eq. (54) to

$$f = (\partial H/\partial L)_{T,p} + T(\partial f/\partial T)_{p,L}. \qquad (57)$$

Another relationship, obtained upon substitution of Eq. (47) in Eq. (43), is

$$f = (\partial U/\partial L)_{T,V} + T(\partial f/\partial T)_{V,L}. \qquad (58)$$

Equations (57) and (58) are two forms of the thermodynamic equation of state for elasticity.

B. Thermoelastic Experiments

The energetic contribution f_e to the elastic force f exhibited by a deformed polymer network is given in the case of uniaxial deformation (elongation or compression) by Eq. (45). A more useful quantity is the fraction f_e/f of the total force that is of energetic origin. Of the different forms of Eq. (49), the following one is generally used:

$$f_e/f = -T[\partial \ln(f/T)/\partial T]_{L,V}. \qquad (59)$$

Thermoelastic measurements have been carried out at constant length and constant volume, but because of experimental difficulties encountered in meeting the latter requirement, most experiments are conducted at constant pressure and either constant length or constant extension or deformation α. The extension α is measured relative to the length L_v^i of the unstretched sample when its volume is fixed at the same volume V as occurs in the stretched state:

$$\alpha = L/L_v^i, \qquad (60)$$

where L is the length of the deformed sample at volume V.

The elastic equation of state for simple elongation, valid for both swollen and unswollen samples, is obtained by using the statistical thermodynamics of random networks. In the case of an affine network (where the cross-links are displaced in proportion to the strain).

$$f\left(\nu_e kT/L_v^i\right)\left(\langle r^2\rangle_v^i/\langle r^2\rangle_0\right)(\alpha - \alpha^{-2}), \qquad (61)$$

where ν_e is the number of effective chains in the network, k Boltzmann's constant, $\langle r^2\rangle_v^i$ the mean-square end-to-end distance for a typical network chain in the isotropic state of volume V, and $\langle r^2\rangle_0$ the corresponding value for the free chain in the absence of constraints imposed by network junctions. The quantity $\langle r^2\rangle_0$ is a characteristic of the polymer chain. It depends on the bond lengths, angles, and potentials hindering rotations about bonds and is therefore a function of temperature. In the case of a phantom network (where the chains can transect one another),

$$f = \left(\xi kT/L_v^i\right)\left(\langle r^2\rangle_u^i/\langle r^2\rangle_0\right)(\alpha - \alpha^{-2}), \qquad (62)$$

where ξ is the cycle rank of the network. These concepts of affine and phantom networks will be developed later in this article. By differentiation of Eq. (61) or (62) at constant length and volume, constant length and pressure, and constant pressure and deformation, respectively, three relationships are obtained:

$$[\partial \ln(f/T)/\partial T]_{V,L} = -d\,\ln\langle r^2\rangle_0/dT, \qquad (63)$$

$$[\partial \ln(f/T)\partial T]_{p,L} = -d\,\ln\langle r^2\rangle_0/dT - b/(\alpha^3 - 1), \qquad (64)$$

and

$$[\partial \ln(f/T)/\partial T]_{p,\alpha} = -d\,\ln\langle r^2\rangle_0/dT + b/3, \qquad (65)$$

where b is the bulk coefficient of thermal expansion:

$$b = (\partial V/\partial T)_p/V. \qquad (66)$$

It was assumed in the derivation of Eqs. (61) through (65) that the stored elastic free energy resides exclusively within the chains of the network. From Eqs. (49), (63), and (65), it can be shown that

$$(\partial f/\partial T)_{p,\alpha} = (\partial f/\partial T)_{V,L} + fb/3. \qquad (67)$$

For measurements at constant length and pressure, an expression for f_e/f is obtained by combination of Eqs. (59), (63), and (64):

$$f_e/f = -T[\partial \ln(f/T)/\partial T]_{L,p} - bT/(\alpha^3 - 1). \qquad (68)$$

Similarly, for measurements at constant pressure and deformation, combining Eqs. (59), (63), and (65) leads to

$$f_e/f = -T[\partial \ln(f/T)/\partial T]_{p,\alpha} + bT/3. \qquad (69)$$

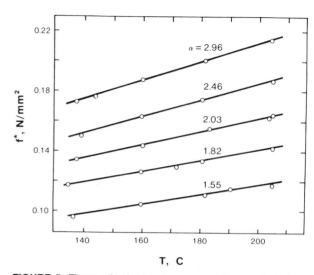

FIGURE 5 Thermoelastic data for an amorphous polyethylene network in the unswollen state, at constant length. The nominal stress f^* is expressed relative to the undeformed cross section at the highest temperature of measurement, and each curve is characterized by the value of the elongation α at the same temperature. [Reprinted with permission from Mark, J. E. (1976). *J. Polymer Sci., Macromol. Rev.* **11**, 135.]

Another fundamental relationship is deduced from Eqs. (59) and (63):

$$f_e/f = T\,d\,\ln\langle r^2\rangle_0/dT. \qquad (70)$$

This equation is particularly important because it ties together thermodynamic and molecular concepts.

A typical apparatus used for thermoelastic studies of networks in elongation is shown in Fig. 4. The use of thermodynamic relationships previously derived requires the measurements of force to be made at equilibrium at each temperature of investigation. A typical series of stress–temperature cures, which have been obtained on amorphous, unswollen polyethylene networks at constant pressure and length, is shown in Fig. 5. Equation (70) relates the thermodynamic quantity f_e/f to its molecular counterpart $d\,\ln\langle r^2\rangle_0/dT$, which can be interpreted in terms of the rotational isomeric state theory of chain configurations. Deformation of a network requires the polymeric chain to switch to less compact conformational states. If these states are of relatively low energy, then f_e/f will be negative. An example of this case is amorphous polyethylene. Its less compact states are the trans (planar) conformations, which are of lowest energy. An example of the opposite behavior is polydimethylsiloxane, in which the less compact states are the gauche (nonplanar) conformations. Since these are of higher energy, f_e/f is found to be positive. Experimental values of f_e/f have been quantitatively interpreted for a wide variety of elastomers.

It has been shown that reliable values of f_e/f may be obtained from measurements at either constant volume or constant pressure. Furthermore, f_e/f does not depend on sample orientation (deformation) prior to cross-linking, or degree of crystallinity, or presence of diluent during cross-linking. Similarly, identical results seem to be obtained for systems cross-linked by chemical means and those cross-linked by high-energy radiation. Also, measurements in elongation are in good agreement with those obtained from other types of deformation. Finally, f_e/f seems to be independent of the extent of deformation and there is no effect of dilution on this ratio. Direct calorimetric measurements carried out on a polymeric network during the deformation process lead to values similar to the thermoelastic results. Typical values of f_e/f for natural rubber, cis-1,4-polybutadiene, and polydimethylsiloxane are, respectively, 0.17, 0.12, and 0.25. A negative value of -0.45 has been reported for polyethylene, in agreement with theory.

A basic postulate of rubberlike elasticity theory is that in the deformation of a polymeric network, the force observed is intramolecular in origin. The elastic response of the network originates within the chains and, not to a significant extent, from interactions between them. Thermoelasticity and neutron scattering results strongly support this assumption. Also, there is no effect of dilution on f_e/f. Furthermore, values of the temperature coefficient $d \ln\langle r^2\rangle_0/dT$ obtained from thermoelastic studies are in good agreement with values obtained from viscosity–temperature measurements on chains of the same polymer dispersed in a solvent at infinite dilution, as can be seen from Table II. Small-angle neutron scattering (SANS) ex-

TABLE III Molecular Dimensions in Amorphous Polymers by Small-Angle Neutron Scattering[a]

Polymer, bulk state	$(\langle s^2\rangle/M)^{1/2}$, (Å/Dalton$^{1/2}$)	
	Dilute solution	Bulk
Polystyrene, glass	0.27	0.27
Polystyrene, glass	0.27	0.28
Polyethylene, melt	0.45	0.46
Polyethylene, melt	0.45	0.45
Polypropylene (isotactic), melt	0.33	0.34
Polymethylmethacrylate, glass	0.30	0.31
Polydimethylsiloxane, melt	0.27	0.25
Polycarbonate, solid	0.43[b]	0.46
Polyvinylchloride	0.35	0.40
Polyisobutylene, melt	0.30	0.31

[a] Reprinted with permission from Flory, P. J. (1984). *Pure Appl. Chem.* **56**, 305.

[b] Calculated.

periments have been performed on amorphous bulk (undiluted) polymers and dilute solutions. It has been shown that chains in a bulk polymer occur in random configurations unperturbed by the neighboring chains with which they are densely packed. Hence, changes in conformations accompanying deformation are not followed by changes in intermolecular energy. The square root of the ratio of the mean-square radius of gyration to the molecular weight, $(\langle s^2\rangle/M)^{1/2}$, as deduced from SANS experiments, is given in Table III for various bulk (undiluted) polymer chains. Its values are compared with data obtained by application of conventional methods (and sometimes of SANS) to dilute solutions of the same polymer in θ solvent. Good agreement is observed.

IV. STATISTICAL THERMODYNAMICS OF RUBBERLIKE ELASTICITY

The mechanical properties of rubber differ from those of other solids, both crystalline and amorphous. Typically, Young's modulus at small extension for rubber is 10^6 N/m^2, whereas those of metal and galss are, respectively, 10^{11} and 5×10^{10}. Similarly, the bulk modulus of rubber is around 3×10^9 N/m^2, wheres those of metal and glass are within an order of magnitude of their Young's moduli. Rubber has a Poisson's ratio of 0.49 compared with 0.3 and 0.25, respectively, for metal and glass. It has the advantage of a low density, 0.9 g/cm^3 (versus approximately 8 for metal and 2.5 for glass), and a high limit of extension without rupture (500–1000%), whereas most substances are no longer elastic after a few percent deformation. It is not surprising to find large departures from

TABLE II Comparison of $d \ln\langle r^2\rangle_0/dT$ Deduced from Thermoelastic Measurements on a Network with Values from Viscosity Measurements on Isolated Chains[a]

Polymer	$10^3 d \ln\langle r^2\rangle_0/dT$	
	$f - T$[b]	$[\eta] - T$
Polyethylene	$-1.05 (\pm 0.10)$	$-1.2 (\pm 0.2)$[c]
		$-1.19 (\pm 0.04)$[d]
		$-0.8 (\pm 0.1)$[e]
Poly(n-pentene-1), isotactic	$0.34 (\pm 0.04)$	$0.52 (\pm 0.05)$[c]
Polyisobutylene	$-0.19 (\pm 0.11)$	$-0.28 (\pm 0.05)$[c]
		$-0.10 (\pm 0.03)$[d]
		$-0.4 (\pm 0.1)$[e]
Polyoxyethylene	$0.23 (\pm 0.02)$	$0.2 (\pm 0.2)$[e]
Polydimethylsiloxane	$0.59 (\pm 0.14)$	$0.52 (\pm 0.20)$[c]

[a] Reprinted with permission from Mark, J. E. (1976). *J. Polymer Sci., Macromol. Rev.* **11**, 135.

[b] Taken from the thermoelastic results present in the literature.

[c] Measurements in an athermal solvent.

[d] Measurements in a series of structurally similar θ solvents.

[e] Approximate results from measurements in a thermodynamically good solvent.

Hooke's law over such a large possible range of extension. Deformation of a metal involves changes in interatomic distances. Since strong valence forces bind atoms together, large energy changes are required. For a polymeric material to exhibit rubberlike elasticity under the usual environmental conditions, it must possess weak intermolecular bonding and have a flexible chain structure. In order to maintain the high segmental mobility for rapid stretching, it must be used well above its glass transition temperature T_g.

There is a thermodynamic analogy between compression of ideal gas and extension of ideal rubber. They both involve entropy changes. The statistical derivation of the force of retraction of rubber is similar to that of the pressure of an ideal gas. When the initial volume V_0 of a gas is reduced to V, the probability Ω that all of the ν molecules of gas, originally in the volume V_0, concentrate in the volume V is

$$\Omega = (V/V_0)^{\nu}. \tag{71}$$

During this compression, the entropy change ΔS is given by the Boltzmann relation

$$\Delta S = k \ln \Omega, \tag{72}$$

where k is Boltzmann's constant. For a perfect gas, the pressure, p is obtained by making use of the equation

$$p = T(\partial S/\partial V)_T. \tag{73}$$

A combination of Eqs. (71) through (73) leads to

$$p = \nu k T/V. \tag{74}$$

A network is an ensemble of chains tied together by cross-linking points. A similar statistical analysis of this network therefore requires a description of a single, isolated polymer chain in an ideal, noninteracting environment.

A. Configurations of Polymer Chains

Macromolecules contain thousands of covalent bonds in linear succession. Rotations occur about these skeletal bonds, and because of Brownian motion, chains assume a great variety of spatial configurations. The unique mechanical and physical properties of polymeric substances are related to the randomness in the conformations of their skeletal bonds that characterizes the random coil. Although the analysis of molecules of such complexity may seem very difficult, one may take advantage of the great number of configurations accessible to the macromolecule by applying statistical methods. Whereas this number is far too great for full enumeration, the vast population can be sampled by randomly selecting a manageable set of

representative configurations. When a high polymer crystallizes, the valence bonds of the chains in the crystalline regions take up regular orientation with respect to their neighbors. For example, the planar zigzag configuration of polyethylene gives the maximum separation between chain ends without distorting the normal bond lengths or angles. However, because of the relative freedom for rotation about single bonds, many other configurations of the chains are intrinsically possible and occur in the amorphous regions of the material. Elastomers are characterized by an absence of strong intermolecular forces and a minimum of geometric interferences to rotation about skeletal bonds. When such molecules are embedded in a medium composed throughout of chemically similar material, many of the configurations that are accessible at constant volume without distorting bond lengths or angles have about the same internal energy and hence about the same *a priori* probability.

A sketch of a chain of four carbon atoms is illustrated in Fig. 6. The bond length l and valence angle θ are fixed quantities for a given chemical structure. The angle ϕ specifies the position of the fourth atom with respect to the other three and defines the conformation. Its minimum energy corresponds to the planar zigzag form, which is the fully extended configuration. The length of the fully extended chain is called the contour length. The quantity most widely used to characterize the configuration of a polymer chain is its end-to-end distance r (displacement length). An average value of r is the root-mean-square $\langle r^2 \rangle^{1/2}$. Another important measure of the effective size of a polymeric molecule is the root-mean-square of the elements of the chain from its center of gravity. This quantity, designated by $\langle s^2 \rangle^{1/2}$, is called the radius of gyration.

1. Freely Jointed Chain

The hypothetical freely jointed chain is considered to consist of n linkages of length l joined in sequence without

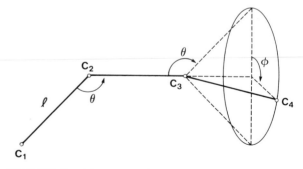

FIGURE 6 Spatial representation of a chain of four carbon atoms. Here l is the bond length, θ the valence angle, and ϕ the rotation angle that specifies the position of atom C_4 on the valence cone generated by the 360° rotation of ϕ.

any restrictions whatever on the angles between successive bonds. The corresponding chain configuration problem is similar to that of random flight. The configuration of the freely jointed chain resembles the path described by a diffusing particle such as a gas molecule. The distribution of end-to-end distances of the chain is first determined in one dimension and then generalized to three dimensions. If the projection of the configuration on the x axis is considered, the average x component of the end-to-end distance is zero since each bond is capable of choosing any direction with equal probability. It can be shown that the root-mean-square value of the length l_x of the projection of a bond is

$$\sqrt{\langle l_x^2 \rangle} = l/\sqrt{3}. \qquad (75)$$

The number n of bonds is large. Thus it can be assumed that each bond makes a contribution along the x axis equal in magnitude to $\sqrt{\langle l_x^2 \rangle}$. Therefore the projection of r on the x axis is given by

$$x = (n_+ - n_-)\sqrt{\langle l_x^2 \rangle} = (n_+ - n_-)l/\sqrt{3}, \qquad (76)$$

where n_+ and n_- are, respectively, the number of bonds making positive and negative contributions upon their projection. The probability $W(x)\,dx$ that x assumes a value between x and $x + dx$ can be calculated under the condition that $|n_+ - n_-| \ll n$ that is, that x is much smaller than the value nl corresponding to full extension of the chain:

$$W(x)\,dx = (\beta/\pi^{1/2}) \exp(-\beta^2 x^2)\,dx, \qquad (77)$$

where

$$\beta = [3/(2nl^2)]^{1/2} \qquad (78)$$

and $W(x)\,dx$ is the Gaussian distribution. In three dimensions, equivalent expressions are obtained for $W(y)$ and $W(z)$, and under the restrictions cited here above, $W(x)$ depends only on x, $W(y)$ on y, and $W(z)$ on z. Thus the probability that the components of r are between x and $x + dx$, and so forth, is the product of the probabilities corresponding to each axis:

$$W(x, y, z)\,dx\,dy\,dz = W(x)W(y)W(z)\,dx\,dy\,dz$$
$$= (\beta/\pi^{1/2})^3 \exp(-\beta^2 r^2)\,dx\,dy\,dz, \qquad (79)$$

where r is the magnitude of the chain end-to-end vector

$$r^2 = x^2 + y^2 + z^2. \qquad (80)$$

Equation (79) gives the probability that if one extremity of vector $\mathbf{r}$ is fixed at the origin of the coordinates, the other lies in the volume $dx\,dy\,dz$ centered around the point (x, y, z). The most probable value of r is zero. The

probability for a chain to have its end in a spherical shell of radius r and thickness dr centered at the origin, irrespective of direction, is the radial distribution

$$W(r)\,dr = (\beta/\pi^{1/2})^3 \exp(-\beta^2 r^2)4\pi r^2\,dr. \qquad (81)$$

The most probable value of r, occurring at the maximum in $W(r)$, is

$$r = 1/\beta. \qquad (82)$$

Thus, this most probable value is not zero. The site of the spherical shell increases with r and is responsible for this difference in the location of the maximum of $W(x, y, z)$ and $W(r)$. The distribution $W(r)$ is represented in Fig. 7 (curve a). The average value of r, calculated by making use of the equation

$$\langle r \rangle = \int_0^\infty r W(r)\,dr \bigg/ \int_0^\infty W(r)\,dr, \qquad (83)$$

is

$$\langle r \rangle = 2/(\pi^{1/2}\beta). \qquad (84)$$

The mean-square end-to-end distance is the second moment of the radial distribution function:

$$\langle r^2 \rangle = \int_0^\infty r^2 W(r)\,dr \bigg/ \int_0^\infty W(r)\,dr, \qquad (85)$$

which gives

$$\langle r^2 \rangle = 3/(2\beta^2). \qquad (86)$$

In the use of the freely jointed chain, β assumes the value given by Eq. (78). Then

$$\langle r^2 \rangle = nl^2. \qquad (87)$$

For linear polymer chains, the mean-square radius of gyration is proportional to $\langle r^2 \rangle$. Specifically,

$$\langle s^2 \rangle = \langle r^2 \rangle/6. \qquad (88)$$

The Gaussian distribution was obtained in the preceding treatment by using the assumption that the chains are far from their full extension. Moreover, the Gaussian distribution predicts zero probability only for $r = \infty$ instead of for all r in excess of that corresponding to full chain extension and does not adequately take into account the significant geometric and conformation differences known to exist among different types of polymer chains. The distribution of the displacement length r for the freely jointed chain may be treated according to methods applied to the orientation of magnetic or electric dipoles by an external field of sufficient strength to produce effects approaching saturation (i.e., complete orientation). A distribution valid for all extensions is expressed by

$$W(r)\,dr = \text{const} \exp\left[-\int_0^r \mathcal{L}^{-1}(r/nl)\,dr/nl\right]4\pi r^2\,dr, \qquad (89)$$

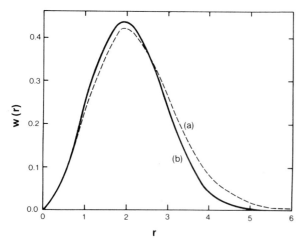

FIGURE 7 Radial distribution function $W(r)$ for a six-link random chain, with $l = 1$: (a) Gaussian limit and (b) inverse Langevin distribution. [Reprinted with permission from Treloar, L. R. G. (1975). "The Physics of Rubber Elasticity," 3rd ed., Clarendon Press, Oxford.]

where $\mathscr{L}^{-1}$ is the inverse of the Langevin function defined by

$$\mathscr{L}(u) = \coth u - 1/u. \tag{90}$$

Equation (89) can be expanded in the series

$$W(r)\,dr = \text{const} \exp(-n\{(\tfrac{3}{2})(r/nl)^2 + (\tfrac{9}{20})(r/nl)^4 + (\tfrac{99}{350})(r/nl)^6 + \cdots\})4\pi r^2\,dr. \tag{91}$$

The Gaussian distribution [Eq. (81)] is recovered for $r \ll nl$.

A representation of the inverse Langevin distribution is given in Fig. 7 (curve b) for a six-link random chain. The approximation of a Gaussian distribution is generally valid for the treatment of rubber elasticity.

2. Real Chains

Real chains have valence angle restrictions. If there is no hindered rotation on the rotational angles (all values of ϕ equally likely), it can be shown that $\langle r^2 \rangle$ is given by

$$\langle r^2 \rangle = nl^2(1 - \cos\theta)/(1 + \cos\theta). \tag{92}$$

However, the angle of rotation ϕ may be restricted by steric interferences between atoms. For n is large and the average value of $\cos\phi$ [i.e., $\langle\cos\phi\rangle$] not too close to unity,

$$\langle r^2 \rangle = nl^2 \left(\frac{1 - \cos\theta}{1 + \cos\theta}\right)\left(\frac{1 + \langle\cos\phi\rangle}{1 - \langle\cos\phi\rangle}\right). \tag{93}$$

3. Equivalent Freely Jointed Chain

Kuhn has shown that real chains of n bonds of length l with fixed bond angles and hindered rotations can be sat-

isfactorily represented by redefined freely jointed chains. The full description of the equivalent freely jointed chain requires two parameters, the number of segments n_e and the length l_e of each. Provided that these segments include a sufficient number of bonds, there will be no angular correlation between the displacement vectors of adjoining segments of even a fairly stiff chain. The equivalent chain obeys the same Gaussian distribution function as the real chain if the number of statistical links n_e is large and if l_e is taken as the root-mean-square length of these links. The parameter β of the distribution function is now

$$\beta = \left[3/(2n_e l_e^2)\right]^{1/2}. \tag{94}$$

The equivalent freely jointed chain is defined as the one that has the same contour length and the same root-mean-square displacement length as the real chain. The quantities n_e and l_e are completely determined by the two equations

$$n_e = n^2 l^2 / \langle r^2 \rangle \tag{95}$$

and

$$l_e = \langle r^2 \rangle / (nl), \tag{96}$$

where $\langle r^2 \rangle$ is the mean-square end-to-end distance of the real chain, which is also equal to the mean-square end-to-end distance $\langle r^2 \rangle_e$:

$$\langle r^2 \rangle = \langle r^2 \rangle_e = n_e l_e^2. \tag{97}$$

The dimensions of the equivalent chain may be derived either by calculation from the known geometry of the real chain or by experiments from the unperturbed chain dimensions. These may be obtained from viscosity or sedimentation measurements carried out in a solvent at the θ temperature.

Monte Carlo simulation has been applied to chain displacement length calculations. Walks are generated on a regular tetrahedral lattice. So that spatial interferences are excluded, all the walks that return to a lattice point that is already occupied are rejected. A free choice of vacant nearest-neighbor sites is allowed for each step so that the paths traced correspond quite closely with those allowed to a singly bonded chain of carbon atoms with three equal potential energy minima in each complete rotation about a chain bond. Such a computer-generated conformation is shown in Fig. 8.

4. Rotational Isomeric State Method

An alternative way to determine configurational averages is the matrix generation technique. The procedure is applicable to chains of any length, to copolymers of any specified composition, and to asymmetric (e.g., vinyl) chains of any stereochemical configuration and sequence, tactic

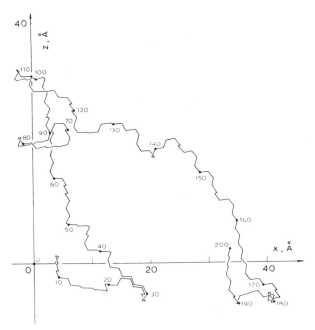

FIGURE 8 Computer-generated conformation of a polymethylene chain of 200 C–C bonds. The projection shown is in the plane of the first two bonds.

or atactic. Spatial configurations of macromolecules are represented in terms of a set of discrete rotational states. Properties of chain molecules that depend on configuration include the dimensions of the spatial configuration measured by the chain displacement vector **r** connecting its ends or by any of various products that may be formed from **r**, the radius of gyration, and, for example, dipole moments and optical anisotropies.

Evaluation of these properties requires summations over contributions from the individual bonds or groups composing the chain. Some applications of the required matrix multiplication technique concerning the chain displacement vector, square of the magnitude of the chain vector, and radius of gyration have been reviewed by Flory and are presented hereafter.

The skeletal bonds of a chain are numbered 1 to n. A Cartesian coordinate system is defined for each bond. The axis x_i for system i is taken along skeletal bond i, the y_i axis in the plane defined by bonds $i-1$ and i, and the z_i axis in the perpendicular direction that completes a right-handed reference frame. A matrix $\mathbf{T}_i$ brings about the transformation between reference frames $i+1$ and i and is a function of angles θ and ϕ (see Fig. 6). The quantities l and θ are fixed by the chemical structure and ϕ by information obtained on structurally related small molecules.

a. Chain displacement vector. In the reference frame of the first bond (defined with the help of an imag-

inary zeroth bond parallel to bond 2), the vector **r** is calculated by

$$\mathbf{r} \equiv \begin{bmatrix} x \\ y \\ z \end{bmatrix} = \mathbf{L}_1 + \mathbf{T}_1\mathbf{L}_2 + \mathbf{T}_1\mathbf{T}_2\mathbf{L}_3 + \cdots$$
$$+ \mathbf{T}_1\mathbf{T}_2 \cdots \mathbf{T}_{n-1}\mathbf{L}_n, \tag{98}$$

where each bond vector $\mathbf{L}_i$ is expressed in its own frame of reference:

$$\mathbf{L}_i = \begin{bmatrix} l_i \\ 0 \\ 0 \end{bmatrix}. \tag{99}$$

A compact form of Eq. (98) is

$$\mathbf{r} \equiv \mathbf{r}_{0,n} = \sum_{j=1}^{n} \mathbf{T}_1^{(j-1)}\mathbf{L}_j, \tag{100}$$

where the serial product $\mathbf{T}_1^{(j-1)}$ is defined by

$$\mathbf{T}_1^{(j-1)} = \mathbf{T}_1\mathbf{T}_2 \cdots \mathbf{T}_{j-1}. \tag{101}$$

In $\mathbf{T}_k^{(h)}$, k is the index of the first matrix and h the number of successive matrices in the product. Serial generation of the sum of terms in Eq. (100) requires performance of all operations pertaining to bond i at step i. At a given step $i-1$, two classes of terms may be distinguished:

$$\mathbf{r}_{0,i-1} = \sum_{h=1}^{i-1} \mathbf{T}_1^{(h-1)}\mathbf{L}_h \tag{102}$$

and

$$\mathbf{T}_1^{(i-1)} = \mathbf{T}_1\mathbf{T}_2 \cdots \mathbf{T}_{i-1}. \tag{103}$$

These two quantities can be gathered into a single matrix

$$\mathbf{M}_{i-1} = \left[\mathbf{T}_1^{(i-1)} \mathbf{r}_{0,i-1} \right]. \tag{104}$$

A generator matrix $\mathbf{A}_i$ may then be formulated to generate $\mathbf{M}_i$:

$$\mathbf{A}_i = \begin{bmatrix} \mathbf{T}_i & \mathbf{L}_i \\ 0 & 1 \end{bmatrix}. \tag{105}$$

Then

$$\mathbf{M}_i = \mathbf{M}_{i-1}\mathbf{A}_i. \tag{106}$$

Let $\mathbf{A}_{[h}$ be defined as the first row of matrix $\mathbf{A}_h$:

$$\mathbf{A}_{[h} = [\mathbf{T}_h\mathbf{L}_h]. \tag{107}$$

The serial product $\mathbf{A}_{[1}\mathbf{A}_2^{(n-2)}$ gives $\mathbf{M}_{n-1}$. Postmultiplication by the last column of $\mathbf{A}_n$, which is denoted by

$$\mathbf{A}_{k]} = \begin{bmatrix} \mathbf{L}_k \\ 1 \end{bmatrix}, \tag{108}$$

with k set equal to n, yields

$$\mathbf{r} = \mathbf{A}_{[1}\mathbf{A}_1^{(n-2)}\mathbf{A}_{n]} \qquad (n \geq 2). \tag{109}$$

The vector **r** is obtained as a serial product of generator matrices. This result admits of immediate generalization to

$$\mathbf{r}_{hk} = \mathbf{A}_{[h+1}\mathbf{A}_{h+2}^{(k-h-2)}\mathbf{A}_{k]} \qquad (k - h \geq 2) \qquad (110)$$

for the vector spanning the sequence of k–h bonds connecting atom h with k.

b. Square of the magnitude of the chain vector.

The square of magnitude of **r** is given by

$$r^2 = \sum_{h=1}^{n} l_h^2 + 2\sum_{h<j} \mathbf{L}_h^T \mathbf{T}_h \mathbf{T}_{h+1} \cdots \mathbf{T}_{j-1}\mathbf{L}_j, \qquad (111)$$

where $\mathbf{L}_h^T$ is the transposed, or row, form of bond vector $\mathbf{L}_h$. For this calculation, the generator matrix is

$$\mathbf{G}_j = \begin{bmatrix} 1 & 2\mathbf{L}_i^T\mathbf{T}_i & l_i^2 \\ 0 & \mathbf{T}_i & 1 \\ 0 & 0 & 1 \end{bmatrix}. \qquad (112)$$

It follows that

$$r^2 = \mathbf{G}_{[1}\mathbf{G}_2^{(n-2)}\mathbf{G}_{n]} \qquad (n \geq 2), \qquad (113)$$

where $\mathbf{G}_{[1}$ is the first row of $\mathbf{G}_1$ and $\mathbf{G}_{n]}$ the final column of $\mathbf{G}_n$.

c. Radius of gyration.

The radius of gyration s for a chain of n bonds is given by

$$s^2 = (n + 1)^{-2} \sum_{0 \leq h < k < n} r_{hk}^2, \qquad (114)$$

the sum being over all pairs of chain atoms. The required generator matrix is

$$\mathbf{S}_i = \begin{bmatrix} 1 & \mathbf{G}_{[i} & l_i^2 \\ 0 & \mathbf{G}_i & \mathbf{G}_{i]} \\ 0 & 0 & 1 \end{bmatrix}. \qquad (115)$$

Thus,

$$s^2 = (n + 1)^{-2}\mathbf{S}_{[1}\mathbf{S}_2^{(n-2)}\mathbf{S}_{n]} \qquad (n \geq 2). \qquad (116)$$

The matrix generation method presented above for evaluating properties of chain molecules in specific configurations is readily elaborated to yield the corresponding statistical mechanical averages over all configurations as represented in terms of a suitable set of rotational isomeric states. The matrix $\mathbf{U}_i$ contains the statistical weight $u_{\zeta\eta,i}$ applicable to rotational isomeric states $\zeta, \eta = \alpha, \beta, \ldots, \nu$, for bonds $i-1$ and i, respectively. The configuration partition function Z for the chain molecule is given by the serial product of these matrices:

$$Z = \mathbf{U}_1^{(n)}, \qquad (117)$$

where

$$U_1 = \text{row}(1, 0, \ldots 0) \qquad (118)$$

and

$$U_n = \text{col}(1, 1, \ldots 1). \qquad (119)$$

The quantity $f = f(\{\phi\})$ is a configuration-dependent molecular property assumed to be expressed as a sum of contributions of each skeletal bond of the chain. For a specific configuration corresponding to the set of angles $\{\phi\}$, the property can be generated by generator matrices $\mathbf{F}_i$ of order s. As an example, if f represents the square of the magnitude of **r**, then $\mathbf{F}_i \equiv \mathbf{G}_i$. In the same way, a generator matrix can be formulated to generate the average $\langle f \rangle$ of f over all configurations:

$$\mathcal{F}_i = \begin{bmatrix} u_{\alpha\alpha}\mathbf{F}(\alpha) & u_{\alpha\beta}\mathbf{F}(\beta)\cdots \\ u_{\beta\alpha}\mathbf{F}(\alpha) & u_{\beta\beta}\mathbf{F}(\beta)\cdots \\ & & \ddots \\ & & & u_{\nu\nu}\mathbf{F}(\nu) \end{bmatrix}_i, \qquad (120)$$

for $1 < i < n$. Terminal matrices have the expressions

$$\mathcal{F}_{[1} = \begin{bmatrix} \mathbf{F}_{[1}\mathbf{00}\cdots\mathbf{0} \end{bmatrix} \qquad (121)$$

and

$$\mathcal{F}_{n]} = \text{col}\big(\mathbf{F}_{n]}\mathbf{F}_{n]} \cdots \mathbf{F}_{n]}\big). \qquad (122)$$

Serial multiplication of matrices $\mathcal{F}_i$ from 1 to n generates the complete set of products $\mathbf{F}_{[1}\mathbf{F}_2 \cdots \mathbf{F}_{n-1}\mathbf{F}_{n]}$, one for each and every configuration of the chain specified by the set of rotational states for all internal bonds. It also generates the product of statistical weights $u_{\zeta\eta}$ that expresses the statistical weight $\Omega(\{\phi\})$ for each configuration. Thus the serial product $\mathcal{F}_{[1}\mathcal{F}_2^{(n-2)}\mathcal{F}_{n]}$ comprises the sum of the complete set of terms $\Omega(\{\phi\})f(\{\phi\})$ for each configuration. Division by the sum Z of the statistical weights $\Omega\{\phi\}$ for every configuration yields $\langle f \rangle$; that is,

$$\langle f \rangle = \mathcal{F}_{[1}\mathcal{F}_2^{(n-2)}\mathcal{F}_{n]}Z^{-1}. \qquad (123)$$

5. Equation of State for a Single-Polymer Chain

If one end of a polymer chain is fixed at the origin of a coordinate system, the other end will fluctuate due to Brownian motion. For a given value of the end-to-end distance r, the chain assumes a large number of configurations, with the probability of each proportional to the particular value of the radial distribution function. If a distance r is imposed, not all of the conformations are still available, and the corresponding decrease of entropy leads to a retractive tension generated by the chain.

According to the first and second laws of thermodynamics,

$$dU = T\,dS - dW, \qquad (124)$$

where W is the work done by the system. At constant temperature, the Helmholtz free energy is

$$dA = dU - T\,dS. \tag{125}$$

Combining Eqs. (124) and (125) gives

$$dA = -dW. \tag{126}$$

The stress–strain work is

$$dW = -f\,dr. \tag{127}$$

The tensile force on a polymer chain at constant temperature and length is obtained from Eqs. (126) and (127):

$$f = (\partial A/\partial r)_T = (\partial U/\partial r)_T - T(\partial S/\partial r)_T. \tag{128}$$

For an ideal chain with no energy barrier hindering the rotation of the segments, the internal energy is the same for all conformations. Thus

$$(\partial U/\partial r)_T = 0. \tag{129}$$

As for an ideal gas, entropy is given by the Boltzmann relation [Eq. (72)]. The number of conformations available to a chain is proportional to $W(x, y, z)$ [Eq. (79)]. The equation of state for a single polymer chain is obtained from Eqs. (72), (79), (128), and (129):

$$f = 2kT\beta^2 r. \tag{130}$$

Force f lies along the direction of vector $\mathbf{r}$. Making use of Eq. (86) yields

$$f = 3kTr/\langle r^2 \rangle. \tag{131}$$

Equation (131) is Hooke's law for a spring with modulus $3kT/\langle r^2 \rangle$. When the displacement length r becomes sufficiently large, the Gaussian distribution function is no longer an adequate representation of $W(x, y, z)$ [Eq. (79)], and it is necessary to use the Langevin equation or its expansion. In this case,

$$f = (kT/l)\mathscr{L}^{-1}(r/nl), \tag{132}$$

where r/nl is the fractional extension of the chain, that is, the ratio of its displacement length to its contour length.

B. Rubberlike Elasticity of a Polymer Network

1. Network Structure

The permanent structure required for elastic recovery is obtained by joining N linear chains by chemical cross-linkages. In a first step, an acyclic structure, or tree, is formed. Topological studies of networks were carried out by Flory, who used graph theory. These are discussed later. A tree is illustrated in Fig. 9. According to graph theory, the chain ends and junctions represented in Fig. 9 are called vertices and the chains that join them are called edges. The

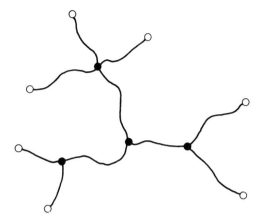

FIGURE 9 Acyclic structure. Chain ends are represented by open circles and junctions by closed circles.

total numbers of vertices and edges are denoted, respectively, by ψ and ν. They are related by

$$\psi = \nu + 1. \tag{133}$$

Flory has shown that the cycle rank ξ defined as the number of independent circuits in the network (or the number of chains that have to be cut to reduce this network to a tree) is a universal measure of network connectivity. Coalescence of ξ pairs of vertices in Fig. 9 introduces ξ-independent circuits and reduces the number of vertices to

$$\psi = \nu - \xi + 1 \simeq \nu - \xi. \tag{134}$$

In the case of a perfect network, the number of vertices equals the number of junctions μ. Additional important relationships are

$$\nu/\mu = \phi/2, \tag{135}$$

$$\xi = (1 - 2/\phi)\nu, \tag{136}$$

and

$$\xi = (\phi/2 - 1)\mu, \tag{137}$$

where ϕ is the functionality, or average functionality, of the junctions.

A real network formed by connecting N linear chains contains ν chains, μ junctions, and $2N$ chain ends. It can be shown by graph theory that

$$\nu = (\phi/2)\mu + N \tag{138}$$

and

$$\psi = \mu + 2N. \tag{139}$$

The cycle rank ξ, obtained by making use of Eq. (134), is

$$\xi = (\phi/2 - 1)\mu - N + 1$$
$$\simeq (\phi/2 - 1)\mu - N. \tag{140}$$

For the network to be formed, it is first necessary to connect chains into a tree. The number of junctions in this acyclic structure, μ^*, calculated by setting $\xi = 0$ in Eq. (140), is

$$\mu^* = N/(\phi/2 - 1), \tag{141}$$

where $\mu - \mu^*$ is the effective number of junctions μ_e. Combination of Eqs. (140) and (141) leads to

$$\xi = (\phi/2 - 1)\mu_e. \tag{142}$$

Equation (135) is still valid when ν and μ are replaced by their effective counterparts ν_e and μ_e. Specifically,

$$\nu_e/\mu_e = \phi/2. \tag{143}$$

Combination of Eqs. (142) and (143) gives a relationship between the cycle rank and the effective number of chains ν_e:

$$\xi = (1 - 2/\phi)\nu_e. \tag{144}$$

Scanlan and Case have defined an active junction as one joined by at least three paths to the gel (network) and an active chain as one terminated by an active junction at both ends. It can be proved that

$$\nu_e - \mu_e = \nu_a - \mu_a, \tag{145}$$

where ν_a and μ_a are, respectively, the active numbers of chains, and junctions. For a perfect network of functionality ϕ,

$$\nu_a/\mu_a = \phi/2. \tag{146}$$

Differences between Flory's and Scanlan and Case's criteria have been discussed in several articles.

The pore or mesh size is a fundamental quantity which characterizes the structure of an insoluble polymer network and can be taken to be the chain molecular weight M_c between two consecutive cross-links.

The following relationships hold for a perfect end-linked network:

$$\nu_e/V_0 = \nu_a/V_0 = \rho/M_c, \tag{147}$$

$$\mu_e/V_0 = \mu_a/V_0 = 2\rho/(\phi M_c), \tag{148}$$

and

$$\xi/V_0 = (1 - 2/\phi)\rho/M_c, \tag{149}$$

where ρ is the polymer density.

Randomly cross-linked networks have a complex topology, and there is no analytical method to determine rigorously the concentration of cross-link units, the junction functionality, or the number of loops and free chains in the network. Several approaches based on probabilistic formations of network structure and soluble fraction measurements have been proposed, but it is difficult to draw

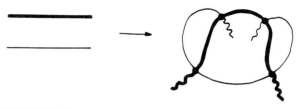

FIGURE 10 First-step network formed by connecting two chains with four tetrafunctional junctions: $N=2$, $\nu=10$, $\psi=84$, $\xi=3$, $\mu(3)=4$, $\nu_1=6$, $M_c = M_n/5$. [Reprinted with permission from J. P. Queslel (1989), *Rubber Chemistry and Technology* **62**, 800.]

reliable conclusions without assumptions on the statistics of network formation. The topology of a random tetra-functionally cross-linked network formed by connecting N primary chains of number-average molecular weight M_n and having no irregularities other than dangling chains has been studied by Queslel and Mark. In this treatment, junctions to which two or three dangling chains are connected are ignored since their number can be considered to be relatively small.

Two steps in the network formation process have been considered. In the first step, chains are connected with a minimum number of tetrafunctional junctions. Four junctions are used to connect two chains, as illustrated in Fig. 10. Six junctions are used to connect three chains by reaction a in Fig. 11. All the junctions are connected by three paths in the first-step network and are therefore active. Their number density $\mu(3)/V_0$ is easily calculated since there is one junction per chain end:

$$\mu(3)/V_0 = 2\rho/M_n. \tag{150}$$

The first-step active chain density ν_1/V_0 can also be calculated since there are $2\rho/M_n$ tetrafunctional junctions which have to be distributed among ρ/M_n chains, each one already having two tetrafunctional junctions. The total number is four tetrafunctional junctions per chain and, thus, three active segments per initial chain (since the chain ends are not active):

$$\nu_1/V_0 = 3\rho/M_n = \mu(3)/V_0 + \rho/M_n. \tag{151}$$

FIGURE 11 Reaction a: First-step network formed by connecting three chains with six junctions: $N=3$, $\nu=15$, $\psi=12$, $\xi=4$, $\mu(3)=6$, $\nu_1=9$, $M_c = M_n/5$. Reaction b: Second-step network formed by adding three new junctions: $\nu=21$, $\psi=15$, $\xi=7$, $\mu(4)=3$, $\mu_a=9$, $\nu_a=15$, $M_c = M_n/7$. [Reprinted with permission from J. P. Queslel (1989), *Rubber Chemistry and Technology* **62**, 800.]

The second-step network is obtained by addition of $\mu(4)$ tetrafunctional junctions which are now connected by four paths to the network. This addition is illustrated by reaction b in Fig. 11. When a tetrafunctional junction is added, the number of active chains is increased by two. The final number density of active chains is then given by

$$\nu_a/V_0 = \mu(3)/V_0 + \rho/M_n + 2\mu(4)/V_0. \qquad (152)$$

It is also the difference between the total number density of chains of molecular weight M_c and the number density of chain ends:

$$\nu_a/V_0 = \rho/M_c - 2\rho/M_n. \qquad (153)$$

The total number density of junctions is

$$\mu/V_0 = \mu(3)/V_0 + \mu(4)/V_0. \qquad (154)$$

The relationship among μ, M_c, and M_n is then obtained by combination of Eqs. (150), (152), (153), and (154):

$$\mu/V_0 = (\rho/2M_c)(1 - M_c/M_n), \qquad (155)$$

where μ is equal to μ_a since all the junctions are active.

The cycle rank ξ is obtained by combination of Eqs. (140) and (155):

$$\xi/V_0 = (\rho/2M_c)(1 - 3M_c/M_n), \qquad (156)$$

since N is equal to ρ/M_n.

A similar treatment has been applied to the topological description of networks formed by random tetrafunctional cross-linking of star polymers having A arms of number-average molecular weight M_n. The first-step network obtained by connecting two three-arm stars is shown in Fig. 12. The treatment leads to the following relationships:

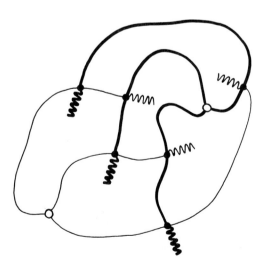

FIGURE 12 First-step network formed by connecting two three-arm stars. [Reprinted with permission from J. P. Queslel and J. E. Mark (1986), *Polymer Journal* **18**, 263.]

$$\mu/V_0 = (\rho/2M_c)[1 - (A - 2)M_c/AM_n] \qquad (157)$$

and

$$\xi/V_0 = (\rho/2M_c)[1 - (A + 2)M_c/AM_n]. \qquad (158)$$

2. Statistical Thermodynamics of Elasticity

a. Affine and phantom networks. The retractive force exhibited by an elastomeric network originates in the tendency of its constituent macromolecules to come back spontaneously to a state of higher entropy. The statistical properties of a single chain have already been described. In particular, it was shown that the chain end-to-end distance r is the quantity of primary importance. A network is an ensemble of such single-polymer chains connected through cross-links, and the statistical analysis of its elasticity consists of calculating the elastic free energy as a function of the macroscopic parameters that characterize the deformation. The following assumptions are made:

1. A chain in a network is defined as that part of a longer chain lying between two successive cross-linking points, and **r** is the vector joining these points.
2. The entropy of the network is the sum of the entropies of the individual chains.
3. The chains are long enough to have a Gaussian distribution of r.
4. The components of the vectorial length of each chain are changed by the deformation in the same ratio as the corresponding dimensions of the sample.

The elastic free energy of this affine network of Gaussian chains relative to the istropic state can be written as

$$\Delta A_{el}(\text{aff}) = (\nu_e kT/2)(I_1 - 3) - (\nu_e - \xi)kT \ln(V/V_0), \qquad (159)$$

where I_1 is the first invariant of the tensor of deformation:

$$I_1 = \lambda_1^2 + \lambda_2^2 + \lambda_3^2. \qquad (160)$$

The quantities λ_1, λ_2, and λ_3 are the principal extension ratios that specify the strain relative to an isotropic state of reference volume V_0. In this state of reference, the mean-square end-to-end distance of chains is assumed to be the corresponding mean-square end-to-end distance for the undeformed chains without cross-links (i.e., $\langle r^2 \rangle_0$). Here V is the volume of the deformed sample. When a polymer is cross-linked in a state of dilution, V_0 is the volume at which cross-linking occurs.

In simple elongation along the axis indexed 1, λ_1 is denoted λ and is the equal to the ratio of length L of deformed specimen of volume V to length L_0 in the isotropic state of volume V_0:

$$\lambda_1 = \lambda = L/L_0. \tag{161}$$

If a change of volume occurs during deformation,

$$\lambda_1\lambda_2\lambda_3 = V/V_0. \tag{162}$$

Because of symmetry of deformation along the 1 axis, $\lambda_2 = \lambda_3$. The elastic equation of state of an affine network has the neo-Hookean form

$$f = (\nu_e kT/L_0)\left[\lambda - V/(V_0\lambda^2)\right], \tag{163}$$

where f is the tensile force.

It is also possible to define an extension ratio α relative to the isotropic state of volume V and of initial length L_v^i by Eq. (60). Here L_v^i is related to L_0 by

$$L_v^i = L_0(V/V_0)^{1/3}. \tag{164}$$

Equation (163) now takes the form

$$f = \left(\nu_e kT/L_v^i\right)(V/V_0)^{2/3}(\alpha - \alpha^{-2}). \tag{165}$$

Equation (61) is recovered since

$$(V/V_0)^{2/3} = \langle r^2\rangle_v^i / \langle r^2\rangle_0. \tag{166}$$

The force per unit cross-sectional area A_0 of the undeformed sample of reference volume V_0 is deduced from Eqs. (164) and (165) to be

$$f/A_0 = (\nu_e kT/V_0)(V_0/V)^{-1/3}(\alpha - \alpha^{-2}). \tag{167}$$

In the deformation of swollen networks, A_0 is the cross-sectional area of the dry sample and V_0/V the volume fraction v_2 of polymer in the swollen system. The extension ratio α is then measured with respect to the isotropic swollen state.

The assumption of affine deformation is rather restrictive. In a different approach, James and Guth studied a statistical model of a network in which chains can pass through one another. Junction fluctuations due to Brownian motion (which are suppressed in the affine model) are large in such a "phantom" network. Then, only the mean positions of the junctions are affine in the strain. The elastic free energy of deformation is now dependent on the cycle rank ξ:

$$\Delta A_{el}(ph) = (\xi kT/2)(I_1 - 3). \tag{168}$$

In uniaxial extension, the elastic equation of state has the form

$$f/A_0 = (\xi kT/V_0)(V_0/V)^{-1/3}(\alpha - \alpha^{-2}). \tag{169}$$

Again Eq. (62) is easily recovered.

b. Experiments. Experimental data obtained in uniaxial extension or compression are usually represented as

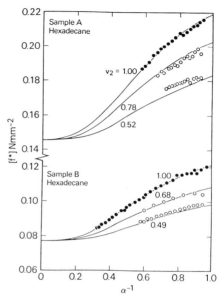

FIGURE 13 Reduced stresses calculated from measurements in uniaxial extension at stress equilibrium. Samples A and B are polydimethylsiloxane cured with 0.2% and 0.1% (w/w) dicumyl peroxide, respectively (25 min at 120°C). Measurements concerning dry and swollen networks are represented by closed and open circles, respectively. The solvent was *n*-hexadecane. The parameters chosen to optimize agreement between theoretical curves and experimental points are for unswollen sample A, $\xi kT/V_0 = 0.146$ N/mm², $\kappa = 5.5$, and $\zeta = 0.05$; and for sample B, 0.077 N/mm², 7.0, and 0.05, respectively. [Reprinted with permission from Erman, B., and Flory, P. J. (1983). *Macromolecules* **16**, 1607.]

the reduced nominal stress $[f^*]$ versus the inverse of deformation α^{-1}. This reduced stress is defined by

$$[f^*] \equiv (f/A_0)v_2^{1/3}/(\alpha - \alpha^{-2}). \tag{170}$$

According to Eqs. (167) and (169), $[f^*]$ is expected to be independent of deformation and swelling for both affine and phantom networks. The elastic behavior of rubbers generally departs from the kinetic theory presented earlier. As an example, measurements in uniaxial extension on dry and swollen dimethylsiloxane networks are reported in Fig. 13. The experimental points are represented by open and closed circles. The reduced stress decreases when deformation or swelling increases. The curves shown in Fig. 13 were calculated from an elasticity theory of real networks that will be discussed later.

c. Non-Gaussian theory. Equation (167) and (169) rely on the Gaussian approximation for the distribution $W(r)$ [Eq. (81)], and this approximation is generally adequate for chains of sufficient length. It is no longer appropriate, however, at very large deformations where chains reach their maximum extensibility. An upturn of the reduced stress occurs in this range of deformation and can be attributed either to this limited chain extensibility or to strain-induced crystallization. Non-Gaussian theories were developed to account for possible departures from the Gaussian distribution function, and they do predict an upturn at high strain. The theory of James and Guth leads to the following force–extension relation for the non-Gaussian affine network in simple extension:

$$f/A_0 = (\tfrac{1}{3})\nu_e kTn^{1/2}[\mathscr{L}^{-1}(\lambda n^{-1/2})$$
$$- \lambda^{-3/2}\mathscr{L}^{-1}(\lambda^{-1/2}n^{-1/2})], \quad (171)$$

where n is the number of randomly jointed links in a chain. The quantity $\mathscr{L}^{-1}$ may be approximated by the series

$$\mathscr{L}^{-1}(t) = 3t + (\tfrac{9}{5})t^3 + (\tfrac{297}{175})t^5 + \cdots. \quad (172)$$

However, the use of the simpler Gaussian theory to study the elastic behavior of real networks has the decisive advantage of providing a good basic molecular picture for rubberlike elasticity.

d. Phenomenological theory.

Continuum mechanics is also used to account for the behavior exhibited by elastomers. The most general form of the strain energy function (which vanishes at zero strain) is the power series

$$\frac{W}{V_0} = \sum_{i,j,k=0}^{\infty} C_{ijk}(I_1 - 3)^i(I_2 - 3)^j(I_3 - 1)^k, \quad (173)$$

where the C_{ijk} coefficients have units of newtons per squre meter. The quantities I_1, I_2, and I_3 are the three strain invariants, I_1 being defined by Eq. (160), and I_2 and I_3 by

$$I_2 = \lambda_1^2\lambda_2^2 + \lambda_2^2\lambda_3^2 + \lambda_3^2\lambda_1^2 \quad (174)$$

and

$$I_3 = \lambda_1^2\lambda_2^2\lambda_3^2. \quad (175)$$

It was found that a simple form of Eq. (173) can reproduce observed dependences. Specifically,

$$W/V_0 = C_1(I_1 - 3) + C_2(I_2I_3^{-1+m/2} - 3) - C_3 \ln I_3, \quad (176)$$

where m is a parameter that can be adjusted to obtain the best fit of the data and C_3/C_2 equals $\tfrac{1}{2}$. For uniaxial extension, the reduced nominal stress is expressed by

$$[f^*] \equiv 2C_1 + 2C_2 v_2^{4/3-m}\alpha^{-1}. \quad (177)$$

The Mooney-Rivlin equation is obtained for $v_2 = 1$ and has been extensively used to fit experimental data in uniaxial extension. Values of m were found to lie between 0 and $\tfrac{1}{2}$. Generally, only a portion of typical experimental curves can be approximated by Eq. (177), which would

be represented by straight lines in Fig. 13. Equation (177) cannot account at all for data obtained in compression.

e. Flory-Erman theory of real networks.

The most nearly complete understanding of the molecular mechanisms governing rubberlike elasticity was achieved by Flory and Erman. Their theory of real networks was successfully tested with dry and swollen networks in different states of strain. In a phantom network, junctions fluctuate around their mean positions due to Brownian motion. The instantaneous distribution of chain vectors **r** is not affine in the strain because it is the convolution of the distribution of the mean vectors $\langle r \rangle$ (which are affine) with the distribution of the fluctuations Δr (which are independent of the strain). In a real network, diffusion of the junctions about their mean positions may be severely restricted by neighboring chains sharing the same region of space. The extreme case is the affine network where the fluctuations are completely suppressed. In this case, the instantaneous distribution of chain vectors is affine in the strain. In the Flory-Erman theory, the fluctuations are dependent on strain and the restrictions on the fluctuations of junctions due to neighboring chains are represented by domains of constraints. These domains, which are due to entanglements with the surrounding chains, can be initially represented as spheres. They are transformed to ellipsoids by the deformation. In uniaxial extension, the main axes of these ellipsoids are along the direction of stretching. Thus fluctuations increase in this direction, along which the stress is measured. The behavior of real networks will tend to that of phantom networks in the limits of infinite deformation. The reduced forces exhibited by an affine network, $[f^*]_{\text{aff}}$, and a phantom network, $[f^*]_{\text{ph}}$, are derived from Eqs. (144), (167), (169), and (170):

$$[f^*]_{\text{aff}} = \nu_e kT/V_0 \quad (178)$$

and

$$[f^*]_{\text{ph}} = \xi kT/V_0$$
$$= (1 - 2/\phi)\nu_e kT/V_0. \quad (179)$$

Since the phantom reduced stress is lower than the affine one, a transition from a real network (where junctions are severely restricted at small strains) to a phantom network at large deformations will result in a decrease of the reduced stress with deformation.

Typical experimental results in uniaxial extension and compression are shown schematically in Fig. 14 and compared there with predictions from affine, phantom, and constrained junction models. Affine and phantom models are inadequate for reproducing these data. A fairly good fit is obtained with the constrained junction model, both in uniaxial extension and compression. The upturn in $[f^*]$ at very high elongations is a non-Gaussian effect and is not

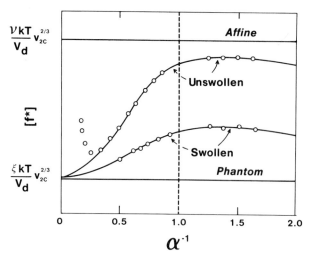

FIGURE 14 Schematic diagram showing the reduced stress $[f^*]$ as a function of reciprocal elongation α^{-1} for both uniaxial extension ($\alpha > 1$) and compression ($\alpha < 1$). The upper and lower horizontal lines represent results from affine and phantom network models, respectively. Circles show representative data from experiments (experimental data obtained in uniaxial extension are similar to those given in Fig. 13). The curves are predictions from the constrained junction theory. [Reprinted with permission from B. Erman and J. E. Mark (1989). *Annu. Rev. Phys. Chem.* **40**, 351.]

accounted for in the constrained junction theory, which is based on the Gaussian approximation. However, network behavior generally follows the Gaussian approximation over a sufficiently large range of elongations that it is possible to apply the constrained junction theory for network characterization under most conditions.

Two parameters are used in this theory: κ is the primary one and measures the severity of the entanglement constraints relative to those imposed by the phantom network connectivity. κ is expected to be proportional to the degree of interpenetration of chains and junctions:

$$\kappa = I\langle r^2\rangle_0^{3/2}(\mu/V_0),\qquad(180)$$

where I is a constant of proportionality generally found to be equal to 0.5, and μ is the number of junctions in the volume V_0 of the state of reference. The ratio μ/V_0 is a measure of the cross-link density and $\langle r^2\rangle_0$ depends on the molecular weight M_c between junctions and the chain flexibility. It can be deduced that κ is proportional to $[f^*]_{ph}^{-1/2}$. Another parameter, ξ, takes into account the nonaffine transformation of the domains of constraints with strain. It is expected to be very small.

The elastic free energy change is written as the sum of the elastic free energy $\Delta A_{el}(ph)$ for a phantom network and a term ΔA_c that accounts for entanglement constraints; $\Delta A_{el}(ph)$ is given in Eq. (168), and

$$\Delta A_{el} = \Delta A_{el}(ph) + \Delta A_c.\qquad(181)$$

The second term can be written as

$$\Delta A_c = \left(\frac{\mu_e kT}{2}\right)\sum_{t=1,2,3}\{(1+g_t)B_t$$
$$-\ln[(B_t+1)(g_t B_t+1)]\},\qquad(182)$$

with

$$B_t = (\lambda_t - 1)\left(1 + \lambda_t - \zeta\lambda_t^2\right)(1+g_t)^{-2}\qquad(183)$$

and

$$g_t = \lambda_t^2\left[\kappa^{-1} + \zeta(\lambda_t - 1)\right].\qquad(184)$$

In the case of uniaxial extension of swollen networks, the reduced nominal stress is given by

$$[f^*] \equiv (\xi kT/V_0)\{1 + (\mu_e/\xi)[\alpha K(\alpha^{-2}v_2^{-2/3})$$
$$-\alpha^{-2}K(\alpha^{-1}v_2^{-2/3})](\alpha - \alpha^{-2})^{-1}\}.\qquad(185)$$

The function K is defined by

$$K(\lambda_1^2) = B_t\left[\dot{B}_t(B_t+1)^{-1}\right.$$
$$\left.+ g_t(\dot{B}_t g_t + \dot{g}_t B_t)(g_t B_t + 1)^{-1}\right],\qquad(186)$$

with

$$\dot{B}_t = B_t\left\{[2\lambda_t(\lambda_t-1)]^{-1} + (1-2\xi\lambda_t)\right.$$
$$\left.\times\left[2\lambda_t\left(1+\lambda_t-\xi\lambda_t^2\right)\right]^{-1} - 2\dot{g}_t(1+g_t)^{-1}\right\}\qquad(187)$$

and

$$\dot{g}_t = \kappa^{-1} - \zeta(1 - 3\lambda_t/2).\qquad(188)$$

The ratio μ_e/ξ is obtained from Eq. (142). The theoretical expression for the reduced force [Eq. (185)] was used to interpret the stress–strain data of Fig. 13. The cycle rank and thus the cross-link density can be deduced from the phantom reduced force, which is obtained by extrapolation of the theoretical curve to infinite deformation ($\alpha^{-1} = 0$). Sample A, cured with 0.2% (w/w) dicumyl peroxide, has a higher phantom reduced force $\xi kT/V_0$ than that of sample B, cured with only 0.1%. Values of $\kappa[f^*]_{ph}^{1/2}$ are 2.10 and 1.94 for samples A and B, respectively. As predicted, this product remains approximately constant and is in agreement with previous results. The cycle rank determined by this procedure permits determination of the cross-link density μ/V_0 of randomly cross-linked networks through Eqs. (155) and (156).

Another illustrative analysis of randomly cross-linked network structures using the constrained junction model concerns *cis*-polyisoprene cured with dicumyl peroxide (dicup). The precursor polymer was an anionically prepared commercial polyisoprene (Shell IR 307) with a high *cis*-1,4 stereochemical structure. Three batches of varying number-average molecular weight, namely, $M_n = 3.60 \times 10^5$, 2.45×10^5, and 1.25×10^5 g/mol, were

TABLE IV Characterization of the *cis*-Polyisoprene Networks by Analysis of Stress–Strain Isotherms in Terms of the Constrained Junction Model[a]

Sample	$10^{-5} M_n$ (g/mol)	Dicup[b]	$[f^*]ph^c$ (MPa)	κ	ζ	$10^{-3} M_c$ (g/mol)	$10^5 \mu/V_0$ (mol/cm^3)
1 IR 360	3.60	1.0	0.2350	4.6	0	4.69	9.58
2 IR 360	3.60	0.8	0.2113	4.9	0.011	5.19	8.63
1 IR 245	2.45	0.8	0.1967	4.9	0.017	5.44	8.17
2 IR 245	2.45	0.6	0.1592	5.4	0.011	6.62	6.68
3 IR 245	2.45	0.35	0.0930	6.8	0	10.72	4.06
1 IR 125	1.25	1.2	0.2452	4.3	0	4.21	10.46
2 IR 125	1.25	0.6	0.1445	5.3	0.035	6.67	6.46
3 IR 125	1.25	0.45	0.0989	6.0	0.010	9.08	4.65
4 IR 125	1.25	0.3	0.0747	6.6	0.016	11.22	3.69

[a] Reprinted with permission from J. P. Queslel (1989), *Rubber Chemistry and Technology* **62**, 800.

[b] Weight percent in bulk (phr).

[c] At $T = 303\ K$.

prepared by degradative working on a two-roll mill. Each precursor was mixed in bulk with several amounts of pure dicup, molded, and cured (30 min at 170°C). The characteristics of the samples are given in Table IV. The equilibrium reduced stresses were measured in uniaxial extension and are shown for four of the samples in Fig. 15. They were fitted by least-squares analysis with theoretical curves calculated through Eqs. (155), (156) (tetrafunctional randomly cross-linked networks), (180), and (185). For *cis*-1,4-polyisoprene, the relationship between $\langle r^2 \rangle_0$ and M_c is

$$\langle r^2 \rangle_0 = 3.8 \times 10^7 M_c/N_A. \tag{189}$$

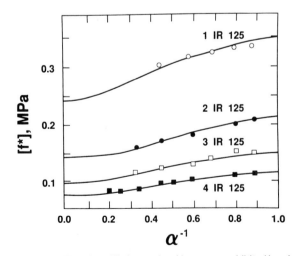

FIGURE 15 Plot of equilibrium reduced stresses exhibited by *cis*-polyisoprene networks versus reciprocal elongation. Continuous curves are predictions of the constrained junction model calculated with parameters listed in Table IV. [Reprinted with permission from J. P. Queslel (1989), *Rubber Chemistry and Technology* **62**, 800.]

The quantities $\langle r^2 \rangle_0$ and M_c are expressed in square centimeters and grams per mole, respectively. N_a is Avogadro's number, and ξ and μ are related to M_c and M_n through Eqs. (155) and (156). κ is related to M_c and M_n through Eqs. (180) and (189). Therefore, the reduced stress given by Eq. (185) depends on only two adjustable parameters, M_c and ζ. These parameters, determined by fitting of the stress–strain isotherms, are given in Table IV. $[f^*]_{ph}$ was then calculated through Eq. (179). The good agreement between experimental data and theoretical predictions confirms the proportionality of parameter κ to the degree of interpenetration of chains and junctions embodied in Eq. (180).

Erman and Monnerie have modified the constrained junction model to include constraints along the chain contours. Analyses of the stress–strain data with the constrained junction and constrained chain models gave similar results. This validates the Flory hypothesis that it is generally sufficient to concentrate the effects of the constraints on the junctions. Nevertheless, the constrained chain model has the advantage of depending on only one parameter instead of two, κ and ζ, for the constrained junction model and gives a more realistic molecular picture.

f. Other theories and models. In the Flory-Erman theory, entanglements are represented as domains of constraints that act on junction fluctuations. It is a mean-field theory since entanglements are seen as averages. Other mean-field theories use the tube concept (in which each network chain is confined within a tube), which was introduced by De Gennes and Doi and Edwards in the viscoelasticity area. In slip-link models, each network chain threads its way through a number of small rings. These discrete theories have the disadvantage of introducing another parameter, the number density of entanglements. Combination of nonlocalization of trapped entanglements and restrictions on the fluctuations of permanent junctions has also been proposed for departures from neo-Hookean behavior in real networks.

Predictions of several entanglement models of rubber elasticity have been compared with experiments performed in uniaxial extension and compression. Both the Flory-Erman theory and the Gaylord tube model fit the stress–strain data over the entire strain range as well as predict the correct position of the maximum in the reduced stress. It is noteworthy that all these theories predict that the entanglement contribution tends to zero in the range of infinite deformation in uniaxial extension. The reason is that the freedom of junctions to fluctuate or of slip-links to slide along the chains increases with deformation in the direction of macroscopic stretching. Another interesting approach of rubber elasticity consists in simulating network chain configurations by a rotational isomeric state technique and deducing a non-Gaussian theory from the distribution of end-to-end distances thus determined. A phenomenological concept of trapped entanglements was introduced by Langley and applied in the range of small deformations in uniaxial extension. During the network formation, a fraction of the entanglements present in the melt are said to be trapped. The contribution of these entanglements to the equilibrium elastic modulus is assumed to be a fraction of the nonequilibrium plateau modulus as determined in viscoelasticity measurements. This fraction is the trapping probability calculated through probability theories that simulate network formation, such as the Macosko-Miller branching theory.

g. Model networks.
Elasticity theories are conveniently tested by using "model" networks. The structural parameters that govern the elasticity behavior of these networks are known from the chemical cross-linking process. Specifically, end-reactive chains of known molecular weight are generally connected by junctions of known functionality. A great variety of network structures can be thus prepared and tested.

h. Swelling of elastomeric networks.
Swelling by an organic solvent is a simple method to study elastomeric structures. Swelling is the result of two thermodynamic phenomena: an increase of entropy of the network–solvent system by introduction of small molecules as diluent and a decrease of entropy of the polymer chains by the isotropic deformation. The Flory-Huggins liquid lattice theory accounts for the first effect, that is, the mixing of polymer and solvent. Use of their relationship and a statistical mechanical expression for the Gibbs-free-energy change with di-

lation led Flory and Rehner to propose a relation between swelling and the degree of cross-linking for an affine network, which has been widely used to characterize a variety of networks. Refined swelling–structure relationships for real networks are obtained by making use of the Flory-Erman theory.

The change ΔA of the Gibbs free energy due to mixing a solvent 1 with a polymer 2 is given by

$$\Delta A = A_M - A_1 - A_2, \qquad (190)$$

where A_M, A_1, and A_2 are the free energies of the system polymer–solvent, pure solvent, and pure polymer, respectively. Chemical potentials of solvent in the un-cross-linked polymer μ_1^u and in the pure state μ_1^0 are obtained by differentiation of ΔA with respect to the number of moles of solvent n_1 at constant n_2 of polymer:

$$(\partial \Delta A / \partial n_1)_{n_2, T, p} = \mu_1^u - \mu_1^0. \qquad (191)$$

The difference $\mu_1^u - \mu_1^0$ is related to the vapor pressures of solvent p_1^u over the solution and p_1^0 of pure solvent by

$$\mu_1^u - \mu_1^0 = RT \ln(p_1^u / p_1^0). \qquad (192)$$

Flory and Huggins used a liquid lattice theory to express $\mu_1^u - \mu_1^0$ as a function of the volume fraction v_2 of polymer in the mixture, the polymer–solvent interaction parameter χ, and the ratio x of molar volume of polymer V_2 and solvent V_1. Specifically,

$$\mu_1^u - \mu_1^0 = RT \left[\ln(1 - v_2) + \chi v_2^2 + v_2 (1 - 1/x) \right], \quad (193)$$

with

$$x = V_2 / V_1 = (M_2 / M_1)(\rho_1 / \rho_2), \qquad (194)$$

where M_1 and M_2 are, respectively, the number-average molecular weights of solvent and polymer and ρ_1 and ρ_2 their densities. The term $1/x$ is small and equal to zero for a network (which is a molecule of infinite molecular weight).

The difference between the chemical potential of solvent in cross-linked polymer μ_1^c and in the pure state μ_1^0 is related to the vapor pressure p_1^c of solvent in the swollen network and over pure solvent p_1^0:

$$\mu_1^c - \mu_1^0 = RT \ln(p_1^c / p_1^0) \qquad (195)$$

Combing Eqs. (192) and (195) at constant volume fraction of solvent leads to

$$\mu_1^c - \mu_1^u = RT \ln(p_1^c / p_1^u). \qquad (196)$$

The difference $\mu_1^c - \mu_1^0$ arises from two contributions: a mixing component $(\mu_1^u - \mu_1^0)_{mix}$ approximately equal to $\mu_1^u - \mu_1^0$ for large x and an elastic component $(\mu_1^c - \mu_1^0)_{el}$. It is generally assumed that these two contributions are separable. Introducing μ_1^0 on the left side of Eq. (196) leads to

$$\left(\mu_1^c - \mu_1^0\right)_{el} = RT \ln\left(p_1^c/p_1^u\right). \tag{197}$$

The quantity $(\mu_1^c - \mu_1^0)_{el}$ results from the elastic deformation (dilation) of the network and is related to the elastic free energy change ΔA_{el} by

$$\left(\mu_1^c - \mu_1^0\right)_{el} = (\partial \Delta A_{el}/\partial n_1)_{T,p}. \tag{198}$$

Combining Eqs. (197) and (198) with the expressions for the activities of solvent over the un-cross-linked polymer $(a_1^u = p_1^u/p_1^0)$ and over the network $(a_1^c = p_1^c/p_1^0)$ at constant volume fraction v_2 of polymer gives

$$RT \ln\left(a_1^c/a_1^u\right) = (\partial \Delta A_{el}/\partial n_1)_{T,p}. \tag{199}$$

If solvent is present in excess, absorption equilibrium is reached when the chemical potential of solvent in the swollen network is equal to that of pure solvent outside the network, that is, $\mu_1^c = \mu_1^0$ Hence,

$$\mu_1^c - \mu_1^0 = \left(\mu_1^c - \mu_1^0\right)_{mix}$$
$$+ \left(\mu_1^c - \mu_1^0\right)_{el} = 0 \tag{200}$$

when $(\mu_1^c - \mu_1^0)_{mix}$ is given by Eq. (193) with $1/x = 0$. The volume fraction of polymer at equilibrium (maximum) swelling is designated v_{2m}. Then,

$$RT\left[\ln(1 - v_{2m}) + \chi v_{2m}^2 + v_{2m}\right]$$
$$+ (\partial \Delta A_{el}/\partial n_1)_{T,p} = 0. \tag{201}$$

The next step is the introduction of an expression for ΔA_{el} in Eq. (201). Continuum mechanics leads to a stored elastic energy given by Eq. (176) with

$$I_3 = (V/V_0)^2$$
$$= v_2. \tag{202}$$

In isotropic swelling, the resulting extension rations are

$$\lambda = \lambda_1 = \lambda_2 = \lambda_3 = [(n_1 V_1 + V_0)/V_0]^{1/3} = v_2^{-1/3}. \tag{203}$$

The stored elastic energy can then be expressed as

$$W = 2C_1 V_0[\lambda^2 - 1 - (\tfrac{1}{2}) \ln \lambda^2]$$
$$+ 3C_2 V_0[(\lambda^2)^{3m/2-1} - 1]. \tag{204}$$

It is necessary to develop the derivative $(\partial \Delta A_{el}/\partial n_1)_{T,p}$ of Eq. (201). Thus,

$$\left(\frac{\partial \Delta A_{el}}{\partial n_1}\right)_{T,p} = \left(\frac{\partial \Delta A_{el}}{\partial \lambda^2}\right)_{T,p} \left(\frac{\partial \lambda^2}{\partial n_1}\right)_{T,p}, \tag{205}$$

with

$$\left(\partial \lambda^2/\partial n_1\right)_{T,p} = 2V_1/3\lambda V_0. \tag{206}$$

Equations (199) and (204) through (206) lead to the isotropic swelling equation

$$\lambda \ln\left(\frac{a_1^c}{a_1^u}\right) = 2\left(\frac{C_1 V_1}{RT}\right)\left[1 - \frac{2}{\lambda^2} - \left(\frac{C_2}{2C_1}\right)(2 - 3m)\lambda^{3m-4}\right]. \tag{207}$$

Use of the Mooney-Rivlin free energy expression [i.e., Eq. (204) with $m = 0$ and without the logarithmic term] leads to the expression

$$\lambda \ln\left(a_1^c/a_1^u\right) = 2(C_1 V_1/RT)\left[1 - C_2/\left(C_1\lambda^4\right)\right]. \tag{208}$$

The Flory-Erman statistical mechanism analysis leads to an elastic free energy change, given by Eq. (181). The derivative of ΔA_{el} as a function of λ^2 is the sum of the corresponding derivatives of ΔA_{el} and ΔA_c:

$$\left(\partial \Delta A_{el}(ph)/\partial \lambda^2\right)_{T,p} = (\tfrac{3}{2})\xi kT \tag{209}$$

and

$$\left(\partial \Delta A_c/\partial \lambda^2\right)_{T,p} = (\tfrac{3}{2})\mu_e kT K(\lambda^2). \tag{210}$$

The final result is

$$\lambda \ln\left(a_1^c/a_1^u\right) = (\xi/V_0)V_1\left[1 + (\mu_e/\xi)K(\lambda^2)\right]. \tag{211}$$

At swelling equilibrium, Eq. (201) gives the relationship

$$\ln(1 - v_{2m}) + \chi v_{2m}^2 + v_{2m}$$
$$= -(\xi/V_0)V_1 v_{2m}^{1/3}\left[1 + (\mu_e/\xi)K\left(v_{2m}^{-2/3}\right)\right], \tag{212}$$

where ξ is expressed in moles. Swelling equations for phantom networks are recovered through $\kappa = 0$ and $K(\lambda^2) = 0$:

$$\lambda \ln\left(a_1^c/a_1^u\right) = (\xi/V_0)V_1 \tag{213}$$

and

$$\ln(1 - v_{2m}) + \chi v_{2m}^2 + v_{2m} = -(\xi/V_0)V_1 v_{2m}^{1/3}. \tag{214}$$

For affine networks, $\kappa = \infty$ and $K(\lambda^2) = 1 - \lambda^{-2}$, so

$$\lambda \ln\left(a_1^c/a_1^u\right) = (\xi/V_0)V_1[1 + (\mu_e/\xi)(1 - \lambda^{-2})] \tag{215}$$

$$\ln(1 - v_{2m}) + \chi v_{2m}^2 + v_{2m}$$
$$= -(\xi/V_0)V_1 v_{2m}^{1/3}\left[1 + (\mu_e/\xi)\left(1 - v_{2m}^{2/3}\right)\right]. \tag{216}$$

The swelling equilibrium experiment consists of determining v_{2m} after completely immersing already-extracted samples in solvent. The interaction parameter χ and the activities of solvent in un-cross-linked and cross-linked polymers can be measured by differential solvent vapor sorption. This technique consists of determination of the amount of solvent absorbed by un-cross-linked and cross-linked polymer samples both placed in a given solvent vapor atmosphere. It was shown that a plot of $\lambda \ln(a_1^c/a_1^u)$ versus λ^2 (where $\lambda = v_2^{-1/3}$) exhibits a maximum that is not

TABLE V Characterization of the *cis*-Polyisoprene Networks by Analysis of Equilibrium Swelling Data in Terms of the Constrained Junction Models[a]

Sample	v_{2m}[b] benezene	κ	$10^{-3} M_c$ (g/mol)	$10^5 \mu_s / V_0$ (mol/cm^3)	v_{2m}[c] cyclohexane	χ_c[d]
1 IR 360	0.183	4.7	4.45	10.12	0.165	0.313
2 IR 360	0.171	5.0	5.07	8.84	0.154	0.315
3 IR 245	0.118	6.9	10.21	4.28	0.107	0.328
1 IR 125	0.183	4.4	4.13	10.64	0.159	0.295
2 IR 125	0.138	5.5	6.70	6.43	0.120	0.307
3 IR 125	0.117	6.2	9.02	4.68	0.105	0.324
4 IR 125	0.097	7.0	11.76	3.51	0.081	0.300

[a] Reprinted with permission from J. P. Queslel (1989), *Rubber Chemistry and Technology* **62**, 800.

[b] Volume fraction of polymer at swelling equilibrium in benzene in 20°C.

[c] Volume fraction of polymer at swelling equilibrium in cyclohexane at 20°C.

[d] Calculated interaction parameter for the system polyisoprene + cyclohexane at 20°C.

predicted by continuum mechanics [Eq. (207)] or phantom and affine models [Eqs. (213) and (215), respectively]. However, qualitative agreement is obtained if the Flory-Erman theory is used [Eq. (211)].

Swelling equilibrium is a simple method to estimate the degree of cross-linking of a network by determination of v_{2m}. If the interaction parameter χ (generally concentration dependent) has already been determined by osmometry, gas–liquid chromatography, freezing point depression of solvent, intrinsic viscosity, or critical solution temperature, then the cycle rank ξ may be calculated through Eqs. (142) and (212). However, Eq. (212) depends also on the two parameters k and ζ, which are not known *a priori*.

Values of v_{2m} in benzene or cyclohexane for the *cis*-polyisoprene networks already characterized by stress–strain isotherms are given in Table V. Network characterization through equilibrium swelling measurements is very sensitive to the elasticity model and interaction parameter used in the calculation. As an example, the swelling result for the 2 IR 360 network in benzene has been interpreted through constrained junction, affine, and phantom models. The total number density of junctions μ_s was calculated from the experimental value of v_{2m} as a function of interaction parameter χ_b. For the constrained junction model, Eqs. (142), (155), (156), (180), (189), and (212) were combined, and the resulting equation was solved with respect to M_c or μ_s. The parameter ζ has the value determined from mechanical measurements and given in Table IV. Equations (214) and (216) were used for phantom and affine calculations, respectively. The plot of μ_s / V_0 versus χ_b is shown in Fig. 16 and illustrates the great sensitivity to choice of elasticity theory and interaction parameter.

Use of an inadequate elasticity model may serve at most for the ranking of series of networks of the same polymer,

on the basis of their cross-link densities. In Fig. 16, the number of junctions calculated using a phantom model is higher than μ_s for an affine one. In a phantom model, junction fluctuations decrease the impact of chain entropy changes. It is therefore necessary to have a phantom network with a higher density of cross-links to counteract this effect and to give the same elastic contribution as in an affine network.

Fortunately, Eichinger and Flory have determined the interaction parameter χ_b for the natural rubber and benzene system by high-pressure osmometry and differential solvent vapor sorption techniques:

$$\chi_b = 0.39 + 0.04 \, v_{2m}. \tag{217}$$

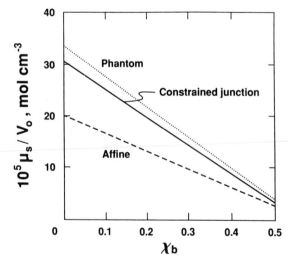

FIGURE 16 Influence of the interaction parameter χ_b and the elasticity model chosen on the determination, by equilibrium swelling, of the number density of junctions μ_s / V_0 of the 2 IR 360 network.

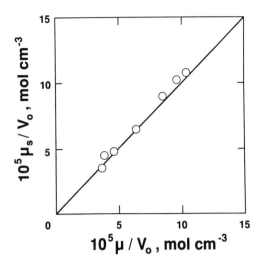

FIGURE 17 Comparison between the number densities of junctions of the polyisoprene networks determined from stress–strain isotherms, μ/V_0, and from equilibrium swelling, μ_s/V_0.

It was thus possible to characterize the seven networks listed in Table V with the suitable values of χ_b. The calculated values of κ, M_c, and μ_s are given in Table V.

Comparison between mechanical and swelling characterizations with the constrained junction model is shown in Fig. 17, where μ_s/V_0 (given in Table V) is plotted versus μ/V_0 previously given in Table IV. The line represents perfect agreement. Fairly good agreement is obtained between these two characterizations. It should be emphasized that this agreement is possible only if an adequate elasticity model and accurate values of the interaction parameter are used.

From the network structures determined previously and from the swelling measurements in cyclohexane reported in Table V, it was possible to calculate the interaction parameter χ_c for the *cis*-polyisoprene and cyclohexane system. The values of χ_c were calculated by solving the combination of Eqs. (142), (156), and (212) while ascribing to κ and M_c the values previously determined by swelling with benzene and reported in Table V. An average value of 0.31 was obtained.

3. Orientation Techniques

When a network is deformed, the distribution of chain end-to-end vectors **r** is no longer isotropic, since chains tend to be aligned in the direction of stretching. Techniques that provide information on the molecular orientation of stretched polymers can be used to test kinetic theories of rubberlike elasticity.

Complete information on the orientation of investigated segments in a deformed network is obtained from knowledge of the orientational distribution function. Mean-square projection of these segments along a given direction represents the second moment of the distribution. Data are generally presented in terms of the orientation function S, defined as the second Legendre polynomial

$$S = (3\langle\cos^2\theta\rangle - 1)/2, \qquad (218)$$

where $\langle\cos^2\theta\rangle$ is the second moment showing mean-square projection of the segment vectors on the laboratory fixed axis, θ being the angle between the segment vector and the axis. For Gaussian networks of freely jointed chains deforming affinely, the expression for S is

$$S = (1/5n)(\lambda^2 - \lambda^{-1}), \qquad (219)$$

where n is the number of segments in a chain. The quantity $1/5n$ is a configurational factor that represents the statistical properties of the chains. The second term, $\lambda^2 - \lambda^{-1}$, is a strain function that relates the orientation to the macroscopic deformation.

In recent theoretical developments, the hypothetical affine deformation condition is replaced by the concept of nonaffine molecular state of deformation first applied to the study of birefringence of real networks. The configurational term is corrected by replacing the freely jointed chain with its real analog, where averaging is performed according to the rotational isomeric state scheme. Calculation of the configuration term is extended to include the effect of orientational correlations of segments in a local steric intermolecular nature with their environments. This field is still actively investigated.

The Flory-Erman theory of real network concerning the relation of stress to strain has been extended by Erman and Flory to birefringence and by Erman and Monnerie to orientation measured by spectroscopic techniques such as infrared dichroism, deuterated nuclear magnetic resonance (NMR), and fluorescence polarization.

The main orientation techniques are SANS (which gives overall dimensions of deuterated chains in a mixture of deuterated and hydrogenated polymers), infrared dichroism (which yields the chain segment orientation averaged over all the chains of the sample), fluorescence polarization (which deals with labeled central or end sequences of labeled chains mixed with normal chains), deuterated NMR, and birefringence.

4. Rupture

Rupture of elastomeric networks is very sensitive to temperature and test conditions. It is possible to characterize the ultimate tensile properties by a "failure envelope" that results from a plot of $\log(\sigma_b T_0/T)$ versus $\log(\lambda_b - 1)$, where T and T_0 are, respectively, the test temperature and a reference temperature, expressed in kelvin; σ_b is the nominal tensile stress at break; and λ_b and is the ultimate extension ratio. The quantities σ_b and λ_b are obtained under conditions of constant strain, constant stress, or constant strain rate. The ultimate properties of rubbers are mainly governed by their viscoelastic characteristics. A change in strain rate or temperature only shifts a point along the failure envelope, which is thus dependent only on the structural characteristics of the elastomer. It has been demonstrated experimentally that the same equivalence of time and temperature applies to rupture properties as one finds for viscoelastic properties, and reduced master curves can be obtained for tensile stress and strain as a function of time to break. The failure process is a nonequilibrium one, which develops with time and involves the consecutive rupture of the molecular chains. Since it is strongly dependent on viscoelastic properties, it can be predicted from creep experiments.

5. Filler Reinforcement

In most of their applications, rubbers are compounded with fillers, mainly carbon black. Abrasian resistance, tear strength, and tensile strength are simultaneously improved. Hystersis, heat build-up, and compression set (permanent deformation) are known to increase as the reinforcing ability of the filler becomes more pronounced. One of the arts of the rubber compounder is the achievement of a proper balance between these desirable and undesirable contributions of a reinforcing filler.

6. Strain-Induced Crystallization

Elongation of a polymer network decreases the entropy of the network chains, and the additional decrease in entropy required for crystallization to occur is therefore relatively small. The melting point $T_m = \Delta H_m / \Delta S_m$, given by the ratio of the enthalpy of fusion to the entropy of fusion, is thus increased, frequently to the extent that considerable amounts of crystallization are induced in the network at the temperature of deformation. This strain-induced crystallization is of great importance with regard to the elastomeric properties of a network. It was found to be the origin of the anomalous increase or upturn in the modulus frequently observed for some polymer networks in the region of very high elongation. Of greater practical importance is the effect of strain-induced crystallization on the ultimate properties and maximum extensibility. Crystallites thus formed act as both physical cross-links and filler particles and, by thus reinforcing the network, greatly increase its ultimate properties. Unfilled polymer networks incapable of the self-reinforcement provided by strain-induced crystallization remain weak, with relatively low ultimate properties.

7. Dynamic Elasticity

Rubberlike materials with permanent cross-linking are typically characterized close to thermodynamic equilibrium, since this facilitates interpretation of elastic properties in terms of molecular structure. Examples of important structural quantities are estimates of the degree of cross-linking or, equivalently, the average molecular weight between cross-links. Dynamic aspects are important even here, however, since junction fluctuations and their suppression by entangling are important in molecular theories such as the Flory-Erman theory described above.

Chain dynamics becomes even more important in the case of an un-cross-linked elastomer. This is due to the fact that such a material will not exhibit a modulus upon stretching, because of the irreversible flow of the chains—rapidly if the chains are short and more slowly if they are long. In the case of such materials, it is important to carry out "dynamic mechanical" studies (creep, stress relaxation, and dynamic shear) to characterize the time-retarded ("viscoelastic") responses in terms of the rubbery plateau region separating the glass-to-rubber transition and the flow region. Establishing the characteristics of this plateau region is very important with regard to the processing and applications of such elastomers. The length of this plateau in plots of the modulus against temperature or time/frequency, for example, depends on the chain structure (molecular, branching, etc.), while its height depends on the number density of interchain entanglements acting as physical cross-links.

Obviously, the degree of such chain entangling is of paramount importance in these situatins. As expected, it varies with the possible presence of diluents or plasticizers and with chain microstructure. The dependences on microstructure are complex, but Graessley and Edwards have made connections between entangling and chain contour lengths and Kuhn step lengths. Specifically, the contour length concentration is defined as the number of main chain bonds per unit volume. It is small in the case of large substituent groups along the chain, since these groups can act as diluting species. The Kuhn step length, on the other hand, increases with chain stiffness, and this stiffness can obviously affect the extent to which a chain can entangle with its neighbors.

8. The Compounding of Rubbers

Natural rubber and its synthetic competitors are generally mixed or "compounded" with a variety of other ingredients prior to being processed into final commercial products. The most important additive of this type is generally a very hard phase for improving an elastomer's mechanical properties. Such "reinforcing" materials can be macroscopic (such as steel or textile cords) or microscopic (small particulate fillers such as carbon black or silica). They are introduced while the rubber is un-cross-linked, with the flow characteristics necessary for the required insertion or blending procedures. Numerous other ingredients are also typically added at this time, including stabilizers (antioxidants and antiozonants), cure ingredients (including catalysts, activators, and accelerators), bonding agents, pigments, waxes, oils, and processing aids. The choices and amounts of the ingredients and the details of their blending are of great importance, since they can bring about vast improvements in properties. Examples are 3-fold increases in stiffness, and 20-fold increases in strength. Such improvements, however, generally come at the cost of some complications, such as increased hysteresis (frictional losses in energy storage due to viscoelastic effects). After such compounding, the mixtures are cured into the final object, by using any of the cross-linking techniques well known to the industry (for example, vulcanization by adding sulfur across double bonds, or free-radical covalent bondings brought about in peroxide thermolyses).

Compounding itself is the art of choosing and mixing the various ingredients to achieve the properties demanded in a specific application. The most important is the selection of the rubber itself, and this choice is typically based on its glass-to-rubber transition temperature; dynamic plateau height; ability to undergo strain-induced crystallization and the effect this has on increasing the modulus and strength; thermal stability; resistance to chemical agents; impermeability; processability; and finally cost. These various properties depend on the structure of the elastomer, and the desire to control them is the basis of the selection of various synthetic rubbers as replacements for natural rubber itself. One result is the commercial availability of wide varieties of specialty synthetic rubbers designed for specific applications.

SEE ALSO THE FOLLOWING ARTICLES

ELASTICITY • PLASTICIZERS • POLYMER PROCESSING • POLYMERS, MECHANICAL BEHAVIOR • RUBBER, NATURAL • RUBBER, SYNTHETIC • THERMODYNAMICS

BIBLIOGRAPHY

Aklonis, J. J., and MacKnight, W. J. (1983). "Introduction to Polymer Viscoelasticity," 2nd ed., Wiley (Interscience), New York.

Erman, B., and Mark, J. E. (1997). "Structures and Properties of Rubberlike Networks," Oxford Univ. Press, New York.

Flory, P. J. (1953). "Principles of Polymer Chemistry," Cornell Univ. Press, Ithaca, N.Y.

Flory, P. J. (1969). "Statistical Mechanics of Chain Molecules," Wiley (Interscience), New York.

Mark, J. E., and Erman, B. (1988). "Rubberlike Elasticity: A Molecular Primer," Wiley (Interscience), New York.

Mark, J. E., and Erman, B. (1992). "Elastomeric Polymer Networks," Prentice Hall, Englewood Cliffs, N.J.

Mark, J. E., Erman, B., and Eirich, F. R., eds. (1994). "Science and Technology of Rubber," 2nd ed., Academic Press, San Diego.

Mark, J. E. *et al.* (1993). "Physical Properties of Polymers," Am. Chem. Soc., Washington, D.C.

Queslel, J.-P. (1989). "Application of elasticity models with constraints on chains and junctions to characterization of randomly cross-linked networks." *Rubber Chem. Technol.* **62,** 800–819.

Queslel, J.-P., and Mark, J. E. (1984). "Molecular interpretation of the moduli of elastomeric polymer networks of known structure." *Adv. Polymer Sci.* **65,** 135–176.

Queslel, J.-P., and Mark, J. E. (1985). "Swelling equilibrium studies of elastomeric network structures." *Adv. Polymer Sci.* **71,** 229–247.

Queslel, J.-P., and Mark, J. E. (1990). "Theoretical equilibrium moduli and swelling extents for elastomer cross-linked in solution." *Rubber Chem. Technol.* **63,** 46–55.

Sperling, L. H. (1992). "Introduction to Physical Polymer Science," 2nd ed., Wiley (Interscience), New York.

Treloar, L. R. G. (1975). "The Physics of Rubber Elasticity," 3rd ed., Clarendon, Oxford.

ISBN 0-12-227414-8

90038 >